GEOGRAPHY

Arthur Getis | Judith Getis | Jerome Fellmann

UNIVERSITY OF ILLINOIS AT URBANA-CHAMPAIGN

GEOG

RAPHY

Macmillan Publishing Co., Inc.
NEW YORK

Collier Macmillan Publishers
LONDON

Macmillan Publishing Co., Inc.
866 Third Avenue, New York, New York 10022

Collier Macmillan Canada, Ltd.

Library of Congress Cataloging in Publication Data

Getis, Arthur (date)
   Geography
   Includes bibliographies and index.
   1. Geography—Text-books—1945-
I. Getis, Judith (date) joint author.
II. Fellmann, Jerome Donald (date) joint
author.   III. Title
G128.G49   910   79-28092
ISBN 0-02-341550-9

Printing: 1 2 3 4 5 6 7 8   Year: 1 2 3 4 5 6 7 8

This book is intended for a one-quarter or one-semester introductory course in geography for college students who have little or no acquaintance with the discipline or awareness of the breadth of its concerns. Our purpose is to convey briefly but incisively the nature of the field and the logical interconnections of its parts.

Although an ability to synthesize large numbers of facts and concepts is a mark of scholarship, most students taking their first geography course are in a weak position to undertake a synthesis of geographic ideas and data. This book divides geographic knowledge into its many subfields so that the content and scope of disciplinary specialties are made clear. With the background thus afforded and with the unifying and interconnected themes of the discipline clarified by the text and the instructor, the student can move on to more advanced courses where the challenge of synthesis is constantly posed.

All geographers share an interest in the nature of the planet we live on. When they study the earth, however, they focus on those aspects that are of particular concern to them. These separate aspects may be thought of as deriving from four broad points of view, each one of which had developed into a school of thought or a "tradition." This book is organized around those four traditions, which are defined in some detail in the Introduction: the earth-science tradition, the culture–environment tradition, the location tradition, and the area-analysis tradition.

Each of the four parts of the book centers on one of these geographic perspectives. Within each part, except that on area analysis, are chapters devoted to the subfields of geography, for example, economic geography or political geography. Each subfield has been placed with the tradition to which we think it belongs; thus, the study of weather and climate is part of the earth-science tradition. Occasionally, our assignment of a topic may appear some-

Preface

what arbitrary, since each tradition contains many emphases and themes and some sub-fields could arguably be attached to more than one tradition. We have made our allocation of topics according to our own perceptions; our rationale for the clustering of chapters is given in the introductions to each part of the text.

The tradition of area analysis—regional geography—is presented in a single chapter that draws upon the preceding traditions and themes. The material and examples it contains are intended to provide the instructor with a springboard for an individualized regional discussion as class or instructor interest may dictate.

A useful textbook must be flexible enough in its organization to permit an instructor to adapt it to the time and subject-matter constraints of a particular course. Although designed with a one-quarter or one-semester course in mind, this text may be used in a full-year introduction to geography when employed as a point of departure for special topics and amplifications introduced by the instructor or when supplemented by additional readings or class projects. Moreover, the chapters are reasonably self-contained and need not be assigned in the sequence here presented. One may deviate from a strict "traditions" approach by rearranging chapters to suit the emphases preferred by the instructor or found to be of greatest interest to the students. With a "current concerns" approach, for example, an instructor might elect to begin with some suitable combination of Chapters 4, 6, 9, 10, and 11 and follow with elaboration of their themes by study of other chapter topics. The instructor of a "human geography" course might decide to use Parts II, III, and IV, the human geography sections, and omit the first four chapters, which focus on the earth-science tradition. The structure of the course may very well reflect a symbiotic relationship between instructor and book rather than being dictated by the book's authors.

Special features of this book are illustrative text inserts and end-of-chapter questions and reading lists. The inserts, which are printed in color, supplement the text presentation of concepts by giving details of a particular instance or amplifying an interesting point made in the text. The "For Review" questions are designed to enable students to check on their understanding of the material and thus reinforce the chapter content. The "Suggested Readings" are intended to facilitate further study and investigation and are therefore confined to easily accessible books. With the exception of a few "classic" works, only recent publications are listed.

Supplementary materials are a Student Study Guide and an Instructor's Manual to accompany this text.

A number of reviewers, generalists as well as specialists, assisted us by providing suggestions for improvement of the manuscript. We want to acknowledge the critical advice of the following: John Allen, Robert Donnell, Ronald Garst, Don R. Hoy, John A. Jakle, George H. Kakiuchi, C. Gregory Knight, J. Kenneth Mitchell, Charles Ogrosky, John Hadley Pierce, Anne Reid, Howard Roepke, and Marilyn Silberfein. We thank them, our Editor Gregory Payne, and our Production Supervisor Elisabeth Belfer for their assistance and concern.

We also wish to express our gratitude to the cartographer James Bier, who designed and drafted the large number of maps and diagrams included in the book. We are indebted to W. D. Brooks and C. E. Roberts, Jr., of Indiana State University for the projection used for many of the maps in this book: a modified van der Grinten. Finally, we extend our thanks to Enid Klass for her assistance in securing many of the photographs that appear in the text.

A. G.
J. G.
J. F.

Contents

vii

viii

GEOGRAPHY

The dawn light showed no change in the towers rising above Pennsylvania's placid Susquehanna River. Their serenity masked an accident that instilled fear, elicited a governmental probe, and would involve millions of dollars to rectify. The event that occurred at the Three Mile Island nuclear power facility on March 28, 1979, called starkly to attention the potential for disaster attendant upon society's apparently insatiable energy demands. Radioactive steam and water and the questioned feasibility of the safe disposition of waste nuclear fuels are dramatic evidence of the pressures being placed by humans upon the environment of which they are a part and whose ultimate limits they cannot extend. Manufactured chemicals and waste products, accelerated erosion through unwise forestry and agricultural practices, the creation of deserts through overgrazing by livestock, the salt poisoning of soils through the development of irrigation—all are evidences of the adverse consequences of human pressures upon natural systems. The social and economic actions of humans occur within the context of the environment and have environmental consequences dangerous to ignore.

The interaction of human and environmental systems works both ways. Human action can, and frequently does, inflict irreparable damage upon the environment; the environment can, and frequently does, exact a frightening toll from the societies that inappropriately exploit it.

November is cyclone season in the Bay of Bengal. At the northern end of the bay lie the islands and the lowlands of the Ganges delta, a vast fertile land, mostly below 30 feet (9 meters) in elevation, made up of old mud, new mud, and marsh. Densely packed by desperately land-needy peoples, this delta area is home to the majority of the population of Bangladesh.

Early in November of 1970 a low-pressure area moved across the Malay Peninsula of

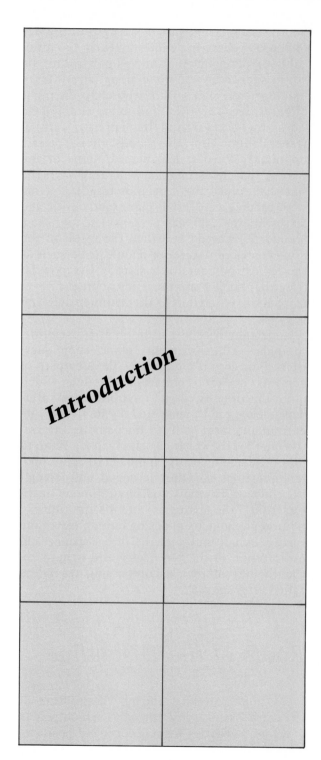

Introduction

1

Southeast Asia and gained strength in the Bay of Bengal, generating winds of nearly 150 miles (240 km) per hour. As it moved northward, the storm sucked up and drew along with it a 20-foot (6-meter) high wall of water. On the night of November 12, with a full moon and highest tides, the cyclone and its battering ram of water slammed into the islands and the deltaic mainland. When it had passed, some of the richest rice fields in Asia were gray with the salt that ruined them, islands totally covered with paddies were left as giant sand dunes, and an estimated 500,000 persons had perished. Should the tragedy be called the result of the blind forces of nature, or should it be seen as the logical outcome of a state of overpopulation that forced human encroachment upon lands more wisely left as the realm of river and sea?

Geography does not attempt, as a discipline, to make value judgments about such questions. Nor does it claim to be the sole path of analysis and interpretation through which questions of cultural and environmental interdependence can be answered. By its history, its current interests, and its unchanging integrative method, however, geography does claim to be a valid and revealing approach to contemporary questions of economic, social, and ecological concern. Humans and environment in interaction; the patterns of distribution of natural phenomena affecting human use of the earth; the cultural patterns of occupance and exploitation of the physical world—these are the themes of that encompassing discipline called *geography*.

Roots of the Discipline

Geography literally means "description of the earth." It is a description in which people and their activities assume a central position.

As a discipline, as a way of thinking about and analyzing the earth's surface, its physical patterns, and human occupance, geography has ancient roots. It grew out of early Greek philosophical concern with "first beginnings" and with the nature of the cosmos. From that broader inquiry, it was refined into a specialized investigation of the physical structure of the earth, including its terrain and its climates and the nature and character of its contrasting inhabited portions. To Strabo (*c*. 64 B.C–A.D. 20), one of the greatest of Greek geographers, the task of the discipline was to "describe the known parts of the inhabited world . . . to write the assessment of the countries of the world [and] to treat the differences between countries." Even earlier, Herodotus (*c*. 484–425 B.C.) had found it necessary to devote much of his book to the lands, peoples, economies, and customs of the various parts of the Persian Empire as necessary background to audience understanding of the causes and course of the Persian wars.

Geographers measured the earth, devised the global grid of latitudes and longitudes, and drew upon this grid surprisingly sophisticated maps of the known world. They explored the apparent latitudinal variations in climate and described in numerous works, both extant and lost, the known worlds of the Mediterranean basin and the more remote, partly rumored lands of northern Europe and of Asia and equatorial Africa. Employing nearly modern concepts, they described river systems, explored cycles of erosion and patterns of deposition, cited the dangers of deforestation, and noted the evident consequences of environmental abuse. Strabo cautioned against the assumption that the nature and actions of humans were determined by the physical environment they inhabited, and he observed that, rather, humans were the active element in a cultural–physical partnership.

So broad and integrated are the concerns of geography, so great the variety of facts, observations, and distributions adduced to guide its analyses, that it has been called the mother of sciences. From its earlier (and still pursued) concerns have sprung such specialized, independent disciplines as meteorology, climatology, cultural anthropology, geology, and a host of other fields of inquiry. From such daughter fields of study, geographers draw the background data that contribute to their own broader investigation of human–environment systems and spatial relationships.

Nature of the Discipline

Although such investigations may serve to give broad definition to geography, it is best understood through the questions geographers ask and the approaches they employ to answer those questions. Of an environmental or cultural phenomenon, they will inquire: What is it? Where is it? Where is it in relation to other physical or social realities that affect it or are affected by it? How did it come to be what it is and where it is? How is it most clearly seen as an integral part of a functional whole? Geography, by the questions it poses, is a survey of distribution, an examination of genesis and process, and a continuing exercise in systems analysis.

So broad a set of questions, so great an assemblage of physical and cultural phenomena inviting examination, has of necessity bred a wide array of specializations and research approaches within the discipline. That very diversity of approaches and topics of inquiry serves to demonstrate an underlying unity, not a divisiveness, in geography. To recognize contrasts between physical and cultural geography is a convenient acknowledgment of differing primary interests in the common objective of the study of human–environment systems in interaction. To set in opposition topical ("sys-tematic") geography (which is a study of preselected classes of phenomena) and regional geography (which seeks to interpret the manner of integration within defined areas of all elements of human–environment systems) artificially obscures the common objective of both approaches.

Geography, then, is a point of view—or, rather, several points of view with a common intellectual objective. It should be seen as having a broad consistency of purpose achieved through the pursuit of a limited number of distinct but closely related traditions. William D. Pattison, who suggested this unifying viewpoint in the mid-1960s, and J. Lewis Robinson, who accepted and expanded Pattison's reasoning in the 1970s (both statements are cited at the end of this introduction), found in the four "traditions" that they recognized an appropriate pluralistic basis for the organization of geographic inquiry. While not all geographic work is confined by the separate themes, Pattison, and later Robinson, maintained that one or more of the traditions were implicit in most geographical studies. Although not labeled precisely, the unifying themes that were recognized—the four traditions within which geographers work—are

1. The earth-science tradition.
2. The culture–environment tradition.
3. The locational tradition.
4. The tradition of area analysis.

We accept these separate but integrated traditions as useful orienting themes for presenting the diverse subject-matter interests of geographers. We have employed them as the device for clustering the several chapters of this book. We trust that the concept of the four traditions, and the centralizing themes they imply, will assist you to recognize the unitary nature of geography while accepting and understanding the diversity of topical interests pursued by those who do geography.

3

Organization of This Book

The content of the earth-science tradition is self-evident: it is that branch of the discipline which addresses itself to the earth as the habitat of humans. It is the tradition that in ancient Greece represented the roots of geography: the description of the physical structure of the earth and of the natural processes that give it detailed form. In modern terms, it is the vital environmental half of the study of human–environment systems, which together constitute geography's subject matter. The earth-science tradition prepares the physical geographer to understand the earth as the common heritage of humankind and to find solutions to the increasingly complex web of pressures placed upon the earth by its expanding, demanding human occupants. By its very nature and content, the earth-science tradition forms the unifying background against which all geographic inquiry is ultimately conducted. Consideration of the elements of the earth-science tradition constitutes Part I of this text.

Part II concerns itself with at least some of the content of the culture–environment tradition. Within this theme of geography, consideration of the earth as a purely physical abstraction gives way to a primary concern with how humans *perceive* the environments they occupy. Its focus is upon culture; the landscapes that are explored and the spatial patterns that are central are those that are cultural in origin and expression. Peoples in their numbers, distributions, and diversity, in their patterns of social and political organization, and in their spatial cognitions and behaviors are the orienting concepts of the culture–environment tradition. The tradition is distinctive in its thrust but unified with the objective of the earth-science tradition by the obvious observation that populations exist, cultures emerge, and behaviors occur within the context of the physical realities and patternings of the earth's surface.

The location tradition—or as it is sometimes called, the spatial tradition—is the subject of the chapters of Part III. It is a tradition that underlies all of geographic inquiry. As Robinson suggested, if we can agree that geology is rocks, that history is time, and that sociology is people, then we can assert that geography is space. The location tradition embodies a reasoned concern with the placement of cultural phenomena or of physical items of significance to human occupance of the earth.

Part, but by no means all, of the location tradition is concerned with distributional patterns. More central are scale, movement, and areal relationships. Map, statistical, geometrical, and systems analysis are among the research tools employed by geographers working within the location tradition. Irrespective, however, of the analytic tools employed or the sets of phenomena studied—economic activities, resource distributions, city systems, or others—the underlying theme is the geometry, or the distribution, of the phenomenon discussed and the flows and interconnections that unite it to related affecting and affected physical and cultural occurrences.

A single-chapter consideration of the area-analysis tradition constitutes Part IV of this introduction to geography and completes our survey of the discipline. Again, the roots of this theme may be traced to antiquity. As Pattison observes, Strabo's *Geography* was addressed to the leaders of Augustan Rome as a summary of the nature of places in their separate characters and differentiations, knowledge deemed useful—indeed, vital—to the statesmen of an empire. Imperial concerns may long since have vanished, but the study of regions and the recognition of their spatial uniformities and of areal differences remain. These uniformities and differences, of course, grow out of the structure of human–environment systems and interrelations that are the study of geography.

The area-analysis tradition, therefore, can be seen not only as a geographic theme in its own right but also as the logical objective and culmination of each of the other traditions. This point of view is further explored in Chapter 12.

The identification of the four traditions of geography is not only an organizational convenience. It is also a recognition that within that diversity of subject matter called geography, unity of interest and objective is ever preserved. The traditions, though recognizably distinctive, are inextricably intertwined and overlapping. The locational tradition, we suggest, considers spatial distributions; but location is as much a perceived as an objective condition, and we consider the perception of area under the theme of culture–environment. We welcome this and other evidences of overlap of concern and approach among the four traditions, for they demonstrate that geography, in Pattison's words, "exhibit[s] a broad consistency, and . . . this essential unity [is] attributable to a small number of distinct but affiliated traditions."

Suggested Readings

HAGGETT, PETER, *Locational Analysis in Human Geography*, ch. 1. Arnold Publishers, London, 1965.

JAMES, PRESTON E., *All Possible Worlds.* The Odyssey Press, Indianapolis, 1972.

PATTISON, WILLIAM D., "The Four Traditions of Geography," *The Journal of Geography*, vol. 63 (1964), pp. 211–16.

ROBINSON, J. LEWIS, "A New Look at the Four Traditions of Geography," *The Journal of Geography*, vol. 75 (1976), pp. 520–30.

TAAFFE, EDWARD J. (ed.), *Geography.* Prentice-Hall, Inc., Englewood Cliffs, NJ, 1970.

6

The Earth-Science Tradition

For nearly a month the mountain had rumbled, emitting puffs of steam and flashes of fire. Within the past week, on its slopes and near its base, deaths had been recorded from floods, mud slides, and falling rock. A little after 8:00 on the morning of May 8, 1902, the climax came—for volcanic Mount Pelée and the thriving port of Saint Pierre on the island of Martinique. To the roar of one of the biggest explosions the world has ever known and to the clanging of church bells aroused in swaying steeples, a fire ball of gargantuan size burst forth from the upper slope of the volcano; lava, ash, steam, and superheated air engulfed the town, and 29,933 people met their death.

More selective but—for those victimized—just as deadly was the sudden "change in weather" that struck central Illinois on December 20, 1836. Within an hour, preceded by winds gusting to 70 miles per hour, the temperature plunged from 40°F to −30°F. On a walk to the post office through the slushy snow, Mr. Lathrop of Jacksonville, Illinois, found, just as he passed the Female Academy, that "the cold wave struck me, and as I drew my feet up the ice would form on my boots until I made a track . . . more like that of a Jumbo than a No. 7 boot." Two young salesmen were found frozen to death, along with their horses; one "was partly in a kneeling position, with a tinderbox in one hand, a flint in the other, with both eyes open as though attempting to light the tinder in the box." Others died, too—inside horses disemboweled and used as makeshift shelter, in fields, woods, and on the road to places both a short distance and an eternity from the traveler's starting point.

Fortunately, few human–environmental encounters are so tragic as these. Rather, the physical world in all its spatial variation provides the constant background against which the human drama is played. It is to that background that physical geographers, acting within the earth-science tradition of the discipline, direct their attention. Their primary concern is with the natural rather than the cultural landscape and with the interplay between the encompassing physical world and the activities of humans.

The interest of those working within the earth-science tradition of geography, therefore, is not in the physical sciences as an end in themselves, but in physical processes as they create landscapes and environments of significance to humankind. The objective is not solely to trace the physical and chemical reactions that have produced a piece of igneous rock or to describe how a glacier has scoured the rock; it is the relationship of the rock to people: geographers are interested in what the rock can tell us about the evolution of the earth as the home of human beings or in what is the significance of certain types of rock for the distribution of mineral resources or of fertile soil.

The questions that physical geographers ask do not usually deal with the catastrophic, though catastrophe is an occasional element of human–earth-surface relationships. Rather, they ask questions that go to the root of understanding the earth as the home and the work place of humankind. What does the earth as our home look like? How have its components been formed? How are they changing, and what changes are likely to occur in the future through natural or human causes? How are environmental features, in their spatially distinct combinations, related to past, present, and prospective human utilization of the earth? These are some of the basic questions that geographers who follow the earth-science tradition attempt to answer.

The four chapters in Part I of our review of geography deal, in turn, with maps, landforms, weather and climate, and human impact upon the environment. It is easy to see why the earth's landforms, weather, and climate come within the earth-science tradition, but it may be less clear why we begin this part with a discussion of maps and end with a consideration of human impact upon the physical world.

Maps are our starting point, for it is impossible to talk of geography or to appreciate its content and lessons without understanding maps. Geography is a spatial science, and since ancient times maps have been used to represent, characterize, and interpret the nature of space. All geographers use maps as basic analytic tools, whether the maps be simple sketches made during field investigations or computer-drawn maps based on data from earth satellites. Mapmakers traditionally sought ways to represent the distribution of land and water features of the earth and to portray the character of the physical landscape. Since the 1700s, they have turned increasing attention to map rendition of cultural and political distributions and to the preparation of thematic maps. In the last few decades, geographers have begun to draw maps based on projections that distort time or space or that are intended to portray "psychological" or "perceived" reality. Whatever the state of their art or the thrust of their attention, mapmakers have been motivated to express the distribution of things central to their interests. As a spatial, distributional science, geography properly begins with a discussion of the nature of maps.

The treatment of so vast a field as landforms (Chapter 2) must of necessity be highly selective within an introductory text. The aim is to summarize the great processes by which landforms are created and to depict the general classes of features resulting from those processes. The objective is to develop an appreciation of the essentials of this facet of the earth-science tradition and to create an awareness of the majestic forces sculpting the earth, without becoming overly involved in scientific reasoning and technical terms.

In Chapter 3, "Physical Geography: Weather and Climate," the major elements of the atmosphere—temperature, precipitation, and air pressure—are discussed, and their regional generalities and associations in patterns of world climates are presented. It is the study of weather and climate that adds coherence to our understanding of the earth as the home of people. Frequent reference will be made to the climatic background of human activities in later sections of our study.

Human actions take place within the context of the physical environment. Equally importantly, those activities have environmental impacts and consequences. Chapter 4, "Human Impact upon the Environment," traces some of those—mainly adverse—consequences and straddles the line between the earth-science and the culture–environment traditions of geography. This chapter emphasizes the effects of human actions on the air, water, soil, and other resources upon which our lives depend; it explores briefly many of the current environmental problems of pollution and of resource management. Thus Chapter 4 can serve as an introduction to the burgeoning field of environmental impact and protection.

The earth-science tradition—like the other traditions of geography—rests upon the distributions of phenomena that affect and are affected by the activities of humans. It begins with the mapmaker's art. It concentrates on the environmental framework within which human occupance of the earth occurs. It leads through a consideration of people's impact upon that environment to a primary focus—in later parts of this book—upon humankind as studied under other traditions of geography.

Imagine that all the maps in the world were destroyed at this moment. Traffic on highways would slow to a crawl. Control towers would be overwhelmed as pilots radioed airports for landing instructions; ships at sea would be forced to navigate without accurately charting their positions. Rangers would be sent to rescue hikers stranded in national parks and forests. Disputes would erupt over political boundaries; military planners would not know how to channel troop movements or where to aim artillery. People would have to rely on verbal descriptions to convey information, and no amount of verbiage can do justice to the detailed information contained on a map of only moderate complexity. Maps are tersely efficient at indicating the location of things relative to one another, and the information void created by their disappearance would have to be filled by volumes of description.

Maps are as fundamental a means of communication as the printed or spoken word or as photographs. In every age, people have produced maps, whether made out of sticks or shells, scratched on clay tablets, drawn on parchment, or printed on paper. Charlemagne even had maps made on solid plates of silver! The map is the most efficient way to portray parts of the earth's surface, to record political boundaries, and to indicate directions to travelers. In addition to their usefulness to the general populace, maps have a special significance to the geographer. In the process of studying the surface of the earth, geographers depend on maps to record, present, and aid them in analyzing the location of points, lines, or objects.

Although the earth scientists of ancient Greece are justly famous for their contributions to mapping—contributions that include the recognition of the spherical form of the earth and the development of map projections and the grid system—modern scientific mapping has its roots in the 17th century. Several developments during the Renaissance gave an impetus to accurate cartography. Among these

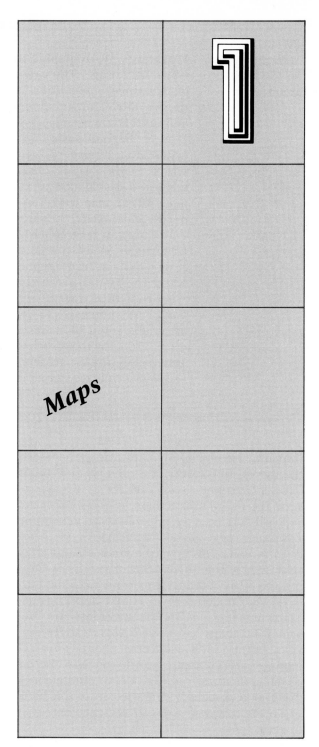

9

Cartography in the Medieval Period

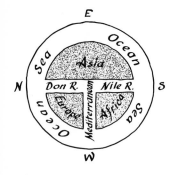

Much of the cartographic tradition of Greece and Rome was lost to medieval Christian Europe. There remained, exclusively in the hands of mariners, charts of coastlines, navigational hazards, and ports of call. World maps ignored even these available data and were based on tradition, supposed biblical authority, and myth. The circular T–O world map was typical of the type. It showed a crudely diagrammatic earth composed of three continents. The land area—centered on Jerusalem—was surrounded by an impassable Ocean Sea, beyond which to the east lay inaccessible Eden and paradise. Place names, if any, marked a casual mix of contemporary and long-gone classical sites. Vacant areas on the map were filled with renderings of fabulous monsters: dog-headed people, monopeds, unicorns.

The Arab geographer Al-Idrisi was directed in 1138 by Roger II, the Christian King of Sicily, to collect all known geographical information and assemble it on a truly accurate representation of the world. An academy of geographers was gathered to assist Al-Idrisi. Books and maps of classical and Islamic origin were consulted, mariners interviewed, and scientific expeditions dispatched. Data collection took 15 years before the final world map was fabricated on a silver disc some 80 inches in diameter, weighing over 300 pounds. Lost to looters in 1160, the map is survived by "Roger's Book," containing the information amassed by the geographers and including a world map, 71 part maps, and 70 sectional itinerary maps.

were the development of printing, the rediscovery of the work of Ptolemy and other Greeks, and the great voyages of discovery. With the rise of nationalism in many European countries, it became imperative to determine and accurately portray boundaries and coastlines as well as to depict the kinds of landforms contained within the borders of a country. During the 17th century, important national surveys were undertaken in France and England. Many conventions in the way data are presented on maps had their origin in these surveys. This early concern about *physical* maps—that is, maps portraying the earth's physical features—allows us to include our discussion of maps within the part of this book exploring topics in the earth-science tradition of geography.

Knowledge of the way in which information is recorded on maps enables us to read and interpret them correctly. To be on guard against drawing inaccurate conclusions or to avoid being swayed by distorted or biased presentations, we must be able to understand and assess the ways in which facts are represented. Of course, all maps are necessarily distorted because of the need to portray the round earth on a flat surface, to use symbols to represent objects, to generalize, and to record features at a different size than they actually are. This distortion of reality is necessary if only because the map is smaller than the things it depicts and because its effective communication depends upon selective emphasis of only a portion of reality. As long as they know the limitations of the commonly used types of maps and understand what relationships are distorted, map readers may easily interpret maps correctly.

Locating Points on a Sphere: The Grid System

In order to visualize the basic system for locating points on the earth, think of the world as a sphere with no markings whatever on it. There would, of course, be no way of describing the exact location of a particular point on the sphere without establishing some system of reference. Thus we use the *grid system,* which consists of a set of imaginary lines drawn across the face of the earth. The key reference points in that system are the North and South poles and the equator, which are given in nature (Figure 1.1), and the prime meridian, which is agreed upon by cartographers.

The North and South poles are the end points of the axis about which the earth spins. The line that encircles the globe halfway between the poles, and that consequently is perpendicular to the axis, is the *equator.* We can describe the location of a point in terms of its distance north or south of the equator measured as an angle at the earth's center. Because a circle contains 360 degrees, the distance between the two poles is 180 degrees and between the equator and each pole, 90 degrees. *Latitude* is the measure of distance north and south of the equator, which is itself, of course, zero degrees. As is evident in Figure 1.1, the lines of latitude, which are parallel, run east–west.

Although in theory an infinite number of parallels of latitude exists, cartographers commonly use only one for each of the 90 degrees between the equator and either pole. The distance between each degree of latitude on the earth is about 69 miles (111 km).* To record the latitude of a place in a more precise way, degrees are divided into minutes and then into seconds, exactly like an hour of time. One minute of latitude is about 1.15 miles (1.85 km), and one second of latitude about 101 feet (31 m). The latitude of the center of Chicago, Illinois, is written 40° 51′ 50″ North. This system of subdivision of the circle is derived from the sexagesimal (base 60) numerical system of the ancient Babylonians, a culture group discussed in Chapter 5.

Because the distance north or south of the equator is not by itself enough to locate a point in space, we need to specify a second coordinate to indicate distance east or west from an agreed-upon reference line. As a starting point for east–west measurement, cartographers in most countries use as the *prime meridian* an imaginary line passing through the Royal Observatory at Greenwich, England. That prime meridian was selected as the zero degree longitude line by an international conference in 1884. Like all meridians, it is a true north–south line connecting the poles of the earth

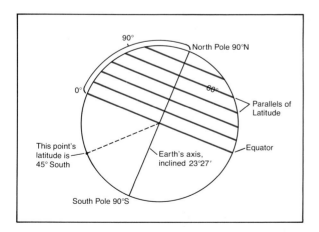

Fig. 1.1 The grid system: parallels of latitude. Note that the parallels become increasingly shorter closer to the poles. The 60th parallel is only one-half as long as the equator.

11

---

* The polar circumference of the earth equals 24,899 miles; thus 1 degree of latitude $= \dfrac{24{,}899}{360} = 69.16$ miles.

Agreement on a Prime Meridian

Latitudinal location relative to the equator, a line given in nature, can be determined by measurement of the angular height of the sun above the horizon at noon on a known date. Measuring devices for this purpose (astrolabes) appear to have been in use by the Babylonians and the Egyptians several thousand years before the Greeks addressed the problem of constructing the globe grid and developing world maps.

The determination of longitude—the other necessity for accurate maps—was more difficult. No reference line equivalent to the equator existed, and no convenient motion of sun or stars indicated the location east or west of a naturally given baseline. This lack was merely annoying when one was traveling on land where surface distances could be measured, or on coasting vessels, from whose decks landfalls could be sighted. It was potentially disastrous when long sea voyages out of sight of land became common in the 16th and 17th centuries; even skilled navigators using dead reckoning could, and did, miss destinations by scores and sometimes hundreds of miles.

Innumerable proposals were made for using water wheels attached to ships' hulls and for other mechanical devices for measuring distance traveled at sea. All were rendered useless by the currents and winds acting upon sailing vessels. The only way longitude could be determined was by the measurement at a given instant of the difference between the time at a known point and the time on a reference line. Since the earth rotates through 360° in 24 hours, each hour of time is the equivalent of 15° of longitude and each degree of longitude equals 4 minutes of time.

Highly accurate pendulum clocks were developed, but were useless on the tossing decks of ships; improved telescopes permitted precise timing of the movement of Jupiter's satellites, but telescopes of the requisite size also required a stable platform. The answer to the longitude problem had to be a small but extremely reliable chronometer (clock) dependent upon a spring-driven movement.

In 1714 an act of the English Parliament established a prize of £20,000 for any device that could determine longitude within 30 minutes of the globe grid, that is, within 2 minutes of time or 34 miles. John Harrison, to whom the prize was eventually awarded, put the chronometer of his invention to an official sea trial on a trip from Portsmouth, England, to Jamaica in the winter of 1761–62. It proved to be in error by only $1\frac{1}{4}$ nautical miles.

The problem of the accurate determination of the longitude of any point at sea or on land was solved, but only with reference to a prime meridian that could be established wherever one wished. For patriotic reasons, most countries located the prime meridian within their own borders, and by the 19th century over a score of different ones were in common use. Recognizing the desirability of establishing a universal prime meridian on which acceptable world maps could be based, scientists held a series of conferences as first suggested by the third International Geographical Congress meeting in Venice in 1881. Delicate negotiations finally produced agreement among 25 nations that 0° longitude would refer to the meridian passing through the Royal Observatory at Greenwich, England. This convention is now almost universally accepted, though one occasionally sees maps based on a different prime meridian.

(Figure 1.2). Meridians are farthest apart at the equator, come closer and closer together as latitude increases, and converge at the North and South poles. Unlike parallels of latitude, all meridians are the same length.

Longitude is the angular distance east or west of the prime meridian. Directly opposite the prime meridian is the 180th meridian, located in the Pacific Ocean. East–west measurements range from 0° to 180°, that is, from the prime meridian to the 180th meridian in each direction. Like parallels of latitude, degrees of longitude can be subdivided into minutes and seconds. However, unlike the degrees of latitude, those of longitude decrease in length away from the equator because of the convergence of meridians at the poles. With the exception of a few Alaskan islands, all places in North and South America are in the area of

west longitude; with the exception of a portion of the Chukchi Peninsula of the U.S.S.R., all places in Asia and Australia have east longitude.

By citing the degrees, minutes, and—if necessary—seconds of longitude and latitude, we can describe the location of any place on the earth's surface. To conclude our earlier example, Chicago is located at 41° 52′ N, 87° 40′ W. Singapore is at 1° 20′ N, 103° 57′ E (Figure 1.3).

Projections and Distortion on Maps

13

The earth can be represented with reasonable accuracy only on a globe, but globes are not as

Fig. 1.2 The grid system: meridians of longitude. These arbitrary but conventional lines, together with the parallels based upon the naturally given equator, constitute the globe grid. Since the meridians converge at the poles, degrees of longitude become shorter the farther one moves away from the equator. Degrees of latitude remain constant throughout the length of every meridian and are equal to the length of a degree measured along the equator.

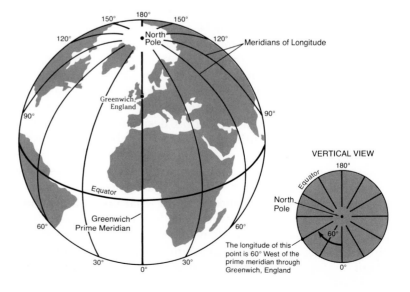

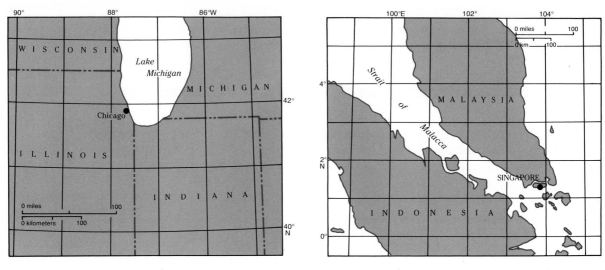

Fig. 1.3 The latitude and longitude of Chicago (41° 52′ N, 87° 40′ W) and of Singapore (1° 20′ N, 103° 57′ E).

14

convenient as flat maps to store or to use, and they cannot depict much detail. For example, if we had a large globe with a diameter of, say, 3 feet we would have to fit the details of nearly 50,000 square miles of earth surface in an area 1 inch on a side. Obviously, a globe of reasonable size cannot show the transportation system of a city or the location of very small towns and villages. In transforming a globe into a map, we cannot flatten the curved surface and keep intact all the properties of the original. Thus, in the drawing of a map, the relationships between points are inevitably distorted in some way. To the geometer, a sphere is a *nondevelopable* surface. It can only be *projected* upon a plane surface; the term *projection* designates the method of representing a curved surface on a flat map.

Because no projection can be entirely accurate, we should know in what respects a particular map correctly reproduces, and in what respects it distorts, earth features. The four main properties of maps—area, shape, distance, and direction—are distorted in different ways and to different degrees by various projections. Figure 1.4 indicates distortion on a popular projection.

Some projections enable the cartographer to represent the *areas* of regions in correct or constant proportion to earth reality. That means that any square inch on the map represents an identical number of square miles (or of similar units) anywhere else on the map. As a result, the shape of the portrayed area is inevitably distorted; a square on the earth, for example, may become a rectangle on the map, but that rectangle has the correct area. Such projections are called *equal-area* or *equivalent.*

Although no projection can provide correct *shapes* for large areas, some accurately portray the shapes of small areas by preserving correct angular relationships. That is, an angle on the globe is rendered correctly on the map. These maps have true shapes, or *conformality.* Parallels and meridians always intersect at right angles on such maps, as they do on a globe. *A map cannot be both equivalent and conformal.*

Distance relationships are nearly always distorted on a map, but some projections do maintain true distances in one direction or along certain selected lines. Others, called *equidistant projections,* show true distance in all directions, but only from one or two central points; see, for example, Figure 1.9.

As is true of distances, *directions* between all points cannot be shown without distortion. *Azimuthal* projections do exist, however, that enable the map user to measure correctly the directions from a single point to any other point; an example is shown in Figure 1.10.

The map user needs to be familiar with the various types of map projections that the cartographer may employ and to understand the ways in which they are accurate or distorted.

While all projections can be described mathematically, some can be thought of as being constructed by geometrical techniques rather than by mathematical formulas. In geometrical projections, the grid system is transformed into a geometrical figure such as a cylinder or a cone, which in turn can be cut and then spread out flat without any stretching or tearing. Such a figure is called *developable.* The selection of the surface to be developed or of the specific mathematical formula to be employed is determined by the properties of the globe grid that one elects to retain. These properties are known to the cartographer and must be recognized by the map reader. As a reminder, globe

Fig. 1.4 Distortion on the Mercator projection. A perfect five-pointed star was drawn on a globe, and the latitude and longitude of the points of the star were transferred to the Mercator map shown above. The manner in which the star is distorted reflects the way the projection distorts land areas. Mercator maps are usually accompanied, as here, by a scale diagram showing the linear values of degrees of longitude at different latitudes.

15

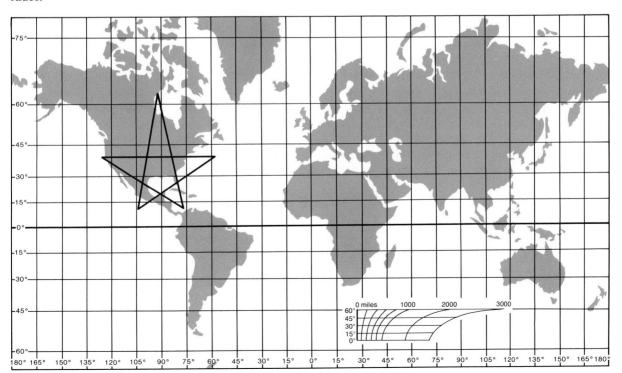

properties are that

1. All meridians are of equal length; each is one-half the length of the equator.
2. All meridians converge at the poles and are true north–south lines.
3. All lines of latitude (parallels) are parallel to the equator and to each other.
4. Parallels decrease in length as one nears the poles.
5. Meridians and parallels intersect at right angles.
6. The scale on the surface of the globe is everywhere the same in every direction.

CYLINDRICAL PROJECTIONS

Three surfaces upon which the grid system of a globe may be projected are the cylinder, the cone, and the plane. As noted, such surfaces are developable: the first two because they can be cut and laid flat without distortion, the third because it is flat at the outset. We shall consider two cases of the cylindrical projection, one useful chiefly for teaching purposes and the other one of the most commonly used (and misused) projections in the world.

Cylindrical Equal-Area Projection. Suppose that we rolled a piece of paper the same height as the globe around the globe, so that it was tangent to the sphere at the equator. Other parts of the cylinder would get farther and farther away from the globe as they approached the North and South poles. If we placed a source of light away from the globe so that the rays of light seemed to be coming from an infinite distance, those rays would be parallel (like the sun's rays striking the earth). The light would project a shadow map upon the cylinder of paper. When we unrolled the cylinder, the map would appear as shown in Figure 1.5.

Note the variance between the grid we have just projected and the true properties of the globe grid. In the cylindrical equal-area projection, meridians do not converge at the poles as they do on a globe, and the poles are not points but lines as long as the equator. The meridians are everywhere equally far apart, and the parallels of latitude have all become the same length. The grid lines do, however, cross each other at right angles, as they do on the globe, and they all lie as straight north–south or east–west lines.

Note also (as one does in evaluating any projection) the ways in which the projection distorts the size and the shape of earth features. The lands around the poles look as if they have been simultaneously flattened in a north–south direction and stretched east–west. Around the equator, the shapes and areas seem to be fairly accurate, but land masses take on a peculiar appearance in the upper latitudes. Although, as its name implies, this is an equal-area projection, it is little used for that property

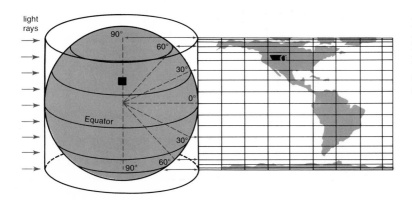

Fig. 1.5 The cylindrical equal-area projection. The North and South poles are as long as the equator, distorting shapes in their vicinities. The shaded area represents a perfect square drawn on the globe.

16

The invention in Europe of cast metal, movable type in the 1440s by John Gutenberg of Mainz, Germany, revolutionized the diffusion of knowledge. Illustrations were particularly appealing in the new printed books, and the maps and text of Claudius Ptolemy's *Geographia* of about A.D. 150 were published in at least seven folio editions before 1500. But a new geographical awareness made Ptolemy's maps less and less acceptable; new regional and world maps were printed in increasing numbers as separate sheets or as accompaniments to texts. About 1550, book publishers switched from woodcut maps to copperplate engravings, and the map trade became concentrated in the Netherlands, noted for the superiority of its engravers.

Gerard Mercator (Latinized from Kremer) was for nearly 60 years the recognized European master of cartography, producing not only the projection and navigational chart bearing his name, but publishing highly accurate regional and world maps based on the latest information and, in his map of Flanders, on his own field surveys. But separate maps were awkward to handle and store, difficult to acquire, and easily lost. What was needed was a compendium of maps of uniform page size, bound in a manageable collection. In 1570, Mercator's friend Abraham Ortels (known as Ortelius) filled this need by publishing his *Theater of the World*, the first modern geographical atlas. Like any present-day world atlas, it contained an introductory general world map, followed by maps of the four continents then known: Europe, Asia, Africa, and America. The remainder of the atlas contained maps of nations and smaller political divisions. At the time of Ortelius's death in 1598, some 28 editions of his atlas had appeared in five languages, and the reference atlas assumed its continuing role as a standard and necessary volume for educated people.

because the shapes are so distorted. Its utility lies solely in the circumstance that it is an easily understood demonstration of how projection from the globe to a flat surface can be accomplished. It also serves as an introduction to one of the most famous of all projections.

The Mercator Projection. We can think of the Mercator projection as also being formed by wrapping a cylinder around a globe, tangent to the equator. Instead of the paper's being the same height as the globe, however, it extends far beyond the poles. The light source is at the center of the globe, rather than off at infinity. As indicated in Figure 1.4, meridians have the same form as on the cylindrical equal-area projection; that is, they are equally spaced parallel vertical lines.

Instead of showing the simultaneous flattening and stretching that occurs near the poles on the cylindrical equal-area projection, the Mercator map balances the spreading of meridians toward the poles by spacing the latitude lines farther and farther apart. On the cylindrical equal-area projection, the distance between the parallels narrows in higher latitudes. On the Mercator projection the opposite occurs; distances between the parallels become increasingly greater and the poles can never be reached. Mathematical tables show the cartog-

Fig. 1.6 The Lambert azimuthal equal-area projection is mathematically derived to display the property of equivalence.

18 rapher how to space the parallels to balance the spreading of meridians so that the result is a conformal projection. On such a map, the shapes of small areas are true, and even large regions are fairly accurately represented.

Note that the Mercator map in Figure 1.4 stops at 80° N and 70° S. A very large map would be needed to show the polar regions, and the sizes of those regions would be enormously exaggerated. The poles themselves can never be shown. In fact, the enlargement of areas with increasing latitude is so considerable that a

Mercator map should not be published without a scale of miles such as that shown.

The Mercator projection has often been misused in atlases and classrooms as a general-purpose world map. Whole generations of schoolchildren have grown to maturity convinced that the state of Alaska is nearly the size of all the lower 48 states put together. To check the distortion of this projection, compare the sizes of Greenland and Canada on Figure 1.4 with those on Figure 1.6, an equal-area projection.

The proper use of the Mercator is for navigation; in fact, it is the standard projection used by navigators because of a peculiarly useful property: a straight line drawn anywhere on the map is a line of constant compass bearing. If such a line, called a *rhumb line,* is followed, a ship's or plane's compass will show that the course is always at a constant angle with respect to geographic north. On no other projection is a rhumb line both straight and true as a direction.

CONIC PROJECTIONS

Of the three developable geometric forms—cylinder, cone, and plane—the cone is the closest in form to one-half of a globe. Conic projections, therefore, are widely used to depict hemispheres or smaller parts of the earth.

A useful projection in this category, and the easiest to understand, is the *simple conic* pro-

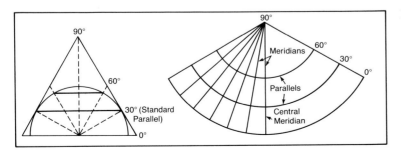

Fig. 1.7 A simple conic projection with one standard parallel. Most conics are adjusted so that the parallels are spaced evenly along the central meridian.

jection. Imagine that a cone is laid over half the globe, as in Figure 1.7, tangent to the globe at the 30th parallel. Distances are true only along the tangent circle, which is called the *standard parallel*. When the cone is developed, of course, the standard parallel becomes an arc of a circle, and all other parallels become arcs of concentric circles. With a central light source, the parallels become increasingly farther apart as they approach the pole, and distortion is accordingly exaggerated.

One can lessen the amount of distortion by shortening the length of the central meridian, spacing the parallels of latitude at equal distances on that meridian, and making the 90th

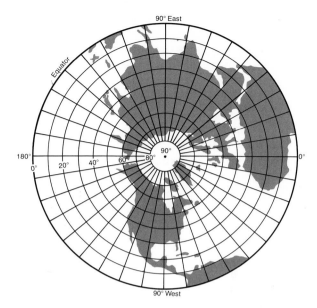

Fig. 1.9 The azimuthal equidistant projection. Parallels of latitude are circles equally spaced on the meridians, which are straight lines. Distances from the center to any other point are true. If the grid is extended to show the southern hemisphere, the South Pole is represented as a circle instead of as a point.

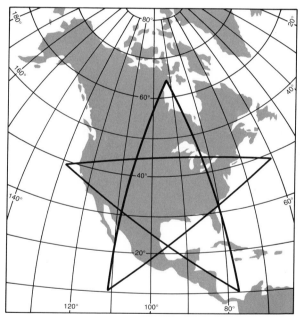

Fig. 1.8 The polyconic projection. The map is produced by bringing together east–west strips from a series of cones, each tangent at a different parallel. This projection differs from the simple conic in that the parallels of latitude are not arcs of concentric circles and the meridians are curved rather than straight lines. While neither equivalent nor conformal, the projection has a high degree of accuracy. Note how closely the star resembles a perfect five-pointed star.

19

parallel (the pole) an arc rather than a point. Most of the conic projections that are in general use employ such mathematical adjustments. Because any projection is most accurate around the standard parallel, a cone that cuts through the globe (a secant cone), and thus is tangent at two parallels, yields a more accurate map than a cone tangent at only one point. When more than one standard parallel is used, a *polyconic* projection results.

Conic projections can be adjusted to achieve desired qualities, and hence they are widely used. They are particularly suited for showing areas in the mid-latitudes that have a greater east–west than north–south extent. Both area and shape can be represented by conic projections without serious distortion (Figure 1.8). Many official map series, such as the topographic sheets and the world aeronautical charts of the U.S. Geological Survey and the Aerospace Center of the Defense Mapping Agency, utilize types of conic projections. They are also used in many atlases.

AZIMUTHAL PROJECTIONS

Azimuths, lines radiating from a central point, are lines of true direction. Azimuthal projections are constructed by placing a plane surface tangent to the globe at a single point. Although the plane may touch the globe anywhere the cartographer wishes, the polar case with the plane centered on either the North or the South Pole is easiest to visualize. It is displayed in Figure 1.9.

Such an azimuthal equidistant projection is useful because it can be centered anywhere, facilitating the correct measurement of distances from that point to all others. For this reason, it is often used to show air navigation routes that originate from a single place. When the plane is centered at places other than the poles, the meridians and the parallels become curiously curved, as is evident in Figure 1.10.

Because they are particularly well suited for showing the arrangement of polar land masses, azimuthal maps are commonly used in atlases. Depending on the particular projection used, either true shape or equal-area or some compromise between them can be depicted.

OTHER PROJECTIONS

Projections can be developed mathematically to show the world, or a portion thereof, in any shape that is desired: ovals, hearts, trapezoids, stars, and so on. One often-used projection is Goode's Homolosine. Usually shown in its interrupted form, as in Figure 1.11, it is actually a

20

Fig. 1.10 An azimuthal equidistant projection centered on Urbana, Illinois. The scale of miles applies only to distances from Urbana or on a line through it. The scale on the rim of the map, representing the antipode of Urbana, is infinitely stretched. (Copyright 1977, Brooks–Roberts; with permission.)

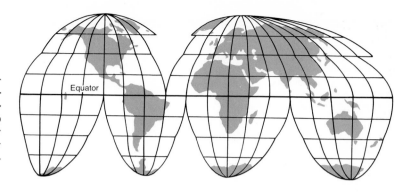

Fig. 1.11 Goode's Homolosine projection, a combination of the sinusoidal and the homolographic projections. This projection can also interrupt the continents to display the ocean areas intact. (Copyright by The University of Chicago Department of Geography.)

combination of two different projections. This equal-area projection also represents shapes well.

Not all maps are equal-area, conformal, or equidistant; many, such as the widely used polyconic (Figure 1.8), are compromises. In fact, some very effective projections are non-Euclidian in origin, transforming space in unconventional ways. Distances may be measured in nonlinear fashion (in terms of time, cost, number of people, or even perception), and maps that show relative space may be constructed from these data. Two examples of such transformations are shown in Figure 1.12.

Mapmakers must be conscious of the properties of the projections they use, selecting the one that best suits their purposes. For navigation by sea or air, a Mercator projection is useful. If numerical data are being mapped, the relative sizes of the areas involved should be correct, and so one of the many equal-area projections is likely to be employed. Most atlases indicate which projection has been used for each map, thus informing the map reader of the properties of the maps and their distortions.

Scale

The scale of a map is the ratio between the measurement of something on the map and the corresponding measurement on the earth. Scale is typically represented in one of three ways: verbally, graphically, or as a representative fraction. A *verbal* scale, as the name implies, is given in words, such as "1 inch to 1 mile." A *graphic* scale is a line or bar marked off in map units but labeled in ground units. A graphic scale of "1 inch to 1 mile" is shown below.

A *representative fraction* (RF) scale gives two numbers, the first representing the map distance and the second the ground distance. The fraction may be written a number of ways. As there are 63,360 inches in 1 mile, the fractional scale of a map at 1 inch to 1 mile could be written

$$1:63,360, \quad \frac{1}{63,360} \quad \text{or} \quad 1/63,360.$$

On the simpler metric scale, 1 centimeter to 1 kilometer is $1:100,000$. The units used in each part of the fractional scale are the same, thus $1:63,360$ could also mean that 1 foot on the map equals 63,360 feet on the ground, or 12 miles—which is, of course, the same as 1 inch equals 1 mile. Numerical scales are the most accurate of all scale statements and can be understood in any language. Figure 1.13 is based on a fractional scale of $1:24,000$.

The terms *large-scale* and *small-scale* are often applied to maps. A large-scale map, such

21

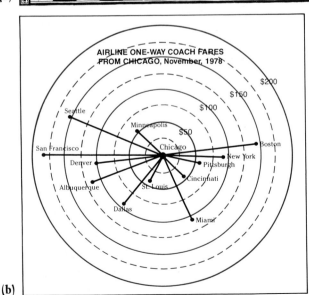

(a)

THE CITY of NEW YORK is unique–it is a NATION. Its inhabitants, of which there are some 7,000,000, are called NEW YORKERS. This MAP is presented, after patient research, as a composite of the NEW YORKERS' ideas concerning THE UNITED STATES · ·

LET THEM SPEAK

We have cousins in the West · They live in Wilmington, Delaware.

He is moving to Dallas so he can be near his little Mother in El Paso

Indiana was an Indian Reservation until just recently, wasn't it?

So you are moving to Indianapolis; you must let me give you a letter to my niece in Minneapolis

Oh yes! he entered the Marathon Swim from Los Angeles to Hawaii · · · ·

A **New Yorker's Idea** of
THE UNITED STATES
OF AMERICA

22

(b)

Fig. 1.12 (a) A New Yorker's Idea of the United States portrays the country not as it actually is but as the artist conceives it. Think of it as a map based on an unknown projection. Distance, area, shape, and direction are all distorted, and a "psychological" rather than a literally true map is achieved. Such "mental maps" are discussed in Chapter 8. (Copyright F. V. Thierfeldt, Milwaukee, WI.) (b) Airline distance from Chicago. A distance-transformation map based upon one-way coach air fares, November 1, 1978.

as a plan of a city, shows a restricted area in considerable detail, and the ratio of map to ground distance is relatively large (for example, 1:7920, or 1 inch equals one-eighth of a mile). Small-scale maps, such as those of countries or continents, have a much smaller ratio. The scale may be 1 inch to 100 or even 1000 miles (1:6,336,000 or 1:63,360,000). By necessity and by design, such maps are very generalized.

Topographic Maps and Terrain Representation

When we speak of the topography of an area, we refer to its terrain. Topographic maps portray the surface features of relatively small areas, often with great accuracy (Figure 1.13). They not only show the elevations of landforms and of streams and other water bodies but also display features that people have added to the natural landscape. These may include transportation routes, buildings, and such land uses as orchards, vineyards, and cemeteries. Boundaries of all kinds, from state boundaries to field or airport limits, may be depicted on topographic maps.

The U. S. Geological Survey (U.S.G.S.), the chief federal agency for topographic mapping in this country, produces several map series, each on a standard scale. A small-scale topographic series, at 1:250,000, or 1 centimeter to 2.5 kilometers, is complete for the United States. Other series are at scales of 1:125,000, 1:62,500, and 1:24,000. A single map in one of these series is called a *quadrangle*. Topographic quadrangles at the scale of 1:24,000 exist for two-thirds of the United States, excluding Alaska.

The principal device used to show elevation on topographic maps is the *contour line,* along which all points are of equal elevation above a datum plane, usually mean sea level. Contours are imaginary lines, perhaps best thought of as the outlines that would occur if a series of progressively higher horizontal slices were made through a vertical feature. Figure 1.14 shows the relationship of contour lines to elevation for an imaginary island.

The contour interval is the vertical spacing between contour lines, and it is normally stated on the map. Contour intervals of 10 and 20 feet are often used, though in relatively flat areas the interval may be only 5 feet. In mountainous areas, the spacing between contours is greater, for graphic convenience. The more irregular the surface, usually, the greater is the number of contour lines that will need to be drawn; the steeper the slope, the closer are the contour lines rendering that slope.

Contour lines are the most accurate method of representing terrain, giving the map reader information about the elevation of any place on the map and the size, shape, and slope of all relief features. They are not truly pictorial, however. To heighten the graphic effect of a topographic map, contours are sometimes supplemented by the use of *shaded relief.* This method of representing the three-dimensional quality of an area is illustrated by Figure 1.15. An imaginary light source, usually in the northwest, can be thought of as illuminating a model of the area, simulating the appearance of sunlight and shadows. Portions that are in the shadow are darkened on the map.

Many symbols besides contour lines appear on topographic maps. The U.S.G.S. produces a sheet listing all the symbols it employs, and some older maps provide legends on the reverse side. On maps of cities, where it would be impossible to locate separately every building, the built-up area is denoted by a special tint, and only streets and public buildings are shown. (See the area of the Watchung School on Figure 1.13.) In the case of running water, separate symbols are used in the legend and on the map to depict perennial streams, intermittent streams, and springs; the location of rapids and falls is also noted. There are three different symbols for dams and three more for types of bridges.

The tremendous amount of information contained on topographic maps makes them useful to engineers, regional planners and land use analysts, and developers, as well as to hikers

23

24

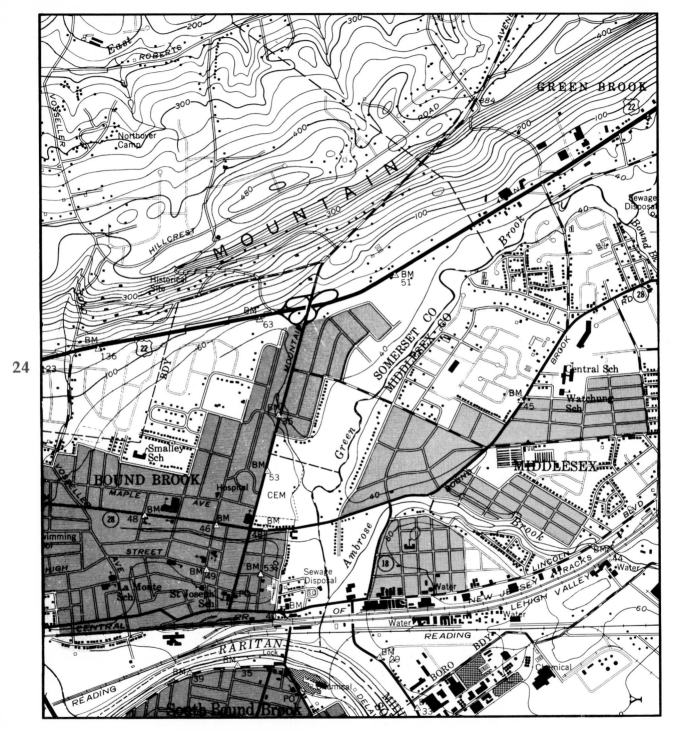

Fig. 1.13 A portion of the Bound Brook, New Jersey, 7.5 minute series of U. S. Geological Survey topographic maps. The fractional scale is 1:24,000.

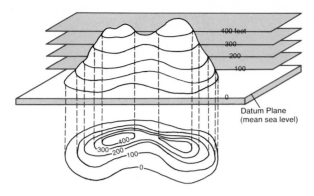

Fig. 1.14 Contours drawn for an imaginary island. The intersection of the landform by a plane held parallel to sea level is a contour representing the height of the plane above sea level.

and casual users. Given such a wealth of information, the experienced map reader can make deductions about both the physical character of the area and the economic and cultural use of the land.

When topographic maps were first developed it was necessary to obtain the data for them through fieldwork, a slow and tedious process. The technological developments that have taken place in aerial photography since the 1930s have made it possible to speed up production and greatly increase the land area represented on topographic maps. Aerial photography is only one of a number of remote sensing techniques now employed.

REMOTE SENSING, LANDSAT IMAGERY, AND THE ORTHOPHOTOMAP

Although *remote sensing* is a new term, coined less than 20 years ago, the process it describes—observation from a remote platform—has been going on for well over a century. Photographs were made from balloons and kites; use was even made of carrier pigeons wearing miniature cameras that made exposures automatically at set intervals. The air-

plane, first used for mapping in the 1930s, provided a platform for the camera and the photographer so that it was possible to take photographs from planned positions. Aerial photographs must, of course, be interpreted through the use of such clues as the size, shape, tone, and color of the recorded objects before maps can be made from them.

Aerial photography employing cameras with returned film is perhaps the most widely used remote sensing technique. Other sensing devices obtain images of an area by responding to wavelengths of light outside the visible spectral region. Infrared systems have proved useful for the classification of vegetation types, laser systems are employed for air pollution mapping, and radar systems can detect ships or aircraft and monitor the weather. Figure 2.13 is a radar image produced by radar equipment mounted on an airplane.

Mapping from the air has certain obvious advantages over surveying from the ground, the most evident being the bird's-eye view that the cartographer obtains. Using stereoscopic devices, the cartographer can determine the exact slope and size of features such as mountains, rivers, and coastlines. Areas that are otherwise hard to survey, such as mountains and deserts, can be mapped easily from the air. Furthermore, millions of square miles can be surveyed in a very short period of time. Maps based on aerial photographs can be made quickly and revised without difficulty, so that they are kept up to date. Thus, aerial photography has meant that the earth can be mapped more accurately, more completely, and more rapidly than ever before.

In the last 20 years, both manned and unmanned spacecraft have supplemented the airplane as the vehicle for terrain photography. Concurrently, many steps have been taken to automate mapping, including the use of electronic mapping techniques, automatic plotting devices, and automatic data processing. Many aerial photographs are now taken either from

25

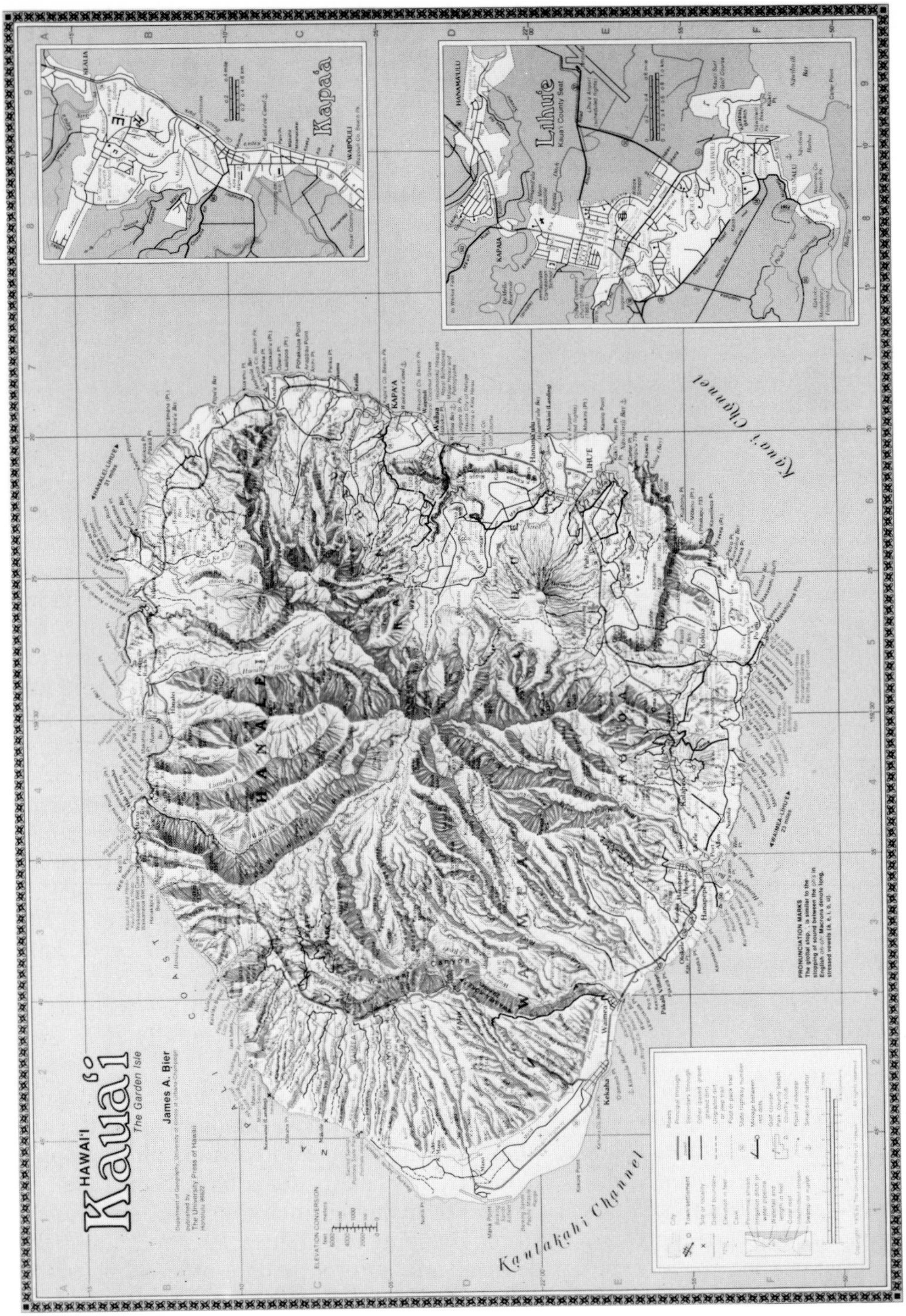

Fig. 1.15 A shaded relief map of the Island of Kaua'i, Hawaii. (Copyright © 1976 by the University Press of Hawaii.)

Every 103 minutes, day in, day out, an American earth resources satellite, **Landsat** Landsat 2, orbits the globe. It travels north to south at an altitude of 568 miles, crossing the equator at roughly the same time on each pass. Its built-in electronic sensors and multilensed cameras provide repetitive coverage of individual spots on earth every 18 days. It has been doing this since January 22, 1975.

Landsat sends back an endless stream of radio data that are converted into photographic images by the National Aeronautical and Space Agency (NASA). Copies of these pictures can be obtained by anybody, anywhere in the world, for a relatively modest fee. Facts about the distribution of vegetation, the progress of various crops, the state of soil moisture, rainfall, snow accumulation, and drainage can be established; fertility can be estimated, threatened floods and erosion factors guarded against. The reservoir buildup of remote headwaters and lakes harnessed to hydroelectric power generation can be swiftly estimated.

Landsat can be consulted by China, for example, if it wants a quick readout on how high the Yangtze River is liable to rise during the next flood season and to check its own surface or aerial reconnaissance observations. A considerable amount of research on potential droughts or floods has already been done from pre-Landsat sightings for the benefit of farmers in the Indus River basin. An earlier United States satellite also located a massive disturbance or fault in the earth's surface in the Central African Republic. Subsequent observations have indicated the existence of an enormous body of iron and possibly uranium ore.

Data from Landsat helped to identify an area of the Niger River basin where the appalling drought conditions afflicting the rest of the Sahel region had been resisted. Field investigators subsequently established that the tribes there practiced managed cattle grazing; communications being poor, their example had never been made known to others.

Landsat 2 has also been used to help forestry. Analysis of aerial color photos and computer readings have enabled lumbermen to select the most promising timber areas in remote sections of the Rocky Mountains. The satellite can tell them where the trees are ripe for cutting and where best to run in their logging trails, and do it quite cheaply. The U. S. Corps of Engineers recently carried out a national dam safety program relying on computer-aided mapping at the low cost of $75 per 1000 square miles.

Condensed by permission from Stephen Barber, "Watching Brief on the World," *Far Eastern Economic Review*, Feb. 25, 1977.

27

continuously orbiting satellites, such as Skylab and Landsat, or from manned spacecraft flights, such as those of Apollo and Gemini. Among the advantages of orbiting spacecraft are the speed of coverage and the fact that views of large regions can be obtained.

28

Fig. 1.16 (opposite) **Fig. 1.16 (opposite) Landsat imagery showing seasonal vegetation change in the Cloud Peak Primitive Area, Bighorn Mountains, in north central Wyoming. May (top) and September (bottom).**

Although remote sensing techniques have yielded maps of many scales, we shall mention just two of the programs sponsored by the federal government—one suitable for small-scale and the other for large-scale maps. The Landsat 1 and 2 satellites (formerly known as *ERTS*), launched in 1972, carry two sets of remote sensing instruments. Data are transmitted to receiving stations in the form of electronic signals. Nearly 80% of the earth's land areas have been imaged by Landsat. The images have been used to produce small-scale maps, which range from 1:250,000 to 1:1,000,000. The images can be adjusted to fit special map projections, thus permitting automated cartographic processing of the data. The Landsat program provides information on seasonal variation in vegetation and on changes in land use in areas as small as 100 acres and can be used to indicate in which areas existing maps need revision. Landsat 3 will reveal significant detail of items 50 meters on a side. Figure 1.16 shows Landsat images of the Cloud Peak, Wyoming, area, taken at different times of the year. A Landsat image of the Mississippi River delta appears in Figure 2.20.

In 1975 the U. S. Department of the Interior adopted the National Mapping Program to improve the collection and analysis of cartographic data and to prepare maps that would assist decision makers who deal with resource and environmental problems. One of the first goals of the program was to achieve complete coverage of the nation with orthophotographic imagery for all areas not already mapped at the scale of 1:24,000. An *orthophotomap* is an aerial photograph that is related to a grid system and to which certain map symbols (contours, boundaries, names) have been added. Note in Figure 1.17 that in contrast to the conventional topographic map, the photograph is the chief means of representing information. Ortho-

photomaps have a wide variety of uses, aiding in forest management, soil surveys, geological investigations, flood hazard and pollution studies, and city planning.

Patterns and Quantities on Maps

The study of the spatial pattern of things, whether people, cows, or traffic flows, is the essence of the locational tradition in geography, a subject explored in Part III of this book. Maps are used to record the location of these phenomena, and different kinds of techniques are employed to depict their presence or numbers at specific points, in given areas, or along lines.

QUANTITIES AT POINTS

A topographic map shows the location at points of many kinds of things, such as churches, schools, and cemeteries. Each symbol counts as one occurrence. Often, however, our interest is in showing the variation in the number of some things that exist at several points, for example, the population of selected cities, or the tonnage handled at certain terminals, or the number of passengers at given airports.

There are two chief means of symbolizing such distributions, as Figures 1.18 and 1.19 indicate. One method is to choose a symbol, usually a dot, to represent a given quantity of the mapped item (such as 50 people) and to repeat that symbol as many times as necessary. Such a map is easily understood because the dots give the map reader a visual impression of the pattern. Sometimes pictorial symbols—for example, human figures or oil barrels—are used instead of dots.

If the range of the data is great, the cartogra-

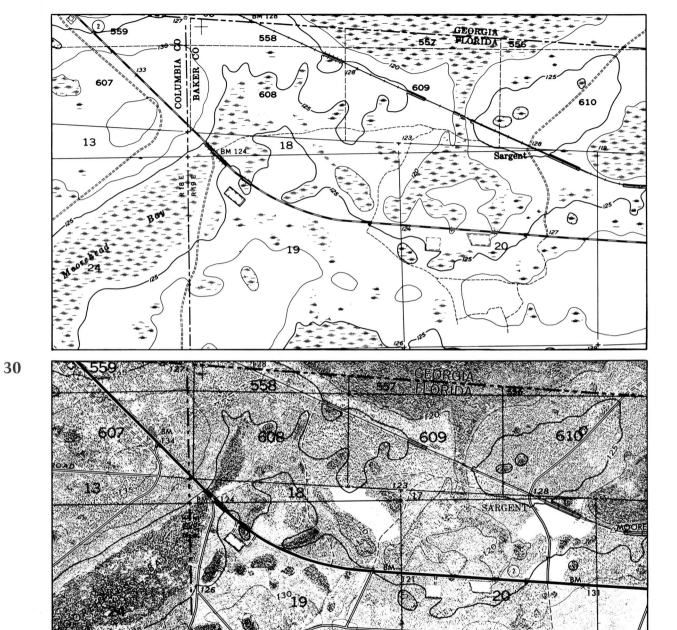

Fig. 1.17 Topographic map (top) and orthophotomap (bottom) of a portion of the Okefenokee Swamp. Orthophotomaps are well suited for portraying swamp and marshlands; flat, sandy terrain; and urban areas.

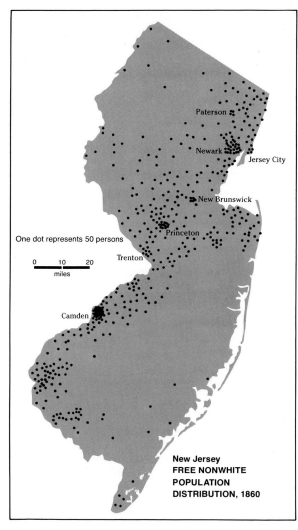

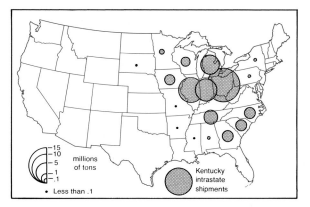

Fig. 1.19 Rail freight shipments of bituminous coal from Kentucky. On this map, the area of the circle is proportional to the tonnage of coal sent.

Fig. 1.18 The dot map provides a visual impression of specific locations and distributions. (Redrawn by permission from *Proceedings* of the Association of American Geographers, vol. 3, 1971, Peter O. Wacker.)

proportional symbols. The size of the symbol is varied according to the quantities represented. Thus, if bars are used, they can be shorter or longer as necessary. If squares or circles are used, the *area* of the symbol is proportional to the quantity shown (Figure 1.19). There are occasions, however, when the range of the data is so great that even circles or squares would take up too much room on the map. In such cases, three-dimensional symbols, usually spheres or cubes, are used; their *volume* is proportional to the data. Unfortunately, many people fail to perceive the added dimension implicit in volume, and most cartographers do not recommend the use of such symbols (Figure 1.20).

31

QUANTITIES IN AREA

Maps showing quantities in area fall into two general categories: those portraying the areas within which a kind of phenomenon occurs, and those showing amounts by area. Atlases contain numerous examples of the first category, such as patterns of religions, languages, political entities, vegetation, or types of rock. Normally, different colors or patterns are used

pher may find it inconvenient to use a repeated symbol. For example, if one port handles 50 or 100 times as much tonnage as another, that many more dots would have to be placed on the map. To circumvent this problem, the cartographer can choose a second method and use

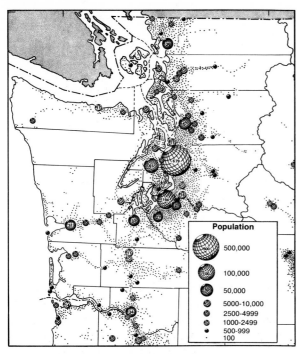

Fig. 1.20 A portion of a map showing the distribution of Washington State population in 1950. The cartographer, J. Michel, used proportional spheres to represent large urban population concentrations.

for different areas, as in Figure 1.21. Among the problems involved in using such maps are that (1) they give the impression of uniformity to areas that may actually contain significant variations; (2) boundaries attain unrealistic precision and significance, implying abrupt changes between areas when, in reality, the changes may be gradual; and (3) unless colors are chosen wisely, some areas may look more important than others.

Variation in the amount of a phenomenon from area to area may be shown by covering subareas with different shades, colors, or patterns. Because the use of different colors or pat-

32

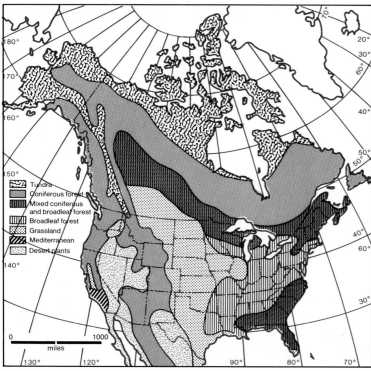

Fig. 1.21 Natural vegetation of North America. Maps such as this one may give the false impression of uniformity within a given area, for example, that grasses and no other vegetation cover the grassland areas. Such maps are intended to represent only the predominant naturally occurring vegetation. (Redrawn with permission from Don R. Hoy, ed., *Geography and Development*, copyright © 1978, Macmillan Publishing Co., Inc.)

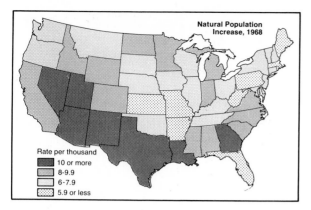

Fig. 1.22 Rate of natural population increase by state, 1968. Quantitative variation by area of a phenomenon is more easily visualized in map than in tabular form. (Redrawn from R. Estall, *A Modern Geography of the United States*, 2nd ed., 1976, Pelican Books, p. 33. © R. Estall, 1972, 1976. Reprinted by permission of Penguin Books Ltd.)

terns may suggest qualitative rather than quantitative differences—that is, differences in kind rather than in degree—the best cartographic practice is to use a single hue and vary its value or intensity. Figure 1.22 is an example of such a map.

An interesting variation of these maps is the

Fig. 1.23 On this cartogram, the areas of states, provinces, and cities have been made proportional to the populations they contained in 1940. (Redrawn from a map prepared by the General Foods Corporation, 1946; with permission.)

cartogram, a map simplified to present a single idea in a diagrammatic way. Railroad and subway route maps commonly take this form. Depending on the idea that the cartographer wishes to convey, the sizes and shapes of areas may be altered, distances and directions may be awry, and contiguity may or may not be preserved. In Figure 1.23 the sizes of the areas represented have been purposely distorted so that they are proportional to the quantities shown. Contiguity has not been totally preserved in the case of one or two mountain states.

LINES ON MAPS

Some lines on maps do not have numerical significance. The lines representing rivers, political boundaries, roads, and railroads, for example, are not quantitative. They are indicated on maps by such standardized symbols as the ones shown here.

33

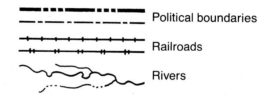

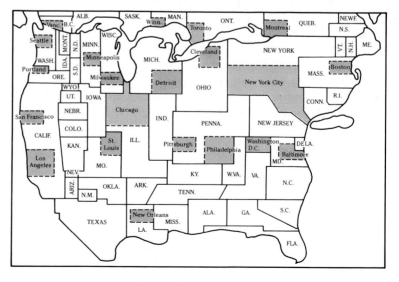

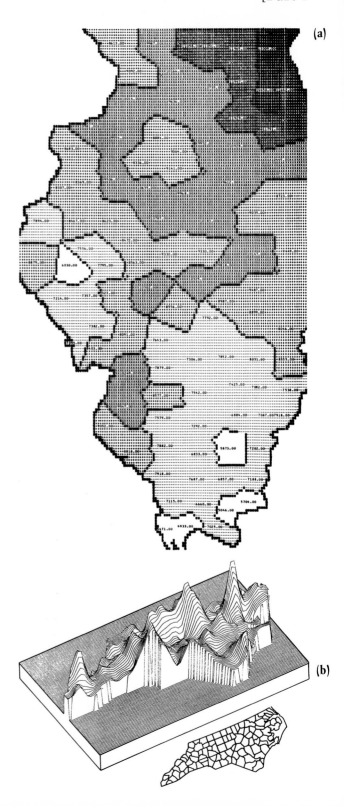

Fig. 1.24 A flow-line map showing annual average interregional migration in the United States, 1955–1965. (Redrawn from R. Estall, *A Modern Geography of the United States*, 2nd ed., 1976, Pelican Books, p. 40. © R. Estall, 1972, 1976. Reprinted by permission of Penguin Books Ltd.)

Often, however, lines on maps do denote specific numerical values. Contour lines that connect points of equal elevation above mean sea level are a kind of *isoline,* or line of constant value. Other examples of isolines are isohyets (equal rainfall), isotherms (equal temperature), and isobars (equal barometric pressure). The implications of isolines as regional boundaries are discussed in Chapter 12.

Flow-line maps are used to portray traffic or commodity flows along a given route, usually a waterway, a highway, or a railway. The location of the route taken, the direction of movement, and the amount of traffic can all be depicted. The amount shown may be either the total or a per-mile figure. In Figure 1.24, the width of the line is proportional to the amount of population movement.

Computer Maps

One of the most recent developments in cartography, and one that holds great promise, is the use of computers to make maps. There are several different methods of programming

34

Fig. 1.25 (opposite) (a) A computer map representation of median family income in 1970 for the State of Illinois. (b) A three-dimensional representation of the number of day-care spaces per 1000 population, 1975, by counties in North Carolina. (a: Spatial Data Analysis Laboratory, Department of Geography, University of Illinois at Urbana-Champaign; b: courtesy of Marc Armstrong.)

computers to draw maps as well as different techniques for the actual production of the maps. All of these methods share the primary advantage of computer cartography—the speed with which maps can be produced. Thousands of facts can be analyzed and a map produced in less than a minute. This characteristic of the computer is especially important for the researcher or planner who needs to analyze many variables simultaneously, often with advanced quantitative techniques. Computer cartography makes possible the production of maps that were virtually impossible to draw 20 years ago. Figure 1.25 shows two examples of computer-drawn maps.

Conclusion

In this chapter, we have not attempted to discuss all aspects of the field of cartography. Mapmaking and map design, systems of land survey, map compilation, and techniques of map reproduction are among the topics that have been deliberately omitted. Our intent has been to introduce those aspects of map study that will aid in map reading and interpretation.

A discussion of cartography is the natural entrance way to the study of geography, for there is no more useful manner of examining the earth than through maps. Indeed, it may properly be said that maps are as indispensable to the geographer as words, photographs, or quantitative techniques of analysis. The number of maps contained in this book amply illustrates this point. As you read the remainder of the book, note their many different uses. For example, notice in Chapter 2 how important maps are to your understanding of the theory of continental drift; in Chapter 6, how maps aid geographers in identifying cultural regions; and in Chapter 8, how behavioral geographers use maps to record people's perceptions of space.

Geographers are not unique in their dependence on maps. People involved in the analysis and solution of many of the problems facing the world also rely on them. Environmental protection, control of pollution, conservation of natural resources, land use planning, and traffic control are just a few of the issues that call for the accurate representation of elements on the earth's surface.

35

For Review

1. List at least five properties of the globe grid. Examine the projections used in Figures 1.4, 1.8, and 1.9; in what ways do each of these projections adhere to or deviate from globe grid properties?

2. What is meant by *map scale*? In what ways may scale be stated?

3. Convert the following map scales into their verbal equivalents.

$$1:1,000,000 \qquad 1:63,360 \qquad 1:12,000$$

4. What is meant by *prime meridian*? What is the purpose of the prime meridian and how is it determined?

5. What happens to the length of a degree of longitude as one approaches the poles? What happens to a degree of latitude between the equator and the poles?

6. From a world atlas, determine, in degrees and minutes, the locations of New York City; Moscow, U.S.S.R.; Sydney, Australia; and your hometown.

7. What is meant by *conformal?* By *equivalent?* By *azimuthal?*

8. What is a *contour line?* What is the contour interval on Figure 1.13? On Figure 1.14?

9. Examine Figure 1.13; imagine that you are on the top floor of the Central School looking in a northwesterly direction. Draw a sketch or write a description of what you would see.

10. The table below gives the gross national product per capita of selected European countries (U. S. $, 1975). On outline maps or as cartograms, represent the data in two different ways.

Austria	4,274	Greece	1,927	Portugal	1,333
Belgium	5,443	Ireland	2,367	Spain	2,080
Denmark	5,877	Italy	2,862	Sweden	7,428
France	5,773	Luxembourg	5,389	Switzerland	6,975
Germany		Netherlands	4,965	United Kingdom	3,583
(Fed. Rep.)	6,144	Norway	5,818		

36

Suggested Readings

BROWN, LLOYD A., *The Story of Maps.* Little, Brown, Boston, 1949.

KEATES, J. S., *Cartographic Design and Production.* Longman Group Ltd., London, 1973.

MONMONIER, MARK S., "Maps, Distortion, and Meaning." Resource Paper No. 75-4, Association of American Geographers, Washington, D.C., 1977.

MUEHRCKE, PHILLIP C., *Map Use—Reading, Analysis and Interpretation.* J P Publications, Madison, WI, 1978.

The National Atlas of the United States of America. U. S. Department of the Interior, Geological Survey, Washington, D.C., 1970.

ROBINSON, ARTHUR H., RANDALL D. SALE, AND JOEL MORRISON, *Elements of Cartography* (4th ed.). Wiley, New York, 1978.

RUDD, ROBERT, *Remote Sensing: A Better View.* Duxbury Press, North Scituate, MA, 1974.

THROWER, NORMAN J. W. (ed.), *Man's Domain, A Thematic Atlas of the World.* McGraw-Hill, New York, 1970.

TYNER, JUDITH, *The World of Maps and Mapping.* McGraw-Hill, New York, 1973.

Before the morning of November 14, 1963, cartographers thought they had finished their work in Iceland. They certainly had been at it long enough—from the island's first appearance on a world map by Eratosthenes some two centuries before Christ. It was a well-mapped area early in the Middle Ages and could be set aside as one of the certainties of the North Atlantic. Then, with a roar, a pillar of steam and fire, and a hail of ash, nature demanded that the cartographers go back to work. Surtsey, the Black One—god of fire and destruction—rose from the sea 20 miles off the Icelandic coast to become new land—1 mile long, 600 feet high, 670 acres of rock, ash, and lava. Cartographers had another job—as would have been their lot had they been around for, say, the last 100 million years. Things just won't stand still. And not only little things, like new islands, or big ones, like mountains rising and being worn low to swampy plains, but even monstrous things: continents that wander about like homeless derelicts, and ocean basins that expand, contract, and split up the middle like worn-out coats. It is a fascinating story, this formation and alteration of the home of humans, which, from their instantaneous view, seems so eternal and unchanging.

Geologic time is long, but the forces that give shape to the land are timeless and constant. Processes of creation and destruction are continually at work to fashion the seemingly eternal structure upon which humankind lives and works. Three types of forces interact to produce those infinite local variations in the surface of the earth called *landforms:* (1) forces that push, pull, move, and raise the earth; (2) forces that scour, scrape, wash, and chip away the land; and (3) forces that deposit soil, pebbles, rocks, and sand in new areas. How long these processes have worked, how they work, and their effects are the subject of this chapter.

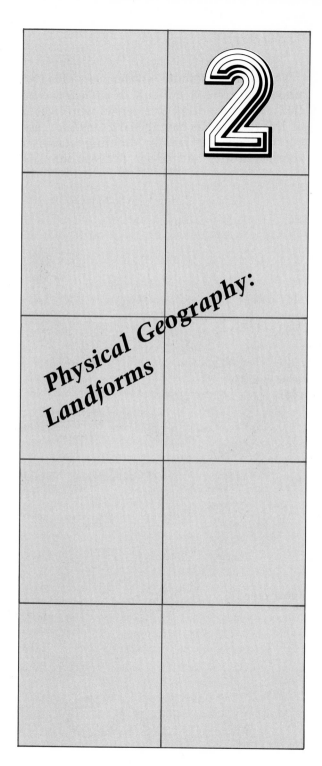

2

Physical Geography: Landforms

Geologic Time

The earth is about 4.7 billion years old. That amount of time is difficult to envisage, especially when we think of a person who lives to be 100 years old as having had a long life, passing through the greatly differing stages of youth, maturity, and old age. Because our usual concept of time is dwarfed when we speak of billions of years, it is useful to compare the ages of the earth with something that is more familiar.

Imagine that the height of the World Trade Center in New York City represents the age of the earth. The twin towers are 1353 feet (412 m) high. Even the thickness of an average piece of paper laid on top of the roof would be

Fig. 2.1 A diagrammatic history of the earth. The sketches depict some of the known characteristics of the named geologic periods. (From *Geologic Time*, U. S. Geological Survey.)

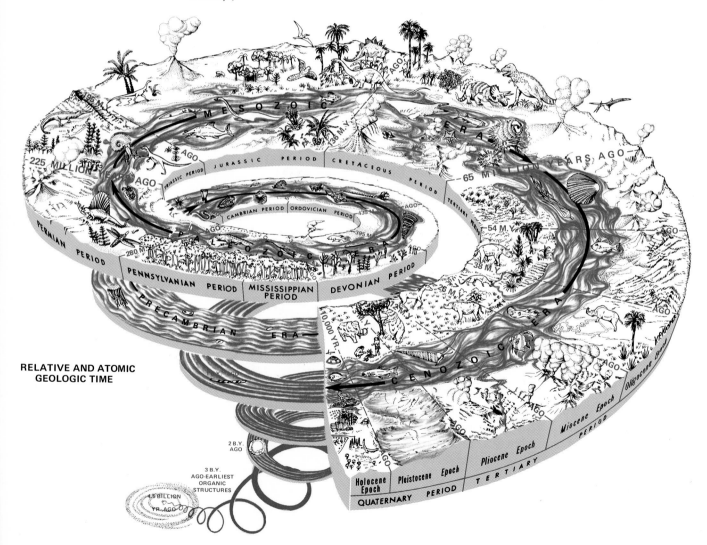

RELATIVE AND ATOMIC
GEOLOGIC TIME

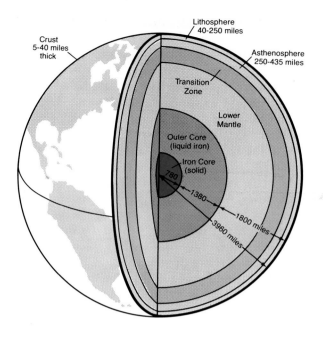

Crust
5-40 miles
thick

Lithosphere
40-250 miles

Asthenosphere
250-435 miles

Transition
Zone

Lower
Mantle

Outer Core
(liquid iron)

Iron Core
(solid)

780
1380
1800 miles
3960 miles

Fig. 2.2 The very thin crust of the earth overlies a layered planetary interior. Zonation of the earth occurred early in its history. Radioactive heating melted the original homogeneous planet. A dense iron core settled to the center, a crust of lighter rock solidified at the surface, and the remnant lower mantle—overlain by a transition zone and the asthenosphere—formed between them. Escaping gases eventually created the atmosphere and the oceans.

too great to represent an average person's lifetime. The height of 6 of the 110 stories would represent the 200 million years that have elapsed since the present ocean basins began to form. The first mammals made their appearance on earth 60 million years ago, or the equivalent of the height of 2 stories. Earth history is so long and involves so many major events that geologists have divided it into a se-

Fig. 2.3 The large lithospheric plates move as separate entities and collide along subduction zones. Assuming the African plate to be stationary, relative plate movements are shown by arrows. Sea floor spreading, the triggering mechanism, takes place along the axes of the ridges. (From "Plate Tectonics" by John F. Dewey. Copyright © 1972 by Scientific American, Inc. All rights reserved.)

39

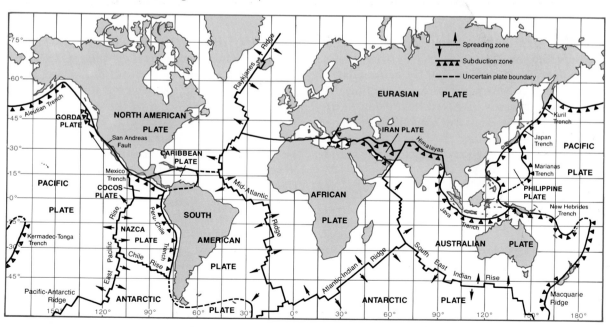

Continental Drift and Plate Tectonics

The idea that the sea floor spreads and continents move has forced us to rethink our conception of the earth as stable and permanent. Only in the last decade has the theory of plate tectonics gained wide acceptance among scientists, although a basic element of the theory—the drifting of continents—was proposed by European geologists more than a century ago.

The man whose name is generally associated with the theory of *continental drift*, Alfred Wegener (1880–1930), believed that the present continents were once united in one supercontinent, to which he gave the name *Pangaea*. In the early decades of the 20th century, he gathered evidence to support his view, but largely because conflicting evidence could be presented and because he and his supporters could not suggest a mechanism to account for the drifting, the idea of continental drift was generally rejected. Indeed, if you were to refer to an earth-science text published before 1960, you would probably find no mention of the theory. In the last decade, however, much evidence has been accumulated to support the even broader theory of *plate tectonics*—evidence related to magnetic patterns, seismic activity, and the age of the sea floor. Additionally, a mechanism—sea-floor spreading resulting from convection currents—has been suggested as the force responsible for the movement of continents. In a short period of time, new evidence and new ways of rethinking old knowledge have led to the wide acceptance by earth scientists of the idea of moving continents.

ries of recognizable, distinctive stages; these are depicted on Figure 2.1

Landforms are ever so slightly being created and destroyed even while you are reading this page. The processes involved have been in operation for hundreds of millions of years; what they have done to shape the landscape you see from your window is very recent by the standards of geologic time.

The Movement of Continents

The landforms that the cartographer maps and on which we live are only the surface features of a thin, floating shell of rock, the earth's crust. Above the core and the lower mantle of the earth, there is a partially molten plastic layer called the *asthenosphere* (Figure 2.2). The asthenosphere supports a thin but strong solid shell of rocks, the *lithosphere.* Resting on the lithosphere is the crust of the earth, which is partly oceanic and partly continental.

The lithosphere is broken into about 10 large plates, each of which, according to the theory of *plate tectonics*, slides or drifts very slowly over the molten asthenosphere. A plate may contain some oceanic and some continental crust. Figure 2.3 shows that the North American plate, for example, contains the North Atlantic Ocean and most, but not all, of North America. (The Caribbean Islands, Mexico, and a bit of California are on other plates.)

Scientists are not certain why the lithospheric plates move. One reasonable theory suggests that molten material raised by heat from the core of the earth is acted upon by the gravitational force of the sun and the moon. In this way, the plates are set in motion. In any case, strong evidence exists that about 200 million years ago the entire continental crust was connected in one supercontinent, to which the name *Pangaea* ("all earth") has been given. Pangaea was broken up as the sea floor began to spread; one crack widened to become the Atlantic Ocean (Figure 2.4).

Fig. 2.4 The drifting of continents. Two hundred million years ago the continents were connected as one large land mass called Pangaea. After they split apart, the continents moved to their present positions. Notice how India broke away from Antarctica and collided with the Eurasian land mass. The Himalayas were formed at the zone of contact.

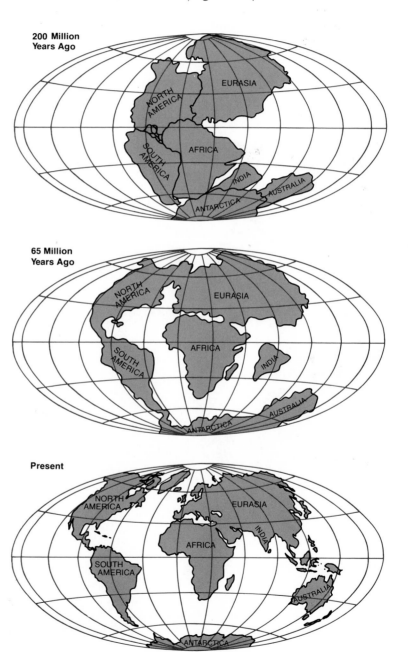

200 Million Years Ago

65 Million Years Ago

Present

41

Materials from the asthenosphere have been rising slowly along the Atlantic Ocean fracture, and as a result, the sea floor has continued to spread. The Atlantic Ocean has widened to about 3000 miles. If it widened by 1 inch per year, as scientists have estimated, one could calculate that the separation of the continents did in fact begin about 200 million years ago.

Notice on Figure 2.5 how the ridge line that makes up the axis of the ocean runs parallel to the eastern coast of North and South America and the western coast of Europe and Africa. Scientists were led to the theory of the *continental drift* of lithospheric plates by the amazing fit of the continents.

Collisions occurred as the lithospheric plates

Fig. 2.5 The configuration of the Atlantic Ocean floor is evidence of the dynamic forces shaping continents and ocean basins. This artist's rendering, based upon deep-sea soundings, is reproduced through the courtesy of the Aluminum Company of America (Alcoa).

42

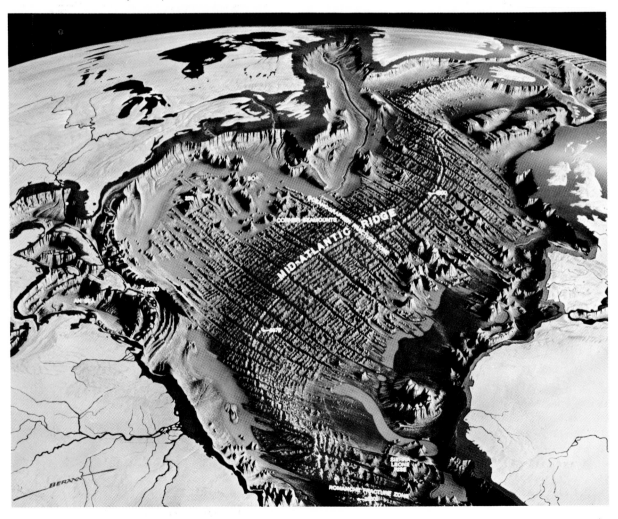

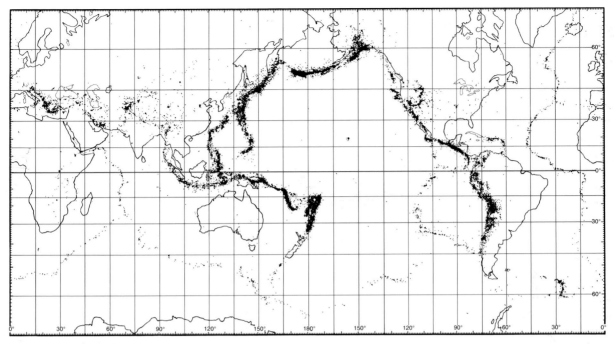

Fig. 2.6 The location of near-surface earthquakes for the period 1962–1967. A comparison of this map with Figure 2.3 shows that most earthquakes occur close to or along plate boundaries. (Locations by U. S. Coast and Geodetic Survey; computer plot by M. Barazangi and J. Dorman, Columbia University.)

moved. The pressure exerted at the intersections of plates resulted in earthquakes, which over periods of many years combined to change the shape and the features of landforms. Figure 2.6 shows the location of near-surface earthquakes for a recent time period. Comparison with Figure 2.3 illustrates that the areas of greatest earthquake activity are at plate boundaries.

The famous San Andreas fault of California is a part of a long fracture separating two lithospheric plates, the North American and the Pacific. Earthquakes occur along faults when the tension and the compression at the junction become so great that only an earth movement can release the pressure. Because the Atlantic Ocean is still widening at the rate of about 1 inch per year, earthquakes must occur from time to time to relieve the stress along the tension zone in the mid-Atlantic and along the other fracture lines, such as the San Andreas fault.

Despite the availability of scientific knowledge about earthquake zones, the general disregard for this danger is a difficult cultural phenomenon with which to deal. Every year there are hundreds, and sometimes hundreds of thousands, of casualties resulting from inadequate preparation for earthquakes (Figure 2.14). In some well-populated areas, the chances that damaging earthquakes will occur are very great. The distribution of earthquakes shown in Figure 2.6 implies the potential dangers to densely settled areas of Japan, the Philippines, parts of Southeast Asia, and the western rim of the Americas.

Movement of the lithospheric plates results in the formation of deep sea trenches and continental-scale mountain ranges, as well as earthquakes. The continental crust is made up of lighter rocks than is the oceanic crust. Thus, where plates with different types of crust at their edges push against each other, there is a tendency for the denser oceanic crust to be forced down into the asthenosphere, causing long and deep trenches to form below the ocean. This type of collision is termed *subduction* (Figure 2.7). The edge of the overriding continental plate is uplifted to form a mountain chain that runs close to, and parallel with, the offshore trench.

Most of the Pacific Ocean is underlain by a plate that, like the others, is constantly pushing and being pushed. The continental crust on adjacent plates is being forced to rise and fracture, making an active earthquake and volcano zone of the rim of the Pacific Ocean. In recent years major earthquakes have occurred in Peru, Central America, Alaska, China, and the Philippines. Many scientists believe that a damaging earthquake will occur along the San Andreas fault in the near future.

Zones around plate intersections not only are susceptible to readjustments in the litho-

sphere, causing earthquakes, but also have areas of weakness where molten material from the asthenosphere may find its way to the surface through a crack or break (fault). The molten material (magma) may explode out of a volcano or ooze out of cracks. Later in this chapter, we will return to the discussion of volcanic and other forces that contribute to the shaping of landforms.

Earth Materials

The rocks of the earth's crust derive their character from their various components. Among the more common varieties of rock are granites, basalts, limestones, sandstones, and slates. Each kind of rock is made up of particles that contain various combinations of such common elements as oxygen, silicon, aluminum, iron, and calcium, together with less-abundant elements. A particular chemical combination that has hardness, density, and its own definite crystal structure is called a *mineral*. Some well-known minerals are quartz, feldspar, and silica. Depending on the nature of the minerals that form them, rocks may be hard or soft, dense or open, one color or another, chemically stable or not. While some rocks resist decomposition, others are very easily broken down.

Although one can classify rocks according to their physical properties, the more common approach is to classify them by the way they evolved. The three main groups of rocks are igneous, sedimentary, and metamorphic.

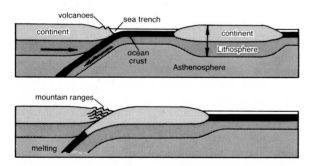

Fig. 2.7 The process of *subduction*. When lithospheric plates collide, the heavier oceanic crust is usually forced beneath the lighter continental material. Deep-sea trenches, mountain ranges, volcanoes, and earthquakes are found along the plate collision lines.

IGNEOUS ROCKS

Igneous rocks are formed by the cooling and hardening of earth material. Weaknesses in the crust give molten material from the asthenosphere an opportunity to find its way into or

44

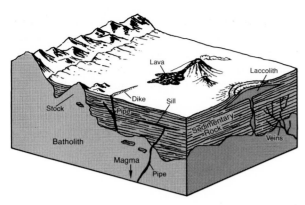

Fig. 2.8 Extrusive and intrusive forms of volcanism. Lava and ejecta (ash and cinders) are extrusions of rock material onto the earth's surface in the form of cones or horizontal flows. *Batholiths* **and** *laccoliths* **are irregular masses of crystalline rock that have cooled slowly below the earth's surface (intrusions).**

onto the crust. When the molten material cools, it hardens and becomes rock. The name for underground molten material is *magma;* above ground it is *lava.* The landforms that are created by hardened magma derive from *intrusive* (below-ground) rock formations, while hardened lava makes up *extrusive* (above-ground) formations (Figure 2.8).

The chemicals making up lava and magma are fairly uniform, but depending on the speed of cooling, different minerals form. Because it is not exposed to the coolness of air, magma hardens slowly, allowing silicon and oxygen to unite and form quartz, a hard, dense mineral. With other components, grains of quartz combine to form the rock called granite.

The lava that oozes out onto the earth's surface, if composed of basic minerals, makes up the rock known as basalt. If, however, the lava erupts from a volcano crater, it may cool very rapidly, allowing air spaces to form in the resulting rock. Some of the rocks formed in this manner are light and angular, such as pumice. Some may be dense, even glassy, as is obsidian.

SEDIMENTARY ROCKS

Sedimentary rocks are made up of particles of gravel, sand, silt, and clay that were eroded from already existing rocks. Surface waters carry the sediment to oceans, marshes, lakes, or tidal basins. Compression of these materials by the weight of additional deposits on top of them, and a cementing process brought on by the chemical action of water and certain minerals, causes sedimentary rock to form.

Sedimentary rocks evolve under water in horizontal beds called *strata.* Usually one type of sediment collects in a given area. If the particles are large—for instance, the size of gravel—a gravelly rock called conglomerate forms. Sand particles are the base for sandstone, while silt and clay form shale or siltstone.

Sedimentary rocks also derive from organic material, such as coral, shells, and marine skeletons. These materials settle into beds in shallow seas and congeal, forming limestone. If the organic material is mainly vegetation, it, too, can develop into a sedimentary rock called coal.

Sedimentary rocks vary considerably in color (from coal black to chalk white), hardness, density, and resistance to chemical decomposition. Large parts of the continents contain sedimentary rocks; nearly the entire eastern half of the United States is overlain with these rocks, for example. Such formations indicate that in the geologic past, seas covered an even larger proportion of the world than they do today.

METAMORPHIC ROCKS

Metamorphic rocks are formed from igneous and sedimentary rocks by earth forces that generate heat, pressure, or chemical reaction. The word *metamorphic* means "changed

45

shape." The internal earth forces that cause the movement and collision of lithospheric plates may be so great that by heat and pressure, the mineral structure of a rock changes, forming new rocks. For example, under great pressure, shale, a sedimentary rock, becomes slate, a rock with different properties. Limestone under certain conditions may become marble, and granite may become gneiss. Materials met-

amorphosed at great depth and exposed only after overlying surfaces have been slowly eroded away are among the very oldest rocks known on earth. Like igneous and sedimentary rocks, however, their formation is a continuing process.

From these raw materials, landforms are shaped by sets of both reinforcing and opposing processes.

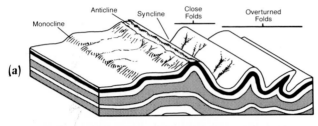

(a)

Fig. 2.9 Degrees of folding vary from slight undulations of strata with little departure from the horizontal to highly compressed or overturned beds. (a) Diagram of stylized forms of folding. (b) Strong asymmetrical folding. The investigator's hand at the left suggests the thickness of the distorted strata. (Photo by C. S. Alexander.)

(b)

Tectonic Forces

Tectonic processes shaping the earth's crust are of two types: *diastrophism* and *volcanism*. Diastrophism is the force that folds, faults, twists, or compresses rocks. Volcanism is the force that transports heated material to or toward the surface of the earth. When continents move, changes occur in the earth's crust. These changes can be as simple as the cracking of a rock or as dramatic as lava exploding from the crater of a volcano.

DIASTROPHISM

In the process of continental drift, pressures build in various parts of the earth's crust, and slowly, over thousands of years, the crust is transformed. By studying rock formations, geologists are able to trace the history of the development of a region. Over geologic time, most continental areas have been subjected to both tectonic and gradational activity—to building up and tearing down. They usually have a complex history of folding, faulting, leveling, deposition of new material, and then more folding, faulting, and leveling. Some flat plains in existence today may hide a history of great mountain development in the past.

Folding. When the pressure caused by moving continents is great, layers of rock (usually sedimentary) are forced to buckle. The result may be a warping or bending effect, and a ridge or a series of parallel ridges or folds may develop. If the stress is pronounced, great wavelike folds form (Figure 2.9). The folds can be thrust upward or laterally many miles. The Appalachian ridges of the eastern United States are at present low parallel mountains (1000–3000 feet above sea level), but the rock evidence suggests that the tops of the present mountains were once the valleys between 30,000-foot (9100-m) crests (Figure 2.10).

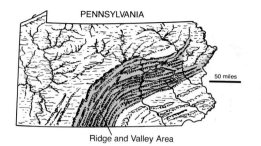

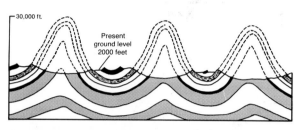

Fig. 2.10 The ridge and valley region of Pennsylvania, now eroded to hill lands, is the relic of 30,000-foot (9100-m) folds that were reduced to form synclinal (downarched) hills and anticlinal (uparched) valleys. The rock in the original troughs, having been compressed, was less susceptible to erosion.

47

Faulting. A fault is a break or fracture in rock. The stress causing a fault results in displacement of the earth's crust along the fracture line. Figure 2.11 depicts diagrammatic examples of fault types. There may be uplift on one side of the fault or downthrust on one side, or uplift on one side with downthrust on the other. In some cases, a steep slope known as a fault *escarpment*, which may be several hundred miles long, is formed. The stress can push one side up over the other side, or a separation away from the fault may cause sinking of the land and create a rift valley (Figure 2.12).

Many faults are merely cracks (called *joints*) with little noticeable movement along them, but in other cases, mountains such as the Sierra Nevada of California have risen as the result of faulting. In some instances, the movement has been horizontal along the surface rather than up or down. The San Andreas fault, pictured in Figure 2.13, is such a case.

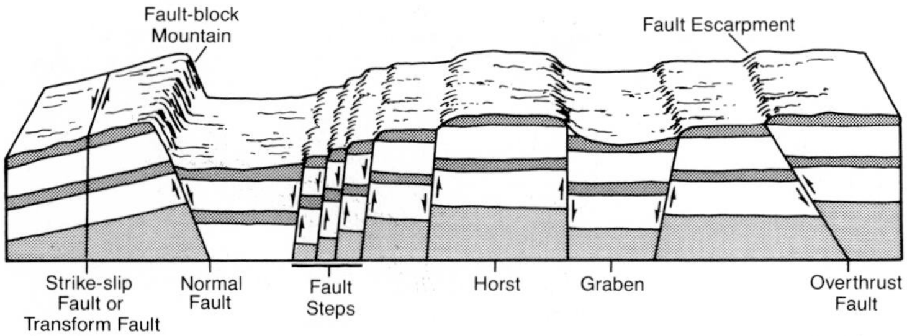

Fault-block Mountain

Fault Escarpment

Fig. 2.11 Faults, in their great variation, are common features of mountain belts where deformation is great. The different forms of faulting are categorized by the direction of movement along the plane of fracture.

Strike-slip Fault or Transform Fault   Normal Fault   Fault Steps   Horst   Graben   Overthrust Fault

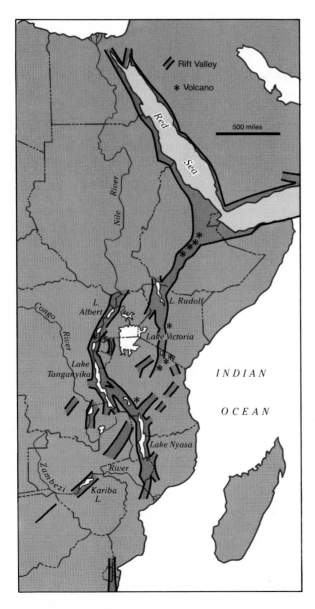

Rift Valley

* Volcano

500 miles

Red Sea

Nile River

Congo River

L. Albert

L. Rudolf

Lake Victoria

Lake Tanganyika

INDIAN

OCEAN

Lake Nyasa

Zambezi River

Kariba L.

48

Earthquakes. Whenever movement occurs along a fault, an earthquake results. The greater the movement, the greater the magnitude of the earthquake. Tension builds in the rock as stress is applied to it, and when after a period of time the critical point is reached, a slippage or earthquake takes place. The earthquake that occurred in Alaska on Good Friday in March 1964 was one of the strongest known. Although the stress point of that earthquake was 75 miles from Anchorage, vibrations called *seismic waves* caused earth movement in the weak clays under the city. Sections of Anchorage literally slid downhill, and part of the business district dropped 10 feet (3 m). Table 2.1 indicates the kinds of effects associated with earthquakes of different magnitudes, and Figure 2.14 depicts various types of earthquake-induced damage.

The seismic waves may generate a sea wave called a *tsunami* (often people use the incorrect name *tidal wave*) that, though not noticeable on the open sea, may become 30 or more feet high as it approaches land. Thus places thousands of miles from the earthquake may be affected by tsunamis. The islands of Hawaii now have a tsunami warning system that was de-

Fig. 2.12 Great fractures in the earth's crust resulted in the creation, through subsidence, of an extensive rift valley system in East Africa. The parallel faults, some reaching more than 2000 feet (610 m) below sea level, are bordered by steep walls of the adjacent plateau, which rises to 5000 feet (1500 m) above sea level and from which the structure dropped.

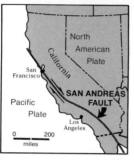

Fig. 2.13 View along the San Andreas fault in California, in the area of the San Francisco peninsula. A transform fault, the San Andreas marks a part of the slipping boundary between the Pacific and the North American plates. The map shows the relative southward movement of the American plate; dislocation averages about 1 cm per year. (N.A.S.A. radar imagery.)

veloped following the devastation at Hilo in 1946.

Earthquakes occur daily in any number of places throughout the world. Most are slight and are noticed only by those observing seismographs, the instruments that record earthquake vibrations, but from time to time there are large-scale stresses, as evidenced by the 1976 earthquakes in Guatemala and China. Most recent earthquakes have taken place on the rim of the Pacific (Figure 2.6), where stress

49

Table 2.1 Richter Scale of Earthquake Magnitude

Magnitude[a]	Characteristic Effects of Earthquakes Occurring near the Earth's Surface[b]
0	not felt
1	not felt
2	not felt
3	felt by some
4	windows rattle
5	windows break
6	poorly constructed buildings destroyed; others damaged
7	widespread damage; steel bends
8	nearly total damage
9	total destruction

[a] Since the Richter scale is logarithmic, each increment of a whole number signifies a 10-fold increase in magnitude. Thus a magnitude 4 earthquake produces a registered effect upon the seismograph 10 times greater than a magnitude 3 earthquake.
[b] The damage levels of earthquakes are presented in terms of the consequences that are felt or seen in populated areas; the recorded seismic wave heights remain the same whether or not there are structures on the surface to be damaged. The actual impact of earthquakes upon humans varies not only with the severity of the quake and such secondary effects as tsunamis or landslides but also with the density of population in the area affected.

Fig. 2.14 Some buildings destroyed by the earthquake in Managua, Nicaragua, in 1972. The quake destroyed or seriously affected several thousand buildings of all sizes. (United Nations/Jerry Frank.) Agua Caliente Bridge (opposite), a link in Guatemala's supply route to the Atlantic, collapsed during the severe earthquake that occurred in February 1976. (United Nations/Sygma/J. P. Laffont.)

from the outward-moving lithospheric plates is greatest. The Aleutian islands of Alaska experience many severe earthquakes each year; fortunately, the islands are only sparsely populated.

The Tsunami   A tsunami follows any submarine earthquake that causes fissures or cracks in the earth's surface. Water rushes in to fill the depression caused by the falling away of part of the ocean bottom; the water then moves outward, building in momentum and rhythm, as swells of tremendous power. The waves that hit Hawaii following the April 1, 1946, earthquake off Dutch Harbor, Alaska, were moving at approximately 400 miles an hour, with a crest-to-crest spacing of some 80 miles.

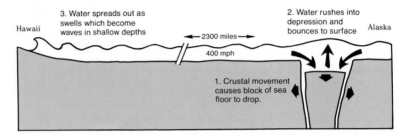

The long swells of a tsunami are largely unnoticed in the open ocean; only when the wave trough scrapes sea bottom in shallow offshore areas does the water pile up into precipitous peaks. The seismic sea waves at Hilo on the exposed northeast part of the island of Hawaii were estimated at between 45 and 100 feet in height. The water smashed into the city, deposited 14 feet of silt in its harbor, left fish stranded in palm trees, and resulted in 173 deaths and $25 million in damages.

VOLCANISM

The second tectonic force is volcanism. The most likely places through which molten material can move toward the surface are at the intersections of plates, but other fault-weakened zones are also subject to volcanic activity. The molten material is called *magma* when it remains below the surface and *lava* when it spews onto the earth's surface.

If sufficient internal pressure forces the magma upward, weaknesses in the crust, or faults, enable molten materials to reach the surface, as indicated in Figure 2.15. The active volcanic belt of the world coincides with the active earthquake and active fault zones. The molten material can either flow smoothly out

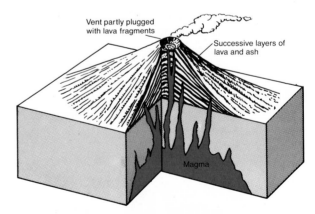

Fig. 2.15 Sudden decompression of gases contained within lavas results in explosions of rock material to form ashes and cinders. Composite volcanoes, such as the one diagramed, are composed of alternating layers of solidified lava and of ash and cinders.

In one of the greatly destructive earthquakes on record, the Lisbon quake of 1755, about 60,000 people were killed when a tsunami 40 feet high swept across much of that city. The Sagami Bay earthquake of 1923 occurred while thousands of housewives in Tokyo and Yokohama were preparing the midday meal. As a side effect, spilled cooking oil ignited, fires broke out everywhere, and over 200,000 people in the two cities were killed.

As these examples illustrate, the effect of earthquakes on populated areas is often due to secondary causes, such as tsunamis, fires, landslides, and collapsing buildings. The examples also suggest that there are two different ways of measuring and comparing the severity of earthquakes: by recording their toll in human lives, structural damage, or monetary loss; or by measurement of the amount of earth movement associated with the quake without reference to its impact on society. Before seismographs were available to record earthquakes, they were rated on their *intensity*—specifically, their effect on people and the amount of damage caused to property. On the Mercalli scale, earthquake intensities were graded from 1 to 12, the higher number denoting complete destruction.

In 1935 C. F. Richter devised a scale of earthquake *magnitude*. An earthquake is really a form of energy expressed as wave motion passing through the surface layer of the earth. Radiating in all directions from the epicenter, seismic waves gradually dissipate their energy at increasing distances from the earthquake focus. On the Richter scale, the amount of energy released during an earthquake is estimated by measurement of the ground motion that occurs. Seismographs record earthquake waves; by a comparison of wave heights, the relative strength of quakes can be determined. Although Richter scale numbers run from 0 to 9, there is no absolute upper limit to earthquake severity; presumably, nature could outdo the magnitude of the most intense earthquakes so far recorded, which reach 8.5–8.6.

Because magnitude, as opposed to intensity, can be measured accurately, the Richter scale has been widely adopted. Nevertheless, it is still only an approximation of the amount of energy released in an earthquake. In addition, the height of the seismic waves can be affected by the rock materials under the seismographic station, and some seismologists believe that the Richter scale underestimates the magnitude of major tremors.

53

of a crater or be shot into the air with great force. The relatively quiet (shield) volcanoes have long gentle slopes, whereas the explosive volcanoes have steep sides. Steam and gases are constantly escaping from the nearly 300 volcanoes active in the world today. When pressure builds up, a crater can become a cauldron of activity, with steam, gas, lava, and ash all billowing out (Figure 2.16).

In many cases the forces beneath the crust are not great enough to allow the magma to reach the surface. Thus the magma hardens

(a)

54

(b)

Fig. 2.16 (a) Steam rising from Halemaumau, the fire pit of Kilauea crater on the slope of Mauna Loa volcano on the island of Hawaii. (b) The near-perfect volcanic cone of Fujiyama, Japan. (Photos by J. A. Bier.)

into a variety of underground formations of igneous rock that do little to affect surface landform features. However, gradational forces may erode overlying rock, so that the igneous rock, which is usually hard and resists erosion, becomes a surface feature. The Palisades, a rocky ridge facing New York City from the west, is such a landform. On other occasions, a

weakness below the earth's surface may allow the growth of a mass of magma that is denied egress to the surface because of firm overlying rock. Through the pressure it exerts, however, the magmatic intrusion may still buckle, bubble, or break the surface rocks, and domes of considerable size may develop.

Evidence from the past shows that some-

(a)

Fig. 2.17 (a) Fluid basaltic lavas created the Columbia Plateau, covering an area of 50,000 square miles (130,000 km²) with a total volume of 25,000 cubic miles (100,000 km³) of lava. Some individual flows were more than 300 feet (100 m) thick and spread up to 40 miles (60 km) from their origin fissures. (b) The jumbled surface of lava flows in south central Idaho. (a: photo by V. E. Livingston, State of Washington Department of Natural Resources; b: photo by M. J. Fellmann.)

(b)

times lava has flowed through fissures or fractures without volcanoes forming. These oozing lava flows have covered large areas to a great depth. The Deccan Plateau of India and the Columbia Plateau of the Pacific Northwest of the United States are examples of this process (Figure 2.17).

Gradational Processes

Gradational processes are responsible for the reduction of the land surface. If a land surface where a mountain once stood is now a low, flat plain, the gradational processes have been at work. The material that has been worn, scraped, or blown away is deposited in new places, and as a result, new landforms are created. Geologically speaking, the Himalayas are a young formation; gradational processes have not yet had time to reduce the huge mountains appreciably.

There are three kinds of gradational processes: weathering, the effect of gravity, and erosion. Weathering processes, both mechanical and chemical, prepare bits of rock for transfer to a different site by means of gravity or erosion. The force of gravity acts to move any higher-lying rock material not held tightly in place, and the agents of running water, moving ice, wind, waves, and currents erode and carry the loose materials to other areas, where landforms are created or changed.

MECHANICAL WEATHERING

Mechanical weathering is the physical disintegration of earth materials at or near the surface. A number of processes cause mechanical weathering, the three most important being frost action, the development of salt crystals, and root action.

If the water that soaks into a rock (between particles or along joints) freezes, ice crystals grow and exert pressure on the rock. If the process is repeated—freezing, thawing, freezing, thawing, and so on—there is a tendency for the rock to begin to disintegrate. Salt crystals act similarly in dry climates, where groundwater is drawn to the surface by *capillary* action (water rising because of surface tension). This action is similar to the process in plants whereby liquid plant nutrients move upward through the stem and leaf system. Evaporation leaves behind salt crystals, which help to disintegrate rocks. Roots of trees and other plants may also find their way into rock joints and, as they grow, break and disintegrate rock. These are all mechanical processes because they are physical in nature and do not alter the chemical composition of the material upon which they act.

CHEMICAL WEATHERING

A number of chemical processes cause rock to decompose rather than to disintegrate—that is, to separate into component parts by chemical reaction rather than to fragment. The three most important are *oxidation*, *hydration*, and *carbonation*. Because each of these processes depends on the availability of water, there is less chemical weathering in dry and cold areas than in moist and warm ones. Chemical reactions are speeded in the presence of moisture and heat.

Oxidation occurs when oxygen combines with rock minerals such as iron to form oxides; as a result, some rock areas in contact with the oxygen begin to decompose. Decomposition also results when water comes into contact with certain rock minerals such as aluminosilicates. The chemical change that occurs is called *hydrolysis*. When carbon dioxide gas from the atmosphere dissolves in water, a weak carbonic acid forms. The action of the acid is particularly evident on limestone; the calcium bicarbonate salt that is created is readily dissolved and removed by ground and surface water.

Weathering, either mechanical or chemical, does not by itself create distinctive landforms. Nevertheless, it acts to prepare rock particles for erosion. After the weathering process decomposes rock, the force of gravity and the erosional forces of running water, wind, and moving ice are able to carry the weathered material to new locations.

GRAVITY

The force of gravity—that is, the attraction of the earth's mass for bodies at or near its surface—is constantly pulling on all materials. Small particles or huge boulders, if not held back by bedrock or other stable material, will fall down slopes. Spectacular acts of gravity include avalanches and landslides, but more widespread are such less noticeable movements as the flow of soil and mud down hillsides.

Especially in dry areas, a common but very

dramatic landform created by the accumulation of rock particles at the base of hills and mountains is the *talus slope,* pictured in Figure 2.18. As pebbles, particles of rock, or even larger stones break away from the exposed bedrock on a mountainside because of weathering, they fall by force of gravity and accumulate, producing large conelike landforms. The larger rocks travel farther than the fine-grained sand particles, which remain near the top of the slope. Thus, talus cones not only are shaped similarly wherever they occur but also contain well-sorted material, with the coarser, heavier particles at the base and the finer ones toward the apex.

EROSIONAL FORCES AND DEPOSITION

Erosional forces such as wind and water carve already existing landforms into new shapes. The material that has been worn, scraped, or blown away is deposited in new places, and new landforms are created. Each agent is associated with a distinctive set of landforms.

57

Fig. 2.18 Mechanical weathering has loosened and broken rock along this butte at Kremmling, Colorado. Simple rockfall has created a pronounced talus slope. (Photo by M. J. Fellmann.)

Running Water. There is no more important erosional force than running water. Water, whether flowing across land surfaces, in stream channels, or underground, plays an enormous role in wearing down and building up landforms.

The ability of running water to erode depends on the amount of precipitation, on the length and steepness of the slope, and on the kind of rock and ground cover. The steeper the slope, the faster the flow and, of course, the more effective the erosion. If the vegetative cover that slows the flow of water is reduced, perhaps because of farming or livestock grazing, erosion can be severe, as shown in Figure 2.19.

Even the impact force of the precipitation—rain or hail—can itself be important. After it dislodges soil and makes the surface more compact, further water cannot penetrate the soil. The result is that more water, prevented from seeping into the ground, becomes available for erosion. Soil and rock particles in the water are carried to streams, leaving behind gullies and small stream channels.

Both the force of water alone and the particles contained in the stream are agents of erosion. The particles act as abrasives, scouring the surface over which they are moved. Abrasion, or wearing away, takes place when the particles strike against stream channel walls and along the stream bed. Large particles such as gravel slide along the stream bed because of the force of the current, grinding rock on the way. Floods and rapidly moving water are responsible for dramatic changes in channel size and configuration, sometimes forming new channels. In cities, where paved surfaces cover soil that could absorb and hold water, surface runoff is so rapid that nearby rivers and streams increase in size and velocity after heavy rains and can become not only flood hazards but also enhanced agents of erosion.

Small particles, such as clay and silt, are sus-

58

Fig. 2.19 Gullying can result from poor farming techniques, including overgrazing by livestock or many years of continuous row crops. Surface runoff removes topsoil easily when vegetation is too thin to protect it. (U.S.D.A. Soil Conservation Service.)

Fig. 2.20 Landsat image of the delta of the Mississippi River. Notice the ongoing deposition of silt and the effect that both river and gulf currents have on the movement of the silt. Note, too, the natural levees that have formed on the riverbanks. (N.A.S.A./U.S.G.S. EROS Data Center.)

pended in water and constitute (together with material dissolved in the water or dragged along the bottom) the *load* of a stream. Rapidly moving floodwaters carry huge loads. As high water or floodwater recedes and stream velocity decreases, the sediment contained within the stream can no longer remain suspended, and the particles begin to settle. Heavy, coarse materials are most quickly dropped; fine particles are carried further. The check in velocity and the resulting deposition are especially pronounced and abrupt when streams meet slowly moving water in bays, oceans, and lakes. Silt and sand accumulate at the intersections, creating *deltas*, as pictured in Figure 2.20. Great rivers such as the Nile, the Mississippi, and the

Yangtze have large, growing deltas (as much as 200 feet per year), but less prominent deltas are found at the mouths of many streams.

In plains adjacent to streams, land is sometimes built up by the deposition of stream load. If the deposited material is rich, it may be a welcome and necessary part of farming activities, as historically it was in Egypt along the Nile. Should the deposition be composed of sterile sands and boulders, however, formerly fertile bottomland may be destroyed. By drowning crops or inundating inhabited areas, the floods themselves, of course, may cause great human and financial loss. More than 900,000 lives were lost in the floods of the Huang He (Yellow River) of China in 1887.

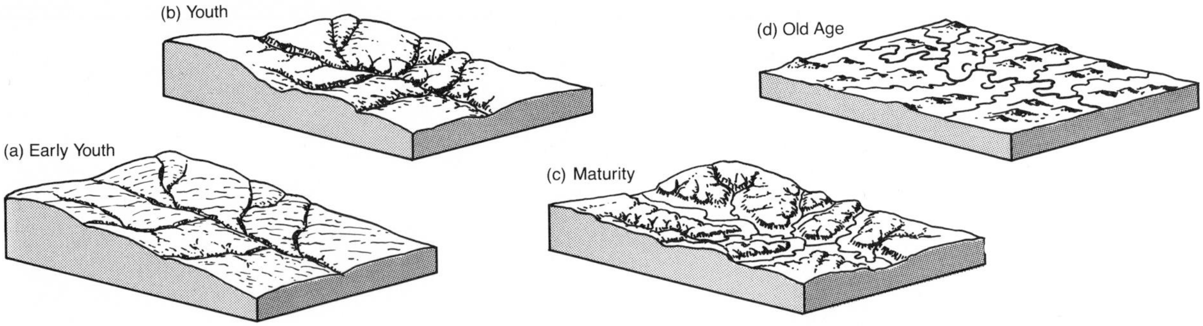

Fig. 2.21 Life cycle of streams and stages of stream-eroded landforms.

Life Cycle of Streams.   Because of erosion, streams themselves have a life cycle. As shown in Figure 2.21, it has recognizable stages from *early youth* to *old age*. As land is uplifted by tectonic force, water flows across the surface seeking lower elevations. Perhaps weak surface material or a depression in the rock allows for the development of a stream channel. In its downhill run at the outset of *early youth*, the stream may flow over precipices, forming falls in the process. The steep downhill gradient allows streams to flow rapidly, cutting narrow V-shaped channels in the rock in humid areas. (In arid regions, a flat-bedded ⊔-shaped water course develops.) Under these conditions, the erosional process is greatly accelerated because stream flow is rapid. In *later youth*, river falls become rapids through headward (upstream) erosion even of resistant rock, and eventually the river channels are incised well below the height of the surrounding landforms, as can be seen at the Grand Canyon of the Colorado River. All of these events mark the youthful stage of a stream.

As the stream reaches *maturity*, erosion tends to level the landforms. The valleys broaden to form flat-bottomed or ⊔-shaped profiles, and landforms become rounded. The stream channel has carved out an area wide enough to be given the name *floodplain*. By *full maturity*, the process of cutting down is over; the stream is actively widening its floodplain. Its course is less direct; the stream meanders, constantly carving out new erosional channels. In fact, the channels left behind as new ones are cut become oxbow-shaped lakes, hundreds of which are found in the Mississippi River floodplain.

By *old age*, nearly flat or very gently sloping landforms are evident. The highest elevations may be at banks of the rivers where *natural levees* are formed by the filtering out of silt at river edges during flood. The lower Mississippi River, in Arkansas and Louisiana, is an example of a river entering the old-age stage of erosion, whereas the nearly flat Amazon River basin is well into old age.

In the course of earth history, the old-age stage of a stream has sometimes been followed by a rejuvenation. The plain may be upraised by crustal movement, and thus the cycle may begin again. This time, the stream cuts through a plain rather than mountains, perhaps creating canyons and plateau-like features in the process. The Colorado Plateau is a good example of an upraised plain. The streams there are now in the youth stage again and have created V-shaped valleys.

A distinction must be made between the results of the stream's erosional cycle in humid areas and the cycle characteristic of arid areas. Water originating in mountainous areas may

60

Fig. 2.22 Alluvial fans are built where the velocity of youthful streams is reduced as they flow out upon the more level land at the base of the mountain slope. The abrupt change in slope and velocity greatly reduces the stream's capacity to carry its load of coarse material. Deposition occurs, choking the stream channel and diverting the flow of water. With the canyon mouth fixing the head of the alluvial fan, the stream sweeps back and forth, building and extending a broad area of deposition. (U. S. Geological Survey)

never reach the sea if the stream channel runs through a desert. In fact, stream channels may be empty except during rainy periods, when water rushes down the hillsides to collect and form a temporary lake. In the process, *alluvium* (sand and mud) builds up in the lakes and at the lower elevations, and alluvial fans are formed along hillsides (Figure 2.22). The fan is produced by the scattering of detritus outward as the stream reaches the lowlands at the base of the slope it traverses. The final stage in the cycle is marked when alluvial deposits bury the eroded mountain masses. It is not unusual to see the remnants of these buried mountains poking through the alluvium as one travels through deserts; the mountain regions of Nevada show many incipient examples.

Because the streams in arid areas may have only temporary existence, their erosional power is less certain than that of the freely flowing streams of humid areas. In some instances, they may barely mark the landscape; in other cases, their swiftly moving flood may carve deep, straight-sided *arroyos*. Often, the water may rush onto an alluvial plain in a complicated pattern resembling a multistrand

braid, leaving in its wake an *alluvial fan*. The channels resulting from this rush of water are called *washes*. The freshet of water, collecting in structural basins, may create temporary lakes called *playas*. The lateral erosional power of unrestricted running water in arid regions is dramatically illustrated by the steep-walled configuration of buttes and mesas such as Mitchell Mesa, Texas, shown in Figure 2.23.

Groundwater. Some of the water supplied by rain and snow sinks underground into the pores and cracks of rock and soil—not in the form of an underground pond or lake, but simply as very wet subsurface material. When it accumulates, a zone of saturation forms. As indicated in Figure 2.24, the upper level of this zone is the *water table;* below it, the soils and rocks are saturated with water. Groundwater moves constantly but very slowly. Most remains underground, seeking the lowest level; however, when the surface of the land dips below the water table, ponds, lakes, and marshes form. Some water is pulled up to the surface by capillary action in the ground or in vegetation. Groundwater, particularly when

61

Fig. 2.23 Mitchell Mesa, south of Alpine, Texas. The resistant caprock of the mesa protects softer, underlying strata from downward erosion. Where the caprock is removed, lateral erosion lowers the surface, leaving the mesa as an extensive and pronounced relic of the former higher-lying landscape. Here the mesa walls have a sloping accumulation of talus. (Photo by J. A. Bier.)

charged with carbon dioxide, can dissolve and transport soluble materials in a process called *solution.*

62   Although groundwater tends to decompose many types of rocks, its effect on limestone is most spectacular. Many of the great caves of the world have been created by the underground movement of water through limestone regions. Water sinking through the overlying rock leaves carbonate deposits as it drips. The deposits hang from cave roofs (stalactites) and build upward from cave floors (stalagmites). In some areas, the uneven effect of groundwater

erosion of limestone leaves a landscape pockmarked by a series of sinkholes, where roofs of caves have collapsed. *Karst* topography refers to a large limestone region marked by sinkholes, caverns, and underground streams. The Dalmatian coast of Yugoslavia, north central Florida, and the Mammoth Cave area of Kentucky are examples of this kind of landform region, depicted in Figure 2.25.

Glaciers.   Another way in which erosion and deposition occur is through the effect of gla-

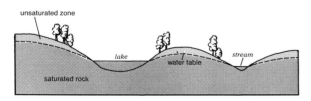

Fig. 2.24 The groundwater table generally follows surface contours, but in a subdued fashion. Water flows slowly through the saturated rock, emerging at earth depressions that are lower than the level of the water table. During a drought, the table is lowered and the stream channel becomes dry.

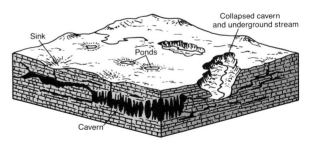

Fig. 2.25 Limestone erodes easily in the presence of water. Karst topography, such as that shown here, occurs in humid areas where limestone in flat beds is at the surface.

ciers. Although they are relatively unimportant today as agents of landform change, glaciers covered a large part of the earth's land area as recently as 8000–15,000 years ago, during the Pleistocene geologic epoch. Many landforms, among them fjords and moraines, were created by the erosional or depositional effects of glaciers.

Only very cold places with short or nonexistent summers are susceptible to snow accumulation, for there the annual snowfall accumulation exceeds the amount of snow that melts and evaporates during the year. The weight of the snow causes it to compact at the base and form ice. When the snowfield reaches a thickness of several hundred feet, the ice at the bottom becomes like plastic and begins slowly to move. A *glacier*, then, is a large body of ice moving slowly down a slope or spreading outward on a land surface (Figure 2.26). Some glaciers appear to be stationary simply because the melting and evaporation at the glacier's edges equal the speed of the ice advance. Glaciers can, however, move as much as several feet per day.

Most theories of glacial formation concern the reasons for the cooling of the earth's climate. Some attribute the ice ages to volcanic dust in the atmosphere. The argument is that the dust, by reducing the amount of solar energy reaching the earth, effectively lowered temperatures at the surface. Another view suggests that widespread mountain building and continental uplift just before and at the start of the Pleistocene brought on colder climates and increased snowfall over certain affected land areas. Yet another theory attributes the ice ages to known changes in the shape, the tilt, and the seasonal positions of the earth's orbit around the sun over the last half-million years. Such changes, it is contended, alter the amount of solar radiation received by the earth as well as its distribution over the earth. Still another view postulates a reduction in atmospheric carbon dioxide, which would inevitably reduce

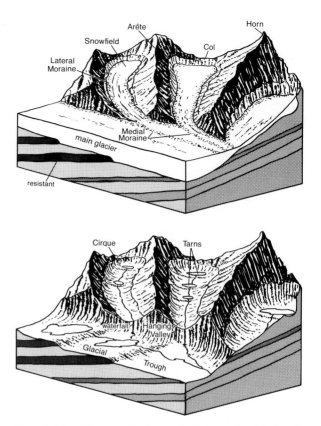

Fig. 2.26 The evolution of alpine glacial landforms. Frost shattering and ice movement carve *cirques*, the irregular bottoms of which may contain lakes *(tarns)* after glacial melt. Where cirque walls adjoin from opposite sides, knifelike ridges called *arêtes* are formed, interrupted by overeroded passes or *cols*. The intersection of three or more arêtes creates a pointed peak, or *horn*. Rock debris falling from cirque walls is carried along by the moving ice. *Lateral moraines* form between the ice and the valley walls; *medial moraines* mark the union of such debris where two valley glaciers join.

63

the earth's surface temperatures and permit the accumulation of ice.

Today, glaciers are found in Antarctica, in Greenland, and in mountainous zones in many other parts of the world. In fact, about 10% of the earth's land area is under ice. During the most recent advance of ice, the continental ice

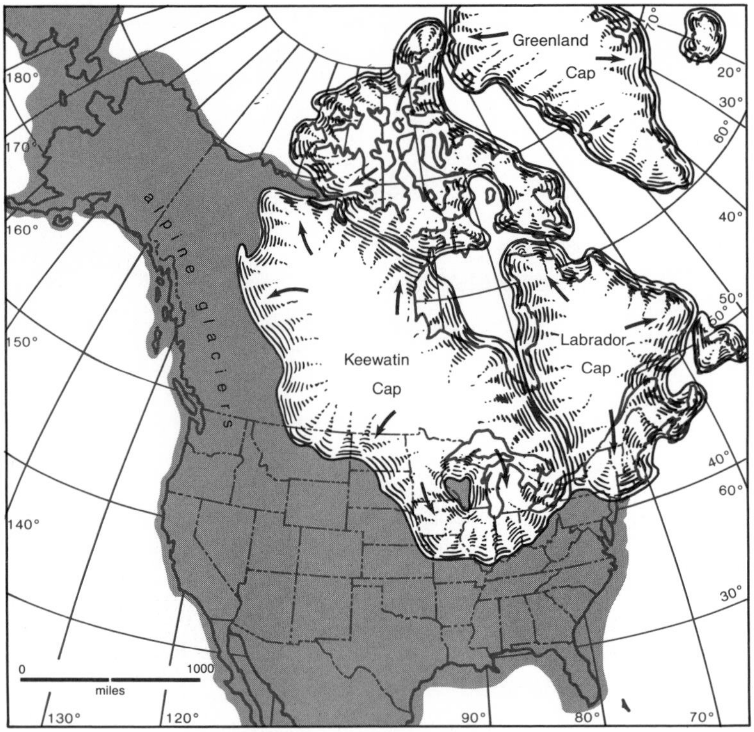

Fig. 2.27 Farthest extent of glaciation in North America. Separate centers of snow accumulation and ice formation developed. Large lakes were created between the western mountains and the advancing ice front; to the south, huge rivers carried away glacial meltwaters. Since large volumes of moisture were trapped as ice on the land, sea levels were lowered and continental margins were extended.

64

of Greenland was part of an enormous glacier that covered nearly all of Canada (Figure 2.27) and the northernmost portions of the United States and Eurasia. These glaciers reached thicknesses of 10,000 feet, enveloping entire mountain systems.

The weight of a glacier helps to break up rock beneath and prepare it for transportation by the moving mass of ice. Consequently, glaciers create landforms by weathering and erosion. Glaciers scour the land as they move, leaving surface scratches on the rocks that remain. Much of eastern Canada has been scoured by glaciers that left little soil but many ice-gouged lakes and streams. The erosional forms created by glacial scourings have a variety of names. A glacial *trough* is a deep, U-shaped valley visible only after the glacier has receded. If the valley is today below sea level, as in Norway or British Columbia, *fjords*, or arms of the sea, are

formed. Some of the landforms created by scouring are depicted in Figure 2.26.

Glaciers also create landforms when they deposit the debris they have transported. These deposits, called *glacial till*, consist of rocks, silt, and sand. As the great tongues of ice move forward, more debris accumulates in some parts of the glacier than in others. The ice that scours valley walls and the ice at the tip of the advancing tongue are particularly filled with this debris (Figure 2.28). As a glacier melts, it leaves behind great hills of glacial till, called *moraines.*

Fig. 2.28 (opposite) Arêtes and lateral and medial moraines mark this photograph of valley glaciers in the Coast Mountains of British Columbia. Note how the lateral moraines join to form a medial moraine. (Courtesy of British Columbia Lands Survey.)

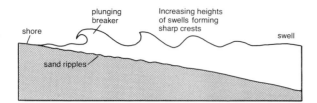

Fig. 2.29 Formation of waves and breakers. As the offshore swell approaches the gently sloping beach bottom, sharp-crested waves form, build up to a steep wall of water, and break forward in plunging surf.

Many other landforms have been made by glaciers. The most important is the *outwash plain*, a gently sloping area in front of the glacier. The melting of the glacier along a broad front sends thousands of small streams running out from the glacier in braided fashion, streams that deposit neatly stratified glacial till. Some of the world's best farmland is on outwash plains, which are essentially great alluvial fans that cover a wide area and provide new, rich parent-material for soil formation. Most of the midwestern part of the United States owes its soil fertility to recent glacial deposition in an area of subsequent luxuriant grass cover and a climate conducive to rapid topsoil development.

Before the end of the most recent ice advance, there were at least three other advances that occurred during the million years of the Pleistocene period. No firm evidence indicates whether we have emerged from the cycle of ice advance and retreat. The factors concerning the earth's changing temperature, which are discussed in the next chapter, must be considered before it is possible to assess the likelihood of a new ice advance. For the first half of this century, the world's glaciers were melting faster than they were building up, although some recent reports indicate that this trend is slowing or coming to a halt.

Waves and Currents. While glacial action is intermittent in earth history, the breaking of ocean waves on continental coasts and on islands is unceasing and causes considerable change in coastal landforms. As waves reach the shallow water close to shore, they are forced to become higher until a breaker is formed, as depicted in Figure 2.29. The uprush of water not only carries sand for deposition but also erodes the landforms at the coast, while the backwash carries the eroded material away. This type of action results in different kinds of landforms, depending on conditions.

If the land at the coast is well above sea level, the wave action causes cliffs to form. The cliffs then erode at a rate dependent on the resistance of the rock to the constant assault from the salt water. During storms, a great deal of power is released by the forward thrust of waves, and much weathering and erosion take place. Landslides are a hazard during coastal storms; they occur particularly in areas where weak sedimentary rock or glacial till exists.

Beaches are formed by the deposition of sand grains contained in the water. The sand originates from the vast amount of coastal erosion and from stream mouths (Figure 2.30). Longshore currents, which move roughly parallel to the shore, transport the sand, forming beaches and sandspits. The more sheltered the area, the better the chance for a beach to build. The

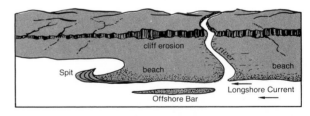

Fig. 2.30 The cliffs behind the shore are eroded by waves during storms and high water. Sediment from the cliff and the river forms the beach deposit; the longshore current moves some sediment downcurrent to form the spit. Offshore bars are created from material removed from the beach and deposited by retreating waves.

66

Fig. 2.31 Sea cliffs, headlands, embayments, and offshore erosional remnants are typical of cliffed shorelines such as this one on the Oregon coast. Beaches occur only near the mouths of rivers or in embayments. (Photo by B. Dahlin.)

backwash of waves, however, takes sand away from the beaches if no longshore current exists. As a result, sandbars develop a short distance away from the shoreline. If the sandbars become large enough, they eventually close off the shore, creating a new coastline that encloses lagoons or inlets. Salt marshes very often develop in and around these areas. Figure 2.31 shows one kind of area partially carved by waves.

Coral reefs, made not from sand but by coral organisms growing in shallow tropical water, are formed by a secretion of lime in the presence of warm water and sunlight. These reefs, consisting of millions of colorful skeletons, develop short distances offshore. When torn loose by wave action, coral fragments may be pulverized and form usual beach features. Off the coast of Australia lies the most famous coral reef, the Great Barrier Reef. Atolls, found in the south Pacific, are reefs formed in shallow water around a volcano that has since been covered or nearly covered by water.

Wind. In dry climates, wind is an agent of weathering, erosion, and deposition, but in humid areas, the effect of wind is confined mainly to sandy beach areas. The limited vegetation in dry areas leaves exposed particles of sand, clay, and silt to movement by wind. Thus many of the sculptured features found in dry areas result from mechanical weathering, that is, from the abrasive action of sand and dust particles as they are blown against rock surfaces. Sand and dust storms occurring in a drought-stricken farm area may make it unusable for agriculture by denuding it of topsoil or by smothering it with sterile sand deposits. Inhabitants of Oklahoma, Texas, and Colorado suffered greatly in the 1930s when their farmlands became the "dust bowl" of the United States.

Several types of landforms are produced by wind-driven sand. Figure 2.32 depicts one of these. Although sandy deserts are much less common than gravelly deserts, their characteristic landforms are better known. Most of the Sahara, the Gobi, and the western United States deserts are covered not with sand but with rocks, pebbles, and gravel. Each also has a small portion (and the Saudi Arabian desert a large area) covered with sand blown by wind into a series of waves or dunes. Unless vegetation stabilizes them, the dunes move as sand is

Fig. 2.32 **The prevailing wind from the left has given these transverse dunes in the Great Sand Dunes of New Mexico a characteristic gentle windward slope and a steep, irregular leeward slope. (Photo by J. A. Bier.)**

68

World areas subject to wind erosion, shifting sands, and arid landform processes are being steadily expanded by the destructive activities of humans. *Desertification*—the extension of desert landscapes by overgrazing, by removal of trees for firewood, and by intensive cropping of agriculturally marginal lands—has by some estimates added an area larger than that of Brazil to the natural deserts of the world. Every year some 14 million additional acres are converted to arid landscapes.

Desertification

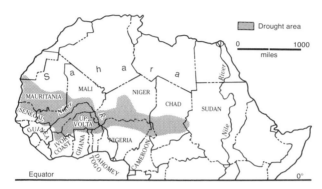

The expansion of the fringes of the Sahara has been well documented, and because of tragic accompanying droughts in the Sahel (Mauritania, Senegal, Mali, Upper Volta, Niger, and Chad) from 1968 through the mid-1970s, it has been called sharply to world attention. In the last 50 years, some 250,000 square miles have been added to the southern Sahara Desert, areas reduced to eroded, stony wasteland and to drifting sand. Elsewhere in Africa, and particularly in the Middle East and India, similar human-induced extension of arid land-scapes—eroded by wind, scarred by infrequent runoff, or subject to moving dunes—gives vivid and catastrophic evidence of cultural impact on landform processes.

69

blown from their windward faces up onto and over their crests. One of the most distinctive sand desert dunes is the crescent-shaped *barchan.* Along seacoasts and inland lake-shores in both wet and dry climates, wind can pile up sand into ridges that may reach a height of 300 feet. Sometimes coastal communities and farmlands are threatened or destroyed by moving sand.

Another kind of wind-deposited material,

silty in texture and pale yellow or buff in color, is called *loess.* Encountered usually in mid-latitude westerly wind belts, it covers extensive areas in the United States (Figure 2.33), central Europe, Central Asia, and Argentina. It has its greatest development in northern China, where it covers hundreds of thousands of square miles, often to depths of more than 100 feet. The wind-borne origin of loess is confirmed by its typical occurrence downwind from exten-

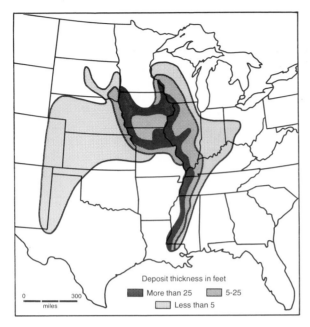

Fig. 2.33 Location of wind-blown silt deposits, including loess, in the United States. The thicker layers found in the upper Mississippi valley area are associated with the wind movement of glacial debris. Farther west, in the Great Plains, wind-deposited materials are sandy in texture, not loessial. (From *Geology of Soils: Their Evolution, Classification, and Uses* by Charles B. Hunt. W. H. Freeman and Company. Copyright © 1972.)

70

sive desert areas, though major deposits are assumed to have resulted from wind erosion of nonvegetated sediment deposited by meltwater from retreating glaciers. Because rich soils usually form from loess deposits, if climatic circumstances are appropriate, these areas are among the most productive agricultural lands in the world.

Conclusion

Landforms are among the most visible elements of the natural stage upon which the drama of human action is played. We have seen

them to be a setting continuously resculpted by the immutable forces that gave them shape. Landforms and the migratory plates on which they develop are more than a mute background observed by humans but unresponsive to our actions. Humans occupying the earth—the study of geography—have in that occupance speeded and thwarted, for good or for ill, the impartial processes of natural land formation. Great dams alter river gradients and impinge upon the cycle of erosion; the filling of tidal marshes and the dredging of reefs destroy the coastlines of nature. The North African granaries of Rome and the lush, fertile hillsides of ancient Greece were turned to deserts and barren crags by the acceleration of erosion that followed human disturbance of their vegetation.

We deal here not with abstractions nor with interesting but remote forces. We deal, in nature, with a stern and always unforgiving arbiter of our actions. They may be actions undertaken for compelling human reasons, but their consequences must be anticipated. The processes we have reviewed are not remote; they are the very stuff of life.

The natural landscape affecting and affected by human action is more than landforms, moving continents, and an occasional earthquake. Except at times of natural disaster, these elements of the physical world are for most of us a quiet, accepted background. More immediately affecting our lives and fortunes are the great patterns of climate that help define the limits of the economically possible at present levels of technology; the daily changes of weather that affect the success of picnic and crop yield alike; and the patterns of vegetation and soils, related in nature to the realms of climate but utilized, altered, and perhaps destroyed by humans. To these elements of the natural environment we turn our attention in the next chapters as a necessary prelude to the consideration of the human side of the human earth-surface equation of geography.

For Review

1. What evidence is there that makes plausible the theory of continental drift?

2. What occurs when moving lithospheric plates collide? What name is given to that process of collision?

3. In what ways may rocks be classified? List three classes of rocks according to origin. In what ways are they distinguished from one another?

4. Explain what is meant by *diastrophism* and *volcanism.*

5. What is meant by *folding*? By *syncline*? By *anticline*? By *overturning*?

6. Draw a diagram indicating as many varieties of fault features as you can.

7. With what earth movements are earthquakes associated? What is a *tsunami* and how does it develop?

8. What is the distinction between chemical and mechanical weathering? Is weathering responsible for landform creation?

9. Describe the stages in the life cycle of streams in humid climates and indicate the landforms associated with each stage of the cycle.

10. How do glaciers form? What landscape characteristics are associated with glacial erosion? With glacial deposition?

11. What landform features can you identify on Figure 1.13? What were the agents of their creation?

12. What processes account for the landform features of the area in which you are now?

Suggested Readings

BUTZER, K. W., *Geomorphology from the Earth.* Harper & Row, New York, 1976.

GOUDIE, A. S., *Environmental Change.* Oxford University Press, New York, 1977.

KOLARS, JOHN F., and JOHN D. NYSTUEN, *Physical Geography: Environment and Man.* McGraw-Hill, New York, 1975.

MULLER, ROBERT A., et al., *Physical Geography Today* (2nd ed.). CRM/Random House, New York, 1978.

PRESS, FRANK, and RAYMOND SIEVER, *Earth*. Freeman, San Francisco, 1974.

STRAHLER, ARTHUR N., *Physical Geography* (4th ed.). Wiley, New York, 1975.

TREWARTHA, GLENN T., ARTHUR H. ROBINSON, and EDWIN H. HAMMOND, *Elements of Geography* (5th ed.). McGraw-Hill, New York, 1967.

It first appeared, as had so many before, as a tiny blip on radar screens; this one was recognized as a small circulation pattern developing off the Yucatán peninsula of Mexico. That was Thursday, June 15, 1972. By late afternoon, growing, it was called a tropical depression and seen to be the possible start of a major storm system in the Caribbean Sea. By the next afternoon, what was to become the most damaging storm in United States history was officially named Agnes, the first hurricane of the 1972 season. By the afternoon of the 17th, when it was 300 miles (480 km) southwest of Key West, Florida, barometric pressure had dropped to 29.12 inches (73.96 cm), and winds increased to between 75 and 95 miles (121 and 153 km) per hour; Agnes measured 250 miles (402 km) in diameter. The hurricane smashed through Florida, spawning 15 tornadoes and two windstorms, with pressure dropping to 28.88 inches (73.36 cm) and rainfall totaling 3–7 inches (7.6–18 cm). Losing energy over land, as expected, Agnes was over Georgia when it was downgraded to a tropical storm by Tuesday, June 20. It began to turn eastward as a normal storm and was expected to die out over the Atlantic. But in Agnes's case a high-pressure system in the Atlantic blocked eastward movement and turned the storm northward while allowing it to gather strength and moisture from coastal waters. Moving up the east coast, Agnes registered record low pressures and deposited unprecedented amounts of rainfall over the major urban industrial areas of the Northeast.

By Tuesday morning, June 27, some two weeks after its beginning, Agnes recorded its death as a weak low near Nova Scotia. It had, in that brief history, caused untold misery: 134 dead, hundreds of thousands of temporary refugees, and devastating flooding from its deposit of 25.5 cubic miles (123 km³) of rainwater—some 28.1 trillion gallons (106 trillion liters). Between $2 billion and $3 billion of property damage was left in the storm's wake.

Physical Geography: Weather and Climate

Agnes was the weather of eastern North America for two weeks; it was one expression of the climatic regime of that same area. Agnes was as well a devastating part of those broad human–earth relations that are the study of geography. This chapter reviews that subsection of physical geography concerned with weather and climate. It deals with normal, patterned phenomena out of which such an abnormality as Agnes may occasionally emerge.

A weather forecaster describes current conditions and predicts tomorrow's and perhaps the following day's weather. If weather conditions at specified moments in time are recorded, such as every hour on the hour, an inventory of weather conditions can be developed. By finding trends in data that have been gathered over an extended period of time, we can speak about average conditions and the likelihood of extreme conditions. These characteristic circumstances describe the *climate* of a region. Thus weather is a moment's view of the lower atmosphere, while climate is a summary of weather conditions in an area or at a place over a period of time. It is with *climate* that the geographer is primarily concerned.

The relationship between geography and climatology goes back to ancient Greeks, now recognized as geographer-philosophers, who pondered the physical nature of the world they knew in the Mediterranean and speculated about what lay to the little-known north and south of their habitable world. They measured variations in angles of the sun's rays at various dates of the year and associated the energy received from the sun with seasonal temperature variations. Reasoning that to the south, where the sun was always high in the sky, it would be intolerably hot and that to the north, where it was always low, it would be uncomfortably cold, they presented the first concept of climatic variation by latitudinal change. Their division of the world into torrid, temperate, and frigid zones, perhaps unfortunately, remains popular to this day.

74

Although meteorology, the study of the earth's atmosphere, has emerged as a discipline in its own right, it grew out of long years of contribution and research by geographers. In recent years, a separate field of atmospheric research has also begun to develop outside of geography, emphasizing the physical relationships responsible for climatic change. Geography, however, has retained its traditional role of analyzing the differences in weather and climate from place to place and of seeking an understanding of how climatic elements, such as precipitation and temperature, affect human occupance of the earth.

In geography, we are particularly interested in the physical environment that surrounds us day to day. That is why it is the *troposphere,* the lowest layer of the earth's atmosphere, that attracts our attention. This layer, extending about 6 miles (10 km) above the ground, contains virtually all of the air, clouds, and precipitation of the earth. In this chapter, we try to answer the questions that are usually raised about the characteristics of that lower atmosphere—the questions people ask about weather and climate. By discussing these questions from the viewpoint of averages or average variations, we are attempting to give a view of the earth's climatic differences, a view held to be most important for understanding the way

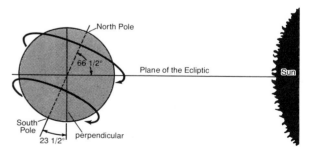

Fig. 3.1 The earth spins on an axis about 66½° away from the plane of the ecliptic, an imaginary line connecting the center of the earth and the center of the sun.

people use the land. Thus climate is a key to explaining, in a broad way, the distribution of world population. People have great difficulty in living in areas that are, on the average, very cold, very hot, very dry, or very wet. They are also negatively affected by huge storms or flooding. The crops they can grow depend partly on climate. Plants, animals, insects, and fish are all affected by climate, and they, in turn, affect people's lives and health. A most obvious day-to-day change, and one recognized by the ancient Greeks as an important contributor to the concept of climate, is temperature.

Air Temperature

Energy from the sun, called *solar energy*, is transformed into heat, primarily at the earth's surface and secondarily in the atmosphere. Of course, not every part of the earth or of its overlying atmosphere receives the same amount of solar energy. At any given place, the amount of solar energy, or *insolation*, available depends on the intensity and duration of radiation from the sun. These are determined by (1) the angle at which the sun's rays strike the earth and (2) the length of the daylight. The earth, as Figure 3.1 indicates, does not spin on an axis that is

A WINTER DAY IN THE NORTHERN HEMISPHERE

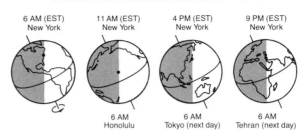

Fig. 3.2 **The process of the 24-hour rotation of the earth on its axis.**

perpendicular to a line connecting the centers of the earth and the sun. Rather, the earth's axis is tilted about $23\frac{1}{2}°$ away from the perpendicular; every 24 hours the earth rotates once on that axis, as shown in Figure 3.2. While rotating, the earth is slowly revolving around the sun in a nearly circular annual orbit (Figure 3.3).

If the earth were not tilted by $23\frac{1}{2}°$ from the perpendicular, the solar energy received at a given latitude would not vary during the course of a year. The rays of the sun would strike the equator most directly, and as the distance away from the equator became greater, the rays would strike the earth at ever-increasing angles, diminishing the intensity of the energy and giving climates a latitudinal stand-

75

Fig. 3.3 **The process of the yearly revolution of the earth about the sun.**

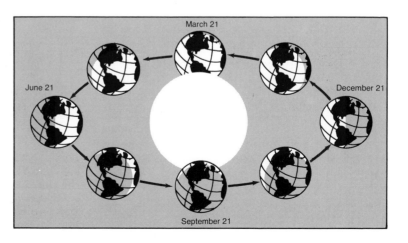

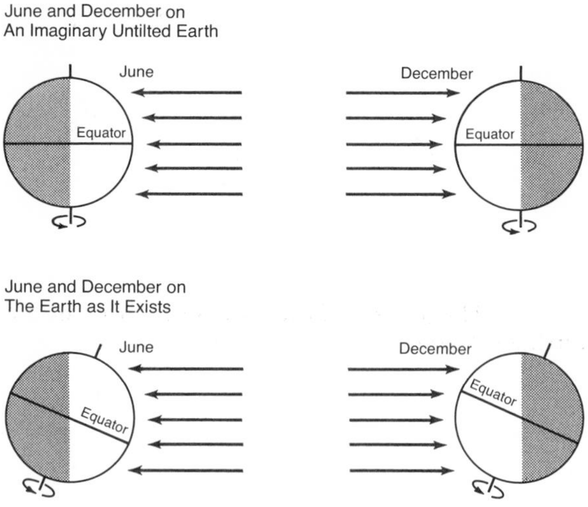

June and December on
An Imaginary Untilted Earth

June   December

June and December on
The Earth as It Exists

June   December

Fig. 3.4 Notice in the lower diagram that as the earth revolves, the north polar area in June is bathed in sunshine for 24 hours, while the south polar areas are dark. The most intense of the sun's rays are felt north of the equator in June and south of the equator in December. None of this is true in the untilted case.

76

ardization (Figures 3.4 and 3.5). Actually, instead of the equator's constantly being the peak target for incoming solar energy, the location of highest incidence varies during the course of the year. When the northern hemisphere is tilted directly toward the sun, the vertical rays are felt as far north as $23\frac{1}{2}°$ N latitude. This position of the earth occurs about June 21, the summer solstice for the northern hemisphere and the winter solstice for the southern. About December 21, when the vertical rays of the sun strike near $23\frac{1}{2}°$ S latitude, it is the beginning of summer in the southern hemisphere and the onset of winter in the northern. During the rest of the year, the position of the earth relative to the sun results in direct rays' migrating from about $23\frac{1}{2}°$ N to $23\frac{1}{2}°$ S and back again. On about March 21 and September 21 (the spring and autumn equinoxes), the vertical rays of the sun strike the equator.

The tilt of the earth also means that the length of day and night varies during the year. One-half of the earth is always illuminated by the sun, but only the equator is constantly lighted for 12 hours each day. As distance away from the equator becomes greater, the hours of daylight or darkness increase, depending on whether the direct rays of the sun are north or south of the equator. In the summer, daylight increases to the maximum of 24 hours in the summer polar region, and during the same period, nighttime finally reaches 24 hours in length in the other polar region.

Because of the 24-hour daylight, it would seem that much solar energy should be available in the summer polar region, but such is not the case. The angle of the sun is so narrow (the sun is low in the sky) that the solar energy is spread over a wide surface. By contrast, the combination of relatively long days and sun angles of close to 90° makes an enormous amount of energy available to areas in the neighborhood of 15° to 30° north and south latitude during each hemisphere's summer.

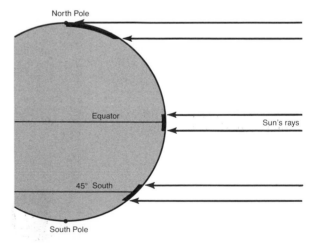

Fig. 3.5 Three equal, imaginary rays from the sun are shown striking the earth at different latitudes at the time of the equinox. As distance increases away from the equator, the rays become more diffused, showing how the sun's intensity is diluted in the high latitudes.

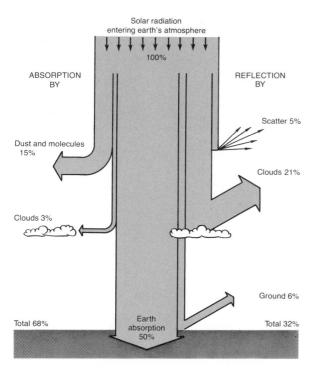

Solar radiation
entering earth's atmosphere

100%

ABSORPTION
BY

REFLECTION
BY

Scatter 5%

Dust and molecules
15%

Clouds 21%

Clouds 3%

Ground 6%

Total 68%

Earth
absorption
50%

Total 32%

Fig. 3.6   Consider the incoming solar radiation as 100%. The portion that is absorbed into the earth (50%) is eventually released to the atmosphere and then reradiated into space. Notice that the outgoing radiation is equal to 100%, showing that there is an energy balance on the earth.

The amount of solar energy at any given place is determined not only by the sun's angle and the day's length but also by the amount of water vapor in the air, by cloud cover, and by the nature of the earth's surface. Much of the shortwave insolation potentially receivable is, in fact, *reflected* back to outer space or diffused in the troposphere. Clouds, which are composed of dust particles and water droplets, reflect a great deal of energy. Light-colored surfaces, especially snow cover, also serve to reflect a very small amount of solar energy.

Energy is lost through reradiation as well as by reflection. As just described, reflection means that the sun's rays are turned back di-

rectly; in the *reradiation* process, the earth acts as a communicator of energy. As indicated in Figure 3.6, the shortwave energy that is absorbed into the land and the water is returned to the atmosphere in the form of longwave terrestrial radiation. On a clear night, when no clouds can block or diffuse the movement, temperatures become lower and lower as the earth reradiates as heat the energy it has received and stored during the course of the day.

Some kinds of earth surface material, especially water, store solar energy more effectively than others. Because water is transparent, solar rays can penetrate a great distance below its surface. If water currents are present, the heat is distributed even more effectively. On the other hand, land surfaces are opaque, and so all of the energy received from the sun is concentrated at the surface. Land, having more heat available at the surface, reradiates its energy faster than does water. Air is heated by this process of reradiation from the earth and not directly by energy from the sun passing through it. Thus, because land heats and cools much more rapidly than water, the extremes of hot and cold temperatures recorded on the earth occur on land and not at sea.

Temperatures are moderated by the presence of large bodies of water near land areas. Note in Figure 3.7 that coastal areas have lower summer temperatures and higher winter temperatures than places at the same distance from the equator that are not seacoasts. Land areas affected by the moderating influences of water are considered marine environments; those areas not affected by nearby water are continental environments.

Temperatures vary in a cyclical way from day to day. In the course of a day, as incoming solar energy exceeds the energy lost through reflection and reradiation, temperatures begin to rise. The ground stores some heat, and temperatures continue to rise until the angle of the sun becomes so narrow that the energy received no longer exceeds that lost by the reflec-

77

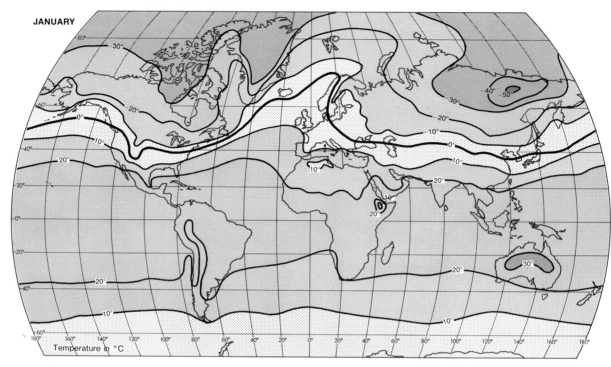

JANUARY

Temperature in °C

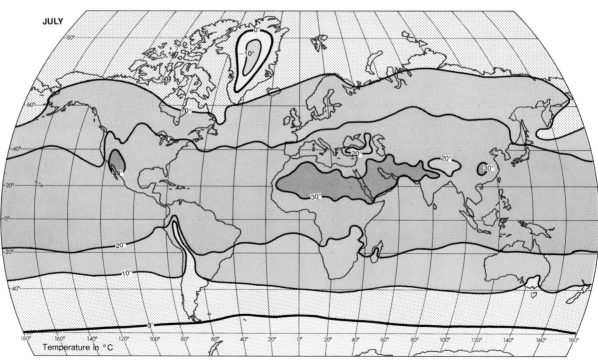

JULY

Temperature in °C

Fig. 3.7 (opposite) At a given latitude, water areas are warmer than land areas in winter and cooler in summer. Isotherms are lines of equal temperature.

tion and reradiation processes. Not all of the heat is lost during the night, but long nights appreciably deplete the stored energy. Because the sun's angle is higher during summer and daylight is longer, the daily high and low temperature ranges are greater than their winter counterparts.

We might think that as we moved vertically away from the earth toward the sun, temperatures would increase; such is not true within the troposphere. The earth is a body absorbing and reradiating heat, and therefore temperatures are usually warmest at the earth's surface and lower as elevation increases. Note on Figure 3.8 that this temperature *lapse rate* (the rate of change of temperature with altitude in the troposphere) averages about 3.5° per 1000 feet (6.4°C per 1000 m). For example, the dif-

ference in altitude between Denver and Pikes Peak is about 9000 feet (2700 m), which would normally result in a 32°F (17°C) difference in temperature. Jet planes flying at an altitude of 30,000 feet (9100 m) are moving through air about 100°F (56°C) colder than ground temperatures.

The normal lapse rate does not always hold, however. Rapid reradiation sometimes causes temperatures to be higher above the earth's surface than at the surface itself. This particular condition, in which air at lower altitudes is cooler than air aloft, is called a *temperature inversion*. An inversion is important because of its effect on air movement. Warm air at the surface, which would normally rise, may be blocked by the still warmer air of the temperature inversion. Thus the surface air is trapped, and if it is filled with automobile exhaust emissions or with smoke, a serious smog condition may develop. Because of the configuration of the nearby mountains, Los Angeles, pictured in Figure 3.9, often experiences temperature inversions, causing the sunlight to be reduced to a dull haze.

79

Air Pressure and Winds

The air we breathe is a gaseous substance whose weight affects air pressure. If it were possible to carve out 1 cubic inch of air at the surface of the earth and weigh it along with all the other cubic inches of air above it, under normal conditions the total weight would come to about 14.7 pounds (6.67 kg) of air as measured at sea level. Actually that is not very heavy when you consider the dimensions of the 1-inch column of air: 1 inch by 1 inch by about 6 miles = 220 cubic feet (6.2 m³). The weight of air 3 miles (4.8 km) above the earth's sur-

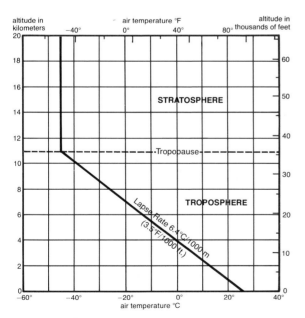

Fig. 3.8 The temperature lapse rate under typical conditions.

Fig. 3.9 Smog in Los Angeles. Subsiding air has forced the temperature up at a height of several hundred feet (chart opposite). Below that level, stagnant air holds increasing amounts of pollutants, caused mainly by automobile exhausts. (EPA-DOCUMERICA.)

face, however, is considerably less than 14.7 pounds because there is correspondingly less air above it. Thus it is clear that air is heavier and air pressure is higher close to the earth's surface. But the weight of air and thus air pressure also vary according to the temperature of

The Donora Valley Tragedy

A heavy fog settled over Donora, Pennsylvania, in late October 1948. Stagnant, moisture-filled air was trapped in the Donora Valley by surrounding hills and by a temperature inversion that held the cooler air, gradually filling with smoke and fumes from the town's zinc works, against the ground under a lid of lighter, warmer upper air. For five days the fog, turned to smog, remained, with sulfur dioxide emitted by the zinc works converted to deadly sulfur trioxide by contact with the air and building in concentration. Both old and young, with and without past histories of respiratory problems, reported to doctors and hospitals difficulty in breathing and unbearable chest pains. Before the rains washed the air clean nearly a week after the smog buildup, 20 were dead and hundreds hospitalized. A normally harmless, water-saturated inversion had been converted to deadly poison by a tragic union of natural weather processes and human activity.

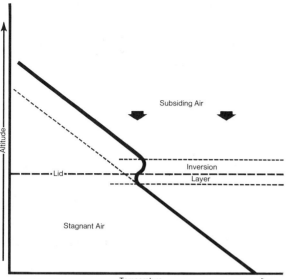

in atmospheric pressure under heated air, and a rise in pressure when cold air is present.

In order to understand the principles of air movements, it is useful to think of air as a liquid made up of a number of fluids of different densities, such as water and gasoline. If the fluids are mixed in a tank and then allowed to separate, the lighter liquid moves to the top. This result represents the vertical motion of air. The heavier liquid spreads out horizontally over the bottom of the tank, so that it is the same thickness everywhere. This flow represents the horizontal movement of air, or the wind. The tendency of air is to attempt to achieve an equilibrium by evening out the pressure imbalances that result from the heating and cooling processes just discussed. Thus the greater the differences in air pressure between places, the greater the wind.

the air. It is a physical law that for equal amounts of cold and hot air, the cold air is heavier. That is why hot-air balloons, filled with lighter air, can rise into the atmosphere.

As early as the 17th century, it was discovered that when a column of mercury was contained within a tube, normal atmospheric pressure at sea level is sufficient to balance the weight of the mercury column to a height of 29.92 inches (76 cm). *Barometers* of varying types are used to record changes in air pressure, and barometric readings in inches of mercury are a normal part, along with recorded temperatures, of every weather report. Since air pressure at a given location changes as surface heating lightens air, barometers record a drop

PRESSURE GRADIENT FORCE

81

Because of differences in the nature of surface, vegetative cover, and other factors affecting energy receipt and retention, there is a strong tendency for zones of high or low air pressure to develop. Sometimes these high- and low-pressure zones cover entire continents, but usually they are considerably smaller—several hundred miles wide. When pressure differences exist between areas, a *pressure gradient* is formed.

Air from the heavier high-pressure areas tends to flow to the low-pressure zones in order to equalize the pressure. The heavy air hugs the earth's surface as it moves, producing winds, and helps to force the warm air to rise. The ve-

Fig. 3.10 If the distance between centers of high pressure and of low pressure is short, the wind is strong. If the pressure differences are great, winds are again strong. The strongest winds are produced by extreme pressure differences over very short distances.

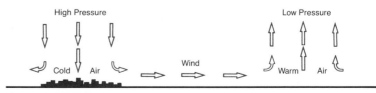

locity, or speed, of the wind is in direct proportion to pressure differences. As depicted in Figure 3.10, winds are caused by pressure differences that tend to induce airflow from points of high to points of low pressure. If distances between high and low pressure are short, pressure gradients are steep and wind velocities are great. The equalization process results in more gentle air movements when zones of different pressures are far apart.

CONVECTION SYSTEM

As an experiment, check the temperature of the room you are in both close to the floor and near the ceiling. Most likely, the temperature is lower near the floor because warm air rises and cool air descends. A convectional wind system results from the flow of air that replaces warm, rising air. Intense surface heating results in rapidly rising air and the rapid movement of replacement air.

A good example of a convectional system is land and sea breezes. Close to a large body of water, the differential in daytime heating between land and water is very great. As a result, the warmer air over the land rises vertically, only to be replaced by cooler air from over the sea. At night, just the opposite occurs. The water is now warmer than the land, which has reradiated much of its heat, and the result is a land breeze toward the sea. These two winds make seashore locations in warm climates particularly comfortable.

MOUNTAIN AND VALLEY BREEZES

Gravitational force causes the heavy cool air that accumulates over snow in mountainous areas to descend into lower valley locations, as suggested on Figure 3.11. Consequently, valleys can become much colder than the slopes, and there is a temperature inversion. Slopes are

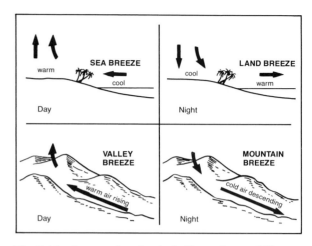

Fig. 3.11 Convectional wind effects due to differential heating and cooling.

the preferred sites for agriculture in some regions because the cold air from the mountain breezes can cause freezing conditions in the valleys. In densely settled narrow valleys where industry is concentrated, air pollution can be particularly dangerous. Mountain breezes usually occur during the night; valley breezes—caused by warm air moving up slopes in mountainous regions—are usually a daytime phenomenon.

CORIOLIS FORCE

In the process of moving from high to low pressure, the wind tends to appear to veer toward the right in the northern hemisphere and toward the left in the southern hemisphere, no matter what the compass direction of the path. This apparent deflection is called the *Coriolis force*. Were it not for this force, winds would move in exactly the direction specified by the pressure gradient.

To illustrate the effect of the Coriolis force upon the winds, a familiar example may be helpful. Think of a line of ice skaters holding hands while skating in a circle, with one of the

82

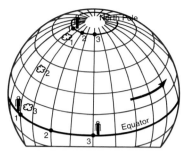 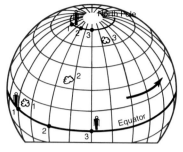

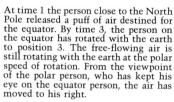

Fig. 3.12   The Coriolis effect.

At time 1 the person close to the North Pole released a puff of air destined for the equator. By time 3, the person on the equator has rotated with the earth to position 3. The free-flowing air is still rotating with the earth at the polar speed of rotation. From the viewpoint of the polar person, who has kept his eye on the equator person, the air has moved to his right.

Air released by the equator person flows faster to the right than the speed of rotation at times 2 and 3 because the release took place at the equator where rotational speed is greatest. From the viewpoint of the equator person, who has kept his eye on the polar person, the air has moved to his right.

skaters nearest the center of the circle. This skater turns slowly, while the skater at the outside of the circle must skate very rapidly in order to keep the line straight. In a similar way, the equatorial regions of the earth are rotating at a much faster rate than the areas around the poles. Next, suppose that the center skater threw a ball toward the skater at the end of the line. By the time the ball arrived, it would pass behind the outside skater. If the skaters are going in a counterclockwise direction, as the earth appears to be moving viewed from the position of the North Pole, the ball passes to the right of the skater. If the skaters are going in a clockwise direction, as the earth appears to be moving viewed from the South Pole position, the deflection is to the left. Because air (like the ball) is not firmly attached to the earth, it, too, will appear to be deflected, as suggested by Figure 3.12. The air maintains its direction of movement, but the earth's surface moves out from under it. Since the position of the air is measured relative to the earth's surface, the air appears to have diverged from its straight path.

The Coriolis force and the pressure gradient force tend to produce spirals rather than simple straight-line patterns of wind, as indicated on Figure 3.13. The spiral of wind is the basic form of the many storms that are so very important to the earth's air circulation system. These storm patterns are discussed later in the chapter.

83

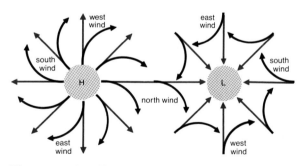

Fig. 3.13   The effect of the Coriolis force on flowing air. The straight green arrows indicate the paths winds would follow flowing out of an area of high (H) pressure, or into one of low (L) pressure, were they to follow the paths dictated by pressure differentials. The black arrows represent the apparent deflecting effect of the Coriolis force. Wind direction—indicated on the diagram for selected black arrows—is always given by the direction from which the wind is coming.

WIND BELTS

Equatorial areas of the earth are zones of low pressure. Intense solar heating in these areas is responsible for a convectional effect. Note on Figure 3.14 how the warm air rises and tends to move away from the *equatorial low* pressure in both a northerly and a southerly direction. As it rises, it cools and eventually becomes heavy. Finally, the heavy air falls, forming zones of high-pressure air. Throughout this

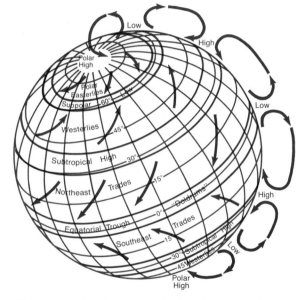

Fig. 3.14 The planetary wind and pressure belts as they would develop on an earth of homogeneous surface. The named high and low pressure belts represent surface pressure conditions; the named wind belts are prevailing surface wind movements responding to pressure gradients and the Coriolis force. In the upper atmosphere conditions are reversed; descending air, creating surface high pressure, represents the contraction of the air column through cooling and the creation of lower pressure aloft. Air rising from surface lows increases upper air pressure at those zones. The upper atmospheric air flows shown on the diagram respond to pressure contrasts at those higher altitudes. Land and water contrasts on the actual earth, particularly evident in the northern hemisphere, create complex distortions of this simplified pattern.

process, there is a tendency for the air to disperse. These areas of *subtropical high* pressure are located at about 30° N and 30° S of the equator.

When this cooled air reaches the earth's surface, it, too, moves in both a northerly and a southerly direction. The Coriolis force, however, modifies the wind direction and creates, in the northern hemisphere, the belts of winds called the *northeast trades* in the tropics and the *westerlies* (really the southwesterlies) in the mid-latitudes. The names refer to the directions from which the winds come. Most of the United States lies within the belt of westerlies; that is, the air usually moves across the country from southwest to northeast. There is also a series of cells of ascending air over the oceans to the north of the westerlies called the *subpolar low*. These areas tend to be cool and rainy. The *polar easterlies* connect the subpolar low areas to the polar high-pressure zones.

This general planetary air-circulation pattern is modified by local wind conditions. It should be clear that these belts move in unison as the vertical rays of the sun change position. For example, equatorial low conditions are evident in the area just north of the equator in the northern-hemisphere summer. The manner in which the seasonal shift takes place has a profound effect on the people of the Indian subcontinent.

The wind, which comes from the southwest in summer in India, reaches the landmass after picking up a great deal of moisture over the warm Indian Ocean. As it crosses the coast mountains and then the foothills of the Himalayas, the monsoon rains begin. *Monsoon* is a term meaning seasonal winds. India's farm economy, and particularly the rice crop, is totally dependent on the summer monsoon rainwater. If, for any of several possible reasons, the wind shift is late or the rainfall is significantly more or less than optimum, crop failure and starvation may result. The undue prolongation of the summer monsoon rains in

1978 caused disastrous flooding and crop and life loss in eastern India and Southeast Asia. The transition to the northeast winter winds occurs gradually across India. First, the dry winds become noticeable in the north of India in September, and by January, most of the subcontinent is dry. Then, beginning in March in the south, the yearly cycle repeats itself.

Ocean Currents

Ocean currents correspond roughly to wind direction patterns because the winds of the world set the ocean currents in motion. In addition, just as differences in air pressure cause wind movements, so differences in the density of water help to force water to move. When water evaporates, the salt and other minerals that will not evaporate are left behind, making the water denser. High-density water exists in areas of high pressure, where descending dry air readily picks up moisture. In areas of low pressure, where rainfall is plentiful, the ocean water is of low density. Wind direction (including the Coriolis effect) and the differences in density cause water to move in wide paths from one part of the ocean to another.

There is an important difference between surface air movements and surface water movements. Landmasses are barriers to water movement, deflecting currents and sometimes forcing them to move in a direction opposite to the main current; air, on the other hand, moves freely over both land and water.

The shape of an ocean basin also has an important effect on ocean current patterns. For

Fig. 3.15 The principal surface ocean currents of the world. Notice how the warm waters of the Gulf of Mexico, the Caribbean, and the tropical Atlantic Ocean drift to northern Europe. (Equa. Count. Cur. = Equatorial Counter Current.)

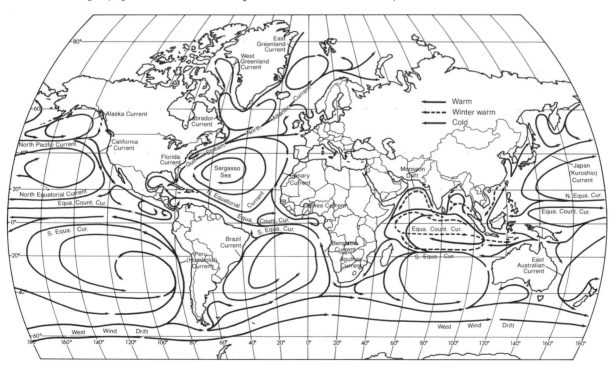

example, the North Pacific current, which moves from west to east, strikes the west coast of Canada and the United States and then is forced to move both north and south (although the major movement is south along the coast of California). In the Atlantic Ocean, however, as Figure 3.15 indicates, the current is deflected in a northeasterly direction by the shape of the coast (Nova Scotia and Newfoundland jut far out into the Atlantic), and then it moves freely across to and past the British Isles and Norway, all the way to the extreme northwest coast of the U.S.S.R. This massive movement of warm water to northerly lands, called the *North Atlantic Drift,* has enormous significance for the inhabitants of those areas. Without it, northern Europe would be much colder than it is.

Ocean currents affect not only the temperature but also the precipitation on the land areas adjacent to the ocean. A cold ocean current near a land area robs the air above the current of its potential moisture supply, thus denying moisture to nearby land. Coastal deserts of the world usually border cold ocean currents. On the other hand, warm ocean currents—such as that off India—bring moisture to the adjacent land area, especially when the prevailing winds are landward.

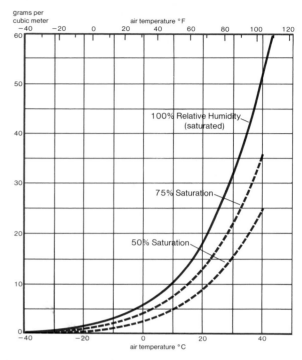

Fig. 3.16 **The water-carrying capacity of air and relative humidity. The actual water in the air (water vapor) divided by the water-carrying capacity ($\times 100$) equals the relative humidity. The solid line represents the maximum water-carrying capacity of air at different temperatures.**

Moisture in the Atmosphere

Air contains water vapor, which is the source of all precipitation. *Precipitation* is water in any form (rain, sleet, snow, hail, dew) deposited on the earth's surface. When the temperature of rising air drops, precipitation occurs. As we know, air pressure is less at higher elevations than at sea level. Therefore, rising air can easily expand because there is less pressure on it. When the heat of the air spreads out through a larger volume, the air

mass becomes cooler. Cool air is less able than warm air to hold water vapor (Figure 3.16). Thus any water vapor in excess of that which the rising, cooling air can possibly hold condenses; that is, it changes from a gas to a liquid. At first, the liquid is in the form of tiny water droplets, usually too light to fall. When there are so many droplets that they coalesce into drops and become too heavy to remain suspended in air, they fall as rain. If temperatures are below the freezing point, the water vapor changes to ice crystals. When enough ice crystals are present, the vapor becomes snow (Figure 3.17).

Rain droplets or ice crystals in large numbers form clouds, which are supported by slight

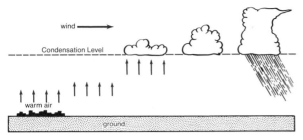

Fig. 3.17 As warm air rises, it cools. As it cools, its water vapor condenses and clouds form. If the water content of the air is greater than the air's capacity to retain moisture, some form of precipitation occurs.

upward movements of air. The form and altitude of clouds depend on the amount of water vapor in the air, the temperature, and the wind movement. Descending air in high-pressure zones usually yields cloudless skies. Whenever warm, moist air rises, clouds form. The most dramatic cloud formation is probably the *cumulonimbus*, pictured in Figure 3.18. This is the anvil-head cloud that often accompanies heavy rain. Low, gray *stratus* clouds appear more often in cooler seasons than in the warmer months. The very high, wispy *cirrus* clouds that may appear in all seasons are made entirely of tiny ice crystals.

The amount of water vapor in the air compared to the maximum amount that the air can possibly hold at a given temperature is called the air's *relative humidity*. The warmer the air, the greater the amount of water vapor it can contain. If the relative humidity is 100%, the air is completely saturated with water vapor. A value of 60% on a hot day means that the air is extremely humid and very uncomfortable. A 60% reading on a cold day, however, tells us that although the air contains relatively large amounts of water vapor, it holds in absolute terms much less than on the hot, muggy day. This example demonstrates the point that relative humidity is meaningful only if we keep the air temperature in mind.

If, one evening, you measured the relative humidity and the air temperature, you could calculate at what temperature condensation would occur (Figure 3.16). If you saw dew or frost on the ground the next morning, you would know that the temperature did in fact drop to the level at which condensation took place. This critical temperature for condensation is the *dew point*. When the air temperature goes below the dew point, condensation occurs. Foggy or cloudy conditions on the earth's surface imply that the dew point has been reached and that the relative humidity is 100%.

TYPES OF PRECIPITATION

When large masses of air rise, precipitation may take place. It can be one of three types: (1) convectional, (2) orographic, or (3) cyclonic or frontal precipitation.

Convectional precipitation results from rising, heated, moisture-laden air. As the air rises, it cools; when its dew point is reached, condensation and precipitation occur, as Figure 3.19 shows. This process is typical of summer storms or showers in tropical and continental climates. Usually the ground is heated during the morning and the early afternoon. The warm air that accumulates begins to rise, and at first, cumulus clouds and then huge cumulonimbus clouds develop. Finally, there is thunder and heavy rainfall, which, since the storm is moving, may affect each part of the ground for only a brief period. It is not unusual for these convectional storms to occur in later afternoon or early evening.

If quickly rising air currents violently circulate the air within the cloud, ice crystals may form near the top of the cloud. These ice crystals may begin to fall when they are large enough, but a new updraft can force them back up, with a subsequent enlargement of the pieces of ice. This process may occur again and

87

(a)

(b)

(c)

Fig. 3.18 Cloud types. (a) cumulonimbus; (b) stratus; (c) cirrus. (National Oceanic and Atmospheric Administration.)

again, until the ice can no longer be sustained by the updrafts and pieces of ice fall to the ground as hail.

Orographic precipitation, depicted in Figure 3.20, occurs when warm air is forced to rise because hills or mountains stand in the way of moisture-laden winds. This type of precipitation is typical of areas where mountains and hills are situated close to oceans or large lakes. Saturated air from over the water blows onshore, rising as the land rises. Again, the process of cooling, condensation, and precipitation takes place. The windward side (the side toward which the winds blow) of the hills and mountains receives the precipitation. The other side, called the *leeward*, and the adjoining regions downwind are very often dry. The air that passes over the mountains or hills de-

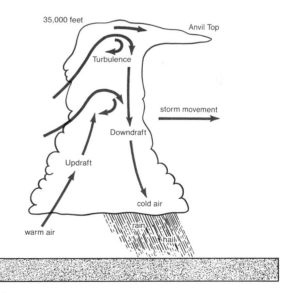

Fig. 3.19 When warm air laden with moisture rises, a cumulonimbus cloud may develop and convectional precipitation occur. The turbulence within the system creates a downdraft of cold upper altitude air.

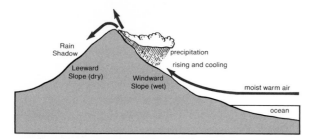

Fig. 3.20 Orographic precipitation. Surface winds may be raised to higher elevations by hills or mountains lying in their paths. If such orographically lifted air is sufficiently cooled, precipitation occurs. Descending air on the leeward side of the upland barrier becomes warmer, its capacity to retain moisture is increased, and water absorption rather than release takes place.

scends and warms, and as we have seen, descending air does not produce precipitation and warming air absorbs moisture from surfaces over which it passes.

Cyclonic or *frontal* precipitation, the third type, is common to the mid-latitudes, where cool and warm air masses meet. It also occurs in the tropics, though less frequently, in the form of hurricanes or typhoons. The most extreme form of cyclonic storm is the tornado, which is much more a windstorm than a rainstorm. In order to understand cyclonic or frontal precipitation, one must first visualize the nature of air masses and the way cyclones develop.

Air masses are large bodies of air with similar temperature and humidity characteristics throughout; they form over a *source region.* Source regions include large areas of uniform surface and relatively consistent temperatures, such as the cold land areas of northern Canada or the north central part of the Soviet Union and the warm tropical water areas in any of the oceans close to the equator. Source regions for North America are shown in Figure 3.21. During a period of a few days or a week, an air mass may form in a source region. For example, in northern Canada, in the fall, when snow has

already covered the vast subarctic landscape, cold, heavy, and dry air develops over the frozen land surface. A further discussion of air masses as regional entities may be found in Chapter 12.

When it is large enough, this continental polar air mass begins to move toward the lighter, warmer air to the south. The leading edge of this tongue of air is called a *front.* The front in this case separates the cold, dry air from whatever air is found in its path. If a warm, moist air mass is in the path of the polar air mass, the heavier cold air hugs the ground and forces the lighter air up above it. The rising moist air condenses, and frontal precipitation occurs. On the other hand, if it is the warm air that is on the move, it slides over the cold air, and again, precipitation occurs. In the first case, when cold air plows into warm air, cumulonimbus clouds form and precipitation is brief and heavy. As the front passes, temperatures drop appreciably, the sky clears, and the air

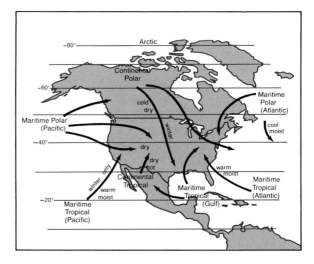

Fig. 3.21 Source regions for air masses in North America. The United States and Canada, lying between major contrasting air-mass source regions, are subject to numerous storms and changes of weather. (After Haynes, U. S. Department of Commerce.)

90

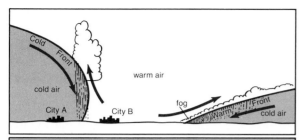

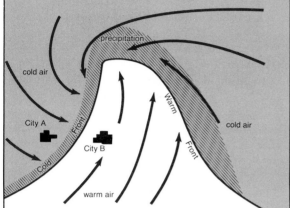

Fig. 3.22 In this diagram, the cold front has recently passed over city A and is heading in the direction of city B. The warm front is moving away from city B.

becomes noticeably drier. In the second case, when warm air is moving over cold air, steel-gray nimbostratus (*nimbo* means "rain") clouds form, and the precipitation is steady and long-lasting. As the front passes, warm, muggy air becomes characteristic of the area. Figure 3.22 summarizes the movement of fronts.

STORMS

When two air masses come into contact, there is the possibility that storms will develop. If the contrasts in temperature and humidity are sufficiently great, or if the wind directions of the two masses that touch are opposites, a wave might develop in the front as

shown in Figure 3.23. Once established, the waves may enlarge and become wedgelike. One wedge is the cooler air moving along the surface, and the other wedge is the warmer air moving up over the cold air. In both wedges, the warm air rises, and a low-pressure center forms. Considerable precipitation is accompanied by counterclockwise winds around the low pressure area. A large system of air circulation centered on a region of low atmospheric pressure such as the one described is called a *cyclone.*

A cyclone may be a weak storm or one of great intensity. The storm is likely to begin over warm tropical waters, far enough from the

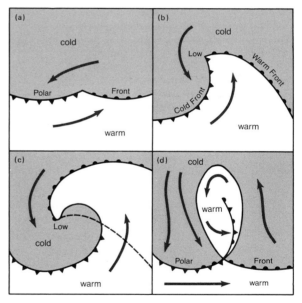

91

Fig. 3.23 When wedges of warm and cold air develop along a low-pressure trough in the mid-latitudes, there is the possibility of cyclonic storm formation. (a) A wave begins to form along the polar front. (b) In the northern hemisphere, cold air begins to turn in a southerly direction, while warm air moves north. The meeting lines of these unlike air masses are called *fronts.* Cold air overtakes (c) and isolates (d) the warm air pocket, removing it from its energy and moisture source. The cyclonic storm dissipates as the polar front is reestablished.

Weather Modification

Not only do people talk about the weather, they have always tried to do something about it. Primitive cultures had ceremonies to induce rainfall or to propitiate the Goddess of Spring so that winter might end. More modern societies have tried technology rather than religion to bend nature to their whim. In the last years of the 19th century, the United States 51st Congress appropriated $9000 for testing the feasibility of blasting rain from clouds by detonating dynamite carried aloft in balloons and rockets. Modern weather modification, largely based on seeding clouds with condensation nuclei to produce ice crystals, began in 1946, when Irving Langmuir and Vincent Shaefer discovered first that dry ice and, later, that any material colder than $-40°C$ $(-40°F)$ would induce condensation. Subsequently, silver iodide was found to be equally satisfactory. The experimental path to scientific weather modification had been opened.

The number and variety of projects designed to modify weather increased greatly after 1960. Project Stormfury, under the direction of the National Hurricane Research Laboratory, was launched in 1962 to explore the possibility of lessening storm intensity through cloud seeding. Although experimental results are varyingly interpreted, the N.H.R.L. claims that the seeding of hurricane Debbie in August of 1969 decreased wind velocities by 30%.

Cloud seeding has been used successfully to disperse cold ground fogs around airports; seeding has increased the snowpack in the Rockies by 15% in some areas, with beneficial effects on river flow, hydroelectric generation, and reservoir storage. Hail suppression through seeding has been a favored objective in agricultural areas. In this case, silver iodide appears to avert large hail formation by causing rapid condensation into raindrops or small hailstones, which melt before reaching the ground. The Russians are active practitioners of hail suppression and claim more than a 70% reduction of crop damage in some experiments. American research suggests decreases of 30%–40% in hail amounts reaching the ground.

Rainmaking, however, has been the most avidly pursued and hotly contested aspect of weather modification. The Florida drought of 1971 is a case in point. Cloud seeding apparently produced some 100,000 acre feet of water, relieving drought in south Florida and simultaneously ruining 3500 acres of tomatoes. Early in 1972, seeding appeared to produce a 12- to 14-inch rainfall west of Rapid City, South Dakota; an ensuing flood and dam rupture killed 230 people and caused more than $150 million in property damage. The cause-and-effect relationship is in dispute. Despite the claimed adverse consequences of rainmaking and the institution of legal action to recover presumed damages, power companies, counties, states, and agricultural associations—particularly in the West and the Midwest—regularly engage in rainmaking attempts. Seeding operations are also actively carried out in the Soviet Union, Australia, Canada, France, Israel, Italy, and Switzerland.

The questions increasingly being raised are: Whose cloud is up there and who has the right to induce a rainfall that may damage someone else's interests? Does rainmaking in one area "steal" moisture from other locations where, by natural causes, it might fall, and if so, what legal liability is involved?

No federal legislation, except a reporting requirement, governs weather modification in the United States, although most states of the West and the Midwest have governing statutes of one form or another. International efforts at control of weather modification, however, are being actively pursued. In response to American cloud seeding for military purposes during the Vietnam war, the United Nations in 1974 called for a draft resolution outlawing weather or environmental modification as a weapon of war. By the late 1970s, the United States was seeking international agreement on cooperation in weather modification to assure minimization of the adverse global consequences of individual national programs.

equator for the Coriolis force to be significant. Here a wave that is not associated with a front may form. If conditions are such that a wedge develops, an intense tropical storm may grow, fed by the energy embodied in the rising moist, warm air. This storm is called a *hurricane* in the Atlantic region (we met Agnes at the start of this chapter) and a *typhoon* in the Pacific area. Figure 3.24 shows the paths of hurricanes in the Atlantic. The winds of these storms whirl about the center, called the *eye*. But within the eye of the storm, there are gentle breezes and relatively clear skies. Over land, these storms lose their warm-water energy source and subside quickly. If they move far into the colder northern waters, they are pushed or blocked by other air masses and also lose their energy source and abate.

The most violent storm of all is the *tornado*, but it is also the smallest (Figure 3.25). A typical tornado is less than 100 feet in diameter. Tornadoes are spawned in the huge cumulonimbus clouds that sometimes travel in advance of a cold front along a squall line. The central part of the United States in spring or fall, when adjacent air masses contrast the most, is the scene of many of these funnel-shaped killer clouds. Although winds are esti-

mated to reach 500 miles per hour, these storms are small and usually travel on the ground for less than a mile, so that they affect only limited areas.

93

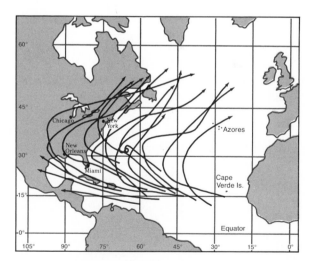

Fig. 3.24 **Tracks of typical Atlantic hurricanes. In the United States, the most vulnerable areas are the Gulf Coast of Florida; Cape Hatteras, North Carolina; Long Island, New York; and Cape Cod, Massachusetts. (After U. S. Navy Oceanographic Office.)**

Climate

We have traced some of the causes of weather changes that occur as air from high-pressure zones flows toward low-pressure areas, fronts pass and waves develop, dew points are reached, and sea breezes arise. As we have noted, in some parts of the world these changes occur more rapidly and more often than in other parts.

Weather conditions from day to day can be explained by the principles we have developed. However, one cannot understand the effect of the weather elements—temperature, precipitation, and air pressure—unless one is conscious of the nature of the earth's surface. Weather forecasters in each location on earth must deal with the elements of weather in the context of the local environment.

The complexities of daily weather conditions may be summarized by statements about climate. The climate of an area is a generalization based on daily and seasonal weather conditions. Are the summers warm on the average? Is there a tendency toward heavy snow in the winter? Are the winds normally from the southeast? Are the climatic averages typical of daily weather conditions, or are the variations from day to day or from week to week so great that one should speak of average variations rather than just of averages? These are the questions we must ask in order to form an intelligent description of the differences in conditions from place to place.

Rather than develop an intricate system for the study of climate, we shall present a straightforward guide to the description of global climates. The two most important elements that differentiate weather conditions are temperature and precipitation. While air pressure is also an important weather element, differences in air pressure are hardly noticeable without the use of a barometer. Thus we may regard warm, moderate, cold, and very cold temperatures as characteristic of a place or region. In addition, high, moderate, and low precipitation are good indicators of the degree of humidity or aridity in a place or region. Thus we shall take these two scales, define the terms more precisely, and map the areas of the world having various combinations of temperatures and precipitation.

Because extreme seasonal changes do occur, two global climatic maps are shown in Figure 3.26, one for winters (a) and one for summers (b). Maps could have been developed for each of four seasons, or for the 12 months, but these two give a rather good, though brief, description of climatic differences. Bear in mind that a summer map of the world is a combination of the climates of the northern hemisphere in July, August, and September and of the southern hemisphere in January, February, and March because the seasons in the two hemispheres are reversed.

In the following paragraphs, the major climates are briefly described, and an indication is given of the important characteristics of their typifying natural vegetation, that is, the plant life that would exist in each area if humans did not directly interfere in the growth process. Natural vegetation, little of which remains today in densely settled portions of the globe, has close interrelationships with soils, landforms, groundwater, elevation, and other features of habitat. The tie between *biochores*—the major structural subdivisions of

95

Fig. 3.25 (opposite) Tornado near the Denver, Colorado, airport, May 1975. In the United States, these violent storms occur most frequently in the central and central southern part of the country (especially in Oklahoma, Kansas, and the Texas Panhandle), where cold polar air very often meets warm, moist Gulf air. (National Oceanic and Atmospheric Administration.)

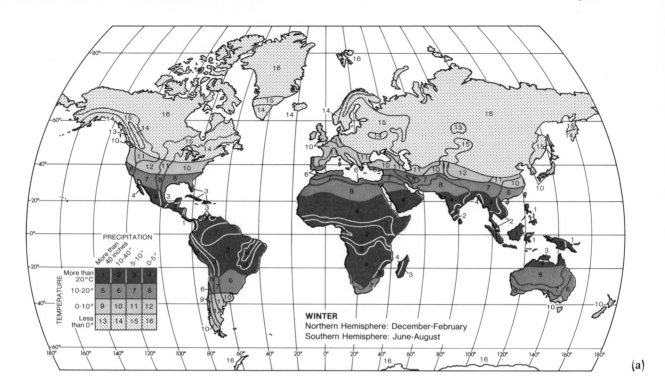

PRECIPITATION

	More than 40 inches	10-40"	5-10"	0-5"
More than 20°C	1	2	3	4
10-20°	5	6	7	8
0-10°	9	10	11	12
Less than 0°	13	14	15	16

TEMPERATURE

WINTER
Northern Hemisphere: December-February
Southern Hemisphere: June-August

(a)

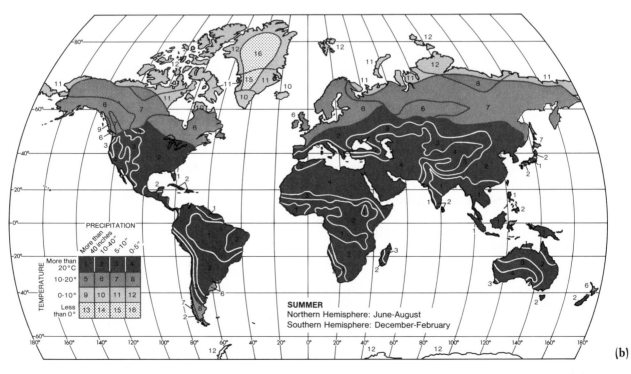

PRECIPITATION

	More than 40 inches	10-40"	5-10"	0-5"
More than 20°C	1	2	3	4
10-20°	5	6	7	8
0-10°	9	10	11	12
Less than 0°	13	14	15	16

TEMPERATURE

SUMMER
Northern Hemisphere: June-August
Southern Hemisphere: December-February

(b)

Fig. 3.26 (opposite) These maps combine temperature and precipitation data to display seasonal variations in the basic components of climate. In reading the maps, remember that the summer climate shown on (b) represents the summer season for both the northern and the southern hemispheres. This means that June, July, and August data were used for the north latitudinal part of the world, and December, January, and February data were used for the south latitudes. As a result, one may obtain a summer view of the world—one of nearly universally warm temperatures and large amounts of precipitation. The winter view (a) shows a world of greatly varying temperatures and small amounts of precipitation.

natural vegetation—and climate is particularly close. Indeed, the earliest maps of world climatic zones were based not upon statistical variation in recorded temperatures, pressures, precipitation, and other measures of the atmosphere but upon observed variation in vegetative regions. Even the descriptive names of the biochores—*forest, savanna, grassland, desert*—are climatic in their implications. The discussion of climates, therefore, benefits from a visualization of regional differences by reference to the type of mature plant communities that, in nature, developed within those climates.

TROPICAL RAIN FOREST (1; 1)*

The areas that straddle the equator are located generally within the equatorial low-pressure zone. They have warm, wet climates in both winter and summer (Figure 3.27). Their rainfall usually comes in the form of daily convectional thunderstorms. Although most days are sunny and hot, by afternoon cumulonimbus clouds form and convectional rain falls.

The natural vegetation is tropical rainforest, which is still present in large areas relatively untouched by humans, such as the Amazon basin of South America and the Congo River basin of Africa. Tall, dense forests of broadleaf trees and heavy vines predominate. Among the

---

* The numbers following the section headings refer to the keys on Figure 3.26. The first figure represents typical winter conditions, and the second is for summer. Each represents idealized conditions. The sample stations and alternates shown on the climagraphs may vary from the norms established in the text.

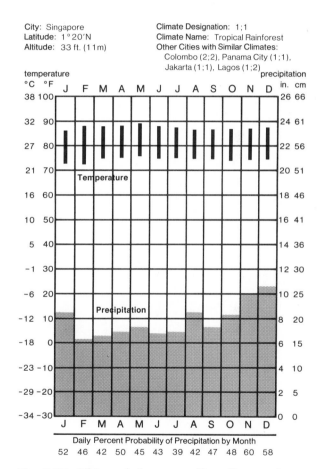

97

City: Singapore
Latitude: 1°20′N
Altitude: 33 ft. (11m)

Climate Designation: 1;1
Climate Name: Tropical Rainforest
Other Cities with Similar Climates:
  Colombo (2;2), Panama City (1;1),
  Jakarta (1;1), Lagos (1;2)

Daily Percent Probability of Precipitation by Month
52 46 42 50 45 43 39 42 47 48 60 58

Fig. 3.27 This and the succeeding climate charts (climagraphs) show average daily high and low temperatures for each month, the average precipitation for each month, and the probability of precipitation on any particular day in a designated month. For Singapore, the average daily high temperature in August is 87°F (30.5°C), the low is 75°F (24°C). The rainfall for the month, on average, is 8.4 inches (21 cm), and on a given day in August, there is a 42% chance of rainfall.

Fig. 3.28 (opposite) Tropical rain forest. The vegetation is characterized by tall, broad-leaf, hardwood trees and vines. (U.S.D.A. photo.)

hundreds of species of trees that are found here, there are dark woods, light woods, and woods of different colors, as well as spongy softwoods such as balsa and hardwoods such as teak and mahogany (Figure 3.28). Rain forests also extend some distance away from the equator along coasts where prevailing winds supply a constant source of moisture to coastal uplands and where the orographic effect provides enough precipitation for heavy vegetation to develop.

SAVANNA (3; 1)

As the sun's vertical rays become more distant from the equator in summer, the equatorial low-pressure zone follows the sun's path. Thus the areas to the north and south of the rain forest are wet only in the summer months. The low-sun months, although still hot, are dry because the moist equatorial low has been replaced by the dry air of the subtropical highs. These areas are known as *savanna lands* be-

Fig. 3.29 The parklike landscape of grasses and trees, characteristic of the tropical savanna, is seen in this view of Ocala National Game Refuge, Florida. (U.S.D.A. Forest Service.)

99

cause of the kind of natural vegetation found there. Although the natural vegetation of savanna areas appears to have been a form of scrub forest, the savanna now is recognized as a grassland with widely dispersed trees. The tendency toward a more forested cover has been discouraged through millennia of periodic firings. Sometimes, savannas seem to have been purposely designed, for they have a parklike look, as indicated by Figure 3.29. The East African region of Kenya and Tanzania contains well-known grasslands and fire-resisting species of trees where large animals like giraffes, lions, and elephants roam. The *campos*

and *llanos* of South America are other huge savanna areas.

There are usually no marked breaks between one kind of climate and another; instead there are zones of transition. These transitional zones are typical of plains and plateaus (they are not so gradual in mountainous regions). Between the tropical rain forest and the savanna, there are other kinds of forests that are less dense than the rain forest itself.

A special case needs mentioning. In Asia, when summer monsoon winds carry water-laden air to the mainland, far more rain falls on the hills, mountains, and adjacent plains than

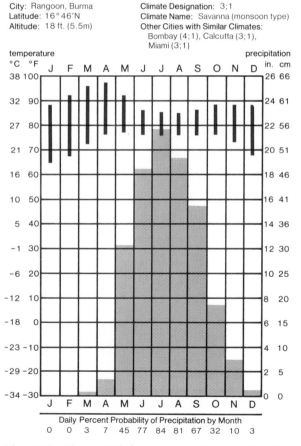

Fig. 3.30 **Climagraph for Rangoon, Burma. (See Figure 3.27 for explanation.)**

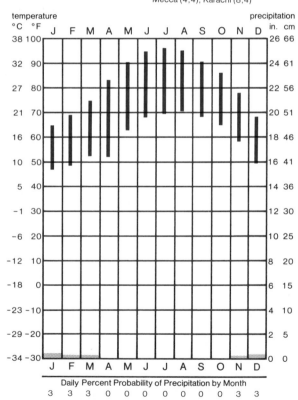

Fig. 3.31 **Climagraph for Cairo, Egypt. (See Figure 3.27 for explanation.)**

100

in the savanna (notice the pattern of precipitation on Figure 3.30). As a result, the vegetation is much denser, even though the winters are dry. Jungle growth and large forests are the natural vegetation here. Much of this vegetation, however, no longer exists because people have been using the land for rice and tea production for many generations.

HOT DESERTS (7; 4)

Eventually the grasses shorten, and desert shrubs become evident on the poleward side of the savannas. Here we approach the belt of subtropical high pressure that brings considerable sunshine and hot summer weather, but very little precipitation. Note the almost total lack of rainfall on Figure 3.31. The precipitation that does fall is of the convectional variety, but it is sporadic. As conditions become drier, there are fewer and fewer drought-resistant shrubs, and in some areas only gravelly and sandy deserts exist, as suggested on Figure 3.32. The great hot deserts of the world, such as the Sahara, the Arabian, the Australian, and the Kalahari deserts, are all the products of high-pressure zones. Often the driest parts of these deserts are on the west coast, where cold ocean currents are found, as Figure 3.26 indicates. Earlier mention was made of the relationship between cold ocean currents and deserts (see page 86).

MEDITERRANEAN CLIMATE (6; 3)

Patterns of climate would be neatly defined, paralleling the lines of latitude, were it not for mountain ranges, warm or cold ocean currents, and particularly land–water configurations. These factors cause the greatest variations in the middle latitudes. Thus the patterns just described, although generally following the earth's pressure belts, do not fit perfectly. There are, however, a number of general rules that help us to understand climates in the complex mid-latitude zone.

Mid-latitude winds are generally from the west in both the northern and the southern

101

Fig. 3.32 Devoid of stabilizing vegetation, desert sands may be constantly rearranged in complex dune formations. (Photo by J. A. Bier.)

hemispheres, and a significant amount of the precipitation is generated by frontal systems. Thus it is important to know whether the water is cold or warm near land areas. Several climatic zones are noticeable in the middle latitudes. They are all marked by warm summer temperatures, except in those areas cooled by westerly winds from the ocean. To the poleward side of the hot deserts, there is a transition zone between the subtropical high and the moist westerlies zones. Here, cyclonic storms bring rainfall only in the winter, when the westerlies shift toward the equator. The summers are dry and hot as the subtropical highs shift slightly poleward (Figure 3.33), and the winters are not cold. These conditions describe the *Mediterranean climate*, which is often found on west coasts of continents in lower-middle latitudes. Southern California, the Mediterranean area itself, western Australia, the tip of South Africa, and central Chile in South America are characterized by such a climate. In these areas, where there is enough precipitation, shrubs and small deciduous trees (trees that lose their leaves once a year) such as scrub oak grow, as shown on Figure 3.34.

The Mediterranean climate area—long and densely settled in Europe, the Near East, and North Africa—has more moisture and a greater variety of vegetation than is found in the desert. Clear, dry air predominates and winters are relatively short and mild. Plants and flowers grow all year around. Even though the summers are hot, the nights are usually cool and clear. Much of the vegetation in this area is now in the form of crops.

MARINE WEST COAST CLIMATE (10; 6)

Closer to the poles, but still within the westerly wind belt, are areas of *marine west coast climate*. Here, cyclonic storms play a relatively large role. In the winter, more rain falls and cooler temperatures prevail than in the Mediterranean zones. Compare the patterns of precipitation in Figures 3.33 and 3.35. Little rain falls in summer in the transitional zone just poleward of the Mediterranean climate, but further toward the poles, rainfall increases appreciably in summer and even more in winter. Marine winds from the west moderate both summer and winter temperatures. Thus summers are pleasantly cool, and winters, though cold, do not normally produce freezing temperatures.

This climate exists somewhere on all continental west coasts that reach sufficient latitude, but it affects relatively small land areas

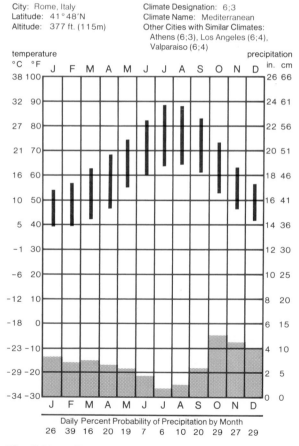

Fig. 3.33 Climagraph for Rome, Italy. (See Figure 3.27 for explanation.)

102

Fig. 3.34 Vegetation typical of an area with a Mediterranean climate. Trees such as scrub oak, pictured above, are short and scattered. (U.S.D.A. Forest Service.)

in all but one region. Because northern Europe has no great mountain belt to thwart the west-to-east flow of moist air, the marine west coast climate stretches well across the continent to Poland. At that point, cyclonic storms originating in the Arctic regions begin to be felt. Northern Europe's moderate climate, then, owes its existence to a lack of mountains and to a relatively warm ocean current whose influence is felt for about 1000 miles, from Ireland to central Europe.

The orographic effect from the mountains in areas such as the northwestern United States, western Canada, and southern Chile produces enormous amounts of precipitation, very often in the form of snow on the windward side. Vast coniferous forests (needle-leaf trees, such as pine, spruce, and fir) cover the mountains' lower elevations. Because the mountains prevent the moist air from continuing to the lee-ward side, mid-latitude deserts are found to the east of these marine west coast areas.

HUMID SUBTROPICAL CLIMATE (6; 2)

On the east coasts of continents, the transition is from the equatorial climate to the *humid subtropical climate.* Here, convectional summer showers and winter cyclonic storms are the sources of precipitation. As illustrated by Figure 3.36, this climate is one of hot, moist summers and moderate, moist winters. In the fall, on occasion, hurricanes strike the coastal areas.

The even distribution of rainfall allows for forests of deciduous, hardwood trees such as oak and maple, whose leaves turn orange and red before falling in autumn. In addition, conifers become mixed with the deciduous trees as a second-growth forest.

The transition poleward to the continental climates is accompanied by increasingly colder winters and shorter summers. In this direction as well, cyclonic storms become more responsible for rainfall than convectional showers,

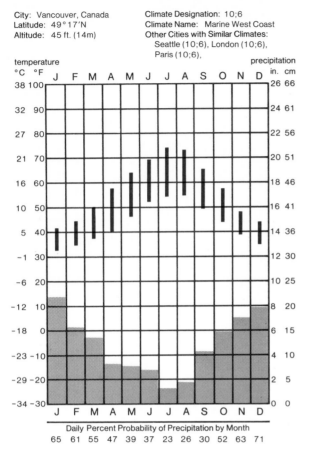

City: Vancouver, Canada
Latitude: 49°17′N
Altitude: 45 ft. (14m)

Climate Designation: 10;6
Climate Name: Marine West Coast
Other Cities with Similar Climates:
Seattle (10;6), London (10;6),
Paris (10;6),

temperature
°C °F

precipitation
in. cm

Daily Percent Probability of Precipitation by Month
65 61 55 47 39 37 23 26 30 52 63 71

Fig. 3.35 Climagraph for Vancouver, Canada. (See Figure 3.27 for explanation.)

City: Charleston, S.C.
Latitude: 32°46′N
Altitude: 16 ft. (5m)

Climate Designation: 6;2
Climate Name: Humid Subtropical
Other Cities with Similar Climates:
Canton (6;2), Sydney (6;2),
New Orleans (6;2)

temperature
°C °F

precipitation
in. cm

Daily Percent Probability of Precipitation by Month
32 32 29 27 26 37 45 45 33 19 23 29

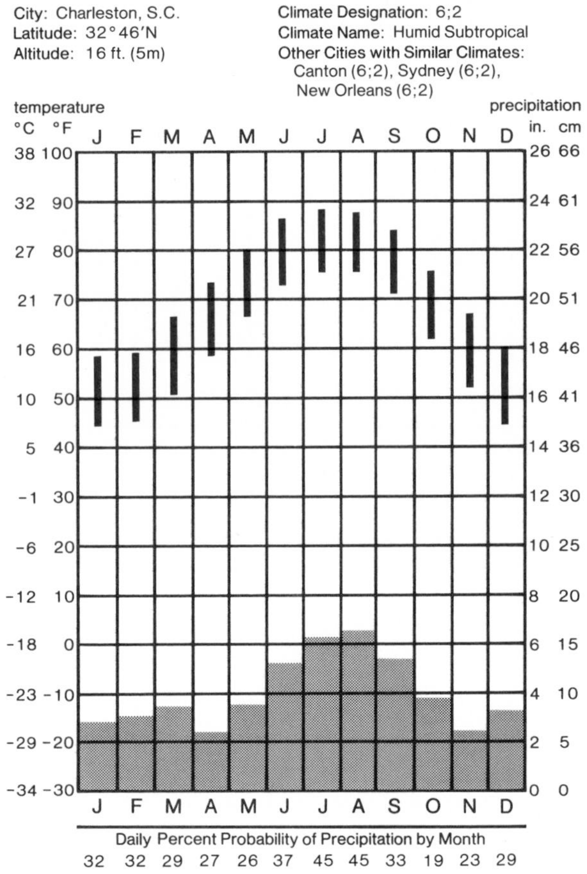

Fig. 3.36 Climagraph for Charleston, South Carolina. (See Figure 3.27 for explanation.)

104

but at this point, the region can no longer be described as humid subtropical; rather it is described as humid continental. Southern Brazil, the southeastern United States, and southern China all have the humid subtropical climate.

HUMID CONTINENTAL CLIMATE (10; 2)

The air masses that drift toward the equator from their origin close to the poles and the air masses that drift toward the poles from the tropics produce frontal precipitation. Whenever warmer air or marine air is in the way of the old continental air masses, or vice versa,

frontal storms develop. We speak of the climates that these air masses influence as *humid continental*. Figures 3.37 and 3.38 show the range and dominance of winter temperatures within this climatic type. The continental climate may be contrasted to the marine west coast climates, the former having prevailing winds from the land, the latter from the sea. Coniferous forests become more plentiful in the direction of the poles, until temperatures become so low that trees are denied an adequate growing season and are therefore stunted (Figure 3.39). The transition from the very cold air masses of the winter to the occasional con-

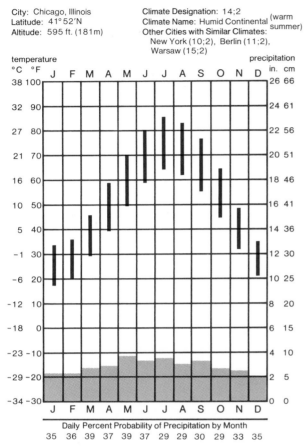

Fig. 3.37 Climagraph for Chicago, Illinois. (See Figure 3.27 for explanation.)

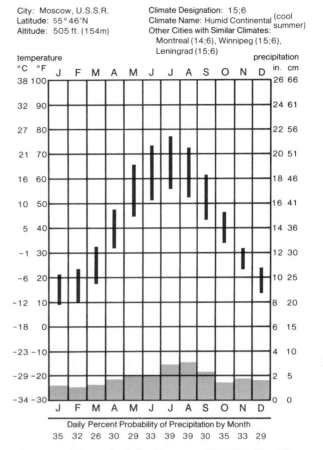

Fig. 3.38 Climagraph for Moscow, U.S.S.R. (See Figure 3.27 for explanation.)

105

vectional storms of the summer means that there are four distinct seasons.

Three huge areas of the world are characterized by the humid continental climate: (1) the north and central United States and southern Canada, (2) most of the European portion of the Soviet Union, and (3) northern China. Because there are no land areas at a comparable latitude in the southern hemisphere, this climate is not represented there. In fact, the only nonmountain cold climate in that hemisphere is the polar climate of Antarctica.

Fig. 3.39 Here, near the source of the Volga River to the north of Moscow, U.S.S.R., the summers are long and warm enough to support a dense coniferous forest. Farther north, growth is less luxuriant. (Novosti Press Agency.)

MID-LATITUDE SEMIDESERTS AND DRY-LAND CLIMATES (10; 4)

In the interior of the continents, behind mountains that block the west winds, or in lands far from the reaches of moist tropical air, there are extensive regions of *semidesert* conditions. Figure 3.40 illustrates typical temperature and precipitation patterns for these areas. Occasionally, a summer convectional storm or a frontal system with some moisture still available occurs. These drier lands are called *steppes*. The natural vegetation is grass, although desert shrubs, pictured in Figure 3.41,

are found in the drier portions of the steppes. Although rain is not plentiful, the soils are rich because the grasses return nutrients to the soil. As a result, the steppes of the United States, Canada, the Soviet Union, and China are among the most productive agricultural regions of the world. These are the wheat, rye, and oats areas of the world. They are also known for their hot, dry summers and for biting winter winds that sometimes bring blizzards.

SUBARCTIC AND ARCTIC CLIMATES (16; 7)

Toward the north and into the interior of the North American and Eurasian land masses, colder and colder temperatures prevail (Figure 3.42). Trees become stunted, and eventually only mosses and other cool-weather plants of the type shown in Figure 3.43 will grow. Because long cold winters predominate, the ground is frozen most of the year. Very often the word *tundra* is used to describe the northern boundary zone beyond the treed subarctic regions. A few cool summer months, with an abundant supply of mosquitoes, break up the monotony of extremely cold, but not very snowy, conditions. Strong easterly winds blow snow, which, combined with ice fogs and little winter sunlight, contributes to a very bleak climate indeed. Alaska, northern Canada, and the northern U.S.S.R. are covered with the stunted trees of the subarctic climate or by the bleak, treeless expanse of the tundra. Antarctica and Greenland, however, are icy deserts.

These thumbnail sketches of climatic conditions throughout the world give us the basic patterns of the larger regions. On any given

106

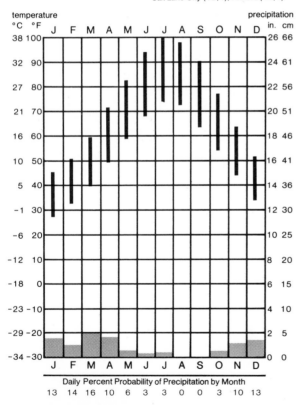

City: Tehran, Iran
Latitude: 35°41′N
Altitude: 4002 ft. (1220m)

Climate Designation: 10;4
Climate Name: Midlatitude Dryland
Other Cities with Similar Climates:
    Salt Lake City (12;4), Ankara (10;4)

Daily Percent Probability of Precipitation by Month
13 14 16 10 6 3 3 0 0 3 10 13

Fig. 3.40   Climagraph for Tehran, Iran. (See Figure 3.27 for explanation.)

Fig. 3.41 (opposite)   Desert shrubs in the mid-latitude drylands of northern Mexico. (United Nations/Jerry Frank)

City: Fairbanks, Alaska
Latitude: 64°51′N
Altitude: 440 ft. (134m)

Climate Designation: 16;6
Climate Name: Subarctic
Other Cities with Similar Climates:
  Yellowknife (15;6), Yakutsk (16;7)

Fig. 3.42 Climagraph for Fairbanks, Alaska. (See Figure 3.27 for explanation.)

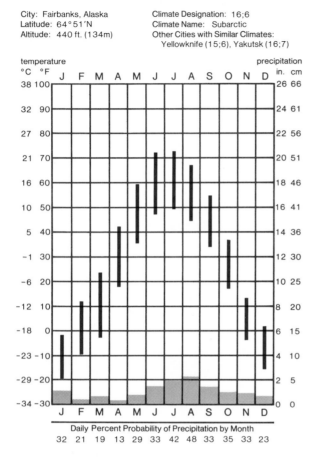

Daily Percent Probability of Precipitation by Month
32 21 19 13 29 33 42 48 33 35 33 23

108

day, conditions may be quite different from those discussed or mapped, but it is the physical, climatological processes in general that concern us. We can deepen our knowledge of climates by applying our understanding of the elements of weather.

Climatic Change

It has been stressed that climates are only averages of perhaps greatly varying day-to-day changes. Figure 3.44 gives an idea of the global variation in precipitation from year to year. Temperatures are less changeable than precipitation on a year-to-year basis, but they too vary. How can we account for these variations? Scientists in research stations all over the world are investigating this question. The data they use range from daily temperature and pre-

Fig. 3.43 Tundra vegetation in the Ruby Range of the Northwest Territories, Canada. (Photo by C. S. Alexander.)

cipitation records to calculations of the position of the earth relative to the sun. Because day-to-day records for most places go back only 50–100 years, scientists look for additional information in rock formations, the chemical composition of earth materials, and astronomical changes.

The fact that a glacier covered most of Canada and about one-third of the United States about 20,000 years ago, as well as at three other times in the last 1 million years, has raised the question of periodic changes in climate. By analyzing layers of fossil microorganisms on the ocean floor, scientists have determined that for the last 400,000 years, there have been climatic cycles of 100,000 years' duration. Some scientists believe that this 100,000-year cycle corresponds to the cyclical changes in the

earth's orbit around the sun and that such changes are responsible for the succession of ice ages. When the orbit is nearly circular, the earth experiences relatively cold temperatures, and when it is elliptical, as it is now, the earth is exposed to more total solar radiation and thus experiences warmer temperatures.

Another cycle of 42,000 years corresponds to the tilt of the earth relative to the orbital plane. The tilt varies from 22.1 degrees to 24.5 degrees. A low tilt position—that is, a more perpendicular position of the earth—is accompanied by periods of colder climate. Cooler summers are thought to be critical in the formation of ice sheets.

Even though the last ice age ended about 11,000 years ago, and the earth is now in one of its warmest periods, scientific evidence sug-

Fig. 3.44 The world pattern of precipitation variability. Note that regions of low total precipitation tend to have high variability. In general, the lower the amount of long-term average annual precipitation, the lower is the probability that the "average" will be recorded in any single year.

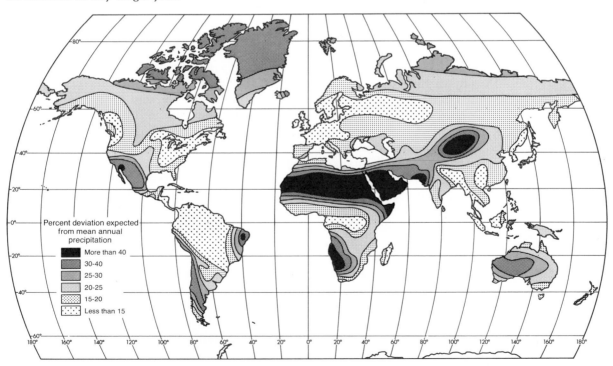

Our Inconstant Climates

In 1125 William of Malmesburg favorably compared the number and productivity of the vineyards of England with those of France; England has almost no vineyards today. In the 10th century, the Vikings established successful colonies in Greenland; by 1250, that island was practically cut off by extensive drift ice, and by the early 15th century, the colonies were forgotten and dead. Many of the California-bound "Forty-niners" reported that a constant danger in crossing the plains was getting separated from companions in a sea of head-high grass; near-desert conditions exist there today.

Climates change, and in the short run, any change may be considered bad since humans and their institutions are carefully adjusted to the weather cycles and climatic conditions under which they developed. Over the longer span of human history, changing climates have been associated with the remarkable spread of humankind over the earth and its astounding growth in technology and numbers. As Chapter 5 will show, the history of cultural development—of agriculture, animal husbandry, and environmental manipulation by humans—began only after the last retreat of major glaciation some 10,000–12,000 years ago. Marine fossil evidence indicative of ancient water temperatures (and thus of gross climatic features) shows that the rise of urban-based civilizations and the creation of modern societies over the past 6000 years have coincided with the warmest temperatures in nearly 100,000 years.

110

But within that long warm spell, frequent and disastrous climatic fluctuations have occurred. Archaeologists and historians are presenting evidence that ancient seats of power—Sumeria, Mycenae, or Mali in Africa—may have fallen not to barbarians but to unfavorable alterations in the climates under which they came to power. Between A.D. 1550 and 1850, a "little ice age" descended on the northern hemisphere; Arctic ice expanded, glaciers advanced, and drier areas of the earth were desiccated. Crop failures were common; Iceland lost one-quarter of its population between 1753 and 1759, and 1816 was the "year without a summer" in New England, when snow fell in June and frost came in July.

A pronounced warming trend began about 1890 and peaked by 1945. During that period, the margin of agriculture was extended northward, the pattern of commercial fishing shifted poleward, and the reliability of crop yields increased. The equable climate, plus medical advances, presumably did much to encourage accelerating growth in the world's population. From 1918 to 1960, a period of explosive expansion in India's numbers, there were far fewer droughts in India than the prior record would have indicated likely.

Now, ample evidence tells us, climate again is changing. Since the 1940s, the mean temperatures of the globe have declined some 2.7°F (1.5°C); the growing season in England is some two weeks shorter; disastrous droughts have occurred in Africa and Asia; unexpected freezes have altered crop patterns in Latin America.

The recent global cooling trend has altered atmospheric circulation and rainfall. The circumpolar vortex, high altitude winds circling the poles from west to east, has remained nearer to the equator than in previous decades. Its stability has occasionally prevented the expected northward movement of monsoonal seasonal winds in the northern hemisphere and has widened and relocated the desert conditions associated with the subtropical high. Whether global cooling and the alteration of wind and pressure systems presage a new "little ice age" cannot yet be foretold. Like all short- and long-term climatic changes of the past, the present shifts from what was for so long the favorable "norm" are certain to bring social adjustments, problems, and—perhaps—conflicts related to environmental change.

gests that there will be substantially more ice on earth 3000 years from now. Climate can change more abruptly than this time scale suggests, however. Relatively small changes in upper-air wind movements can change a climate significantly in decades. Today, polar air is more dominant than tropical air. Changes in the amounts of precipitation or in the reliability of precipitation, as well as temperature changes, can have an enormous impact on agriculture and patterns of human settlement. In India, monsoon winds have fluctuated greatly in intensity recently, and in North Africa, summer rains have periodically failed for a number of years.

Great volcanic eruptions that spread dust particles around the world may also produce climatic change because the dust blocks the sun's rays to some extent. For three years after the volcano Krakatoa near Java erupted in 1883, there was a noticeable decline in temperatures throughout the world.

Finally, some scientists are concerned about the effect of human civilization on climatic change. There is evidence that industrial and automobile emissions are changing the gaseous mixture in the atmosphere and that supersonic aircraft affect the ozone layer. These changes are among the topics discussed in Chapter 4, dealing with environmental concerns.

111

For Review

1. What is the difference between *weather* and *climate*?

2. What determines the amount of insolation received at a given point? Does all solar energy potentially receivable actually reach the earth? If not, why?

3. How is the atmosphere heated? What is the *lapse rate*? What does it indicate about the atmospheric heat source? What is a *temperature inversion*?

4. What is the relationship between atmospheric pressure and surface temperatures? What is a *pressure gradient,* and of what concern is it in weather forecasting?

5. In what ways do land and water areas respond differently to equal insolation? How are these responses related to atmospheric temperatures and pressures?

6. Draw and label a diagram of the planetary wind and pressure system. Account for the occurrence and character of each wind and pressure belt. Why are the belts latitudinally ordered?

7. What is *relative humidity?* How is it affected by changes in air temperature? What is the *dew point?*

8. What are the three types of large-scale precipitation? How does each occur?

9. What are *air masses?* What is a *front?* Describe the development of a cyclonic storm, showing how it is related to air masses and fronts.

10. What factors were chiefly responsible for today's weather?

11. Name the nine climatic regimes discussed in this chapter. Summarize the distinguishing characteristics of each.

Suggested Readings

112

ANTHES, R. A., H. A. PANOFSKY, J. J. CAHIR, and A. RANGO, *The Atmosphere* (2nd ed.). Merrill, Columbus, OH 1978.

BLAIR, THOMAS A., and ROBERT C. FITE, *Weather Elements: A Text in Elementary Meteorology* (5th ed.). Prentice-Hall, Englewood Cliffs, NJ, 1976.

HAYS, J. D., J. IMBRIE, and N. J. SHACKLETON, "Variations in the Earth's Orbit: Pacemaker of the Ice Ages," *Science,* vol. 194 (Dec. 10, 1976), pp. 1121–32.

MILLER, ALBERT, *Meteorology* (3rd ed.). Merrill, Columbus, OH 1976.

NEIBURGER, M., J. G. EDINGER, and W. D. BONNER, *Understanding Our Atmospheric Environment.* Freeman, San Francisco, 1973.

STEWART, GEORGE R., *Storm.* Random House, New York, 1941.

STRAHLER, ARTHUR N., *Physical Geography* (4th ed.). Wiley, New York, 1975.

TREWARTHA, GLENN T., ARTHUR H. ROBINSON, and EDWIN H. HAMMOND. *Elements of Geography* (5th ed.). McGraw-Hill, New York, 1967.

Weatherwise. A periodical issued six times a year by Weatherwise, Inc., 230 Nassau St., Princeton, NJ.

Life Science Products Company was remarkably casual in its handling of Kepone. A close relative of DDT and aldrin—potent insecticides banned for general use in the United States—Kepone was produced by Life Science (in an abandoned gas station in Hopewell, Virginia) under contract to Allied Chemical Corporation. Although research evidence was available to indicate the potential health and environmental hazards of careless handling of the pesticide, no safety precautions were imposed in the manufacture or handling of Kepone or in the disposal of its residue.

Working without protective clothing or masks, and eating on tables covered with Kepone dust, employees inhaled and ingested the powder, unaware of the dangers involved. Within a year of the start of manufacture, a number of workers reported serious health problems, including tremors, loss of equilibrium, sterility, and liver lesions. When blood samples showed that all workers and many of their family members—exposed to Kepone-coated work clothes—had far higher than allowable concentrations of the pesticide in their systems, the plant was shut down in July 1975.

Tragic as were the personal consequences, the environmental effects of Kepone were even more widespread and devastating. Although water-quality officials in Virginia knew that Kepone had destroyed digester bacteria in Hopewell's sewage system, they permitted the pesticide to be dumped untreated into the James River, where an estimated 100,000 pounds still rest in river muds. The poison was carried by the James into Chesapeake Bay, where the $50-million-a-year commercial fishing industry was preserved by the simple device of raising the allowable Kepone-contamination level in fish. Pollution has also spread upstream from Hopewell and to tributary streams by the tidal transport of contaminated river sediment.

Two hundred yards from the Life Science plant, an air pollution monitoring station col-

Human Impact upon the Environment

lected data sufficient to indicate that dangerous levels of Kepone were in the atmosphere. It did not occur to anyone, however, to test for Kepone until the plant was closed. A year after manufacture ceased, Kepone levels were still unacceptably high in the atmosphere, and serious consideration was being given to removal or burial of contaminated soil around the plant site.

Although fines totaling nearly $18 million were levied against Allied Chemical, Life Science Products, and their officers, the long-term costs in human suffering, fishery damage, and environmental degradation may exceed these amounts many times over. The story of Kepone is one of many that could have been selected to illustrate how humans can affect the quality of the water, air, and soil upon which their existence depends.

Terrestrial features and ocean basins, elements of weather and characteristics of climate, flora and fauna comprise the building blocks of that complex mosaic called the *environment*—an overworked word that means the totality of things that in any way may affect an organism. Humans exist within a natural environment—the sum of the physical world—which they have modified by their individual and collective actions. Forests have been cleared, grasslands plowed, dams built, and cities constructed. Upon the natural environment, then, has been erected a cultural environment, modifying, altering, or destroying the balance of nature that existed before human impact was expressed. This chapter is concerned with the interface between humans and the natural environment that they have so greatly altered and endangered.

Humans, certainly since the dawn of agriculture, have changed the face of the earth, have distorted delicate balances and interplays of nature, and, in the process, have endangered (and, perhaps, enhanced) the societies and economies erected in the process of these alterations. The essentials of the natural balance and the ways in which humans have altered it

are not only our topics here but are also matters of social concern that rank among the principal domestic and international issues of our times. For the geographer, however, human impact upon the natural environment is more than a social issue involving legislation, disputation, and public and private expenditures; it represents as well a vital bridge between the physical and the cultural traditions of the discipline, the area in which human–earth-surface relationships are most obvious, immediate, and extensive.

The starting point is energy. The basis for life on earth originates 93 million miles away, with the sun. Solar energy, which through photosynthesis is converted to chemical energy in plants, is responsible for our survival and the survival of all other living things. We eat plants, and in the process, the chemical energy of plants is transformed into energy of motion. Some living things bypass plants as a source of food, but the energy they obtain by eating animals, insects, and fish can be traced back to plants. Only plants have the ability to change solar energy into a form of energy that can be used by animals. Plants, one estimate suggests, use 100 times more energy than all of man's machines combined. Some inedible energy is stored in rocks such as coal or in liquids such as petroleum, but these too have their origin in plants.

The transfer of energy from one living thing to another is never completely efficient. Some energy is always lost. A typical carnivore absorbs less than 10% of the energy entering the plant-eating animal upon which it dines. The other 90% is stored, used for respiration, or lost as heat. Figure 4.1 depicts the stages in a simple process through which heat energy is added to the environment.

Human technologies have altered the simple, natural energy exchange on the earth; to normal heat loss must be added that heat which is generated and dissipated in the conversion of latent energy to forms usable by humankind. The higher the technological level

114

Humans have erected their advanced societies through the utilization of inanimate energy resources. That utilization has a profound and irreversible impact upon the environment.

Energy—the ability to do work—exists in two forms: *potential* and *kinetic.* Potential energy is expressed through the relation of an object to a force that may act upon it. If you drop this book from waist height, its potential energy state is lowered to zero with respect to the floor. But no energy has been lost; it has simply been passed from the object to the environment. The first law of thermodynamics assures us that energy is not created or destroyed in any process, though it may be converted from one form to another. The falling book transformed its potential energy into the kinetic energy of motion of the book and of the surrounding air molecules. Since molecular motion is measured by temperature, the book's loss of potential energy is expressed by the transfer of heat to the environment. Part of the book's potential energy, of course, could be expended in work; a string could be tied to it and a lighter object lifted. Useful energy conversions, however—for example, from coal to electricity—are never complete; some potential energy is always converted to heat and dissipated to the surroundings.

This is another way of saying that disorder is the end product of natural processes. Heat passes from a warm to a cold object; if they remain in contact, both eventually assume the same temperature. Ordered molecules of dye color the entire bowl of water in which they dissolve; the orderly garden, left untended, becomes disordered and overgrown. The measure of relative disorder is called *entropy.*

115

Order can, of course, be created, but only at the price of expenditure of energy. The ice cube can be frozen, the molecule of dye can be manufactured, the garden can be hoed. But these nonspontaneous processes also increase system and environmental entropy by the loss of heat they involve. The second law of thermodynamics guarantees that a closed thermodynamic system tends to maximize its entropy, or reach a state of equilibrium. Any system plus its surroundings tends toward a state of increasing disorder or entropy.

Since we cannot repeal the first and second laws, we must accept their implications. Anything that is done to increase order in a system requires the expenditure of energy. But part of that energy input is lost in heat. The effort spent to bring order to a part of nature results in an increase in environmental disorder exceeding the order created. Nature and all of its parts are systems that are continually running down. The presumed human "control of nature" accelerates the process.

of a society, the greater its dependence upon energy consumption, and the greater the amount of heat loss. At a minimum, humans must consume about 2000 calories of energy

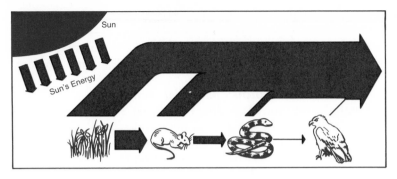

Fig. 4.1 Loss of heat in energy transformations. With each link in the food chain—here, from plant to mouse to snake to bird—there is a decrease of efficiency in the transfer of energy. A general rule is that only 10% of the net energy of a lower link can survive to the next higher link in the food chain. The remainder, consumed in sustaining life processes, is lost as heat into the environment.

per day to sustain life at reasonable levels of efficiency; if they rear livestock and use energy to cook food, their energy consumption increases to some 10,000 calories per person per day. When they smelt steel, build and operate engines and automobiles, generate electricity, and produce and use the myriad goods and services of which a high-technology society is composed, energy consumption and consequent heat loss undergo a staggering increase. **116** The average American now uses directly or indirectly about 200,000 calories per day. The United States, comprising some 6% of the world's population, consumes about 35% of the world's energy—though it simultaneously produces a correspondingly large proportion of the world's food, manufactured goods, and services.

The energy needs of the earth's populations cannot be satisfied, under present technologies, by the capture of radiant solar energy; we must rely upon the utilization, and depletion, of the stored energy of fossil fuels. The consumption of those fuels, the conversion through them of raw materials into useful commodities, and the disposition of the wastes of the technological societies we have created constitute destructive alterations of the *biosphere* or *ecosphere*—the thin film of air, water, and earth within which we live.

Ecosystems

This biosphere is composed of three interrelated parts: (1) the atmosphere, some 6–7 miles (9.5–11.25 km) thick; (2) surrounding and subsurface waters in oceans, rivers, lakes, glaciers, or ground water—much of it locked in ice or earth and not immediately available for use; (3) the upper reaches of the earth's crust—a few thousand feet at most—containing the soils supporting plant life, the minerals that plants and animals require to exist, and the fossil fuels and ores that humans exploit. It is an intricately interlocked system, containing all that is needed for life, all that life has to utilize,

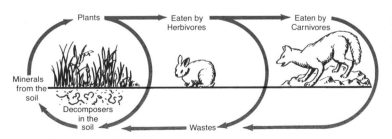

Fig. 4.2 A simplified example of an ecosystem illustrating the interdependence of organisms and the physical environment.

and, presumably, all that ever will be available. The ingredients of the thin ecosphere must be, and are, constantly recycled and renewed in nature: plants purify the air; the air helps to purify the water; the water and minerals are used by plants and animals and are returned for reuse.

The biosphere, therefore, consists of two inextricable components: (1) a nonliving outside (solar) energy source and requisite chemicals and (2) a living world of plants and animals. In turn, the biosphere may be subdivided into specific *ecosystems*, which are made up of the populations of organisms existing together in a particular area and of the necessary base environment of energy, air, water, soil, and chemicals. Figure 4.2 depicts one kind of ecosystem and in crude form suggests the most important principle concerning them: everything is interconnected. Any intrusion, any interruption, in the balance that has been naturally achieved inevitably results in dislocations and generally unexpected and perhaps undesirable effects elsewhere in the system. Each organism occupies a specific and essential *niche*, or place, within an ecosystem. In the energy-exchange system, each organism plays a definite role; individual organisms survive because of other organisms that also live in that environment. The problem lies not in recognizing the niches but in anticipating the chain of causation and the readjustments of the system consequent upon disturbing the occupants of a particular niche.

In the tundra of North America, wolves eat the sick, old, weak caribou, controlling their population and indirectly saving the land from overgrazing. Since wolves have been viewed as an unpleasant and dangerous predator, hunting them has been freely permitted and even encouraged by the granting of bounties on proof of their death. Reduction of the predator population means an increase in the caribou population, with resultant overgrazing, range destruction, and the weakening and starvation of

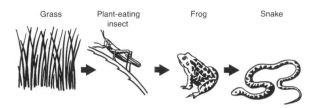

Fig. 4.3　**A simple food chain. Although energy is lost in biomass conversion, nutrients move in cycles. When an organism dies and decays, its nutrients are released into soil or water, where they can be reabsorbed by plants and animals.**

the herbivores of all types in the tundra *biome* (ecosystem). In this instance, the causal chain is short and obvious; in most instances of biome tampering, the results of human action may be so subtle or so remote from first causes that it is nearly impossible to anticipate them in advance or fully to appreciate their distant severity.

FOOD CHAINS AND FOOD WEBS

117

Life depends on the energy flowing through an ecosystem. The transfer of energy from one organism to another is one link in a *food chain*, defined as the sequence of such transfers within a biome. Some food chains have only two links, as when human beings eat rice; most have only three or four. The caribou–wolf relationship is a link in a chain consisting of plants → caribou → wolves. Figure 4.3 shows a chain with four links.

Since the ecosystem in nature is in a continuous cycle of integrated operation, there is no "start" or "end" to a food chain; there are, however, nutritional transfer stages—called *trophic levels*—in which each lower level transfers part of its contained energy to the next-higher-level consumer. Each step in the food chain involves energy loss as heat is dissipated into the environment; on average, only some 10% of the energy stored in living tissue is transferred to the next higher trophic level.

While the food chain is a meaningful way to illustrate the flow of energy from organism to organism, it does not accurately indicate the complexity of the actual operation of energy transfer in an ecosystem. Thus there are many animals that feed upon the grass in a meadow, and many other animals eat them, live on them (e.g., fleas), or live inside them (bacteria). *Food web* is a better way to describe the total flow of energy in an ecosystem. Hundreds of food chains exist in even a moderately complicated food web.

The decomposers pictured in Figure 4.2 are essential to our survival. They cause the disintegration of organic matter—animal carcasses and droppings, dead vegetation, trash, and so on. In the process of decomposition, the chemical nature of the material is changed and the nutrients become available for reuse by plants or animals. Nutrients, the minerals and other elements that organisms need for growth, are never destroyed; they keep moving from living to nonliving things and back again. Our bodies contain nutrients that were once part of other organisms, perhaps a fox, a hawk, or an oak tree.

RENEWAL CYCLES

The idea of a cycle is important in furthering our understanding of the natural renewal of the elements essential to life, which include carbon, oxygen, hydrogen, and phosphorus. Since the supply of these materials is fixed, they must be continuously recycled through food chains. Figures 4.4 and 4.5 show two important cycles, the phosphorus cycle and the hydrologic cycle.

Phosphorus is an element essential to the survival of many species of plants and animals. It is an important component of DNA and of bones and teeth. Plants absorb phosphate from the soil; other organisms acquire it by eating plants. It is returned to the soil through animal excretions or when plants and animals decay. When phosphate that has been used as an agricultural fertilizer is carried to rivers and eventually to the sea through runoff, only a small percentage is returned to the land through fish catches and the excretions of guano birds, which eat fish. Most of it remains in deep ocean waters until geologic processes cause uplifting of the ocean floor.

Food chains, systems of energy transfer, and chemical element recycling achieve natural balance and sequence within undisturbed biomes. Humans, however, the one universal dominant species, have had a long record of altering the smooth flow of these cycles. The impact of humans upon ecosystems—small, at first, with low population, energy consumption, and technological levels—has increased

118

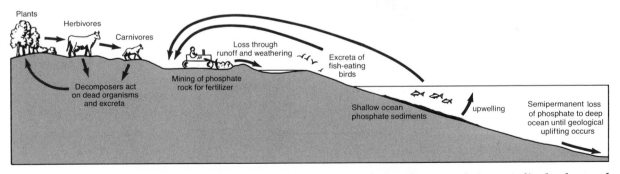

Fig. 4.4 The phosphorus cycle. The natural phosphorus cycle is exceedingly slow and has been upset by human activity. Currently, phosphorus is being washed into the sea faster than it is being returned to the land, primarily because it is being mined for use as a fertilizer.

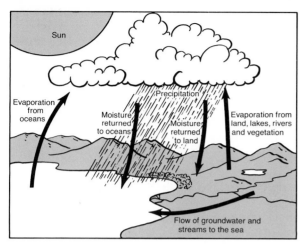

Fig. 4.5 The hydrologic cycle. The sun provides energy for the evaporation of fresh and ocean water. The water is held in the air until it becomes supersaturated. Atmospheric moisture is returned to the earth's surface as solid or liquid precipitation to complete the cycle. Since precipitation is not uniformly distributed, moisture is not necessarily returned to areas in the same quantity as has evaporated from those areas. The continents receive more water than they lose; the excess returns to the seas as surface water or ground water. A global water balance, however, is always maintained.

so rapidly and pervasively as to present us with widely recognized and varied ecological crises. Some of the effects of humans on the natural environment are the topic of the remainder of this chapter.

Impact on Water

The supply of water is constant; the system by which it is continuously circulated through the biosphere is called the *hydrologic cycle*, shown in Figure 4.5. In that cycle, water may change form and composition, but under natural environmental circumstances it is marvelously purified in the recycling process and again made available with appropriate proper-

ties to the ecosystems of the earth. Evaporation and transpiration (the emission of water vapor from plants) are the mechanisms by which water is redistributed. Water vapor collects in clouds, condenses under conditions reviewed in Chapter 3, then falls again to earth. There it is reevaporated and retranspired, only to fall once more as precipitation occurs.

Humans have a discernible impact upon the hydrologic cycle by affecting the speed of water recirculation and by hastening the return of water to the sea. Massive withdrawals of water from surface supplies by thermal power plants or by industry for cooling purposes can abbreviate the normal cycle by prematurely converting water from a liquid to its vaporous state. Comparable accelerated vaporization occurs through evaporation from reservoirs, particularly in arid regions. The amount of usable water available at some point in the cycle is thereby reduced.

But the primary adverse human impact is felt in the area of water quality. People withdraw water from lakes, rivers, or underground deposits to use for drinking, bathing, agriculture, industry, and a hundred other purposes. Although the water that is withdrawn returns to the cycle, it is not always returned in the same condition as at the time of withdrawal. Water, like other segments of the ecosystem, is subject to serious problems of pollution.

THE MEANING OF WATER POLLUTION

As a general definition, *environmental pollution* by humans means the introduction into the biosphere of wastes that, because of their volume or their composition or both, cannot be readily disposed of by natural recycling processes. In the case of water, the central idea is that pollution exists when water composition has been so modified by the presence of one or more substances that either it cannot be used for a specific purpose or it is less suitable for that use than it was in its natural state. Pollu-

119

**Human
Alteration
of Nature**

The human costs of tampering with nature's balance are high and frequently unforeseen. The planned extermination of the hippopotamus and the crocodile in parts of Africa led to the silting of streams, to the loss of fish populations upon which humans depend for protein, and to the spread of the debilitating disease schistosomiasis. Indonesian farms were supplied with an insecticide to control the rice borer, but the poison also killed the fish living in the rice paddies, which had fertilized the rice, acted as a control on caterpillars, and served as a source of protein and as a cash crop. Rice yields and income fell; rural and urban diets deteriorated.

Malaria seemed almost conquered in the decade following World War II, but the resistance of the mosquitoes to pesticides and the adjustment of the malarial parasites to common drugs have lately resulted in an increase in the incidence of the disease to at least 200 million cases worldwide. In 1976, the World Health Organization estimated 1.5 million malaria deaths. The agricultural development projects deemed so essential to third-world countries are contributing to its spread. Frequent draining of irrigation canals in water projects in the Gezira region of the Sudan initially kept malaria at bay; the introduction of rice and other crops demanding constant water availability has increased crop productivity but, as well, has added a vast new area of malaria infestation to the African continent.

With a new awareness of the complexity and the number of possible consequences of human alterations of nature, developmental agencies and lending institutions have, since the early 1970s, increasingly required both environmental and economic impact statements in an attempt to anticipate the full consequences of programs designed to address apparently limited objectives. Although they are a useful recognition of the complexity of nature's web, such statements must inevitably be incomplete and may frequently be biased to favor the "progress" of regions and nations faced with pressing and immediate human needs.

120

tion is brought about by the discharge into water of substances that cause unfavorable changes in its chemical or physical nature or in the quantity and quality of organisms living in the water. Pollution is a relative term. Water that is not suitable for drinking may be completely satisfactory for cleaning streets. Water that is too polluted for fish may provide an acceptable environment for certain water plants.

Human activity is not the only cause of water pollution. Leaves that fall from trees and decay, animal wastes, oil seepages, and other natural phenomena may affect water quality. There are natural processes, however, to take care of such pollution. Organisms in water are able to degrade, assimilate, and disperse such substances in the amounts in which they naturally occur. Only in rare instances do natural

Fig. 4.6 (a) Pollutants flowing out of an eroded sewage pipe into the Mississippi River are especially noticeable during periods of low water, as seen here. (b) Contaminants do not always remain confined to their point of occurrence. The drawing suggests the damage to groundwater and the poisoning of trees at a distance from storage piles of highway salt in Illinois. Contamination was discovered when well water, charged with salt that had been dissolved and carried into the groundwater supply, would no longer freeze in the ice machine of the inn. (a: U.S.D.A. Soil Conservation Service, b: Data from James P. Gibb, Illinois State Water Survey.)

(b)

anisms, or that take a very long time to break down.

As long as there are people on earth, there will be pollution. Thus the problem is one not of eliminating pollution but of controlling it. This is particularly true in the technologically advanced countries. The more developed and affluent the country, the more resources it uses and the more it pollutes.

In the United States, about half of the water used daily is used by industry. Another 40% is used for agricultural purposes, chiefly irrigation, and the remainder is used for domestic purposes. Each type of usage alters the water in some way before returning it to the environment. Table 4.1 shows the kinds of water pollutants associated with different sources.

POLLUTION BY DUMPING

Many industries dump organic and inorganic wastes into bodies of water. These may be

pollutants overwhelm the cleansing abilities of recipient waters. What is happening now is that the quantities of wastes discharged by humans often exceed the ability of a given body of water to purify itself (Figure 4.6). In addition, humans are introducing pollutants such as metals or inorganic substances that cannot be broken down at all by natural mech-

Table 4.1 Sources and Types of Water Pollutants

Source of Waste	Kind of Pollutant
Municipalities and residences	Human wastes, detergents, garbage, trash
Urban drainage	Suspended sediment, fertilizers, pesticides, road salt
Agriculture Crop production	Fertilizers, herbicides, pesticides, erosion sediment
Irrigation return flow	Mineral salts and erosion sediment
Industry	The widest possible range of pollutants, including biodegradable wastes in the paper and food-processing industries, heat discharge in a variety of activities, nondegradable wastes in the chemical and iron and steel industries, and radioactivity
Mining	Acids, sediment, metal wastes, culm (coal dust)
Electric power production	Heat and nuclear wastes
Recreation and navigation	Human wastes, garbage, fuel wastes

Source: Lawrence G. Hines, *Environmental Issues*, W. W. Norton & Co., New York, 1973, p. 197, Table 9.1. Reprinted by permission.

122

acids, highly toxic minerals such as mercury or arsenic, or, in the case of petroleum refineries, toxic organic chemicals. Organisms that are not adapted to living in water thus contaminated die, the water may become unsuitable for domestic use or irrigation, or the wastes may reenter the food chain with deleterious effects on humans. A particularly unfortunate example brought to the attention of the public in 1953 was of mercury poisoning in Japan. A chemical plant in Minamata Bay that used mercury chloride in its manufacturing process discharged the waste mercury into the bay, where it settled with the mud. Fish that fed upon organisms in the mud absorbed the mercury and concentrated it; the fish were in turn eaten by humans. Over 100 people suffered death, deformity, or other permanent disability. Many people fear that the discharge of radioactive isotopes from nuclear power plants may lead to similar contamination of the water supply.

More recently, attention in this country has focused on polychlorinated biphenyls (PCBs), compounds that have been used in capacitors, devices that increase the efficiency of electric current in a wide range of products. During the manufacturing process, companies have dumped PCBs into rivers, from which they have entered the food chain. Several states have banned commercial fishing in lakes and rivers where fish have higher levels of PCBs than are considered safe. A rock bass containing 355 parts per million of PCBs was caught in the Hudson River; had that fish been eaten, the consumer would have taken in one-third the lifetime PCB limit as set by the Food and Drug Administration. Although not all of the effects of PCBs on human health are known, skin eruptions, excessive eye discharges, and possibly cancer of the liver are thought to be linked to them.

The herbicides and pesticides used in agriculture eventually find their way into the water supply, a problem discussed in more detail on following pages. The long-term effects of such usage are not always immediately known. DDT, for example, was used for many years before people discovered its effect on birds, fish, and plant life in the water. One of the dilemmas we are facing is that we have not yet discovered how to balance our need for increased agricultural yields in the short run—and the manipulation of the environment that it appears to entail—with what is in our best interest in the long run.

Domestic sewage can also pollute water, depending on how well it is treated before being discharged. In general, this is less of a problem in countries such as the United States than in countries where such treatment is either not practiced or is not thorough. This country is not without such sources of water

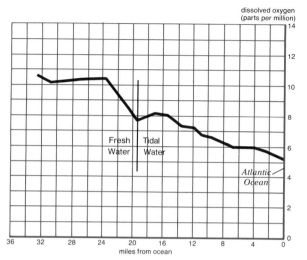

Fig. 4.8 Dissolved-oxygen profile of the Raritan River, New Jersey. As the Raritan flows toward the Atlantic Ocean, it receives so much industrial waste that aquatic life can barely survive in the estuarine zone. Fish can live with less than 5 parts per million of dissolved oxygen, but only under duress. Raritan Bay, rich in nutrients, is the source of the "red tides" that occur along the New Jersey coast in summer. The red algae bloom in the bay and are pushed out by prevailing currents. The red tides, fatal to some marine life and a cause of discomfort to swimmers, can be as much as 3 miles wide and 50 miles long. (From George Carey et al., *Urbanization, Water Pollution, and Public Policy*, Center for Urban Policy Research, Rutgers University, New Brunswick, NJ, 1972.)

123

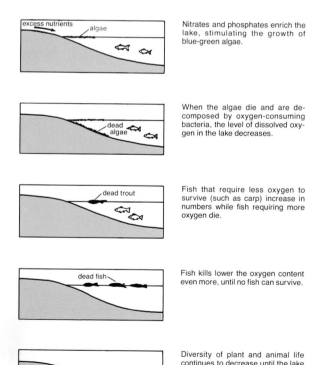

Nitrates and phosphates enrich the lake, stimulating the growth of blue-green algae.

When the algae die and are decomposed by oxygen-consuming bacteria, the level of dissolved oxygen in the lake decreases.

Fish that require less oxygen to survive (such as carp) increase in numbers while fish requiring more oxygen die.

Fish kills lower the oxygen content even more, until no fish can survive.

Diversity of plant and animal life continues to decrease until the lake is suitable only for bacteria that do not require oxygen.

Fig. 4.7 Eutrophication is hastened by artificial sources of nutrients.

pollution, however. Where antipollution laws are not well enforced, pollution from domestic sewage occurs. The city of Houston, Texas, among many others, has been allowed to dump raw, untreated human waste into nearby water bodies for years—waste containing viruses responsible for polio, hepatitis, spinal meningitis, and other diseases.

EUTROPHICATION

Agriculture is a chief contributor of excess nutrients to water bodies. Pollution occurs when nitrates and phosphates that have been used in fertilizers and that are present in animal manure drain into streams and rivers, eventually accumulating in ponds, lakes, and estuaries. The nutrients hasten the process of *eutrophication,* or the increase of nutrients in a water body. Eutrophication occurs naturally when nutrients in the surrounding area are washed into the water, but when the sources of enrichment are artificial, as is true of fertilizers, the body of water may become overloaded with nutrients. The end result may be an oxygen deficiency in the water.

Figure 4.7 illustrates one form that overfertilization of a water body can take. Algae and other plants are stimulated to grow abundantly, crowding out other species. When the algae die, the level of dissolved oxygen in the water decreases, primarily because of the bacteria acting on the dead and decomposing vegetation. Fish and plants that cannot tolerate the poorly oxygenated water are eliminated. Figure 4.8 illustrates the results of a study of the variation in the amount of dissolved oxygen in the

Raritan River in New Jersey. Scientists have estimated that as many as one-third of the medium- and large-sized lakes in the United States have been affected by accelerated eutrophication.

THERMAL POLLUTION

Many industrial processes require the use of water as a coolant (Figure 4.9). *Thermal pollution* occurs when water that has been heated is returned to the environment and has adverse effects on the plants and animals in the water body. Many plants and fish cannot survive changes of even a few degrees in the water temperature. They either die or migrate; the species that depend on them for food must also either die or migrate. Thus the food chain has been disrupted. In addition, the higher the temperature of the water, the less oxygen it contains, which means only lower-order plants and animals can survive.

24

Fig. 4.9 Cooling towers at the Watts Bar Nuclear Plant near Spring City, Tennessee. (Tennessee Valley Authority.)

One of Egypt's proudest achievements was the completion in 1970 of the Aswân High Dam. The dam had two primary purposes: water storage and flood control, and hydroelectric power supply. By controlling the annual flood of the Nile River and releasing water slowly for irrigation, the dam was intended to enable farms to change from basin to perennial irrigation. Water would be available for the establishment of a four-crop rotation system along the flood-plain from Aswân to Cairo. In addition, the planners expected that between 1 million and 2 million acres of land would become available for cultivation, an increase in arable area of about 20%.

While the dam has accomplished the purposes for which it was built, there have been unexpected side effects. The dam has modified the ecology of the Nile River valley, a silent witness to the principle that one cannot take an action in the environment without triggering a chain reaction. In terms of environmental disruption, the costs of the dam may in the end come to out-weigh its benefits.

The Nile River used to flood annually, depositing silt that enriched the soil, carrying nutrients to fish and shellfish, and flushing away salts in the river bottom. The fertile silt now accumulates in Lake Nasser, the huge reservoir behind the dam. The fertility of the agricultural land in the lower valley has decreased to such an extent that farmers must now apply costly artificial fer-tilizers. Salts, no longer washed away, accumulate in the river and the irrigation canals, threatening crops.

Some
Environmental
Effects of
the Aswân
High Dam

Construction of the Aswan High Dam. (United Nations.)   125

The Nile delta is experiencing a loss in the amount of land suitable for agriculture. Erosion of the shoreline in the Mediterranean Sea is occurring because sediment is no longer deposited by floodwaters. Expensive seawalls may have to be built to protect the northern delta from further erosion, from inundation by seawater, and from the harmful effects of sea spray on crops.

The sardine fishery industry, which used to add millions of dollars to the Egyptian economy each year, has suffered a disastrous decline since the completion of the dam. With the elimination of the nutrients that the floodwaters formerly carried, the southeastern Mediterranean no longer provides a suitable environment for the fish. The development of freshwater fishing on Lake Nasser may balance the loss of the sardine industry, but the outcome will not be known until environmental conditions and the fish population of the lake stabilize.

Finally, the quiet waters of the irrigation canals have become breeding grounds for disease-carrying organisms. In particular, a large increase in the incidence of schistosomiasis has occurred. A debilitating disease that is almost impossible to cure, it is transmitted to humans by snails. The eggs are carried by the snails; larvae emerge from the snails in stagnant waters and penetrate humans, where they grow into worms. Because floods no longer clean the canals, snails thrive. Schistosomiasis has spread so rapidly that in some areas, over half of the human population is affected. Cattle and sheep contract a similar disease, fascioliasis, also borne by snails. Attempts to eradicate the snails through the use of such solutions as copper sulfate have their own secondary effect: massive fish kills downstream.

126

Impact on Air

The troposphere, the thin layer of air just above the earth's surface, contains all the air that we breathe. Every day thousands of tons of pollutants are discharged into the air by cars and incinerators, factories and airplanes. Air that contains substances that have a harmful effect on living things is polluted.

AIR POLLUTANTS

Truly clean air has probably never existed. Just as there are natural sources of water pollution, so are there substances that pollute the air without the aid of humans. Dust from volcanic eruptions, marsh gases, smoke from forest fires, and wind-blown dust are natural sources of air pollution.

Normally these pollutants are of low volume and they are widely dispersed throughout the atmosphere. On occasion, a major volcanic eruption may produce so much dust that the atmosphere is temporarily altered. The eruption in 1883 of a volcano on the island of Krakatoa, west of Java, produced dust that lingered in the area for over a year. In general, however, the natural sources of air pollution do not have a significant, long-term effect on air, which like water is able to cleanse itself.

Far more important than naturally occurring pollutants are the substances that people cause

Fig. 4.10 The burning of fossil fuels in motor vehicles results in the discharge of millions of tons of carbon monoxide, sulfur and nitrogen gases, hydrocarbons, and particulates each year. (Courtesy of Association of American Railroads.)

to be discharged into the air. These pollutants are primarily the waste products that result from the burning of fossil fuels (coal and oil) and other materials. Fossil fuels are burned in power plants that generate electricity, in many industrial plants, in home furnaces, and in cars and trucks, buses and airplanes, as pictured in Figure 4.10. Scientists estimate that about three-quarters of all air pollutants come from the burning of fossil fuels. Industrial processes other than fuel burning, the incineration of solid wastes, forest and agricultural fires, and the evaporation of solvents account for most of the remaining pollutants.

Perhaps the best known of air pollutants and the most dangerous to health are *sulfur oxides,* which come primarily from the burning of fossil fuels. They cause industrial smog, which is associated with respiratory ailments in humans. When unusual weather conditions cause the smog to remain over an area for several days, the result may be a disaster. For example, in 1952, London, England, experienced an acute episode of such pollution as a result of which 4000 more people died than normal.

Sulfur oxides are also responsible for a phenomenon known as *acid rains.* When sulfur dioxide is absorbed into water vapor in the atmosphere, it becomes sulfuric acid. This acid, when washed out of the air by rain or snow, damages vegetation and other materials with which it comes into contact. In parts of Europe, one can see its corrosive effects on marble and limestone sculptures and buildings and on metals such as iron and bronze, as suggested by Figure 4.11. In North America and northern Europe, atmospheric acid has measurably reduced rates of forest growth; in many areas, crop patterns have been altered as plants particularly susceptible to acid rain have had to be eliminated from the agricultural system. Figure 4.12 depicts the major sources and types of air pollutants.

Many of the fuel-burning processes already mentioned discharge solid particles of soot and dust, sulfate and flouride, asbestos and metallic particles (beryllium, arsenic, cadmium, lead) into the air. Other particulates are formed in the air from gases emitted during combustion. Human production of particulates accounts for a significant proportion of the total volume, although natural sources of particulate pollution exist.

Large and medium-sized particulates, those greater than 1 micron in diameter, are heavy and tend to wash out of the air rapidly, discoloring buildings, clothing, and cars. The technology is available to collect most of these particles in industrial and power plants before they are discharged into the air. The emission

127

Fig. 4.11 The deleterious effect of sulfur oxides on statuary may be seen on the famous bronze horses atop St. Mark's Basilica in Venice. (Courtesy of Italian Government Travel Office.)

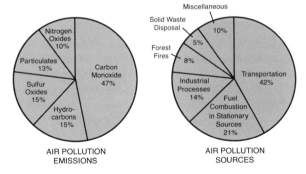

AIR POLLUTION
EMISSIONS

Nitrogen
Oxides
10%

Particulates
13%

Sulfur
Oxides
15%

Carbon
Monoxide
47%

Hydro-
carbons
15%

AIR POLLUTION
SOURCES

Miscellaneous

Solid Waste
Disposal

Forest
Fires

10%

5%

8%

Industrial
Processes
14%

Transportation
42%

Fuel
Combustion
in Stationary
Sources
21%

Fig. 4.12 Air pollution emissions in the United States in percentages by weight. If pollution from vehicles were eliminated, emissions would be reduced by 42%. (National Air Pollution Control Administration, U. S. Department of Health, Education, and Welfare.)

of very fine particulates is more difficult to control. Because they are light, they can remain suspended in the air for days or even weeks. Many are known to have an adverse effect on human health; lung cancers, for example, have been linked to the inhalation of asbestos fibers. The possibility also exists that the increase in particulates in the atmosphere may have a long-term effect on climate by reducing the amount of solar radiation that reaches the earth's surface.

Many people fear that the discharge of radioactive pollutants from nuclear power plants and nuclear weapons may lead to the contamination of both air and water, a problem of particular concern because of the long time it takes for some isotopes to decay. The *half-life* of an isotope is the time required for one-half of the atomic nuclei to decay. After one half-life has elapsed, half of the nuclei remain; after another half-life, half of the remainder (one-quarter of the original nuclei) remain, and so on. The half-life of the isotope indicates whether the problem posed by a particular radioactive pollutant is a short- or a long-term one. The half-life of uranium-238, which is used in breeder reactors, is 4.5 billion years. Other isotopes have half-lives measured in fractions of a second.

FACTORS AFFECTING AIR POLLUTION

Many factors affect the type and degree of air pollution found at a given place. Those over which people have relatively little control are climate, weather, wind patterns, and topography. These determine whether pollutants will be blown away or whether a buildup is likely to occur. Thus a city on a flat plain is less likely to experience a buildup than is a city in a valley near the coast.

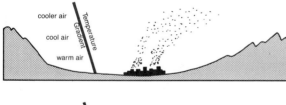

Fig. 4.13 Under normal conditions (top), air temperature decreases as altitude increases. During a temperature inversion (bottom), there is a mass of warmer air between surface and upper altitude air. A temperature inversion intensifies the effect of air pollution; instead of being dispersed, the pollutants are held close to the ground.

Unusual weather occurrences can alter the normal patterns of pollutant dispersal. A *temperature inversion*, depicted in Figure 4.13 (see also Figure 3.9), magnifies the effects of air pollution. Under normal circumstances, air temperature decreases away from the earth's surface. If, however, there is a stationary layer of warm, dry air over a region, the normal rising and cooling of air from below is prevented.

The air becomes stagnant; pollutants accumulate in the lowest layer instead of being blown away. The air becomes more and more contaminated. Normally, inversions last for only a few hours, although certain areas experience them much of the time. Los Angeles is characterized in summer by a temperature inversion most of the time. If an inversion lingers long enough—for example, for several days—it can seriously affect human health (Figure 4.14). Temperature inversions have been associated with increases in mortality in Pennsylvania and New York, as well as in the episode in London mentioned above.

129

Some areas are particularly likely to suffer from *photochemical smog,* which is created when oxides of nitrogen react with oxygen present in water vapor in the air to form nitrogen dioxide. In the presence of sunlight, nitrogen dioxide reacts with hydrocarbons from automobile exhausts and industry to form new compounds, such as *ozone.* This complex reaction sequence illustrates the principle of *synergy;* in this case, the reaction of the pollutants with one another and with other components of air produces an effect that the substances could not have had singly. In the United States, Los Angeles, Salt Lake City, and Denver have the kinds of climate and topography that favor the formation of photochemical smog.

The air pollutants generated in one place may have their most serious effect in areas 100 or 200 miles away. Thus the worst effects of the photochemical air pollution that originates in New York City are felt in Connecticut and parts of Massachusetts. The chemical reaction

Fig. 4.14 (opposite) Because of its location and prevailing wind patterns, New York City often avoids the buildup of pollutants, but when adverse weather conditions occur (low wind speeds and temperature inversions), sulfur dioxide and smoke concentrations reach peak values. The city experienced major air pollution episodes in 1953, 1963, and 1966, causing hundreds of deaths above statistical expectations. (EPA-DOCUMERICA—Chester Higgins.)

that produces ozone takes a few hours, by which time air currents have carried the pollutants away from the city. In a similar fashion, New York City is the recipient of pollutants produced in other places, notably New Jersey.

Other factors that affect the type and degree of air pollution at a given place are the levels of urbanization and industrialization. Population densities, traffic densities, the type and density of industries, and home-heating practices all help to determine the kinds of substances discharged into the air at a single point. In general, the more urbanized and industrialized a place is, the more responsible it is for pollution. The United States may contribute as much as one-third of the world's air pollution, a figure roughly equivalent to the portion of the world's resources consumed in this country.

EFFECTS OF AIR POLLUTION

In recent years researchers have begun to realize that air pollution has an effect on the major factors that control the temperature of the earth's surface, but how much of an effect is not yet known. The pollution processes that send particulates into the upper atmosphere,

Fig. 4.15 Researchers are trying to determine the long-run effect of the cloud cover produced by contrails from jet aircraft. (National Oceanic and Atmospheric Administration.)

131

and the cloud cover produced by contrails from jet planes (Figure 4.15), have an "icebox" effect on the earth. The particulates and the clouds reflect incoming sunlight back into space before it reaches the earth. The result is a cooler earth than normal.

On the other side of the heat balance sheet is the "greenhouse" effect. The burning of fossil fuels adds carbon dioxide (CO_2) to the atmosphere. CO_2 slows down the reradiation of heat from earth back into space, acting as glass in a greenhouse does to heat up the climate. The more CO_2 there is in the atmosphere, the higher temperatures on earth should be, and increasing industrial activity over the next one or two centuries is likely to raise the carbon dioxide content of the atmosphere many times. It should be noted that there is considerable scientific controversy over this presumed consequence of rising carbon dioxide levels; some authorities suggest that a net cooling—not warming—of the earth's atmosphere results from increases in its CO_2 content. In any event, it is evident than modern, or technological, society is capable of producing significant heating or cooling effects that can transform climates.

The effect of air pollution on plant life has been well documented. Chemicals washed out of the air by rain and snow contaminate soil and water. Acid rains and smog have caused extensive crop damage and the reduction of yields in some areas; for example, citrus groves in the vicinity of Florida phosphate plants have been devastated. Long-term changes in climate induced by air pollution would, of course, have important secondary effects on vegetation.

The short-term effect of air pollution on human health has been evidenced on a number of occasions. Highly polluted air can be lethal, as disasters in London, England, and Donora, Pennsylvania, as well as a number of other cities have shown. The long-term effect of most pollutants is less well understood. It is possible that constant exposure to low levels of polluted air damages lungs, increasing the incidence of such respiratory ailments as pneumonia, emphysema, and asthma. Increases in certain cancers and in heart disease may also turn out to be related to air pollution.

Impact on Land and Soils

People have affected the earth wherever they have lived. Whatever we do, or have done in the past, to satisfy our basic needs has affected the landscape. To provide food, clothing and shelter, transportation and defense, we have cleared the land and replanted it, rechanneled waterways, and built roads, fortresses, and cities. We have mined the earth's resources, logged entire forests, terraced mountainsides, even reclaimed land from the sea. The nature of the changes made in any single area depends on what was there to begin with and how people have used the land. Some of these changes are examined in the following sections.

PROBLEMS IN THE COASTAL ZONE

Human impact on the land has been particularly strong in areas where land and water meet. In many places the actual shape of the coastline has been culturally altered, either deliberately or inadvertently. In the Netherlands, for example, millions of acres of land have been reclaimed from the sea by the building of dikes and canals to enclose polders. Farming practices in river valleys have had significant effects on deltas, with increased sedimentation often extending the area of land into the sea.

Many cities are built upon coasts, and human occupance of coastal areas has altered them significantly. *Estuaries* are the zones where fresh water from rivers and streams meets the sea; land areas adjacent to estuaries are called *wetlands*. As the name suggests, these may be marshes, swamps, or tidal flats, but in any case, they are extraordinarily productive, as are freshwater wetlands (Figure 4.16).

Wetlands are marked by a complex ecosystem, in which fish, birds, and plant life are intricately interrelated. People have tended to destroy the wetlands, draining, filling, and building upon them. We have built upon sand dunes, rather than leaving them unused as natural barriers between human occupance and the sea. To stabilize beaches, we have constructed seawalls and jetties. All of these acts result in extensive changes in coastal areas.

SURFACE MINING

A practice that has both a direct, immediate impact on the environment and slower-acting and less visible effects is *surface mining*. It involves the removal of vegetation, topsoil, and rock from the earth's surface in order to get at the resources underneath. Open-pit mining and strip mining are the methods most commonly used.

Open-pit mining is used primarily to obtain iron and copper, sand, gravel, and stone. As Figure 4.17 indicates, an enormous pit remains after the mining has been completed because most of the material has been removed for processing. Strip mining is being increasingly employed in this country as a source of coal; more coal per year now comes from strip mines than from underground mines. Phosphate is also mined in this way. A trench is dug, the material is excavated, and another trench is dug, the soil and waste rock being deposited in the empty first trench, and so on. The result is a ridged landscape.

Landscapes marred by vast open pits or unevenly filled trenches are one of the most visi-

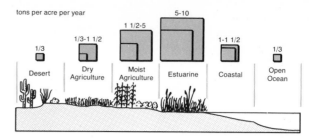

Fig. 4.16 Tons per acre of biomass production. Estuaries, and particularly salt marshes, are among the most productive of ecosystems in the generation of biomass. The influx of nutrients from surface runoff, their concentration at the meeting zone of salt and fresh water, and the efficiency of higher plants and phytoplankton (minute floating plant organisms) in converting solar energy to potential animal food make estuarine zones rich larders for marine and terrestrial animals, including humans. Marsh vegetation captures about 6% of the available sunlight; a stand of growing corn catches only 2% of the sun's energy available to it. Salt marshes produce up to 10 tons per acre of organic matter; highly productive hay lands yield no more than 4 tons. (From *Life and Death of the Salt Marsh* by J. and M. Teal, by permission of Little, Brown and Co. in association with the Atlantic Monthly Press. Illustrations copyright © 1969 by Richard G. Fish.)

133

ble results of surface mining. Thousands of square miles of land have been affected, with the prospect of thousands more to come as the amount of surface mining increases. Damage to the aesthetic value of an area is not the only liability of surface mining. If the area is large, wildlife habitats are disrupted, and surface and subsurface drainage patterns are disturbed. In recent years, concern over the effect of strip mining has prompted federal legislation to increase regulation and stop the worst abuses. Strip-mining companies are now expected to restore mined land to its original contours and to replant vegetation.

Surface mining also leaves tons of wastes. In fact, in terms of tonnage, mining is the single greatest contributor to solid wastes, with about 2 billion tons per year left to be disposed of in

(a)

this country alone. The normal practice is to dump waste rock and mill tailings in huge heaps near the mine sites. Carried by wind and water, dust from the wastes pollutes the air, while dissolved minerals pollute nearby water sources. Occasionally the wastes cause greater damage, as happened in Wales in 1966, when slag heaps from the coal mines slid onto the village of Aberfan, burying over 140 schoolchildren beneath them. Such tragedies call attention to the need for less destructive ways of disposing of mine wastes.

Fig. 4.17 (a) Aerial view of the Bingham Canyon open-pit copper mine in Utah. The pit is over 1500 feet deep; the operations cover more than 1000 acres. (b) About 150 square miles of land surface in the United States is lost each year to the strip mining of coal and other resources. (a: Kennecott Copper Corporation photo by Don Green; b: Illinois State Geological Survey.)

(b)

Fig. 4.18 Subsidence in the Long Beach, California, harbor area. Lines of equal subsidence are shown in feet. The settling of the land in this area occurred over a period of decades as oil was withdrawn from the Wilmington oil field below the harbor. In recent years, subsidence has been checked in part by pumping water into deep wells to maintain subsurface fluid pressure. (Courtesy of the Port of Long Beach.)

EXTRACTION OF SUBSURFACE FLUIDS

The extraction of fluids from beneath the ground brings about different changes than surface mining. The withdrawal of large amounts of oil, gas, or groundwater from underground deposits has caused *subsidence* in areas around the world. Mexico City, Houston (Texas), the Po delta in Italy, the Los Angeles–Long Beach area, and the environs of Lake Maracaibo in Venezuela are among the places that have experienced this phenomenon. Recent evidence indicates that the withdrawal of trillions of gallons of water from a 4500-square-mile area of Arizona has resulted in the subsidence of parts of the area of more than 7 feet since 1952. When the fluids are removed, sediments com-

pact and the land surface sinks. The subsidence in Long Beach has been more than 20 feet, as indicated in Figure 4.18.

As one might expect, subsidence damages structures built upon the land, including buildings, roads, and sewage lines. In coastal areas only a few feet above sea level to begin with, it brings the danger of flooding. A dramatic example of the effects of subsidence occurred in Los Angeles in 1963, when it caused the dam at the Baldwin Hills Reservoir to crack; in less

Fig. 4.19 (opposite) Subsidence caused cracking and emptying of the Baldwin Hills Reservoir of Los Angeles in December 1963. (Courtesy of California Department of Water Resources.)

than 2 hours, the water emptied into the city, resulting in millions of dollars worth of property damage (Figure 4.19).

AGRICULTURE, VEGETATION, AND SOILS

Another human activity that affects both the landscape and the nature of the soil is agriculture. People, by design or by accident, have brought about many changes in the physical, chemical, and biochemical nature of the soil; some of these changes have made the soil less productive. The exact nature of the changes in any area depends on what the past practices have been as well as on the nature of the land to begin with. Irrigation, for example, not only alters drainage patterns but may also lead to excessive salinity of the soil. Because the water tends to move slowly and thus to evaporate more rapidly, the salts carried in the water are absorbed into the ground, in time affecting the productivity of the land. In Iraq and Iran, thousands of once fertile acres have had to be abandoned; the Central Valley of California is being similarly affected by salinization.

Other agricultural activities, such as overgrazing and deforestation, accelerate erosion, which in turn can turn vast areas into useless

Fig. 4.20 Part of the Montfort feed lots in Greeley, Colorado. The sanitary disposal of organic wastes generated by such concentrations of animals is a problem only recently addressed by environmental protection agencies. (Courtesy of Montfort of Colorado, Inc.)

138

Fig. 4.21 Open refuse dumps, while an easy means of waste disposal, often harbor disease-carrrying organisms and contribute to the pollution of soil, water, and air. U.S.D.A. Soil Conservation Service.)

dust bowls through processes of *desertification*. As much as 5% of the world's deserts may have been caused not by climate but by human abuse of the land. Where population pressures impel millions of people to use marginal lands for cultivation or animal grazing, the natural vegetation is destroyed. The gathering of wood for fuel also causes deforestation. People have known for several years that the Sahara Desert is spreading southward; less well known is the fact that hundreds of thousands of acres of land become desert each year on the Sahara's northern edge, as well as in India, the Middle East, and eastern Africa.

Countries where animals are raised intensively face the problem of disposing of the wastes—chiefly manure and animal carcasses. This is a problem particularly in feedlots, where animals are crowded together at maxi-mum densities to be fattened before slaughter. Large feedlots, such as the one pictured in Figure 4.20, may produce as much waste as would a large city. The manure pollutes both soil and water with infectious agents and excess nutrients.

SOLID WASTE DISPOSAL

Modern technologies and the societies that have developed them produce enormous amounts of solid wastes of which disposition must be made. The rubbish heaps of past cultures suggest that humankind has always been faced with the problem of ridding itself of materials no longer needed or of the waste products of its consumption system. The problem for advanced societies, with their ever-greater variety, amount, and durability of refuse, is ever more serious: how to dispose of the solid wastes produced by residential, commercial, and industrial processes. Although these account for much less tonnage than the wastes produced by mining or agriculture, they are everywhere, a problem with which each individual and each municipality must deal. They are the newspapers and the beer cans, the toothpaste tubes and the old television sets, the broken refrigerators and the rusted cars—as well as chemicals and radioactive wastes. The amount of waste per person has increased rapidly in recent years, to its current level in the United States of about 5 pounds per person per day.

Each method of disposing of such wastes has its own impact on the environment. Loading wastes onto barges and dumping them in the sea, long a practice for coastal communities, inevitably pollutes the ocean. Damage to sea life follows, particularly in estuaries. Open dumps on the land of the type pictured in Figure 4.21 are a menace to public health, for they harbor disease-carrying rats and insects. When combustibles are burned, either at dumps or in private or municipal incinerators, chemicals and particulates are discharged into the air. In extreme cases, gases and chemicals may be a health hazard. For example, when polyvinyl chloride (PVC), a component of plastic, is burned, hazardous hydrogen chloride is released.

Sanitary landfills occupy many thousands of acres in this country. Under this method of solid waste disposal, the refuse is deposited in a pit or a trench and compacted, then covered with soil, as depicted in Figure 4.22. There may be several layers of refuse sandwiched between layers of soil. The wastes are then left to decompose. While less of a public health hazard than an open dump, some landfills have had to be shut down when the rain and snow percolating through them have washed out chemicals that then contaminated nearby water supplies.

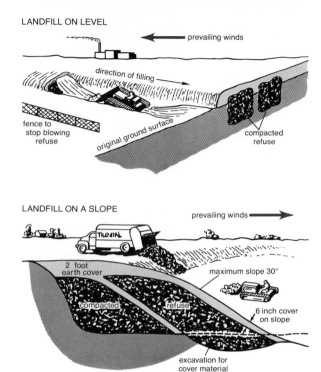

Fig. 4.22 How well-planned sanitary landfills are constructed. (From *Our Precarious Habitat*, Revised Edition, by Melvin A. Benarde, by permission of W. W. Norton & Company, Inc. Copyright © 1973, 1970 W. W. Norton & Company, Inc.)

In addition, the building of houses or other structures on landfills has proved to be of questionable wisdom, as the uneven settling of the soil causes structural damage and the possibility of explosions from unsafe levels of gas.

The disposal of radioactive wastes generated by nuclear fission in reactors that produce electricity for commercial use and in government-owned reactors that produce plutonium for military purposes has proved especially troublesome. The escape of the material into land, air, or water could pose a hazard for thousands, millions, or even billions of years, given the long half-life of some of the isotopes.

No method has yet been devised to ensure

that leakages will not occur. Some wastes have been sealed in protective tanks and dumped at sea. However, the tanks may be moved from the original dumping site by strong currents and may be crushed by water pressure, causing leakage of the wastes. Over 47,000 drums of low-level radioactive wastes were dumped into the ocean 35 miles west of San Francisco in the period 1946–1970. Recent investigations have shown that some of the drums had been crushed, contaminating sediment in the area with plutonium. Even without such physical damage, the life expectancy of the containers must be presumed to be far shorter than the half-life of their radioactive contents.

Other radioactive wastes have been placed in

Fig. 4.23   Two of the thirty storage tanks at the U. S. Department of Energy's Savannah River Plant in South Carolina. Built to contain high-level radioactive wastes, the tanks are shown before they were encased in concrete and buried underground. (E. I. du Pont de Nemours & Co., SC, from U. S. Department of Energy.)

Disposal of Toxic Wastes

Thousands of industrial dumps exist across the nation and, Environmental Protection Agency officials estimate, well over 1000 of them are potential man-made environmental disasters.

One of these, perhaps a warning of counterpart problems elsewhere, was created—quite legally at the time—by Hooker Chemical and Plastics Corporation, which in 1942 began to bury toxic wastes in metal drums in the receding waters and bottom muds of the unused Love Canal in Niagara Falls, New York. The dump site was sold in 1953 to the city's board of education, which built a school on one part of it and sold the rest to a developer for private home construction.

As long as the containers remained intact—nearly 20 years—no surface evidence of contamination appeared. Unusually heavy rains during the 1970s, however, began to carry chemicals leaking from the corroding drums to the surface, where a devil's brew of more than 80 chemicals has been identified; at least 10 of these chemicals have been found seeping into homes bordering the canal. Of these, 7 are proved carcinogens in animals. Accompanying the backyard puddles of mixed toxic substances and the infusion of chemical odors in neighborhood homes have been increases in the incidence of birth defects, mental retardation in children, and a miscarriage rate 50% above normal. In August 1978, the New York State health commissioner declared the area unsafe.

State purchase of private properties at fair market value, undertaken late in 1978, provided monetary compensation, though not physical or mental relief, for the property owners. The larger concern, however, remains: How many other ticking time bombs exist in the form of old industrial dumps filled before recent stringent state and federal safeguard controls, and what ecological disasters of even greater magnitude are yet to befall?

142

tanks and buried in the earth. By 1976, 75 million gallons of liquid high-level waste and 51 million cubic feet of low-level waste were stored at nine locations in the United States, one of which is shown in Figure 4.23. At least one of these storage areas, the one at Hanford, Washington, has experienced leakages, with seepage of one-half million gallons of high-level waste into the surrounding soil.

Another method of waste disposal, used for chemicals as well as for radioactive wastes, has been to inject them into deep wells. For example, waste chemical fluids and radioactive wastes from chemical weapons manufactured at the Rocky Mountain Arsenal of the U. S. Army were injected into a well over 2 miles deep during the period 1962–1966. The practice was halted when it was shown to be triggering earthquakes in the area around Denver, Colorado.

Solid waste will cease to be a problem only when ways are found to reuse the resources it contains. Several methods of reuse are being explored, to different degrees in different countries. Conversion of the wastes into fuels or into compost, a low-grade fertilizer, is feasible

Fig. 4.24 Rabbits converging on a water hole in Australia during a drought. Deliberately imported to Australia, the rabbit quickly became an economic burden, competing with sheep for grazing land and stimulating soil erosion. (Courtesy of Australian Information Service.)

if there is a market for those products. Likewise, many materials in the waste, excluding plastics or textiles, can be recycled and probably will be once the costs are shown to be economical. Until then, current methods of waste disposal will continue to pollute soil, air, and water.

Impact on Plants and Animals

People have modified plant and animal life on the earth through several different processes. One is by deliberately or inadvertently introducing a plant or an animal into an area where it did not previously exist. If conditions in the new area favor the species, it may multiply rapidly, often with damaging and unforeseen consequences.

INTRODUCTION OF NEW SPECIES

The rabbit, for example, was purposely introduced to Australia in 1859. The original dozen pairs multiplied to a population in the thousands only a few years later and, despite programs of control, to an estimated 1 billion a hundred years later, when the photograph in Figure 4.24 was taken. Inasmuch as five rabbits eat about as much as one sheep, a national problem had been created. Much of the grassland on which sheep could graze has been denuded by the rabbits.

Many similar stories could be told. For in-

Fig. 4.25 Whitefish with lamprey attached. (INTERIOR U. S. Fish and Wildlife Service.)

stance, goats were imported to the island of St. Helena by the Portuguese, who could not have foreseen that in time the goats would completely destroy the indigenous vegetation. Overgrazing by grass-eating animals has altered vegetative patterns in many parts of the world, a notable example being the Sahel area of Africa. The sea lamprey, a parasite that attaches itself to a fish and sucks on its blood until the fish dies, found its way into the Great Lakes when the Welland Canal was deepened in 1932 (Figure 4.25). Within a few years, catches of lake trout and whitefish declined precipitously, all but putting an end to these fish as commercial crops.

Plants as well as animals can alter vegetative patterns. The Asiatic chestnut blight, for instance, has destroyed most of the native American chestnut trees in the United States—trees with significant commercial as well as aesthetic value. The cause was the importation of some chestnut trees from China to the United States. They carried a fungus fatal to the American chestnut tree but not to the Asiatic variety, which was largely immune to it. The water hyacinth entered this country in 1884 when bulbs were given as souvenirs at an exposition in New Orleans. Before the end of the century, the spread of the plant had become a matter of concern. In one growing season, a single plant can produce over 60,000 offshoots. To the consternation of sailors and fishermen, swamps, lakes, and canals in the southern United States have become clogged with hyacinths, which also affect fish and plankton.

These examples are just a few of the many that could have been selected to illustrate that plant and animal life are so interrelated that when people introduce a new species to a region, whether by choice or by chance, there may be untold and far-reaching consequences.

DESTRUCTION OF PLANTS AND ANIMALS

Another process by which humans have affected plants and animals is by their deliberate destruction. When people clear an area to plant crops, or spread defoliants in a war, or use herbicides, they are destroying vegetation. Herbicides are chemical weed killers designed to increase agricultural yields in the short run; as illustrated by Figure 4.26, they may have other consequences in the long run.

Whenever we modify the vegetation in an area, we affect plant and animal populations and soils. Thus soil in the tropics exposed by defoliants to wind and sun may become less productive. If herbicides kill microorganisms in the soil, the soil ecology is affected. We are learning that the food chain is part of an interlocking web; to touch one strand of the web is to set off reverberations in the rest.

Humans have caused the extinction of some animals and have placed many others on the

144

Fig. 4.26 A crippled calf. Some live-stock disorders are due to the ingestion of agricultural biocides with feed. (U.S.D.A. photo.)

endangered-species list. We have overhunted and overfished, for food, fur, or sport. Jaguars and buffaloes, passenger pigeons and whales, and hundreds of other animals have been thoughtlessly exploited.

We have also destroyed animals by disrupting the habitats in which they live. When we drain swamps. fell forests, dig mines, or build cities, we are affecting the ecosystem of the region. Often the animals can simply move elsewhere, but not always. The whooping crane, for example, has been virtually eliminated in this country, partly because the marshes in which it lived were drained; its comeback, sought by breeding programs in the United States and Canada, is still uncertain.

POISONING AND CONTAMINATION

Humans have also affected plant and animal life by poisoning or contamination. In the last several years we have become acutely conscious of the effect of insecticides and herbicides, known collectively as *biocides*. The best-known and most widely used has been DDT, although there are now thousands of compounds in use.

DDT was first used during World War II to kill insects that carried diseases such as malaria and yellow fever. In the years following the war, tons of DDT and other biocides were used, sometimes to combat disease, sometimes to increase agricultural yields. Insects, after all, can destroy a significant percentage of a given crop, either when it is in the field or after it has been harvested.

In the last 10 years some of the side effects of these biocides have been well enough documented for us to question their indiscriminate use. Once used, a biocide settles into the soil, where it may remain or may be washed into a body of water. In either case, it is absorbed by organisms living in the soil or mud. Through a

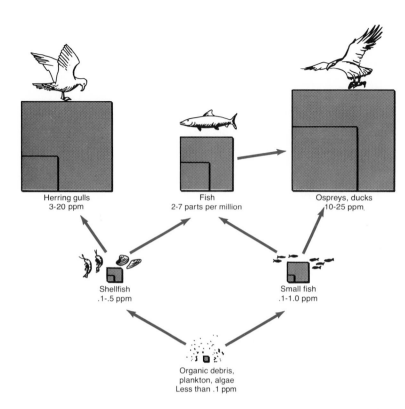

Herring gulls
3-20 ppm

Fish
2-7 parts per million

Ospreys, ducks
10-25 ppm

Shellfish
.1-.5 ppm

Small fish
.1-1.0 ppm

Organic debris,
plankton, algae
Less than .1 ppm

Fig. 4.27 A simplified example of *biological magnification.* **Although the level of DDT in the water and mud may be low, the impact on organisms at the top of the food chain can be significant. In this example, birds at the top of the chain have concentrations of residues as much as 250 times greater than their prey, and one-half million times greater than the concentration in the water. Radioisotopes such as strontium-90 and cesium-137 undergo magnification in the food chain just as insecticides do.**

146

process known as *biological magnification,* illustrated in Figure 4.27, the biocide accumulates and is concentrated at progressively higher levels in the food chain. By remaining as residue in fatty tissues, a very small amount of a biocide produces unexpected effects, with predators accumulating larger amounts than their prey.

Thus the concentration of DDT in the mud at the bottom of a lake may be on the order of 0.01 parts per million (ppm); small crustaceans eating algae or other organisms may exhibit concentrations of DDT of 0.5 ppm; fish feeding on them may show 4 ppm; at the highest level of that food chain, gulls or other birds that feed on those fish may have concentrations of 20 ppm. The higher the level of an organism in the food chain, the greater the concentration of DDT will be, and that concentration may be lethal. Robins and other small birds die when

they eat earthworms that have ingested DDT that has settled into the earth after being sprayed on trees.

DDT has also been shown to cause a decrease in the thickness of the eggshells of some of the larger birds, causing a greater number of eggs to break than normally would. Declines in the population of ospreys, peregrine falcons, and the bald eagle have been traced to this disruption of the reproductive process.

DDT is so long-lived and has been so widely used that it is present everywhere. Even penguins and cormorants in the Antarctic, hundreds of miles from the nearest point of DDT use, have detectable levels of the insecticide. In the United States, DDT and other biocides are present in most foods, including human milk.

Although the use of DDT has declined as its effects have become apparent, other chlorinated hydrocarbon compounds have been developed and are in wide use. PCBs have, like DDT, entered the food chain on a worldwide basis. They, too, are long-lived, accumulate in fatty tissues, and display biological magnification in the food chain. As with DDT, the long-term effect of low concentrations of PCBs in humans is unknown.

By altering the natural processes that determine which insects in a population will survive, we have inadvertently spurred the development of resistant species. If all but 5% of the mosquito population in an area is killed by an insecticide, the ones that survive are the most resistant individuals, and it is they who produce succeeding generations. There are now insects that are totally resistant to DDT, which has led some scientists to conclude that the entire process of insecticide development may be self-defeating.

Conclusion

The intricately interconnected systems of the ecosphere—the atmosphere, the hydrosphere, the lithosphere, and their contained biomes—have been subjected to profound, frequently unwittingly destructive, alteration by humans. The effects of human impact upon the environment are complex and are never isolated. An external action that impinges upon any part of the web of nature inevitably triggers chain reactions the ultimate impacts of which appear never to be fully anticipated.

Human disruption of the fragile structure of the environment is increasing at an increasing rate. World population expansion and the ecological pressures it implies, increases in per capita energy consumption, and the creation of ever more grandiose schemes to alter nature to satisfy felt developmental needs are among the obvious causes of that disruption and are indications of the human conviction that we alone stand above the imperatives of nature. Other species may perish in a changing environment; only humans have sought to divorce themselves from it.

Of course, it is now accepted as an abstract truism that humans are part of the natural environment and depend—literally, for their lives—on the water, air, food, and energy resources that the biosphere contains. But for many, it is an acceptance outside of the realm of daily concern. Ever more forceful, however, is the evidence that humans cannot manipulate, distort, pollute, or destroy any part of the ecosystem without diminishing its quality or disrupting its structure. The increasing frequency of actual environmental danger and disaster has converted the abstraction to a present reality.

Alteration of the environment by human action is as old as the history of humans themselves. It has marked in increasing degree their rise from primitive clans to organized societies, their shift from rural to urban life, their incredible growth in numbers, and their increasing ability through advanced technology to work their will upon the environments of the world. To these matters we turn our attention in the following chapters.

147

For Review

1. What is meant by the *biosphere*? Of what parts is it composed?

2. What are *food chains*? What is meant by *trophic levels*? What is the concept of *entropy* and how is it related to food chains?

3. Draw a diagram of or briefly describe the *hydrologic cycle.* In what ways do humans have an impact upon the hydrologic cycle? What effect does urbanization have upon it?

4. What is meant by *pollution*? By *pollutant*? Is all pollution the consequence of human action alone?

5. The increasing use of fossil fuels over the past 200 years has had a measurable effect upon the environment. List those effects of which you are aware. What measures are advocated or imposed to minimize them?

6. What steps have coal companies and the government taken to guard against environmental damage from strip mining? All reclamation efforts involve economic cost, and all costs must ultimately be borne by the consumer. The U. S. Office of Surface Mining has estimated that its reclamation requirements would add 25 cents per ton to the price of coal mined on the level terrain of the Midwest, and perhaps as much as $2.16 a ton on the steep slopes of the Appalachian fields. Industry sources say that these estimates are from three to eight times too low. In a paragraph, discuss whether these additional costs—magnified as they are incorporated into the prices of electricity, manufactured goods, services, and so on in the total economy—should be willingly assumed or should be resisted by the public.

7. How is solid waste disposed of in your community? What ecological problems, actual or potential, are implicit in this means of disposal?

8. What is the *nitrogen cycle* and what have people done to disrupt it? What are the short-term and long-term consequences of that disruption?

9. What is meant by *biological magnification*? What does it imply at the human trophic level?

Suggested Readings

COMMITTEE ON GEOLOGICAL SCIENCES, NATIONAL ACADEMY OF SCIENCES, *The Earth and Human Affairs.* Canfield Press, San Francisco, 1972
DETWYLER, THOMAS R., *Man's Impact on Environment.* McGraw-Hill, New York, 1971.
FARVAR, M. TAGHI, AND JOHN P. MILTON (eds.), *The Careless Technology.* Natural History Press, Garden City, NY, 1972.

148

FOIN, THEODORE C., *Ecological Systems and the Environment.* Houghton Mifflin, Boston, 1976.

MILLER, G. TYLER, *Living in the Environment.* Wadsworth, Belmont, CA, 1975.

STRAHLER, ARTHUR N., AND ALAN H. STRAHLER. *Geography and Man's Environment.* Wiley, New York, 1977.

WAGNER, RICHARD H., *Environment and Man.* Norton, New York, 1971.

150

The Culture–Environment Tradition

The Crow country. The Great Spirit put it exactly in the right place; while you are in it, you fare well; whenever you get out of it, whichever way you travel, you fare worse. . . . The Crow country is in exactly the right place. It has snowy mountains and sunny plains; all kinds of climates and good things for every season. When the summer heats scorch the prairies, you can draw up under the mountains, where the air is sweet and cool. . . . In the autumn when your horses are fat and strong from the mountain pastures, you can go down on the plains and hunt the buffalo or trap beaver on the streams. And when winter comes on, you can take shelter in the woody bottoms along the rivers.

The Crow country is exactly in the right place. Everything good is found there. There is no country like the Crow country.

Such was the opinion of Arapoosh, Chief of the Crows, speaking of the Big Horn basin country of Wyoming in the early 19th century. In the 1860s, General Raynolds reported to the Secretary of War that the basin was "repelling in all its characteristics, surrounded on all sides by mountain ridges [and presenting] but few agricultural advantages."

In the four chapters of Part I, our primary concern was with the physical landscape. But while the physical environment may be described by process and data, it takes on human meaning only through the filter of culture. Arapoosh and Raynolds viewed the same landscape, but from the standpoints of their separate cultures and conditionings. Culture is like a piece of tinted glass, affecting and distorting our view of the earth. Culture conditions the way people think about the land, the way they use and alter the land, and the way they interact with one another upon the land.

Such conditioning is a focus of the culture–environment tradition of geography, a tradition still concerned with the landscape but not in the physical science sense of Part I. In Part II, humans are the focus; yet the physical environment is always in our minds as we develop the notion of cultural difference and reality. Our landscapes become human rather than purely physical.

The four chapters in this part explore the environment–culture–people relationship. In Chapter 5, "Historical Cultural Geography," we go back in time to the origin of people, outlining the process by which cultures evolved and dwelling on the meaning of culture. An important question is how cultures became regionally differentiated from each other when humans are believed to have had a common origin in East Africa.

Chapter 6, "Population Geography and Cultural Diversity," focuses on those variables that help to account for differences in patterns of behavior over the face of the earth. Those aspects of population that affect the social organization of a group of people are examined, and some population patterns are described. The remainder of the chapter deals with key variables in defining a culture and in producing variations in culture from place to place.

One of these variables, politics, is explored in detail in Chapter 7, "Political Geography." Political systems and processes strongly influence the form and distribution of many elements of culture. Economic and transportation systems coincide with national boundaries. Political regulations, whether detailed zoning codes or broad environmental protection laws, have a marked effect on the cultural landscape. In some countries, religion, literature,

151

music, and the fine arts are affected by the government, which may encourage certain forms of expression and reject others as reactionary or decadent. Even sports are subject to political influence; for example, the relationship between nationalism and the Olympics is noted in the chapter.

Patterns of human spatial behavior and the factors that account for the manner in which people use space are the subject of Chapter 8, "Behavioral Geography." The way people view the environment is important, for human actions are guided as much by how things are perceived to be arranged as by the objective reality of their location. The individual and group behaviors of humans create the cultural landscapes—those of production, resource utilization, and urban settlement—that are the topics of Part III, "The Location Tradition." Chapter 8, therefore, may be seen as an entity in itself and as a bridge to other subfields of the discipline.

Our theme in Part II, however, is people and the collective and personal cultural landscapes they create or envision. Let us begin with a consideration of the origins and the cultural variations of humankind.

Culture is as much a part of the regional differentiation of the earth as are the topography, the climate, and other aspects of the physical environment. The visible and invisible evidences of culture—buildings and farming patterns, language and political organization—are elements in the diversity that invites and is subject to geographic inquiry. Cultures—the learned ways of life of a people—result in human landscapes with variations as subtle as the difference in "feel" of Paris, Moscow, and New York or as obvious as the contrasts of rural Thailand and the American Midwest.

Since such differences exist, cultural geography exists. Since they have existed in various forms for millennia, one branch of cultural geography addresses the question of "Why?" Why, since humankind constitutes a single species, are cultures so diverse? What were the origins of the different culture regions we now observe? How, from whatever limited areas in which single culture traits and amalgams developed, were they diffused over a wider portion of the globe? How did people who had roughly similar origins come to display significant areal differences in technology, social structure, and ideology? These questions are the concern of the present chapter.

Human Beginnings

Perhaps 3–4 million years ago, the ancestors of modern humans were making primitive stone tools in East Africa. Since that time, people have made astonishing discoveries and inventions. They have created great civilizations and they have also plundered them. They have achieved sophisticated control of their environment, created materials that do not appear in nature, and harnessed sources of energy unknown to their ancestors, although not always safely. The stone tools created so long ago were the beginning of a prolonged developmental

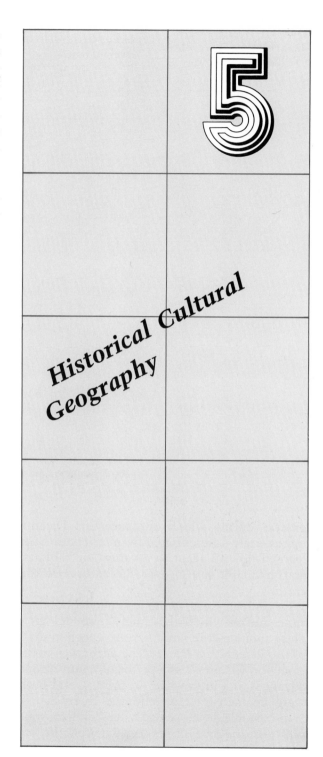

5

Historical Cultural Geography

Table 5.1 Key Dates in the Evolution of Culture

Recent fossil discoveries make uncertain the early sequence of hominid development.

Cultural Period	Approximate Number of Years Ago[a]	Chief Developments
	10–15 million	Evolution of hominids
	5–3.75 million	Emergence of *Homo* and *Australopithecus*
Paleolithic	1.5 million	Evolution of *Homo erectus*, migration to Asia and Europe; use of fire and crude tools
	250,000–11,000	*Homo sapiens* develops and disperses across world; hunting-and-gathering economy; variety of tools; artwork; burial rituals; retreat of last glaciers
Mesolithic	11,000–9000	Domestication of plants and animals, some production of food; semipermanent settlements; further development of tools
Neolithic	9000–5500	Systems of agriculture; use of animals for work; specialization in occupation
	5500 to present	Growth of culture hearths, cities, city-states, and empires; continuous development in all systems of culture

[a]Dates apply to East Africa and the Old World. Societies in other parts of the world passed through many of these stages at later dates.

process during which our ancestors became increasingly differentiated from other animals, learned how to adapt to a variety of habitats, and gradually spread over the surface of the earth.

In geologic terms, the extended period of time during which these developments took place includes the late *Pliocene* epoch and all of the *Pleistocene* epoch. The latter was mentioned in Chapter 2 as a period of recurrent ice advances and interglacial warmings, and both were characterized by climatic fluctuations. While people developed advanced civilizations only during a recent interglacial period, the origins of humans go back far beyond the onset of the glacial epoch. The incompleteness of the fossil record renders the exact details and sequence of human evolution uncertain. The frequency of new discoveries subjects even that uncertain record to changing interpretations. Some approximate key dates on which there is general agreement are given in Table 5.1. By any standards, mankind's biological and cultural evolution has traversed an immense time span.

Perhaps 10–15 million years ago, *hominids*, or "manlike" creatures, diverged from their apelike progenitors. Not accepted by all paleo-

The developmental history of early humans is undergoing constant revision as new fossil evidence is discovered, particularly in East Africa, and as new theories and lines of evidence are proposed. The temporal sequence summarized in the text stands, as of 1980, as accepted but challenged wisdom.

> **The Emergence of Humans: An Alternate View**

One serious challenge comes from the biochemists, who have rejected as impossible the anthropologists' suggestion that the hominid–ape split came as early as 10–15 million years ago. Biochemically, the great apes and humans are closely related, and in some ways, chimpanzees are closer to humans than to gorillas. On the basis of biochemical similarities, a date of 3–7 million years B.P. for the hominid revolution would appear more logical. Recent fossil evidence from Ethiopia seems to support the biochemists' timetable: some 3-million-year-old remains have been found of the "Afar ape-man," which show evidence of divergence into both human and *Australopithecus* descendants. Considering other fossil evidence as well as that of the Afar ape-man, the emergence of true humans as recently as 7 million B.P. is now considered likely. The still tentative recorded and conjectural sequence is diagrammed below.

Orthodox evolutionary texts hold that *Dryopithecus* was the ancestor of modern apes; *Ramapithecus* was either itself hominid or gave rise to hominids and thus to humans. (Adapted by permission from *The Economist*, London, June 10, 1978, p. 81.)

155

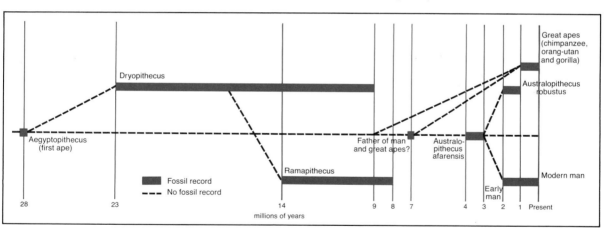

anthropologists as a hominid, *Ramapithecus* appears to have moved from the trees to a more open savanna environment. It may represent a link between the early apes and the two (and perhaps more) genera of later hominids, undoubted members of the human family, which began to develop approximately 5 million years ago.

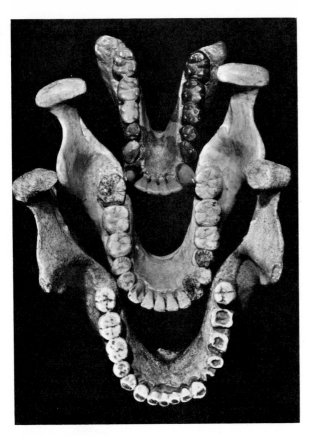

156

Fig. 5.1 Models of the lower jaws of (from top to bottom) a hominid, *Australopithecus*, and *Homo erectus*. Over time, variations in the size of teeth may have been partially determined by the types of food consumed and the extent to which teeth were used for gripping and chewing nonfood items, such as animal hides. (Courtesy of The American Museum of Natural History.)

One of these was *Australopithecus*, "near man." Walking erect and with limb and tooth structures that were clearly hominid, small-brained *Australopithecus* may or may not—authorities differ—have given rise to the genus *Homo*, or "true man" (Figure 5.1). It does appear, however, that *Australopithecus* was contemporary with early humans, arrived at an evolutionary dead end, and became extinct perhaps 1 million years ago.

The date of the appearance of *Homo* is much in doubt, ranging back by some disputed estimates to 3.75–4 million years B.P. (before present). Little different physically from the related australopiths, the first humans showed clear evidence of the beginnings of culture: tool and fire making, the use of language, and the organization of groups into linguistically based bands.

The emergence of these cultural essentials and their later regional refinements are roughly associated in time with the Pleistocene epoch, beginning approximately 2 million years B.P. For the purpose of a general differentiation of the stages of cultural development—not necessarily related to the progress of individual social groups—anthropologists recognize the series of cultural-technological periods we employ here and in Table 5.1.

THE PALEOLITHIC PERIOD

By 1.5 million years ago the hominid form called *Homo erectus* had evolved and migrated from Africa to Asia and southern Europe. Its stone tools, more carefully shaped and differentiated than those made by its predecessors, show clear evidence of being worked to standardized designs. These tools indicate, too, that societies had an established division of labor between men and women and between adults of the same sex. A hunter and gatherer, *Homo erectus* maintained home bases with fireplaces and had begun the construction of shelters. *Homo sapiens*, the only species of the genus *Homo* alive today, is thought to have evolved no later than 250,000 years ago.

Neanderthal and, later, Cro-Magnon people, perhaps the best known of the early *Homo sapiens*, inhabited Western Europe and the lands around the Mediterranean Sea from about 100,000 to approximately 10,000 years ago. They possessed a brain as large as our own; indeed, in the case of *Homo neanderthalensis*,

Fig. 5.2 **Cave art of the Cro-Magnon culture: a woolly rhinoceros. (Reproduced by permission of The University Museum, University of Pennsylvania.)**

the cranial capacity exceeded ours. They made many kinds of tools, used animal skins and furs for clothing, and were able to hunt big game as well as the smaller animals. Art and religion became detectable parts of their culture. Beautiful drawings have been discovered in some of the caves in which Cro-Magnon people lived in southern France and Spain (Figure 5.2). Evidence of religious ritual is marked by obvious ceremonial burial, even by the earlier Neanderthals, with tools, food, ornaments, and even flowers laid with the body in dug graves.

The term commonly employed to describe the stage of human culture during this period is *Paleolithic*; it is also known as Old Stone Age culture. It began when people first used stone tools several million years ago; the end of this period coincided in much of the Old World with the retreat of the last glaciers in Europe about 12,000 years ago. During the last very few thousand years of that time span, the rapidity and frequency of climatic change put new stresses on human populations by altering

their environment and requiring an adjustment of their techniques of food acquisition to changing flora and fauna. The same changes gave rise to sharp regional contrasts and accelerated the differentiation of culture that for so many millennia previously had evolved comparably and very slowly among all human groups.

Various lines of evidence tell us that during the late Pleistocene period, beginning about 40,000 years B.P., the unglaciated sections of western, central, and northeastern Europe were covered with tundra vegetation (Figure 5.3). Southeastern Europe and southern Russia had forest, tundra, and grasslands, while the Mediterranean area had forest cover. Gigantic herds of herbivores—reindeer, bison, mammoth, horses—browsed, bred, and migrated throughout the tundra and grasslands; an abundant animal life filled the forests.

By the end of the Paleolithic period, humans had spread across the earth to all the continents but Antarctica. The settlement of the

157

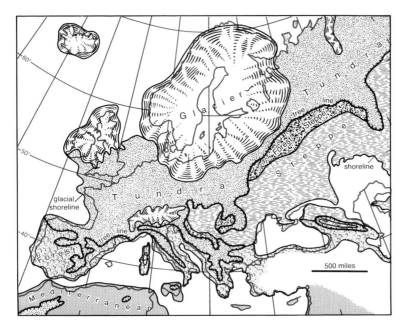

Fig. 5.3 The extent of glaciation and the distribution of vegetation during the Upper Paleolithic period. (After *Descriptive Paleoclimatology*, A. E. M. Nairn, ed. Copyright © 1961 by John Wiley & Sons. Reprinted by permission of John Wiley & Sons, Inc.)

158

lands bordering the Pacific Ocean is shown in Figure 5.4. While spreading, populations also increased, benefiting in growing numbers from the more abundant food supply. Estimates of the number of humans range from 5 to 10 million by about 9000 B.C. Human tool-working

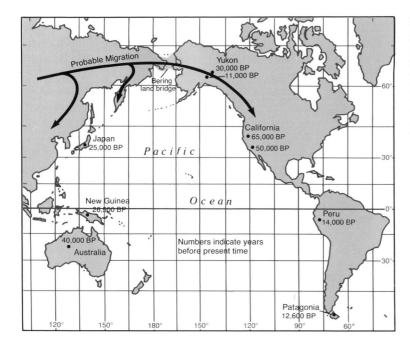

Fig. 5.4 Settlement of the Pacific region. Recent archaeological evidence indicates that the land areas surrounding the Pacific Ocean were settled at least 25,000 years ago. Paleolithic hunters and gatherers reached the American continent across a land bridge in the Bering Strait area. The land bridge existed during several ice ages, so that there may have been repeated migrations, which would account for the variously dated remains from the Yukon.

techniques had become much more sophisticated and diverse, and the variations in tool assemblages characteristic of the different population groups steadily increased (Figure 5.5). Technology thus greatly extended the range of possibilities—not perceived by more primitive peoples—in the use of locally available raw materials. The result was more efficient and extensive exploitation of the environment.

Hunting, gathering, and, locally, fishing formed the basis of existence. Fruits, nuts, tubers, and edible plants joined birds, small and large game, and fish in the diet. The proportion of hunted to gathered food, of meat to vegetable matter, depended on their relative availability and on the technological (tool-making) level of spatially separated populations. Recurring periods of climatic change, altering the regional floral and faunal constitution and abundance, presumably encouraged the diversification of adaptive strategies and rewarded innovation— or at least accelerated the adoption of innovations—within populations reduced by a sudden depletion of food.

We may assume, then, that the Upper Paleolithic period was marked by a continuous process of learning how to cope with a world changing in environment and in potentiality for exploitation. What was learned—in tool making, hunting strategy, language, religion, art—was transmitted within the culture group. The increasing variety of adaptive strategies and technologies and the diversity of noneconomic creations in art, religion, and custom meant an inevitable cultural variation of humankind. The differentiation among societies so evident today began during the Paleolithic.

CULTURE AND RACE

Those specialized behavioral patterns, responses, cognitions, and adaptations that are called *culture* summarize the way of life of a group of people. Culture is both the totality and the details of patterns of *learned* behavior

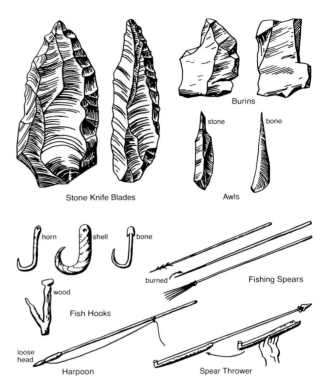

Fig. 5.5 Representative tools of the Paleolithic period in Europe. The basis of many Upper Paleolithic tools was the stone blade, formed by new techniques of flaking stone that permitted efficient manufacture of special-purpose implements. Among these was the *burin*, a chisel-edged blade for working wood, bone, and antlers. The wooden or bone spearthrower (called the *atlal* after its Aztec name) permitted more reliable harvesting of large game. Development of bone and antler fish hooks and harpoons made possible a sedentary life based on fishing; clothing and tents could be fashioned with bone awls, needles, and fasteners.

transmitted to successive generations of a social group through imitation and through that distinctively human capability, speech. We shall return to a consideration of the concept and the content of culture and to the innovation and diffusion of its elements later in this chapter.

The spread of human beings over the earth and their adaptation to different environments were accompanied by the development of those physical variations in the basic human stock that are commonly called racial. In very general terms, a *race* may be defined as a breeding population whose members have in common some hereditary biological characteristics that set them apart physically from other human groups. Since the reference is solely to *genes*, the term *race* cannot be applied meaningfully to national, linguistic, religious, or other culturally based classifications of populations.

Although some anthropologists argue that race is a meaningless concept, there exists a common understanding that populations are distinctively grouped in accordance with differences in pigmentation, hair characteristics, and other evidences largely related to variations in soft tissue. Subtle skeletal differences among peoples also exist, on the basis of which the anthropologist Carlton S. Coon considered Cro-Magnon remains to be Caucasoid (or "white"), some African finds of the Upper Paleolithic to display Negroid traits, and comparably aged Chinese finds to be recognizably Mongoloid. Certainly, racial differentiation is old in humankind and can reasonably be dated at least to the Paleolithic spread and isolation of groups of people. In the minds of many, the problem of race as a useful scientific concept lies in the lack of agreement on an acceptable classification system. Caucasoid, Mongoloid, Negroid, Congoid, Australoid, Polynesian, Amerindian, and innumerable alternate and additional major and minor groupings have been proposed.

The question here is not the utility or the applicability of the concept of race. Rather, it is the visible variation of populations as a distributional characteristic of human development. If all humankind belongs to a single species, *Homo sapiens,* that can freely interbreed and produce fertile offspring, how did such observable differentiation occur? Among several causative factors commonly discussed, two appear to be most important: *genetic drift* and

adaptation. Genetic drift refers to a heritable trait that appears by chance in one group and becomes accentuated through inbreeding. If two populations are spatially too far apart for much interaction to occur (isolation), a trait may develop in one but not in the other. Adaptation to particular environments is thought to account for the rest of the racial differentiation, although it should be recognized that we are only guessing here. One cannot say with certainty that any racial characteristic is the result of such evolutionary changes. Genetic studies, however, seem to indicate considerable relationship between, for example, solar radiation and blood type and between temperature and body size.

THE MESOLITHIC PERIOD

The retreat of the last glaciers about 12,000 years ago marked the end of the Paleolithic period and the beginning of the Mesolithic period (the Middle Stone Age) in cultural development in the Old World. With glacial recession came climatic, vegetational, and faunal changes that imposed upon humans new ecological conditions to which adaptation was required. The weather became warmer. Forests began to appear on the open plains and tundra of Europe and northern China. In the Near East, where plant and animal domestication would later occur, dry steppes were replaced by savanna vegetation. The large grazing animals—reindeer, mammoth, buffalo, and others—retreated to the north or disappeared. Their passing was reflected by a sudden absence of animal depiction in cave art.

As the food and ecological base altered, so did human technologies. Fishing gear—hooks and seine nets, spears and harpoons, dugout canoes and skin boats—reflects an increase in fish and shellfish in the diet of Europeans. Notched bows, arrows with refined heads, and spears were employed in harvesting the more elusive forest game. The spread of forests led to an increased use of wood, to the appearance of hafted axes of antler, bone, or stone, and to wooden shelters. Importantly, a broadly based rather than a narrowly based subsistence economy came into being, utilizing a variety of food sources within an area and replacing the former reliance upon a single food base such as large game. The broadened and more reliable food base made sedentary village life a possible alternative to nomadism. The domestication of both plants and animals, which began during this period, led subsequently to full-blown agricultural societies and, eventually, to the development of cities, city-states, and empires. The Mesolithic period—from about 11,000 to 5000 B.C. in Europe—was crucial for the transition from the simple collection of food to its production.

There is no agreement on whether the domestication of animals preceded or followed that of plants. It is quite likely that the sequence was different in different areas. Certainly, animal domestication—which means, simply, the successful breeding of species that are dependent on human beings—was in the main begun during the Mesolithic epoch and was apparently independently developed in several world regions. Radiocarbon dates suggest the occurrence of the domestication of goats in the Near East as early as 8000 B.C., of sheep in the Near East or Turkey by about 7500 B.C., and of cattle and pigs in both Greece and the Near East about 7000 B.C. The horse underwent domestication around 4000 B.C. in the grasslands of the Ukraine. Recently dated fossils from Iraq indicate that dogs had been domesticated by 12,000 B.C. To date, this is the earliest evidence of animal domestication.

As with animals, so plant domestication appears to have occurred independently in more than one area of settlement, as shown on Figure 5.6. One source region, and possibly the oldest, was in the Middle East in present-day Iraq and Iran. Others were in south and east Asia, north China, and north and possibly west Africa. In the Americas, Mexico, the Yucatán, and Peru appear to have been separate sites of plant and animal domestication. Plant domestication, therefore, was a cultural development undertaken by many hunting–gathering peoples who found in their separate home territories appropriate wild plant stock and favorable cultural and environmental preconditions. A distinguished geographer, Carl O. Sauer, summarized those preconditions as

1. A population already well fed and able to afford the time investment implicit in plant selection, propagation, and improvement.
2. Occupancy of areas of marked plant diversity.
3. The existence of wooded areas, devoid of heavy sod, and of hilly terrain away from river valleys subject to flood.
4. A sedentary population, workers of wood, and possessors of forest-clearing traditions and tools.

The Mesolithic regions of the origins of agriculture all show the gradual introduction of tools—mortars and pestles for grinding seeds, flint blades for harvesting—that indicated a growing awareness of and dependence upon wild grains in the food base. The domestication of plants, which implies simply the deliberate planting by humans of seeds, roots, or shoots of selected stock, was initiated during the Mesolithic period, but in its refined form, it marks the onset of the Neolithic period. It appears to have taken place at the outer margins of the most favorable areas of growth of wild plant stock. Usually it involved the movement of selected species to ecological niches for which they were not ideally adapted, thus hastening mutation. The human cultivators, too, adapted: to sedentary residence to protect the planted forest clearings from animal, insect, and human predators; to greater labor specialization; and to a more formalized and expansive

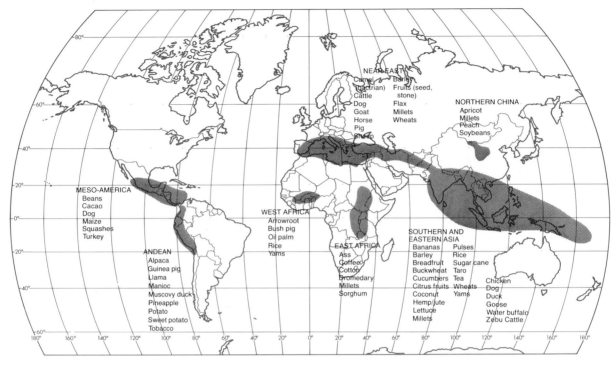

Fig. 5.6 Chief centers of plant and animal domestication. The southern and southeastern Asian hearth was characterized by the domestication of plants, such as taro, that are propagated by the division and replanting of existing plants. Reproduction by the planting of seeds (e.g., maize and wheat) was more characteristic of Meso-American and the Near East. (After J. O. M. Broek and J. W. Webb, *A Geography of Mankind*, McGraw-Hill Book Company, New York, 1968.)

religious structure in which fertility and harvest rites became important elements. The regional contrasts between hunter–gatherer and sedentary agricultural societies increased.

NEOLITHIC PERIOD

Neolithic (or New Stone Age) designates a stage of cultural development, not a specific span of time. The term implies the creation of an advanced set of tools and technologies to deal with conditions and needs encountered by an expanding, sedentary population whose economy was based upon the agricultural management of the environment. Not all peoples in all areas of the earth made the same cultural transition at the same time. In the Near East, from which most of our knowledge of this late prehistoric period comes, the Neolithic period lasted from approximately 8000 to 3500 B.C. There, as elsewhere, it brought complex and revolutionary changes in human life. Culture began to alter at an accelerating pace, and change itself became a way of life. In an interconnected adaptive web, technological and social innovations came with a speed and a genius surpassing all developmental progress of the preceding millions of years.

Humans learned the arts of spinning and weaving plant and animal fibers; learned to use the potter's wheel and to fire clay and make utensils; developed techniques of brick making, mortaring, and building construction; and discovered the skills of mining, smelting, and casting metals. On the foundation of such technical advancements appeared formal villages and the new social environment they created. A more complex exploitative culture engendered a stratified society to replace the rough equality of adults in hunting–gathering economies. Special local advantages in resources or products promoted the development of long-distance trading connections, which the invention of the sailboat helped to maintain.

By the end of the Neolithic epoch, certain spatially restricted groups had created a food-producing—rather than a foraging—society. In so doing, they undertook the purposeful restructuring of their environment through plant and animal modification; by management of soil, terrain, water, and mineral resources; and by utilization of animal energy to supplement that of humans. Metal was used to make refined tools and superior weapons—first, pure copper, and later, the alloy of tin and copper that produced the harder, more durable bronze. Humans had moved from adopting and shaping to the art of creating.

As people gathered together in larger communities, new and more formalized rules of conduct and control emerged, especially important where the use of land was involved. We see the beginnings of governments to enforce laws and specify punishments for wrongdoers. The protection of private property—so much greater in amount and variety than that carried by the nomad—demanded more complex legal codes. Religions became increasingly formalized. Early people saw themselves as subject to the whim of uncontrollable forces outside themselves, forces such as the wind and the rain, floods and droughts, the wanderings of game animals and the dangers of the hunt. For the hunter, religion could be individualistic,

and his worship was concerned with personal health and safety. The collective concerns of agriculturalists were based on the calendar: the cycle of rainfall, the seasons of planting and harvesting, the rise and fall of waters to irrigate the crops. This religion became more formalized, with seasonal rituals and an established priesthood that stood not only as intermediaries between people and the forces of nature but also as authenticators of the timing and structure of the rituals.

These developments occurred over several thousand years. They led to increasing specialization in occupations. Metalworkers, potters, sailors, priests, and, in some areas, warriors complemented the work of farmers and hunters.

Culture Hearths

The social and technical revolutions that characterized the Neolithic period were, at their origin, spatially confined. The new technologies, the new ways of life, and the new social structures diffused from those points of origin and were selectively adopted by peoples not a party to their creation. The term *culture hearth* is used to describe advanced centers of culture systems that exerted an influence on surrounding regions.

In the classical Near East, we may speak of an explosion in knowledge in the years that followed the Neolithic period. Large cities, city-states, and empires were founded. Political systems and religious ideas became highly organized. Rules were established to govern land ownership, the payment of taxes, trade, and other activities. Technology advanced rapidly.

In increasingly complex societies of laws, formal commercial transactions, and involved religious traditions, permanent record keeping was needed. Writing appeared in Mesopotamia and Egypt in the last quarter of the fourth millennium B.C., as cuneiform in the former and as hieroglyphics in the latter. The separate forms of writing have suggested to some that they were of independent origin in separate hearths; others maintain that the *idea* of writing originated in Mesopotamia and diffused outward to Egypt, the Indus Valley, Crete, and perhaps even to China. It is usually assumed that the systems of writing or record keeping developed in New World hearths were not related to Old World origins. Also accepted is the idea that the simplest, most efficient writing system, the alphabetic system, was invented only once, by a Semitic people in southwestern Asia around 1500 B.C.

Predictive sciences of astronomy and mathematics were developed, stimulated by the calendrical demands of a regulated agricultural year and its associated religious ceremonies and by the measurement needs of an intricate land tenure system. Writing, in order to preserve records of observations, was essential to astronomy, as it was to the advancement of mathematical knowledge. Art, much of it concerned with worship, warfare, and rulers, blossomed. Architecture, associated with the structural needs of complex societies and including massive constructions for ceremonial purposes, flourished. The level of technological and cultural development was, by definition, higher in these centers of innovation and invention than elsewhere, and the new ways of doing things and thinking about things spread outward from them to people in adjoining territories.

Four major culture hearths thrived in the years that followed the Neolithic period. These centers of early creativity were located in Egypt, Mesopotamia, the Indus Valley, and northern China. Other hearths were located in the Americas, west Africa, Crete, and Syria (Figure 5.7). These developed later than the hearths mentioned above, and the latter two appear to have been stimulated by contact with outside groups.

These hearths of advanced culture are generally agreed to have reached the status of *civilizations*. The definition of that term is not precise, but indicators of its achievement are commonly assumed to be writing, metallurgy, established trade connections, predictive sciences, class stratifications and labor specialization, formalized governmental systems, and a structured urban culture.

The difficulty in limiting the definition to a single one of these characteristics, as is occasionally done, is that they emerged gradually and in different degrees and sequences in individual culture regions. Such is the case with cities and city-dominated regions. In some hearth areas, as Mesopotamia and Egypt, the transition from settled village to urban form was gradual and prolonged; Minoan Crete never developed a true urban culture. In the Indus Valley, the concept of *city* appears to have been imported, and the chief centers of Mohenjo-Daro and Harappa had from the start an advanced structural form.

The association of city with civilization is inevitable and historically accurate. The association of *city* with *state* is also accepted, for in the classical cities was found the concentration of administrative control, priesthood and temple, artisan and merchant, defensive ramparts and lay nobility that, together, represented the centralization of nonagricultural functions essential to the organized control of territory.

To some, the emergence of the state and its visible functional center, the city, was the culminating contribution of ancient culture hearths, separating Neolithic societies from predecessor cultures. But what circumstances demanded the creation of the state and of its epitome, the capital city? Karl A. Wittfogel has seen as a common thread the needs of increasingly populous societies to control and render more productive environments in which water was a scarce and periodic resource. He calls the ancient civilizations of Mesopotamia, Egypt, India, China, and lowland Peru "hydraulic societies." Each arose in roughly comparable

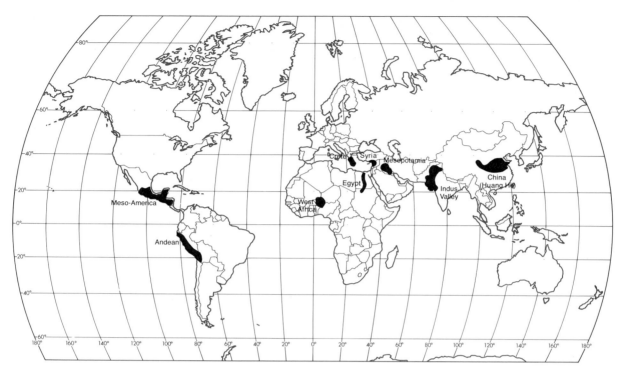

Fig. 5.7   **Early culture hearths of the Old World and the New World.**

The characterizing elements of the advanced civilizations that apparently suddenly flowered in the Near East and in North Africa some 5000–6000 years B.P. had their roots in the dim reaches of prehistory. Each feature of sophisticated cultures such as the Sumerian and the Egyptian—formal systems of writing, a great store of astronomical knowledge, and accurate calendars—owed its origin not to spontaneous generation but to the slow accumulation of knowledge and skills among predecessor populations, particularly the Cro-Magnon.

Far from the hulking, brutish cave dwellers of popular conception, the Cro-Magnon peoples lived in Eurasia as hunters and gatherers from about 40,000 to 10,000 years B.P. The first of modern humans, they left behind no enduring structures and wrote nothing in a currently comprehensible language. They did leave, however—as drawings in the caves they decorated but did not inhabit, in sculpture, in markings on bone and stone, and in discards of their creativity—ample evidence that they, too, were inheritors, developers, and transmitters of still more ancient, sophisticated cultural traits.

Cro-Magnon cave art, dated as 17,000–20,000 years old, was first discovered at Altamira, Spain, in 1879. Subsequent finds elsewhere, occurring at the

The Cro-Magnon People

rate of one or two a year, quickly established that the artists not only controlled advanced knowledge of pigments, binders, and painting tools and techniques but also adhered to recognizable, regionally specialized "schools" of drawing. Different cultural traditions had obviously developed across Europe during the Pleistocene. Analyses of their pigments show that Cro-Magnon artists utilized perhaps half a dozen different colored minerals, grinding them into fine powders and mixing them with binders presumably made from animal fat, egg white, blood, plant juices, and fish glue. Paint jars, stone sketch pads, animal hair and frayed twig brushes, and stone palette knives all document the care and preparation that attended the renderings of animals and vegetation upon the walls and ceilings of deep cave recesses. Even older than the drawings of late Cro-Magnon times was a tradition of sculpture marked by widespread finds of voluptuous "Venus" figurines of 20,000–27,000 years B.P. and by the oldest known sculpture, a 32,000-year-old image of a horse in mammoth ivory that, according to some art historians, ranks with the finest works attributable to any culture in any age.

That Cro-Magnon art was closely related to religious observations is generally accepted. More debatable, but still persuasively argued, is that indecipherable complex markings on bone and stone obviously not used as tools represent a form of record keeping, of notation of the seasons and the phases of the moon that records astronomical observations of the type to be carried still further in such observatories as that of Stonehenge, England.

These and other lines of evidence led increasing numbers of prehistorians to postulate that Cro-Magnons—and their predecessors—possessed the linguistic and conceptual sophistication and had developed the rudiments of science and craft that were the requisite prologue to the "spontaneous" emergence of advanced civilizations.

physical settings; except for Peru, each was in a semidesert through which ran a major stream. The Nile, which gets its water from the mountains of Ethiopia, cuts through the desert in Egypt on its way to the Mediterranean Sea, while the Indus rises in the Himalayas of Asia, flows through the Thar Desert, and empties into the Indian Ocean. Mesopotamia centered on the arid lowland area through which pass the Tigris and Euphrates rivers. The Wei River of China, a tributary of the Huang He (Yellow River), which was an East Asian culture hearth, also flows through a semiarid region. In Peru several (rather than a single) snow-fed rivers cut across an arid coastal plain.

In each instance a limited floodplain environment for which water supplies could be controlled and directed efficiently by a highly organized society prompted the sequence of events culminating in the emergence of the state, of the central dominating city, and of civilization. Other authors have proposed other circumstances as causative agents: group rivalry over limited land resources; the increase in utilizable (animal) energy as a necessary prelude to cultural advance; and the inevitable need for more restrictive formalized control of populations whose increase precluded informal social organization and demanded regulated and reliable food production.

It has been traditional to regard food surpluses from intensive agriculture as the stimulus to population growth. The surpluses, and the greater labor efficiency implied, presumably made possible the support of the people required for the development of structured religions, the arts, and science. Recently this sequence has been criticized as a reversal of the logical chain of causation. The contrasting view now offered is that it is most unlikely that food production was expanded beyond expected family needs in order to provide a surplus to be consumed by populations that did not exist. Rather, it is felt, population growth created pressure for increased productivity, a pressure reinforced and made effective by the demands of a centralized government. Thus, the argument concludes, population growth and increasing density led to the formation of the state and all of its discerned cultural trappings, with the state demanding obedience and efforts on its behalf from those it governed.

The contrasting views concerning the nature of the causal chain leading to civilization are beyond the scope of our present discussion. It is important to note, however, that there exists a wide diversity of opinion on such a basic cultural event as the rise of civilization. Our knowledge of culture history—of historical cultural geography—is far from complete; the number of questions that are recognized greatly exceed the number of answers that are accepted. The scope of the problem may be glimpsed in the similarities and differences of the two well-studied civilizations that are described below.

Fig. 5.8   The ancient Near East. Mesopotamia is the name given to the area around and between the Tigris and Euphrates rivers. The larger area that includes Mesopotamia but extends in a semicircle as far west as the coastal areas of Palestine and Syria is often referred to as the *Fertile Crescent*.

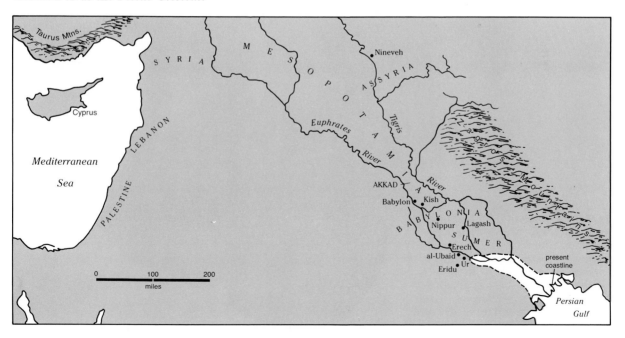

MESOPOTAMIA

Mesopotamia, the area associated with the Tigris and Euphrates rivers largely within the confines of present-day Iraq, was at a crossroads of human migration (Figure 5.8). Lacking clear natural boundaries, it was subject time and again to the invasion of tribes that lived to the east, north, and west. The history of the region during the period of time with which we are concerned was a succession of upheavals followed by relatively short periods of stability.

Three main eras may be distinguished. About 3000 B.C. the central part of the region was inhabited by Akkadians, a Semitic people who spoke a language different than that of the Sumerians, who lived to the south. People lived in a number of independent city-states, each with its local deities. The period 2900–2400 B.C. was a prosperous one, particularly for the Sumerians. Cuneiform writing on clay tablets developed, the arch was invented, and an extensive system of waterways and canals was built to protect people from the often violent annual floods (Figure 5.9).

The second period began about 2400 B.C., when Sargon of Akkad brought the land under a single rule. The empire he created, which lasted for approximately 200 years, included all of Mesopotamia, from the Persian Gulf in the south to Syria and Lebanon in the west. The empire collapsed under the pressure of invasions from the Guti, who lived in the Zagros Mountains in the east. In 2100 B.C. Mesopo-

Fig. 5.9 A portion of an alabaster relief from the Neo-Assyrian period (circa 700 B.C.), late in Mesopotamian history, shows organized work teams engaged in the development work of the nation. (Reproduced by permission of The University Museum, University of Pennsylvania.)

Fig. 5.10 Gold artifacts from the royal graves of Ur. The tombs also yielded finely wrought objects of copper, silver, and semiprecious stones such as alabaster, carnelian, and lapis lazuli. (Reproduced by permission of The University Museum, University of Pennsylvania.)

tamia was briefly reunited by Ur-Nammu. Although the dynasty he founded lasted only a century, it appears to have been extremely active intellectually. Figure 5.10 pictures some of the artifacts from royal tombs, which indicate astonishing wealth and creativity. The city of Ur was destroyed when nomadic tribes from the east and the west invaded the region, and once again the many small city-states became autonomous.

During the 18th century B.C. Mesopotamia was reunified by Hammurabi, the ruler of Babylon. He made that city the capital of the Babylonian Empire, which extended westward as far as the Mediterranean Sea and northward into the territory of the Assyrians. He is perhaps best known for his achievements in the field of law; under his direction laws and usages that went far back in time were codified. For a brief time, Babylon was the commercial center of a rich empire. Its temple was one of the seven wonders of the ancient world. A century after Hammurabi's death, Babylonia was invaded by tribes who plundered the cities and burned the crops.

EGYPT

Unlike the civilization of open and exposed Mesopotamia, the Egyptian civilization that developed along the banks of the Nile was isolated and protected by deserts to the east and the west and by difficult terrain and the Nubian desert to the south. The Nile River is the most significant fact about the geography of Egypt. Viewed from the air, the fertile strip along the river is a dark ribbon against the desert sands. As the historian Herodotus recognized, Egypt is "the gift of the river." Figure 5.11 indicates the degree to which settlement was concentrated around the Nile in ancient

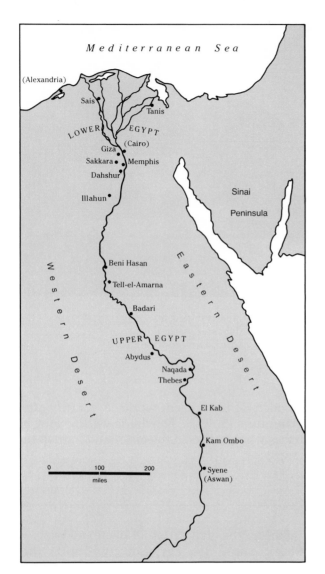

Fig. 5.11 Ancient Egypt and its urban network. Alexandria, Cairo, and the modern name Aswan have been added for reference.

170

the next 1500 years a wealthy and powerful country under the rule of a succession of single kings. Memphis, founded by Menes, remained the capital for the entire period.

A well-developed political system emerged during the years that ensued. The country was divided into provinces, each under the control of a governor appointed by and responsible to the king. Each provincial governor depended in turn on a staff of officials and village mayors. The vizier—an office established in the Fourth Dynasty—acted as a minister between the king and various central administrators, such as the treasurer and the minister of agriculture.

The division of labor became increasingly complex. The agricultural surpluses were great enough to support a large number of people who were freed from the necessity of growing their own food. Priests, scribes, soldiers, artisans, and musicians are among those depicted in tomb paintings (Figure 5.12). The pyramids, the monumental tombs of the rulers, testify to the ability of the government to organize people for special tasks.

An elaborate system of taxation was developed. Skill in hydraulic engineering increased; a canal built to link the Nile Valley with the Red Sea is but one example. A solar calendar of 365 days was adopted. Beautiful metalwork, fine carpentry, and lovely art have been discovered in some of the pyramids.

The Egyptians carried on trade with outlying regions. Gold, ivory, and ebony came from Nubia, to the south. Copper and malachite came from mines in the Sinai. From Lebanon came cedar, and from Syria and Palestine, olive oil and wine.

COMPARISON AND CONTRAST

The civilizations that emerged in Mesopotamia and Egypt during the period 3200–1700 B.C. were alike in a number of important respects. Both had political and religious hierarchies that administered a given region, and

times. Two regions were settled during Neolithic times: Lower Egypt, around the delta, and Upper Egypt, the area south of the delta.

About 3200 B.C. Egypt was united by Menes, who took as one of his titles "King of Upper and Lower Egypt." With the exception of a 200-year period of weakness, Egypt was for

Fig. 5.12 Portion of an Egyptian tomb painting from about 1400 B.C. Various agricultural activities are depicted, among them cutting, threshing, and winnowing grain. Scribes note the yield of the harvest. (The Metropolitan Museum of Art, New York.)

both had social systems involving a complex division of labor. Both developed systems of writing and contributed in significant ways to the development of literature, science, and the arts. Religious buildings and other public works were supported by a system of taxation.

Yet differences as well as similarities characterized the two civilizations. The physical isolation of Egypt in comparison with Mesopotamia has already been mentioned. Egypt was for nearly 1500 years a single kingdom with a common language, while except for relatively short periods Mesopotamia was composed of autonomous city-states, each with its own local customs. For 1500 years there was only one capital in Egypt, Memphis, while Mesopotamia experienced a constant shift of power. Although Memphis extended for 10 miles along the banks of the Nile, Egypt was basically a semiurban country, where a number of small cities served chiefly as market towns for the agricultural produce of the countryside. In Mesopotamia, on the other hand, a number of large urban centers were the focuses of the pop-

ulation and of social and intellectual activity. Other differences appeared: in the concept of the divinity of the king, in the uses to which writing was put, in the purposes of the monumental buildings, and in the content and function of art. As the years elapsed, the two civilizations became not only increasingly complex but also increasingly different from one another. From similar origins, but by separate adaptations and innovations, distinctive cultures emerged.

The System of Culture

Culture is the totality of learned behaviors and attitudes transmitted within a society to succeeding generations by imitation, instruction, and example. It has nothing to do with instinct or with gene transmissions, topics both debatable and divorced from our concerns. Culture implies the acquisition by group members of an integrated and cohesive body of

171

Disappearance of the Indus Valley Culture	Just as the activities of humans may alter or destroy their environment, so may spontaneous changes in the environment bring low the works of organized society. Recent research suggests that one of the world's great early cultures, that of the Indus Valley of Pakistan, was destroyed not by fire or sword but by drowning in a sea of mud. The great cities of that civilization, among them Harappa and Mohenjo-Daro, underwent an abrupt downfall about 2000 B.C. Excavations showing deep accumulations of mud, erosional collapse of city walls, and recurring attempts at rebuilding city structures document a losing battle against rising water backed up behind a dam created by earthquakes, rock shifts, and mudslides. The Indus was converted from a free-flowing stream to a huge, swampy lake, trapping mud formerly carried to the sea. As the lake rose, forests were destroyed and agriculture was ruined; dikes and brick walls were erected to shield the cities, but their food bases and very life were choked. The dam was cut through by the river and replugged at least five times, each recurrence taking its toll of urban life. The degeneration of pottery and other evidences of cultural decline document an ultimately lost battle with an altering environment.

172

behavioral patterns, environmental and social perceptions, and knowledge of existing technologies. Naturally, although one of necessity learns the culture in which one is born and reared, one need not—indeed, may not—learn its totality. Age, sex, status, or occupation may dictate the facets of the cultural whole in which one becomes fully indoctrinated.

Culture is an interlocked web of such complexity and pervasiveness that it cannot be grasped and in fact may be misapprehended by being glimpsed or generalized from limited, obvious traits. Language, eating utensils, the use of gestures, or the nature of religious ceremonies may sum and stereotype a culture for the casual observer. These are, however, only individually insignificant parts of a whole that can be understood and weighted only when the whole is envisaged. Understanding a culture fully is, perhaps, impossible for one who is not part of it. For analytical purposes, however, elements of culture—its building blocks and expressions—may be grouped and examined as clustered subsets of the whole. Following the anthropologist Leslie White, we may divide culture into three major subsystems: the technological, the sociological, and the ideological. Together, they comprise the system of culture as a whole. But they are integrated; each reacts upon the others and is affected by them in turn.

The *technological subsystem* is composed of the material objects, together with the techniques for their use, by means of which people are able to live. The material objects are the tools and other instruments that enable us to feed, clothe, house, defend, and amuse ourselves. We must have food, we must be protected from the elements, and we must be able to defend ourselves against enemies.

Cultural diversity among societies, ancient and modern, reflects in part the retention and utilization of technological innovations that have assisted groups to adapt successfully to their environments. Technological development in human history has been characterized by changes in the raw materials utilized and in

(a)

Fig. 5.13 (a) An Indian farmer working with an ox-drawn plow exemplifies the low technological levels of subsistence economies. (b) Power implements represent the control of inanimate energy for the use of humankind. The size of the power shovel may be estimated by comparison with the toylike object beside it—a bulldozer. (a: United Nations; b: Illinois State Geological Survey.)

(b)

173

the techniques and efficiency of tool creation, and by an increasing functional specificity and diversity of tools. As we have seen, in the earliest periods of human history technologies and tools were essentially the same among all groups. Later, regional differences appeared, and by the Upper Paleolithic period, a great proliferation of technological traditions was evident.

Beginning with metalworking, a reverse trend—toward commonality of technology—began, so that today advanced societies are nearly indistinguishable in the tools and techniques at their command. Those differences in technological traditions that still exist between developed and underdeveloped societies reflect in part national and personal wealth, stage of economic advancement and complexity, and, importantly, the level and type of energy used (Figure 5.13). This latter consideration is discussed in Chapter 9.

174

Changes in its technological base inevitably meant a chain of reactions and readjustments in the other subsystems of culture. Plant and animal domestication led to new complexities of life that demanded behavioral changes. These included a sedentary rather than a wandering life, permanent contact with larger numbers of the same people, cooperative defense and worship, and organizational change from band to tribe. Similarly, the harnessing of fossil fuels and the introduction of the factory system created new behavioral patterns appropriate to fixed hours of repetitive cooperative labor and to the new impersonality of employer–supervisor–worker relationships.

The *sociological subsystem* of a culture is the sum of those expected and accepted patterns of interpersonal relations that find their outlet in economic, political, military, religious, kinship, and other associations. These define the social organization of a culture. They define how the individual functions relative to the group, whether it be family, church, or state.

There are no "givens" as far as patterns of interaction in any of these associations are concerned, except that most cultures possess all these ways of structuring behavior. The form of behavior that is expected and the relative importance to the society of the differing behavior sets vary among and constitute obvious differences of cultures. Differing patterns of behavior are learned and are transmitted from one generation to the next (Figure 5.14).

Rejection of the sociological patterns appropriate to the culture constitutes, beyond narrow limits, unacceptable, deviant behavior subject to formal or informal rebuke, rejection,

Fig. 5.14 All societies prepare children for membership in the culture group. In each of these settings, certain values, beliefs, skills, and proper ways of acting are being transmitted to the youngsters. (a: United Nations; b: photo by Kao Hung, China Photo Service/Eastphoto; c: United Press International photo.)

(a)

or punishment. Such censure, like the education of the young, may be seen as the socialization of the untrained or of the deviant, imparting or reinforcing the culture's sociological system.

The *ideological subsystem* is the summation and abstraction of the sociological system. It consists of ideas, beliefs, and knowledge and of the ways we express these things in our speech or other forms of communication. Mythologies and theologies, legend, literature, philosophy, folk wisdom, and commonsense knowledge make up this category. Passed on from generation to generation, these statements tell us what we ought to believe, what we should value, and how we ought to act. Beliefs form the basis of the socialization process.

Often we know—or think we know—what the beliefs of a group are from written sources. Sometimes, however, we depend on the actions or objectives of a group to tell us what its true ideas and values are. "Do as I say, not as I do" is a cynical recognition of the fact that actions and words do not always coincide. The values of a group cannot be deduced from the written record alone.

Changes in the ideas held by a society may affect the sociological and technological systems just as, for example, changes in technology force changes in the social system. The abrupt alteration of the ideological structure of Russia from a monarchical, agrarian, capitalistic system to an industrialized, communistic society involved sudden, interrelated alteration in all facets of the culture system formerly observed within that nation. The interlocking nature of all aspects of a culture is termed *cultural integration*.

Culture traits enable us to differentiate one culture from another by considering the way these three systems manifest themselves in the behavior of groups of people. How can the technological, sociological, and ideological systems of different groups be identified? What traits in a culture indicate significant differentiation from other culture complexes? We may

176

reject as superficial and misleading generalizations derived from trivialities, the foods people eat for breakfast, for example, or the kinds of eating implements they use.

This rejection is a reflection of the kinds of understanding we seek and the level of generalization we desire. There is no single most appropriate way to designate or recognize a culture or to delimit a culture region. As geographers concerned with world systems, we are interested in those aspects of culture that vary over extensive regions of the world and differentiate societies in a broad, summary fashion. With equal validity, a social anthropologist may study, categorize, and map variations in kinship networks, power structures, or class stratifications in smaller culture groups distributed over restricted areas. For our purposes, the way energy is used, the form social experience takes, the ideas that people hold sacred, and the expression of all of these in language are broad elements that, when categorized, may be considered culture traits of significance. In the next chapter, we discuss some of these indicators of cultural differences and present a map (Figure 6.21) that shows one geographer's view of the world's culture regions.

Interaction of People with Their Environment

Culture develops in a physical environment that, in its way, contributes to differences among people. The acquisition of food, shelter, and clothing, all parts of culture, depends on the utilization of the natural resources at hand. The reaction of people to the environment of a given area and their impact on that environment are two interwoven themes of cultural geography. From at least the time of the ancient Greeks, we have evidence of concern with the environment–culture relationship.

While schools of thought have come and

Environmental determinism has, since the early Greek geographer-historians, been a favored explanation of human behavior and of the observed differences between cultures. Elevated to the status of a "scientific" examination of causation during the 19th and the early 20th centuries, determinism rested upon the basic and unquestioned assumption that the physical environment provided the sole set of stimuli to which humans must respond. More sophisticated cultural–geographic and anthropological research, revealing the complex strategies by which "primitive" peoples utilize their environment, renders unacceptable such simplistic views as were expressed in a high school textbook published in 1941:

Environmental Determinism

> Environment determines, to a large extent, what man uses for food, shelter, and clothing. In the Tropics he does not have to exert himself to any great degree. There, clothing may be scant, shelter is easily provided, and food may be obtained without much labor. But in the colder Temperate Zones climate makes much greater demands on human energy. There, environment is a definite challenge to human effort. Man has to labor to obtain protection from severe changes of weather, to secure food, to build a shelter that will provide him with a means of warmth during winter months. Thus work has been developed as more of a permanent habit by peoples living in Temperate Zones than by those dwelling in warmer climates.*

*From Louise I. Capen, *Across the Ages.* American Book Co., New York, 1941, pp. 306–307.

gone, some form of *environmental determinism* has had enduring popularity. Proponents of this theory have held that the physical environment shapes humans, their actions, and their thoughts. Indeed, in extreme form, the theory went beyond the notion of environmental shaping or leading of human action: it postulated an inevitability of predetermined human responses to physical circumstances. Further, in early versions, environmental determinism assured Europeans, already confident of the superiority of their culture and their stage of advancement, that the contrasts they observed between their technologies and those of "inferior" native populations in colonial territories were the will of the Creator, who had ordained the environmental differences that engendered the observed cultural contrasts.

Although the physical environment consists of the range of landforms, soils, vegetation, and weather and climate found on the earth, environmental determinists usually singled out climate as having the most direct influence on people. There are, according to determinists, climates that stimulate us, others that make us easygoing and carefree, and still others that have a positively deleterious effect on our character. Environmental determinism was a widely held and widely taught notion in this country in the early years of this century.

Modern geographers reject environmental determinism, believing it to be too simplistic a

177

description of the very complex environment—culture relationship. Environmental factors alone cannot account for the cultural variations that occur around the world. Levels of technology, systems of organization, and ideas about what is true and right have no obvious relationship to environmental circumstances.

The environment does place certain limitations on the human use of territory, although such limitations must be seen not as absolute, enduring restrictions but as relative to technologies, cost considerations, national aspirations, and linkages with the larger world. Human choices in the use of landscapes are affected by group perception of the feasibility and desirability of their occupance and exploitation. These are not circumstances inherent in the land. Mines, factories, and cities were (and are being) created in the formerly nearly unpopulated tundra and forests of Siberia, not in response to recent environmental improvement but as a reflection of Soviet developmental programs, of rising internal and foreign interest in the resources of the territory, and of pressing domestic and international political considerations. Similarly, even with the availability of current technology, many potentially productive areas of the world remain sparsely settled, including much of South America, parts of Australia, sections of Southeast Asia, and portions of Anglo-America.

Map evidence, of course, suggests that the environment has placed certain limitations on people. The vast majority of the world's population is concentrated on less than one-half of the earth's land surface. In some large world regions, such as Australia, the concentration is even greater (Figure 5.15). Areas with relatively mild climatic conditions that offer a supply of fresh water, fertile soil, and abundant mineral resources are densely settled, reflecting in part the different carrying capacities of the land under earlier technologies. Under past conditions of human perceptions, and to a large extent even today, the Polar regions, high and rugged mountains, deserts, and some hot and

178

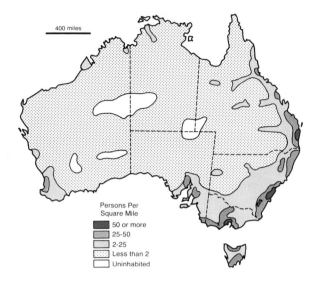

Fig. 5.15 Distribution of population in Australia. The country is very unevenly settled, with the bulk of the population living in coastal areas. Although an extreme case, Australia's settlement pattern reflects the differences in population-supporting capacities of lands throughout the world.

humid areas contain very few people. The physical environment, then, must be perceived as offering us the possibility of feeding, clothing, and housing ourselves. If such resources are lacking in an area, or if our perceptions do not discern them, there is no inducement for people to be there.

Environments that contain such resources provide the framework within which a culture operates. Geographers believe that any of these environments offer a range of possibilities; the needs, traditions, and level of technology of a culture affect how these possibilities are perceived and what choices are made regarding them. Each society uses natural resources in accordance with its culture. Changes in a group's technical abilities or objectives bring about changes in its perceptions of the usefulness of the land.

Coal, oil, and natural gas have been in their present locations for many thousands of years, but they were of no major use to Neolithic people for energy purposes and hence did not con-fer any discernible advantage on a particular site. Not until the advent of the Industrial Rev-olution did these deposits gain importance and come to influence the location of such great industrial complexes as the Midlands in Eng-land, the Ruhr in Germany, and the Pittsburgh steel-making district of the United States. American Indians made one use of the envi-ronment around Pittsburgh, while settlers that came later made quite another.

People are also able to modify their environ-ment, and this is the other half of the relation-ship mentioned above. Cultural geography, we said, is concerned both with the reactions of people to the physical environment and with their impact on that environment. We modify our environment through the material objects we place on the landscape: cities, farms, roads, airports, and so on (Figure 5.16). The form these

take is the product of the kind of culture group in which we live.

The *cultural landscape* (the surface of the earth as modified by human action) is of partic-ular significance to the historical–cultural ge-ographer. It is a tangible, physical record of a given culture. House types, kinds of public buildings, and the size and distribution of set-tlements are among the indicators of the use that humans have made of the land.

The more technologically advanced and complex the culture, the greater its impact on the environment. In sprawling urban–indus-trial societies, the cultural landscape has come to outweigh the natural physical environment in its impact on people's daily lives. It inter-poses itself between "nature" and humans, and residents of such cities can go through life with very little contact with or concern about the physical environment.

Fig. 5.16 **The physical and cultural landscapes in juxtaposition. Advanced societies are capable of so altering the circumstances of nature that the cultural landscapes they create become controlling environments. (United Nations.)**

179

Culture Change

A recurring theme of historical cultural geography is change. No culture is, or has been, characterized by a fixed set of material objects, systems of organization, or even ideologies. Cultures are constantly changing; they are always in a state of flux. This theme has been emphasized in the course of this chapter. Some cultural changes are major and pervasive. Thus the transition from hunter-gatherer to sedentary farmer affected markedly every facet of the cultures experiencing that change; so, too, did the Industrial Revolution and the associated urbanization of the societies it has touched. Other changes are so slight individually as to go, at their inception, almost unnoticed, though cumulatively they can substantially alter the affected culture. Think of how the culture of the United States in the 1980s differs from that of the 1940s. Notice how the role of cars and planes has changed. Television is now an integral part of life and a means of mass communication effectively unknown until well after World War II. Social mores have altered dramatically. Not all change is progress, of course, but progress can come about only through change.

INNOVATION AND DIFFUSION

The two mechanisms that induce change within a culture are innovation and diffusion. *Innovation* implies alterations to a culture that result from ideas created within the social group itself. The novelty may be an invented change in material technology, as the bow and arrow or the jet engine. It may involve the development of nonmaterial forms of social structure and interaction: feudalism, for example, or Christianity. *Diffusion* can take several forms; these are discussed in Chapter 8. Those varying forms, each different in its impact on social groups, may be summarized as two basic types. Either people move, for any of a number of reasons, to a new area and take their culture

Fig. 5.17 A street in Guatemala City. In modern society, advertising is a potent force for diffusion. Advertisements over radio and television, in newspapers and magazines, and on billboards and signs communicate information about many different products and innovations. (United Nations/Jerry Frank.)

(a)

Fig. 5.18 Outward appearance is not the only similarity between Egyptian (a) and Mayan (b) pyramids. In both societies, the pyramids were royal burial sites, goods were placed inside the tombs to accompany the corpse on its journey to eternity, and pains were taken to hide the entrance to the tomb. (a: United Press International photo; b: Courtesy of Mexican National Tourist Council.)

(b)

with them (e.g., immigrants to the American colonies), or information about an innovation (e.g., barbed wire) may spread throughout a culture (Figure 5.17).

Together, innovation and diffusion suggest that cultures change because social groups either think of a new way of doing something or borrow that new way from another culture group. A very few innovations and inventions have led to revolutionary changes in a culture, for example, the domestication of plants and the first smelting of metals. Most inventions are small improvements on already existing artifacts or techniques; the automobile is a combination of elements that had existed for many years before they were put together to form a car.

It is not always possible to determine whether the existence of a culture trait in two different areas is the result of diffusion or of independent invention. The pyramids of Egypt

and of the Mayan civilization, pictured in Figure 5.18, are thought by some to have developed in each place independently and by others to be the result of pre-Columbian voyages from Egypt to the Americas. Sometimes, however, and particularly when historical records exist, diffusion can be documented. We know, for example, how Christianity, first formulated in the Middle East, spread across much of the world.

When one culture group undergoes a major modification by adopting many of the characteristics of another, dominant culture group, *acculturation* has occurred. This may be the result of conquest, when one society overcomes another and occupies its territory. Very often the loser is forced to acculturate or does so voluntarily, overwhelmed by the superiority in numbers or the technical level of the conqueror. This kind of acculturation was experienced by the primitive Europeans after Roman

182

Fig. 5.19 Baseball is one of the most popular sports in Japan, attracting millions of spectators annually. (Consulate General of Japan, New York.)

conquest, by native populations in the face of the Slavic conquest of Siberia, and by Amerindians during the European settlement of North America. In a different fashion, it is evident in changes in Japanese political organization and philosophy imposed by occupying Americans after World War II or by Japanese adoption of some more frivolous aspects of American life (Figure 5.19). On occasion, of course, the invading group is assimilated into the older conquered culture. For example, Chinese culture prevailed over the conquering tribes of Mongols. Acculturation may also be the result of commercial expansion. The relationship of a mother country to its colony may result in permanent changes in the culture of the colonized society.

CONTACT BETWEEN REGIONS

Each of the culture hearths in existence at the end of the Neolithic period was a core area from which influence extended outward. The hearths were centers of both innovation and diffusion, and thus centers of change. Culture diffused from these centers, first to regions under their direct control and then to peripheral areas. In time, secondary centers developed in Europe, Africa, Southeast Asia, Mongolia, and the Americas. Sometimes these secondary centers eclipsed the primary centers, as happened, for example, in western Europe. There the center of control moved from the Near East to Greece, then to Italy, and then northward to France and England.

Recent changes in technology permit us to travel farther than ever before, with greater safety and with greater speed. This acceleration has resulted in the rapid spread of goods and ideas. Large commercial and industrial organizations market their goods throughout the world, something undreamed of even a century ago. One can purchase Coca Cola in Katmandu, Nepal, as well as in the company's headquarters city of Atlanta, Georgia. The worldwide communication of ideas and information is equally rapid. The Voice of America beams broadcasts across national frontiers, a U.S.S.R. cultural magazine is specifically designed for English-speaking readers, and so on.

Barriers to diffusion are any elements that hinder either the flow of information or the movement of people and thus retard or prevent the acceptance of an innovation. Because of the *friction of distance*, generally the farther two areas are from one another, the less likely it is that there will be interaction. All other things being equal, the amount of interaction decreases as the distance between two areas increases. Of course, all other things are not equal, and interregional contact can also be hindered by the physical environment. Oceans and rugged mountains are physical barriers to diffusion. The effects of two different kinds of environments on interaction were noted in the cases of early Egypt and Mesopotamia.

The more similar two cultural areas are to one another, the greater the likelihood of the adoption of an innovation, for diffusion is a selective process. The receiver culture may adopt some goods or ideas from the donor culture and may reject others. The decision to adopt is governed by the receiving group's own culture. Political restrictions, religious taboos, and other social customs are cultural barriers to diffusion. The French Canadians, although close geographically to many centers of diffusion (Toronto, New York, and Boston) are only minimally influenced by such centers. There are many reasons for the selectivity of the French Canadians, but the language barrier is an important one.

On a larger scale, Japan until the mid-19th century and Nepal until the 1950s kept their doors closed to foreigners. Burma and mainland China are only now beginning to let in visitors on a restricted basis after a number of years of self-imposed isolation. None of these countries wanted its way of life contaminated by foreign influence. Such cases are the excep-

183

tion rather than the rule, however, and when the barriers to diffusion break down, a country may be subject to nearly overwhelming outside influence.

People are agents of cultural diffusion. They travel for many reasons: to trade, to explore, to conquer, to find a better place to live. They have often been motivated by religious reasons, and we find that religions have been a major factor in diffusion. Islam swept across northern Africa and into southern Europe in one direction and eastward across South Asia in the other. Christianity has found a home in Europe and the Americas, but missionaries continue to try to convert peoples in other parts of the world. Buddhism spread eastward from its origin in India, where it is now a minor religion, to embrace a large percentage of the people of Asia.

REGIONALLY DIFFERENTIATED CULTURES

184

Humans spread over much of the earth in prehistoric times and very early displayed a cultural differentiation that is evident in artifacts dating from the Paleolithic period. Some groups were more technologically advanced than others. Some had contact with other peoples; some were more isolated. Some were nomadic and roamed over hundreds of miles, while others led a relatively sedentary existence.

By Mesolithic, and particularly Neolithic, times pronounced variations existed in the technology used by different human groups. For their houses, some people used bricks, others stone, others a kind of plaster, still others wood. Some groups used fire for cooking inside the house, while others kept it outside. The plants and animals available in different areas varied. Many kinds of weapons were used: hunting tools included axes, spears, clubs, and bow and arrow. The kinds of fish available varied from region to region, and nets, hooks, and spears were all used for fishing. Based on the

kind of crops they raised, and the kinds of territories they occupied, people used different tools for cultivation and different systems of cultivation. Thus, as early as 7000 B.C., significant variations in culture existed from region to region.

Certain aspects of the cultural differentiation we have been describing have continued down to the present day. Nontechnological differences in religion, kinship patterns, political organization, language, and social customs obviously still exist. It is equally apparent that there are also sharp differences between societies in the general availability of or access to modern technology: an Indian peasant may carry a burden to market; an American farmer uses a motortruck.

At the same time, however, the development of a common, worldwide technology has in important ways reduced formerly great cultural contrasts between colonizing and colonial or developed and developing societies. The ease of acquisition and the success of employment of transistor radios and sewing machines by peoples through the world, for example, suggest a general convergence of cultures.

THE INTERDEPENDENCE OF CULTURES

Two major influences—one that had impact some time ago and one at work now—counteract the trend toward differentiation begun in prehistory. They have acted and are acting to bring about similarities rather than differences. The first was European imperialism and colonialism, which began in the 16th century and continued for over 400 years. Colonialism resulted in the exportation of both advanced technologies and culture-specific ideas about social, political, and economic organization to all parts of the world. The era was unique; never before had so many of the world's peoples been exposed to and affected by a single kind of culture.

The second major factor promoting cultural

convergence is technological advances in communication. These have ensured that no group exists today that has not felt the impact of modern industrial civilization. Easy communication has assured the rapid diffusion of innovations—or, at least, an awareness of them—to even the most physically and socially isolated of cultures. Communication advances also promote interdependence among nations. Telephone and telegraph lines, railroad lines, shipping routes, gas and oil pipelines, airlines, and communication satellites have united all countries into one enormous communications network. While nationalism and isolationism act as barriers to convergence, advances in communication tend irreversibly to make the world more uniform and more interdependent, though significant diversities, as Chapter 6 will show, still remain.

Conclusion

From its first appearance, the genus *Homo* developed and transmitted by language and example the tools, skills, behavior patterns, and beliefs that constitute culture. With the areal spread of populations, with isolation, with new environmental circumstances to face, and with new solutions to those circumstances, humankind not only established itself throughout the world but created a mosaic of racial, linguistic, technological, and habit patterns that differentiated it into distinct culture groupings. Culture hearths—areas of new solutions, new knowledge, and new techniques of environmental manipulation—emerged; hunters and gatherers became villagers, and sedentary populations were the rootstock of complex urban civilizations.

The pervasive, interlocking web of culture developed not in response to the imperatives of environment but as a reflection of the invention and the diffusion of ideas, techniques, and knowledge. The emergence and the spread of new traits prohibited cultural stagnation within most societies and assured that change would remain the recurring theme of cultural geography.

The rate of change was accelerated as population numbers increased and intergroup contacts became easier, although significant world patterns of cultural variation remain to the present as reminders of the heritage of the past. It is to population growth and cultural diversity that we turn our attention in Chapter 6.

185

For Review

1. What is meant by the term *culture*? What are the subdivisions of a system of culture? What is the content of each of those subdivisions?

2. What aspects of cultural development are associated with the Paleolithic period? With the Mesolithic? With the Neolithic?

3. Consider the culture of which you are a part and a product. Briefly outline the essentials of that culture, employing the concepts you developed in Question 1.

4. What is meant by *race*? Can the term be employed as a synonym of *culture*?

5. How did different races, however recognized or classified, develop from a presumably common ancestral stock?

6. What is meant by *domestication*? When and where did the domestication of plants and animals occur? What preconditions for plant domestication have been recognized?

7. What impact upon culture and population numbers did plant domestication have? Animal domestication?

8. What is a *culture hearth*? Identify the major and secondary hearths that emerged during the post-Neolithic period. What new traits of culture developed in all of them? In only some of them?

9. What is meant by *innovation*? By *diffusion*? Discuss the role played by innovation and diffusion in altering the cultural web from that known to your great-grandparents to that in which you participate.

Suggested Readings

BUTZER, KARL, *Early Hydraulic Civilization in Egypt: A Study in Cultural Ecology.* University of Chicago Press, Chicago, 1976.

FRANKFORT, HENRI, *The Birth of Civilization in the Near East.* Doubleday Anchor Books, Garden City, NY, 1956.

GLACKEN, CLARENCE, *Traces on the Rhodian Shore: Nature and Culture in Western Thought from Ancient Times to the End of the Eighteenth Century.* University of California Press, Berkeley, 1967.

LYNCH, KEVIN, *What Time Is This Place?* M.I.T. Press, Cambridge, MA, 1972.

NANCE, JOHN, *The Gentle Tasaday.* Harcourt Brace Jovanovich, New York, 1975.

POUNDS, NORMAN J. G., *An Historical Geography of Europe, 450* B.C.–A.D. *1330.* Cambridge University Press, Cambridge, England, 1973.

SAUER, CARL, *Agricultural Origins and Dispersals.* American Geographical Society, New York, 1952.

SPENCER, J. E., and W. L. THOMAS, *Introducing Cultural Geography.* Wiley, New York, 1973.

VANCE, JAMES E., *This Scene of Man.* Harper's College Press, New York, 1977.

WHITE, LESLIE A., *The Science of Culture.* Grove, New York, 1949.

186

A respected geographer, Glenn T. Trewartha, once observed that "numbers, densities, and qualities of the population provide the essential background for all geography. Population is the point of reference from which all other elements are observed." While not all geographers agree with the absolute primacy given by Trewartha to population studies, they recognize that humankind in all its diversity demands special emphasis in the study of the surface of the earth and of the manner in which people occupy and utilize that surface.

A number of subdisciplines have emerged focusing upon restricted aspects of this concern with humans in their numbers, distributions, activities, and cultural diversity. Some of these, labeled *political, economic, urban,* or *behavioral geography,* are considered in separate chapters in this book. They represent arbitrary subdivisions of study made useful by the great complexity of humans in their infinite activities and characters. Underlying those subdivisions, however, are common fundamental concerns, among which are the evolution of human societies, the largely quantitative assessment of human population, and the principal cultural traits that tend to differentiate humankind into identifiable societies occupying significant portions of the earth's surface.

The first, historical cultural geography, was considered in Chapter 5. The second and third—population geography and world regional cultural traits—are discussed here both for the further elucidation of the *human* component of the geographer's "human–earth's surface" equation and as a prelude to specialized systematic views of human geography contained in subsequent chapters.

Population Geography

Population geography is concerned with the growth, composition, and distribution of

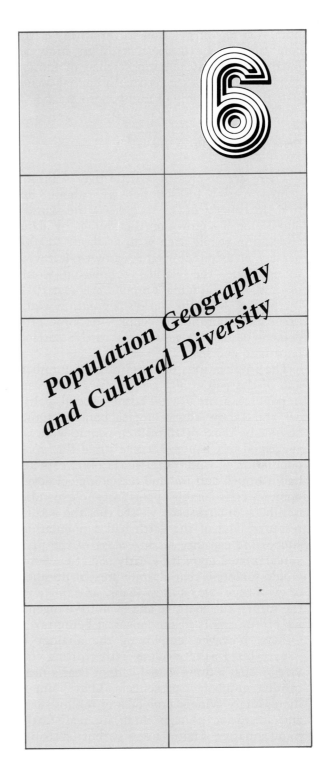

Population Geography and Cultural Diversity

187

human groups in relation to variations in the conditions of earth space. It differs from demography, the systematic science of human population, in its concern with spatial analysis—the relationship of numbers to area.

POPULATION GROWTH

There are at present a little over 4 billion people on earth. In 1960 there were about 3 billion. In other words, since 1960 the world's population has grown on the average by about 65 million people each year, or about 200,000 per day, an astounding rate of growth. If it were to continue, the world's population would reach the 8-billion mark around the year 2010. By the year 3000, the world's population would exceed a billion billion people—or some 1700 persons per square yard of the earth's surface, both land and sea.

The implications of those present numbers and of the potentially astronomical increases in population are of vital current social, political, and—above all—ecological concern. Quite obviously, the 5 or 10 million people who, we suggested in Chapter 5, constituted the world population of *Homo sapiens* 11,000 years ago had, through cultural and technological development, considerable opportunity to expand in numbers. In retrospect, we see that the natural resource base of the earth had a population-supporting capacity far in excess of the pressures exerted upon it by early societies. Some would maintain that, despite present numbers or even those we can reasonably anticipate for the future, the adaptive and exploitive ingenuity of humans is in no danger of being taxed. Others, however, employing the analogy of "Spaceship Earth," declare with chilling conviction that a finite vessel cannot bear a near-infinite number of passengers. They point to increasingly evident problems of malnutrition and starvation (though these are realistically more a matter of failures of distribution than of

inability to produce enough foodstuffs). They cite dangerous conditions of air and water pollution, the nearing exhaustion of mineral resources and of fossil fuels, particularly petroleum, and other evidences of strains on world resources as foretelling the discernible outer limits of population growth.

Why are we suddenly confronted with what seems to many an insoluble problem—the apparently unremitting propensity of humankind to increase its numbers? On a worldwide basis, populations grow only one way: the number of births in a given period must exceed the number of deaths. Ignoring for the moment regional population changes resulting from migration, we can conclude that the observed and projected dramatic increases in population must result from the failure of natural controls to limit the number of births or to increase the number of deaths, or from the success of human ingenuity in circumventing such controls when they exist. We shall return to these considerations after defining some terms important in the study of population geography.

SOME DEFINITIONS

Birth and death rates and rates of natural increase are among the simplest measures used to analyze population. The *birth rate* (technically, the crude birth rate) is the annual number of live births per 1000 population. A country with a population of 2 million and with 40,000 births a year would have a birth rate of 20 per 1000.

$$\frac{40,000}{2,000,000} = 20 \text{ per } 1000$$

Today, birth rates over 30 per 1000 are considered high, yet over two-thirds of the world's population live in countries with birth rates that are that high or higher. Birth rates less than 20 per 1000 are low and are confined to some European countries, the United States,

188

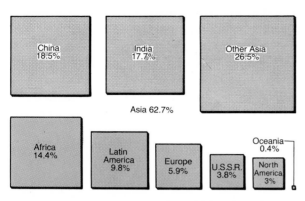

Fig. 6.1 In the mid-1970s, three-fifths of world population growth came in Asia. That continent as a whole recorded a birth rate of 31 per 1000 persons in 1974. That rate, however, marked a pronounced decline from the 39 per 1000 registered in 1965. (Data from United Nations and U.S. Bureau of the Census.)

Canada, and Japan. Regional variations in percentage contributions to world population growth are summarized in Figure 6.1.

Crude birth rates may display such regional variability because of variations in age and sex composition or disparities in births among reproductive-age, rather than total, population. They are therefore not of much use for regional comparative or predictive purposes. To overcome these deficiencies in the crude birth rate, the population geographer can, with greater effort or data input, make more sophisticated calculations.

The crude *death rate* is calculated the same way as the crude birth rate: the annual number of deaths per 1000 population. Like crude birth rates, crude death rates are meaningful for comparative purposes only when identically structured populations are studied. To overcome that limitation, death rates can be calculated for specific age groups. The *infant mortality rate* is the ratio of deaths of infants aged one year or under per 1000 live births.

$$\frac{\text{deaths, age 1 year or less}}{1000 \text{ live births}}$$

Infant mortality rates are significant because that is where the greatest declines in mortality have occurred, largely as a result of the increased availability of health services. The drop in infant mortality accounts for a large part of the decline in the general death rate in the last few decades, for mortality during the first year of life usually is greater than in any other year.

Two centuries ago it was not uncommon for 200–300 infants per 1000 to die in their first year. There are still countries with infant mortality rates over 100 per 1000. In contrast, in all Western European countries except Portugal, the rate is well under 30 per 1000 live births. The time and space variations in infant mortality and in crude death rates are obviously not due to differences in the innate ability of different groups to survive. They are essentially the result of the achievement, during the 1940s, of significant controls of important causes of death. Penicillin and other antibiotics, DDT and other pesticides protected, preserved, and prolonged lives that, before their introduction, would have been lost. Modern medicine and sanitation have altered life expectancy and the age-old relationships between birth and death rates. The availability and employment of these modern methods, however, have varied regionally, and the developed countries have been most able—financially, administratively, and educationally—to benefit from them.

The *rate of natural increase* is obtained by subtracting the crude death rate from the crude birth rate. *Natural* signifies that increases or decreases due to migration are not included. If a country had a birth rate of 22 per 1000 and a death rate of 12 per 1000 for a given year, the growth rate would be 10 per 1000. This rate is usually expressed as a percentage, that is, as a rate per 100 rather than per 1000. In the example above, the annual increase would be 1%.

The rate of increase can be related to the time it takes for a population to double. Table 6.1 shows that it would take 70 years for a pop-

189

Table 6.1 Number of Years Required for a Population to Double at Various Rates of Increase

Annual Percentage Increase	Doubling Time (Years)
0.5	140
1.0	70
2.0	35
3.0	24
4.0	17
5.0	14
10.0	7

Table 6.3 Population Growth and Approximate Doubling Times Since the Year A.D. 1

Year	Estimated Population	Doubling Time (Years)
1	250 million	
1650	500 million	1650
1850	1 billion	200
1930	2 billion	80
1975	4 billion	45
2010	8 billion	35

ulation with a rate of increase of 1% (approximately the present rate of growth in the United States) to double. A 2% rate of increase means that the population will double in only 35 years. How could adding only 20 persons per 1000 cause a population to grow so quickly? The principle is the same as compounding interest in a bank. Table 6.2 shows the number yielded by a 2% rate of increase at the end of successive five-year periods.

For the world as a whole, rates of increase have risen over the span of human history. Therefore the "doubling time" has decreased.

Table 6.2 Population Growth Yielded by a 2% Rate of Increase
The population doubles in 35 years.

Year	Population
0	1000
5	1104
10	1219
15	1345
20	1485
25	1640
30	1810
35	2000

Note in Table 6.3 how the population of the world has doubled in successively shorter periods of time. The last figure in Table 6.3 indicates that the world will reach a population of 8 billion by the year 2010 if the present rate of growth continues. In countries with high rates of increase (Figure 6.2), the doubling time is even less than the 35 years projected for the world as a whole.

Here, then, lies the answer to the question posed earlier. We are dealing with geometric rather than arithmetic growth. The ever-increasing base population has reached such a size that each additional doubling results in an astronomical increase in the total. One author, G. Tyler Miller, offers a graphic illustration of the inevitable consequences of such doubling, or J-curve, growth. He asks that we mentally fold in half a page of this book (assume that the page has a thickness of about $\frac{1}{254}$ of an inch— $\frac{1}{100}$ of a centimeter). We have, of course, doubled its thickness. We can—mentally, though certainly not physically—continue the doubling process several times without great effect, but 12 such doublings would yield a thickness of one foot (30.5 cm) and 20 doublings would result in a thickness of nearly 260 feet (79 m). From then on, the results of further doubling are astounding. Doubling the page only about 52 times, Miller reports, would give a thickness reaching from the earth to the sun.

Recently recorded trends in world population increase suggest that both the family and the economic planning programs adopted by the developing nations are taking effect. An international demographic transition at numbers far lower than earlier forecasts now seems likely. In mid-1978 the United Nations reported that the rate of world population growth was slowing, though not all nations were exhibiting success in limiting explosive increases.

Decline in Fertility Rates

The growth rate in Indonesia declined from 2.5% in 1970 to 1.7% by 1978. Its earlier projected population in the year 2000 was cut from 300 million to 190 million. Since "explosive" population growth is reckoned to exist at and above 2.1% annually, Indonesia's rate of increase has dropped below the presumed danger point.

Similar major declines to near or below the explosive growth level have been noted for other developing nations, including China and India. Nevertheless, growth rates in the less-developed countries, comprising some three-quarters of the total world population, were 2.3% in 1975; if continued, they promised a doubling of population in 30 years. In Latin America, the increase has been at an annual rate of 2.8%. "Developed" nations averaged less than a 1% annual increase during the late 1970s.

Recently recorded declines in fertility rates give hope that the most dire implications of the J-curve may yet be avoided. The success of the family planning programs now operating in some 70 developing countries is evident in demographers' revised projections of future world population—to, perhaps, 5.8 billion by the year 2000, in contrast to earlier United Nations estimates of 6.3 billion.

191

Rounding the bend on the J-curve, which world population has done (see Figure 6.3), poses problems and implications for human occupance of the earth of a vastly greater order of magnitude than ever faced before.

THE DEMOGRAPHIC CYCLE

The most extreme consequences of exponential population growth cannot be realized. A braking mechanism, the substantial lowering of birth rates through what is known as the *demographic transition*, may result in the voluntary control of totally unregulated population growth. Should that fail, involuntary controls of an unpleasant nature will be set in motion.

As long as births only slightly exceed deaths, even when the rates of both are high, the population will grow only slowly. This was the case for most of human history until about A.D. 1750 (Figure 6.3). Demographers think that it took from approximately A.D. 1 to A.D. 1650 for the population to increase from 250 million to 500 million, a doubling time of more than a millennium and a half. Growth was not steady, of course. Wars, famine, and other disasters took heavy tolls. For example, the bubonic plague (the Black Death), which swept across

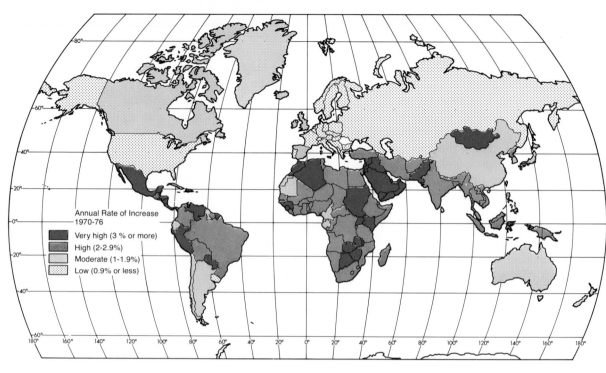

Fig. 6.2   Annual rate of natural population increase, 1970–1976.

Europe in the 14th century, is estimated to have killed over one-third of the population of that continent. High birth and high but fluctu- ating death rates represent the first stage in the demographic cycle (Figure 6.4).

In a few areas of the world, notably parts of

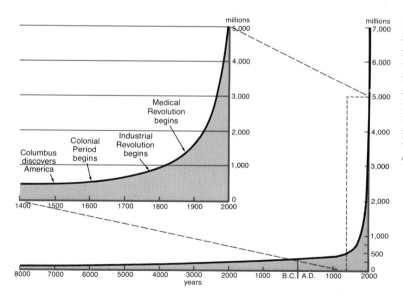

Fig. 6.3   World population growth, 8000 B.C. to A.D. 2000. At the current rate of increase, the population of the world will double in 35 years. Notice that the bend in the J-curve begins in about the mid-1700s, when industrial- ization started to provide new means to support population growth. Im- provements in medical science served to reduce death rates near the opening of the 20th century.

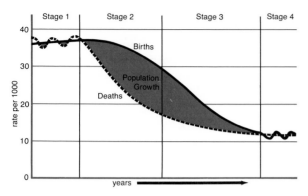

Fig. 6.4 Stages in the population cycle. During the first stage, birth and death rates are both high and population grows slowly. When the death rate drops and birth rates remain high, there is a rapid increase in numbers. During the third stage, birth rates decline and population growth is less rapid. The fourth stage is marked by low birth and death rates and, consequently, by a low rate of natural increase.

Africa and Southeast Asia, high birth and death rates still prevail (Figure 6.5). Women have many children, but life expectancy is low. There is a real need for many children just so that society can survive. In addition, large families are considered advantageous. Children contribute to the family by starting to work at an early age and by supporting their parents in their old age.

With the industrialization that began about 1750, Europe entered a second stage of the population cycle, some effects of which have been felt worldwide even without conversion to an industrial economy. This second stage is marked by declining death rates accompanied by continuing high birth rates. Rapidly rising populations result from increases in life expectancy. Falling death rates are due to advances in medical and sanitation practices, improved mechanisms for foodstuff storage and distribution, a rising per capita income, and rapid urbanization. Birth rates do not fall as soon as death rates; ingrained cultural patterns change more slowly than technologies. Many countries in South America and parts of South-

east Asia are at this stage in the population cycle (Figure 6.4). The annual rates of increase in such countries are about 30 per 1000; this means that their populations will double in a little over 20 years. Such rates, of course, do not mean that the full impact of the Industrial Revolution has been worldwide; they do mean that the underdeveloped societies have been beneficiaries of the life preservation techniques associated with it.

The next stage follows when birth rates decline as people begin to control the number of births. The advantages that having many children had in an agrarian society are not so evident in urbanized, industrialized cultures. In fact, children become economic liabilities rather than assets. Housing is crowded, rising incomes mean that the family can rely on a single earner, and children are sent to school instead of to work. When the birth rate falls, and death rates remain low, population begins to level off.

A few countries have entered what appears to be the final stage of the demographic cycle, which is characterized by very low birth and

193

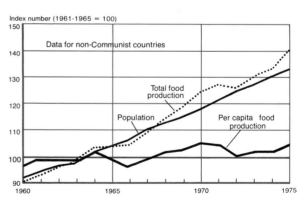

Fig. 6.5 Recent food and population trends in developing countries. The general upward trend in total food production in developing countries has been matched by population growth in those same nations. In most years, per capita food production has remained at about its 1960 level. (Reprinted with permission from *World Population Growth and Response, 1965–75*, Population Reference Bureau, Washington, D.C., 1976.)

Effect of Infectious Disease on Population Size

As long as populations were small and dispersed, and while separately developing societies were effectively isolated one from another, infectious diseases were specific to limited areas and restricted peoples that developed acceptable tolerance to them. The numerous diseases endemic to the tropical areas of human development were not present in the middle and upper latitudes of subsequent settlement. There, other diseases, which were frequently associated with the domestication of animals and to which comparable adaptation was made, were encountered. Isolated groups in a sparsely populated world developed disease tolerances not shared by societies elsewhere.

As culture hearths developed and as populations expanded, became concentrated in village and city, and came into increasing contact, the spread of diseases for which no tolerance or immunity had been acquired was inevitable and devastating. Susceptible new hosts in numbers sufficient to maintain the chain of infection are required for a massive and deadly incidence of transmissible disease. A common childhood infection demonstrates the pattern: a modern city population of some 500,000 with a recurring susceptible population of at least 7000 children results in a wavelike recurrence of outbreaks of measles at about two-year intervals.

Adaptations between human hosts and parasites are gradually acquired, and where populations are sufficiently large, reservoirs of disease are created. They remained unique to separate environments and cultures before, perhaps, 3000 B.C. After that time migration, trade, and conquest brought formerly isolated societies into contact with previously unknown forms of infection.

The expansion of north Chinese settlement to the fertile Yangtze Valley was marked by a millennium of halting advance as devastating diseases—probably malaria and dengue fever—were encountered and slowly, through acquired tolerances, overcome. The advances of Aryan conquerors into tropical India were slowed and dissipated more by encounter with deadly diseases unknown to the invaders than by the armed resistance of the native inhabitants. Athens was dealt a crippling blow from which it never fully recovered by an unidentified disease described by Thucydides. Killing one-fourth of the Athenian army, the infection had spread from Ethiopia to Egypt and thence, by ship, to Piraeus, the port of Athens. Within a single season, it had burnt itself out, leaving no traces except its presumed effect upon the course of the Peloponnesian War.

Recurring devastating plagues struck the expanding Roman Empire as trade and conquest brought the population into inevitable contact with previously isolated diseases. The plague bacillus *Pasteurella pestis,* the Black Death of medieval Europe, was brought to Western rodents—and thus to humans—from the Himalayas by expanding caravan trade and Mongol conquest. Worldwide sea commerce, with its uninvited cargo of rats, spread the rodent carriers to all continents.

The Americas, prior to European penetration, were remarkably free of in-

194

fectious diseases. The complex of parasites and acquired immunities characteristic of Eurasia were unknown in the western hemisphere, perhaps because domesticable animals in the New World were not themselves hosts to transferable parasitism. An almost totally disease-susceptible population succumbed, in Central and Andean America, not to numerical or military superiority but to transmitted microorganisms. While the Aztecs were successfully driving Cortes's tiny band from Mexico City, an imported smallpox epidemic was raging, quickly paralyzing the native population. The same disease promptly spread southward, devastating the Inca Empire and opening it to easy conquest by Pizarro. The fate of a continent was determined by unseen and unknown organisms.

---

death rates, and hence very slight percentage increases in population. The United States, Canada, Japan, and many northwest European countries have entered this stage. Birth rates vary, of course, but are characteristically low and are affected by conscious decisions related to the business cycle and attitudes toward abortion and contraception.

The population cycle just described recounts the experience of the northwest European countries as they went from rural, agrarian societies to urban, industrial ones, but it is also displayed in other regions of comparable economic development. Some doubt its applicability to other parts of the world.

In some countries, the death rate can be—and has been—lowered quickly and dramatically without modernization of the economy. The use of insecticides, vaccinations, and better public health measures has accomplished in a few years what it took Europe 50 or 100 years to experience. Sri Lanka, for example, sprayed extensively with DDT to combat malaria; life expectancy jumped from 44 years in 1946 to 60 only 8 years later. With similar public health programs, India has experienced a steady decline in its death rate since 1947. The resulting rapid population growth often hinders economic development because of the strain put on the economy. When population doubles

each generation, it is hard to achieve social and economic progress; just keeping up with the new mouths to feed is a major accomplishment. In India, for example, grain production has doubled since the 1940s, but so has the population (Figure 6.6).

The rapid increase in population that accompanied Europe's industrialization was alleviated by the emigration of millions of people to America, Australia, New Zealand, and other areas. This solution is not as possible today. There are few areas of the world into which millions of people could move easily, and in any case most countries limit by quotas the number of immigrants they will receive. International migrations, then, cannot significantly ease population pressures.

Another factor that makes the new demographic changes unlike those experienced earlier is that many countries are currently at very high levels of population. Even low rates of increase mean enormous additions to the population. India's population of over 600 million (1976) increases by about 13 million people per year, or roughly the population of the New York metropolitan area. European countries at the start of industrialization did not have to contend with this kind of population growth; they had not yet reached the critical bend on the J-curve.

195

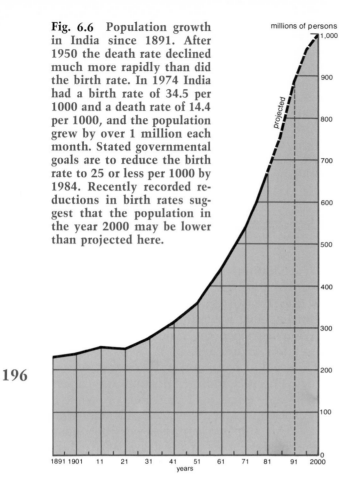

Fig. 6.6 Population growth in India since 1891. After 1950 the death rate declined much more rapidly than did the birth rate. In 1974 India had a birth rate of 34.5 per 1000 and a death rate of 14.4 per 1000, and the population grew by over 1 million each month. Stated governmental goals are to reduce the birth rate to 25 or less per 1000 by 1984. Recently recorded reductions in birth rates suggest that the population in the year 2000 may be lower than projected here.

World Population Distribution

Where are the more than 4 billion people of the earth at the present time? How are they distributed? The most striking feature of the population map presented inside the back cover of this book is the very unevenness of the placement of the world's peoples. Some areas are nearly uninhabited, others are sparsely settled, and yet others contain dense agglomerations of people.

The virtually empty regions include extremely cold areas, deserts, and high mountainous zones. Sparsely settled areas, which are found on all continents, include dry and cold plateaus, fringes of deserts, and tropical rainforests. Climate and landforms make some areas less conducive to settlement than others. Usually it is the land's capability of being farmed successfully that separates the habitable from the uninhabitable land areas. There are notable exceptions, however, especially where some special circumstance, such as the occurrence of a valuable mineral, attracts people to regions of low agricultural carrying capacity. Wherever people have developed an industrial base successful enough to allow for the importation of foodstuffs, the farming factor is less significant; the United Kingdom and Japan are cases in point.

The areas of dense population include (1) the fertile river valleys of the warm lands of Africa and south and east Asia; (2) the coastal plains of moderate and warm climates; and (3) the urbanized, industrialized portions of Europe and North America. In addition, areas benefiting from special circumstances of high accessibility to industrial and food raw materials and world markets, such as in South Africa and eastern South America, should be mentioned. Perhaps the most anomalous situation is the dense settlement in the Andes mountains of South America and the plateau of Mexico. Here Indians found temperate conditions away from the dry coastal regions and the inhospitable Amazon basin, and the fertile plateaus have served a large population for more than a thousand years.

Density

Population density figures are useful—if sometimes misleading—statements summarizing regional variations of human distribution. The *crude density of population* is the commonest and least satisfying expression of that variation; it is the calculation of the number of people per unit area of land within the boundaries of a political entity. It can be misleading because a nation may contain extensive regions that are only sparsely populated and, perhaps, largely undevelopable. In general, the larger the political unit for which crude population density is calculated, the less useful is the figure. This figure is, however, commonly employed simply because the data are easy to obtain.

Various modifications may be made to refine density as a meaningful abstraction of distribution. Its descriptive precision is improved if the area in question can be broken down into comparable regions or units. Thus it is more revealing to know that New Jersey has a density of 841 and Nevada of 4 people per square mile than to know only that the figure for the United States is 70 per square mile. The calculation may also be modified to provide density distinctions between classes of population, rural versus urban, for example. Rural densities in the United States rarely exceed 300 per square mile, while portions of an urban area such as New York City can have as many as 250,000 people per square mile.

A favored and revealing refinement of density calculations is to relate population not simply to total national territory but to that area of a country that is or may be cultivated. These figures have been called *physiological densities* and are, in a sense, an expression of the population pressure differentially exerted upon agricultural land. Note the difference between the United States and the People's Republic of China in Table 6.4. Although the crude density calculation shows a ratio of 1:3.74 and indicates China's significantly higher population density, it is the physiological density figures that point up the startling differences in actual densities in the inhabited parts of the two countries.

THE FUTURE

We began the discussion of population by pointing out what could happen if population continued to increase at the present rate. However, we cannot predict with any certainty what in fact will happen to the growth rate. Even small changes in the rate have tremendous consequences. Increases or decreases of a fraction of a percent involve millions of people.

Population pressures do not come from the amount of space humans occupy. It has been calculated, for example, that the entire human race could stand in an area slightly larger than the island of Manhattan. The problems stem from the food, energy, and other resources necessary to support the population and from the impact on the environment of the increasing demands and the technologies required to meet them. In addition, the rates of growth currently prevailing in many countries make it nearly impossible for them to achieve the kind of social and economic development they would like. The equilibrium between high birth rates and high death rates has been upset; at the present time, the human population is growing at a rate that will be impossible to sustain.

197

	China	United States
Crude population density (people per square mile)	260	70
Physiological density (people per square mile of cultivated land)	2119	316

Table 6.4 Comparison of Population Density Figures for the United States and Mainland China (based on estimated 1971 population figures)

Clearly, at some point population will have to stop increasing as fast as it has been. That is, either there will inevitably be imposed the self-induced limitations on expansion implicit in the demographic transition or an equilibrium between population and resources will be established in more dramatic fashion.

Recognition of this eventuality is not new. In a treatise published in 1798, Thomas Robert Malthus, an English economist and demographer, put the problem succinctly: all biological populations have a potential for increase that exceeds the actual rate of increase, and the resources required for the support of increase are limited. In later publications, Malthus amplified his thesis by noting that

1. Population is inevitably limited by the means of subsistence.
2. Populations invariably increase with increase in the means of subsistence unless prevented by powerful checks.
3. The checks that inhibit the reproductive capacity of populations and keep it in balance with means of subsistence are either "private" (moral restraint, celibacy, and chastity) or "destructive" (war, poverty, pestilence, and famine).

The chilling consequences of Malthus's dictum that unchecked population increases geometrically while food production can increase only arithmetically have been evident throughout human history, as they are today. Starvation, the ultimate expression of resource depletion, is no stranger to the past or the present. By conservative estimate, some 50 persons will starve to death during the two minutes it takes you to read this page; half will be children under five. They will, of course, be more than replaced numerically by new births during the same two minutes. Losses always are. It has been estimated that all battlefield casualties, some 35 million in all of mankind's wars over the last 500 years, equal only a six-month replacement period at present rates of population growth.

198

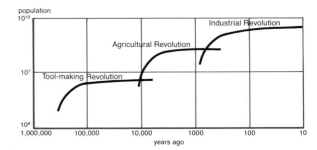

Fig. 6.7 The steadily higher homeostatic plateaus (states of equilibrium) achieved by humans are evidence of their ability to increase the carrying capacity of the land through technological advance. Each new plateau represents the conversion of the J–curve into an S–curve. The diagram is suggestive of the process; it does not represent actual world population figures, which are shown here on a logarithmic scale. (From "The Human Population" by Edward S. Deevey, Jr. Copyright © 1960 by Scientific American, Inc. All rights reserved.)

Yet, inevitably—following the logic of Malthus, the evidence of history, and our observations of animal populations—equilibrium must be achieved between numbers and support resources. When overpopulation of any species occurs, a population dieback is inevitable. The madly ascending leg of the J-curve is bent by external forces to the horizontal, and the J-curve is converted to an S-curve. It has happened before in human history, as Figure 6.7 summarizes. The S-curve represents a population size consistent with, and supportable by, the exploitable resource base. The population is said to have reached a *homeostatic plateau*, or to be equivalent to the *carrying capacity* of the occupied area.

Fig. 6.8 (opposite) In the 1970s, the Indian government undertook an extensive campaign to check rapid population growth. It encouraged sterilization, liberalized abortion laws, penalized civil servants who did not restrict the number of their children to two or three, and rewarded those who did by providing access to housing, medical care, and education. (United Nations/Farkas.)

In animals, overcrowding and environmental stress apparently release an automatic physiological suppressant of fertility. In humans, no such natural control is discernible; population limitation must be either brutish or self-imposed. Increasingly, governmental attitudes over much of the world indicate acceptance of the need for family planning to achieve a steady state where birth and death rates are equal and zero population change is maintained.

In some countries government leaders have made family planning part of their programs for national economic development. And governments can do much to increase people's incentives to restrict fertility as well as their abilities to control the number of births. For example, Japan legalized sterilization and abortions and encouraged contraception in 1947. In 10 years, the birth rate dropped from 34 per 1000 to 17 per 1000. Mainland China appears to have birth control programs aimed at a zero growth rate by the year 2000. Impelled by a sense of urgency, India adopted an extensive birth control campaign (Figure 6.8).

On the other hand, not all governments or cultural traditions favor conscious limitations on population growth. Some nations follow the Marxian concept that overpopulation is simply an expression of a calculated maldistribution of the means of subsistence, not of their shortage. Some religious groups oppose voluntary population limitation as immoral and destructive of the family basis of society. Cultural ideas about ideal family size, a belief in the value of male as opposed to female offspring, and attitudes toward abortion and contraception are deeply ingrained. In technologically less-developed societies, the main restraint on population growth in the years to come could very well be a rise in the death rate, brought about by famine and disease. The malnutrition and undernourishment that already exist can only be aggravated as population pressure on the food supply increases.

Finally, because of the age composition of many societies, it will be a long time before their populations stop growing *even if* each couple simply replaces itself (i.e., has only two children). When a high proportion of the population is young, larger and larger numbers enter the childbearing ages each year. Between 35% and 40% of the people now on earth are under 15 years old. In developing countries the figure may approach 45%. The consequences of the fertility of these young people are yet to be realized. Inevitably, even the most stringent national policies limiting growth must be affected by their reproductive potentials. In the United States, for example, with its large percentage of people under 20 (Figure 6.9), population would increase by about 40% even if, starting today, couples produced just enough children to ensure that each generation would exactly replace itself.

200

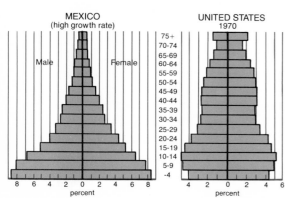

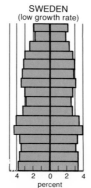

MEXICO (high growth rate) — Male / Female

UNITED STATES 1970

SWEDEN (low growth rate)

Age groups: 75+, 70-74, 65-69, 60-64, 55-59, 50-54, 45-49, 40-44, 35-39, 30-34, 25-29, 20-24, 15-19, 10-14, 5-9, -4

percent

Fig. 6.9 Age–sex pyramids for selected countries. These pyramids show the age distribution of population in a rapidly growing country (Mexico), in one with a low rate of growth (Sweden), and in the United States. Over the next several decades, the age distribution of population in the United States is expected to become more and more similar to that of Sweden. The proportion of elderly people will increase, the proportion of young people will decrease, and the median age of the population will rise. This change will affect many aspects of American life, including housing, education, politics, and the economy.

While international dismay is expressed over rising world populations, some countries are finding their own population stability or decline a matter of serious domestic concern. Four European countries—Austria, East Germany, Luxembourg, and West Germany—have constant or decreasing populations. Belgium, Britain, and France should reach the same condition by the early 1980s. By 1985 Denmark, the Netherlands, Norway, Sweden, and Switzerland are expected to have balanced birth and death rates, and the whole of Europe, if present trends continue, will experience population decline by the year 2000.

Zero Population Growth (ZPG)

There are national social and economic consequences of population stability or reduction not always perceived by those who advocate "zero population growth" as a world objective. An exact equation of births and deaths means an increasing proportion of older citizens, a reduced working-age population, and higher taxes on fewer producers to pay the pensions so common in developed nations. To maintain economic strength, more foreign workers may be required, increasing cultural conflict and loss of national and ethnic values. Absolute decline implies the eventual disappearance of the affected culture group. The German Demographic Society, projecting West German population trends of 1978—when the nation displayed the lowest birth rate of any major country in the world—predicts that the country's population will be reduced from 60 million to 40 million in 50 years, to 20 million in 100 years, with extinction foreseeable by the year 2500.

201

Considerations of financial solvency, national power, and self-preservation are creating demands in European nations that present population trends be halted and reversed. Proposals have been made in France to increase monthly allowances and to award additional electoral votes to parents in proportion to the number of offspring produced. East Germany assures long, paid leaves of absence from work for new mothers and provides monthly stipends for children; the result of this encouragement has been a 24% increase in the birth rate.

Control of human numbers may appear to be rational, if not imperative, on a world scale, but on a national level, it raises serious questions, requiring a definition of goals and an achievement of consensus.

Cultural Diversity

Knowledge of patterns of population is basic to an understanding of cultural geography. Patterns and trends in fertility and mortality; the size, age, and sex distribution of the population; the density of settlement; the rates of population growth; the stage in the population cycle—all of these aspects of population affect the social organization of a culture. They aid us in understanding how the people in a given area live, how they may interact with one another, how they use and affect the land, what pressure there is on resources, and what the future may bring.

However, population factors alone are not sufficient to enable us to make sense of billions of individual lives or to explain variations in culture. We must look to other pattern-producing variables to see order in human settlement. Although there are many such variables, an understanding of a few will help us enormously in characterizing cultures. These are technology, language, religion, and ethnicity.

At the same time, we must recognize the limitations of any description of a culture group. No discussion will allow one to know exactly how a single individual will behave. Generalizations are based on the collective traits of many individuals in a particular society.

TECHNOLOGY

In tracing human evolution, in assessing variations in culture, and in differentiating the states and kinds of human existence from place to place, answers are sought to a series of commonplace questions: How do the people in an area make a living? What resources and what tools do they use to feed, clothe, and house themselves? Is a larger percentage of the population engaged in agriculture than in manufacturing? Do people travel to work in cars, on bicycles, or on foot? Do they shop for food or grow their own?

In a broad sense, these questions concern the different adaptive strategies used in getting food, in "making a living." For most of human history, people lived by hunting and gathering, taking the bounty of nature with only minimal dependence on weaponry, implements, and the controlled use of fire. While their adaptive skills were great, their technological level—possession and use of specialized tools, ability to exploit a range of potential resources, control of external forces of energy—was low. Consequently, the homeostatic plateau was low.

As we saw in Chapter 5, some 10,000 years ago a major technological revolution—the domestication of plants and animals—occurred, the human carrying capacity of the earth increased, and populations grew. The increased size and reliability of the food base was the cause, of course, but those increases were expressions of the sudden (in human historical terms) development of new technologies that made possible a more intensive and extensive utilization of resources. Together with tools and engineering skills, the new technology depended on the control and use of energy, especially that of animals, to supplement human efforts.

The Industrial Revolution, beginning in England in the 18th century, represented a second major change in people's control of energy. For many, it led to life divorced from their own agricultural efforts, to new forms of social organization, to urbanization and trade, and to new carrying capacities of the earth. This second energy revolution began with reliance upon water power, but it quickly progressed to the use of more dependable inanimate sources, primarily coal, oil, and natural gas.

The Industrial Revolution wrought enormous changes in society, changes that diffused to other parts of the world, chiefly through European colonization and trade. Europeans who migrated to the Americas, Australia and New Zealand, and parts of Africa took their technologically more developed culture to those regions. These are among the areas of the world that today can be described as "industrialized" or "technologically advanced."

The Soviet Union, which also fits into this category, owes its development, first, to a conscious effort by Czar Peter the Great in the 18th century to open Russia to Western ideas, and second, to the efforts of Soviet leaders in the 20th century to industrialize the country. In the second half of the 19th century, Japan began a very rapid and successful attempt at modernization when its leaders assumed that unless they did so, Japan would not be able to maintain itself against the West.

202

Many aspects of a society are related to its level of technological development. The term *standard of living* brings to mind some of these, such as personal income, level of education, food consumption, life expectancy, and availability of health care. The complexity of the occupational structure, the degree of specialization in jobs, the ways natural resources are used, and the degree of industrialization are also intimately related to the technology of a given society.

In technologically advanced countries, many people are employed in manufacturing or allied service trades. Per capita incomes tend to be high, as do levels of education. These countries wield great economic and political power. In contrast, technologically less-advanced countries have a high percentage of people engaged in agricultural production, with much of the

agriculture at a subsistence level. The gross national products (the GNP equals the total value of all goods and services produced by a country during a year) of these countries are much lower than those of industrialized nations. Per capita incomes tend to be low, and illiteracy rates high.

Labels such as *advanced–less advanced, developed–underdeveloped,* or *industrial–nonindustrial* can mislead us into thinking in terms of a dichotomy. They may also lead the unthinking into the belief that they mean, in general cultural terms, "superior/inferior." This belief is totally improper, since the terms are related solely to economic and technological circumstances and bear no qualitative relationship to such vital aspects of culture as music, art, or religion. Improperly used, these technological comparative terms imply a static

Table 6.5 Per Capita Energy Consumption, 1976, for Countries with More Than 10 Million Inhabitants (in kilograms of coal equivalent)

203

Country	Per Capita Consumption[a]	Country	Per Capita Consumption[a]	Country	Per Capita Consumption[a]
United States	11,554	Venezuela	2,838	Morocco	273
Canada	9,950	Spain	2,399	India	218
Czechoslovakia	7,397	Yugoslavia	2,016	Indonesia	218
East Germany	6,789	Argentina	1,804	Pakistan	181
Australia	6,657	Iran	1,490	Kenya	152
Netherlands	6,224	Mexico	1,227	Sudan	143
West Germany	5,922	South Korea	1,020	Vietnam	124
United Kingdom	5,268	Turkey	743	Sri Lanka	106
U.S.S.R.	5,259	Brazil	731	Nigeria	94
Poland	5,253	Algeria	729	Tanzania	68
France	4,380	China	706	Zaire	62
Romania	4,036	Colombia	685	Burma	49
Japan	3,679	Peru	642	Afghanistan	41
Hungary	3,553	Malaysia	578	Bangladesh	32
Italy	3,284	Egypt	473	Ethiopia	27
North Korea	3,072	Philippines	329	Nepal	11
South Africa	2,985	Thailand	308		

[a]Does not include energy derived from the human body or from animals.

Source of data: *Statistical Yearbook, 1977.* United Nations, Department of Economic and Social Affairs.

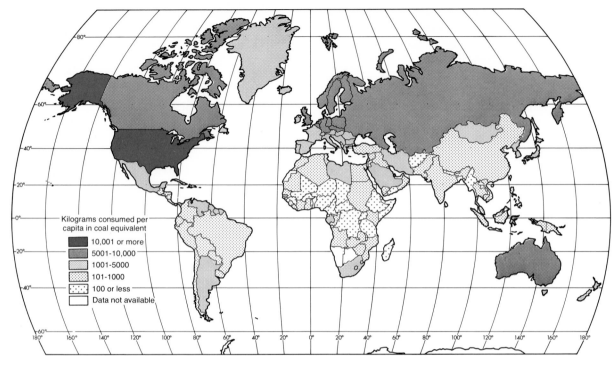

Fig. 6.10   National levels of energy consumption, 1976.

rather than a dynamic situation; many countries are in the process of altering their level of technological development. We can most usefully think of the nations of the world arrayed along a continuum. Table 6.5 lists (and Figure 6.10 maps) larger countries by the amount of energy consumed per person, a common measure of technological advancement. Other measures that could also be used to illustrate technological level are national product per capita, percentage of workers engaged in agricultural production, or transportation facilities per person.

It is important to recognize that all countries include areas that are at different levels of development. In a highly developed, industrialized country such as the United States, there are still areas that have not reached the developmental level of the rest of society; Appalachia is one such region. To take the reverse

case, many countries that are characterized as technologically less advanced contain regions that are highly developed, with important industries, well-developed transport networks, and people earning relatively high incomes.

LANGUAGE

Among all the culture traits, the two of which we are perhaps most immediately made aware are nationality and language. In all societies, the foreigner is in greater or lesser degree a person apart, a member of another culture and a citizen of a different nation. The foreigner is presumed to embody at least some of the attitudes, customs, and political stances we associate with his or her country of citizenship. We will discuss aspects of nationality and other political affiliations as a special part of cultural differentiation—political geography—in Chapter 7. The other inescapable trait that cannot usually or easily be disguised by adoption of the dress, customs, or citizenship of another country is language, dialect, or accent.

Language makes possible the understandings and shared behavior patterns called culture. In addition, like all other aspects of culture, it is transmitted to successive generations by learning and imitation. Some anthropologists argue that the language of a society structures the perceptions of its speakers; by the words that it contains and the concepts it can formulate, language is said to determine the attitudes, understandings, and responses of the society to which it belongs. Language, therefore, may be both a cause and a symbol of cultural differentiation.

If that conclusion is true, one aspect of cultural heterogeneity may be easily understood. The roughly 4 billion people on earth speak about 3000 languages. From our limited experience, we may find it difficult to name more than 30, or 1%, of the languages currently in use; but being informed that as many as 1000 languages are spoken in Africa, we are immediately able to understand the political and social instability in that continent. Europe alone has more than 100 languages, and, at least in part, language differences were a basis for national delimitation in the political restructuring of Europe after World War I. Language, then, is the hallmark of cultural diversity (Figure 6.11).

Some languages obviously have many more speakers than others; Table 6.6 lists those that are each spoken by more than 30 million people. Even this level of diversity may be reduced if we group languages into families. A linguistic family is a group of languages thought to

Fig. 6.11　World pattern of languages.

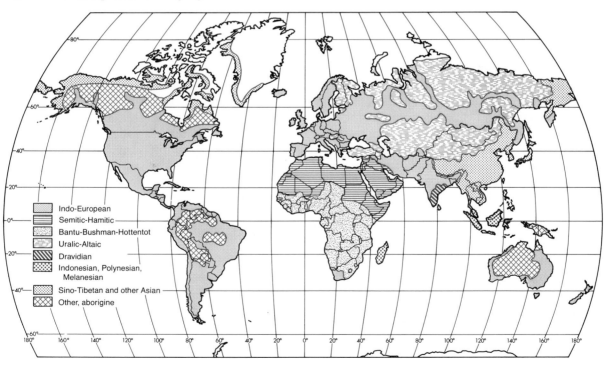

Indo-European
Semitic-Hamitic
Bantu-Bushman-Hottentot
Uralic-Altaic
Dravidian
Indonesian, Polynesian, Melanesian
Sino-Tibetan and other Asian
Other, aborigine

Table 6.6 Languages Spoken by More than 30 Million People (1976)

Language	Millions of Speakers	Language	Millions of Speakers
Mandarin (China)	660	Telegu (India)	54
English	363	Tamil (Sri Lanka, India)	54
Russian	240	Korean	53
Spanish	219	Marathi (India)	52
Hindi	213	Cantonese	48
Arabic	130	Javanese	45
Portuguese	129	Wu (China)	43
Bengali (Bangladesh, India)	127	Ukrainian	42
German	120	Turkish	40
Japanese	112	Min (China)	39
Malay-Indonesian	96	Vietnamese	37
French	92	Polish	36
Italian	60	Thai	31
Urdu (Pakistan, India)	58	Gujarati (India)	30
Punjabi (Pakistan, India)	57		

have a common origin. For example, the Indo-European family of languages includes, among many others, English, Greek, Hindi, and Russian (Table 6.7). All told, languages in the Indo-European family are spoken by about half the world's peoples. Linguists think that these languages derive from a common ancestor language called *proto-Indo-European,* spoken by tribes in east central Europe (though some believe the southern Russian plains eastward to the Caspian Sea to be the more likely site of origin) several thousand years ago. As these groups spread across Europe and Asia and settled in areas isolated from one another, their languages evolved differently.

Within a linguistic family, we can distinguish subfamilies. The Romance languages (e.g., French, Spanish, and Italian) and the Germanic languages (e.g., English, German, and Dutch) are subfamilies of Indo-European. Languages in a subfamily often show similarities in sounds, grammatical structure, and vocabulary, such as English *daughter,* German *Tochter,* and Swedish *dotter.*

Linguists also make a distinction between the language as it is spoken by perhaps millions of people and the standard or official version—that is, the form carrying official governmental or educational sanction. In Arab countries, for example, classical Arabic is the language of the mosque, of education, and of the newspapers, while colloquial Arabic is used at home, in the street, and at the market. The United States, Canada, and the United Kingdom all have different forms of standard English. In England, standard English is commonly thought of as that used by the announcers of the British Broadcasting Corporation.

Dialects are the regional variations of a language. Vocabulary, pronunciation, spelling, and the speed at which the language is spoken may set groups of speakers apart from one another. In George Bernard Shaw's play *Pygmalion,* on which the musical *My Fair Lady* was based, Professor Henry Higgins is able to deduce the place of origin of a flower girl by listening to her vocabulary and accent. An American spe-

Table 6.7   **Major Languages in the Indo-European Family**

Almost half the people in the world speak an Indo-European language.

Subfamily	Modern Language	Where Spoken
Albanian	Albanian	Albania
Baltic	Latvian	Latvian S.S.R.
	Lithuanian	Lithuanian S.S.R.
Armenian	Armenian	Armenian S.S.R.
Greek	Greek	Greece, Cyprus
Indic	Hindi	India
	Urdu	Pakistan, India
	Bengali	Bangladesh, India
	Marathi	India
	Punjabi	Pakistan, India
Iranian	Persian	Iran, Afghanistan
Romance	Italian	Italy, Switzerland
	Romanian	Romania, Balkans
	Spanish	Spain, Latin America
	Portuguese	Portugal, Brazil, parts of Africa
	French	France, Belgium, Switzerland, parts of Africa
Celtic	Gaelic	Ireland, parts of Scotland
	Welsh	Wales
	Breton	Brittany
Germanic	Dutch	Netherlands
	English	Anglo-America, Great Britain, Australasia, parts of Africa and Asia
	German	Germany, Switzerland, Austria
	Scandinavian languages	Norway, Sweden, Denmark
Slavic	Russian	U.S.S.R.
	Ukrainian	Ukrainian S.S.R.
	Polish	Poland
	Czech	Czechoslovakia
	Slovak	Czechoslovakia

cialist in linguistics could probably do the same with a native speaker of American English. Figure 6.12 indicates the variation in usage associated with just one phrase. Southern English and New England speech are among the dialects spoken in the United States that are most easily recognized from distinctive accents. There may be so much variation that some dialects are almost unintelligible to other speakers of the language. Effort is re-quired for Americans to understand Australian English or that spoken in Liverpool, England, or in Glasgow, Scotland.

Language is the most important medium by which culture is transmitted. It is what helps make it possible for parents to teach their children what the world they live in is like and what they must do to become functioning members of society. A common language fosters unity among people. It promotes a feeling

Pidgin and Creole Languages

Outside of the formally recognized families of languages lie "pidgin" and "creole" speech, which nevertheless are important, even primary, means of intra- and intergroup communication among millions of people throughout the world. A *pidgin* of whatever form is an auxiliary language not learned as a mother tongue by any of its users. Derived with reduced vocabulary and simplified structure from other languages, a pidgin is generally restricted to specific functions such as trade, communication with visitors, administration, or work supervision. A *creole* is a language in the common meaning, evolved from a pidgin and developed to become the first language of a society. It has the full range of functions and complexity of structure necessary for expressive intra-group communication.

Known pidgins were first developed in the early years of European colonization and are now largely found in Asia, the Pacific, and Africa. They are not just impromptu distortions of a European tongue, but real languages with specific vocabularies and rules of structure and are not necessarily understood by speakers of other versions of pidgin. Used daily by several million persons throughout the world, pidginized languages may evolve into more complex primary media of communication. In central Africa, a pidginized (with French) Sango is spreading, while the true Sango is disappearing. Pidgin English has emerged as a powerful and needed integrative tool in linguistically chaotic Papua, New Guinea, and gives evidence of becoming the national language of that country.

208

Creoles are spoken by some 6 million persons of the Caribbean area; by a number of groups in Sierra Leone, Cameroon, Ivory Coast, Guinea, and elsewhere in West Africa; and in the Asian areas of India, Macao, and the Philippines. Afrikaans and Bahasa Indonesian may also be considered creoles. Gullah, once widely spoken in Georgia, South Carolina, and the Sea Islands, is a United States example. A creolized Swahili has been made the national language of administration and education in Tanzania. Creoles, as fully developed languages, are emerging in several parts of the world as symbols of national and cultural identity; perhaps their development reflects part of the process of prehistoric linguistic diversification.

Adapted by permission of the author from Dell Hymes, "Pidginization and Creolization of Languages, Their Social Contexts," *Items*, Social Science Research Council, Vol. 22, No. 2, June 1968.

for a region; if it is spoken throughout a country, it fosters nationalism. This is why languages sometimes gain political significance and serve as a focus of opposition to what is perceived as foreign domination. Although people in Wales speak English, many also want

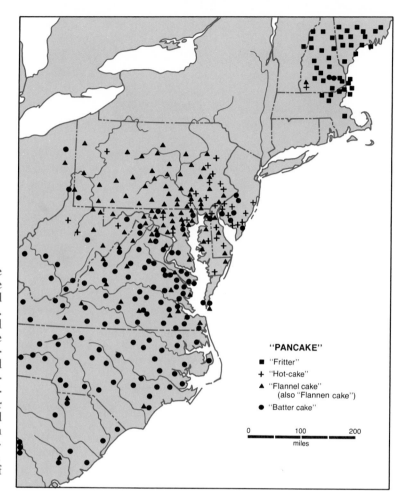

Fig. 6.12 Maps such as this one are used to record variations over space and among social classes in word usage, accent, and pronunciation. Differences are due not only to initial settlement patterns but also to more recent large-scale movements of people, from rural to urban areas and from the South to the North. Television and radio counteract regional differences by promoting a "general" or "standard" American accent and usage. (Reprinted with permission from Hans Kurath, *A Word Geography of the Eastern United States.* Copyright © by The University of Michigan Press, 1970.)

"PANCAKE"

■ "Fritter"
+ "Hot-cake"
▲ "Flannel cake"
  (also "Flannen cake")
● "Batter cake"

0   100   200
              miles

to preserve Welsh because they consider it an important aspect of their culture. They think that if the language is forgotten, their entire culture may also be threatened. French Canadians have asked for and received government recognition of their language and now have imposed it as the official language of Quebec. In India, where over 100 languages are spoken (Figure 6.13), serious riots were caused in 1965 by people expressing opposition to the imposition of Hindi as the single official language.

The classification of countries by language is complicated tremendously by bilingualism or multilingualism. Areas are considered bilingual if more than one language is spoken by a significant proportion of the population. In some countries—for example, Belgium—there is more than one official language. In many others, such as the United States, only one language may have official government sanction, while several others are spoken. Speakers of one of these may be concentrated in a few areas (e.g., most speakers of French in Canada live in Quebec); less often, they may be distributed fairly evenly throughout the country. There may be roughly equal numbers of speakers of

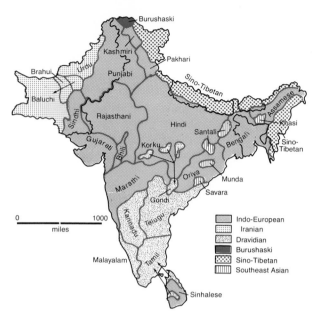

Fig. 6.13 Principal languages of the Indian subcontinent. Many additional tongues further complicate this pattern of major languages. The many mutually unintelligible languages led, in the interest of national unity, to a constitutional provision that Hindi would become the official language of India. At the same time, recognition of the practical and cultural differentiation of the nation implicit in linguistic regionalism led to a political subdivision of the nation into states based largely on language differences. (Reprinted with permission of the author from Rhoads Murphey, *Introduction to Geography*, 3rd ed., Rand McNally, Chicago, 1971.)

210

each language, or one group may be significantly in the majority. When this is the case, the minority group may have to learn the language of the majority to receive an education, obtain certain types of employment, or fill a position in government. In some countries, the language in which instruction, commercial transactions, and government business take place is neither a native language nor the majority language, but yet another language. This is the position of English in many former colonies (e.g., Nigeria, Ghana, and Bangladesh).

RELIGION

Language is one of the most fiercely defended elements of a culture. The official Committee for the Defense of the French Language of the French Academy is charged by the government with the elimination from the mother tongue of foreign words and phrases—even those in popular use. Spanish Americans demand the right of instruction in their own language, and Basques wage civil war to achieve a linguistically based separatism.

Yet language yields to religion as a cultural rallying point. French Catholics and French Huguenots (Protestants) freely slaughtered each other in the name of religion in the 16th century; English Roman Catholics were hounded from the nation after the establishment of the Anglican church; religious enmity between Muslims and Hindus forced the partition of the Indian subcontinent after the departure of the British. The Crusades were fought with each side calling upon its own God; and the treatment of Jews by medieval and modern Christian Europe is an unremitting tale of hatred and religious bigotry.

Religion may intimately affect all facets of a culture, openly or indirectly. Most Hindus do not eat any meat; Muslims do not eat pork; Roman Catholics until recently ate no meat on Fridays. Christians regard Sunday as the day of rest; for Jews it is Saturday; for Muslims, Friday. These are small examples of the control that religious beliefs exert over people's daily lives. The list could be expanded many times.

There are other important ways in which religions shape believers' lives and thoughts. Religions are an integral part of a culture. They are formalized views about the relation of the individual to the world and to the hereafter. Each carries a distinct conception of the meaning and value of this life, and most contain strictures about what must be done to achieve salvation (Figure 6.14). These rules become interwoven with the traditions of a culture.

Fig. 6.14 Worshipers praying inside the Royal Mosque in Lahore, Pakistan. Many rules concerning daily life are given in the Koran, the holy book of the Muslims. All Muslims are expected to observe the five pillars of the faith: (1) repeated saying of the basic creed; (2) prayers five times daily, facing Mecca; (3) a month of daytime fasting (Ramadan); (4) almsgiving; and (5) if possible, a pilgrimage to Mecca. (United Nations.)

One cannot understand India without a knowledge of Hinduism, or Israel without an understanding of Judaism.

Even economic patterns may be intertwined with religion. Religious restrictions on food and drink affect the economic structure of an area, very strongly in some places, less so in others. The kinds of animals that are raised or avoided, the crops that are grown, and the importance of those crops in the daily diet are often determined by religious beliefs. To take another example, the frequently used term *Protestant ethic* connotes a presumed attitude toward economic progress that is quite different than that of, say, Hinduism. Individualism, hard work, and the acquisition of material wealth are said to characterize the former, while in Hinduism, presumably, there is considerably less concern about the accumulation of material wealth in this life.

In many countries there is a state religion. Buddhism, for example, is the official state religion in Burma, Thailand, Laos, and Cambo-dia. In a few countries government leaders are church officials; in others political parties have religious affiliations that are reflected in the voting patterns of the electorate, as in Northern Ireland. Laws, too, may reflect religious beliefs. The penalty attached to inadvertently killing a cow, for example, is very high in Nepal where the cow is a sacred animal.

In a few countries political ideologies have a quasi-religious role. They have many of the elments of a religion, including a set of beliefs, ethical standards, revered leaders, an organization, and a body of literature akin to scrip-

Fig. 6.15 People lined up to visit Lenin's tomb in Red Square, Moscow. V. I. Lenin was a leader of the Communist revolution of 1917, which marked the end of the rule of the czars and the establishment of the Union of Soviet Socialist Republics. Lenin's body is preserved in a glass-covered coffin inside the tomb, which is visited by thousands of people each day. Millions also make a pilgrimage to Lenin's birthplace on the Volga. The tomb and Lenin's birthplace may be said to be sacred shrine for Soviet Communists. (United Nations.)

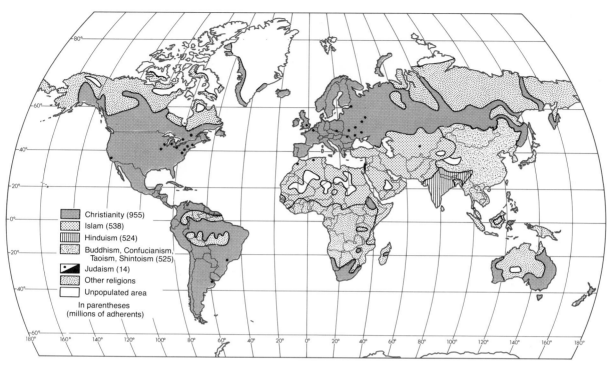

Fig. 6.16 The distribution of major religions. Although there are hundreds of religions, those shown on the map include about half the people of the world.

ture (Figure 6.15). Such ideologies can exert great influence on people's lives. In addition, adherents may display an almost religious fervor in their desire to proselytize (convert nonbelievers) and to root out heretical beliefs and practices. In this fashion, communism may be viewed as the state religion of the U.S.S.R. It combines a belief in the redeeming role of the proletariat with an ethos of work and sacrifice.

Settlement patterns are also linked to religion. Members of a religion tend to be areally concentrated, a fact readily apparent from Figure 6.16. Note the predominance of Hinduism in India and of Buddhism in most of Southeast Asia. Although no map distinction is made between subsets of Christians, Protestantism predominates in the United States and Australia, while South America is a Roman Catholic area. Within countries, members of a particular religion may also display areal concentration. Thus in the United States, Jews and Unitarians tend to live in urban areas, and Mennonites are found mainly in rural settings. Figure 6.17 suggests the regional concentration of American Lutherans. More Baptists live in the South than elsewhere, while Mormons are found chiefly in Utah and portions of the adjoining states.

Many other instances of religion as an important strand in the fabric of culture could be cited. Religious ideas affect the role of women in a society, marriage and divorce customs and taboos, and the importance attached to education (Figure 6.18). Our ideas about the significance of this life, whether we have only one life or many, what we should strive for in life—all of these are molded in part by religious precepts.

Religion is a strong force for promoting cul-

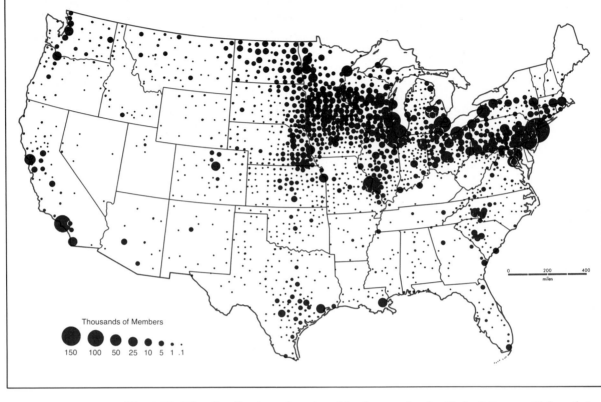

Thousands of Members

150　100　50　25　10　5　1　.1

Fig. 6.17　The distribution of reported Lutherans in the United States. Although Lutheran church members may be found in every state, their great preponderance is in the upper Midwest and the urbanized Northeast. The pattern has not changed significantly since 1952, the year these data were gathered. (Redrawn by permission from the *Annals of the Association of American Geographers*, vol. 51, 1961, Wilbur Zelinsky.)

tural stability. Through rituals and scripture, time-honored ways of doing things are fostered. Religions in general act to retard cultural change. Much of the diversity within a given religion has occurred as believers have been faced with changing times or conditions. Church leaders have had to ask themselves whether they should insist upon strict adherence to religious doctrine or adjust to new ways of life.

All religions contain such diversity. Most Americans are familiar with at least the names of many of the various sects of Protestantism:

Lutheran, Unitarian, Baptist, Presbyterian, and Episcopalian, to name only a few. Each is characterized by beliefs or practices that distinguish it from the rest. Similarly, Islam contains several sects, among them the Sunnites, the Shiites, and the Wahhabites. Judaism has spawned Orthodox, Conservative, and Reform branches. Hinduism contains a variety of creeds and practices. Often this diversity has been caused by different interpretations of religious doctrine, sometimes by differences that evolved when a major religion came into conflict with native cultures.

Fig. 6.18 Monks at the Temple of the Emerald Buddha in Bangkok, Thailand. In some countries, such as Burma and Thailand, all males are expected to spend at least a year of their lives as Buddhist monks. These men, with shaved heads and dressed in their distinctive saffron-colored robes, are a familiar sight on the streets of Bangkok. Emphasis on monastic training has led to a high rate of lieracy among males, an example of the effect of religion on other facets of culture. (United Nations.)

The study of religion aids us in understanding the diverse cultural patterns we see on the face of the earth. Religions both promote and hinder contact between groups; people of different religions may feel so hostile to one another that wars erupt. The Crusades and the wars of the Reformation are just two examples of the religious wars that have marked history. At the same time, religions have promoted contact between regions by inspiring believers to convert others, through either conquest or peaceful missionary efforts. Many of the major religions have thus acted as unifying forces, bringing large numbers of people formerly more diverse to a belief in a common faith.

ETHNICITY

Any discussion of cultural diversity would be incomplete without a mention of ethnicity. The root word *ethnos* means "nation"; the term is usually used to refer to the ancestry of a particular people. Members of an ethnic group have some common characteristic traits or customs that are tied in with their ancestry; relgion and language may be among those traits. Ethnicity is a composite of the kinds of cultural indices we have been discussing (Figure 6.19). Normally, one refers to ethnic groups when they are in a minority position in a society. Thus Japanese in Japan would not be identified as an ethnic group because theirs is the dominant culture, but Japanese living in Korea would constitute an ethnic group.

Immigrants to a country typically have one of two choices. They may hope to become assimilated, to give up many of their past cultural traits and move into the mainstream of the dominant culture, or they may try to retain their cultural heritage. In either case, they usually settle initially in an area where other members of their ethnic group live. Bound as they are to the group by ties of ancestry, religion, and perhaps language or ways of earning their livelihood, this is an area where they feel secure—in fact, "at home." Those who have a desire to assimilate may move out after a generation or two, and their place may be taken by other, newly arrived immigrants. Sometimes settlers have no desire to assimilate or are not

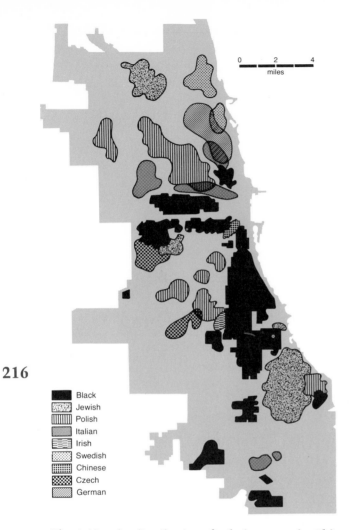

216

Black
Jewish
Polish
Italian
Irish
Swedish
Chinese
Czech
German

Fig. 6.19 The distribution of ethnic groups in Chicago, 1957. Most large cities of the United States are ethnically heterogeneous, particularly in the industrial North, to which came successive waves of immigrants. The heterogeneity is usually expressed by a well-defined series of homogeneous neighborhoods rather than by a smooth intermixture of peoples. On this map, black areas are defined as neighborhoods with 25% or more black inhabitants; they have expanded significantly in size since 1957. (Fig. 21-e from the Appendix, "Black Metropolis 1961," in *Black Metropolis*/Vol. 1. © 1962 by St. Clair Drake and Horace R. Cayton. Reprinted by permission of Harper & Row, Publishers, Inc.)

allowed to assimilate, so that they and their descendants form a more or less permanent subculture in the larger society. The Chinese in Malaysia belong to this category. Ethnicity in the context of nationality is discussed more fully in Chapter 7, "Political Geography."

OTHER ASPECTS OF DIVERSITY

Culture is the sum total of the way of life of a society. We have isolated only a few elements—technology, language, religion, and ethnicity—as particularly evident characteristics differentiating culture groups. Other suggestive but less basic elements exist. Architectural styles in public and private buildings are evocative of region of origin even when they are encountered in indiscriminate juxtaposition in American cities. The Gothic and New England churches, the neoclassical bank, and the sky-

scraper office building suggest not only the functions they house but the culturally and regionally variant design solutions that gave them form. The Spanish, Tudor, French Provincial, or ranch-style residence may not reveal the ethnic background of its American occupant, but it does constitute a culture statement of the area and the society from which it diffused.

Music, food, games, and other evidences of the joys of life, too, are cultural indicators associated with particular world or national areas. Music is an emotional form of communication found in all societies, but, being culturally patterned, it varies among them. Instruments, scales, and types of compositions are technical forms of variants; the emotions aroused and the responses evoked among peoples to musical cues are learned pyschological responses. The Christian hymn means nothing emotionally to a pagan New Guinea clan; the music of

Table 6.8 Selected Foods and Beverages That Are Consumed at Home at a Rate at Least 8% Greater Than the Regional Expectation Based on National Consumption Patterns   **217**

Northeast	North Central	South	West
Macaroni, spaghetti, noodles	Butter	Flour, cornmeal	Rice
Cakes, pies, pastry	Pie mix and fillings	Fresh snap beans	Nonwhite bread
Fish and seafood	Fresh berries	Frozen lima beans	Cottage, other soft cheese
Cream	Doughnuts, sweet rolls, coffee cake	Lard	Diced fruits (prunes, raisins, and others)
Fresh juices	Potato chips	Cola drinks	Frozen lemonade
Canned pineapple juice	Icings and fudge mixes	Bacon, poultry	Canned and bottled apple juice
Frozen green beans, spinach		Sugar, syrup, molasses, honey	Frozen peas
Frozen French-fried potatoes		Mayonnaise	Frozen prepared dinner
Ginger ale		Coffee in bags	Frozen fruit, berry, cream pies
Alcoholic beverages, especially wine and whiskey		Leavening agents	Noncarbonated drinks
Butter		Chewing tobacco	Wine

Source of data: Fabian Linden, ed., *Market Profiles of Consumer Products*. National Industrial Conference Board, New York, 1967.

(a)

(b)

(c)

Fig. 6.20 Sports and other recreational pursuits vary from culture to culture. Shown here are hurling (Ireland), platform tennis (eastern United States), Thai boxing (Thailand), and sumo wrestling (Japan). (a: Irish Tourist Board; b: American Platform Tennis Association; c: Thailand Tourist Organization; d: Japan Airlines photo.)

(d)

a Chinese opera may be simply noise to the European ear; the ceremonial music of the Plains Indians is vital to their culture, but irrelevant to outsiders. Where there is sufficient similarity between musical styles and instrumentation, blending (syncretism) and transferral may occur. American jazz represents a blend; calypso and flamenco music have been transferred to the North American scene.

Other transferrals are the foods of culture areas. Taco stands, pizza parlors, and Chinese, French, Greek, Armenian, and other ethnic food dispensaries suggest not only areal variation in preferred and familiar foods, but the diffusion of food variants through an assimilating American culture region. Even within American cuisine, which the Chinese writer Lin Yutang dismissed as "dull and insipid and extremely limited in variety," recognizable regional variants abound (Table 6-8). Recreational pursuits, too, vary markedly from culture to culture (Figure 6.20), but the Olympic Games and the universal appeal of soccer indicate the ease of diffusion of sports.

These are additional minor statements of the variety and the intricate interrelationships of that human mosaic called culture. Individually and collectively such indicators are, in their areal variations, the subject matter of the cultural geographer.

Culture Realms

The world maps of the culture traits of technology, language, and religion presented in this chapter show the subdivision of earth space into special-purpose regions. A region is an area possessing certain specified characteristics that are different from those of surrounding or adjacent areas. For cultural geographers, these characteristics are in some way related to the attributes or activities of people: the political organizations they devise, the beliefs of their religion, the form of economy they pursue. There are as many cultural regions as there are cultural features.

Individual places may be grouped into something called a region only after we have defined the purpose of the study and selected the criteria. Suppose that interest centers on the spatial extent of Islam. We must choose the time period(s) that we are concerned with and then define what number of members of the Muslim religion must reside at a place for it to be considered part of the Islamic realm. A certain proportion of believers relative to the total population might be the criterion, or the percentage of members in a political unit who are Muslims. Only after having chosen our criteria may we plot the data on a map and begin to draw a border around the region.

The easiest kinds of regions to delimit are *single-factor* regions, that is, areas defined by the possession of a single characteristic such as religious affiliation. More often than not, we are interested in *multiple-factor* regions—areas marked by the possession of a set of interrelated characteristics. Such regions might be displayed on a world map of developed and underdeveloped economies, for example. In determining which nations are "developed," we might consider their achievement of such established criteria as national product per capita, energy consumption per capita, the density of the transportation network, the level of urbanization, and the percentage of the population that is literate. In general, the more factors used, the harder it will be to delineate an area where all the factors defining a specific mix of characteristics are uniformly achieved. There will be an area over which these factors coincide, but there will also be areas where only one or two of the criteria are met. These fringe areas are often zones of transition between one region and the next.

Two elements that are crucial in delineating regions are purpose and scale. Culture is composed of so many elements that their patterns of distribution never coincide neatly in any one area. Therefore any single study must be limited to those culture traits whose distribution is of particular significance. Scale is important because any single place is a member of many different regions. A whole hierarchy of regions exists, from neighborhoods in a city to entire continents. Large regions may be subdivided until we eventually reach individual places or districts within them. The scale at which we choose to work depends on the purpose of our study. These and other aspects of regionalization are treated in Chapter 12.

Our present purpose is to consolidate the many individual mosaics of culture traits into a world view of multifactor culture realms. Ideally, the amalgam of characteristics within such a realm should be distinctively different from the combination of traits displayed by other culture regions. There should be an underlying consistency, unity, and integration of significant elements in each that clearly sets it off from all other depicted realms.

Clearly, our present data base is inadequate to the task; political structure, economic orientation, patterns of behavior, and levels of urbanization—all aspects of culture—are yet to be considered. A preliminary cultural regionalization may be attempted, however, on the basis of the fundamental differentiating characteristics already discussed. Language, religion, and ethnicity are ingrained and persistent attributes of populations that color their acceptance of and regulate their adoption of modern economies and technologies. "Developing" realms are modernizing in ways distinctively their own, taking on the trappings of an integrated world community, but remaining

Fig. 6.21 Culture realms of the modern world. This is just one of many possible subdivisions of the world into multifactor cultural regions.

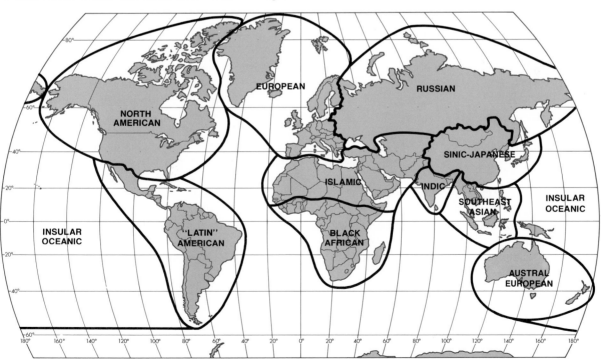

clearly separated in underlying culture traits.

Weakening the validity of this claim that uniform and distinct culture realms may be distinguished is the inevitable cultural heterogeneity resulting from past and present migrations of unlike peoples. Earlier European colonization and emigration and the more recent massive refugee relocations of the 20th century are striking departures from the slow migrations and assimilations of earlier human history. The "melting pot" of North America has produced not a culture uniform in all respects but a collection of peoples whose ethnic and cultural differences are increasingly a matter of pride and emphasis. Similar confused and indistinct mixtures cloud the picture in the Soviet Union, in South America, in Southeast Asia, and in Australia. Nevertheless, overriding the ethnic and even the linguistic and religious differences within a culture realm are adherences to adopted or imposed behaviors, technologies, and aspirations characteristic of the larger society in which minority groups participate. Based in part on such presumed distinguishing commonalities, Figure 6.21 is offered as a world subdivision into culture realms, that is, into regionally discrete areas that are more alike internally than they are like other realms.

Conclusion

The large and ever-growing populations of the world pose a crisis related to the earth's carrying capacity and ecological structure not encountered in earlier periods of human history. That a new homeostatic plateau will be reached is certain; whether it will be attained by human action or imposed by nature is less sure. The unevenly distributed populations of the earth exert unremitting and increasing demands upon that small fraction of the globe that can, through natural endowment, house and sustain them.

Within that limited area, the boundaries of which are steadily and destructively being extended, the world's population is differentiated into culture realms by mosaics of social characteristics and outlooks. In this chapter, it has been suggested that major traits distinguishing those separate realms—which are more alike internally than they are similar to other culture regions—are technology, language, religion, and ethnicity. Each of the principal traits may serve as evidence of related aspects of cultural variation from place to place.

Closely akin to states of technological development are the ways people can make a living, their level of income and education, and their attitudes toward change. The way that people view the meaning and purpose of life and their attitudes toward other cultures and societies are conditioned by their religion and, sometimes, their ethnicity. Language, the chief means of communicating culture, can be a strong indicator and shaper of cultural differences and culture systems.

Of nearly comparable importance as a determinant of societal identity at all levels of generalization are the political affiliations entered into or imposed upon individuals and groups. Formal political organization, the allocation and exercise of power it implies, and the role of nationality as a prime element of cultural variation are topics dealt with by political geography, to which we next turn our attention.

222

For Review

1. What is meant by *rate of natural increase*? How is the rate expressed? What rate of natural increase would double a population in 35 years?

2. If zero population growth (ZPG) becomes a reality in the United States, what kinds of changes are likely to occur in American society over the next 50 years?

3. What is the *demographic transition*? Where has it been achieved? What are the population implications of the terms J-*curve* and S-*curve*?

4. Contrast *crude population densities* and *physiological densities*. Under what circumstances might each be usefully employed?

5. What was Malthus's underlying assumption concerning the relationship between population growth and food supply? What is the *homeostatic plateau*? How did Malthus suggest it would be reached?

6. In what ways is the technological level achieved by a society an indicator of culture, as the term is employed in this chapter? Briefly discuss why technology is unsatisfactory as a prime trait of cultural differentiation.

7. *Hoagy (hoagie), submarine (sub), grinder,* and *hero* are various names for the same kind of sandwich. Name at least five other terms that vary regionally. (Hint: Is a creek always a creek?)

8. In what way may religion affect other culture traits of a society?

9. What is meant by *culture realm*? What, in your opinion, is the validity (if any) of the concept in the modern world?

223

Suggested Readings

BROEK, JAN, and JOHN WEBB, *A Geography of Mankind.* McGraw-Hill, New York, 1978.

HART, JOHN F., *The Look of the Land.* Prentice-Hall, Englewood Cliffs, NJ, 1975.

JORDAN, TERRY G., and LESTER ROWNTREE, *The Human Mosaic.* Canfield Press, San Francisco, 1976.

KARIEL, HERBERT G., and PATRICIA E. KARIEL, *Explorations in Social Geography.* Addison-Wesley, Reading, MA, 1972.

MEINIG, DONALD W. (ed.), *The Interpretation of Ordinary Landscapes.* Oxford University Press, New York, 1979.

PETERS, GARY L., and ROBERT P. LARKIN, *Population Geography: Problems, Concepts, and Prospects.* Kendall/Hunt, Dubuque, IA, 1979.

WARD, DAVID, *Geographic Perspectives on America's Past.* Oxford University Press, New York, 1979.

World Population Growth and Response: 1965–75. Population Reference Bureau, Washington, DC, 1976.

ZELINSKY, WILBUR, *The Cultural Geography of the United States.* Prentice-Hall, Englewood Cliffs, NJ, 1973.

On December 4, 1977, Jean-Bedel Bokassa crowned himself Bokassa I, Emperor of the Central African Empire. As reported by the press, the ceremony combined elements of Africa's tribal past with those of its more recent status as a French colony. The emperor rode to the coronation in a carriage similar to one Napoleon used, through wooden triumphal arches and past plaster pillars. Tribal dancers performed in a sports stadium built on the edge of the jungle. Tight security measures were taken during the ceremony; squadrons of mounted lancers escorted the carriage, troops armed with submachine guns guarded the stadium, and Soviet-made helicopters circled overhead.

Festive though the occasion may have been, the emperor faced an awesome task. He had somehow to forge a nation out of a patchwork of 80 tribes of diverse customs and languages. The country's 3 million inhabitants are scattered over an area about the size of France. Most are illiterate, most are poor. The country is landlocked, and until a rail or truck route can be built to the ocean, the country's mineral resources cannot be economically exploited. Less than two years after his coronation, Bokassa I was overthrown and the country was renamed the Central African Republic. The problems that faced Bokassa I and that now face his successor differ in degree but not in kind from those of every governmental leader in the world.

Nationality is a basic element in cultural variations between societies. It is one aspect of politics, which deals with the management of public affairs, the organized subdivision and control of territory, and the formalized distribution of power. Political geographers analyze the organization and distribution of political phenomena in their areal expression; they seek to relate those concerns to such other spatial evidences of culture as religion, language, and ethnicity. Traditionally, political geography has had a primary interest in nations: in the spatial patterns reflecting the exercise of cen-

7

Political Geography

225

tral governmental control, in questions of boundary delimitation and effect, and in the economic and organizational viability of states. Political geographers have also turned their attention to smaller areas of organizational control and response: to the voting behaviors of ethnic or socioeconomic groups, to the implications and consequences of political-district boundary changes, and the like.

The nation is only one geographic expression of the legally bounded spaces within which humans organize themselves—or have organization imposed upon them—to advance their individual and collective interests. Since these interests characteristically have spatial roots reflecting different levels of administrative concern, from the local school district or voting precinct to supranational economic unions or military alliances, political geographic inquiry takes place in a framework defined by scale. In this chapter, we consider some of the problems of defining political jurisdictions, asking why political power is fragmented and seeking the elements that lend cohesion to a political entity. We begin with local administrative organization and end with international political systems.

Political Systems

The ways in which we live our lives, conduct our daily affairs, and make plans for our futures are affected by decisions made by public agencies. Where children go to school, during which hours it costs least to make long-distance telephone calls, and how much we pay in taxes are examples of such decisions. In these American instances, the controlling decisions are made by many different organizations, ranging from the local school board through state and federal agencies to the Congress of the United States. Each system or organization operates within a carefully defined

geographic area. Congress can make laws for the entire country, but a school board has the power to operate only within its school district.

Even these limited examples suggest that populations are associated with, or are effectively controlled by, two different classes of organizational entities. On the one hand are those like the Congress or the state legislatures, county governments, or city councils that establish general laws or objectives that apply to the legally bounded territories under their control. In turn, legislatively enacted general bodies of law are converted into specific programs and detailed actions by the second class of control, elected or appointed administrative or regulatory agencies, such as the Interstate Commerce Commission, state departments of conservation, or local sanitary districts.

Particularly at the lower political-administrative levels, jurisdictional boundaries are largely determined by the stated objectives of the organizational units, and at all levels they reflect spatial expressions of differing common interests. Political organizations are given the responsibility of identifying people's goals, which may be economic or social, educational, or military, and are endowed with the power to pursue them. If the goal is attained, such as the building of a dam, the organization may cease to exist. More often, however, goals are long-run and may even be unattainable—for example, "to insure domestic tranquillity"—in which case the controlling political mechanism constitutes a continuing effort.

Political systems operate within specific geographic areas. The limits of those areas usually do not exist on the landscape. Boundaries of wards in a city, city limits, and even most state boundaries are not physical entities that can be seen or touched, and yet people have emotional feelings about many territorial units. In the United States most people identify strongly with their town and their state.

226

Most people have a feeling that they are in a "foreign," or at least a different, area when they cross from their own state into another.

Political entities purposely try to instill feelings of allegiance in their constituents, for such feelings give the system strength. People who have such allegiance are likely to accept the rules governing behavior in the area and to contribute to the decision-making process. The forces that promote cohesiveness in regions are examined later in this chapter. Let us look now at the problems encountered in delimiting boundaries on the local level.

Local Political Organization: The Districting Problem

There are over 48,000 local governmental units in the United States. This figure includes only municipalities, townships, and counties;

it does not include the thousands of school districts, water control districts, and other special-purpose districts that also exist. Around each of these districts boundaries have been drawn. While the number of districts does not change greatly from year to year, many boundary lines are redrawn in any single year.

For example, rulings of the U. S. Supreme Court have led to the redrawing of thousands of attendance boundaries of school districts. The court's decision in 1954 in *Brown* v. *Board of Education of Topeka, Kansas,* held that the doctrine of "separate but equal" school systems for blacks and whites was unconstitutional because separate educational institutions were inherently unequal. Subsequent rulings broadened the decision to mean that no racially segregated school system is allowed, whether it occurs by design or inadvertently as the result of *de facto* segregation in housing. Boundaries of school attendance zones have had to be adjusted to ensure that schools will be racially integrated.

Electoral districts within states and cities

227

Fig. 7.1 Plans for Congressional redistricting in the State of Washington. The plans of the House and the Senate, controlled by the Republicans and the Democrats, respectively, were influenced by the desire of each party to maximize its influence. The court approved the plan of Richard Morrill, a geographer. Among the criteria Morrill employed to delimit boundaries were (1) to make the districts as compact as possible while ensuring that they had roughly equal numbers of people (2) and to adhere, where possible, to existing county lines. (From *Geography and Politics in America* by Stanley D. Brunn: Fig. 6.10 "Congressional Redistricting Plans for Washington" [after Richard Morrill], p. 159. Copyright © 1974 by Stanley D. Brunn. Reprinted by permission of Harper & Row, Publishers, Inc.)

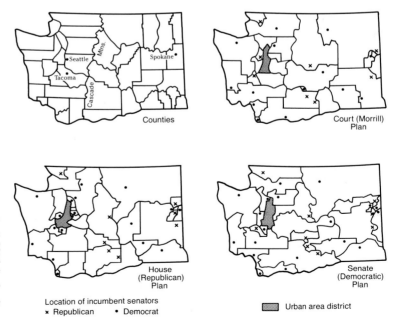

often need to be adjusted so that they contain roughly equal numbers of voters. Reapportionment is occasioned by shifts in population, as areas gain or lose people. *Gerrymandering* is the term used to describe the division of an area into voting districts in such a way as to give one political party an unfair advantage in elections, to fragment voting blocks, or to achieve other nondemocratic objectives (Figure 7.1). However, the drawing of boundaries is tremendously complicated, and it is not always clear what is "unfair."

The way boundaries are delimited depends on the desired ends to be achieved. The power of a group of people may be maximized, minimized, or effectively diluted by the way the lines are drawn. Assume that X and Y represent two groups with different policy preferences. In Figure 7.2a, the X's are concentrated in one district and have one representative out of four; in Figure 7.2b, the X's are dispersed, are not a majority in any district, and stand the chance of not electing any representative at all. The power of the X's is maximized in Figure 7.2c, where they control two of the four districts.

Boundary drawing is never easy, particularly when political groups want to maximize their representation. Furthermore, the boundaries we may want for one set of districts may *not* be those that we want for another. For example,

228

sewage districts must take natural drainage features into account, while police districts may be based on the distribution of population or the number of miles of street to be patrolled, and school districts must take into account the numbers of school-aged children and the capacities of individual schools.

Political boundary making, ideally, involves the practical application of geographic principles of regionalization, and geographers have been professionally involved in resolving delimitation problems. They were important contributing members of President Woodrow Wilson's staff after World War I, giving the American representatives at the Paris Peace Conference advice on boundary making in Eastern Europe, where new nations were created from the defeated Austro-Hungarian Empire and from the western reaches of the former Russian Empire. A geographer was appointed Special Master by the federal district court in Seattle, Washington, in the early 1970s to reapportion congressional and legislative districts in the State of Washington when the two political parties in the state legislature were unable to agree upon new district boundaries. These are just two examples of the real-life application of the skills of political and regional geographers.

Regional Organization: Political Fragmentation in the United States

The United States is subdivided into great numbers of political administrative units whose areas of control and influence are spatially limited. The 50 states are partitioned into more than 3000 counties ("towns" in New England and "parishes" in Louisiana), which in their turn contain some 21,000 townships, each of which has a still lower level of spatial

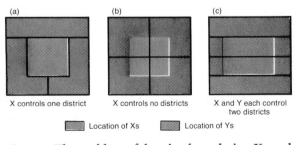

(a) (b) (c)

X controls one district   X controls no districts   X and Y each control two districts

Location of Xs   Location of Ys

Fig. 7.2 The problem of drawing boundaries. Xs and Ys might represent Republicans and Democrats, urban and rural voters, blacks and whites, and so on. (Redrawn by permission of the author from Edward F. Bergman, *Modern Political Geography*, Wm. C. Brown Co., Dubuque, Iowa, 1975.)

influence and governing power. This political fragmentation is further increased by the existence, within both urban and rural areas, of nearly innumerable special-purpose districts whose boundaries rarely coincide with the standard major and minor civil divisions of the nation, or even with each other. Each district represents a form of political allocation of territory to achieve a specific aim of local need or legislative intent.

Most Americans live in large and small cities. These, too, are subdivided—not only into wards or precincts for voting purposes, but also into special districts for such functions as fire and police protection, water and electricity supply, education, recreation, and sanitation. These districts almost never coincide with one another, and the larger the urban area, the greater the proliferation of small, special-purpose governing and taxing units.

Our largest recognized urban areas are the some 250 Standard Metropolitan Statistical Areas (SMSAs), which usually extend over several municipalities (the "central city" and its suburbs) and, often, several counties. Figure 7.3 shows one such extended "city." Some, such as the New York metropolitan area, even straddle state lines. In the New York–New Jersey SMSA, there are over 300 municipal governments alone; counting all of the general- and special-purpose legal units, the Chicago metropolitan area has over 1000 governments.

The proliferation of governmental units in small areas creates a complex mosaic of controls that often hinders achievement of the very aim behind their creation. The theory is that if control is kept at the local level, the citizens are assured a voice in government. Yet even the best-informed citizens cannot follow the workings of all of the many boards and agencies that are supposed to represent them. They may be able to keep up with the local transit and planning commissions, but have to neglect the city council, the board of health, and the school board, to name only a few. Thus

true understanding and extensive participation are impossible, one reason for the low turnout of voters in most local elections.

The existence of such a great number of districts in metropolitan areas causes inefficiency in public services and hinders the orderly use of space. Zoning ordinances, for example, are determined by each municipality. They are intended to allow citizens to decide how land is to be used and thus are a clear example of the effect of political decisions on the division and development of space. Zoning policies dictate where industries will be located, and whether they will be light or heavy industries; the sites of parks and other recreational areas; the location of business districts; what kind of housing is allowed, and where it can be built. Unfortunately, in large urban areas, the efforts of one community may be hindered by the practices of neighboring communities. Thus land zoned for an industrial park or perhaps a slaughterhouse in one city may abut on land zoned for single-family residences in an adjoining municipality. Each district pursues its own interests, which may not coincide with those of the larger region.

Inefficiency and duplication of effort characterize not just zoning but many of the services provided by local governments. The efforts of one community to avert air and water pollution may be, and often are, counteracted by the rules of other towns in the region, although state and national environmental protection standards are now reducing such potential conflicts. The provision of health care facilities, electricity and water, transportation, and recreational space affects the whole region and, many think, should be under the control of a single metropolitan government.

Such consolidation of control has long been proposed, but its introduction is prohibited by law in some states and in all instances involves the surrender of jealously protected local authority and autonomy to a larger unit. In the form of relatively complete "metro" govern-

229

230

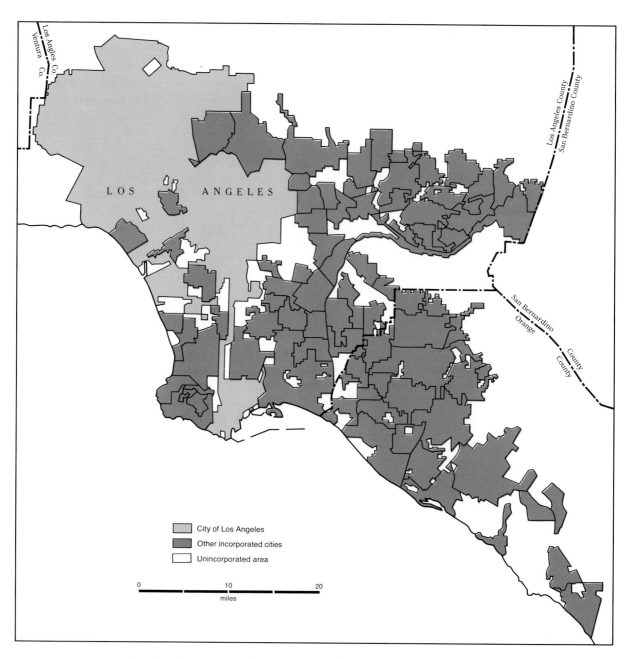

Los Angeles Co.
Ventura Co.

L O S   A N G E L E S

Los Angeles County
San Bernardino County

San Bernardino
County

Orange
County

City of Los Angeles

Other incorporated cities

Unincorporated area

0   10   20
miles

Fig. 7.3 The complex intermixture of incorporated and unincorporated areas in the Los Angeles metropolitan area hinders effective areawide planning.

ment, Dade County (Miami), Florida, and To-ronto, Canada, have had successful experiences

ences in region-wide administration. Figure 7.4 shows one proposal to reorganize the states of

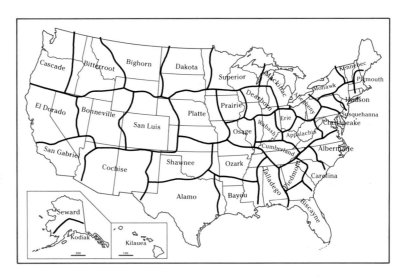

Fig. 7.4 **Pearcy's proposed 38-state United States.** Named after prominent physical or cultural features, the new states, where possible, have at their centers a major metropolitan area that serves as their focus. The new states are designed to reflect the economic and cultural realities of contemporary America. Relief, compactness of shape, and the pattern of transport routes replace the surveyor's lines and river courses of present state boundaries. (Copyright © 1973 by Plycon Press, P.O. Box 220, Redondo Beach, CA 90277, and G. Etzel Pearcy.)

the United States with metro-centered regions.

Problems of local governmental control are not unique to the United States. In every metropolitan area decisions must be made about what services are to be provided, how and where they will be provided, and who is to do the providing. In most countries, however, these decisions are not left to the smallest of local units. Central authorities have the power to decide when reorganization of the political system is necessary and to redraw the boundaries of regions.

In Great Britain, a royal commission was appointed in 1974 to study the problem of the relationship of the over 1000 local districts that existed. Each of these was directly responsible to the central government. The commission redrew county lines and gave control over certain functions—such as education, planning, and transportation—to the counties. Within the counties, district councils were established that were responsible for providing other services, including housing and garbage collection. Believing that certain functions demanded a higher level of organization, the commission created a limited number of regional authorities with control over health services and water supply. The commission's

plan attempted to strike a balance between centralization and decentralization of responsibility and control, based on the recognition that some functions should be controlled and provided locally and others areawide.

231

The Nation-State

Although local control of government in the United States creates areal differences in the behavior of citizens, the most profound contrasts in culture tend to occur between rather than within nations. The border between the United States and Canada has more impact both on peoples' lives and on the cultural landscape than do state boundaries in the United States. Similarly, people separated by the Mexican–United States border lead vastly different lives. San Diego resembles Los Angeles more than it does Tijuana, Mexico, although Tijuana is also a large city, and a mere 15 miles (25 km) away. Most people work in their own country, share a common language, and even telephone within their own country more often than to other countries. (Figures 7.5 and 7.6).

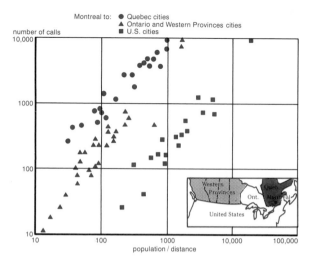

Fig. 7.5 Telephone calls from Montreal to selected cities. The graph depicts the number of calls placed from Montreal to cities in Quebec, in Ontario and the western provinces, and in the United States over a period of 10 days. United States cities are telephoned less frequently than would be expected if distance alone were the controlling factor. The international boundary inhibits interaction. (Reprinted with permission from J. Ross McKay, "The Interactance Hypothesis and Boundaries in Canada," *Canadian Geographer*, volume 11, 1958, p. 5.)

232

THE IDEA OF THE NATION-STATE

One of the most significant elements in cultural geography is the nearly complete division of the earth's land surface into separate national units as shown on the world political map inside the front cover. Even Antarctica is subject to the rival territorial claims of seven nations, although these claims, with their implication of sovereign control, have not been pressed because of the Antarctic Treaty of 1959 (Figure 7.7). A second element is that this division into national units is relatively recent. Although nation-states have existed since the days of early Egypt and Mesopotamia, only in the last century has the world been almost completely divided into independent governing entities. Now people everywhere accept the idea of the state, and its claim to sovereignty within its borders, as normal.

The idea of the modern state was developed by political philosophers in the 18th century. The concept that people owe allegiance to a country—to a state and the people it represents—rather than to its leader, such as a

Fig. 7.6 The discontinuous railway network between Canada and the United States. Note how the political boundary reduces interaction. Many branch lines approach the border, but only eight cross it. In fact, for over 300 miles (480 km), no railway bridges the boundary line.

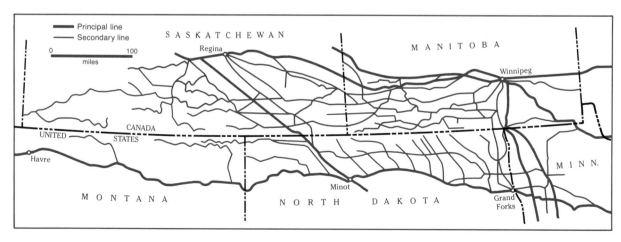

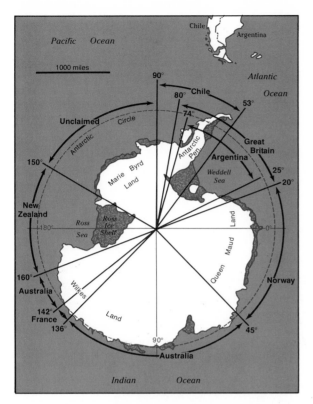

Fig. 7.7 Territorial claims upon Antarctica. Although seven nations claim sovereignty over portions of Antarctica, with three of the claims overlapping, the continent has no permanent inhabitants. In 1959, 19 governments, including the 7 claimants, signed the Antarctic Treaty, which outlaws military activity from the continent and designates it as a theater of scientific research. The discovery of mineral deposits and offshore food and fuel resources may test the strength of the treaty.

king or a feudal lord, is called *nationalism*. It coincided in France with the French Revolution and spread over Western Europe, to England, Spain, and Germany. The term *nation* implies a feeling of loyalty and attachment to a country and a willingness to abide by its rules. Many of the countries that were created in Europe after World War I resulted from an attempt to redraw boundaries that coincided with such feelings.

Many states are the result of European ex-

pansion during the 17th, 18th, and 19th centuries, when much of Africa, Asia, and the Americas was divided into colonies. Usually these colonial claims were given fixed and described boundaries where earlier none had been formally defined. Of course, precolonial native populations had relatively fixed home areas of control claimed by themselves (and recognized by neighboring groups), within which there was recognized dominance and border defense and from which there were, perhaps, raids of plunder or conquest of neighboring "foreign" territories. Beyond understood tribal territories, great empires arose, again with recognized outer limits of influence or control: Mogul and Chinese; Benin and Zulu; Inca and Aztec. Upon them where they still existed, and upon the less formally organized spatial patterns of effective tribal control, European colonizers imposed their arbitrary new administrative divisions of the land. In fact, tribes that had little in common were joined in the same colony (Figure 7.8). The new divisions, therefore, were not usually based on meaningful cultural or physical lines. Instead, the boundaries simply represented the limits of the colonizing empire's power.

As these former colonies have gained political independence, they have retained the idea of the state. They have generally accepted—in the case of Africa by conscious decision to avoid precolonial territorial or ethnic claims that could lead to war—the borders established by their former European rulers. The problem that many of the new countries face is to develop feelings of nationalism and loyalty to the state among their arbitrarily associated citizens. Zaire, the former Belgian Congo, contains more than 250 frequently antagonistic tribes. Only if past tribal animosities can be converted into an overriding spirit of national cohesion will countries such as Zaire truly be nation-states.

233

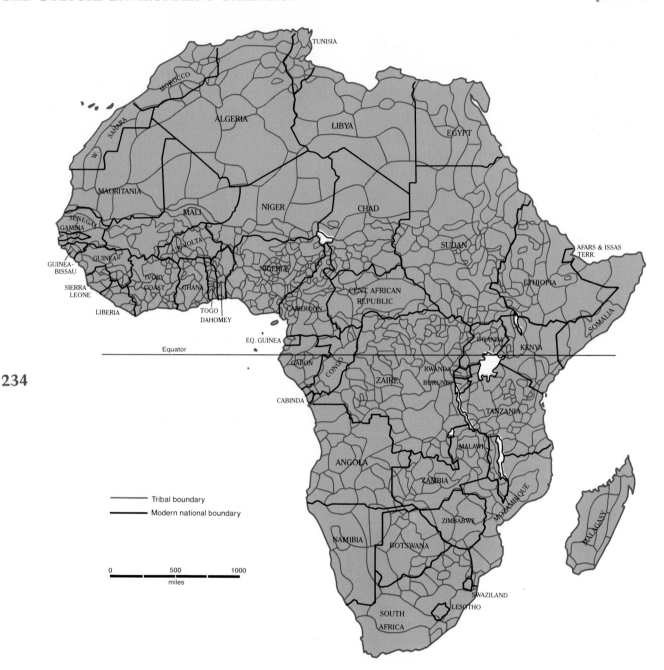

234

Fig. 7.8 The discrepancies between tribal and national boundaries in Africa. Tribal boundaries were ignored by European colonial powers; the result has been significant ethnic diversity in nearly all African countries. (Fig. 5, p. 248, in *World Regional Geography: A Question of Place* by Paul Ward English, with James Andrew Miller. Copyright © 1977 by Harper & Row, Publishers, Inc. Redrawn by permission.)

STATE COHESIVENESS

What are those forces that make a state cohesive—that enable it to function and give it strength? We have already mentioned one of them: a feeling of collective distinction from all other peoples and lands, that is, an identification with the state and an acceptance of national goals that is called *nationalism*. A sense of unity binding the people of a state together is necessary to overcome the divisive forces present in most societies.

The concepts of *nation* and *state* are not always synonymous. The nation—in the sense of an ethnically distinct population with its own traditions, language, literature, and sense of identity—may exist within or extend beyond the administrative bounds of the political state. The constitution of Somalia—a state in East Africa—recognizes a Somali population living outside of the country's territorial limits in Ethiopia and Kenya; it commits the state to the reunification of the politically divided "nation." Here, obviously, nationality as recognized by those who feel themselves a separate people exceeds the areal limits of the state. Conversely, in the constitutional structure of the Soviet Union, one division of the legislative branch of the government is termed the *Soviet of Nationalities*. It is composed of stated numbers of representatives from civil divisions of the Soviet Union populated by groups of officially recognized "nations": Ukrainians, Kazakhs, Estonians, and others. In this instance, the concept of nationality is territorially less than, and subservient to, the extent and control of the state.

National Symbols. *Iconography* is a word that has been used to denote the study of symbols that bind a people together. National anthems and other patriotic songs, flags, national flowers and animals, colors, and rituals are all developed as symbols of a state in order to attract allegiance (Figure 7.9). They ensure that all citizens, no matter how diverse the population may be, will have at least these symbols in common. They impart a sense of belonging to a political entity called, for example, Japan or Canada. In some countries, certain documents, such as the Magna Charta or the Declaration of Independence, serve the same purpose. Royalty may fill the need: in Sweden, Japan, and Great Britain, the monarchy functions as the symbolic focus of people's allegiance, while real power lies elsewhere.

Fig. 7.9 The ritual of the pledge of allegiance is just one way in which schools seek to instill a sense of nationalism in students. (United Press International Photo.)

Nationalism and the Olympics

The following excerpt is taken from an article written at the close of the 1976 summer Olympics:

> As the curtain drops on the Games of the XXI Olympiad, this image emerges: Remove nationalism from the Olympics and there can be no Olympics. Let the purists moralize and the editorial writers agonize, but the sad fact is—for better or for worse—nationalism is what the Olympics are all about.
>
> The truth of this was written in every quivering rafter of the Forum on Tuesday night when the United States team won the basketball gold medal by defeating Yugoslavia. American flags waved against a backdrop of faces screaming "U-S-A, U-S-A" and the scene was one of rampant nationalism.
>
> There were voices and flags for the Yugoslavs, too, and earlier for the Russians and the Canadians. Had these been teams without a country there would have been neither cheers nor flags—because there would have been no people.
>
> The people were there to cheer not individuals and individual performances but to vent the strong feelings of patriotism and to see their national team win. . . .
>
> It is fashionable and chic and one-worldish to expound the joys of competition for its own sake, but the human spirit needs something more to cling to.
>
> Eliminate the majesty of the opening Olympic ceremony, the parade of nations, the flags, the ethnic costumes. Let the athletes march only as individuals, under one generic flag with only an Olympic anthem to sing. Should these things happen, the Olympic movement would soon die of financial malnutrition.

236

National legends and heroes perform the same function. Immigrant children in the United States, whether from China or Central Europe, have been taught about George Washington and the cherry tree and Benjamin Franklin and his kite. In the U.S.S.R. Marx and Lenin are revered. Symbols such as these are significant insofar as ideologies and beliefs are an important aspect of culture. When a culture is very heterogeneous, composed of people with different customs, religions, and languages, belief in the nation can help weld them together.

Institutions That Promote Nationalism. Schools, particularly elementary schools, are among the most important of the institutions that promote national feelings. Children learn the history of their own country and relatively little about other countries. Schools are expected to inculcate the society's goals, values, and traditions; allegiance to the state is accepted as the norm. On the whole, schools teach youngsters to identify with their country rather than with the world or with humanity as a whole.

The concept of the state and the sense of nationalism that overrides small-group consciousness have suffered erosion and have been subject to compromise in all parts of the world in recent years. Since the end of World War II, ethnic-group awareness and identification with region rather than with the state have created currents of unrest and division, disturbing established national patterns of cohesion.

Language is the most obvious and emotive minority rallying point, serving as the immediate identification of groups outside of, and feeling somehow oppressed by, the remainder of the state to which they are attached. That the regional minorities waging language wars, particularly in Europe, are usually located outside of the economic heartlands of their states adds the frustrations of underdevelopment to their sense of diminished identity and status. Demands for linguistic recognition may also be seen as regional rejections of strongly centralized governments of both Western and Eastern Europe; decentralized Switzerland, with four official languages, has no linguistic conflict.

Dialect and language regions of France. (Reproduced by permission of *The Economist*, London, Sept. 16, 1978.)

237

Highly centralized France, Spain, Britain, Belgium, Yugoslavia, and the Soviet Union are experiencing such language revolts. Until 1970, the spoken regional languages and dialects of France were ignored and denied recognition by the state. Even though the ban on teaching regional tongues was dropped in that year, the Association for the Defense and Promotion of the Languages of France has protested to the United Nations about the imposition of the French language throughout the country; Bretons have taken more direct and violent paths to assertion of their separate identity.

Post-Franco Spain has relaxed the earlier total rejection of Basque and Catalan as regional languages, has granted modest local autonomy to those who speak them, and has given state support to instruction in them. In Britain, Welsh and Gaelic have been symbols of regional rejections of centralized English domination. Parliamentary debates concerning greater regional autonomy ("devolution") in the United Kingdom have resulted in bilingual road and informational signs in Wales and, beginning in 1982, there will be a publicly supported Welsh-language television channel. Belgium, which has been divided between Dutch- and French-speakers since its founding in 1830, has the most serious and potentially destructive language conflict. There, the majority (56%) Dutch-speaking Flemish have struggled constantly to establish their language as equal with the French of the minority Walloons. Impermanent legal compromises and the overturn of governments have marked the conflict in recent years. Seeking linguistic neutrality, the Belgian navy has adopted English as the language of command.

The nationality question based upon ethnic–linguistic differentiation is a continuing concern in parts of Eastern Europe as well, and particularly in the U.S.S.R. There, the Great Russian (Slavic) inheritors of a czarist empire encompassing more than 100 nationality groups face a maze of emotional and practical problems. Most of the 14 non-Russian Union Republics of the country had long histories of independence and cultural achievement prior to coming under Great Russian control. For many, pride in their history, their culture, and their language remains great, and attempts at russification are deeply resented. The Russian language, taught in primary and secondary schools throughout the U.S.S.R. as a means of unification, is rarely used by non-Russians who remain in their own homelands. Adherence to regional, rather than state, loyalties was apparent in the 1978 mass demonstrations in Tbilisi, the capital of the Georgian Republic, against the omission in a proposed new constitution of a statement that Georgian was the official language of the region. The government acquiesced and granted the same recognition to the Armenian language in Soviet Armenia. The problem of maintaining unquestioned Great Russian control of the state is apt to increase rather than diminish. Because of continuing high birth rates among non-Slavic—particularly Central Asian—peoples, the proportion of Russians, more than 53% of the total population in the census of 1970, fell below that figure in the 1979 census returns.

The concept of State—the adherence to a national identity irrespective of regional or ethnic differentiation—has been a periodically enforced organization of peoples. That its validity is currently so widely challenged is a reflection of the durability of the cultural geographic variables explored in previous chapters.

238

Schools promote nationalism in another way, by teaching students a common language.

The importance of language as an expression of culture was discussed in Chapter 6. Linguistic

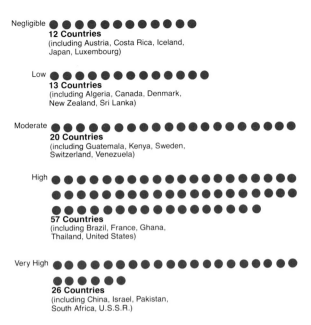

Negligible
12 Countries
(including Austria, Costa Rica, Iceland,
Japan, Luxembourg)

Low
13 Countries
(including Algeria, Canada, Denmark,
New Zealand, Sri Lanka)

Moderate
20 Countries
(including Guatemala, Kenya, Sweden,
Switzerland, Venezuela)

High

57 Countries
(including Brazil, France, Ghana,
Thailand, United States)

Very High

26 Countries
(including China, Israel, Pakistan,
South Africa, U.S.S.R.)

Fig. 7.10 National commitment to military expenditures. Each dot represents one of the 128 nations
that were classified according to their governmental
expenditures (per capita) for military purposes in
contrast to spending on health and education. For
example, Iceland in the "negligible" category spends
considerably more on health and education than it
does for military purposes. Israel in the "very high"
category spends more for military purposes than for
health and education.

diversity tends to hamper national cohesiveness; linguistic homogeneity fosters it. By authorizing the use of a single language for instruction in the schools, a state facilitates
communication and promotes a national identity.

Other institutions that promote nationalism
are the armed forces, political parties, and,
sometimes, a state church. The military organization fulfills a primary state goal: the provision of security, both internal and external. A
high percentage of most states' budgets is spent
to secure such protection (Figure 7.10). The
armed forces are of necessity taught to identify
with the state. They see themselves as protecting the state's welfare from what are perceived

to be its enemies. In some countries the armed
forces are used as much to guarantee internal
stability as to protect the country from outside
danger.

Political parties are a less obvious but still a
significant force for promoting cohesion in a
state. In one-party nations, the single official
political party serves as the apparent channel
through which popular support of governmental programs is expressed. In multiparty states,
the political party is the instrument though
which separate and perhaps quite divergent
economic and social interests are presented for
mass support. In their appeals for adherents to
their goals and leadership, political parties
strive to promote national cohesion under
their banners. Whether or not they are successful in advancing their separate programs, political parties serve to increase participation in
the political process and to arouse popular interest in the affairs of government.

In some countries the religion of the majority of the people may be designated a state
church. When this occurs, the church becomes
a force for cohesion, helping to unify the population. This is true of the Roman Catholic
church in the Republic of Ireland, Islam in Pakistan, and Judaism in Israel. In countries such
as these, the religion and the church are so
identified with the state that belief in one is
transferred to allegiance to the other.

A final institution that brings about cohesion may be the state itself, through its control
of the channels of communication. These include radio and television, newspapers, and
magazines. Control may take the form of ownership by the state, licensing, or censorship.
What is communicated—whether it is outright
propaganda or the teaching of the alphabet—is
perhaps not as important as the mere fact that
there is communication. Citizens, though they
might be physically remote, are not isolated if
they are within reach of the communications
media.

The schools, the armed forces, the political
parties, the church, and the media are institu-

239

tions that teach people what it is like to be members of a nation-state. These institutions operate primarily on the level of the ideological system. By themselves, they are not enough to give cohesion and thus strength to a state.

Organization. A second necessity for cohesion —in addition to the feeling of nationalism— is effective organization of the state. Can it provide security from external aggression and internal conflict? Are its resources distributed and allocated in such a way as to promote the economic welfare of its citizens? Are there institutions that encourage consultation and the peaceful settlement of disputes? How firmly established are the rule of law and the power of the courts? Is the system of decision-making responsive to the people's needs?

Let us look at just one set of these questions, that dealing with the state and the economy. Today as never before, the promotion of economic well-being is seen as a task of government. People expect the government to act in such a way as to enhance their welfare. Although the degree to which the state is involved in the economy varies from country to country—from the public ownership of means of production to private ownership—in no country is there an absence of regulations. Tariff barriers restrict trade, protecting home industries from foreign competition; direct aid is given to certain industries in many countries; some nations may try to attract foreign investment, while others strictly control or discourage it. Governments extend subsidies to those engaged in certain kinds of agriculture. They may sponsor programs aimed at increasing agricultural yields or at supporting basic industry.

The best efforts of a government, however, are conditioned by certain elements over which it has little control. One of these elements is the physical makeup of the territory of the nation. The area that the state occupies may be large, as is true of China, or small, as is true of Liechtenstein. The larger in area the state is, the greater is the chance that it will have resources such as fertile soil and minerals from which it can benefit. On the other hand, a very large country may have vast areas that are inaccessible, sparsely populated, and hard to govern. Thus size alone is not critical in determining a country's stability and strength. Potentially important is a nation's relative location. If it is located on major trade routes, it will benefit materially. It will also be exposed to the diffusion of new ideas and new ways of doing things.

A country's shape may affect patterns of organization. The most efficient form would be a circle, assuming no major topographic barriers, with the capital located in the center. In such a country, all places could be reached from the center in a minimum amount of time with the least expenditure for roads, railway lines, and so on. It would also have the shortest possible borders to defend. Uruguay and Libya have roughly circular shapes.

The least efficient shape is represented by countries such as Norway and Chile, which are long and narrow. In such states, the parts of the country far from the capital are likely to be isolated because great expenditures are required to link them to the core. These countries are also likely to encompass more diversity of climate, resources, and peoples than compact states, perhaps to the detriment of national cohesion or, perhaps, to the promotion of economic strength.

A third class of shapes is called *fragmented.* This class includes countries composed entirely of islands (e.g., the Philippines and Indonesia), countries that are partly on islands and partly on the mainland (Italy, Malaysia), and those chiefly on the mainland but whose territory is separated by another state (the United States). Pakistan was a fragmented nation until 1971, when the eastern part of the country broke away from the west and declared itself the independent state of Bangladesh. Examples

240

Brazil exemplifies a state's determination to achieve full development by planned programs of capital city relocation, highway and railroad construction, and subsidized shifts of population and economic activity to underutilized sections of the nation.

Development of Brazil's Interior

Larger than the conterminous United States (excluding Alaska and Hawaii), with the eighth largest population in the world and with vast resources, the country is unevenly settled. Most Brazilians live along the eastern seacoast, particularly in the urban complexes of Rio de Janiero and São Paulo, a settlement pattern reminiscent of the concentration of population along the Atlantic seaboard at an earlier time in the United States. The eastern seacoast is the core of Brazil, the heart of economic and cultural activities. The isolation of the rest of the country from the developed coastal core region left the bulk of the nation underdeveloped and its vast mineral and agricultural potential underutilized.

241

In the 1950s, government leaders decided to encourage settlement in and the economic development of the extensive interior of the country. The first step was to move the capital from Rio de Janiero to a new site over 500 miles (800 km) away, a distance comparable to that from Washington, D.C., to Indianapolis. A highway was constructed, the site was cleared, and in less than five years, Brasília had been built. The capital is now a modern metropolis with a population over 1 million.

A program of road building was also planned. The 1000-mile (1600-km) highway from Brasília to Belém represented the first overland connection between the central part of the country and the north. The Trans-Amazon Highway, about 3500 miles (5600 km) long, connects the east coast with the western states. Just as the building of railroads opened up the interior of the United States to settlement, these highways were designed to promote development of the Brazilian interior.

Plans were made to attract landless peasants from the northeast coastal zone into the Amazon basin by giving them land and government aid. Projects got under way to mine bauxite, iron ore, manganese, and other minerals. The government encouraged businesses to locate in the interior by giving them tax relief on profits invested in the Amazon, by establishing a favorable loan program, and by declaring Manaus a duty-free zone. The Itaipú Dam, under construction in the south of Brazil, will be the largest hydroelectric dam in the world when completed, generating five times as much electricity as the Aswân Dam in Egypt.

Although the Brazilian experiment represents a significant, planned approach to the development of a country, it has not achieved all that it set out to. The soil of the Amazon basin loses its fertility rapidly once the rain forest canopy is cleared. The peasants attracted to the area were not experienced in farming under the environmental conditions of the new territories and were not given adequate technical assistance. The government has found it costly to maintain highways in good condition, and some are deteriorating. Government attention is also diverted by the same kinds of problems that other countries face, such as the high cost of energy, a high rate of population growth, and a large body of poor people. Nevertheless, the Brazilian attempt shows the importance of political integration and how it is related to transportation, communication, and patterns of settlement.

242

of differing national shapes are shown on Figure 7.11.

The location of the capital city, the seat of central authority, sometimes becomes a matter of concern. All other things being equal, a capital located in the center of the nation provides equal access to the government, facilitates communication to and from the capital, and enables the government to exert its authority easily. Many capital cities, like Washington, D.C., were centrally located when they were designated as capitals but lost their centrality as the state expanded. Some capital cities have been relocated, at least in part to achieve the presumed advantages of centrality: in the U.S.S.R., the capital was moved to Moscow from St. Petersburg (now Leningrad); in Turkey, to Ankara from Istanbul; in Brazil, to Brasília from Rio de Janeiro.

Centrality was sought in a number of states within the United States where the capital city is now at or near the geographic center. In Pennsylvania, Ohio, Michigan, and Illinois, the capitals were moved from previous sites to more central locations. Alaskans have approved a proposal to move the capital from Juneau to Willow, a small town near Anchorage (Figure 7.12).

Transportation and Communication. A state's transportation network fosters political integration. When transportation brings about in-

Fig. 7.11 Shapes of nations. The national territories are not drawn to scale.

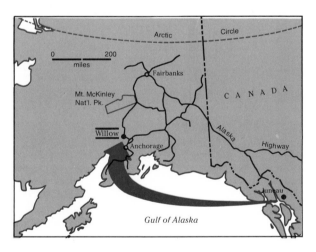

Fig. 7.12 The proposed shift of the Alaskan capital. A referendum approved by voters in 1974 called for the capital to be moved from Juneau, which is 600 miles (1000 km) away from Anchorage, the center of population and commerce. The voters chose Willow, a small town about 70 miles (110 km) north of Anchorage, as the site of the new political center.

teraction between areas, the people may be tied economically and socially to each other. The role of a transportation network in uniting a country has been recognized since ancient times. The saying that all roads lead to Rome had its origin in the impressive system of roads that linked Rome to the rest of the empire. Centuries later a similar network was built in France, linking Paris to the various departments of the country. Often the capital city is better connected to other cities than outlying cities are to one another. In France, for example, it can take less time to travel from one city to another by way of Paris than by direct route.

Roads and railroads have played a historically significant role in promoting political integration. In the United States and Canada they not only opened up new areas for settlement but increased interaction between rural and urban areas. Comparable benefits of economic, social, and political interaction and territorial development are anticipated by the Russians upon the completion through Siberia of the Baikal–Amur Mainline Railroad (BAM) by the middle 1980s. The $9–$15 billion cost of the 2000-mile (3200-km) project (Figure 7.13) is expected to yield great returns by tapping some of the world's richest deposits of minerals and timber, by opening up vast new areas of settlement, and by providing a militarily secure link between the western and the extreme far-eastern sections of the U.S.S.R.

Transportation and communication are encouraged within a state and are curtailed or at least controlled between states. Mechanisms of control include restrictions on trade through tariffs or embargoes, legal barriers to immigration and emigration, and limitations on travel through passports and visa requirements. A pointed illustration is the Australian railroad system (Figure 7.14). Until their federation in 1901, the states of Australia were independent of one another. As competitors for British markets, they had no desire to foster mutual interaction. Different rail gauges rather than a single, standard gauge were used, so that people

243

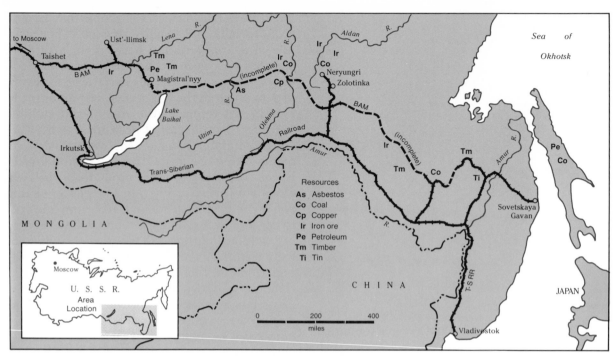

Fig. 7.13 The Baikal–Amur Mainline Railroad, under construction, is a vital link in the long-term development of Siberian resources undertaken by the Soviet Union.

and goods traveling between states had to change trains at the border.

Transportation systems play a major role in a state's economic development. In general, the more economically advanced a country is, the more extensive is its transport network. By fostering interdependence between regions, a transport network keeps regions from having to be self-sufficient. Each can specialize in the production of the goods and services for which it is best suited. Specialization results in an overall increase in productivity. The concept of comparative advantage, to be discussed in Chapter 9, helps to explain how this system works. At the same time, the higher the level of development, the more money there is to be spent on building transport routes. In other words, the two feed on one another. Towns and cities that are near an interstate highway route

and are connected with it by an interchange benefit, while those that are bypassed tend to suffer economically.

The Seas: Open or Closed?

We have seen that boundaries play an important role in defining those areas in which certain activities may or may not be allowed to take place. However, we have considered only boundaries on land. Since water covers about two-thirds of the earth's surface and the seas have been used by people since ancient times, it is not surprising that boundaries across water are also important.

A basic question of increasing international

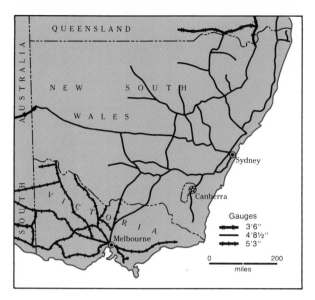

Fig. 7.14 Railway gauges in eastern Australia. Gauges of different widths were established to hinder interaction between the states of Australia, which until federation owed primary allegiance to Great Britain rather than to one another. (Fig. 7.13, "Railway networks . . ." [After Oxford Atlas for New Zealand, 1960] p. 266, in *Geography as Social Science* by Gordon J. Fielding. Copyright © 1974 by Gordon J. Fielding. Redrawn by permission of Harper & Row, Publishers, Inc., and Oxford University Press.)

concern involves the right of nations to use water and the resources it contains. The waters in a country, such as rivers and lakes, have traditionally been regarded as being within the sovereignty of that country. Oceans, however, are not within a single nation's borders. Should they then be open to all nations to use, or may a country claim sovereignty over them, closing them to use by other countries?

The question was debated in the 17th and 18th centuries in European nations whose economic well-being depended materially on the use of the seas. The result was a general agreement adhered to through the 19th century, that a coastal country could claim sovereignty over a continuous belt 3 or 4 nautical miles wide (a

nautical mile equals 1.15 statute miles or 1.85 km). The remainder of the ocean was open to use by all.

The period since World War II has seen many countries extend their claims to 6, 12, and even 200 nautical miles (11, 22, and 370 km) of offshore waters. The United States and several other countries claim exclusive fishing rights up to 200 miles offshore. Many countries, including the United States, claim mineral rights on the continental shelf, again up to 200 miles. Some African countries claim absolute sovereignty over both mineral and fishing rights for 100 miles (185 km) or more offshore.

International disputes have resulted from these extended claims (Figure 7.15). Britain first challenged Iceland's claim to exclusive fishing rights in a 50-mile (93-km) zone by sending warships to protect British fishing boats and then retaliated by establishing a 200-mile fishing limit around the British Isles. Unlicensed United States fishermen have been arrested for fishing within the 200-mile limit claimed by Peru, while the United States has protested the presence of Soviet and Japanese ships off its coasts. North Koreans captured the United States ship *Pueblo* on the basis that it was within that country's territorial waters. Arguments over sovereignty are becoming increasingly frequent as countries use the oceans more heavily as a source of food, minerals, and offshore oil and gas. As competition for those resources has become keener, the need for a less ambiguous definition of sovereignty has increased.

There is also a need for countries to agree on how the resources of the ocean will be used. Modern fishing techniques make it possible to overfish—that is, to take in fish faster than they can reproduce and consequently to decrease the number available in future years. Several species have become extinct, and it is more than likely that others, including some types of whales, will become extinct in the very near future.

245

Fig. 7.15 A U. S. Coast Guard cutter escorts a Spanish fishing vessel to port after seizing it for violation of the Fisheries Conservation and Management Act. In 1976, the United States extended its claim of jurisdiction over fishery resources from 12 to 200 nautical miles (22 to 370 km) from its shoreline. Congress was impelled by a need to conserve seafood stocks before they were depleted to the point of exhaustion. In 1973, foreign fleets, primarily Japanese and Soviet, harvested 7.9 billion pounds (3.6 billion kg) of seafood within American boundary waters, while United States fishermen landed only 1.4 billion pounds (630 million kg). By 1977, the foreign catch was reduced to 3.7 billion pounds (1.68 billion kg). (Courtesy of *National Fisherman* Magazine.)

International agreements to limit the amount of fish taken in a specified period of time help alleviate the danger of overfishing, but the political mechanisms for making and enforcing such agreements are inadequate. The Japanese have agreed, by treaty, to limit their taking of North Pacific salmon to waters west of 175° W longitude, but new techniques for open-sea capture of that species continue to endanger North American stocks. The International Whaling Commission, given the power to regulate that industry in the southern hemisphere since 1946, has regularly set the quotas on desirable species so high as to fail to guarantee their survival; indeed, the blue whale has been declared officially extinct. A 1978 thirteen-nation draft convention to conserve "Antarctic marine living resources," notably krill (a tiny shrimplike crustacean), has been reached, in this case before serious depletion of a potentially invaluable marine source of protein. As with many other such agreements, however, the treaty group is a self-selecting society whose arrangements do not have the force of international law.

Countries do not have to sign agreements if they do not want to, and nonsignatory countries can make useless the treaties that others observe. Sometimes shipowners transfer their vessels to foreign registry to avoid agreements binding upon their own country. The quotas set in the treaties may be so high as to permit overfishing anyway. In addition, such agreements are hard to police; individual fishermen may think that it is not in their own best interest to observe quotas that they have not agreed to. Finally, there is no international body, not even the International Court of Justice, that

has the power to enforce the recommendations it makes. Until such a body exists to make decisions that member countries must adhere to, and until suitable policing mechanisms are developed, there is little hope of avoiding overfishing.

In the years to come, competing claims to the mineral resources contained in the ocean floor may occasion more disputes than fishing does. Petroleum and natural gas are in increasing demand as energy supplies. Many countries claim sovereignty over the continental shelf where such resources are known to exist and are relatively accessible. However, who should have the right to exploit the resources in the ocean floor—internationally licensed companies, or countries contiguous to those waters? When natural gas was discovered under the North Sea, in waters not claimed by any country, the contiguous nations agreed to a parti-

Fig. 7.16 These landlocked nations are among those that advocate international control of deep seabed natural resources.

247

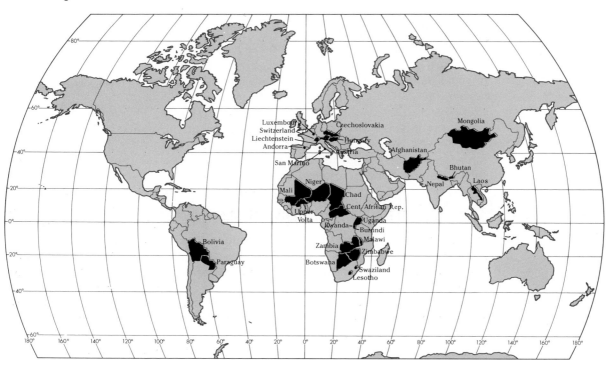

**The Law
of the Sea**

For most of human history, the two-thirds of the earth's surface covered by the oceans remained effectively outside individual national control or international jurisdiction. The seas were a common highway for those daring enough to venture upon them, an inexhaustible larder for fishermen, and a bottomless refuse pit for the muck of civilization. By the end of the 19th century, however, the claims of most coastal countries to a 3-mile sovereignty off their shores were generally recognized. Such sovereignty, though recognizing the rights of others to innocent passage, permitted national protection of coastal fisheries, allowed the enforcement of quarantine and customs regulations, and made claims of neutrality effective during other people's wars. The primary concern was with security and unrestricted commerce. No separately codified law of the sea existed, however, and none seemed needed until after World War I.

A League of Nations Conference for the Codification of International Law, convened in 1930, inconclusively discussed maritime legal matters and served to identify areas of concern that were to become increasingly pressing after World War II. Important among these was an emerging shift from interest in commerce and national security to a preoccupation with the resources of the seas, an interest fanned by the Truman Proclamation of 1945, which claimed American control over seabed and subsoil resources of the continental shelf contiguous to its coasts. Other nations, many claiming even broader areas of control, quickly followed suit. By the late 1970s nearly a third of all the oceans had been arbitrarily appropriated by some 60 coastal states, and seabed claims were being extended even further.

The rush to annex marine resources led to a realization that the established principles of international law needed to be buttressed by specific agreement upon a law of the sea. A United Nations conference, convened in 1958, resolved some issues but left others—the width of territorial waters and the creation of exclusive fishing zones, for example—unsettled. In 1967 an explosive new issue was injected into international debate on the control of maritime resources: the concept, advocated by the underdeveloped and the landlocked nations, that the seabed and the resources it contained should be recognized as the "common heritage of mankind."

Between 1950 and 1970 the world's annual fish catch quadrupled, and the regenerative abilities of many food species were being endangered; merchant shipping—particularly oil tanker tonnage—also quadrupled, and coast and marine damage from oil spills and other man-made pollution was threatening the self-cleansing capability of such enclosed waters as the Mediterranean and even that of the open ocean. Such exploitation and destruction of the marine environment was felt to be of concern to all peoples, not just to developed maritime nations. Of equal or greater interest was the vast mineral treasure in the form of manganese nodules strewn across the ocean bottoms. Potato-shaped crumbly accretions containing up to 27 separate elements, the nodules have 4 minerals of particular interest: nickel, copper, cobalt, and manganese. Their potential value in the late 1970s was conservatively estimated at more than $3 trillion. It is particularly this treasure that has captured the attention of nations that are technologically unable to exploit it or are denied access to it because they are landlocked.

Various sessions of the United Nations Conference on the Law of the Sea, first called in 1973, wrestled inconclusively with conflicting national claims, the "common heritage" doctrine, and the desires of private American corporations to extract the nodule resources that they had the technology to exploit. In the face of an often-expressed sense of urgency to codify a law of the sea meeting all nations' requirements, interests, and expectations, total international agreement seems distant and threatened by the unilateral *de facto* claims and controls increasingly imposed on the watery two-thirds of the globe.

tioning of the sea. Such a peaceful solution may prove to be the exception rather than the rule, and in any case, it begs the question whether such resources are available for use by all or only by those favored by an accident of location (Figure 7.16).

The United Nations Conference on the Law of the Sea, held periodically since 1973, represents an attempt by 156 nations to achieve consensus on how the oceans and their resources should be shared. The most difficult issue has been who should control the exploration and exploitation of the deep seabeds, which are rich in minerals such as nickel and copper, cobalt and manganese. Countries have aligned themselves in various ways on this question.

Industrialized countries possess the technology and the capital necessary to mine the deep seabeds and want to be able to do so through private investment. In contrast, developing countries argue that the resources belong to all nations and that exploration, mining, and the distribution of profits should be controlled by an international authority. The industrialized nations fear that the developing countries would by their numbers dominate such an authority.

Landlocked countries and those with brief shorelines want transit rights over nearby coastal states and special privileges in the waters they control. Many of the coastal states, however, claim exclusive rights to seabed minerals. Failure to reach agreement on this issue will result in the intensification of national rivalries.

249

Aspects of International Political Systems

One of the most significant trends in recent years has been the rapid increase in the number of nation-states. At the beginning of World War II, in 1939, there were about 70 independent countries. By 1970 the number had more than doubled, and at the end of 1979, 151 independent states had membership in the United Nations.

THE DISSOLUTION OF EMPIRES

Most of the newly formed states were once colonies in one or another of the European empires. Expanding political consciousness in Asia, Africa, and South America brought a demand for change and, subsequently, an end to the colonial era. Faced with strong nationalistic feelings on the part of the inhabitants and sometimes recognizing the legitimacy of those feelings, convinced at other times that the colonies were no longer assets either economically or politically, or responding to the pressures of world opinion, the European countries relinquished their control.

From the former British Empire there have come, just in the period since World War II, the independent countries of India, Pakistan, Bangladesh, Malaysia, and Singapore in Asia; Ghana, Nigeria, Kenya, Uganda, Tanzania, Malawi, Botswana, Zimbabwe, and Zambia in

Africa. Even this extensive list is not complete. These and other now independent but former British colonies are shown in Figure 7.17. A similar process has occurred with most of the former colonies of France, the Netherlands, Spain, and Portugal.

Empires have been created—and have fallen—throughout history: the Roman, the Ottoman, and the Austro-Hungarian, to name only three. Each was, for a while, very strong; each eventually dissolved. What is there, then, that is significant about the dissolution of empires in this century?

The answer lies in the speed of the process, in the areal extent over which it has occurred, and in what has happened to the former colonies. The modern European empires, compris-

Fig. 7.17 Some 43 former possessions, colonies, and protectorates of Great Britain (black areas) have achieved independent statehood since 1945. This map show how large and far-flung was the territory formerly controlled by a single small European nation.

250

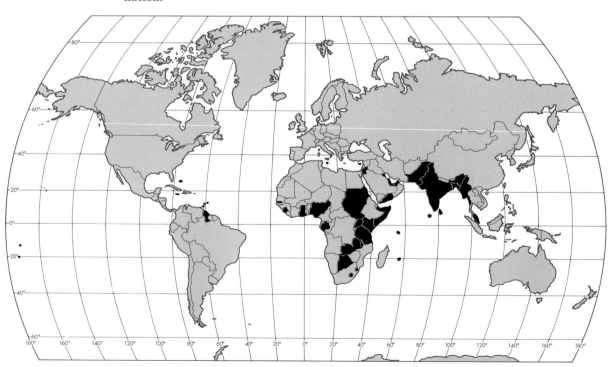

Fig. 7.18 The World Health Organization is the agency of the United Nations that directs international health work. Because human and animal diseases ignore international boundaries, international agencies such as W.H.O. provide a level of control and coordination above that of the nation-state. W.H.O. workers help to eradicate certain diseases, improve nutrition and living conditions, and aid developing countries in strengthening their health services. Pictured is a W.H.O. team in a Burmese village. (United Nations.)

ing vast parts of the globe, crumbled almost simultaneously in a few years' span after World War II. The colonies of those empires have now become independent states, not a consequence when previous empires broke apart before the concept of nationalism became dominant. Nationalism is now a truly universal phenomenon; there will soon be very few people left who are not citizens of independent states. It is in this that the uniqueness of the present period lies.

In many ways countries now are weaker than ever before. Many are economically weak, others are politically unstable, and some are both. Strategically, no country is safe from military attack, for technology now enables us to shoot weapons halfway around the world. Some people believe that no national security is possible in the atomic age. The recognition that a country cannot by itself guarantee either its prosperity or its own security has led to increased cooperation among states. In a sense, these cooperative ventures are replacing the empires of yesterday. They are proliferating quickly, and they involve countries everywhere.

NEW INTERNATIONAL SYSTEMS

Worldwide. The United Nations (U.N.) is the only organization that tries to be universal, and even it is not all-inclusive. The United Nations is the most ambitious attempt ever undertaken to bring together the world's nations in international assembly and to promote world peace. It is stronger and more representative than its predecessor, the League of Nations. It provides a forum where countries may meet to discuss international problems and a mechanism, admittedly weak but still significant, for forestalling disputes or, when necessary, for ending wars. The United Nations also sponsors a number of agencies aimed at fostering international cooperation with respect to specific goals (Figure 7.18). Among these are the World Health Organization (W.H.O.), the Food and Agriculture Organization (F.A.O.), and the United Nations Educational, Scientific, and Cultural Organization (U.N.E.S.C.O.).

The United Nations would be stronger if countries were willing to surrender enough of their sovereignty to make it a true world government—specifically, their sovereignty with respect to the ability to wage war. The weakness of the United Nations stems partly from its legal inability to make and enforce a world law. There is little world law, and there is no permanent world police force. Although there is recognized international law adjudicated by the International Court of Justice, rulings by this body are sought only by nations agreeing beforehand to abide by its arbitration. The United Nations has no authority over the military forces of individual countries, it cannot take prompt and effective action to counter aggression, and thus there is no guarantee of peace.

Countries have shown themselves to be more willing to surrender some of their sovereignty to participate in smaller multinational systems. These systems may be economic, military, or political, and many have been formed since 1945.

Regional Alliances. Cooperation in the economic sphere seems to come more easily to states than does political or military coopera-

252

tion. Among the most powerful and far-reaching of the economic alliances are those that have evolved in Europe, particularly the Common Market. It had several forerunners.

Shortly after the end of World War II, the Benelux countries (Belgium, the Netherlands, and Luxembourg) formed an economic union to create a common set of tariffs and to eliminate import licenses and quotas. At about the same time were formed the Organization for European Economic Cooperation, which coordinated the distribution and use of Marshall Plan funds, and the European Coal and Steel Com-

Fig. 7.19 The original "Inner Six" and "Outer Seven" of Europe.

munity, which integrated the development of that industry in the member countries. A few years later, in 1957, the European Economic Community, or Common Market, was born, composed at first of only six countries: France, Italy, West Germany, and the Benelux countries.

To counteract these Inner Six, as they were called, other countries joined in the European Free Trade Association (E.F.T.A.). Known as the Outer Seven, they were the United Kingdom, Norway, Denmark, Sweden, Switzerland, Austria, and Portugal (Figure 7.19). In 1973 the United Kingdom, Denmark, and Ireland applied for and were granted membership in the Common Market. Through the workings of the E.F.T.A. and the Common Market, a free-trade zone has been established over much of Western Europe. Not only have tariff barriers among member states been eliminated, but there is free movement of labor and services as well. In addition, there is some coordination of planning with respect to transportation and agriculture.

We have traced this developmental process not because it is important to remember all the forerunners of the Common Market but to illustrate the rapidity with which regional alliances are made. Countries come together in an alliance, some drop out, and others join. New treaties are made, and new alliances emerge. It seems safe to predict that while the alliances themselves will change, the idea of economic associations is permanently added to the political and military alliances that are as old as the concept of nations.

Three further points about economic unions are worth noting. The first—and it applies to military and political alliances as well—is that the formation of an alliance in one area often stimulates the creation of another alliance by countries left out of the first. Thus the union of the Inner Six gave rise to the treaty between the Outer Seven. Similarly, a counterpart to the Common Market is the Council of Mutual

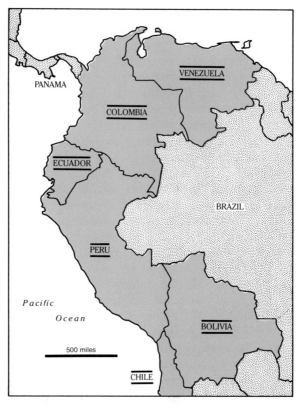

Fig. 7.20 **The Andean Group nations as originally constituted. The Andean pact economic union was formed by contiguous nations in 1969 to promote freer trade among member countries and to pool capital and technical resources.**

253

Economic Assistance, also known as Comecon, which links the Communist countries of Eastern Europe and the U.S.S.R. through trade agreements.

Second, the new economic unions tend to be composed of contiguous states (Figure 7.20). This was not the case with the recently dissolved empires, which included far-flung territories. Contiguity facilitates the movement of people and goods. Communication and transportation are simpler and more effective among adjoining countries than among those far removed from one another.

Finally, it does not seem to matter whether countries are homogeneous or heterogeneous in their economies, as far as joining in economic unions is concerned. There are examples of both. If countries are dissimilar, they may complement each other. This is the basis for the European Common Market. Dairy products and furniture from Denmark are sold in France, freeing that country to specialize in the production of machinery and clothing, although the Common Market is faced by expensive gluts of, for example, butter and wine, which are produced in more than one country and are basic products of politically potent national farm groups. On the other hand, countries that produce the same raw materials find that by joining together in an economic alliance, they are able to enhance their control of markets and prices for their products. The Organization of Petroleum Exporting Countries (O.P.E.C.) is a particularly significant case in point. Less successful attempts at the formation of commodity cartels and price agreements between producing and consuming nations are represented by the International Tin Agreement, International Coffee Agreement, and others.

Countries form alliances for other than economic reasons. Strategic, political, and cultural considerations may also foster cooperation. Military alliances are based on the principle that in unity there is strength. Such pacts usually provide for mutual assistance in the case of aggression. Once again, action breeds reaction when such an association is created. The formation of the North Atlantic Treaty Organization (N.A.T.O), a defensive alliance of many European countries and the United States, was countered by the establishment of the Warsaw Treaty Organization, which joins the U.S.S.R. and its satellite countries of Eastern Europe. Both pacts allow member states to base armed forces in one another's territories—a relinquishment of a certain degree of sovereignty unique to this century.

Military alliances depend on the perceived common interests and political goodwill of the countries involved. As political realities change, so, too, do the strategic alliances. N.A.T.O. was altered in the 1960s as a result of policy disagreements between France and the United States. The conflict over Cyprus between Greece and Turkey, both N.A.T.O. members, and their response to American reaction to that dispute altered the original cohesiveness of that alliance. The organization similar to N.A.T.O. in Asia, S.E.A.T.O. (Southeast Asia Treaty Organization), was disbanded in response to changes in the policies of its member countries.

All international alliances recognize communities of interest. In economic and military associations, common objectives are clearly seen and described, and joint actions are agreed upon with respect to the achievement of those objectives. More generalized common concerns or appeals to historical interests may be the basis for alliances that are primarily political. Such alliances tend to be rather loose, not requiring members to yield much power to the union. Examples are the British Commonwealth, composed of many former British colonies, and the Organization of American States, both of which offer economic benefits as well as political ones. There are many examples of abortive political unions that have foundered precisely because the individual countries could not agree on questions of policy and were unwilling to subordinate individual interests to make the union succeed. The United Arab Republic, the Central African Federation, the Federation of Malaysia and Singapore, and the Federation of the West Indies fall within this category.

Observers of the world scene have wondered about the prospect that "superstates" will emerge from the many international alliances that now exist. Will a United States of Europe, for example, evolve out of the many pacts that now bind the countries of Western Europe economically, militarily, politically, and cultur- ally? Will unification in one sector trigger uni- fication in the others, eventually creating ties of mutual interest so strong that a common government is the logical next step? No one really knows, but as long as the individual state is regarded as the highest form of political and social organization (as it is now) and as the body in which sovereignty rests, such unifica- tion is unlikely.

However, Western Europe is the area to watch in this regard. If a successful interna- tional government is going to be formed, it will probably be there. It is there that unification, including partial monetary unification, has proceeded so far and in so many different sec- tors that eventual political union is a possibil- ity. In fact, a quasi-parliament, the Council of Europe, has existed since 1949, but as now con- stituted, it lacks the power to make and en- force laws.

Conclusion

The political subdivision of the world into nation-states and lower orders of territorial ju- risdiction constitutes an expression of cultural separation and identity as pervasive as that inherent in language, religion, or ethnicity. Indeed, political concerns impinge more force- fully upon the affairs of peoples and societies than do the more subtle expressions of cultural differentiation, as the reading of any newspaper will show. It is also apparent that control over our lives is tending to shift to increasingly higher levels of political organization. The cul- mination of this process is seen in the many international systems that are evolving. At the same time, as nations become interlocked in their economic systems and are tied together by a global communications network, there is increasing though sometimes reluctant recog- nition of the fact that developments in one part of the world affect the rest of it and that certain elements of sovereignty must, in the national and the international interest, be surrendered.

That changing pattern of political control necessarily affects the behavior not only of the nations directly involved but also of the indi- vidual citizens of affected societies. Behavioral geography, the pattern of individual and group action against the background of their culture, is our next topic of concern.

255

For Review

1. What reasons can you suggest for the great political fragmentation of the United States? Since few other nations have such a multiplicity of local gov- erning and administrative units, is such subdivision—in your opinion— evidence of internal weakness? Why?

2. What is meant by *metro government*? What would appear to be its advan- tages? Its disadvantages?

3. Why, in your opinion, were the culture realms suggested on Figure 6.21 gen- erally drawn so that they were composed of whole, rather than parts of, nation-states? Comment on the ways that political jurisdiction reinforces other cultural contrasts between the United States and Mexico.

4. What are some of the devices by which national cohesion and identity are achieved? What internal cultural characteristics might serve to disrupt or destroy the cohesiveness that nation-states attempt to foster?

5. What are some of the pressing issues facing the nations attempting to codify a law of the sea? Do you feel that the interests of your nation would be best served by tight international control of maritime resources? Why or why not?

6. What made the post-1945 dissolution of empires unique in the political history of the world?

7. What international organizations and alliances can you name? What were the purposes of their establishment? Have they, in your opinion, served to increase or to reduce cooperation, communication, and understanding within the world community?

Suggested Readings

BERGMAN, EDWARD F., *Modern Political Geography*. Wm. C. Brown, Dubuque, IA, 1975.

DE BLIJ, HARM J., *Systematic Political Geography*. Wiley, New York, 1967.

BRUNN, STANLEY D., *Geography and Politics in America*. Harper & Row, New York, 1974.

COPPOCK, J. T., and W. R. D. SEWELL (eds.), *Spatial Dimensions of Public Policy*. Pergamon, New York, 1976.

JOHNSTON, R. J., *Political, Electoral and Spatial Systems: An Essay in Political Geography*. Clarendon Press, Oxford, England, 1979.

KASPERSON, ROGER E., and JULIAN V. MINGHI (eds.), *The Structure of Political Geography*. Aldine, Chicago, 1969.

PRESCOTT, J. R. V., *The Geography of Frontiers and Boundaries*. Hutchinson University Library, London, 1967.

Daniel Boone first heard of the Dark and Bloody Ground, of Kentucky, in 1755, as a young volunteer in Braddock's army. The land's dense forests and fertile soils, and particularly its richness in game, were described by those who themselves had ventured down the Ohio River or who, as campfire storytellers, were relating the tales of others. Knowledge of game, especially of deer, was important in a society that measured income in the money equivalent of "bucks". When Daniel Boone returned home to the Yadkin Valley of North Carolina, he began hunting and trapping both east and west of the Alleghenies, trying, but never quite succeeding, to rid himself of the debts that his farm and growing family somehow kept building.

In 1769, by now a competent woodsman, Boone eagerly accepted the suggestion of an old army acquaintance, John Findley, to join in an extended hunting trip to Kentucky. Findley had himself traded along the Ohio. Although death and privation attended the journey, Boone learned firsthand of the wealth and promise of the western lands. Enlisting both neighbors and relatives, Boone set out with a small band of families and their livestock westward along the Wilderness Trace, no more than a faint trail along which one walked or rode mounted, for no wagon could pass. Although terrorized by Indian attacks and delayed by defections and border warfare, Boone's party had by spring of 1775 established the new settlement of Boonesborough, Kentucky, and added their group to the growing number of westward migrants.

Boone's story is unique only in its particulars. As did his predecessors back to the beginnings of humankind, Boone acted in space and over space on the basis of acquired information and awareness of opportunity. By imparting his knowledge to others, he contributed to a group decision to migrate. His story summarizes the content of this chapter, a survey of one of the most current and exciting thrusts of geographic

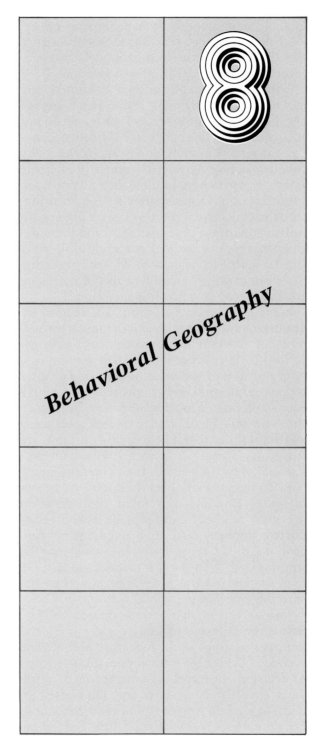

Behavioral Geography

inquiry: how individuals make spatial behavioral decisions and how those separate decisions may be summarized by models and generalizations to explain collective actions.

Behavioral geography deals with choices and movements of individuals about which we can develop descriptive models. That is, we do not anticipate what one person will do but summarize the results of individually and collectively coherent action. Physicists cannot predict the action of separate molecules; they rely on such summary observations as Boyle's law to anticipate the collective response of gases under controlled conditions of temperature and pressure. Our survey of behavioral geography follows a similar summary approach. We seek to define the ways in which individual spatial decisions are aggregated into patterns of action.

In preceding chapters within this section we discussed group behavior and the resultant patterns of cultural development and political action; in later chapters, we will review economic structure and urban life. In this chapter we pose a question basic to the themes we have been exploring: What considerations influence the way individual human beings use space and act within it? Implicit in this question are subsidiary analytical concerns: How do individuals view (perceive) their environment? How do those views affect their separate action in space? How are movement and migration decisions reached? How is information transmitted through space and acted upon? And how might all these individual decisions be summarized, as in the law of gases, so that we may understand the order that underlies the seeming chaos of individual action?

Two aspects of that action concern us. The first is the daily or temporary use of space—the journeys to stores or to work or to school. The second is the longer-term commitment related to decisions to travel, to migrate, or to settle away from the home territory. This latter aspect suggests an implicit additional concern of behavioral geography: that of time. Spatial actions of humans are not instantaneous; they operate over time as an expression of process or system. Elements of both aspects are embodied in the way individuals perceive space and act within it.

Individual Activity Space

We have seen in Chapter 7 that groups and nations draw boundaries around themselves and divide space into territories that are, if necessary, defended. The concept of *territoriality*—the fierce defense of home ground—has been seen by some as a root explanation of much of human action and response. It is true that some collective activity appears to be governed by territorial defense responses: the conflict between street groups in claiming and protecting "turf" (and their fear for their lives when venturing beyond it) and the sometimes violent rejection by strong ethnic urban neighborhoods of an encroaching black, Hispanic, or other population group. But for most, our personal sense of territoriality is a tempered one; homes and property are regarded as defensible private domains but are opened to innocent visitors, known or unknown, or to those on private or official business.

Nor do we confine our activities so exclusively within controlled home territories as street-gang members do within theirs. Rather, we have a more or less extended home range, an *activity space*, within which we move freely on our rounds of regular activity, sharing that space with others also about their daily affairs. Figure 8.1 suggests a probable activity space for a suburban family of five for one day. Note that the activity space for each individual for one day is rather limited, even though two members of the family use automobiles. If one week's activity were shown, more paths would

258

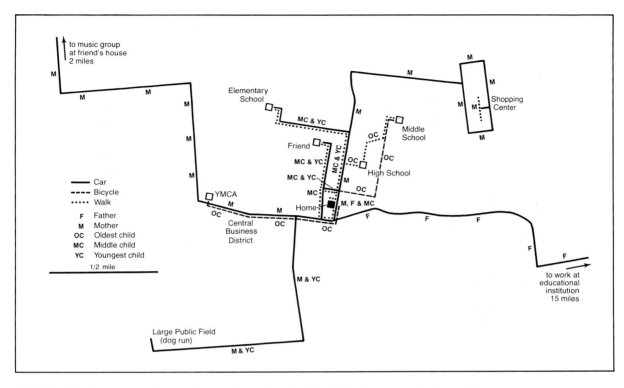

Fig. 8.1 Activity space for each member of a family of five for a typical weekday. Routes of regular movement and areas recurrently visited help to foster a sense of territoriality and to color one's perceptions of space.

have to be added to the map, and in a year's time, several long trips would probably have to be noted. Since long trips are taken irregularly, we will confine our idea of activity space to often-visited places. That activity is more varied and frequent than we might usually recollect or record. A recent study of 119 children in Kansas concluded that the subjects were involved in more than 36 million behavioral actions in the span of one year, though, of course, most of those did not involve movement outside the home.

The kind of activities individuals engage in can be classified according to type of trip: journeys to work, to school, to shops, for recreation, and so on. People in nearly all parts of the world make these same types of journeys,

though the spatially variable requirements of culture and economy dictate their frequency, duration, and significance in the time budget of an individual. Figures 8.2, 8.3, and 8.4 illustrate this point. Figure 8.2 depicts variations in travel patterns for two different culture groups in rural midwestern Canada. It suggests that "modern" rural Canadians, who want to take advantage of the variety of goods offered in the regional capital, are willing to travel longer distances than are people of a traditionalist culture who have different tastes in clothing and consumer goods and whose demands are satisfied in local settlements. Figures 8.3 and 8.4 suggest the importance of the journey to work among urban populations in two different North American cities.

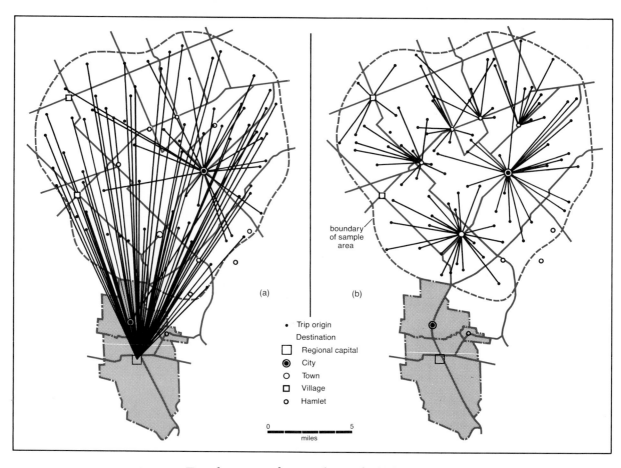

Fig. 8.2 Travel patterns for purchase of clothing and yard goods of (a) rural cash-economy Canadians and (b) Canadians of the old-order Mennonite sect. These strikingly different travel behaviors demonstrate the great differences that may exist in the action spaces of different culture groups occupying the same territory. (Redrawn with permission from R. A. Murdie, "Cultural Differences in Consumer Travel," *Economic Geography*, Volume 41, 1965.)

260

The types of trips individuals make and thus the extent of their activity space depend on at least three variables: their stage in the life cycle; the means of mobility at their command; and the demands or opportunities implicit in their daily activities. The first, stage in the life cycle, refers to membership in a specific age group. Preschoolers stay close to home unless they accompany parents. School-aged children usually travel short distances to lower

schools and longer distances to upper-level schools. After-school activities tend to be limited to walking or bicycle trips to nearby locations. High school students are usually more mobile and take part in more activities than do younger children. Adults responsible for household duties make shopping trips and trips related to child care as well as journeys away from home for social, cultural, or recreational purposes. Wage-earning adults usually travel

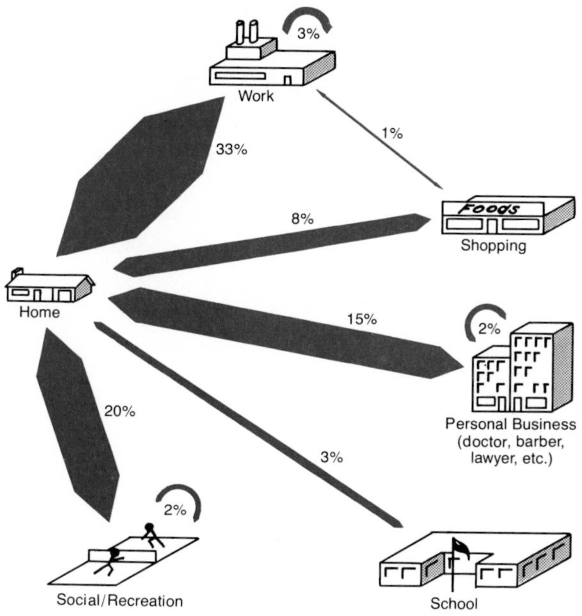

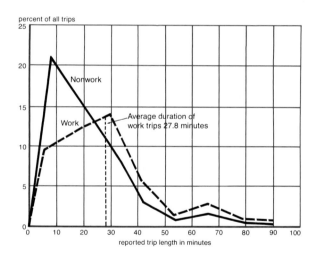

Fig. 8.3 Chicago travel patterns. The numbers are the percentages of all urban vehicular trips taken in Chicago. By far the greatest single movement is the journey to and from work. Over 87% of all trips are represented on the diagram. (Source: *Chicago Area Transportation Study*, vol. 1, 1959.)

means of personal conveyance, daily non-emergency action space may be limited to the shorter range afforded by the bicycle or by walking. Obviously, both intensity of purpose and the condition of the roadway affect the execution of movement decisions.

The mobility of individuals in countries or in sections of countries with high incomes is relatively great; activity space horizons are broad. These horizons, however, are not limitless. There are a fixed number of hours in a day, most of them consumed in performing work, preparing and eating food, and sleeping. In addition, there are a fixed number of road, rail, and air routes, so that even the most mobile individuals are constrained in the amount of activity space they can use. No one can easily claim the world as his or her activity space.

A third factor limiting activity space is the

farther distances from home than other family members. Elderly people normally do not find it feasible or desirable to have extended activity spaces. These age-specific movement patterns were illustrated by a study of the home ranges within which populations normally moved in the town of Midwest, Kansas, and in Yoredale, Yorkshire, England, summarized in Table 8.1.

The second variable that affects the extent of activity space is mobility, or the ability to travel. An informal consideration of the cost and effort required to overcome the friction of distance is implicit. Where incomes are high, automobiles available, and the cost of fuel reckoned minor in the family budget, mobility may be great and individual action space large. In societies where cars are not a standard

Fig. 8.4 The frequency distribution of work and nonwork trip lengths in minutes in Toronto. Work trips are usually longer than other internal city journeys. For North America, 20 minutes seems to be the upper limit of commuting time before people begin to consider time seriously in their choice of home site. Most nonwork trips are very short. (Source: *Metropolitan Toronto and Region Transportation Study*, The Queen's Printer, Toronto, 1966, Fig. 42.)

Table 8.1 Population (P), Territorial Range (TR), and Territorial Index (TI) of Age Groups in Midwest and Yoredale[a]

The activity space of the older members of society is materially lower than that of working-age populations and most closely resembles that of adolescents.

Age Group (years)	Midwest			Yoredale		
	P	TR	TI	P	TR	TI
Aged (65 and over)	162	462	80	178	332	67
Adult (18–64)	375	578	99	770	491	99
Adolescent (12–17)	50	464	80	107	329	67
Older School (9–11)	26	389	67	51	274	55
Younger school (6–8)	28	359	62	72	251	51
Preschool (2–5)	50	363	63	81	214	43
Infant (under 2)	24	329	57	41	125	25
All ages	715	579	100	1300	494	100

[a] Population (P) equals the number of individuals in the age group sample; territorial range (TR) is the calculated total number of different behavior settings participated in by the age group population; territorial index (TI) is the group's percentage of the "all ages" territorial range.

Reprinted by permission from Barker & Barker, "The Psychological Ecology of Old People in Midwest, Kansas, and Yoredale, Yorkshire," *Journal of Gerontology,* Vol. 16 (April 1961), p. 146.

262

individual assessment of the availability of possible activities or opportunities. In the subsistence economies discussed in Chapter 9, the needs of daily life are satisfied at home; the impetus for journeys away from the residence is minimal. If there are no stores, schools, factories, or roads, one's expectations and opportunities are limited, and the activity space is therefore reduced. In impoverished nations or neighborhoods, low incomes limit the inducements, opportunities, destinations, and necessity of travel.

Distance and Human Interactions

People make many more short-distance trips than long ones. If we drew a boundary line around our activity space, it would be evident that trips to the boundary are taken much less often than short-distance trips around home. Think of activity space as more intensively used near one's home place or base and as declining in use with increasing distance from the base. This is the principle of *distance decay*—the exponential decline of an activity or function with increasing distance from its point of origin. The tendency is for the frequency of trips to fall off very rapidly beyond an individual's *critical distance.* Figure 8.5 illustrates this principle with regard to journeys from the home site. The critical distance is the distance beyond which cost, effort, and means play an overriding role in our willingness to travel.

For example, a small child will make many trips up and down its block but is inhibited by parental admonitions from crossing the street. Different but equally effective constraints control adult behavior. Daily or weekly shopping may be within the critical distance of an indi-

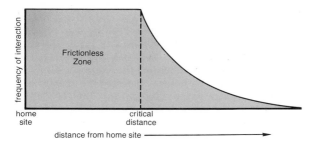

Fig. 8.5 **This general diagram indicates how distance is observed by most people. For each activity, there is a distance beyond which the intensity of contact declines. This is called the *critical distance*, if distance alone is being considered, or the *critical isochrone*, if time is the measuring rod. For the distance up to the critical distance, a frictionless zone is identified in which time or distance considerations do not effectively figure in the trip decision.**

vidual, and little thought may be given to the cost or the effort involved, but shopping for special goods is relegated to infrequent trips, and cost and effort are considered.

Spatial Interaction and the Accumulation of Information

The critical distance for each person is different. The variables of stage in life cycle, mobility, and opportunity together with an individual's interests and demands help to define how much and how far a person will travel. On the basis of these variables, we can make inferences about the likelihood that a person will gain more or less information about his or her activity space and the space beyond.

We gain information about the world from many sources (Figure 8.6). Although information obtained from radio, television, and newspapers is important to us, face-to-face contact is assumed to be the most effective means of communication. Daniel Boone convinced his neighbors to migrate without the aid of additional exhortatory devices.

If we combine the ideas of activity space and distance decay, we see that as the distance away from the home place increases, the number of possible face-to-face contacts usually decreases. We expect more human interactions at short distances than at long distances. Because more information is available to those with larger activity spaces, there are just that many more possible chances for face-to-face contact. Where population densities are high, such as in cities (particularly central business districts during business hours), the spatial interaction between individuals can be at a very high level, which is one reason that these centers of commerce and entertainment are often also centers for the development of new ideas.

The fact that the possible number of interactions is high, however, does not necessarily mean that the effective occurrence of interactions will be high. In crowded areas people commonly set psychological barriers around themselves so that only a limited number of interactions take place. The barriers are raised in defense against information overload and for psychological well-being. We must have a sense of privacy in order to filter out the information that does not directly concern us. As a result, individuals tend to reduce their interests to a narrow range when they find themselves in crowded situations, allowing their wider interests to be satisfied by use of the communications media.

263

SPATIAL INTERACTION AND INNOVATION

The probability that new ideas will be generated out of old ideas is a function of the number of available old ideas in contact with one another. People who specialize in a particular field of interest seek out others with whom they wish to interact. The gathering place has a

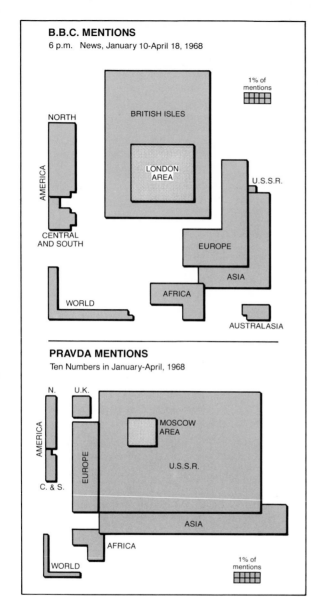

B.B.C. MENTIONS
6 p.m. News, January 10–April 18, 1968

1% of mentions

NORTH

AMERICA

CENTRAL
AND SOUTH

WORLD

BRITISH ISLES

LONDON
AREA

U.S.S.R.

EUROPE

ASIA

AFRICA

AUSTRALASIA

PRAVDA MENTIONS
Ten Numbers in January–April, 1968

264

N.   U.K.

AMERICA

EUROPE

C. & S.

MOSCOW
AREA

U.S.S.R.

ASIA

AFRICA

WORLD

1% of mentions

Fig. 8.6 View of the world from the British Broadcasting Company and *Pravda*. Pick up a newspaper and see how much of it is devoted to local, national, and international news. You will discover that in general, as distance increases, the information available about places decreases. Perhaps the reason that London news is greater in amount on the B.B.C. than is Moscow news in *Pravda* is that London contains a greater proportion of the population of the British Isles than does Moscow of the U.S.S.R. (Redrawn with permission from J. P. Cole and P. Whysall, "Places in the News, A Study of Geographical Information," *Bulletin of Quantitative Data for Geographers*, Nottingham University, 1968.)

movements are usually generated in circumstances of high spatial interaction. An exception, of course, is the case of highly traditional societies—China before World War II, perhaps— where the culture rejects innovation and clings steadfastly to customary ideas and methods.

The culture hearths of an earlier day were the most densely settled, high-interaction centers of the world. At the present time the great national and regional capital cities attract people who want or need to interact with others in special-interest fields. The association of population concentrations and the expression of human ingenuity has long been noted. The home addresses recorded for patent applicants by the U. S. Patent Office over the last century indicate that inventors were typically residents of major urban centers, presumably closely in contact and exchanging ideas with those in shared fields of interest. It still appears that the great national and regional capital cities of the world attract those who are young and ambitious, and that face-to-face contact is important in the creation of new ideas and products.

The revolution in communications initiated early in the 20th century by the development of the vacuum tube and the extension of the invention to the telephone, television, and data-processing equipment have suggested to some that the traditional importance of cities as collectors of creative talent may be past. The

higher density than previously. Crowded central cities are characteristically composed of specialists in very narrow fields of interest. Consequently, under short-distance–high-density circumstances, the old ideas are given a hearing and new ideas are generated by the interaction. New inventions and new social

A skilled, long-term observer of rural and village life in underdeveloped countries suggests that in recent years, resistance to innovation has dramatically decreased in many of them. In Asia, particularly, the spatial diffusion of new ideas concerning contraception, population control, and agricultural methods has resulted in a transformation of peasant cultures and has broken down earlier adamant resistance to change. What were strange and often unwelcome Western innovations are now coming to be taken as the desirable village cultural norm: two-child families, multiple cropping, and all kinds of new scientific farming techniques.

Acceptance of Cultural Innovation

Since these new ideas did not reach Asian villages until the late 1960s, what seem to be epic historical changes took only about a decade in areas culturally constant for preceding millennia. Among the evidences of acceptance of innovation are

1. Adoption of contraception and plummeting birth rates.
2. Increases in rice production even in Asia's poorest regions (in Java by 4%, in Bangladesh by 5%), associated with the rapid spread of high-yield, fertilizer-intensive rice, multiple cropping, and irrigation.
3. A pattern of reverse migration in cities such as Jakarta, Calcutta, and Bombay, as for the first time since World War II, significant numbers of peasants are returning to newly prospering villages.

National economies are beginning to reflect the new rural prosperities through rising exports and falling deficits.

Acceptance of cultural innovation on the Asian model has not yet occurred in black Africa, in the Latin Roman Catholic world (including its Asian outpost, the Philippines), or in the Moslem countries from Morocco to Pakistan (though Egyptian and Southeast Asian Muslims have changed in the general Asian fashion). Economies in these resisting areas are sluggish to faltering, and traditional cultural values and production methods still hold sway in the villages.

In the Latin American and Moslem countries, there is a strong belief, unique to these village cultures, in male superiority and the equation of many children with a man's worth and masculinity. This macho emphasis is a main obstacle to cultural adjustment and reduction of birth rates, and it is reinforced by the active resistance to innovation from cultural leaders. Roman Catholic peasants are affected by the Vatican's opposition to contraception. Muslims are influenced by a growing militant minority movement opposing change and dedicated to retaining traditional values.

265

Adapted by permission from "Epic Change in Asia's Villages," by Richard Critchfield, *The Christian Science Monitor*, December 15, 1978. © 1978 The Christian Science Publishing Society. All rights reserved.

change in educational levels in the central city and its reduced role as a corporate headquarters, noted in Chapter 11, suggest that the formerly vital large-city residence is of declining significance in the creative arts.

SPATIAL DIFFUSION

When a substance or a concept is dispersed outward from a center of origin, the process is known as *spatial diffusion.* Ideas generated in a center of activity will remain there unless some process is available for their spread. Innovations—ideas that change people's behavior—spread in various ways. Some new concepts are so obviously advantageous that they are quickly put to use by those who can profit from their employment; a new development in petroleum extraction may promise such material reward as to lead to its quick adoption by all of the major petroleum companies, irrespective of their distance from the point of introduction. The new strains of wheat and rice that were part of the "Green Revolution" (see Chapter 9) were quickly made known to agronomists in all cereal-producing countries, but they were more slowly taken up by the farmers of those countries to whom the advantages and the production techniques connected with the new grains had to be taught by example. Obviously, many innovations are of little consequence by themselves, but sometimes the widespread adoption of seemingly inconsequential innovations brings about large changes when viewed over a period of time.

A new tune, "adopted" by a few people, may lead many individuals to adopt that tune plus others of a similar sound, which in turn may have a bearing on dance routines, which in turn may have a bearing on clothes selection, which in turn may have a bearing on retailers' advertising campaigns and on consumers' patterns of expenditures. Eventually a new cultural form will be identified that may have an important impact on the thinking processes of the adopters and on those who come into contact with the adopters. Notice that a broad definition of innovation is used, but notice also that what is important is whether or not innovations are adopted.

In spatial terms, we may identify several different diffusion processes. These are distance-affected diffusion, population-density–affected diffusion, and hierarchical diffusion.

Distance-Affected Diffusion. Let us suppose that an automobile enthusiast develops a gasoline additive that noticeably improves the performance of his or her car. Suppose further that the enthusiast shows friends and associates the invention and that they in turn tell others. This process is similar to the spread of ripples in calm water. The innovation will continue to diffuse until barriers are met (that is, people not interested in adopting the new idea) or until the area is saturated (that is, no more people can be contacted because the boundaries of the island or the country or the activity spaces are reached). This step-by-step diffusion process follows the rules of distance decay at each step; short-distance contacts are more likely than long-distance contacts. But, over time, the idea may have spread far from the original site, as indicated by Figure 8.7.

A number of characteristics of this kind of diffusion are worth noting. If the idea has merit in the eyes of potential adopters, the number of contacts of adopters with potential adopters will compound. Consequently, the innovation will spread slowly at first and then more and more rapidly, until there is saturation or a barrier is reached. The incidence of adoption is diagrammed by the S-shaped curve in Figure 8.8 and is mapped in Figure 8.9. In a similar manner, the area in which the adopters are located will at first be small, and then the area will enlarge at a faster and faster rate. The spreading process will slow as the available areas and/or people decrease. Precisely the

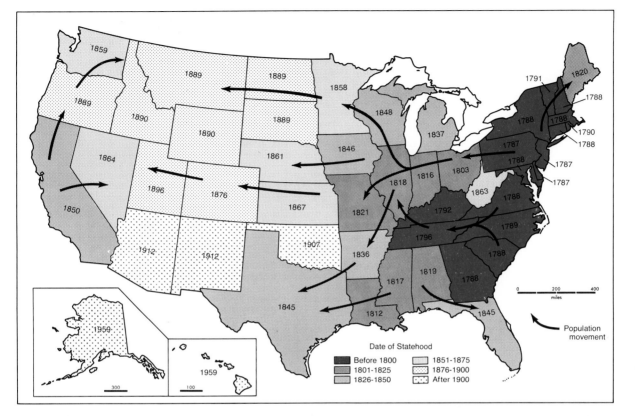

Fig. 8.7 The diffusion of statehood. The achievement of statehood was, of course, preceded by the development of territorial settlements. The process of statehood was a distance-affected diffusion phenomenon.

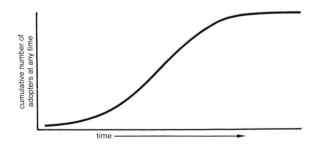

Fig. 8.8 The diffusion of innovations over time. The number of adopters of an innovation increases at an increasing rate until the point at which about one-half of the total who ultimately decide to adopt the innovation have made the decision. At this point, the number of adopters increases at a decreasing rate.

same process occurs as with the spread of a contagious disease.

Barriers to adoption, besides the obvious ones of physical boundaries (mountains, oceans, rivers) and cultural obstacles (different religions, languages, ideologies, political systems), include psychological barriers. Within any population, there are those who will not innovate, that is, will not accept new ideas. In some societies—particularly traditional, dispersed agricultural societies—the number of nonadopters for any innovation may be very high indeed. In other societies where innovation and change are given high status, and

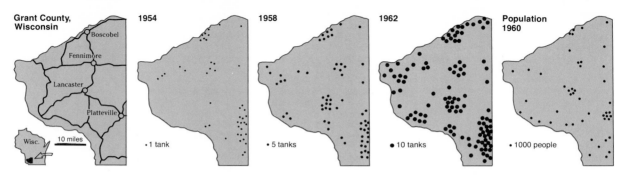

Fig. 8.9 Propane tanks to store fuel for heating and cooling were introduced to Grant County, Wisconsin, in 1951 at Platteville, the main town of the county. Lancaster was slow to accept this innovation, but distance from Platteville in general played an important role in the early diffusion of the innovation. By 1962, the pattern of acceptances was similar to the distribution of population. While the number of early adoptions increased rapidly, by 1962 the rate of increase per year began to slow. (Data from L. A. Brown, "The Diffusion of Innovation: A Markov Chain-Type Approach," *Discussion Paper*, No. 3, Department of Geography, Northwestern University, 1963.)

where the innovations are not risky, ideas spread much more rapidly.

268 **Population-Density-Affected Diffusion.** If an inventor's idea fell into the hands of a commercial distributor, the diffusion process would follow a somewhat different course than that discussed above. The distributor might "force" the idea into the minds of individuals by using the mass media. If the media are local in impact, such as newspapers, then the pattern of adoptions will be similar to that described above. If, however, a nationwide television or magazine advertising campaign were undertaken, the innovation would become known in numbers roughly corresponding to population density. Where more people live, there will, of course, be more potential adopters. Economic or other barriers may also affect the diffusion. One immediately sees, however, why large TV markets are so valuable.

This type of diffusion process may act together with the distance-decay process. Many of those who accept the innovation after learning of it in the mass media will tell others, so that the distance-decay effect will begin to take

over soon after the original contact is made. Each type of medium has its own level of effectiveness. Advertisers have found that they must repeat messages time and again before they are accepted as important information. This fact says something about the effectiveness of mass media as opposed to, say, face-to-face contact. It may also be a comment on the type of innovations advertised in the mass media.

Hierarchical Diffusion. The third way innovations are spread combines some aspects of the first two plus the inclusion of a new element: the hierarchy. A hierarchy is a classification of objects into categories so that each category is increasingly complex or has increasingly higher status. Hierarchies are found in many systems of organization, such as government offices (the organization chart) or universities (instructors, professors, deans, and so on). In this discussion, we refer to varying levels of city complexity as the *hierarchy of cities.*

Large cities are much more complex than small ones and occupy different levels in the urban hierarchy. Furthermore, the way cities of similar size are structured in terms of transport systems, available opportunities, land uses, and so on is similar, especially within one country or major region. Small cities do not offer many of the activities found in the large cities and are therefore less complex. We may add as many levels to the hierarchy of cities as there are different types of structural elements that make cities what they are. For convenience, urban areas may be clustered within a five-level hierarchy: huge metropolises, regional centers, cities, towns, and villages.

For many types of innovations, diffusion is affected by neither distance-decay nor density; it is hierarchical in nature. Let us take as an example the diffusion of a new medical discovery reported in a medical journal that is widely read by doctors. The adoption of the innovation may be dependent on the purchase of expensive equipment. If so, only large hospitals serving large populations will be in a position to obtain the new equipment. Such hospitals are usually found in large metropolitan areas; thus the innovation will spread within the highest level of the urban or hospital hierarchy. When purchasing costs for the equipment become lower or when the demand for the service increases, there is a possibility that the innovation will be accepted at a lower level in the hierarchy.

A hypothetical scheme showing how a four-level hierarchy might be connected for the flow of information is presented in Figure 8.10. Note that the lowest-level centers are connected to higher-level centers but not to each other. Note, too, that connections might bypass intermediate levels and link only with the highest-level center.

Many times, hierarchical diffusion takes place simultaneously with the other two kinds of diffusion. One might expect variations when the density of high-level centers is great and when the distances between centers are short.

A quick and inexpensive way to spread an idea is to communicate information about it at high-order hierarchical levels. Then the three types of diffusion processes may be used most effectively; even while an idea is diffusing through a high level in the hierarchy, it is also spreading outward from the high-level centers. Consequently, low-level centers that are near high-level centers may be apprised of the innovation before medium-level centers. People living in suburbs and small towns near a large city are privy to much that is new in the large city, as are individuals in other large cities half a continent away.

All these forms of diffusion operate in the spread of culture. They add their impact to the enlargements of activity space and the consequences are the spatial interaction and innovation discussed earlier in this chapter. We should also recall, from Chapters 5 and 6, that migration, invasions, selective cultural adoptions, and cultural transference aid the diffusion of innovation. These broader movements and exchanges represent interactions of people beyond their usual activity spaces and the creation within them of awarenesses not normally likely when they are confined to a restricted home range without the benefit of modern communication. It took a war to inform Daniel Boone about the opportunities of Kentucky, and it took a far-roving army friend to induce him to act upon that knowledge. For a further understanding of these interactions, movements, and migrations, it is useful to consider how individuals think of areas that are essentially unfamiliar to them.

269

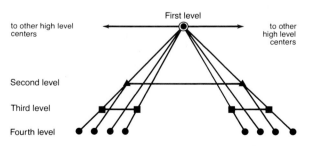

Fig. 8.10 A four-level communication hierarchy.

Perception of the Environment

The term *environmental perception* refers to our awareness, as individuals, of home and distant places and the beliefs we have about them. It involves our feelings, reasoned or irrational, about the complex of natural and cultural characteristics of an area. Whether our view accords with that of others or truly reflects the "real" world seen in abstract descriptive terms is not the major concern. Our perceptions are the important thing, for the decisions people make about the use of their lives are based not necessarily on reality but upon their perceptions of reality.

Psychologists and geographers are interested in determining how we arrive at our environmental perceptions. We shall focus attention on perceptions of places beyond our normal activity space, although the question of how we view areas with which we are quite familiar is also of interest.

In technologically advanced societies, television and radio, magazines and newspapers, books and lectures, travel brochures and hearsay all combine to help us develop a mental picture of unfamiliar places. Again, however, the most effectively transmitted information seems to come from word-of-mouth reports. These may be in the form of letters or visits from relatives, friends, and associates. Probably the strongest lines of attachment to relatively unknown areas develop through the information supplied by family members and friends.

As we know, our knowledge of close places is greater than our knowledge of far places. But barriers to information flow give rise to *directional biases.* Not having friends or relatives in one part of a country may represent a barrier to individuals, so that interest in and knowledge of the area beyond the "unknown" region are sketchy. In the United States, both northerners

270

Fig. 8.11 Mental maps do not necessarily aid communication. (Drawing by Stevenson; © 1976 The New Yorker Magazine, Inc. Reproduced by permission.)

and southerners tend to be less well informed about each other's areas than about the western part of the country. Traditional communication lines in the United States follow an east–west rather than a north–south direction, the result of early migration patterns, business connections, and the pattern of the development of high-order cities.

When information about a place is sketchy, blurred pictures develop. These influence the impression we have of places and cannot be discounted. Many important decisions are made on the basis of incomplete information or biased reports—decisions such as to visit or not, to migrate or not, to hate or not, even to make war or not.

We might say that each individual has a mental map of the world. No single person, of course, has a true and complete image of the world; therefore there can be no completely accurate mental map. In fact, the best mental map that most individuals have is that of their own activity space.

MENTAL MAPS

No one can reproduce on paper an exact replica of the mental image he or she might have of an area. The study of mental maps must by necessity be indirect. If we want to know how particular people envisage their town, we must either ask them questions about the town or ask them to draw sketch maps. While the result will not be a completely accurate picture of what they have in mind, it will be suggestive of the mental map.

Whenever individuals think about a place or how to get to a place, they produce a mental map (Figure 8.11). Unnecessary details are left out, and only the important elements are included. Those elements usually include awareness that the object or the destination does indeed exist, some conception of the distances separating the starting point and the named object(s), and a feeling for the directional rela-

tionships between points. A mental route map might also include reference points to be encountered on the chosen path of connection or on alternate lines of travel. Although mental maps are highly personalized, people with similar experiences tend to give similar answers to

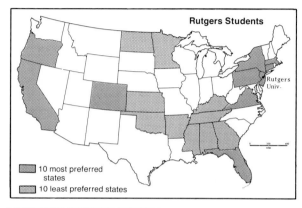

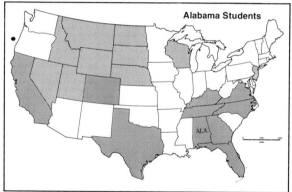

Fig. 8.12 These maps are based on data collected from samples of college students at Rutgers University in New Jersey and at the University of Alabama. The students were asked, "In what order do you rank the 48 conterminous states with regard to residential desirability?" The maps show the 10 most preferred states and the 10 least preferred states for each set of students. Note that both groups agreed on California, Colorado, Virginia, and Florida as desirable places to live. They also agreed that North Dakota was undesirable, but they disagreed on Alabama, New Jersey, Kentucky, Tennessee, and Georgia. (Alabama data from Peter Gould, "On Mental Maps," Inter-University Community of Mathematical Geographers, Ann Arbor, 1965.)

271

questions about the environment and to produce roughly comparable sketch maps.

Awareness of places is usually accompanied
by opinions about them, but there is no necessary relationship between the depth of knowledge and the perceptions held. In general, the
more familiar we are with a locale, the more
sound will be the factual basis of our mental
image of it. But individuals form firm impres-
sions of places totally unknown by them per-
sonally, and these may color travel or migration decisions.

One way to ascertain how individuals envisage the environment is to ask them what they
think of various places. For instance, they may
be asked to rate places according to desirability—perhaps residential desirability—or to
make a list of the 10 best and the 10 worst cities in the United States. With the exception of
certain glamour spots, such as Aspen, Colorado, a certain regularity appears in such studies. Figure 8.12 presents some residential desirability data as elicited from college students in
New Jersey and Alabama. These and comparable mental maps suggest that near places are
preferred to far places unless much information is available about the far places. Places
with similar cultural forms are preferred, as are
places with high standards of living. Individuals tend to be indifferent to unfamiliar places,
and to dislike unfamiliar areas that have competing cultural interests (such as disliked political and military activities). People tend mentally to increase the size of the familiar and to
decrease the size of all else. They tend to place
their own location in a central position and
increase the size of things nearby. Also, as we
grow older our perspectives change, as shown
in Figure. 8.13.

272

PERCEPTION OF NATURAL HAZARDS

Mental maps of home areas do not generally
include as an overriding concern an acknowl-

edgment of potential natural dangers. An in-
triguing area of research to which geographers
have addressed themselves deals with the way
people perceive *natural hazards*, defined as
processes or events in the physical environment not caused by humans but with consequences that are harmful to them. Most climatic (e.g., hurricanes, cyclones, blizzards) and
geological (earthquakes, volcanic eruptions)
hazards cannot be prevented; their consequences, as touched upon in Chapters 2 and 3,
may be disastrous. The cyclone that struck the
delta area of Bangladesh on November 12,
1970, left dead at least 500,000 people; the July
28, 1976, earthquake in the Tangshan area of
China devasted a major urban-industrial complex, with casualties presumed to lie between
700,000 and 1 million. These were major and
exceptional natural hazards, but more common occurrences are experienced and apparently discounted by those in their likely areas
of effect. Johnstown, Pennsylvania, has suffered recurrent floods, and yet residents rebuild; violent storms strike the Gulf and East
coasts of the United States, and people remain
or return.

Why do people choose to settle in high-
hazard areas, in spite of the potential threat
to their lives and property? Why do hundreds of
thousands of people live along the San Andreas
fault in California, build houses in Atlantic
coastal areas known to experience hurricanes
(Figure 8.14), or farm in flood-prone areas adjacent to the Mississippi River? What is it that
makes the risk worth taking?

There are many reasons that hazardous areas
are perceived differently by people and thus
that some people choose to settle in them. Of
major import is the fact that specific hazards
are relatively rare occurrences. Many people
think that the likelihood of an earthquake or a
flood or other natural calamity is sufficiently
remote so that it is not economically feasible
to protect themselves against it. They are also
influenced by the fact that scientists who

Fig. 8.13 Three young people, aged 6, 10, and 13, who lived in the same house, were asked to draw maps of their neighborhood. No further instructions were given. Notice how perspectives broaden and neighborhoods expand with age.

For the 6-year-old, the "neighborhood" consists of the houses on either side of her own.

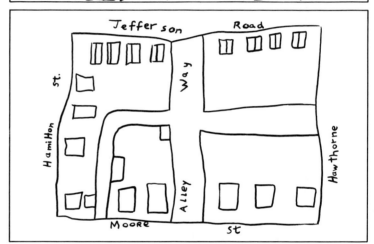

The square block on which she lives is the neighborhood for the 10-year-old.

The wider activity space of the 13-year-old is reflected in this drawing. The square block that the 10-year-old drew is shaded in this sketch.

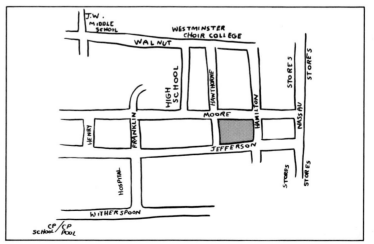

study such hazards may themselves differ on the probability of an event or on the potential damage that it might inflict. And in fact, the prediction of hazards is not an exact science, being based on the calculation of the probability of occurrence of what are uncommon events. The predictions are subject to uncertainty.

People are also influenced by their past experiences in high-hazard areas. If they have not suffered much damage in the past, they may be optimistic about the future. If, on the other hand, past damage has been great, they may think that the probability of similar occurrences in the future is low (Table 8.2). People's memories can be short. In the years following an earthquake, for example, a sense of security grows, building codes or their interpretation are relaxed, and population in the area increases.

High-hazard areas are often sought out not because they pose risks but because they possess desirable topography or scenic views, as do, for instance, the Atlantic and California

Fig. 8.14 Damaged houses on Virginia Beach, Virginia, after a storm and high tide. (U.S.D.A. photo.)

Table 8.2 Common Responses to the Uncertainty of Natural Hazards

Eliminate the Hazard	
Deny or Denigrate Its Existence	**Deny or Denigrate Its Recurrence**
"We have no floods here, only high water."	"Lightning never strikes twice in the same place."
"It can't happen here."	"It's a freak of nature."
Eliminate the Uncertainty	
Make It Determinate and Knowable	**Transfer Uncertainty to a Higher Power**
"Seven years of great plenty. . . . After them seven years of famine."	"It's in the hands of God."
"Floods come every five years."	"The government is taking care of it."

Reprinted with permission from Ian Burton and Robert Kates, "The Perception of Natural Hazards in Resource Management," *Natural Resources Journal*, Vol. 3, 1964, p. 435, published by the University of New Mexico School of Law, Albuquerque, NM 87131.

coasts. Once people have purchased property in a known hazard area, they may be unable to sell it for a reasonable price even if they so desire. They think that they have no choice but to remain and protect their investment. The cultural hazard—loss of livelihood and investment—appears to be more serious than whatever natural hazards there may be.

Migration

In the $1\frac{1}{2}$ to 2 million years that have elapsed since *Homo erectus* is presumed to have emerged in East Africa, his descendants have populated the earth. Each individual's action space and awareness space were limited, but because of pressures of numbers, availability of food, and migration of game, those spaces were collectively enlarged to encompass the world. An important aspect of human history has been the migration of peoples, the evolution of their separate cultures, and the diffusions of those cultures and their components by in-

terchange and communication. Portions of that story have been touched upon in Chapters 5 and 6.

These past collective movements and their intergroup and landscape consequences have, of course, their more modern counterparts. The settlement of North America, Australia, and New Zealand involved great forced and free long-distance movement of peoples and the impact of their cultures upon largely natural landscapes. The flight of refugees from past and recent wars, the settlement of Jews in Israel, the forced urbanization of Soviet peasants, and innumerable other examples of mass movement come quickly to mind. In all cases, societies were intermixed, ideas diffused, and history altered. The vast numbers who suffered the hardships of movement to the American West were the logical followers of Daniel Boone's equally hazardous, but shorter, journeys to Kentucky.

In the study of those individual actions that result in the collective responses summarized in behavioral geography, two types of long-distance movement are recognized. The first is

275

simply the planned two-way trip: the exciting, perhaps historically decisive, daring journey of exploration about which we read in history, or the more common contemporary vacation or business or social trip in which modern societies indulge so freely. The latter sort of trip, of course, enhances the mental maps and enlarges the awareness space of the participants. It may also contribute to the diffusion of information through cultural contact, but its individual impact is small because it is so transitory.

Much more important is migration: a relocation of both residential environment and activity space. Naturally, the length of the move and its degree of disruption of normal household activities raise distinctions important in the study of migration. A change of residence from the central city to the suburbs certainly changes the activity space of schoolchildren and of adults in many of their nonworking activities, but the workers may still retain the city—indeed, the same place of employment there—as an action space. On the other hand, immigration from Europe to the United States and the massive farm-to-city movements of rural Americans late in the last and early in the present century meant total change of all aspects of behavioral patterns.

THE DECISION TO MIGRATE

The decision to move is a cultural and temporal variable. Nomads fleeing famine and spreading deserts in the Sahel obviously are motivated by different considerations than are the executive receiving a job transfer to Chicago, the resident of Appalachia seeking factory employment in the city, or the retired couple searching for sun and sand. For the modern American, decisions to migrate have been summarized into a limited number of categories that are not mutually exclusive. They include (1) changes in life cycle (e.g., getting married, having children, getting a divorce, or

needing less dwelling space when the children leave home); (2) changes in the career cycle (e.g., getting a first job or a promotion, receiving a career transfer, or retiring); (3) forced migrations associated with urban development, construction projects, and the like; (4) neighborhood changes from which there is flight, perhaps pressures from new and unwelcome ethnic groups, building deterioration, street gangs, and similar rejected alterations in activity space; and (5) changes of residence associated with individual personality. Some people simply seem to move often for no discernible reason, while others, *stayers,* settle into a community permanently. Of course, for a nation such as the Soviet Union, with its prohibitions on emigration, a tightly controlled system of internal passports, restriction of job changes, and severe housing shortages, a totally different set of summary migration factors would be present.

The factors that contribute to mobility tend to change over time. There is a group, however,

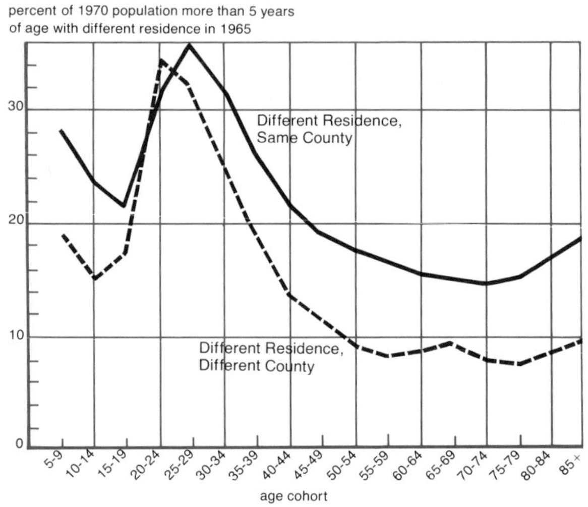

percent of 1970 population more than 5 years
of age with different residence in 1965

Fig. 8.15 Young adults figured most prominently in both short- and long-distance moves between 1965 and 1970. Mobility is disinctly age-related. (Reproduced by permission from the Association of American Geographers, Resource Paper 77-2, 1977, C. C. Roseman.)

An example of a structure developed for making a rational decision whether or not to migrate may be found in the case of the selection of a job by a hypothetical unmarried male who is a recent college graduate. This person partitioned his consideration into four major aspects: monetary compensation, geographical location, travel requirements, and the nature of the work. Each of these aspects was subdivided, and some were then subdivided again, as illustrated in the schema.

Rational Decision Making: An Example

			Relative Worth Placed on Each Category	*1* Large National Company	*2* Regional Office	*3* Local Job
					Ratings on Job Opportunities	
Nature of work	Current and future work features	Management training program	.040	× 95 = 3.80	80 = 3.20	50 = 2.00
		Variety of work	.059	× 80 = 4.72	75 = 4.43	70 = 4.13
		Technical challenge	.049	× 80 = 3.92	80 = 3.92	95 = 4.66
	Immediate work features		.132	× 70 = 9.24	80 = 10.56	90 = 11.88
Travel requirements	Long trips away from office	Trip lengths	.082	× 90 = 7.38	70 = 5.74	60 = 4.92
		Proportion time away	.054	× 80 = 4.32	70 = 3.78	60 = 3.24
	Daily commuting characteristics		.034	× 50 = 1.70	70 = 2.38	100 = 3.40
Geographical location		Climate	.034	× 80 = 2.72	70 = 2.38	60 = 2.04
		Degree of urbanity	.068	× 90 = 6.12	80 = 5.44	60 = 4.08
		Proximity to relatives	.068	× 60 = 4.08	70 = 4.76	100 = 6.80
Monetary compensation	Future salary prospects	10-year increase	.035	× 90 = 3.15	80 = 2.80	80 = 2.80
		3-year increase	.064	× 60 = 3.84	70 = 4.48	80 = 5.12
	Immediate prospects — Fringe benefits	Retirement	.009	× 95 = 0.86	70 = 0.63	60 = 0.54
		Insurance	.014	× 95 = 1.33	70 = 0.98	60 = 0.84
		Starting salary	.209	× 70 = 14.63	75 = 15.68	100 = 20.90
			1.000	71.81	71.16	77.35

After Howard Raiffa, Harvard University.

277

Each category was given a value on a scale from 0 to 1.0, according to the importance the potential migrant placed on it. The restriction that the total of values must add up to 1.0 enabled the graduate to multiply across any path to find the true relative worth of the 15 categories. In the schema, we see that the graduate placed the greatest emphasis on starting salary and the kind of work to be done at the outset. Retirement and insurance benefits were least important.

Once the graduate had a clearer idea of his values, he could take each of his job opportunities and evaluate them according to the best information available about each subcategory. For each possible job, he assigned a value to every category according to the system 90–100 is excellent, 80– 90 is good, 70–80 is fair, and so on. By multiplying each of these by the relative worth of each category and summing for each column, the graduate was able to make an objective decision. For example, for job opportunity No. 1, $95 \times .040 = 3.80$; $80 \times .059 = 4.72$; and so on. For job opportunity No. 1, the score was 71.8; for No. 2, the value was 71.2; and for No. 3 it was 77.3.

This decision was based on the subject's own value system. Each person's categories and weighting schemes would be different. In the case illustrated, the graduate finally ranked the local opportunity ahead of the national and regional opportunities. This ranking led to a decision not to migrate at the present time.

that in most societies has always been the most mobile: young adults (Figure 8.15). They are the members of society who are launching a career and making their first decisions about occupation and location. They have the fewest responsibilities of all adults and thus are not as strongly tied to family and institutions as older people. Most of the major voluntary migrations have been composed primarily of young people who suffered from lack of opportunities in the home area and who were easily able to take advantage of opportunities elsewhere.

The concept of *place utility* helps us understand the decision-making process that potential migrants undergo. It is the value an individual places on each potential migration site. The decision to migrate is a reflection of the appraisal by a potential migrant of the current home site as opposed to other potential sites. The individual may adjust to conditions at the home site and thus decide not to migrate.

In the evaluation of the place utility of each potential site, the decision maker considers not only present relative place utility but also expected place utility. The evaluations are matched with the individual's aspiration level, that is, the level of accomplishment or ambition that the individual sees for himself or herself. Aspirations tend to be adjusted to what an individual considers attainable. If one is satisfied with present circumstances, search behavior is not initiated. If, on the other hand,

there is dissatisfaction, a utility is assigned to each of the possible new sites. The utility is based on past or expected future rewards at the various sites. Since the new places are unfamiliar to the individual, the information received about them acts as a substitute for the personal experience of the home site. The decision maker can do no more than sample information about the new sites, and, of course, there may be sampling errors.

One goal of the potential migrant is to minimize uncertainty. Most decision makers either elect not to migrate or postpone the decision unless uncertainty can be lowered sufficiently. Most migrants reduce uncertainty by imitating the successful procedures followed by others. We see that the decision to migrate is not a perfunctory, spur-of-the-moment reaction to information. It is usually a long, drawn-out process based on a great deal of sifting and evaluation of information.

An example of some of these observations and generalizations in action can be seen in the case of the large numbers of young Eastern European males who migrated to the United States in the 1890s and the early 1900s. Faced with rural poverty and overpopulation at home, they had to consider the place utility there to be minimal. Their space-searching ability was, however, limited by lack of money and by lack of alternatives in the lands of their birth. With a willingness to work and with as-

278

pirations for success—perhaps wealth—in the United States, they were eager listeners to representatives of American steel and coal companies seeking a cheap, docile, and numerous labor supply not easily then available in this nation. Thousands, presented with the glittering prospect of a certain job, quickly placed high utility on relocation to the United States, and acting on assurances given them by company representatives, they accepted employment, passage money, and a sign—to be worn about their necks at disembarkation—indicating company affiliation and city of destination. They were the last great wave of European immigrants to this country, acting on information received; whether or not they experienced sampling error, only they could report. At least, their arrival indicated their assignment of higher utility to the new site than to the old.

BARRIERS TO MIGRATION

Migration depends on a knowledge of opportunities in other areas. People with a limited knowledge of opportunities elsewhere are less likely to migrate than are better-informed individuals. Other barriers to migration, besides limited knowledge, are the cost of moving, ties to individuals and institutions in the original activity space, and government regulations.

The cost factor limits long-distance movement, but the larger the differential between present circumstances and perceived opportunities, the more individuals are willing to spend on moving. Figure 8.16 is an example of the effect of distance on relocation. For many people, especially older people, the differentials must be extraordinarily high for movement to take place.

Family, religious, ethnic, and community relationships defy the principle of differential opportunities. Many people will not migrate under any but the most pressing of circumstances. Even if ties are not strong, people may

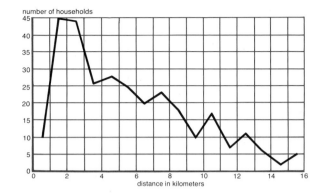

Fig. 8.16 Distances between old and new residences in the Asby area of Sweden. Notice how the number of movers decreases with increasing distance. (Data from T. Hägerstrand, *Innovation Diffusion as a Spatial Process*, University of Chicago Press, 1967.)

think that a move would be too disruptive.

For many reasons, most people migrate within their own country. Many governments frown upon movements into or outside of their own borders. Countries where per capita incomes are high or are perceived as high are generally the most desired international destinations. In order to protect themselves against overwhelming migration streams, such countries have restrictive policies on immigration. Other countries, recognizing that emigration might be economically or politically disadvantageous, restrict out-migration.

On the other hand, countries suffering from an excess of workers often encourage emigration. The huge migration of people to the Americas in the late 19th and the early 20th centuries is a good example of perceived opportunities for economic gain far greater than in the home country. Many European countries were overpopulated, and their political and economic systems stifled domestic economic opportunity at a time when people were needed by American entrepreneurs hoping to increase their wealth in virtually untapped resource-rich areas.

279

In the 20th century, nearly all countries have experienced a movement of population to cities from their agricultural areas. The migration has presumably paralleled the number of perceived opportunities within cities and convictions of absence of place utility in the rural districts. Perceptions, of course, do not necessarily accord with reality, as we have seen; the enormous influx of rural folk to major urban areas in the underdeveloped and developing countries and the economic destitution of many of those in-migrants suggest recurrent gross misperceptions, faulty information, and sampling error.

PATTERNS OF MIGRATION

Several geographic concepts deal with patterns of migration. The first of these is the *migration field.* For any single place, the origin of its in-migrants and the destination of its out-migrants remain fairly stable spatially over

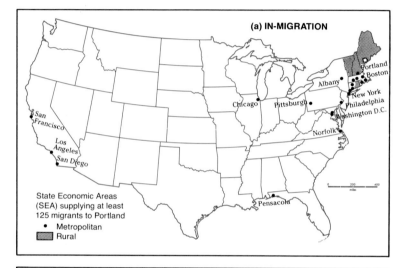

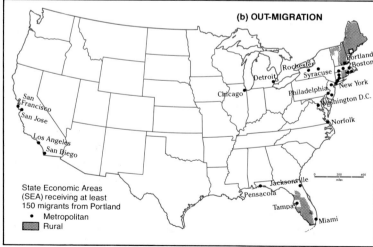

Fig. 8.17 The in-migration field (a) and the out-migration field (b) of Portland, Maine. Except for the retirement center of Florida, the in- and out-migration patterns for Portland are roughly similar. Long-distance migration is mainly to and from large cities, understandable if we keep in mind how information is transmitted. (From *Human Spatial Behavior* by Jakle et al. © 1976 by Wadsworth Pub. Co., Belmont, CA 94402. Reprinted by permission of the publisher, Duxbury Press.)

time. Areas that dominate a locale's in- and out-migration patterns constitute the migration fields for the place in question, as shown in Figure 8.17. As we would expect, areas near the point of origin make up the largest part of the migration field. However, places far away, especially large cities, may also be prominent. These characteristics of migration fields are functions of the hierarchical movement to larger places (discussed below) and the fact that there are so many people in large metropolitan areas that one might expect some migration into and out of them from most areas within a country.

Migration fields do not conform exactly to the diffusion concepts mentioned earlier. As shown by Figure 8.18, some migration fields reveal a distinctly *channelized* pattern of flow. The channels link areas that are socially and economically tied to one another by past migration patterns, by economic trade considerations, or by some other affinity. As a result, flows of migration along these channels are greater than would otherwise be the case. The movement of blacks from the southern United States to the North, of Scandinavians to Minnesota and Wisconsin, of Mexicans to such border states as Texas and New Mexico, of retirees to Florida and Arizona are all examples of channelized flows.

Return migration is a term used to refer to those who soon after migrating decide to return to their point of origin. If freedom of movement is not restricted, it is not unusual for as many as 25% of all migrants to return to their place of origin. Unsuccessful migration is sometimes due to an inability to adjust to the new environment; more often, it is a result of false expectations based on distorted mental images of the destination at the time of the move. Myths, secondhand information, lies, and people's own exaggerations contribute to what turn out to be mistaken decisions to move. While return migration often represents the unsuccessful adjustment of individuals to a

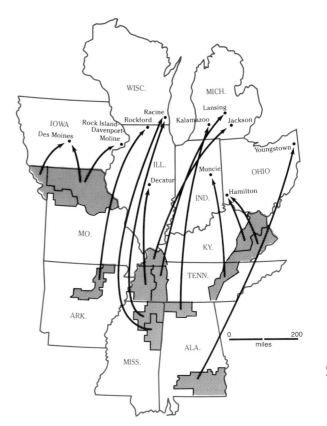

Fig. 8.18 Channelized migration flows from the rural South to midwestern cities of medium size. Distance is not necessarily the main determinant of flow direction. Perhaps through family and friendship links, the rural southern areas are tied to particular midwestern destinations. (Reproduced by permission from *Proceedings* of the Association of American Geographers, vol. 3, 1971, C. C. Roseman.)

281

new environment, it does not necessarily mean that negative information about a place returns with the migrant. It usually means a reinforcement of the channel, since communication lines between the unsuccessful migrant and would-be migrants take on added meaning and understanding.

In addition to channelization, the influence of large cities causes migration fields to deviate from the distance-decay pattern. The concept

**Recent
Residential
Preferences
in the
United States**

Since the late 1960s, migration patterns in the United States have appeared to be running contrary to the tradition of hierarchical movement, that is, of relocation from small places to larger ones. Metropolitan areas between 1970 and 1978 grew at a much slower rate than in previous postwar periods (many even declined in population), and central cities continued a relatively recent pattern of absolute population loss. In contrast, a substantially increased rate of growth was recorded in nonmetropolitan centers. The general reversal of hierarchical growth patterns, associated with the alteration of earlier migration streams, has also been observed in Western Europe and in Japan. Underlying the migration flow reversal is some combination of a change in the locational preferences of people and their changing abilities to act upon those preferences.

Several national surveys were conducted in the United States during the 1970s, seeking expressions of residential preference by size of community and by type of area. The results showed that a majority of the respondents were attracted to just those community and areal locations that had been losing population up to the late 1960s: smaller towns in nonmetropolitan districts. One survey revealed that although 20% of those sampled lived in cities of above 500,000 population, only 9% preferred such locations; 24% lived in communities of between 50,000 and 500,000, but only 16% wanted to be there. Among all the respondents, however, although a clear majority preferred residence in smaller centers than they currently inhabited, a similar majority wished to be located in areas accessible to large urban complexes. Residential preferences, then, were clearly in favor of small towns outside of, but close to, metropolitan concentrations.

These may always have been the locational wishes of Americans; no early comparable surveys exist to prove the point either way. What does exist is direct evidence that changing patterns of employment opportunities and altering stages in life cycles now provide the population with the ability to act upon their expressed preferences.

Industrial and commercial decentralization within metropolitan areas has created more employment centers outside of central cities, employment that can be reached from residences in counties immediately outside of standard metropolitan areas. A small-town domicile near—but not in—major urban agglomerations is increasingly possible, satisfying the residential desires of most survey respondents. In addition, a general decentralization of American industry is reflected in the greater relative growth of small rather than larger metropolitan areas, encouraging migration down rather than up the urban hierarchy. Industrial dispersion has also expanded the economic base of both the nonmetropolitan and the urban Southeast, encouraging a trend—first noticed in the late 1970s—toward reverse migration of blacks from northern cities to their former home areas.

Finally, lower retirement ages, increasing pension incomes, and a steady increase in the number and proportion of older citizens permit large numbers of

Americans (and their European and Japanese counterparts) to act upon their residential preference for resort or retirement-area homes in lower-cost regions outside metropolitan centers. The South—including, particularly, Florida—the Ozark district, the northern Midwest, and the interior West have experienced growth based upon their amenities. To the extent that their growth reflects a return to the home areas of the metropolitan-bound migrants of 30–40 years ago, an element of return migration as well as reverse hierarchical movement is evident.

---

of *hierarchical* migration assists us in understanding the nature of migration fields. Earlier we noted that sometimes information diffuses according to a hierarchical rule, that is, from city to city at the highest level in the hierarchy and then to lower levels. Hierarchical migration, in a sense, is a response to that flow. The tendency is for individuals in domestic relocations to move up the level in the hierarchy, from small places to larger ones. Very often, levels are skipped on the way up; only in periods of general economic decline is there considerable movement down the hierarchy. Suburbs of large cities are considered part of metropolitan areas, so that movement from a town to a small suburb would be considered moving up the hierarchy. From this pattern, we can envisage information flowing to a place via a hierarchical routing. Once the information is digested, some respond by migrating to the area from whence the information came.

Conclusion

In this discussion of behavioral geography, we have emphasized the factors that influence the way individuals use space. At the beginning of the chapter, we posed a number of questions, used as our theme. In considering how individuals view their environment, we spoke of the nature of the information availa-ble and of the age of people, their past experiences, and their values. The question of differences in the extent of space used led us to the concept of activity space. The age of an individual, the degree of mobility, and the availability of opportunities all play a role in defining the limits of individual activity space. We saw, however, that as distance increases, our familiarity with our environment decreases. The concept of the critical distance is helpful to a recognition of the distance beyond which the decrease in familiarity with places begins to be significant.

The way space is used is a function of all of the factors mentioned, but certain factors having to do with the diffusion of innovations indicate what opportunities exist for individuals living in various places. Distance, density, and hierarchical diffusion influence the geographic direction that cultural change will take. Finally, we examined how our ability to make well-reasoned, meaningful decisions is a function of the utility we assign to places and the opportunities at those places.

Behavioral geographers view individual action in a special way. There is a strong emphasis on information flow and the frictional effects of distance. In many respects, it is an interdisciplinary field in that the combined ideas from geographers, psychologists, economists, and other social scientists are crucial to an understanding of the spatial behavior of humans.

283

For Review

1. What is meant by *activity space*? What factors affect the areal extent of the activity space of an individual?

2. Recall the places you have visited in the past week. In your movements, were the *distance-decay* and *critical-distance* rules operative? What variables affect an individual's critical distance?

3. Briefly distinguish between distance-affected diffusion, population-density–affected diffusion, and hierarchical diffusion. In what ways, if any, were these forms of diffusion in operation in the culture hearths discussed in Chapter 5?

4. On a blank piece of paper, and without any maps to guide you, draw a map of the United States, putting in state boundaries wherever possible; this is your mental map of the nation. Compare it with a standard atlas map. What conclusions can you reach?

5. What considerations affect a decision to migrate? What is *place utility,* and how does its perception induce or inhibit migration? What common barriers to migration exist?

6. Define the term *migration field.* Some migration fields show a channelized flow of people. Select a particular channelized migration flow (such as the movement of Scandinavians to the United States, or people from the Great Plains to California, or southern blacks to the North) and explain why a channelized flow developed.

284

Suggested Readings

ABLER, RONALD, JOHN S. ADAMS, and PETER GOULD, *Spatial Organization.* Prentice-Hall, Englewood Cliffs, NJ, 1971.

DOWNS, ROGER M., and DAVID STEA, *Maps in Mind: Reflections on Cognitive Mapping.* Harper & Row, New York, 1977.

GOULD, PETER, and RODNEY WHITE, *Mental Maps.* Penguin Books, Harmondsworth, Middlesex, England, 1974.

JAKLE, JOHN A., STANLEY BRUNN, and CURTIS C. ROSEMAN, *Human Spatial Behavior.* Duxbury Press, North Scituate, MA, 1976.

LOWE, JOHN C., and S. MORYADAS, *The Geography of Movement.* Houghton Mifflin, Boston, 1975.

PORTEOUS, J. DOUGLAS, *Environment and Behavior: Planning and Everyday Life.* Addison-Wesley, Reading, MA, 1977.

SAARINEN, THOMAS F., *Environmental Planning: Perception and Behavior.* Houghton Mifflin, Boston, 1976.

TUAN, YI-FU, *Topophilia: A Study of Environmental Perception, Attitudes, and Values.* Prentice-Hall, Englewood Cliffs, NJ, 1974.

WOLPERT, JULIAN, "Behavioral Aspects of the Decision to Migrate," *Papers* of the Regional Science Association, vol. 15, 1965, pp. 159–69.

285

286

The Location Tradition

The Scythia, the Thule, the Britain, the Germany, and the Gaul which the Roman writers describe in such forbidding terms, have been brought almost to rival the native luxuriance and easily won plenty of Southern Italy; and, while the fountains of oil and wine that refreshed old Greece and Syria and Northern Africa have almost ceased to flow, and the soils of those fair lands are turned to thirsty and inhospitable deserts, the Hyperborean regions of Europe have conquered, or rather compensated, the rigors of climate, and attained to a material wealth and variety of product that, with all their natural advantages, the granaries of the ancient world can hardly be said to have enjoyed.

So, in the lavish prose of the 19th century, did George Perkins Marsh capsulize the changing fortunes of peoples and places. Natural endowment, cultural development, and changing patterns of land use are all implicit in his remarks, as they are in the location tradition of geography.

A theme that has run like a thread through the first eight chapters of this book may aptly be termed the *theme of location.* Distributions of climates, landforms, culture hearths, and religions were points of interest in our examinations of physical and cultural landscapes. In Part III of our study, the analysis of location becomes our central concern rather than just one strand among many, and the location tradition of geography is brought to the fore.

In the study of economic, resource, or urban geography, a central question has to do with *where* certain types of human activities take place, not just in absolute terms but also in relation to one another. In studying the distribution of a given activity, such as the making of iron and steel, the concern is with identifying the locational pattern, analyzing it to see why it is arranged the way it is, and searching for underlying principles of location. Because the elements of culture are integrated, such a study often must take into consideration the way in which the location of one element affects the locations of others. For example, certain characteristics of the labor pool have helped to alter the distribution of the textile industry in America, and this alteration has had ramifications for both the New England area and the South, as well as affecting the location of other economic activities.

In Chapter 9, "Economic Geography," our attention is directed to the location of economic activities, as we seek to answer the question of why they are distributed as they are. What forces operate to make some regions extremely productive and others less so? Broad types of economic systems, such as commercial and planned, are analyzed, and a close look is taken at the locational characteristics of some important manufacturing and agricultural industries.

The resources upon which humankind draws and depends for its sustenance and development are not distributed uniformly in quantity or quality over the earth. Chapter 10, "The Geography of Natural Resources," centers on the relationship between the demands people place upon the spatially varying environment—demands that are culture-bound—and the ability of the environment to sustain those demands. In an increasingly integrated economic and social world of growing consumption demands, control of resources and their wise and productive use loom ever more importantly in human affairs. The understanding of resources in both their distributive and their exploitative characteristics is a theme of growing concern in the location tradition of geography.

The increasing urbanization of the world's population represents another stage of the processes of cultural development, economic change, and environmental control, aspects of which have been the topics of separate chapters. In Chapter 11, urban areas are viewed in two ways. First, the focus is upon cities as points in space with patterns of distribution and specializations of functions that invite analysis. Second, cities are considered as landscape entities with specialized arrangements of land uses resulting from recognizable processes of urban growth and development. Within their confines, and by means of the interconnected functional systems that they form, cities summarize a present stage of economic patterning and resource utilization of humankind. The consideration of urban geography thus logically concludes our examination of the location tradition of geography.

288

The crop bloomed luxuriantly that summer of 1846. The disaster of the preceding year seemed over, and the potato, the sole sustenance of some 8 million Irish peasants, would again yield in the bounty needed. Yet within a week, wrote Father Mathew, "I beheld one wide waste of putrefying vegetation. The wretched people were seated on the fences of their decaying gardens . . . bewailing bitterly the destruction that had left them foodless." Colonel Gore found that "every field was black," and an estate steward noted that "the fields . . . look as if fire has passed over them." The potato was irretrievably gone for a second year; famine and pestilence were inevitable.

Within five years, the settlement geography of the most densely populated country in Europe was forever altered. The United States received a million immigrants, who provided the cheap labor needed for the canals, railroads, and mines that it was creating in its rush to economic development. New patterns of commodity flows were initiated as American maize for the first time found an Anglo-Irish market—as part of Poor Relief—and then entered a wider European market that had also suffered general crop failure in that bitter year.

In an instant, a microscopic organism, the cause of the potato blight, had altered the economic and human geography of two continents. It had done so by a chain of interlocking causes and consequences dramatically demonstrating the unitary nature of those physical and cultural patterns which make up that world view called geography. Central among those patterns are those that the economic geographer isolates for special study.

Simply stated, economic geography is the study of how people earn their living, how livelihood systems vary by area, and how economic activities are interrelated and linked. It is, of course, vastly complex, as many lengthy volumes attest. In fact, we cannot really comprehend the totality of the economic pursuits of 4 billion human beings. We cannot examine

Economic Geography

289

the infinite variety of productive activities found on each unit area of the earth's surface; nor can we trace all their innumerable interrelationships, linkages, and flows. Even if that level of understanding were possible, it would be valid for only a fleeting instant of time.

The economic geographer seeks consistencies. The attempt is to discover generalizations that will aid in the comprehension of the maze of economic variations characterizing human existence. The search may be regional or topical—contrasts between rain forest and grassland or between agricultural and manufacturing industries. The search may be, as it is here, directed to a recognition of the types of economic systems identified with generalized patterns of production and consumption and an examination of the location of economic activities within those systems. Ultimately, as geographers, economic geographers address the how and the why of variations in the spatial patterns of economic activity.

In the compass of a single chapter, we can hope only to suggest the breadth of economic geographic interests and their centrality to all human geographic inquiry. More will be left out than included of the range of concerns, contributions, and conclusions of economic geographers. Some matters, particularly those dealing with the ways people earn their living in cities, are deferred to a later chapter. The intent is introduction and the creation of an interest and awareness that will excite further inquiry. The potato blight, apparently of concern only to a small island, ultimately affected the economies of continents; the depletion of America's petroleum resources is altering international alliances, the relative wealth of nations, and more. Economic geographic understanding is essential to the comprehending citizen.

Types of Economic Systems

Broadly speaking, there are three major types of economic systems: subsistence, commercial, and planned. In a *subsistence* economic system, goods and services are created for the use of the producers and their kinship group. There is, therefore, little exchange of goods and only limited need for markets. In a *commercial* economy, as the term is used here, the producers or their agents market the goods and services, and the laws of supply and demand determine their price and quantity. Market competition is the primary force shaping the location patterns involved. Trading, bargaining, and bidding are characteristic of the system. A *planned* economy is one in which the producers or their agents market the goods and services, usually to a government agency that controls their supply and price. The quantities produced and the locational patterns of production are carefully programmed by the planning agency.

In actuality, few people are members of only one of these systems. A farmer in India may produce rice and vegetables privately for the family's consumption, but very often the family will save some of the produce to sell. Members of the family may make cloth or other handicrafts that are marketed either by the farmer or by a trader in a local marketplace. With the money derived from the sale of the goods the Indian peasant is able to buy, among other things, clothes for the family, tools, or fuel. Thus, the farmer is a member of at least two systems: subsistence and commercial.

A Polish farmer may produce a specified amount of grain for a government food agency and another amount for the family and then sell the surplus in a nearby marketplace. Thus that farmer is a member of all three types of systems.

290

In the United States, government controls on the production of various types of goods and services (such as wheat, alcohol, nuclear power plants, and truck routes) mean that the United States does not have a purely commercial economy; to a limited extent, it is a planned one. Example after example would show that there are very few people in the world who are members of only one type of economic system.

In a given country, however, one of the three forms of economic system is usually dominant, and it is relatively easy to classify countries by them. Much more difficult is equating these separate systems with the degree of technological advancement displayed by nations so classified. Although it is true that subsistence-dominated countries are usually less developed technologically than either commercial or planned economies, in recent years there has been a strong movement among them to in-

crease the level of administrative control over their economies. More and more they are adopting advanced technologies and simultaneously becoming either more commercial, more planned, or, as is usually the case, a combination of both.

Figure 9.1 shows the pattern of domination of the three economic systems. For nations that are now dominated by commercial or planned economies but that still have the majority of their people at the subsistence level, light stripes have been placed within the boundaries. For example, China has a governmentally controlled economic system, but with so many of the Chinese producing agricultural goods for their immediate working groups, one must recognize the degree of subsistence economy in the nation.

Japan, Western Europe, and the United States are among those areas characterized by com-

Fig. 9.1   Generalized economic systems of the world.   **291**

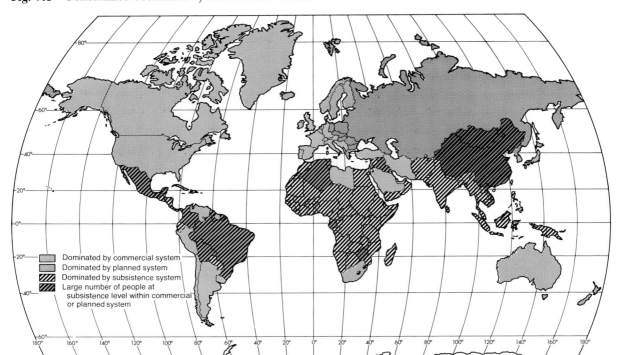

Dominated by commercial system
Dominated by planned system
Dominated by subsistence system
Large number of people at subsistence level within commercial or planned system

mercial or market economies. Although the Western European countries are increasingly developing government-specified production goals, land, mineral wealth, farms, and factories have traditionally been owned by private individuals. Only 75 years ago, Japan was dominated by subsistence agricultural activities, but mainly through government support for commercial production, the country developed into one of the leading urban-industrial economies of the world, and few people now engage in subsistence farming.

In Eastern Europe and the Soviet Union, government programs for economic growth have been used to change subsistence-dominated economies into technologically advanced, planned systems. In these and several other countries shown on Figure 9.1, the means of production—land, minerals, farms, and factories—are owned by the state, although some

variant ownership forms (private and cooperative) are permitted in a few nations such as Poland and Yugoslavia.

Most of the countries of South America, Africa, Asia, and the Middle East are dominated by subsistence economies. These areas, however, are undergoing rapid change. In fact, a most striking aspect of recent history is the extraordinary change in economic system and development characteristic of many of these countries. It is convenient to separate the nations of these areas into two groups: those with abundant resources per capita—for example, the Middle Eastern countries ("the have countries")—and the nations with poor resource bases ("the have-not countries"). Although there is a uniformity in the nature of the subsistence economies, the forces that are promoting change within them differ.

Fig. 9.2 Nomadic herding, supporting relatively few people, occupies large parts of the dry and cold world.

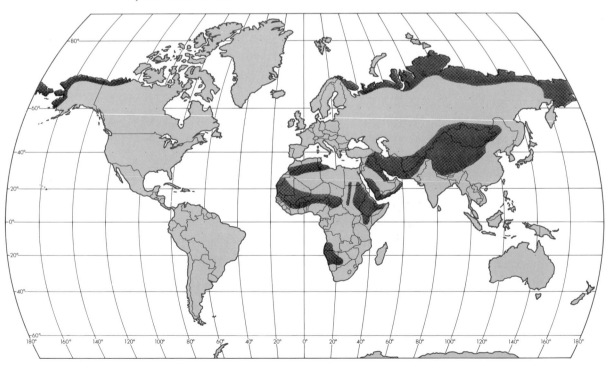

Subsistence Economic Systems

Since, by definition, a subsistence economic system involves nearly total self-sufficiency on the part of its members, production for exchange is minimal and each family or close-knit social group relies upon itself for its food and other most essential requirements. Subsistence systems are, then, overwhelmingly agricultural—more precisely, rural—and technologically "underdeveloped." Within that generalized description, two chief types of subsistence economic systems may be recognized: extensive and intensive. The variants within each type are several; the essential contrast between the two types, however, is realizable yield per acre utilized and, therefore, population-supporting potential. Extensive subsistence agriculture, including livestock specialization, involves large areas of land and minimal labor input. Both product per acre and population densities are low. Intensive subsistence farming, in which livestock rearing is less significant, involves the cultivation of small land holdings through the expenditure of great amounts of labor. Yields per unit area and population densities are both high.

EXTENSIVE SUBSISTENCE SYSTEMS

Of the several types of extensive subsistence systems—varying one from another in their intensities of land use—two are of particular importance.

Nomadic herding—the wandering but controlled movement of livestock solely dependent upon natural forage—is the most extensive type of land-use system. Over large portions of the Asiatic semidesert and desert areas and on the fringes of and within the Sahara Desert, there is a relatively small number of people who graze animals not for market sale but for consumption by the herder group (Figure 9.2). Sheep, goats, and camels are most common, while cattle, horses, and yaks are locally important; the reindeer of Lapland are part of the same system. Whatever the animals involved, their common characteristics are hardiness, mobility, and an ability to subsist on sparse forage. The animals provide a variety of products: milk, cheese, and meat for food; fibers and skins for clothing; skins for shelter; excrement for fuel; and bones for tools. For the herder, they represent total subsistence. Nomadic movement is tied to sparse and seasonal rainfall or to cold temperature regimes and to the areally varying appearance and exhaustion of forage. Extended stays in a given location are neither desirable nor possible.

As a type of economic system, nomadic herding is declining. More and more the governments of countries in which herders roam are becoming uneasy about their inability to govern the nomads, whose tradition is the warlike independence associated vividly with the Mongol hordes of Genghis Khan and his successors. Nomads generally disregard the boundaries of countries as irrelevant to the needs of herds, an attitude that is troublesome to those who wish to maintain the territorial integrity of their respective states. But probably of most significance is the availability of better opportunities as industrialization comes to many of the less technologically advanced countries. For example, the oil industry of the Middle East is attracting many Bedouin tribesmen to jobs in the oil fields or in the growing cities. A negative factor, too, is the growing nomad population, which demands larger and larger herds to sustain itself in traditional ways, putting intolerable pressure on a fragile environment.

A much differently based and distributed form of extensive subsistence agriculture is found in all of the warm, wet, tropical areas of the world. There, largely in the tropical rainforest environments discussed in Chapter 3, many people engage in a kind of nomadic farm-

293

ing. Because the drenched soil comes to be leached of many of its nutrients, the farmers need to move on after harvesting several crops. This type of *shifting cultivation* has a number of names, the most common of which is *slash and burn*. With sharp knives, the farmers hack down the natural vegetation, usually trees and vines, burn the cuttings, and then plant such crops as maize (corn), millet (a cereal grain), rice, starchy manioc or cassava, yams, and sugar (Figure 9.3). Initial yields—the first and second crops—may be very high in biomass (the quantity of living organic matter in a given area) in comparison with other world agricultural forms, but because of rapid soil exhaustion and the swift return of native vegetation,

the yields quickly become lower with each planting on the same plot. Wet lowlands are less favored for farm clearings than the better drained hillsides where erosion exposes fresh soil. Wherever sited, plots are soon invaded by native vegetation and must be relocated.

As indicated by Figure 9.4, large areas of the world are given over to this kind of system. Population densities are low, for much land is needed to support few people, but here as elsewhere, population density must be considered a relative term. In actuality, although overall density is low, persons per square mile of cultivated land may be high. Shifting cultivation may be seen as a transitional agricultural form in that it is labor-intensive on shifting plots. It

Fig. 9.3 Slash-and-burn agriculture in the Amazon Basin. Land that was once jungle has been partially cleared and burned. Rice is growing between the stumps. (United Nations.)

294

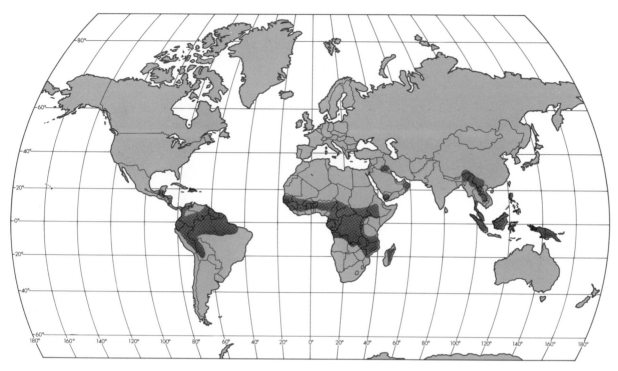

Fig. 9.4 Areas where shifting cultivation and some livestock herding are practiced. These are essenitally subsistence agricultural regions.

is found in Indonesia, and on the islands of Borneo, New Guinea, and Sumatra; it is part of the economy of the uplands of South Asia in Vietnam, Thailand, Burma, and the Philippines. Nearly the whole of Central and West Africa away from the coasts, Brazil's Amazon Basin, and large portions of Central America are all noted for this type of extensive subsistence agriculture. These areas are among the most isolated in the world. Roads, railroads, and telephones are largely absent. As modern technology is introduced, these cultivators will inevitably be reduced in number. It may be argued that shifting cultivation is a highly efficient cultural adaptation when the measuring rod is the ratio of energy input to food output. Fertilizers, of course, would help sustain initial high yields, but would also increase the ratio. In a world short of energy, progress brings its penalties.

INTENSIVE SUBSISTENCE SYSTEMS

Over one-half of the people of the world are engaged in intensive subsistence agriculture, which predominates in the areas shown on Figure 9.5. Hundreds of millions of Indians, Chinese, Pakistanis, Banglas, and Indonesians plus further millions in other Asian, African, and Latin American nations are small-plot rice and wheat farmers. Most live in monsoon Asia, and we will devote our attention to that area.

Intensive subsistence farmers are concentrated in such major river valleys and deltas as the Ganges and the Yangtze, and in smaller valleys close to coasts—level areas with fertile alluvial soils. These warm, moist districts are well suited to the production of rice, a crop that under ideal conditions can provide large amounts of food per unit of land. Rice also requires a great deal of time and attention, for

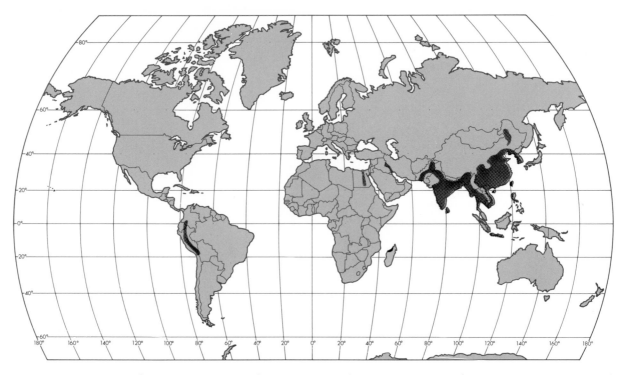

Fig. 9.5 Intensive subsistence agriculture. Large parts of Asia support millions of people engaged in sedentary subsistence cultivation, with rice and wheat the chief crops.

planting rice shoots by hand in standing fresh water is a tedious art (Figure 9.6). In the cooler and drier portions of Asia, wheat is grown intensively, along with millet and upland rice.

Intensive subsistence farming is characterized by large inputs of labor per unit of land, by the promise of high yields in good years, by small plots, and by the intensive use of fertilizers—mostly animal manure. For food security and dietary custom, there is some diversification of products; vegetables and some livestock are part of the agricultural system. Cattle are raised for labor and for food. Swine, ducks, and chickens are used for food, but since Muslims eat no pork, hogs are absent in their areas of settlement. Hindus generally eat no meat, but the large number of cattle in India are vital for labor, as a source of milk, and as producers of fertilizer and fuel.

The introduction of improved health care in this century has strongly influenced population growth rates in these countries. Lowered infant and crude death rates have helped to increase population enormously since birth rates have only recently shown decrease. Rising population, of course, puts increasing pressure on the land; the response has been to increase the intensity of production. Lands formerly considered unsuitable for agriculture by reason of low fertility, inadequate moisture, difficulty of clearing and preparation, isolation from settlement areas, and other factors have been brought into cultivation. To till those additional lands, a price must be paid. In any economic activity, there is an additional (marginal) cost incurred—in labor, capital, or other unit of expenditure—to bring into existence an added unit of production. When the value of

The village of Nanching is in subtropical southern China on the Pearl River (now the Xi River) delta near Canton. Its pre-Communist subsistence agricultural system is described by a field investigator, whose account is here condensed. The system is found in its essentials in other rice-oriented societies.

The Economy of a Chinese Village

In this double-crop region, rice was planted in March and August and harvested in late June or July and again in November. March to November was the major farming season. Early in March the earth was turned with an iron-tipped wooden plow pulled by a water buffalo. The very poor who could not afford a buffalo used a large iron-tipped wooden hoe for the same purpose.

The plowed soil was raked smooth, fertilizer was applied, and water was let into the field, which was then ready for the transplanting of rice seedlings. Seedlings were raised in a seedbed, a tiny patch fenced off on the side or corner of the field. Beginning from the middle of March, the transplanting of seedlings took place. The whole family was on the scene. Each took the seedlings by the bunch, ten to fifteen plants, and pushed them into the soft inundated soil. For the first thirty or forty days the emerald green crop demanded little attention except keeping the water at a proper level. But after this period came the first weeding; the second weeding a month later. This was done by hand, and everyone old enough for such work participated. With the second weeding went the job of adding fertilizer. The grain was now allowed to stand to "draw starch" to fill the hull of the kernels. When the kernels had "drawn enough starch," water was let out of the field, and both the soil and the stalks were allowed to dry under the hot sun.

Then came the harvest, when all the rice plants were cut off a few inches above the ground with a sickle. Threshing was done on a threshing board. Then the grain and the stalks and leaves were taken home with a carrying pole on the peasant's shoulder. The plant was used as fuel at home.

As soon as the exhausting harvest work was done, no time could be lost before starting the chores of plowing, fertilizing, pumping water into the fields, and transplanting seedlings for the second crop. The slack season of the rice crop was taken up by chores required for the vegetables which demanded continuous attention, since every peasant family devoted a part of the farm to vegetable gardening. In the hot and damp period of late spring and summer eggplant and several varieties of squash and beans were grown. The green-leafed vegetables thrived in the cooler and drier period of fall, winter and early spring. Leeks grew the year round.

When one crop of vegetables was harvested, the soil was turned and the clods broken up by a digging hoe and leveled with an iron rake. Fertilizer was applied, and seeds or seedlings of a new crop were planted. Hand weeding was a constant job; watering with the long-handled wooden dipper had to be done an average of three times a day, and in the very hot season when evaporation was rapid, as frequently as six times a day. The soil had to be cultivated with the hoe

297

frequently as the heavy tropical rains packed the earth continuously. Instead of the two applications of fertilizer common with the rice crop, fertilizing was much more frequent for vegetables. Besides the heavy fertilizing of the soil at the beginning of a crop, usually with city garbage, additional fertilizer, usually diluted urine or a mixture of diluted urine and excreta, was given every ten days or so to most vegetables.

Adapted from *A Chinese Village in Early Communist Transition* by C. K. Yang by permission of the M.I.T. Press, Cambridge. Massachusetts. Copyright 1959.

that added (marginal) production at least equals the marginal cost, the effort may be undertaken. In past periods of lower population pressure, there was no incentive to extend cul-

tivation to less productive or more expensive unneeded lands; now, circumstances are altered.

There are limits to possible agricultural in-

Fig. 9.6 Transplanting rice seedlings requires arduous hand labor by all members of the family. The newly flooded paddies, previously plowed and fertilized, will have their water level maintained until the grain is ripe. This photograph was taken in Cambodia; the scene is repeated wherever subsistence paddy-rice agriculture is practiced. (United Nations.)

tensification and expansion, and new population-supporting mechanisms are being eagerly sought, particularly in China, India, Pakistan, Bangladesh, and Indonesia. At least partial conversion of national economies to a stronger emphasis on industrial development is viewed as indispensable both to provide employment opportunities divorced from agriculture and to secure to the nation the modernization and the release from subsistence that such development presumably promises. To guarantee the type of industrialization deemed socially and economically desirable, and because the private sector cannot offer the capital for the large-scale schemes envisioned, these subsistence economies are moving toward at least partially administered economic systems, with central planning and state control of production.

The same forces for change exist in the other countries of Asia. All would like to be more self-sufficient and not to have to import huge amounts of food and machinery from abroad. Improvements in crop varieties, better farming practices, and the use of chemical fertilizers are increasing crop yields, but the pressure on the land is still great. The promise of the "Green Revolution"—a term indicating the huge yield increases to be realized as improved strains of wheat and rice became adopted by peasant farmers—is only gradually being fulfilled. Failures in central planning, in capital attraction, and in the promotion of national unity have been obstacles to many developmental efforts. Perhaps most important is the fact that in crowded, traditional societies, people's day-to-day survival depends so much on the land and on traditional modes of action that they find it difficult to look to and provide for a different tomorrow.

There are variations on the systems just described. For example, there are still some tribes of hunters and gatherers in the Arctic north and in the desert and rain forest south. Subsistence fishing villages abound along the coasts of South Pacific islands and in Southeast Asia.

Wherever commercial economies have had an impact, these subsistence systems have been modified.

Commercial Economic Systems

Modifications of subsistence systems have inevitably made them more complex by imbuing them with at least some of the heterogeneity and linkages of activity that mark the commercial, or Western, economic system. The subsistence economies individually tend toward uniformity of product mix and lack of labor specialization. They lack those linkages between production stages and between producer and consumer that require transport and marketing functions. The commercial system—that of advanced economies—possesses the complex specializations and linkages that the subsistence economies lack. The results of specialized labor inputs are hierarchically and spatially linked in interdependent patterns of production, exchange of goods and services, and ultimate consumption. By its very complexity, the commercial economic system has invited the most serious attention of economists and economic geographers. The body of theory and model designed to explain the system's formal and informal mechanisms is vast and far beyond the scope of this introduction. The purpose here is merely to outline the system's central elements.

Three important characteristics of commercial systems are specialization, profit, and interdependence. Individuals, industries, and even areas specialize, allowing producers to be efficient, so that their unit costs of making and handling goods are as low as possible. Low prices usually bring higher sales and greater profits. Interdependence occurs because producers must cooperate with each other to ensure efficient marketing procedures and in-

299

creased profits. Businesspeople recognize, of course, that profits allow them to invest more capital in production and thereby reap even greater rewards. The profit motivation of commercialists can result in savings, investment, and increased production. From the geographer's point of view, specialization, interdependence, and profit imply that businesspeople select efficient locations for their activities, locations that are the generating points of trade.

Clearly some sites are much more advantageous than others for commercial activities. Location at a favorable site allows for low production costs and perhaps large-scale production. Some locations have both raw material resources and markets nearby; water and a food supply may be present, too, as well as a productive labor force. Transportation is a key variable; in fact, Western economies cannot flourish without well-connected transport systems (Figure 9.7). Outmoded vehicles or silted harbors, potholed roads or expensive turnpikes, all put restraints on trade by raising transportation costs. The advantages of a site may be so great that an industry or a specialized form of agriculture can overcome these restraints, but all site costs are considered very carefully by profit-motivated commercialists.

Not only is there specialization at sites, but countries themselves specialize. If one country can gain an advantage in a particular enterprise

Fig. 9.7 Accessibility is a key measure of economic development and of the degree to which a world region can participate in interconnected market activities. Isolated areas of nations with advanced economies suffer price disadvantage because of high transportation costs. Lack of accessibility in subsistence economic areas slows their modernization and hinders their participation in the world market. (From *Comparative World Atlas 1971.* Copyright permission—Hammond Incorporated Maplewood, NJ.)

300

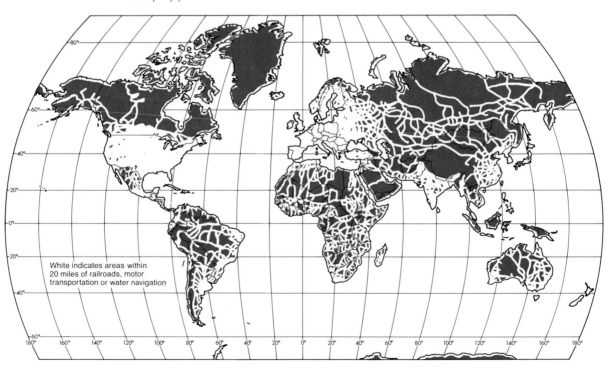

White indicates areas within 20 miles of railroads, motor transportation or water navigation

because of low costs, it is motivated to increase its degree of specialization. The concept of *comparative advantage* is helpful to an understanding of specialization and trade. Assume that two countries both have a need for, and are able successfully to produce domestically, two commodities. Further assume that there is no transport cost consideration. No matter what its cost of production of either commodity, Country A will choose to specialize in only one of them if, by that specialization and through exchange with Country B for the other, A stands to gain more than it loses. The key to comparative advantage is the utilization of resources in such a fashion as to gain, by specialization, a volume of production and a selling price that permit exchange for a needed commodity at a cost level that is below that of the domestic production of both.

At first glance, the concept of comparative advantage may at times seem to defy logic. For example, Japan may be able to produce airplanes and appliances more cheaply than the United States, thereby giving it an apparent advantage in both goods. But it benefits both countries if they specialize in the good in which they have a comparative advantage. In this instance, Japan's lower labor costs make it more profitable for Japan to specialize in the volume production of appliances and to buy airplanes from the United States. Resource-poor countries tend to develop specializations that other, more well-endowed countries are less inclined to pursue. In this way, poorer nations enter into international trade.

Often the thought of a particular country stimulates us to think of products associated with it. Usually these are the "specialized" goods that enter into international trade. French wines, Scotch whisky, teakwood products from Thailand, shoes from Italy, coffee from Brazil, and television sets from Japan are all examples of products that have become specializations in the nations named.

Transportation costs are central to the spatial patterning of production, explaining the location of a large variety of economic activities. Water-borne transportation is nearly always cheaper than any other mode of conveyance, and the enormous amount of commercial activity that takes place on coasts or on navigable rivers leading to coasts is an indication of that cost advantage. When railroads were developed and the commercial exploitation of inland areas could begin, coastal sites continued to be important as more and more goods were transferred there between low-cost water and land media. Resources at many potential sites have not been developed simply because the costs of moving the goods to markets would be too high. Every increase or decrease in transportation charges has an effect on commercial location.

In the rare instance when transportation costs become a negligible factor in production and marketing, an economic activity is said to be *footloose*. Some manufacturing activities are located without reference to raw materials; for example, the raw materials for electronic products such as computers are so valuable, light, and compact that transportation costs have little bearing on where production takes place.

AGRICULTURE

Agriculture fully shares with industry the basic characteristics of commercial economic systems: *specialization*—by enterprise (farm), by area, and even by nation; *profit*—rather than self-sufficiency and subsistence; and *interdependence*—with suppliers and buyers through the market mechanism. Like industrialists, commercial farmers attempt to reap the highest profits possible from their labors. To a greater extent than in industry, perhaps, elements of seeming irrationality may affect their decisions and thwart the maximization of net return that they seek. Among those elements are a less certain knowledge of the economic factors affecting production and marketing

301

decisions, less complete control over the conditions of production (e.g., weather and field conditions), and—for the smaller, if not the corporate, farmer—a greater influence of tradition or personal preference on production decisions.

The major difficulty that farmers face is that when they prepare their fields for planting, they do not know what the market price will be for their produce when the crops are ready for harvesting. That price reflects conditions over which they individually have no control, which are not related to their production costs, and which, indeed, may reflect not just national but international circumstances. In selling, the farmer inevitably stands at a disadvantage. The producers are many and small, the buyers are few and large and better informed. At the market, the latter obviously have an advantage: the price is set with no room for bargaining. Nonetheless, as part of the commercial system, farmers strive for as much information as they can get in order to make sound economic judgments. Thus the crop or the mix of crops and livestock that individual farmers produce is a result of a careful appraisal of profit possibilities. Markets must be assessed, the physical nature of farmland and the possible weather conditions must be evaluated, and the costs of production (fuel, fertilizer, capital equipment, labor) must be reckoned. Farmers do not necessarily grow what they want. Of the hundreds of possible crops that may be grown on a given farm, only a handful can bring profit to the producer.

The market for any good in a commercial society is basically a function of supply and demand, though this idealization is only partially realized in modern economies, with their strong infusion of governmentally administered constraints. If there is a glut of wheat on the market, for example, the price per ton will come down and the area sown will diminish. It will also diminish if governments, responding to economic or political considerations, impose acreage controls. If, on the other hand, an extended dry spell reduces the amount and the quality of the wheat produced, the market price for wheat will increase, and more farmers may be tempted to grow it. Again, this expected response to market conditions may be subverted by such government-imposed constraints as price controls and crop agreements. To a limited extent, farmers may withold their product from the market in the hope of improved prices. However, this option lasts for only a short while for livestock farmers, whose ongoing feed costs might prove ruinous; it does not exist at all for the producer of perishable fruits and vegetables.

Heinrich von Thünen observed early in the last century that lands of apparently identical physical properties (the term *isotropic* is used to connote uniformity) were utilized for different agricultural purposes. Seeking regularities governing farmers' individual land-use decisions, von Thünen deduced that the uses to which parcels were put was a function of the differing values (rents) placed upon seemingly identical lands. Those differences, he determined, reflected the cost of overcoming the distance separating a given farm from a central market town. The greater the distance, the higher was the production cost of the farmer, since transport charges had to be added to the producer's other expenses. A simple exchange model emerged: the greater the transportation cost, the lower was the rent that could be paid for land to remain competitive in the market. This understood, a rational patterning of crop decisions was revealed that was based not on land quality but on land cost. Von Thünen's model, diagramed in Figure 9.8, helps explain the changing crop patterns and farm sizes evident on the landscape at increasing distances from major cities. Farmland close to markets takes on high value, is used *intensively* for high-value crops, and is subdivided into relatively small units. Land far from markets is used *extensively* and in larger units. The effect

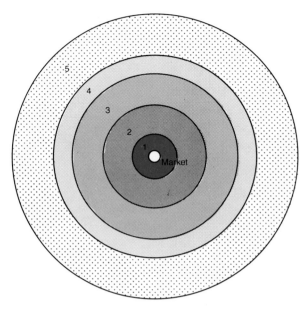

Fig. 9.8 Von Thünen's schema. Recognizing that as distance from the market increases, the value of land decreases, von Thünen developed a descriptive model of intensity of land use that holds up reasonably well in practice. The most intensively produced agricultural crops are found on the land close to the market; the less intensively produced commodities are located at more distant points. Compare this diagram with Figure 9.9. In the diagram, circle 1 is an area of dairying and truck gardening (horticulture); circles 2, 3, and 4 represent various crop-rotation farming systems, each of which includes grain. Circle 4 is farmed less intensively than are circles 3 or 2. Circle 5 represents the area of extensive stock raising. As the market at the center increases in size, the agricultural specialty areas are displaced outward, but the relative position of each is retained.

of these relationships on a national basis is seen on Figure 9.9.

Intensive Commercial Agriculture. Farmers who apply large amounts of capital (for machinery and fertilizers, for example) and/or labor per unit of land engage in intensive agriculture. The crops that are characterized by high yields and high market value per unit of

land are fruits, vegetables, and dairy products, all of which are highly perishable. Near most medium-sized and large cities, there are truck farms that produce a wide range of vegetables, and dairy farms. The fact that the produce is perishable increases transport costs because of the special handling that is needed, such as refrigerated trucks and special packaging. This is another reason for locations close to market. Note the location of truck farming in Figure 9.10.

Livestock-grain farming involves the growing of grain to be fed on the producing farm to livestock, which constitutes the cash product of the farm. Livestock-grain farmers also work their land intensively, but the value of their product per unit of land is usually less than that of the truck farm. Consequently livestock-grain farms are farther from the main markets than are truck and dairy farms.

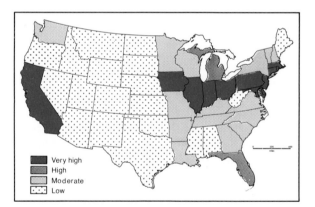

303

Fig. 9.9 Relative 1978 value per acre of farmland and buildings. In a generalized way, per acre valuations support von Thünen's model. Major metropolitan markets of the Northeast, the Midwest, and California are in part reflected by high rural property valuations, while fruit and vegetable production along the Gulf Coast increases land values there. National and international agricultural goods markets, soil productivity, climate, and terrain characteristics are also reflected in the map pattern. (Data from *Statistical Abstract of the United States,* 1979.)

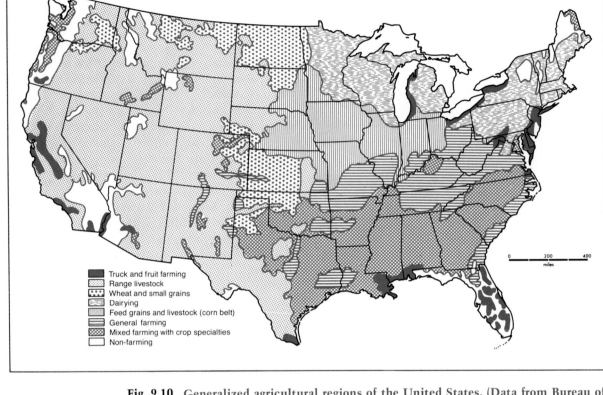

Fig. 9.10 Generalized agricultural regions of the United States. (Data from Bureau of Agricultural Economics.)

Normally the profits for marketing livestock (chiefly hogs and beef cattle in the United States) per pound are greater than those for selling corn or other feed, such as alfalfa and clover. As a result, farmers convert their corn into meat on the farm by feeding it to the livestock, efficiently avoiding the cost of buying grain. Where land is too expensive, especially near cities, feed must be shipped to the farm. The grain–livestock belts of the world are close to the great coastal and industrial zone markets. The "corn belt" of the United States and the livestock region of Western Europe are two examples. Lower-cost land traditionally allowed livestock to be shipped to the large European market from as far away as Argentina and New Zealand.

Extensive Commercial Agriculture. Further from the market, on less expensive land, there is less need to use the land intensively. Cheaper land gives rise to larger farm units. Extensive agriculture is typified by large wheat farms and livestock ranching.

There are, of course, limits to the land-use explanations attributable to von Thünen's theory. While it is evident from Figure 9.9 that farmland values decline westward with increasing distance from the northeastern market of the United States, they show no corresponding increase with increasing proximity to the massive West Coast market region. The western lands are characterized by extensive agriculture, but as a consequence of physical, not distance, considerations. Climatic regionalization, discussed in Chapter 3, obviously affects the productivity and the potential agricultural use of an area, as do associated soils regions and topography. In the United States, of course, increasing distance westward from eastern markets is fortuitously associated with increasing aridity and the beginning of mountainous terrain. In general, rough terrain and subhumid climates, rather than simple distance from market, underlie the widespread occurrence of extensive agriculture under all economic systems.

Livestock ranching differs significantly from livestock-grain farming. On ranches, young cattle or sheep are allowed to graze over thousands of acres. When the cattle have gained enough weight so that weight loss in shipping will not be a problem, they are sent to live-

stock-grain farms for accelerated fattening or to feedlots near the slaughterhouses for the same purpose. Ranching can be an economic activity only where alternative land uses are nonexistent. Consequently, ranching is usually found in mountainous areas, on dry lands, or in areas very far from markets. In the United States, ranching is important in the Rocky Mountain area and in the Southwest.

Wheat farming requires large capital inputs for planting and harvesting machinery, but the inputs per unit of land are low; wheat farms are very large. Wheat may be stored, if care is taken, so that wheat prices reflect not only current crop conditions but also the amount available in grain elevators. World trade in wheat is considerable. Figure 9.11 shows that the United States, Canada, Australia, and Argen-

Fig. 9.11 Wheat-growing areas and world trade in wheat. The arrows mark characteristic origins and flows of wheat in international trade. The volume of movements as well as the origin and destination nations, alters from year to year with variations in climatic conditions and political circumstances.

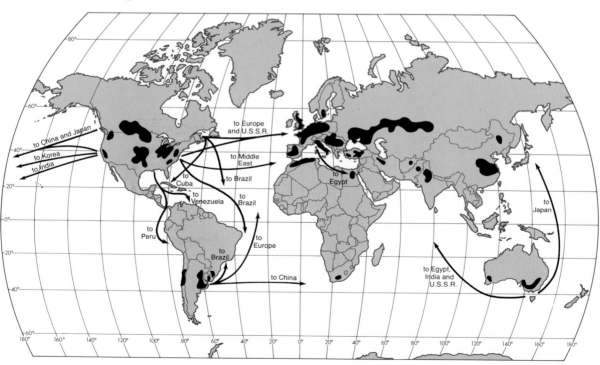

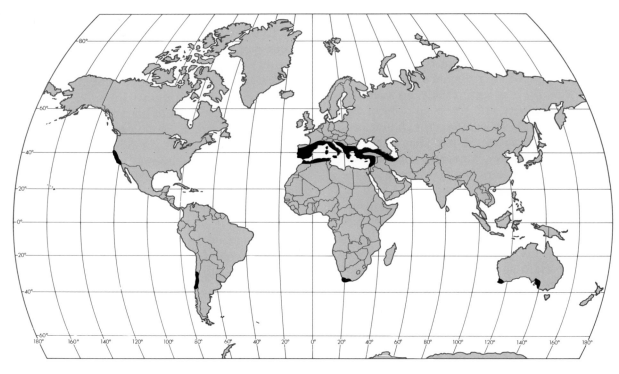

Fig. 9.12 Areas of Mediterranean agriculture. These areas have roughly similar climates and specialize in similar agricultural commodities, such as grapes, oranges, olives, peaches, and vegetables.

tina are usually wheat exporters, while the U.S.S.R., India, China, and the Western European countries are frequent wheat importers.

Special Crops. Proximity to the market does not guarantee the intensive production of high-value crops, should terrain or climatic circumstances preclude it. Nor does great distance from the market inevitably determine that extensive agriculture on low-priced land will be the sole agricultural option. Special circumstances, most often climatic, make some places far from markets intensively developed agricultural areas. Two special cases are agriculture in Mediterranean climates and in plantation areas.

Mediterranean agriculture is known for grapes, olives, oranges, figs, raisins, and other commodities that cannot be grown easily in and around the great industrial zones, which generally are found in areas of cooler climate. These crops need warm temperatures all year round and a great deal of sunshine in the summer. Mediterranean agriculture exists primarily in southern Europe and North Africa, California, Chile, Australia, and the Republic of South Africa, as indicated by Figure 9.12. These are among the most productive agricultural lands in the world. Farmers can regulate output in sunny areas such as these because storms and other inclement weather problems are infrequent. Also, the rainfall regime of Mediterranean climate areas, as shown in Chapter 3, lends itself to the controlled use of water. Of course, much capital must be spent for the irrigation systems—another reason for the intensive use of the land.

Climate is also the vital element in the production of plantation crops, which are particularly well adapted to the heavy rainfall of the tropics. There is little demand for these crops within the tropics, but they have a large market elsewhere. Such crops as tea, rubber, cacao, cane sugar, and bananas cannot be grown outside the tropics unless costly hothouse techniques are used. Commercialists in Western countries such as England, France, the Netherlands, and the United States became interested in the tropics partly because they afforded these countries the opportunity to set up plantations. Rather than depend on local suppliers, Westerners ran the farms themselves. In many cases, they handled the protection of their interests simply by making the territory a colony. Thus the commercial purpose of the plantation system was to place on the world market those goods produced by Western countries in non-Western areas.

For ease of access to shipping, most of the

plantations were developed along or near coasts. The major plantation crops and the areas where they are produced, illustrated by Figure 9.13, are tea (India and Sri Lanka); jute (India and Bangladesh); rubber (Malaysia and Indonesia); cacao (Ghana and Nigeria); cane sugar (Cuba and the Caribbean area, Brazil, Mexico, India, and the Philippines); coffee (Brazil and Colombia); and bananas (Central America).

Until recently the plantation crop system dominated the economies of the countries just named. The wealth produced by the system did not flow back into those countries, and some people believe that they were therefore prevented from developing more diversified, healthier economies. Nationalist feeling became very strong, particularly after World War II, and the colonial ties were first loosened and then cut. The influence of the United States remains strong in the Caribbean area but has

Fig. 9.13 Plantation agriculture regions.   307

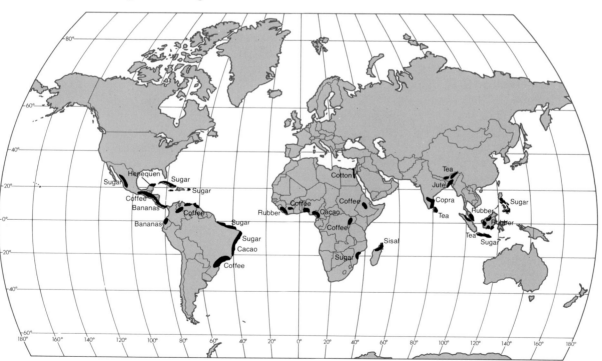

weakened elsewhere, while European economic power has weakened rapidly in Africa and Southeast Asia.

The newly independent countries, although attempting to free themselves of political domination by the Western countries, remain centers of plantation agriculture. Although some farms are still owned and operated by Europeans and Americans, many have been nationalized by the new governments; all depend on cheap local labor to offset the high transportation costs to European and American markets. There has been a strong and successful movement in recent years, however, to break up the plantations into small tracts distributed to former plantation workers.

Cotton is a subtropical plantation crop that requires a long growing season. Together with tobacco, it was raised as a plantation crop in the southeastern United States until the Civil War. Cotton growing was located outside the belt for truck and dairy farming but could displace livestock-grains. Now agriculture in the South has diversified; foreign competition in cotton and competing farm commodities, such as livestock-grain, are reducing the importance of cotton there.

Agriculture shares with all economic–geographic phenomena the characteristic of constant, frequently unpredictable change. As with all such changes, the consequences for other human patterns may be great. A case in point has been the enormous impact of changes in agricultural technology upon patterns of population. Significant migrations occur as people, released from the land, search for jobs in industrial zones. In the early years of this century in the United States, the introduction of increasingly efficient farm machinery made redundant millions of farm workers, who, in their turn, joined the urban labor force in the industrial Northeast. After World War II, the improved cotton picker released great numbers of southern blacks, who also sought new employment in northern cities, altering the urban ethnic mix there. In Europe, recent mechanization and farm consolidation have induced migrant laborers to move from the south and the east to the northern industrial zones, frequently changing their nation of residence.

POWER AND ORES

The primacy of power and mineral raw materials in their economies is a clear measure of the contrast between the developed and the underdeveloped economies. Of course, there is no meaningful distinction between the commercial and planned systems of advanced technology, though there are contrasts in their respective reactions to cost considerations. Commercial systems are, relatively, more sensitive to changes in the market price of these important components of production.

Except for the brief and localized importance of waterpower, coal was the form of inanimate energy upon which the Industrial Revolution was founded and developed. The volume of its reserves—estimated to be sufficient for 2000 years at present rates of extraction—assure coal's future importance, particularly as supplies of petroleum and natural gas approach exhaustion. At the time of the Industrial Revolution, it became clear that it was more costly to transport coal than manufactured goods. Coalfields became efficient locations for production because they offered low total transportation costs. As a result, most of the great industrial belts of the world are located on or very near coal deposits. The Westphalian (Ruhr) and English Midlands fields, shown in Figure 9.14, and the Pittsburgh–Cleveland region in the United States are examples of early major industrial zones. They not only fed the coastal cities with manufactured products but became major markets themselves because of their attraction for labor. Industries that were related not directly to coal but to the coal-

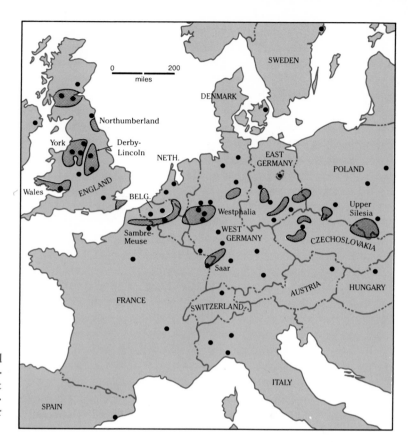

Fig. 9.14 Coalfields in Europe and major manufacturing cities. Although a number of cities are distant from coalfields, the majority of leading industrial centers are at or near them.

309

related industries began to cluster in these zones; industrial agglomerations of great size and complex linkages developed. The associated large industrial cities represented an attraction for a further increase in market-oriented manufacturing activity. Coal still plays a role in industrial development, and though no longer dominant, it still accounts for nearly one-half of the world's inanimate energy. As petroleum and natural gas become less plentiful, there are indications that coal resources unutilized as yet because of their remoteness are becoming more attractive as sites for industry.

Petroleum became a major power source only early in this century. The rapidity of its adoption as both a favored energy resource and a raw material, along with the limited size and the speed of depletion of known and probable reserves, augurs its near disappearance as an industrial fuel by the close of the century. With its high energy value in relation to its bulk, petroleum has tended to be transported to, and refined at, consuming areas already established (Figure 9.15). Relatively few industrial complexes of major size have developed with petroleum deposits serving as the prime attraction. Where there has been such a spatial relationship—on or near producing fields or along major pipelines—the attraction has been to petroleum as a raw material, not as a fuel.

The petroleum-producing countries located outside of the commercial system (e.g., Saudi Arabia, Kuwait, and Venezuela) were treated

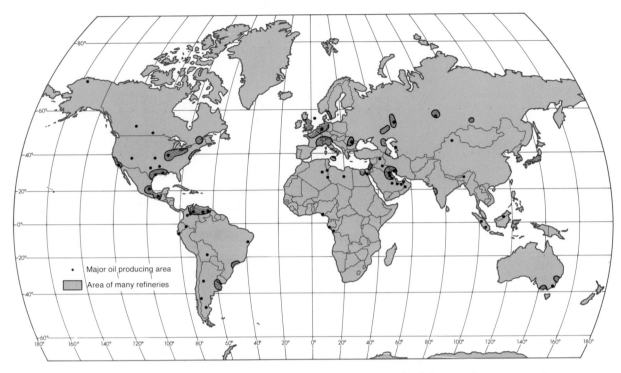

Fig. 9.15 Major sources of oil and regions in which large refinery operations are conducted. Refineries are generally located in established industrial zones, while petroleum is exploited wherever it is found. The result, of course, is an enormous trade in petroleum, as Figure 10.6 illustrates.

310

very much like colonies. The resources were owned and operated by Western interests, with a portion of the royalties going to the local governments. Now that these countries have gained control of their own resources, manufacturing complexes will most likely develop.

There is a strong desire in the Western countries, particularly in the United States, to develop alternate power supplies. Nuclear and solar energy are two candidates for the next phase in the development of industrialized societies. At present, safety factors or development costs detract from their widespread use. If safety, cost, and potential environmental damage were not factors, nuclear energy could now be better established within the already-existing industrial zones. Because of the high

cost, solar energy schemes have only recently become reasonably practicable, as discussed in Chapter 11.

Hydroelectric power is developed where sites for dams are available, though, of course, not all such sites can or will be developed. More often than not, relatively rugged mountain areas contain the preferred physiographic conditions for the building of dams. Unfortunately, hydroelectric power sites are not plentiful, and as Figure 9.16 shows, developable potential is often remote from existing markets. Most of the best sites have already been developed in the technically advanced countries. The potential is in Africa and Asia, where rainfall is plentiful and where great rivers run through undammed gorges, but where indus-

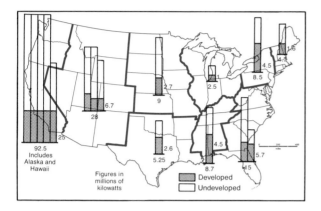

Fig. 9.16 Waterpower, developed and undeveloped, in the United States. The greatest potential for hydroelectric power development is in the Mountain and the Pacific states. Industries may take greater advantage of this potential in the future, although dam construction increasingly meets objections raised by environmental protection groups and alternate users of the water in question. (Data from Federal Power Commission.)

trial demands and available capital are limited.

Transportation costs play a great role in determining where low-value minerals will be mined. Minerals such as gravel, limestone for cement, and aggregate are found in such abundance that they have value only when they are near the site where they are to be used. For example, gravel for road building has value if it is at the road-building site, not otherwise. Transporting gravel hundreds of miles is an unprofitable activity.

The production of other minerals, especially metallic minerals such as copper, lead, and iron ore, is affected by a balance of three forces: the quantity available, the richness of the ore, and the distance to markets. When the ore is rich in metallic content, it is profitable to ship it directly to the market for refining. But, of course, the highest-grade ores tend to be mined first. Consequently, the demand for lower-grade ores has been increasing in recent years. Low-grade ores are often refined or upgraded by various types of separation treatments at the

mine site to avoid the cost of transportation of waste materials not wanted at the market. In the case of copper, concentration takes place near the mines (Figure 9.17) and refining takes place near areas of consumption. The large percentage of waste in copper (more than 99% of the ore) and most other industrially significant ores should not be considered the mark of an unattractive deposit. Indeed, the opposite may be true. Many higher-content ores are left unexploited—because of the cost of extraction or the smallness of the reserves—in favor of the utilization of large deposits of even very-low-grade ore. The attraction of the latter is a size of reserve sufficient to justify the long-term commitment of developmental capital and, simultaneously, to assure a long-term source of supply. At one time, high-grade magnetite iron ore was mined and shipped from the Mesabi area of Minnesota; these deposits are now exhausted. Yet immense amounts of capital have been invested in the mining and processing into high-grade iron ore pellets of the virtually unlimited supplies of low-grade iron-bearing rock (taconite) still remaining.

311

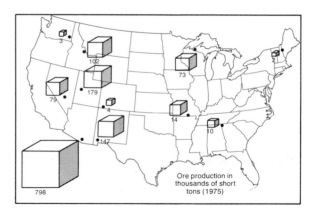

Fig. 9.17 Copper mining in the United States. Most copper ore domestically mined contains less than 1% copper. Before most of it is sent to the east coast for refining, it is concentrated near the mines to avoid transporting waste materials. (Data from _Commodities Yearbook,_ Commodity Research Bureau, New York, 1976.)

**Industrial
Location
Consultants**

In the years after World War II, a sizable new profession emerged, that of industrial location consultants, of whom many are geographers. The numerous independent consultants to industry have been more than matched by professional locaters attached to state governments, municipal development associations, utility companies, railroads, and other organizations that hope to develop arguments sufficiently convincing to "sell" relocating firms on their locales.

The new professional groups approach the question of industrial location in a highly sophisticated manner, taking the Weberian cost-minimization analysis as a philosophical starting point, but refining and expanding the factor considerations and data acquisition to book-filling proportions. As expensive site searches testify, few management decisions promise such rewards or threaten such losses as those putting at risk perhaps millions of dollars in land, building, and equipment outlays for new plant construction. The task of the location

**Costs Associated
with a Site Can Be
Classified by Area**

312

Area of Occurrence	Type of Cost
National or regional	Outbound freight costs Inbound freight costs
National, with many regional and local variations	Labor Power Fuel Climate (heating and air conditioning) Construction costs
State	Business taxes Business organization taxes; sales tax on equipment and materials Air and water Pollution control Financing programs
Local	Building site Water Real estate taxes Air and water Pollution control Financing programs

[a] Some consideration frequently must be given to other influences related to governmental procedures. Examples are tax forgiveness to new industry for a specified period, right-to-work laws, and transportation regulations.

Modified from Robert A. Will, "Finding the Best Plant Location," *Chemical Engineering*, Vol. 72, March 1, 1965.

consultant is to minimize those risks by evaluating, within guidelines established by the client, all possible alternate locations in order to determine a best possible recommendation.

Questionnaires, checklists, or outlines for research have been devised to assist clients in establishing their locational criteria and consultants in pursuing their research. The abbreviated table opposite suggests the categories of important considerations used in making locational analyses. In practice, each of the types of factors may be divided and further subdivided to secure data deemed essential to the ultimate site decision.

Expensive mistakes may still be made. Increasingly, however, entrepreneurs in commercial economies exert themselves to make locational decisions based upon the clear realities of cost considerations.

MANUFACTURING

In commercial economies, entrepreneurs use their knowledge to locate industries at the lowest-cost sites. In order to assess the advantages of one location over another, industrialists must evaluate the most important variable costs. They subdivide their total costs into categories and note how each cost will vary from place to place. In different industries, transportation cost, labor cost, power cost, plant cost, the cost of money, or the cost of raw materials may be the major variable cost. The industrialist must look at each of these and by a process of elimination eventually select the lowest-cost site. If the producer then determines that a large enough market can be reached cheaply enough, the location promises to be profitable. Because of a changing mix of input costs, production techniques, and marketing activities, most such initially profitable locations do not remain advantageous. Migrations of population, technological advances, and changes in the demand for products affect industrialists and industrial locations greatly. The abandoned mills and factories of New England or of the steel towns of Pennsylvania are testimony to the impermanence of the "best" locations.

The concern with variable costs as a deter-minant in industrial location decisions has elicited an extensive theoretical literature. Most of it is based upon, and extends ideas presented by, the distinguished German location economist Alfred Weber. Figure 9.18 gives a picture of his analytic technique, which is employed in the following brief discussions of a few of the characteristic industries in advanced commercial economies.

Iron and Steel Industry. The metal-making and metalworking industries reach their highest level of importance in the most advanced economies; the foundation stone of industrial power is frequently seen to be the iron and steel industry. Small wonder that grandiose steel mill projects appear so frequently and often so irrationally in the development plans of nations barely emerging from subsistence economies. The steel mill serves as the ultimate mark of achievement and national pride.

A primary raw material in steelmaking is, of course, iron ore, but equally important are coke (a by-product of coal), limestone, and ferroalloys, such as nickel and tungsten. The iron ore, limestone, and coke are heated to yield iron, which is then treated in open-hearth furnaces or by the basic oxygen process to remove excess carbon. To the resulting steel, in

313

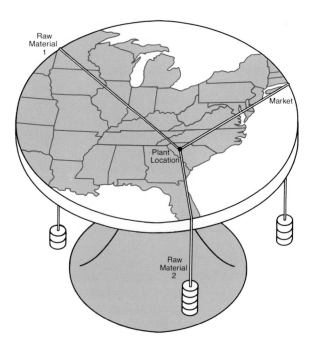

Fig. 9.18 Plane table solution to a plant location problem. By allowing a weight to represent the "pull" of raw material and market locations, an equilibrium point is found on the plane table. That point is the location at which all forces balance each other. It represents the least-cost location—the place where, theory suggests, the plant should be located. Alfred Weber developed the analytic techniques for determining least-cost locations for such simple cases and for instances where costs other than transportation charges—for example, labor and power costs—must be taken into account.

314 its liquid form, one or more ferroalloys are added to give the steel the specific qualities required for its ultimate use. In recent years, scrap metal has often been used in the production process in place of, or in combination with, iron and the ferroalloys.

Plant, labor, and capital are not major variable costs affecting location in the iron and steel industry because they do not vary markedly within the bounds of a single country. We might assume, therefore, that the industrialist would attempt to locate close to raw materials. This is not quite the case, since the buyers of steel must pay the transportation costs required for delivering it. Logically, then, the buyers would want to locate close to the steel mills. But the demand for a product usually precedes the development of the capacity to supply it. Thus the steel industrialists take careful account of the location of their customers before they invest in huge plants. If they should locate too far from the consumers, the delivered price would be high and the demand

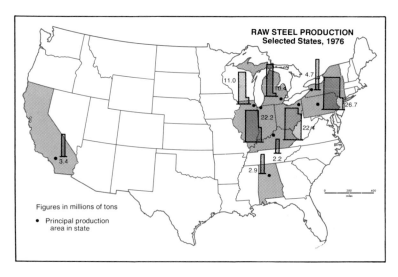

Fig. 9.19 Steel production in the United States is still largely concentrated in the older industrialized Northeast. The ready market and the low-cost transportation of raw materials through the Great Lakes and from overseas ore sources favored concentration there. Regional markets and transportation costs have encouraged some production in the western states. More than 82% of American raw steel production in 1976 came from the states for which output totals are indicated on the map. (Data from American Iron and Steel Institute.)

lowered. Therefore, the earliest steel plant locations were tradeoffs between the transportation costs for raw materials and the transport costs for delivering the finished product to market (Figure 9.19).

Aluminum Industry.   The major costs in the aluminum industry are raw materials, power, plant, labor, and capital. The question is: Which of these is a major variable cost? To answer that question we need first to sketch how aluminum is made. It is ultimately derived from the mineral bauxite. Because the bauxite contains industrial waste as well as aluminum,

it is changed—by leaching under high temperatures and pressures—to a material called *alumina,* which is an aluminum–oxygen compound. To extract the aluminum, massive charges of electricity are sent through the alumina while it is in solution. Electrical power is the major variable cost within a single country in the production of aluminum. Plant, labor, and capital costs for this industry do not vary significantly from place to place within a country.

The source of electrical power is irrelevant in the locational decision; it need only be available in great amounts at low cost. The familiar

Fig. 9.20   **Location of primary aluminum capacity in the United States, and the relative cost of electric power. Aluminum manufacturers generally choose sites where low-cost power is available. Long-term, large-volume contracts for power may reduce electricity rates to mills below the industrial average for their area. (Data from U. S. Department of the Interior and Federal Power Commission.)**

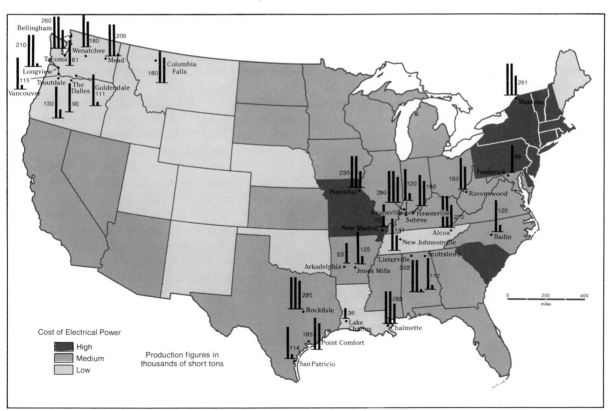

mental association of aluminum plants and waterpower sites, though sometimes accurate, is not necessarily valid. Extreme examples of the influence of power costs to the exclusion of other cost considerations are the Kitimat plant on the west coast of Canada and the Bratsk plant near Lake Baikal in eastern Siberia. Both are far from bauxite supplies and from markets, but both are adjacent to huge hydroelectric power sources. Yet well over one-half of the aluminum refined in the United States uses power from other than hydroelectric sources (Figure 9.20).

Transportation costs for alumina are about as important as those for iron ore in the steel industry. Direct-shipping iron ore averages above 50% iron content, while enriched pellets average 65% iron; alumina averages about 50% aluminum content. However, finished aluminum per unit weight is much more valuable than an equal amount of steel. As a consequence, customer location is not as important for the aluminum as for the steel industry.

Textile Industry. The major costs in the textile, or cloth-making, industry are plant, labor, capital, and raw materials. The raw material is cotton, wool, or any of a number of artificial fibers. There is little waste; these are not ores, so the transportation costs of raw materials and finished goods, while important, are not a major variable. Among the cost factors, therefore, that of labor is the locational determinant.

A great deal of labor is necessary to turn the raw materials into a finished product, but it need not be skilled labor. In the United States, nearly all of the cotton and wool textile industries were located in New England until the turn of the century. The industrialists took advantage of a surplus unskilled labor supply, much of it immigrant, in the cities and towns of that area. The industry first began to decline in New England when coal (which the area did not have) replaced the waterpower upon which the industry was originally based. The decline accelerated as plants became obsolete and as unions, which increased wage levels, became more powerful. When it became clear that a surplus of cheap labor was available in the Piedmont district of the South, the industry quickly moved in that direction. Today, as Figure 9.21 shows, the United States textile industry is concentrated in the Piedmont.

In many countries newly emerging from subsistence economies, such as Thailand, India, and Taiwan, the textile industry is seen as a near-ideal employer of an abundant labor force. In such areas, the growth of the industry to significant proportions is creating unwanted and resisted competition to textile manufacturers in established commercial economies. The advantage now enjoyed by emerging nations may not endure, as it has not in former world centers of the textile industry. As more employment opportunities are created by accelerating industrialization, and as labor skills and educational levels rise, so, too, will the wage rates whose original minimal levels gave the new areas their comparative advantage.

The Fashion Industry. The conversion of the finished product of the textile industry into clothing is loosely known as the *apparel industry*. Like textile manufacturing, it finds its most profitable location in areas of cheap labor, such as Taiwan, South Korea, and Mississippi. Repetitious, limited-skill, assembly-line operations are requisite for volume production of clothing for a mass, highly price-competitive market. A special branch of the apparel trade, the *fashion* industry, however, departs from the mass-produced, low-quality norm of the general clothing trade. Relatively higher labor skills are required, and the value of the finished product permits the payment of the wage rates needed to retain competent workers. Because fashions change quickly, are design- rather than price-competitive, and require close collaboration between creator and producer, the

316

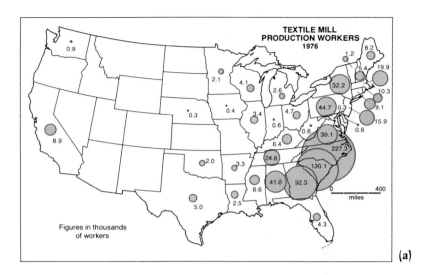

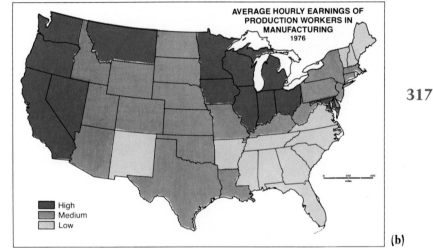

Fig. 9.21 (a) Location of textile-mill production workers in the United States and (b) relative hourly earnings in the manufacturing industries. Areas of low labor rates in the South show the largest concentrations of textile-industry workers. The average low labor rates shown for New England do not reflect the wages there of highly unionized textile workers.

317

industry must be well informed about rapidly changing market conditions and buyer preferences; it benefits from spatial concentration.

This industry tends to cluster in the recognized fashion capitals of the world: London, Paris, Rome, Los Angeles (for fashion sports wear), and New York. In New York, the skilled craftsmen and their showrooms are concentrated in a small area of Manhattan (Figure 9.22). Even in this industry, so closely associated traditionally with New York, changes are

occurring in response to cost considerations. Labor rates, rents, taxes, suburbanization of the labor force, and many similar factors are inducing relocation from the Manhattan concentration to northeastern Pennsylvania, still near the showrooms and designers, but with a more favorable cost structure.

The Automobile Industry. This industry typifies manufacturing in a highly advanced economy; the varied outputs of a great many indus-

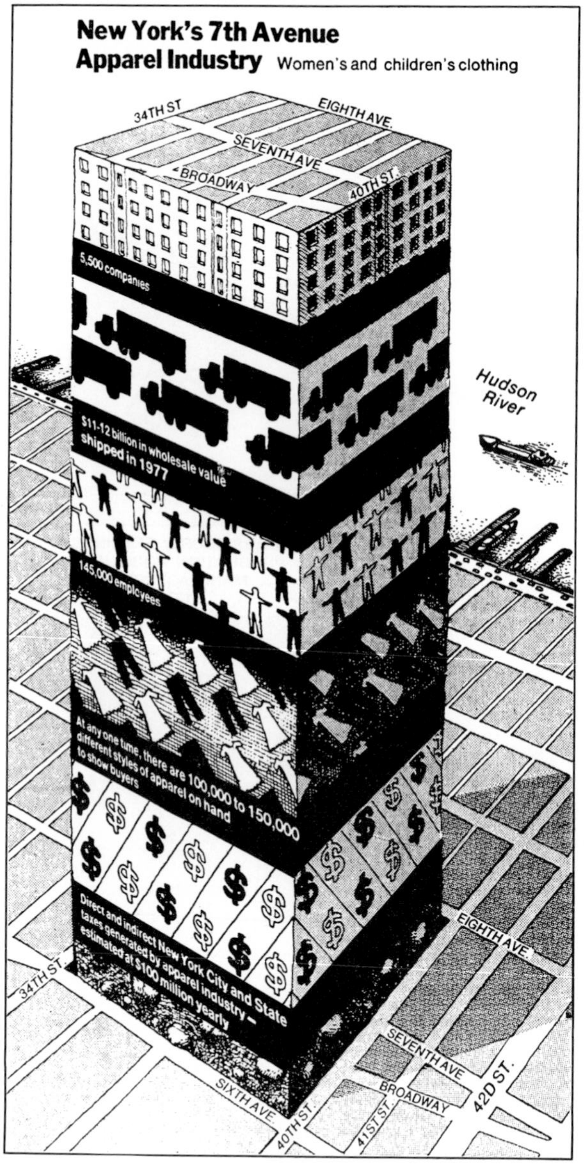
New York's 7th Avenue
Apparel Industry Women's and children's clothing
34TH ST
EIGHTH AVE.
SEVENTH AVE
BROADWAY
40TH ST.
5,500 companies
Hudson River
$11-12 billion in wholesale value shipped in 1977
145,000 employees
At any one time, there are 100,000 to 150,000 different styles of apparel on hand to show buyers
Direct and indirect New York City and State taxes generated by apparel industry – estimated at $100 million yearly
EIGHTH AVE.
34TH ST.
SEVENTH AVE
BROADWAY
42D ST.
SIXTH AVE
40TH ST.
41ST ST.

Fig. 9.22 The apparel industry is still New York City's largest employer. The individual small firms, beset by rising costs, are often displaced by the conversion of their loft-building workshops to apartments and to artists' studios. (The New York *Times*/Bob Gale. © 1978 by The New York Times Company. Reprinted by permission.)

tries are assembled to make a single high-value product. In any given country, labor and plant costs are uniformly high; most auto workers are represented by strong labor unions, and tooling costs are enormous. The high transportation costs for the *finished* product affect location the most. Since this is the case, we would expect the industry to locate in the major markets of the commercial world. This is generally true for final assembly plants, but not for the earlier stages of manufacturing.

The industry is so huge that it creates a market itself. That is, the myriad industries that supply the auto industry locate close to the big plants, resulting in the development of large industrial complexes or agglomerations. The enormous labor supply that is needed helps to create cities, which become, as do all population concentrations, centers of demand for the automobiles.

In the United States, the auto industry got its start in many locations throughout the Northeast and the Midwest. Essentially, these were the locations that were already the market centers of the country in 1900. In the Midwest, however, the distance between towns was greater than in the Northeast. There was also flat land and plenty of gravel for roads. Wood was important to the early auto industry for body parts, as were the gasoline engines already used for motorboating on the Great Lakes. These factors, together with the reluctance of eastern bankers to invest in the fledgling industry and the fact that Michigan was the home of Henry Ford, led to the first large-scale production of autos in the Detroit area. This was an accident of history; it could have happened anywhere along the south shores of the Great Lakes.

Today the centralization of the industry in the southern Michigan, Ohio, Indiana, and Wisconsin area is altered by the existence of car assembly plants built as close to the large markets as possible. This dispersal is evident

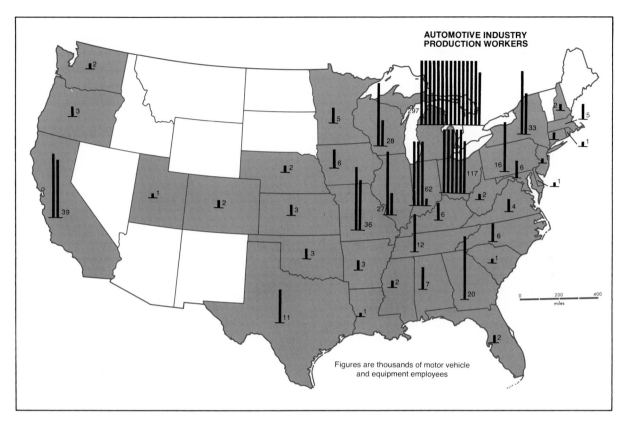

Fig. 9.23 Distribution of production workers in the motor vehicle and equipment industries. The automobile industry is one of the most concentrated manufacturing activities in the United States. (Data from U. S. Bureau of the Census, *Census of Manufacturers, 1977.*)

in Figure 9.23. The parts are shipped to assembly plants, which sell the cars in the local markets.

Imposed Considerations.   These few briefly discussed examples of the cost factors affecting the location of individual industries have concentrated on the locational controls cited by Weber and other location economists and geographers. Both theory and observation suggest that in a pure, competitive commercial economy, the costs of material, transportation, labor, and plant should control locational decisions. Obviously, neither in the United States

nor in any other nation assigned to the "commercial economy" category do the idealized conditions exist. Other constraints—some representing cost considerations, others political or social impositions—also affect, perhaps decisively, the locational decision process. Land-use and zoning controls, environmental quality standards, governmental area-development inducements, local tax-abatement provisions or developmental bond authorizations, noneconomic pressures on quasi-governmental corporations, and other considerations constitute attractions or repulsions for industry outside of the context and consideration of pure theory. If these noneconomic forces become overwhelmingly compelling, the assumptions of the commercial economy classification no longer obtain, and the transition to a controlled mixed or wholly planned economy has occurred.

TERTIARY ACTIVITIES

Primary activities are those involved in raw material and basic foodstuff production; secondary activities process raw materials and convert them into demanded products (the manufacturing and construction industries). A separate and essential sector of commercial economies, and the one with which we most often come into contact—tertiary activities—consists of those business and labor specializations that provide the vital link between producer and consumer. They are the wholesaling and retailing activities that fulfill the exchange function of advanced economies and provide market availability to satisfy the consumption requirements of individual members of highly interdependent societies. In commercial economies, tertiary activities also provide vitally needed information to manufacturers: the knowledge of market demand without which economically justifiable production decisions are impossible.

The retailing component of tertiary occupations is totally demand-oriented. The spatial distribution of retailing, therefore, is completely controlled by the spatial distribution of effective demand, that is, by wants made meaningful by purchasing power. A considerable theoretical and empirical literature exists reviewing locational patterns, hierarchies, and land-use arrangements of individual stores, store clusters, and communities functioning as retailing centers. Although the topic is beyond the scope of this chapter, some aspects of it are considered in Chapter 11, "Urban Geography."

Wholesaling and retailing are joined in the general tertiary-occupations category by some individuals and businesses that are, in a sense, producers. They produce *services* sold and purchased as intangible commodities: personal and professional services and financial, administrative, and governmental activities. They, too, represent a mark of contrast between advanced and subsistence societies, and the greater their proliferation, the greater the interdependence of the society of which they are a part.

Planned Economies

As the name implies, planned economies have a degree of centrally directed control of resources and of key sectors of the economy that permits the attainment of governmentally determined objectives. All advanced economies have elements of planned control, at the very least through the allocation of resources to social welfare programs, mass-transit systems, a defense establishment with associated industries, and the like. However, a clear distinction may be made between those economies categorized as planned and those categorized as commercial. In the former, the preponderant control of resource allocation is centrally administered; in the latter, the allocation of resources results primarily from a response to market forces.

Just as there are variations within individual commercial economies in their degrees of realization of the abstract ideal, so, too, do planned economies differ in the extent to which centralized control is exerted upon the society. The usual distinction made is between socialism and communism. Both derive from the teachings of Karl Marx, but they have taken basically different and mutually exclusive paths to attain the objectives he espoused.

Under socialism, the less extreme form of the planned economy, planning does not presuppose public ownership of *all* the means of production. In fact, according to a declaration adopted by the Socialist International in 1951,

320

planning is "compatible with the existence of private ownership in important fields, for instance in agriculture, handicraft, retail trade, and small and middle-sized industries." What is involved is the concept of the replacement of capitalism by a system of democratically planned production that would achieve socially accepted goals. This aim would be accomplished by public ownership and operation, at various governmental levels, of key industries. The remainder of the economy would be retained in the hands of cooperative organizations and private owners under such public regulation as necessary to reach democratically agreed-upon objectives. Public ownership would not be a goal in itself, but the means of the achievement of goals.

Under this definition and set of assumptions, many Western nations may be classified as having planned economies on the socialist model. For example, the government of Great Britain owns much of the transportation system of the country and controls the output of the coal and steel industries as well as the development of most housing. It also controls the location choices of major industrial investments. Sweden has developed a similar system with even more central planning. Strong government participation in, as well as actual ownership of, major industries, transportation systems, and banks is found in Italy and France. Many economically developing countries have central authorities that control the use of resources and dictate output requirements.

Communism is the extreme example of the planned economy. Whether under the name of Marxism–Leninism, Maoism, or some other designation, it is a theory of economic organization based upon total state control of the means and tools of production. Since one of the tools of production is labor, the ideology translates itself into planned direction of all facets of individual and collective social action.

Communism in its idealized form seeks totally to abolish private capitalism, the free market, and the informational guidance of the free price system. In their place is state ownership of all enterprise, including agriculture, and of every personal or professional service. Production is divorced from market control; irrespective of costs, prices are established to achieve predetermined goals.

Self-declared Communist nations are many and growing in number; they are found in Europe, Asia, Africa, and the Americas. In Eastern Europe, communism was imposed by Russian occupation following World War II. In the developing nations outside Europe, militant Communist parties have achieved power through the support of disparate groups dissatisfied with the political and social conditions under which they lived. Of course, the forerunner of them all is the Union of Soviet Socialist Republics, which may serve as an example of an advanced planned economy and as a contrast to the theoretical pure commercial and subsistence economies earlier discussed.

The U.S.S.R.: Planned Economy in Operation. The Communist revolution of 1917 was followed by a confused period of civil war, a New Economic Policy of limited capitalism, and finally, after the assumption of total authority by Stalin, the establishment of the principles and mechanisms of the complete state control of economy and society that—with modifications—remain in force today.

Economic development of the then largely peasant society was to proceed, in the form of successive five-year plans, under the direction of the State Planning Agency (Gosplan). Predetermined maximum and minimum production levels for every sector of the economy were forecast, and their achievement was assigned to ministries responsible for the management of designated sectors: heavy industry, agriculture, and so on. Party-established guiding principles of development shaped the details of the plan: (1) emphasis upon capital goods, heavy industry, and armaments; (2) full development of the resources of the vast country, no matter

321

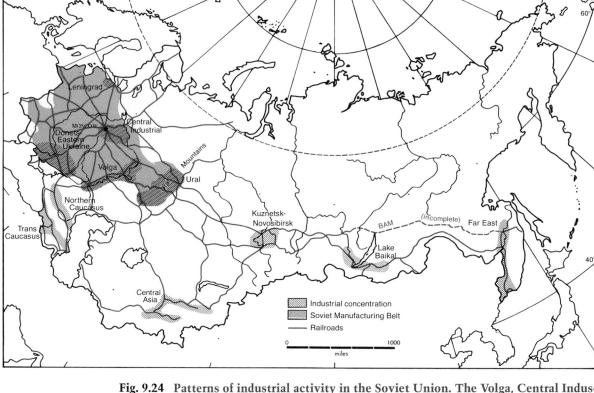

Fig. 9.24 Patterns of industrial activity in the Soviet Union. The Volga, Central Industrial, and Leningrad concentrations within the Soviet manufacturing belt are not dependent upon local raw materials. All other U.S.S.R. industrial concentrations have a strong orientation to materials and were developed, by plan, despite their distance from the population and markets of European Russia.

how remote from existing European concentrations of population and industry; (3) the creation of population centers and industry at the sites of resources; (4) investment capital accumulation by a set of administered prices that would extract from the agricultural and consumer sectors the funds needed for industrial growth; and (5) the creation of a set of physically separate, independent centers of production of basic commodities, without regard to location of markets, transportation costs, or the other economic constraints accepted in commercial economies.

In the industrial sector, the application of these policies has resulted in the development of the world's second largest producer of manufactured goods overall and, in many areas of heavy industry—steel, for example—its leading performer. New, populous, integrated centers of industry have been established in the eastern reaches of the country—in the Urals, in the Kuznetsk Basin of western Siberia, in Kazakhstan, in Central Asia, and in the Soviet Far East (Figure 9.24). Since the primary market for most commodities remains in the heavily settled and industrialized west, the price of delivered goods from the high-cost eastern producers must be discounted to fit within the

Six hundred miles east of Moscow in the isolated Tatar Republic, the world's largest industrial project, the Kama River truck plant, was constructed during the 1970s. It was as much a symbol of Soviet industrial location philosophies as it was a significant addition to the manufacturing capabilities of the nation.

The Kama River Truck Plant

Kama is big even by Russian standards. The plant alone covers some 40 square miles and is made up of six separate installations, giants in their own right: foundry, forge, pressing and stamping, tooling and repair, engines and transmissions, and final assembly. Additional factories to produce building materials were added to the plant complex, for the isolation of the assemblage required self-sufficiency in construction.

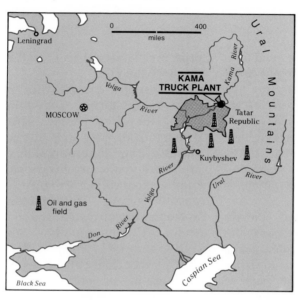

323

Certain practical, economic considerations dictated the location of the truck plant. The broad Kama River, a tributary of the Volga, provides cheap and easy transportation of materials and generation of electric power. The largely rural Tatar Republic contained underemployed farm populations that could be turned into factory workers. Proximity to the Ural Mountains and to the Ural–Volga oil and gas fields permitted easy access to raw materials and fuels.

These are considerations comprehensible to industrial locaters in commercial economies. Not so understandable to private entrepreneurs was a decisive Soviet consideration: the determination, at whatever cost, to develop fully every section of the nation and to bring urbanization and industrialization equally to all regions.

The decision to advance that objective through the Kama River plant entailed investment commitments far beyond the estimated $5 billion cost of the plant itself. The isolated, undeveloped character of the area demanded the construction of an entire city to house nearly 100,000 workers and their families. Since nothing existed before, not only apartment housing but the full range of urban facilities had to be built: schools, hospitals, stores, restaurants, sports arenas, community centers, and a hotel. To supply food, nine collective farms were organized, and local plants were built to produce vodka, ice cream, bread, and other necessities.

The creation of a single-purpose gigantic company town is fully consistent with Russian developmental and industrialization concepts originated at the start of the Stalinist period. The fully planned economies can envision objectives and allocate resources beyond the interests and capacities of entrepreneurs in commercially oriented societies.

budget allocations of the receiving establishments (Figure 9.25). In part, the government subsidizes industries by ignoring the full cost of transportation services, and in part, by absorbing the unrecorded loss within the industry sector budget. The concept of optimal location in the commercial economy sense, although recognized and argued, has characteristically been secondary to the achievement of other planning objectives.

Those objectives are discerned clearly enough in broad outline; it is the necessary detailed planning—budget, production goals, material sources, product destination, labor force size, wage rates, and every other facet of plant operation—for some quarter million industrial enterprises that has proved most complicated. An enormous work force—estimated at more than 15 million—is engaged in the planning process, though not all, of course, confined to the industrial sector. The planners, lacking the feedback information available in commercial economies through the market price system, must draw up, for enterprises, economic sectors, planning regions, and the nation, enormous balance sheets with all the supplies for the economy entered on one side and planned uses on the other. Complex and time-consuming bargaining sessions follow, with planners seeking to minimize allocations to enterprises and maximize quotas, and enter-

prise managers and sector ministries aiming at the opposite goal.

Comparable complexities of management exist in agriculture. Governmental control is basic throughout Soviet farming, including ownership of the land, direction of the crop economy, and determination of inputs. There are three main organizational forms in the rural sector: collective farms, state farms, and private holdings. The approximately 500 million acres of cultivated land in the "socialized" sector are about equally divided between collective and state farms. The latter are generally much larger than collectives. Some 15 million acres are in "private" plots.

Collectives were created in the late 1920s and the early 1930s in a massive program of forced consolidation of separate land holdings into cooperatives, with individual farm families jointly working the allotted communal rent-free land. Brigades of workers are assigned specific tasks during the crop year. Originally they received remuneration only as a share in farm profit, if any was achieved, but increasingly they are paid by a wage system. Crops to be grown and quotas to be met are established by regional agricultural planners. Supervision is under a farm manager ostensibly elected by the members of the collective, but in practice selected by the Communist Party.

Collectivization released millions of farm

324

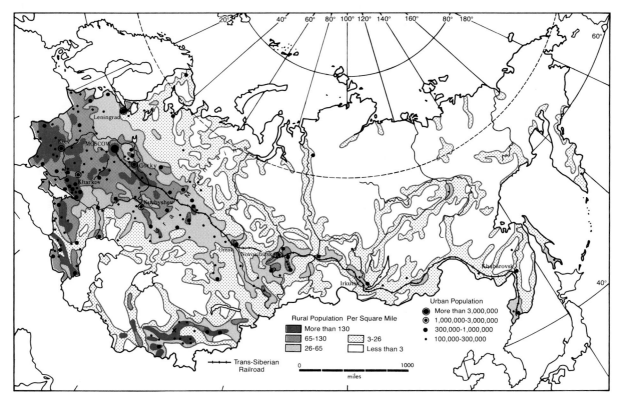

Fig. 9.25 Population distribution in the U.S.S.R. The vast majority of the Soviet Union's 262 million people live in the European part of the country to the west of the Ural Mountains (to the west of the 60° East longitude line). The Trans-Siberian Railroad is the main connecting link between the Pacific coast in the east through Siberia to the Urals and the industrial center at Moscow. Along it lie the developed areas and the higher population densities of Siberia.

workers for urban employment during the industrialization program of the U.S.S.R. and so achieved, at great human cost, one of its primary objectives. A second was also achieved: the generation of investment capital for the industrial sector. By paying low prices for forced delivery of farm products and disposing of them in the urban market at much higher prices, the government in effect imposed a massive tax upon the agricultural sector that provided funds for industrial development.

The *state farm* is a government enterprise operated by employees of the state, which provides all inputs and claims all yields. It is a rural factory in every administrative way equivalent to its urban counterpart. The state farm tends to be larger than the collective and more specialized in its output. For ideological reasons, increasing numbers of collectives have been placed under full state control in recent years.

Private plots, up to a maximum of about $1\frac{1}{4}$ acres, are labor-intensive holdings granted primarily to collective farm families. They depend upon them for home-produced food—including a limited number of animals—and for cash income through sale in urban farmers' markets where state price controls do not apply. According to some reports, these carefully tended personal plots produce 30% or more of the agricultural output of the country.

A success of the planned economy in achieving national goals of increased grain production is to be found in the Virgin and Idle Lands program. It spread the pattern of farming from established, primarily European, agricultural regions eastward to the semiarid midlatitude steppes of the southern Urals, southwestern Siberia, and northern Kazakhstan (Figure 9.26). Launched in 1954, the program directed the cultivation of some 70 million acres (28 mil-

lion hectares)—ultimately to rise to 115 million acres (47 million hectares)—of new land comparable in physical character to the Dust Bowl area of the United States. Increases in output have been impressive but erratic because of climatic conditions.

Comparable achievements can be found in all sectors of the Soviet planned economy, for they represent the pursuit, regardless of cost considerations, of a national objective. That they may represent a misallocation of resources and unbalanced development is not the point. The planned economy is predicated on

Fig. 9.26 The Virgin and Idle Lands program extended grain production, primarily spring wheat, eastward into marginal land. Wheat constitutes nearly 90% of total Soviet food grain production and 50% of all grains grown. Although the Virgin Lands program, begun in the 1950s, added significantly to the established agricultural regions of the nation, variability in weather in the new grain areas greatly affects Soviet wheat supplies and world patterns of wheat trade.

326

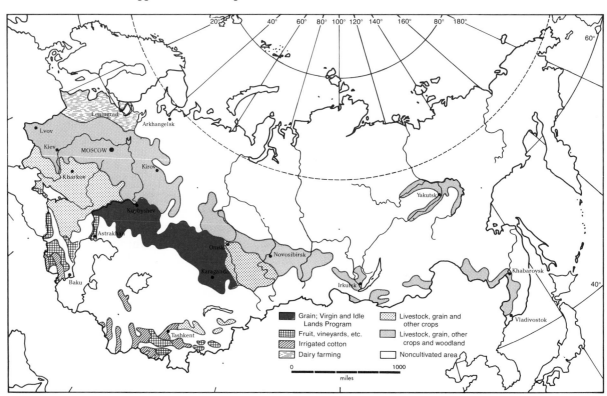

the desirability of the achievement of goals of national priority by the assignment of funds and efforts to their fulfillment.

Conclusion

Very few people live in pure subsistence, commercial, or planned economies, although planned economies tend to be more pervasive in their countries of occurrence than do the other two forms in theirs. The trends in the world seem clear: subsistence economies are giving way to planned and commercial organization. Commercial economies are introducing more and more government control and planning in their operations. Planned economies, seeking ways to overcome at least some of their cumbersome rigidities and inefficiencies, are turning to some of the controls and allocation methods characteristic of commercial economies.

It is also clear that economic activities, patterns, and events are the very stuff of life. As they are pivotal to all facets of human existence and experience, so they are central to geographic understanding. World patterns of economic activity join together the physical and cultural sides of the discipline. They are rooted in the realities of climates, soils, mineral deposits, navigable waters, and other natural phenomena. The economies pervasively influence, and are influenced by, the organizational forms of society, religion, custom, and way of life.

One final reminder is needed: an economy in isolation no longer exists. The world network of transportation, communication, trade, and interdependence is too complete to allow totally separated national economies. Events affecting one ramify to affect all. Crop failure in the Soviet Union influences world prices of wheat, patterns of shipping, balances of payments, and even international pacts and agreements.

The potato blight in an isolated corner of Europe a century and one-third ago still holds its message: despite differences in language, culture, or ideology, we are inextricably a single people economically.

327

For Review

1. What are the distinguishing characteristics of the economic systems labeled *subsistence, commercial, and planned*?

2. What are the ecological consequences of the different forms of *extensive subsistence* land use? In what world regions are such systems found? What, in your opinion, are the prospects of these land uses for the future?

3. How is *intensive subsistence* agriculture distinguished from extensive subsistence cropping? Why, in your opinion, have such different land-use forms developed in separate areas of the warm, moist tropics?

4. What is the concept of *comparative advantage*? How is it related to world trade patterns? How does it help us understand the developmental choices open to underdeveloped nations?

5. Briefly summarize the assumptions and dictates of von Thünen's agricultural model. The model suggests that concentric circles of agricultural specialization develop around a solitary, isolated market. What changes in that proposed pattern would you anticipate if a railroad, highway, or navigable waterway were located along a single radius of those circles? Diagram the new pattern you think would emerge.

6. What role do prices play in the allocation of resources in commercial economies? What role do they play in planned economies? What differences in locational patterns of industry are implicit in the different treatments of costs in the two economic systems?

7. What kinds of constraints other than pure cost considerations affect industrial location and other forms of economic development in commercial economies? In your opinion, are such constraints appropriate in a commercial economy? Why or why not?

8. Contrast the roles of government in economic organization and control suggested by the terms *socialism* and *communism*. Recognizing that developing countries seem to choose some form of central guidance and control in their development programs, why have some, like China, chosen communism and others, like India, opted for modified socialism?

328

Suggested Readings

ALEXANDER, JOHN W., and L. J. GIBSON, *Economic Geography* (2nd ed.). Prentice-Hall, Englewood Cliffs, NJ, 1979.

BERRY, BRIAN J. L., EDGAR C. CONKLING, and D. MICHAEL RAY, *The Geography of Economic Systems*. Prentice-Hall, Englewood Cliffs, NJ, 1976.

CONKLING, EDGAR C., and MAURICE YEATES, *Man's Economic Environment*. McGraw-Hill, New York, 1976.

DE SOUZA, ANTHONY R., and J. BRADY FOUST, *World Space-Economy*. Merrill, Columbus, OH, 1979.

LLOYD, PETER, and PETER DICKEN, *Location in Space* (2nd ed.). Harper & Row, New York, 1977.

MORRILL, RICHARD L., and JACQUELINE M. DORMITZER, *The Spatial Order*. Duxbury Press, North Scituate, MA, 1979.

Oxford Economic Atlas of the World (4th ed.). Oxford University Press, London, 1972.

THOMAN, RICHARD S., and PETER B. CORBIN, *The Geography of Economic Activity* (3rd ed.). McGraw-Hill, New York, 1974.

The economic integration of the world's people and nations, a basic theme of Chapter 9, is rooted in a common dependence upon the finite resources of the earth. Long lines at service stations, rapidly escalating coal and petroleum prices, natural gas shortages during frigid winters, balance-of-payment difficulties associated with increasing oil imports, the establishment of the U. S. Department of Energy, legislation mandating greater vehicular fuel efficiency, proposals for tax rebates for home insulation—all these and more were evidence in the United States of the 1970s of a growing national awareness of the energy shortages so long predicted but so little believed.

Wise management of energy supplies is an obvious first step in addressing the problem posed by their changing price and availability. Conservation measures include the reduction of lighting, heating, and air-conditioning loads in homes, offices, and industry; experimentation with rate reductions during off-peak hours; cost penalties rather than rewards for high-volume energy users; the introduction of co-generation facilities (using industrial boiler heat for in-plant electrical generation); and the extraction of energy from urban and feedlot refuse.

The second step is the development of alternate energy supplies to compensate for the declining availability and the rising cost of traditional fuels. Their development, however, is a long-term solution to the world's energy needs, and no matter what alternate sources are developed, they will be exploited only at economic costs reflecting the conversion of energy from a cheap resource to be used prodigally to a scarce resource to be used as conservatively as possible.

Energy resources are one type of the resources upon which human beings depend; shortages in their supply are merely an indication of the growing scarcity of all. Resource depletion affects not just the United States, not just the technologically advanced economies,

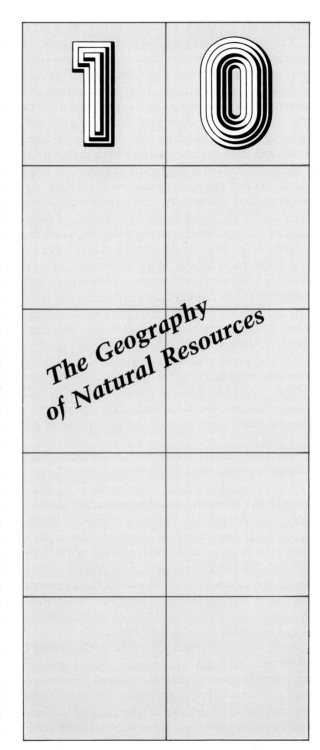

The Geography
of Natural Resources

but, in varying ways, all countries of the world.

The cultural and technological advance of humankind has increased our range of resource exploitation. Growing population numbers and economic development have magnified the extent and the intensity of human depletion of the treasures that were part of the original biosphere. Political subdivisions of the earth have endowed the separate nations with differential control of unevenly distributed resources. Wars have been fought over access to those resources, but ultimately they are the common heritage of all. Since resources of land, of ores, and of energy are finite, but the resource demands of an expanding, economically advancing population appear to be insatiable, an imbalance between resource availability and utilization has emerged as a critical concern in the human–earth-surface equation of geography. Because resources of all kinds are unevenly distributed in kind, amount, and quality over the earth and do not match comparably uneven distributions of population and demand, a consideration of natural resources and their conservation falls within the locational–distributional tradition of the discipline of geography.

Concern with conservation—which may simply be defined as the wise use of resources, not the cessation of use—and interest in environmental protection and ecosystem maintenance, discussed in Chapter 4, represent an intertwined awareness of human dependence upon the delicate and limited fabric of the biosphere. The overall conservation question may be stated rather simply: How can the demands placed on the natural environment be satisfied without disruption of that environment in ways injurious to it and to humans?

Save for subsistence agriculture, the world's economic systems are geared toward growth and development, that is, toward continued and accelerated pressures upon the ecosystems and resources of the earth. The logic of unceasing economic growth, however, is being chal-

lenged, and an irreconcilable conflict arises between two groups. One, fearing uncontrolled resource depletion and environmental disaster, advocates the immediate worldwide adoption of "no-growth" policies. The second is composed of those who, not having achieved the levels of development and of living attained in advanced nations, see "no growth" as forever holding them in second-class economic circumstances. Although this conflict is not assessed directly here, the effect of rising incomes and a growing population on the finite supply of the earth's raw materials is a real concern. Before we address that concern, it will be useful to define some commonly employed terms and concepts.

Definitions

By *conservation* we mean resource use with as little waste and environmental degradation as possible. The term implies a deliberate policy of trying to assure society as a whole as much future benefit from the exploitation of natural resources as is now experienced. In all respects, conservation is a human-centered concern; it seeks to maximize human benefit by maintaining the supply and the quality of natural resources at levels sufficient to meet present and future needs.

Resources are the items that a population, at any given state of economic development and technological awareness, perceives to be neces-

Fig. 10.1 (opposite) The original hardwood forest covering these West Virginia hills was removed by settlers who saw greater resource value in the underlying soils. The soils, in their turn, were stripped away for access to the still more valuable coal deposits. Resources are as a culture perceives them, though their exploitation may consume them and destroy the potential of an area for alternative uses. (U.S.D.A. Soil Conservation Service.)

sary and useful to its maintenance and well-being. To be considered a resource, a given commodity must first of all be *understood* to be a resource. This is a cultural, not purely a physical, circumstance. Native Americans may have viewed the resource base of, say, Pennsylvania largely as composed of forests for shelter and fuel and as the habitat of the game resources upon which their food supply was based. European settlers viewed the forests as the unwanted covering of the resource they perceived to be of value: soil for agriculture. Still later, industrialists appraised the underlying coal deposits, ignored or unrecognized as a resource by earlier occupants, as the landscape item of value for exploitation. A resource exists, then, only when it is perceived to exist (Figure 10.1). Even when perception is present, accessibility colors resource appraisal. If the rock in the ocean deeps cannot be mined, it is not considered a resource. If manganese nodules on the sea bed can be mined, but mining is prohibited by international legal complications, the appraised resource is not an effective one.

Resources are both human and physical. Willing, healthy, and skilled workers constitute a valuable resource, but without physical resources such as petroleum, iron ore, or fertile soil, the human resources are ineffective. Because of this secondary position of people in the chain of resource status, we shall devote attention to physically occurring resources, or as they are more commonly called, *natural resources*. In discussions centering on the theme of conservation, natural resources are usually recognized as falling into one of two broad classes: nonrenewable and renewable.

NONRENEWABLE RESOURCES

Nonrenewable resources, in their original forms, exist in finite amounts. Although the elements of which these resources are composed cannot be destroyed, they can be altered to less useful or available forms. They are subject to depletion. The enery stored in a unit volume of the fossil fuels—coal, oil, natural gas—may have taken eons to concentrate in usable form; it can be converted to heat in an instant and effectively lost forever.

Fortunately, many minerals of the earth can be recycled even though they cannot be replaced. If they are not chemically destroyed—that is, if they retain their original chemical composition—they are potentially reusable. Aluminum, iron, lead, zinc, and the other metallic resources plus many of the nonmetallics, such as diamonds and petroleum by-products, can be used time and time again. However, many of these materials are used in small amounts in any given manufactured object, so that recouping them is economically unfeasible. In addition, many materials are now being used in manufactured products, making them unavailable for recycling unless the product is destroyed. Consequently, the term *reusable resource* must be employed carefully. At present, all mineral resources are being mined much faster than they are being recycled.

RENEWABLE RESOURCES

These are resources that can both be consumed and restored after use. The hydrologic cycle assures that water, no matter how often used or how much abused, will return over and over to the land for further exploitation. Of course, groundwater extracted beyond the replacement rate in arid areas may be as permanently dissipated as if it were a nonrenewable ore (Figure 10.2). Soils can be continuously used productively and even improved by proper management and fertilization practices; they can also be lost by mismanagement that leads to total erosion from rocky slopes, as the agricultural history of the Mediterranean basin amply attests. Forests, wildlife, and fish re-

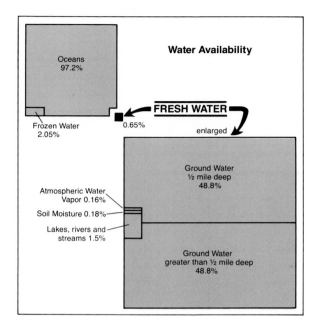

Oceans
97.2%

Water Availability

FRESH WATER

0.65%

Frozen Water
2.05%

enlarged

Ground Water
½ mile deep
48.8%

Atmospheric Water
Vapor 0.16%

Soil Moisture 0.18%

Lakes, rivers and
streams 1.5%

Ground Water
greater than ½ mile deep
48.8%

Fig. 10.2 Less than 1% of the world's water supply is available for human use in freshwater lakes and rivers and from wells. An additional 2% is, despite exotic proposals to tow Antarctic ice to Middle Eastern deserts, effectively locked in glaciers and polar ice caps.

sources are also naturally renewable, provided exploitation to extinction has not occurred.

Sometimes the renewal cycle takes hundreds of years. For example, soils depleted of their nutrients can usually regain them if conditions are right, but the process takes a long time. Humans can speed up the processes by adding chemical nutrients in the case of soil or by planting trees in the case of forest resources or by cleansing water in the case of that resource.

For practical purposes, a renewable resource cannot be classified as such unless the process takes place over a short period of time. Consequently, we should not say that forests are a renewable resource unless humans are planting at least as much wood as is being cut. Even this sustained-yield approach to renewable resources is not quite good enough. With rising

expectations and growing populations, the renewed resource may not be sufficient to satisfy future demands. A good example of this problem is the world's fish resources. For many years, people thought that natural increases in fish supplies would more than offset increasing demands for fish products. Now it has become clear that the world's fish supply may be called a renewable resource only if measures are taken to conserve it.

Some resources, such as ocean tides and solar energy, are not truly renewable. They are there in fixed amounts over time, and their supply does not increase or diminish. If they are not used, their potential is lost, but they remain available for future use.

While the classification of resources into categories labeled *nonrenewable* and *renewable* is not totally satisfying, it sheds some light on the sensitivity of earth materials to human use.

333

Energy Resources

The key to income growth and high standards of living is energy production (Figure 10.3). Over the 300 or so years since the beginnings of the use of fossil fuels for industrial purposes, the employment of those fuels has increased about twice as fast as the rate of population growth (4% versus 2% per year). It is the control of energy, particularly that derived from fossil fuels, that has made possible the population increases discussed in Chapter 6 and has given population-supporting capacity to areas far in excess of what would be possible without inanimate energy resources. Modern technologies and societies have become so dependent upon the constant availability and flow of energy that any variation in supply or disruption of availability has profound impact. The New York City blackouts of 1965 and 1977, though traceable to combinations of mechanical and

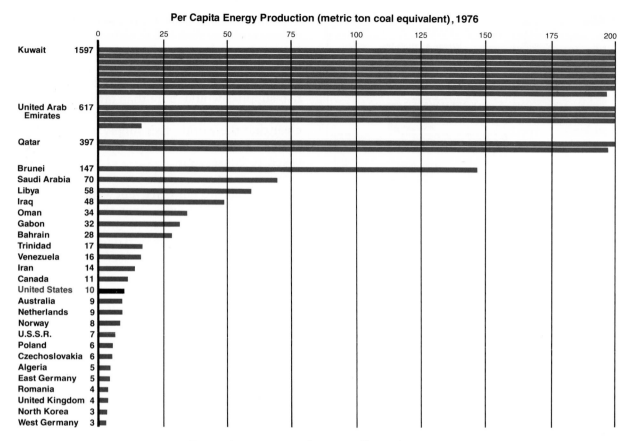

Fig. 10.3 Per capita energy production reflects the unequal world distribution of energy resources, the application of foreign or domestic capital to their exploitation, the population of producing nations, and the existence of international markets for energy supplies. It has no necessary relationship to national energy consumption or to the degree of economic development within producing nations. Note the prominence of oil-rich, lesser developed nations of the Middle East on this chart, and the differences between the chart and the record of energy consumption shown on Figure 6.10. (Data from United Nations, *Statistical Yearbook, 1977.*)

human error, were a chilling reminder of the absolute dependence by modern cultures upon uninterrupted energy supplies.

Energy can be produced in a number of ways. Humans themselves are energy producers, acquiring their fuel from the energy contained in food. Of course, as we saw in Chapter 4, it is solar energy that is stored in plants, from which our food is derived. In fact, nearly all energy sources are really storehouses for energy originally derived from the sun; among them are wood, water, fossil fuels (coal, petroleum, natural gas), ocean tides, wind, and the sun itself.

Each of these energy sources has been harnessed to a greater or lesser degree by humans. For the present, world societies are primarily dependent upon the fossil fuels, and the conse-

"The poor man's energy crisis" is a phrase increasingly applicable to the rising demand for, and the decreasing supply of, wood for fuel in the developing nations. It is a different kind of crisis than that faced by the industrialized nations, which are encountering rising prices and diminished supplies of petroleum and natural gas. The crisis of the underdeveloped societies involves cooking food and keeping warm, not running machines, cooling theaters, or burning lights.

The United Nations Food and Agriculture Organization (F.A.O.) estimates that wood accounts for 90% of the fuel used in the developing countries, and demand has been expected to rise by 50% between 1978 and 1994. The agency reports that wood accounts for nearly 60% of all energy consumed in Africa, more than 40% in the Far East (excluding China), 20% in Latin America, and 14% in the Near East.

Increasing demand and declining supplies are having serious human and natural consequences. In such widely scattered areas as Nepal and Haiti, the wood shortage is forcing families to change their diets to primary dependence upon less nutritious foods that need no cooking. Reports of whole villages reduced to only one cooked meal a day are common. With the average villager requiring a ton of wood per year, an increasing proportion of labor must be expended to secure even minimal supplies of fuel, to the detriment of food or income-producing activities. The F.A.O. reports that in parts of Tanzania in East Africa, between 250 and 300 man-days of labor are needed to fill the yearly firewood requirements of a single household.

Obvious environmental consequences of the increasing demand for fuel wood involve total deforestation and the uprooting of shrubs and grasses over wide areas. Soil erosion and desertification are inevitable and are increasing in regions already poor and overpopulated. A less apparent but nonetheless real consequence of expanding energy needs is the substitution of animal dung for unavailable wood for fuel. The result is loss of manure for soil enrichment. An estimated 400 million tons of potential fertilizer are burned each year in Africa and Asia, lowering the productivity of already marginal lands.

The Energy Crisis in Less Developed Countries

335

quences are made clear by repeated warnings of early depletion of oil and natural gas supplies and by their steadily rising prices. For most of human history, wood was the predominant source of fuel, and even today at least one-half of the world's people depend partly or largely on fuel wood for cooking and heating. The growing populations of the underdeveloped nations dependent upon wood have so depleted supplies that the cost and the expenditure of human effort to travel ever farther from home to secure fuel have both increased seriously, as have the soil erosion and the spread of deserts associated with deforestation. Waterpower was briefly important in manufacturing early in the Industrial Revolution and still remains significant in the generation of electrical energy. The other sun-derived sources and atomic energy,

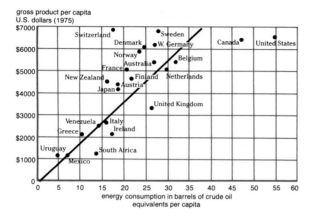

Fig. 10.4 Energy consumption rises with increase in gross national product. Since the internal combustion engine accounts for a large share of national energy consumption, this graph is a statement both of economic development and of the roles of mass transportation, automotive efficiency, and mechanization (including that of agriculture) in different national economies. (Data from United Nations, *Statistical Yearbook, 1975.*)

336

despite their potential, still play only minor roles in the total world energy budget.

Energy consumption goes hand-in-hand with industrial production and with increases in per capita income (Figure 10.4). By the application of energy, the conversion of materials into commodities and the performance of services far beyond the capabilities of any single individual are made possible. Further, the application of energy can overcome deficiencies in the material world that humans exploit; high-quality iron ore may be depleted, but by massive applications of energy, very low-quality iron-bearing rocks—taconite—can have their contained iron extracted and concentrated for industrial use. Because of the association of energy and economic development, a basic conflict between societies is made clear. Nations that can afford high levels of energy production and consumption continue to expand their economies and increase their levels of living; those without access to energy or unable to afford it see the gap between their economic

prospects and those of the developed nations grow ever greater.

In exploitation of materials, including fuels, humans turn first to the sources of the highest quality, the greatest amount, the greatest accessibility, and the cheapest extraction within the context of existing awareness and technology. Only as these are depleted are more difficult, less desirable supplies exploited, with an ever-greater cost in energy input. Other things being equal, the closer the fuels are located to industries, the cheaper they are. As discussed in Chapter 9, the high cost of the transportation of coal early in the industrial period induced the development of major heavy industrial centers directly on coalfields, for example, Pittsburgh, the Ruhr, Silesia, the English Midlands, and the Donets district of the Ukraine.

The substitution of petroleum for coal as the favored industrial energy source did little to alter earlier patterns of manufacturing and population concentration. Because petroleum is easier and cheaper to transport than coal, it was moved to existing centers of consumption

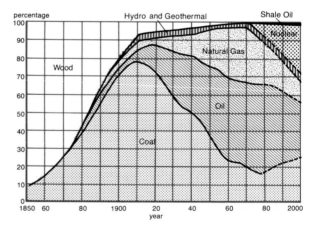

Fig. 10.5 Sources of energy, actual and projected, in the United States, 1850–2000. First, it was wood that declined as the leading source of energy, then coal. Energy from nuclear fuels is just beginning to make a significant impact. Projections were based on supplies, prices, technologies, and potential investments in the mid-1970s.

and did little to induce a relocation of industry to petroleum deposits. Petroleum also rose to importance because of its more complete combustion characteristics and its adaptability as a concentrated energy source for powering moving vehicles.

Since the distribution of petroleum supplies differs markedly from that of the coal deposits upon which urban-industrial markets developed, intricate and extensive national and international systems of petroleum transportation were created. Transfer costs were reduced by high-capacity pipelines and by the construction of ever-larger tankers. The efficiency of these transport media and the energy/cost ratio advantages of petroleum created a world dependence on that fuel even though coal was still generally and cheaply available. Figure 10.5 illustrates past energy consumption patterns and projections of the importance of oil and natural gas in the United States to the year 2000.

American dependence upon oil, the importance of petroleum to other advanced economies, and the spatial separation of petroleum deposits from consuming centers and nations have led to an enormous world trade in oil. Petroleum from a variety of production centers flows, primarily by water, to industrially advanced countries, as shown in Figure 10.6. The United States, which was nearly self-sufficient in oil as late as the 1960s, now depends upon foreign sources for a large and growing proportion of its supplies. Increases in worldwide demand and the depletion of many older fields have given an emergency cast to the continuing and expensive search for new deposits—in the wastes of Siberia, on the Arctic coast of North America, and off the coasts of the world's continents.

THE ENERGY CRISIS: PETROLEUM AND NATURAL GAS

The word *crisis* properly implies a decisive moment, a turning point, or the culmination of a period of prosperity and success; in that sense, it is appropriately applied to the world's energy situation. Nations such as the United States have come to the end of a period in which energy was a cheap and prodigally used commodity, a freely available and lavishly expended base for remarkable economic advancement. Now circumstances have altered, and economies and societies are forced to make adjustments unanticipated by the majority of people a few years ago.

Energy takes many forms, no one of which can satisfy all contemporary needs. Petroleum is being depleted, coal so far cannot power automobiles, and nuclear energy as of now cannot produce all the needed electricity. Because of our need for energy in diverse forms, at least two energy "crises" confront society. One is the crisis of absolute depletion. Despite variant estimates of world petroleum or natural gas reserves (or those of any other chemically bound fossil storage of solar energy), the eventual exhaustion of supplies at currently feasible expenditures of money and energy to recover is inevitable.

A proviso of the first crisis implies the second: acceptable cost. The investments necessary to maintain a constant—let alone the necessarily expanding—energy supply increase exponentially as earlier good, large, accessible, and cheap deposits are exhausted. Tremendous petroleum reserves are locked in oil shales and tar sands. No one seriously considered exploiting them until world oil prices had risen to a point that was considered astronomical a generation ago but that is now seen to be at a level that begins to justify the capital investment necessary to develop the shales and sands. The glowing promise of nuclear energy has been dimmed by the capital costs of the installed electrical generating capacity of nuclear-powered plants, by questions raised about their operational safety, by the attendant expense and the uncertainty of disposal of radioactive wastes. Whatever the energy source now to be developed or exploited, its cost will contribute

337

1956

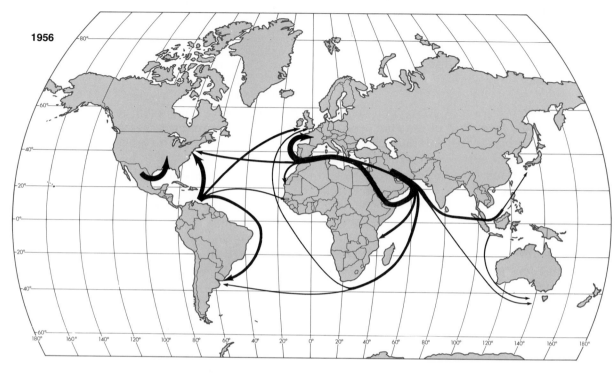

338

1976

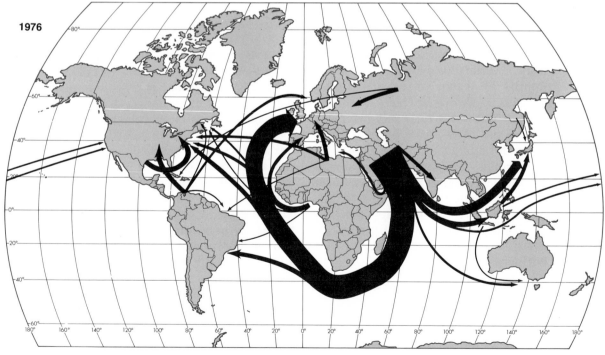

Rising costs of fuel have accelerated the search for means to use it more efficiently. Research into magnetohydrodynamics (MHD), jointly sponsored by the United States and the Soviet Union, gives promise of producing far more electricity from a unit of hydrocarbon fuel than can be obtained by existing thermal electric plants. In the MHD process, coal or natural gas is burned at temperatures up to 5000°F. The resulting ionized gas is directed at bullet speed down a channel through an intense magnetic field, thus producing electricity. The hot gas may then be used to heat water for the steam operation of standard turbine generators. The two-step process increases fuel efficiency from a conventional 35% in the conversion of coal to electricity to about 65%; a theoretical maximum efficiency of 75% might be approached. Plants with a power of 1–2 million kilowatts are envisioned by the early 1990s.

Magneto-hydrodynamics

to the increased price of energy and of everything consuming energy.

These development-cost considerations have been compounded by political moves. In recent years, the Middle Eastern countries and the other nonindustrial countries producing oil have managed to gain control over their own resources. The international cartel called the Organization of Petroleum Exporting Countries (O.P.E.C.) has raised prices and manipulated supplies. Each price rise, besides adding to the cost of energy, has created balance-of-payment problems for importing nations, problems serious for developed nations and

Fig. 10.6 (opposite) The international sea-borne flow of crude oil and petroleum products underwent great changes between 1956 and 1976. World consumption increased, exports from the Middle East expanded rapidly, and the United States became very much more dependent upon foreign sources. The closing of the Suez Canal in 1967 and the growing use of "supertankers" too large to use it when it was reopened reduced flows through the Mediterranean. The line widths are proportional to average daily movements; the same scale is employed on both maps. (Redrawn from The British Petroleum Company Limited, *BP Statistical Review of the World Oil Industry, 1976.*)

devastating for third-world countries that do not produce their own oil or compensating items of export value.

To these direct costs must be added those of environmental protection and degradation. Expenditures mandated in the United States by the Environmental Protection Agency to control particulate matter and chemical emissions from power stations, to reduce the temperature of thermally polluted cooling waters, and to guarantee the suppression of radioactive leakage are all chargeable against each unit of energy produced. Disastrous oil spills from runaway offshore wells or from major and minor tanker wrecks and leakages have cleanup, scenic, and wildlife costs that are not easily measured but that are societal costs of energy nonetheless (Figure 10.7). Other charges—including those represented by taxation to discourage consumption, by expenditures on pipelines in remote and difficult Arctic areas, and by other additions to actual extraction costs—are also part of the problem.

Natural gas came to importance later than petroleum and does not move as freely in international trade. When found in conjunction with oil, gas was early "flared off" (burned) as

339

340

Fig. 10.7 (opposite) Results of an August 1979 oil-well blowout off the coast of Mexico. Each day thousands of barrels of oil poured into the Gulf of Mexico; currents carried the oil 500 miles to the Texas coast, where it fouled island beaches such as the one shown here. (United Press International photo.)

an unwanted by-product of the oil industry, as it still is in oil fields without appropriate capturing equipment. Natural gas, like petroleum, flows easily and cheaply by pipeline, but water movement involves costly equipment for liquefaction and for vessels that can contain the liquid under appropriate temperature conditions. Natural gas is therefore largely a commodity of domestic or international pipeline movement rather than of water-borne commerce.

COAL AND OIL SHALE PROBLEMS

Although coal is a nonrenewable resource, world supplies are so great that resource life expectancy may be measured in centuries, not in the tens of years usually cited for petroleum and natural gas. Of an original estimated world reserve of coal on the order of 17,000 billion tons, only some 130 billion tons have so far been used. Coal is therefore seen as a likely substitute for petroleum, an ironic circumstance in view of the rapid rise of oil as a substitute for coal. Coal is not a resource of constant quality, accessibility, or cost. It ranges from lignite (barely compacted from the original peat) through bituminous (soft coal) to anthracite (hard coal), each class representing increased carbon content and heat value. Good-quality bituminous coals with the caloric content and the physical properties suitable for producing coke for the steel industry are decreasingly available readily and are of increasing cost. Anthracite, formerly a dominant fuel for home heating, finds no ready industrial market at its high mining costs.

Vast quantities of coal suitable for industrial use and thermal electric generation are available. Much can be mined relatively cheaply by open-pit (surface) techniques, in which huge shovels strip off surface material and remove the exposed seams. Much, however, is available only by expensive and dangerous shaft mining, as in Appalachia and most of Europe. Even where cheap production methods may be employed, variations in coal quality have cost implications. The thermal coals of the eastern United States have relatively high sulfur content, and costly techniques for the removal of sulfur from stack gases (and for the removal of other noxious by-products of burning) are now required by most industrial nations, including the United States. The mutilation of the original surface and the acid contamination of water associated with strip mining are in the United States at least partially controlled by environmental protection laws, which add to energy costs. Western United States coals, now attractive because of their low sulfur content, require expensive transportation to markets or high-cost transmission lines if they are used to generate electricity for distant consumers (Figure 10.8).

Technologies are developing to use coal as a raw material for the production of oils, alcohols, and gasolines that can substitute for petroleum, and the gasification of coal promises a substitute for natural gas. Although possible—and in some countries long employed—these techniques become economically feasible only as the costs of petroleum and natural gas increase, so that coal conversion becomes competitive.

A similar situation colors the prospects of the extraction of oil from oil shale, a tremendous potential reserve of hydrocarbon energy. In fact, the rocks involved are not shales but calcium and magnesium carbonates, and the hydrocarbon is not oil but a waxy substance called *kerogen* that adheres to the grains of carbonate material. Covering, in the United States, large areas of northeastern Utah, northwestern

341

Colorado, and southwestern Wyoming, shales are found containing astronomical potential quantities of oil. Even the amount that appears to be recoverable under proven technology at foreseeable energy prices equals some 80 billion barrels. But cost and environmental damage are serious considerations. Only recently has oil shale, because of world petroleum prices, become a feasible resource, and the costs of moving from pilot plant to full-production capability are tremendous. If the shale is mined by surface extraction, vast contour disruption, landscape scars, and detritus heaps will result despite the best reclamation efforts. Ecologically more promising are techniques for burning the rock in place in underground mines and pumping the oil to the surface. In either case, inputs of energy, equipment, and money assure high product prices.

Fig. 10.8 Long-distance transportation adds to the cost of the low-sulfur western coals that are demanded by environmental protection laws but are remote from eastern United States markets. To minimize these costs, unit trains carrying only coal engage in a continuous shuttle movement between western strip mines and eastern utility companies. (Courtesy of National Coal Association.)

Beneath the wilderness bogs of northern Alberta, Canada, lie a potential but expensive 300 billion barrels of petroleum in the world's largest known single deposit. Although only slightly smaller than the proven Middle East oil reserves, the Athabasca tar sands have for over 100 years defied all efforts to exploit their potential at an economic cost. Now, with high and rising world petroleum prices, with the development of appropriate technologies, and with massive capital investments, the oil wealth of the sands begins to justify the cost of its development.

Actually the tar sands are misnamed, for they are really saturated not with tar but with bitumen, a viscous, high-carbon petroleum thicker than molasses, containing both sulfur and traces of vanadium. Each grain of tar sand is coated with a thin film of water surrounded by bitumen. This physical separation of the petroleum and the sand makes the extraction of oil, although difficult, at least possible.

Deposits that can be mined from the surface lie 200 feet below overlying sand and shale topped by some 20 feet of muskeg (thick peat accumulations), which must be bulldozed away during the months of winter freeze; in summer, heavy equipment simply disappears into the unstable bog. With the overburden of sand and shale removed, huge drag lines dig the sticky tar sands, whose abrasive qualities drastically shorten the useful life of all mining equipment.

In the refinery, the tar sand is mixed with hot water, steam, and a bit of sodium hydroxide. The resultant mixture is aerated and transferred to holding vessels, where the bitumen rises to the surface to be skimmed off for further processing. The sand-free bitumen is next heated and separated ("cracked") into solid coke and both a light gas and a heavy gas oil. Sulfur is extracted from the two individually treated oils, and a high-quality, light crude oil is produced by a recombination of the original gas oils. The cost of these operations makes crude oil from the tar sands perhaps the world's most expensive to produce. Rising world oil prices and declining alternate cheaper sources, however, make the Athabasca tar sands potentially profitable for the first time.

But the cost is more than monetary. The delicate muskeg ecosystem is destructively disturbed by the vast open-pit mining operations, refineries, and road network. Plans are to return the cleansed sands to the pits, mix them with muskeg, fertilize them, and plant them. Reestablishment of Arctic vegetative cover, however, is extremely difficult, and even if successful, it involves a great many years in areas of short growing season and harsh climate. Even if revegetation is successful, the inevitable sulfur dioxide emissions to the air and the necessary deposit of alkaline processing water in immense holding ponds assure serious ecological disturbance in a largely pristine area.

The price of satisfying energy needs is high and varied.

343

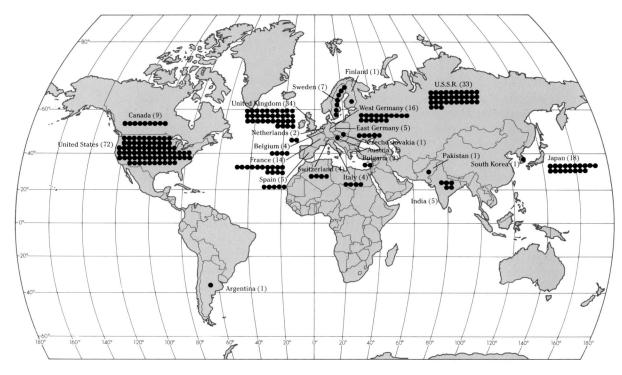

Fig. 10.9 The location of nuclear power plants, including those operational and under construction in 1977. The United States leads with 44% of the world's capacity in megawatts. (Data from International Atomic Energy Agency.)

NUCLEAR FUELS

Nuclear energy has been heralded as a major long-term solution to the energy crisis. Within a single gram of uranium-235 is the energy equivalence of nearly 14 barrels of oil. To tap that energy, some 240 nuclear power plants had already been built around the world by 1977 in both developed and developing nations. By the same year, 72 plants were in operation in the United States alone, though their construction was increasingly opposed by environmental groups and many scientists. Those in operation had a total rated capacity of some 53,000 megawatts, or about 20% of the nation's installed electrical generating capacity. Because of design problems and environmental concerns, not all were operating at design capacity (Figure 10.9).

Basically, energy can be created from the atom in two ways: nuclear fission and nuclear fusion. *Nuclear fission* for power production involves the controlled "splitting" of an atomic nucleus of uranium-235, a raw material itself in short supply. When U-235 atoms are split, about one thousandth of the original mass is converted to heat. The released heat is transferred through a heat exchanger to create steam, which drives turbines to generate electricity. If fissionable U-235 is wrapped in a blanket of U-238, a more abundant but nonfissionable isotope, the U-235 both generates electricity and converts its wrapping to fissionable plutonium. This is the breeder reactor, which produces more fuel than it consumes, but the fuel produced, plutonium, is one of the most toxic substances known. Problems of assuring complete safety in any fission reactor have not been solved (Figure 10.10). Even more troublesome is the matter of disposal of the great amount of radioactive waste produced by either fission method of power production. With half-lives in the hundreds and thousands of years, radioactive wastes—deadly contaminants all—represent a potential for ecological and human disaster on an unimaginable scale.

Nuclear fusion represents the same kind of energy generation that is carried on in the sun and the stars. The process involves forcing two atoms of deuterium, an isotope of hydrogen, to fuse into a single atom of helium, with a consequent release of large amounts of energy. The process is accomplished in an uncontrolled fashion in the hydrogen bomb. The problem,

Fig. 10.10 A geological fault, with possible earthquake hazard, is located only 2.5 miles from Diablo Canyon nuclear power plant near San Luis Obispo, California. Although the plant's structural stability in the event of an earthquake up to 7.5 on the Richter scale was affirmed by the Nuclear Regulatory Agency in October 1979, the possibility of earthquake danger has been cited in the cases of other nuclear installations in the eatern United States. (Courtesy of Pacific Gas and Electric Company, San Francisco, CA.)

Fig. 10.11 A solar house in Harrisville, New Hampshire. Several thousand buildings in the United States are now heated by solar energy. Federal regulations direct that a portion of all new military housing include supplementary solar energy systems, and actual or prospective state and federal tax concessions encourage new home builders to do the same. Retrofitting of some 85 million existing houses with auxiliary solar systems is unlikely. (Goosebrook Solar Home designed and built in 1977 by Total Environmental Action, Inc., in Harrisville, NH.)

not yet solved for commercial use, is to control fusion for a usable release of its energy. Should that control prove feasible, human energy requirements would presumably be satisfied from this source alone for millions of years. Each cubic meter of fresh or salt water contains deuterium atoms equivalent to some 270 tons of coal energy. The technical problems involve the heating of deuterium to 180 million °F (82.2 million °C) and containing a substance of that temperature. If the developmental problems are solved, fusion offers the advantages of the total absence of radioactive wastes and greater inherent safety in energy generation than fission plants.

Solar Energy

Each year the earth intercepts solar energy equivalent to many thousands of times the energy currently used by humans. An inexhaustible, nonpolluting energy source, solar energy is the ultimate origin from which most forms of utilized energy were derived: fossil fuels and current plant life, waterpower, and windpower. It is, however, the direct capture of solar energy that has excited recent interest. The technology for developing solar energy is well known, especially for the heating and cooling of homes and small office buildings (Figure 10.11).

The most widely employed method of utilizing solar energy involves the installation of black, glass-covered panels to collect solar heat. The heat is transferred to water- or air-storage tanks by tubes that run through the collector panels. Pumps or fans distribute the heated water or air around the small buildings for which the system is adapted. Present technology cannot maintain usable heat levels for more than a few days at a time, so that solar heating must be supplemented by heat from conventional sources during cloudy periods or in cool areas.

Reflecting mirror systems can concentrate solar rays, developing heat levels sufficient to drive turbines with heated fluids or gases for generating electricity. Serious proposals have been advanced for utilizing the reflecting mirror technology to satisfy all United States electrical needs. Assuming 30% efficiency of operation, the system would involve covering some 31,000 square miles (80,500 sq. km) of the southwestern part of the nation with mirror reflector power installations. If feasible, the system could be important in any world area with a high annual number of hours of sunshine.

Electricity can also be generated directly from solar rays by photovoltaic, or energy, cells. A variety of crystalline substances are capable of making such a direct conversion, and solar cells of silicon have been used in spacecraft. Costs, however, are still high for energy cells and efficiency is relatively low. Again, large land areas would be required to generate significant amounts of energy; as with other means of solar generation of power, storage of electricity for the hours of darkness poses problems. It has also been proposed that

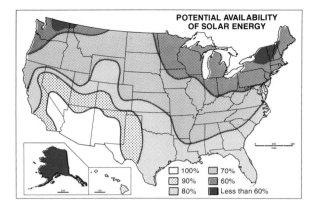

Fig. 10.12   Solar energy availability. The incidence of solar energy in the United States is a climate- and latitude-induced variable. The North and Northeast, with the greatest heating demand, are at least able to satisfy energy requirements through insolation. (© 1977 by the New York Times Company. Reprinted by permission.)

orbiting space stations equipped with solar cells could send generated power to earth as microwave beams. The harnessing of solar rays is seen by many as the best hope to satisfy a large proportion of future energy needs with minimal environmental damage and maximum conservation of earth resources (Figure 10.12).

OTHER ENERGY POSSIBILITIES

Hydropower, solar sea power, wind power, tidal power, and geothermal power are other energy sources cited as partial solutions to the energy crisis.

Although in the United States, at least, most significant hydroelectric sites are already developed, some modest power additions could be gained by damming, or utilizing already dammed, smaller streams in the humid areas of the country. Solar sea power involves the utilization of the temperature differential between warm surface water and cold deep water in tropical oceans. One estimate suggests that

347

solar heat input into the warm water layers is 10,000 times the energy requirements of humankind. Ammonia in a heat exchanger would be liquid at the lower temperatures of deep water and would be converted to gas at the upper levels; the expanding gas would drive a turbine and generate electricity.

Wind power has been used for centuries to pump water, grind grain, and drive machinery. Now it is proposed that banks of giant windmills appropriately sited could generate useful amounts of electrical power; both private and federal funds in the United States are being invested to explore the feasibility of this energy source.

Tidal energy is represented by the ebb and flow of water in response to the gravitational pulls of the sun and the moon; the useful contribution by tidal power to the world's energy needs appears to be small, however. Sites with appropriate tidal ranges and geographical features are few. One plant, the world's first, was completed in 1966 on the Rance River in France to utilize the tidal flows of the English Channel. In the United States, plans for the Passamaquoddy project between Maine and Canada have not yet reached fruition.

When surface water—either seeping naturally or pumped downward—is converted to steam in hot regions of the earth's crust, geothermal power is developed as the expanding steam issues from the earth. Geothermal energy is already being utilized for household heat or for electrical generation in Iceland, New Zealand, the United States (California), the Soviet Union, and other countries. Some 24 nations reportedly have geothermal potential, though its contribution to world energy needs

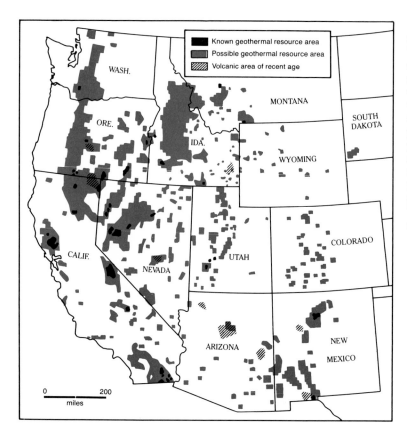

Fig. 10.13 Known and potential geothermal resources in the western United States. Recent exploratory drilling by petroleum companies has revealed several major western reservoirs of water superheated from past volcanic activity. Although the environmental problems are minor compared with those of coal-fired electric plants, geothermal water tends to be mineral-laden and is usually returned to the ground via injection wells to prevent the contamination of local streams. (© 1977 by The New York Times Company. Reprinted by permission.)

would undoubtedly be small even if all potential sites were exploited (Figure 10.13).

Minerals

Metallic and nonmetallic minerals are, like the fossil fuels, nonrenewable resources upon which modern economies make insatiable demands. Usable mineral deposits are finite in the number and the size of reserves, with predictable rates and dates of effective exhaustion. Table 10.1 gives one estimate of the "years remaining" for important metals, though it should be taken as suggestive rather than definitive, as new discoveries and new recovery techniques may amend the predictions.

Although human societies began to use metals as early as 3500 B.C., world demand remained small until the Industrial Revolution of the 18th and 19th centuries. It was not until

Table 10.1 **World Mineral Reserves**

	Years Remaining[a]
Bauxite (aluminum)	100
Coal	2300
Copper	32
Gold	31
Iron	114
Lead	17
Manganese	123
Mercury	19
Natural gas	32
Petroleum	25
Silver	11
Tin	34
Zinc	27

[a] Number of years the world's reserves (known supplies) will last from 1975, based on the expected rate of mineral consumption between the years 1974 and 2000.

Source: U. S. Bureau of Mines, *Mineral Facts and Problems, 1975.* Government Printing Office, Washington, DC, 1976.

after World War II that increasing shortages and rising prices—and in the United States, increasing dependence upon foreign sources—began to impress themselves upon the general consciousness. Worldwide technological development has established ways of life in which minerals are the essential constituent. A 1972 United Nations Conference on the Environment estimated that if the whole world were to begin using minerals at the 1970 United States rate, the annual world production of tin would have to increase 250 times, of copper 100 times, and of iron ore 75 times. A minerals crisis every bit as foreseeable and serious as the energy crisis is close at hand.

In workable deposits, minerals constitute only a tiny fraction of the earth's crust—far less than 1%. That industrialization has proceeded so rapidly and so cheaply is the direct result of the earlier ready availability of rich and accessible deposits of the requisite materials. Economies grew fat by skimming the cream. It has been suggested that should some catastrophe occur to return human cultural levels to the Stone Age, it would be extremely unlikely that humankind ever again could move along the road of industrialization with the resources left at its disposal.

All workable mineral deposits are the results of geologic processes that have concentrated one or more desired materials in particular locations. Although additional techniques of concentration may be necessary after mining, as preparation for further processing, ultimately the result of the utilization of any mineral or fuel resource is to disperse the material used. In the case of fuels, the loss is in the form of heat. In the case of metals, the dispersion occurs as fabricated materials rust, corrode, or abrade; are dispersed as fractional alloys in other metals; or are dissipated in water, in soil, or in the air, as is the metal content of leaded fuels.

Our successes in exploiting metallic minerals have been achieved not only at the expense of usable world reserves but at increasing mon-

349

Fig. 10.14 The average standard-sized American automobile of the early 1970s had approximately the following material content, in pounds:

Iron and Steel (frame, engine, and body)	2775
Aluminum (components and some engines)	100
Copper (radiators and electrical systems)	50
Lead (battery)	25
Zinc (castings, galvinizing, and tires)	50
Glass (windshields and windows)	100
Rubber and plastics (tires and fittings)	250
Miscellaneous (fabric, insulation, etc.)	150
	3500

Nearly all of these inputs, including the plastic and synthetic rubber, derive from minerals. Assuming an annual production run of some 10 million cars and commercial vehicles, the above unit demands for vehicle manufacturing alone would have to be multiplied 10 million times, with additional allowance made for unavoidable material loss in processing activities. Later model compact cars, such as those pictured, contain proportionately less materials. (From John D. Morgan, Jr., "Future Use of Minerals: The Question of 'Demand'," in Eugene N. Cameron, ed., *The Mineral Position of the United States, 1975–2000.* Madison: University of Wisconsin Press, 1973. Copyright Board of Regents of the University of Wisconsin System. Photo courtesy of General Motors.)

etary costs as high-grade deposits are depleted. The quick national assets represented by the richest, most accessible ores are soon exhausted. Costs increase as more advanced energy-consuming technologies must be applied to extract the desired materials from ever greater depths or from new deposits of lesser mineral content. Consequent environmental costs increase greatly.

THE SUBSTITUTION PROBLEM

The increasing costs and the declining availability of metals encourage the search for substitutes. As Figure 10.14 suggests, the range and quantity of material going into such a common product as a full-sized American car indicate the potential need for multiple forms of substitution. The fact that industrial chemists and metallurgists have been so successful in the search for new materials to substitute for (or perform better than) the traditional resources has tended to allay fears of possible resource depletion. But it must be understood that for some minerals, no adequate replacements have been found; other substitutes are frequently synthetics, often employing increasingly scarce and costly hydrocarbons in their production. Many in their use or disposal constitute environmental hazards, and all have their own high and increasing price tags.

Obviously, then, both conservation in use and recycling of minerals to as great an extent as possible constitute clearly desirable approaches to the problem of mineral resource depletion. Current fuel efficiency requirements for cars sold in the United States have resulted in vehicle size and weight reductions, achieving at a stroke conservation of both fuel and minerals. Such gains are not easily made and are offset by rising populations even when per capita consumption remains constant or is slightly reduced. Much more promising is the recycling of already extracted and utilized minerals. Old cars can be made into new ones; old aluminum cans may, at 10% of the energy input of new aluminum production, be made

Fig. 10.15 The per capita consumption of new minerals in the United States in pounds for 1975. More conservative mineral use, the substitution of alternatives for scarce minerals, and recycling are seen as devices for reducing some demands for new minerals. (Data from U. S. Bureau of Mines, *Mineral Facts and Problems.*)

351

APPROXIMATELY 38,000 POUNDS OF NEW MINERAL MATERIALS . . .

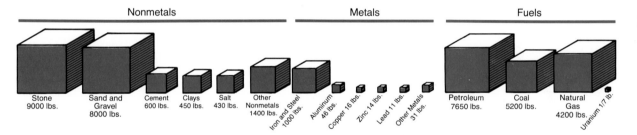

INCLUDING ENERGY EQUIVALENT TO 300 PERSONS WORKING AROUND THE CLOCK . . .

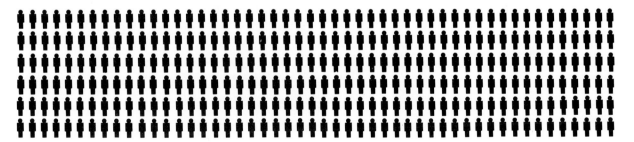

ARE REQUIRED ANNUALLY FOR EACH U.S. CITIZEN

Fig. 10.16 (opposite) The some 140 million tons of refuse thrown away by Americans each year contain an estimated 11.5 million tons of steel and iron, 870,000 tons of aluminum, some 435,000 tons of other metals, principally copper, and more than 13 million tons of glass. Burnable organic matter (including paper) equals some 60 million tons—the equivalent of 150 million barrels of crude oil in energy content. The photographs show one method of recycling industrial scrap to fuel steam power boilers. Solid waste is collected at the plant (upper left), shredded (upper right), and then nonburnable materials are removed (lower left). The remaining scrap is stored in a huge silo (lower right), later to be burned with coal to produce steam power. (GMC Truck and Coach Public Relations.)

into new cans; and so on. At present in the United States, 38,000–40,000 pounds of new raw materials are consumed per capita per year (Figure 10.15), and of this amount about 20,000 pounds become solid waste. Only about 1000–2000 pounds are recycled. Figure 10.16 suggests other benefits of recycling.

Table 10.2 Leading Producers of Critical Raw Materials by Percent of World Total

	Bauxite (Aluminum Ore)		Copper		Iron Ore	
	1. Australia	25.2	1. U.S.A.	21.8	1. U.S.S.R.	25.0
	2. Jamaica	19.2	2. Canada	11.4	2. U.S.A.	10.3
	3. Surinam	11.5	3. Chile	10.3	3. Australia	9.8
	4. U.S.S.R.	6.1	4. Zambia	9.9	4. China	7.6
	5. Guyana	4.5	5. U.S.S.R.	9.8	5. Brazil	6.7
Biggest user	→ U.S.A.	42.0	U.S.A.	33.0	U.S.A.	28.0
	Lead		Manganese		Tin	
	1. U.S.A.	15.5	1. U.S.S.R.	36.1	1. Malaysia	30.6
	2. U.S.S.R.	13.3	2. S. Africa	18.9	2. Bolivia	12.8
	3. Australia	11.5	3. Brazil	9.7	3. U.S.S.R.	12.5
	4. Canada	10.9	4. Gabon	8.7	4. Indonesia	9.5
	5. Peru	5.6	5. India	6.9	5. Thailand	8.9
Biggest user	→ U.S.A.	25.0	U.S.S.R.	NA[a]	U.S.A.	24.0
	Coal		Crude Oil		Phosphate Rock (for Fertilizer)	
	1. U.S.S.R.	23.6	1. U.S.A.	16.5	1. U.S.A.	38.2
	2. U.S.A.	17.2	2. U.S.S.R.	15.2	2. U.S.S.R.	23.0
	3. China	13.6	3. S. Arabia	13.6	3. Morocco	17.1
	4. E. Germany	7.7	4. Iran	10.5	4. Tunisia	3.5
	5. W. Germany	6.8	5. Venezuela	6.0	5. China	2.3
Biggest user	→ U.S.S.R.	NA[a]	U.S.A.	29.9	U.S.A.	NA

[a]NA: data not available.

Source: U. S. Bureau of Mines, *Minerals Yearbook, 1973.* Government Printing Office, Washington, DC.

353

THE MATTER OF LOCATION

Because usable mineral deposits are the result of geologic accident, it follows that the larger the nation, the more probable it is that such accidents will have occurred within national territory. And in fact, the U.S.S.R., Canada, China, the United States, and Brazil are the leading mining countries. As Table 10.2 indicates, these nations produce the bulk of the metals (such as iron, manganese, nickel, and tungsten) and of nonmetals (such as potash and

sulfur). The table reflects conditions in the early 1970s; for some commodities—notably crude oil—new rankings have emerged.

But not even these large nations have domestic reserves of all needed minerals. Some, like the United States, which were bountifully supplied by nature, have spent their assets and depend upon foreign sources. Although the United States was virtually self-sufficient in mineral supplies in the 1940s and 1950s, it is not so today. Because of its past history of utilization of domestic reserves and its continually

Fig. 10.17 Imports supply significant percentages of the minerals and metals consumed in the United States, as these 1975 data show. Reliance on foreign suppliers for most listed materials continues to increase, posing problems of national security and balance of payments. (Data from U. S. Bureau of Mines, *Mineral Facts and Problems.*)

354

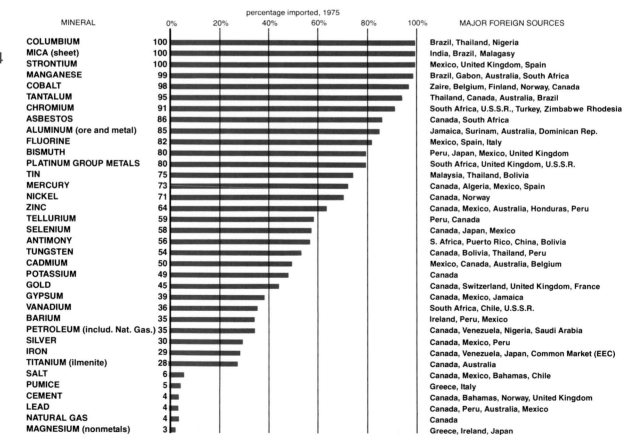

MINERAL	percentage imported, 1975	MAJOR FOREIGN SOURCES
COLUMBIUM	100	Brazil, Thailand, Nigeria
MICA (sheet)	100	India, Brazil, Malagasy
STRONTIUM	100	Mexico, United Kingdom, Spain
MANGANESE	99	Brazil, Gabon, Australia, South Africa
COBALT	98	Zaire, Belgium, Finland, Norway, Canada
TANTALUM	95	Thailand, Canada, Australia, Brazil
CHROMIUM	91	South Africa, U.S.S.R., Turkey, Zimbabwe Rhodesia
ASBESTOS	86	Canada, South Africa
ALUMINUM (ore and metal)	85	Jamaica, Surinam, Australia, Dominican Rep.
FLUORINE	82	Mexico, Spain, Italy
BISMUTH	80	Peru, Japan, Mexico, United Kingdom
PLATINUM GROUP METALS	80	South Africa, United Kingdom, U.S.S.R.
TIN	75	Malaysia, Thailand, Bolivia
MERCURY	73	Canada, Algeria, Mexico, Spain
NICKEL	71	Canada, Norway
ZINC	64	Canada, Mexico, Australia, Honduras, Peru
TELLURIUM	59	Peru, Canada
SELENIUM	58	Canada, Japan, Mexico
ANTIMONY	56	S. Africa, Puerto Rico, China, Bolivia
TUNGSTEN	54	Canada, Bolivia, Thailand, Peru
CADMIUM	50	Mexico, Canada, Australia, Belgium
POTASSIUM	49	Canada
GOLD	45	Canada, Switzerland, United Kingdom, France
GYPSUM	39	Canada, Mexico, Jamaica
VANADIUM	36	South Africa, Chile, U.S.S.R.
BARIUM	35	Ireland, Peru, Mexico
PETROLEUM (includ. Nat. Gas.)	35	Canada, Venezuela, Nigeria, Saudi Arabia
SILVER	30	Canada, Mexico, Peru
IRON	29	Canada, Venezuela, Japan, Common Market (EEC)
TITANIUM (ilmenite)	28	Canada, Australia
SALT	6	Canada, Mexico, Bahamas, Chile
PUMICE	5	Greece, Italy
CEMENT	4	Canada, Bahamas, Norway, United Kingdom
LEAD	4	Canada, Peru, Australia, Mexico
NATURAL GAS	4	Canada
MAGNESIUM (nonmetals)	3	Greece, Ireland, Japan

expanding economy, the United States now depends on a number of countries for supplies of such things as tin (Malaysia) and bauxite, the ore for aluminum (Jamaica), and mercury (Canada), among others, as Figure 10.17 suggests. Partially to protect the nation from interruption of foreign supplies, after World War II the country established a strategic stockpile of commodities for which domestic self-sufficiency was not assured. Although primarily reserved for military needs, minerals from that stockpile have on occasion been sold for commercial use as a device to hold down rapidly inflating prices during times of acute market shortage.

Agricultural Resources

Efforts at the conservation or recycling of nonrenewable resources only postpone the day at which, inevitably, they must no longer be available or affordable. Such resources can be preserved only by cessation of use, in which case they no longer are meaningfully resources. With proper management techniques, however, the agricultural resource base of traditional areas of intensive farming could remain renewably productive for all time. The farmlands and terraced slopes of Southeast Asia have been continuously cropped for thousands of years with no destruction of the resource base and no diminution of yield. Unfortunately, increasing population pressures are putting destructive strains upon the world's agricultural resources. Humans are fully capable of dissipating even essential "renewable" resources.

One of the most critical problems facing the world is the race between people and food. As we have seen (Chapter 6), the world's population is increasing with frightening rapidity. We can reasonably expect that the world's population will grow from its present 4 billion to perhaps 6 billion or more by the year 2000. As Fig-

ure 10.18 indicates, many nations are already faced with problems of malnutrition, although international aid programs and increased world trade in foodstuffs have, through more adequate distribution systems, alleviated the situation in recent years. Local climatic catastrophes, the inadequacy of storage facilities and road systems in impoverished nations, and problems of credits and trade imbalances make access to world supplies unsure for many undernourished peoples.

For everyone in the world to have an adequate diet, food production should rise in pace with population increases. Ideally, foodstuff supplies should exceed population growth to provide the grain reserves needed to improve diets above subsistence levels. One can speak, of course, only of average levels of production, for crop yields vary greatly from year to year. A series of poor monsoon rains during the early 1970s made India a massive importer of the grains that it could not produce domestically. Another series of excellent rainy seasons in the later 1970s gave India grain surpluses to carry forward against future need. The variations in wheat production in the Soviet Union are reported annually in American newspapers, as the U.S.S.R. switches from self-sufficiency to world market purchases with variations in growing-season weather.

The secondary effects of chronic or periodic food shortages are of concern both to individual nations and to the international community. When malnutrition is the rule, not just starvation but low resistance to infectious diseases, lethargy, high child mortality rates, mental damage, social disorder, and political unrest or upheaval are likely consequences. The interconnections of the world's peoples mean that food supply problems are not simply domestic concerns in the seemingly remote areas of their occurrence but in some form have impact on all societies.

Some 70% of the world's 36.6 billion acres (14.8 billion hectares) of land is not suitable for intensive human use; therefore essentially all

355

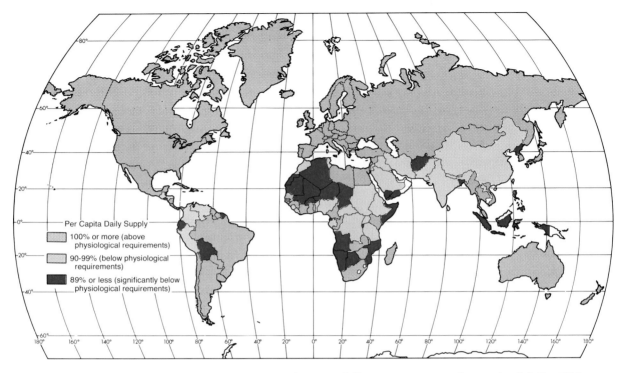

Fig. 10.18 The percentage of required dietary energy supply received daily. This map shows what percentage of the required food supply is received on average by the population of a given country. Data are lacking for areas left blank. (Data from United Nations, Food and Agriculture Organization, *Population, Food Supply, and Development.*)

356

activities must be concentrated on the remaining 30%, or some 11 billion acres (4.5 billion hectares). On this limited total area must be accommodated essentially all the cultural uses of land: roads, airports, cities, industries, and agriculture. One of these uses may be expanded only at the expense of the others. Of those 11 billion acres, only some 3.5 billion (1.4 billion hectares)—or less than 32% of the inhabitable land area—are actually used in intensive food production, and total arable (potentially farmable) land amounts to only 6 billion acres (2.6 billion hectares), according to the United Nations Food and Agricultural Organization (Figure 10.19). On a world basis, the 4 billion inhabitants of the earth have on average nine-

tenths of an acre (0.36 hectare) of food-producing land; when we reach a population of 6 billion, each will have slightly over one-half acre. It is not a speculation but a certainty that within your lifetime dietary levels and food varieties available to even the most affluent of nations will undergo drastic reductions, while food expenditures as a percentage of per capita income will increase greatly.

EXPANSION OF CULTIVATED AREAS

One way to increase food production is to expand the areas being cultivated. Estimates of the amount of land suitable for agricultural use

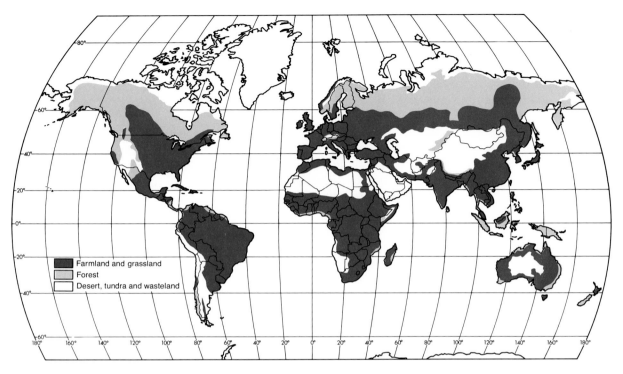

Fig. 10.19 **The more than two-thirds of the world's land area in forest and various forms of wasteland is effectively excluded from agricultural use through climatic, terrain, and soil limitations. A growing world population means ever smaller per capita food-producing areas. (© 1977 by the New York Times Company. Reprinted by permission.)**

range from 4.5 to 7 billion acres (1.8 to 2.8 billion hectares), depending on definitions of "suitable" and "agricultural use." Clearly, some land is more suitable than other land for regular, sustained food production, and most experts agree that most of the land in the world that is well suited for agriculture is already so used.

Let us accept the United Nation figures and assume that an additional 2.5 billion acres (1 billion hectares) could be brought into at least marginal production. Should that occur, of course, the natural environment of every habitable acre on earth would be forever destroyed, and food production would still not have kept pace with potential population increases. The

fact is, however, that we regularly take more land *out* of production than we add. In the United States alone, more than 2 million acres (800,000 hectares)—almost all of it prime agricultural land—are lost to expanding cities, highways, and other nonagricultural uses each year. Additional millions of acres are lost annually through soil erosion and salinization and to the spread of deserts by overgrazing and deforestation.

Obviously, any attempt to expand the area of agricultural land will be made at the expense of environments that all previous human experience has adjudged to be marginal, unproductive, or unsuitable under technologies formerly or currently economically available. Most of the land deemed potentially suitable for cultivation is in Africa and South America. Although population pressures and food needs are great in those areas, the uncultivated land that they contain have remained so because of their marginal quality and their unsuitability for the systems of agriculture developed in those continents. Desert soils need vast amounts of irrigation water to sustain plant growth, and its provision is expensive. Furthermore, when supplied, it may result in resource degradation through salinization of the soil, as noted in Chapter 4. Soils in the tropical rain forest are delicate, are not particularly responsive to fertilizers, and, because they need time to recover their fertility after cropping, are not well suited to supporting sustained yields under conditions of high agricultural density. Steppe lands normally receive an undependable rainfall of below 20 inches (50 cm) per year, sufficient for some agricultural purposes but too unreliable or insufficient for assured yields of most desirable crops. The great fluctuations in grain yields in the Soviet Union are a reflection of the significant areal expansion of agricultural land (the Virgin and Idle Lands program) at the expense of dry steppe territories. Instances of great crop success in years of favorable weather are offset by years of partial or total failures when, in some locales, not even the investment in seed is recouped.

The cultivation of lands only marginally suited to agriculture, then, cannot be viewed as the solution to increasing world foodstuff demands and may, indeed, render unfit for any use land that is improperly added to the agricultural base (Figure 10.20). In the case of all marginal lands, the economic value of the materials, energy, and equipment necessary to make them usefully productive would generally exceed that of the expected product. At the same time, the elimination of pastureland would remove rich animal protein bases that would have to be replaced by less concentrated plant proteins and more carbohydrates.

358

Clearly, then, increasing the amount of cultivated land will not alone solve the world's food problem.

INCREASING YIELDS

The second chief method of expanding agricultural output is by increasing the yields per acre, that is, getting more food from a given piece of land than before. This increase is most often accomplished by multiple cropping (growing two or three crops in rotation in one year on a single piece of land); by using new, high-yielding varieties of grains; and/or by the increased use of fertilizers. A particularly productive strategy that has been used in the United States is to apply fertilizers to fertile and not just to marginal land. Farm mechanization is also a tool for increasing yields, but heavy farm machines are expensive and not well suited to use on small farms, in areas where holdings are fragmented, or where average farm incomes are low.

Two problems of the yield-increase strategy are immediately apparent. First, the most dramatic production increments are potentially obtainable in the less-developed countries, which are least able financially to make the conversion from labor-intensive to capital-intensive agriculture. While American farmers produce, on average, some 400 food calories for each calorie of human effort expended, they do so by virtue of very high costs in equipment, fertilizers, and energy. Underdeveloped nations, according to one estimate, garner no more than 10 produced food calories for each

Fig. 10.20 (opposite) A drought-stricken area in Upper Volta, one of six countries in the Sahel. The cultivation of marginal land, overgrazing by livestock, and climatic change during the 1970s have led to the destruction of native vegetation, to erosion, and to the expansion of the deserts of the world. (United Nations/F.A.O. Photo by F. Botts.)

calorie of human energy invested. By United States standards, subsistence and semisubsistence farmers are, therefore, inefficient producers. To raise their productivity in caloric terms to Western standards would, however, require capital investment levels so far totally unattainable for agriculture in the poorer nations.

Second, many of the proposals for yield increases carry the possibility, even the probability, of ecological and human disaster. Nitrogen fertilizers consume great quantities of natural gas in their manufacture, using a scarce resource and contributing to atmospheric heat. More immediately, the pollution of surface waters by fertilizer runoff has been demonstrated, and damage to the atmosphere by the nitrogen oxide destruction of the ozone layer has been alleged. Scientists have estimated that if commercial fertilizers were used in India at the rate they are in the Netherlands,

Fig. 10.21 Fertilization can dramatically increase plant growth and crop yields, but not without monetary cost and environmental hazard. Investment in fertilizers, herbicides, or pesticides is beyond the resources of most of the world's farmers. Runoff from fields carries chemical pollutants to nearby streams and lakes. At experimental plots, the low yields of strips in continuous corn cultivation are contrasted with production from strips where lime, nitrogen, phosphorus, and potassium have been added (center), and where crop rotation and manuring have been practiced with only minor chemical fertilization. (Courtesy of College of Agriculture, University of Illinois.)

360

India would need nearly one-half of the present world output of fertilizer. Assuming the validity of the claims of pollution, and knowing that the production of 1 million tons of fertilizer requires 1 million tons of steel and about 5 million tons of mostly fossil fuels, the consequences of worldwide expansion of yields by heavy fertilizer applications are obvious and serious.

Further, even assuming a no-cost, no-damage situation, there still is a point beyond which crops will not respond to higher and higher levels of fertilizer application (Figure 10.21). During the 1950s, each additional million tons of fertilizer used annually in the United States was accompanied by a 10-million-ton rise in the grain harvest; during the early 1960s, the grain result of that million tons of input declined to 8.2 million tons, then to 7.2 million tons by the late 1960s, and to 5.8 million tons by the early 1970s. Comparable arguments may be offered to demonstrate the great ecological dangers and the reduced cost-efficiency of the massive utilization of herbicides and pesticides.

High-yield agricultural techniques have human implications beyond those inherent in their ecologically destructive characteristics. Double cropping increases the dangers of soil erosion and soil structure degradation; the expansion of farming into marginal land destroys the original ecological balance, creating dust bowls, deserts, or sterile, eroded lands. The famine that awaits those who come to depend upon agricultural lands extended beyond reasonable environmental limits is inevitable.

The Green Revolution—promising such great results in vastly increased yields of special strains of wheat, rice, and some tubers—endangers populations by encouraging dependence upon *monoculture* (total emphasis upon a single crop). Discounting for the moment the required management skills, water supplies, fertilizer applications, and other inputs that make the Green Revolution more attractive in experimental plots than on the farms of the less-developed countries, monoculture is inherently dangerous. It demands total dependence for human existence in marginally fed countries upon a single crop with a wide variety of essential inputs. Should the crop fail from disease, insect damage, or input shortage, disaster is unavoidable. Perhaps unwittingly, but wisely, subsistence agriculture, dependent solely upon natural conditions for success, is characterized by the production of a great variety of foodstuffs. Diversity is the subsistence farmer's insurance policy.

OTHER FOOD RESOURCE BASES

Since expanding cultivated acreage and intensifying efforts to increase yields are obviously both dangerous and inadequate responses to human foodstuff demands, the expansion or creation of alternate resource bases has attracted interest and hope. As always when humankind moves from the easy and plentiful to the difficult and scarce, cost penalties are severe. Further, the more exotic the new food resource, the greater may be the expected resistance to the requisite dietary change.

The direct synthesis of solar radiation—bypassing photosynthesis to create starches and fats from carbon dioxide, water, and trace elements—is one such exotic approach. Monetary cost, lack of developed technology, and need for plant capacity far in excess of the synthetic organic industry of the United States just to keep pace with annual population growth suggest the difficulties in realizing this theoretical approach to food needs.

It is, however, protein—not starch or fat—that is the pressing nutritional need in much of the world; the amino acids contained in protein are the essential building blocks of the body. Many animal products alone and some vegetable products in combination are complete proteins, supplying all the needed amino

361

acids. Various formulated foods compounded of protein sources and the less nutritive cereals could supply human needs, though perhaps with some loss of aesthetic and taste appeal.

Totally new protein sources of promise include the culture of microbiological organisms to produce single cell protein (SCP). Through a fermentation process, strains of yeast can convert constituents of crude petroleum into the protein of the rapidly reproducing yeast cells themselves. One estimate suggests that only 2.5% of annual petroleum production could yield 22 million tons of protein, about equal to the present total production of animal protein. In addition, bacteria have been developed that can ferment methane, yielding (when dried) SCP of about 50% protein content. Theoretically, 1000 pounds (455 kg) of appropriate fermenting microorganisms could produce 1 million pounds of protein per day. Since methane can be produced from garbage and manure, no particular added resource pressure is evident, although the acceptability of SCP for human consumption may be in doubt from an aesthetic standpoint.

The sea has been seen as the great protein reservoir for extended exploitation, but if it is to be that, it must be more wisely handled than in the past. Between 1950 and 1970, fish supplied an expanding part of human diets as modern technology was applied to harvesting food fish. But in 1970, the trend was interrupted, and the productivity of fishing areas drastically declined as catches above reproduc-

tion rates dangerously reduced regenerative capacity (Figure 10.22). Since that time, world population growth has led to an 11% decline in per capita catch and, of course, to rising prices.

Two somewhat hopeful signs have appeared. First, recent extensions of territorial waters and fishing zones by maritime nations have reduced fishing pressures and have given promise of permitting replacement stocks to develop. Within its claimed 200-mile fishing limit, the United States controls nearly 20% of the most productive world fishing grounds. Beginning in 1977, it has imposed a million-ton reduction in catch, to a maximum of 2 million tons taken annually. Second, there remain, largely unfished, vast supplies of unconventional protein. "Junk fish" can be consumed directly or, more probably, by being used as a fish-meal protein extender of grains. Perhaps an additional 10 million tons of squid could be taken annually, although few cultures now use them as food. Estimates suggest that 50–100 million tons of krill—a nutritious, shrimplike crustacean of no present market acceptance— could be taken from Antarctic waters annually, though perhaps with disastrous results for the Antarctic food chain.

363

AGRICULTURAL RESOURCE MANAGEMENT

In the future, laboratory foods may prove to be useful supplements to the foods that are more traditionally produced. They will be, however, additional nutrient sources won at a high, perhaps intolerable, price. Any industrial process, particularly on a worldwide scale, implies massive inputs of energy, material, and money. The resultant pollution of air and water and the increase of atmospheric heat from constructing and operating the vast number of processing plants that would be required for major dependence upon synthetic food would further strain the fragile biosphere to

Fig. 10.22 (opposite) A factory ship with attendant fishing vessels. The increased efficiency of commercial fishermen using factory ships for preparing and freezing the catch of smaller vessels, the use of sonar to locate schools of food species, and the employment of more efficient nets and tackle result in "clean sweeps" of fishing grounds. Serious depletion of fish stocks endangers a food source once thought inexhaustible. (TASS from Sovfoto.)

which so much harm has already been done.

Exotic food sources may become necessary, but far more vital is the maintenance of the resource base that makes traditional agriculture possible. Humans have been notably remiss in this maintenance. The verdant North African fields of grain that were the foundation of Roman power are replaced by sandy or rocky waste; paved Roman roads stand high above the lands with which they were once level. The soil removed by erosion was an unrecognized nonrenewable resource. Once stripped from the land, it cannot be replaced by nature within any time frame meaningful to contemporary society.

Fig. 10.24 (opposite) Strip cropping is a method of erosion control. Crops that do little to protect the soil, such as corn, are planted in alternating strips with grass or leaf crops that give better protection. (U.S.D.A. Forest Service.)

Two forms of abuse of the world's soil resource have diminished the amount and the productivity of this most vital agricultural base: depletion and erosion. Depletion simply means that the soil has lost some or all of its vital nutrients through crop removal. The soil has been "mined," crop yields drop, plants are weakened, and their nutritive value is reduced. The same soils that produced cotton and to-

Fig. 10.23 Soil erosion in the United States. In all parts of the nation, agricultural alteration of the environment has resulted in soil resource stress and depletion. (Data from U.S.D.A. Soil Conservation Service.)

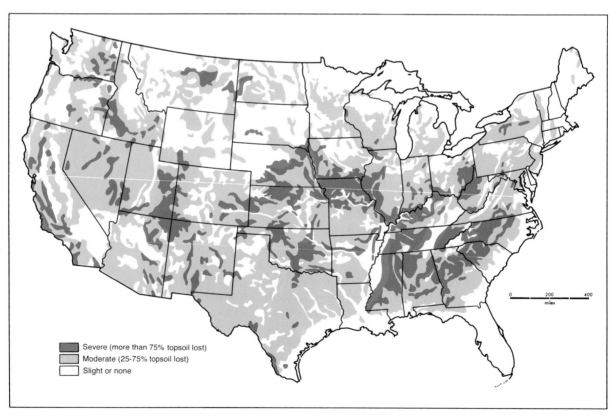

Severe (more than 75% topsoil lost)
Moderate (25-75% topsoil lost)
Slight or none

0   200   400
miles

bacco so bountifully in colonial days in the southeastern United States soon were abandoned as useless. We are told that in 1834, a visitor to Washington's Mount Vernon saw it as a "widespread and perfect agricultural ruin."

Plant nutrients can be replaced by fertilization. What is more difficult to repair in damaged soil is the deterioration in physical structure resulting from depletion, particularly, of the organic matter (humus) essential to hold water in the soil. From that loss comes the desertification that destroyed North Africa and earlier destroyed the Near Eastern civilizations of antiquity. The same process continues today; one expert estimates that man-made deserts now cover nearly 7% of the earth's land area.

The depletion of humus, the removal of forest and grass cover, the plowing of hillsides, and other acts of land resource exploitation prepare the thin layer of productive soil for the erosive forces of wind and running water. Gullying, sheet wash, and wind erosion remove the fragile resource and destroy the hopes and the livelihood of the culprits, their inheritors, and all of society (Figure 10.23; see also Figure 2.19).

Techniques such as contour plowing, terracing, the erection of windbreaks, strip cropping, and crop rotation (Figure 10.24) can be employed to hold soils in place and keep them productive. Instead of sole reliance on commercial fertilizers, the nutrients in animal and agricultural wastes can be returned to the soil,

Soil Loss in the United States

366

We are accustomed to hearing baleful reports of encroaching deserts in Africa and Asia and to being warned of disastrous declines in the fertility of as much as one-fifth of the world's cropland. However, Americans, the comforting feeling is, can be assured that our bountiful crops year after year are as much an evidence of the undiminished quality of our agricultural soils as they are a witness to our splendid climate and skillful farmers. Certainly, it is widely believed, the excellent work of the U. S. Soil Conservation Service in the 1930s and 1940s has repaired past and prevented future erosion problems by introducing contour plowing, terracing, grass waterways, check dams, and other soil retention practices.

Not so, both the Worldwatch Institute and the Soil Conservation Service reported in 1978. Soil erosion in the United States was reported to be at an all-time high and conservation practices noticeably poorer than in the previous generation. In 1977 alone, more than 2 billion tons of topsoil was washed away from croplands, and an additional 3 billion tons was lost through wind erosion on cropland and rangeland.

The hardest-hit areas were not in the southwestern Dust Bowl of terrible memory but in the moist, rolling-hill regions of western Mississippi, western Tennessee, and Missouri. Each of these areas lost more than 10 tons of topsoil per acre of cropland that year. Even fertile Iowa, that recognized model of American agriculture, was reported to be losing over 200 million tons of topsoil per year, and wind was removing some 10 tons per acre per year from the cotton and wheat lands of northwest Texas and from the feed-grain lands of New Mexico and Arizona.

The loss of 5 tons per acre is equivalent to about one-thirtieth of an inch of dirt. Since good cropland may have 10–14 inches of fertile topsoil, the danger may not appear to be great or immediate. There are, however, consequences of soil loss that are serious, though unrelated to crop yields. Streams and reservoirs experience accelerated siltation when farm-soil loss increases. Urban water supplies are reduced and more farmland is removed from cultivation as new and larger reservoirs are constructed. Flood danger is increased as bottomlands fill with silt, and the costs of maintaining navigation channels grow.

The short supplies and the increasing costs of gasoline since 1974 have brought growing interest in the conversion of agricultural residues to alcohol to produce motor fuel. The implication is that stalks, cobs, and other organic wastes from food production can be turned to useful service, not left idly in the fields. Yet, under normal weather conditions, such residue left on the land can replace as much as 5 tons per acre of soil loss. Gasohol may succeed at the price of still further destruction of the nation's precious soil resource.

restoring both fertility and vital humus. To avoid the deleterious side effects of herbicides and pesticides, specific biocides (biological pest control) can be more widely used. The necessity in the face of rising populations and food requirements is to balance the desire for high productivity in the short run with the long-run goal of maintaining the resource base upon which all depend.

Forests

Soil is only one of the presumed renewable or sustainable resources in danger of irreparable damage by human action. Forests, wetlands, and wildlife—and the scenic resource they imply—all are similarly endangered.

The adage about not being able to see the forest for the trees is applicable to those who view forests only for the commercial value of the trees they contain. Forests are more than trees, and timbering is only one purpose that forests serve. Chief among the other purposes are soil and watershed conservation, the provision of a habitat for wildlife, and recreation. Nevertheless, timbering is important. In the United States, the forest products industry—which includes timber processing and the manufacture, transportation, and marketing of wood products—employs several million workers and accounts for about 5% of the gross national product.

Environmentalists have long recognized that forests are not a resource to be thoughtlessly exploited and abused, and some evidence indicates that the general public is beginning to accept this view (Figure 10.25). The widely adopted concept of multiple-use management states that forests serve a variety of purposes; the kind of management techniques employed in any one area depend on the particular use(s) to be emphasized. Thus, if the goal is to maintain a diversity of native plant species in order to provide a maximum number of ecological niches for wildlife, the forest will be managed differently than if it is designed for public recreation or the protection of watersheds. Even if the use to be emphasized is timber production, different management approaches may be taken. Logging techniques for the production

367

Fig. 10.25 Destructive logging practices result in eroded barren slopes unable to support a renewed forest cover. (U.S.D.A. Forest Service.)

of plywood or wood chips, for example, differ from those used for the production of high-quality lumber.

In the United States, 17% of the forest land is in national forests owned by the public and managed by the U. S. Forest Service (Figure 10.26). Logging by private companies is permitted, but *sustained-yield* techniques are practiced, at least in theory. Sustained yields result when harvesting is balanced with new growth, thus ensuring that there will be forests in the years to come. Sound management dictates that the emphasis should be on the long-run goal of perpetual crops rather than on the maximization of short-term profits (Figure 10.27).

Need we be concerned about the state of the world's forest resources? After all, forests cover roughly one-third of the earth's land surface. However, not all forested lands are available or extensive enough for any of the purposes described above, and we can reasonably expect no significant increase in the amount of land devoted to forests. Furthermore, there has been an increasing demand for wood and wood products. As with many other resources, the United States is a large consumer. With only 6% of the world's population, we consume about 40% of the world's total forest resource output. Already this country must import wood from abroad (notably Canada) to meet the demand. If we are to continue to satisfy our needs without endangering future supplies, wise management must be combined with a decrease in wasteful usage of wood products and an increased emphasis on recycling. Some European forests have significantly higher yields per acre than those in the United States, indicating the value of better care and management techniques.

Fig. 10.26 National forests of the United States. More than one-half of America's softwood for papermaking and home construction grows in the national forests. These forests provide annually about one-quarter of all timber used in the United States. Stands under federal control are principally Douglas fir, Western hemlock, pine, and spruce. (Data from U.S.D.A. Forest Service.)

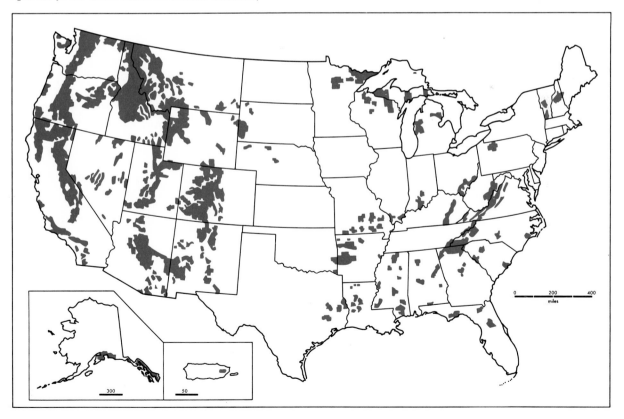

(a)

(b)

Fig. 10.27 (a) Sustained-yield forestry includes *block cutting* when stands of valuable species are approximately of the same age and size class. A seed tree is left standing to encourage regeneration of the forest. (b) *Selective cutting* is practiced in mixed-forest stands where different tree ages, sizes, and species are represented. Older, mature specimens are removed at first cutting; younger trees are left for later harvesting. (U.S.D.A. Forest Service.)

The Wetlands

Coastal wetlands are defined as marshes, swamps, tidal flats, or other zones where fresh and saltwater mix on shelves extending back from the sea. Techniques for their management and conservation differ from those for forests in two important respects. First, they occupy much less area. In the United States, less than 0.1% of land can be so classified. Second, the ecosystem they support is very delicately balanced. People can easily and irrevocably affect that balance.

Wetlands and estuarine zones, where fresh water from streams and rivers meets saltwater from the oceans, are extraordinarily productive. Trapping the silt and the other organic matter that rivers bring downstream, they provide shelter and food for a variety of fish and waterfowl. In addition, they are homes or spawning grounds for many species of fish and shellfish. Not only are these areas highly pro-

ductive themselves, but they also contribute to the continued productivity of a much larger area. Other fish feed on the life that flows from estuaries and wetlands into the sea. Adjoining the estuarine zones is the neritic zone (Figure 10.28), the relatively shallow part of the sea lying above the continental shelf. Because it is not very deep, sunlight can penetrate the neritic zone. It also receives the nutrients flowing into oceans from streams and rivers, so that vegetation can flourish. In general, this zone is the most productive of all ocean waters, supporting the greatest variety of aquatic life. However, the neritic zone depends to a considerable extent on the continued functioning of the estuarine zone.

Unfortunately, wetlands and estuaries are in danger. Scientists have estimated that as much as one-half of such areas in the United States has been misused or polluted (Figure 10.29). The lands have been drained, built upon, mined for phosphate, and used for waste disposal, as well as polluted by chemicals, excess

372

Fig. 10.28 The outflow of fresh water from rivers and streams and the action of tides and wind serve to mix deep ocean waters with surface waters in estuaries, contributing to their biological productivity. The saline content of estuaries is lower than that of the open sea. Many fish and shellfish require water of low salinity at some point in their life cycles.

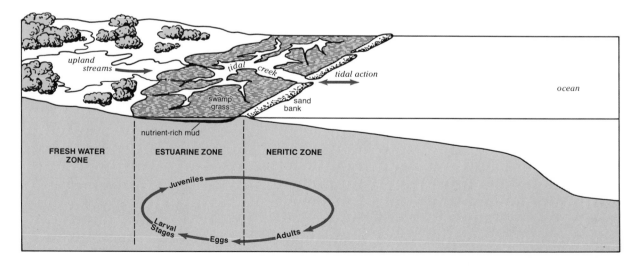

nutrients, and other wastes that the streams and rivers carry downstream. Habitat destruction or alteration inevitably affects the existing intricate ecosystems. At present, the only way to halt this disruption seems to be to preserve the wetlands as they are, as ecological reserves. Should we fail to do so, there will be negative consequences not only for the water life they sustain but also for that in the adjoining oceans.

Wildlife Resources

Wildlife resources have an unusual position in the classification of resources as renewable or nonrenewable, for while they are commonly classified as renewable resources, once a species becomes extinct, it is nonrenewable. There is no way that it can be re-created in the laboratory. The chief problem we are facing with respect to wildlife is the increasing number of species that are becoming extinct or are in danger of doing so. In the United States, 15% of the approximately 400 native mammals are on the Rare and Endangered Species list—a list maintained by the national government and updated yearly.

The picture is bleaker in some parts of the world, less bleak in others. Many people fear that as countries in Africa and South America become more industrialized and more urbanized and expand their areas under cultivation, there will be an increasingly stronger negative impact on wildlife. The reason is that one of the main causes of extinction in other countries has been the loss or alteration of habitats for wildlife. By clearing forest land, draining wetlands, extending farm land, and building cities, people modify or destroy the habitats in which animals have lived. Since the habitat needs of each species vary, alteration makes an area more suitable for some species and less suitable for others. Skyscrapers, TV towers,

and lighthouses affect birds just as dams, canals, and irrigation ditches affect fish, favorably for some species, negatively for others. In this country, people have begun to invade with snowmobiles and dune buggies what used to be relatively safe (at least in certain seasons) sanctuaries for certain species. As noted in Chapter 4, other causes of extinction include direct killing, pollution, and the introduction of competing or predatory species (Figure 10.30).

Conclusion: The Limits to Growth

Much current thinking about resource development is based on projected needs for the next 20 or 30 years. Only a few attempts have been made to project supplies and demands to the year 2100. The researchers for the Club of Rome, an international group of concerned academics and industrialists, foretell impending disaster.

Taking the available knowledge of the way that economic and demographic variables are intertwined, they developed a computer model designed to predict for the years up to 2100 the world levels of pollution, food intake, industrial output, population, and resource availability. Under vastly different assumptions about what might happen in the future, all of the models pointed to the same disastrous results: low worldwide food intake levels by the year 2100. Some predictions are shown in Figure 10.31. Resource depletion is the culprit in some of the computer runs, but even where it is assumed that resources will not be substantially depleted and that population will be controlled, industrial output reaches awesome heights, resulting in environmental degradation and eventually low agricultural output.

There are those who dispute these results. In response to criticisms, a new version of the

373

Fig. 10.29 Tidal marshlands have been subjected to dredging and filling for residential and industrial development. The loss of such areas reduces the essential habitat of waterfowl, fish, crustaceans, and mollusks. Many waterfowl breed and feed in coastal marshes and use them for rest during long migrations. (Photo by Mary M. Thacher. National Audubon Society Collection/Photo Researchers, Inc.)

(a)

Fig. 10.30 (a) *European Starling.* The introduction of alien species into a new habitat can have far-reaching effects. Sparrows and starlings, deliberately introduced into this country in the last century, have largely driven out native songbirds such as wrens and bluebirds. (b) *Coyote.* Much killing of wildlife to protect crops and livestock is simplistic. The killing of coyotes, which occasionally eat young sheep, has resulted in an increase in the rabbit population on western ranges. Rabbits eat grasses that the sheep need for forage. (c) *Canadian Harp Seal Pups.* Active debate continues over whether clubbing and skinning harp seal pups constitute harvesting a natural resource and a necessary part of Newfoundland's economy or should be condemned as inhumane slaughter. (d) *Deer in Winter.* Overprotection of a species can lead to increase in the population that the habitat cannot support. These deer may face starvation for lack of food. (a: INTERIOR—U. S. Fish and Wildlife Service; b: INTERIOR—Sport Fisheries and Wildlife; c: N.F.B. Photothèque photo by D. Wilkinson; d: N.F.B. Photothèque photo by Jeanne White.)

(b)

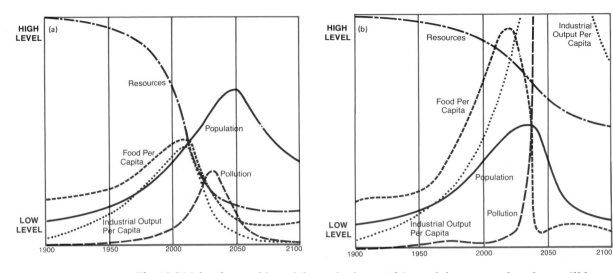

Fig. 10.31 (a) The world model standard run. This model assumes that there will be no major changes in physical, economic, or social relationships; that is, it is a continuation of patterns that have obtained since A.D. 1900. Food, industrial output, and population grow exponentially until the depletion of the resource base causes a slowdown in industrial growth and, eventually, a halt in population growth. (b) The world model with unlimited resources, pollution controls, and increased agricultural productivity. In this model, the assumption is that unlimited nuclear energy has solved the resource problem, that there are strict pollution controls, and that food yields per acre have doubled. With the removal of these constraints to growth, population and industrial output reach such high levels that a pollution crisis develops in spite of the pollution control policy, and further growth is halted. (*The Limits to Growth: A Report for the Club of Rome's Project on the Predicament of Mankind*, by Donella H. Meadows, Dennis L. Meadows, Jorgen Randers, and William W. Behrens III. A Potomac Associates book published by Universe Books, New York, 1972, 1974. Graphics by Potomac Associates.)

378

models was presented in *Mankind at the Turning Point* in 1974. The outlook is still bleak. This time, the moral and survival aspects of large-scale nuclear reactors scattered over the globe are considered. People with more faith in technological control of the environment believe that investment in new technology will follow from the threat of lowered standards of living.

The problem that all countries are now facing or will soon face is best viewed not as a resource problem but as a cultural problem. At some point, we will have to make the transition from a philosophy of exponential growth in rates of energy use and consumption to a philosophy of limited growth or no growth. This change will raise many questions about human values, about the rights of the individual versus devotion to the common good, about the kinds of political and economic systems that seek to avoid crises and confrontations, and about the objectives of the developed versus the developing countries and the obligations of the former to the latter. Realistic appraisals will have to be made of the tradeoffs between societal costs and societal benefits vis-à-vis resource extraction and utilization in the effort to develop systems that allow for the long-run maintenance of acceptable quality-of-life standards for all people.

For Review

1. Since the first law of thermodynamics (see Chapter 4) tells us that energy is neither created nor destroyed, how is it possible to speak of an energy crisis related to coal or petroleum?

2. Does *conservation of resources* imply cessation of use? If not, in your opinion by whom or by what means should levels of resource consumption be determined? What groups might reasonably be expected to object to the authority for or the means of consumption control you suggest?

3. What is the basic distinction between a renewable and a nonrenewable resource? Does the utilization of nonrenewable bauxite pose the same problem of resource depletion as the consumption of nonrenewable petroleum? Why or why not?

4. Since agricultural resources are deemed renewable, why is there concern about their exploitation? Make a list of conservation issues of which you are aware that have direct bearing upon agriculture. Discuss each briefly from the standpoint of (a) the immediate interests of those creating the issue of concern; (b) those who are concerned but not directly involved; (c) posterity.

5. What effects would a "no-growth" policy have on the social, economic, and political structure of a technologically advanced society? Write a brief essay clearly stating your own views on the proposition: I would, as a member of an economically advanced society, accept the consequences of a national "no-growth" policy.

6. What is meant by *multiple-use* forest management? What differing objectives are sought by those advocating such a management policy? In your opinion, is the multiplication of goals in forest management consistent with the efficient realization of any one of them?

7. What is the *neritic zone*? What is its functional role in the maintenance of marine life? In what ways is the ecology of the neritic zone being assaulted? Any such assault is usually justified in the name of progress or economic benefit. How, in your opinion, should benefit be weighed against ecological cost?

379

Suggested Readings

COMMONER, BARRY, *The Poverty of Power: Energy and the Economic Crisis.* Knopf, New York, 1976.

Cook, Earl, *Man, Energy, Society*. Freeman, San Francisco, 1976.

Dasmann, Raymond F., *Environmental Conservation* (3rd ed.). Wiley, New York, 1972.

Ehrlich, Paul R. and Anne H., *Population, Resources and Environment*. Freeman, San Francisco, 1970.

Enthoven, Alain C., and A. Myrick Freeman III, *Pollution, Resources and the Environment*. Norton, New York, 1973.

Greenwood, N. J., and J. M. B. Edwards, *Human Environments and Natural Systems* (2nd ed.). Duxbury Press, North Scituate, MA, 1979.

Griggs, G. B., and J. A. Gilchrist, *The Earth and Land-Use Planning*. Duxbury Press, North Scituate, MA, 1977.

Heilbroner, Robert L., *An Inquiry into the Human Prospect*. Norton, New York, 1974.

Meadows, Dennis, et al., *The Limits to Growth*. Universe Books, New York, 1972.

Mesarović, Mihailo, and Eduard Pestel, *Mankind at the Turning Point: The Second Report to the Club of Rome*. Dutton, New York, 1974.

Miller, G. Tyler, *Living in the Environment*. Wadsworth, Belmont, CA, 1975.

U. S. Bureau of Mines, *Minerals Yearbook*. U. S. Government Printing Office, Washington, D.C.

380

The royal edict, carefully incised in wet clay in cuneiform, might have read:

[Here let my city be raised; let it be surrounded in the form of a rectangle with mighty walls of burnt brick and with towers and crenelations. Within, lay out a quadratic grid of streets each 11 cubits wide, save a main thoroughfare of 15 cubits, and let all intersect at right angles. Let the walls be pierced with mighty gates, each to its allotted purpose: one for the metalworkers and other artisans, one to take from the city the refuse thereof, one for the flocks and food-stuffs, and one leading to the spring. And where needful, let the market for commodities be in open spaces within the prescribed gate. Outside the walls, let there be placed the port of the city and therein the manufactories of metal and cloth, and elsewhere without, the reed and mud huts of the lowly and of the laborers, and the fields and pastures and the necessary canals to bring them water. Within, place the brick houses of landowners, citizens, and officials. Build a separate set of walls and enclose therein the temple and my own palace. And let the whole be maintained by levies of labor and by the wealth of conquest, the tribute of the temple, and the profits of trade.]

Of course, no such tablet of 2500 B.C. has been, or will be, found. Mesopotamia, discussed in Chapter 5 as a major culture hearth of antiquity, lay open, was crossed and conquered by many peoples, and was the site of innumerable settlements from villages to major cities. These settlements had myriad forms, planned and unplanned, but our imaginary edict summarizes some recurring features as recorded on tablets and discovered by archaeologists. More importantly, it suggests the early appearance of complex structural and functional agglomerations—cities—as the mark of the social and economic organization of territories larger than the agglomerations themselves. The allocation of land uses, the

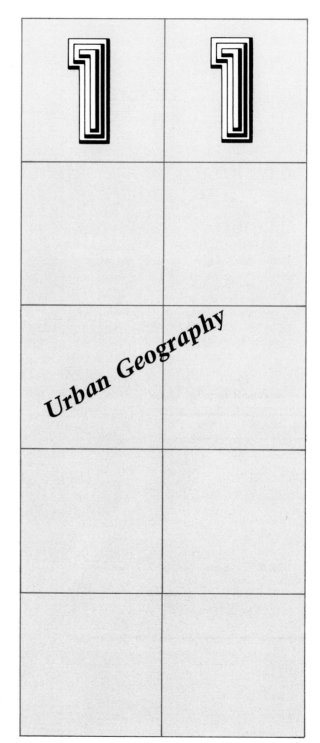

Fig. 11.1 The difference in size, density, and land-use complexity between New York City and a small Illinois farm town is immediately obvious. Although both places are *urban*, the class names *city* and *town* help to identify their respective roles in a national system of urban units. (a: Port Authority of New York and New Jersey; b: U.S.D.A. Agricultural Stabilization and Conservation Service.)

differentiation of streets, and the functional diversity of the urban unit even in early Mesopotamian times were just the forerunners of the urban complexities yet to appear and to be made part of the city landscape in changing form to the present day.

Cities are among the oldest marks of civilization and among the newest experiences of the world's population. Only in this century have cities become the home of the majority of people in industrialized countries and the inhospitable place of refuge for uncounted millions in modernizing subsistence economies. In all their incarnations, ancient and modern, recurring themes and regularities appropriate to their time and place of origin are evident within cities. All perform functions—have an economic base—generating the income necessary to support themselves and to sustain their contained population. None exists in a vacuum, but each is part of a larger urban and nonurban society and economy with which it has essential reciprocal connections. Each has a more or less orderly internal arrangement of land uses, social groups, and economic functions partially planned and controlled and partially determined by individual decisions, competition for sites, and operational efficiencies. All, large or small, ancient or modern, have or have had maladjustments of land use, recurring and serious social problems and conflicts, environmental inadequacies, and concerns of administration and orderly development. Yet cities remain, though flawed, the capstone of our cultures, the organizing focuses of modern societies and economies. In this chapter, we explore why this should be so, observe how cities function, and seek explanation of why they sometimes operate in ways that harm at least some of their inhabitants. The emphasis is on North American cities—typical of advanced commercial economies; the implications are worldwide, and brief reference will be made to the patterns and problems characteristic of cities in newly changing subsistence economies and in tightly controlled planned economies.

384

The Functions of Cities

Cities exist for the efficient performance of functions required by the economy and the society that create them, functions that cannot be adequately carried out in dispersed locations. Cities reflect the savings of time, energy, and money that the agglomeration of people and activities implies. The nearer the producer to the consumer, the worker to the work place, the lawyer or doctor to the client, or the machine shop to the factory it serves, the more efficient is the performance of their separate activities, and the more effective is the integration of the urban economy. Since all urban functions and people cannot be located at a single point, cities themselves must take up space, and land uses and populations must have room within them. Since interconnection is essential, there must be a transportation and communication system to foster the enormous amount of interaction that takes place. The more advanced the transportation technology and the more creatively the transportation facilities are used, the more efficiently the activities of large cities are performed. The totality of people, functions, and transportation facilities constitutes a distinctive cultural landscape subject to geographic analysis.

SOME DEFINITIONS

The city landscape is not of a single type, structure, or size. In those nucleated, nonagricultural settlements called *urban*, we deal with a population and functional continuum. Its terminal points are, at the low end, the smallest of hamlets with perhaps one or two stores or a grain elevator and, at the opposite end, the complex, multifunctional metropolitan area, supercity, or "megalopolis" (Figure 11.1). While words designating urban units in their different size classes are employed in common speech, they are not uniformly applied by all users. What is recognized as a *city* by a resident of North Dakota, our least urbanized state, might not at all be afforded that name and status by an inhabitant of New Jersey, our most urbanized.

It is therefore necessary to agree upon the meanings of terms commonly employed but varyingly interpreted. Here, the general term *urban* means a nonagricultural, structured agglomeration. The word *city* is reserved for nucleated settlements with 10,000 or more people, multifunctional in character, and of a size presumably requiring and displaying orderly internal land-use organization, including an established central business district and both residential and nonresidential land uses. The word *town* is used to mean urban places that have lesser size and functional complexity but still have a nuclear business concentration and are functionally complete—that is, with varied economic activities and much of the population necessary for their performance. *Suburb* denotes a subsidiary area, a functionally specialized segment of a larger urban complex. It may be dominantly or exclusively residential, industrial, or commercial, but by the specialization of its land uses and functions, the suburb cannot have independent urban status.

Some or all of these urban types may be spatially and functionally associated into larger landscape units. The *urbanized area* refers to a continuously built-up landscape defined by building and population densities with no reference to the political boundaries that limit the legal city of which it is an extension. It may be viewed as the *physical* city and may contain a central city and many contiguous cities, towns, suburbs, and unincorporated urban tracts. The *metropolitan area*, on the other hand, refers to a large-scale *functional* entity, perhaps discontinuously built up but nonetheless operating as a coherent and integrated economic whole. The United States census recognizes these latter units—urbanized area and Standard Metropolitan Statistical Area—as well as smaller, politically separate units, in

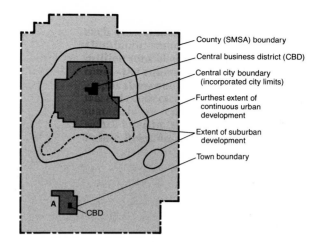

Fig. 11.2   A hypothetical spatial arrangement of urban units within a Standard Metropolitan Statistical Area. Sometimes city limits are very extensive and contain areas commonly thought of as suburban or even rural. Older eastern American cities more often have restricted limits and contain only part of the high-density land uses associated with them. In the diagram, A may be thought of as suburban as well as urban.

385

the recording of urban data for this country. A schematic spatial relationship of these various units is shown on Figure 11.2.

THE LOCATION OF CITIES

The American geographer Mark Jefferson once observed, "Cities do not grow up of themselves. Countrysides set them up to do tasks that must be performed in central places." That trenchant comment reminds us of two essential facts. First, the city is part of a larger, partially nonurban economic whole, with which it has close reciprocal relationships constituting the reason for the city's existence. Although a consumer of food, a processor of materials, and an accumulator and dispenser of goods and services, the city must depend upon the "countryside" that established it for its essential supplies and as a market for its activities. Second, in order to perform adequately the tasks that support it and to add new functions as demanded by the larger economy, the city must be efficiently located. That efficiency may be marked by spatial or transportation centrality to the area served. It may derive from location at points whose physical characteristics permit the performance of assigned functions. Or placement may be related to the resources, productive regions, and transportation network of the nation, so that effective performance of a wide array of activities is pos-

sible without reference to the requirements of the immediate local region.

In discussing urban location, geographers frequently differentiate between *site* and *situation*. Site is usually thought of as the locale of the city in physical terms: it lies on an outwash plain, at the head of navigation, or on a deep-water harbor. As Figure 11.3 suggests, transportation—particularly water transportation—was an important localizing factor when the major American cities were established.

If *site* suggests absolute location, *situation* indicates relative location, the placement of

Fig. 11.3 The 25 most populous Standard Metropolitan Statistical Areas in the United States (1975 estimate). Notice the general association of principal cities and navigable water; before the advent of the railroads in the middle of the 19th century, all major cities were associated with waterways.

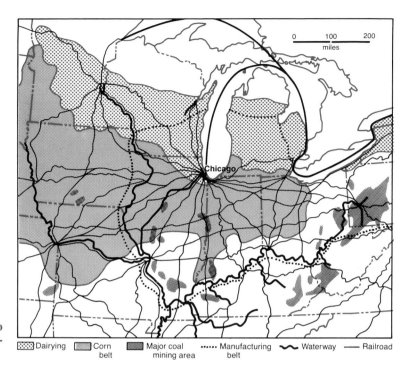

Fig. 11.4 The "situation" of Chicago helps suggest the reasons for its functional diversity.

Dairying   Corn belt   Major coal mining area   ⋯⋯ Manufacturing belt   Waterway   Railroad

387

the urban unit in relation to the physical and human characteristics of surrounding areas: to raw material distributions, market areas, agricultural regions, and the like. By combining these concepts, one can gain an understanding of the location of cities not conveyed, say, by reciting coordinates of latitude and longitude. One can locate Chicago, as was done in Chapter 1, at 41° 52′ N, 87° 40′ W. Much more revealing analytically is its description as lying on a lake plain at the deepest penetration of the Great Lakes system into the interior of the country, astride the Great Lakes–Mississippi waterways, and near the western margin of the manufacturing belt, the northern boundary of the corn belt, and the southeastern reaches of a major dairy region. Reference to railroads, coal deposits, and ore fields would amplify its situational characteristics (Figure 11.4). By this description, functional implications relating to market, to raw materials, and to transportation centrality can be drawn.

The site or situation that originally gave rise to an urban unit may not long remain the essential ingredient in its functional success or basic to its economic structure. Agglomerations originally successful for whatever reason may by their success attract people and activities totally unrelated to the initial localizing forces (Figure 11.5). By what has been called a process of "circular and cumulative causation," a successful urban unit may acquire new populations and functions attracted by the already existing markets, labor force, and urban facilities.

THE ECONOMIC BASE

Part of the employed population of an urban unit is engaged in the production of goods or the performance of services for areas and people outside the city itself. They are workers engaged in "export" activities, whose efforts

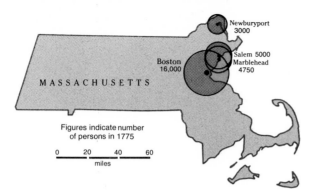

Fig. 11.5 The importance of maritime activities, situation advantages, terrain discontinuities, and diversity of functions early gave rise to agglomerations of urban units in eastern Massachusetts. Their development and success encouraged continuing urbanization despite changing economic conditions. The cities shown are those known to have contained at least 3000 residents in 1775. (Source: *Atlas of Early American History*, Princeton University Press, 1976.)

388

result in money flowing into the community. Collectively, they constitute the *basic sector* of the city's total economic structure. Other workers support themselves by producing things for residents of the urban unit itself. Their efforts, necessary to the well-being and the successful operation of the city, do not generate new money for it but comprise a *service,* or *nonbasic, sector* of its economy. These people are responsible for the internal functioning of the city. They are crucial to the continued operation of its stores, professional offices, city government, local transit media, and school systems.

The total economic structure of a city equals the sum of basic and nonbasic activities. In actuality, it is the rare urbanite who can be classified as belonging entirely to one sector or another. Some part of the work of most people involves financial interaction with residents of other areas. Doctors, for example, may have mainly local patients and thus are members of the nonbasic sector; but the moment they pro-

vide a service for someone from outside the community, they bring new money into the city and become part of the basic sector.

Variations in basic employment structure among urban units characterize the specific functional role played by individual cities within the national economy and provide a means for a classification and comparison of cities. Some cities, particularly smaller ones, may be termed *unifunctional;* their basic sector is composed entirely of a single activity type. They are the small farm-market towns, the coal-mining or lumbering towns, the county seats or university towns without other basic activities. Most cities, however, perform many export functions, and the larger the urban unit, the more multifunctional it becomes. Nonetheless, even in cities with a diversified economic base, one or a very small number of export activities tend to dominate the structure of the community and to identify its operational purpose within a national system of cities.

Such functional specialization permits the classification of cities into categories: manufacturing, retailing, wholesaling, transportation, government, and so on. Such specialization may also evoke images when the city is named: Detroit, Michigan, or Gary, Indiana, as manufacturing centers; Tulsa, Oklahoma, for oil production; Hartford, Connecticut, as an insurance capital; Washington, D.C., in public administration; and so on. Certain large regional, national, or world capitals—as befits major multifunctional concentrations—call up a whole series of mental associations, such as New York with banking, the stock exchange, entertainment, the fashion industry, port activities, and the like. Some functionally specialized and the very large diversified cities of the United States are categorized on Table 11.1.

Assuming it were possible to divide with complete accuracy the employed population of a city into totally separate basic and nonbasic components, a ratio between the two employ-

Notice that the most diversified cities tend to be the largest. This follows from the fact that the concentration of many basic functions in them prevents any one activity from attaining absolute dominance. Note that some metropolitan areas contain proportions of basic workers sufficient to qualify the centers as functionally specialized in more than one activity.

Table 11.1 Functional
Specialization of Some
U. S. Metropolitan Areas[a]

Agriculture

1. Fresno, CA
2. Bakersfield, CA
3. Lancaster, PA
4. Orlando, FL
5. Phoenix, AZ
6. San Bernardino, CA

Minerals

1. Duluth, MN (iron ore)
2. Johnstown, PA (coal)
3. Tulsa, OK (oil)
4. Bakersfield, CA (oil)
5. Wilkes-Barre, PA (coal)

Construction

1. Fort Lauderdale, FL
2. Orlando, FL
3. Honolulu, HI
4. Albuquerque, NM
5. Tampa–Saint Petersburg, FL

Durable Manufacturing

1. Flint, MI
2. Youngstown, OH
3. Gary, IN
4. Bridgeport, CT
5. Canton, OH
6. Detroit, MI

Nondurable Manufacturing

1. Akron, OH
2. Wilmington, DE
3. Wilkes-Barre, PA
4. Beaumont, TX
5. Reading, PA
6. Allentown, PA

Transportation

1. Omaha, NB
2. Duluth, MN
3. New Orleans, LA
4. Jersey City, NJ
5. Huntington, WV

Wholesaling

1. Charlotte, NC
2. Spokane, WA
3. Salt Lake City, UT
4. Memphis, TN
5. Portland, OR
6. Houston, TX

Retailing

1. Fort Lauderdale, FL
2. Tampa–Saint Petersburg, FL
3. El Paso, TX
4. San Antonio, TX
5. Spokane, WA
6. Tucson, AZ

Finance

1. Hartford, CT
2. Des Moines, IA
3. Jacksonville, FL
4. Fort Lauderdale, FL
5. Omaha, NB
6. Richmond, VA
7. New York, NY

Entertainment

1. Fort Lauderdale, FL
2. Los Angeles, CA
3. Miami, FL
4. Honolulu, HI
5. New York, NY

389

(table continues)

Table 11.1 *(cont.)*

Public Administration (incl. Military)	Most Diversified
1. Washington, DC	1. Los Angeles, CA
2. Harrisburg, PA	2. New York, NY
3. Sacramento, CA	3. Philadelphia, PA
4. San Antonio, TX	4. San Francisco, CA
5. Mobile, AL	5. Chicago, IL
	6. Boston, MA
	7. Saint Louis, MO

[a] Within each category, the cities are listed in order of the degree to which employment is specialized in the activity named. Thus, Fort Lauderdale has a higher proportion of its employees engaged in retailing than does any other city.

ment groups could be established. With some exception for high-income communities, this basic/nonbasic ratio is roughly similar for cities of similar size irrespective of their functional specializations. Further, research has shown that as a city increases in size, the number of nonbasic personnel grows faster than the number of new basic workers. Thus, in cities with a population of 1 million, the ratio is about 2 nonbasic workers for every basic worker; the addition of 10 new basic employees implies the expansion of the labor force by 30 (10 basic, 20 nonbasic) and an increase in total population equal to the added workers plus their dependents. A multiplier effect thus exists, associated with economic growth. The term *multiplier effect* implies the addition of nonbasic workers and dependents to a city's total employment and population as an expected, normal supplement of new basic employment; the size of the effect is determined by the city's demonstrated basic/nonbasic ratio. It is an effect much less prominent in smaller centers. Towns of 10,000 or so show a basic/nonbasic ratio of, perhaps, 2:1, since the lesser total population and purchasing power of the smaller unit cannot support a wide range of service activities. These changing numerical relationships are shown in Figure 11.6.

The changing numerical relationships are understandable when we consider how cities add functions and grow in population. A new industry selling goods to other communities requires new workers, who thus increase the basic work force. These new employees, in turn, demand certain goods and services, such as clothing, food, and medical assistance, which are provided locally. Those who perform such services must themselves have services available to them. For example, a grocery clerk must also buy groceries. The more nonbasic

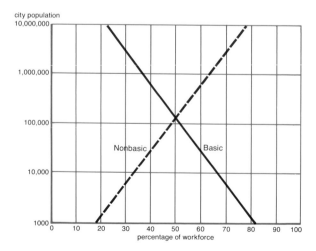

Fig. 11.6 The proportion of the work force engaged in basic and nonbasic activities by city size. As cities become larger, a greater proportion of their work force is employed in nonbasic activities. Larger cities are therefore more self-contained.

workers a city has, the more nonbasic workers are needed to support them, and the application of the multiplier effect becomes obvious.

We have also seen that the growth of cities may be self-generating—"circular and cumulative" in a way related not to the development of basic industry but to the attraction of what would be classified as *service* in an economic base study. Banking and legal services, a sizable market, a diversified labor force, extensive public services, and the like may attract or generate additions to the labor force not classically *basic* by definition. The multiplier effect in larger metropolitan centers is discerned in many guises.

SYSTEMS OF CITIES

The economic structure displayed by an individual city is reflected not only in the size of that city but also in its location and in its relationships with other units in the larger city system of which all are a part.

In a crude but revealing classification, from a functional standpoint urban units may be designated as primarily either transportation centers, special-function cities, or central places. The distributional pattern of transportation-based cities is that of alignment—along seacoasts, major and minor rivers, canals, or railways. Routes of communication form the orienting axes along which urbanization has occurred and upon which at least initial economic success depended (Figure 11.7). Special-function cities are those engaged in mining, manufacturing, or other functions, the localizing forces of which are not dependent upon an immediately surrounding market area but are related to raw materials, agglomeration economies, or the self-generating attractions of constantly growing market and labor concentrations. Special-function cities show a clustered pattern of development, epitomized by such major metropolitan concentrations as New York–northeastern New Jersey, Chicago–

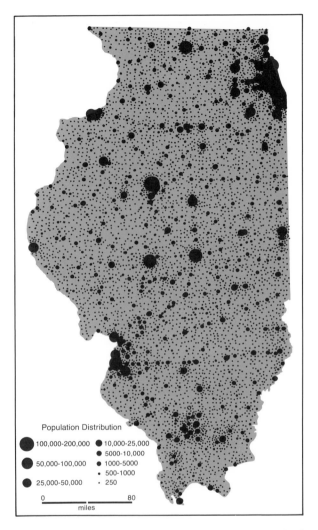

Fig. 11.7 Railroads preceded settlement in much of the continental interior, and urban centers were developed—frequently by the railroad companies themselves—as collecting and distributing points expected to grow as farm populations increased. Located at constant 5–6-mile intervals in Illinois, the rail towns were the focal points of an expanding commercial agriculture. The linearity of the town pattern of Illinois in 1940, at the peak of railroad influence, marks the rail routes unmistakably.

northwestern Indiana, and the San Francisco or Los Angeles multi-city clusters.

The third category, with size classes and

spacing patterns peculiar to it alone, is composed of "central places." They are nodal points for the distribution of goods and services to surrounding nonurban populations and, as well, vital links in a system of interdependent cities. This functional system of urban units was subjected to special study by the German geographer Walter Christaller, who laid the foundations of central-place theory.

In 1933, Christaller attempted to explain the size and the location of settlements. While he was not totally successful, he performed a valuable service by developing a framework for understanding much about urban interdependence. Christaller first decided that his theory could best be developed in rather idealized circumstances. He assumed that towns that provide the surrounding countryside with such fundamental goods as groceries and clothing would develop in a plain where farmers specialized in commercial agricultural production. He assumed also that the farm population would be dispersed in an even pattern and that the characteristics of the people would be uniform; that is, they would possess similar tastes, demands, and incomes. Christaller knew full well that such assumptions are hardly realistic, but he recognized that his theory would have to be infinitely more complicated if such assumptions were not used. They are, of course, similar to the isotropic plain assumptions of von Thünen, discussed in Chapter 9.

Christaller further assumed that each kind of product or service supplied to the dispersed population had its own *threshold*, which is the minimum market (number of consumers or volume of purchasing power) needed to support the supply of a product. For example, a shoe store can prosper in an area only if there is sufficient demand for shoes and money to buy shoes. There will be stores offering products with high thresholds (requiring high purchasing power) only if there is a large market area. High-threshold goods, such as diamonds or fur

392

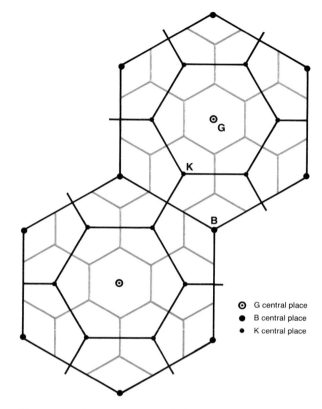

Fig. 11.8 The two G central places are the largest on the above diagram of one of Christaller's models. The B central places offer fewer goods and services for sale and serve only the areas of the intermediate-sized hexagons. The many K central places, which are considerably smaller and more closely spaced, serve still smaller hinterlands. The goods offered in the K places are also offered in the B and G places, but the latter offer considerably more, and higher-order, goods. Notice that places of the same size are equally spaced. (A. Getis and J. Getis, "Christaller's Central Place Theory," *Journal of Geography*, 1966. Reprinted by permission of the *Journal of Geography*, National Council for Geographic Education.)

coats, are either expensive or not in great demand. On the other hand, bread, a low-threshold product, can be marketed from many locations since demand is high and the product is relatively inexpensive. A smaller *hinterland* (market area) can support an enterprise selling bread and other low-order goods.

Christaller assumed that the businesspeople of the agricultural plain would divide the area into noncompeting markets where each entrepreneur would have exclusive rights to the sale of a particular product. This idea follows from a set of assumptions about the efficient use of transportation by consumers—they will travel to the nearest opportunity (store)—and the efficient division of market areas by those supplying goods and services. When all of the assumptions and definitions are considered simultaneously, the result is a series of hexagonal market areas that cover the entire plain, as shown in Figure 11.8. The hexagonal market area is a necessary postulate; the hexagon is the most compact geometric form into which the entire hypothetical plain can be subdivided, leaving no unserved hinterland and no areas unassigned to a closest central place. The largest hexagonal market areas have as their centers central places that supply goods and services of the highest and lowest orders, that is, *all* goods and services supportable by the demand and the purchasing power of the entire area. Contained within or at the edge of the largest market areas are central places serving a smaller population and offering fewer goods and services to their respectively smaller hinterlands.

Two important conclusions were reached by Christaller. First, towns of the same size will be evenly spaced and larger towns will be farther apart than smaller ones. This means that there will be many more small than large towns. In the region under consideration, the ratio of the number of small towns to towns of the next larger size will be constant. In Figure 11.8, the ratio in a completed network would be 3 to 1. This distinct steplike series of towns in size classes differentiated by both size and function is called a *hierarchy of central places.*

Second, the system of towns is interdependent. If one town were eliminated, the entire

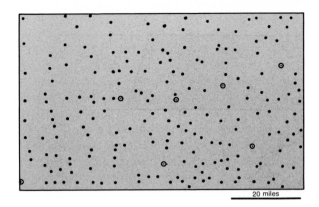

Fig. 11.9 The pattern of hamlets, towns, and cities in a portion of Indiana. This map represents an area 82 by 51 miles just north of Indianapolis. Cities containing more than 10,000 people are circled. Notice that the pattern is remarkably even and includes a number of linear arrangements that correspond to highways and railroads.

system would have to readjust. Consumers need a variety of products, each of which has a different threshold. The towns containing many goods and services become regional retailing centers, while the small central places serve just the people immediately in their vicinity. The higher the threshold of a desired product, the farther, on average, the consumer must travel to purchase it.

These conclusions have been shown to be generally valid in widely differing areas within the commercial world. When varying incomes, cultures, landscapes, and transportation systems are taken into consideration, the results of Christaller hold up rather well. They are particularly applicable to agricultural areas, especially with regard to the size and spacing of cities and towns, as Figure 11.9 suggests. One has to stretch things a bit to see the model operate in highly industrialized areas. However, if we combine a Christaller-type approach with the ideas that help us understand industrial location and transportation alignments, we have a fairly good understanding of the location of the majority of cities and towns.

393

INFLUENCE

The theme of central-place theory is that each urban unit has a surrounding area over which it exercises dominating influence for the class of functions it contains. By extension, each unit—as part of a larger system of cities—has some impact, however modest, upon all other elements of that system. Again (by the dictates of central-place theory and the logic of the increasing significance to the city system of the larger, more complex urban units), it follows that (1) as the distance away from a city increases, its influence on the surrounding countryside declines and (2) the sphere of influence of a city is usually proportional to its size.

In both instances, we see the application of a gravity or interaction model. Based on concepts of Newtonian physics, the *gravity model* simply states that the interaction between two bodies varies directly with their mass and inversely with some function (usually assumed to be the square) of the distance separating them. In the case of cities, mass is measured by population size. The influence zone of an urban unit and its interaction with all other

Fig. 11.10 Influence zones based upon correspondent banking linkages between major and lower-order centers. The map suggests the influence of principal cities on regions beyond their metropolitan areas. The areas not within one of the zones represent districts of competition for influence between principal cities. (Reproduced by permission from the *Annals* of the Association of American Geographers, vol. 62, 1972, John R. Borchert.) The *distance-decay* relationship of urban influence is shown in the lower left corner of the map.

394

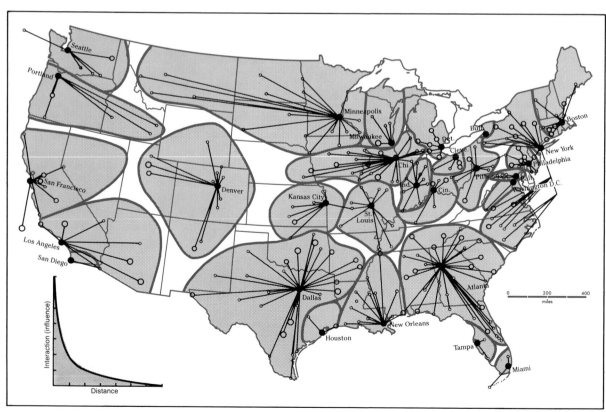

cities within the system are thus seen to be a consequence of the size of the city in question and its distance away from every other city. With respect to its hinterland dominance, a city is said to display a distance–decay relationship; that is, the influence of the urban unit decreases in a negative exponential fashion (decreases at a decreasing rate) with distance from the city. Figure 11.10 illustrates aspects of these relationships.

In reality, of course, influence zones within the system of cities are much more involved than these basic models suggest. The hierarchical ordering of urban functional specializations suggests complex overlaps and controls between cities irrespective of distance or size considerations. Linkages between cities reflect the existence of *complementarity* within the overall pattern of urban functional specializations. Complementarity can be said to form the basis of potential interaction between two cities when one produces a supply of a good or service for which there is demand in the other. The individual functional specializations of cities assure the existence of a widespread complementarity of activities and thus of linkages. The functional concentration represented by very large cities or by those with unique and necessary goods or services assures a hierarchical ordering of influence patterns.

A small city may influence a local region of, say, 25 square miles by having its newspaper delivered to that region. Beyond that region, another city may be the dominant influence. A large city located 100 miles away from the small city may influence the small city and other small cities through its banking services, TV station, and large shopping areas. Consequently, influence areas are very much like the market areas of central-place theory. There is an overlapping, hierarchical arrangement, and the influence of the largest cities is felt over the widest areas.

More intricate interrelationships and hierarchies are common. Consider Grand Forks, North Dakota, which for local market pur-

poses dominates the rural area immediately surrounding it. However, Grand Forks is influenced by political decisions made in the capital, Bismarck. For a variety of cultural and higher-order commercial and banking activities, Grand Forks feels the influence of Minneapolis–Saint Paul. A center of wheat production, Grand Forks is subordinate to the grain market established by the Chicago Board of Trade and the grain companies' dependence upon the New York–centered financial community. The pervasive agricultural controls administered from Washington, D.C., are exerted without reference to its distance from Grand Forks or to the size of either city. The commercial economy of the United States involves complex urban interrelationships and patterns of influence that cannot be fully described by limited-assumption theories.

Inside the City

Yet theory, the search for regularity and consistency, is essential to our understanding not only of the general structure of the system of cities but also of the land-use arrangements and population patterns we find within cities.

It is a common observation that a recurring pattern of land-use arrangements and population densities exists within older central cities and their suburban fringes. It is deduced that those regularities have been created by the circumstances of differential accessibility, the impersonal operations of a free market in land, and the collective consequences of individual residential locational decisions. The whole urban structure is, of course, also being modified by land-use controls, obsolescence, transportation alterations, and governmental redevelopment programs.

THE COMPETITIVE BIDDING FOR LAND

For its effective operation, the city requires close spatial association of the functions it

houses and of the populations carrying out those functions. As long as functions were few and population small, pedestrian movement and pack-animal haulage were sufficient for the effective integration of the urban community. With the addition of large-scale manufacturing and the accelerated urbanization of the economy during the 19th century, however, functions and populations—and therefore city area—grew beyond the interaction capabilities of pedestrian movement alone. Increasingly efficient, and costly, mass-transit systems were installed. Even with their introduction, however, only land within walking distance of mass-transit routes or terminals could successfully be incorporated into the expanding urban structure. Usable land, therefore, was a scarce commodity, and by its scarcity, it assumed high market value and demanded intensive, high-density utilization. The industrial city of the mass-transit era, because of its limited supply of usable land, was compact, was characterized by high residential and structural densities, and showed a sharp break on its margins between urban and nonurban uses. The older central cities of, particularly, the northeastern United States and southeastern Canada were of that vintage and pattern.

397

Within the city, parcels of land were theoretically allocated among alternate potential users on the basis of the relative ability of those users to outbid competitors for a chosen site. There was, in gross generalization, a continuous open auction in land in which users

Fig. 11.11 In older cities the central business district, localized by converging mass-transit lines, had a superior accessibility sought by many functions. Competition for space and resulting high land values demanded high-density use of a restricted area. Districts farther from the peak mass-transit focus had less attraction and lower values and were developed at lower densities. Courtesy of Chicago Association of Commerce and Industry.)

would locate, relocate, or be displaced in accordance with "rent-paying ability." The attractiveness of a parcel, and therefore the price at auction that it could command, was a function of its accessibility. Ideally, the most desirable and efficient location for all the functions and the people of a city would be at the single point at which the maximum possible interchange could be achieved.

Such total coalescence of activity is obviously impossible. Since uses must therefore arrange themselves spatially, the attractiveness of a parcel is rated by its relative accessibility to all other land uses of the city. Store owners wish to locate where they can easily be reached by potential customers; factories need a convenient assembly of their workers; residents desire easy connection with their jobs, stores, and schools; and so forth. Within the older central city, the radiating mass-transit lines established the elements of the urban land-use structure by freezing in the landscape a clear-cut pattern of differential accessibility. The convergence of that system upon the city core gave that location the highest accessibility, the highest desirability, and, hence, the highest land values of the entire built-up area. Similarly, transit junction points were accessible to larger segments of the city than locations along single traffic routes; these latter were more desirable than parcels lying between the radiating lines (Figure 11.11).

Society deems certain functions desirable without regard to their economic competitiveness. Churches, schools, parks, and public buildings are assigned space without being participants in the auction for land. Other uses, by the process of that auction, are assigned spaces by market forces. Merchants with the highest-order goods and the largest threshold requirements bid most for, and occupy, parcels within the central business district (CBD), which is localized at the convergence of mass-transit lines. Successful bidders for slightly less accessible CBD parcels are the developers of the tall

398

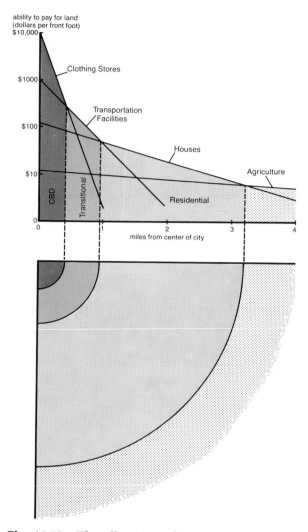

Fig. 11.12 The allocation of space in accordance with rent-paying ability. The inner zone of most accessible land is claimed by commercial users, here typified as "clothing stores." Central business fringe (transitional) uses, residences, and finally nonurban activities complete the generalized land-use arrangement. Each outer use is excluded from more central locations by inability to meet the rent bids of the occupants who can pay more for the benefits of centrality. (Fig. 9.8, "Different activities . . . of accessibility," p. 211 in *The North American City*, 2nd ed., by Maurice Yeates and Barry Garner. Copyright © 1971, 1976 by Maurice H. Yeates and Barry J. Garner. Reprinted by permission of Harper & Row, Publishers, Inc.)

office buildings of major cities, the principal hotels, and similar land uses. Comparable, but lower-order, commercial aggregations develop at outlying intersections—transfer points—of the mass-transit systems. Industry takes control of parcels adjacent to essential cargo routes: rail lines, waterfronts, rivers, or canals. Strings of stores, light industries, and high-density apartment structures can afford and benefit from location along high-volume transit routes. The least-accessible locations within the city are left for the least-competitive users—low-density residences. A diagrammatic summary of this repetitive allocation of space among competitors for urban sites is shown in Figure 11.12.

LAND VALUES AND POPULATION DENSITY

Theoretically, two separate but related distance-decay patterns should be the consequence of the open land auction, one related to land values and the other to population density. The land value surface of the central city, if viewed as a topographic map such as Figure 11.13, with hills representing high valuations and depressions showing low prices, would display a series of peaks, ridges, and valleys representing the differentials in accessibility marked by the pattern of mass-transit lines, their intersections, and the unserved interstitial areas. Dominating these local variations, however, is an overall decline of valuations with increasing distance away from the peak value intersection, the most accessible and costly parcel of the central business district. As we would expect in a distance-decay pattern, the drop in valuation is precipitous within a short linear distance from that point, and then the valuation declines at a lesser rate to the margins of the built-up area.

With one important variation, the population density pattern of the central city shows a comparable distance-decay arrangement, as

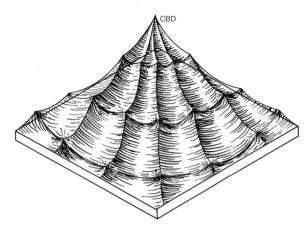

Fig. 11.13 **Generalized pattern of land values, Chicago, 1961. The "topography" represented on this diagram shows the major land-value peak of the central business district. The ridges radiating from that point mark mass-transit lines and major business thoroughfares. Regularly occurring minor peaks indicate commercial agglomerations at intersections of mass-transit lines or main traffic routes—points of accessibility superior to those of interstitial troughs. (Reproduced with permission from B. J. L. Berry,** *Commercial Structure and Commercial Blight,* **Research Paper 85, Department of Geography Research Series, The University of Chicago, 1963.)**

399

suggested by Figure 11.14. The hollow at the center, the central business district, represents the total inability of residential uses to compete for space against alternative occupants of supremely accessible parcels. Yet accessibility has its residential attractions and brings its penalty in high land prices. The result is the high-density residential occupance of parcels near to the center of the city—by those who are too poor to afford a long-distance journey to work, who are consigned by their poverty to the high-density, obsolescent slum tenements near the heart of the inner city, or who are self-selected occupants of the high-density, high-rent apartments made necessary by the price of land. Other urbanites, if financially able, may opt to trade off higher commuting

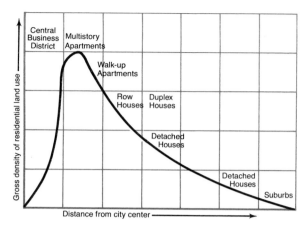

Fig. 11.14 A generalized population density curve. As distance from the area of multistory apartments increases, population density declines. (After M. Clawson, R. Held, and C. Stoddard, *Land for the Future*. A Resources for the Future Book published by The Johns Hopkins University Press, 1960. Reproduced by permission.)

MODELS OF URBAN LAND-USE STRUCTURE

Generalized models of urban growth and land-use patterning have been proposed to summarize the observable results of these organizing forces and controls. The starting point of them all is the distinctive central business district possessed by every older central city. The *core* of this area is characterized by intensive land development: tall buildings, many stores and offices, and crowded streets. Just outside the core is the *frame* of the CBD; as indicated by Figure 11.16, this is an area of wholesaling activities, transportation terminals, warehouses, new-car dealers, furniture stores, and even light industries. Just beyond the central business district is the beginning of residential land uses.

The land-use models depicted on Figure 11.17 diverge in their summarization of patterns outside of the CBD. The *concentric zone* model is an urban application of von Thünen's agricultural zonation, discussed in Chapter 9, and is based upon the same premise of an iso-

400 costs for lower-priced land and may reside on larger parcels away from high-accessibility, high-congestion locations. Residential density declines with increasing distance from the city center as this option is exercised.

As a city grows in population size, peak densities no longer increase, and the pattern of population distribution becomes more uniform. Secondary centers begin to compete with the CBD for customers and industry, and residential areas become less associated with the city center and more dependent upon high-speed transportation arteries. Peak densities in the inner city decline, and peripheral areas increase in population concentration.

The validity of these generalizations may be seen on Figure 11.15, a time-series graph of population density patterns for Montreal over a 30-year period. Within about 2 miles of the CBD, densities increased from 1941 to 1961; after that date, as the city expanded, density decreased. At distances greater than 4 miles from the center, population density has been increasing.

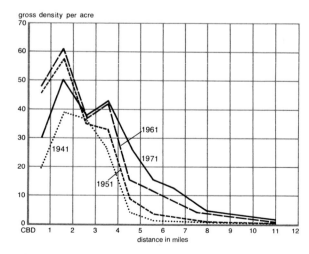

Fig. 11.15 Population density–distance curves for Montreal, 1941–1971. (Redrawn with permission of the author from Maurice Yeates, *Main Street: The Windsor–Quebec City Urban Axis*, Macmillan of Canada, Toronto, 1975, p. 87, Fig. 3.2A.)

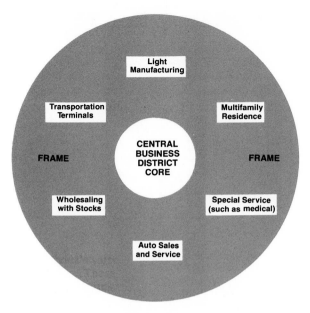

Fig. 11.16 The core–frame concept. Commercial land is used intensively at the core of the central business district; the frame contains less intensive commercial uses. Some of the activities of the frame zone are related to each other and to the core. (Redrawn with permission from E. Horwood and R. Boyce, *Studies of the Central Business District and Urban Freeway Development,* University of Washington Press, Seattle, 1959.)

tropic plain. Developed to explain the observed sociological patterning of American cities, the model recognizes five concentric circles of mostly residential diversity at increasing distance in all directions from the core.

1. The high-density CBD with wholesaling, warehousing, light industry, and transport depots at its margins.
2. A zone in transition marked by the deterioration of old residential structures abandoned, as the city expanded, by the former, more wealthy occupants and now containing high-density, low-income slums, rooming houses, and, perhaps, ethnic ghettos.
3. A zone of "independent workingmen's homes" occupied by industrial workers, per-

haps second-generation Americans able to afford modest but older homes on small lots.
4. A zone of better residences, single-family homes, or high-rent apartments occupied by those wealthy enough to exercise choice in housing location and to afford the longer, more costly journey to CBD employment.
5. A commuters' zone of low-density, isolated residential suburbs, just beginning to emerge when this model was proposed in the 1920s.

The model is dynamic; it imagines the continuous expansion of inner zones at the expense of the next outer developed circles and suggests a ceaseless process of invasion, succession, and population segregation by income level (Figure 11.18).

The *sector* model (Figure 11.17b) also concerns itself with patterns of housing and wealth, but it arrives at the conclusion that high-rent residential areas dominate expansion patterns and grow outward from the center of

401

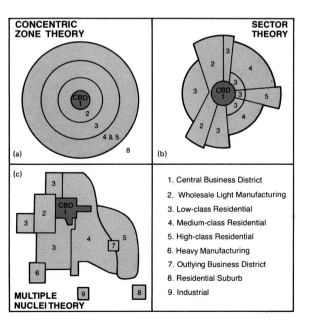

Fig. 11.17 Three classic models of the internal structure of cities.

Fig. 11.18 The end of the line in the process of invasion, succession, and abandonment. A derelict slum building in Chicago. (E.P.A.-DOCUMERICA.)

the city along defined sectoral paths, new housing for the wealthy being added as an outward extension of existing high-rent axes as the city grows. Middle-income housing sectors lie adjacent to the high-rent areas, and low-income residents occupy the remaining sectors of growth. There tends to be a "filtering-down" process as older areas are abandoned by the outward movement of their original inhabitants, with the lowest-income populations (closest to the center of the city and farthest from the current location of the wealthy) the dubious beneficiaries of the least-desirable va-

cated areas. The accordance of the sector model with the actual pattern of Calgary, Canada, is suggested on Figure 11.19.

These "single-node" models of growth and patterning are countered by a *multiple-nuclei* concept (Figure 11.17c), which postulates that large cities develop by peripheral spread not from one but from several nodes of growth. Individual nuclei of special function—commercial, industrial, port, residential—are originally developed in response to the benefits accruing from the spatial association of like activities. Peripheral expansion of the separate

As urban governments and state and federal authorities attempt to find ways to revitalize the ailing older American cities, a potentially significant process is occurring: *gentrification,* or the movement of middle-class people into the inner city. This movement does not yet counterbalance the exodus to the suburbs that has taken place since the end of World War II, but it marks an interesting reversal in what had seemed to be the inevitable decline of cities. According to an estimate of the Urban Land Institute, 70% of all sizable American cities are experiencing a significant renewal of deteriorated areas. Gentrification is especially noticeable in the major cities of the North and the East, from Boston down the Atlantic Coast to Charleston, South Carolina, and Savannah, Georgia.

During the early years of suburbanization, there was also a major movement of low-income nonwhites into central cities. That migration stream has now slowed to a trickle, and attention is centered on the movement of young, affluent, tax-paying professionals into neighborhoods close to the city center. These formerly depressed areas are being rehabilitated and made attractive by the new residents. The prices of houses and apartments in former slums have soared. New, trendy restaurants and boutiques open daily.

The reasons for the upsurge of interest in urban housing reflect to some extent the recent changes in American family structure. Just a decade ago, the suburbs, with their green spaces, were a powerful attraction for young married couples who considered single-family houses with ample yards ideal places to raise children. Now that the proportion of single people (whether never or formerly married) in the American population has increased, the attraction of the jobs and the recreational activities offered in cities outweighs the perceived need for green space.

In addition, as gasoline prices have risen, a city location enables workers to reduce commuting costs significantly. Perhaps most important, however, is the cost of urban housing. During the outflow of people to the low-priced land of the suburbs, many economic forces kept housing costs high in the city. These forces have weakened considerably; before housing prices in a renewed area are bid up, urban housing is relatively cheaper than suburban housing.

Though heartening to those who believe that cities need "saving," gentrification may well bring its own problem: the displacement of the poor, the nonwhite, and the jobless. Some observers of the urban scene fear that these people will be crowded out of the inner city to bleak lower-class suburbs. If so, the urban problem will not have been solved; it will have been simply switched around as a suburban crisis replaces the urban crisis.

nuclei eventually leads to coalescence and the juxtaposition of incompatible land uses along the lines of juncture. The urban land-use pattern, therefore, is not regularly structured from

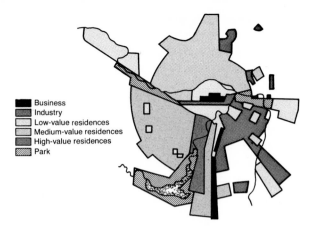

Business
Industry
Low-value residences
Medium-value residences
High-value residences
Park

Fig. 11.19 The land-use pattern of Calgary, Alberta, Canada, in 1961. The circular arrangement of uses suggested in Figure 11.12 results if a city develops on a flat surface. In reality, hills, rivers, railroads, and highways affect land uses in uneven ways. Physical and cultural barriers and the evolution of cities over time tend to result in a sectoral pattern of similar land uses. Calgary's central business district is the focus for many of the sectors. (Redrawn with permission from P. J. Smith, "Calgary: A Study in Urban Pattern," *Economic Geography*, vol. 38, 1962.)

404

a single center in a sequence of circles or a series of sectors, but based upon separate expanding clusters of contrasting activities.

SOCIAL AREAS OF CITIES

Although too simplistic to be fully satisfactory, these classical models of urban structure receive some confirmation from modern interpretations of small-area census data detailing place-to-place variation in the social indicators of urban populations: income level, age structure, family status, ethnicity, and housing-stock characteristics.

The more complex cities are economically and socially, the stronger is the tendency for city residents to segregate themselves into groups based on social status, stage in the life cycle, and ethnicity. In a large metropolitan region, this territorial behavior may be a de-

fense against the unknown or the unwanted. Most people feel more secure when they are near those with whom they can easily identify. In traditional societies, these groups are the families and tribes. In modern society, people group according to income or occupation (social status), age or family characteristics (life cycle status), and language or race (ethnic characteristics). Many of these groupings are fostered by the size and the value of available housing. Land developers, especially in cities, produce homes of similar quality in specific areas. Of course, as time elapses, there is change in the quality of houses, and new groups may replace old groups. In any case, neighborhoods of similar social characteristics evolve.

Social Status. The social status of an individual or a family is determined by income, education, occupation, and home value. In North America, high income, a college education, a professional or managerial position, and high home value constitute high status. *High home value* can mean an expensive rented apartment as well as a large house with extensive grounds.

A good housing indicator of social status is persons per room. A low number of persons per room tends to indicate high status. Low-status characterizes people with low-income jobs living in low-value housing. There are many levels of status, and people tend to filter out into neighborhoods where most of the heads of households are of similar rank.

In most cities, people of similar social status are grouped in sectors whose points are in the innermost urban residential areas. The pattern in Chicago is illustrated in Figure 11.20. If the number of people within a given social group increases, they tend to move away from the central city along an arterial connecting them with the old neighborhood. Major transport routes leading to the city center are the usual migration routes out from the center. Social-status patterning agrees with sector theory.

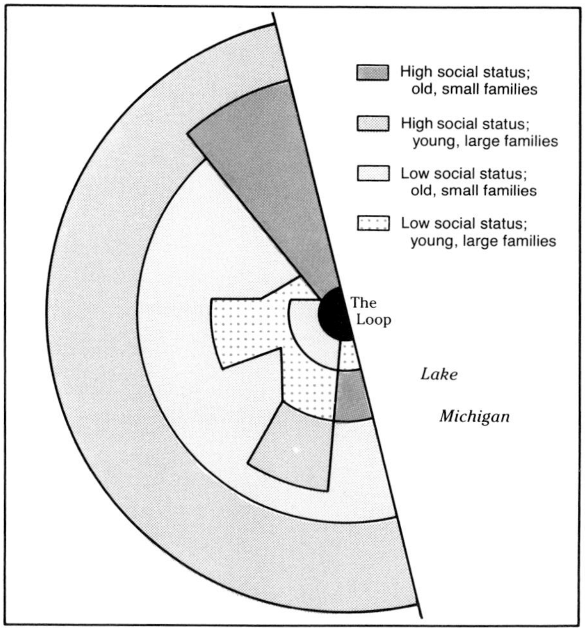

Fig. 11.20 A diagrammatic representation of the major social areas of the Chicago area. The central business district is known as "The Loop." (Redrawn from Philip Rees, "The Factorial Ecology of Metropolitan Chicago," M.A. thesis, University of Chicago, 1968.)

Family Status. As distance from the center of each sector increases, the average age of the head of the household declines, or the size of the family increases, or both. Within a particular sector—say, that of high status—older people whose children do not live with them or young professionals without families tend to live close to the city center. Between these will be the older families who lived at the outskirts of the city in an earlier period. The young families seek space for child rearing, while older people covet more the accessibility to the cultural and business life of the city. Where inner-city life is unpleasant, there is a tendency for older people to migrate to the suburbs or to retirement communities.

Within the lower-status sectors, the same pattern tends to emerge. Transients and single people are housed in the inner city, while fami-

lies, if they find it possible or desirable, live further from the center. The arrangement that emerges is a concentric-circle patterning according to family status.

Ethnicity. Groups for whom ethnicity is a residential locational determinant more important than social or family status appear in the social geography of cities as separate clusters or nuclei, reminiscent of the multiple-nuclei theory of urban structure. For some ethnic groups, cultural segregation is both sought and vigorously defended, even in the face of pressures for neighborhood change exerted by potential competitors for housing space. The durability of "Little Italies" and "Chinatowns," and of Polish, Greek, Armenian, and other ethnic neighborhoods in many American cities is evidence of the persistence of self-maintained segregation.

Certain ethnic or racial groups, especially blacks, have had segregation in nuclear communities forced upon them. Every city has one or more black areas, which in many respects may be considered cities within a city. Figure 11.21 illustrates the concentration of blacks and of Puerto Ricans in certain portions of Brooklyn, New York. The barriers to movement outside of the area have always been high. White society has consistently blocked blacks from gaining the social status that would allow them a greater choice in neighborhood selection. In most American cities, the poorest residents are the blacks, who are relegated to the lowest-quality housing in the least-desirable areas of the city. Similar restrictions have been placed on Hispanics and American Indians.

INSTITUTIONAL CONTROLS

Within this century, and particularly since World War II, institutional controls have strongly influenced the land-use arrangements and growth patterns of most cities. Land-use

405

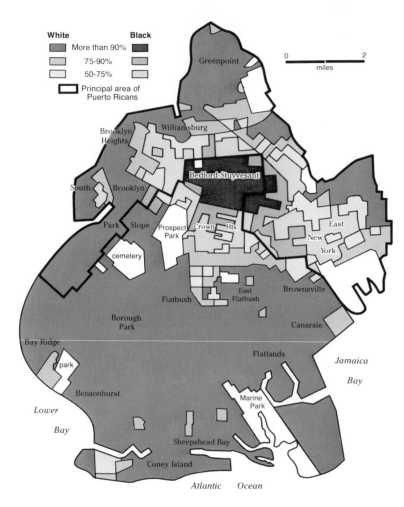

Fig. 11.21 Black and Puerto Rican areas of Brooklyn, New York, 1970. The main black area is in the Bedford-Stuyvesant district, not far from Brooklyn's commercial center, which is just northwest of Prospect Park. In most North American cities, the black ghetto is in the most densely populated area close to the CBD. (From Bergman–Phol: *A Geography of the New York Metropolitan Region,* Kendall/Hunt Publishing Company, Dubuque, Iowa, 1975. Reprinted with permission.)

plans have been adopted, and subdivision control and zoning ordinances have been enacted, to realize those plans. Building, health, and safety codes have been adopted to assure publicly accepted and acceptable urban development and maintenance. All are ultimately based upon broad applications of the police powers of municipalities and their rights to assure public health, safety, and well-being even when traditional private-property rights are infringed. These nonmarket controls on land use presuppose the possibility of achieving an ideal arrangement of uses, minimizing incom-

patibilities (residences adjacent to heavy industry, for example), maximizing all real estate values, and providing for the creation in appropriate locations of all public and private uses needed for and conducive to a balanced, orderly community. Presumably such careful planning would preclude the emergence of slums, so often the result of undesirable adjacent uses, and would stabilize neighborhoods by reducing market-induced pressures for land-use change.

Such controls, particularly zoning ordinances and subdivision control regulations specifying acre or larger residential building lots and very large house floor areas, have been adopted as devices to exclude from upper-income areas lower-income populations or those who would choose to build or occupy other forms of residences: apartments, special housing for the aged, and so forth. Bitter court battles have been waged, with mixed results, over "exclusionary" zoning practices that in the view of some serve to separate rather than to unify the total urban structure and to maintain or increase diseconomies of land-use development.

Zoning ordinances and land-use plans have also been criticized as being unduly restrictive and unresponsive to changing land-use needs, patterns of economic development, and public objectives. Some argue that they freeze existing land uses and congeal the city into outdated forms reflecting the wisdom of the past rather than the opportunities of the present.

Suburbanization

The North American mass-transit city of repetitive land use and social area structure, controlled in its development by market forces and by fixed and focused lines of traffic and rigid patterns of accessibility, reached its climax by the late 1920s. Although the Great Depression of the 1930s and the national priorities of World War II effectively halted urban expansion and land-use replacement, the intervening years saw the creation of technological, physical, and institutional structures that resulted, after the war, in a sudden and massive alteration of past urban forms.

The improvement of the automobile increased its reliability, range, and convenience, freeing its owner from dependence upon fixed-route public transit for access to home, work, or shopping. The road-building and improvement projects undertaken by all levels of government to relieve urban unemployment dur-

ing the Depression opened up vast new acreages of nonurban land for urban development, with the automobile—not mass transit—as the means of access. The enactment, as part of the New Deal programs under President Franklin D. Roosevelt, of a maximum 40-hour work week guaranteed millions of Americans the time for a commuting journey not possible when workdays of 10 or more hours were common. And finally, to stimulate the economy by the ripple effect associated with a prosperous construction industry, the Federal Housing Administration was established to guarantee creditors the security of their mortgage loans, thus reducing down-payment requirements and lengthening mortgage repayment periods. Although the Depression and the war precluded most Americans from acting upon the new opportunities, home ownership became an eventual possibility for additional millions of families of middle- and lower-middle-income levels.

Demands for housing, pent up by years of economic depression and wartime restrictions, were loosed in a flood after 1945, and a massive suburbanization of populations and urban functions altered the existing pattern of urban America. Between 1950 and 1970, the two most prominent patterns of population growth were the metropolitanization of people and, within metropolitan areas, their suburbanization. Growth patterns for the Chicago area are shown on Figure 11.22. Since 1970, as Chapter 8 disclosed, new residential movements have emerged, again based upon the automobile and also detrimental to large-central-city vitality.

Residential land uses led the rush to suburbanize, with a resultant pattern of individually uniform, spatially discontinuous additions beyond the political limits of most older central cities. This unfocused (because not tied to mass-transit lines and the accessibility structure they impose) residential sprawl represented a massive relocation of purchasing power, to which, in accordance with the man-

407

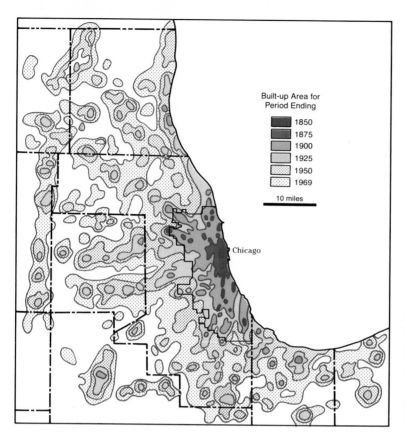

Built-up Area for
Period Ending

1850
1875
1900
1925
1950
1969

10 miles

Chicago

408

Fig. 11.22 Urban sprawl. In Chicago, as in most large North American cities, the size of the urbanized area has increased dramatically during the last 100 years, particularly since World War II. (Redrawn with permission from *Chicago: Transformation of an Urban System*, Copyright 1976, Ballinger Publishing Co.)

dates of central-place theory, retail merchants were quick to respond. Following the prototypes established by Shopper's World outside of Framingham, Massachusetts, and by Northgate in suburban Seattle, Washington, opened, respectively, in 1949 and 1950 (though Country Club Plaza, opened in 1923 in Kansas City, Missouri, served—as did a very few others—as earlier models), the planned regional shopping center spread to expanding metropolitan areas throughout the nation. The consequences for the metropolitan area pattern of commercial locations and the impact upon the central-city retailing dominance is shown on Figure 11.23 for the Chicago area. Faced with a newly suburbanized labor force, new space requirements for car-driving employees, new indus-

trial technologies demanding a horizontal rather than a vertical flow of material through plants, and the new locational flexibility permitted by the motor truck and the expressway, industry followed the outward move. This trend added more low-density uses to the suburban zone and further eroded the employment and economic base of the older central city. One summarization of the new patterns is shown on Figure 11.24.

Outside North America, the dependence on the automobile for work and shopping trips is less well established (Figure 11.25). As a result, land-use patterning is more compact. Densities are higher, there is less low-density urban sprawl, less land is vacant or undeveloped, and more people live in apartments as opposed to

Fig. 11.23 The changing pattern of the retailing business in Chicago, 1948–1958. The suburbanization of population and purchasing power shown on Figure 11.22 provides the rationale for the postwar changes in the pattern of commercial activities displayed on this map. (Redrawn with permission from B. J. L. Berry and F. E. Horton, *Geographic Perspectives on Urban Systems*, © 1970 Prentice-Hall, Inc.)

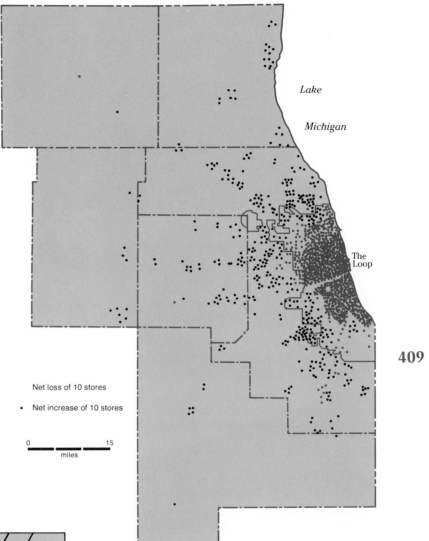

Lake

Michigan

The Loop

409

Fig. 11.24 Generalized urban-area land-use patterns. The peripheral zone separating the high-density inner city from the lower-density suburbs emerged as a district of shopping centers, industrial parks, and both single-family and apartment housing developments. The radial suburbs are tied to commuter transportation lines, including expressways, which connect with the central business district. (After Edward Taaffe et al., *The Peripheral Journey to Work*, Northwestern University Press, 1963. Reproduced by permission of Northwestern University Transportation Center. All rights reserved.)

Net loss of 10 stores

• Net increase of 10 stores

0 ——————— 15
miles

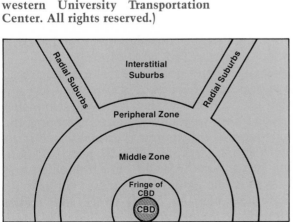

single-family houses. The city centers are much more of a focus for urban activities than is true in North America. Very often, the majority of the rich or the powerful live close to the center of the city in apartments and town houses. However, they may have summer homes or estates far from the city. If there are large numbers of urban poor, as is the case in Asia, Africa, and South America, they are

(a)

(b)

Fig. 11.25 (a) A congested city street in the old section of New Delhi, India. Notice the large variety of conveyances. (b) Tokyo is a congested city with a highly efficient and dense network of commuter trains. At peak hours, "pushers" are employed to get as many people as possible into the trains. (c) In Rio de Janiero, Brazil, many of the wealthy live in apartments in and near the center of the city. (a: United Nations/J. P. Lafonte; b, c: United Nations.)

found in densely settled shanty towns on the periphery of the city. Mass-transit facilities figure prominently in the journey to work.

In time, in the United States, an established social and functional pattern of suburban land use emerged, giving evidence of a lower-density, more extensive repetition of the models of land use developed to describe the structure of the central city. Multiple nuclei of specialized land uses were created, expanded, and coalesced. Sectors of high-income residential use continued their outward extension beyond the central-city limits, usurping the most scenic and the most desirable suburban areas and segregating them by price and zoning restrictions. Middle-, lower-middle-, and lower-income groups found their own income-segregated portions of the fringe; and ethnic minorities, frequently relegated to the older

411

(c)

**The
Self-sufficient
Suburbs**

The suburbanization of the American population and of the commerce and industry dependent upon its purchasing power and labor has led to more than the simple physical separation of the suburbanite from the central city. It has created a metropolitan area, many of whose inhabitants have no connection with the core city, feel no ties to it, and find satisfaction of all their needs within the peripheral zone.

A *New York Times* "suburban poll," conducted in the summer of 1978,

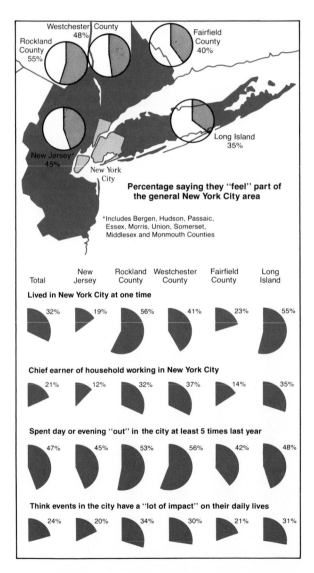

412

revealed a surprising lack of interest in New York City on the part of those usually assumed to be intimately involved in the economic, cultural, and social life of the "functional city," or the metropolitan area dominated by New York. The poll discovered that in 80% of suburban households, the principal wage earner did not work in New York City. The concept of the *commuter zone,* implicit or explicit in the classic urban models shown as Figure 11.17, obviously no longer has validity if *commuting* implies, as it formerly did, a daily journey to work to the central city.

Just as employment ties have been broken with majority of suburbanites working in or near their outlying areas of residence, so, too, have the cultural and service ties been severed that were traditionally thought to bind the metropolitan area into a single unit dominated by the core city. The *Times* survey found that one-half of the suburbanites polled made fewer than five nonbusiness visits to New York City each year; one-quarter said they never went there. Even the presumed concentration in the city of high-order goods and services—those with the largest service areas—did not constitute an attraction for most suburbanites. Only 20% of the respondents thought that they would journey to New York City to see a medical specialist, and only 7% or 8% for a lawyer, an accountant, or to make a major purchase. Even needs at the upper reaches of the central-place hierarchy were satisfied within the fringe.

The suburbs, as the *Times* survey documented, have outgrown their former role as bedroom communities and have emerged as a chain of independent multinucleated urban developments. Together, they are largely self-sufficient and divorced from the central city and have little feeling of subordination to it or dependence upon it.

413

industrial suburbs, found theirs. A rigid social geography was developed, with intermingling minimized and even shopping centers oriented in goods and prices to restricted submarkets of the suburban population. With increasing suburban sprawl and the rising costs implicit in the ever-greater spatial separation of the functional segments of the fringe, the limits of feasible expansion were reached, the supply of developable land was reduced (with corresponding increases in its price), and the intensity of land development grew. Changing lifestyles and cost constraints have resulted in a proliferation of suburban apartment complexes and the disappearance of open land. The maturation and the coalescence of urban land uses have resulted in the emergence of coherent metropolitan-area cities that are suburban in traditional name only, containing the central business district and mix of land uses that the designation *city* implies.

Functional Relocation and Political Fragmentation

The structure, the functional base, and the financial stability of the central city have been grievously damaged by the process of suburbanization. In earlier periods of growth, as new settlement areas or functional nodes developed beyond the political margins of the city, annexation absorbed new growth within the corporate boundaries of the expanding older city. The additional tax base and employment centers became part of the municipal whole. But the city is the creature of the state government, and where the states recognized the right of separate incorporation of new growth areas—particularly in the eastern part of the United States—the ability of the city to continue to expand was restricted. Lower suburban residential densities made septic tanks and private wells adequate substitutes for expensive city-sponsored sewers and water mains; lower structural densities and lower crime rates meant less felt need for high-efficiency city fire and police protection. Where possible, suburbanites opted for separation from the central city and for aloofness from the costs, the deterioration, and the adversities associated with it. Their homes, jobs, shopping, schools, and recreation all existed outside the confines of the city from which they had divorced them-

selves, and the close-knit fabric of urban life was rent.

In many western states of more recent urbanization, legislatures adopted a more permissive annexation stance. Newer cities—Los Angeles, Houston, Oklahoma City, and others—that grew to sprawling maturity by automobile rather than the mass-transit movement of people, were legally able to expand their corporate boundaries not only to include new growth areas as they emerged but also to annex in anticipation of their development vast tracts of totally rural land in their environs (Figure 11.26).

The redistributions of population attendant upon the process of suburbanization resulted not only in the spatial but also in the political segregation of social groups of the metropolitan area. The upwardly mobile residents of the city—younger, wealthier, and better educated—took advantage of the automobile and the freeway to leave the central city. The poorer, older, least-advantaged urbanites were left behind. In many cases, the desire of these latter groups to move to suburban homes has been thwarted

414

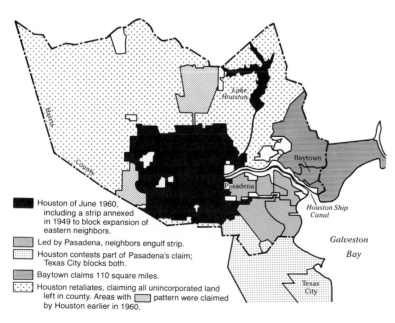

Led by Pasadena, neighbors engulf strip.

Houston of June 1960, including a strip annexed in 1949 to block expansion of eastern neighbors.

Houston contests part of Pasadena's claim; Texas City blocks both.

Baytown claims 110 square miles.

Houston retaliates, claiming all unincorporated land left in county. Areas with ⧄ pattern were claimed by Houston earlier in 1960.

Fig. 11.26 In 1960, in response to a series of "blocking" annexations initiated by neighboring municipalities and to acts of incorporation of communities that sought annexation powers under Texas law, Houston took for itself all the remaining unincorporated area of Harris County. (Reprinted from the July 23, 1960, issue of *Business Week* by special permission, © 1960 by McGraw-Hill, Inc., all rights reserved.)

Governmental chaos in major metropolitan areas has reached such proportions that new approaches to administrative control and the integration of those areas are being increasingly proposed and adopted. The aims of all plans of metropolitan government are to create or preserve coherence in the management of metropolitan area concerns and to assure that both the problems and the benefits of growth will be shared irrespective of their jurisdictional location. Two approaches—unified government and predevelopment annexation—have been followed to achieve the elimination of jurisdictional overlap and rivalry, the abolition of fiscal inequities between municipalities, and the removal of the fragmented approach to areawide problems, and to secure unified administrative control of extensive urbanized areas.

The North American model of unified government—sometimes called *Unigov* or *Metro*—is found in the Province of Ontario, Canada, and, particularly, in metropolitan Toronto. The provincial government proposed and is carrying out a reorganization of Ontario's 900-plus cities, towns, and villages into some 30 two-tiered systems of metropolitan government. The upper tier of consolidation will provide broad-scale direction and fiscal control; the lower tier will assure some measure of local autonomy. The Metropolitan Government of Toronto, established in the mid-1950s, was the prototype. The central city combined with five surrounding boroughs to create the 240-square-mile metropolitan unit that assumed responsibility for problems of public transportation, water and sewerage, centralized police protection, unified capital borrowing, and centralized land use and capital planning. Consolidation on the Ontario model is intended to spread growth and tax revenues more equitably by putting all communities within Metro jurisdiction upon the same footing, although most taxes collected by the separate municipalities are passed upward to the Metro government. To preserve a sense of identity, and to free Metro government from concern with problems best handled locally, municipalities retain their separate names and local officials and perform such housekeeping tasks as maintaining local roads and sidewalks, providing fire protection, and collecting garbage.

Unigov has won many adherents and some successes in the United States, where the usual pattern is the consolidation of the central city with the county of which it is a part. City-county governments have been successfully operated in Indianapolis/Marion County, Indiana; Miami/Dade County and Jacksonville/Duval County, Florida; Nashville/Davidson County, Tennessee; Portland/Multnomah County, Oregon; and Seattle/King County, Washington. A similar and equally successful approach to metropolitan-area control is marked by the Twin Cities Metropolitan Council centered in Minneapolis–Saint Paul. Here, Minnesota's two largest cities, some 130 smaller surrounding municipalities, seven counties, and great numbers of special-purpose agencies—containing about half the state's population—were partially subordinated to a metropolitan council with functions limited to those few activities that de-

415

manded centralized, areawide administration. These include sewerage, water supply, airport location, highway routes, and the preservation of open space. To the local communities were left their own forms of government and the control of fire and police services, schools, street maintenance, and other local functions.

The preservation of the tax base and the retention for the central city of expansion room are the driving forces behind the predevelopment annexations permitted under the laws of some, particularly western, states. Cities such as Oklahoma City (650 square miles) and Houston (510 square miles) have taken advantage of such permissive state regulation to assure that they will not be constricted by a ring of incorporated suburbs that would syphon off growth, population, jobs, and income generated by the presence of the central city itself. Since such annexation rights over contiguous unincorporated areas are granted to all incorporated communities (generally of a specific minimum size), unseemly "land grabs" have been initiated by rival communities, to the detriment of both undeveloped rural areas and the rational control of logically single metropolitan areas.

416

by bankers, real estate operators, and white social pressure, as well as by their inability to pay for suburban homes or commuting costs. Large areas within the city's boundaries now contain only the poor and ethnic minorities, a population little able to pay the rising costs of social services that their numbers, neighborhoods, and condition require.

The services needed to support the poor include welfare payments, social workers, extra police and fire protection, health delivery systems, and subsidized housing. Central cities are unable to raise by themselves the taxes needed to support such an array and intensity of social service now that they have lost the tax bases represented by suburbanized commerce, industry, and upper-income residential uses. Lost, too, are the job opportunities formerly a part of the central-city structure. Increasingly, the poor and the minorities are trapped in a central city without the possibility of nearby employment and isolated by distance, immobility, and unfamiliarity from the few remaining low-skill jobs, which are now largely suburbanized.

The separation of tax bases and tax needs by the arbitrary political barriers fragmenting the metropolitan area poses a fearful question for the maintenance of the central city. Metropolitan government, discussed in Chapter 7, has been offered as a solution to the need for a unified economic base and tax support for the central city and the suburban zone alike, but only in the regions with the least depressed central cities do suburbanites seem willing to entertain such a plan.

The abandonment of the central city by people and functions has nearly destroyed the traditional active open auction for urban land, which led to the replacement of obsolescent uses and inefficient structures in a continuing process of urban modernization. In the vacuum left by the departure of private investors, the federal government, particularly since the landmark Housing Act of 1949, has initiated urban renewal programs with or without provi-

At Stalin's death in 1953, a housing shortage of monumental proportions existed in the Soviet Union. Urban housing construction had been very low on the Stalinist list of investment priorities during the prewar five-year plans. The massive rural-to-urban population shifts demanded by the rapid industrialization of the country and the wartime destruction in the urbanized western sections of the nation compounded the pressures upon the prerevolutionary housing stock and its meager postrevolutionary additions. Popular unrest, intolerable overcrowding, and the need to redevelop and expand the cities of the U.S.S.R. changed housing from a low to a high priority from the mid-1950s to the present time.

Speed was essential if the minimum expectations of a demanding population were to be met and the promise of rapid rectification of past deficiencies was to gain credence. The solutions seized upon were component prefabrication and standardization of building and project design. Both had the advantage of minimizing cost and labor inputs.

Thus were created, during the first phase of the housing program, the ubiquitous microdistricts of uniform five-story apartment blocks. By law, buildings taller than five stories required elevators, a money and maintenance expenditure to be avoided if possible. In 1966, by which time the drab sameness of community layout and building design was being questioned on sociological and economic grounds, the microdistrict was described by two Soviet architects.

417

> The usual microdistrict—the design is the same, East, West, North or South, in big or small cities—covers some 25–30 hectares, has a population of 8000–10,000 living in five-story units of 100 or 120 apartments each, and contains nursery schools, kindergartens, a school, shops, restaurants, a club-house, motion picture theater, bath house, department store and out-patient clinic. The present standards provide that the school shall be no more than 500 meters from the farthest dwelling, the nursery schools and kindergartens no more than 200 meters from the residences served by them. The food shops and department store are in the heart of the district, no more than 500 meters from the outermost apartment house.*

Whatever the advantages of the adopted construction and design program in providing the rapid relief of a serious housing problem, the microdistrict did consume vast amounts of land, contributing to the sprawl of the larger cities, to increased expense in extending transportation and utility lines, and to the absorption under concrete of excessive amounts of agricultural land.

By the 1970s, greater flexibility in housing design was being achieved to meet the needs of different-sized urban centers. In 1971, the Director of the Construction Bank's Housing Department commented, "The big-city builders' bias in favor of the five-story apartment house as the most economical type of residential building has proved mistaken in the light of [the excessive population growth of the largest cities]." He pointed out that where the housing need was greatest and the space most limited, the costs per square meter of nine-story buildings made them a more acceptable response.

*"Reflections on the Microdistrict," E. Levina and Ye. Syrkina, *Zvezda*, No. 10, October, 1966, pp. 150–156.

sions for a partnership with private housing and redevelopment investment. Under a wide array of programs instituted and funded since the late 1940s, slum areas have been cleared, public housing built, cleared land conveyed at subsidized cost to private developers for middle-income housing construction, cultural complexes and industrial parks created, and city centers reconstructed. With the continuing erosion of the urban economic base and the disadvantageous restructuring of the central-city population base, the hard-fought governmental battle to maintain or revive the central city is generally judged to be a losing one (Figure 11.27).

Cities in Noncommercial Economies

A great acceleration of city formation and expansion and a great increase in urbanization have affected all the world's nations and economic systems. In other economies and societies, the impact of urbanization and the responses to it differ from the patterns and problems observable in the United States.

Developing countries, emerging from formerly dominantly subsistence economies, have experienced disproportionate population concentrations, particularly in national and regional capitals. Lacking or relatively undeveloped is the substructure of maturing, functionally complex smaller and medium-sized centers characteristic of more advanced and diversified economies. Cairo contains 22% of Egypt's total population; 19% of all Nicaraguans live in Managua; and Greater Lima contains 18% of the Peruvian populace. The low-income surplus rural populations have fled to developed seats of wealth and political centrality as to places of refuge. Finding the jobs they seek nonexistent, they often must settle for a barely sustainable life of beggary. While attention may be lavished on creating urban cores on the skyscraper model of Western cities, the new urban multitudes pack themselves into squatter shanty communities on the fringes of the city, isolated from the sanitary facilities, the public utilities, and the job opportunities found only at the center. Such impoverished squatter districts are found around most major cities in Africa, Asia, and Latin America; an example is shown in Figure 11.28.

Such planned economies as that of the Soviet Union have pursued urban formation and expansion under controlling conditions totally different from the mechanisms operating in commercial economies. In Communist nations, urban land is treated as a noncost commodity to be allocated to uses in accordance with planned developmental patterns. Those patterns are based upon primary depend-

418

ence on the mass-transit movement of people, high-density apartment complexes, the close spatial association of industrial employment nuclei and the working population serving within them, and the development of outlying, self-contained satellite cities within growing metropolitan areas. Cross-commuting is prohibited in the fringe zones and discouraged to or from the central city.

The *microdistrict* is the basic residential planning unit, a superblock apartment complex containing the schools, stores, and recreational and public facilities necessary for the minimum requirements of the local occupants, reducing the need for more than occasional travel beyond the confines of the home district. To reduce travel requirements further, new housing complexes are frequently built immediately adjacent to main employment centers, with residential populations spatially segregated by place of employment rather than primarily on income or ethnic grounds, as in the United States. City centers, the focuses of mass-transit lines, are not taken over by commercial functions but reserved for public buildings, parks, and a large paved square for parades and public assembly.

Conclusion

The city so carefully planned by our imaginary Mesopotamian potentate probably developed as many problems, conflicts, diseconomies, and disappointments as have all successor cities to the present, irrespective of the culture or the system of economic organization under which they developed.

The city is the essential functional node in the chain of linkages and interdependencies that marks any society or economy advanced beyond the level of primitive subsistence. The more complex the society, the greater the degree of specialization and exchange it demands for its maintenance, and the more urbanized it becomes, that is, the more concentrated in functional agglomerations are the activities and people of which the society is composed.

In the United States, at least, cities are the ultimate expression of a way of life and an organization of society and economy to which, apparently, we have irrevocably committed ourselves. Urban areas are the home of more than three-fourths of our population, represent investments of tens of trillions of dollars, and constitute by every measure both our primary national resource and our principal, unavoidable preoccupation. The provision of services, physical and social; the administration of urban territories and populations; the control of the city's growth; the organization of its structure; the movement of its people; the replacement or preservation of its forms and structures; the education of its young; the maintenance of its elderly; and the sustenance and housing of its poor—all are topics of continuing national and local debate, public decision, tax expenditure, and legal opinion. As citizens, we pay for and are individually affected by it all. An understanding of the city and of the system of cities in which we live is an essential first step toward a responsible public and personal response to the urban-centered discourse to which we are constantly subjected. The study of urban geography and the understanding of its conclusions and propositions can make that discourse comprehensible.

The Mesopotamian king had the easy part: he only ordered the city built. People have to live in it.

419

421

Fig. 11.27 Some elaborate governmental projects for urban redevelopment, slum clearance, and rehousing of the poor have been failures. The demolition of the Pruitt–Igoe project in St. Louis was a recognition, resulting from soaring vandalism and crime rates, that public high-rise developments intended to revive the central city did not always meet the housing and social needs of their inhabitants. (The Pulitzer Publishing Company/St. Louis Post-Dispatch.)

422

Fig. 11.28 As many as one-half million people live in squatter slums such as this one in New Delhi, India. This pattern is repeated in most of the large cities of Asia, Africa, and Latin America. (United Press International photo.)

For Review

1. Consider the urban area in which you live or attend school or with which you are most familiar. In a brief paragraph, discuss that community's *site* and *situation;* point out the connection, if any, between its site and situation and the basic functions that it earlier performed or now performs.

2. Describe the *multiplier effect* as it relates to the functional and the population growth of urban units.

3. What is a *central place* in the classification of cities? What area does it serve, and what kinds of functions does it perform? If an urban system were composed solely of central places, what summary statements could you make about the spatial distribution and the urban size hierarchy of that system?

4. How might the *interaction model* help explain the kinds of shopping trips your family characteristically makes? Does the concept of *threshold* add to that explanation?

5. Is there a hierarchy of retailing activities in the community with which you are most familiar? Of how many and what kinds of levels is that hierarchy composed? What localizing forces affect the distributional pattern of retailing within that community?

6. Briefly describe the urban land-use patterns predicted by the *concentric circle,* the *sector,* and the *multiple-nuclei* models of urban development. Which one, if any, best corresponds to the growth and land-use pattern of the community most familiar to you?

7. In what ways do *social status, family status,* and *ethnicity* affect the residential choices of households? What expected distributional patterns of urban *social areas* are associated with each? Does the social geography of your known community conform to the predicted pattern?

Suggested Readings

Association of American Geographers Comparative Metropolitan Analysis Project.
Vol. 1, *Contemporary Metropolitan America: Twenty Geographical Vignettes.* Ballinger, Cambridge, MA, 1976.
Vol. 2, *Urban Policymaking and Metropolitan Dynamics: A Comparative Geographical Analysis.* Ballinger, Cambridge, MA, 1976.

Vol. 3, *A Comparative Atlas of America's Great Cities: Twenty Metropolitan Regions*. University of Minnesota Press, Minneapolis, 1976.

BERRY, BRIAN J. L., and FRANK E. HORTON (eds.), *Geographic Perspectives on Urban Systems*. Prentice-Hall, Englewood Cliffs, NJ, 1970.

BERRY, BRIAN J. L., AND JOHN D. KASARDA, *Contemporary Urban Ecology*. Macmillan, New York, 1977.

ELIOT-HURST, MICHAEL E. (ed.), *Transportation Geography*. McGraw-Hill, New York, 1974.

HARVEY, DAVID, *Social Justice and the City*. Edward Arnold, London, 1973.

KING, LESLIE, and REGINALD G. GOLLEDGE, *Cities, Space, and Behavior: The Elements of Urban Geography*. Prentice-Hall, Englewood Cliffs, NJ, 1978.

RUGG, DEAN S., *Spatial Foundations of Urbanism* (2nd ed.). Wm. C. Brown, Dubuque, IA, 1979.

YEATES, MAURICE, and BARRY GARNER, *The North American City* (3rd ed.). Harper & Row, New York, 1980.

425

IV

426

The Area-Analysis Tradition

Julius Caesar began his account of his transalpine campaigns by observing that all of Gaul was divided into three parts. With that spatial summary, he gave to every schoolchild an example of geography in action.

Caesar's report to the Romans demanded that he convey to an uninformed audience a workable mental picture of place. He was able to achieve that aim by aggregating spatial data, by selecting and emphasizing what was important to his purpose, and by submerging or ignoring what was not. He was pursuing the geographic tradition of area analysis, a tradition that is at the heart of the discipline and focuses on the recognition of spatial uniformities and the elucidation of their significance.

The tradition of area analysis is commonly associated with the term *regional geography*, the study of particular portions of the earth's surface. As did Caesar, the regional geographer attempts to view a particular area and to summarize what is spatially significant about it. One cannot, of course, possibly know everything about a region, nor would "everything" contribute to our understanding of its essential nature. Regional geographers, however, approach a preselected earth space—a continent, a nation, or some other division—with the intent of making it as fully understood as possible in as many facets of its nature as they deem practicable. It is from this school of area analysis that "regional geographies" of, for example, Africa, the United States, or the Pacific Northwest emanate.

Because the scope of their inquiries is so broad, regional geographers must become thoroughly versed in all of the topical subfields of the discipline, such as those discussed in this book. Only in that way can area specialists select those phenomena that give insight into the essential unity and diversity of their region of study. In their approach, and based upon their topical knowledge, regional geographers frequently seek to delimit and study regions defined by one or a limited number of criteria.

The three topical traditions already discussed have often misleadingly been contrasted to the tradition of area analysis. In actuality, practitioners of both topical and regional geography inevitably work within the tradition of area analysis. Both seek an organized view of earth space. The one asks what are the regional units that evolve from the consideration of a particular set of preselected phenomena; the other asks how, in the study of a region, its varied content may best be elucidated.

Chapter 12 is devoted to the area-analysis tradition. Its introductory pages explore the nature of regions and of methods of regional analysis. The body of the chapter consists of a series of regional vignettes, each based upon a theme introduced in one of the preceding chapters of the book. Since those chapters were topically organized, the separate studies that follow demonstrate the tradition of area analysis in the context of topical (sometimes called "systematic") geography. They are examples of the ultimate regionalizing objectives of geographers who seek an understanding of their data's spatial expression. The step from these limited topical studies to the broader, composite understandings sought by the area analyst is both short and intellectually satisfying.

To say "Kent State" is for many not only to recall a single indelibly etched moment of localized tragedy, but also to arouse memories of a prolonged period of national division, student protest, political action, and unpopular war. "Paleolithic" summarizes an ill-defined but generally understood period of preagricultural human development when hunter-gatherers, working with nonmetallic tools, spread their domain throughout the inhabitable world. And "Roaring Twenties" suggests a temporally restricted period of American life characterized by a set of manners, music, and morals that passed from the scene at the end of a decade.

Such shorthand references are in reality complex constructs employed in both casual and professional discussion in the expectation that a mosaic of interrelated phenomena will be visualized by both speaker and listener. The references are more than that; they represent, as well, recognition of a need to aggregate into useful and revealing segments the broad sweep of human history and to place in perspective the details of daily events.

As in history (for time), so in geography (for space), there is a need to generalize and to categorize that is felt alike in common speech and professional study. The region is the geographer's equivalent of the historian's epoch: a device that segregates into acceptable component parts the total reality that is the surface of the earth.

All of us have a general idea of the meaning of *region,* and all of us employ the regional concept in everyday speech and action. We visit "the old neighborhood" or "go downtown"; we plan to vacation or retire in the "sun belt"; or we speculate on the effects of weather conditions in the Middle West or the "corn belt" on next year's food prices. In each instance, we individually have a mental image of the area of reference based upon criteria of delimitation or on acceptable bounds of generalization that seem important to us and, presumably, to our listeners. We have engaged in an informal place

12 The Regional Concept

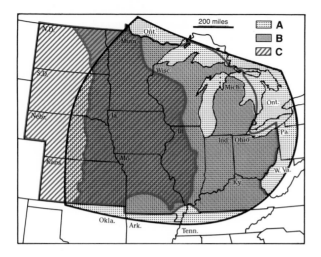

Fig. 12.1 The Middle West as seen by different professional geographers. Agreement upon the need to recognize spatial order and to define regional units does not imply unanimity in the selection of boundary criteria. Each of the sources concurs in the significance of the Middle West as a regional entity in the spatial structure of the United States and agrees upon its core area. These sources differ, however, in their assessments of its limiting characteristics. (*A* from John H. Garland, ed., *The North American Midwest*, Wiley, New York, 1955; *B* from John R. Borchert and Jane McGuigan, *Geography of the New World*, Rand McNally, Chicago, 1961; *C* from Otis P. Starkey and J. Lewis Robinson, *The Anglo-American Realm*, McGraw-Hill, New York, 1969.)

430

classification to convey an array of spatial, organizational, or content ideas.

Geography, as primarily a spatial discipline, accepts as one of its charges the arrangement of its data into defined regional units (Figure 12.1). It applies the regional concept to bring order to the infinite heterogeneity of the earth's surface. But neither the concept nor the unit areas it seeks to classify or clarify are simple. Only the purpose is simple: to create understanding out of the seeming chaos of the physical and cultural variation of the earth we inhabit. The recognition of regional units—the process of regionalization—makes possible the structured view of space necessary to the com-

prehension and the analysis of the world around us.

The region is an areal grouping of phenomena that contributes to the analysis of the intellectual problem at hand. Although the region may appear to be self-evident and unmistakable—as a lush, irrigated agricultural area set in the midst of barren desert—in reality it is a concept of the intellect, a created entity. Its purpose is the selection for special study of items or characteristics that contribute to the understanding of a specific topic or areal problem. All other variables are disregarded as irrelevant. In our example, the irrigated agricultural district has regional import only if our topic of concern is the economic patterning or the human use of arid lands. If our interest is the distribution and classification of the world's desert areas, the intermittent occurrence of irrigation agriculture is immaterial.

Since regions are spatial expressions of mental constructs, they share—no matter the criteria upon which they are based—certain common elements related to space.

1. They have location, often expressed in the applied regional name: the Middle West, the Near East, North Africa, or Southeast Asia. It is apparent from this form of terminology that the defined region is located relative to some larger understood territorial unit or land mass and that it is usually and conveniently less than continental in dimension. It is also evident that such locational designations contain no necessary specification of the content or the nature of the region. Even when the regional name involves no areal reference, uniqueness of location is implicit. We may not at the outset of a discussion of agricultural areas of the United States know exactly where the cotton belt lies, but by its specification, we understand it to be distinct from, say, the corn belt or the dairy belt.

2. Since regions are defined spatial realities, they have spatial extent. As we have seen, regions are recognized on the basis of criteria se-

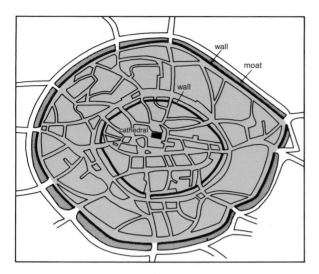

Fig. 12.2 Aachen in 1649. The regional units of contemporary concern are rarely as precisely and visibly demarked as the walled medieval city. Its sprawled modern counterpart may be more difficult to define, but the boundary significance of the concept of *urban* remains.

lected as relevant for the purposes of study. They imply an areal variation of phenomena and the existence on the earth's surface of districts where the phenomena of interest are present or dominant and of locales where they are absent or subordinate.

3. Regions derive from the way geographers discern reality; the acceptance of regional extent implies the recognition of regional boundaries. At some defined point, *urban* is replaced by *nonurban* (Figure 12.2), the Middle West ends and the Plains begin, or the rain forest ceases and the savanna emerges. Since regions are purposeful intellectual constructs based upon specific criteria, their boundaries are drawn where the criteria defining them no longer occur or dominate.

4. It is, of course, the content of the region that gives it definition and determines the criteria of its delimitation. Although there are as many individual regions as the objectives of

spatial study and understanding demand, generalized *types* of regions are recognized.

A *formal* region is one throughout which a single or limited combination of phenomena are of such uniformity as to permit an areal generalization based upon their existence. Formal regions are areas of a homogeneity that may be physical or cultural in nature. Your home state is a formal political region within which uniformity of law and administration is expressed. The Columbia Plateau or the humid subtropical climatic zone is an expression of uniformity of physical characteristics that make up formal natural regions. In previous chapters, we encountered homogeneous cultural regions, areas of the earth's surface in which standardized characteristics of language, religion, ethnicity, or livelihood existed. Whatever may be the criteria of its content, the formal region is the largest area over which a valid generalization of static uniformity may be made. Whatever is stated about one part of it holds true for its remainder.

The *functional* region, in contrast, may be visualized as a spatial system; its parts are interdependent, and throughout its extent, the functional region operates as a dynamic, organizational unit. Such a region has unity not in the sense of static content but in the manner of its operational integration. It may be conceived of as having a node, or core area, of influence and control, centered upon which is the total region defined by the type and extent of control specified. The "central place" recognized in Chapter 11 has a trade area or a complementary region focused upon it by connections of trade and limits of influence. Such regional capitals as Chicago, Atlanta, or Minneapolis exercise financial, administrative, wholesaling, and retailing centrality for the areas of the nation that they serve as organizational nodes.

Either the formal or the functional regional type is a recognition of significant differences in the characteristics of earth space. Interest therefore centers upon the cores of regions

431

where defined differences are best expressed and most easily recognized. Boundaries usually simply separate different regions and mark the outermost limit of the area over which the minimum essential criteria of regional character may be noted. For formal regions, the boundary may be abrupt and unmistakable, as the escarpment marking the margin of highlands, or the guarded border of a nation. It may also be a precisely defined division along a continuum, as an average warmest month temperature of 50°F or 60% of cultivated land in spring wheat. For functional regions, boundaries are drawn where the influences of two competing core regions are equal, opposite, and least distinct. Regional limits marked on the map usually recognize by line or zone the least, not the greatest, areas of interest.

Since regions are a product of perception-with-a-purpose, they vary in scale, type, and degree of generalization. They are, inevitably, hierarchically arranged. On a formal regional scale of progression, the Delmarva Peninsula of the eastern United States may be seen as part of the Atlantic Coastal Plain, which is in turn a portion of the eastern North American humid continental climatic region, a hierarchy that changes criteria of delimitation as the level and purpose of generalization alter (Figure 12.3). The central business district of Chicago is one land-use complex in the functional regional hierarchy that encompasses the variety of nodal influences of the city of Chicago and of the metropolitan region of which it is the core. Each recognized regional entity may stand alone and, simultaneously, exist as a subset of a larger, equally coherent, territorial unit.

The recognition of regions, the determination of their limits, the definition of their type and their hierarchical relationships are subordinate to the fact that regions are created as perceptual generalizations to serve a purpose. They focus our attention upon spatial uniformities and bring clarity to the seeming confusion of the infinitely variegated natural and cultural

432

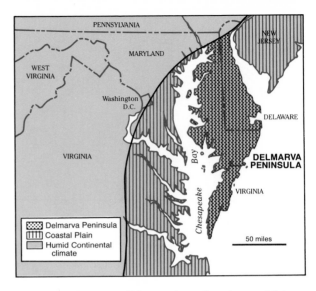

Fig. 12.3 One possible nesting of regions within a regional hierarchy defined by differing criteria. Each regional unit has internal coherence; the recognition of each adds comprehension of the nature of its subdivisions.

world around us. They provide the framework for the purposeful organization of spatial data. In the pages that follow, a variety of approaches to regional recognition will be explored. Each is based upon the content of one of the preceding chapters of this book, and each is meant to suggest the diversity of ways, determined by interest and purpose, in which regionalization and the regional concept may be employed.

Regions in the Earth-Science Tradition

The simplest of all regions to define, and generally the easiest to recognize, is the formal region based upon a single readily apparent component or characteristic. The island is land, not water, and its unmistakable boundary is naturally given where the one element passes to the other. The terminal moraine may mark the transition from the rich, black soils of recent formation to the particolored clays of earlier generation; the dense forest may break dramatically upon the glade or the open prairies. The nature of change is singular and apparent.

The physical geographer, although concerned with the whole array of earth sciences that elucidate the natural environment, deals at the outset with single factor formal regions. Many of the phenomena of concern to physical geography, of course, do not exist in static, clearly defined units; they require for regionalization the application of bounding criteria. A particular amount of received precipitation, the presence of certain primary soil characteristics, the dominance in nature of particular plant associations—all must be arbitrarily selected as regional limits, and all such limits are subject to change through time or by purpose of recognition.

LANDFORMS AS REGIONS

The landform region exists in a more sharply defined fashion than such transitional physical features as soil, climate, or vegetation for which boundary criteria must be selected and defended. The landform region arises, visible and apparently immutable, from nature itself, independent of human influence and unaffected by time on the human scale. Landforms constitute basic, naturally defined regions of physical geographic concern. The existence of major landform regions—mountains, lowlands, plateaus—is unquestioned in popular recognition or scientific definition; their influence on climates, vegetational patterns, even upon the primary economies of subsistence populations has been noted in earlier portions of this book. The following discussion of a distinct landform region, describing its constitu-

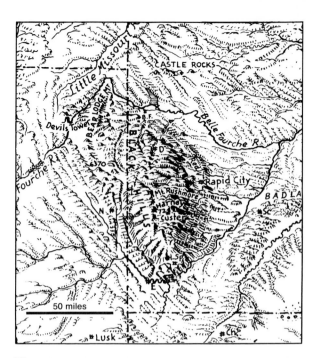

Fig. 12.4 The Black Hills physiographic province. (Reproduced by permission of Mrs. E. Raisz from *Landforms of the United States*, by Erwin Raisz, 6th revised edition, 1957.)

433

tion and its relationship to other physical and cultural phenomena, is adapted from a classic study by Wallace W. Atwood.

The Black Hills Province.* The Black Hills rise abruptly from the surrounding plains (Figure 12.4). The break in topography at the margin of this province is obvious to the most casual observer who visits that part of our country. Thus the boundaries of this area, based on contrasts in topography, are readily determined.

One who looks more deeply into the study of the natural environment may recognize that in the neighboring plains the rock formations lie

* From *The Physiographic Provinces of North America* by Wallace W. Atwood, © Copyright, 1940, by Ginn and Company (Xerox Corporation). Used with permission.

in a nearly horizontal position. They are sandstones, shales, conglomerates, and limestones. In the foothills those same sedimentary formations are bent upward and at places stand in a nearly vertical position. Precisely where the change in topography occurs, we find a notable change in the geologic structure and thus discover an explanation for the variation in relief.

The Black Hills are due to a distinct upwarping, or doming, of the crustal portion of the earth. Subsequent removal by stream erosion of the higher portions of that dome and the dissection of the core rocks have produced the present relief features. As erosion has proceeded, more and more of a complex series of ancient metamorphic rocks has been uncovered. Associated with the very old rocks of the core and, at places, with the sedimentary strata, there are a number of later intrusions which have cooled and formed solid rock. They have produced minor domes about the northern margin of the Black Hills.

434 With the elevation of this part of our country there came an increase in rainfall in the area, and with the increase in elevation and rainfall came contrasts in relief, in soils, and in vegetation.

As we pass from the neighboring plains, where the surface is monotonously level, and climb into the Black Hills area, we enter a landscape having great variety in the relief. In the foothill belt, at the southwest, south, and east, there are hogback ridges interrupted in places by *water gaps,* or gateways, that have been cut by streams radiating from the core of the Hills. Between the ridges there are roughly concentric valley lowlands. On the west side of the range, where the sedimentary mantle has not been removed, there is a plateau-like surface; hogback ridges are absent. Here erosion has not proceeded far enough to produce the landforms common to the east margin. In the heart of the range we find deep canyons, rugged intercanyon ridges, bold mountain forms, craggy knobs, and other picturesque features

(Figure 12.5). The range has passed through several periods of mountain growth and several stages, or cycles, of erosion.

The rainfall of the Black Hills area is somewhat greater than that of the brown, seared, semiarid plains regions, and evergreen trees survive among the hills. We leave a land of sagebrush and grasses to enter one of forests. The dark-colored evergreen trees suggested to early settlers the name Black Hills. As we enter the area, we pass from a land of cattle ranches and some seminomadic shepherds to a land where forestry, mining, general farming, and recreational activities give character to the life of the people. In color and form, in topography, climate, vegetation, and economic opportunities, the Black Hills stand out conspicuously as a distinct geographic unit.

DYNAMIC REGIONS IN WEATHER AND CLIMATE

The unmistakable clarity and durability of the Black Hills landform region and the precision with which its boundaries may be drawn are rarely echoed in other types of formal physical regions. Most of the natural environment, despite its appearance of permanence and certainty, is dynamic in nature. Vegetations, soils, and climates change through time by natural process or by the action of humans. Boundaries shift, perhaps abruptly, as witness the recent migration southward of the Sahara; the core characteristics of whole provinces change as marshes are drained or forests are replaced by cultivated fields.

That complex of physical conditions that we recognize locally and briefly as weather and summarize as climate displays particularly clearly the transitory nature of much of the natural environment that surrounds us. Yet even in the turbulent change of the atmosphere, discrete regional entities exist with definable boundaries and internally consistent

Fig. 12.5 The "Needles" of the Black Hills result from erosion along vertical cracks and crevices in granite. (Courtesy of South Dakota Division of Tourism.)

horizontal and vertical properties. "Air masses" and the consequences of their encounters constitute a major portion of contemporary weather analysis and prediction. Air masses further fulfill all the criteria of multi-factor formal regions, though their dynamic quality and their patterns of movement obviously mark them as being of a nature distinctly different from such static physical entities as landform regions, as the following extract from *Climatology and the World's Climates* by George R. Rumney makes clear.

Air Masses. * An air mass is a portion of the atmosphere having a uniform horizontal distribution of certain physical characteristics, especially of temperature and humidity. These

* Copyright George R. Rumney, 1968 (Macmillan, New York). Adapted from pp. 64–65 with permission of the author.

qualities are acquired when a mass of air stagnates or moves very slowly over a large and relatively unvaried surface of land or sea. Under these circumstances surface air gradually takes on properties of temperature and moisture approaching those of the underlying surface, and there then follows a steady, progressive transmission of properties to greater heights, resulting finally in a fairly clearly marked vertical transition of characteristics. Those parts of the earth where air masses acquire their distinguishing qualities are called source regions.

The height to which an air mass is modified depends upon the length of time it remains in its source region and also upon the difference between the initial properties of the air when it first arrived and those of the underlying surface. If, for exampe, an invading flow of air is cooler than the surface beneath as it comes to virtual rest over a source region, it is warmed from below, and convective currents are formed, rapidly bearing aloft new characteristics of temperature and moisture to considerable heights. If, on the other hand, it is warmer than the surface of the source region, cooling of its surface layers takes place, vertical thermal currents do not develop, and the air is modified only in its lower portions. The process of modification may be accomplished in just a few days of slow horizontal drift, although it often takes longer, sometimes several weeks. Radiation, convection, turbulence, and advection are the chief means of bringing it about.

The prerequisite conditions for these developments are very slowly migrating, outward spreading, and diverging air and a very extensive surface beneath that is fairly uniform in nature. Light winds and relatively high barometric pressure characteristically prevail. Hence, most masses form within the great semipermanent anticyclonic regions of the general circulation, where calms, light variable winds, and overall subsidence of the atmosphere are typical.

Four major types of source regions are recog-

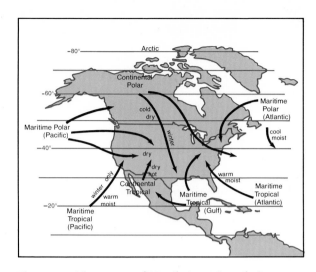

Fig. 12.6 Air masses of North America, their source regions, and their paths of movement. (After Haynes, U. S. Department of Commerce.)

nized: continental polar, maritime polar, continental tropical, and maritime tropical. Polar air masses are continental when they develop over land or ice surfaces in high latitudes; these are cold and dry. They are maritime when they form over the oceans in high latitudes. An air mass from these sources is cold and moist. Similarly, tropical air is continental when it originates along the Tropics of Cancer and Capricorn over northern Africa and northern Australia and is therefore warm and dry. It is maritime when it forms along the Tropics over the oceans, where it develops as a mass of warm, moist air. A single air mass usually covers thousands of square miles of the earth's surface when fully formed.

An air mass is recognizable chiefly because of the uniformity of its primary properties—temperature and humidity—and the vertical distribution of these. Secondary qualities, such as cloud types, precipitation, and visibility, are also taken into account. These qualities are retained for a remarkably long time, often for several weeks, after an air mass has traveled far

436

from its source region, and they are thus the means of distinguishing it from other masses of air.

The principal air masses of North America, their source regions, and seasonal movements are shown in Figure 12.6.

Air masses, intangible and ephemeral as they may appear, adhere to the requirements for recognition as special-type geographic regions. They have location—at their origin and on their paths of movement to ultimate dissipation. They may be assigned regional names indicative of their origin locations, their characteristics, or both. They have extent, a surface area over which their properties are present and dominant. And they have definable boundaries—called *fronts,* as we saw in Chapter 3—where different sets of temperature and humidity properties are encountered.

ECOSYSTEMS AS REGIONS

A traditional, simplistic definition of geography as "the study of the areal variation of the surface of the earth" suggested that the discipline centered on the classification of areas and on the subdivision of the earth into its constituent regional entities. Considerations of organization and function were secondary and even unnecessary to the purpose of regionalization: the definition of cores and the designation of boundaries of static assemblages of physical or cultural properties.

Newer research approaches stress the need for the study of spatial relationships from the standpoint of systems analysis, which emphasizes the organization, structure, and functional dynamics within an area and provides for the quantification of the linkages between the components of space. The ecosystem or biome, introduced in Chapter 4, provides a systems-analytic concept of great flexibility, permitting the study of form, structure, and the integration of physical, biological, and human phenomena in any desired spatial dimension.

The ecosystem—an interacting functional system composed of organisms and their effective environment—brings together in a single framework environment, humans, and the biological realm, permitting an analysis of the interrelationships between these components. Since that relationship is structured, structure rather than spatial uniformity attracts attention and leads to a new understanding of the region. Since ecosystems can be conceived of at different scales and levels of complexity, they can serve as modeled frameworks for the delimitation and study of regions of varying size. They can particularly serve as the vehicle for investigating the complex consequences of human impact upon the natural environment, the theme of Chapter 4.

Note how, in the following description drawn from an article by William J. Schneider, the ecosystem concept is employed in the recognition and analysis of regions and subregions of varying size, complexity, and nature, all from the point of view of the functional integration of content. Introduced, too, is the concept of *ecotone,* or zone of ecological stress, in this case induced by human pressures upon the natural system.

437

The Everglades. * The Everglades is a river. Like the Hudson or the Mississippi, it is a channel through which water drains from higher to lower ground as it moves to the sea. It extends in a broad, sweeping arc from the southern end of Lake Okeechobee in central Florida to the tidal estuaries of the Gulf Coast and Florida Bay. As much as 70 miles wide, but generally averaging 40, it is a large, shallow slough that weaves tortuously through acres of saw grass and past "islands" of trees, its waters, even in the wet season, rarely deeper than 2 feet. But, again like the Hudson or the Missis-

* "Water and the Everglades" by William J. Schneider. With permission from *Natural History,* November, 1966, pp. 32–40. Copyright American Museum of Natural History, 1966.

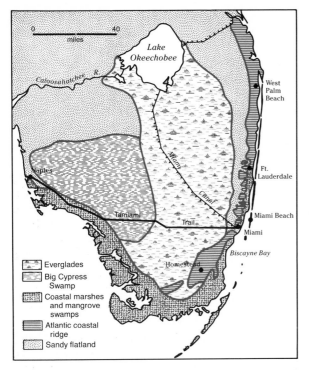

Fig. 12.7 The Everglades is part of a complex of eco-systems stretching southward in Florida from Lake Okeechobee to the sea. Drainage and water-control systems have altered its natural condition.

sippi, the Everglades bears the significant imprint of civilization.

Since the close of the Pleistocene, 10,000 years ago, the Everglades has been the natural drainage course for the periodically abundant overflow of Lake Okeechobee. As the lake filled during the wet summer months or as hurricane winds blew and literally scooped the water out of the lake basin, excess water spilled over the lake's southern rim. This overflow, together with rainfall collected en route, drained slowly southward between Big Cypress Swamp and the sandy flatlands to the west and the Atlantic coastal ridge to the east, sliding finally into the brackish water of the coastal marshes (Figure 12.7).

Water has always been the key factor in the life of the Everglades. Three-fourths of the annual average 55 inches (140 cm) of rainfall occurs in the wet season, June through October, when water levels rise to cover 90% of the land area of the Everglades. In normal dry seasons in the past, water covered no more than 10% of the land surface. Throughout much of the Everglades, and prior to recent engineering activities, this seasonal rain cycle caused fluctuations in water levels that averaged 3 feet. Both occasional severe flooding and prolonged drought accompanied by fire imposed periodic stress upon the ecosystem. It may be that randomly occurring ecologic trauma is vital to the character of the Everglades.

Three dominant biological communities—open water, saw grass, and woody vegetation—reflect small, but consistent, differences in the surface elevation of peat soils that cover the Everglades (Figure 12.8). The open-water areas occur at the lower soil elevations; inundated much of the year, they contain both sparse, scattered marsh grasses and a mat of algae. The saw-grass communities develop on a soil base only a few inches higher than that in the surrounding open glades. The soil base is thickest under the tree islands. The few inches' difference in soil depth apparently governs the species composition of these three communities.

Today, the Everglades is no longer precisely a natural river. Much of it has been altered by an extensive program of water management, including drainage, canalization, and the building of locks and dams. Large withdrawals of groundwater for municipal and industrial use have depleted the underlying aquifer and permitted the landward penetration of sea water through the aquifer and through the surface canals. Thousands of individual water-supply wells have been contaminated by encroaching saline water; large biotic changes have taken place in the former freshwater marshes south of Miami. Mangroves—indicators of salinity—have extended their habitat inland, and fires rage across areas that were formerly much wet-

Fig. 12.8   Open water, saw grasses, and tree islands constitute separate biomes of the Everglades. (Photo by Kit and Max Hunn. National Audubon Society Collection/Photo Researchers, Inc.)

ter. The ecotone—the zone of stress between dissimilar adjacent ecosystems—is altering as a consequence of these human-induced modifications of the Everglades ecosystem.

The organization, structure, and functional dynamics of the Everglades ecosystems are thus undergoing change. The structured relationships of its components—in nature affected and formed by stress—are being subjected to distortions by humans in ways not yet fully comprehended.

Regions in the Culture–Environment Tradition

The earth-science tradition of geography imposes certain distinctive parameters upon area analysis. However defined, the regions that may be drawn derive from the circumstances of nature and do not rest upon the vagaries of human action. The culture–environment tradition, however, introduces to regional geography the infinite variations of human occupation and organization of space. There is a corresponding multiplication of topical concepts of regional cores and of criteria for the recognition of regional boundaries.

Despite the differing orientation of emphasis and interest between the physical and the cultural geographer, one element of study is common to their concerns, that of process. The "becoming" of an ecosystem, of a cultural landscape, or of the pattern of exchanges in an economic system is an integral and implicit part of nearly all geographic study. Evidence of the past adduced as an aid to understanding of the present is inevitably basic to the genetic approach to geographical investigation, which recognizes that distributional patterns or qualities of regions mark a merely transitory stage in a continuing process of change.

HISTORICAL CULTURAL REGIONS

While the recognition of past formal or functional regions may be considered an implied element in most geographic study, it becomes a primary purpose for the historical geographer, who is concerned with the reconstruction of a sequence of change or the depiction of a past stage of reality. Regional recognition and delimitation, of course, increase in difficulty and decrease in precision as one moves backward in time and loses the advantage of contemporary observation and documentation. Studies of tree rings, lake sediments, or pollen may help in the reconstruction of past climatic or vegetation regions, but absolute certainty in drawing boundaries of or in assigning dates to spatial change is impossible. Similarly, the consideration of past cultural landscapes and patterns is made easier by the discovery of datable artifacts, by reference to historical records, or by lines of evidence of physical environmental change. Again, however, boundary delimitation lacks the certainty made possible by observation of the present landscape or by use of contemporary data. This lack of precision is nowhere more evident than in attempts to isolate regions of the origins of agriculture or to trace the paths of diffusion of the concepts of cultivation and of animal domestication.

Near Eastern Agricultural Origins and Dispersals. The most important cultural advance of humankind occurred thousands of years ago, without benefit of documentation and with little currently discernible impact upon the landscape. After undetermined millennia as hunter-gatherers, humans settled in villages and began to domesticate animals and cultivate crops. The how, why, and when of the agricultural transformation—and, for the cultural geographer, the where—remain incompletely answered despite a vast array of new investigative techniques now being brought to bear on this most basic of cultural questions.

For the Near East, the general location and sequence of the food producing revolution now appear clear. The Mesopotamian culture hearth, discussed in Chapter 5, existed as an early irrigated-agriculture civilization that, if not itself the developer of its basic grains and domesticated animals, must surely have been very near in space to their actual points of origin. This assumption was given strength by early discoveries in Near Eastern caves of evidence of habitation dating from early in the Ice Age. Among that evidence, though not among the earliest artifacts, were tool assemblages that included primitive sickles. Elsewhere in the same area, traces of flourishing agricultural villages were unearthed by archeologists, although such settlements were datably later habitations than were the caves. To resolve the question of agricultural origins, given these sparse locational and temporal data, is an ongoing task involving a vast array of talent and techniques from many sciences. Now, the outlines of the origins and dispersals of a cultural revolution are emerging and are lending themselves to the regional delimitations necessary to the historical cultural geographer.

The Near Eastern region of agricultural origin was the Zagros Mountains of northern Iraq, forming part of the frame of the more famous, but later, "Cradle of Civilization" of Mesopotamia. Investigation in these grassy uplands, at elevations of from 1250 to 3000 feet, revealed certain basic facts localizing the cultural revolution there. First, it was proved that plants and animals capable of domestication had lived in the area long before humans undertook to tame them. Wild cereal grass pollens 16,000 years old and bones of wild sheep and goats more than 40,000 years old were identified. Second, both pollen analysis and the shells of land snails still common to the area showed that no climatic revolution had occurred in the region as a necessary precondition of the agricultural way of life. The Zagros Mountains had been deforested since the Ice Age.

With these facts as starting points, researchers turned to identifying early gathered wild grains and their conversion to cultivated forms, and to tracing the sequence of animal domestication. Imprints of grain retained in pottery vessels demonstrated what had to

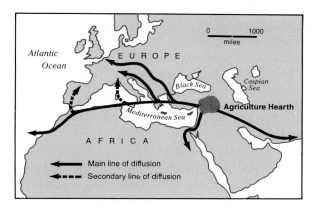

Fig. 12.9 From a known core region whose boundaries are uncertain, the concept of the domestication of grains and of sedentary agriculture spread from the Near East eastward to South Asia and westward through North Africa and Europe.

occur in an area of agricultural origin: primitive wild cereal varieties coexisted with more advanced cultivated forms, showing a conscious selection and adaptation of desirable plant characteristics. Discoveries of caches of purely wild grain at open village sites rendered it certain that humans were sedentary gatherers of grain—barley and wheat—before they became skilled sowers and farmers. Finally, the possibility of mutation of wild into domesticated grains was demonstrated when laboratory X-ray exposure of primitive two-row barley converted it into the six-row form that was a major food crop of Mesopotamia. Further research seems to prove that the sheep was the earliest domesticated animal and that its taming preceded that of cereals.

The sequence in the food-producing revolution began to emerge. At about 10,000 B.C., animal domestication was undertaken on the grasslands of northern Syria and Iraq. Around 9000 B.C., there appeared villages of seed collectors and hunters, followed, by 7500 B.C., by village farming communities. From that origin area came, inevitably, dispersion of the concept and techniques of the agricultural

441

transformation. Early farming villages in Greece and the Balkans dating from the sixth millennium provide evidence of the movement of the new way of life through Anatolia to Europe, whence it spread up the Danube, reaching the Atlantic some 3000 years later. Main and subsidiary lines of dispersion are generalized on Figure 12.9.

The cultural geographer, drawing upon lines of evidence developed by archeology, pollen analysis, genetic research, and other scientific fields and techniques, is now able with more accuracy than ever before to draw boundaries and trace flows related to the cultural transition from primitive to advanced society.

LANGUAGE AS REGION

The great culture realms of the world outlined in Chapter 6 constitute historical and contemporary summarizations of significant composites of peoples. They are not rigorously identified with nation, language, religion, or technology, but with all these and more in varying amalgams. Culture realms are therefore multifactor regions that obscure more than they illumine the distinctions between peoples that are so fundamental to the human mosaic of the earth. Basic to cultural geography is the recognition of small regions of single-factor homogeneity that give character to their areas of occurrence and that collectively provide a needed balance to the sweeping generalizations of the culture realm.

Language provides an example of such small-area variation, one partially explored in Chapter 6. The language families shown on Figure 6.11 conceal the identity of and the distinctions among the divergent official tongues of nation-states; these in turn ignore or submerge the language forms of minority populations,who may base their own sense of proud identity upon their regional linguistic separateness. In scale and recognition even below these ethnically identified regional languages

are those local speech variants frequently denied status as identifiable languages and cited as proof of the ignorance and the cultural deprivation of their speakers. Yet such a limited-area, limited-population tongue contains all the elements of the classic culturally based region; its area is defined, its boundaries easily drawn; it represents homogeneity and majority behavior among its members; and it summarizes, by a single culture trait, a collection of areally distinctive outlooks and orientations.

Gullah as Language. Isolation is a key element in the retention or the creation of distinctive and even externally unintelligible languages. The isolation of the ancestors of the some quarter-million present-day speakers of Gullah—themselves called Gullahs—was nearly complete. Held by the hundreds as slaves on the offshore islands and in the nearly equally remote low country along the southeastern United States coast from South Carolina to the Florida border (Figure 12.10), the blacks retained the speech patterns of the African languages—Ewe, Fanti, Bambara, Twi, Wolof, Ibo, Malinke, Yoruba, Efik—native to the slave groups. Forced to use English words for minimal communication with their white overseers, but modifying, distorting, and interjecting African-based substitute words into that unfamiliar language, the Gullahs kept intonations and word and idea order in their spoken common speech that made it unintelligible to white masters or to more completely integrated mainland blacks. Because the language was not understood, its speakers were considered ignorant, unable to master the niceties of English. Because ignorance was ascribed to the Gullahs, they learned to be ashamed of themselves, of their culture, and of their tongue, which even they themselves did not recognize as a highly structured and sophisticated separate language.

In common with many linguistic minorities, the Gullahs are losing their former sense of inferiority and gaining pride in their cultural

442

Fig. 12.10 Gullah speakers are concentrated on the Sea Islands and the coastal mainland of South Carolina and Georgia. The isolation that promoted their linguistic distinction is now being eroded.

heritage and in the distinctive tongue that represents it. Out of economic necessity, standard English is being taught to their schoolchildren, but an increasing scholarly and popular interest in the structure of their language and in the nature of their culture has caused Gullah to be rendered as a written language studied as a second language, and translated into English.

In both the written and the spoken versions, Gullah betrays its African syntax patterns, particularly in its employment of terminal locator words: "Where you goin' at?" The same African origins are revealed by the absence in English translation of distinctive tenses: "I be tired" conveys the concept that "I have been tired for a period of time." Though tenses exist in the African root languages, they are noted more by inflection than by special words and structures.

"He εn gut no morratater fer mak no pie wid" may be poor English, but it is good Gullah. Its translation—"He has no more sweet potatoes for making pie"—renders it intelligible to

ears attuned to English but loses the musical lilt of original speech and, more importantly, obscures the cultural identity of the speaker, a member of a regionally compact group of distinctive Americans whose territorial extent is clearly marked by its linguistic dominance.

POLITICAL REGIONS

The most rigorously defined formal cultural region is the nation-state. Its boundaries are presumably carefully surveyed and are, perhaps, marked by fences and guard posts. There is no question of an arbitrarily divided transition zone or of diminution toward the periphery of the basic quality of the regional core. The rigidity of a nation's boundaries, its unmistakable placement in space, and the trappings—flag, anthem, government, army—that are uniquely its own give to the nation-state an appearance of permanence and immutability not common in other, more fluid, cultural regions. But its stability is often more imagined than real. Political boundaries are not necessarily permanent. They are subject to change, sometimes violent change, as a result of internal and external pressures. The Indian subcontinent illustrates the point.

Political Regions in the Indian Subcontinent. The history of the subcontinent since about 400 B.C. has been one of the alternating creation and dissolution of empires, of the extension of central control based upon the Ganges Basin, and of resistance to that centralization by the marginal territories of the peninsula. British India, created largely unintentionally by 1858, was only the last, though perhaps the most successful, attempt to bring under unified control the vast territory of incredibly complex and often implacably opposed racial, religious, and linguistic groupings.

A common desire for independence and freedom from British rule united the subcontinent's disparate populations at the end of

443

World War II. That common desire, however, was countered by the mutual religious antipathies felt by Muslims and Hindus, each dominant in separate regions of the colony and each unwilling to be affiliated with or subordinated to the other. When the British surrendered control of the subcontinent in 1947, they recognized these apparently irreconcilable religious differences and partitioned the subcontinent into the second and seventh most populous nations on earth. The independent nation of India was created out of the largely Hindu areas constituting the bulk of the former colony. Separate sovereignty was granted to most of the Muslim-majority area under the name of Pakistan. Even so, the partition left boundaries, notably in the Vale of Kashmir, dangerously undefined or in dispute.

An estimated 1 million persons died in the religious riots that accompanied the partition decision. In perhaps the largest short-term mass migration in history, some 10 million Hindus moved from Pakistan to India, and 7.5 million Muslims left India for Pakistan, "The Land of the Pure."

Unfortunately, the purity resided only in common religious belief, not in spatial coherence or in shared language, ethnicity, customs, food, or economy. During its 23 years of existence as originally conceived, Pakistan was a sorely divided nation. The partition decision created an eastern and a western component separated by more than 1000 miles of foreign territory and united only by a common belief in Mohammed (Figure 12.11). West Pakistan, as large as Texas and Oklahoma combined, held 55 million largely light-skinned Punjabis with Urdu language and strong Middle Eastern cultural ties. Some 70 million Bengali-speakers, making up East Pakistan, were crammed into an Iowa-sized portion of the delta of the Ganges and Brahmaputra rivers. The western segment of the nation was part of the semiarid world of western Asia; the eastern portion of Pakistan was joined to humid, rice-producing Southeast Asia.

Beyond the affinity of religion, little else united the awkwardly separated nation. East Pakistan felt itself exploited by a dominating western minority that sought to impose its lan-

444

Fig. 12.11 The sequence of political change on the Indian subcontinent. British India was transformed in 1947 to the nations of India and Pakistan, the latter a Muslim state with a western and an eastern component. In 1970 Pakistan itself was torn by civil war based on ethnic and political contrasts, and the eastern segment became the new nation of Bangladesh.

guage and its economic development, administrative objectives, and military control. Rightly
or wrongly, East Pakistanis saw themselves as
aggrieved and abused; they complained of a per
capita income level far below that of their
western compatriots, claimed discrimination
in the allocation of investment capital, found
disparities in the pricing of imported foods, and
asserted that their exports of raw materials—
particularly jute—were supporting a national
economy in which they did not share proportionally. They argued that their demands for
regional autonomy, voiced since nationhood,
had been denied.

When, in November of 1970, East Pakistan
was struck by a cyclone and tidal wave that
took an estimated 500,000 lives, the limit of
eastern patience was reached. Resentful over
what they saw as a totally inadequate West
Pakistani effort of aid in the natural disaster
that had befallen them, the East Pakistanis
were further incensed by the refusal of the
central government to convene on schedule a
national assembly to which they had won an
absolute majority of delegates. Civil war resulted, and the separate nation-state of Bangladesh was created. The sequence of political
change in the subcontinent is traced in Figure
12.11. That further boundary change through
war or forced annexation will occur seems
implicit in the structure of divergent national
claims and cultural affiliations and diversities
of the subcontinent.

The nation-state so ingrained in our consciousness and so firmly defined, as displayed
on the map inside the front cover, is both a recent and an ephemeral creation of the cul-
tural-regional landscape. It rests upon a claim,
more or less effectively enforced, of a monopoly of power and allegiance resident in a government and superior to the communal, linguistic, ethnic, or religious affiliations that
preceded it or that claim loyalties overriding it.
As the violent recent history of the Indian subcontinent demonstrates, nationalism may be

sought, but its maintenance is not assured by
the initiating motivations.

MENTAL REGIONS

The regional entities so far employed as examples, and the methods of regionalization by
which they were derived, have a concrete reality. They are formal or functional regions
of specified, measurable content; they have
boundaries drawn by some objective criterion
of change or alteration of content; they have
location upon an accurately measured global
grid.

Individuals and whole cultures may operate—and successfully—with a much less formalized and less precise picture of the nature
of the world and of the structure of its parts.
The mental maps discussed in Chapter 8 represent personal views of regions and regionalization. The private world views they embody are,
as we also saw, colored by the culture of which
their holders are individual members. The
T–O map shown in Chapter 1 is a shared world
view of Christian Europe in early post-Roman
times. It provided a rational, mapped regionalization based upon observation and church doctrine. That it depicts imaginary regions—as an
unknowable Paradise in the East—made it no
less a valid and operational regional delinea-
tion. Primitive societies, particularly, have distinctive world views by which they categorize
what is familiar and satisfactorily account for
what is not. The Yurok Indians of the Klamath
River area of northern California were no exception. Their geographic concepts were reported by T. T. Waterman from whose paper,
"Yurok Geography," the following summary is
drawn.

The Yurok World View.* The Yurok imagines

445

* *University of California Publications in American
Archaeology and Ethnology*, Vol. 16, No. 5 (1920). Selection from pp. 189–193 (with permission).

himself to be living on a flat extent of land-scape, which is roughly circular and surrounded by ocean. By going far enough up the river, it is believed, "you come to salt water again." In other words, the Klamath River is considered, in a sense, to bisect the world. This whole earth-mass, with its forests and mountains, its rivers and sea cliffs, is regarded as slowly rising and falling, with a gigantic but imperceptible rhythm, on the heaving, primeval flood. The vast size of the "earth" causes you not to notice this quiet heaving and settling. This earth, therefore, is not merely surrounded by the ocean but floats upon it. At about the central point of this "world" lies a place which the Yurok call qe'nek, on the southern bank of the Klamath, a few miles below the point where the Trinity River comes in from the south. In the Indian concept, this point seems to be accepted as the center of the world.

446   At this locality also the sky was made. Above the solid sky there is a sky-country, wo'noiyik, about the topography of which the Yurok's ideas are almost as definite as are his ideas of southern Mendocino county, for instance. Downstream from qe'nek, at a place called qe'nek-pul ("qe'nek-downstream"), is an invisible ladder leading up to the sky-country. The ladder is still thought to be there, though no one to my knowledge has been up it recently. The sky-vault is a very definite item in the Yurok's cosmic scheme. The structure consisting of the sky dome and the flat expanse of landscape and waters that it encloses is known to the Yurok as ki-we'-sona (literally "that which exists"). This sky, then, together with its flooring of landscape, constitutes "our world." I used to be puzzled at the Yurok confusing earth and sky, telling me, for example, that a certain gigantic redwood tree "held up the world." Their ideas are of course perfectly logical, for the sky is as much a part of the "world" in their sense as the ground is.

The Yurok believe that passing under the sky

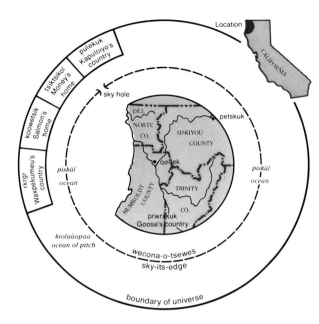

Fig. 12.12 The world view of the Yuroks as pieced together by T. T. Waterman during his anthropological study of the tribe. Qe'nek, in the center of the diagram, marks the center of the world in Indian belief. (Redrawn with permission from T. T. Waterman, "Yurok Geography," *University of California Publications in American Archaeology and Ethnology,* **vol. 16, 1920.)**

edge and voyaging still outward you come again to solid land. This is not our world, and mortals ordinarily do not go there; but it is good, solid land. What are breakers over here are just little ripples over there. Yonder lie several regions. To the north (in our sense) lies pu'lekuk, downstream at the north end of creation. South of pu'lekuk lies tsi'k-tsik-ol ("money lives") where the dentalium-shell, medium of exchange, has its mythical abode. Again, to the south there is a place called kowe'tsik, the mythical home of the salmon, where also all have a "house." About due west of the mouth of the Klamath lies rkrgr', where lives the culture-hero wo'xpa-ku-mä ("across-the-ocean that widower").

Still to the south of rkrgr' there lies a broad

sea, kiolaaopa'a, which is half pitch—an Algonkian myth idea, by the way. All of these solid lands just mentioned lie on the margin, the absolute rim of things. Beyond them the Yurok does not go even in imagination. In the opposite direction, he names a place pe'tskuk ("up-river-at)", which is the upper "end" of the river but still in this world. He does not seem to concern himself much with the topography there.

The Yurok's conception of the world he lives in may be summed up in the accompanying diagram (Figure 12.12).

Regions in the Location Tradition

While location, as we have seen, is a primary attribute of all regions, regionalization in the location tradition of geography implies far more than a named delimitation of earth space. The central concern is with the distribution of human activities and of the resources upon which those activities are based.

In this sense, world regionalization of agriculture and of the soils and climates with which it is related is within the location tradition; for practical and accepted reasons, the underlying physical patterns have been included under the earth-science tradition. But the point is made: the location tradition of geography emphasizes the "doing" in human affairs, and "doing" is not an abstract thing but an interrelated mosaic of life and the resources upon which life is dependent.

The location tradition, therefore, encourages the recognition and the definition of a far wider array of regional types than do the earth-science or culture–environment traditions. Any single pattern of economic activity or of resource distribution invites the recognition of definable formal regions. The interchange of

commodities, the control of urban market areas, the flows of capital, or the collection and distribution activities of ports are just a few examples of the infinite number of analytically useful functional regions that one may recognize.

ECONOMIC REGIONS

Economic regionalization is among the most frequent, familiar, and useful employments of the regional method. By it, the geographer is enabled to make spatially meaningful isolations of activities and resources, making them the objects of the special investigation, recognition of cores and boundaries, and understanding of interrelationships and flows that the complexities of the contemporary world demand.

The economic region, examples of which were explored in Chapter 9, should be seen as potentially more than a device for recording what *is* in either a formal or a functional sense. It has increasingly become a device for examining what might or should be. The concept of the, primarily, economic region as a tool for planning and a framework—in the location tradition—for the manipulation of the peoples, resources, and economic structure of a composite region first took root in the United States during the Depression years of the 1930s.

The key element in the planning region is the public recognition of a major territorial unit in which economic change or decline is seen as the remediable root-cause of a galaxy of such interrelated problems as regional outmigration, isolation, cultural deprivation, underdevelopment, and poverty.

Appalachia. Until the early 1960s, "Appalachia" was for most a loose reference to the complex physiographic province of the eastern

447

Fig. 12.13 (opposite) **A low-income farm area in Perry County, Kentucky. In the 1950s and 1960s, large numbers of people left this area for the industrial cities to the north. (U.S.D.A. photo)**

United States associated with the Appalachian mountain chain. If thought of at all, it was apt to be visualized as rural, isolated, and tree-covered; as an area of coal mining, hillbillies, and folksongs (Figure 12.13).

During the 1950s, however, the economic stagnation and the functional decline of the area became increasingly noticeable in the national context of economic growth, rising personal incomes, and growing concern with the elimination of the poverty and the deprivation of every group of citizens. Less dramatically but just as decisively as the Dust Bowl or the Tennessee Valley of an earlier era, Appalachia became simultaneously a popularly recognized economic and cultural region and a governmentally determined planning region.

Evidences of poverty, underdevelopment, and social crisis were obvious to a nation committed to recognize and eradicate such conditions within its own borders. By 1960, per capita income within Appalachia was $1400 when the national average was $1900. In the decade of the 1950s, mine employment fell 60% and farm jobs declined 52%; the rest of the country lost only 1% of mining jobs and 35% of agricultural employment. Rail employment fell with the drop in coal mining. Massive outmigration occurred among the young adults, with such cities as Chicago, Detroit, Dayton, Cleveland, and Gary the targets. Even with these departures, unemployment among those who remained in Appalachia averaged 50% higher than the national rate. Because of the departures, the remnant population—only 47% of whom lived, in 1960, in or near cities, as against 70% for the entire United States— was distorted in age structure. The very young and the old were disproportionally represented; the productive working-age groups were, at least temporarily, emigrants.

When these and other socioeconomic indicators were plotted by counties and by state economic areas, an elongated but regionally coherent and clearly bounded Appalachia as newly understood was revealed by maps (Figure 12.14). It extended through 12 states from Alabama to New York, covered some 182,000 square miles, and contained 17 million people, 93% of them white.

By 1963, awareness of the problems of the area at federal and state levels passed beyond recognition of a multifactor economic region to the establishment of a planning region. A joint federal–state Appalachian Regional Commission was created to develop a program designed to meet the perceived needs of the entire area. The approach chosen was one of limited investment in a restricted number of highly localized developments, with the expectation that these would spark economic growth supported by private funds.

In outline, the plan was (1) to ignore those areas of poverty and unemployment that were in isolated, inaccessible "hollows" throughout the region; (2) to designate "growth centers" where development potential was greatest and concentrate all spending for economic expansion there; the regional growth potential in targeted expenditure was deemed sufficient to overcome charges of aiding the prosperous and depriving the poor; and (3) to create a new network of roads so that the isolated jobless could commute to the new jobs expected to form in and near the favored growth centers; road construction would also, of course, open inaccessible areas to tourism and strengthen the economic base of the entire planning region.

In the years since the Appalachian Regional Commission was established, and in ways not anticipated, the economic prospects of Appalachia have altered. National energy crises have revived mining and transportation employment; new industrial jobs have multiplied as

449

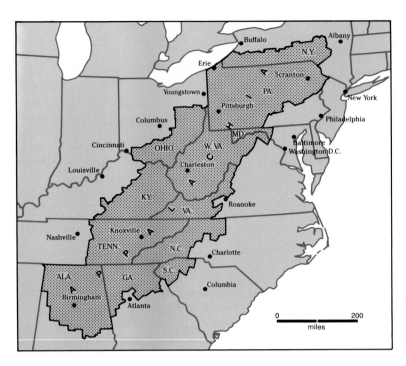

Fig. 12.14 "Appalachia" as defined by the Appalachian Regional Commission. The boundary criteria were social and economic conditions, not topography.

450

manufacturing has relocated to, or has been newly developed in, the Appalachia portion of the sun belt. New employment opportunities have exceeded local labor pools, and out-migrants have returned home from cities outside the region, bringing again a more balanced population pyramid.

REGIONS OF NATURAL RESOURCES

The unevenly distributed resources upon which humankind depends for existence are logical topics of concern within the integrative location tradition of geography. Resource regions are mapped, material qualities and quantities are discussed, areal relationships to industrial concentrations, flows to markets, and the exploitative impacts upon alternate uses of areas of occurrences are typical interests in resource geography and in the definition of resource regions.

Those regions, however, are usually treated as if they were expressions of observable surface phenomena; as if, somehow, reality was confounded and an oilfield were as exposed and two-dimensional as a soil region or a manufacturing belt. What is ignored is that most mineral resources are three-dimensional regions beneath the ground. In addition to the characteristics of an area that may form the basis for regional delimitations and descriptions of surface phenomena, regions beneath the surface add their own particularities to the problem of regional definition. They have, for example, upper and lower boundaries in addition to the circumferential bounds of surface features; they may have an internal topography of concern divorced from the visible landscape. The relationship of the underground region to surface natural features and human developments invites comment and analysis, but subsurface relationships, though hidden, may be even more basic for these specialized, but real, re-

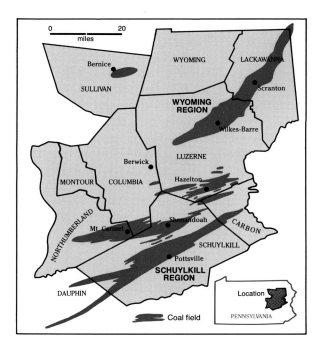

Fig. 12.15 The anthracite regions of eastern Pennsylvania are sharply defined by the geologic events that created them.

The Schuylkill Anthracite Region. Nothing in the wild surface terrain of the anthracite country suggested the existence of an equally rugged subterranean topography of coal beds and interstratified rock, slate, and fire clays forming a total vertical depth of 3000 feet at greatest development. Yet the creation of the surface landscape was an essential determinant of the areal extent of the Schuylkill district, of the contortions of its bedding, and of the nature of its coal content. A county history of the area reports, "The physical features of the anthracite country are wild. Its area exhibits an extraordinary series of parallel ridges and deep valleys, like long, rolling lines of surf which break upon a flat shore." Both the surface and the subsurface topographies reflect the strong folding of strata after the coal seams were deposited; the anthracite (hard) coal resulted from metamorphic carbonization of the original bituminous beds. Subsequent river and glacial erosion removed as much as 95% of the original anthracite deposits and gave to those that remained discontinuous existence in sharply bounded fields like the Schuylkill (Figure 12.15), a discrete areal entity of 181 square miles.

The irregular topography of the underground Schuylkill region means that the interbedded coal seams, the most steeply inclined of all the anthracite regions (Figure 12.16), outcrop visibly at the surface on hillsides and along stream

451

gions—for example, mineral distribution and exploitability in relation to its enclosing rock or to groundwater amounts and movement. An illustration drawn from the Schuylkill field of the anthracite region of northeastern Pennsylvania may help illustrate the nature of regions beneath the surface.

Fig. 12.16 The deep folding of the Schuylkill coal seams made them costly to exploit. The Mammoth Seam runs as deep as 1500–2000 feet below the surface.

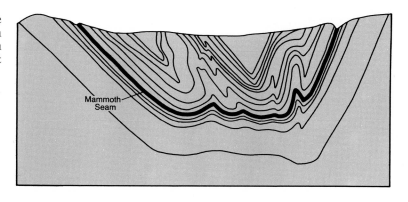

valleys. The outcrops made the presence of coal known as early as 1770, but not until 1795 did Schuylkill anthracite find its first use by local blacksmiths. Reviled as "stone coal" or "black stone" that would not ignite, anthracite found no ready commercial market, although it was used in wire and rolling mills located along the Schuylkill River before 1815 and to generate steam in the same area by 1830.

The resources of the subterranean Schuylkill region affected human patterns of surface regions only after the Schuylkill Navigation Canal was completed in 1825 (Figure 12.17), providing a passage to rapidly expanding exter-

nal markets for the fuel and for the output of industry newly located atop the region. Growing demand induced a boom in coal exploration, an exhaustion of the easily available outcrop coal, and the beginning of the more arduous and dangerous underground mining.

The early methods of mining were simple: merely quarrying the coal from exposed outcrops, usually driving on a slight incline to permit natural drainage. Deep shafts were unnecessary and, indeed, not thought of, since the presence of anthracite at depth was not suspected. Later, when it was no longer possible to secure coal from a given outcrop, a small pit

Fig. 12.17 The Schuylkill Navigation Canal was the outlet to market for the area's anthracite coal resource after the canal's completion in 1825. (Photo by J. A. Jakle.)

was sunk to a depth of 30–40 feet; when the coal and the water that accumulated in the pit could no longer safely be brought to the surface by windlass, the pit was abandoned and a new one was started. Shaft mining, in which a vertical opening from the surface provides penetration to one or several coal beds, eventually became a necessity; with it came awareness of the complex interrelationships between seam thickness, the nature of interstratified rock and clays, the presence of gases, and the movement of subsurface water.

The subterranean Schuylkill region has a three-dimensional pattern of utilization. The configuration and the variable thickness of the seams demands concentration of mining activities. Minable coal is not uniformly available along any possible vertical or horizontal cross section because of the interstratification and the extreme folding of the beds. Mining is concentrated further by the location of shafts and the construction of passages, in their turn determined by both patterns of ownership and thickness of seam. In general, no seam less than 2 feet is worked, and the absolute thickness—50 feet—is found only in the Mammoth seam of the Schuylkill region.

Friable interstratified rock increases the danger of coal extraction and raises the costs associated with cave-in prevention. Although the Schuylkill mines are not gassy, the possibility of gas release from the collapse of coal pillars left as mine supports makes necessary systems of ventilation even more elaborate than those minimally required to provide adequate air to miners. Water is ever present in the anthracite workings, and constant pumping or draining is necessary for mine operation. The collapse of strata underlying a river may result in sudden disastrous flooding.

The Schuylkill subterranean anthracite region presents a pattern of complexity in distribution of physical and cultural features and of interrelationships between phenomena every bit as great and inviting of geographical analysis as any purely surficial region.

URBAN REGIONS

Urban geography represents a climax stage in the location tradition of geography. Modern integrated, interdependent society on a world basis is urban-centered. Cities are the indispensable functional focuses of production, exchange, and administration. They exist individually as essential elements in interlocked hierarchical systems of cities; internally, they display complex but repetitive spatial patterns of land uses and functions.

Because of the multifaceted character of urbanism, cities are particularly susceptible to the application of regional methodology. They are themselves, of course, formal regions; in the aggregate, their distributions give substance to formal regions of urban concentration. Cities are the cores of functional regions of varying types and hierarchical orders. Their internal heterogeneity of functional, land-use, and socioeconomic patterning invite the regional approach to analysis. The employment of that approach in both its formal and its functional modes is classically displayed by Jean Gottmann, who, in examining the data and the landscapes of the eastern United States at mid-century, recognized and analyzed Megalopolis. The following is taken from his study.

Megalopolis.* The northeastern seaboard of the United States is today the site of a remarkable development—an almost continuous stretch of urban and suburban areas from southern New Hampshire to northern Virginia and from the Atlantic shore to the Appalachian foothills (Figure 12.18). The processes of urbanization, rooted deep in the American past, have worked steadily here, endowing the region with unique ways of life and of land use. No other section of the United States has such a large concentration of population, with such a

453

* Jean Gottmann, *Megalopolis: The Urbanized Northeastern Seaboard of the United States.* Copyright © 1961 by the Twentieth Century Fund, Inc. Reprinted by permission.

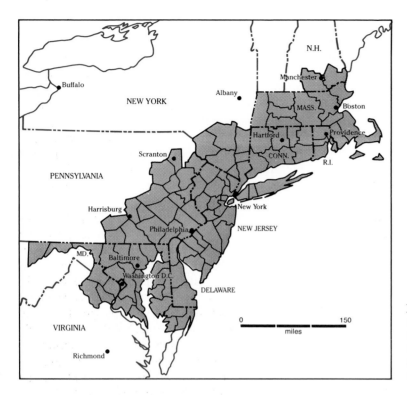

Fig. 12.18 Megalopolis in 1960. The region was composed of counties that, by United States census definition, were "urban" in population characteristics. Much of the area is distinctly "rural" in land use.

high average density, spread over such a large area. And no other section has a comparable role within the nation or a comparable importance in the world. Here has been developed a kind of supremacy, in politics, in economics, and possibly even in cultural activities, seldom before attained by an area of this size.

Great, then, is the importance of this section of the United States and of the processes now at work within it. And yet it is difficult to single this area out from surrounding areas, for its limits cut across established historical divisions, such as New England and the Middle Atlantic states, and across political entities, since it includes some states entirely and others only partially. A special name is needed, therefore, to identify this special geographical area.

This particular type of region is new, but it is the result of age-old processes, such as the growth of cities, the division of labor within a civilized society, the development of world resources. The name applied to it should, therefore, be new as a place name but old as a symbol of the long tradition of human aspirations and endeavor underlying the situations and problems now found here. Hence the choice of the term *Megalopolis*, used in this study.

As one follows the main highways or railroads between Boston and Washington, D.C., one hardly loses sight of built-up areas, tightly woven residential communities, or powerful concentrations of manufacturing plants. Flying this same route one discovers, on the other hand, that behind the ribbons of densely occupied land along the principal arteries of traffic, and in between the clusters of suburbs around the old urban centers, there still remain large areas covered with woods and brush alternating with some carefully cultivated patches of farmland (Figure 12.19). These green spaces, however, when inspected at closer range, appear stuffed with a loose but immense scattering of buildings, most of them residential but some of industrial character. That is, many of these sections that look rural actually function largely as suburbs in the orbit of some city's downtown. Even the farms, which occupy the larger tilled patches, are seldom worked by people whose only occupation and income are properly agricultural.

Thus the old distinctions between rural and urban do not apply here any more. Even a quick look at the vast area of Megalopolis reveals a revolution in land use. Most of the people living in the so-called rural areas, and still classified as "rural population" by recent censuses, have very little, if anything, to do with agricul-

Fig. 12.19 The "Pine Barrens" of New Jersey is still a largely undisturbed natural enclave in the heart of Megalopolis. Its preservation is a subject of dispute between environmentalists and developers. (Photo by Patricia J. Baxter, New Jersey Conservation Foundation.)

ture. In terms of their interests and work they are what used to be classified as "city folks," but their way of life and the landscapes around their residences do not fit the old meaning of urban.

In this area, then, we must abandon the idea of the city as a tightly settled and organized unit in which people, activities, and riches are crowded into a very small area clearly separated from its nonurban surroundings. Every city in this region spreads out far and wide around its original nucleus; it grows amidst an irregularly colloidal mixture of rural and suburban landscapes; it melts on broad fronts with other mixtures, of somewhat similar though different texture, belonging to the suburban neighborhoods of other cities.

Thus an almost continuous system of deeply interwoven urban and suburban areas, with a total population of about 37 million people in 1960, has been erected along the Northeastern Atlantic seaboard. It straddles state boundaries, stretches across wide estuaries and bays, and encompasses many regional differences. In fact, the landscapes of Megalopolis offer such variety that the average observer may well doubt the unity of the region. And it may seem to him that the main urban nuclei of the seaboard are little related to one another. Six of its great cities would be great individual metropolises in their own right if they were located elsewhere. This region indeed reminds one of

456

Aristotle's saying that cities such as Babylon had "the compass of a nation rather than a city."

Conclusion

The region is a mental construct, a created entity whose sole function is the purposeful organization of spatial data. The scheme of that organization, the selection of data to be analyzed, and the region resulting from these decisions are reflections of the intellectual problem posed.

This chapter has not attempted to explore all the philosophical and methodological aspects of regionalism and of the regional method. It has endeavored, by example, only to document its basic theme: the geographer's regions are arbitrary but deliberately conceived devices for the isolation in space of phenomena, patterns, interrelations, and flows that invite geographic analysis. In this sense, all geographers are regional geographers, and a sharp distinction between regional and systematic geography is not useful.

As the region represents the application and the culmination of the geographer's craft, so these examples and this chapter complete our survey of the four traditions of geography.

For Review

1. What do geographers seek to achieve when they recognize or define regions? On what basis are regional boundaries drawn? Are regions concrete entities whose dimensions and characteristics are agreed upon by all who study the same general segment of earth space? Ask three fellow students not participants in this course for their definition of the "South." If their answers differed, what implicit or explicit criteria of regional delimitation were they employing?

2. What are the spatial elements or the identifying qualities shared by all regions?

3. What is the identifying characteristic of a formal region? How are its boundaries determined? Name three different examples of formal regions drawn from any earlier chapters of this book. How was each defined and what was the purpose of its recognition?

4. How are functional regions defined? What is the nature of their bounding criteria? Give three or four examples of functional regions that were defined earlier in the text.

5. The ecosystem was suggested as a viable device for regional delimitation. What regional geographic concepts are suggested in ecosystem recognition? Is an ecosystem identical to a formal region? Why or why not?

6. National, linguistic, historical, planning, and other regions have been recognized in this chapter. With what other regional entities do you have acquaintance in your daily affairs? Are fire protection districts, police or voting precincts, or zoning districts regional units identifiable with "geographer's" regions as discussed in this chapter or elsewhere in this book? How influenced are you in your private life by your, or others', regional delimitations?

457

Suggested Readings

FREEMAN, T. W., *A Hundred Years of Geography*, ch. 2, "The Regional Approach." Aldine, Chicago, 1961.

MINSHULL, ROGER, *Regional Geography*, Aldine, Chicago, 1967.

WHITTLESEY, DERWENT, in PRESTON E. JAMES and CLARENCE F. JONES (eds.), *American Geography: Inventory and Prospect*. Syracuse University Press (for the Association of American Geographers), Syracuse, NY, 1954.

WOOLDRIDGE, S. W., and W. G. EAST, *The Spirit and Purpose of Geography*, ch. VIII, "Regional Geography and the Theory of Regions." Capricorn Books, New York, 1967.

accessibility The relative ease with which a destination may be reached from other locations.

acculturation Cultural modification or change resulting from one culture group or individual adopting traits of a more advanced or dominant society; cultural development through "borrowing."

acid rain Precipitation containing sulfuric acid, caused by sulfur dioxide dissolving in water vapor in the atmosphere.

activity space The area within which people move freely on their rounds of regular activity.

adaptation A presumed modification of heritable traits through response to environmental stimuli.

agglomeration economies The savings that result from the spatial grouping of activities, such as industries or retail stores.

air mass A large body of air with little horizontal variation in temperature, pressure, and humidity.

alluvial fan A fan-shaped accumulation of alluvium deposited by a stream at the base of a hill or mountain.

alluvium Sediment carried by a stream and deposited in a floodplain or delta.

arroyo A steep-sided, flat-bottomed gully, usually dry, carved out of desert land by rapidly flowing water.

asthenosphere A partially molten, plastic layer above the core and lower mantle of the earth.

Australopithecus ("near man") A hominid of African origin, *Australopithecus* was a terrestrial biped who lived from 5 million to 1 million years B.P.

azimuthal map projection Any of several map projections based on the projection of the grid system onto a plane; lines from a single central point have true compass direction.

barchan A crescent-shaped sand dune; the horns of the crescent point downwind.

Glossary

459

basic sector Those products of an urban unit that are exported outside the city itself.

biochore A major structural subdivision of natural vegetation.

biocide A chemical used to kill pests and disease organisms.

biological magnification The accumulation of a chemical in the fatty tissue of an organism and its concentration at progressively higher levels in the food chain.

biome *See* ecosystem.

biosphere (*syn:* ecosphere) The thin film of air, water, and earth within which we live, including the atmosphere, surrounding and subsurface waters, and the upper reaches of the earth's crust.

birth rate The ratio of the number of live births during one year to the total population, usually at the midpoint of the same year, expressed as the number of births per annum per 1000 population.

carrying capacity The population numbers that can be adequately supported by the known and utilized resources—usually agricultural—of an area.

cartogram A map that has been simplified to present a single idea in a diagrammatic way; the base is not normally true to scale.

central business district (CBD) The center or "downtown" of an urban unit, where retail stores, offices, and cultural activities are concentrated and where land values are high.

central place A nodal point for the distribution of goods and services to a surrounding hinterland population.

channelized migration The tendency for migration to flow between areas that are socially and economically allied by past migration patterns, by economic trade considerations, or by some other affinity.

chemical weathering The decomposition of earth materials due to chemical reactions that include oxidation, hydration, and carbonation.

circumpolar vortex High-altitude winds circling the poles from west to east.

climate A summary of weather conditions in a place or region over a period of time.

collective farm In the Soviet planned economy, the term refers to the cooperative operation of an agricultural enterprise under State control of production and market, but without full status or support as a State enterprise.

commercial economy The production of goods and services for exchange in competitive markets where price and availability are determined by supply and demand forces.

comparative advantage A region's profit potential for a productive activity compared to alternate areas of production of the same good or to alternate uses of the region's resources.

complementarity A term used to describe the actual or potential relationship between two units that each produce different goods or services for which the other has a demand, resulting in an exchange between the units.

concentric zone One of a series of circular belts of land uses around the central business district, each belt having distinct functions.

conformal map projection One on which the shapes of small areas are accurately portrayed.

conic map projection One based on the projection of the grid system onto a cone.

conservation The preservation of natural resources so as to maintain supplies and qualities at levels sufficient to meet present and future needs.

continental drift The hypothesis that an original single land mass (Pangaea) broke apart and that the continents have moved very slowly over the asthenosphere to their present locations.

contour line A map line along which all points are of equal elevation above or below a datum plane, usually mean sea level.

460

convectional precipitation Rain produced when heated, moisture-laden air rises and then cools below the dew point.

coral reef A rocklike landform in shallow tropical water composed chiefly of compacted coral and other organic material.

core In urban geography, that part of the central business district characterized by intensive land development.

coriolis force A fictitious force used to describe motion relative to a rotating earth; specifically, the force that tends to deflect a moving object or fluid to the right (clockwise) in the northern hemisphere and to the left (counterclockwise) in the southern hemisphere.

Creole A language developed from a pidgin to become the native tongue of a society.

critical distance The distance beyond which cost, effort, and/or means play an overriding role in the willingness of people to travel.

Cro-Magnon Extinct species of *Homo sapiens* of Western Europe. Traces of their advanced culture are found in exquisite cave art and carefully worked tools.

cultural landscape The surface of the earth as modified by human action, including housing types, settlement patterns, and agricultural land use.

culture The totality of learned behaviors and attitudes transmitted within a society to succeeding generations by imitation, instruction, and example.

culture hearth A region within which an advanced, distinct set of culture traits developed and from which there was diffusion of distinctive technologies and ways of life.

culture trait A single distinguishing feature of regular occurrence within a culture, such as the use of chopsticks or the observance of a particular caste system.

cyclone A type of atmospheric disturbance in which masses of air circulate rapidly about a region of low atmospheric pressure.

DDT A chlorinated hydrocarbon that is among the most persistent of the biocides in general use.

death rate A mortality index usually calculated as the number of deaths per annum per 1000 population.

delta A triangular-shaped deposit of mud, silt, or gravel created by a stream where it flows into a body of standing water.

demographic transition A model of the effect of economic development on population growth. A first stage involves both high birth and death rates; the second phase displays high birth rates and falling mortality rates and population increases. Phase three shows reduction in population growth as birth rates decline to the level of death rates. The final, fourth, stage implies again a population stable in size but larger in numbers than at the start of the transition cycle.

demography The science of population study, with particular emphasis upon quantitative aspects.

density of population Crude density is a gross measurement of the numbers of persons per unit area of land within predetermined limits, usually political or census boundaries. *See also* physiological density.

desertification Extension of the desert landscape as a result of overgrazing, destruction of the forests, or other human-induced changes.

developable surface A geometric surface such as a cylinder or cone that may be spread out flat without tearing or stretching.

dew point The temperature at which air becomes saturated with water vapor.

dialect A regional variation of a more widely spoken language.

diastrophism The earth force that folds, faults, twists, and compresses rock.

diffusion The gradual spread over space from an origin of new ideas or practices.

directional bias The tendency for people to

461

have greater knowledge of places in some directions than in others as a result of barriers to the flow of information.

distance-decay The decline of an activity or function with increasing distance from its point of origin.

domestication The successful transformation of plant or animal species from a wild state to a condition of dependency upon human management, usually with distinct physical change from wild forebears.

ecosphere *See* biosphere.

ecosystem (*syn*: biome) A population of organisms existing together in a particular area and the energy, air, water, soil, and chemicals upon which it depends.

ecotone The zone of stress between dissimilar, adjacent ecosystems.

environment Surroundings; the totality of things that in any way may affect an organism, including both physical and cultural conditions; a region characterized by a certain set of physical conditions.

462

environmental determinism The theory that the physical environment, particularly climate, molds human behavior.

environmental perception The way people observe and interpret, and the ideas they have about, near or distant places.

equator An imaginary line that encircles the globe halfway between the North and South poles.

equidistant map projection One on which true distances in all directions can be measured from one or two central points.

equivalent map projection One on which the areas of regions are represented in correct or constant proportion to earth reality; also called equal-area.

estuary The lower course or mouth of a river where tides cause fresh water and salt water from the sea to mix.

ethnicity Social status afforded to, usually, a minority group within a national population. Recognition is based primarily upon culture

traits such as religion, distinctive customs, or native or ancestral national origin.

eutrophication The increase of nutrients in a body of water; the nutrients stimulate the growth of algae, whose decomposition decreases the dissolved oxygen content of the water.

extensive agriculture Crop or livestock system in which land quality or extent is more important than capital or labor inputs in determining output.

extrusive rock Rock solidified from molten material that has issued out onto the earth's surface.

fault A break or fracture in rock produced by stress or the movement of lithospheric plates.

fault escarpment A steep slope formed by the vertical movement of the earth along a fault.

filtering In urban geography, a process whereby individuals of one income group replace residents of a portion of an urban area who are of another income group.

fjord A glacial trough the lower end of which is filled with sea water.

folding The buckling of rock layers under pressure of moving lithospheric plates.

food chain A sequence of nutritional energy transfers in an ecosystem accomplished when organisms at one trophic level feed upon those at a lower level.

food web The total flow of energy in an ecosystem.

footloose A descriptive term applied to manufacturing activities for which the cost of transporting material or product is not important in determining location or production.

formal region An earth area throughout which a single phenomenon or limited combination of phenomena is of such uniformity that it can serve as basis for an areal generalization.

fossil fuels Nonrenewable energy resources occurring as sediments or contained within sedimentary rock, specifically either coal, petroleum, or natural gas. They represent the remains or the derivatives of remains of ancient plants or animals.

frame In urban geography, that part of the central business district characterized by such low-intensity land uses as warehouses and automobile dealers.

friction of distance A measurement indicating the effect of distance upon the extent of interaction between two points. Generally, the greater the distance, the less the interaction or exchange or the greater the cost of achieving the exchange.

front The line or zone of separation between two air masses of different temperatures and humidities.

frontal precipitation Rain or snow produced when moist air of one air mass is forced to rise over the edge of another air mass.

functional region An earth area recognized as an operational unit based upon defined organizational criteria.

genetic drift A chance modification of gene composition occurring in an isolated population and becoming accentuated through inbreeding.

gentrification The movement into the inner portions of American cities of middle-class people who replace low-income populations and rehabilitate structures.

geothermal power Energy developed when surface water is converted to steam in hot regions of the earth's crust.

gerrymander To divide an area into voting districts in such a way as to give one political party an unfair advantage in elections, to fragment voting blocks, or to achieve other nondemocratic objectives.

glacial till Deposits of rocks, silt, and sand left by a glacier after it has receded.

glacial trough A deep, U-shaped valley or trench formed by glacial erosion.

glacier A huge mass of slowly moving land ice.

gradation The process responsible for the gradual reduction of the land surface.

gravity model A mathematical prediction of the interaction between two bodies as a function of their size and of the distance separating them.

Green Revolution Term suggesting the great increases in food production, primarily in subtropical areas, accomplished through the introduction of very high-yielding grain crops, particularly wheat and rice.

"greenhouse" effect Heating of the earth's surface as shortwave solar energy passes through an atmosphere transparent to it but opaque to reradiated longwave terrestrial energy. Also refers to increasing the opacity of the atmosphere through addition to it of increased amounts of carbon dioxide from burning of fossil fuels.

grid system The set of imaginary lines of latitude and longitude that intersect at right angles to form a system of reference for locating points on the earth.

gross national product (GNP) The total value of all goods and services produced by a nation per year.

groundwater Subsurface water that accumulates beneath the water table in the pores and cracks of rock and soil; occupies the saturated zone.

growth center An urban center that, if its economy was stimulated, would serve to generate economic expansion in surrounding areas.

half-life The time required for one-half of the atomic nuclei of an isotope to decay.

hierarchical migration The tendency for individuals to move from small places to larger ones.

hierarchy of central places The steplike series of urban units in classes differentiated by both size and function.

hinterland The market area or region served by an urban unit.

homeostatic plateau The equilibrium level of

463

population that can be supported adequately by available resources. Equivalent to carrying capacity.

hominid "Manlike", refers to any humanlike creature, including the early ancestors of the human species.

Homo erectus First clearly recognizable human species (about 1.5 million–250,000 years B.P.) with a bone structure largely indistinguishable from modern humans. With a wide geographical range, *H. erectus* had well-developed tools and knowledge of fire.

Homo sapiens Modern humans; the only contemporary species of the genus *Homo*. Evolved perhaps 250,000 years B.P.

hydrologic cycle The system by which water is continuously circulated through the biosphere.

"icebox" effect The tendency for certain kinds of air pollutants to lower temperatures on earth by reflecting incoming sunlight back into space, preventing it from reaching the earth.

iconography In political geography, a term denoting the study of symbols that unite a nation.

ideological subsystem The complex of ideas, beliefs, knowledge, and means of their communication that characterize a culture.

igneous rock Rock formed as molten earth materials cool and harden either above or below the earth's surface.

Industrial Revolution The term applied to the rapid economic and social changes in agriculture and manufacturing that followed the introduction of the factory system to the textile industry of England in the last quarter of the 18th century.

infant mortality rate A refinement of the death rate to specify the ratio of deaths of infants age one year or less per 1000 live births.

innovation An alteration of custom or culture that originates within the social group itself.

insolation The amount of radiant energy received from the sun at a given place.

intensive agriculture The application of large amounts of capital and/or labor per unit of cultivated land to increase output.

interaction model *See* gravity model.

intrusive rock A landform created by hardening of magma beneath the earth's surface.

isoline A map line connecting points of constant value.

isotropic plain A hypothetical portion of the earth's surface where, it is assumed, the land is everywhere the same and the characteristics of the people are everywhere similar.

karst A limestone region marked by sinkholes, caverns, and underground streams.

lapse rate The rate of change of temperature with altitude in the troposphere; the average lapse rate is about 3.5°F per 1000 feet (6.4°C per 1000 m).

latitude A measure of distance north or south of the equator, given in degrees.

lava Molten material that has emerged onto the earth's surface.

linguistic family A group of languages thought to have a common origin.

lithosphere Solid shell of rocks resting on the asthenosphere.

loess A deposit of wind-blown silt.

longitude A measure of distance east or west of the prime meridian, given in degrees.

longshore current A current that moves roughly parallel to the shore and transports the sand that forms beaches and sand spits.

magma Underground molten material.

malnutrition Food intake insufficient in quantity or deficient in quality to sustain life at optimal conditions of health.

Malthusianism Doctrine proposed by Thomas R. Malthus (1766–1834), that unless checked by self-control, war, or natural disaster, population will inevitably increase faster than will the food supplies needed to sustain it.

map projection A method of transferring the

grid system from the earth's curved surface to the flat surface of a map.

map scale The ratio between length or distance on a map and the corresponding measurement on the earth. Scale may be represented verbally, graphically, or as a fraction.

marginal cost The additional cost incurred to produce an additional unit of output.

mechanical weathering The physical disintegration of earth materials, commonly by frost action, root action, or the development of salt crystals.

mental map A map drawn to represent the mental image(s) a person has of an area.

Mercator projection A true conformal cylindrical projection first published in 1569, useful for navigation.

meridian A north–south line of longitude; on the globe, all meridians are of equal length and converge at the poles.

Mesolithic period Middle Stone Age; the transitional period between the more clearly recognized Paleolithic and Neolithic periods, during which refined and specialized tools were developed and early domestication of plants and animals occurred.

metamorphic rock Rock transformed from igneous and sedimentary rocks by earth forces that generate heat, pressure, or chemical reaction.

microdistrict The basic residential planning unit, usually a superblock, characteristic of new urban developments in planned economies.

migration The movement of people or other organisms from one region to another.

migration field An area that sends major migration flows to a given place or the area that receives major flows from a place.

mineral A natural inorganic substance that has a definite chemical composition and characteristic crystal structure, hardness, and density.

monoculture Agricultural system dominated by a single crop.

monsoon A wind system that reverses direction seasonally, producing wet and dry seasons; used especially to describe the wind system of South, Southeast, and East Asia.

moraine Any of several types of landforms composed of debris transported and deposited by a glacier.

mountain wind The downward flow of heavy, cool air at night from mountainsides to lower valley locations.

multiple-nuclei theory The postulate that large cities develop by peripheral spread not from one but from several nodes of growth and that hence there are many centers of the various land-use functions in an urban area.

multiplier effect The expected addition of nonbasic workers and dependents to a city's total employment and population that accompanies new basic employment.

nationalism A sense of unity binding the people of a state together; devotion to the interests of a particular nation; an identification with the state and an acceptance of national goals.

natural hazard A process or event in the physical environment that has consequences harmful to humans.

natural levee An embankment on the sides of a meandering river formed by deposition of silt during floods.

natural resource A physically occurring item that a population perceives to be necessary and useful to its maintenance and well-being.

Neanderthal Early *Homo sapiens* (approximately 100,000–10,000 years B.P.).

Neolithic period New Stone Age, during which advanced tools and technologies were developed by sedentary populations engaged in production as distinguished from hunting and gathering.

465

niche The place an organism or species occupies in an ecosystem.

nomadic herding Migratory but controlled movement of livestock solely dependent upon natural forage.

nonbasic sector Those economic activities of an urban unit that service the resident population.

nonrenewable resource A natural resource that exists in a finite amount.

North Atlantic drift The massive movement of warm water in the Atlantic Ocean from the Caribbean Sea and Gulf of Mexico in a northeasterly direction to the British Isles and the Scandinavian peninsula.

North and South poles The endpoints of the axis about which the earth spins.

nuclear fission The controlled splitting of an atom to release energy.

nuclear fusion The combining of two atoms of deuterium into a single atom of helium in order to release energy.

nutrient A mineral or other element an organism requires for normal growth and development.

orographic precipitation Rain or snow caused when warm, moisture-laden air is forced to rise over hills or mountains in its path and is thereby cooled.

orthophotomap An aerial photograph to which a grid system and certain map symbols have been added.

outwash plain A gently sloping area in front of a glacier composed of neatly stratified glacial till carried out of the glacier by meltwater streams.

Paleolithic period Old Stone Age; the earliest stage of human culture, it was characterized by hunting and gathering economies and by use of fire and worked tools.

Pangaea The name given to the supercontinent that is thought to have existed 200 million years ago.

parallel of latitude An east–west line indicating distance north or south of the equator.

466

peak value intersection The most accessible and costly parcel of land in the central business district and, therefore, in the entire urbanized area.

photochemical smog A form of polluted air produced by the interaction of hydrocarbons and oxides of nitrogen in the presence of sunlight.

physiological density The number of persons per unit area of agricultural land. *See also* density of population.

pidgin An auxiliary language derived, with reduction of vocabulary and simplification of structure, from other languages. Not a native tongue, it is employed to provide a mutually intelligible vehicle for limited transactions of trade or administration.

place utility The perceived attractiveness of a place in its social, economic, or environmental attributes.

planned economy Production of goods and services, usually consumed or distributed by a governmental agency, in quantities and at prices determined by governmental program.

plantation A large agricultural holding, frequently foreign-owned, devoted to the production of a single export crop.

plate tectonics The theory that the lithosphere is divided into plates that slide or drift very slowly over the asthenosphere.

playa A temporary lake or a lake bed found in a desert environment.

Pleistocene The geological epoch dating from 2 million to 10 thousand years B.P. during which four stages of continental glaciation occurred.

Pliocene The geologic period dating from 13 million to 2 million years B.P. during which the early evolution of hominids occurred.

pollution The introduction into the biosphere of materials that, because of their quantity, chemical nature, or temperature, have a negative impact on the ecosystem or that cannot be readily disposed of by natural recycling processes.

pressure gradient force Differences in air pressure between areas that induce air to flow from areas of high to areas of low pressure.

primary activities Those parts of the economy involved in making natural resources available for use or further processing; includes mining, agriculture, forestry, fishing or hunting, grazing.

prime meridian An imaginary line passing through the Royal Observatory at Greenwich, England, serving by agreement as the zero degree line of longitude.

Ramapithecus A genus accepted by many as an early hominid dated at 10 or more million years B.P.

region In geography, the term applied to an earth area that displays a distinctive grouping of physical or cultural phenomena.

relative humidity A measure of the relative dampness of the atmosphere; the ratio between the amount of water vapor in the air and the maximum amount that it could hold at the same temperature, given in percent.

remote sensing Any of several techniques of obtaining images of an area without having the sensor in direct physical contact with it.

renewable resource A natural resource, such as water, that can be consumed and restored after use in a relatively short period of time. *See also* sustained yield.

reradiation A process by which the earth returns solar energy to space; some of the shortwave solar energy that is absorbed into the land and water is returned to the atmosphere in the form of longwave terrestrial radiation.

rhumb line A line of constant compass bearing; it cuts all meridians at the same angle.

Richter scale A logarithmic scale used to express the magnitude of an earthquake.

sand bar An offshore shoal of sand created by the backwash of waves.

savanna A tropical grassland characterized by widely dispersed trees and experiencing pronounced yearly wet and dry seasons.

secondary activities Those parts of the economy involved in the processing of raw materials derived from primary activities; includes manufacturing, construction, power generation.

sector theory A model used to describe wedge-shaped sectors of diverse land uses outward from the central business district.

sedimentary rock Rock formed from particles of gravel, sand, silt, and clay that were eroded from already existing rocks.

seismic waves Vibrations within the earth set off by earthquakes.

shaded relief A method of representing the three-dimensional quality of an area by use of continuous graded tone to simulate the appearance of sunlight and shadows.

shifting cultivation (*syn:* slash and burn agriculture) Crop production on forest clearings kept in cultivation until their quickly declining fertility is lost. Cleared plots are then abandoned and new sites are prepared.

site The place where something is located; the immediate surroundings.

situation The location of something in relation to the physical and human characteristics of a larger region.

slash and burn agriculture *See* shifting cultivation.

sociological subsystem The totality of expected and accepted patterns of interpersonal relations common to a culture or subculture.

source region In climatology, an area where an air mass forms.

spatial diffusion The outward spread of a substance, concept, or population from its point of origin.

standard parallel The tangent circle, usually a parallel of latitude, in a conic projection; along the standard line the scale is as stated on the map.

state farm Under Soviet and other planned economies, a government agricultural enterprise operated with paid employees.

stream load Eroded material carried by a

467

stream in one of three ways, depending on the size and composition of the particles: (1) in dissolved form, (2) suspended by the water, (3) rolled along the stream bed.

subduction The process by which one lithospheric plate is forced down into the asthenosphere as a result of collision with another plate.

subsidence Settling or sinking of a portion of the land surface, sometimes as a result of the extraction of fluids such as oil or water from underground deposits.

subsistence agriculture Any of several farm economies in which most crops are grown for food, nearly exclusively for local consumption.

subsistence economy A system in which goods and services are created for the use of producers or their immediate families. Market exchanges are limited and of minor importance.

sustained yield The practice of balancing harvesting with growth of new stocks so as to avoid depletion of the resource and ensure a perpetual supply.

syncretism The development of a new form of, for example, religion or music, through the coalescence of distinctive parental elements.

systems analysis An approach to the study of large systems through (1) segregation of the entire system into its component parts, (2) investigation of the interactions between system elements, and (3) study of inputs, outputs, flows, interactions, and boundaries within the system.

talus cone A landform composed of rock particles that have accumulated at the base of a cliff, hill, or mountain.

technological subsystem The complex of material objects together with the techniques of their use by means of which people carry out their productive activities.

technology An integrated system of knowledge and skills developed within a culture to carry out successfully purposeful and productive tasks.

temperature inversion The condition caused by rapid reradiation in which air at lower altitudes is cooler than air aloft.

territoriality Persistent attachment of most animals to a specific area; the behavior associated with the defense of the home territory.

tertiary activities Those parts of the economy that fulfill the exchange function and that provide market availability of commodities; includes wholesale and retail trade and associated transportational, governmental, and informational services.

thermal pollution The introduction of heated water into the environment, with consequent adverse effects on plants and animals.

threshold In economic geography, the minimum market needed to support the supply of a product or service.

T–O map A type of world map drawn during the Middle Ages, showing a circular world divided into three parts by a T-shaped partition. The T consists of three principal waterways; the continents are Asia, Europe, and Africa.

topographic map One that portrays the surface features of a relatively small area, often in great detail.

trophic level A nutritional transfer stage in an ecosystem.

troposphere The lowest layer of the atmosphere, extending to about 6 miles (10 km) above the earth's surface.

tsunami A seismic sea wave generated by an earthquake or volcanic eruption.

tundra A treeless area lying between the tree line of arctic regions and the permanently ice-covered region.

Unigov The unified government of a city and all or parts of its metropolitan region.

valley wind The flow of air up mountain slopes during the day.

variable costs In economic geography, costs of

468

production inputs that display place to place differences.

volcanism The earth force that transports heated material to or toward the surface of the earth.

washes The dry, braided channels in deserts that remain after a rush of water.

water table The upper limit of the saturated zone and hence of groundwater.

weathering Mechanical and chemical proc- esses that fragment and decompose rock materials.

wetland A lowland area that is saturated with moisture, such as a marsh or tidal flat.

zero population growth (ZPG) A term suggest- ing a population in equilibrium, with birth rates and death rates nearly identical.

zoning Designating by ordinance areas in a municipality for particular types of land use.

469

(Italicized page numbers refer
to illustrations and maps)

A

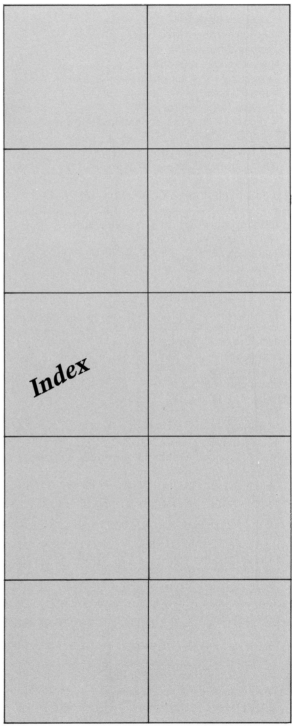

Index

471

472

473

475

477

479

H

I

485

487

490

O

P

495

497

499

Persons Per
Square Mile

Persons Per
Square Kilometer

500 or more
125-500
25-125
2-25
1-2
Sparsely
populated

200 or more
50-200
10-50
2-10
1-2
Sparsely
populated

• Urbanized area of more
than 1,000,000

WORL

International Editions

Multinational Business Finance has been used throughout the world to teach students of international finance. Our books are published in a number of foreign languages including Chinese, French, Spanish, Indonesian, Portuguese, and Ukrainian.

Acknowledgments

The authors are very thankful for the many detailed reviews of previous editions and suggestions from a number of colleagues. The final version of *Multinational Business Finance*, Fourteenth Edition, reflects most of the suggestions provided by these reviewers. The survey reviewers were anonymous, but the detailed reviewers were:

Jennifer Foo, *Stetson University*

John Gonzales, *University of San Francisco*

Delroy M. Hunter, *University of Southern Florida*

Chee K. Ng, *Fairleigh Dickinson University*

Richard L. Patterson, *Indiana University, Bloomington*

Sanjiv Sabherwal, *University of Texas at Arlington*

Tzveta Vateva, *Kent State University*

Special thanks are extended to the reviewers and survey participants of the previous editions:

Otto Adleberger
Essen University, Germany

Alan Alford
Northeastern University

Stephen Archer
Willamette University

Bala Arshanapalli
Indiana University Northwest

Hossein G. Askari
George Washington University

Robert T. Aubey
University of Wisconsin at Madison

David Babbel
University of Pennsylvania

James Baker
Kent State University

Morten Balling
Arhus School of Business, Denmark

Arindam Bandopadhyaya
University of Massachusetts at Boston

Ari Beenhakker
University of South Florida

Carl Beidleman
Lehigh University

Robert Boatler
Texas Christian University

Gordon M. Bodnar
Johns Hopkins University

Nancy Bord
University of Hartford

Finbarr Bradley
University of Dublin, Ireland

Tom Brewer
Georgetown University

Michael Brooke
University of Manchester, England

Robert Carlson
Assumption University, Thailand

Kam C. Chan
University of Dayton

Chun Chang
University of Minnesota

Sam Chee
Boston University Metropolitan College

Kevin Cheng
New York University

It-Keong Chew
University of Kentucky

Frederick D. S. Choi
New York University

Jay Choi
Temple University

Nikolai Chuvakhin
Pepperdine University

Mark Ciechon
University of California, Los Angeles

J. Markham Collins
University of Tulsa

Alan N. Cook
Baylor University

Kerry Cooper
Texas A&M University

Robert Cornu
Cranfield School of Management, U.K.

Roy Crum
University of Florida

Steven Dawson
University of Hawaii at Manoa

David Distad
University of California, Berkeley

Gunter Dufey
University of Michigan, Ann Arbor

Mark Eaker
Duke University

Rodney Eldridge
George Washington University

Imad A. Elhah
University of Louisville

Vihang Errunza
McGill University

Cheol S. Eun
Georgia Tech University

Mara Faccio
University of Notre Dame

Larry Fauver
University of Tennessee

Joseph Finnerty
*University of Illinois
at Urbana-Champaign*

William R. Folks, Jr.
University of South Carolina

Lewis Freitas
University of Hawaii at Manoa

Anne Fremault
Boston University

Fariborg Ghadar
George Washington University

Ian Giddy
New York University

Martin Glaum
*Justus-Lievig-Universitat Giessen,
Germany*

Deborah Gregory
University of Georgia

Robert Grosse
Thunderbird

Christine Hekman
Georgia Tech University

Steven Heston
University of Maryland

James Hodder
University of Wisconsin, Madison

Alfred Hofflander
*University of California, Los
Angeles*

Janice Jadlow
Oklahoma State University

Veikko Jaaskelainen
*Helsinki School of Economics
and Business Administration*

Benjamas Jirasakuldech
University of the Pacific

Ronald A. Johnson
Northeastern University

Fred Kaen
University of New Hampshire

John Kallianiotis
University of Scranton

Charles Kane
Boston College

Robert Kemp
University of Virginia

W. Carl Kester
Harvard Business School

Seung Kim
St. Louis University

Yong Kim
University of Cincinnati

Yong-Cheol Kim
*University of
Wisconsin-Milwaukee*

Gordon Klein
*University of California,
Los Angeles*

Steven Kobrin
University of Pennsylvania

Paul Korsvold
*Norwegian School
of Management*

Chris Korth
University of South Carolina

Chuck C. Y. Kwok
University of South Carolina

John P. Lajaunie
Nicholls State University

Sarah Lane
Boston University

Martin Laurence
William Patterson College

Eric Y. Lee
Fairleigh Dickinson University

Yen-Sheng Lee
Bellevue University

Donald Lessard
*Massachusetts
Institute of Technology*

Arvind Mahajan
Texas A&M University

Rita Maldonado-Baer
New York University

Anthony Matias
Palm Beach Atlantic College

Charles Maxwell
Murray State University

Sam McCord
Auburn University

Jeanette Medewitz
University of Nebraska at Omaha

Robert Mefford
University of San Francisco

Paritash Mehta
Temple University

Antonio Mello
*University of Wisconsin
at Madison*

Eloy Mestre
American University

Kenneth Moon
Suffolk University

Gregory Noronha
Arizona State University

Edmund Outslay
Michigan State University

Lars Oxelheim
Lund University, Sweden

Jacob Park
Green Mountain College

Yoon Shik Park
George Washington University

John Petersen,
George Mason University

Harvey Poniachek
New York University

Yash Puri
*University of Massachusetts
at Lowell*

R. Ravichandrarn
*University of Colorado
at Boulder*

Scheherazade Rehman
George Washington University

Jeff Rosenlog
Emory University

David Rubinstein
University of Houston

Alan Rugman
Oxford University, U.K.

R. J. Rummel
University of Hawaii at Manoa

Mehdi Salehizadeh
San Diego State University

Michael Salt
San Jose State University

Roland Schmidt
*Erasmus University,
the Netherlands*

Lemma Senbet
University of Maryland

Alan Shapiro
University of Southern California

Hany Shawky
State University of New York, Albany

Hamid Shomali
Golden Gate University

Vijay Singal
Virginia Tech University

Sheryl Winston Smith
University of Minnesota

Luc Soenen
California Polytechnic State University

Marjorie Stanley
Texas Christian University

Joseph Stokes
University of Massachusetts-Amherst

Jahangir Sultan
Bentley College

Lawrence Tai
Loyola Marymount University

Kishore Tandon
CUNY—Bernard Baruch College

Russell Taussig
University of Hawaii at Manoa

Lee Tavis
University of Notre Dame

Sean Toohey
University of Western Sydney, Australia

Norman Toy
Columbia University

Joseph Ueng
University of St. Thomas

Gwinyai Utete
Auburn University

Rahul Verma
University of Houston-Downtown

Harald Vestergaard
Copenhagen Business School

K. G. Viswanathan
Hofstra University

Joseph D. Vu
University of Illinois, Chicago

Mahmoud Wahab
University of Hartford

Masahiro Watanabe
Rice University

Michael Williams
University of Texas at Austin

Brent Wilson
Brigham Young University

Bob Wood
Tennessee Technological University

Alexander Zamperion
Bentley College

Emilio Zarruk
Florida Atlantic University

Tom Zwirlein
University of Colorado, Colorado Springs

**Industry
(present or former affiliation)**

Paul Adaire
Philadelphia Stock Exchange

Barbara Block
Tektronix, Inc.

Holly Bowman
Bankers Trust

Payson Cha
HKR International, Hong Kong

John A. Deuchler
Private Export Funding Corporation

Kåre Dullum
Gudme Raaschou Investment Bank, Denmark

Steven Ford
Hewlett Packard

David Heenan
Campbell Estate, Hawaii

Sharyn H. Hess
Foreign Credit Insurance Association

Aage Jacobsen
Gudme Raaschou Investment Bank, Denmark

Ira G. Kawaller
Chicago Mercantile Exchange

Kenneth Knox
Tektronix, Inc.

Arthur J. Obesler
Eximbank

I. Barry Thompson
Continental Bank

Gerald T. West
Overseas Private Investment Corporation

Willem Winter
First Interstate Bank of Oregon

A note of thanks is also extended to our accuracy reviewer, Dev Prasad, of the University of Massachusetts Lowell.

We would also like to thank all those with Pearson Education who have worked so diligently on this edition: Kate Fernandes, Kathryn Dinovo, and Meredith Gertz. In addition, Gillian Hall, our outstanding project manager at The Aardvark Group, deserves much gratitude.

Finally, we would like to dedicate this book to our parents, the late Wilford and Sylvia Eiteman, the late Harold and Norma Stonehill, and Bennie Ruth and the late Hoy K. Moffett, who gave us the motivation to become academicians and authors. We thank our wives, Keng-Fong, Kari, and Megan, for their patience while we were preparing *Multinational Business Finance*, Fourteenth Edition.

Pacific Palisades, California   D.K.E.

Honolulu, Hawaii   A.I.S.

Glendale, Arizona   M.H.M.

About the Authors

David K. Eiteman. David K. Eiteman is Professor Emeritus of Finance at the John E. Anderson Graduate School of Management at UCLA. He has also held teaching or research appointments at the Hong Kong University of Science & Technology, Showa Academy of Music (Japan), the National University of Singapore, Dalian University (China), the Helsinki School of Economics and Business Administration (Finland), University of Hawaii at Manoa, University of Bradford (U.K.), Cranfield School of Management (U.K.), and IDEA (Argentina). He is a former president of the International Trade and Finance Association, Society for Economics and Management in China, and Western Finance Association.

Professor Eiteman received a B.B.A. (Business Administration) from the University of Michigan, Ann Arbor (1952); M.A. (Economics) from the University of California, Berkeley (1956); and a Ph.D. (Finance) from Northwestern University (1959).

He has authored or co-authored four books and twenty-nine other publications. His articles have appeared in *The Journal of Finance*, *The International Trade Journal*, *Financial Analysts Journal*, *Journal of World Business*, *Management International*, *Business Horizons*, *MSU Business Topics*, *Public Utilities Fortnightly*, and others.

Arthur I. Stonehill. Arthur I. Stonehill is a Professor of Finance and International Business, Emeritus, at Oregon State University, where he taught for 24 years (1966–1990). During 1991–1997 he held a split appointment at the University of Hawaii at Manoa and Copenhagen Business School. From 1997 to 2001 he continued as a Visiting Professor at the University of Hawaii at Manoa. He has also held teaching or research appointments at the University of California, Berkeley; Cranfield School of Management (U.K.); and the North European Management Institute (Norway). He was a former president of the Academy of International Business, and was a western director of the Financial Management Association.

Professor Stonehill received a B.A. (History) from Yale University (1953); an M.B.A. from Harvard Business School (1957); and a Ph.D. in Business Administration from the University of California, Berkeley (1965). He was awarded honorary doctorates from the Aarhus School of Business (Denmark, 1989), the Copenhagen Business School (Denmark, 1992), and Lund University (Sweden, 1998).

He has authored or co-authored nine books and twenty-five other publications. His articles have appeared in *Financial Management*, *Journal of International Business Studies*, *California Management Review*, *Journal of Financial and Quantitative Analysis*, *Journal of International Financial Management and Accounting*, *International Business Review*, *European Management Journal*, *The Investment Analyst* (U.K.), *Nationaløkonomisk Tidskrift* (Denmark), *Sosialøkonomen* (Norway), *Journal of Financial Education*, and others.

Michael H. Moffett. Michael H. Moffett is Continental Grain Professor in Finance at the Thunderbird School of Global Management at Arizona State University, where he has been since 1994. He also has held teaching or research appointments at Oregon State University (1985–1993); the University of Michigan, Ann Arbor (1991–1993); the Brookings Institution,

Washington, D.C.; the University of Hawaii at Manoa; the Aarhus School of Business (Denmark); the Helsinki School of Economics and Business Administration (Finland), the International Centre for Public Enterprises (Yugoslavia); and the University of Colorado, Boulder.

Professor Moffett received a B.A. (Economics) from the University of Texas at Austin (1977), an M.S. (Resource Economics) from Colorado State University (1979), an M.A. (Economics) from the University of Colorado, Boulder (1983), and Ph.D. (Economics) from the University of Colorado, Boulder (1985).

He has authored, co-authored, or contributed to a number of books, articles, case studies, and other publications. He has co-authored two books with Art Stonehill and David Eiteman, *Fundamentals of Multinational Finance*, and this book, *Multinational Business Finance*. His articles have appeared in the *Journal of Financial and Quantitative Analysis, Journal of Applied Corporate Finance, Journal of International Money and Finance, Journal of International Financial Management and Accounting, Contemporary Policy Issues, Brookings Discussion Papers in International Economics*, and others. He has contributed to a number of collected works including the *Handbook of Modern Finance*, the *International Accounting and Finance Handbook*, and the *Encyclopedia of International Business*. He is also co-author of two books in multinational business with Michael Czinkota and Ilkka Ronkainen, *International Business* (7th Edition) and *Global Business* (4th Edition), and *The Global Oil and Gas Industry: Strategy, Finance, and Management*, with Andrew Inkpen.

Brief Contents

Contents

Global Financial Environment

CHAPTER 1

Multinational Financial Management: Opportunities and Challenges

The objects of a financier are, then, to secure an ample revenue; to impose it with judgment and equality; to employ it economically; and, when necessity obliges him to make use of credit, to secure its foundations in that instance, and for ever, by the clearness and candor of his proceedings, the exactness of his calculations, and the solidity of his funds.

—Edmund Burke, *Reflections on the Revolution in France*, 1790, p. 467.

LEARNING OBJECTIVES

- Understand the complexity of risks associated with financial globalization
- Explore how global capital markets are critical for the exchange of products, services, and capital in the execution of global business
- Consider how the theory of comparative advantage establishes the foundations for the justification for international trade and commerce
- Discover what is different about international financial management, and which market imperfections give rise to the multinational enterprise
- Examine how imperfections in global markets translate into opportunities for multinational enterprises
- Consider how the globalization process moves a business from a purely domestic focus in its financial relationships and composition to one truly global in scope

The subject of this book is the financial management of multinational enterprises (MNEs)—multinational financial management. MNEs are firms—both for-profit companies and not-for-profit organizations—that have operations in more than one country and conduct their business through *branches*, foreign subsidiaries, or joint ventures with host country firms. That conduct of business comes with challenges as suggested by the following news release from Procter & Gamble Co. (P&G), an American multinational consumer goods company:

> *"The October–December 2014 quarter was a challenging one with unprecedented currency devaluations," said Chairman, President and Chief Executive Officer A.G. Lafley. "Virtually every currency in the world devalued versus the U.S. dollar, with the Russian Ruble leading the way. While we continue to make steady progress on the strategic*

transformation of the company—which focuses P&G on about a dozen core categories and 70 to 80 brands, on leading brand growth, on accelerating meaningful product innovation and increasing productivity savings—the considerable business portfolio, product innovation, and productivity progress was not enough to overcome foreign exchange."

—P&G News Release, January 27, 2015.

P&G is not alone. New MNEs are appearing all over the world every day, while many of the older and established ones (like P&G) are struggling to survive. Businesses of all kinds are seeing a very different world than in the past. Today's MNEs depend not only on the emerging markets for cheaper labor, raw materials, and outsourced manufacturing, but also increasingly on those same emerging markets for sales and profits. These markets—whether they are emerging, less developed, or developing, or are *BRIC* (Brazil, Russia, India, and China), BIITS (Brazil, India, Indonesia, Turkey, South Africa, which are also termed the Fragile Five), or MINT (Mexico, Indonesia, Nigeria, Turkey)—represent the majority of the earth's population and, therefore, the majority of potential customers. And adding market complexity to this changing global landscape is the risky and challenging international macroeconomic environment, both from a long-term and short-term perspective. The global financial crisis of 2008–2009 is already well into the business past, and capital is flowing again—although both into and out of economies—at an ever-increasing pace.

How to identify and navigate these risks is the focus of this book. These risks may all occur on the playing field of the global financial marketplace, but they are still a question of management—of navigating complexity in pursuit of the goals of the firm.

Financial Globalization and Risk

Back in the halcyon pre-crisis days of the late 20th and early 21st centuries, it was taken as self evident that financial globalization was a good thing. But the subprime crisis and eurozone dramas are shaking that belief. . . . what is the bigger risk now—particularly in the eurozone—is that financial globalization has created a system that is interconnected in some dangerous ways.

—"Crisis Fears Fuel Debate on Capital Controls,"
Gillian Tett, *Financial Times*, December 15, 2011.

The theme dominating global financial markets today is the complexity of risks associated with financial globalization—far beyond whether it is simply good or bad, but how to lead and manage multinational firms in the rapidly moving marketplace. The following is but a sampling of this complexity of risks.

- The international monetary system, an eclectic mix of floating and managed fixed exchange rates, is under constant scrutiny. The rise of the Chinese *renminbi* is changing much of the world's outlook on currency exchange, reserve currencies, and the roles of the dollar and the euro (see Chapter 2).

- Large fiscal deficits, including the current eurozone crisis, plague most of the major trading countries of the world, complicating fiscal and monetary policies, and ultimately, interest rates and exchange rates (see Chapter 3).

- Many countries experience continuing balance of payments imbalances, and in some cases, dangerously large deficits and surpluses—whether it be the twin surpluses enjoyed by China, the current account surplus of Germany amidst a sea of eurozone deficits, or the continuing current account deficit of the United States, all will inevitably move exchange rates (see Chapter 3).

■ Ownership, control, and governance vary radically across the world. The publicly traded company is not the dominant global business organization—the privately held or family-owned business is the prevalent structure—and goals and measures of performance vary dramatically between these business models (see Chapter 4).

■ Global capital markets that normally provide the means to lower a firm's cost of capital, and even more critically, increase the availability of capital, have in many ways shrunk in size and have become less open and accessible to many of the world's organizations (see Chapter 2).

■ Today's emerging markets are confronted with a new dilemma: the problem of first being the recipients of capital inflows, and then of experiencing rapid and massive capital outflows. Financial globalization has resulted in the ebb and flow of capital into and out of both industrial and emerging markets, greatly complicating financial management (Chapters 5 and 8).

This first chapter is meant only as an introduction and a taste of the complexity of risks associated with financial globalization. The Mini-Case at the end of this first chapter, *Crowdfunding Kenya*, is intended to push you in your thinking about how and why money moves across the globe today.

The Global Financial Marketplace

Business—domestic, international, global—involves the interaction of individuals and individual organizations for the exchange of products, services, and capital through markets. The global capital markets are critical for the conduct of this exchange. The global financial crisis of 2008–2009 served as an illustration and a warning of how tightly integrated and fragile this marketplace can be.

Assets, Institutions, and Linkages

Exhibit 1.1 provides a map of the global capital markets. One way to characterize the global financial marketplace is through its assets, institutions, and linkages.

Assets. The assets—financial assets—at the heart of the global capital markets are the debt securities issued by governments (e.g., U.S. Treasury Bonds). These low-risk or risk-free assets form the foundation for the creation, trading, and pricing of other financial assets like bank loans, corporate bonds, and equities (stock). In recent years, a number of additional securities have been created from existing securities—derivatives, the value of which is based on market value changes of the underlying securities. The health and security of the global financial system relies on the quality of these assets.

Institutions. The institutions of global finance are the central banks, which create and control each country's money supply; the commercial banks, which take deposits and extend loans to businesses, both local and global; and the multitude of other financial institutions created to trade securities and derivatives. These institutions take many shapes and are subject to many different regulatory frameworks. The health and security of the global financial system relies on the stability of these financial institutions.

Linkages. The links between the financial institutions—the actual fluid or medium for exchange—are the interbank networks using currency. The ready exchange of currencies in the global marketplace is the first and foremost necessary element for the conduct of financial trading, and the global currency markets are the largest markets in the world. The exchange of currencies, and the subsequent exchange of all other securities globally via currency, is the

EXHIBIT 1.1   **Global Capital Markets**

The global capital market is a collection of institutions (central banks, commercial banks, investment banks, not-for-profit financial institutions like the IMF and World Bank) and securities (bonds, mortgages, derivatives, loans, etc.), which are all linked via a global network—the *Interbank Market*. This interbank market, in which securities of all kinds are traded, is the critical pipeline system for the movement of capital.

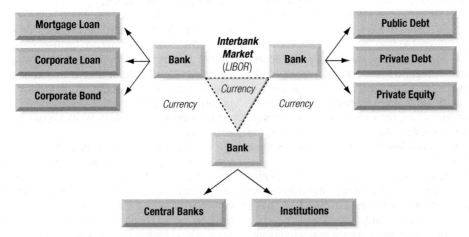

The exchange of securities—the movement of capital in the global financial system—must all take place through a vehicle—currency. The exchange of currencies is itself the largest of the financial markets. The interbank market, which must *pass-through* and exchange securities using currencies, bases all of its pricing through the single most widely quoted interest rate in the world—LIBOR (the London Interbank Offered Rate).

international interbank network. This network, whose primary price is the London Interbank Offered Rate (LIBOR), is the core component of the global financial system.

The movement of capital across currencies and continents for the conduct of business has existed in many different forms for thousands of years. Yet, it is only within the past 50 years that these capital movements have started to move at the pace of an electron in the digital marketplace. And it is only within the past 20 years that this market has been able to reach the most distant corners of the earth at any moment of the day. The result has been an explosion of innovative products and services—some for better and some for worse. And as illustrated by *Global Finance in Practice 1.1*, conditions and markets can change—often quickly.

The Market for Currencies

The price of any one country's currency in terms of another country's currency is called a foreign currency exchange rate. For example, the exchange rate between the U.S. dollar ($ or USD) and the European euro (€ or EUR) may be stated as "1.0922 dollar per euro" or simply abbreviated as $1.0922/€. This is the same exchange rate as when stated "EUR1.00 = USD1.0922." Since most international business activities require at least one of the two parties in a business transaction to either pay or receive payment in a currency that is different from their own, an understanding of exchange rates is critical to the conduct of global business.

Currency Symbols. As noted, USD and EUR are often used as the symbols for the U.S. dollar and the European Union's euro. These are the computer symbols (ISO-4217 codes) used today on the world's digital networks. The field of international finance, however, has a rich history of using a variety of different symbols in the financial press, and a variety of different

GLOBAL FINANCE IN PRACTICE 1.1

The Rocketing Swiss Franc

The Swiss franc has been fighting its appreciation against the European *euro* for years. While it is not a member of the European Union and while its currency has been one of the world's most stable for over a century, Switzerland is an economy and a currency completely encased within the Eurozone.

In 2011, in an attempt to stop the Swiss franc from continuing to grow in value against the euro (to stop its *appreciation*), the Swiss Central Bank announced a "floor" on its value against the euro of 1.20 Swiss francs to one euro. To preserve this value, the Bank would intervene in the market by buying euros with Swiss francs anytime the market exchange rate threatened to hit the floor. As illustrated in Exhibit A, the Bank had to intervene only a few select times in the past three years—until early 2015.

In early 2015, the markets continued to apply upward pressure on the Swiss franc's value against the euro (which means pushing its exchange value to less than 1.20 Swiss francs per euro). The Swiss Central Bank continued to intervene, buying euros with Swiss francs and accumulating more and more euros in its reserves of foreign currency. The Bank also set central bank interest rates at negative

levels—yes, *negative*. This meant that the Bank charged depositors to hold Swiss franc deposits, an effort to dissuade investors from exchanging any currency, including the euro, for Swiss francs.

But the European Union's economies continued to struggle in 2014, and early reports of economic activity in 2015 were showing further slowing. Investors wished to exit the euro fearing its future fall in value. The European Central Bank added to investor anxiety when it announced that it would be undertaking expansionary government debt purchases—quantitative easing—(expansionary monetary policy) to kick-start the sluggish EU economy.

On the morning of January 15, 2015, the Swiss Central Bank shocked the markets by announcing that it was abandoning the 1.20 floor and cutting interest rates further (more negative). It had concluded that with the forthcoming monetary expansion from the ECB, there was no longer any way to keep the flood gates closed. The Swiss franc, as illustrated in Exhibit B, appreciated against the euro within minutes. For two of the world's major currencies, it was a very eventful day.

EXHIBIT A   Swiss Franc–Euro Exchange Rate

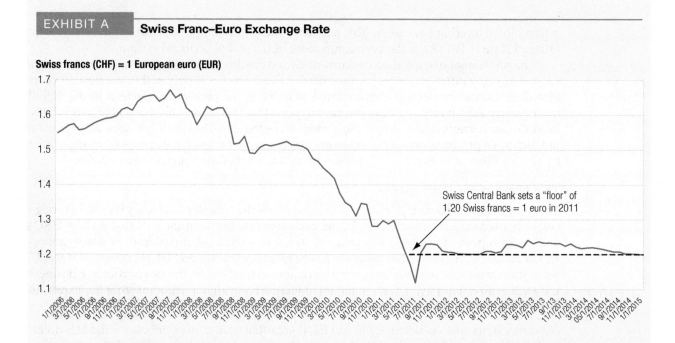

Swiss francs (CHF) = 1 European euro (EUR)

Swiss Central Bank sets a "floor" of 1.20 Swiss francs = 1 euro in 2011

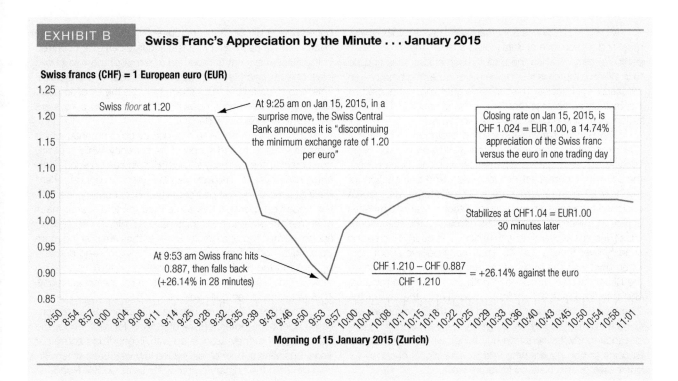

EXHIBIT B   **Swiss Franc's Appreciation by the Minute . . . January 2015**

Swiss francs (CHF) = 1 European euro (EUR)

Swiss *floor* at 1.20

At 9:25 am on Jan 15, 2015, in a surprise move, the Swiss Central Bank announces it is "discontinuing the minimum exchange rate of 1.20 per euro"

Closing rate on Jan 15, 2015, is CHF 1.024 = EUR 1.00, a 14.74% appreciation of the Swiss franc versus the euro in one trading day

Stabilizes at CHF1.04 = EUR1.00 30 minutes later

At 9:53 am Swiss franc hits 0.887, then falls back (+26.14% in 28 minutes)

$$\frac{\text{CHF } 1.210 - \text{CHF } 0.887}{\text{CHF } 1.210} = +26.14\% \text{ against the euro}$$

Morning of 15 January 2015 (Zurich)

abbreviations are commonly used. For example, the British pound sterling may be £ (the pound symbol), GBP (Great Britain pound), STG (British pound sterling), ST£ (pound sterling), or UKL or UK£ (United Kingdom pound). This book uses both the simpler common symbols—the $ (dollar), the € (euro), the ¥ (yen), the £ (pound)—and the three letter ISO codes throughout. And as *Global Finance in Practice 1.2* describes, this would include BTC, the three-letter digital code for Bitcoin.

GLOBAL FINANCE IN PRACTICE 1.2

Bitcoin-Cryptocurrency or Commodity?

The difference is that established fiat currencies—ones where the bills and coins, or their digital versions, get their value by dint of regulation or law—are underwritten by the state which is, in principle at least, answerable to its citizens. Bitcoin, on the other hand, is a community currency. It requires self-policing on the part of its users. To some, this is a feature, not a bug. But, in the grand scheme of things, the necessary open-source engagement remains a niche pursuit.

   —"Bits and bob," *The Economist*, June 13, 2011.

Bitcoin is an open-source, peer-to-peer, digital currency. It is a *cryptocurrency*, a digital currency that is created and managed using advanced encryption techniques known as cryptography. And it may be the world's first completely decentralized digital-payments system. But is Bitcoin a true currency?

Bitcoin was invented in 2009 by a man calling himself Satoshi Nakamoto. Nakamoto published, via the Internet, a nine-page paper outlining how the Bitcoin system would work. He also provided the open-source code needed to both produce the digital coins (*mine* in Bitcoin terminology) and trade Bitcoins digitally as money. (Nakamoto is not thought to be a real person, likely being a *nome de plume* for some relatively small working group.)

Mining of Bitcoins is a mathematical process. The *miner* must find a sequence of data (called a *block*) that produces a particular pattern when the Bitcoin hash algorithm is applied to it. When a match is found, the miner obtains a bounty—an *allocation*—of Bitcoins. This repetitive guessing, conducted by increasingly complex computers, is called hashing. The motivation for mining is clear: to make money.

The Bitcoin software system is designed to release a 25-coin reward to the miner in the worldwide network who succeeds in solving the mathematical problem. Once solved, the solution is broadcast network-wide, and competition for the next 25-coin reward begins. The system's protocol is designed to release a new block of Bitcoins every 10 minutes until all 21 million are released. The difficulty of the search continually increases over time with mining. This causes increasing scarcity over time, similar to what many believe about gold when gold was the basis of currency values. But ultimately the Bitcoin system is limited in both time (every 10 minutes) and total issuance (21 million). Theoretically the last of the 21 million Bitcoins would be mined in 2140. Once mined, Bitcoins are considered a *pseudonymous*—nearly anonymous—cryptocurrency. Owners can buy things with Bitcoins or sell Bitcoins to non-miners who wish to use digital currency for purchases or speculate on its future value.

Ownership of each and every coin is verified and registered through a digital chain timestamp across the thousands of network nodes. Like cash, this prevents double spending, since every Bitcoin exchange is authenticated across the decentralized Bitcoin network (currently estimated at 20,000 nodes). Unlike cash, every transaction that has ever occurred in the Bitcoin system is recorded in terms of the two public keys (the transactors, the Bitcoin addresses) in the system. This record, called the *block chain*, includes the time, amount, and the two near-anonymous IP addresses (public keys are not tied to any person's identity).

Traditional currencies are issued by governments, which regulate the growth and supply of the currency, while implicitly guaranteeing the currency's value. Bitcoin has no such guarantor, no insurer, and no *lender-of-last-resort*. A gold standard like that used in the first part of the twentieth century, is a system based on *specie*; it has some fixed link to a scarce metal of some intrinsic value. Bitcoins have no intrinsic value; they are not composed of a precious metal; they are nothing more than digital code. Their value reflects the supply and demand by those in the marketplace who believe in their value. This makes Bitcoin a *fiat currency* similar to major currencies today. Their value has been quite volatile.

The ability of Bitcoin to bypass authorities has led to concerns about the potential use of Bitcoin for illicit trade, the laundering of money associated with illegal drugs and other illegal business activity globally. One low was seen when Bitcoin became the primary currency for sales on Silk Road, an underground Web site for illegal drug trafficking. Although eventually shut down by the U.S. government, Bitcoin's potential use for illegal activities has impacted the public's perception of its potential. Others, however, see promise.

Exchange Rate Quotations and Terminology. Exhibit 1.2 lists currency exchange rates for December 31, 2014, as would be quoted in New York or London. The exchange rate listed is for a specific country's currency—for example, the Argentina peso is Peso 8.7851 = 1.00 U.S. dollar, is Peso 9.5990 = 1.00 Euro, and Peso 13.1197 = 1.00 British pound. The rate listed is termed a "mid-rate" because it is the middle or average of the rates at which currency traders buy currency (*bid rate*) and sell currency (*offer rate*).

The U.S. dollar has been the focal point of most currency trading since the 1940s. As a result, most of the world's currencies have been quoted against the dollar—Mexican pesos per dollar, Brazilian real per dollar, Hong Kong dollars per dollar, etc. This quotation convention is also followed against the world's major currencies, as listed in Exhibit 1.2. For example, the Japanese yen is commonly quoted against the dollar, euro, and pound, as in ¥119.765/$, ¥130.861/€, and ¥178.858/£.

Quotation Conventions. Several of the world's major currency exchange rates, however, follow a specific quotation convention that is the result of tradition and history. The exchange rate between the U.S. dollar and the euro is always quoted as "dollars per euro" or $/€. For example, $1.0926 listed in Exhibit 1.2 under "United States." Similarly, the exchange rate between the U.S. dollar and the British pound is always quoted as "dollars per pound" or $/£. For example, $1.4934 listed under "United States" in Exhibit 1.2. In addition, countries that were formerly members of the British Commonwealth will often be quoted against the U.S. dollar, as in U.S. dollars per Australian dollar or U.S. dollars per Canadian dollar.

| EXHIBIT 1.2 | Selected Global Currency Exchange Rates |

March 23, 2015 Country	Currency	Symbol	Code	Currency Equal to 1 Dollar	Currency Equal to 1 Euro	Currency Equal to 1 Pound
Argentina	peso	Ps	ARS	8.7851	9.5990	13.1197
Australia	dollar	A$	AUD	1.2744	1.3924	1.9032
Brazil	real	R$	BRL	3.1610	3.4538	4.7206
Canada	dollar	C$	CAD	1.2516	1.3676	1.8691
Chile	peso	$	CLP	625.98	683.97	934.84
China	yuan	¥	CNY	6.2160	6.7919	9.2830
Czech Republic	koruna	Kc	CZK	25.0240	27.3424	37.3710
Denmark	krone	Dkr	DKK	6.8243	7.4566	10.1915
Egypt	pound	£	EGP	7.6301	8.3369	11.3948
Euro	euro	€	EUR	0.9152	1.0000	1.3668
India	rupee	Rs	INR	62.2650	68.0336	92.9870
Indonesia	rupiah	Rp	IDR	13,019.50	14,225.70	19,443.41
Israel	shekel	Shk	ILS	3.9655	4.3329	5.9221
Japan	yen	¥	JPY	119.765	130.861	178.858
Kenya	shilling	KSh	KES	91.90	100.41	137.24
Malaysia	ringgit	RM	MYR	3.6950	4.0373	5.5181
Mexico	new peso	$	MXN	14.9583	16.3441	22.3387
New Zealand	dollar	NZ$	NZD	1.3120	1.4335	1.9593
Nigeria	naira	₦	NGN	199.700	218.201	298.233
Norway	krone	NKr	NOK	7.8845	8.6149	11.7747
Philippines	peso	₱	PHP	44.7900	48.9396	66.8897
Poland	zloty	—	PLN	3.7600	4.1084	5.6153
Russia	ruble	R	RUB	58.7450	64.1875	87.7302
Singapore	dollar	S$	SGD	1.3671	1.4938	2.0416
South Africa	rand	R	ZAR	11.9100	13.0134	17.7865
South Korea	won	W	KRW	1,114.65	1,217.92	1,664.63
Sweden	krona	SKr	SEK	8.5016	9.2892	12.6963
Switzerland	franc	Fr.	CHF	0.9669	1.0565	1.4440
Taiwan	dollar	T$	TWD	31.4350	34.3473	46.9452
Thailand	baht	B	THB	32.5450	35.5601	48.6029
Turkey	lira	YTL	TRY	2.5477	2.7837	3.8048
United Kingdom	pound	£	GBP	0.6696	0.7316	1.0000
Ukraine	hrywnja	—	UAH	22.6500	24.7484	33.8257
Uruguay	peso	$U	UYU	25.5300	27.8952	38.1267
United States	dollar	$	USD	1.0000	1.0926	1.4934
Venezuela	Bolivar fuerte	Bs	VEB	6.2935	6.8766	9.3988
Vietnam	dong	d	VND	21,525.00	23,519.23	32,145.66
Special Drawing Right	—	—	SDR	0.7199	0.7867	1.0752

Note that a number of different currencies use the same symbol (for example, both China and Japan have traditionally used the ¥ symbol, yen or yuan, meaning round or circle). All quotes are mid-rates, and are drawn from the *Financial Times*.

Eurocurrencies and LIBOR

One of the major linkages of global money and capital markets is the eurocurrency market and its interest rate, LIBOR. *Eurocurrencies* are domestic currencies of one country on deposit in a second country for a period ranging from overnight to more than a year or longer. Certificates of deposit are usually for three months or more and in million-dollar increments. A eurodollar deposit is not a *demand deposit*—it is not created on the bank's books by writing loans against required fractional reserves, and it cannot be transferred by a check drawn on the bank having the deposit. Eurodollar deposits are transferred by wire or cable transfer of an underlying balance held in a *correspondent bank* located within the United States. In most countries, a domestic analogy would be the transfer of deposits held in nonbank savings associations. These are transferred when the association writes its own check on a commercial bank.

Any *convertible currency* can exist in "euro" form. Note that this use of "euro" prefix should not be confused with the European currency called the euro. The eurocurrency market includes eurosterling (British pounds deposited outside the United Kingdom); euroeuros (euros on deposit outside the eurozone); euroyen (Japanese yen deposited outside Japan) and eurodollars (U.S. dollars deposited outside the U.S.). Eurocurrency markets serve two valuable purposes: (1) eurocurrency deposits are an efficient and convenient money market device for holding excess corporate liquidity; and (2) the eurocurrency market is a major source of short-term bank loans to finance corporate working capital needs, including the financing of imports and exports. Banks in which eurocurrencies are deposited are called eurobanks. A *eurobank* is a financial intermediary that simultaneously bids for time deposits and makes loans in a currency other than that of its home currency. Eurobanks are major world banks that conduct a eurocurrency business in addition to all other banking functions. Thus, the eurocurrency operation that qualifies a bank for the name eurobank is, in fact, a department of a large commercial bank, and the name springs from the performance of this function.

The modern eurocurrency market was born shortly after World War II. Eastern European holders of dollars, including the various state trading banks of the Soviet Union, were afraid to deposit their dollar holdings in the United States because those deposits might be attached by U.S. residents with claims against communist governments. Therefore, Eastern European holders deposited their dollars in Western Europe, particularly with two Soviet banks: the Moscow Narodny Bank in London and the Banque Commerciale pour l'Europe du Nord in Paris. These banks redeposited the funds in other Western banks, especially in London. Additional dollar deposits were received from various central banks in Western Europe, which elected to hold part of their dollar reserves in this form to obtain a higher yield. Commercial banks also placed their dollar balances in the market because specific maturities could be negotiated in the eurodollar market. Such companies found it financially advantageous to keep their dollar reserves in the higher-yielding eurodollar market. Various holders of international refugee funds also supplied funds.

Although the basic causes of the growth of the eurocurrency market are economic efficiencies, many unique institutional events during the 1950s and 1960s contributed to its growth.

- In 1957, British monetary authorities responded to a weakening of the pound by imposing tight controls on U.K. bank lending in sterling to nonresidents of the United Kingdom. Encouraged by the Bank of England, U.K. banks turned to dollar lending as the only alternative that would allow them to maintain their leading position in world finance. For this they needed dollar deposits.

- Although New York was "home base" for the dollar and had a large domestic money and capital market, international trading in the dollar centered in London because of that city's expertise in international monetary matters and its proximity in time and distance to major customers.

■ Additional support for a European-based dollar market came from the balance of payments difficulties of the U.S. during the 1960s, which temporarily segmented the U.S. domestic capital market.

Ultimately, however, the eurocurrency market continues to thrive because it is a large international money market relatively free from governmental regulation and interference.

Eurocurrency Interest Rates. The reference rate of interest in the eurocurrency market is the *London Interbank Offered Rate,* or *LIBOR.* LIBOR is the most widely accepted rate of interest used in standardized quotations, loan agreements, or financial derivatives valuations. The use of interbank offered rates, however, is not confined to London. Most major domestic financial centers construct their own interbank offered rates for local loan agreements. Examples of such rates include *PIBOR* (Paris Interbank Offered Rate), *MIBOR* (Madrid Interbank Offered Rate), *SIBOR* (Singapore Interbank Offered Rate), and *FIBOR* (Frankfurt Interbank Offered Rate), to name just a few.

The key factor attracting both depositors and borrowers to the eurocurrency loan market is the narrow interest rate *spread* within that market. The difference between deposit and loan rates is often less than 1%. Interest spreads in the eurocurrency market are small for many reasons. Low lending rates exist because the eurocurrency market is a wholesale market where deposits and loans are made in amounts of $500,000 or more on an unsecured basis. Borrowers are usually large corporations or government entities that qualify for low rates because of their credit standing and because the transaction size is large. In addition, overhead assigned to the eurocurrency operation by participating banks is small.

Deposit rates are higher in the eurocurrency markets than in most domestic currency markets because the financial institutions offering eurocurrency activities are not subject to many of the regulations and reserve requirements imposed on traditional domestic banks and banking activities. With these costs removed, rates are subject to more competitive pressures, deposit rates are higher, and loan rates are lower. A second major area of cost avoided in the eurocurrency markets is the payment of deposit insurance fees (such as the Federal Deposit Insurance Corporation, FDIC) and assessments paid on deposits in the United States.

The Theory of Comparative Advantage

The *theory of comparative advantage* provides a basis for explaining and justifying international trade in a model world assumed to enjoy free trade, perfect competition, no uncertainty, costless information, and no government interference. The theory's origins lie in the work of Adam Smith, and particularly with his seminal book, *The Wealth of Nations*, published in 1776. Smith sought to explain why the division of labor in productive activities, and subsequently international trade of those goods, increased the quality of life for all citizens. Smith based his work on the concept of *absolute advantage*, with every country specializing in the production of those goods for which it was uniquely suited. More would be produced for less. Thus, with each country specializing in products for which it possessed absolute advantage, countries could produce more in total and trade for goods that were cheaper in price than those produced at home.

In his work, *On the Principles of Political Economy and Taxation*, published in 1817, David Ricardo sought to take the basic ideas set down by Adam Smith a few logical steps further. Ricardo noted that even if a country possessed absolute advantage in the production of two goods, it might still be relatively more efficient than the other country in one good's production than the production of the other good. Ricardo termed this comparative advantage. Each country would then possess comparative advantage in the production of one of the two products, and both countries would then benefit by specializing completely in one product and trading for the other.

Although international trade might have approached the comparative advantage model during the nineteenth century, it certainly does not today, for a variety of reasons. Countries do not appear to specialize only in those products that could be most efficiently produced by that country's particular factors of production. Instead, governments interfere with comparative advantage for a variety of economic and political reasons, such as to achieve full employment, economic development, national self-sufficiency in defense-related industries, and protection of an agricultural sector's way of life. Government interference takes the form of *tariffs, quotas*, and other non-tariff restrictions.

At least two of the factors of production—capital and technology—now flow directly and easily between countries, rather than only indirectly through traded goods and services. This direct flow occurs between related subsidiaries and affiliates of multinational firms, as well as between unrelated firms via loans and license and management contracts. Even labor flows between countries, such as immigrants into the United States (legal and illegal), immigrants within the European Union, and other unions.

Modern factors of production are more numerous than in this simple model. Factors considered in the location of production facilities worldwide include local and managerial skills, a dependable legal structure for settling contract disputes, research and development competence, educational levels of available workers, energy resources, consumer demand for brand name goods, mineral and raw material availability, access to capital, tax differentials, supporting infrastructure (roads, ports, and communication facilities), and possibly others.

Although the *terms of trade* are ultimately determined by supply and demand, the process by which the terms are set is different from that visualized in traditional trade theory. They are determined partly by administered pricing in oligopolistic markets.

Comparative advantage shifts over time as less-developed countries become more developed and realize their latent opportunities. For example, over the past 150 years, comparative advantage in producing cotton textiles has shifted from the United Kingdom to the United States, to Japan, to Hong Kong, to Taiwan, and to China. The classical model of comparative advantage also did not really address certain other issues such as the effect of uncertainty and information costs, the role of differentiated products in imperfectly competitive markets, and economies of scale.

Nevertheless, although the world is a long way from the classical trade model, the general principle of comparative advantage is still valid. The closer the world gets to true international specialization, the more world production and consumption can be increased, provided that the problem of equitable distribution of the benefits can be solved to the satisfaction of consumers, producers, and political leaders. Complete specialization, however, remains an unrealistic limiting case, just as perfect competition is a limiting case in microeconomic theory.

Global Outsourcing of Comparative Advantage

Comparative advantage is still a relevant theory to explain why particular countries are most suitable for exports of goods and services that support the global supply chain of both MNEs and domestic firms. The comparative advantage of the twenty-first century, however, is one that is based more on services, and their cross-border facilitation by telecommunications and the Internet. The source of a nation's comparative advantage, however, still is created from the mixture of its own labor skills, access to capital, and technology.

For example, India has developed a highly efficient and low-cost software industry. This industry supplies not only the creation of custom software, but also call centers for customer support, and other information technology services. The Indian software industry is composed of subsidiaries of MNEs and independent companies. If you own a Hewlett-Packard computer and call the customer support center number for help, you are likely to reach a call center in India. Answering your call will be a knowledgeable Indian software engineer or programmer who will "walk you through" your problem. India has a large number of well-educated,

English-speaking technical experts who are paid only a fraction of the salary and overhead earned by their U.S. counterparts. The overcapacity and low cost of international telecommunication networks today further enhances the comparative advantage of an Indian location.

The extent of global outsourcing is already reaching out to every corner of the globe. From financial back-offices in Manila, to information technology engineers in Hungary, modern telecommunications now take business activities to labor rather than moving labor to the places of business.

What Is Different about International Financial Management?

Exhibit 1.3 details some of the main differences between international and domestic financial management. These component differences include institutions, foreign exchange and political risks, and the modifications required of financial theory and financial instruments.

Multinational financial management requires an understanding of cultural, historical, and institutional differences such as those affecting corporate governance. Although both domestic firms and MNEs are exposed to foreign exchange risks, MNEs alone face certain unique risks, such as political risks, that are not normally a threat to domestic operations.

MNEs also face other risks that can be classified as extensions of domestic finance theory. For example, the normal domestic approach to the *cost of capital*, sourcing debt and equity, capital budgeting, *working capital management*, taxation, and credit analysis needs to be modified to accommodate foreign complexities. Moreover, a number of financial instruments that are used in domestic financial management have been modified for use in international financial management. Examples are foreign currency options and futures, interest rate and currency swaps, and letters of credit.

The main theme of this book is to analyze how an MNE's financial management evolves as it pursues global strategic opportunities and new constraints emerge. In this chapter, we will take a brief look at the challenges and risks associated with Ganado Corporation (Ganado), a company evolving from domestic in scope to becoming truly multinational. The discussion

EXHIBIT 1.3   What Is Different about International Financial Management?

Concept	International	Domestic
Culture, history, and institutions	Each foreign country is unique and not always understood by MNE management	Each country has a known base case
Corporate governance	Foreign countries' regulations and institutional practices are all uniquely different	Regulations and institutions are well known
Foreign exchange risk	MNEs face foreign exchange risks due to their subsidiaries, as well as import/export and foreign competitors	Foreign exchange risks from import/export and foreign competition (no subsidiaries)
Political risk	MNEs face political risk because of their foreign subsidiaries and high profile	Negligible political risks
Modification of domestic finance theories	MNEs must modify finance theories like capital budgeting and the cost of capital because of foreign complexities	Traditional financial theory applies
Modification of domestic financial instruments	MNEs utilize modified financial instruments such as options, forwards, swaps, and letters of credit	Limited use of financial instruments and derivatives because of few foreign exchange and political risks

GLOBAL FINANCE IN PRACTICE 1.3

Corporate Responsibility and Corporate Sustainability

Sustainable development is development that meets the needs of the present without compromising the ability of future generations to meet their own needs.
— Brundtland Report, 1987, p. 54.

What is the purpose of the corporation? It is accepted that the purpose of the corporation is to certainly create profits and value for its stakeholders, but the responsibility of the corporation is to do so in a way that inflicts no costs on society, including the environment. As a result of globalization, this growing responsibility and role of the corporation in society has added a level of complexity to the leadership challenges faced by the multinational firm.

This developing controversy has been somewhat hampered to date by conflicting terms and labels — corporate goodness, corporate responsibility, corporate social responsibility (CSR), corporate philanthropy, and corporate sustainability, to list but a few. Confusion can be reduced by using a guiding principle — that sustainability is a goal, while responsibility is an obligation. It follows that the obligation of leadership in the modern multinational is to pursue profit, social development, and the environment, all along sustainable principles.

The term sustainability has evolved greatly within the context of global business in the past decade. A traditional primary objective of the family-owned business has been the "sustainability of the organization" — the long-term ability of the company to remain commercially viable and provide security and income for future generations. Although narrower in scope, the concept of environmental sustainability shares a common core thread — the ability of a company, a culture, or even the earth, to survive and renew over time.

will include constraints that a company will face in terms of managerial goals and governance as it becomes increasingly involved in multinational operations. But first we need to clarify the unique value proposition and advantages that the MNE was created to exploit. And as noted by *Global Finance in Practice 1.3*, the objectives and responsibilities of the modern multinational have grown significantly more complex in the twenty-first century.

Market Imperfections: A Rationale for the Existence of the Multinational Firm

MNEs strive to take advantage of imperfections in national markets for products, factors of production, and financial assets. Imperfections in the market for products translate into market opportunities for MNEs. Large international firms are better able to exploit such competitive factors as economies of scale, managerial and technological expertise, product differentiation, and financial strength than are their local competitors. In fact, MNEs thrive best in markets characterized by international oligopolistic competition, where these factors are particularly critical. In addition, once MNEs have established a physical presence abroad, they are in a better position than purely domestic firms to identify and implement market opportunities through their own internal information network.

Why Do Firms Become Multinational?

Strategic motives drive the decision to invest abroad and become an MNE. These motives can be summarized under the following categories:

1. *Market seekers* produce in foreign markets either to satisfy local demand or to export to markets other than their home market. U.S. automobile firms manufacturing in Europe for local consumption are an example of market-seeking motivation.

2. *Raw material seekers* extract raw materials wherever they can be found, either for export or for further processing and sale in the country in which they are found — the host country. Firms in the oil, mining, plantation, and forest industries fall into this category.

3. *Production efficiency seekers* produce in countries where one or more of the factors of production are underpriced relative to their productivity. Labor-intensive production of electronic components in Taiwan, Malaysia, and Mexico is an example of this motivation.

4. *Knowledge seekers* operate in foreign countries to gain access to technology or managerial expertise. For example, German, Dutch, and Japanese firms have purchased U.S. electronics firms for their technology.

5. *Political safety seekers* acquire or establish new operations in countries that are considered unlikely to expropriate or interfere with private enterprise. For example, Hong Kong firms invested heavily in the United States, United Kingdom, Canada, and Australia in anticipation of the consequences of China's 1997 takeover of the British colony.

These five types of strategic considerations are not mutually exclusive. Forest products firms seeking wood fiber in Brazil, for example, may also find a large Brazilian market for a portion of their output.

In industries characterized by worldwide oligopolistic competition, each of the above strategic motives should be subdivided into proactive and defensive investments. Proactive investments are designed to enhance the growth and profitability of the firm itself. Defensive investments are designed to deny growth and profitability to the firm's competitors. Examples of the latter are investments that try to preempt a market before competitors can get established in it, or capture raw material sources and deny them to competitors.

The Globalization Process

Ganado is a hypothetical U.S.-based firm that will be used as an illustrative example throughout the book to demonstrate the globalization process—the structural and managerial changes and challenges experienced by a firm as it moves its operations from domestic to global.

Global Transition I: Ganado Moves from the Domestic Phase to the International Trade Phase

Ganado is a young firm that manufactures and distributes an array of telecommunication devices. Its initial strategy is to develop a sustainable competitive advantage in the U.S. market. Like many other young firms, it is constrained by its small size, competitors, and lack of access to cheap and plentiful sources of capital. The top half of Exhibit 1.4 shows Ganado in its early domestic phase.

Ganado sells its products in U.S. dollars to U.S. customers and buys its manufacturing and service inputs from U.S. suppliers, paying U.S. dollars. The creditworth of all suppliers and buyers is established under domestic U.S. practices and procedures. A potential issue for Ganado at this time is that although Ganado is not international or global in its operations, some of its competitors, suppliers, or buyers may be. This is often the impetus to push a firm like Ganado into the first phase of the globalization process—into international trade. Ganado was founded in Los Angeles by James Winston in 1948 to make telecommunications equipment. The family-owned business expanded slowly but steadily over the following 40 years. The demands of continual technological investment in the 1980s, however, required that the firm raise additional equity capital in order to compete. This need led to its initial public offering (IPO) in 1988. As a U.S.-based publicly traded company on the New York Stock Exchange, Ganado's management sought to create value for its shareholders.

As Ganado became a visible and viable competitor in the U.S. market, strategic opportunities arose to expand the firm's market reach by exporting products and services to one or

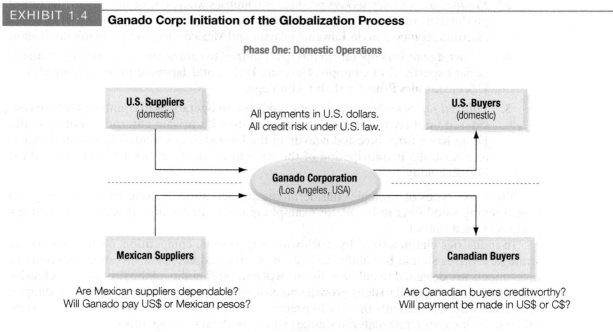

EXHIBIT 1.4   Ganado Corp: Initiation of the Globalization Process

Phase One: Domestic Operations

U.S. Suppliers (domestic)

All payments in U.S. dollars.
All credit risk under U.S. law.

U.S. Buyers (domestic)

Ganado Corporation
(Los Angeles, USA)

Mexican Suppliers

Canadian Buyers

Are Mexican suppliers dependable?
Will Ganado pay US$ or Mexican pesos?

Are Canadian buyers creditworthy?
Will payment be made in US$ or C$?

Phase Two: Expansion into International Trade

more foreign markets. The *North American Free Trade Agreement (NAFTA)* made trade with Mexico and Canada attractive. This second phase of the globalization process is shown in the lower half of Exhibit 1.4. Ganado responded to these globalization forces by importing inputs from Mexican suppliers and making export sales to Canadian buyers. We define this phase of the globalization process as the International Trade Phase.

Exporting and importing products and services increases the demands of financial management over and above the traditional requirements of the domestic-only business in two ways. First, direct foreign exchange risks are now borne by the firm. Ganado may now need to quote prices in foreign currencies, accept payment in foreign currencies, or pay suppliers in foreign currencies. As the values of currencies change from minute to minute in the global marketplace, Ganado will increasingly experience significant risks from the changing values associated with these foreign currency payments and receipts.

Second, the evaluation of the credit quality of foreign buyers and sellers is now more important than ever. Reducing the possibility of non-payment for exports and non-delivery of imports becomes a key financial management task during the international trade phase. This credit risk management task is much more difficult in international business, as buyers and suppliers are new, subject to differing business practices and legal systems, and generally more challenging to assess.

Global Transition II: The International Trade Phase to the Multinational Phase

If Ganado is successful in its international trade activities, the time will come when the globalization process will progress to the next phase. Ganado will soon need to establish foreign sales and service affiliates. This step is often followed by establishing manufacturing operations abroad or by licensing foreign firms to produce and service Ganado's products.

EXHIBIT 1.5   **Ganado's Foreign Direct Investment Sequence**

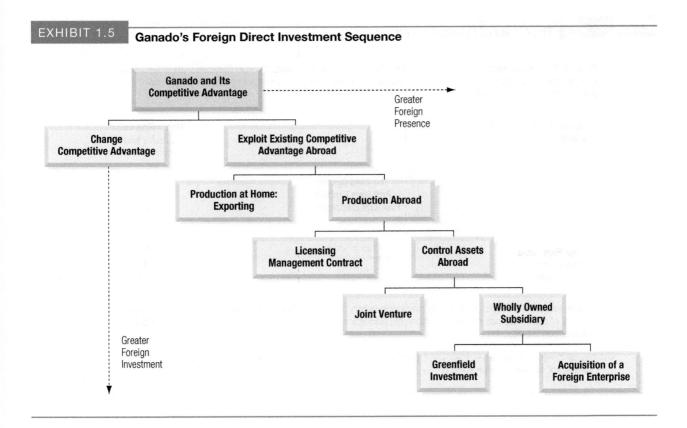

The multitude of issues and activities associated with this second larger global transition is the real focus of this book.

Ganado's continued globalization will require it to identify the sources of its competitive advantage, and with that knowledge, expand its intellectual capital and physical presence globally. A variety of strategic alternatives are available to Ganado—the foreign direct investment sequence—as shown in Exhibit 1.5. These alternatives include the creation of foreign sales offices, the licensing of the company name and everything associated with it, and the manufacturing and distribution of its products to other firms in foreign markets. As Ganado moves further down and to the right in Exhibit 1.5, the degree of its physical presence in foreign markets increases. It may now own its own distribution and production facilities, and ultimately, may want to acquire other companies. Once Ganado owns assets and enterprises in foreign countries it has entered the multinational phase of its globalization.

The Limits to Financial Globalization

The theories of international business and international finance introduced in this chapter have long argued that with an increasingly open and transparent global marketplace in which capital may flow freely, capital will increasingly flow and support countries and companies based on the theory of comparative advantage. Since the mid-twentieth century, this has indeed been the case as more and more countries have pursued more open and competitive markets. But the past decade has seen the growth of a new kind of limit or impediment to financial globalization: the growth in the influence and self-enrichment of organizational insiders.

EXHIBIT 1.6   **The Limits of Financial Globalization**

There is a growing debate over whether many of the insiders and rulers of organizations with enterprises globally are taking actions consistent with creating firm value or consistent with increasing their own personal stakes and power.

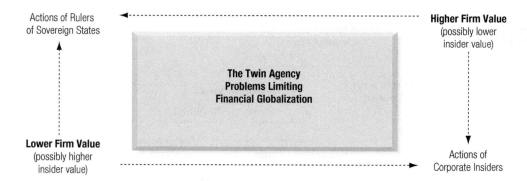

If these influential insiders are building personal wealth over that of the firm, it will indeed result in preventing the flow of capital across borders, currencies, and institutions to create a more open and integrated global financial community.

Source: Constructed by authors based on "The Limits of Financial Globalization," Rene M. Stulz, *Journal of Applied Corporate Finance*, Vol. 19, No. 1, Winter 2007, pp. 8–15.

One possible representation of this process can be seen in Exhibit 1.6. If influential insiders in corporations and sovereign states continue to pursue the increase in firm value, there will be a definite and continuing growth in financial globalization. But, if these same influential insiders pursue their own personal agendas, which may increase their personal power and influence or personal wealth, or both, then capital will not flow into these sovereign states and corporations. The result is the growth of financial inefficiency and the segmentation of globalization outcomes—creating winners and losers. As we will see throughout this book, this barrier to international finance may indeed be increasingly troublesome.

This growing dilemma is also something of a composite of what this book is about. The three fundamental elements—financial theory, global business, and management beliefs and actions—combine to present either the problem or the solution to the growing debate over the benefits of globalization to countries and cultures worldwide.

We close this chapter—and open this book—with the simple words of one of our colleagues in a recent conference on the outlook for global finance and global financial management.

> *Welcome to the future. This will be a constant struggle. We need leadership, citizenship, and dialogue.*

> —Donald Lessard, in *Global Risk, New Perspectives and Opportunities*, 2011, p. 33.

SUMMARY POINTS

- The creation of value requires combining three critical elements: (1) an open marketplace; (2) high-quality strategic management; and (3) access to capital.

- The theory of comparative advantage provides a basis for explaining and justifying international trade in a model world of free and open competition.

- International financial management requires an understanding of cultural, historical, and institutional differences, such as those affecting corporate governance.

- Although both domestic firms and MNEs are exposed to foreign exchange risks, MNEs alone face certain unique risks, such as political risks, that are not normally a threat to domestic operations.

- MNEs strive to take advantage of imperfections in national markets for products, factors of production, and financial assets.

- The decision whether or not to invest abroad is driven by strategic motives and may require the MNE to enter into global licensing agreements, joint ventures, cross-border acquisitions, or greenfield investments.

- If influential insiders in corporations and sovereign states pursue their own personal agendas, which may increase their personal power, influence, or wealth, then capital will not flow into these sovereign states and corporations. This will, in turn, create limitations to globalization in finance.

MINI-CASE

Crowdfunding Kenya[1]

The concept of crowdfunding has a number of parallels in traditional Kenyan culture. Harambee is a long-used practice of collective fundraising for an individual obligation like a travel or medical expense. Another Kenyan practice, chama, involves group fundraising for loans or investments by private groups. In either case, they have strong links to the fundamental principle of a community. In the case of crowdfunding, it is an online community.

Crowdfunding is an Internet-enabled method of raising capital for business startups without going through the arduous, costly, and time-consuming process of traditional equity capital fundraising. The rapid growth in crowdfunding over recent years has been based primarily in the major industrial country markets of North America and Western Europe where there is a highly organized, developed, and deep financial sector, but a sector that often shuts out the small, innovative, non-traditional entrepreneur.

The concept of raising funds from a large crowd or group is not new. It is a technique that has been employed by individuals, organizations, and even governments for centuries. Beethoven and Mozart both raised funds for their work through pre-creation subscriptions. The United States and France both used a form of crowdfunding fundraising to construct the Statue of Liberty. But crowdfunding's real potential may now lie in funding new business startups in emerging markets—markets where the capital sources and institutions available to small and medium enterprises (SMEs) within the country may be limited. If crowdfunding can provide access to capital that many entrepreneurs need, tapping a larger more affordable cross-border financial ecosystem, business and economic and social development in the emerging markets may be able to take a great step forward. Kenya is one country attempting to pilot the effort.

The Capital Lifecycle

The ability of a startup business to access affordable capital through the early stages of its lifecycle has been the focus of a multitude of financial innovations in the past two decades. But the *capital lifecycle*, the institutions and sources of capital available to an enterprise as it evolves, has until recently possessed a number of gaps, putting many startup businesses at risk.

Exhibit A illustrates the capital lifecycle of a for-profit enterprise. An entrepreneur—the founder—puts up his own money in the first stage, the *proof of concept*. This is followed by further *pre-seed* capital typically funded

[1]Copyright © 2015 Thunderbird School of Global Management at Arizona State University. All rights reserved. This case was prepared by Professor Michael H. Moffett for the purpose of classroom discussion only. The author would like to thank Sherwood Neiss of Crowdfunding Capital Advisors for helpful comments; any errors are, however, the sole responsibility of the author.

EXHIBIT A   **The Capital Lifecycle**

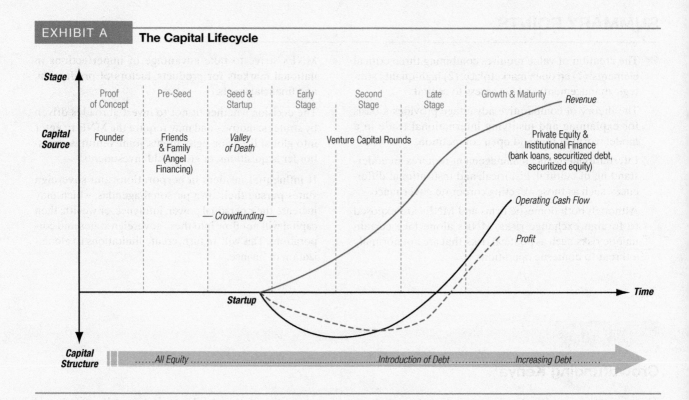

from friends and family, or in some cases, *angel financing* from angel investors. *Angel investors* are individuals or small groups of professional investors who invest at the earliest stages of business development, playing the role of a "guardian angel." The principle is to provide the capital to move the business opportunity further along while still protecting the interests of the entrepreneurial owners. This is often referred to as the pre-seed stage of business development.

It is immediately after this, in the seed stage, that many firms fail to advance in their development due to a vacuum of available capital and capital providers. This vacuum, often referred to as the *Valley of Death*, is a critical period in which the firm is building and moving toward operational launch. But without operating activities and therefore revenues and cash flows, additional investors and access to capital is scarce. It is this vacuum which crowdfunding has filled in many industrial country markets.

Following their launch, promising businesses often then find they have added sources of capital for the financing of rapid growth—the venture capital rounds. *Venture capitalists* (VCs) are investment firms focused on taking an equity position in new businesses that are showing revenue results, but may not yet be positive in terms of cash flow or

profitability. VCs focus their attention on businesses that are considered to have high growth potential but need capital—now—to acquire the scale and assets needed to pursue the growth opportunity.

The final stage of the capital lifecycle is that of the growing and maturing company. It is only now that the business possesses the track record of sales, profits, and cash flow that assure bank lenders of the creditworthiness of the firm. Bank loan-based debt is now accessible. It is also at this time that the firm may consider an *initial public offering* (IPO), to issue equity and raise capital in the marketplace. Firms now gain access to debt, bank loans, after operations have commenced and firms demonstrate operating cash flow capability. However, debt service obligations are not desirable in rapid growth businesses trying to retain as much capital as humanly possible in these early growth stages.

If the business appears to have true value growth prospects, it may catch the eye of *private equity* (PE). Private equity firms invest in greater amounts at later stages of business development. PE investors provide capital to businesses that are fully established and successful, but are in need of capital for growth and business strategy fulfillment. They rarely invest in startup business, searching instead for investment opportunities that will yield higher rates of return than traditional investments in public companies.

Crowdfunding Principles

I believe that crowdfunding may have the potential to help catalyze existing efforts to create entrepreneurial cultures and ecosystems in developing nations. Development organizations like the World Bank and other institutions will play an ongoing role to act as "trusted third parties" in creating these new models of funding and providing mentorship, capacity building as well as ongoing monitoring and reporting.

—Steve Case, Chairman and CEO, Revolution, and Founder, America Online[2]

Crowdfunding began as an online extension of the pre-seed stage in which traditional financing relies upon friends and family to pool funds to finance business development. It seeks to connect an extended group of interested investors, still based on friends and family—the so-called *crowdfunding ecosystem*—directly with startups in need of seed capital. It attempts to open up these funding channels by bypassing the traditional regulatory and institutional barriers, restrictions, costs, and burdens, that capital raising carries in every country around the globe.

Crowdfunding structures typically fall into any one of four categories: donation-based, rewards-based, loan or debt-based, or equity-based.[3]

1. **Donation-based.** Non-profit foundations often employ crowdfunding methods to raise funds for causes of all kinds. Contributors receive nothing in return for their gifts other than positive emotional and intellectual gratification.

2. **Rewards-based.** In rewards-based crowdfunding efforts contributors receive a perk, a benefit, a T-shirt, a ticket, a back-stage pass, some small form of reward. One highly successful platform using this structure is Kickstarter, a U.S.-based arts and project-based fund raiser. As with donation-based funds, there is no guarantee of the project's execution or success, and no return on the investment other than a small reward, perk, or token benefit.

3. **Debt-based.** Debt-based or lending-based crowdfunding efforts provide capital to individuals and organizations in need of growth capital in return for repayment of principal. Micro-finance organizations like Grameen Bank have long used this structure successfully to fund entrepreneurial efforts particularly in emerging markets. The investor is typically promised repayment of principal, but often—as is the case of kiva.org, no payment of interest is made by the borrower or paid to the "investor."

4. **Equity-based.** Investors gain a share of ownership in the project or company. These are enterprise funding efforts to support for-profit business development, the investor receiving voting rights and the possibility (but not the promise) of a return on their capital. This is an investment, not a gift, and although the investors may be drawn from interested or like-minded groups, returns on investment are expected and therefore the business plan and prospects are evaluated critically.

The last two are fundraising efforts focused on business development, and categorically referred to as investment-geared crowdfunding (IGCF). For longer-term sustainable market-based economies, it is category four—equity-based crowdfunding—that is thought to offer the greatest potential for economic development and employment.

Critical Requirements

There are at least three critical components to a successful equity-based crowdfunding initiative: (1) a well-defined and capable crowdsourcing ecosystem; (2) a defined solid business plan and competitive analysis; and (3) a motivated, capable, and committed entrepreneur.

Crowdfunding's true singular strength is the ability of a potential investment to reach an extended crowdfunding ecosystem—a linked crowd accessible via the Internet and therefore not limited by geography, currency, or nationality. It is based on the digital reach of the Internet via social networks and viral marketing, rather than on the traditional institutional structure of the financial and investment sectors in countries. However, given that the object of the investment is a for-profit business that is resident in a difficult to fund or finance marketplace, a successful ecosystem will still be defined by some commonality of experience, culture, ethnicity, or diaspora.[4] As many in the crowdfunding sector will note, when you are raising funds for a for-profit investment anywhere in the world, relationships and linkages play a critical role in moving from a token "gift for a good cause" to an investment in a business.

[2]*Crowdfunding's Potential for the Developing World*, infoDev/The World Bank, by Jason Best, Sherwood Neiss, and Richard Swart, Crowdfunding Capital Advisors (CCA), 2013.

[3]"Issue Brief: Investment-Geared Crowdfunding," CFA Institute, March 2014.

[4]*Crowdfunding Investing for Dummies*, Sherwood Neiss, Jason W. Best, Zak Cassady-Dorion, John Wiley & Sons, Inc., 2013.

Secondly, a business plan must be defined. Crowdfunding is not based on the madness of crowds, but rather their strength in numbers, knowledge, and will. If enough small individual investors collectively support a startup enterprise, anywhere in the world, they can fund the development and growth of the business. But to even reach the proposal stage at which point a crowdfunding platform will entertain discussions, the entrepreneur will need to have refined a business plan. This must include prospective profitability, financial forecasts, and competitive analysis. Any business anywhere, needs a plan to generate sales, control costs, and compete if it is to eventually make a profit.

Finally, as it has been since the beginning of time, success will only come from a truly capable and committed founder—the entrepreneur. Even a business which is well-funded, well-defined, and exceptionally innovative will fail without an entrepreneur who is willing to roll up his sleeves, day after day after day, to go the extra mile (or kilometer) to achieve success. Whether that entrepreneur is named Rockefeller, Gates, Jobs, or Zuckerburg, ultimately—commitment, passion, hunger—has to be ingrained in his DNA.

Kenyan Challenge

Kenya was not all that different from many other major emerging markets when it came to business startups: a shortage of capital and institutions and interest in funding new business development. Funding startups, particularly SMEs, was always challenging, even in the largest and most developed industrial countries. Gaining access to affordable capital in a country like Kenya, even with a burgeoning domestic economy, was extremely difficult.

After a series of successive rounds of evaluation and competition, four crowdfunding projects had been identified for its pilot program by infoDev of the World Bank Group, working through its Kenya Climate Innovation Center (KICC) with the support of Crowdfund Capital Advisors (CCA).

- **Lighting Up Kenya.** *Join the Solar Generation creation in Kenya. Help us extinguish kerosene lamps and improve lives.* Co-Founder of Skynotch Energy Africa, Patrick Kimathi is trying to bring clean-lighting solutions (solar lamps) for off-grid indoor lighting.
- **Wanda Organic.** *Nurture the Soil—Climate Smart Agriculture. Help us improve access to bio-organic fertilizer and biotechnology for farmers in Kenya.* Marion Moon, founder of Wanda Organic, wants to enable Kenyan farmers to produce more, increase profitability and family income, improve nutrition, and create new employment in rural economies, while restoring and strengthening the health of Kenya's soil.
- **Briquette Energy Drive.** *Biomass Briquettes are made from agricultural plant waste and are a replacement for fossil fuels such as oil or coal, and burn hotter, cleaner, and longer.* Allan Marega is the managing director of Global Supply Solutions, whose goal is to make briquettes the preferred replacement to charcoal and wood fuel.
- **iCoal Concepts.** James Nyaga, Director of Strategy and Innovation at iCoal Concepts, wants to use recycled charcoal dust to make briquettes that are denser, burn longer, odorless, and smokeless, ultimately reducing indoor air pollution.

The Kenyan projects are among a number of pilot programs testing crowdfunding applications in emerging markets. Only time and experience will tell if crowdfunding delivers sustainable financial development for the global economy.

Mini-Case Questions

1. Where does crowdfunding fit in the capital lifecycle of business development?
2. Is crowdfunding really all that unique? What does it offer that traditional funding channels and institutions do not?
3. What is likely to differentiate successes from failures in emerging market crowdfunding programs?

QUESTIONS

These questions are available in MyFinanceLab.

1. **Globalization Risks in Business.** What are some of the risks that come with the growing globalization of business?

2. **Globalization and the Multinational Enterprise (MNE).** The term globalization has become widely used in recent years. How would you define it?

3. **Assets, Institutions, and Linkages.** Which assets play the most critical role in linking the major institutions that make up the global financial marketplace?

4. **Currencies and Symbols.** What technological change is even changing the symbols we use in the representation of different country currencies?

5. **Eurocurrencies and LIBOR.** Why have eurocurrencies and LIBOR remained the centerpiece of the global financial marketplace for so long?

6. **Theory of Comparative Advantage.** Define and explain the theory of comparative advantage.

7. **Limitations of Comparative Advantage.** Key to understanding most theories is what they say and what they don't. Name four key limitations to the theory of comparative advantage.

8. **International Financial Management.** What is different about international financial management?

9. **Ganado's Globalization.** After reading the chapter's description of Ganado's globalization process, how would you explain the distinctions between international, multinational, and global companies?

10. **Ganado, the MNE.** At what point in the globalization process did Ganado become a multinational enterprise (MNE)?

11. **Role of Market Imperfections.** What is the role of market imperfections in the creation of opportunities for the multinational firm?

12. **Why Go.** Why do firms become multinational?

13. **Multinational Versus International.** What is the difference between an international firm and a multinational firm?

14. **Ganado's Phases.** What are the main phases that Ganado passed through as it evolved into a truly global firm? What are the advantages and disadvantages of each?

15. **Financial Globalization.** How do the motivations of individuals, both inside and outside the organization or business, define the limits of financial globalization?

PROBLEMS

These problems are available in MyFinanceLab.

1. **Comparing Cheap Dates Around the World.** Comparison of prices or costs across different country and currency environments requires translation of the local currency into a single common currency. This is most meaningful when the comparison is for the identical or near-identical product or service across countries. Deutsche Bank has recently started publishing a comparison of cheap dates—an evening on the town for two to eat at McDonald's, see a movie, and drink a beer. Once all costs are converted to a common currency, the U.S. dollar in this case, the cost of the date can be compared across cities relative to the base case of a cheap date in USD in New York City.

   After completing the table below, answer the following questions.

   a. Which city in the table is truly the cheapest date?

   b. Which city in the table is the most expensive cheap date?

   c. If the exchange rate in Moscow on the Russian ruble (RUB) was 0.04200, instead of 0.0283, what would be the USD price?

   d. If the exchange rate in Shanghai was CNY6.66 = 1 USD, what would be its cost in USD and relative to a cheap date in New York City?

Country	City	Cheap Date in Local Currency	Exchange Rate Quote	Exchange Rate 7 April 2014	USD Price	Relative to NYC
Australia	Sydney	AUD111.96	USD = 1 AUD	0.9290	104.01	112%
Brazil	Rio de Janeiro	BRL135.43	USD = 1 BRL	0.4363	_____	_____
Canada	Ottawa	CAD78.33	USD = 1 CAD	0.9106	_____	_____
China	Shanghai	CNY373.87	USD = 1 CNY	0.1619	_____	_____
France	Paris	EUR75.57	USD = 1 EUR	1.3702	_____	_____
Germany	Berlin	EUR76.49	USD = 1 EUR	1.3702	_____	_____
Hong Kong	Hong Kong	HKD467.03	USD = 1 HKD	0.1289	_____	_____
India	Mumbai	INR1,379.64	USD = 1 INR	0.0167	_____	_____
Indonesia	Jakarta	IDR314,700	USD = 1 IDR	0.0001	_____	_____
Japan	Tokyo	JPY10,269.07	USD = 1 JPY	0.0097	_____	_____
Malaysia	Kuala Lumpur	MY 117.85	USD = 1 MYR	0.3048	_____	_____
Mexico	Mexico City	MX 423.93	USD = 1 MXN	0.0769	_____	_____
New Zealand	Auckland	NZD111.52	USD = 1 NZD	0.8595	_____	_____
Phillipines	Manila	PHP1,182.88	USD = 1 PHP	0.0222	_____	_____

(continued)

Country	City	Cheap Date in Local Currency	Exchange Rate Quote	Exchange Rate 7 April 2014	USD Price	Relative to NYC
Russia	Moscow	RUB2,451.24	USD = 1 RUB	0.0283	_____	_____
Singapore	Singapore	SGD77.89	USD = 1 SGD	0.7939	_____	_____
South Africa	Cape Town	ZAR388.58	USD = 1 ZAR	0.0946	_____	_____
United Kingdom	London	GBP73.29	USD = 1 GBP	1.6566	_____	_____
United States	New York City	USD93.20	1 USD	1.0000	93.20	100%
United States	San Francisco	USD88.72	1 USD	1.0000	_____	_____

Source: Data drawn from *The Random Walk, Mapping the World's Prices 2014*, Deutsche Bank Research, 09 May 2014, Figures 30 and 32, with author calculations. "Relative to NYC" is calculated as = (Cheap Date in USD)/(93.20).

Note: The *cheap date* combines the local currency cost of a cab ride for two, two McDonald's hamburgers, two soft drinks, two movie tickets, and two beers. In 2013 Deutsche Bank had included sending a bouquet of roses in the date, but did not include that in the 2014 index, making the two years not directly comparable.

2. **Blundell Biotech.** Blundell Biotech is a U.S.-based biotechnology company with operations and earnings in a number of foreign countries. The company's profits by *subsidiary*, in local currency (in millions), are shown in the table (a) below for 2013 and 2014.

   The average exchange rate for each year, by currency pairs, was as shown in table (b) below. Use this data to answer the following questions.

   a. What were Blundell Biotech's consolidated profits in U.S. dollars in 2013 and 2014?
   b. If the same exchange rates were used for both years, what was the change in corporate earnings on a "constant currency" basis?
   c. Using the results of the constant currency analysis in part b, is it possible to separate Blundell's growth in earnings between local currency earnings and foreign exchange rate impacts on a consolidated basis?

(a)

Net Income	Japanese Subsidiary	British Subsidiary	European Subsidiary	Chinese Subsidiary	Russian Subsidiary	United States Subsidiary
2013	JPY1,500	GBP100.00	EUR204.00	CNY168.00	RUB124.00	USD360.00
2014	JPY1,460	GBP106.40	EUR208.00	CNY194.00	RUB116.00	USD382.00

(b)

Exchange Rate	JPY = 1 USD	USD = 1 GBP	USD = 1 EUR	CNY = 1 USD	RUB = 1 USD	USD
2013	97.57	1.5646	1.3286	6.1484	31.86	1.0000
2014	105.88	1.6473	1.3288	6.1612	38.62	1.0000

Comparative Advantage

Problems 3–7 illustrate an example of trade induced by comparative advantage. They assume that China and France each have 1,000 production units. With one unit of production (a mix of land, labor, capital, and technology), China can produce either 10 containers of toys or 7 cases of wine. France can produce either 2 cases of toys or 7 cases of wine. Thus, a production unit in China is five times as efficient compared to France when producing toys, but equally efficient when producing wine. Assume at first that no trade takes place. China allocates 800 production units to building toys and 200 production units to producing wine. France allocates 200 production units to building toys and 800 production units to producing wine.

Problem 8.

Fixed Rmb Pricing of the PT350 Plasma Cutting Torch

Year	Cost (Rmb)	Margin (Rmb)	Price (Rmb)	Margin (percent)	Average Rate (Rmb/US$)	Price (US$)	Percent Change in US$ Price
2007	16,000	2,000	18,000	11.1%	7.61	2,365	—
2008	15,400	_____	_____	_____	6.95	_____	_____
2009	14,800	_____	_____	_____	6.83	_____	_____
2010	14,700	_____	_____	_____	6.77	_____	_____
2011	14,200	_____	_____	_____	6.46	_____	_____
2012	14,400	_____	_____	_____	6.31	_____	_____
2013	14,600	_____	_____	_____	6.15	_____	_____
2014	14,800	_____	_____	_____	6.16	_____	_____
Cumulative							_____

3. **Production and Consumption.** What is the production and consumption of China and France without trade?

4. **Specialization.** Assume complete specialization, where China produces only toys and France produces only wine. What would be the effect on total production?

5. **Trade at China's Domestic Price.** China's domestic price is 10 containers of toys equals 7 cases of wine. Assume China produces 10,000 containers of toys and exports 2,000 to France. Assume France produces 7,000 cases of wine and exports 1,400 cases to China. What happens to total production and consumption?

6. **Trade at France's Domestic Price.** France's domestic price is 2 containers of toys equals 7 cases of wine. Assume China produces 10,000 containers of toys and exports 400 containers to France. Assume France in turn produces 7,000 cases of wine and exports 1,400 cases to China. What happens to total production and consumption?

7. **Trade at Negotiated Mid-Price.** The mid-price for exchange between France and China can be calculated as follows. What happens to total production and consumption?

Assumptions	Toys (containers/ unit)	Wine (cases/unit)
China — output per unit of production input	10	7
France — output per unit of production input	2	7
China — total production inputs	1,000	
France — total production inputs	1,000	

8. **Peng Plasma Pricing.** Peng Plasma is a privately held Chinese business. It specializes in the manufacture of plasma cutting torches. Over the past eight years it has held the Chinese renminbi price of the PT350 cutting torch fixed at Rmb 18,000 per unit. Over that same period it has worked to reduce costs per unit, but has struggled of late due to higher input costs. Over that same period the renminbi has continued to be revalued against the U.S. dollar by the Chinese government. After completing the table — assuming the same price in renminbi for all years — answer the following questions.

a. What has been the impact of Peng's pricing strategy on the US$ price? How would you expect their U.S. dollar-based customers to have reacted to this?

b. What has been the impact on Peng's margins from this pricing strategy?

9. **Santiago Pirolta's Compensation Agreement.** Santiago Pirolta has accepted the Managing Director position for Vitro de Mexico's U.S. operations. Vitro is a Mexico-based manufacturer of flat and custom glass products. Much of its U.S. sales are based on a variety of bottle products, both mass market (e.g., glass bottles for soft drinks and beer) as well as specialty products (high-end cosmetic bottles with rare metal coloring and quality). He will live and work in the United States (Dallas, Texas) and wishes to be paid in U.S. dollars. Vitro has agreed that his base salary of USD350,000 will be paid in U.S. dollars, but Vitro wishes to tie his annual performance bonus (potentially 10% to 30% above his base salary) to the Mexican peso value of U.S. sales since Vitro consolidates all final results for reporting to stockholders in Mexican pesos (MXN).

Santiago, however, is a bit uncertain on having his bonus based on the Mexican peso values of U.S. sales. As a close friend and colleague, what advice would you give him based on your completion of the table below?

Year	(million USD)	Change	MXN = 1 USD	(million MXN)	Change
2011	USD820		12.80	MXN _____	
2012	USD842	____%	13.30	MXN _____	____%
2013	USD845	____%	12.70	MXN _____	____%
2014	USD860	____%	13.40	MXN _____	____%

Americo Industries—2010

Problems 10–14 are based on Americo Industries. Americo is a U.S.-based multinational manufacturing firm with wholly owned subsidiaries in Brazil, Germany, and China, in addition to domestic operations in the United States. Americo is traded on the NASDAQ. Americo currently has 650,000 shares outstanding. The basic operating characteristics of the various business units is as follows:

Business Performance (000s)	U.S. Parent (US$)	Brazilian Subsidiary (reais, R$)	German Subsidiary (euros, €)	Chinese Subsidiary (yuan, ¥)
Earnings before taxes (EBT)	$4,500	R$6,250	€4,500	¥2,500
Corporate income tax rate	35%	25%	40%	30%
Average exchange rate for the period	—	R$1.80/$	€0.7018/$	¥7.750/$

10. **Americo Industries' Consolidate Earnings.** Americo must pay corporate income tax in each country in which it currently has operations.
    a. After deducting taxes in each country, what are Americo's consolidated earnings and consolidated earnings per share in U.S. dollars?
    b. What proportion of Americo's consolidated earnings arise from each individual country?
    c. What proportion of Americo's consolidated earnings arise from outside the United States?

11. **Americo's EPS Sensitivity to Exchange Rates (A).** Assume a major political crisis wracks Brazil, first affecting the value of the Brazilian reais and, subsequently, inducing an economic recession within the country. What would be the impact on Americo's consolidated EPS if the Brazilian reais were to fall in value to R$3.00/$, with all other earnings and exchange rates remaining the same?

12. **Americo's EPS Sensitivity to Exchange Rates (B).** Assume a major political crisis wracks Brazil, first affecting the value of the Brazilian reais and, subsequently, inducing an economic recession within the country. What would be the impact on Americo's consolidated EPS if, in addition to the fall in the value of the reais to R$3.00/$, earnings before taxes in Brazil fell as a result of the recession to R$5,800,000?

13. **Americo's Earnings and the Fall of the Dollar.** The dollar has experienced significant swings in value against most of the world's currencies in recent years.
    a. What would be the impact on Americo's consolidated EPS if all foreign currencies were to appreciate 20% against the U.S. dollar?
    b. What would be the impact on Americo's consolidated EPS if all foreign currencies were to depreciate 20% against the U.S. dollar?

14. **Americo's Earnings and Global Taxation.** All MNEs attempt to minimize their global tax liabilities. Return to the original set of baseline assumptions and answer the following questions regarding Americo's global tax liabilities:
    a. What is the total amount—in U.S. dollars—that Americo is paying across its global business in corporate income taxes?
    b. What is Americo's *effective tax rate* (total taxes paid as a proportion of pre-tax profit)?
    c. What would be the impact on Americo's EPS and global effective tax rate if Germany instituted a corporate tax reduction to 28%, and Americo's earnings before tax in Germany rose to €5,000,000?

INTERNET EXERCISES

1. **International Capital Flows: Public and Private.** Major multinational organizations (some of which are listed next) attempt to track the relative movements and magnitudes of global capital investment. Using these Web pages and others you may find, prepare a two-page executive briefing on the question of whether capital generated in the industrialized countries is finding its way to the less-developed and emerging markets. Is there some critical distinction between "less-developed" and "emerging"?

The World Bank	www.worldbank.org
OECD	www.oecd.org
European Bank for Reconstruction and Development	www.ebrd.org

2. **External Debt.** The World Bank regularly compiles and analyzes the external debt of all countries globally. As part of their annual publication on World Development Indicators (WDI), they provide summaries of the long-term and short-term external debt obligations of selected countries online like that of Poland shown here. Go to their Web site and find the decomposition of external debt for Brazil, Mexico, and the Russian Federation.

The World Bank	www.worldbank.org/data

3. **World Economic Outlook.** The International Monetary Fund (IMF) regularly publishes its assessment of the prospects for the world economy. Choose a country of interest and use the IMF's current analysis to form your own expectations of its immediate economic prospects.

IMF Economic Outlook	www.imf.org/external/index.htm

4. ***Financial Times* Currency Global Macromaps.** The *Financial Times* provides a very helpful real-time global map of currency values and movements online. Use it to track the movements in currency.

Financial Times	markets.ft.com/research/Markets/Currencies

CHAPTER 2

The International Monetary System

*The price of every thing rises and falls from time to time and place to place;
and with every such change the purchasing power of money changes so far
as that thing goes.*

—Alfred Marshall.

LEARNING OBJECTIVES

- Explore how the international monetary system has evolved from the days of the gold standard to today's eclectic currency arrangement
- Detail how the International Monetary Fund categorizes the many different exchange rate regimes operating across the globe today
- Examine how the choice of fixed versus flexible exchange rate regimes is made by a country in the context of its desires for economic and social independence and openness
- Explain the dramatic choices the creation of a single currency for Europe—the euro—required of the European Union's member states
- Study the complexity of exchange rate regime choices faced by many emerging market countries today
- Describe the detailed strategy being deployed by China in the gradual globalization of the Chinese renminbi

This chapter begins with a brief history of the international monetary system, from the days of the classical gold standard to the present time. The first section describes contemporary currency regimes and their construction and classification, fixed versus flexible exchange rate principles, and what we would consider the theoretical core of the chapter—the attributes of the ideal currency and the choices nations must make in establishing their currency regime. The second section describes the many different exchange rate regimes at work today, following the IMF's classification system. The third section details the differences between fixed and flexible exchange rate systems, leading to the fourth section's description of the creation and development of the euro for European Union participating countries. The fifth section then details the difficult currency regime choices faced by many emerging market countries. The sixth and final section traces the rapid globalization of the Chinese renminbi now underway. The chapter concludes with the Mini-Case, *Iceland—A Small Country in a*

Global Crisis, which examines the rather classic case of how Iceland was confronted with the theoretical choices a country must make in defining its currency—described throughout the chapter—the *Impossible Trinity*.

History of the International Monetary System

Over the centuries, currencies have been defined in terms of gold, silver, and other items of value, all within a variety of different agreements between nations to recognize these varying definitions. A review of the evolution of these systems—or *eras* as we refer to them in Exhibit 2.1—provides a useful perspective against which to understand today's rather eclectic system of fixed rates, floating rates, crawling pegs, and others, and helps us to evaluate weaknesses in and challenges for all enterprises conducting global business.

The Gold Standard, 1876–1913

Since the days of the pharaohs (about 3000 B.C.), gold has served as a medium of exchange and a store of value. The Greeks and Romans used gold coins, and they passed on this tradition through the mercantile era to the nineteenth century. The great increase in trade during the free-trade period of the late nineteenth century led to a need for a more formalized system for settling international trade balances. One country after another set a par value for its currency in terms of gold and then tried to adhere to the so-called "rules of the game." This later came to be known as the *classical gold standard*. The gold standard, as an international monetary system, gained acceptance in Western Europe in the 1870s. The United States was something of a latecomer to the system, not officially adopting the gold standard until 1879.

Under the gold standard, the *rules of the game* were clear and simple: Each country set the rate at which its currency unit (paper or coin) could be converted to a given weight of gold. The United States, for example, declared the dollar to be convertible to gold at a rate of $20.67 per ounce (this rate was in effect until the beginning of World War I). The British pound was

---

EXHIBIT 2.1   **The Evolution of the Global Monetary System**

	Classical Gold Standard	Inter-War Years	Fixed Exchange Rates	Floating Exchange Rates	Emerging Era
	1870s — 1914	1923 — 1938	1944 — 1973	1973 — 1997	1997 —

Timeline: 1860, 1880, 1900, 1920, 1940, 1960, 1980, 2000, 2020

World War I (around 1914–1920); World War II (around 1938–1945)

Impact on Trade	Trade dominated capital flows	Increased barriers to trade & capital flows	Capital flows begin to dominate trade	Capital flows dominate trade	Selected emerging nations open capital markets
Impact on Economies	Increased world trade with limited capital flows	Protectionism & nationalism	Expanded open economies	Industrial economies increasingly open; emerging nations open slowly	Capital flows drive economic development

pegged at £4.2474 per ounce of gold. As long as both currencies were freely convertible into gold, the dollar/pound exchange rate was

$$\frac{\$20.67/\text{Ounce of Gold}}{£4.2474/\text{Ounce of Gold}} = \$4.8665/£$$

Because the government of each country on the gold standard agreed to buy or sell gold on demand at its own fixed parity rate, the value of each individual currency in terms of gold, and therefore exchange rates between currencies, was fixed. Maintaining reserves of gold that were sufficient to back its currency's value was very important for a country under this system. The system also had the effect of implicitly limiting the rate at which any individual country could expand its money supply. Growth in the money supply was limited to the rate at which official authorities (government treasuries or central banks) could acquire additional gold.

The gold standard worked adequately until the outbreak of World War I interrupted trade flows and the free movement of gold. This event caused the main trading nations to suspend operation of the gold standard.

The Interwar Years and World War II, 1914–1944

During World War I and through the early 1920s, currencies were allowed to fluctuate over fairly wide ranges in terms of gold and in relation to each other. Theoretically, supply and demand for a country's exports and imports caused moderate changes in an exchange rate about a central equilibrium value. This was the same function that gold had performed under the previous gold standard. Unfortunately, such flexible exchange rates did not work in an equilibrating manner. On the contrary: international speculators sold the weak currencies short, causing them to fall further in value than warranted by real economic factors. *Selling short* is a speculation technique in which an individual speculator sells an asset, such as a currency, to another party for delivery at a future date. The speculator, however, does not yet own the asset and expects the price of the asset to fall before the date by which the speculator must purchase the asset in the open market for delivery.

The reverse happened with strong currencies. Fluctuations in currency values could not be offset by the relatively illiquid forward exchange market, except at exorbitant cost. The net result was that the volume of world trade did not grow in the 1920s in proportion to world gross domestic product. Instead, it declined to a very low level with the advent of the Great Depression in the 1930s.

The United States adopted a modified gold standard in 1934 when the U.S. dollar was devalued to $35 per ounce of gold from the $20.67 per ounce price in effect prior to World War I. Contrary to previous practice, the U.S. Treasury traded gold only with foreign central banks, not private citizens. From 1934 to the end of World War II, exchange rates were theoretically determined by each currency's value in terms of gold. During World War II and its chaotic aftermath, however, many of the main trading currencies lost their convertibility into other currencies. The dollar was one of the few currencies that continued to be convertible.

Bretton Woods and the International Monetary Fund, 1944

As World War II drew to a close in 1944, the Allied Powers met at Bretton Woods, New Hampshire, to create a new postwar international monetary system. The Bretton Woods Agreement established a U.S. dollar–based international monetary system and provided for two new institutions: the International Monetary Fund and the World Bank. The *International Monetary Fund* (IMF) aids countries with balance of payments and exchange rate problems. The *International Bank for Reconstruction and Development* (IBRD or the *World Bank*) helped fund postwar reconstruction and has since supported general economic development. *Global Finance in Practice 2.1* provides some insight into the debates at Bretton Woods.

GLOBAL FINANCE IN PRACTICE 2.1

Hammering Out an Agreement at Bretton Woods

The governments of the Allied powers knew that the devastating impacts of World War II would require swift and decisive policies. In the summer of 1944 (July 1–22), representatives of all 45 allied nations met at Bretton Woods, New Hampshire, for the United Nations Monetary and Financial Conference. Their purpose was to plan the postwar international monetary system. It was a difficult process, and the final synthesis was shaded by pragmatism.

The leading policymakers at Bretton Woods were the British and the Americans. The British delegation was led by Lord John Maynard Keynes, known as "Britain's economic heavy weight." The British argued for a postwar system that would be more flexible than the various gold standards used before the war. Keynes argued, as he had after World War I, that attempts to tie currency values to gold would create pressures for deflation in many of the war-ravaged economies.

The American delegation was led by the director of the U.S. Treasury's monetary research department, Harry D. White, and the U.S. Secretary of the Treasury, Henry Morgenthau, Jr. The Americans argued for stability (fixed exchange rates) but not a return to the gold standard itself. In fact, although the U.S. at that time held most of the gold of the Allied powers, the U.S. delegates argued that currencies should be *fixed in parities*, but that redemption of gold should occur only between official authorities like central banks.*

On the more pragmatic side, all parties agreed that a postwar system would be stable and sustainable only if there was sufficient credit available for countries to defend their currencies in the event of payment imbalances, which they knew to be inevitable in a reconstructing world order.

The conference divided into three commissions for weeks of negotiation. One commission, led by U.S. Treasury Secretary Morgenthau, was charged with the organization of a fund of capital to be used for exchange rate stabilization. A second commission, chaired by Lord Keynes, was charged with the organization of a second "bank" whose purpose would be for long-term reconstruction and development. A third commission was to hammer out details such as what role silver would have in any new system.

After weeks of meetings, the participants came to a three-part agreement—the Bretton Woods Agreement. The plan called for: (1) fixed exchange rates, termed an "adjustable peg," among members; (2) a fund of gold and constituent currencies available to members for stabilization of their respective currencies, called the International Monetary Fund (IMF); and (3) a bank for financing long-term development projects (eventually known as the World Bank). One proposal resulting from the meetings, which was not ratified by the United States, was the establishment of an international trade organization to promote free trade.

*Fixed in parities is an old expression in this field, which means that the value of currencies should be set or fixed at rates that equalize their value, typically purchasing power.

The IMF was the key institution in the new international monetary system, and it has remained so to the present day. The IMF was established to render temporary assistance to member countries trying to defend their currencies against cyclical, seasonal, or random occurrences. It also assists countries having structural trade problems if they promise to take adequate steps to correct their problems. If persistent deficits occur, however, the IMF cannot save a country from eventual devaluation. In recent years, the IMF has attempted to help countries facing financial crises, providing massive loans as well as advice to Russia, Brazil, Greece, Indonesia, and South Korea, to name but a few.

Under the original provisions of Bretton Woods, all countries fixed the value of their currencies in terms of gold but they were not required to exchange their currencies for gold. Only the dollar remained convertible into gold (at $35 per ounce). Therefore, each country established its exchange rate vis-à-vis the dollar, and then calculated the gold par value of its currency to create the desired dollar exchange rate. Participating countries agreed to try to maintain the value of their currencies within 1% (later expanded to 2.25%) of par by buying or selling foreign exchange or gold as needed. Devaluation was not to be used as a competitive trade policy, but if a currency became too weak to defend, devaluation of up to 10% was allowed without formal approval by the IMF. Larger devaluations required IMF approval. This became known as the *gold-exchange standard*.

An additional innovation introduced by Bretton Woods was the creation of the *Special Drawing Right* or SDR. The SDR is an international reserve asset created by the IMF to supplement existing foreign exchange reserves. It serves as a unit of account for the IMF and other international and regional organizations. It is also the base against which some countries peg the exchange rate for their currencies. Initially defined in terms of a fixed quantity of gold, the SDR is currently the weighted average of four major currencies: the U.S. dollar, the euro, the Japanese yen, and the British pound. The weight assigned to each currency is updated every five years by the IMF. Individual countries hold SDRs in the form of deposits in the IMF. These holdings are part of each country's international monetary reserves, along with its official holdings of gold, its foreign exchange, and its reserve position at the IMF. Member countries may settle transactions among themselves by transferring SDRs.

Fixed Exchange Rates, 1945–1973

The currency arrangement negotiated at Bretton Woods and monitored by the IMF worked fairly well during the postwar period of reconstruction and rapid growth in world trade. However, widely diverging national monetary and fiscal policies, differential rates of inflation, and various unexpected external shocks eventually resulted in the system's demise. The U.S. dollar was the main reserve currency held by central banks and was the key to the web of exchange rate values. Unfortunately, the U.S. ran persistent and growing deficits in its balance of payments. A heavy capital outflow of dollars was required to finance these deficits and to meet the growing demand for dollars from investors and businesses. Eventually, the heavy overhang of dollars held by foreigners resulted in a lack of confidence in the ability of the U.S. to meet its commitments in gold.

This lack of confidence came to a head in the first half of 1971. In a little less than seven months, the United States suffered the loss of nearly one-third of its official gold reserves as global confidence in the value of the dollar plummeted. Exchange rates between most major currencies and the U.S. dollar began to float, and thus indirectly, their values relative to gold. A year and a half later, the U.S. dollar once again came under attack, thereby forcing a second devaluation in February 1973; this time by 10% to $42.22 per ounce of gold. By late February 1973, a fixed-rate system no longer appeared feasible given the speculative flows of currencies. The major foreign exchange markets were actually closed for several weeks in March 1973. When they reopened, most currencies were allowed to float to levels determined by market forces.

The Floating Era, 1973–1997

Since March 1973, exchange rates have become much more volatile and less predictable than they were during the "fixed" exchange rate period, when changes occurred infrequently. Exhibit 2.2 illustrates the wide swings exhibited by the nominal exchange rate index of the U.S. dollar since 1964. Clearly, volatility has increased for this currency measure since 1973.

Exhibit 2.2 notes some of the most important shocks in recent history: the creation of the European Monetary System (EMS) in 1979; the run-up and peak of the U.S. dollar in 1985; the EMS crisis of 1992; the Asian crisis of 1997; the launch of the European euro in 1999; the rise of the dollar in 2014 and 2015.

The Emerging Era, 1997–Present

The period following the Asian Crisis of 1997 has seen growth in both breadth and depth of emerging market economies and currencies. We may end up being proven wrong on this count, but the final section of this chapter argues that the global monetary system has already begun embracing—for over a decade now—a number of major emerging market currencies, beginning with the Chinese renminbi. Feel free to disagree.

EXHIBIT 2.2   **The BIS Exchange Rate Index of the Dollar**

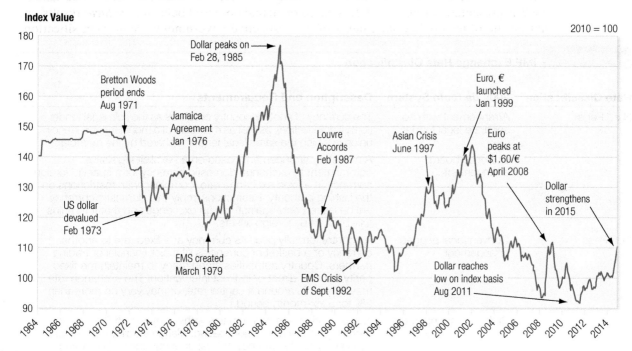

Source: BIS.org. Nominal exchange rate index (narrow definition) for the U.S. dollar.

IMF Classification of Currency Regimes

The global monetary system—if there is indeed a singular "system"—is an eclectic combination of exchange rate regimes and arrangements. Although there is no single governing body, the International Monetary Fund (IMF) has at least played the role of "town crier" since World War II. We present its current classification system of currency regimes here.

Brief Classification History

The IMF was for many years the central clearinghouse for exchange rate classifications. Member states submitted their exchange rate policies to the IMF, and those submissions were the basis for its categorization of exchange rate regimes. However, that all changed in 1997–1998 with the Asian Financial Crisis. During the Asian Financial Crisis, many countries began following very different exchange rate practices than those they had committed to with the IMF. Their actual practices—their *de facto* systems—were not what they had publicly and officially committed to—their *de jure* systems.

Beginning in 1998 the IMF changed its practice and stopped collecting regime classification submissions from member states. Instead, it confined its regime classifications and reports to analysis performed in-house. (This included the cessation of publishing its Annual Report on Exchange Arrangements and Exchange Restrictions, a document on which much of the world's financial institutions depended on for decades.) As a global institution, which is in principle apolitical, the IMF's analysis today is focused on classifying currencies on the basis of an *ex post* analysis of how the currency's value was based in the recent past. This analysis focuses on observed behavior, not on official government policy pronouncements.

The IMF's *de facto* System

The IMF's methodology of classifying exchange rate regimes today, in effect since January 2009, is presented in Exhibit 2.3. It is based on actual observed behavior, *de facto* results, and not on the official policy statements of the respective governments, *de jure* classification.[1]

EXHIBIT 2.3   IMF Exchange Rate Classification

Rate Classification	2009 *de facto* System	Description and Requirements
Hard Pegs	Arrangement with no separate legal tender	The currency of another country circulates as the sole legal tender (formal dollarization), as well as members of a monetary or currency union in which the same legal tender is shared by the members.
	Currency board arrangement	A monetary arrangement based on an explicit legislative commitment to exchange domestic currency for a specific foreign currency at a fixed exchange rate, combined with restrictions on the issuing authority. Restrictions imply that domestic currency will be issued only against foreign exchange and that it remains fully backed by foreign assets.
Soft Pegs	Conventional pegged arrangement	A country formally pegs its currency at a fixed rate to another currency or a basket of currencies of major financial or trading partners. Country authorities stand ready to maintain the fixed parity through direct or indirect intervention. The exchange rate may vary ±1% around a central rate, or may vary no more than 2% for a six-month period.
	Stabilized arrangement	A spot market rate that remains within a margin of 2% for six months or more and is not floating. Margin stability can be met by either a single currency or basket of currencies (assuming statistical measurement). Exchange rate remains stable as a result of official action.
	Intermediate pegs:	
	Crawling peg	Currency is adjusted in small amounts at a fixed rate or in response to changes in quantitative indicators (e.g., inflation differentials).
	Crawl-like arrangement	Exchange rate must remain with a narrow margin of 2% relative to a statistically defined trend for six months or more. Exchange rate cannot be considered floating. Minimum rate of change is greater than allowed under a stabilized arrangement.
	Pegged exchange rate within horizontal bands	The value of the currency is maintained within 1% of a fixed central rate, or the margin between the maximum and minimum value of the exchange rate exceeds 2%. This includes countries that are today members of the Exchange Rate Mechanism II (ERM II) system.
Floating Arrangements	Floating	Exchange rate is largely market determined without an ascertainable or predictable path. Market intervention may be direct or indirect, and serves to moderate the rate of change (but not targeting). Rate may exhibit more or less volatility.
	Free floating	A floating rate is freely floating if intervention occurs only exceptionally, and confirmation of intervention is limited to at most three instances in a six-month period, each lasting no more than three business days.
Residual	Other managed arrangements	This category is residual, and is used when the exchange rate does not meet the criteria for any other category. Arrangements characterized by frequent shifts in policies fall into this category.

Source: "Revised System for the Classification of Exchange Rate Arrangements," by Karl Habermeier, Anamaria Kokenyne, Romain Veyrune, and Harald Anderson, IMF Working Paper WP/09/211, International Monetary Fund, November 17, 2009.

[1] "Revised System for the Classification of Exchange Rate Arrangements," by Karl Habermeier, Annamaria Kokenyne, Romain Veyrune, and Harald Anderson, Monetary and Capital Markets Department, IMF Working Paper 09/211, November 17, 2009. The system presented is a revision of the IMF's 1998 revision to a *de facto* system.

The classification process begins with the determination of whether the exchange rate of the country's currency is dominated by markets or by official action. Although the classification system is a bit challenging, there are four basic categories.

Category 1: Hard Pegs. These countries have given up their own sovereignty over monetary policy. This category includes countries that have adopted other countries' currencies (e.g., Zimbabwe's *dollarization*—its adoption of the U.S. dollar), and countries utilizing a currency board structure that limits monetary expansion to the accumulation of foreign exchange.

Category 2: Soft Pegs. This general category is colloquially referred to as *fixed exchange rates*. The five subcategories of *soft peg regimes* are differentiated on the basis of what the currency is fixed to, whether that fix is allowed to change—and if so under what conditions, what types, magnitudes, and frequencies of intervention are allowed/used, and the degree of variance about the fixed rate.

Category 3: Floating Arrangements. Currencies that are predominantly market-driven are further subdivided into free floating with values determined by open market forces without governmental influence or intervention, and simple floating or floating with intervention, where government occasionally does intervene in the market in pursuit of some rate goals or objectives.

Category 4: Residual. As one would suspect, this category includes all exchange rate arrangements that do not meet the criteria of the previous three categories. Country systems demonstrating frequent shifts in policy typically make up the bulk of this category.

Exhibit 2.4 provides a glimpse as to what these major regime categories translate into in the global market—fixed or floating. The vertical dashed line, the *crawling peg*, is the zone some currencies move into and out of depending on their relative currency stability. Although the classification regimes appear clear and distinct, the distinctions are often more difficult

---

EXHIBIT 2.4   Taxonomy of Exchange Rate Regimes

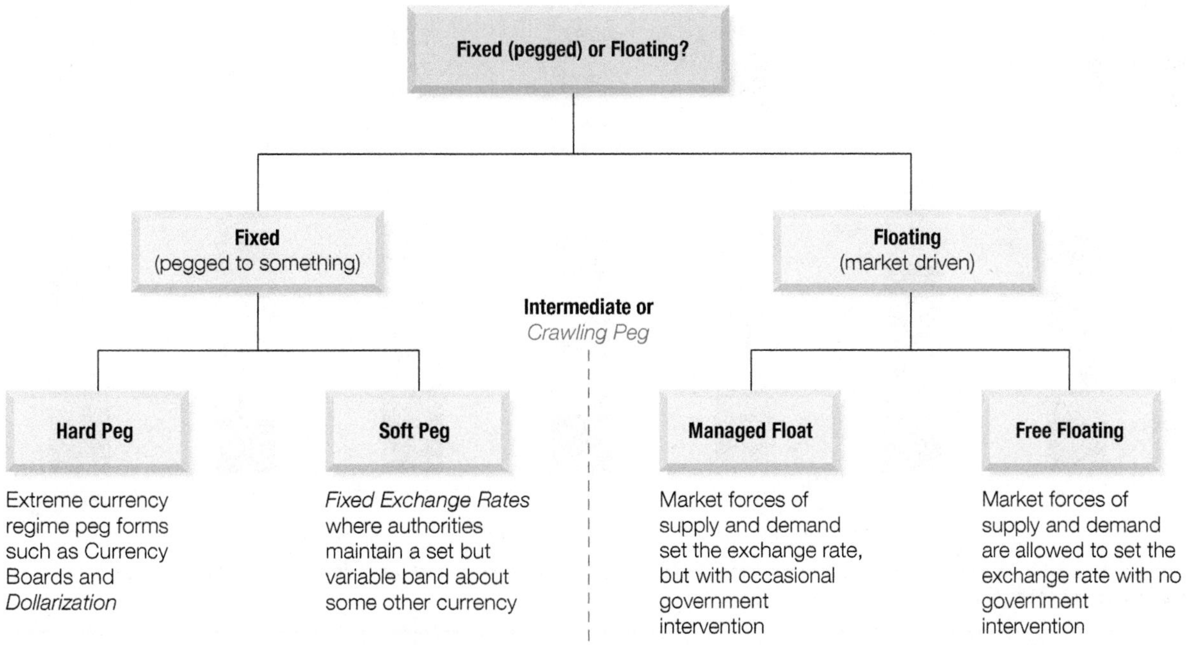

Hard Peg	Soft Peg	Managed Float	Free Floating
Extreme currency regime peg forms such as Currency Boards and *Dollarization*	*Fixed Exchange Rates* where authorities maintain a set but variable band about some other currency	Market forces of supply and demand set the exchange rate, but with occasional government intervention	Market forces of supply and demand are allowed to set the exchange rate with no government intervention

to distinguish in practice in the market. For example, in January 2014, the Bank of Russia announced it would no longer conduct intervention activities with regard to the value of the ruble and that it planned to allow the ruble to trade freely, with no intervention.

A Global Eclectic

Despite the IMF's attempt to lend rigor to regime classifications, the global monetary system today is indeed a global eclectic in every sense of the term. As Chapter 5 will describe in detail, the current global market in currency is dominated by two major currencies, the U.S. dollar and the European euro, and after that, a multitude of systems, arrangements, currency areas, and zones.

The euro itself is an example of a rigidly fixed system, acting as a single currency for its member countries. However, the euro is also an independently floating currency against all other currencies. Other examples of rigidly fixed exchange regimes include Ecuador, Panama, and Zimbabwe, which use the U.S. dollar as their official currency; the Central African Franc (CFA) zone, in which countries such as Mali, Niger, Senegal, Cameroon, and Chad among others use a single common currency (the franc, which is tied to the euro); and the Eastern Caribbean Currency Union (ECCU), a set of countries that use the Eastern Caribbean dollar.

At the other extreme are countries with independently floating currencies. These include many of the most developed countries, such as Japan, the United States, the United Kingdom, Canada, Australia, New Zealand, Sweden, and Switzerland. However, this category also includes a number of unwilling participants—emerging market countries that tried to maintain fixed rates but were forced by the marketplace to let them float. Among these are Korea, the Philippines, Brazil, Indonesia, Mexico, and Thailand.

As illustrated by Exhibit 2.5, the proportion of IMF member countries (188 reporting in 2014) with floating regimes (managed floats and free floats) has been holding at about 34%. Soft pegs continue to dominate, at 43.5% of all member countries in 2014. Although the

---

EXHIBIT 2.5   IMF Membership Exchange Rate Regime Choices

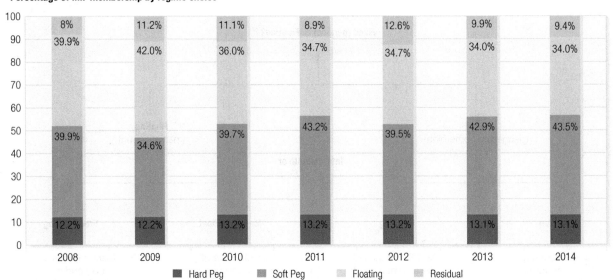

Percentage of IMF membership by regime choice

Source: Data drawn from *Annual Report on Exchange Arrangements and Exchange Restrictions 2014*, International Monetary Fund, 2014, Table 3, Exchange Rate Arrangements 2008–2014.

contemporary international monetary system is typically referred to as a "floating regime," it is clearly not the case for the majority of the world's nations.

Fixed versus Flexible Exchange Rates

A nation's choice as to which currency regime to follow reflects national priorities about all facets of the economy, including inflation, unemployment, interest rate levels, trade balances, and economic growth. The choice between fixed and flexible rates may change over time as priorities change. At the risk of overgeneralizing, the following points partly explain why countries pursue certain exchange rate regimes. They are based on the premise that, other things being equal, countries would prefer fixed exchange rates.

- Fixed rates provide stability in international prices for the conduct of trade. Stable prices aid in the growth of international trade and lessen risks for all businesses.
- Fixed exchange rates are inherently anti-inflationary, requiring the country to follow restrictive monetary and fiscal policies. This restrictiveness, however, can often be a burden to a country wishing to pursue policies to alleviate internal economic problems such as high unemployment or slow economic growth.

Fixed exchange rate regimes necessitate that central banks maintain large quantities of international reserves (hard currencies and gold) for use in the occasional defense of the fixed rate. As international currency markets have grown rapidly in size and volume, increasing reserve holdings has become a significant burden to many nations.

Fixed rates, once in place, may be maintained at levels that are inconsistent with economic fundamentals. As the structure of a nation's economy changes, and as its trade relationships and balances evolve, the exchange rate itself should change. Flexible exchange rates allow this to happen gradually and efficiently, but fixed rates must be changed administratively—usually too late, with too much publicity, and at too large a one-time cost to the nation's economic health.

The terminology associated with changes in currency values is also technically specific. When a government officially declares its own currency to be worth less or more relative to other currencies, it is termed a *devaluation* or *revaluation*, respectively. This obviously applies to currencies whose value is controlled by government. When a currency's value is changed in the open currency market—not directly by government—it is called a *depreciation* (with a fall in value) or *appreciation* (with an increase in value).

The Impossible Trinity

If the ideal currency existed in today's world, it would possess the following three attributes, illustrated in Exhibit 2.6, often referred to as the *impossible trinity*.

1. **Exchange rate stability.** The value of the currency is fixed in relationship to other major currencies, so traders and investors could be relatively certain of the foreign exchange value of each currency in the present and into the near future.

2. **Full financial integration.** Complete freedom of monetary flows would be allowed, so traders and investors could easily move funds from one country and currency to another in response to perceived economic opportunities or risks.

3. **Monetary independence.** Domestic monetary and interest rate policies would be set by each individual country to pursue desired national economic policies, especially as they might relate to limiting inflation, combating recessions, and fostering prosperity and full employment.

These qualities are termed the *impossible trinity* (also referred to as the *trilemma of international finance*) because the forces of economics do not allow a country to simultaneously

EXHIBIT 2.6	**The Impossible Trinity**

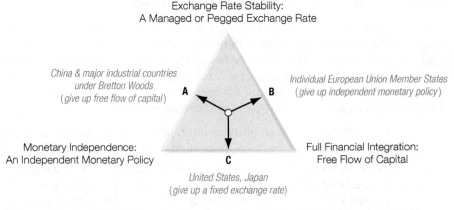

Nations must choose in which direction to move from the center—toward points A, B, or C. Their choice is a choice of what *to pursue* and what *to give up*—that of the opposite point of the pyramid. Marginal compromise is possible, but only marginal.

achieve all three goals: monetary independence, exchange rate stability, and full financial integration. For example a country like the United States has knowingly given up having a fixed exchange rate—moving from the center of the pyramid toward point C—because it wishes to have an independent monetary policy, and it allows an extremely high level of freedom in the movement of capital into and out of the country.

China today is a clear example of a nation that has chosen to continue to control and manage the value of its currency and to conduct an independent monetary policy—moving from the center of the pyramid toward point A—while continuing to restrict the flow of capital into and out of the country. To say it has "given up" the free flow of capital probably would be inaccurate, as China has allowed no real freedom of capital flows in the past century.

The consensus of many experts is that the force of increased capital mobility has been pushing more and more countries toward full financial integration in an attempt to stimulate their domestic economies and to feed the capital appetites of their own MNEs. As a result, their currency regimes are being "cornered" into being either purely floating (like the United States) or integrated with other countries in monetary unions (like the European Union). *Global Finance in Practice 2.2* drives this debate home.

GLOBAL FINANCE IN PRACTICE 2.2

Who Is Choosing What in the Trinity/Trilemma?

The global financial crisis of 2008–2009 sparked much debate over the value of currencies—in some cases invoking what some termed *currency wars*. With most of the non-Chinese world suffering very slow economic growth, and under heavy pressure to stimulate their economies and alleviate high unemployment rates, more and more arguments and efforts for a weak or undervalued currency arose. Although sounding logical, the impossible trinity makes it very clear that each economy must choose its own medicine. Here is what many argue are the choices of three of the major global economic players:

	Choice #1	Choice #2	Implied Condition #3
United States	Independent monetary policy	Free movement of capital	Currency value floats
China	Independent monetary policy	Fixed rate of exchange	Restricted movement of capital
Europe (EU)	Free movement of capital	Fixed rate of exchange	Integrated monetary policy

The choices made by the European Union (EU) are clearly the more complex. As a combination of different sovereign states, the EU has pursued integration of a common currency, the euro, and free movement of labor and capital. The result, according to the impossible trinity, is that EU member states had to give up independent monetary policy, replacing individual central banks with the *European Central Bank* (ECB). The recent fiscal deficits and near-collapses of government debt issuances in Greece, Portugal, and Ireland have raised questions over the efficacy of the arrangement.

A Single Currency for Europe: The Euro

Beginning with the Treaty of Rome in 1957 and continuing with the Single European Act of 1987, the Maastricht Treaty of 1991–1992, and the Treaty of Amsterdam of 1997, a core set of European countries worked steadily toward integrating their individual countries into one larger, more efficient, domestic market. However, even after the launch of the 1992 Single Europe program, a number of barriers to true openness remained, including the use of different currencies. The use of different currencies required both consumers and companies to treat the individual markets separately. Currency risk of cross-border commerce still persisted. The creation of a single currency was seen as the way to move beyond these last vestiges of separated markets.

The original 15 members of the EU were also members of the *European Monetary System* (EMS). The EMS formed a system of fixed exchange rates amongst the member currencies, with deviations managed through bilateral responsibility to maintain rates at ±2.5% of an established central rate. This system of fixed rates, with adjustments along the way, remained in effect from 1979–1999. Its resiliency was seriously tested with exchange rate crises in 1992 and 1993, but it held and moved onward.

The Maastricht Treaty and Monetary Union

In December 1991, the members of the EU met at Maastricht, the Netherlands, and concluded a treaty that changed Europe's currency future. The Maastricht Treaty specified a timetable and a plan to replace all individual EMS member currencies with a single currency—eventually named the euro. Other aspects of the treaty were also adopted that would lead to a full European Economic and Monetary Union (EMU). According to the EU, the EMU is a single-currency area within the singular EU market, now known informally as the *eurozone*, in which people, goods, services, and capital are allowed to move without restrictions.

The integration of separate country monetary systems is not, however, a minor task. To prepare for the EMU, the Maastricht Treaty called for the integration and coordination of the member countries' monetary and fiscal policies. The EMU would be implemented by a process called *convergence*.

Before becoming a full member of the EMU, each member country was expected to meet a set of convergence criteria in order to integrate systems that were at the same relative performance levels: (1) nominal inflation should be no more than 1.5% above the average for the three members of the EU that had the lowest inflation rates during the previous year; (2) long-term interest rates should be no more than 2% above the average of the three members with the lowest interest rates; (3) individual government budget deficits—fiscal deficits—should be no more than 3% of gross domestic product; and (4) government debt outstanding should be no more than 60% of gross domestic product. The convergence criteria were so tough that few, if any, of the members could satisfy them at that time, but 11 countries managed to do so just prior to 1999 (Greece was added two years later).

The European Central Bank (ECB)

The cornerstone of any monetary system is a strong, disciplined, central bank. The Maastricht Treaty established this single institution for the EMU, the European Central Bank (ECB), which was established in 1998. (The EU created the European Monetary Institute (EMI) in 1994 as a transitional step in establishing the European Central Bank.) The ECB's structure and functions were modeled after the German Bundesbank, which in turn had been modeled after the U.S. Federal Reserve System. The ECB is an independent central bank that dominates the activities of the individual countries' central banks. The individual central banks continue to regulate banks resident within their borders, but all financial market intervention and the issuance of the single currency is the sole responsibility of the ECB. The single most important mandate of the ECB is its charge to promote price stability within the European Union.

The Launch of the Euro

On January 4, 1999, 11 member states of the EU initiated the EMU. They established a single currency, the euro, which replaced the individual currencies of the participating member states. The 11 countries were Austria, Belgium, Finland, France, Germany, Ireland, Italy, Luxembourg, the Netherlands, Portugal, and Spain. Greece did not qualify for EMU participation at the time, but joined the euro group later, in 2001. On December 31, 1998, the final fixed rates between the 11 participating currencies and the euro were put into place. On January 4, 1999, the euro was officially launched.

The United Kingdom, Sweden, and Denmark chose to maintain their individual currencies. The United Kingdom has been skeptical of increasing EU infringement on its sovereignty, and has opted not to participate. Sweden, which has failed to see significant benefits from EU membership (although it is one of the newest members), has also been skeptical of EMU participation. Denmark, like the United Kingdom, Sweden, and Norway has so far opted not to participate. (Denmark is, however, a member of ERM II, the Exchange Rate Mechanism II, which effectively allows Denmark to keep its own currency and monetary sovereignty, but fixes the value of its currency, the krone, to the euro.)

The official currency symbol of the euro, EUR, was registered with the International Standards Organization. The official symbol of the euro is €. According to the EU, the € symbol was inspired by the Greek letter epsilon (ε), simultaneously referring to Greece's ancient role as the source of European civilization and recalling the first letter of the word Europe.

The euro would generate a number of benefits for the participating states: (1) Countries within the eurozone enjoy cheaper transaction costs; (2) Currency risks and costs related to exchange rate uncertainty are reduced; and (3) All consumers and businesses both inside and outside the eurozone enjoy price transparency and increased price-based competition. The primary "cost" of adopting the euro, the loss of monetary independence, would be a continuing challenge for the members for years to come.

On January 4, 1999, the euro began trading on world currency markets. Its introduction was a smooth one. The euro's value slid steadily following its introduction, however, primarily as a result of the robustness of the U.S. economy and U.S. dollar, and sluggish economic sectors in the EMU countries. Beginning in 2002, the euro appreciated versus the dollar. Since that time, as illustrated in Exhibit 2.7, it had remained in a range of roughly $1.20 to $1.50 per euro. It has, however, demonstrated significant volatility.

The use of the euro has continued to expand since its introduction. As of January 2012, the euro was the official currency for 17 of the 27 member countries in the European Union, as well as five other countries (Montenegro, Andorra, Monaco, San Marino, and the Vatican) that may eventually join the EU. The 17 countries that currently use the euro—the so-called eurozone— are Austria, Belgium, Cyprus, Estonia, Finland, France, Germany, Greece, Ireland, Italy, Luxembourg, Malta, the Netherlands, Portugal, Slovakia, Slovenia, and Spain. Although all members of

EXHIBIT 2.7	**The U.S. Dollar–European Euro Spot Exchange Rate**

U.S. dollars (USD) to = 1 euro (EUR)

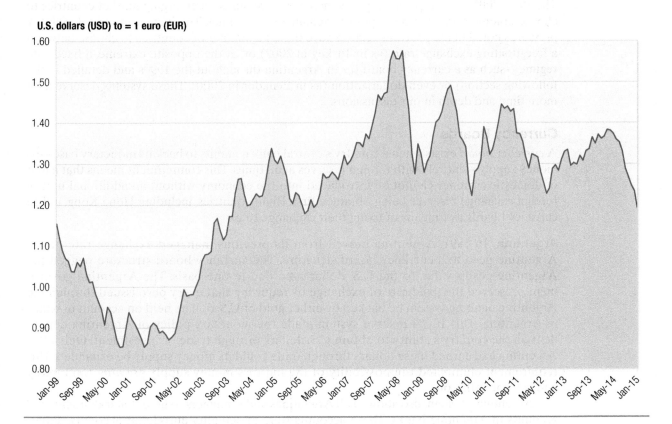

the EU are expected eventually to replace their currencies with the euro, recent years have seen growing debates and continual postponements by new members in moving toward full euro adoption. The continuing issues with European sovereign debt, as discussed in *Global Finance in Practice 2.3*, also continue to pose serious challenges to further euro expansion.

GLOBAL FINANCE IN PRACTICE 2.3

The Euro and the Greek/EU Debt Crisis

The European Monetary Union is a complex organism compared to the customary country structure of fiscal and monetary policy institutions and policies described in a typical Economics 101 course. The members of the EU do not have the ability to conduct independent monetary policy. When the EU moved to a single currency with the adoption of the euro, its member states agreed to use a single currency (exchange rate stability), allow the free movement of capital in and out of their economies (financial integration), but give up individual control of their own money supply (monetary independence). Once again, a choice was made among the three competing dimensions of the impossible trinity; in this case, to form a single monetary policy body—the European Central

Bank (ECB)—to conduct monetary policy on behalf of all EU members.

But fiscal and monetary policies are still somewhat intertwined. Government deficits that are funded by issuing debt to the international financial markets still impact monetary policy. Proliferating sovereign debt—debt issued by Greece, Portugal, and Ireland, for example—may be euro-denominated, but it is the debt obligation of each individual government. However, if one or more of these governments flood the market with debt, this may result in increased cost and decreased availability of capital to other member states. In the end, if monetary independence is not preserved, then one or both of the other elements of the impossible trinity may fail—capital mobility or exchange rate stability.

Emerging Markets and Regime Choices

The 1997–2005 period specifically saw increasing pressures on emerging market countries to choose among more extreme types of exchange rate regimes. The increased capital mobility pressures noted in the previous section have driven a number of countries to choose between a free-floating exchange rate (as in Turkey in 2002) or, at the opposite extreme, a fixed-rate regime—such as a currency board (as in Argentina throughout the 1990s and detailed in the following section) or even dollarization (as in Ecuador in 2000). These systems deserve a bit more time and depth in our discussions.

Currency Boards

A *currency board* exists when a country's central bank commits to back its monetary base—its money supply—entirely with foreign reserves at all times. This commitment means that a unit of domestic currency cannot be introduced into the economy without an additional unit of foreign exchange reserves being obtained first. Eight countries, including Hong Kong, utilize currency boards as a means of fixing their exchange rates.

Argentina. In 1991, Argentina moved from its previous managed exchange rate of the Argentine peso to a currency board structure. The currency board structure pegged the Argentine peso's value to the U.S. dollar on a one-to-one basis. The Argentine government preserved the fixed rate of exchange by requiring that every peso issued through the Argentine banking system be backed by either gold or U.S. dollars held on account in banks in Argentina. This 100% reserve system made the monetary policy of Argentina dependent on the country's ability to obtain U.S. dollars through trade or investment. Only after Argentina had earned these dollars through trade could its money supply be expanded. This requirement eliminated the possibility of the nation's money supply growing too rapidly and causing inflation.

Argentina's system also allowed all Argentines and foreigners to hold dollar-denominated accounts in Argentine banks. These accounts were in actuality eurodollar accounts, dollar-denominated deposits in non-U.S. banks. These accounts provided savers with the ability to choose whether or not to hold pesos.

From the very beginning there was substantial doubt in the market that the Argentine government could maintain the fixed exchange rate. Argentine banks regularly paid slightly higher interest rates on peso-denominated accounts than on dollar-denominated accounts. This interest differential represented the market's assessment of the risk inherent in the Argentine financial system. Depositors were rewarded for accepting risk—for keeping their money in peso-denominated accounts. In January 2002, after months of economic and political turmoil and nearly three years of economic recession, the Argentine currency board was ended. The peso was first devalued from Peso1.00/$ to Peso1.40/$, then it was floated completely. It fell in value dramatically within days. The Argentine decade-long experiment with a rigidly fixed exchange rate was over.

Dollarization

Several countries have suffered currency devaluation for many years, primarily as a result of inflation, and have taken steps toward dollarization. Dollarization is the use of the U.S. dollar as the official currency of the country. Panama has used the dollar as its official currency since 1907. Ecuador, after suffering a severe banking and inflationary crisis in 1998 and 1999, adopted the U.S. dollar as its official currency in January 2000. One of the primary attributes of dollarization was summarized well by *BusinessWeek* in a December 11, 2000, article entitled "The Dollar Club":

> *One attraction of dollarization is that sound monetary and exchange-rate policies no longer depend on the intelligence and discipline of domestic policymakers. Their monetary policy becomes essentially the one followed by the U.S., and the exchange rate is fixed forever.*

The arguments for dollarization follow logically from the previous discussion of the impossible trinity. A country that dollarizes removes any currency volatility (against the dollar) and would theoretically eliminate the possibility of future currency crises. Additional benefits are expectations of greater economic integration with other dollar-based markets, both product and financial. This last point has led many to argue in favor of regional dollarization, in which several countries that are highly economically integrated may benefit significantly from dollarizing together.

Three major arguments exist against dollarization. The first is the loss of sovereignty over monetary policy. This is, however, the point of dollarization. Second, the country loses the power of *seigniorage*, the ability to profit from its ability to print its own money. Third, the central bank of the country, because it no longer has the ability to create money within its economic and financial system, can no longer serve the role of lender of last resort. This role carries with it the ability to provide liquidity to save financial institutions that may be on the brink of failure during times of financial crisis.

Ecuador. Ecuador officially completed the replacement of the Ecuadorian sucre with the U.S. dollar as legal tender in September 2000. This step made Ecuador the largest national adopter of the U.S. dollar, and in many ways it made Ecuador a test case of dollarization for other emerging market countries to watch closely. This was the last stage of a massive depreciation of the sucre in a brief two-year period.

During 1999, Ecuador suffered a rising rate of inflation and a falling level of economic output. In March 1999, the Ecuadorian banking sector was hit with a series of devastating "bank runs," financial panics in which all depositors attempted to withdraw all of their funds simultaneously. Although there were severe problems in the Ecuadorian banking system, the truth was that even the healthiest financial institution would fail under the strain of this financial drain. Ecuador's president immediately froze all deposits (this was termed a *bank holiday* in the United States in the 1930s when banks closed their doors). The value of the Ecuadorian sucre plummeted in early March, inducing the country to default on more than $13 billion in foreign debt in 1999 alone. Ecuador's president moved quickly to propose dollarization to save the Ecuadorian economy.

By January 2000, when the next president took office (after a rather complicated military coup and subsequent withdrawal), the sucre had fallen in value to Sucre 25,000/$. The new president continued the dollarization initiative. Although unsupported by the U.S. government and the IMF, Ecuador completed its replacement of its own currency with the dollar over the next nine months. The results of dollarization in Ecuador are still unknown. Today, many years later, Ecuador continues to struggle to find both economic and political balance with its new currency regime.

Currency Regime Choices for Emerging Markets

There is no doubt that for many emerging markets the choice of a currency regime may lie somewhere between the extremes of a hard peg (a currency board or dollarization) or free-floating. However, many experts have argued for years that the global financial marketplace will drive more and more emerging market nations toward one of these extremes. As shown in Exhibit 2.8, there is a distinct lack of middle ground between rigidly fixed and free-floating extremes. But is the so-called *bi-polar choice* inevitable?

EXHIBIT 2.8   **Regime Choices for Emerging Markets**

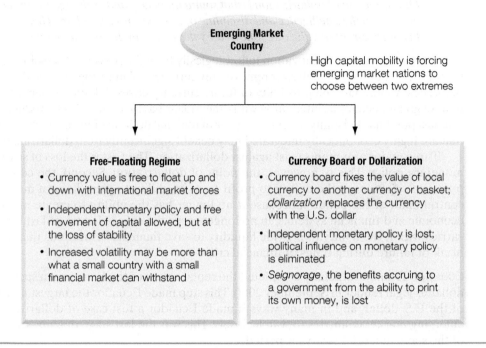

Emerging Market Country

High capital mobility is forcing emerging market nations to choose between two extremes

Free-Floating Regime
- Currency value is free to float up and down with international market forces
- Independent monetary policy and free movement of capital allowed, but at the loss of stability
- Increased volatility may be more than what a small country with a small financial market can withstand

Currency Board or Dollarization
- Currency board fixes the value of local currency to another currency or basket; *dollarization* replaces the currency with the U.S. dollar
- Independent monetary policy is lost; political influence on monetary policy is eliminated
- *Seignorage*, the benefits accruing to a government from the ability to print its own money, is lost

---

There is general consensus that three common features of emerging market countries make any specific currency regime choice difficult: (1) weak fiscal, financial, and monetary institutions; (2) tendencies for commerce to allow currency substitution and the denomination of liabilities in dollars; and (3) the emerging market's vulnerability to sudden stoppages of outside capital flows. Calvo and Mishkin may have said it best:[2]

> *Indeed, we believe that the choice of exchange rate regime is likely to be one second order importance to the development of good fiscal, financial and monetary institutions in producing macroeconomic success in emerging market countries. Rather than treating the exchange rate regime as a primary choice, we would encourage a greater focus on institutional reforms like improved bank and financial sector regulation, fiscal restraint, building consensus for a sustainable and predictable monetary policy and increasing openness to trade.*

In anecdotal support of this argument, a poll of the general population in Mexico in 1999 indicated that 9 out of 10 people would prefer dollarization to a floating-rate peso. Clearly, many in the emerging markets have little faith in their leadership and institutions to implement an effective exchange rate policy.

Globalizing the Chinese Renminbi

> *Logically, it would be reasonable to expect China to make the RMB fully convertible before embarking on the ultimate goal of internationalizing the currency. But China appears to have put "the horse before the cart" by creating an offshore market to promote*

---

[2]"The Mirage of Exchange Rate Regimes for Emerging Market Countries," Guillermo A. Calvo and Frederic S. Mishkin, *The Journal of Economic Perspectives*, Vol. 17, No. 4, Autumn 2003, pp. 99–118.

the currency's use in international trade and investments first. And this offshore trade has taken the lead over the onshore market.

— "RMB to Be a Globally Traded Currency by 2015,"
John McCormick, RBS, in the May 3, 2013, China Briefing.

The Chinese renminbi (RMB) or yuan (CNY) is going global.[3] Although trading in the RMB is closely controlled by the People's Republic of China (PRC)—with all trading inside China between the RMB and foreign currencies (primarily the U.S. dollar) being conducted only according to Chinese regulations—its reach is spreading. The RMB's value, as illustrated in Exhibit 2.9, has been carefully controlled but allowed to gradually revalue against the dollar. It is now quickly moving toward what most think is an inevitable role as a true international currency.

Two-Market Currency Development

The RMB continues to develop along a segmented onshore/offshore two-market structure regulated by the PRC, as seen in Exhibit 2.10. The onshore market (carrying the official ISO code for the Chinese RMB, CNY) is a two-tier market, with retail exchange and an interbank wholesale exchange. The currency has, since mid-2005, been a managed float regime. Internally,

---

EXHIBIT 2.9   **The Revaluation of the Chinese Yuan (1994–2015)**

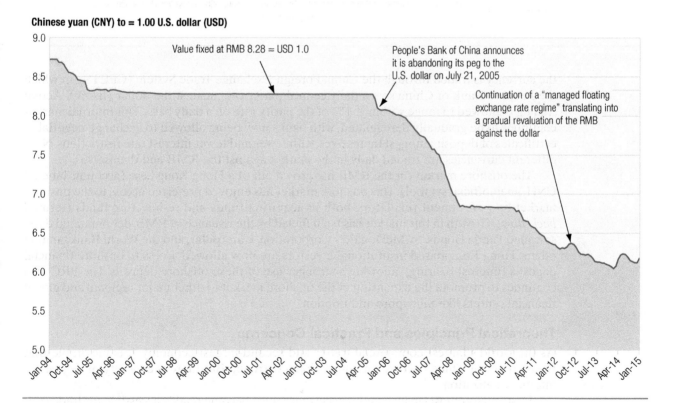

Chinese yuan (CNY) to = 1.00 U.S. dollar (USD)

Value fixed at RMB 8.28 = USD 1.0

People's Bank of China announces it is abandoning its peg to the U.S. dollar on July 21, 2005

Continuation of a "managed floating exchange rate regime" translating into a gradual revaluation of the RMB against the dollar

---

[3]The People's Republic of China officially recognizes the terms *renminbi* (RMB) and *yuan* (CNY) as names of its official currency. *Yuan* is used in reference to the unit of account, while the physical currency is termed the *renminbi*.

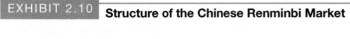

EXHIBIT 2.10 **Structure of the Chinese Renminbi Market**

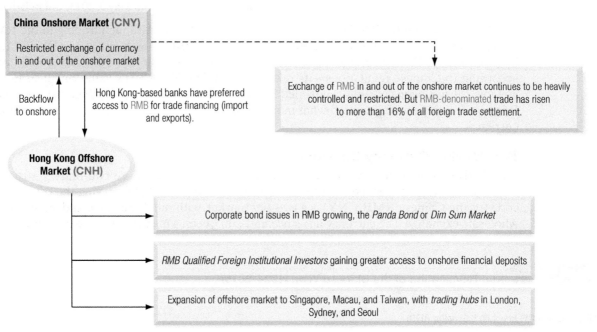

the currency is traded through the China Foreign Exchange Trade System (CFETS), in which the People's Bank of China sets a daily central parity rate against the dollar (fixing). Actual trading is allowed to range within ±1% of the parity rate on a daily basis. This internal market continues to be gradually deregulated, with banks now being allowed to exchange negotiable certificates of deposit amongst themselves, with fewer and fewer interest rate restrictions. Nine different currencies are traded daily in the market against the RMB and themselves.

The offshore market for the RMB has grown out of a Hong Kong base (accounts labeled CNH, an unofficial symbol). This offshore market has enjoyed preferred access to the onshore market by government regulators, both in acquiring funds and re-injecting funds (termed *back-flow*). Growth in this market has been fueled by the issuance of RMB-denominated debt, so-called Panda Bonds, by McDonald's Corporation, Caterpillar, and the World Bank, among others. Hong Kong-based institutional investors are now allowed access to onshore financial deposits (interest bearing), allowing a stronger use of these offshore deposits. The PRC also continues to promote the expansion of the offshore market to other major regional and global financial centers like Singapore and London.

Theoretical Principles and Practical Concerns

As the world's largest commercial trader and second-largest economy, it is inevitable that the currency of China become an international currency. But there is a variety of degrees of internationalization.

First and foremost, an international currency must become readily accessible for trade (this is technically described as Current Account use, to be described in detail in the next chapter). As noted in Exhibit 2.11, it is estimated that more than 16% of all Chinese trade is now denominated in RMB, which although small, is a radical increase from just 1% a mere four years ago.

EXHIBIT 2.11   **Exchange Rate Regime Tradeoffs**

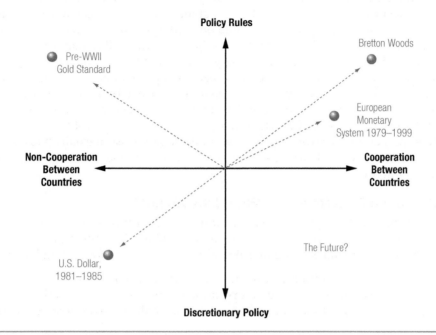

A Chinese exporter was typically paid in U.S. dollars, and was not allowed to keep those dollar proceeds in any bank account. Exporters were required to exchange all foreign currencies for RMB at the official exchange rate set by the PRC, and to turn them over to the Chinese government (resulting in a gross accumulation of foreign currency reserves). Now, importers and exporters are encouraged to use the RMB for trade denomination and settlement purposes.

A second degree of internationalization occurs with the use of the currency for international investment—capital account/market activity. This is an area of substantial concern and caution for the PRC at this time. The Chinese marketplace is the focus of many of the world's businesses, and if they were allowed free and open access to the market and its currency there is fear that the value of the RMB could be driven up, decreasing Chinese export competitiveness. Simultaneously, as major capital markets like the dollar and euro head into stages of rising interest rates, there is a concern that large quantities of Chinese savings could flow out of the country in search of higher returns—capital flight.

A third degree of internationalization occurs when a currency takes on a role as a reserve currency (also termed an anchor currency), a currency to be held in the foreign exchange reserves of the world's central banks. The continued dilemma of fiscal deficits in the United States and the European Union has led to growing unease over the ability of the dollar and euro to maintain their value over time. Could, or should, the RMB serve as a reserve currency? Forecasts of the RMB's share of global reserves vary between 15% and 50% by the year 2020.

The Triffin Dilemma. One theoretical concern about becoming a reserve currency is the *Triffin Dilemma* (or sometimes called the *Triffin Paradox*).[4] The Triffin Dilemma is the potential conflict in objectives that may arise between domestic monetary and currency policy objectives

[4]The theory is the namesake of its originator, Belgian-American economist Robert Triffin (1911–1963), who was an outspoken critic of the Bretton Woods Agreement, as well as a strong advocate and collaborated in the development of the European Monetary System (EMS).

and external or international policy objectives when a country's currency is used as a reserve currency. Domestic monetary and economic policies may on occasion require both contraction and the creation of a current account surplus (balance on trade surplus).

If a currency rises to the status of a global reserve currency, in which it is considered one of the two or three key stores of value on earth (possibly finding its way into the IMF's Special Drawing Right [SDR] definition), other countries will require the country to run current account deficits, essentially dumping growing quantities of the currency on global markets. This means that the country needs to become internationally indebted as part of its role as a reserve currency country. In short, when the world adopts a currency as a reserve currency, demands are placed on the use and availability of that currency, which many countries would prefer not to deal with. In fact, both Japan and Switzerland both worked for decades to prevent their currencies from gaining wider international use, partially because of these complex issues. The Chinese RMB, however, may eventually find that it has no choice—the global market may choose.

Exchange Rate Regimes: What Lies Ahead?

All exchange rate regimes must deal with the tradeoff between rules and discretion, as well as between cooperation and independence. Exhibit 2.11 illustrates the tradeoffs between exchange rate regimes based on rules, discretion, cooperation, and independence.

1. Vertically, different exchange rate arrangements may dictate whether a country's government has strict intervention requirements (rules) or if it may choose whether, when, and to what degree to intervene in the foreign exchange markets (discretion).

2. Horizontally, the tradeoff for countries participating in a specific system is between consulting and acting in unison with other countries (cooperation) or operating as a member of the system, but acting on their own (independence).

Regime structures like the gold standard required no cooperative policies among countries, only the assurance that all would abide by the "rules of the game." Under the gold standard, this assurance translated into the willingness of governments to buy or sell gold at parity rates on demand. The Bretton Woods Agreement, the system in place between 1944 and 1973, required more in the way of cooperation, in that gold was no longer the "rule," and countries were required to cooperate to a higher degree to maintain the dollar-based system. Exchange rate systems, like the European Monetary System's (EMS) fixed exchange rate band system used from 1979 to 1999, were hybrids of these cooperative and rule regimes.

The present international monetary system is characterized by no rules, with varying degrees of cooperation. Although there is no present solution to the continuing debate over what form a new international monetary system should take, many believe that it will succeed only if it combines cooperation among nations with individual discretion to pursue domestic social, economic, and financial goals.

SUMMARY POINTS

- Under the gold standard (1876–1913), the "rules of the game" were that each country set the rate at which its currency unit could be converted to a weight of gold.

- During the inter-war years (1914–1944), currencies were allowed to fluctuate over fairly wide ranges in terms of gold and each other. Supply and demand forces determined exchange rate values.

- The Bretton Woods Agreement (1944) established a U.S. dollar-based international monetary system. Under the original provisions of the Bretton Woods Agreement, all countries fixed the value of their currencies in terms of gold but were not required to exchange their currencies for gold. Only the dollar remained convertible into gold (at $35 per ounce).

- A variety of economic forces led to the suspension of the convertibility of the dollar into gold in August 1971. Exchange rates of most of the leading trading countries were then allowed to float in relation to the dollar and thus indirectly in relation to gold.

- If the ideal currency existed in today's world, it would possess three attributes: a fixed value, convertibility, and independent monetary policy. However, in both theory and practice, it is impossible for all three attributes to be simultaneously maintained.

- Emerging market countries must often choose between two extreme exchange rate regimes: a free-floating regime or an extremely fixed regime, such as a currency board or dollarization.

- The members of the European Union are also members of the European Monetary System (EMS). This group has tried to form an island of fixed exchange rates among themselves in a sea of major floating currencies. Members of the EMS rely heavily on trade with each other, so the day-to-day benefits of fixed exchange rates between them are perceived to be great.

- The euro affects markets in three ways: (1) Countries within the eurozone enjoy cheaper transaction costs; (2) Currency risks and costs related to exchange rate uncertainty are reduced; and (3) All consumers and businesses both inside and outside the eurozone enjoy price transparency and increased price-based competition.

MINI-CASE

Iceland—A Small Country in a Global Crisis[5]

There was the short story, and the longer more complex story. Iceland had seen both. And what was the moral of the story? Was the moral that *it's better to be a big fish in a little pond*, or was it *once burned twice shy*, or something else?

Iceland was a country of only 300,000 people. It was relatively geographically isolated, but its culture and economy were heavily intertwined with that of Europe, specifically northern Europe and Scandinavia. A former property of Denmark, it considered itself both independent and yet Danish. Iceland's economy was historically driven by fishing and natural resource development. Although not flashy by any sense of the word, they had proven to be solid and lasting industries, and in recent years, increasingly profitable. At least that was until Iceland discovered "banking."

The Icelandic Crisis: The Short Story

Iceland's economy had grown very rapidly in the 2000 to 2008 period. Growth was so strong and so rapid that inflation—an ill of the past in most of the economic world—was a growing problem. As a small, industrialized and open economy, capital was allowed to flow into and out of Iceland with economic change. As inflationary pressures rose, the Central Bank of Iceland had tightened monetary policy, interest rates rose. Higher interest rates attracted capital from outside Iceland, primarily European capital, and the banking system was flooded with capital. The banks in turn

invested heavily in everything from real estate to Land Rovers (or *Game Overs* as they became known).

Then September of 2008 happened. The global financial crisis, largely originating in the United States and its real estate-securitized-mortgage-debt-credit-default-swap crisis brought much of the international financial system and major industrial economies to a halt. Investments failed—in the U.S., in Europe, in Iceland. Loans to finance those bad investments fell delinquent. The Icelandic economy and its currency—the *krona*—collapsed. As illustrated in Exhibit A, the Krona fell more than 40% against the euro in roughly 30 days, more than 50% in 90 days. Companies failed, banks failed, unemployment grew, and inflation boomed. A long, slow, and painful recovery began.

The Icelandic Crisis: The Longer Story

The longer story of Iceland's crisis has its roots in mid-1990s, when Iceland—like many other major industrial economies—embraced privatization and deregulation. The financial sector, once completely owned and operated by government, was privatized and largely deregulated by 2003. Home mortgages were deregulated in 2003; new mortgages required only a 10% down payment. Investment—foreign direct investment (FDI)—flowed into Iceland rapidly. A large part of the new investment was in aluminum production, an energy-intensive process that

[5]Copyright © 2015 Thunderbird School of Global Management at Arizona State University. All rights reserved. This case was prepared by Professor Michael H. Moffett for the purpose of classroom discussion only, and not to indicate either effective or ineffective management.

EXHIBIT A   **The Icelandic Short Story—Fall of the Krona**

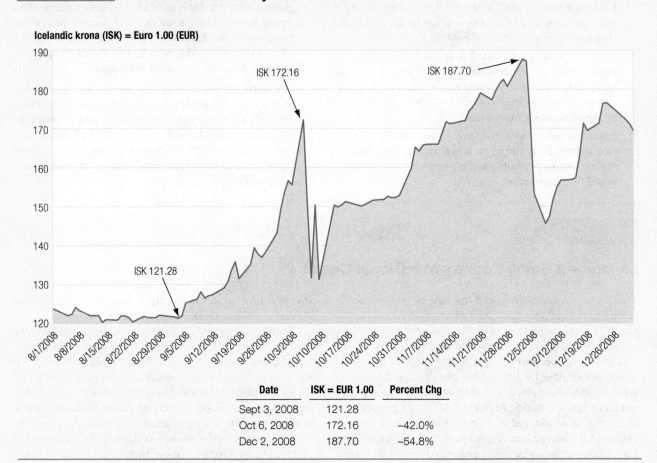

Date	ISK = EUR 1.00	Percent Chg
Sept 3, 2008	121.28	
Oct 6, 2008	172.16	−42.0%
Dec 2, 2008	187.70	−54.8%

could utilize much of Iceland's natural (*natural* after massive damn construction) hydroelectric power. But FDI of all kinds also flowed into the country, including household and business capital.

The new Icelandic financial sector was dominated by three banks—Glitnir, Kaupthing, and Landsbanki Islands. Their opportunities for growth and profitability seemed unlimited, both domestically and internationally. Iceland's membership in the European Economic Area (EEA) provided the Icelandic banks a *financial passport* to expand their reach throughout the greater European marketplace.

As capital flowed into Iceland rapidly in 2003–2006, the krona rose, increasing the purchasing power of Icelanders but raising concerns with investors and government. Gross domestic product (GDP) had grown at 8% in 2004, 6% in 2005, and was still above 4% by 2006. While the average unemployment rate of the major

economic powers was roughly 6%, Iceland's overheating economy had only 3% unemployment. But rapid economic growth in a small economy, as happens frequently in economic history, stoked inflation. And the Icelandic government and central bank then applied the standard prescription: slow money supply growth to try to control inflationary forces. The result—as expected—was higher interest rates.

A financial crash in Iceland snowballed yesterday, setting off a series of tremors as far afield as Brazil and South Africa. At one point the Icelandic krona was down 4.7 per cent at a 15-month low of IKr69.07 to the dollar, having fallen a further 4.5 per cent on Tuesday, its biggest one-day slide in almost five years. The krona's collapse meant carry trade investors who borrowed in euros to gain exposure to Reykjavik's 10 per cent interest rate, saw one-and-a-half years' worth of carry trade profit wiped out in less than two days.

The collapse, which was sparked by Fitch downgrading its outlook on Iceland, citing fears over an "unsustainable" current account deficit and drawing parallels with the imbalances evident before the 1997 Asian crisis, led to a generalised sell-off in Icelandic assets . . .

— "Iceland's Collapse Has Global Impact,"
Financial Times, Feb 23, 2006, p. 42.

Lessons Not Learned

Brennt barn forðast eldinn (A burnt child keeps away from fire)

— Icelandic proverb

The mini-shock suffered by Iceland in 2006 was short lived, and investors and markets quickly shook off its effects. Bank lending returned, and within two years the Icelandic economy was in more trouble than ever.

In 2007 and 2008 Iceland's interest rates continued to rise—both market rates (like bank overnight rates) and central bank policy rates. Global credit agencies rated the major Icelandic banks AAA. Capital flowed into Icelandic banks, and the banks in turn funneled that capital into all possible investments (and loans) domestically and internationally. Iceland's banks created *Icesave*, an Internet banking system to reach out to depositors in Great Britain and the Netherlands. It worked. Iceland's bank balance sheets grew from 100% of GDP in 2003 to just under 1,000% of GDP by 2008.

Iceland's banks were now more international than Icelandic. (By the end of 2007 their total deposits were 45% in British pounds, 22% Icelandic krona, 16% euro, 3% dollar, and 14% other.) Icelandic real estate and equity prices boomed. Increased consumer and business spending resulted in the growth in merchandise and service imports, while the rising krona depressed exports. The merchandise, service, and income balances in the current account all went into deficit. Behaving like an emerging market country that had just discovered oil, Icelanders dropped their fish hooks, abandoned their boats, and became bankers. Everyone wanted a piece of the pie, and the pie appeared to be growing at an infinite rate. Everyone could become rich.

Then it all stopped, suddenly, without notice. Whether it was caused by the failure of Lehman Brothers in the U.S., or was a victim of the same forces, it is hard to say. But beginning in September 2008 the krona started falling and capital started fleeing. Interest rates were increased even further to try and entice (or 'bribe') money to stay in Iceland and in krona. None of it worked. As illustrated by Exhibit B, the krona's fall was large, dramatic, and somewhat permanent. In retrospect, the 2006 crisis had been only a small rain shower; 2008 proved a *tsunami*.

Now those same interest rates, which had been driven up by both markets and policy, prevented any form of renewal—mortgage loans were either impossible to get or impossible to afford, business loans were too expensive given the new limited business outlook. The international interbank market, which had largely frozen-up during the midst of the crisis in September and October 2008, now treated the Icelandic financial sector like a leper. As illustrated by Exhibit C, interest rates had a long way to fall to reach earth (the Central Bank of Iceland's overnight rate rose to well over 20%).

Aftermath: The Policy Response

There is a common precept observed by governments and central banks when they fall victim to financial crises: *save the banks*. Regardless of whether the banks and bankers were considered the cause of the crisis, or complicit (one Icelandic central banker termed them *the usual suspects*), it is common belief that all economies need a functioning banking system in order to have any hope for business rebirth and employment recovery. This was the same rule used in the U.S. in the 1930s and across South Asia in 1997 and 1998.

But the Icelandic people did not prescribe to the usual medicinal. Their preference: *let the banks fail*. Taking to the streets in what was called the *pots and pans revolution*, the people wanted no part of the banks, the bankers, the bank regulators, or even the Prime Minister. The logic was some combination of "allow free markets to work" and "I want some revenge." (This is actually quite similar to what many analysts have debated over what happened in the U.S. at the same time when the U.S. government *let Lehman go*.)

In contrast to the bank bailouts in the United States in 2008 following the onset of the financial crisis undertaken under the mantra of "too big to fail," Iceland's banks were considered "too big to save." Each of the three major banks, which had all been effectively nationalized by the second week of October in 2008, was closed. As illustrated in Exhibit D, although Iceland's bank assets and external liabilities were large, and had grown rapidly, Iceland was not alone. Each failed bank was reorganized by the government into a *good bank* and a *bad bank* in terms of assets, but not combined into singular good banks and bad banks.

The governing authorities surviving in office in the fall of 2008 undertook a thee-point emergency plan: (1) stabilize the exchange rate; (2) regain fiscal sustainability; and (3) rebuild the financial sector. The primary tool was *capital controls*. Iceland shut down the borders and the Internet lines for moving capital into or out of the country. The most immediate problem was the exchange rate.

EXHIBIT B   **The Icelandic Krona—European Euro Spot Exchange Rate**

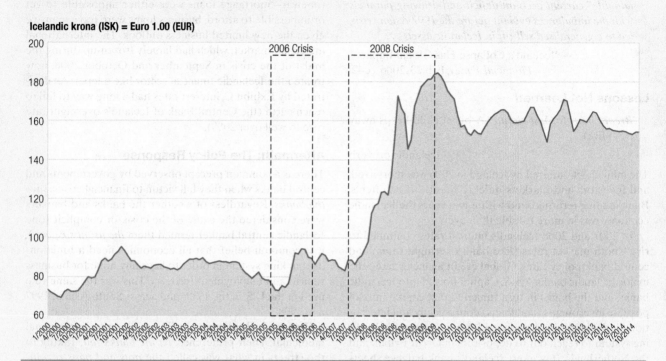

EXHIBIT C   **Icelandic Central Bank Interest Rates**

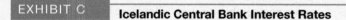

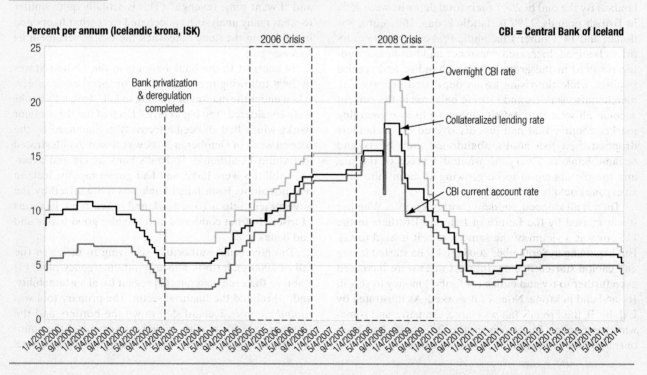

| EXHIBIT D | Icelandic Banks Compared to Others in Potential Crisis |

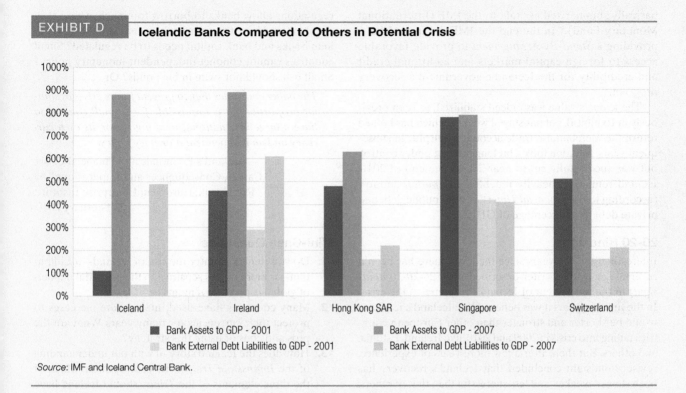

- Bank Assets to GDP - 2001
- Bank External Debt Liabilities to GDP - 2001
- Bank Assets to GDP - 2007
- Bank External Debt Liabilities to GDP - 2007

Source: IMF and Iceland Central Bank.

The falling krona had decimated purchasing power, and the rising prices of imported goods were adding even more inflationary pressure.

> *Given the substantial macroeconomic risks, capital controls were an unfortunate but indispensable ingredient in the policy mix that was adopted to stabilise the króna when the interbank foreign-exchange market was restarted in early December 2008.*
>
> *— Capital Control Liberalisation,* Central Bank of Iceland, August 5, 2009, p. 2.

The bank failures (without bailout) raised serious and contentious discussions between Iceland and other authorities in the United Kingdom, the EU, the Netherlands, and elsewhere. Because so many of the deposits in Iceland banks were from foreign depositors, home-country authorities wanted assurance that their citizens' financial assets would be protected. In Iceland, although the government guaranteed domestic residents that their money was insured (up to a limit), foreign depositors were not. Foreign residents holding accounts with Icelandic financial institutions were prohibited from pulling the money out of Iceland and out of the krona.

Capital controls were introduced in October — upon the recommendation of the IMF — and then altered and magnified in revisions in November and December 2008 and again in March of 2009.

> *Payments linked to current account transactions and inward FDI were released after a short period of time. Thus, transactions involving actual imports and exports of goods and services are allowed and so are interest payments, if exchanged within a specified time limit. Most capital transactions are controlled both for residents and non-residents; that is, their ability to shift between ISK and FX is restricted. Króna-denominated bonds and other like instruments cannot be converted to foreign currency upon maturity. The proceeds must be reinvested in other ISK instruments. Furthermore, the Rules require residents to repatriate all foreign currency that they acquire.*
>
> *— Capital Control Liberalisation,* Central Bank of Iceland, August 5, 2009, p. 2–3.

It also turned out that the crisis itself was not such a big surprise. The Central Bank of Iceland had approached the European Central Bank (ECB), the Bank of England, and the U.S. Federal Reserve in the spring of 2008 (months before the crisis erupted), hoping to arrange foreign exchange swap agreements in case its foreign exchange reserves proved inadequate. The answer was *no,* basically summarized as "talk to the IMF (International Monetary Fund)." In the end the IMF did indeed help, providing a *Stand-By Arrangement* to provide favorable access to foreign capital markets and additional credit and credibility for the Icelandic government's recovery program.

The krona's value was indeed stabilized, as seen previously in Exhibit B, but has stayed weaker, which has helped return the merchandise trade account to surplus in subsequent years. Inflation took a bit longer to get under control, but was successfully cut to near 2% by the end of 2010. Iceland remains a heavily indebted *Lilliputian* country (according to the *Financial Times*), in both public debt and private debt as a percentage of GDP.

20-20 Hindsight

Interestingly, in the years since the crisis, there has been a reversal (or as one writer described it, *20-20-20-20 hindsight*) in the assessment of Iceland's response to the crisis. In the first few years it was believed that Iceland's recovery would be shorter and stronger than other European countries falling into crisis in 2009 and 2010, like Ireland, Estonia, and others. But then, after a few more years of experience, revised hindsight concluded that Iceland's recovery has been slower, weaker, and less successful than that of others, partly a result of allowing the banks to fail, partly a result of the country's "addiction" to capital controls.

And the lessons? What are the lessons to be taken from the Icelandic saga? Deregulation of the financial system is risky? Banks and bankers are not to be trusted? Cross-border banking is risky? Inadequate cross-border banking

regulations allow banks to borrow too much, where they shouldn't, and invest too much where they shouldn't? Bank loan books and bank capital needs to be regulated? Small countries cannot conduct independent monetary policy? Small fish should not swim in big ponds? Or . . .

> *The paper concludes that, to prevent future crises of similar proportions, it is impossible for a small country to have a large international banking sector, its own currency and an independent monetary policy.*

> — "Iceland's Economic and Financial Crisis: Causes, Consequences and Implications," by Rob Spruk, European Enterprise Institute, 23 February 2010.

Mini-Case Questions

1. Do you think a country the size of Iceland—a Lilliputian—is more or less sensitive to the potential impacts of global capital movements?

2. Many countries have used interest rate increases to protect their currencies for many years. What are the pros and cons of using this strategy?

3. How does the Iceland story fit with our understanding of the *Impossible Trinity*? In your opinion, which of the three elements of the *Trinity* should Iceland have taken steps to control more?

4. In the case of Iceland, the country was able to sustain a large current account deficit for several years, and at the same time have ever-rising interest rates and a stronger and stronger currency. Then one day, it all changed. How does that happen?

QUESTIONS

These questions are available in MyFinanceLab.

1. **The Rules of the Game.** Under the gold standard, all national governments promised to follow the "rules of the game." What did this mean?

2. **Defending a Fixed Exchange Rate.** What did it mean under the gold standard to "defend a fixed exchange rate," and what did this imply about a country's money supply?

3. **Bretton Woods.** What was the foundation of the Bretton Woods international monetary system, and why did it eventually fail?

4. **Technical Float.** What specifically does a floating rate of exchange mean? What is the role of government?

5. **Fixed versus Flexible.** What are the advantages and disadvantages of fixed exchange rates?

6. *De facto* **and** *de jure.* What do the terms *de facto* and *de jure* mean in reference to the International Monetary Fund's use of the terms?

7. **Crawling Peg.** How does a crawling peg fundamentally differ from a pegged exchange rate?

8. **Global Eclectic.** What does it mean to say the international monetary system today is a global eclectic?

9. **The Impossible Trinity.** Explain what is meant by the term *impossible trinity* and why it is, in fact, "impossible."

10. **The Euro.** Why is the formation and use of the euro considered to be of such a great accomplishment? Was it really needed? Has it been successful?

11. **Currency Board or Dollarization.** Fixed exchange rate regimes are sometimes implemented through a currency board (Hong Kong) or dollarization (Ecuador). What is the difference between the two approaches?

12. **Argentine Currency Board.** How did the Argentine currency board function from 1991 to January 2002 and why did it collapse?

13. **Special Drawing Rights.** What are Special Drawing Rights?

14. **The Ideal Currency.** What are the attributes of the ideal currency?

15. **Emerging Market Regimes.** High capital mobility is forcing emerging market nations to choose between free-floating regimes and currency board or dollarization regimes. What are the main outcomes of each of these regimes from the perspective of emerging market nations?

16. **Globalizing the Yuan.** What are the major changes and developments that must occur for the Chinese yuan to be considered "globalized"?

17. **Triffin Dilemma.** What is the Triffin Dilemma? How does it apply to the development of the Chinese yuan as a true global currency?

18. **China and the Impossible Trinity.** What choices do you believe that China will make in terms of the Impossible Trinity as it continues to develop global trading and use of the Chinese yuan?

PROBLEMS

These problems are available in MyFinanceLab.

1. **Chantal DuBois in Brussels.** Chantal DuBois lives in Brussels. She can buy a U.S. dollar for €0.7600. Christopher Keller, living in New York City, can buy a euro for $1.3200. What is the foreign exchange rate between the dollar and the euro?

2. **Quartzite Inc.** The spot rate for Mexican pesos is Ps12.42/$. If the U.S.-based company Quartzite Inc. buys Ps500,000 spot from its bank on Monday, how much must Quartzite pay and on what day?

3. **Gilded Question.** Before World War I, $20.67 was needed to buy one ounce of gold. If, at the same time, one ounce of gold could be purchased in France for FF410.00, what was the exchange rate between French francs and U.S. dollars?

4. **Golden Rule.** Under the gold standard, the price of an ounce of gold in U.S. dollars was $20.67, while the price

of that same ounce in British pounds was £3.7683. What would be the exchange rate between the dollar and the pound if the U.S. dollar price had been $42.00 per ounce of gold?

5. **Toyota Exports to the United Kingdom.** Toyota manufactures in Japan most of the vehicles it sells in the United Kingdom. The base platform for the Toyota Tundra truck line is ¥1,650,000. The spot rate of the Japanese yen against the British pound has recently moved from ¥197/£ to ¥190/£. How does this change the price of the Tundra to Toyota's British subsidiary in British pounds?

6. **Loonie Parity.** If the price of former Chairman of the U.S. Federal Reserve Alan Greenspan's memoir, The Age of Turbulence, is listed on Amazon.ca as C$26.33, but costs just US$23.10 on Amazon.com, what exchange rate does that imply between the two currencies?

7. **Mexican Peso Changes.** In December 1994, the government of Mexico officially changed the value of the Mexican peso from 3.2 pesos per dollar to 5.5 pesos per dollar. What was the percentage change in its value? Was this a depreciation, devaluation, appreciation, or revaluation? Explain.

8. **Hong Kong Dollar and the Chinese Yuan.** The Hong Kong dollar has long been pegged to the U.S. dollar at HK$7.80/$. When the Chinese yuan was revalued in July 2005 against the U.S. dollar from Yuan8.28/$ to Yuan8.11/$, how did the value of the Hong Kong dollar change against the yuan?

9. **Chinese Yuan Revaluation.** Many experts believe that the Chinese currency should not only be revalued against the U.S. dollar as it was in July 2005, but also be revalued by 20% or 30%. What would be the new exchange rate value if the yuan were revalued an additional 20% or 30% from its initial post-revaluation rate of Yuan8.11/$?

10. **Ranbaxy (India) in Brazil.** Ranbaxy, an India-based pharmaceutical firm, has continuing problems with its cholesterol reduction product's price in one of its rapidly growing markets, Brazil. All product is produced in India, with costs and pricing initially stated in Indian rupees (Rps), but converted to Brazilian reais (R$) for distribution and sale in Brazil. In 2009, the unit volume was priced at Rps21,900, with a Brazilian reais price set at R$895. But in 2010, the reais appreciated in value versus the rupee, averaging Rps26.15/R$. In order to preserve the reais price and product profit margin in rupees, what should the new rupee price be set at?

11. **Vietnamese Coffee Coyote.** Many people were surprised when Vietnam became the second largest coffee producing country in the world in recent years, second only to Brazil. The Vietnamese dong, VND or d, is managed against the U.S. dollar but is not widely traded. If you were a traveling coffee buyer for the wholesale market (a "coyote" by industry terminology), which of the following currency rates and exchange commission fees would be in your best interest if traveling to Vietnam on a buying trip?

Currency Exchange	Rate	Commission
Vietnamese bank rate	d19,800	2.50%
Saigon Airport exchange bureau rate	d19,500	2.00%
Hotel exchange bureau rate	d19,400	1.50%

12. **Chunnel Choices.** The Channel Tunnel or "Chunnel" passes underneath the English Channel between Great Britain and France, a land-link between the Continent and the British Isles. One side is therefore an economy of British pounds, the other euros. If you were to check the Chunnel's rail ticket Internet rates you would find that they would be denominated in U.S. dollars (USD). For example, a first class round trip fare for a single adult from London to Paris via the Chunnel through RailEurope may cost USD170.00. This currency neutrality, however, means that customers on both ends of the Chunnel pay differing rates in their home currencies from day to day. What is the British pound and euro denominated prices for the USD170.00 round trip fare in local currency if purchased on the following dates at the accompanying spot rates drawn from the *Financial Times*?

Date of Spot Rate	British Pound Spot Rate (£/$)	Euro Spot Rate (€/$)
Monday	0.5702	0.8304
Tuesday	0.5712	0.8293
Wednesday	0.5756	0.8340

13. **Barcelona Exports.** Oriol D'ez Miguel S.R.L., a manufacturer of heavy-duty machine tools near Barcelona, ships an order to a buyer in Jordan. The purchase price is €425,000. Jordan imposes a 13% import duty on all products purchased from the European Union. The Jordanian importer then re-exports the product to a Saudi Arabian importer, but only after imposing its own resale fee of 28%. Given the following spot exchange rates on April 11, 2010, what is the total cost to the Saudi Arabian importer in Saudi Arabian riyal, and what is the U.S. dollar equivalent of that price?

INTERNET EXERCISES

1. **International Monetary Fund's Special Drawing Rights.** Use the IMF's Web site to find the current weights and valuation of the SDR.

International Monetary Fund	www.imf.org/external/np/tre/sdr/ sdrbasket.htm

2. **Malaysian Currency Controls.** The institution of currency controls by the Malaysian government in the aftermath of the Asian currency crisis is a classic response by government to unstable currency conditions. Use the following Web site to increase your knowledge of how currency controls work.

International Monetary Fund	www.imf.org/external/pubs/ft/ bl/rr08.htm

3. **Personal Transfers.** As anyone who has traveled internationally learns, the exchange rates available to private retail customers are not always as attractive as those accessed by companies. The OzForex Web site possesses a section on "customer rates" that illustrates the difference. Use the site to calculate what the percentage difference between Australian dollar/U.S. dollar spot exchange rates are for retail customers versus interbank rates.

OzForex	www.ozforex.com.au/exchange-rate

4. **Exchange Rate History.** Use the Pacific Exchange Rate database and plot capability to track value changes of the British pound, the U.S. dollar, and the Japanese yen against each other over the past 15 years.

Pacific Exchange Rate Service	fx.sauder.ubc.ca

CHAPTER 3

The Balance of Payments

The sort of dependence that results from exchange, i.e., from commercial transactions, is a reciprocal dependence. We cannot be dependent upon a foreigner without his being dependent on us. Now, this is what constitutes the very essence of society. To sever natural interrelations is not to make oneself independent, but to isolate oneself completely.

—Frederic Bastiat.

LEARNING OBJECTIVES

- Learn how government and multinational enterprise management uses balance of payments accounts and accounting in decision-making
- Examine how the primary accounts of the balance of payments reflects fundamental economic and financial activities across borders
- Explore how changes in the balance of payments affect key macroeconomic rates like exchange rates and interest rates
- Analyze how exchange rate changes affect the prices and competitiveness of international trade
- Evaluate how governments have responded to the globalization of capital markets in their use of capital restrictions in an effort to hinder capital mobility

The measurement of all international economic transactions that take place between the residents of a country and foreign residents is called the *balance of payments* (BOP). This chapter provides a sort of navigational map to aid in interpreting the balance of payments and the multitude of economic, political, and business issues that it involves. But our emphasis is far from descriptive, as a deep understanding of trade and capital flows is integral to the management of multinational enterprises. In fact, the second half of the chapter emphasizes a more detailed analysis of how elements of the balance of payments affect trade volumes and prices, as well as how capital flows, capital controls, and capital flight alter the cost and ability to do business internationally. The chapter concludes with a Mini-Case, *Global Remittances*, a sector only recently explored in depth by governments as they try to monitor and control capital flows across their borders.

Home-country and host-country BOP data, and their subaccounts, are important to business managers, investors, consumers, and government officials because the data simultaneously influences and is influenced by other key macroeconomic variables, such as gross domestic product (GDP), employment levels, price levels, exchange rates, and interest rates. Monetary and fiscal policy must take the BOP into account at the national level. Business managers and investors need BOP data to anticipate changes in host-country economic policies that might be driven by BOP events. BOP data is also important for the following reasons:

- The BOP is an important indicator of pressure on a country's foreign exchange rate, and thus of the potential for a firm trading with or investing in that country to experience foreign exchange gains or losses. Changes in the BOP may predict the imposition or removal of foreign exchange controls.

- Changes in a country's BOP may signal the imposition or removal of controls over payment of dividends and interest, license fees, royalty fees, or other cash disbursements to foreign firms or investors.

- The BOP helps to forecast a country's market potential, especially in the short run. A country experiencing a serious trade deficit is not as likely to expand imports, as it would be if running a surplus. It may, however, welcome investments that increase its exports.

Fundamentals of BOP Accounting

BOP accounting is saddled with terminology from corporate accounting that has different meanings in this instance. The word "balance" creates a false image of a corporate balance sheet. A BOP statement is a statement of cash flows over an interval of time more in accord with a corporate income statement, but on a cash basis. It also uses the terms *debit* and *credit* in its own unique way. A BOP *credit* is an event, such as the export of a good or service, that records foreign exchange earned—an inflow of foreign exchange to the country. A *debit* records foreign exchange spent, such as payments for imports or purchases of services—an outflow of foreign exchange.

International transactions take many forms. Each of the following examples is an international economic transaction that is counted and captured in the U.S. balance of payments:

- A U.S.-based firm, Fluor Corporation, manages the construction of a major water treatment facility in Bangkok, Thailand.

- The U.S. subsidiary of a French firm, Saint Gobain, pays profits back to its parent firm in Paris.

- An American tourist purchases a small Lapponia necklace in Finland.

- The U.S. government finances the purchase of military equipment for its military ally, Norway.

- A Mexican lawyer purchases a U.S. corporate bond through an investment broker in Cleveland.

The BOP provides a systematic method for classifying these transactions. A rule of thumb always aids the understanding of BOP accounting: *Follow the cash flow.*

The BOP has three major sub-accounts—the *current account*, the *capital account*, and the *financial account*.

The BOP must balance. If it does not, something has not been counted or has been counted improperly. Therefore, it is incorrect to state that "the BOP is in disequilibrium." It cannot be. The supply and demand for a country's currency may be imbalanced, but that is not the same thing as for the whole BOP. A sub-account of the BOP, such as the *balance on goods and services* (a sub-account of any country's current account), may be imbalanced (in surplus or deficit), but the entire BOP of a single country is always balanced.

Exhibit 3.1 illustrates that the BOP does indeed balance, in this case for the United States from 2005 through 2012. The five balances listed in Exhibit 3.1—*current account, capital*

| EXHIBIT 3.1 | The U.S. Balance of Payments Accounts, Summary | | | | | | | | |

Balance	2005	2006	2007	2008	2009	2010	2011	2012	2013
Current Account Balance	−740	−807	−719	−687	−381	−444	−459	−461	−400
Capital Account Balance	13	−2	0	6	0	0	−1	7	0
Financial Account Balance	687	807	617	735	283	439	532	428	368
Net Errors and Omissions	26	−1	101	−50	150	7	−55	30	30
Reserves and Related	14	2	0	−5	−52	−2	−16	−4	3
Sum or Total	0	0	0	0	0	0	0	0	0

Source: International Monetary Fund, *Balance of Payments Statistics Yearbook, 2013.*

account, financial account, net errors and omissions, and *reserves and related items*—do indeed sum to zero.

There are three main elements of the actual process of measuring international economic activity: (1) identifying what is and is not an international economic transaction; (2) understanding how the flow of goods, services, assets, and money creates debits and credits to the overall BOP; and (3) understanding the bookkeeping procedures for BOP accounting.

Defining International Economic Transactions

Identifying international transactions is ordinarily not difficult. The export of merchandise— goods such as trucks, machinery, computers, telecommunications equipment, and so forth—is obviously an international transaction. Imports such as French wine, Japanese cameras, and German automobiles are also clearly international transactions. But this merchandise trade is only a portion of the thousands of different international transactions that occur in the United States and other countries each year.

Many other international transactions are not so obvious. The purchase of a good, like a glass figure in Venice, Italy, by a U.S. tourist is classified as a U.S. merchandise import. In fact, all expenditures made by U.S. tourists around the globe for services provided by, for example, restaurants and hotels are recorded in the U.S. balance of payments as imports of travel services in the current account.

The BOP as a Flow Statement

As noted above, the BOP is often misunderstood because many people infer from its name that it is a *balance sheet*, whereas in fact it is a *cash flow statement*. By recording all international transactions over a period of time such as a year, the BOP tracks the continuing flows of purchases and payments between a country and all other countries. It does not add up the value of all assets and liabilities of a country on a specific date like a balance sheet does for an individual firm (that is, in fact, the *net international investment position* (NIIP) of a country, described in a later section). Two types of business transactions dominate the balance of payments:

1. **Exchange of *real assets*.** The exchange of goods (e.g., automobiles, computers, textiles) and services (e.g., banking, consulting, and travel services) for other goods and services (barter) or for money

2. **Exchange of *financial assets*.** The exchange of financial claims (e.g., stocks, bonds, loans, and purchases or sales of companies) for other financial claims or money

Although assets can be identified as real or financial, it is often easier to think of all assets as goods that can be bought and sold. The purchase of a hand-woven area rug in a shop in

Bangkok by a U.S. tourist is not all that different from a Wall Street banker buying a British government bond for investment purposes.

BOP Accounting

The measurement of all transactions in and out of a country is a daunting task. Mistakes, errors, and statistical discrepancies will occur. The primary problem is that double-entry bookkeeping is employed in theory, but not in practice. Individual purchase and sale transactions should—in theory—result in financing entries in the balance of payments that match. In reality, current, capital, and financial account entries are recorded independently of one another, not together as double-entry bookkeeping would prescribe. Thus, there will be discrepancies (to use a nice term for it) between debits and credits.

The Accounts of the Balance of Payments

The balance of payments is composed of three major sub-accounts: the *current account*, the *capital account*, and the *financial account*. In addition, the *official reserves account* tracks government currency transactions, and a fifth statistical sub-account, the *net errors and omissions account*, is produced to preserve the balance in the BOP. The word "net" in account titles means that payments and receipts, i.e., debits and credits, are netted within that account.

The Current Account

The current account includes all international economic transactions with income or payment flows occurring within the year, the current period. The current account consists of four subcategories:

1. **Goods trade.** The export and import of goods is known as the goods trade. Merchandise trade is the oldest and most traditional form of international economic activity. Although many countries depend on both imports and exports of goods, most countries seek to preserve either a balance or surplus on goods trade.

2. **Services trade.** The export and import of services is known as the services trade. Common international services are financial services provided by banks to foreign importers and exporters, travel services of airlines, and construction services of domestic firms in other countries. For the major industrial countries, this sub-account has shown the fastest growth in the past decade.

3. **Income.** This is predominantly current income associated with investments made in previous periods. If a U.S. firm created a subsidiary in South Korea to produce metal parts in a previous year, the proportion of net income that is paid back to the parent company in the current year (the dividend) constitutes current investment income. Additionally, wages and salaries paid to nonresident workers are also included in this category.

4. **Current transfers.** Financial settlements associated with the change in ownership of real resources or financial items are called current transfers. Any transfer between countries that is one-way—a gift or grant—is termed a current transfer. For example, funds provided by the U.S. government to aid in the development of a less-developed nation are a current transfer. Transfer payments made by migrant or guest workers back to their home countries, *global remittances,* are an example of current transfers.

All countries possess some amount of trade, most of which is merchandise. Many less-developed countries have little in the way of service trade, or items that fall under the income or transfers sub-accounts. The current account is typically dominated by the first component described above, the export and import of merchandise. For this reason, the *balance of trade* (BOT) that is so widely quoted in the business press refers to the balance of exports and imports of goods trade only. If the country is a larger industrialized country, however, the BOT is somewhat misleading, in that service trade is not included.

Exhibit 3.2 presents the two major components of the U.S. current account for the 2000–2012 period: (1) goods trade and (2) services trade and investment income. The first and most striking message is the magnitude of the goods trade deficit. The balance on services and income, although not large in comparison to net goods trade, has run a small but consistent surplus over the past two decades.

Merchandise trade is the original core of international trade. The manufacturing of goods was the basis of the industrial revolution and the focus of the *theory of comparative advantage in international trade*. Manufacturing is traditionally the sector of the economy that employs most of a country's workers. Declines in the U.S. BOT attributed to specific sectors, such as steel, automobiles, automotive parts, textiles, and shoe manufacturing, caused massive economic and social disruption.

Understanding merchandise import and export performance is much like understanding the market for any single product. The demand factors that drive both are income, the economic growth rate of the buyer, and price of the product in the eyes of the consumer after passing through an exchange rate. U.S. merchandise imports reflect the income level of U.S. consumers and growth of industry. As income rises, so does the demand for imports. Exports follow the same principles, but in the reverse. U.S. manufacturing exports depend not on the incomes of U.S. residents, but on the incomes of buyers of U.S. products in all other countries around the world. When these economies are growing, the demand for U.S. products is growing.

As illustrated in Exhibit 3.2, the United States has consistently run a surplus in services trade income. The major categories of services include travel and passenger fares; transportation services; expenditures by U.S. students abroad, foreign students studying in the U.S.; telecommunications services; and financial services.

EXHIBIT 3.2   **U.S. Trade Balances on Goods and Services**

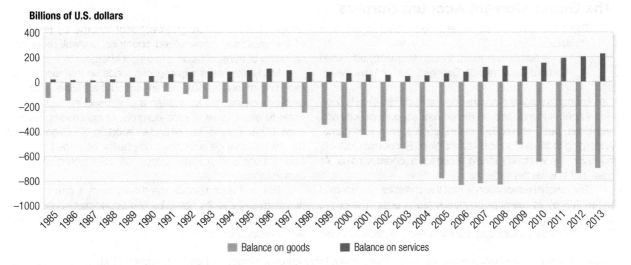

Source: Balance of Payments Statistics Yearbook, International Monetary Fund, December 2014, p. 1,060.

The Capital and Financial Accounts

The capital and financial accounts of the balance of payments measure all international economic transactions of financial assets. The capital account is made up of transfers of financial assets and the acquisition and disposal of nonproduced/nonfinancial assets. This account has been introduced as a separate component in the IMF's balance of payments only recently. The magnitude of capital transactions covered by the Capital Account is relatively minor, and we will include it in principle in all of the following discussions of the financial account. But as noted in *Global Finance in Practice 3.1*, some mysteries in global accounts remain!

Financial Account

The financial account consists of four components: direct investment, portfolio investment, net financial derivatives, and other asset investment. Financial assets can be classified in a number of different ways, including by the length of the life of the asset (its maturity) and the nature of the ownership (public or private). The financial account, however, uses *degree of control* over assets or operations to classify financial assets. *Direct investment* is defined as investment that has a long-term life or maturity and in which the investor exerts some explicit degree of control over the assets. In contrast, *portfolio investment* is defined as both short-term in maturity and as an investment in which the investor has no control over the assets.

Direct Investment. This investment measure is the net balance of capital dispersed from and into a country like the United States for the purpose of exerting control over assets. If a U.S. firm builds a new automotive parts facility in another country or purchases a company in another country, this is a direct investment in the U.S. balance of payments accounts. When the capital flows out of the U.S., it enters the balance of payments as a negative cash flow. If, however, a foreign firm purchases a firm in the U.S., it is a capital inflow and enters the balance of payments positively.

Foreign resident purchases of assets in a country are always somewhat controversial. The focus of concern over foreign investment in any country, including the United States, is on

---

GLOBAL FINANCE IN PRACTICE 3.1

The Global Current Account Surplus

There are three kinds of lies: lies, damned lies and statistics.

　　　　—Author unknown, though frequently attributed to Lord Courtney, Sir Charles Dilke, or Mark Twain.

One country's surplus is another country's deficit. That is, individual countries may and do run current account deficits and surpluses, but it should be, theoretically, a zero sum game. According to the IMF's most recent World Economic Outlook, however, the world is running a current account surplus. At least that is what the statistics say.*

The rational explanation is that the statistics, as reported to the IMF by its member countries, are in error. The errors are most likely both accidental and intentional. The IMF believed for many years that the most likely explanation was under-reporting of foreign investment income by residents of the wealthier industrialized countries, as well as under-reporting of transportation and freight charges.

Many alternative explanations focus on intentional mis-reporting of international current account activities. Over- or under-invoicing has long been a ploy used in international trade to avoid taxes, capital controls, or purchasing restrictions. Other arguments, like under-reporting of foreign income for tax avoidance and the complexity of intra-company transactions and transfer prices, all offer potential partial explanations.

But in the end, while the theory says it can't be, the numbers say it is. As noted by *The Economist*, planet Earth appears to be running a current account surplus in its trade with extraterrestrials.**

World Economic Outlook: Slowing Growth, Rising Risks, International Monetary Fund, September 2011.
**"Economics Focus, Exports to Mars," *The Economist*, November 12, 2011, p. 90

two issues: control and profit. Some countries place restrictions on what foreigners may own in their country. This rule is based on the premise that domestic land, assets, and industry in general should be owned by residents of the country. The U.S., however, has traditionally had few restrictions on what foreign residents or firms can own or control in the country (with the exception of national security concerns). Unlike the case in the traditional debates over whether international trade should be free, there is no consensus on international investment.

The second major focus of concern over foreign direct investment is who receives the profits from the enterprise. Foreign companies owning firms in the U.S. will ultimately profit from the activities of those firms—or to put it another way, foreign companies will profit from the efforts of U.S. workers. In spite of evidence that indicates foreign firms in the U.S. reinvest most of their profits in their U.S. businesses (in fact, at a higher rate than do domestic firms), the debate on possible profit drains has continued. Regardless of the actual choices made, workers of any nation feel that the profits of their work should remain in their own hands in their own country.

The choice of words used to describe foreign investment can also influence public opinion. If these massive capital inflows are described as "capital investments from all over the world showing their faith in the future of U.S. industry," the net capital surplus is represented as decidedly positive. If, however, the net capital surplus is described as resulting in "the United States being the world's largest debtor nation," the negative connotation is obvious. Both are essentially spins on the same economic principles at work.

Capital, whether short-term or long-term, flows to where the investor believes it can earn the greatest return for the level of risk. And although in an accounting sense this is "international debt," when the majority of the capital inflow is in the form of direct investment, a long-term commitment to jobs, production, services, technological, and other competitive investments, the impact on the competitiveness of industry located within a country is increased. Net direct investment cash flows for the U.S. are shown in Exhibit 3.3.

Portfolio Investment. This is the net balance of capital that flows into and out of a country but that does not reach the 10% ownership threshold of direct investment. If a U.S. resident purchases shares in a Japanese firm but does not attain the 10% threshold, we define the purchase as a portfolio investment (and an outflow of capital). The purchase or sale of debt securities (like U.S. Treasury bills) across borders is also classified as portfolio investment, because debt securities by definition do not provide the buyer with ownership or control.

Portfolio investment is capital invested in activities that are purely profit-motivated (return), rather than activities to control or manage the investment. Purchases of debt securities, bonds, interest-bearing bank accounts, and the like are intended only to earn a return. They provide no vote or control over the party issuing the debt. Purchases of debt issued by the U.S. government (U.S. Treasury bills, notes, and bonds) by foreign investors constitute net portfolio investment in the United States. It is worth noting that most U.S. debt purchased by foreigners is U.S. dollar-denominated in the currency of the issuing country (dollars). Much of the foreign debt issued by countries such as Russia, Brazil, and Southeast Asian countries is also U.S. dollar-denominated, and is therefore the currency of a foreign country. The foreign country must then earn dollars to repay its foreign-held debt, typically through exports.

As illustrated in Exhibit 3.3, portfolio investment has shown much more volatile behavior than net foreign direct investment has over the past decade. Many U.S. debt securities, such as U.S. Treasury securities and corporate bonds, are consistently in high demand by foreign investors of all kinds. The motivating forces for portfolio investment flows are always the same: return and risk. These same debt securities have also been influential in a different measure of international investment activity, as described in *Global Finance in Practice 3.2*.

EXHIBIT 3.3	The U.S. Financial Accounts, 1985–2013

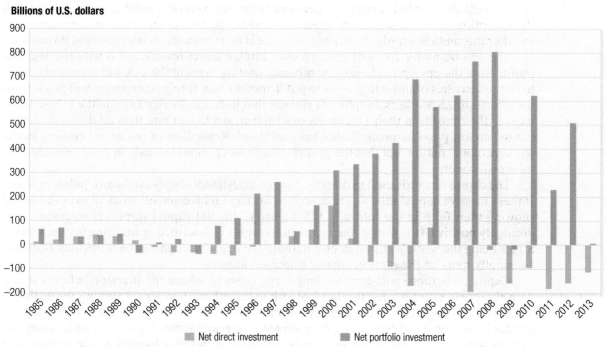

Source: *Balance of Payments Statistics Yearbook,* International Monetary Fund, 2013.

GLOBAL FINANCE IN PRACTICE 3.2

A Country's Net International Investment Position (NIIP)

The net international investment position (NIIP) of a country is an annual measure of the assets owned abroad by its citizens, its companies, and its government, less the assets owned by foreigners public and private in their country. Whereas a country's balance of payments is often described as a country's international *cash flow statement*, the NIIP may be interpreted as the country's international *balance sheet*. NIIP is a country's stock of foreign assets minus its stock of foreign liabilities.

The NIIP, in the same way company cash flows are related to a company's balance sheet, is based upon and categorized by the same capital and financial accounts used in the balance of payments: direct investment, portfolio investment, other investment and reserve assets. As international capital has found it easier and easier to move between currencies and cross borders in recent years, ownership of assets and securities has clearly boomed.

One common method of putting a country's NIIP into perspective is to measure it as a percentage of the total economic size of the nation—the Gross Domestic Product (GDP) of the country. As illustrated here, the NIIP of the U.S. has clearly seen a dramatic increase since 2005, now averaging 25% of U.S. GDP.

Although some observers have seen this growing percentage as a risk to the U.S. economy (such as calling the U.S. the world's largest debtor nation), these investments in assets of all kinds in many ways represent the faith foreign investors have in the future of the nation and its economy. A large part of this investment is the purchase of U.S. government securities, Treasury notes and bonds, issued in part to finance the U.S. government's growing deficits. These foreign purchasers have therefore aided in the financing of the U.S. government's budget deficit.

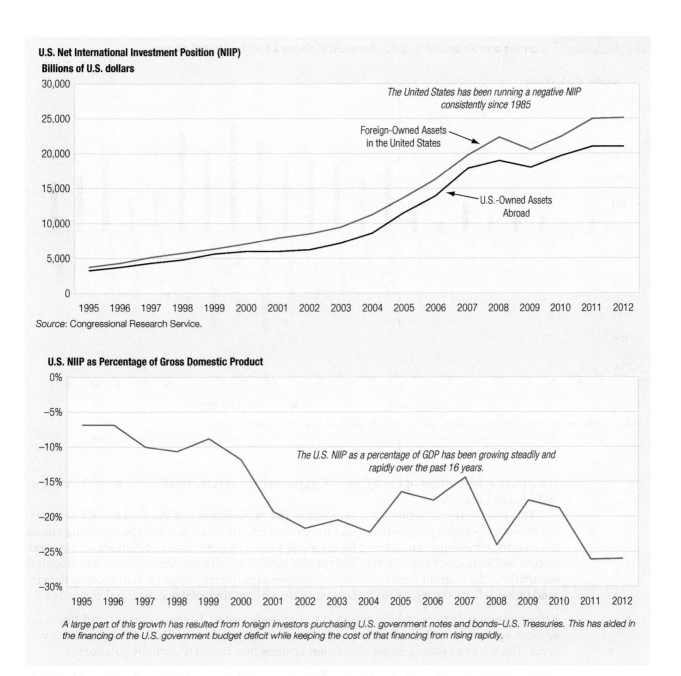

U.S. Net International Investment Position (NIIP)
Billions of U.S. dollars

The United States has been running a negative NIIP consistently since 1985

Foreign-Owned Assets in the United States

U.S.-Owned Assets Abroad

Source: Congressional Research Service.

U.S. NIIP as Percentage of Gross Domestic Product

The U.S. NIIP as a percentage of GDP has been growing steadily and rapidly over the past 16 years.

A large part of this growth has resulted from foreign investors purchasing U.S. government notes and bonds–U.S. Treasuries. This has aided in the financing of the U.S. government budget deficit while keeping the cost of that financing from rising rapidly.

Other Asset Investment. This final component of the financial account consists of various short-term and long-term trade credits, cross-border loans from all types of financial institutions, currency deposits and bank deposits, and other accounts receivable and payable related to cross-border trade.

Net Errors and Omissions and Official Reserves Accounts

Exhibit 3.4 illustrates the current account balance and the capital/financial account balances for the United States over recent years. The exhibit shows one of the basic economic and

EXHIBIT 3.4   **Current and Financial/Capital Account Balances for the United States**

Billions of U.S. Dollars

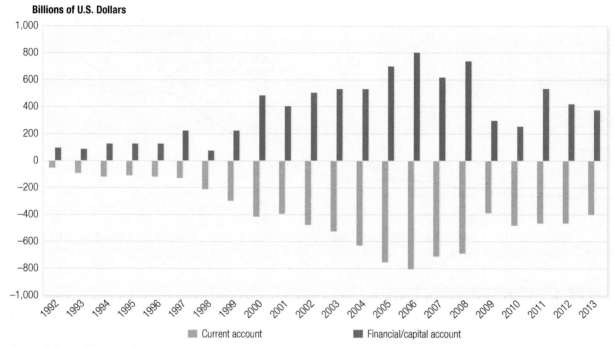

Current account   Financial/capital account

Source: Balance of Payments Statistics Yearbook, International Monetary Fund, December 2013, p. 1,032.

accounting relationships of the balance of payments: the inverse relation between the current and financial accounts.

This inverse relationship is not accidental. The methodology of the balance of payments, double-entry bookkeeping requires that the current and financial accounts be offsetting unless the country's exchange rate is being highly manipulated by governmental authorities. The next section on China describes one very high profile case in which government policy has thwarted economics—the twin surpluses of China. Countries experiencing large current account deficits fund these deficits through equally large surpluses in the financial account, and vice versa.

Net Errors and Omissions Account. As previously noted, because current and financial account entries are collected and recorded separately, errors or statistical discrepancies will occur. The net errors and omissions account ensures that the BOP actually balances.

Official Reserves Account. The Official Reserves Account is the total reserves held by official monetary authorities within a country. These reserves are normally composed of the major currencies used in international trade and financial transactions (so-called "hard currencies" like the U.S. dollar, European euro, and Japanese yen; gold; and special drawing rights, SDRs).

The significance of official reserves depends generally on whether a country is operating under a fixed exchange rate regime or a floating exchange rate system. If a country's currency is fixed, the government of the country officially declares that the currency is convertible into a fixed amount of some other currency. For example, the Chinese yuan was fixed to the U.S. dollar for many years. It was the Chinese government's responsibility to maintain this fixed rate, also called parity rate. If for some reason there was an excess supply of yuan on the currency market, to prevent the value of the yuan from falling, the Chinese government would have to support the yuan's value by purchasing yuan on the open market (by spending its hard currency reserves) until the excess supply was eliminated. Under a floating rate system, the Chinese government possesses no such responsibility and the role of official reserves is diminished. But as described in the following section, the Chinese government's foreign exchange reserves are now the largest in the world, and if need be, it probably possesses sufficient reserves to manage the yuan's value for years to come.

Breaking the Rules: China's Twin Surpluses

Exhibit 3.5 documents one of the more astounding BOP behaviors seen globally in many years—the twin surplus balances enjoyed by China in recent years. China's surpluses in both the current and financial accounts—termed the *twin surplus* in the business press—are highly unusual. Ordinarily, for example, in the cases of the United States, Germany, and Great Britain, a country will demonstrate an inverse relationship between the two accounts. This inverse relationship is not accidental, and typically illustrates that most large, mature, industrial countries "finance" their current account deficits through equally large surpluses in the financial account. For some countries like Japan, it is the inverse; a current account surplus is matched against a financial account deficit.

China, however, has experienced a massive current account surplus and a sometimes sizable financial account surplus simultaneously. This is rare, and an indicator of just how exceptional the growth of the Chinese economy has been. Although current account surpluses of this magnitude would ordinarily create a financial account deficit, the positive prospects of the Chinese economy have drawn such massive capital inflows into China in recent years

EXHIBIT 3.5   **China's Twin Surplus, 1998–2013**

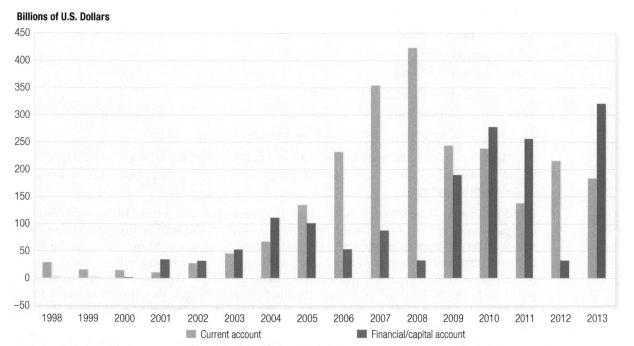

Billions of U.S. Dollars

Legend: Current account / Financial/capital account

Source: Balance of Payments Statistics Yearbook, International Monetary Fund, December 2013.

that the financial account too is in surplus. Note that the financial account surplus fell sizably in 2012, only to rise to a record level in 2013—as a result of continuing deregulation of capital inflows into the country's economy.

The rise of the Chinese economy has been accompanied by a rise in its current account surplus, and subsequently, its accumulation of foreign exchange reserves. China's foreign exchange reserves increased by a factor of 16 between 2001 and 2013—rising from $200 billion to nearly $3.7 trillion. There is no real precedent for this build-up in foreign exchange reserves in global financial history. These reserves allow the Chinese government to manage the value of the Chinese yuan and its impact on Chinese competitiveness in the world economy. The magnitude of these reserves will allow the Chinese government to maintain a relatively stable managed fixed rate of the yuan against other major currencies like the U.S. dollar as long as it chooses.

BOP Impacts on Key Macroeconomic Rates

A country's balance of payments both impacts and is impacted by the three macroeconomic rates of international finance: exchange rates, interest rates, and inflation rates.

The BOP and Exchange Rates

A country's BOP can have a significant impact on the level of its exchange rate and vice versa, depending on that country's exchange rate regime. The relationship between the BOP and exchange rates can be illustrated by using a simplified equation that summarizes BOP data:

Current Account Balance		Capital Account Balance		Financial Account Balance		Reserve Balance		Balance of Payments
$(X - M)$	$+$	$(CI - CO)$	$+$	$(FI - FO)$	$+$	FXB	$=$	BOP

X = exports of goods and services,

M = imports of goods and services,

CI = capital inflows,

CO = capital outflows,

FI = financial inflows,

FO = financial outflows,

FXB = official monetary reserves such as foreign exchange and gold.

The effect of an imbalance in the BOP of a country works somewhat differently depending on whether that country has fixed exchange rates, floating exchange rates, or a managed exchange rate system.

Fixed Exchange Rate Countries. Under a fixed exchange rate system, the government bears the responsibility to ensure that the BOP is near zero. If the sum of the current and capital accounts do not approximate zero, the government is expected to intervene in the foreign exchange market by buying or selling official foreign exchange reserves. If the sum of the first two accounts is greater than zero, a surplus demand for the domestic currency exists in the world. To preserve the fixed exchange rate, the government must then intervene in the foreign exchange market and sell domestic currency for foreign currencies or gold in order to bring the BOP back to near zero.

If the sum of the current and capital accounts is negative, an excess supply of the domestic currency exists in world markets. Then the government must intervene by buying the domestic currency with its reserves of foreign currencies and gold. It is obviously important for a government to maintain significant foreign exchange reserve balances, sufficient to allow it to intervene effectively. If the country runs out of foreign exchange reserves, it will be unable to buy back its domestic currency and will be forced to devalue its currency.

Floating Exchange Rate Countries. Under a floating exchange rate system, the government of a country has no responsibility to peg its foreign exchange rate. The fact that the current and capital account balances do not sum to zero will automatically—in theory—alter the exchange rate in the direction necessary to obtain a BOP near zero. For example, a country running a sizable current account deficit and a capital and financial accounts balance of zero will have a net BOP deficit. An excess supply of the domestic currency will appear on world markets. Like all goods in excess supply, the market will rid itself of the imbalance by lowering the price. Thus, the domestic currency will fall in value, and the BOP will move back toward zero.

Exchange rate markets do not always follow this theory, particularly in the short to intermediate term. This delay is known as the *J-curve* (detailed in an upcoming section). The deficit gets worse in the short run, but moves back toward equilibrium in the long run.

Managed Floats. Although still relying on market conditions for day-to-day exchange rate determination, countries operating with managed floats often find it necessary to take action to maintain their desired exchange rate values. They often seek to alter the market's valuation of their currency by influencing the motivations of market activity, rather than through direct intervention in the foreign exchange markets.

The primary action taken by these governments is to change relative interest rates, thus influencing the economic fundamentals of exchange rate determination. In the context of the equation presented earlier, a change in domestic interest rates is an attempt to alter the capital account balance $(CI - CO)$, especially the short-term portfolio component of these capital flows, in order to restore an imbalance caused by the deficit in the current account.

The power of interest rate changes on international capital and exchange rate movements can be substantial. A country with a managed float that wishes to defend its currency may choose to raise domestic interest rates to attract additional capital from abroad. This step will alter market forces and create additional market demand for the domestic currency. In this process, the government signals to the markets that it intends to take measures to preserve the currency's value within certain ranges. However, process also raises the cost of local borrowing for businesses, so the policy is seldom without domestic critics.

The BOP and Interest Rates

Apart from the use of interest rates to intervene in the foreign exchange market, the overall level of a country's interest rates compared to other countries has an impact on the financial account of the balance of payments. Relatively low real interest rates should normally stimulate an outflow of capital seeking higher interest rates in other country currencies. However, in the case of the United States, the opposite effect has occurred. Despite relatively low real interest rates and large BOP deficits on the current account, the U.S. BOP financial account has experienced offsetting financial inflows due to relatively attractive U.S. growth rate prospects, high levels of productive innovation, and perceived political safety. Thus, the financial account inflows have helped the United States to maintain its lower interest rates and to finance its exceptionally large fiscal deficit. However, it is beginning to appear that the favorable inflow on the financial account is diminishing while the U.S. balance on the current account is worsening.

The BOP and Inflation Rates

Imports have the potential to lower a country's inflation rate. In particular, imports of lower-priced goods and services place a limit on what domestic competitors can charge for comparable goods and services. Thus, foreign competition substitutes for domestic competition to maintain a lower rate of inflation than might have been the case without imports.

On the other hand, to the extent that lower-priced imports substitute for domestic production and employment, gross domestic product will be lower and the balance on the current account will be more negative.

Trade Balances and Exchange Rates

A country's import and export of goods and services is affected by changes in exchange rates. The transmission mechanism is in principle quite simple: changes in exchange rates change relative prices of imports and exports, and changing prices in turn result in changes in quantities demanded through the price elasticity of demand. Although the theory seems straightforward, real global business is more complex.

Trade and Devaluation

Countries occasionally devalue their own currencies as a result of persistent and sizable trade deficits. Many countries in the not-too-distant past have intentionally devalued their currencies in an effort to make their exports more price-competitive on world markets. These competitive devaluations are often considered self-destructive, however, as they also make imports relatively more expensive. So what is the logic and likely results of intentionally devaluing the domestic currency to improve the trade balance?

The J-Curve Adjustment Path

International economic analysis characterizes the trade balance adjustment process as occurring in three stages: (1) the currency contract period; (2) the pass-through period; and (3) the quantity adjustment period. These three stages are illustrated in Exhibit 3.6. Assuming that the trade balance is already in deficit prior to the devaluation, a devaluation at time t_1 results initially in a further deterioration in the trade balance before an eventual improvement. The path of adjustment, as shown, takes on the shape of a flattened "j."

In the first period, the *currency contract period*, a sudden unexpected devaluation of the domestic currency has a somewhat uncertain impact, simply because all of the contracts for exports and imports are already in effect. Firms operating under these agreements are required to fulfill their obligations, regardless of whether they profit or suffer losses. Assume that the United States experienced a sudden fall in the value of the U.S. dollar. Most exports were priced in U.S. dollars but most imports were contracts denominated in foreign currency. The result of a sudden depreciation would be an increase in the size of the trade deficit at time t_1 because the cost to U.S. importers of paying their import bills would rise as they spent more dollars to buy the foreign currency they needed, while the revenues earned by U.S. exporters would remain unchanged. There is little reason, however, to believe that most U.S. imports are denominated in foreign currency and most exports in dollars.

The second period of the trade balance adjustment process is termed the *pass-through period*. As exchange rates change, importers and exporters eventually must pass these exchange rate changes through to their own product prices. For example, a foreign producer selling to the U.S. market after a major fall in the value of the U.S. dollar will have to cover its own domestic costs of production. This need will require the firm to charge higher dollar prices in order to earn its own local currency in large enough quantities. The firm must raise its prices in the U.S. market. U.S. import prices then rise, eventually passing the full exchange rate changes through to prices. Similarly, the U.S. export prices are now cheaper compared to foreign competitors' because the dollar is cheaper. Unfortunately for U.S. exporters, many of the inputs for their final products may actually be imported, dampening the positive impact of the fall of the dollar.

EXHIBIT 3.6   **Trade Adjustment to Exchange Rates: The J-Curve**

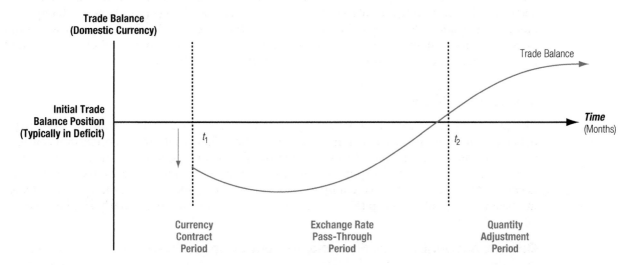

If export products are predominantly priced and invoiced in domestic currency, and imports are predominantly priced and invoiced in foreign currency, a sudden devaluation of the domestic currency can possibly result—initially—in a deterioration of the balance on trade. After exchange rate changes are passed through to product prices, and markets have time to respond to price changes by altering market demands, the trade balance will improve. The currency contract period may last from three to six months, with pass-through and quality adjustment following for an additional three to six months.

The third and final period, the *quantity adjustment period*, achieves the balance of trade adjustment that is expected from a domestic currency devaluation or depreciation. As the import and export prices change as a result of the pass-through period, consumers both in the United States and in the U.S. export markets adjust their demands to the new prices. Imports are relatively more expensive; therefore the quantity demanded decreases. Exports are relatively cheaper; therefore the quantity demanded increases. The balance of trade—the expenditures on exports less the expenditures on imports—improves.

Unfortunately, these three adjustment periods do not occur overnight. Countries like the U.S. that have experienced major exchange rate changes, have also seen this adjustment take place over a prolonged period. Empirical studies have concluded that for industrial countries, the total time elapsing between time t_1 and t_2 can vary from 3 to 12 months. To complicate the process, new exchange rate changes often occur before the adjustment is completed.

Trade Balance Adjustment Path: The Equation

A country's trade balance is essentially the net of import and export revenues, where each is a multiple of price—$P_x^\$$ and P_m^{fc}—the prices of exports and imports, respectively. Export prices are assumed to be denominated in U.S. dollars, and import prices are denominated in foreign currency. The quantity of exports and the quantity of imports are denoted as Q_x and Q_m, respectively. Import expenditures are then expressed in U.S. dollars by multiplying the foreign currency denominated expenditures by the spot exchange rate $S^{\$/fc}$. The U.S. trade balance, expressed in U.S. dollars, is then expressed as follows:

$$\text{U.S. Trade Balance} = P_x^\$ Q_x) - (S^{\$/fc} P_M^{fc} Q_M)$$

The immediate impact of a devaluation of the domestic currency is to increase the value of the spot exchange rate $S^{\$/fc}$, resulting in an immediate deterioration in the trade balance (*currency contract period*). Only after a period in which the current contracts have matured, and new prices reflecting partial to full pass-through of the exchange rate change, will improvement in the trade balance be evident (*pass-through period*). In the final stage, in which the price elasticity of demand has time to take effect (*quantity adjustment period*), is the actual trade balance expected to rise above where it started in Exhibit 3.6.

Capital Mobility

The degree to which capital moves freely cross-border is critically important to a country's balance of payments. We have already seen how the U.S. has suffered a deficit in its current account balance over the past 20 years while running a surplus in the financial account, and how China has enjoyed a surplus in both the current and financial accounts over the last decade. But these are only two country cases, and may not reflect the challenges that changing balances in trade and capital may mean for many countries, particularly smaller ones or emerging markets.

Current Account versus Financial Account Capital Flows

Capital inflows can contribute significantly to an economy's development. Capital inflows can increase the availability of capital for new projects, new infrastructure development, and productivity improvements. These, in turn, may stimulate general economic growth and job creation. For domestic holders of capital, the ability to invest outside the domestic economy may reap greater investment returns, portfolio diversification, and extend the commercial development of domestic enterprises.

That said, the free flow of capital in and out of an economy can potentially destabilize economic activity. Although the benefits of free capital flows have been known for centuries, so have the negatives. For this very reason, the creators of the Bretton Woods system were very careful to promote and require the free movement of capital for current account transactions — foreign exchange, bank deposits, money market instruments — but they did not require such free transit for capital account transactions — foreign direct investment and equity investments.

Experience has shown that current account-related capital flows can be more volatile, with capital flowing in and out of an economy and a currency on the basis of short-term interest rate differentials and exchange rate expectations. This volatility is somewhat compartmentalized, not directly impacting real asset investments, employment, or long-term economic growth. Longer-term capital flows reflect more fundamental economic expectations, including growth prospects and perceptions of political stability.

The complexity of issues, however, is apparent when you consider the plight of many emerging market countries. Recall the *impossible trinity* from Chapter 2—the theoretical structure that states that no country can maintain a fixed exchange rate, allow complete capital mobility (both in and out of the country), and conduct independent monetary policy—simultaneously. Many emerging market countries have continued to develop by maintaining a near-fixed (soft peg) exchange rate regime—a strictly independent monetary policy—while restricting capital inflows and outflows. With the growth of current account business activity (exports and imports of goods and services), more current account-related capital flows are deregulated. If, however, the country experiences significant volatility in these short-term capital movements, capital flows potentially impacting either exchange rate pegs or monetary policy objectives, authorities are often quick to reinstitute capital controls.

The growth in capital openness over the past 30 years resulted in a significant increase in political pressures for more countries to open up more of their financial account sectors to international capital. But the devastation of the Asian Financial Crisis of 1997/1998 brought much of that to a halt. Smaller economies, no matter how successful their growth and development may have been under export-oriented trade strategies, found themselves still subject to sudden and destructive capital outflows in times of economic crisis and financial contagion.

Historical Patterns of Capital Mobility

Before leaving our discussion of the balance of payments, we need to gain additional insights into the history of capital mobility and the contribution of capital outflows—capital flight—to balance of payments crises. Has capital always been free to move in and out of a country? Definitely not. The ability of foreign investors to own property, buy businesses, or purchase stocks and bonds in other countries has been controversial.

Exhibit 3.7, first introduced in Chapter 2, provides a way of categorizing historical eras of capital mobility over the last 150 years. The exhibit divides economic history into five distinct exchange rate eras and their associated implications for capital mobility (or lack thereof). These exchange rate eras obviously reflect the exchange rate regimes we discussed and detailed in Chapter 2, but also reflect the evolution of political economy beliefs and policies of both industrialized and emerging market nations over this period.

The Gold Standard (1860–1914). Although an era of growing capital openness in which trade and capital began to flow more freely, it was an era dominated by industrialized nation economies that were dependent on gold convertibility to maintain confidence in the system.

The Inter-War Years (1914–1945). This was an era of retrenchment in which major economic powers returned to policies of isolationism and protectionism, thereby restricting trade and nearly eliminating capital mobility. The devastating results included financial crisis, a global depression, and rising international political and economic disputes that drove nations into a second world war.

EXHIBIT 3.7   **The Evolution of the Global Monetary System**

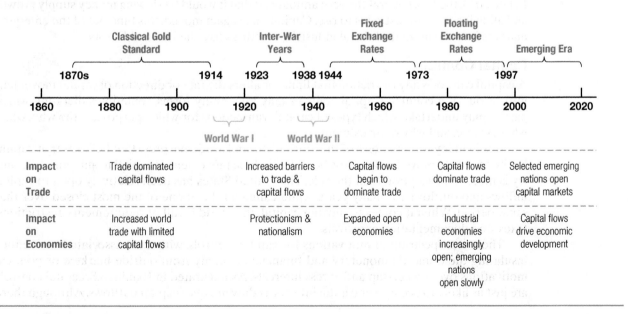

The Bretton Woods Era (1945–1971). The dollar-based fixed exchange rate system under Bretton Woods gave rise to a long period of economic recovery and growing openness of both international trade and capital flows in and out of countries. Many researchers (for example Obstfeld and Taylor, 2001) believe it was the rapid growth in the speed and volume of capital flows that ultimately led to the failure of Bretton Woods—global capital could no longer be held in check.

The Floating Era (1971–1997). The Floating Era, saw the rise of a growing schism between the industrialized and emerging market nations. The industrialized nations (primary currencies) moved to—or were driven to—floating exchange rates by capital mobility. The emerging markets (secondary currencies), in an attempt to both promote economic development and maintain control over their economies and currencies, opened trade but maintained restrictions on capital flows. Despite these restrictions, the era ended with the onslaught of the Asian Financial Crisis in 1997.

The Emerging Era (1997–Present). The emerging economies, led by China and India, attempt to gradually open their markets to global capital. But, as the impossible trinity taught the industrial nations in previous years, the increasing mobility of capital now requires that they give up either the ability to manage their currency values or to conduct independent monetary policies. The most challenging dimension in this current era is that a number of emerging market currencies are now being buffered by the magnitude of non-current account capital flows—portfolio capital or "hot money" flows as it has been termed—and their currencies now suffer larger swings in appreciation or depreciation as capital flows grow in magnitude.

The 2008–2014 period reinforced what some call the double-edged sword of global capital movements. The credit crisis of 2008–2009, beginning in the United States, quickly spread to the global economy, pulling and pushing down industrial and emerging market economies alike. But in the post credit crisis period, global capital now flowed toward the emerging markets. Although funding and fueling their rapid economic recoveries, it came—in the words of one journalist—"with luggage." The increasing pressure on emerging market currencies to appreciate is partially undermining their export competitiveness. But then, just as suddenly as the capital came, it went. In late 2013, the U.S. Federal Reserve announced that it would be slowing money supply growth and allowing U.S. interest rates to rise. Capital once again moved; this time out of the emerging markets into the more traditional industrial countries like the U.S. and Europe.

Capital Controls

A capital control is any restriction that limits or alters the rate or direction of capital movement into or out of a country. Capital controls may take many forms, sometimes dictating which parties may undertake which types of capital transactions for which purposes—the who, what, when, where, and why of investment.

It is in many ways the bias of the journalistic and academic press that believes that capital has been able to move freely across boundaries. Free movement of capital in and out of a country is more the exception than the rule. The United States has been relatively open to capital inflows and outflows for many years, while China has been one of the most closed over that same period. When it comes to moving capital, the world is full of requirements, restrictions, taxes, and documentation approvals.

There is a spectrum of motivations for capital controls, with most associated with either insulating the domestic monetary and financial economy from outside markets or political motivations over ownership and access interests. As illustrated in Exhibit 3.8, capital controls are just as likely to occur over capital inflows as they are over capital outflows. Although there

EXHIBIT 3.8	**Purposes of Capital Controls**		

Control Purpose	Method	Capital Flow Controlled	Example
General Revenue/ Finance War Effort	Controls on capital outflows permit a country to run higher inflation with a given fixed-exchange rate and also hold down domestic interest rates.	Outflows	Most belligerents in WWI and WWII
Financial Repression/ Credit Allocation	Governments that use the financial system to reward favored industries or to raise revenue, may use capital controls to prevent capital from going abroad to seek higher returns.	Outflows	Common in developing countries
Correct a Balance of Payments Deficit	Controls on outflows reduce demand for foreign assets without contractionary monetary policy or devaluation. This allows a higher rate of inflation than otherwise would be possible.	Outflows	U.S. interest equalization tax 1963–1974
Correct a Balance of Payments Surplus	Controls on inflows reduce foreign demand for domestic assets without expansionary monetary policy or revaluation. This allows a lower rate of inflation than would otherwise be possible.	Inflows	German Bardepot Scheme 1972–1974
Prevent Potentially Volatile Inflows	Restricting inflows enhances macroeconomic stability by reducing the pool of capital that can leave a country during a crisis.	Inflows	Chilean *encaje* 1991–1998
Prevent Financial Destabilization	Capital controls can restrict or change the composition of international capital flows that can exacerbate distorted incentives in the domestic financial system.	Inflows	Chilean *encaje* 1991–1998
Prevent Real Appreciation	Restricting inflows prevents the necessity of monetary expansion and greater domestic inflation that would cause a real appreciation of the currency.	Inflows	Chilean *encaje* 1991–1998
Restrict Foreign Ownership of Domestic Assets	Foreign ownership of certain domestic assets— especially natural resources—can generate resentment.	Inflows	Article 27 of the Mexican Constitution
Preserve Savings for Domestic Use	The benefits of investing in the domestic economy may not fully accrue to savers so the economy as a whole can be made better off by restricting the outflow of capital.	Outflows	—
Protect Domestic Financial Firms	Controls that temporarily segregate domestic financial sectors from the rest of the world may permit domestic firms to attain economies of scale to compete in world markets.	Inflows and Outflows	—

Source: "An Introduction to Capital Controls," Christopher J. Neely, *Federal Reserve Bank of St. Louis Review*, November/December 1999, p. 16.

is a tendency for a negative connotation to accompany capital controls (possibly the bias of the word "control" itself), the impossible trinity requires that capital flows be controlled if a country wishes to maintain a fixed exchange rate and an independent monetary policy.

Capital controls may take a variety of forms that mirror restrictions on trade. They may simply be a tax on a specific transaction, they may limit the quantity or magnitude of specific capital transactions, or they may prohibit transactions altogether. The controls themselves have tended to follow the basic dichotomy of the balance of payments current account transactions versus financial account transactions.

In some cases capital controls are intended to stop or thwart capital outflows and currency devaluation or depreciation. The case of Malaysia during the Asian Crisis of 1997–1998 is one example. As the Malaysian currency came under attack and capital started to leave the Malaysian economy, the government imposed capital controls to stop short-term capital movements, in or out, but not hinder nor restrict long-term inward investment. All trade-related requests for access to foreign exchange were granted, allowing current account-related capital flows to continue. But access to foreign exchange for inward or outward money market or capital market investments were restricted. Foreign residents wishing to invest in Malaysian assets—real assets not financial assets—had open access.

Capital controls can be implemented in the opposite case, in which the primary fear is that large rapid capital inflows will both cause currency appreciation (and therefore harm export competitiveness) and complicate monetary policy (capital inflows flooding money markets and bank deposits). Chile in the 1990s provides an example. Newfound political and economic soundness started attracting international capital. The Chilean government responded with its *encaje* program, which imposed taxes and restrictions on short-term (less than one year) capital inflows, as well as restrictions on the ability of domestic financial institutions to extend credits or loans in foreign currency. Although credited with achieving its goals of maintaining domestic monetary policy and preventing a rapid appreciation in the Chilean peso, this program came at substantial cost to Chilean firms, particularly smaller ones.

A similar use of capital controls to prevent domestic currency appreciation is the so-called case of *Dutch Disease*. With the rapid growth of the natural gas industry in the Netherlands in the 1970s, there was growing fear that massive capital inflows would drive up the demand for the Dutch guilder and cause a substantial currency appreciation. A more expensive guilder would harm other Dutch manufacturing industries, causing them to decline relative to the natural resource industry. This is a challenge faced by a number of resource-rich economies of relatively modest size and with relatively small export sectors in recent years, including oil and gas development in Azerbaijan, Kazakhstan, and Nigeria, to name but a few.

An extreme problem that has raised its head a number of times in international financial history is *capital flight*, one of the problems that capital controls are designed to control. Although defining capital flight is a bit difficult, the most common definition is the rapid outflow of capital in opposition to or in fear of domestic political and economic conditions and policies. Although it is not limited to heavily indebted countries, the rapid and sometimes illegal transfer of convertible currencies out of a country poses significant economic and political problems. Many heavily indebted countries have suffered significant capital flight, compounding their problems of debt service.

There are a number of mechanisms used for moving money from one country to another, some legal, some not. Transfers via the usual international payments mechanisms (regular bank transfers) are obviously the easiest and lowest cost, and are legal. Most economically healthy countries allow free exchange of their currencies, but of course for such countries capital flight is not a problem. The opposite, transfer of physical currency by bearer (the proverbial smuggling out of cash in the false bottom of a suitcase) is more costly and, for transfers out of many countries, illegal. Such transfers may be deemed illegal for balance of payments reasons or to make difficult the movement of money from the drug trade or other illegal activities.

And there are other more creative solutions. One is to move cash via collectibles or precious metals, which are then transferred across borders. Money laundering is the cross-border purchase of assets that are then managed in a way that hides the movement of money and its ownership. And finally, false invoicing of international trade transactions occurs when capital is moved through the under-invoicing of exports or the over-invoicing of imports, where the difference between the invoiced amount and the actual agreed upon payment is deposited in banking institutions in a country of choice.

Globalization of Capital Flows

Notwithstanding these benefits, many EMEs [emerging market economies] are concerned that the recent surge in capital inflows could cause problems for their economies. Many of the flows are perceived to be temporary, reflecting interest rate differentials, which may be at least partially reversed when policy interest rates in advanced economies return to more normal levels. Against this backdrop, capital controls are again in the news.

A concern has been that massive inflows can lead to exchange rate overshooting (or merely strong appreciations that significantly complicate economic management) or inflate asset price bubbles, which can amplify financial fragility and crisis risk. More broadly, following the crisis, policymakers are again reconsidering the view that unfettered capital flows are a fundamentally benign phenomenon and that all financial flows are the result of rational investing/borrowing/lending decisions. Concerns that foreign investors may be subject to herd behavior, and suffer from excessive optimism, have grown stronger; and even when flows are fundamentally sound, it is recognized that they may contribute to collateral damage, including bubbles and asset booms and busts.

— "Capital Inflows: The Role of Controls," Jonathan D. Ostry, Atish R. Ghosh, Karl Habermeier, Marcos Chamon, Mahvash S. Qureshi, and Dennis B.S. Reinhardt, IMF Staff Position Note, SPN/10/04, February 19, 2010, p. 3.

Traditionally, the primary concern over capital inflows is that they are short-term in duration, may flow out with short notice, and are characteristics of the politically and economically unstable emerging markets. But as described in the preceding quote, two of the largest capital flow crises in recent years have occurred within the largest, most highly developed, mature capital markets—the United States and Western Europe.

In both the 2008 global credit crisis, which had the United States as its core, and the ensuing European sovereign debt crisis, crisis befell markets that have long been considered some of the most mature, the most sophisticated, and the "safest."

SUMMARY POINTS

- The BOP is the summary statement—a cash flow statement—of all international transactions between one country and all other countries over a period of time, typically a year.

- The two sub-accounts of the BOP that receive the most attention are the current account and the financial account. These accounts summarize the current trade and international capital flows of the country, respectively.

- The current account and financial account are typically inverse on balance, one in surplus and the other in deficit. China, however, has been consistently enjoying a surplus of both.

- Monitoring the various sub-accounts of a country's BOP activity is helpful to decision-makers and policymakers, on all levels of government and industry, in detecting the underlying trends and movements of fundamental economic forces driving a country's international economic activity.

- Changes in exchange rates affect relative prices of imports and exports, and changing prices in turn result in changes in quantities demanded through the price elasticity of demand.

- A devaluation results initially in a further deterioration of the trade balance before an eventual improvement—the path of adjustment taking on the shape of a flattened "j."

- The ability of capital to move instantaneously and massively cross-border has been one of the major factors in the severity of recent currency crises. In cases such as Malaysia in 1997 and Argentina in 2001, the national governments concluded that they had no choice but to impose drastic restrictions on the ability of capital to flow.

- Although not limited to heavily indebted countries, the rapid and sometimes illegal transfer of convertible currencies out of a country poses significant economic problems. Many heavily indebted countries have suffered significant capital flight, which has compounded their problems of debt service.

Global Remittances[1]

One area within the balance of payments that has received intense interest in the past decade is that of remittances. The term *remittance* is a bit tricky. According to the International Monetary Fund (IMF), remittances are international transfers of funds sent by migrant workers from the country where they are working to people, typically family members, in the country from which they originated. According to the IMF, a *migrant* is a person who comes to a country and stays, or intends to stay, for a year or more. As illustrated by Exhibit A, it is estimated that nearly $600 billion was remitted across borders in 2014.

Remittances make up a very small, often negligible cash outflow from sending countries like the United States. They do, however, represent a more significant volume, for example as a percent of GDP, for smaller receiving countries, typically developing countries, sometimes more than 25%. In many cases, this is greater than all development capital and aid flowing to these same countries. And although the historical record on global remittances

is short, as illustrated in Exhibit A, it has shown dramatic growth in the post-2000 period. Its growth has been rapid and dramatic, falling back only temporarily with the global financial crisis of 2008–2009, before returning to its rapid growth path once again from 2010 on.

Remittances largely reflect the income that is earned by migrant or guest workers in one country (source country) and then returned to families or related parties in their home countries (receiving countries). Therefore it is, not surprising that although there are more migrant worker flows between developing countries, the high-income developed economies remain the main source of remittances. The global economic recession of 2009 resulted in reduced economic activities like construction and manufacturing in the major source countries; as a result, remittance cash flows fell in 2009 but rebounded slightly in 2010.

Most remittances occur as frequent small payments made through wire transfers or a variety of informal

Global Remittance Inflows, 1970–2014 (millions of U.S. dollars)

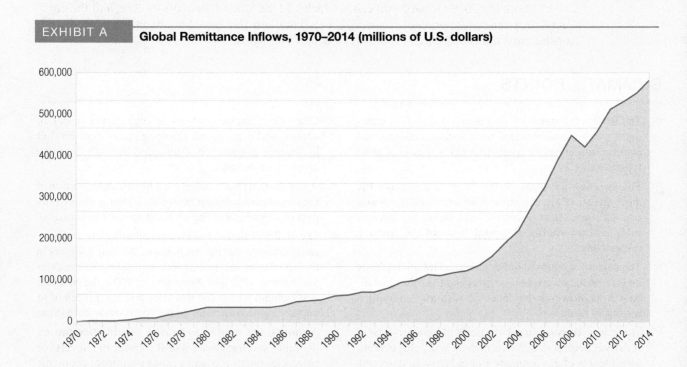

channels (some even carried by hand). The United States Bureau of Economic Analysis (BEA), which is responsible for the compilation and reporting of U.S. balance of payments statistics, classifies migrant remittances as "current transfers" in the current account. Wider definitions of remittances may also include capital assets that migrants take with them to host countries and similar assets that migrants bring back with them to their home countries. These values, when compiled, are generally reported under the capital account of the balance of payments. However, discerning exactly who is a "migrant," is also an area of some debate. Transfers back to their home country made by individuals who may be working in a foreign country (for example, an expat working for a multinational organization) but who are not considered "residents" of that country, may also be considered global remittances under current transfers in the current account.

Remittance Prices

Given the development impact of remittance flows, we will facilitate a more efficient transfer and improved use of remittances and enhance cooperation between national and international organizations, in order to implement the recommendations of the 2007 Berlin G8 Conference and of the Global Remittances Working Group established in 2009 and coordinated by the World Bank. We will aim to make financial services more accessible to migrants and to those who receive remittances in the developing world. We will work to achieve in particular the objective of a reduction of the global average costs of transferring remittances from the present 10% to 5% in five years through enhanced information, transparency, competition and cooperation with partners, generating a significant net increase in income for migrants and their families in the developing world.

— The G8 Final Declaration on Responsible
Leadership for a Sustainable Future, paragraph 134.

A number of organizations have devoted significant effort in the past five years to better understanding the costs borne by migrants in transferring funds back to their home countries. The primary concern has been excessive remittance charges—the imposition of what many consider exploitive charges related to the transfer of these frequent small payments.

The G8 countries launched an initiative in 2008 entitled "5 × 5", to reduce transfer costs from a global average of 10% to 5% in five years (by 2014). The World Bank

supported this initiative by creating Remittance Prices Worldwide (RPW), a global database to monitor remittance price activity across geographic regions.[2] It was hoped that, through greater transparency and access to transfer cost information, market forces would drive these costs down. Although the global average cost had fallen to a low of 7.90% in the fall 2014, the program was still clearly far from its goal of 5%. Funds remitted from the G8 countries themselves fell to 7.49% in 2014, 7.98% for the G20 countries in the same period. This was particularly relevant given that these are the source countries of a large proportion of all funds remitted.

Little was known of global remittance costs until the World Bank began collecting data in the RPW database. The database collects data on the average cost of transactions conducted along a variety of country corridors globally (country pairs). Exhibit B provides one sample of what these cost surveys look like. This corridor transaction, the transfer of ZAR 1370 (South African Rand, equivalent to about USD 200 at that time) from South Africa to Malawi was the highest cost corridor in the RPW.

Remittance costs shown in Exhibit B are of two types: (1) a transaction fee, which in this case ranges between ZAR 43 and 390; and (2) an exchange rate margin, which is an added cost over and above the organization's own cost of currency. The resulting total cost per transaction can be seen to rise as high as 30.6% for this specific corridor. Given that most transfers are by migrant or guest workers back to their home countries and families, and they are members usually of the lowest income groups, these charges—30%—are seen as exploitive.

It should also be noted that these are charges imposed upon the sender, at the origin. Other fees or charges may occur to the receiver at the point of destination. It is also obvious from the survey data in Exhibit B that fees and charges may differ dramatically across institutions. Hence the objective of the program—to provide more information that is publicly available to people remitting funds thereby adding transparency to the process—is clear.

Other results from the RPW cost survey initiative include the following.

- China is the most expensive country in the G20 to send money to, while South Africa continues to be the mostly costly G20 country to send money from.
- South Asia is the least costly region to send money to, while Sub-Saharan Africa continues to be the most expensive region to send money to globally.

[2]Latest corridor cost surveys are available at the RPW Web site at http://remittanceprices.worldbank.org.

Remittance Price Comparison for Transfer from South Africa to Malawi

ZAR 1,370.00

Firm	Firm Type	Product	Fee	Exchange Rate Margin (%)	Total Cost Percent (%)	Total Cost (currency)	Net Transfer (currency)
MoneyGram	MTO	Branch	149.60	2.10	13.02	178.37	1,191.63
Mukuru	MTO	Branch	123.30	6.76	15.76	215.91	1,154.09
Mukuru	MTO	Branch, call-center	123.30	6.76	15.76	215.91	1,154.09
Western Union	MTO	Branch, call-center	194.84	1.70	15.92	218.13	1,151.87
Nedbank	Bank	Branch, call-center	228.00	6.06	22.70	311.02	1,058.98
ABSA	Bank	Branch, call-center	193.80	9.39	23.54	322.44	1,047.56
Standard	Bank	Bank Branch, call-center	235.00	10.35	27.50	376.80	993.21
Bidvest	Bank	Bank Branch, call-center	356.00	2.10	28.09	384.77	985.23
Bank of Athens	Bank	Branch, call-center	390.00	1.96	30.43	416.85	953.15
FNB of South Africa	MTO	Branch, call-center	235.00	19.45	36.60	501.47	868.54
South African Post Office*	Post	Branch	43.10	0.00	3.15	43.10	1,326.90
Bank Average			280.56	5.97	26.45	362.38	1,007.62
Money Transfer Average			165.21	7.35	19.41	265.96	1,104.04
Post Office Average			43.10	0.00	3.15	43.10	1,326.90
Total Average			206.54	6.06	21.13	289.52	1,080.48

MTO: Money transfer operator.

Fee: Currency fee charged customer per transaction of ZAR 1370 (South African Rand), equal to USD 200.00.

Exchange Rate Margin: Additional margin charged customer over and above the interbank exchange rate (percentage difference).

Total Cost (%): Total cost to customer of a single transaction including the transaction fee and exchange rate margin.

Total Cost (currency): Total cost in ZAR of a single ZAR 1370.00 transaction × total cost in percent.
Net Transfer (currency): Net remittance after total costs (ZAR1370.00−total cost in ZAR).

* The South African Post Office is not transparent; it does not disclose the exchange rate used prior to executing the transaction, hence it is not zero.

Source: World Bank, "Sending money from South Africa to Malawi," remittanceprices/worldbank.org/en/corridor, data collected by World Bank on November 11, 2014, and author calculations.

- The five highest cost corridors (always available on the RPW Web site) continue to be intra-Africa.

- In 2013, India received foreign exchange remittances worth $70 billion from its migratory workforce to retain the top spot in the world amid a broad slow-down caused by regulatory hindrances on both move-ment of people and capital.

- The top 10 remittance recipient nations in 2013 following India were China ($60 billion), the Philippines ($25 billion), Mexico ($22 billion), Nigeria ($21 billion), Egypt ($17 billion), Pakistan ($15 billion), Bangladesh ($14 billion), Vietnam ($11 billion), and Ukraine ($10 billion).

Product Types and Innovation

RPW tracks a number of different data dimensions to this growing industry, including costs by service provider types—commercial banks, money transfer operators (MTOs), and post offices—and product types classified by cash/account transactions. According to the World Bank commercial banks continue to be the most costly, with MTOs on average—although it does differ dramatically across region and corridors—being the cheapest.

Exhibit C provides a breakdown of the types of transactions conducted in the global remittance mar-ket in 2013. Not surprisingly, nearly 50% of the remit-tances are still cash-to-cash, reflecting both the source and use of the funds being remitted. Online services have been growing, and appear to be some of the lowest cost. Account-to-account services are by far the most expensive, although arguably they represent the most organizationally formal transactions (bank-to-bank accounts held by customers).

This global segment, however, should soon be fruitful ground for product innovation, such as the use of Bitcoin, the cryptocurrency that has gained increasing use globally. A number of companies are now attempting to build cross-border remittance platforms using digital currencies like Bitcoin, yet no single company or platform seems to have yet established dominance. One of the continuing barriers is regulatory, as many governments restrict access to financial service linkages. As a result, some platforms, like HelloBit, are attempting to form sub-platforms that use already existing regulatory access, but allow the use of Bitcoin rather than traditional national currencies.

Growing Controversies

With the growth in global remittances has come a growing debate as to what role they do or should play in a country's balance of payments, and more importantly, economic development. In some cases, like India, there is growing resistance from the central bank and other banking institutions to allow online payment services like PayPal to process remittances. In other countries, like Honduras, Guatemala, and Mexico, there is growing debate on whether the remittances flow to families, or are actually payments made to a variety of Central American human-trafficking smugglers.

In Mexico for example, remittances now make up the second largest source of foreign exchange earnings, second only to oil exports. The Mexican government has increasingly viewed remittances as an integral component of its balance of payments, and in some ways, a "plug" to replace declining export competition and dropping foreign direct investment. But there is also growing evidence that remittances flow to those who need it most, the lowest income component of the Mexican population, and therefore mitigate poverty and support consumer spending. Former President Vicente Fox was quoted as saying that Mexico's workers in other countries remitting income home to Mexico are "heroes." Mexico's own statistical agencies also disagree on the size of the funds remittances received, as well as to whom the income is returning (family or non-family interests).

Mini-Case Questions

1. Where are remittances across borders included within the balance of payments? Are they current or financial account components?
2. Under what conditions—for example, for which countries currently—are remittances significant contributors to the economy and overall balance of payments?
3. Why is the cost of remittances the subject of such intense international scrutiny?
4. What potential do new digital currencies—cryptocurrencies like Bitcoin—have for cross-border remittances?

EXHIBIT C   **Remittance Product Use and Cost**

Product Types	Percent of Transactions	Average Cost
Cash to cash	45%	7.0%
Account to account (to any bank)	19%	12.5%
Online	17%	5.9%
Cash to account	8%	5.6%
Account to cash	4%	7.8%
Account to account (within same bank)	2%	7.9%
Mobile	1%	6.5%
Prepaid card	1%	8.4%
Other	3%	9.5%
	100%	

Source: *Remittance Prices Worldwide*, Issue No. 11, September 2014, Figures 11 and 12, p. 7.

QUESTIONS

These questions are available in MyFinanceLab.

1. **Balance of Payments Defined.** What is the balance of payments?

2. **BOP Data.** What institution provides the primary source of similar statistics for balance of payments and economic performance worldwide?

3. **Importance of BOP.** Business managers and investors need BOP data to anticipate changes in host-country economic policies that might be driven by BOP events. From the perspective of business managers and investors, list three specific signals that a country's BOP data can provide.

4. **Flow Statement.** What does it mean to describe the balance of payments as a flow statement?

5. **Economic Activity.** What are the two main types of economic activity measured by a country's BOP?

6. **Balance.** Why does the BOP always "balance"?

7. **BOP Accounting.** If the BOP were viewed as an accounting statement, would it be a balance sheet of the country's wealth, an income statement of the country's earnings, or a funds flow statement of money into and out of the country?

8. **Current Account.** What are the main component accounts of the current account? Give one debit and one credit example for each component account for the United States.

9. **Real versus Financial Assets.** What is the difference between a real asset and a financial asset?

10. **Direct versus Portfolio Investments.** What is the difference between a direct foreign investment and a portfolio foreign investment? Give an example of each. Which type of investment is a multinational industrial company more likely to make?

11. **Net International Investment Position.** What is a country's net international investment position and how does it differ from the balance of payments?

12. **The Financial Account.** What are the primary subcomponents of the financial account? Analytically, what would cause net deficits or surpluses in these individual components?

13. **Classifying Transactions.** Classify each of the following as a transaction reported in a sub-component of the current account or of the capital and financial accounts of the two countries involved:

    a. A U.S. food chain imports wine from Chile.
    b. A U.S. resident purchases a euro-denominated bond from a German company.
    c. Singaporean parents pay for their daughter to study at a U.S. university.
    d. A U.S. university gives a tuition grant to a foreign student from Singapore.
    e. A British Company imports Spanish oranges, paying with eurodollars on deposit in London.
    f. The Spanish orchard deposits half its proceeds in a eurodollar account in London.
    g. A London-based insurance company buys U.S. corporate bonds for its investment portfolio.
    h. An American multinational enterprise buys insurance from a London insurance broker.
    i. A London insurance firm pays for losses incurred in the United States because of an international terrorist attack.
    j. Cathay Pacific Airlines buys jet fuel at Los Angeles International Airport so it can fly the return segment of a flight back to Hong Kong.
    k. A California-based mutual fund buys shares of stock on the Tokyo and London stock exchanges.
    l. The U.S. army buys food for its troops in South Asia from local vendors.
    m. A Yale graduate gets a job with the International Committee of the Red Cross in Bosnia and is paid in Swiss francs.
    n. The Russian government hires a Norwegian salvage firm to raise a sunken submarine.
    o. A Colombian drug cartel smuggles cocaine into the United States, receives a suitcase of cash, and flies back to Colombia with that cash.
    p. The U.S. government pays the salary of a foreign service officer working in the U.S. embassy in Beirut.
    q. A Norwegian shipping firm pays U.S. dollars to the Egyptian government for passage of a ship through the Suez Canal.
    r. A German automobile firm pays the salary of its executive working for a subsidiary in Detroit.
    s. An American tourist pays for a hotel in Paris with his American Express card.
    t. A French tourist from the provinces pays for a hotel in Paris with his American Express card.
    u. A U.S. professor goes abroad for a year on a Fulbright grant.

14. **The Balance.** What are the main summary statements of the balance of payments accounts and what do they measure?

15. **Twin Surpluses.** Why are China's twin surpluses—surpluses in both the current and financial accounts—considered unusual?

16. **Capital Mobility—United States.** The U.S. dollar has maintained or increased its value over the past 20 years despite running a gradually increasing current account deficit. Why has this phenomenon occurred?

17. **Capital Mobility—Brazil.** Brazil has experienced periodic depreciation of its currency over the past 20 years despite occasionally running a current account surplus. Why has this phenomenon occurred?

18. **BOP Transactions.** Identify the correct BOP account for each of the following transactions:
    a. A German-based pension fund buys U.S. government 30-year bonds for its investment portfolio.
    b. Scandinavian Airlines System (SAS) buys jet fuel at Newark Airport for its flight to Copenhagen.
    c. Hong Kong students pay tuition to the University of California, Berkeley.
    d. The U.S. Air Force buys food in South Korea to supply its air crews.

    e. A Japanese auto company pays the salaries of its executives working for its U.S. subsidiaries.
    f. A U.S. tourist pays for a restaurant meal in Bangkok.
    g. A Colombian citizen smuggles cocaine into the United States, receives cash, and smuggles the dollars back into Colombia.
    h. A U.K. corporation purchases a euro-denominated bond from an Italian MNE.

19. **BOP and Exchange Rates.** What is the relationship between the balance of payments and a fixed or floating exchange rate regime?

20. **J-Curve Dynamics.** What is the J-Curve adjustment path?

21. **Evolution of Capital Mobility.** Has capital mobility improved steadily over the past 50 years?

22. **Restrictions on Capital Mobility.** What factors seem to play a role in a government's choice to restrict capital mobility?

23. **Capital Controls.** Which do most countries control, capital inflows or capital outflows? Why?

24. **Globalization and Capital Mobility.** How does capital mobility typically differ between industrialized countries and emerging market countries?

PROBLEMS

These problems are available in MyFinanceLab.

Australia's Current Account

Use the following balance of payments data for Australia from the IMF to answer Problems 1–4.

Assumptions (million US$)	2000	2001	2002	2003	2004	2005	2006	2007	2008	2009	2010	2011	2012	2013
Goods, credit (exports)	64,052	63,676	65,099	70,577	87,207	107,011	124,913	142,421	189,057	154,777	213,782	271,719	257,950	254,164
Goods, debit (imports)	−68,865	−61,890	−70,530	−85,946	−105,238	−120,383	−134,509	−160,205	−193,972	−159,216	−196,303	−249,238	−270,136	−249,774
Services, credit (exports)	18,677	16,689	17,906	21,205	26,362	31,047	33,088	40,496	45,240	40,814	46,968	51,653	53,034	53,344
Services, debit (imports)	−18,388	−16,948	−18,107	−21,638	−27,040	−30,505	−32,219	−39,908	−48,338	−42,165	−51,313	−61,897	−65,405	−67,399
Primary income: credit	8,984	8,063	8,194	9,457	13,969	16,445	21,748	32,655	37,320	27,402	35,711	47,852	47,168	45,910
Primary income: debit	−19,516	−18,332	−19,884	−24,245	−35,057	−44,166	−54,131	−73,202	−76,719	−65,809	−84,646	−102,400	−88,255	−83,618
Secondary income: credit	2,622	2,242	2,310	2,767	3,145	3,333	3,698	4,402	4,431	4,997	5,813	7,510	7,271	7,206
Secondary income: debit	−2,669	−2,221	−2,373	−2,851	−3,414	−3,813	−4,092	−4,690	−4,805	−5,799	−7,189	−9,723	9,635	9,390

Note: The IMF has recently adjusted their line item nomenclature. *Exports* are all now noted as *credits*, imports as *debits*.

1. What is Australia's balance on goods?

2. What is Australia's balance on services?

3. What is Australia's balance on goods and services?

4. What is Australia's current account balance?

India's Current Account

Use the following balance of payments data for India from the IMF to answer Problems 5–9.

Assumptions (millions of US$)	2000	2001	2002	2003	2004	2005	2006	2007	2008	2009	2010	2011	2012	2013
Goods, credit (exports)	43,247	44,793	51,141	60,893	77,939	102,175	123,876	153,530	199,065	167,958	230,967	307,847	298,321	319,110
Goods, debit (imports)	−53,887	−51,212	−54,702	−68,081	−95,539	−134,692	−166,572	−208,611	−291,740	−247,908	−324,320	−428,021	−450,249	−433,760
Services, credit (exports)	16,684	17,337	19,478	23,902	38,281	52,527	69,440	86,552	106,054	92,889	117,068	138,528	145,525	148,649
Services, debit (imports)	−19,187	−20,099	−21,039	−24,878	−35,641	−47,287	−58,514	−70,175	−87,739	−80,349	−114,739	−125,041	−129,659	−126,256
Primary income: credit	2,521	3,524	3,188	3,491	4,690	5,646	8,199	12,650	15,593	13,733	9,961	10,147	9,899	11,230
Primary income: debit	−7,414	−7,666	−7,097	−8,386	−8,742	−12,296	−14,445	−19,166	−20,958	−21,272	−25,563	−26,191	−30,742	−33,013
Secondary income: credit	13,548	15,140	16,789	22,401	20,615	24,512	30,015	38,885	52,065	50,526	54,380	62,735	68,611	69,441
Secondary income: debit	−114	−407	−698	−570	−822	−869	−1,299	−1,742	−3,313	−1,764	−2,270	−2,523	−3,176	−4,626

5. What is India's balance on goods?

6. What is India's balance on services?

7. What is India's balance on goods and services?

8. What is India's balance on goods, services, and income?

9. What is India's current account balance?

China's (Mainland) Balance of Payments

Use the following balance of payments data for China (Mainland) from the IMF to answer Problems 10–14.

Assumptions (million US$)	2000	2001	2002	2003	2004	2005	2006	2007	2008	2009	2010	2011	2012	2013
A. Current account balance	20,518	17,401	35,422	45,875	68,659	134,082	231,844	353,183	420,569	243,257	237,810	136,097	215,392	182,807
B. Capital account balance	−35	−54	−50	−48	−69	4,102	4,020	3,099	3,051	3,938	4,630	5,446	4,272	3,052
C. Financial account balance	1,958	34,832	32,341	52,774	110,729	96,944	45,285	91,132	37,075	194,494	282,234	260,024	−36,038	323,151
D. Net errors and omissions	−11,748	−4,732	7,504	17,985	10,531	15,847	3,502	13,237	18,859	−41,181	−53,016	−13,768	−87,071	−77,628
E. Reserves and related items	−10,693	−47,447	−75,217	−116,586	−189,849	−250,975	−284,651	−460,651	−479,554	−400,508	−471,658	−387,799	−96,555	−431,382

10. Is China experiencing a net capital inflow or outflow?

11. What is China's total for Groups A and B?

12. What is China's total for Groups A through C?

13. What is China's total for Groups A through D?

14. Does China's BOP balance?

Russia's (Russian Federation's) Balance of Payments

Use the following balance of payments data for Russia (Russian Federation) from the IMF to answer Problems 15–19.

Assumptions (million US$)	2000	2001	2002	2003	2004	2005	2006	2007	2008	2009	2010	2011	2012	2013
A. Current account balance	46,839	33,935	29,116	35,410	59,512	84,602	92,316	72,193	103,935	50,384	67,452	97,274	71,282	34,141
B. Capital account balance	10,676	−9,378	−12,396	−993	−1,624	−12,764	291	−10,641	−104	−12,466	−41	130	−5,218	−395
C. Financial account balance	−34,295	−3,732	921	3,024	−5,128	1,025	3,612	97,108	−139,705	−28,162	−21,526	−76,115	−25,675	−44,983
D. Net errors and omissions	−9,297	−9,558	−6,078	−9,179	−5,870	−7,895	11,248	−9,732	−3,045	−6,392	−9,135	−8,651	−10,370	−10,842
E. Reserves and related items	−13,923	−11,266	−11,563	−28,262	−46,890	−64,968	−107,466	−148,928	38,919	−3,363	−36,750	−12,638	−30,020	22,078

15. Is Russia experiencing a net capital inflow?

16. What is Russia's total for Groups A and B?

17. What is Russia's total for Groups A through C?

18. What is Russia's total for Groups A through D?

19. Does Russia's BOP balance?

Euro Area Balance of Payments

Use the following balance of payments data for the Euro Area from the IMF to answer Problems 20–24.

Assumptions (billion US$)	2000	2001	2002	2003	2004	2005	2006	2007	2008	2009	2010	2011	2012	2013
A. Current account balance	−81.8	−19.7	44.5	24.9	81.2	19.2	−0.3	24.9	−195.9	−12.5	12.2	16.0	171.4	305.4
B. Capital account balance	8.4	5.6	10.3	14.3	20.5	14.2	11.7	5.4	14.2	11.2	7.3	14.0	7.3	27.9
C. Financial account balance	50.9	−41.2	−15.3	−47.6	−122.9	−71.4	−28.5	−3.4	175.3	73.8	−12.2	−103.2	−213.6	−345.2
D. Net errors and omissions	6.4	38.8	−36.5	−24.4	5.6	15.0	19.6	−21.3	11.2	−12.6	−7.2	38.4	33.7	1.8
E. Reserves and related items	16.2	16.4	−3.0	32.8	15.6	23.0	−2.6	−5.7	−4.9	−59.7	−0.1	34.8	1.3	10.1

20. Is the Euro Area experiencing a net capital inflow?

21. What is the Euro Area's total for Groups A and B?

22. What is the Euro Area's total for Groups A through C?

23. What is the Euro Area's total for Groups A through D?

24. Does the Euro Area's BOP balance?

25. Trade Deficits and J-Curve Adjustment Paths. Assume the United States has the following import/export volumes and prices. It undertakes a major "devaluation" of the dollar, say 18% on average against all major trading partner currencies. What is the pre-devaluation and post-devaluation trade balance?

Initial spot exchange rate ($/fc)	2.00
Price of exports, dollars ($)	20.0000
Price of imports, foreign currency (fc)	12.0000
Quantity of exports, units	100
Quantity of imports, units	120
Percentage devaluation of the dollar	18.00%
Price elasticity of demand, imports	−0.90

INTERNET EXERCISES

1. World Organizations and the Economic Outlook. The IMF, World Bank, and United Nations are only a few of the major world organizations that track, report, and aid international economic and financial development. Using these Web sites and others that may be linked, briefly summarize the economic outlook for the developed and emerging nations of the world. For example, Chapter 1 of the World Economic Outlook published annually by the World Bank is available through the IMF's Web page.

International Monetary Fund	www.imf.org/
United Nations	www.unsystem.org/
The World Bank Group	www.worldbank.org/
Europa (EU) Homepage	europa.eu/
Bank for International Settlements	www.bis.org/

2. St. Louis Federal Reserve. The Federal Reserve Bank of St. Louis provides a large amount of recent open-economy macroeconomic data online. Use the following addresses to track down recent BOP and GDP data.

Recent international economic data	research.stlouisfed.org/ publications/iet/
Balance of payments statistics	research.stlouisfed.org/fred2/ categories/125

3. U.S. Bureau of Economic Analysis. Use the following Bureau of Economic Analysis (U.S. government) and the Ministry of Finance (Japanese government) Web sites to find the most recent balance of payments statistics for both countries.

Bureau of Economic Analysis	www.bea.gov/international/
Ministry of Finance	www.mof.go.jp/

4. World Trade Organization and Doha. Visit the WTO's Web site and find the most recent evidence presented by the WTO on the progress of talks on issues including international trade in services and international recognition of intellectual property at the WTO.

World Trade Organization	www.wto.org

5. Global Remittances Worldwide. The World Bank's Web site on global remittances is a valuable source for new and developing studies and statistics on cross-border remittance activity.

World Bank	http://remittanceprices.worldbank.org/

CHAPTER 4

Financial Goals and Corporate Governance

Yogi Berra never went to Harvard Business School, but he did understand one important principle. To quote Yogi, "In theory there is no difference between theory and practice. In practice, there is."

— Peter Rose, CEO and Chairman, Expeditors International, 8k, November 20, 2006, p. 4.

LEARNING OBJECTIVES

- Examine the different ownership structures for businesses globally and how this impacts the separation between ownership and management—the agency problem
- Explore the different goals of management—stockholder wealth maximization versus stakeholder capitalism
- Analyze how financial management differs between the public traded and the privately held firm
- Evaluate the multitude of goals, structures, and trends in corporate governance globally

This chapter examines how legal, cultural, political, and institutional differences affect a firm's choice of financial goals and corporate governance. The owner of a commercial enterprise, and his or her specific personal and professional interests, has a significant impact on the goals of the corporation and its governance. We therefore examine business ownership, goals, and corporate governance in turn. We then explore how governance failures have led to different approaches around the world to improve governance, by both regulatory and other means. The chapter concludes with the Mini-Case, *Luxury Wars—LVMH vs. Hermès*, the recent struggle by Hermès of France to remain family controlled.

Who Owns the Business?

We begin our discussion of corporate financial goals by asking two basic questions: (1) Who owns the business? and (2) Do the owners of the business manage the business themselves? In global business today the ownership and control of organizations varies dramatically across countries and cultures. To understand how and why those businesses operate, one must first understand the many different ownership structures.

Types of Ownership

The terminology associated with the ownership of a business can be confusing. A business owned by a government, the state, is a public enterprise. A business that is owned by a private individual, a private company, or simply a non-government entity, is a private enterprise.

A second distinction on ownership clouds the terminology. A business owned by a private party, or a small group of private individuals, a private enterprise, is termed privately held. If those owners, however, wish to sell a portion of their ownership in the business in the capital markets, for example by listing and trading the company's shares on a stock exchange, the firm's shares are then publicly traded. It is therefore important to understand that shares in a publicly traded firm can be purchased and owned by private parties. Exhibit 4.1 provides a brief overview of these ownership distinctions.

Ownership can be held by a variety of different groups or organizations as well. A business may be owned by a single person (sole proprietorship), two or more people (partnership), a family (family-owned business), two other companies (joint venture), thousands of individuals (publicly traded company), a government (state-owned enterprise), or some combination.

The following three multinational enterprises are examples of how ownership differs in global business, as well as how it may evolve within any single enterprise over time.

- Petróleo Brasileiro S.A., or Petrobras, is the national oil company of Brazil. Founded in 1953, it was originally 100% owned by the Brazilian government, and was therefore a state-owned enterprise. Over time, however, it sold portions of its ownership to the public, becoming publicly traded on the Sao Paulo stock exchange. Today the Brazilian government owns approximately 64% of the shares of Petrobras, with the remaining 36% in the hands of private investors—shareholders—all over the world.

EXHIBIT 4.1   **Business Ownership**

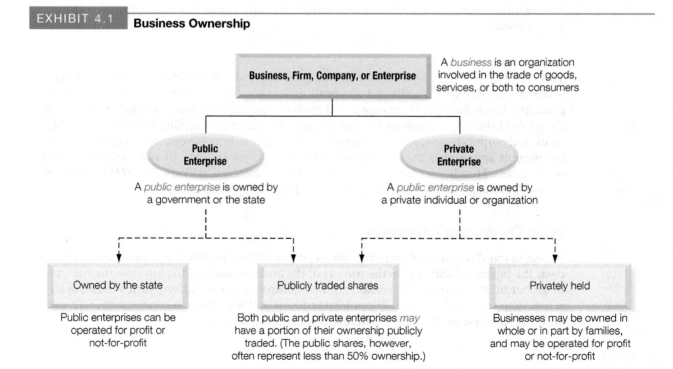

■ Apple was founded in 1976 as a partnership of Steve Jobs, Steve Wozniak, and Ronald Wayne. On January 3, 1977, Apple was incorporated in the United States, with Ronald Wayne selling his ownership interest to his two partners. In 1980 Apple sold shares to the public for the first time in an initial public offering (IPO), with its shares listed (traded) on the NASDAQ Stock Market. Today Apple has roughly 900 million shares outstanding, and is considered "widely held" as no single investor holds 5% of its shares. In recent years Apple has periodically been the world's most valuable publicly traded company, as calculated by market capitalization (shares outstanding multiplied by the share price).

■ Hermès International is a French multinational producer of luxury goods. Founded by Thierry Hermès in 1837, it has been owned and operated by the Hermès family for most of its history, making it a family-owned business. In 1993 the company "went public" for the first time, selling 27% of its interest to the general public. The family, however, retained 73% and therefore the control of the company. (The Mini-Case at the end of this chapter details the battle fought by the family in 2010 to retain its control.)

Once the ownership of the business is established, it is then easier to understand where control lies, as ownership and control are separate concepts. Petrobras is a publicly traded Brazilian business that is controlled by the Brazilian government. Hermès International is a publicly traded family-controlled French-based business. Apple is a publicly traded and widely held business, so control rests with its Board of Directors and the senior leadership team hired by the Board to run the company. Individual investors who hold shares in Apple may vote on issues presented to them on an annual basis, so they have a degree of high-level influence, but the daily strategy, tactics, operations, and governance of Apple is under the control of the senior management team and the Board.

Any business, whether initially owned by the state, a family, or a private individual or institution, may choose to have a portion of its ownership traded as shares in the public marketplace, as noted in Exhibit 4.1. (Note that we say a portion, as a 100% publicly traded firm can no longer be either state-owned or privately held by definition.) For example many SOEs are also publicly traded. China National Petroleum Corporation (CNPC), the state-owned parent company of PetroChina, is one example, having shares traded on stock exchanges in Shanghai, Hong Kong, and New York, yet majority ownership and control still rests with the government of China.

If a firm's ownership decides to sell a portion of ownership to the public market, it conducts an initial public offering, or IPO. Typically only a relatively small percentage of the company is initially sold to the public, anywhere from 10% to 20%, resulting in a company that may still be controlled by a small number of private investors, a family, or a government, but now with a portion of its ownership traded publicly. Over time, a company may sell more and more of its equity interest into the public marketplace, eventually becoming totally publicly traded. Alternatively, a private owner or family may choose to retain a major share, but may not retain control. It is also possible for the controlling interest in a firm to reverse its pubic share position, reducing the number of shares outstanding by repurchasing shares.

The acquisition of one firm by another demonstrates yet another way ownership and control can change. For example, in 2005 a very large private firm, Koch Industries (U.S.), purchased all outstanding shares of Georgia-Pacific (U.S.), a very large publicly traded company. Koch took Georgia-Pacific private.

Even if a firm is publicly traded, it may still be controlled by a single investor or by a small group of investors, including major institutional investors. This means that the control of a publicly traded company is much like that of a privately held company, reflecting the interests and goals of the controlling individual investor or family. A continuing characteristic of many emerging markets is the dominance of family-controlled firms, although many are simultaneously publicly traded. Many family-controlled firms may outperform publicly traded firms.

As discussed later in this chapter, there is another significant implication of an initial sale of shares to the public: The firm becomes subject to many of the increased legal, regulatory, and reporting requirements related to the sale and trading of securities. In the U.S., for example, going public means the firm must disclose a sizable degree of financial and operational detail, publish this information at least quarterly, comply with Securities and Exchange Commission (SEC) rules and regulations, and comply with all the specific operating and reporting requirements of the specific exchange on which its shares are traded.

Separation of Ownership from Management

One of the most challenging issues in the financial management of the enterprise is the possible separation of ownership from management. Hired or professional management may be present under any ownership structure, however it is most often observed in SOEs and widely held publicly traded companies. This separation of ownership from management raises the possibility that the two entities may have different business and financial objectives. This is the so-called principal agent problem or simply the agency problem. There are several strategies available for aligning shareholder and management interests, the most common of which is for senior management to own shares or share options. What is then good for the managers' own personal wealth is similar to that of general shareholders.

The United States and United Kingdom are two country markets characterized by widespread ownership of shares. Management may own some small portion of stock in their firms, but largely management is a hired agent that is expected to represent the interests of shareholders. In contrast, many firms in many other global markets are characterized by controlling shareholders, such as government, institutions (e.g., banks in Germany), family (e.g., in France, Italy, and throughout Asia and Latin America), and consortiums of interests (e.g., keiretsus in Japan and chaebols in South Korea). A business that is owned and managed by the same entity does not suffer the agency problem.

In many of these cases, control is enhanced by ownership of shares with dual voting rights, interlocking directorates, staggered election of the board of directors, takeover safeguards, and other techniques that are not used in the Anglo-American markets. However, the recent emergence of huge equity funds and hedge funds in the United States and the United Kingdom has led to the privatization of some very prominent publicly traded firms around the world.

The Goal of Management

As companies become more deeply committed to multinational operations, a new constraint develops—one that springs from divergent worldwide opinions and practices as to just what the firms' overall goal should be from the perspective of top management, as well as the role of corporate governance.

> *What do investors want? First, of course, investors want performance: strong predictable earnings and sustainable growth. Second, they want transparency, accountability, open communications and effective corporate governance. Companies that fail to move toward international standards in each of these areas will fail to attract and retain international capital.*
>
> —"The Brave New World of Corporate Governance," *LatinFinance*, May 2001.

An introductory course in finance is usually taught based on the assumption that maximizing shareholder wealth is the goal of management. In fact, every business student memorizes the concept of maximizing shareholder wealth sometime during his or her college education. Maximizing shareholder wealth, however, has at least two major challenges: (1) It is not necessarily the accepted goal of management across countries to maximize the wealth of shareholders—other stakeholders may have substantial influence; and (2) It is extremely difficult to carry out. Creating shareholder wealth is—like so many lofty goals—much easier said than done.

Although the idea of maximizing shareholder wealth is probably realistic both in theory and in practice in the Anglo-American markets, it is not always exclusive elsewhere. Some basic differences in corporate and investor philosophies exist between the Anglo-American markets and those in the rest of the world.

The Shareholder Wealth Maximization Model

The Anglo-American markets have a philosophy that a firm's objective should follow the *shareholder wealth maximization model* (SWM). More specifically, the firm should strive to maximize the return to shareholders, as measured by the sum of capital gains and dividends, for a given level of risk. Alternatively, the firm should minimize the risk to shareholders for a given rate of return.

The SWM theoretical model assumes, as a universal truth, that the stock market is efficient. This means that the share price is always correct because it captures all the expectations of return and risk as perceived by investors. It quickly incorporates new information into the share price. Share prices, in turn, are deemed the best allocators of capital in the macro economy.

The SWM model also treats its definition of risk as a universal truth. Risk is defined as the added probability of varying returns that the firm's shares bring to a diversified portfolio. The *operational risk*, the risk associated with the business line of the individual firm, can be eliminated through portfolio diversification by investors. Therefore the *unsystematic risk*, the risk of the individual security, should not be a prime concern for management unless it increases the possibility of bankruptcy. On the other hand, *systematic risk*, the risk of the market in general, cannot be eliminated through portfolio diversification and is risk that the share price will be a function of the stock market.

Agency Theory. The field of agency theory is the study of how shareholders can motivate management to accept the prescriptions of the SWM model.[1] For example, liberal use of stock options should encourage management to think like shareholders. Whether these inducements succeed is open to debate. However, if management deviates from the SWM objectives of working to maximize shareholder returns, then it is the responsibility of the board to replace management. In cases where the board is too weak or ingrown to take this action, the discipline of equity markets could do it through a takeover. This discipline is made possible by the one-share-one-vote rule that exists in most Anglo-American markets.

Long-Term versus Short-Term Value Maximization. During the 1990s, the economic boom and rising stock prices in most of the world's markets exposed a flaw in the SWM model, especially in the United States. Instead of seeking long-term value maximization, several large U.S. corporations sought short-term value maximization (e.g., the continuing debate about meeting the market's expected quarterly earnings). This strategy was partly motivated by the overly generous use of stock options to motivate top management.

[1]Michael Jensen and W. Meckling, "Theory of the Firm: Managerial Behavior, Agency Costs, and Ownership Structure," *Journal of Financial Economics*, No. 3, 1976, and Michael C. Jensen, "Agency Cost of Free Cash Flow, Corporate Finance and Takeovers, *American Economic Review*, 76, 1986, pp. 323–329.

This short-term focus sometimes created distorted managerial incentives. In order to maximize growth in short-term earnings and to meet inflated expectations by investors, firms such as Enron, Global Crossing, Health South, Adelphia, Tyco, Parmalat, and WorldCom undertook risky, deceptive, and sometimes dishonest practices for the recording of earnings and/or obfuscation of liabilities, which ultimately led to their demise. It also led to highly visible prosecutions of their CEOs, CFOs, accounting firms, legal advisers, and other related parties.

This, sometimes destructive, short-term focus on the part of both management and investors has been correctly labeled *impatient capitalism*. This point of debate is also sometimes referred to as the firm's *investment horizon* in reference to how long it takes the firm's actions, its investments and operations, to result in earnings. In contrast to impatient capitalism is patient capitalism, which focuses on long-term shareholder wealth maximization. Legendary investor Warren Buffett, through his investment vehicle Berkshire Hathaway, represents one of the best of the patient capitalists. Buffett has become a billionaire by focusing his portfolio on mainstream firms that grow slowly but steadily with the economy, such as Coca Cola.

The Stakeholder Capitalism Model

In the non-Anglo-American markets, controlling shareholders also strive to maximize long-term returns to equity. However, they are more constrained by other powerful stakeholders. In particular, outside the Anglo-American markets, labor unions are more powerful and governments interfere more in the marketplace to protect important stakeholder groups, such as local communities, the environment, and employment. In addition, banks and other financial institutions are more important creditors than securities markets. This model has been labeled the *stakeholder capitalism model* (SCM).

Market Efficiency. The SCM model does not assume that equity markets are either efficient or inefficient. It does not really matter because the firm's financial goals are not exclusively shareholder-oriented since they are constrained by the other stakeholders. In any case, the SCM model assumes that long-term "loyal" shareholders, typically controlling shareholders, should influence corporate strategy rather than the transient portfolio investor.

Risk. The SCM model assumes that total risk, that is, operating risk, does count. It is a specific corporate objective to generate growing earnings and dividends over the long run with as much certainty as possible, given the firm's mission statement and goals. Risk is measured more by product market variability than by short-term variation in earnings and share price.

Single versus Multiple Goals. Although the SCM model typically avoids a flaw of the SWM model, namely, impatient capital that is short-run oriented, it has its own flaw. Trying to meet the desires of multiple stakeholders leaves management without a clear signal about the tradeoffs. Instead, management tries to influence the tradeoffs through written and oral disclosures and complex compensation systems.

The Scorecard. In contrast to the SCM model, the SWM model requires a single goal of value maximization with a well-defined scorecard. According to the theoretical model of SWM described by Michael Jensen, the objective of management is to maximize the total market value of the firm.[2] This means that corporate leadership should be willing to spend or invest more if each additional dollar creates more than one dollar in the market value of the company's equity, debt, or any other contingent claims on the firm.

Although both models have their strengths and weaknesses, in recent years two trends have led to an increasing focus on the shareholder wealth form (SWM). First, as more of

[2]Michael C. Jensen, "Value Maximization, Stakeholder Theory, and the Corporate Objective Function," *Journal of Applied Corporate Finance*, Fall 2001, Volume 14, No. 3, pp. 8–21, p. 12.

the non-Anglo-American markets have increasingly privatized their industries, a shareholder wealth focus is seemingly needed to attract international capital from outside investors, many of which are from other countries. Second, and still quite controversial, many analysts believe that shareholder-based MNEs are increasingly dominating their global industry segments. Nothing attracts followers like success.

Operational Goals

It is one thing to state that the objective of leadership is to maximize shareholder value, but it is another to actually do it. The management objective of maximizing profit is not as simple as it sounds, because the measure of profit used by ownership/management differs between the privately held firm and the publicly traded firm. In other words, is management attempting to maximize current income, capital appreciation, or both?

The return to a shareholder in a publicly traded firm combines current income in the form of dividends and capital gains from the appreciation of share price:

$$\text{Shareholder Return} = \frac{D_2}{P_1} + \frac{P_2 - P_1}{P_1}$$

where the initial price, P_1, is the beginning price, the initial investment by the shareholder, P_2, is the price of the share at the end of period, and D_2 is the dividend paid at the end of the period. The shareholder theoretically receives income from both components. For example, in the United States in the 1990s a diversified investor may have received an average annual return of 14%, 2% from dividends and 12% from capital gains. This "split" between dividend and capital gain returns, however, differs dramatically across the world's major markets over time.

Management generally believes it has the most direct influence over the first component—the *dividend yield*. Management makes strategic and operational decisions that grow sales and generate profits. Then it distributes those profits to ownership in the form of dividends. *Capital gains*—the change in the share price as traded in the equity markets—is much more complex, and reflects many forces that are not in the direct control of management. Despite growing market share, profits, or any other traditional measure of business success, the market may not reward these actions directly with share price appreciation. Many top executives believe that stock markets move in mysterious ways and are not always consistent in their valuations. In the end, leadership in the publicly traded firm typically concludes that it is its own growth—growth in top-line sales and bottom-line profits—that is its great hope for driving share price upward. This can, however, change over time with continued success as seen in *Global Finance in Practice 4.1.*

GLOBAL FINANCE IN PRACTICE 4.1

Why Did Apple Start Paying a Dividend and Raising Debt?

In March 2012, Apple announced it would end a 17-year period of no dividend payments. This was followed by a similarly shocking announcement in April 2013 that it would raise nearly $17 billion in debt, although the company had an enormous cash balance. Both financial policy changes were surprising to the market and uncommon in the tech sector. So why?

One way to understand the financial policy change is to consider whether Apple had grown and evolved from a *growth*

firm to a *value firm*. *Growth firms* are small- to medium-sized firms in the relatively rapid growth stages of their businesses. Their value is growing rapidly in the public markets, their share prices rising creating substantial capital gains for their shareholders. At the same time, since they are in such rapid growth stages they need all the capital they can get their hands on, so arguably, they need the dividend cash flow more than their shareholders. Although they could raise debt during this stage, debt is often seen as an obligation that slows the ability of corporate leadership to respond quickly to changing customer and technology market needs.

Value firms are large, more mature businesses, which find it increasingly difficult to create material value changes in their business. Once an ExxonMobil or Microsoft gets so large, major new business developments and successes can rarely change the company's financial earnings and results significantly. Share price movements become slower, and more subtle, over time. Yet the companies continue to generate massive amounts of cash flow and earnings. In this case, shareholders—often in the persona of activist investors like Carl Icahn—pressure management to take on more and more debt and pay out greater and greater dividends. The debt requires corporate management to stay sharp in keeping up debt service obligations while the dividend distributions provide added and sometimes significant returns to shareholders who are seeing capital appreciation slowing or stagnating.

In Apple's case, there is an additional issue driving its financial policy change: U.S. taxes. Although the company held massive cash balances, this cash was largely offshore—outside the United States—and would result in major increased tax payments if the cash was repatriated to the U.S. in order to pay the newly declared cash dividends.

A privately held firm has a much simpler shareholder return objective function: maximize current and sustainable financial income to its owners. The privately held firm does not have a share price (it does have a value, but this is not a definitive market-determined value in the way in which we believe markets work). It therefore simply focuses on generating current financial income, dividend income (as well as salaries and other forms of income provided owners), to generate the returns to its ownership. If the privately held ownership is a family, the family may also place a great emphasis on the ability to sustain those earnings over time while maintaining a slower rate of growth, which can be managed by the family itself. Without a share price, "growth" is not of the same strategic importance in the privately held firm. It is therefore critical that ownership and ownership's specific financial interests be understood from the very start if we are to understand the strategic and financial goals and objectives of management.

The privately held firm may also be less aggressive (take fewer risks) than the publicly traded firm. Without a public share price, and therefore the ability of outside investors to speculate on the risks and returns associated with company business developments, the privately held firm—its owners and operators—may choose to take fewer risks. This may mean that it will not attempt to grow sales and profits as rapidly, and therefore may not require the capital (equity and debt) needed for rapid growth. A recent study by McKinsey found that privately held firms consistently used significantly lower levels of debt (averaging 5% less debt-to-equity) than publicly traded firms. Interestingly, these same private firms also enjoyed a lower cost of debt, roughly 30 basis points lower based on corporate bond issuances.[3]

Exhibit 4.2 provides an overview of the variety of distinctive financial and managerial differences between state-owned, publicly traded, and privately held firms.

Operational Goals for MNEs. The MNE must be guided by operational goals suitable for various levels of the firm. Even if the firm's goal is to maximize shareholder value, the manner in which investors value the firm is not always obvious to the firm's top management. Therefore, most firms hope to receive a favorable investor response to the achievement of operational goals that can be controlled by the way in which the firm performs, and then hope—if we can use that term—that the market will reward their results. The MNE must determine the proper balance between three common operational financial objectives:

  1. Maximization of consolidated after-tax income
  2. Minimization of the firm's effective global tax burden

---

[3]"The five attributes of enduring family businesses," Christian Caspar, Ana Karina Dias, and Heinz-Peter Elstrodt, *McKinsey Quarterly*, January 2010, p. 6.

EXHIBIT 4.2	Public versus Private Ownership		
Organizational Characteristic	**State-Owned**	**Publicly Traded**	**Privately Held**
Entrepreneurial	No	No; stick to core competencies	Yes; do anything the owners wish
Long-term or short-term focus	Long-term focus; political cycles	Short-term focus on quarterly earnings	Long-term focus
Focused on profitable growth	No	Yes; growth in earnings is critical	No; needs defined by owners earnings need
Adequately financed	Country-specific	Good access to capital and capital markets	Limited in the past but increasingly available
Quality of leadership	Highly variable	Professional; hiring from both inside and outside	Highly variable; family-run firms are lacking
Role of earnings (Profits)	Earnings may constitute funding for government	Earnings to signal the equity markets	Earnings to support owners and family
Leadership are owners	Caretakers, not owners	Minimal interests; some have stock options	Yes; ownership and mgmt often one and the same

3. Correct positioning of the firm's income, cash flows, and available funds as to country and currency

These goals are frequently incompatible, in that the pursuit of one may result in a less desirable outcome of another. Management must make decisions daily in business about the proper tradeoffs between goals (which is why companies are managed by people and not computers).

Consolidated Profits. The primary operational goal of the MNE is to maximize consolidated profits, after-tax. Consolidated profits are the profits of all the individual units of the firm, originating in many different currencies, and expressed in the currency of the parent company. This is not to say that management is not striving to maximize the present value of all future cash flows. It is simply the case that most of the day-to-day decision-making in global management is about current earnings. The leaders of the MNE, the management team who are implementing the firm's strategy, must think far beyond current earnings.

For example, foreign subsidiaries have their own set of traditional financial statements: (1) a statement of income, summarizing the revenues and expenses experienced by the unit over the year; (2) a balance sheet, summarizing the assets employed in generating the unit's revenues, and the financing of those assets; and (3) a statement of cash flows, summarizing those activities of the unit that generate and then use cash flows over the year. These financial statements are expressed initially in the local currency of the unit for tax and reporting purposes to the local government, but they must also be consolidated with the parent company's financial statements for reporting to shareholders.

Public/Private Hybrids. The global business environment is, as one analyst termed it, "a messy place," and the ownership of companies of all kinds, including MNEs, is not necessarily purely public or purely private. One recent study of global business found that fully one-third of all companies in the S&P 500 were technically family businesses. And this was not just the case

| EXHIBIT 4.3 | **The Superior Performance of Family** |

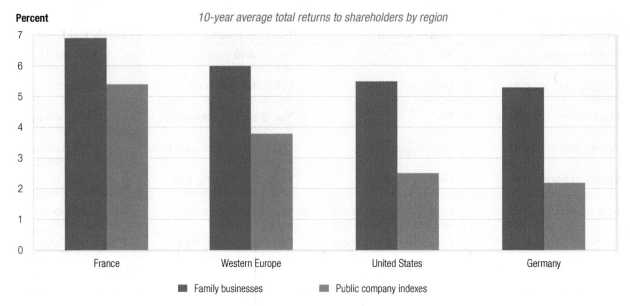

Source: Authors' presentation based on data presented in "The Five Attributes of Enduring Family Businesses," Christian Caspar, Ana Karina Dias, and Heinz-Peter Elstrodt, *McKinsey Quarterly*, January 2010, p. 7. Index of public companies by region: France, SBF120; Western Europe, MSCI Europe; United States, S&P500; Germany, HDAX.

for the U.S.; roughly 40% of the largest firms in France and Germany were heavily influenced by family ownership and leadership.

In other words, the firm may be publicly traded, but a family still wields substantial power over the strategic and operational decisions of the firm. This may prove to be a good thing. As illustrated in Exhibit 4.3, the financial performance of publicly traded family-controlled businesses (as measured by total returns to shareholders) in five different regions of the globe were superior to their nonfamily publicly traded counterparts.

Why do family-influenced businesses seemingly perform better than others? According to Credit Suisse, there are three key catalysts for the performance of *stocks with significant family influence* (SSFI): (1) management with a longer-term focus; (2) better alignment between management and shareholder interests; and (3) stronger focus on the core business of the firm.

Publicly Traded versus Privately Held: The Global Shift

Is the future of the publicly traded firm really at risk, or is it just that U.S.-based publicly traded shares are on the decline? Exhibit 4.4 provides a broad overview of global equity listings, separating the number of listings between those on U.S. exchanges and all others.

Exhibit 4.4, based on listings data from the World Federation of Exchanges, raises a number of questions about trends and tendencies across the global equity markets:

| EXHIBIT 4.4 | **Global Equity Share Listings** |

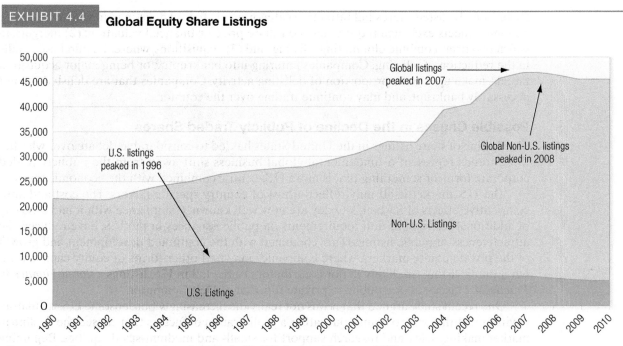

Source: Derived by authors from statistics collected by the World Federation of Exchanges (WFE), www.world-exchanges.org.

- Global equity listings grew significantly over the past 20 years, but they peaked in 2008. And although the true residual impact of the 2008–2009 global financial crisis is yet unknown, it is clear that the crisis, amid other factors, has seemingly slowed the growth of public share listings.

- The U.S. share of global equity listings has declined dramatically and steadily since the mid-1990s. At the end of 2010, of the 45,508 equities listed on 54 stock exchanges globally, U.S. listings comprised 11.0% of the total. That was a dramatic decline from 1996 when the U.S. comprised 33.3%.

- U.S. public share listings fell by 3,767 (from 8,783 in 1996 to 5,016 in 2010), 42.9% over the 14-year period since its peak. Clearly, the attraction of being a publicly traded firm on a U.S. equity exchange had declined dramatically.

Listings Measurement

New listings on a stock exchange are the net result of *listing additions*, new companies joining the exchange, reduced by *delistings*, companies exiting the exchange.

Listing Additions. Stock exchange listing additions arise from four sources: (1) initial public offerings (IPOs); (2) movements of share listings from one exchange to another; (3) spinouts from larger firms; and (4) new listings from smaller non-exchanges such as bulletin boards. Since movements between exchanges typically are a zero sum within a country, and spinouts and bulletin board movements are few in number, real growth in listings comes from IPOs.

Delistings. Delisted shares fall into three categories: (1) forced delistings, in which the equity no longer meets exchange requirements on share price or financial valuation; (2) mergers, in which two firms combine eliminating a listing; and (3) acquisitions, where the purchase results in the reduction of a listing. Companies entering into bankruptcy, or being major acquisition targets, make up a great proportion of delisting activity. Companies that are delisted are not necessarily bankrupt, and may continue trading over the counter.

Possible Causes in the Decline of Publicly Traded Shares

The decline of share listings in the United States has led to considerable debate over whether these trends represent a fundamental global business shift away from the publicly traded corporate form, or something that is more U.S.-centric combined with the economic times.

The U.S. market itself may reflect a host of country specific factors. The cost and anti-competitive effects of Sarbanes-Oxley are now well known. Compliance with it and a variety of additional restrictions and requirements on public issuances in the U.S. have reduced the attractiveness of public listings. This, combined with the continued development and growth of the private equity markets, where companies may find other forms of equity capital without a public listing, are likely major contributors to the fall in U.S. listings. *Global Finance in Practice 4.2* discusses one public to private flip, that of Dell Computer.

One recent study argued that it was not really the increasingly burdensome U.S. regulatory environment that was to blame, but rather a proliferation of factors that caused the decline in market making, sales, and research support for small- and medium-sized equities. Beginning with the introduction of online brokerage in 1996 and online trading rules in 1997, more and more equity trading in the United States shifted to ECNs, electronic communication networks, which allowed all market participants to trade directly with the exchange order books, and not through brokers or brokerage houses. Although this increased competition reduced

---

GLOBAL FINANCE IN PRACTICE 4.2

Why Did Dell Go Private?

In November 2013, Dell Computer, one of the true publicly traded equity stars of the 1990s, went private. Dell's founder, Michael Dell, together with the capital of the hedge fund Silver Lake Partners and nearly $15 billion in debt, engineered a $24.4 billion purchase of Dell, taking it private. But why? What could Michael Dell and his team do differently as a private firm that they could not accomplish as a publicly traded firm?

Strategically, Michael Dell had been trying to transform Dell Computer from a PC company to an enterprise company for nearly five years. The equity markets, however, still saw Dell as only a PC company, and as a PC company, it was failing. At one time the market leader of the PC industry, it had fallen to third and was continuing to slide. From Michael Dell's perspective, the publicly traded markets were too impatient, too short-term in focus, too simplistic in their understanding of

strategic change, to allow Dell to successfully transform itself. Michael Dell believed that Dell needed to take more risk, to make long-term investments that would create a new long-term future, but would require sacrificing short-term financial returns. But he had run head-on into a number of activist investors like Carl Icahn, who had taken positions in the company and demanded the firm raise debt for greater dividend distributions, not for longer-term investments in support of a new corporate strategic future.

As Michael Dell noted in a *Wall Street Journal* opinion piece one year after taking the company private: *Shareholders increasingly demanded short-term results to drive returns; innovation and investment too often suffered as a result. Shareholder and customer interests decoupled.*[*] Time will tell whether a longer-term perspective will result in a newly transformed and competitive Dell.

[*]"Going Private Is Paying Off for Dell: A year later, we're able to focus on customers and the long term, rather than activist investors," by Michael Dell, *Wall Street Journal*, Nov. 24, 2014.

transaction costs dramatically, it also undermined the profitability of the retail brokerage houses, which had always supported research, market making, and sales and promotion of the small- to medium-sized equities. Without this financial support, the smaller stocks were no longer covered and promoted by the major equity houses. Without that research, marketing, promotion and coverage, their trading volumes and values fell.

Corporate Governance

Corporate governance is the system of rules, practices, and processes by which an organization is directed and controlled. Although the governance structure of any company—domestic, international, or multinational—is fundamental to its very existence, this subject has become the lightning rod of political and business debate in the past few years as failures in governance in a variety of forms has led to corporate fraud and failure. Abuses and failures in corporate governance have dominated global business news in recent years. Beginning with the accounting fraud and questionable ethics of business conduct at Enron culminating in its bankruptcy in the fall of 2001, failures in corporate governance have raised issues about the ethics and culture of business conduct.

The Goal of Corporate Governance

The single overriding objective of corporate governance in the Anglo-American markets is the optimization over time of the returns to shareholders. In order to achieve this, good governance practices should focus the attention of the board of directors of the corporation on this objective by developing and implementing a strategy for the corporation that ensures corporate growth and equity value creation. At the same time, it should ensure an effective relationship with stakeholders. A variety of organizations, including the *Organization for Economic Cooperation and Development* (OECD), have continued to refine their recommendations regarding five primary areas of governance:

1. **Shareholder rights.** Shareholders are the owners of the firm, and their interests should take precedence over other stakeholders.
2. **Board responsibilities.** The board of the company is recognized as the individual entity with final full legal responsibility for the firm, including proper oversight of management.
3. **Equitable treatment of shareholders.** Equitable treatment is specifically targeted toward domestic versus foreign residents as shareholders, as well as majority and minority interests.
4. **Stakeholder rights.** Governance practices should formally acknowledge the interests of other stakeholders—employees, creditors, community, and government.
5. **Transparency and disclosure.** Public and equitable reporting of firm operating and financial results and parameters should be done in a timely manner, and should be made available to all interests equitably.

These principles obviously focus on several key areas—shareholder rights and roles, disclosure and transparency, and the responsibilities of boards, which we will discuss in more detail.

The Structure of Corporate Governance

Our first challenge is to understand what people mean when they use the expression "corporate governance." Exhibit 4.5 provides an overview of the various parties and their responsibilities associated with the governance of the modern corporation. The modern corporation's actions and behaviors are directed and controlled by both internal and external forces.

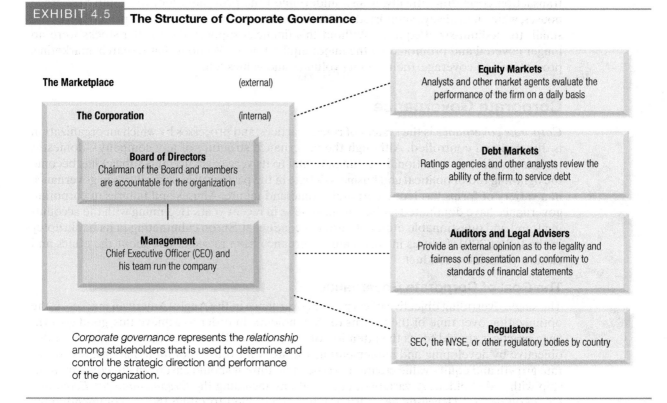

EXHIBIT 4.5   **The Structure of Corporate Governance**

Corporate governance represents the *relationship* among stakeholders that is used to determine and control the strategic direction and performance of the organization.

The *internal forces*, the officers of the corporation (such as the chief executive officer or CEO) and the board of directors of the corporation (including the chairman of the board), are those directly responsible for determining both the strategic direction and the execution of the company's future. But they are not acting within a vacuum; the internal forces are subject to the constant prying eyes of the *external forces* in the marketplace, that question the validity and soundness of the decisions and performance of internal forces. External forces include the equity market (stock exchanges) on which the company's shares are traded, the investment banking analysts who cover and critique the company shares, the creditors of the companies, the credit rating agencies that assign credit ratings to the company's debt or equity securities, the auditors and legal advisers who testify to the fairness and legality of the company's financial statements, and the multitude of regulators who oversee the company's actions — all in an attempt to assure the validity of information presented to investors.

Board of Directors. The legal body that is accountable for the governance of the corporation is its board of directors. The board is composed of both employees of the organization (inside members) and senior and influential nonemployees (outside members). Areas of debate surrounding boards include the following: (1) the proper balance between inside and outside members; (2) the means by which board members are compensated; and (3) the actual ability of a board to monitor and manage a corporation adequately when board members are spending few days a year in board activities. Outside members, often the current or retired chief executives of other major companies, may bring with them a healthy sense of distance and impartiality, which, although refreshing, may also result in limited understanding of the true issues and events within the company.

Management. The senior officers of the corporation—the chief executive officer (CEO), the chief financial officer (CFO), and the chief operating officer (COO)—are not only the most knowledgeable of the business, but also the creators and directors of its strategic and operational direction. The management of the firm is, according to theory, acting as a contractor—as an agent—of shareholders to pursue value creation. These officers are positively motivated by salary, bonuses, and stock options and negatively motivated by the risk of losing their jobs. They may, however, have biases of self-enrichment or personal agendas, which the board and other corporate stakeholders are responsible for overseeing and policing. Interestingly, in more than 80% of the companies in the Fortune 500, the CEO is also the chairman of the board. This is, in the opinion of many, a conflict of interest and not in the best interests of the company and its shareholders.

Equity Markets. The publicly traded company, regardless of country of residence, is highly susceptible to the changing opinion of the marketplace. The equity markets themselves, whether they are the New York Stock Exchange or the Mexico City Bolsa, should reflect the market's constant evaluation of the promise and performance of the individual company. The analysts are experts employed by the many investment banking firms that also trade in the client company shares. These analysts are expected (sometimes naively) to evaluate the strategies, plans for execution of the strategies, and financial performance of the firms on a real-time basis, and they depend on the financial statements and other public disclosures of the firm for their information.

Debt Markets. Although the debt markets (banks and other financial institutions providing loans and various forms of securitized debt like corporate bonds) are not specifically interested in building shareholder value, they are indeed interested in the financial health of the company. Their interest, specifically, is in the company's ability to repay its debt in a timely manner. Like equity markets, they must rely on the financial statements and other disclosures (public and private in this case) of the companies with which they work.

Auditors and Legal Advisers. Auditors and legal advisers are responsible for providing an external professional opinion as to the fairness, legality, and accuracy of corporate financial statements. In this process, they attempt to determine whether the firm's financial records and practices follow what in the United States is termed generally accepted accounting principles (GAAP) in regard to accounting procedures. But auditors and legal advisers are hired by the very firms they are auditing, putting them in the rather unique position of policing their employers.

Regulators. Publicly traded firms in the United States and elsewhere are subject to the regulatory oversight of both governmental organizations and nongovernmental organizations. In the United States, the Securities and Exchange Commission (SEC) is a careful watchdog of the publicly traded equity markets, both of the behavior of the companies themselves in those markets and of the various investors participating in those markets. The SEC and other similar authorities outside of the United States require a regular and orderly disclosure process of corporate performance so that all investors may evaluate the company's investment value with adequate, accurate, and fairly distributed information. This regulatory oversight is often focused on when and what information is released by the company, and to whom.

A publicly traded firm in the United States is also subject to the rules and regulations of the exchange upon which they are traded (New York Stock Exchange/Euronext, American Stock Exchange, and NASDAQ are the largest). These organizations, typically categorized as self-regulatory in nature, construct and enforce standards of conduct for both their member companies and themselves in the conduct of share trading.

Comparative Corporate Governance

The origins of the need for a corporate governance process arise from the separation of ownership from management, and from the varying views by culture of who the stakeholders are and of their significance.[4] This assures that governance practices will differ across countries and cultures. As seen in Exhibit 4.6, corporate governance regimes may be classified by the evolution of business ownership over time.

Market-based regimes—like those of the United States, Canada, and the United Kingdom—are characterized by relatively efficient capital markets in which the ownership of publicly traded companies is widely dispersed. Family-based regimes, like those characterized in many of the emerging markets, Asian markets, and Latin American markets, not only started with strong concentrations of family ownership (as opposed to partnerships or small investment groups that are not family-based), but have also continued to be largely controlled by families even after going public. Bank-based and government-based regimes are those reflecting markets in which government ownership of property and industry has been a constant over time, resulting in only marginal public ownership of enterprise, and even then, subject to significant restrictions on business practices.

All exchange rate regimes are therefore a function of at least four major factors in the evolution of corporate governance principles and practices globally: 1) the financial market development; 2) the degree of separation between management and ownership; 3) the concept of disclosure and transparency; and 4) the historical development of the legal system.

Financial Market Development. The depth and breadth of capital markets is critical to the evolution of corporate governance practices. Country markets that have had relatively slow growth, as in the emerging markets, or have industrialized rapidly utilizing neighboring capital markets (for example, Western Europe), may not form large public equity market systems. Without significant public trading of ownership shares, high concentrations of ownership are preserved and few disciplined processes of governance are developed.

Separation of Management and Ownership. In countries and cultures where the ownership of the firm has continued to be an integral part of management, agency issues and failures have been less problematic. In countries like the United States, in which ownership has become

---

EXHIBIT 4.6   Comparative Corporate Governance Regimes

Regime Basis	Characteristics	Examples
Market-based	Efficient equity markets; Dispersed ownership	United States, United Kingdom, Canada, Australia
Family-based	Management and ownership is combined; Family/majority and minority shareholders	Hong Kong, Indonesia, Malaysia, Singapore, Taiwan,
Bank-based	Government influence in bank lending; Lack of transparency; Family control	Korea, Germany
Government-affiliated	State ownership of enterprise; Lack of transparency; No minority influence	China, Russia

Source: Based on "Corporate Governance in Emerging Markets: An Asian Perspective," by J. Tsui and T. Shieh, in *International Finance and Accounting Handbook*, Third Edition, Frederick D.S. Choi, editor, Wiley, 2004, pp. 24.4–24.6.

---

[4]For a summary of comparative corporate governance see R. La Porta, F. Lopez-de-Silanes and A. Schleifer, "Corporate Ownership Around the World," *Journal of Finance*, 54, 1999, pp. 471–517. See also A. Schleifer and R. Vishny, "A Survey of Corporate Governance," *Journal of Finance*, 52, 1997, pp. 737–783, and the *Journal of Applied Corporate Finance*, Vol. 19, No. 1, Winter 2007.

largely separated from management (and widely dispersed), aligning the goals of management and ownership is much more difficult.

Disclosure and Transparency. The extent of disclosure regarding the operations and financial results of a company vary dramatically across countries. Disclosure practices reflect a wide range of cultural and social forces, including the degree to which ownership is public, the degree to which government feels the need to protect investor's rights versus ownership rights, and the extent to which family-based and government-based business remains central to the culture. Transparency, a parallel concept to disclosure, reflects the visibility of decision-making processes within the business organization.

Historical Development of the Legal System. Investor protection is typically better in countries where English common law is the basis of the legal system, compared to the codified civil law that is typical in France and Germany (the so-called Code Napoleon). English common law is typically the basis of the legal systems in the United Kingdom and former colonies of the United Kingdom, including the United States and Canada. The Code Napoleon is typically the basis of the legal systems in former French colonies and the European countries that Napoleon once ruled, such as Belgium, Spain, and Italy. In countries with weak investor protection, controlling shareholder ownership is often a substitute for a lack of legal protection. Note that we have not mentioned ethics. All of the principles and practices described so far have assumed that the individuals in roles of responsibility and leadership pursue them truly and fairly. That, however, has not always been the case.

Family Ownership and Corporate Governance

Although much of the discussion about corporate governance concentrates on the market-based regimes (see Exhibit 4.6), family-based regimes are arguably more common and more important worldwide. For example, in a study of 5,232 corporations in 13 Western European countries, family-controlled firms represented 44% of the sample compared to 37% that were widely held.[5] *Global Finance in Practice 4.3* highlights some of the history of family power, the family cartel that controlled Italy for nearly 60 years.

GLOBAL FINANCE IN PRACTICE 4.3

Italian Cross—Shareholding and the End of the Salatto Buono

Italy, in the years following World War II, was a country teetering on collapse. In an effort to stabilize industrial activity, the powerful families of the north—the Agnellis (of Fiat fame), the Pesentis, the Pirellis, the Ligrestis, and later the Benettons—formed salotto buono—"the fine drawing room"—to control Italian finance, industry, and media, through relatively small stakes. At the core of the relationship was that each family business held significant ownership and control in the other in an interlocking or cross-shareholding structure that assured that no outsiders could gain ownership or influence.

The creator of salotto buono was Enrico Cuccia, the founder of Mediobanca, the Milan-based investment bank.

One man in particular, Cesare Geronzi, rose to the top of Italian finance. And every step of the way, he took three scarlet chairs with him. The chairs sat in his waiting room at Mediobanca and eventually Generali, Italy's largest financial group. Geronzi rose to the pinnacle of power despite twice being the target of major financial and accounting fraud cases, including Parmalat. Over the following half-century anyone wishing to gain influence had to pass through the "three chairs," the salatto buono.

But, alas, the global financial crisis of 2008–2009 broke down many of the world's last bastions of private power. One such casualty was salotto buono, as more and more of its vested families fell further into debt and bankruptcy.

---

[5]Mara Faccio and Larry H.P. Lang, "The Ultimate Ownership of Western European Corporations," *Journal of Financial Economics*, 65 (2002), p. 365. See also: Torben Pedersen and Steen Thomsen, "European Patterns of Corporate Ownership," *Journal of International Business Studies*, Vol. 28, No. 4, Fourth Quarter, 1997, pp. 759–778.

Failures in Corporate Governance

Failures in corporate governance have become increasingly visible in recent years. The Enron scandal in the United States is well known. In addition to Enron, other firms that have revealed major accounting and disclosure failures, as well as executive looting, are WorldCom, Parmalat, Tyco, Adelphia, and HealthSouth. In each case, prestigious auditing firms, such as Arthur Andersen, missed the violations or minimized them possibly because of lucrative consulting relationships or other conflicts of interest. Moreover, security analysts and banks urged investors to buy the shares and debt issues of these and other firms that they knew to be highly risky or even close to bankruptcy. Even more egregious, most of the top executives who were responsible for the mismanagement that destroyed their firms walked away (initially) with huge gains on shares sold before the downfall, and even overly generous severance packages.

Good Governance and Corporate Reputation

Does good corporate governance matter? This is actually a difficult question, and the realistic answer has been largely dependent on outcomes historically. For example, as long as Enron's share price continued to rise, questions over transparency, accounting propriety, and even financial facts were largely overlooked by all of the stakeholders of the corporation. Yet, eventually, the fraud, deceit, and failure of the multitude of corporate governance practices resulted in bankruptcy. It not only destroyed the wealth of investors, but the careers, incomes, and savings of so many of its own employees. Ultimately, good governance should matter.

One way in which companies may signal good governance to the investor markets is to adopt and publicize a fundamental set of governance policies and practices. Nearly all publicly traded firms have adopted this approach, as becomes obvious when visiting corporate Web sites. This has also led to a standardized set of common principles, as described in Exhibit 4.7, which might be considered a growing consensus on good governance practices. Those practices—board composition, management compensation structure and oversight, corporate auditing practices, and public disclosure—have been widely accepted.

EXHIBIT 4.7   **The Growing Consensus on Good Corporate Governance**

Although there are many different cultural and legal approaches used in corporate governance worldwide, there is a growing consensus on what constitutes good corporate governance.

- **Composition of the Board of Directors.** A board of directors should have both internal and external members. More importantly, it should be staffed by individuals of true experience and knowledge of not only their own rules and responsibilities, but of the nature and conduct of the corporate business.

- **Management Compensation.** A management compensation system should be aligned with corporate performance (financial and otherwise) and have significant oversight from the board and open disclosure to shareholders and investors.

- **Corporate Auditing.** Independent auditing of corporate financial results should be conducted on a meaningful real-time basis. An audit process with oversight by a Board committee composed primarily of external members would be an additional significant improvement.

- **Public Reporting and Disclosure.** There should be timely public reporting of both financial and nonfinancial operating results that may be used by investors to assess the investment outlook. This should also include transparency and reporting around potentially significant liabilities.

A final international note of caution: The quality and credibility of all internal corporate practices on good governance are still subject to the quality of a country's corporate law, its protection of both creditor and investor rights (including minority shareholders), and the country's ability to provide adequate and appropriate enforcement.

In principle, the idea is that good governance (at both the country and corporate levels) is linked to cost of capital (lower), returns to shareholders (higher), and corporate profitability (higher). An added dimension of interest is the role of country governance, as it may influence the country in which international investors may choose to invest. Curiously, however, not only have corporate rankings been highly uncorrelated across ranking firms, but a number of academic studies have also indicated little linkage between a firm's corporate governance ranking and its future likelihood of restating earnings, shareholder lawsuits, return on assets, and a variety of measures of stock price performance.

An additional way to signal good corporate governance, in non-Anglo-American firms, is to elect one or more Anglo-American board members. This was shown to be true for a select group of Norwegian and Swedish firms in a study by Oxelheim and Randøy.[6] The firms had superior market values. The Anglo-American board members suggested a governance system that would show better monitoring opportunities and enhanced investor recognition.

Corporate Governance Reform

Within the United States and the United Kingdom, the main corporate governance problem is the one addressed by agency theory: With widespread share ownership, how can a firm align management's interest with that of the shareholders? Since individual shareholders do not have the resources or the power to monitor management, the U.S. and U.K. markets rely on regulators to assist in the monitoring of agency issues and conflicts of interest. Outside the U.S. and the U.K., large controlling shareholders (including Canada) are in the majority. They are able to monitor management in some ways better than regulators. However, controlling shareholders pose a different agency problem. It is extremely difficult to protect the interests of minority shareholders (investors holding small numbers of share and therefore little voting power) against the power of controlling shareholders, whether the controlling shareholders are major institutions, large wealthy private investors, or even controlling families.

In recent years, reform in the United States and Canada has been largely regulatory. Reform elsewhere has been largely focused on the adoption of principles rather than stricter legal regulations. The principles approach is softer, less costly, and less likely to conflict with other existing regulations.

Sarbanes-Oxley Act. The U.S. Congress passed the *Sarbanes-Oxley Act* (SOX) in July 2002. Named after its two primary congressional sponsors, SOX had four major requirements: (1) CEOs and CFOs of publicly traded firms must vouch for the veracity of the firm's published financial statements; (2) corporate boards must have audit and compensation committees drawn from independent (outside) directors; (3) companies are prohibited from making loans to corporate officers and directors; and (4) companies must test their internal financial controls against fraud.

The first provision — the so-called *signature clause* — has already had significant impacts on the way in which companies prepare their financial statements. The provision was intended to instill a sense of responsibility and accountability in senior management (and to therefore reduce management explanations of "the auditors signed off on it"). The companies themselves have pushed the same procedure downward in their organizations, often requiring business unit managers and directors at lower levels to sign their financial statements. Regardless of the form of corporate government reform, as discussed in *Global Finance in Practice 4.4*, the definition of good governance is still under debate.

---

[6]Lars Oxelheim and Trond Randøy, "The Impact of Foreign Board Membership on Firm Value," *Journal of Banking and Finance*, Vol. 27, No. 12, 2003, pp. 2,369–2,392.

Is Good Governance Good Business Globally?

The term "good governance" is, in many instances, a highly politically charged term. When talking to the press, many directors and executives argue that the pursuit of good governance practices is good for business globally. However, those same officials may also argue that stringent reporting and disclosure requirements, like those imposed by the United States under Sarbanes-Oxley, harm business competition and growth and, ultimately, the attractiveness of listing and trading their securities in the United States. In the end, the devil may indeed be in the detail.

One size may not fit all, however. Culture has an enormous impact on business conduct, and many countries are finding their own way without necessarily following U.S. or European practices. For example, a number of Japanese leaders note that the Japanese corporate governance system differs from the Western system and has evolved while preserving Japanese culture and history. They argue that cultural background and history should not be ignored when developing and implementing global standards, regulations, and oversight.

Poor performance of management usually requires changes in management, ownership, or both. Exhibit 4.8 illustrates some of the alternative paths available to shareholders when they are dissatisfied with firm performance. Depending on the culture and accepted practices, it is not unusual for many investors to—for an extended time—remain quietly disgruntled regarding share price performance. A more active response is to sell their shares. It is with the third and fourth possible actions, changing management and initiating a takeover, that management hears a much louder dissatisfied "voice."

Corporate Responsibility and Sustainability

Sustainable development is development that meets the needs of the present without compromising the ability of future generations to meet their own needs.

—Brundtland Report, 1987, p. 54.

EXHIBIT 4.8   Potential Responses to Shareholder Dissatisfaction

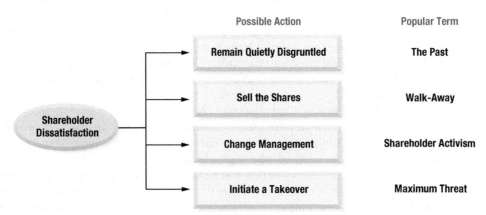

What counts is that the management of a publicly quoted company, and its board of directors, know that the company can become the subject of a hostile takeover bid if they fail to perform. The growth of equity and hedge funds in the United States and elsewhere in recent years has strengthened this threat as leveraged buyouts are once again common.

What is the purpose of the corporation? It is increasingly accepted that the *purpose* of the corporation is to certainly create profits and value for its stakeholders, but the *responsibility* of the corporation is to do so in a way that inflicts no costs on the environment and society. As a result of globalization, this growing responsibility and role of the corporation in society has added an additional level of complexity to the modern corporation never seen before.

The discussion has been somewhat hampered to date by a lot of conflicting terms and labels—*corporate goodness*, *corporate responsibility*, *corporate social responsibility* (CSR), *corporate philanthropy*, and *corporate sustainability*, to list but a few. To simplify, *sustainability* is often described as a goal, while *responsibility* is an obligation of the corporation. The obligation is to pursue profit, social development, and the environment—but to do so along sustainable principles.

Nearly two decades ago a number of large corporations began to refine their publicly acknowledged corporate objective as "the pursuit of the triple bottom line." This triple bottom line—*profitability*, *social responsibility*, and *environmental sustainability*—was considered an enlightened development of modern capitalism. What some critics referred to as a softer and gentler form of market capitalism, was a growing acceptance on the part of the corporation for doing something more than generating a financial profit. One way to explain this development of an expanded view of corporate responsibilities divides the arguments along two channels, the *economic channel* and the *moral channel*.

The *economic channel* argues that by pursuing corporate sustainability objectives the corporation is actually still pursuing profitability, but is doing so with a more intelligent longer-term perspective—"enlightened self-interest." It has realized that a responsible organization must assure that its actions over time, whether or not required by law or markets, do not reduce future choices. Alternatively, the *moral channel* argues that since the corporation has the rights and responsibilities of a citizen, including the moral responsibility to act in the best interests of society, regardless of its impacts on profitability. And you thought the management of a company was simple!

SUMMARY POINTS

- Most commercial enterprises have their origins with either entrepreneurs (private enterprise) or governments (public enterprise). Regardless of origin, if they remain commercial in focus, they may over time choose to go public (in whole or in part) via an initial public offering (IPO).

- When a firm becomes widely owned, it is typically managed by hired professionals. Professional managers' interests may not be aligned with the interests of owners, thus creating an agency problem.

- The Anglo-American markets subscribe to a philosophy that a firm's objective should follow the shareholder wealth maximization (SWM) model. More specifically, the firm should strive to maximize the return to shareholders, the sum of capital gains and dividends, for a given level of risk.

- As do shareholders in Anglo-American markets, controlling shareholders in non-Anglo-American markets

strive to maximize long-term returns to equity. However, they also consider the interests of other stakeholders, including employees, customers, suppliers, creditors, government, and community. This is known as stakeholder capitalism.

- The return to a shareholder in a publicly traded firm combines current income in the form of dividends and capital gains from the appreciation of share price. A privately held firm tries to maximize current and sustainable income since it has no share price.

- The MNE must determine for itself the proper balance between three common operational objectives: maximization of consolidated after-tax income; minimization of the firm's effective global tax burden; and correct positioning of the firm's income, cash flows, and available funds as to country and currency.

- The relationship between stakeholders that is used to determine the strategic direction and performance of an organization is termed corporate governance. Dimensions of corporate governance include agency theory; composition and control of boards of directors; and cultural, historical and institutional variables.

■ A number of initiatives in corporate governance practices in the United States, the United Kingdom, and the European Union—including board structure and

compensation, transparency, auditing, and minority shareholder rights—are spreading to a number of today's major emerging markets.

■ Corporate governance practices are seen, in some countries and cultures, as overly intrusive and occasionally are viewed as damaging to the competitive capability of the firm. The result is an increasing reluctance to go public in selective markets.

MINI-CASE

Luxury Wars—LVMH vs. Hermès[7]

The basic rule is to be there at the right moment, at the right place, to seize a promising opportunity in an environment guaranteeing sufficient longer-term growth.

—Bernard Arnault, Chairman and CEO, LVMH.

Patrick Thomas focused intently on not letting his hands shake as he quietly ended the call. He had been riding his bicycle in rural Auvergne, in south-central France, when his cell phone buzzed. He took a long deep breath and tried to think. He had spent most of his professional life working at Hermès International, SA and had assumed the position of CEO in 2006 after the retirement of Jean-Louis Dumas. The first nonfamily CEO to run the company was now facing the biggest threat to the family-controlled company in its 173-year history.

The LVMH Position

The man on the other end of the phone had been none other than Bernard Arnault, Chairman and CEO of LVMH (Moët Hennessy Louis Vuitton), the world's largest luxury brand company, the richest man in France, and a major competitor. Arnault was calling to inform Thomas that LVMH would be announcing in two hours that they had acquired a 17.1% interest in Hermès. Thomas had simply not believed Arnault for the first few minutes, thinking it impossible that LVMH could have gained control of that significant a stake without his knowledge. Arnault assured Thomas it was no joke and that he looked forward to participating in the company's continued success as a shareholder. Patrick Thomas began assessing the potential threat, if it was indeed a threat.

Hermès International. Hermès International, SA is a multibillion-dollar French company that makes and sells luxury goods across a number of different product categories including women's and men's apparel, watches, leather goods, jewelry, and perfume. Thierry Hermès, who was known for making the best saddles and harnesses in Paris, founded the company in 1837. The company's reputation soared as it began to provide its high-end products to nobility throughout Europe, North Africa, Russia, Asia, and the Americas. As the years passed, the company began to expand its product line to include the finest leather bags and silk scarves on the market, all while passing the company down through generations and maintaining family control.

Despite going public in 1993, roughly 60 direct descendants of Thierry Hermès, comprising the 5th and 6th generations, still controlled approximately 73% of the company. In 2006, the job of CEO was assumed, for the first time, by a nonfamily member, Patrick Thomas.

Bernard Arnault

Arnault is a shrewd man. He has reviewed his portfolio and sees what he is missing—a company that still produces true luxury—and he is going after it.

—Anonymous luxury brand CEO speaking on the LVMH announcement.

Bernard Arnault had made a very profitable career out of his penchant for taking over vulnerable family-owned businesses (earning him the colorful nickname of "the wolf in cashmere"). Born in Roubaix, France, to an upper class family, Arnault excelled as a student and graduated from France's prestigious engineering school, Ecole Polytechnique, before working as an engineer and taking over his family's construction business. When the French government began looking for someone to acquire the bankrupt company, Boussac (and its luxury line, Christian Dior), Arnault promptly bought the company. It proved the first step in building what would eventually become the luxury titan, LVMH, and propelling Arnault to the title of France's wealthiest man.

From that point on, Arnault began assembling what his competitors referred to as "the evil empire," by preying on susceptible family-owned companies with premium names. His takeover of Louis Vuitton was said to have gotten so personal and vicious that, after the last board meeting, the Vuitton family packed their belongings and left the building in tears. In addition to Louis Vuitton, Arnault had spent the last three decades forcibly acquiring such family-owned luxury brands as Krug (champagne), Pucci (fashion), Chateau d'Yquem (vineyard), and Celine (fashion), among others.

Arnault's only failure had been his attempted takeover of Gucci in 1999, when he was beaten by Francois Pinault, whose company PPR served as the white knight for Gucci and stole the deal out from under Arnault. It marked the one time in LVMH's history that it had failed in a takeover bid.

Autorité des Marchés Financiers (AMF). Arnault's announcement of LVMH's ownership stake in Hermès came as a shock to both the fashion industry and the family shareholders of Hermès. Exhibit A is Hermès public response to LVMH's initiative. The French stock market regulator, the Autorité des Marchés Financiers (AMF), required any investor gaining a 5% or greater stake in a publicly traded company to file their ownership percentage publicly, as well as a document of intent. But no such notice had yet been filed.

In the days following the October 23 press release, LVMH confirmed that the company had complied with all current rules and regulations in the transactions, and that it would file all the necessary documentation within the allotted time. The AMF announced that it would investigate LVMH's acquisition of the Hermès stock. This, however, was little consolation to Thomas and the Hermès family. Even if violations were found, the only penalty LVMH could suffer would be the loss of voting rights for two years.

Equity Swaps. LVMH had, it turns out, acquired its large ownership position under the radar of the Hermès family, company management, and industry analysts, through the use of equity swaps. Equity swaps are derivative contracts where two parties agree to swap future cash flows at a preset date. The cash flows are referred to as "legs" of the

---

EXHIBIT A   **Hermès Response via Press Release: October 24, 2010**

Hermès has been informed that LVMH has acquired a 17% stake in the Company. In 1993, the shareholders of Hermès International, all descendants from Emile Hermès, decided to enlist the Company on the Paris stock exchange. This decision was made with two objectives in mind: 1) support the long term development of the Company; 2) make shares easier to trade for the shareholders.

Over the last 10 years, the Hermès group has delivered an average annual growth rate of 10% of its net result and currently holds a very strong financial position with over M€ 700 of free cash. Today, Hermès Family shareholders have a strong majority control of nearly 3/4 of the shares. They are fully united around a common business vision. Their long term control of Hermès International is guaranteed by its financial status as limited partnership by shares and the family shareholders have confirmed that they are not contemplating any significant selling of shares. The public listing of shares, allows investors who want to become minority shareholders to do so. As a Family Company Hermès has treated and will always treat its shareholders with utmost respect.

The Executive Management,
Sunday October 24th, 2010

Source: Hermes.com.

swap. In most equity swaps, one leg is tied to a floating rate like LIBOR (the floating leg), and the other leg is tied to the performance of a stock or stock index (the equity leg). Under current French law, a company must acknowledge when they attain a 5% or more equity stake in another company, or the rights to purchase a 5% or more stake via derivatives like equity swaps.

But there was a loophole. The swaps could be structured so that their value was tied to the equity instrument only; at closeout the contract may be settled in cash, not shares. Using this structure, the swap holder is not required to file with the AMF, since they will never actually own the stock.

The LVMH Purchase. It was widely known that Arnault had long coveted Hermès as a brand. In fact, Mr. Arnault had previously owned 15% of Hermès when he first took over LVMH in the 1990s. At the time, Mr. Arnault had his hands full with reorganizing and redirecting LVMH after his takeover of the company, so he agreed to sell the shares to then Hermès CEO Jean-Louis Dumas when Dumas wanted to take the company public.

But things had changed for LVMH and Arnault since then. Mr. Arnault had grown his company to the largest luxury conglomerate in the world, with over $55 billion in annual sales. He accomplished this through organic growth of brands and strategic purchases. Known for his patience and shrewd business acumen, when he saw an opportunity to target a long coveted prize, he took advantage.

The attack on Hermès shares was one of Arnault's most closely kept secrets, with only three people in his empire aware of the equity swap contracts. Arnault began making his move in 2008 when three blocks of Hermès shares—totaling 12.8 million shares—were quietly placed on the market by three separate French banks. The origins of these shares were unknown, but many suspected they had come from Hermès family members.

It is believed that Arnault was contacted by the banks and was given 24 hours to decide whether he would like to purchase them or not. Arnault was hesitant to take such a large ownership stake in Hermès, particularly one requiring registration with the AMF. Arnault and the banks then developed the strategy whereby he would hold rights to the shares via equity swaps, but only as long as he put up the cash. At contract maturity, LVMH would realize the profit/loss on any movement in the share price. As part of the agreement with the banks, however, LVMH would have the option to take the shares instead. Had the contracts required share settlement, under French law LVMH would have had to publicly acknowledge its potential equity position in Hermès.

The design of the contracts prevented LVMH from actually holding the shares until October 2010, when they publicly announced their ownership stake in Hermès. During this period Hermès share price floated between €60 and €102. This explained how LVMH was able to acquire its shares in Hermès at an average price of €80 per share, nearly a 54% discount on the closing price of €176.2 on Friday, October 22.

LVMH could have actually held its swap contracts longer and postponed settlement and disclosure, but the rapid rise in Hermès share price over the previous months forced the decision (which many analysts attributed to market speculations of an LVMH takeover plot). If LVMH had postponed settlement, it would have had to account for €2 billion in paper profit, earned on the contracts when publishing their year-end accounts in February 2011.

The Battle Goes Public

Although the original press release by LVMH made it clear that the company had no greater designs on controlling Hermès, Hermès management did not believe it, and moved quickly. After a quick conference call amongst Hermès leadership, Hermès CEO Thomas and Executive Chairman Puech gave an interview with *Le Figaro* on October 27.

It's clear his [Mr. Arnault] intention is to take over the company and the family will resist that.

—Patrick Thomas, CEO Hermès, *Le Figaro*, October 27, 2010.

We would like to convince him [Mr. Arnault] that this is not the right way to operate and that it's not friendly. If he entered in a friendly way, then we would like him to leave in a friendly way.

—Mr. Puech, Executive Chairman of Emile Hermès SARL, *Le Figaro*, October 27, 2010.

Arnault wasted no time in responding in an interview given to the same newspaper the following day: I do not see how the head of a listed company can be qualified to ask a shareholder to sell his shares. On the contrary he is supposed to defend the interests of all shareholders.

—Bernard Arnault, CEO LVMH, *Le Figaro*, October 28, 2010.

Pierre Godé, Vice President LVMH. On November 10, after much speculation regarding LVMH's intent, Pierre Godé, an LVMH Vice President, gave an interview with *Les Echos* newspaper (itself owned by LVMH) to discuss how and why the transactions took place the way they did, as well as to dispel media speculation about a potential hostile takeover attempt from LVMH. In the interview, Godé was questioned about why LVMH chose to purchase the equity swap contracts against Hermès in the first place, and why LVMH chose to close those contracts in Hermès shares rather than in cash.

Godé confided that LVMH had begun looking at Hermès in 2007 when the financial crisis started and the stock exchange began to fall. LVMH was looking for financial investments in the luxury industry—as that is where their expertise lies—and concluded that Hermès would weather the financial crisis better than other potential investments. It was for this reason alone that LVMH chose to purchase equity swaps with Hermès shares as the equity leg.

Godé argued that equity swaps with cash payment and settlement were trendy at that time, and virtually every bank offered this derivative. Even though LVMH already had just under a 5% stake of Hermès stock at the time the derivatives were being set up, Godé stated that LVMH never even considered the possibility of closing out the swaps in shares. For one thing, it was something they could not do contractually (according to Godé), nor did LVMH want to ask the banks for equity settlements. But by 2010, the situation had changed, prompting LVMH to reassess their Hermès equity swap contracts. The contracts themselves were running out, and LVMH had a premium of nearly €1 billion on them. According to Godé, the banks that had covered their contracts with LVMH were now tempted to sell the shares, which represented 12% of Hermès' capital.

Godé explained that selling the shares, in and of itself, did not concern LVMH. What LVMH did worry about however, was where the shares might end up. Godé stressed that at that time there were rumors that both a "powerful group from another industry" and Chinese investment funds were interested in the Hermès shares. LVMH management felt the rising share price of Hermès lent support to these rumors. Additionally, the market had been improving and LVMH had the financial means to be able to pay for the contracts and settle in shares. As a result, LVMH spoke with the banks to assess their position, and after, several weeks of talks, LVMH reached an agreement with the banks in October for part of the shares.

At this time, Godé explained, "the Board had to choose between receiving a considerable amount on the equity swaps or take a minority participation in this promising group but where our power would be very limited as the family controlled everything. There was an intense debate and finally the Board chose to have share payments." Godé completed the interview by stating that LVMH was surprised by the strong negative response from Hermès, especially considering that LVMH had owned a 15% stake in the company in the early 1990s.

Evolution of Hermès International and Its Control. Hermès was structured as a Société en Commandite, the French version of a limited partnership in the United States. In the case of Hermès, this structure concentrates power in the hands of a ruling committee, which is controlled by the family.

In addition to the Société en Commandite structure of the company, former Hermès CEO Jean-Louis Dumas established a partner company, Hermès SARL, in 1989. This company represented the interests of family shareholders (only direct Hermès descendants could be owners), and was the sole authority to appoint management and set company strategy. This unusual structure provided the Hermès family with the ability to retain decision-making power even if only one family member remained as a shareholder. The structure had been adopted as protection against a hostile takeover after Dumas saw the way Bernard Arnault had dealt with the Vuitton family when he acquired their company.

In a further attempt to placate family members and minimize family infighting, Dumas listed 25% of Hermès SA on the French stock market in 1993. This was done to provide family members with a means to value their stake in the company as well as partially cash out if they felt their family dividends were not enough (several family members were known to live large, and Dumas feared their lifestyles might exceed their means). Dumas believed—at least at that time—that his two-tier structure would insulate Hermès from a potential hostile takeover.

However, analysts were now speculating that Hermès SARL might only provide protection through the 6th generation, and that with just a 0.1% stake in the company being worth approximately €18 million at current market prices, there was reason to fear some family members "defecting." This concern was made all the more real when it became known through AMF filings that Laurent Mommeja, brother to Hermès supervisory board member Renaud fe, sold €1.8 million worth of shares on October 25 at a share price of €189 per share.

After considerable debate, the Hermès family decided to consolidate their shares into a trust in the form of a holding company that would ensure their 73% ownership stake would always vote as one voice and ultimately secure the family's continued control of the company. On December 21, LVMH announced that it had raised its total stake in Hermès to 20.21% and that it had filed all required documents with AMF upon passing the 20% threshold. LVMH also reiterated that it had no intention of taking control of Hermès or making a public offer for its shares. Under French law, once LVMH reached one-third share ownership it would have to make a public tender for all remaining shares.

Mini-Case Questions

1. Hermès International was a family-owned business for many years. Why did it then list its shares on a public market? What risks and rewards come from a public listing?
2. Bernard Arnault and LVMH acquired a large position in Hermès shares without anyone knowing. How did they do it and how did they avoid the French regulations requiring disclosure of such positions?
3. The Hermès family defended themselves by forming a holding company of their family shares. How will this work and how long do you think it will last?

QUESTIONS

These questions are available in MyFinanceLab.

1. **Business Ownership.** What are the predominant ownership forms in global business?

2. **Business Control.** How does ownership alter the control of a business organization? Is the control of a private firm that different from a publicly traded company?

3. **Separation of Ownership and Management.** Why is the separation of ownership from management so critical to the understanding of how businesses are structured and led?

4. **Corporate Goals: Shareholder Wealth Maximization.** Explain the assumptions and objectives of the shareholder wealth maximization model.

5. **Corporate Goals: Stakeholder Capitalism Maximization (SCM).** Explain the assumptions and objectives of the stakeholder capitalization model.

6. **Management's Time Horizon.** Do shareholder wealth maximization and stakeholder capitalism have the same time-horizon for the strategic, managerial, and financial objectives of the firm? How do they differ?

7. **Operational Goals.** What should be the primary operational goal or goals of an MNE?

8. **Financial Returns.** How do shareholders in a publicly traded firm actually reap cash flow returns from their ownership? Who has control over which of these returns?

9. **Dividend Returns.** Are dividends really all that important to investors in publicly traded companies? Aren't capital gains really the point or objective of the investor?

10. **Ownership Hybrids.** What is a hybrid? How may it be managed differently?

11. **Corporate Governance.** Define corporate governance and the various stakeholders involved in corporate governance. What is the difference between internal and external governance?

12. **Governance Regimes.** What are the four major types of governance regimes and how do they differ?

13. **Governance Development Drivers.** What are the primary drivers of corporate governance across the globe? Is the relative weight or importance of some drivers increasing over others?

14. **Good Governance Value.** Does good governance have a "value" in the marketplace? Do investors really reward good governance, or does good governance just attract a specific segment of investors?

15. Shareholder Dissatisfaction. If shareholders are unhappy with the current leadership of a firm—its actual management and control—what are their choices?

16. Emerging Markets Corporate Governance Failures. It has been claimed that failures in corporate governance have hampered the growth and profitability of some prominent firms located in emerging markets. What are some typical causes of these failures in corporate governance?

17. Emerging Markets Corporate Governance Improvements. In recent years emerging market MNEs have improved their corporate governance policies and become more shareholder-friendly. What do you think is driving this phenomenon?

PROBLEMS

These problems are available in MyFinanceLab.

Use the following formula to answer problems on shareholder returns, where P_t is the share price at time t, and D_t is the dividend paid at time t.

$$\text{Shareholder Return} = \frac{D_2}{P_1} + \frac{P_2 - P_1}{P_1}$$

1. Emaline Returns. If the share price of Emaline, a New Orleans-based shipping firm, rises from $12 to $15 over a one-year period, what is the rate of return to the shareholder given each of the following:
   a. The company paid no dividends
   b. The company paid a dividend of $1 per share
   c. The company paid the dividend and the total return to the shareholder is separated into the dividend yield and the capital gain

2. Vaniteux's Returns (A). Spencer Grant is a New York-based investor. He has been closely following his investment in 100 shares of Vaniteux, a French firm that went public in February 2010. When he purchased his 100 shares at €17.25 per share, the euro was trading at $1.360/€. Currently, the share is trading at €28.33 per share, and the dollar has fallen to $1.4170/€.
   a. If Spencer sells his shares today, what percentage change in the share price would he receive?
   b. What is the percentage change in the value of the euro versus the dollar over this same period?

   c. What would be the total return Spencer would earn on his shares if he sold them at these rates?

3. Vaniteux's Returns (B). Spencer Grant chooses not to sell his shares at the time described in Problem 2. He waits, expecting the share price to rise further after the announcement of quarterly earnings. His expectations are correct, and the share price rises to €31.14 per share after the announcement. He now wishes to recalculate his returns. The current spot exchange rate is $1.3110/€.

4. Vaniteux's Returns (C). Using the same prices and exchange rates as in Problem 3, Vaniteux (B), what would be the total return on the Vaniteux investment by Laurent Vuagnoux, a Paris-based investor?

5. Microsoft's Dividend. In January 2003, Microsoft announced that it would begin paying a dividend of $0.16 per share. Given the following share prices for Microsoft stock in the recent past, how would a constant dividend of $0.16 per share per year have changed the company's average annual return to its shareholders over this period?

First Trading Day	Closing Share Price	First Trading Day	Closing Share Price
1998 (Jan 2)	$131.13	2001 (Jan 2)	$43.38
1999 (Jan 4)	$141.00	2002 (Jan 2)	$67.04
2000 (Jan 3)	$116.56	2003 (Jan 2)	$53.72

6. Carty's Choices. Brian Carty, a prominent investor, is evaluating investment alternatives. If he believes an individual equity will rise in price from $59 to $71 in the coming one-year period, and the share is expected to pay a dividend of $1.75 per share, and he expects at least a 15% rate of return on an investment of this type, should he invest in this particular equity?

7. Fashion Acquisitions. During the 1960s, many conglomerates were created by firms that were enjoying a high price/earnings ratio (P/E). These firms then used their highly valued stock to acquire other firms that had lower P/E ratios, usually in unrelated domestic industries. Conglomerates went out of fashion during the 1980s when they lost their high P/E ratios, thus making it more difficult to find other firms with lower P/E ratios to acquire.

Problem 7.

Company	P/E Ratio	Number of Shares	Market Value per Share	Earnings	EPS	Total Market Value
ModoUnico	20	10,000,000	$20.00	$10,000,000	$1.00	$200,000,000
Modern American	40	10,000,000	$40.00	$10,000,000	$1.00	$400,000,000

During the 1990s, the same acquisition strategy was possible for firms located in countries where high P/E ratios were common compared to firms in other countries where low P/E ratios were common. Consider the hypothetical firms in the pharmaceutical industry shown in the table at the top of the page.

Modern American wants to acquire ModoUnico. It offers 5,500,000 shares of Modern American, with a current market value of $220,000,000 and a 10% premium on ModoUnico's shares, for all of ModoUnico's shares.

a. How many shares would Modern American have outstanding after the acquisition of ModoUnico?

b. What would be the consolidated earnings of the combined Modern American and ModoUnico?

c. Assuming the market continues to capitalize Modern American's earnings at a P/E ratio of 40, what would be the new market value of Modern American?

d. What would be the new earnings per share of Modern American?

e. What would be the new market of a share of Modern American?

f. How much would Modern American's stock price increase?

g. Assume that the market takes a negative view of the acquisition and lowers Modern American's P/E ratio to 30. What would be the new market price per share of stock? What would be its percentage loss?

8. **Corporate Governance: Overstating Earnings.** A number of firms, especially in the United States, have had to lower their previously reported earnings due to accounting errors or fraud. Assume that Modern American (Problem 7) had to lower its earnings to $5,000,000 from the previously reported $10,000,000. What might be its new market value prior to the acquisition? Could it still do the acquisition?

9. **Yehti Manufacturing (A).** Dual classes of common stock are common in a number of countries.

Assume that Yehti Manufacturing has the following capital structure at book value. The A-shares have ten votes per share and the B-shares have one vote per share.

Yehti Manufacturing	Local Currency (millions)
Long-term debt	200
Retained earnings	300
Paid-in common stock: 1 million A-shares	100
Paid-in common stock: 4 million B-shares	400
Total long-term capital	1,000

a. What proportion of the total long-term capital has been raised by A-shares?

b. What proportion of voting rights is represented by A-shares?

c. What proportion of the dividends should the A-shares receive?

10. **Yehti Manufacturing (B).** Assume all of the same debt and equity values for Yehti Manufacturing in Problem 9, with the sole exception that both A-shares and B-shares have the same voting rights—one vote per share.

a. What proportion of the total long-term capital has been raised by A-shares?

b. What proportion of voting rights is represented by A-shares?

c. What proportion of the dividends should the A-shares receive?

11. **Lantau Beer (A): European Sales.** Lantau Beer is a Hong Kong-based brewery and files all of its financial statements in Hong Kong dollars (HK$). The company's European sales director, Phillipp Bosse, has been criticized for his performance. He disagrees, arguing that sales in Europe have grown steadily in recent years. Who is correct?

Problem 11.

	2010	2011	2012
Total net sales, HK$	171,275	187,500	244,900
Percent of total sales from Europe	48%	44%	39%
Total European sales, HK$	_____	_____	_____
Average exchange rate, HK$/€	11.5	11.7	10.3
Total European sales, euros	_____	_____	_____
Growth rate of European sales	_____	_____	_____

Problem 12.

	2010	2011	2012
Annual yen payments on debt agreement (¥)	12,000,000	12,000,000	12,000,000
Average exchange rate, ¥/HK$	12.3	12.1	11.4
Annual yen debt service, HK$	_____	_____	_____

12. **Lantau Beer (B): Japanese Yen Debt.** Lantau Beer of Hong Kong borrowed Japanese yen under a long-term loan agreement several years ago. The company's new CFO believes, however, that what was originally thought to have been relatively "cheap debt" is no longer true. What do you think?

13. **Mattel's Global Performance.** Mattel (U.S.) achieved significant sales growth in its major international regions between 2001 and 2004. In its filings with the United States Security and Exchange Commission (SEC), it reported both the amount of regional sales and the percentage change in those sales resulting from exchange rate changes.
   a. What was the percentage change in sales, in U.S. dollars, by region?
   b. What was the percentage change in sales, by region, net of currency change impacts?
   c. What impact did currency changes have on the level and growth of consolidated sales between 2001 and 2004?

Problem 13.

Mattel's Global Sales

(thousands of US$)	2001 Sales ($)	2002 Sales ($)	2003 Sales ($)	2004 Sales ($)
Europe	$ 933,450	$ 1,126,177	$ 1,356,131	$ 1,410,525
Latin America	471,301	466,349	462,167	524,481
Canada	155,791	161,469	185,831	197,655
Asia Pacific	119,749	136,944	171,580	203,575
Total International	$ 1,680,291	$ 1,890,939	$ 2,175,709	$ 2,336,236
United States	3,392,284	3,422,405	3,203,814	3,209,862
Sales Adjustments	(384,651)	(428,004)	(419,423)	(443,312)
Total Net Sales	$ 4,687,924	$ 4,885,340	$ 4,960,100	$ 5,102,786

	Impact of Change in Currency Rates		
Region	2001–2002	2002–2003	2003–2004
Europe	7.0%	15.0%	8.0%
Latin America	−9.0%	−6.0%	−2.0%
Canada	0.0%	11.0%	5.0%
Asia Pacific	3.0%	13.0%	6.0%

Source: Mattel, Annual Report, 2002, 2003, 2004.

14. **Chinese Sourcing and the Yuan.** Harrison Equipment of Denver, Colorado, purchases all of its hydraulic tubing from manufacturers in mainland China. The company has recently completed a corporate-wide initiative in six sigma/lean manufacturing. Completed oil field hydraulic system costs were reduced 4% over a one-year period, from $880,000 to $844,800. The company is now worried that all of the hydraulic tubing that goes into the systems (making up 20% of their total costs) will be hit by the potential revaluation of the Chinese yuan—if some in Washington get their way. How would a 12% revaluation of the yuan against the dollar impact total system costs?

15. **S&P Equity Returns History.** The U.S. equity markets have delivered very different returns over the past 90 years. Use the following data arranged by decade to answer the following questions about these U.S. equity investment returns.

   a. Which period shown had the highest total returns? The lowest?

   b. Which decade had the highest dividend returns? When were dividends clearly not a priority for publicly traded companies?

   c. The 1990s was a boom period for U.S. equity returns. How did firms react in terms of their dividend distributions?

   d. How has the 2000s period fared? How do you think publicly traded companies have started changing their dividend distribution habits as a result?

Problem 15.

S&P 500 Equity Returns, 1926–2014 (average annual return, percent)

Period	1930s	1940s	1950s	1960s	1970s	1980s	1990s	2000s	1926 to 2014
Capital appreciation	−5.3%	3.0%	13.6%	4.4%	1.6%	12.6%	15.3%	−2.7%	5.9%
Dividend yield	5.4%	6.0%	5.1%	3.3%	4.2%	4.4%	2.5%	1.8%	4.0%
Total return	0.1%	9.0%	18.7%	7.7%	5.8%	17.0%	17.8%	−0.9%	9.9%

Source: Data drawn from "JP Morgan Guide to the Markets, 2015," JP Morgan Asset Management.

INTERNET EXERCISES

1. **Multinational Firms and Global Assets/Income.** The differences among MNEs is striking. Using a sample of firms such as the following, pull from their individual Web pages the proportions of their incomes that are earned outside their countries of incorporation. (Note how Nestlé calls itself a "transnational company.")

Walt Disney	disney.go.com
Nestlé S.A.	www.nestle.com
Intel	www.intel.com
Mitsubishi	www.mitsubishi.com
Nokia	www.nokia.com
Royal Dutch/Shell	www.shell.com

   Also note the way in which international business is now conducted via the Internet. Several of the home pages listed allow the user to choose the language of the presentation viewed.

2. **Corporate Governance.** There is no hotter topic in business today than corporate governance. Use the following site to view recent research, current events and news items, and other information related to the relationships between a business and its stakeholders.

Corporate Governance Net	www.corpgov.net

3. **Fortune Global 500.** Fortune magazine is relatively famous for its listing of the Fortune 500 firms in the global marketplace. Use Fortune's Web site to find the most recent listing of the global firms in this distinguished club.

Fortune	www.fortune.com/fortune

4. **Financial Times.** The *Financial Times*, based in London—the global center of international finance—has a Web site with a wealth of information. After going to the home page, go to "Markets" and then to the "Markets Data" page. Examine the recent stock market activity around the globe. Note the similarity in movement on a daily basis among the world's major equity markets.

Financial Times	www.ft.com

Foreign Exchange Theory and Markets

CHAPTER 5

The Foreign Exchange Market

The best way to destroy the capitalist system is to debauch the currency. By a continuing process of inflation, governments can confiscate, secretly and unobserved, an important part of the wealth of their citizens.

—John Maynard Keynes.

LEARNING OBJECTIVES

- Explore the multitude of functions of the foreign exchange market
- Detail how the structure of the global currency market has been changing
- Describe the financial and operational transactions conducted in the foreign exchange market
- Examine how the size of the global currency market has changed with global economics
- Learn the forms of currency quotations used by currency dealers, financial institutions, and agents of all kinds when conducting foreign exchange transactions

The *foreign exchange market* provides the physical and institutional structure through which the money of one country is exchanged for that of another country—the rate of exchange between currencies is determined and foreign exchange transactions are physically completed. *Foreign exchange* means the money of a foreign country; that is, foreign currency bank balances, banknotes, checks, and drafts. A *foreign exchange transaction* is an agreement between a buyer and seller that a fixed amount of one currency will be delivered for some other currency at a specified rate. This chapter describes the following features of the foreign exchange market:

- Its three main functions
- Its participants
- Its immense daily transaction volume
- Its geographic extent
- Types of transactions, including spot, forward, and swap transactions
- Exchange rate quotation practices

The chapter concludes with the Mini-Case, *The Venezuelan Bolivar Black Market*, which describes a businessman's challenge in accessing hard currency in a restricted exchange market.

Functions of the Foreign Exchange Market

Money is an object that is accepted as payment for goods, services, and in some cases, past debt. There are typically three functions of money: as a unit of account, as a store of value, and as a medium of exchange. The foreign exchange market is the mechanism by which participants transfer purchasing power between countries by exchanging money, obtain or provide credit for international trade transactions, and minimize exposure to the risks of exchange rate changes.

- The transfer of purchasing power is necessary because international trade and capital transactions normally involve parties living in countries with different national currencies. Usually each party wants to deal in its own currency, but the trade or capital transaction can be invoiced in only one currency. Hence, one party must deal in a foreign currency.

- Because the movement of goods between countries takes time, inventory in transit must be financed. The foreign exchange market provides a source of credit. Specialized instruments, such as bankers' acceptances and letters of credit are available to finance international trade.

- The foreign exchange market provides "hedging" facilities for transferring foreign exchange risk to someone else who is more willing to carry risk.

Structure of the Foreign Exchange Market

The foreign exchange market has, like all markets, evolved dramatically over time. Beginning with money changing hands in stalls on the streets of Florence and Venice, to the trading rooms in London and New York in the twentieth century, the market is based on supply and demand, market information and expectations, and negotiating strength.

The global foreign exchange market today is undergoing seismic change. That change involves every dimension of the market—the time, the place, the participants, the purpose, and the instruments. The forces driving change in the foreign exchange market are fundamental: electronic trading platforms, algorithmic trading programs and routines, and the increasing role of currencies as an asset class. These forces and other facilitators have combined to expand the depth, breadth, and reach of the foreign exchange market.

Time of Day and Currency Trading

The foreign exchange market spans the globe, with prices moving and currencies trading— somewhere—every hour of every business day. As illustrated in Exhibit 5.1, the world's trading day starts each morning in Sydney and Tokyo; moves west to Hong Kong and Singapore; passes on to the Middle East; shifts to the main European markets of Frankfurt, Zurich, and London; jumps the Atlantic to New York; continues west to Chicago; and ends in San Francisco and Los Angeles. Many large international banks operate foreign exchange trading rooms in each major geographic trading center in order to serve both their customers and themselves (so-called *proprietary trading*) on a 24-hour-a-day basis.

Although global currency trading is indeed a 24-hour-a-day process, there are segments of the 24-hour day that are busier than others. Historically, the major financial centers of the 19th and 20th centuries dominated—London and New York. But as is the case with much of global commerce today, the Far East, represented by Tokyo and Hong Kong, are now threatening that dominance. When these city-based trading centers overlap, the global currency markets exhibit the greatest depth and liquidity.

EXHIBIT 5.1   **Global Currency Trading: The Trading Day**

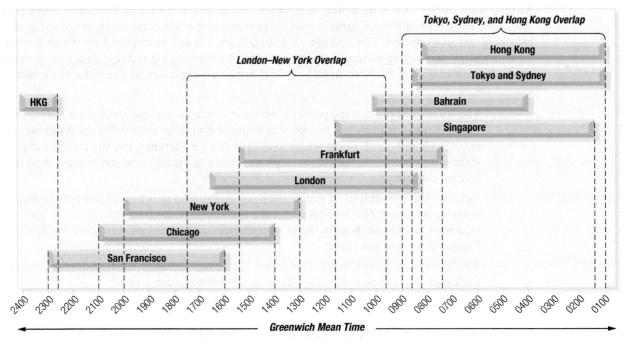

The currency trading day literally extends 24 hours per day. The busiest time of the day, which used to be when London and New York overlapped, has now started shifting "further East," to the Tokyo–Hong Kong dominated part of the day.

Trading Platforms and Processes

Currency trading is conducted in a variety of ways including individual-to-individual personal transactions, on official trading floors by open bidding, and increasingly by electronic platforms. Although continuous trading is indeed possible and increasingly prevalent, a "closing price" is often needed for a variety of record keeping and contractual purposes. These closing prices are often published as the official price, or "fixing," for the day, and certain commercial and investment transactions are based on this official price. Business firms in countries with exchange controls, like mainland China, often must surrender foreign exchange earned from exports to the central bank at the daily fixing price.

Currency traders are connected by highly sophisticated telecommunications networks, with which dealers and brokers exchange currency quotes instantaneously. A growing part of the industry is automated trading, electronic platforms in which corporate buyers and sellers trade currencies through Internet-based systems provided or hosted by major trading institutions. Although some of the largest currency transactions are still handled by humans via telephone, the use of computer trading has grown dramatically in recent years. The largest traditional providers of foreign exchange rate information and trading systems—Reuters, Telerate, EBS, and Bloomberg—are still substantial, but there has been a host of new service providers flooding the foreign exchange markets in recent years. But, despite all of the technology, the human element is still present as described in *Global Finance in Practice 5.1*.

Market Participants

One of the biggest changes in the foreign exchange market in the past decade has been its shift from a two-tier market (the interbank or wholesale market, and the client or retail market)

GLOBAL FINANCE IN PRACTICE 5.1

FX Market Manipulation: Fixing the Fix

Following the turmoil surrounding the setting of LIBOR rates in the interbank market, similar allegations arose over the possible manipulation of benchmarks in the foreign exchange markets in 2013 and 2014.

Much of the focus was on the *London fix*, the 4 P.M. daily benchmark rate used by a multitude of institutions and indices for marking value. Market analysts had noted steep spikes in trading just prior to the 4 P.M. fix, spikes that were not sustained in the hours and days that followed. Traders were alleged to be exchanging emails, using social networking sites, and even phone calls, to collaborate on market movements and price quotes at key times. After-hours personal trading by currency traders, an area of only marginal concern before, was also under review. Moving from voice-trading (telephone) to electronic trading was thought to be one possible fix, but the currency markets had long been something of an enigma when it came to electronic trading. *FICC trading*—fixed income, currencies, and commodities—was one of the first sectors to adopt electronic

trading in the mid-1990s, but the market had proven slow to change.

But change had finally come. By 2014, nearly 75% of all currency trading was electronic. The logic of the fix was simple: computer algorithms were less likely to pursue fraudulent trading for fixing. And research had shown strong evidence that electronic trading was more stabilizing than voice-trading, as most of the algorithmic codes were based on reversion to the mean—to the market averages—over time.

But as is the case with many technological fixes, the fix did not eliminate the problem, it had possibly just changed it. E-trading might still facilitate market manipulation, just of a more sophisticated kind. For example, there was a rumor of software in development that could detect mouse movements by other currency traders on some of the largest electronic platforms, allowing one computer (human attached) to detect the trader's mouse hovering over the bid or ask button prior to execution. Alas, it appeared there would always be the human element in trading, for better or for worse.

to a single-tier market. Electronic platforms and the development of sophisticated trading algorithms have facilitated market access by traders of all kinds and sizes.

Participants in the foreign exchange market can be simplistically divided into two major groups, those trading currency for commercial purposes, *liquidity seekers*, and those trading for profit, *profit seekers*. Although the foreign exchange market began as a market for liquidity purposes, facilitating the exchange of currency for the conduct of commercial trade and investment purposes, the exceptional growth in the market has been largely based on the expansion of profit-seeking agents. As might be expected, the profit seekers are typically much better informed about the market, looking to profit from its future movements, while liquidity seekers simply wish to secure currency for transactions. As a result, the profit seekers generally profit from the liquidity seekers.

Five broad categories of institutional participants operate in the market: (1) bank and nonbank foreign exchange dealers; (2) individuals and firms conducting commercial or investment transactions; (3) speculators and arbitragers; (4) central banks and treasuries; and (5) foreign exchange brokers.

Bank and Nonbank Foreign Exchange Dealers

Bank and nonbank traders and dealers profit from buying foreign exchange at a *bid* price and reselling it at a slightly higher *ask* (also called an *offer*) price. Competition among dealers worldwide narrows the spread between bids and offers and so contributes to making the foreign exchange market "efficient" in the same sense as are securities markets.

Dealers in the foreign exchange departments of large international banks often function as "market makers." Such dealers stand willing at all times to buy and sell those currencies in which they specialize and thus maintain an "inventory" position in those currencies. They trade with other banks in their own monetary centers and with other centers around the world in order to maintain inventories within the trading limits set by bank policy. Trading limits are important because foreign exchange departments of many banks operate as profit centers, and individual dealers are compensated on a profit incentive basis.

Currency trading is quite profitable for many institutions. Many of the major currency-trading banks in the United States derive on average between 10% and 20% of their annual net income from currency trading. Currency trading is also very profitable for the bank's traders who typically earn a bonus based on the profitability to the bank of their individual trading activities.

Small- to medium-size banks and institutions are likely to participate but not to be market makers in the interbank market. Instead of maintaining significant inventory positions, they often buy from and sell to larger institutions in order to offset retail transactions with their own customers or to seek short-term profits for their own accounts. *Global Finance in Practice 5.2* provides some insight into one "newbie's" experience on a currency trading desk.

Individuals and Firms Conducting Commercial and Investment Transactions

Importers and exporters, international portfolio investors, multinational corporations, tourists, and others use the foreign exchange market to facilitate execution of commercial or investment transactions. Their use of the foreign exchange market is necessary, but incidental, to their underlying commercial or investment purpose. Some of these participants use the market to hedge foreign exchange risk as well.

Speculators and Arbitragers

Speculators and arbitragers seek to profit from trading within the market itself. True profit seekers, they operate in their own interest, without a need or obligation to serve clients or to ensure a continuous market. Whereas dealers seek profit from the spread between bids and offers in addition to what they might gain from changes in exchange rates, speculators seek

---

GLOBAL FINANCE IN PRACTICE 5.2

My First Day of Foreign Exchange Trading

For my internship I was working for the Treasury Front and Back Office of a major investment bank's New York branch on Wall Street. I was, for the first half of my internship, responsible for the timely input and verification of all foreign exchange, money market, securities, and derivative products. The second half was more interesting. I received training in currency trading.

I started on the spot desk, worked there for two weeks, and then moved to the swap desk for the remaining three weeks of my internship. From the first day I knew I would have to stay on my toes. The first two weeks of my training I was assigned to the spot desk where my supervisor was a senior trader who was very young (only 23 years old) and extremely ambitious.

On the very first day, about 11 A.M., she bet on the rise of the Japanese yen after the elections of the new Japanese Prime Minister. She had a long position on the yen and was short on the dollar. Unfortunately she lost $700,000 in less than 10 minutes. It is still unclear to me why she made such a bet. *The Wall Street Journal* and the *Financial Times* (both papers were used heavily in the trading room) were very negative regarding the new Prime Minister's ability to reverse the financial crisis in Japan. It was clear that her position was based purely on emotions, instinct, savvy—anything but fundamentals.

To understand the impact of a $700,000 loss, you must understand that every trader on a spot desk has to make eight times his or her wage in commission. Let's say that my supervisor was making $80,000 a year. She would then need to make $640,000 in commission on spreads during that year to keep her job. A loss of $700,000 put her in a very bad position, and she knew it. But to her credit, she remained quite confident and did not appear shaken.

But after my first day I was pretty shaken. I understood after this experience that being a trader was not my cup of tea. It is not because of the stress of the job—and it is obviously very stressful. It was more that most of the skills of the job had nothing to do with what I had been learning in school for many years. And when I saw and experienced how hard these people partied up and down the streets of New York many nights—then trading hundreds of millions of dollars in minutes the following day—well, I just did not see this as my career track.

Source: Reminiscences of an anonymous intern.

all of their profit from exchange rate changes. Arbitragers try to profit from simultaneous exchange rate differences in different markets.

Traders employed by those banks conduct a large proportion of speculation and arbitrage on behalf of major banks. Thus, banks act both as exchange dealers and as speculators and arbitragers. (Banks seldom admit to speculating, instead characterizing themselves as "taking an aggressive position.")

Central Banks and Treasuries

Central banks and treasuries use the market to acquire or spend their country's foreign exchange reserves as well as to influence the price at which their own currency is traded, a practice known as *foreign exchange intervention*. They may act to support the value of their own currency because of national policies or because of commitments to other countries under exchange rate relationships or regional currency agreements. Consequently, the motive is not to earn a profit as such, but rather to influence the foreign exchange value of their currency in a manner that will benefit the interests of their citizens. In many instances, they do their job best when they willingly take a loss on their foreign exchange transactions. As willing loss takers, central banks and treasuries differ in motive and behavior from all other market participants.

Foreign Exchange Brokers

Foreign exchange brokers are agents who facilitate trading between dealers without themselves becoming principals in the transaction. They charge a small commission for this service. They maintain instant access to hundreds of dealers worldwide via open telephone lines. At times, a broker may maintain a dozen or more such lines to a single client bank, with separate lines for different currencies and for spot and forward markets.

It is a broker's business to know at any moment exactly which dealers want to buy or sell any currency. This knowledge enables the broker to find an opposite party for a client without revealing the identity of either party until after a transaction has been agreed upon. Dealers use brokers to expedite the transaction and to remain anonymous, since the identity of participants may influence short-term quotes.

Continuous Linked Settlement

In 2002, the *Continuous Linked Settlement* (CLS) system was introduced. The CLS system eliminates losses if either party of a foreign exchange transaction is unable to settle with the other party. It links the number of settlement systems, which operate on a real-time basis, and is expected to eventually result in same-day settlement, replacing the traditional two-day transaction period.

The CLS system should help counteract fraud in the foreign exchange markets as well. In the United States, the responsibility for regulating foreign exchange trading is assigned to the *U.S. Commodity Futures Trading Commission* (CFTC).

Transactions in the Foreign Exchange Market

Transactions in the foreign exchange market can be executed on a spot, forward, or swap basis. A broader definition of the market, one including major derivatives, would include foreign currency options, futures, and swaps.

Spot Transactions

A *spot transaction* in the interbank market is the purchase of foreign exchange with delivery and payment between banks taking place normally on the second following business day. The Canadian dollar settles with the U.S. dollar on the first following business day. Exhibit 5.2

EXHIBIT 5.2   **Foreign Exchange Transactions and Settlement**

Foreign exchange operations are defined by the timing—*the future date*—set for delivery. There are in principle three major categories of over-the-counter transactions categorized by future delivery: *spot* (which may be *overnight*), *forward* (including *outright forward*), and *swap transactions*.

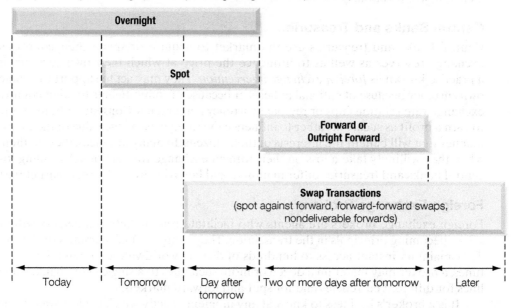

provides a time-map of the three major types of over-the-counter currency transactions typically executed in the global foreign exchange market: *spot transactions, forward transactions,* and *swap transactions*. Although there are a number of deviations on these types, some of which are described in the section that follows, all transactions are defined by their future date for delivery. (Note that we are not including futures transactions here; they parallel the time footprint of forwards, but are not executed over-the-counter.)

The date of settlement is referred to as the *value date*. On the value date, most dollar transactions in the world are settled through the computerized *Clearing House Interbank Payments System* (CHIPS) in New York, which calculates net balances owed by any one bank to another and which facilitates payment of those balances by 6:00 P.M. that same day in Federal Reserve Bank of New York funds. Other central banks and settlement services providers operate similarly in other currencies around the world.

A typical spot transaction in the interbank market might involve a U.S. bank contracting on a Monday for the transfer of £10,000,000 to the account of a London bank. If the spot exchange rate were $1.8420 to each British pound (£), the U.S. bank would transfer £10,000,000 to the London bank on Wednesday, and the London bank would transfer $18,420,000 to the U.S. bank at the same time. A spot transaction between a bank and its commercial customer would not necessarily involve a wait of two days for settlement.

Outright Forward Transactions

A *forward transaction* (or more formally, an *outright forward transaction*) requires delivery at a future value date of a specified amount of one currency for a specified amount of another currency. The exchange rate is established at the time of the agreement, but payment and delivery are not required until maturity. Forward exchange rates are normally quoted for value dates of one, two, three, six, and twelve months. Although the heaviest demand is for maturities of one year or less, forwards today are often quoted as far as 20 years into the future. According to the IMF, in 2014 there were 127 countries with forward markets.

Payment on forward contracts is on the second business day after the even-month anniversary of the trade. Thus, a two-month forward transaction entered into on March 18 will be for a value date of May 20, or the next business day if May 20 falls on a weekend or holiday.

Note that as a matter of terminology we can speak of "buying forward" or "selling forward" to describe the same transaction. A contract to deliver dollars for euros in six months is both "buying euros forward for dollars" and "selling dollars forward for euros."

Swap Transactions

A *swap transaction* in the interbank market is the simultaneous purchase and sale of a given amount of foreign exchange for two different value dates. Both purchase and sale are conducted with the same counterparty. There are several types of swap transactions.

Spot Against Forward. The most common type of swap is a *spot against forward*. The dealer buys a currency in the spot market (at the *spot rate*) and simultaneously sells the same amount back to the same bank in the forward market (at the forward exchange rate). Since this is executed as a single transaction with just one counterparty, the dealer incurs no unexpected foreign exchange risk. Swap transactions and outright forwards combined made up more than half of all foreign exchange market activity in recent years.

Forward-Forward Swaps. A more sophisticated swap transaction is called a *forward-forward swap*. For example, a dealer sells £20,000,000 forward for dollars for delivery in, say, two months at $1.8420/£ and simultaneously buys £20,000,000 forward for delivery in three months at $1.8400/£. The difference between the buying price and the selling price is equivalent to the interest rate differential, which is the interest rate parity described in Chapter 6, between the two currencies. Thus, a swap can be viewed as a technique for borrowing another currency on a fully collateralized basis.

Nondeliverable Forwards (NDFs). Created in the early 1990s, the *nondeliverable forward* (NDF), is now a relatively common derivative offered by the largest providers of foreign exchange derivatives. NDFs possess the same characteristics and documentation requirements as traditional forward contracts, except that they are settled only in U.S. dollars; the foreign currency being sold forward or bought forward is not delivered. The dollar-settlement feature reflects the fact that NDFs are contracted offshore, for example in New York for a Mexican investor, and so are beyond the reach and regulatory frameworks of the home country governments (Mexico in this case). NDFs are traded internationally using standards set by the *International Swaps and Derivatives Association* (ISDA). Although originally envisioned to be a method of currency hedging, it is now estimated that more than 70% of all NDF trading is for speculation purposes.

NDFs are used primarily for emerging market currencies or currencies subject to significant exchange controls, like Venezuela. Emerging market currencies often do not have open spot market currency trading, liquid money markets, or quoted Eurocurrency interest rates. Although most NDF trading focused on Latin America in the 1990s, many Asian currencies—including the Chinese renminbi—have been very widely traded in recent years. In general, NDF markets normally develop for country currencies having large cross-border capital movements, but still being subject to convertibility restrictions.

Pricing of NDFs reflects basic interest differentials, as with regular forwards, plus some additional premium charged by the bank for dollar settlement. If, however, there is no accessible or developed money market for interest rate setting, the pricing of the NDF takes on a much more speculative element. Without true interest rates, traders may price NDFs based on what they believe spot rates may be in the future.

NDFs are traded and settled outside the country of the subject currency, and therefore are beyond the control of the country's government. In the past, this has created a difficult situation, in which the NDF market serves as something of a gray market in the trading of that currency. For example, in late 2001, Argentina was under increasing pressure to abandon its fixed exchange rate regime of one peso equaling one U.S. dollar. The NDF market began quoting rates of ARS1.05/USD and ARS1.10/USD, in effect a devalued peso, for NDFs settling within the next year. This led to increasing speculative pressure against the peso (to the ire of the Argentine government).

NDFs, however, have proven to be something of an imperfect replacement for traditional forward contracts. The problems with the NDF typically involve its "fixing of spot rate on the fixing date," the spot rate at the end of the contract used to calculate the settlement. In times of financial crisis, for example with the Venezuelan bolivar in 2003, the government of the subject currency may suspend foreign exchange trading in the spot market for an extended period. Without an official fixing rate, the NDF cannot be settled. In the case of Venezuela, the problem was compounded when a new official "devalued bolivar" was announced, but was still not traded.

Size of the Foreign Exchange Market

The *Bank for International Settlements* (BIS), in conjunction with central banks around the world, conducts a survey of currency trading activity every three years. The most recent survey, conducted in April 2013, estimated daily global net turnover in the foreign exchange market to be $5.3 trillion. The BIS data for surveys between 1989 and 2013 is shown in Exhibit 5.3.

EXHIBIT 5.3	Global Foreign Exchange Market Turnover, 1989–2013

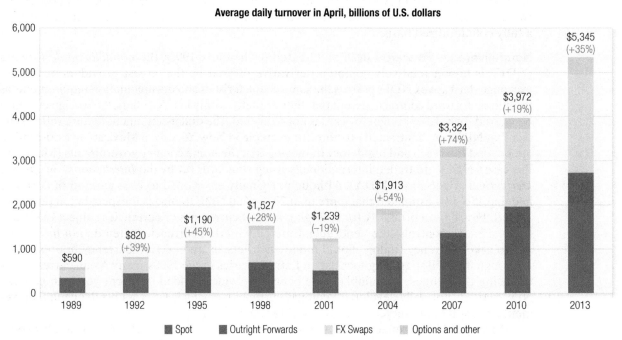

Source: Bank for International Settlements, "Triennial Central Bank Survey: Foreign Exchange and Derivatives Market Activity in April 2013: Preliminary Results," December 2013, www.bis.org.

Global foreign exchange turnover in Exhibit 5.3 is divided into the three categories of currency instruments discussed previously (spot transactions, forward transactions, and swap transactions) plus a fourth category of options and other variable-value foreign exchange derivatives. Growth has been dramatic; since 1989, the foreign exchange market has grown at an average annual rate of 9.6% per year.

As of 2013 (daily trading in April), trading in the foreign exchange market was at an all-time high of $5.3 trillion per day. Although the global recession in 2000–2001 clearly dampened market activity, the global financial crisis of 2008–2009 did not. According to the BIS, the organization that collects and deciphers this data, the primary driver of rapid foreign exchange growth in recent years is the increasing profit seeker activity facilitated by electronic trading and access to the greater market.

Geographical Distribution

Exhibit 5.4 shows the proportionate share of foreign exchange trading for the most important national markets in the world between 1992 and 2013. (Note that although the data is collected and reported on a national basis, the "United States" and "United Kingdom" should largely

EXHIBIT 5.4   **Top 10 Geographic Trading Centers in the Foreign Exchange Market, 1992–2013**

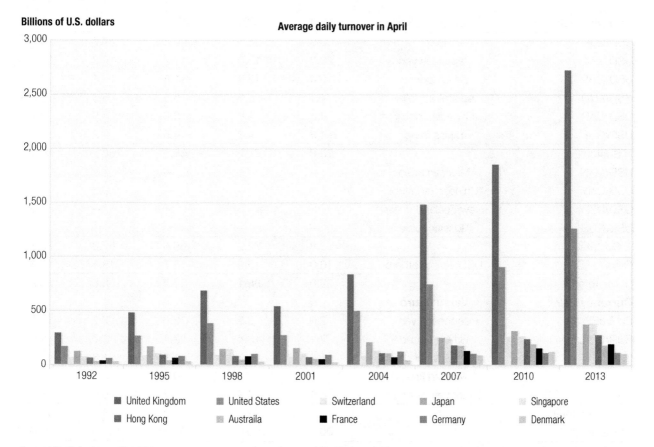

Source: Bank for International Settlements, "Triennial Central Bank Survey: Foreign Exchange and Derivatives Market Activity in April 2013: Preliminary Results," December 2013, www.bis.org.

be interpreted as "New York" and "London," respectively, because the majority of foreign exchange trading takes place in each country's major financial city.)

The United Kingdom (London) continues to be the world's major foreign exchange market in traditional foreign exchange market activity with 40.9% of the global market. The United Kingdom is followed by the United States with 18.9%, Singapore with 5.7%, Japan (Tokyo) with 5.6%, Switzerland with 3.2%, and Hong Kong now reaching 4.1% of global trading. Indeed, the United Kingdom and United States together make up nearly 60% of daily currency trading. The relative growth of currency trading in Asia versus Europe over the past 15 years is pronounced, as the growth of the Asian economies and markets has combined with the introduction of the euro to shift currency exchange activity.

Currency Composition

The currency composition of trading, as shown in Exhibit 5.5, also indicates significant global shifts. Because all currencies are traded against some other currency pairs, all percentages shown in Exhibit 5.5 are for that currency versus another. The dollar continues to dominate

EXHIBIT 5.5 **Daily FX Trading by Currency Pair** (percent of total)

Currency Pair	Versus Dollar	2001	2004	2007	2010	2013
USD/EUR	Euro	30.0	28.0	26.8	27.7	24.1
USD/JPY	Japanese yen	20.2	17.0	13.2	14.3	18.3
USD/GBP	British pound	10.4	13.4	11.6	9.1	8.8
USD/AUD	Australian dollar	4.1	5.5	5.6	6.3	6.8
USD/CAD	Canadian dollar	4.3	4.0	3.8	4.6	3.7
USD/CHF	Swiss franc	4.8	4.3	4.5	4.2	3.4
Subtotal		73.8	72.2	65.5	66.2	65.1
USD/MXN	Mexican peso	—	—	—	—	2.4
USD/CNY	Chinese renminbi	—	—	—	0.8	2.1
USD/NZD	New Zealand dollar	—	—	—	—	1.5
USD/RUB	Russian ruble	—	—	—	—	1.5
Subtotal		73.8	72.2	65.5	67.0	72.6
Other/USD	USD versus others	16.0	15.9	16.7	18.8	13.3
Dollar Total		89.8	88.1	82.2	85.8	85.9
Currency Pair	Versus Euro					
EUR/JPY	Japanese yen	2.9	3.2	2.6	2.8	2.8
EUR/GBP	British pound	2.1	2.4	2.1	2.7	1.9
EUR/CHF	Swiss franc	1.1	1.6	1.9	1.8	1.3
EUR/SEK	Swedish krona	—	—	0.7	0.9	0.5
Other	Other versus other	4.1	4.7	11.2	6.9	8.1
Non-dollar total		10.2	11.9	17.8	14.2	14.1
Global Total		100.0	100.0	100.0	100.0	100.0

Source: Constructed by authors based on data presented in Table 3, p. 11, of Triennial Central Bank Survey, Foreign exchange turnover in April 2013: preliminary global results, Bank for International Settlements, Monetary and Economic Department, September 2013.

global trading, ultimately involved in 85.9% of all currency trading. The USD/EUR makes up 24.1% of trading, followed by the USD/JPY with 18.3%, the USD/GBP with 8.8%, and the USD/AUD at 6.8%. According to the BIS, the "big three" (dollar, euro, and yen) continue to dominate global currency trades, totaling roughly 92% of all trading surveyed.

There is, however, growing awareness of the rapid development of several major emerging market currencies, namely the Mexican peso, the Chinese renminbi, and the Russian ruble. It may not be long before several of these (most analysts are betting on the renminbi) become prominent currencies in the global market.

Foreign Exchange Rates and Quotations

A *foreign exchange rate* is the price of one currency expressed in terms of another currency. A foreign exchange quotation (or quote) is a statement of willingness to buy or sell at an announced rate. As we delve into the terminology of currency trading, keep in mind basic pricing, say the pricing of an orange. If the price is $1.20/orange, the "price" is $1.20, the "unit" is the orange.

Currency Symbols

Quotations may be designated by traditional currency symbols or by ISO codes. ISO—the International Organization for Standardization—is the world's largest developer of voluntary standards. ISO 4217 is the International Standard for currency codes, most recently updated in ISO 4217:2008.

The ISO codes were developed for use in electronic communications. Both traditional symbols and currency codes are given in full at the end of this book, but the major ones used throughout this chapter are the following:

Currency	Traditional Symbol	ISO 4217 Code
U.S. dollar	$	USD
European euro	€	EUR
Great Britain pound	£	GBP
Japanese yen	¥	JPY
Mexican peso	Ps	MXN

Today, all electronic trading of currencies between institutions in the global marketplace uses the three-letter ISO codes. Although there are no hard and fast rules in the retail markets and in business periodicals, European and American periodicals have a tendency to use the traditional currency symbols, while many publications in Asia and the Middle East have embraced the use of ISO codes. The paper currency (banknotes) of most countries continues to be represented using the country's traditional currency symbol. As illustrated in *Global Finance in Practice 5.3*, some countries like Russia are trying to return to traditional symbol use.

Exchange Rate Quotes

Foreign exchange quotations follow a number of principles, which at first may seem a bit confusing or nonintuitive. Every currency exchange involves two currencies, currency 1 (CUR1) and currency 2 (CUR2):

$$\textbf{CUR1/CUR2}$$

The currency to the left of the slash is called the *base currency* or the *unit currency*. The currency to the right of the slash is called the *price currency* or *quote currency*. The quotation

GLOBAL FINANCE IN PRACTICE 5.3

Russian Symbolism

During an era in which currencies have increasingly been identified by their 3-digit ISO codes, the Russian government has decided that it is time for the Russian ruble (or rouble depending on your preference) to have its own symbol.

 In December 2013, the Bank of Russia had a contest, a popular vote, to choose among five different symbolic choices to be the new face of the currency. The winner (at left), with 61% of the vote, was the Russian letter R. The new symbol, in the words of the Governor of the Bank of Russia, "embodied the stability and reliability of the currency."

When asked if the new ruble symbol may end up being confused with the Latin letter P, the Governor said it wasn't a problem because the dollar sign looks like the letter S.

The ruble's new symbol now joins that historical list of symbols—the $ (U.S. dollar), £ (British pound sterling), ¥ (Japanese yen), and the relatively youthful € (European Union euro)—as a declaration of currency value. More and more countries of late have promoted their own currency symbols in a show of nationalistic pride. India adopted a new symbol for the rupee in 2010 (₹) and Turkey's lira got its own new symbol in 2012 (₺).

always indicates the number of units of the price currency, CUR2, required in exchange for receiving one unit of the base currency, CUR1.

For example, the most commonly quoted currency exchange is that between the U.S. dollar and the European euro. For example, a quotation of

$$\text{EUR/USD1.2174}$$

designates the euro (EUR) as the *base currency*, the dollar (USD) as the *price currency*, and the exchange rate is USD 1.2174 = EUR 1.00. If you can remember that the currency quoted on the left of the slash is always the base currency, and always a single unit, you can avoid confusion. Exhibit 5.6 provides a brief overview of the multitude of terms often used around the world to quote currencies, through an example using the European euro and U.S. dollar.

Market Conventions

The international currency market, although the largest financial market in the world, is steeped in history and convention.

European Terms. European terms, the quoting of the quantity of a specific currency per one U.S. dollar, has been market practice for most of the past 60 years or more. Globally, the base

---

EXHIBIT 5.6   Foreign Currency Quotations

European terms Foreign currency price of one dollar (USD)	American terms U.S. dollar price of one euro (EUR)
USD/EUR 0.8214 or **USD 1.00 = EUR 0.8214**	**EUR/USD 1.2174** or **EUR 1.00 = USD 1.2174**
USD is the base or unit currency *EUR is the quote or price currency*	*EUR is the base or unit currency* *USD is the quote or price currency*

$$\frac{1}{\text{EUR } 0.8214 \text{ / USD}} = \text{USD } 1.2714 \text{ / EUR}$$

currency used to quote a currency's value has typically been the U.S. dollar. Termed *European terms*, this means that whenever a currency's value is quoted, it is quoted in terms of number of units of currency to equal one U.S. dollar.

For example, if a trader in Zurich, whose home currency is the Swiss franc (CHF), were to request a quote from an Oslo-based trader on the Norwegian krone (NOK), the Norwegian trader would quote the value of the NOK against the USD, not the CHF. The result is that most currencies are quoted per U.S. dollar — Japanese yen per U.S. dollar, Norwegian krone per U.S. dollar, Mexican pesos per U.S. dollar, Brazilian real per U.S. dollar, Malaysian ringgit per U.S. dollar, Chinese renminbi per U.S. dollar, and so on.

American Terms. There are two major exceptions to this rule of using European terms: the euro and the U.K. pound sterling (the pound sterling for historical tradition). Both are normally quoted in *American terms* — the U.S. dollar price of one euro and the U.S. dollar price of one pound sterling. Additionally, Australian dollars and New Zealand dollars are normally quoted on American terms.

For centuries, the British pound sterling consisted of 20 shillings, each of which equaled 12 pence. Multiplication and division with the nondecimal currency were difficult. The custom evolved for foreign exchange prices in London, then the undisputed financial capital of the world, to be stated in foreign currency units per pound. This practice remained even after sterling changed to decimals in 1971.

The euro was first introduced as a substitute or replacement for domestic currencies like the Deutsche mark and French franc. To make the transition simple for residents and users of these historical currencies, all quotes were made on a "domestic currency per euro" basis. This held true for its quotation against the U.S. dollar; hence, "U.S. dollars per euro" is the common quotation used today.

American terms are also used in quoting rates for most foreign currency options and futures, as well as in retail markets that deal with tourists and personal remittances. Again, this is largely a result of established practices that have been perpetuated over time, rather than some basic law of nature or finance.

Currency Nicknames. Foreign exchange traders may also use nicknames for major currencies. "Cable" means the exchange rate between U.S. dollars and U.K. pounds sterling, the name dating from the time when transactions in dollars and pounds were carried out over the Transatlantic telegraph cable. A Canadian dollar is a "loonie," named after the water fowl on Canada's one dollar coin. "Kiwi" stands for the New Zealand dollar, "Aussie" for the Australian dollar, "Swissie" for Swiss francs, and "Sing dollar" for the Singapore dollar.

Direct and Indirect Quotations. A *direct quote* is the price of a foreign currency in domestic currency units. An *indirect quote* is the price of the domestic currency in foreign currency units.

In retail exchange in many countries (such as currency exchanged in hotels or airports), it is common practice to quote the home currency as the price and the foreign currency as the unit. A woman walking down the Avenue des Champs-Elysèes in Paris might see the following quote:

EUR 0.8214 = USD 1.00

Since in France the *home currency* is the euro (the price) and the *foreign currency* is the dollar (the unit), in Paris this quotation is a *direct quote on the dollar* or a *price quote on the dollar*. She might say to herself, "0.8214 euros per dollar," or "it will cost me 0.8214 euros to get one dollar." These are European terms.

At the same time a man walking down Broadway in New York City may see the following quote in a bank window:

USD 1.2174 = EUR 1.00

Since in the U.S. the *home currency* is the dollar (the price) and the *foreign currency* is the euro (the unit), in New York this would be a *direct quote on the euro* (the home currency price of one unit of foreign currency) and an *indirect quote on the dollar* (the foreign currency price of one unit of home currency). The man might say to himself, "I will pay $1.2174 dollars per euro." These are American terms. The two quotes are obviously equivalent (at least to four decimal places), one being the reciprocal of the other:

$$\frac{1}{\text{EUR } 0.8214 \text{ / USD}} = \textbf{USD 1.2174 / EUR}$$

Bid and Ask Rates. Although a newspaper or magazine article will state an exchange rate as a single value, the market for buying and selling currencies, whether it be retail or wholesale, uses two different rates, one for buying and one for selling. Exhibit 5.7 provides a sample of how these quotations may be seen in the market for the dollar/euro.

A *bid* is the price (i.e., exchange rate) in one currency at which a dealer will buy another currency. An *ask* is the price (i.e., exchange rate) at which a dealer will sell the other currency. Dealers bid (buy) at one price and ask (sell) at a slightly higher price, making their profit from the spread between the prices. The *bid-ask spread* may be quite large for currencies that are traded infrequently, in small volumes, or both.

Bid and ask quotations in the foreign exchange markets are superficially complicated by the fact that the bid for one currency is also the offer for the opposite currency. A trader seeking to buy dollars with euros is simultaneously offering to sell euros for dollars. Closing rates for 47 currencies (plus the SDR) as quoted by *The Wall Street Journal* are presented in Exhibit 5.8.

EXHIBIT 5.7   Bid, Ask, and Mid-Point Quotation

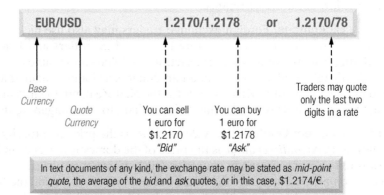

For example, the *Wall Street Journal* would quote the following currencies as follows:

	Last Bid		Last Bid
Euro (EUR/USD)	1.2170	Brazilian Real (USD/BRL)	1.6827
Japanese Yen (USD/JPY)	83.16	Canadian Dollar (USD/CAD)	0.9930
U.K. Pound (GBP/USD)	1.5552	Mexican Peso (USD/MXN)	12.2365

EXHIBIT 5.8	**Exchange Rates: New York Closing Snap shot**

February 18, 2015

Country	Currency	Symbol	Code	USD equivalent	Currency per USD
Americas					
Argentina	peso	Ps	ARS	0.1151	8.6904
Brazil	real	R$	BRL	0.3529	2.8338
Canada	dollar	C$	CAD	0.8029	1.2455
Chile	peso	$	CLP	0.001614	619.6
Colombia	peso	Col$	COP	0.0004114	2430.74
Ecuador	U.S. dollar	$	USD	1	1
Mexico	new peso	$	MXN	0.0673	14.8638
Peru	new sol	S/.	PEN	0.3241	3.0855
Uruguay	peso	$U	UYU	0.04062	24.62
Venezuela	boliviar fuerte	Bs	VND	0.15885497	6.2951
Asia-Pacific					
Australia	dollar	A$	AUD	0.7812	1.2801
China	yuan	¥	CNY	0.1598	6.2564
Hong Kong	dollar	HK$	HKG	0.1289	7.7586
India	rupee	₹	INR	0.01611	62.0568
Indonesia	rupiah	Rp	IDR	0.0000778	12857
Japan	yen	¥	JPY	0.00842	118.79
Malaysia	ringgit	RM	MYR	0.2764	3.6175
New Zealand	dollar	NZ$	NZD	0.7545	1.3254
Pakistan	rupee	Rs.	PKR	0.00986	101.45
Philippines	peso	₱	PHP	0.0226	44.2375
Singapore	dollar	S$	SGD	0.7376	1.3558
South Korea	won	W	KRW	0.000902	1108.62
Taiwan	dollar	T$	TWD	0.03159	31.65
Thailand	baht	B	THB	0.0307	32.57
Vietnam	dong	d	VND	0.00004688	21333
Europe					
Czech Republic	koruna	Kc	CZK	0.04182	23.912
Denmark	krone	Dkr	DKK	0.1531	6.5326
Euro area	euro	€	EUR	1.1398	0.8774
Hungary	forint	Ft	HUF	0.00372537	268.43
Iceland	krona	kr	ISK	0.007569	132.11
Norway	krone	NKr	NOK	0.1328	7.5279
Poland	zloty	—	PLN	0.2724	3.671
Romania	leu	L	RON	0.2562	3.9039
Russia	ruble	₽	RUB	0.01624	61.579
Sweden	krona	SKr	SEK	0.1196	8.3619
Switzerland	franc	Fr.	CHF	1.061	0.9425
Turkey	lira	₺	TRY	0.4093	2.4433
United Kingdom	pound	£	GBP	1.5438	0.6478
Middle East/Africa					
Bahrain	dinar	—	BHD	2.6524	0.377
Egypt	pound	£	EGP	0.1312	7.6241
Israel	shekel	Shk	ILS	0.2595	3.8542
Kuwait	dinar	—	KWD	3.3835	0.2956
Saudi Arabia	riyal	SR	SAR	0.2665	3.7518
South Africa	rand	R	ZAR	0.0862	11.5995
United Arab Emirates	dirham	—	AED	0.2723	3.673

Note: SDR from the International Monetary Fund; based on exchange rates for U.S., British, and Japanese currencies.

Quotes based on trading among banks of $1 million and more, as quoted at 4 P.M. ET by Reuters. Rates are drawn from *The Wall Street Journal* online on February 19, 2015.

The Wall Street Journal gives American terms quotes under the heading "USD equivalent" and European terms quotes under the heading "Currency per USD." Quotes are given on an outright basis for spot, with forwards of one, three, and six months provided for a few select currencies. Quotes are for trading among banks in amounts of $1 million or more, as quoted at 4 P.M. EST by Reuters. The *Journal* does not state whether these are bid, ask, or midrate (an average of the bid and ask) quotations.

The order of currencies in quotations used by traders can be confusing (at least the authors of this book think so). As noted by one major international banking publication: *The notation EUR/USD is the system used by traders, although mathematically it would be more correct to express the exchange rate the other way around, as it shows how many USD have to be paid to obtain EUR 1.*

This is why the currency quotes shown previously in Exhibit 5.7—like EUR/USD, USD/JPY, or GBP/USD—are quoted and used in business and the rest of this text as $1.2170/€, ¥83.16/$, and $1.5552/£. International finance is not for the weak of heart!

Cross Rates

Many currency pairs are only inactively traded, so their exchange rate is determined through their relationship to a widely traded third currency. For example, a Mexican importer needs Japanese yen to pay for purchases in Tokyo. Both the Mexican peso (MXN or the old peso symbol, Ps) and the Japanese yen (JPY or ¥) are commonly quoted against the U.S. dollar (USD or $). Using the following quotes from Exhibit 5.8,

		Currency per USD
Japanese yen	USD/JPY	118.79
Mexican peso	USD/MXN	14.8638

the Mexican importer can buy one U.S. dollar for MXN14.8638, and with that dollar can buy JPY118.79. The *cross rate* calculation would be as follows:

$$\frac{\text{Japanese Yen} = 1 \text{ U.S. Dollar}}{\text{Mexican Peso} = 1 \text{ U.S. Dollar}} = \frac{¥118.79/\$}{Ps14.8638/\$} = ¥7.9919/Ps$$

The cross rate could also be calculated as the reciprocal, with the USD/MXN rate divided by the USD/JPY rate, yielding Ps0.1251/¥.

Cross rates often appear in various financial publications in the form of a matrix to simplify the math. Exhibit 5.9 calculates a number of key cross rates from the quotes presented in Exhibit 5.8, including the Mexican peso/Japanese yen calculation just described (the Ps0.1251/¥ rate).

Intermarket Arbitrage

Cross rates can be used to check on opportunities for intermarket arbitrage. Suppose the following exchange rates are quoted:

Citibank quotes U.S. dollars per euro	USD1.3297 = 1 EUR
Barclays Bank quotes U.S. dollars per pound sterling	USD1.5585 = 1 GBP
Dresdner Bank quotes euros per pound sterling	EUR1.1722 = 1 GBP

The euro-pound sterling cross rate, derived from the Citibank and Barclays Bank quotes, is

$$\frac{USD1.5585/GBP}{USD1.3297/EUR} = EUR\ 1.721/GBP$$

EXHIBIT 5.9	Key Currency Cross-Rate Calculations for February 18, 2015

				Calculated			
	Dollar	**Euro**	**Pound**	**SFranc**	**Peso**	**Yen**	**CdnDlr**
Canada	1.2455	1.4195	1.9227	1.3215	0.0838	0.0105	—
Japan	118.786	135.384	183.368	126.033	7.9916	—	95.372
Mexico	14.8638	16.9407	22.9450	15.7706	—	0.1251	11.9340
Switzerland	0.9425	1.0742	1.4549	—	0.0634	0.0079	0.7567
U.K.	0.6478	0.7383	—	0.6873	0.0436	0.0055	0.5201
Euro	0.8774	—	1.3544	0.9309	0.0590	0.007	0.704
U.S.	—	1.1397	1.5437	1.0610	0.0673	0.0084	0.8029

Note: Cross-rates are calculated from the quotes presented in the first column "Dollar."

Note that the calculated cross rate is *not the same* as Dresdner Bank's quotation of EUR1.1722/ GBP, so an opportunity exists to profit from arbitrage between the three markets. Exhibit 5.10 shows the steps in what is called *triangular arbitrage*.

A market trader at Citibank New York, with USD1,000,000, can sell that sum spot to Barclays Bank for British pounds sterling, and then in turn, sell these pounds to Dresdner Bank for euros. In the third and final simultaneous sale, the trader can sell the euros to Citibank for USD1,000,112.

The profit on one such "turn" is a risk-free USD112: $1,000,112 − $1,000,000. We know, that's not much, but it's digital! Such triangular arbitrage can continue until exchange rate equilibrium is reestablished. "Reestablished" in this case means when the calculated cross rate equals the actual quotation, less any tiny margin for transaction costs.

EXHIBIT 5.10	Triangular Arbitrage by a Market Trader

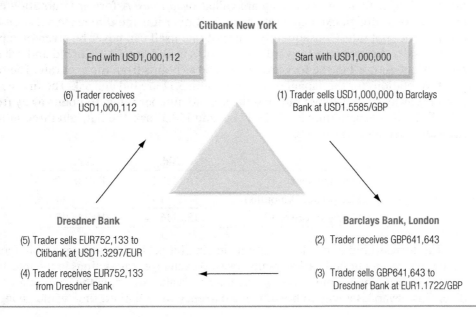

Citibank New York

End with USD1,000,112   Start with USD1,000,000

(6) Trader receives
USD1,000,112

(1) Trader sells USD1,000,000 to Barclays
Bank at USD1.5585/GBP

Dresdner Bank

(5) Trader sells EUR752,133 to
Citibank at USD1.3297/EUR

(4) Trader receives EUR752,133
from Dresdner Bank

Barclays Bank, London

(2) Trader receives GBP641,643

(3) Trader sells GBP641,643 to
Dresdner Bank at EUR1.1722/GBP

Percentage Change in Spot Rates

Assume that the Mexican peso has recently changed in value from USD/MXN 10.00 to 11.00. Your home currency is the U.S. dollar. What is the percent change in the value of the Mexican peso? The calculation depends upon the designated home currency.

Foreign Currency Terms. When the foreign currency price (the price, Ps) of the home currency (the unit, $) is used, Mexican pesos per U.S. dollar in this case, the formula for the percent change (%Δ) in the foreign currency becomes

$$\% \Delta = \frac{\text{Beginning Rate} - \text{Ending Rate}}{\text{Ending Rate}} \times 100 = \frac{\text{Ps10.00/\$} - \text{Ps11.00/\$}}{\text{Ps11.00/\$}} \times 100 = -9.09\%$$

The Mexican peso fell in value 9.09% against the dollar. Note that it takes more pesos per dollar, and the calculation resulted in a negative value, both characteristics of a fall in value.

Home Currency Terms. When the home currency price (the price) for a foreign currency (the unit) is used—the reciprocals of the numbers above—the formula for the percent change in the foreign currency is:

$$\% \Delta = \frac{\text{Ending Rate} - \text{Beginning Rate}}{\text{Beginning Rate}} \times 100 = \frac{\$0.09091/\text{Ps} - \$0.1000/\text{Ps}}{\$0.1000/\text{Ps}} \times 100 = -9.09\%$$

The calculation yields the identical percentage change, a fall in the value of the peso by 9.09%. Although many people find the second calculation, the home currency term calculation, to be the more "intuitive" because it reminds them of many percentage change calculations, one must be careful to remember that these are exchanges of currency for currency, and the currency that is designated as home currency is significant.

Forward Quotations

Although spot rates are typically quoted on an outright basis (meaning all digits expressed), forward rates are, depending on the currency, typically quoted in terms of points or *pips*, the last digits of a currency quotation. Forward rates of one year or less maturity are termed *cash rates*; for longer than one-year they are called *swap rates*. A forward quotation expressed in points is not a foreign exchange rate as such. Rather it is the *difference* between the forward rate and the spot rate. Consequently, the spot rate itself can never be given on a points basis.

Consider the spot and forward point quotes in Exhibit 5.11. The bid and ask spot quotes are outright quotes, but the forwards are stated as points from the spot rate. The three-month points quotations for the Japanese yen in Exhibit 5.11 are bid and ask. The first number refers to points away from the spot bid, and the second number refers to points away from the spot ask. Given the outright quotes of 118.27 bid and 118.37 ask, the outright three-month forward rates are calculated as follows:

	Bid	Ask
Outright spot	JPY118.27	JPY118.37
Plus points (3 months)	1.43	1.40
Outright forward	JPY116.84	JPY116.97

The forward bid and ask quotations in Exhibit 5.11 of two years and longer are called swap rates. As mentioned earlier, many forward exchange transactions in the interbank market involve a simultaneous purchase for one date and sale (reversing the transaction) for another date. This "swap" is a way to borrow one currency for a limited time while giving up the use

| EXHIBIT 5.11 | Spot and Forward Quotations for the Euro and Japanese Yen |

| | | Euro: Spot and Forward ($/€) | | | | Japanese Yen: Spot and Forward (¥/$) | | | |
| | | Bid | | Ask | | Bid | | Ask | |
	Term	Points	Rate	Points	Rate	Points	Rate	Points	Rate
	Spot		1.0897		1.0901		118.27		118.37
Cash	1 week	3	1.0900	4	1.0905	−10	118.17	−9	118.28
rates	1 month	17	1.0914	19	1.0920	−51	117.76	−50	117.87
	2 months	35	1.0932	36	1.0937	−95	117.32	−93	117.44
	3 months	53	1.0950	54	1.0955	−143	116.84	−140	116.97
	4 months	72	1.0969	76	1.0977	−195	116.32	−190	116.47
	5 months	90	1.0987	95	1.0996	−240	115.87	−237	116.00
	6 months	112	1.1009	113	1.1014	−288	115.39	−287	115.50
	9 months	175	1.1072	177	1.1078	−435	113.92	−429	114.08
	1 year	242	1.1139	245	1.1146	−584	112.43	−581	112.56
Swap	2 years	481	1.1378	522	1.1423	−1150	106.77	−1129	107.08
rates	3 years	750	1.1647	810	1.1711	−1748	100.79	−1698	101.39
	4 years	960	1.1857	1039	1.1940	−2185	96.42	−2115	97.22
	5 years	1129	1.2026	1276	1.2177	−2592	92.35	−2490	93.47

of another currency for the same time. In other words, it is a short-term borrowing of one currency combined with a short-term loan of an equivalent amount of another currency. The two parties could, if they wanted, charge each other interest at the going rate for each of the currencies. However, it is easier for the party with the higher-interest currency to simply pay the net interest differential to the other. The swap rate expresses this net interest differential on a points basis rather than as an interest rate.

Forward Quotations in Percentage Terms. The percent per annum deviation of the forward from the spot rate is termed the *forward premium*. However, as with the calculation of percentage changes in spot rates, the forward premium—which may be either a positive (a *premium*) or negative value (a *discount*)—depends upon the designated home (or base) currency.

Assume the following spot rate for our discussion of *foreign currency terms* and *home currency terms*.

	Foreign currency (price)/ home currency (unit)	Home currency (price)/ foreign currency (unit)
Spot rate	¥ 118.27/$	USD/JPY0.0084552
3-month forward	¥ 116.84/$	USD/JPY0.0085587

Foreign Currency Terms. Using the foreign currency as the price of the home currency (the unit), JPY/USD spot and forward rates, and 90 days forward, the forward premium on the yen (f^{JPY}) is calculated as follows:

$$f^{JPY} = \frac{\text{Spot} - \text{Forward}}{\text{Forward}} \times \frac{360}{90} \times 100 = \frac{118.27 - 116.84}{116.84} \times \frac{360}{90} \times 100 = +4.90\%$$

The sign is positive indicating that the Japanese yen is selling forward at a premium of 4.90% against the U.S. dollar.

Home Currency Terms. Using the home currency (the dollar) as the price for the foreign currency (the yen) and the reciprocals of the spot and forward rates from the previous calculation, the forward premium on the yen (f^{JPY}) is calculated as follows:

$$f^{\text{JPY}} = \frac{\text{Spot} - \text{Forward}}{\text{Forward}} \times \frac{360}{90} \times 100 = \frac{\dfrac{1}{116.84} - \dfrac{1}{118.27}}{\dfrac{1}{118.27}} \times \frac{360}{90} \times 100$$

such that

$$f^{\text{JPY}} = \frac{\text{Forward} - \text{Spot}}{\text{Spot}} \times \frac{360}{90} \times 100 = \frac{0.0085587 - 0.0084552}{0.0084552} \times \frac{360}{90} \times 100 = +4.90\%$$

Again, the result is identical to the previous premium calculation: a positive 4.90% premium of the yen against the dollar.

SUMMARY POINTS

- The three functions of the foreign exchange market are to transfer purchasing power, provide credit, and minimize foreign exchange risk.

- One of the biggest changes in the foreign exchange market in the past decade has been in its shift from a two-tier market (the interbank or wholesale market, and the client or retail market) to a single-tier market.

- Electronic platforms and the development of sophisticated trading algorithms have facilitated market access by traders of all kinds and sizes.

- Geographically the foreign exchange market spans the globe, with prices moving and currencies traded somewhere every hour of every business day.

- A foreign exchange rate is the price of one currency expressed in terms of another currency. A foreign exchange quotation is a statement of willingness to buy or sell currency at an announced price.

- Transactions within the foreign exchange market are executed either on a spot basis, requiring settlement

two days after the transaction, or on a forward or swap basis, which requires settlement at some designated future date.

- European terms quotations are the foreign currency price of a U.S. dollar. American terms quotations are the dollar price of a foreign currency.

- Quotations can also be direct or indirect. A direct quote is the home currency price of a unit of foreign currency, while an indirect quote is the foreign currency price of a unit of home currency.

- Direct and indirect are not synonyms for American and European terms, because the home currency will change depending on who is doing the calculation, while European terms are always the foreign currency price of a dollar.

- A cross rate is an exchange rate between two currencies, calculated from their common relationships with a third currency. When cross rates differ from the direct rates between two currencies, intermarket arbitrage is possible.

The Venezuelan Bolivar Black Market[1]

It's late afternoon on March 10th, 2004, and Santiago opens the window of his office in Caracas, Venezuela. Immediately he is hit with the sounds rising from the plaza—cars honking, protesters banging their pots and pans, street vendors hawking their goods. Since the imposition of a new set of economic policies by President Hugo Chávez in 2002, such sights and sounds had become a fixture of city life in Caracas. Santiago sighed as he yearned for the simplicity of life in the old Caracas.

Santiago's once-thriving pharmaceutical distribution business had hit hard times. Since capital controls were implemented in February of 2003, dollars had been hard to come by. He had been forced to pursue various methods—methods that were more expensive and not always legal—to obtain dollars, causing his margins to decrease by 50%. Adding to the strain, the Venezuelan currency, the bolivar (Bs), had been recently devalued (repeatedly). This had instantly squeezed his margins as his costs had risen directly with the exchange rate. He could not find anyone to sell him dollars. His customers needed supplies and they needed them quickly, but how was he going to come up with the $30,000—the hard currency—to pay for his most recent order?

Political Chaos

Hugo Chávez's tenure as President of Venezuela had been tumultuous at best since his election in 1998. After repeated recalls, resignations, coups, and reappointments, the political turmoil had taken its toll on the Venezuelan economy as a whole, and its currency in particular. The short-lived success of the anti-Chávez coup in 2001, and his nearly immediate return to office, had set the stage for a retrenchment of his isolationist economic and financial policies.

On January 21st, 2003, the bolivar closed at a record low—Bs1853/$. The next day President Hugo Chávez suspended the sale of dollars for two weeks. Nearly instantaneously, an unofficial or black market for the exchange of Venezuelan bolivars for foreign currencies (primarily U.S. dollars) sprouted. As investors of all kinds sought ways to exit the Venezuelan market, or simply obtain the hard-currency needed to continue to conduct their businesses (as was the case for Santiago), the escalating capital flight caused the black market value of the bolivar to plummet to Bs2500/$ in weeks. As markets collapsed and exchange values fell, the Venezuelan inflation rate soared to more than 30% per annum.

Capital Controls and CADIVI

To combat the downward pressures on the bolivar, the Venezuelan government announced on February 5th, 2003, the passage of the 2003 Exchange Regulations Decree. The Decree took the following actions:

1. Set the official exchange rate at Bs1596/$ for purchase (bid) and Bs1600/$ for sale (ask);

2. Established the Comisin de Administracin de Divisas (CADIVI) to control the distribution of foreign exchange; and

3. Implemented strict price controls to stem inflation triggered by the weaker bolivar and the exchange control-induced contraction of imports.

CADIVI was both the official means and the cheapest means by which Venezuelan citizens could obtain foreign currency. In order to receive an authorization from CADIVI to obtain dollars, an applicant was required to complete a series of forms. The applicant was then required to prove that they had paid taxes the previous three years, provide proof of business and asset ownership and lease agreements for company property, and document the current payment of Social Security.

Unofficially, however, there was an additional unstated requirement for permission to obtain foreign currency: authorizations would be reserved for Chávez supporters. In August 2003 an anti-Chávez petition had gained widespread circulation. One million signatures had been collected. Although the government ruled that the petition was invalid, it had used the list of signatures to create a database of names and social security numbers that CADIVI utilized to cross-check identities on hard currency requests. President Chávez was quoted as saying "Not one more dollar for the putschits; the bolivars belong to the people."[2]

[1]Copyright © 2004 Thunderbird School of Global Management. All rights reserved. This case was prepared by Nina Camera, Thanh Nguyen, and Jay Ward under the direction of Professor Michael H. Moffett for the purpose of classroom discussion only and not to indicate either effective or ineffective management. Names of principals involved in the case have been changed to preserve confidentiality.
[2]"Venezuela Girds for Exchange Controls," *The Wall Street Journal* (Eastern edition), February 5, 2003, p. A14.

Santiago's Alternatives

Santiago had little luck obtaining dollars via CADIVI to pay for his imports. Because he had signed the petition calling for President Chávez's removal, he had been listed in the CADIVI database as anti-Chávez, and now could not obtain permission to exchange bolivar for dollars.

The transaction in question was an invoice for $30,000 in pharmaceutical products from his U.S.-based supplier. Santiago intended to resell these products to a large Venezuelan customer who would distribute the products. This transaction was not the first time that Santiago had been forced to search out alternative sources for meeting his U.S. dollar-obligations. Since the imposition of capital controls, his search for dollars had become a weekly activity for Santiago. In addition to the official process—through CADIVI—he could also obtain dollars through the gray or black markets.

The Gray Market: CANTV Shares

In May 2003, three months following the implementation of the exchange controls, a window of opportunity had opened up for Venezuelans—an opportunity that allowed investors in the Caracas stock exchange to avoid the tight foreign exchange curbs. This loophole circumvented the government-imposed restrictions by allowing investors to purchase local shares of the leading telecommunications company CANTV on the Caracas' bourse, and to then convert those shares into dollar-denominated American Depositary Receipts (ADRs) traded on the NYSE.

The sponsor for CANTV ADRs on the NYSE was the Bank of New York, the leader in ADR sponsorship and management in the U.S. The Bank of New York had suspended trading in CANTV ADRs in February after the passage of the Decree, wishing to determine the legality of trading under the new Venezuelan currency controls. On May 26th, after concluding that trading was indeed legal under the Decree, trading resumed in CANTV shares. CANTV's share price and trading volume both soared in the following week.[3]

The share price of CANTV quickly became the primary method of calculating the implicit gray market exchange rate. For example, CANTV shares closed at Bs7945/share on the Caracas bourse on February 6, 2004. That same day, CANTV ADRs closed in New York at $18.84/ADR. Each New York ADR was equal to seven shares of CANTV in Caracas. The implied gray market exchange rate was then calculated as follows:

$$\frac{\text{Implicit Gray}}{\text{Market Rate}} = \frac{7 \times \text{Bs7945/Share}}{\$18.84/\text{ADR}} = \text{Bs2952/\$}$$

The official exchange rate on that same day was Bs1598/$. This meant that the gray market rate was quoting the bolivar about 46% weaker against the dollar than what the Venezuelan government officially declared its currency to be worth. Exhibit A illustrates both the official exchange rate and the gray market rate (calculated using CANTV shares) for the January 2002 to March 2004 period. The divergence between the official and gray market rates beginning in February 2003 coincided with the imposition of capital controls.[4]

The Black Market

A third method of obtaining hard currency by Venezuelans was through the rapidly expanding black market. The black market was, as is the case with black markets all over the world, essentially unseen and illegal. It was, however, quite sophisticated, using the services of a stockbroker or banker in Venezuela who simultaneously held U.S. dollar accounts offshore. The choice of a black market broker was a critical one; in the event of a failure to complete the transaction properly there was no legal recourse.

If Santiago wished to purchase dollars on the black market, he would deposit bolivars in his broker's account in Venezuela. The agreed upon black market exchange rate was determined on the day of the deposit, and usually was within a 20% band of the gray market rate derived from the CANTV share price. Santiago would then be given access to a dollar-denominated bank account outside of Venezuela in the agreed amount. The transaction took, on average, two business days to settle. The unofficial black market rate was Bs3300/$.

In early 2004 President Chávez had asked Venezuela's Central Bank to give him "a little billion"—millardito—of its $21 billion in foreign exchange reserves. Chávez argued that the money was actually the people's, and he wished to invest some of it in the agricultural sector. The Central Bank refused. Not to be thwarted in its search for funds, the Chávez government announced on February 9, 2004, another devaluation. The bolivar was devalued 17%, falling in official value from Bs1600/$ to Bs1920/$ (see Exhibit A).

[3]In fact CANTV's share price continued to rise over the 2002 to 2004 period as a result of its use as an exchange rate mechanism. The use of CANTV ADRs as a method of obtaining dollars by Venezuelan individuals and organizations was typically described as "not illegal."
[4]Morgan Stanley Capital International (MSCI) announced on November 26, 2003, that it would change its standard spot rate for the Venezuelan bolivar to the notional rate based on the relationship between the price of CANTV Telefonos de Venezuela D in the local market in bolivars and the price of its ADR in U.S. dollars.

| EXHIBIT A | Official and Gray Market Exchange Rates for the Venezuelan Bolivar |

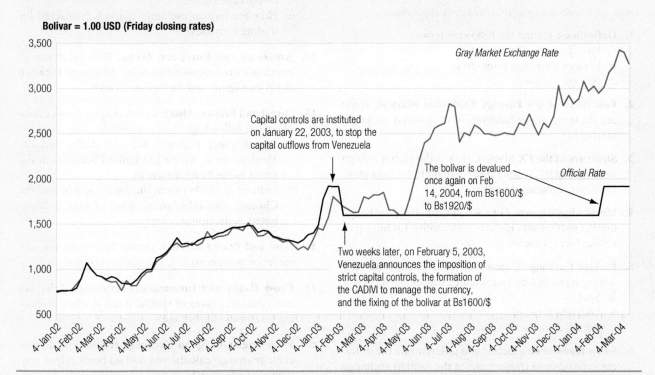

Bolivar = 1.00 USD (Friday closing rates)

Gray Market Exchange Rate

Capital controls are instituted on January 22, 2003, to stop the capital outflows from Venezuela

The bolivar is devalued once again on Feb 14, 2004, from Bs1600/$ to Bs1920/$

Official Rate

Two weeks later, on February 5, 2003, Venezuela announces the imposition of strict capital controls, the formation of the CADIVI to manage the currency, and the fixing of the bolivar at Bs1600/$

With all Venezuelan exports of oil being purchased in U.S. dollars, the devaluation of the bolivar meant that the country's proceeds from oil exports grew by the same 17% as the devaluation itself.

The Chávez government argued that the devaluation was necessary because the bolivar was "a variable that cannot be kept frozen, because it prejudices exports and pressures the balance of payments" according to Finance Minister Tobias Nobriega. Analysts, however, pointed out that Venezuelan government actually had significant control over its balance of payments: oil was the primary export, the government maintained control over the official access to hard currency necessary for imports, and the Central Bank's foreign exchange reserves were now over $21 billion.

Time Was Running Out

Santiago received confirmation from CADIVI on the afternoon of March 10th that his latest application for dollars was approved and that he would receive $10,000 at the official exchange rate of Bs1920/$. Santiago attributed his good fortune to the fact that he paid a CADIVI insider

an extra 500 bolivars per dollar to expedite his request. Santiago noted with a smile that "the Chávistas need to make money too."

The noise from the street seemed to be dying with the sun. It was time for Santiago to make some decisions. None of the alternatives were bonita, but if he was to preserve his business, bolivars—at some price—had to be obtained.

Mini-Case Questions

1. Why does a country like Venezuela impose capital controls?
2. In the case of Venezuela, what is the difference between the gray market and the black market?
3. Create a financial analysis of Santiago's choices. Use it to recommend a solution to his problem.

Post Script. Although President Chávez died in 2013, and the Venezuelan bolivar has been devalued repeatedly and renamed the *bolivar fuerte* since the time of this case, it remains a currency that is overvalued by its government and restricted in its exchange, and therefore continues to lead a double life—officially and unofficially.

QUESTIONS

These questions are available in MyFinanceLab.

1. **Definitions.** Define the following terms:
   a. Foreign exchange market
   b. Foreign exchange transaction
   c. Foreign exchange

2. **Functions of the Foreign Exchange Market.** What are the three major functions of the foreign exchange market?

3. **Structure of the FX Market.** How is the global foreign exchange market structured? Are digital telecommunications replacing people?

4. **Market Participants.** For each of the foreign exchange market participants, identify their motive for buying or selling foreign exchange.

5. **Foreign Exchange Transaction.** Define each of the following types of foreign exchange transactions:
   a. Spot
   b. Outright forward
   c. Forward-forward swaps

6. **Swap Transactions.** Define and differentiate the different type of swap transactions in the foreign exchange markets.

7. **Nondeliverable Forward.** What is a nondeliverable forward and why does it exist?

8. **Foreign Exchange Market Characteristics.** With reference to foreign exchange turnover in 2013, rank the following:
   a. The relative size of spot, forwards, and swaps
   b. The five most important geographic locations for foreign exchange turnover
   c. The three most important currencies of denomination

9. **Foreign Exchange Rate Quotations.** Define and give an example of the following:
   a. Bid rate quote
   b. Ask rate quote

10. **Reciprocals.** Convert the following indirect quotes to direct quotes and direct quotes to indirect quotes:
    a. Euro: €1.22/$ (indirect quote)
    b. Russia: Rbl30/$ (indirect quote)
    c. Canada: $0.72/C$ (direct quote)
    d. Denmark: $0.1644/DKr (direct quote)

11. **Geography and the Foreign Exchange Market.** Answer the following:
    a. What is the geographical location of the foreign exchange market?

    b. What are the two main types of trading systems for foreign exchange?
    c. How are foreign exchange markets connected for trading activities?

12. **American and European Terms.** With reference to interbank quotations, what is the difference between American terms and European terms?

13. **Direct and Indirect Quotes.** Define and give an example of the following:
    a. Direct quote between the U.S. dollar and the Mexican peso, where the United States is designated as the home country.
    b. Indirect quote between the Japanese yen and the Chinese renminbi (yuan), where China is designated as the home country.

14. **Base and Price Currency.** Define base currency, unit currency, price currency, and quote currency.

15. **Cross Rates and Intermarket Arbitrage.** Why are cross currency rates of special interest when discussing intermarket arbitrage?

16. **Percentage Change in Exchange Rates.** Why do percentage change calculations end up being rather confusing on occasion?

PROBLEMS

These problems are available in MyFinanceLab.

1. **Isaac Díez.** Isaac Díez Peris lives in Rio de Janeiro. While attending school in Spain he meets Juan Carlos Cordero from Guatemala. Over the summer holiday Isaac decides to visit Juan Carlos in Guatemala City for a couple of weeks. Isaac's parents give him some spending money, R$4,500. Isaac wants to exchange it for Guatemalan quetzals (GTQ). He collects the following rates:

Spot rate on the GTQ/€ cross rate   GTQ10.5799 = €1.00

Spot rate on the €/R$ cross rate   €0.4462 = R$1.00

   a. What is the Brazilian reais/Guatemalan quetzal cross rate?
   b. How many quetzals will Isaac get for his reais?

2. **Victoria Exports.** A Canadian exporter, Victoria Exports, will be receiving six payments of €12,000, ranging from now to 12 months in the future. Since the company keeps cash balances in both Canadian dollars and U.S. dollars, it can choose which currency to exchange the euros for at the end of the various

periods. Which currency appears to offer the better rates in the forward market?

Period	Days Forward	C$/euro	US$/euro
spot	–	1.3360	1.3221
1 month	30	1.3368	1.3230
2 months	60	1.3376	1.3228
3 months	90	1.3382	1.3224
6 months	180	1.3406	1.3215
12 months	360	1.3462	1.3194

3. **Yen Forward.** Use the following spot and forward bid-ask rates for the Japanese yen/U.S. dollar (¥/$) exchange rate from September 16, 2010, to answer the following questions:
   a. What is the mid-rate for each maturity?
   b. What is the annual forward premium for all maturities?
   c. Which maturities have the smallest and largest forward premiums?

Period	¥/$ Bid Rate	¥/$ Ask Rate
spot	85.41	85.46
1 month	85.02	85.05
2 months	84.86	84.90
3 months	84.37	84.42
6 months	83.17	83.20
12 months	82.87	82.91
24 months	81.79	81.82

4. **Credit Suisse Geneva.** Andreas Broszio just started as an analyst for Credit Suisse in Geneva, Switzerland. He receives the following quotes for Swiss francs against the dollar for spot, 1 month forward, 3 months forward, and 6 months forward.

Spot exchange rate:
Bid rate	SF1.2575/$
Ask rate	SF1.2585/S
1 month forward	10 to 15
3 months forward	14 to 22
6 months forward	20 to 30

   a. Calculate outright quotes for bid and ask and the number of points spread between each.
   b. What do you notice about the spread as quotes evolve from spot toward 6 months?
   c. What is the 6-month Swiss bill rate?

5. **Munich to Moscow.** On your post-graduation celebratory trip you decide to travel from Munich, Germany, to Moscow, Russia. You leave Munich with 15,000 euros in your wallet. Wanting to exchange all of them for Russian rubles, you obtain the following quotes:

Spot rate on the dollar/euro cross rate	$1.3214/€
Spot rate on the ruble/dollar cross rate	Rbl30.96/$

   a. What is the Russian ruble/euro cross rate?
   b. How many rubles will you obtain for your euros?

6. **Moscow to Tokyo.** After spending a week in Moscow you get an email from your friend in Japan. He can get you a very good deal on a plane ticket and wants you to meet him in Tokyo next week to continue your post-graduation celebratory trip. You have 450,000 rubles left in your money pouch. In preparation for the trip you want to exchange your Russian rubles for Japanese yen so you get the following quotes:

Spot rate on the rubles/dollar cross rate	Rbl30.96/$
Spot rate on the yen/dollar cross rate	¥84.02/$

   a. What is the Russian ruble/yen cross rate?
   b. How many yen will you obtain for your rubles?

7. **Asian Pacific Crisis.** The Asian financial crisis that began in July 1997 wreaked havoc throughout the currency markets of East Asia.
   a. Which of the following currencies had the largest depreciations or devaluations during the July to November period?
   b. Which seemingly survived the first five months of the crisis with the least impact on their currencies?

Country and Currency	July 1997 (per US$)	Nov 1997 (per US$)
China yuan	8.40	8.40
Hong Kong dollar	7.75	7.73
Indonesia rupiah	2,400	3,600
Korea won	900	1,100
Malaysia ringgit	2.50	3.50
Philippines peso	27	34
Singapore dollar	1.43	1.60
Taiwan dollar	27.80	32.70
Thailand baht	25.0	40.0

Problem 8.

Currency	USD	EUR	JPY	GBP	CHF	CAD	AUD	HKD
HKD	7.7736	10.2976	0.0928	12.2853	7.9165	7.6987	7.6584	—
AUD	1.015	1.3446	0.0121	1.6042	1.0337	1.0053	—	0.1306
CAD	1.0097	1.3376	0.0121	1.5958	1.0283	—	0.9948	0.1299
CHF	0.9819	1.3008	0.0117	1.5519	—	0.9725	0.9674	0.1263
GBP	0.6328	0.8382	0.0076	—	0.6444	0.6267	0.6234	0.0814
JPY	83.735	110.9238	—	132.3348	85.2751	82.9281	82.4949	10.7718
EUR	0.7549	—	0.009	1.193	0.7688	0.7476	0.7437	0.0971
USD	—	1.3247	0.0119	1.5804	1.0184	0.9904	0.9852	0.1286

8. **Bloomberg Currency Cross Rates.** Use the table at the top of the page from Bloomberg to calculate each of the following:
   a. Japanese yen per U.S. dollar?
   b. U.S. dollars per Japanese yen?
   c. U.S. dollars per euro?
   d. Euros per U.S. dollar?
   e. Japanese yen per euro?
   f. Euros per Japanese yen?
   g. Canadian dollars per U.S. dollar?
   h. U.S. dollars per Canadian dollar?
   i. Australian dollars per U.S. dollar?
   j. U.S. dollars per Australian dollar?
   k. British pounds per U.S. dollar?
   l. U.S. dollars per British pound?
   m. U.S. dollars per Swiss franc?
   n. Swiss francs per U.S. dollar?

9. **Dollar/Euro Forwards.** Use the following spot and forward bid-ask rates for the U.S. dollar/euro (US$/€) from December 10, 2010, to answer the following questions:
   a. What is the mid-rate for each maturity?
   b. What is the annual forward premium for all maturities based on the mid-rates?
   c. Which maturities have the smallest and largest forward premiums based on the mid-rates?

Period	Bid Rate	Ask Rate
spot	1.3231	1.3232
1 month	1.3230	1.3231
2 months	1.3228	1.3229
3 months	1.3224	1.3227
6 months	1.3215	1.3218
12 months	1.3194	1.3198
24 months	1.3147	1.3176

10. **Swissie Triangular Arbitrage.** The following exchange rates are available to you. (You can buy or sell at the stated rates.) Assume you have an initial SF12,000,000. Can you make a profit via triangular arbitrage? If so, show the steps and calculate the amount of profit in Swiss francs (Swissies).

Mt. Fuji Bank	¥ 92.00/$
Mt. Rushmore Bank	SF1.02/$
Mt. Blanc Bank	¥ 90.00/SF

11. **Aussie Dollar Forward.** Use the following spot and forward bid-ask rates for the U.S. dollar/Australian dollar (US$ = A$1.00) exchange rate from December 10, 2010, to answer the following questions:
   a. What is the midrate for each maturity?
   b. What is the annual forward premium for all maturities based on the mid-rates?
   c. Which maturities have the smallest and largest forward premiums based on the mid-rates?

Period	Bid Rate	Ask Rate
spot	0.98510	0.98540
1 month	0.98131	0.98165
2 months	0.97745	0.97786
3 months	0.97397	0.97441
6 months	0.96241	0.96295
12 months	0.93960	0.94045
24 months	0.89770	0.89900

12. **Transatlantic Arbitrage.** A corporate treasury working out of Vienna with operations in New York simultaneously calls Citibank in New York City and Barclays in London. The banks give the following quotes on the euro simultaneously.

Citibank NYC	Barclays London
$0.7551–61 = €1.00	$0.7545–75 = €1.00

Using $1 million or its euro equivalent, show how the corporate treasury could make geographic arbitrage profit with the two different exchange rate quotes.

13. **Venezuelan Bolivar (A).** The Venezuelan government officially floated the Venezuelan bolivar (Bs) in February 2002. Within weeks, its value had moved from the pre-float fix of Bs778/$ to Bs1025/$.
    a. Is this a devaluation or a depreciation?
    b. By what percentage did the value change?

14. **Venezuelan Bolivar (B).** The Venezuelan political and economic crisis deepened in late 2002 and early 2003. On January 1, 2003, the bolivar was trading at Bs1400/$. By February 1, its value had fallen to Bs1950/$. Many currency analysts and forecasters were predicting that the bolivar would fall an additional 40% from its February 1 value by early summer 2003.
    a. What was the percentage change in January?
    b. What is the forecast value for June 2003?

15. **Indirect on the Dollar.** Calculate the forward premium on the dollar (the dollar is the home currency) if the spot rate is €1.3300/$ and the 3-month forward rate is €1.3400/$.

16. **Direct on the Dollar.** Calculate the forward discount on the dollar (the dollar is the home currency) if the spot rate is $1.5800/£ and the 6-month forward rate is $1.5550/£.

17. **Mexican Peso–European Euro Cross Rate.** Calculate the cross rate between the Mexican peso (Ps) and the euro (€) from the following spot rates: Ps12.45/$ and €0.7550/$.

18. **Pura Vida.** Calculate the cross rate between the Costa Rican colón (₡) and the Canadian dollar (C$) from the following spot rates: ₡500.29/$ and C$1.02/$.

19. **Around the Horn.** Assuming the following quotes, calculate how a market trader at Citibank with $1,000,000 can make an intermarket arbitrage profit.

Citibank quotes U.S. dollar per pound	$1.5900 = £1.00
National Westminster quotes euros per pound	€1.2000 = £1.00
Deutschebank quotes U.S. dollar per euro	$0.7550 = €1.00

20. **Great Pyramids.** Inspired by his recent trip to the Great Pyramids, Citibank trader Ruminder Dhillon wonders if he can make an intermarket arbitrage profit using Libyan dinars (LYD) and Saudi riyals (SAR). He has USD1,000,000 to work with so he gathers the following quotes. Is there an opportunity for an arbitrage profit?

Citibank quotes U.S. dollar per Libyan dinar	$1.9324 = LYD1.00
National Bank of Kuwait quotes Saudi riyal per Libyan dinar	SAR1.9405 = LYD1.00
Barclay quotes U.S. dollar per Saudi riyal	$0.2667 = SAR1.00

INTERNET EXERCISES

1. **Bank for International Settlements.** The Bank for International Settlements (BIS) publishes a wealth of effective exchange rate indices. Use its database and analyses to determine the degree to which the dollar, the euro, and the yen (the "big three currencies") are currently overvalued or undervalued.

Bank for International Settlements	www.bis.org/statistics/eer/index.htm

2. **Bank of Canada Exchange Rate Index (CERI).** The Bank of Canada regularly publishes an index of the Canadian dollar's value, the CERI. The CERI is a multilateral trade-weighted index of the Canadian dollar's value against other major global currencies relevant to the Canadian economy and business landscape. Use the CERI from the Bank of Canada's Web site to evaluate the relative strength of the loonie in recent years.

Bank of Canada exchange rates	www.bankofcanada.ca/rates/exchange/ceri/

3. **Forward Quotes.** FXStreet foreign exchange services provides representative forward rates on a multitude of currencies online. Use the following Web site to search out forward exchange rate quotations on a variety of currencies.

FXStreet	www.fxstreet.com/rates-charts/forward-rates/

4. **Federal Reserve Statistical Release.** The United States Federal Reserve provides daily updates of the value of the major currencies traded against the U.S. dollar on its Web site. Use the Fed's Web site to determine the relative weights used by the Fed to determine the index of the dollar's value.

Federal Reserve	www.federalreserve.gov/releases/h10/update/

5. **Daily Market Commentary.** Many different online currency trading and consulting services provide daily assessments of global currency market activity. Use the following GCI site to find the market's current assessment of how the euro is trading against both the U.S. dollar and the Canadian dollar.

GCI Financial Ltd.   www.gcitrading.com/fxnews/

6. **Pacific Exchange Rate Service.** The Pacific Exchange Rate Service Web site, managed by Professor Werner Antweiler of the University of British Columbia, possesses a wealth of current information on currency exchange rates and related statistics. Use the service to plot the recent performance of currencies that have recently suffered significant devaluations or depreciations, such as the Argentine peso, the Venezuelan bolivar, the Turkish lira, and the Egyptian pound.

Pacific Exchange Rate Service   fx.sauder.ubc
                                 .ca/plot.html

International Parity Conditions

. . . if capital freely flowed towards those countries where it could be most profitably employed, there could be no difference in the rate of profit, and no other difference in the real or labour price of commodities, than the additional quantity of labour required to convey them to the various markets where they were to be sold.

—David Ricardo, *On the Principles of Political Economy and Taxation*, 1817, Chapter 7.

LEARNING OBJECTIVES

- Examine how price levels and price level changes (inflation) in countries determine the exchange rates at which their currencies are traded
- Show how interest rates reflect inflationary forces within each country and currency
- Explain how forward markets for currencies reflect expectations held by market participants about the future spot exchange rate
- Analyze how, in equilibrium, the spot and forward currency markets are aligned with interest differentials and differentials in expected inflation

What are the determinants of exchange rates? Are changes in exchange rates predictable? Managers of MNEs, international portfolio investors, importers and exporters, and government officials must deal with these fundamental questions every day. This chapter describes the core financial theories surrounding the determination of exchange rates. Chapter 8 will introduce two other major theoretical schools of thought regarding currency valuation and combine the three different theories in a variety of real-world applications.

The economic theories that link exchange rates, price levels, and interest rates are called international parity conditions. In the eyes of many, these international parity conditions form the core of the financial theory that is considered unique to the field of international finance. These theories do not always work out to be "true" when compared to what students and practitioners observe in the real world, but they are central to any understanding of how multinational business is conducted and funded in the world today. And, as is often the case, the mistake is not always in the theory itself, but in the way it is interpreted or applied in practice. This chapter concludes with a Mini-Case, *Mrs. Watanabe and the Japanese Yen Carry Trade*, which demonstrates how both the theory and practice of international parity conditions sometimes combine to form unusual opportunities for profit—for those who are willing to bear the risk!

Prices and Exchange Rates

If identical products or services can be sold in two different markets, and no restrictions exist on the sale or transportation of product between markets, the product's price should be the same in both markets. This is called the *law of one price*.

A primary principle of competitive markets is that prices will equalize across markets if frictions or costs of moving the products or services between markets do not exist. If the two markets are in two different countries, the product's price may be stated in different currency terms, but the price of the product should still be the same. Comparing prices would require only a conversion from one currency to the other. For example,

$$P^{\$} \times S = P^{¥},$$

where the price of the product in U.S. dollars ($P^{\$}$), multiplied by the spot exchange rate (S, yen per U.S. dollar), equals the price of the product in Japanese yen ($P^{¥}$). Conversely, if the prices of the two products were stated in local currencies, and markets were efficient at competing away a higher price in one market relative to the other, the exchange rate could be deduced from the relative local product prices:

$$S = \frac{P^{¥}}{P^{\$}}.$$

Purchasing Power Parity and the Law of One Price

If the law of one price were true for all goods and services, the *purchasing power parity* (PPP) exchange rate could be found from any individual set of prices. By comparing the prices of identical products denominated in different currencies, one could determine the "real" or PPP exchange rate that should exist if markets were efficient. This is the absolute version of purchasing power parity. *Absolute purchasing power parity* states that the spot exchange rate is determined by the relative prices of similar baskets of goods.

The "Big Mac Index," as it has been christened by *The Economist* (see Exhibit 6.1) and calculated regularly since 1986, is a prime example of the law of one price. Assuming that the Big Mac is indeed identical in all countries listed, it serves as one form of comparison of whether currencies are currently trading at market rates that are close to the exchange rate implied by Big Macs in local currencies.

For example, using Exhibit 6.1, in China a Big Mac costs Yuan 17.2 (local currency), while in the United States the same Big Mac costs $4.79. The actual spot exchange rate was Yuan 6.2115 = $1 at this time. The price of a Big Mac in China in U.S. dollar terms was therefore

$$\frac{\text{Price of Big Mac in China in Yuan}}{\text{Yuan/\$ Spot Rate}} = \frac{\text{Yuan17.2}}{\text{Yuan6.2115/\$}} = \$2.77.$$

This is the value in column 3 of Exhibit 6.1 for China. We then calculate the implied *purchasing power parity rate of exchange* using the actual price of the Big Mac in China (Yuan17.2) over the price of the Big Mac in the United States in U.S. dollars ($4.79):

$$\frac{\text{Price of Big Mac in China in Yuan}}{\text{Price of Big Mac in the U.S. in \$}} = \frac{\text{Yuan17.2}}{\$4.79} = \text{Yuan3.591/\$}.$$

This is the value in column 4 of Exhibit 6.1 for China. In principle, this is what the Big Mac Index is saying the exchange rate between the Yuan and the dollar should be according to the theory.

| EXHIBIT 6.1 | Selected Rates from the Big Mac Index | | | | |

Country and Currency		(1) Big Mac Price in Local Currency	(2) Actual Dollar Exchange Rate January 2015	(3) Big Mac Price in Dollars	(4) Implied PPP of the Dollar	(5) Under/ Overvaluation Against Dollar**
United States	$	4.79	—	4.56	—	—
Britain	£	2.89	1.5115*	4.37	1.6574*	−8.8%
Canada	C$	5.70	1.2286	4.64	1.190	−3.1%
China	Yuan	17.2	6.2115	2.77	3.591	−42.2%
Denmark	DK	34.5	6.4174	5.38	7.203	12.2%
Euro area	€	3.68	1.1587*	4.26	1.302*	−11.0%
India	₹	116.3	61.615	1.89	24.269	−60.6%
Japan	¥	370	117.77	3.14	77.244	−34.4%
Mexico	Peso	49.0	14.6275	3.35	10.230	−30.1%
Norway	kr	48.0	7.6225	6.30	10.021	31.5%
Peru	Sol	10.0	3.008	3.32	2.088	−30.6%
Russia	P	89.0	65.227	1.36	18.580	−71.5%
Switzerland	SFr	6.50	0.86165	7.54	1.357	57.5%
Thailand	Baht	99.0	32.605	3.04	20.668	−36.6%

*These exchange rates are stated in US$ per unit of local currency, $/£ and $/€.

**Percentage under/over valuation against the dollar is calculated as (Implied − Actual)/(Actual), except for the Britain and Euro area calculations, which are (Actual − Implied)/(Implied).

Source: Data for columns (1) and (2) drawn from "The Big Mac Index," *The Economist*, January 22, 2015.

Now comparing this implied PPP rate of exchange, Yuan 3.591/$, with the actual market rate of exchange at that time, Yuan6.2115/$, the degree to which the yuan is either *undervalued* (−%) or *overvalued* (+%) versus the U.S. dollar is calculated as follows:

$$\frac{\text{Implied Rate} - \text{Actual Rate}}{\text{Actual Rate}} = \frac{\text{Yuan}3.591/\$ - \text{Yuan}6.2115/\$}{\text{Yuan}6.2115/\$} \approx -42.2\%.$$

In this case, the Big Mac Index indicates that the Chinese yuan is *undervalued* by 42.2% versus the U.S. dollar as indicated in column 5 for China in Exhibit 6.1. *The Economist* is also quick to note that although this indicates a sizable undervaluation of the managed value of the Chinese yuan versus the dollar, the theory of purchasing power parity is supposed to indicate where the value of currencies should go over the long-term, and not necessarily its value today.

It is important to understand why the Big Mac may be a good candidate for the application of the law of one price and measurement of under or overvaluation. First, the product itself is nearly identical in each market. This is the result of product consistency, process excellence, and McDonald's brand image and pride. Second, and just as important, the product is a result of predominantly local materials and input costs. This means that its price in each country is representative of domestic costs and prices and not imported ones, which would be influenced by exchange rates themselves. The index, however, still possesses limitations. Big Macs cannot be traded across borders, and costs and prices are influenced by a variety of other factors in each country market, such as real estate rental rates and taxes.

A less extreme form of this principle would be that in relatively efficient markets the price of a basket of goods would be the same in each market. Replacing the price of a single product with a price index allows the PPP exchange rate between two countries to be stated as

$$S = \frac{PI^{¥}}{PI^{\$}},$$

where $PI^{¥}$ and $PI^{\$}$ are price indices expressed in local currency for Japan and the United States, respectively. For example, if the identical basket of goods cost ¥1,000 in Japan and $10 in the United States, the PPP exchange rate would be

$$\frac{¥1000}{\$10} = ¥100/\$ \quad \text{or} \quad ¥100 = \$1.00.$$

Just in case you are starting to believe that PPP is just about numbers, *Global Finance in Practice 6.1* reminds you of the human side of the equation.

Relative Purchasing Power Parity

If the assumptions of the absolute version of PPP theory are relaxed a bit, we observe what is termed *relative purchasing power parity*. Relative PPP holds that PPP is not particularly helpful in determining what the spot rate is today, but that the relative change in prices between two

The Immiseration of the North Korean People—The "Revaluation" of the North Korean Won

The principles of purchasing power are not just theoretical, they can also capture the problems, poverty, and misery of a people. The devaluation of the North Korean won (KPW) in November 2009 was one such case.

The North Korean government has been trying to stop the growth and activity in the street markets of its country for decades. For many years the street markets have been the sole opportunity for most of the Korean people to earn a living. Under the communist state's stewardship, the quality of life for its 24 million people has continued to deteriorate. Between 1990 and 2008, the country's infant mortality rate had increased 30%, and life expectancy had fallen by three years. The United Nations estimated that one in three children under the age of five suffered malnutrition. Although most of the working population worked officially for the government, many were underpaid (or in many cases not paid at all). They often bribed their bosses to allow them to leave work early to try to scrape out a living in the street markets of the underground economy.

But it was this very basic market economy that President Kim Jong-il (now deceased) and the governing regime wished to stamp out. On November 30, 2009, the Korean government made a surprise announcement to its people: a new, more valuable Korean won would replace the old one. "You have until the end of the day to exchange your old won for new won." All old 1,000 won notes would be replaced with 10 won notes, knocking off two zeros from the officially recognized value of the currency. This meant that everyone holding old won, their cash and savings, would now officially be worth 1/100th of what it was previously. Exchange was limited to 100,000 old won. People who had worked and saved for decades to accumulate what was roughly $200 or $300 in savings outside of North Korea were wiped out; their total life savings were essentially worthless. By officially denouncing the old currency, the North Korean people would be forced to exchange their holdings for new won. The government would indeed undermine the underground economy.

The results were devastating. After days of street protests, the government raised the 100,000 ceiling to 150,000. By late January 2010, inflation was rising so rapidly that Kim Jong-il apologized to the people for the revaluation's impact on their lives. The government administrator who had led the revaluation was arrested, and in February 2010, executed "for his treason."

countries over a period of time determines the change in the exchange rate over that period. More specifically:

> *If the spot exchange rate between two countries starts in equilibrium, any change in the differential rate of inflation between them tends to be offset over the long run by an equal but opposite change in the spot exchange rate.*

Exhibit 6.2 shows a general case of relative PPP. The vertical axis shows the percentage change in the spot exchange rate for foreign currency, and the horizontal axis shows the percentage difference in expected rates of inflation (foreign relative to home country). The diagonal parity line shows the equilibrium position between a change in the exchange rate and relative inflation rates. For instance, point P represents an equilibrium point at which inflation in the foreign country, Japan, is 4% lower than in the home country, the United States. Therefore, relative PPP predicts that the yen will appreciate by 4% per annum with respect to the U.S. dollar. If current market expectations led to either point W or S in Exhibit 6.2, the home currency would be considered either *weak* (point W) or *strong* (point S), and the market would not be in equilibrium.

The logic behind the application of PPP to changes in the spot exchange rate is that if a country experiences inflation rates higher than those of its main trading partners, and its exchange rate does not change, its exports of goods and services become less competitive with comparable products produced elsewhere. Imports from abroad become more price-competitive with higher-priced domestic products. These price changes lead to a deficit on the current account in the balance of payments unless offset by capital and financial flows.

Empirical Tests of Purchasing Power Parity

There has been extensive testing of both the absolute and relative versions of purchasing power parity and the law of one price. These tests have, for the most part, not proved PPP to

---

EXHIBIT 6.2   **Relative Purchasing Power Parity**

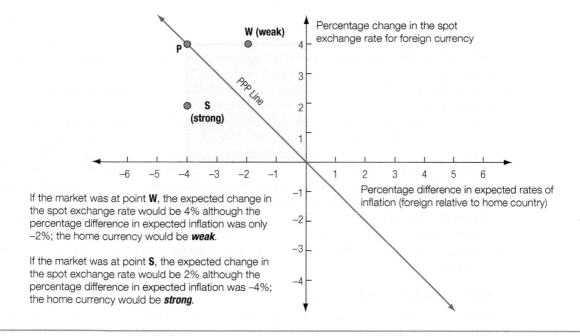

If the market was at point **W**, the expected change in the spot exchange rate would be 4% although the percentage difference in expected inflation was only –2%; the home currency would be **weak**.

If the market was at point **S**, the expected change in the spot exchange rate would be 2% although the percentage difference in expected inflation was –4%; the home currency would be **strong**.

be accurate in predicting future exchange rates. Goods and services do not in reality move at zero cost between countries, and in fact many services are not "tradable," for example, for haircuts. Many goods and services are not of the same quality across countries, reflecting differences in the tastes and resources of the countries of their manufacture and consumption.

Two general conclusions can be made from these tests: (1) PPP holds up well over the very long run but poorly for shorter time periods; and (2) The theory holds better for countries with relatively high rates of inflation and underdeveloped capital markets.

Exchange Rate Indices: Real and Nominal

Because any single country trades with numerous partners, we need to track and evaluate its individual currency value against all other currency values in order to determine relative purchasing power. The objective is to discover whether a country's exchange rate is "overvalued" or "undervalued" in terms of PPP. One of the primary methods of dealing with this problem is the calculation of *exchange rate indices*. These indices are formed through trade—by weighting the bilateral exchange rates between the home country and its trading partners.

The *nominal effective exchange rate index* uses actual exchange rates to create an index, on a weighted average basis, of the value of the subject currency over time. It does not really indicate anything about the "true value" of the currency or anything related to PPP. The nominal index simply calculates how the currency value relates to some arbitrarily chosen base period, but it is used in the formation of the *real effective exchange rate index*. The real effective exchange rate index indicates how the weighted average purchasing power of the currency has changed relative to some arbitrarily selected base period. Exhibit 6.3 plots the real effective exchange rate indexes for Japan, the euro area, and the U.S. for the 1980–2012 period.

The real effective exchange rate index for the U.S. dollar, $E_R^\$$, is found by multiplying the nominal effective exchange rate index, $E_N^\$$, by the ratio of U.S. dollar costs, $C^\$$ over foreign currency costs, C^{FC}, both in index form:

$$E_R^\$ = E_N^\$ \times \frac{C^\$}{C^{FC}}.$$

If changes in exchange rates just offset differential inflation rates—if purchasing power parity holds—all the real effective exchange rate indices would stay at 100. If an exchange rate strengthened more than was justified by differential inflation, its index would rise above 100. If the real effective exchange rate index were above 100, the currency would be considered "overvalued" from a competitive perspective, and vice versa.

Exhibit 6.3 shows how the real effective exchange rate of the U.S. dollar, Japanese yen, and the European euro have changed over the past 35 years. The dollar's index value was substantially above 100 in the early 1980s (overvalued), falling below 100 during the 1988–1996 period (undervalued), then rising above 100 again. The dollar has been only slightly overvalued since 2012. While the euro has not strayed far from "proper valuation" since 2009, the Japanese yen has been clearly undervalued since mid-2012.

Apart from measuring deviations from PPP, a country's real effective exchange rate is an important tool for management when predicting upward or downward pressure on a country's balance of payments and exchange rate, as well as an indicator of whether producing for export in that country could be competitive.

Exchange Rate Pass-Through

Exchange rate pass-through is a measure of the response of imported and exported product prices to changes in exchange rates. When that pass-through is *partial*, meaning the full percentage change in the exchange rate is not reflected in prices, a country's real effective

EXHIBIT 6.3   **Real Effective Exchange Rate Indexes (base year 2010 = 100)**

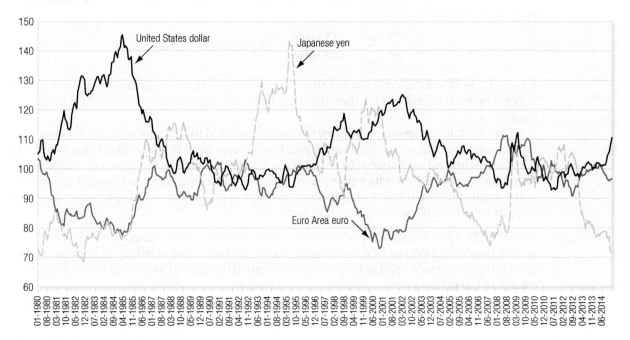

Source: Bank for International Settlements, www.bis.org/statistics/eer/.

exchange rate index can deviate from its PPP-equilibrium level of 100. Although PPP implies that all exchange rate changes are passed through by equivalent changes in prices to trading partners, empirical research in the years following the growth in floating-rate currencies questioned this long-held assumption.

Complete versus Partial Pass-Through. To illustrate exchange rate pass-through, assume that Volvo produces an automobile in Belgium and pays all production expenses in euros (Volvo currently has no manufacturing in North America). The price of this specific model is €50,000. When the firm exports the auto to the United States, the price of the Volvo in the U.S. market should simply be the euro value converted to U.S. dollars at the spot exchange rate:

$$P^\$_{\text{Volvo}} = P^{\text{€}}_{\text{Volvo}} \times S^{\$/\text{€}}$$

where $P^\$_{\text{Volvo}}$ is the Volvo price in dollars, $P^{\text{€}}_{\text{Volvo}}$ is the Volvo price in euros, and $S^{\$/\text{€}}$ is the spot exchange in number of dollars per euro. If the euro were to appreciate 20% versus the U.S. dollar—from \$1.00/€ to \$1.20/€—the price of the Volvo in the U.S. market should theoretically rise to \$60,000. If the price in dollars increases by the same percentage change as the exchange rate, then there has been *complete pass-through* (or 100%) of changes in exchange rates.

$$\frac{P^\$_{\text{Volvo,2}}}{P^\$_{\text{Volvo,1}}} - \frac{\$60,000}{\$50,000} = 1.20 \text{ or a } 20\% \text{ increase}$$

However, if Volvo worried that a price increase of this magnitude in the U.S. market would severely decrease sales volumes, it might work to prevent the dollar price of this model from rising the full amount in the U.S. market. If the price of this same Volvo model rose to only $58,000 in the U.S. market, the percentage increase would be less than the 20% appreciation of the euro versus the dollar.

$$\frac{P^\$_{Volvo,2}}{P^\$_{Volvo,1}} - \frac{\$58,000}{\$50,000} = 1.60 \text{ or a } 16\% \text{ increase}$$

If the price in U.S. dollars rises by less than the percentage change in exchange rates (as is often the case in international trade), then there has been only *partial pass-through* of exchange rate changes.

For example, components and raw materials imported to Belgium cost less in euros when the euro appreciates versus the currency of foreign suppliers. It is also likely that some time may pass before all exchange rate changes are finally reflected in the prices of traded goods, including the period over which previously signed contracts are delivered upon. It is obviously in the interest of Volvo to do what it can to prevent appreciation of the euro from raising the price of its automobiles in major export markets.

Price Elasticity of Demand. The concept of *price elasticity of demand* is useful when determining the desired degree of pass-through. Recall that the price elasticity of demand for any good is the percentage change in quantity of the good demanded as a result of the percentage change in the good's price:

$$\text{Price elasticity of demand} = e_p = \frac{\%\Delta Q_d}{\%\Delta P}$$

where Q_d is quantity demanded and P is product price. If the absolute value of e_p is less than 1.0, then the good is relatively "inelastic." If it is greater than 1.0, the good is relatively "elastic."

A Belgian product that is relatively price-inelastic—meaning that the quantity demanded is relatively unresponsive to price changes—may often demonstrate a high *degree of pass-through*. This is because a higher dollar price in the United States market would have little noticeable effect on the quantity of the product demanded by consumers. Dollar revenue would increase, but euro revenue would remain the same. However, products that are relatively price-elastic would respond in the opposite way. If the 20% euro appreciation resulted in 20% higher dollar prices, U.S. consumers would decrease the number of Volvos purchased. If the price elasticity of demand for Volvos in the United States were greater than one, total dollar sales revenue of Volvos would decline.

Pass-Through and Emerging Market Currencies. A number of emerging market countries have chosen in recent years to change their objectives and choices as described in the impossible trinity (introduced and detailed previously in Chapter 2). These countries have changed from choosing a pegged exchange rate and independent monetary policy over the free flow of capital (point A in Exhibit 6.4) to policies allowing more capital flows at the expense of a pegged or fixed exchange rate (toward point C in Exhibit 6.4).

This change in focus has also now introduced exchange rate pass-through as an issue in these same emerging markets. With changing exchange rates and increased trade and financial product movements in and out of these countries, prices are changing. Although price volatility alone is a source of growing concern, price changes contributing to inflationary pressure is even more unsettling. The root cause of these problems lies not with choices made by the emerging market nations, but rather with the interest rate choices made by the major industrial countries with which they trade.

Since 2009, all the major industrial country currency markets—the dollar, the euro, the yen—have been characterized by extremely low interest rates, as concerns over economic

EXHIBIT 6.4   **Pass-Through, the Impossible Trinity, and Emerging Markets**

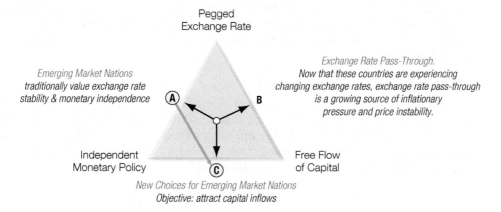

Pegged
Exchange Rate

Emerging Market Nations
traditionally value exchange rate
stability & monetary independence   (A)   B

Exchange Rate Pass-Through.
Now that these countries are experiencing
changing exchange rates, exchange rate pass-through
is a growing source of inflationary
pressure and price instability.

Independent
Monetary Policy   (C)

Free Flow
of Capital

New Choices for Emerging Market Nations
Objective: attract capital inflows

Many emerging market countries have chosen to move from Point A to Point C, exchanging fixed exchange rates for the chance of attracting capital inflows. The result is that these countries are now the subject to varying levels of exchange rate pass-through.

growth and employment have dominated. Select emerging market countries have then experienced appreciating currencies in some cases (because their interest rates are higher than industrial country currencies). Those exchange rate changes have led to exchange rate pass-through of imported products—rising prices—contributing to inflationary pressures.

Interest Rates and Exchange Rates

We have already seen how prices of goods in different countries should be related through exchange rates. We now consider how interest rates are linked to exchange rates.

The Fisher Effect

The Fisher effect, named after economist Irving Fisher, states that nominal interest rates in each country are equal to the required real rate of return plus compensation for expected inflation. More formally, this is derived from $(1 + r)(1 + \pi) - 1$ as

$$i = r + \pi + r\pi,$$

where i is the nominal rate of interest, r is the real rate of interest, and π is the expected rate of inflation over the period of time for which funds are to be lent. The final compound term, $r\pi$ is frequently dropped from consideration due to its relatively minor value. The Fisher effect then reduces to (approximate form)

$$i = r + \pi.$$

The Fisher effect applied to the United States and Japan would be as follows:

$$i^\$ = r^\$ + \pi^\$; \quad i^\yen = r^\yen + \pi^\yen.$$

where the superscripts $\$$ and $\yen$ pertain to the respective nominal (i), real (r), and expected inflation (π) components of financial instruments denominated in dollars and yen, respectively. We need to forecast the future rate of inflation, not what inflation has been. Predicting the future is, well, difficult.

Empirical tests using *ex post* national inflation rates have shown that the Fisher effect usually exists for short-maturity government securities such as Treasury bills and notes. Comparisons based on longer maturities suffer from the increased financial risk inherent in fluctuations of the market value of the bonds prior to maturity. Comparisons of private sector securities are influenced by unequal creditworthiness of the issuers. All the tests are inconclusive to the extent that recent past rates of inflation are not a correct measure of future expected inflation.

The International Fisher Effect

The relationship between the percentage change in the spot exchange rate over time and the differential between comparable interest rates in different national capital markets is known as the *international Fisher effect*. "Fisher-open," as it is often termed, states that the spot exchange rate should change in an equal amount but in the opposite direction to the difference in interest rates between two countries. More formally,

$$\frac{S_1 - S_2}{S_2} \times 100 = i^\$ - i^¥,$$

where $i^\$$ and $i^¥$ are the respective national interest rates, and S is the spot exchange rate using indirect quotes (an indirect quote on the dollar is, for example, ¥ = \$1.00) at the beginning of the period (S_1) and the end of the period (S_2). This is the approximation form commonly used in industry. The precise formulation is as follows:

$$\frac{S_1 - S_2}{S_2} = \frac{i^\$ - i^¥}{1 + i^¥}.$$

Justification for the international Fisher effect is that investors must be rewarded or penalized to offset the expected change in exchange rates. For example, if a dollar-based investor buys a 10-year yen bond earning 4% interest, instead of a 10-year dollar bond earning 6% interest, the investor must be expecting the yen to appreciate vis-à-vis the dollar by at least 2% per year during the 10 years. If not, the dollar-based investor would be better off remaining in dollars. If the yen appreciates 3% during the 10-year period, the dollar-based investor would earn a bonus of 1% higher return. However, the international Fisher effect predicts that, with unrestricted capital flows, an investor should be indifferent to whether his bond is in dollars or yen—because investors worldwide would see the same opportunity and compete it away.

Empirical tests lend some support to the relationship postulated by the international Fisher effect, although considerable short-run deviations occur. A more serious criticism has been posed, however, by recent studies that suggest the existence of a foreign exchange risk premium for most major currencies. Also, speculation in uncovered interest arbitrage creates distortions in currency markets. Thus, the expected change in exchange rates might consistently be greater than the difference in interest rates.

The Forward Rate

A *forward rate* (or *outright forward* as described in Chapter 5) is an exchange rate quoted today for settlement at some future date. A forward exchange agreement between currencies states the rate of exchange at which a foreign currency will be "bought forward" or "sold forward" at a specific date in the future (typically after 30, 60, 90, 180, 270, or 360 days).

The forward rate is calculated for any specific maturity by adjusting the current spot exchange rate by the ratio of euro currency interest rates of the same maturity for the two subject currencies. For example, the 90-day forward rate for the Swiss franc/U.S. dollar exchange rate ($F^{SF/\$}$) is found by multiplying the current spot rate ($S^{SF/\$}$) by the ratio of the 90-day euro-Swiss franc deposit rate (i^{SF}) over the 90-day eurodollar deposit rate.

$$F_{90}^{SF/\$} = S^{SF/\$} \times \frac{\left[1 + \left(i^{SF} \times \frac{90}{360}\right)\right]}{\left[1 + \left(i^{\$} \times \frac{90}{360}\right)\right]}$$

Assuming a spot rate of SF1.4800/$, a 90-day euro Swiss franc deposit rate of 4.00% per annum, and a 90-day eurodollar deposit rate of 8.00% per annum, the 90-day forward rate is SF1.4655/$.

$$F_{90}^{SF/\$} = SF1.4800/\$ \times \frac{\left[1 + \left(0.0400 \times \frac{90}{360}\right)\right]}{\left[1 + \left(0.0800 \times \frac{90}{360}\right)\right]} = SF1.4800/\$ \times \frac{1.01}{1.02} = SF1.4655/\$$$

The forward premium or discount is the percentage difference between the spot and forward exchange rate, stated in annual percentage terms. When the foreign currency price of the home currency is used, as in this case of SF/$, the formula for the percent-per-annum premium or discount becomes

$$f^{SF} = \frac{Spot - Forward}{Forward} \times \frac{360}{Days} \times 100.$$

Substituting the SF/$ spot and forward rates, as well as the number of days forward (90),

$$f^{SF} = \frac{SF1.4800/\$ - SF1.4655/\$}{SF1.4655/\$} \times \frac{360}{90} \times 100 = +3.96\% \text{ per annum.}$$

The sign is positive, indicating that the Swiss franc is *selling forward* at a 3.96% per annum premium over the dollar (it takes 3.96% more dollars to get a franc at the 90-day forward rate).

As illustrated in Exhibit 6.5, the forward premium on the eurodollar forward arises from the differential between eurodollar and Swiss franc interest rates. Because the forward rate for any particular maturity utilizes the specific interest rates for that term, the forward premium or discount on a currency is visually obvious—the currency with the higher interest rate (in this case the U.S. dollar)—will sell forward at a discount, and the currency with the lower interest rate (here the Swiss franc) will sell forward at a premium.

The forward rate is calculated from three observable data items—the spot rate, the foreign currency deposit rate, and the home currency deposit rate—and is not a forecast of the future spot exchange. However, the forward rate is frequently used by managers as a forecast, yielding mixed results, as the following section describes.

Interest Rate Parity (IRP)

The theory of *interest rate parity* (IRP) provides the link between the foreign exchange markets and the international money markets. The theory states:

> *The difference in the national interest rates for securities of similar risk and maturity should be equal to, but opposite in sign to, the forward rate discount or premium for the foreign currency, except for transaction costs.*

EXHIBIT 6.5	**Currency Yield Curves and the Forward Premium**

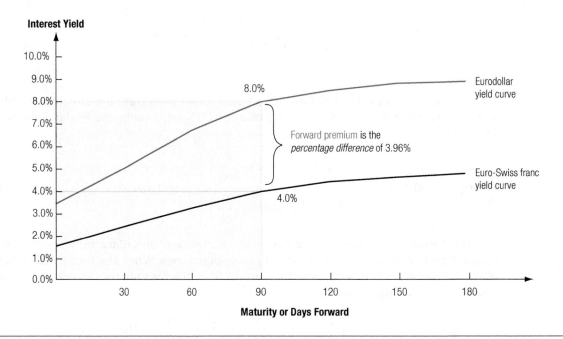

Exhibit 6.6 shows how the theory of interest rate parity works. Assume that an investor has $1,000,000 and several alternative but comparable Swiss franc (SF) monetary investments. If the investor chooses to invest in a dollar money market instrument, the investor would earn the dollar rate of interest. This results in $(1 + i^\$)$ at the end of the period, where $I^\$$ is the dollar rate of interest in decimal form.

EXHIBIT 6.6	**Interest Rate Parity (IRP)**

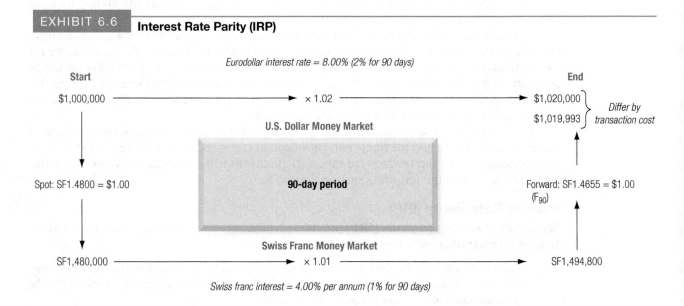

The investor may, however, choose to invest in a Swiss franc money market instrument of identical risk and maturity for the same period. This action would require that the investor exchange the dollars for francs at the spot rate, invest the francs in a money market instrument, sell the francs forward (in order to avoid any risk that the exchange rate would change), and at the end of the period convert the resulting proceeds back to dollars.

A dollar-based investor would evaluate the relative returns of starting in the top-left corner and investing in the dollar market (straight across the top of the box) compared to investing in the Swiss franc market (going down and then around the box to the top-right corner). The comparison of returns would be

$$(1 + i^\$) = S^{SF/\$} \times (1 + i^{SF}) \times \frac{1}{F^{SF/\$}},$$

where S = spot rate of exchange and F = the forward rate of exchange. Substituting in the spot rate (SF1.4800/\$) and forward rate (SF1.4655/\$) and respective interest rates from Exhibit 6.6, the interest rate parity condition is

$$(1 + 0.02) = 1.4800 \times (1 + 0.01) \times \frac{1}{1.4655}.$$

The left-hand side of the equation is the gross return the investor would earn by investing in dollars. The right-hand side is the gross return the investor would earn by exchanging dollars for Swiss francs at the spot rate, investing the franc proceeds in the Swiss franc money market, and simultaneously selling the principal plus interest in Swiss francs forward for dollars at the current 90-day forward rate.

Ignoring transaction costs, if the returns in dollars are equal between the two alternative money market investments, the spot and forward rates are considered to be at IRP. The transaction is "covered," because the exchange rate back to dollars is guaranteed at the end of the 90-day period. Therefore, as shown in Exhibit 6.6, in order for the two alternatives to be equal, any differences in interest rates must be offset by the difference between the spot and forward exchange rates (in approximate form):

$$\frac{F}{S} = \frac{(1 + i^{SF})}{(1 + i^\$)}, \text{ or } \frac{SF1.4655/\$}{SF1.4800/\$} = \frac{1.01}{1.02} = 0.9902 \approx 1\%.$$

Covered Interest Arbitrage (CIA)

The spot and forward exchange markets are not constantly in the state of equilibrium described by interest rate parity. When the market is not in equilibrium, the potential for "riskless" or arbitrage profit exists. The arbitrager who recognizes such an imbalance will move to take advantage of the disequilibrium by investing in whichever currency offers the higher return on a covered basis. This is called *covered interest arbitrage* (CIA).

Exhibit 6.7 describes the steps that a currency trader, most likely working in the arbitrage division of a large international bank, would implement to perform a CIA transaction. The currency trader, Fye Hong, may utilize any of a number of major eurocurrencies that his bank holds to conduct arbitrage investments. The morning conditions indicate to Fye Hong that a CIA transaction that exchanges 1 million U.S. dollars for Japanese yen, invested in a six-month euroyen account and sold forward back to dollars, will yield a profit of \$4,638 (\$1,044,638 − \$1,040,000) over and above the profit available from a eurodollar investment. Conditions in the exchange markets and euromarkets change rapidly however, so if Fye Hong waits even a few minutes, the profit opportunity may disappear.

| EXHIBIT 6.7 | **Covered Interest Arbitrage (CIA)** |

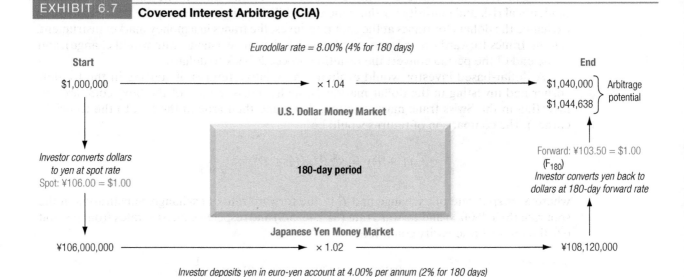

Fye Hong now executes the following transaction:

Step 1: Convert $1,000,000 at the spot rate of ¥106.00/$ to ¥106,000,000 (see "Start" in Exhibit 6.7).

Step 2: Invest the proceeds, ¥106,000,000, in a euroyen account for six months, earning 4.00% per annum, or 2% for 180 days.

Step 3: Simultaneously sell the future yen proceeds (¥108,120,000) forward for dollars at the 180-day forward rate of ¥103.50/$. This action "locks in" gross dollar revenues of $1,044,638 (see "End" in Exhibit 6.7).

Step 4: Calculate the cost (opportunity cost) of funds used at the eurodollar rate of 8.00% per annum, or 4% for 180 days, with principal and interest then totaling $1,040,000. Profit on CIA ("End") is $4,638 ($1,044,638 − $1,040,000).

Note that all profits are stated in terms of the currency in which the transaction was initialized, but that a trader may conduct investments denominated in U.S. dollars, Japanese yen, or any other major currency. All that is required to make a covered interest arbitrage profit is for interest rate parity not to hold. Depending on the relative interest rates and forward premium, Fye Hong would have started in Japanese yen, invested in U.S. dollars, and sold the dollars forward for yen. The profit would then end up denominated in yen. But how would Fye Hong decide in which direction to go around the box in Exhibit 6.7?

Rule of Thumb. The key to determining whether to start in dollars or yen is to compare the differences in interest rates to the forward premium on the yen (the cost of cover). For example, in Exhibit 6.7, the difference in 180-day interest rates is 2.00% (dollar interest rates are higher by 2.00%). The premium on the yen for 180 days forward is as follows:

$$f^{¥} = \frac{\text{Spot} - \text{Forward}}{\text{Forward}} \times \frac{360}{180} \times 100 = \frac{¥106.00/\$ - ¥103.50/\$}{¥103.50/\$} \times 200 = 4.8309\%$$

In other words, by investing in yen and selling the yen proceeds forward at the forward rate, Fye Hong earns more on the combined interest rate arbitrage and forward premium than if he continues to invest in dollars.

> *Arbitrage Rule of Thumb:* If the difference in interest rates is greater than the forward premium (or expected change in the spot rate), invest in the higher interest yielding currency. If the difference in interest rates is less than the forward premium (or expected change in the spot rate), invest in the lower interest yielding currency.

Using this rule of thumb should enable Fye Hong to choose in which direction to go around the box in Exhibit 6.7. It also guarantees that he will always make a profit if he goes in the right direction. This rule assumes that the profit is greater than any transaction costs incurred.

This process of CIA drives the international currency and money markets toward the equilibrium described by interest rate parity. Slight deviations from equilibrium provide opportunities for arbitragers to make small riskless profits. Such deviations provide the supply and demand forces that will move the market back toward parity (equilibrium).

Covered interest arbitrage opportunities continue until interest rate parity is reestablished, because the arbitragers are able to earn risk-free profits by repeating the cycle as often as possible. Their actions, however, nudge the foreign exchange and money markets back toward equilibrium for the following reasons:

1. The purchase of yen in the spot market and the sale of yen in the forward market narrows the premium on the forward yen. This is because the spot yen strengthens from the extra demand and the forward yen weakens because of the extra sales. A narrower premium on the forward yen reduces the foreign exchange gain previously captured by investing in yen.

2. The demand for yen-denominated securities causes yen interest rates to fall, and the higher level of borrowing in the United States causes dollar interest rates to rise. The net result is a wider interest differential in favor of investing in the dollar.

Uncovered Interest Arbitrage (UIA)

A deviation from covered interest arbitrage is *uncovered interest arbitrage* (UIA), wherein investors borrow in countries and currencies exhibiting relatively low interest rates and convert the proceeds into currencies that offer much higher interest rates. The transaction is "uncovered," because the investor does not sell the higher yielding currency proceeds forward, choosing to remain uncovered and accept the currency risk of exchanging the higher yield currency into the lower yielding currency at the end of the period. Exhibit 6.8 demonstrates the steps an uncovered interest arbitrager takes when undertaking what is termed the "yen carry-trade."

The "yen carry-trade" is an age-old application of UIA. Investors, from both inside and outside Japan, take advantage of extremely low interest rates in Japanese yen (0.40% per annum) to raise capital. Investors exchange the capital they raise for other currencies like U.S. dollars or euros. Then they reinvest these dollar or euro proceeds in dollar or euro money markets where the funds earn substantially higher rates of return (5.00% per annum in Exhibit 6.8). At the end of the period—a year, in this case—they convert the dollar proceeds back into Japanese yen in the spot market. The result is a tidy profit over what it costs to repay the initial loan.

The trick, however, is that the spot exchange rate at the end of the year must not change significantly from what it was at the beginning of the year. If the yen were to appreciate significantly against the dollar, as it did in late 1999, moving from ¥120/$ to ¥105/$, these "uncovered" investors would suffer sizable losses when they convert their dollars into yen to repay the yen they borrowed. Higher return at higher risk. The Mini-Case at the end of this chapter details one of the most frequent carry trade structures, the Australian dollar/Japanese yen cross rate.

EXHIBIT 6.8   Uncovered Interest Arbitrage (UIA): The Yen Carry Trade

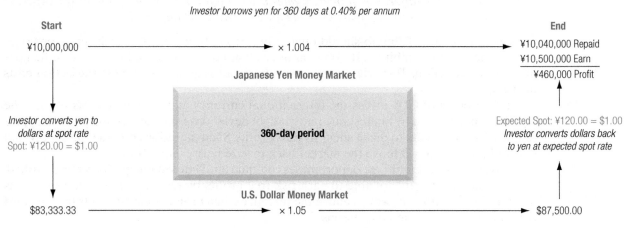

Investor borrows yen for 360 days at 0.40% per annum

Start		End
¥10,000,000	× 1.004	¥10,040,000 Repaid
		¥10,500,000 Earn
		¥460,000 Profit

Japanese Yen Money Market

Investor converts yen to dollars at spot rate
Spot: ¥120.00 = $1.00

360-day period

Expected Spot: ¥120.00 = $1.00
Investor converts dollars back to yen at expected spot rate

U.S. Dollar Money Market

$83,333.33 — × 1.05 — $87,500.00

Investor deposits dollars in U.S. dollar money market at 5.00% per annum

Equilibrium between Interest Rates and Exchange Rates

Exhibit 6.9 illustrates the conditions necessary for equilibrium between interest rates and exchange rates. The vertical axis shows the difference in interest rates in favor of the foreign currency, and the horizontal axis shows the forward premium or discount on that currency. The interest rate parity line shows the equilibrium state, but transaction costs cause the line to be a band rather than a thin line.

EXHIBIT 6.9   Interest Rate Parity and Equilibrium

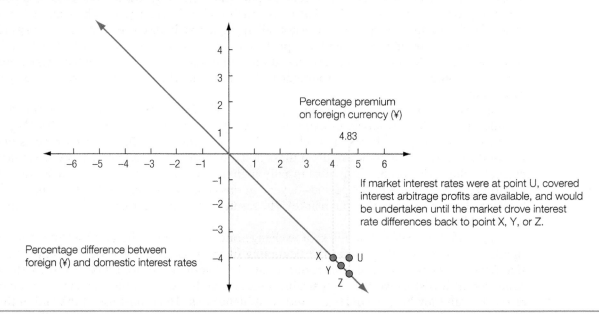

Percentage premium on foreign currency (¥)

4.83

Percentage difference between foreign (¥) and domestic interest rates

If market interest rates were at point U, covered interest arbitrage profits are available, and would be undertaken until the market drove interest rate differences back to point X, Y, or Z.

Transaction costs arise from foreign exchange and investment brokerage costs on buying and selling securities. Typical transaction costs in recent years have been in the range of 0.18% to 0.25% on an annual basis. For individual transactions, like Fye Hong's covered interest arbitrage (CIA) activities illustrated in Exhibit 6.7, there is no explicit transaction cost per trade; rather, the costs of the bank in supporting Fye Hong's activities are the transaction costs. Point X in Exhibit 6.9 shows one possible equilibrium position, where a 4% lower rate of interest on yen securities would be offset by a 4% premium on the forward yen.

The disequilibrium situation, which encouraged the interest rate arbitrage in the previous CIA example of Exhibit 6.7, is illustrated in Exhibit 6.9 by point U. Point U is located off the interest rate parity line because the lower interest on the yen is 4% (annual basis), whereas the premium on the forward yen is slightly over 4.8% (annual basis). Using the formula for forward premium presented earlier, we calculate the forward premium on the Japanese yen as follows:

$$\frac{¥106.00/\$ - 103.50/\$}{¥103.50/\$} \times \frac{360 \text{ Days}}{180 \text{ Days}} \times 100 = 4.83\%$$

The situation depicted by point U is unstable, because all investors have an incentive to execute the same covered interest arbitrage. Except for a bank failure, the arbitrage gain is virtually risk-free.

Some observers have suggested that political risk does exist, because one of the governments might apply capital controls that would prevent execution of the forward contract. This risk is fairly remote for covered interest arbitrage between major financial centers of the world, especially because a large portion of funds used for covered interest arbitrage is in eurodollars. The concern may be valid for pairings with countries not noted for political and fiscal stability.

The net result of the disequilibrium is that fund flows will narrow the gap in interest rates and/or decrease the premium on the forward yen. In other words, market pressures will cause point U in Exhibit 6.9 to move toward the interest rate parity band. Equilibrium might be reached at point Y, or at any other locus between X and Z, depending on whether forward market premiums are more or less easily shifted than interest rate differentials.

Uncovered interest arbitrage takes many forms in global financial markets today, and opportunities do exist for those who are willing to bear the risk (and potentially pay the price). *Global Finance in Practice 6.2* describes one such speculation, foreign-currency home mortgages in Hungary, which as described, can turn an innocent homeowner into a foreign currency speculator.

GLOBAL FINANCE IN PRACTICE 6.2

Hungarian Mortgages

No one understands the linkage between interest rates and currencies better than Hungarian homeowners. Given the choice of taking mortgages out in local currency (Hungarian forint) or foreign currency (Swiss francs for example), many chose francs because the interest rates were lower.

But regardless of the actual interest rate itself, the fall in the value of the forint against the franc by more than 40% resulted in radically increasing mortgage debt service payments in Hungarian forint. These borrowers are now trying to have their own mortgages declared "unconstitutional" in order to get out from under these rising debt burdens.

(continued)

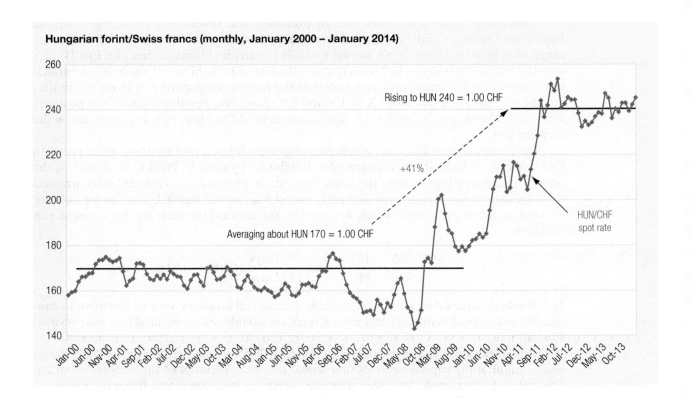

Hungarian forint/Swiss francs (monthly, January 2000 – January 2014)

Rising to HUN 240 = 1.00 CHF

+41%

Averaging about HUN 170 = 1.00 CHF

HUN/CHF
spot rate

Forward Rate as an Unbiased Predictor of the Future Spot Rate

Some forecasters believe that foreign exchange markets for the major floating currencies are "efficient" and forward exchange rates are unbiased predictors of future spot exchange rates.

Exhibit 6.10 demonstrates the meaning of "unbiased prediction" in terms of how the forward rate performs in estimating future spot exchange rates. If the forward rate is an unbiased predictor of the future spot rate, the expected value of the future spot rate at time 2 equals the present forward rate for time 2 delivery, available now, $E_1 (S_2) = F_{1,2}$.

Intuitively, this means that the distribution of possible actual spot rates in the future is centered on the forward rate. The fact that it is an unbiased predictor, however, does not mean that the future spot rate will actually be equal to what the forward rate predicts. Unbiased prediction simply means that the forward rate will, on average, overestimate and underestimate the actual future spot rate in equal frequency and degree. The forward rate may, in fact, never actually equal the future spot rate.

The rationale for this relationship is based on the hypothesis that the foreign exchange market is reasonably efficient. Market efficiency assumes that (1) All relevant information is quickly reflected in both the spot and forward exchange markets; (2) Transaction costs are low; and (3) Instruments denominated in different currencies are perfect substitutes for one another.

Empirical studies of the efficient foreign exchange market hypothesis have yielded conflicting results. Nevertheless, a consensus is developing that rejects the efficient market hypothesis. It appears that the forward rate is not an unbiased predictor of the future spot rate and that it does pay to use resources to attempt to forecast exchange rates.

If the efficient market hypothesis is correct, a financial executive cannot expect to profit in any consistent manner from forecasting future exchange rates, because current quotations in the forward market reflect all that is presently known about likely future rates. Although future exchange rates may well differ from the expectation implicit in the present forward market

EXHIBIT 6.10   **Forward Rate as an Unbiased Predictor of Future Spot**

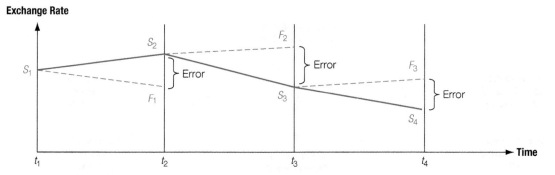

The forward rate available "today" (*t*) for delivery at a future time (*t* + 1) is used as a forecast or *predictor* of the spot rate at time *t* + 1. The difference between the spot rate which then occurs and the forward rate is the forecast error. When the forward rate is termed an "unbiased predictor of the future spot rate" it means that the errors are normally distributed around the mean future spot rate (the sum of the errors equal zero).

quotation, we cannot know today in which way actual future quotations will differ from today's forward rate. The expected mean value of deviations is zero. The forward rate is therefore an "unbiased" estimator of the future spot rate.

Tests of foreign exchange market efficiency, using longer time periods of analysis, conclude that either exchange market efficiency is untestable or, if it is testable, that the market is not efficient. Furthermore, the existence and success of foreign exchange forecasting services suggest that managers are willing to pay a price for forecast information even though they can use the forward rate as a forecast at no cost. The "cost" of buying this information is, in many circumstances, an "insurance premium" for financial managers who might get fired for using their own forecast, including forward rates, when that forecast proves incorrect. If they "bought" professional advice that turned out wrong, the fault was not in their forecast!

If the exchange market is not efficient, it is sensible for a firm to spend resources on forecasting exchange rates. This is the opposite conclusion to the one in which exchange markets are deemed efficient.

Prices, Interest Rates, and Exchange Rates in Equilibrium

Exhibit 6.11 illustrates all of the fundamental parity relations simultaneously, in equilibrium, using the U.S. dollar and the Japanese yen. The forecasted inflation rates for Japan and the United States are 1% and 5%, respectively—a 4% differential. The nominal interest rate in the U.S. dollar market (1-year government security) is 8%—a differential of 4% over the Japanese nominal interest rate of 4%. The spot rate is ¥104/$, and the 1-year forward rate is ¥100/$.

Relation A: Purchasing Power Parity (PPP). According to the relative version of purchasing power parity, the spot exchange rate one year from now, S_2, is expected to be ¥100/$:

$$S_2 = S_1 \times \frac{1 + \pi^{¥}}{1 + \pi^{\$}} = ¥104/\$ \times \frac{1.01}{1.05} = ¥100/\$.$$

This is a 4% change and equal, but opposite in sign, to the difference in expected rates of inflation (1% − 5%, or −4%).

| EXHIBIT 6.11 | **International Parity Conditions in Equilibrium (Approximate Form)** |

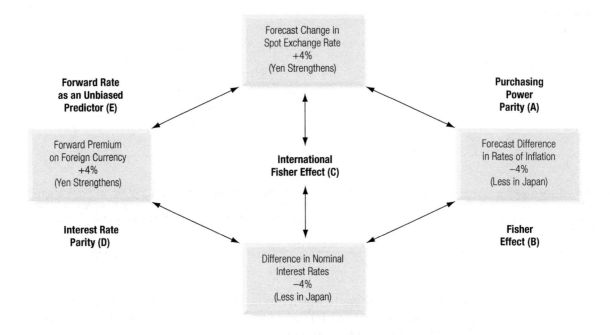

Relation B: The Fisher Effect. The real rate of return is the nominal rate of interest less the expected rate of inflation. Assuming efficient and open markets, the real rates of return should be equal across currencies.

Here, the real rate is 3% in U.S. dollar markets ($r = i - \pi = 8\% - 5\%$) and in Japanese yen markets (4% − 1%). Note that the 3% real rate of return is not in Exhibit 6.11, but rather the Fisher effect's relationship—that nominal interest rate differentials equal the difference in expected rates of inflation, −4%.

Relation C: International Fisher Effect. The forecast change in the spot exchange rate, in this case 4%, is equal to, but opposite in sign to, the differential between nominal interest rates:

$$\frac{S_1 - S_2}{S_2} \times 100 = i^{\yen} - i^{\$} = -4\%$$

Relation D: Interest Rate Parity (IRP). According to the theory of interest rate parity, the difference in nominal interest rates is equal to, but opposite in sign to, the forward premium. For this numerical example, the nominal yen interest rate (4%) is 4% less than the nominal dollar interest rate (8%):

$$i^{\yen} - i^{\$} = 1\% - 5\% = -4\%$$

and the forward premium, $f^{\yen}$, is a positive 4%:

$$f^{\yen} = \frac{S_1 - F}{F} \times 100 = \frac{\yen 104/\$ - \yen 100/\$}{\yen 100/\$} \times 100 = 4\%.$$

Relation E: Forward Rate as an Unbiased Predictor. Finally, the 1-year forward rate on the Japanese yen, F, if assumed to be an unbiased predictor of the future spot rate, also forecasts ¥100/$.

SUMMARY POINTS

- Parity conditions have traditionally been used by economists to help explain the long-run trend in an exchange rate.

- Under conditions of freely floating rates, the expected rate of change in the spot exchange rate, differential rates of national inflation and interest, and the forward discount or premium are all directly proportional to each other and mutually determined. A change in one of these variables has a tendency to change all of them with a feedback on the variable that changes first.

- If the identical product or service can be sold in two different markets, and there are no restrictions on its sale or transportation costs of moving the product between markets, the product's price should be the same in both markets. This is called the law of one price.

- The absolute version of purchasing power parity states that the spot exchange rate is determined by the relative prices of similar baskets of goods.

- The relative version of purchasing power parity states that if the spot exchange rate between two countries starts in equilibrium, any change in the differential rate of inflation between them tends to be offset over the long run by an equal but opposite change in the spot exchange rate.

- The Fisher effect, named after economist Irving Fisher, states that nominal interest rates in each country are equal to the required real rate of return plus compensation for expected inflation.

- The international Fisher effect, "Fisher-open" as it is often termed, states that the spot exchange rate should change in an equal amount, but in the opposite direction, to the difference in interest rates between two countries.

- The theory of interest rate parity (IRP) states that the difference in the national interest rates for securities of similar risk and maturity should be equal to, but opposite in sign to, the forward rate discount or premium for the foreign currency, except for transaction costs.

- When the spot and forward exchange markets are not in equilibrium, as described by interest rate parity, the potential for "riskless" or arbitrage profit exists. This is called covered interest arbitrage (CIA).

- Some forecasters believe that for the major floating currencies, foreign exchange markets are "efficient" and forward exchange rates are unbiased predictors of future spot exchange rates.

MINI-CASE

Mrs. Watanabe and the Japanese Yen Carry Trade[1]

At more than ¥1,500,000bn (some $16,800bn), these savings are considered the world's biggest pool of investable wealth. Most of it is stashed in ordinary Japanese bank accounts; a surprisingly large amount is kept at home in cash, in tansu savings, named for the traditional wooden cupboards in which people store their possessions. But from the early 2000s, the housewives—often referred to collectively as "Mrs. Watanabe," a common Japanese surname—began to hunt for higher returns.

> —"Shopping, Cooking, Cleaning Playing the Yen Carry Trade," *Financial Times*, February 21, 2009.

Over the past 20 years, Japanese yen interest rates have remained extremely low by global standards. For years the monetary authorities at the Bank of Japan have worked tirelessly fighting equity market collapses, deflationary pressures, liquidity traps, and economic recession, all by keeping yen-denominated interests rates hovering at around 1% per annum or lower. Combined with a sophisticated financial industry of size and depth, these low interest rates have spawned an international financial speculation termed the *yen carry trade*.

In the textbooks, this trading strategy is categorized more formally, as *uncovered interest arbitrage* (UIA). It is a fairly simple speculative position: borrow money where it is cheap and invest it in a different currency market with higher interest returns. The only real trick is to time the market correctly so that when the currency in the high-yield market is converted back to the original currency, the exchange rate has either stayed the same or moved in favor

---

of the speculator. "In favor of" means that the high-yielding currency has strengthened against the borrowed currency. And as Shakespeare stated, "ay, there's the rub."

Yen Availability

But why the focus on Japan? Aren't there other major currency markets in which interest rates are periodically low? Japan and the Japanese yen turn out to have a number of uniquely attractive characteristics to investors and speculators pursuing carry trade activities.

First, Japan has consistently demonstrated one of the world's highest savings rates for decades. This means that an enormous pool of funds has accumulated in the hands of private savers, savers who are traditionally very conservative. Those funds, whether stuffed in the mattress or placed in savings accounts, earn little in return. (In fact, given the extremely low interest rates offered, there is little effective difference between the mattress and the bank.)

A second factor facilitating the yen carry trade is the sheer size and sophistication of the Japanese financial sector. Not only is the Japanese economy one of the largest industrial economies in the world, it is one that has grown and developed with a strong international component. One only has to consider the size and global reach of Toyota or Sony to understand the established and developed infrastructure surrounding business and international finance

in Japan. The Japanese banking sector, however, has been continuously in search of new and diverse investments with which to balance the often despondent domestic economy. It has therefore sought out foreign investors and foreign borrowers who are attractive customers. Multinational companies have found ready access to yen-denominated debt for years—debt, which is, once again, available at extremely low interest.

A third expeditor of the yen carry trade is the value of the Japanese yen itself. The yen has long been considered the most international of Asian currencies, and is widely traded. It has, however, also been exceedingly volatile over time. But it is not volatility alone, as volatility itself could undermine interest arbitrage overnight. The key has been in the relatively long trends in value change of the yen against other major currencies like the U.S. dollar, or as in the following example, the Australian dollar.

The Australian Dollar/Japanese Yen Exchange Rate

Exhibit A illustrates the movement of the Japanese yen/Australian dollar exchange rate over a 13-year period, from 2000 through 2013. This spot rate movement and long-running periodic trends have offered a number of extended periods in which interest arbitrage was highly profitable.

EXHIBIT A	The Trending JPY and AUD Spot Rate

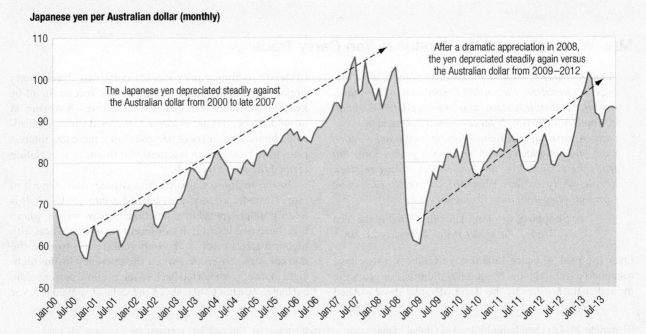

Japanese yen per Australian dollar (monthly)

The Japanese yen depreciated steadily against the Australian dollar from 2000 to late 2007

After a dramatic appreciation in 2008, the yen depreciated steadily again versus the Australian dollar from 2009–2012

The two periods of Aussie dollar appreciation are clear—after-the-fact. During those periods, an investor who was short yen and long Aussie dollars (and enjoying relatively higher Aussie dollar interest) could and did enjoy substantial returns.

But what about shorter holding periods, say a year, in which the speculator does not have a crystal ball over the long-term trend of the spot rate—but only a guess? Consider the one-year speculation detailed in Exhibit B. An investor looking at the exchange rate in January 2009 (Exhibit A) would see a yen that had reached a recent historical "low"—a strong position against the Aussie dollar. Betting that the yen would likely bounce, weakening once again against the Aussie dollar, she could borrow JPY50 million at 1.00% interest per annum for one year. She could then exchange the JPY50 million yen for Australian dollars at JPY60.91 = AUD1.00, and then deposit the AUD820,883 proceeds for one year at the Australian interest rate of 4.50% per annum. The investor could even have rationalized that even if the exchange rate did not change, she would earn a 3.50% per annum interest differential.

As it turned out, the spot exchange rate one year later, in January 2010, saw a much weaker Japanese yen against the Aussie dollar, JPY83.19 = AUD1.00. The one-year Aussie-Yen carry trade position would then have earned a very healthy profit of JPY20,862,296.83 on a one-year investment of JPY50,000,000, a 41.7% rate of return.

Post 2009 Financial Crisis

The global financial crisis of 2008–2009 has left a marketplace in which the U.S. Federal Reserve and the European Central Bank have pursued easy money policies. Both central banks, in an effort to maintain high levels of liquidity and to support fragile commercial banking systems, have kept interest rates at near-zero levels. Now global investors who see opportunities for profit in an anemic global economy are using those same low-cost funds in the U.S. and Europe to fund uncovered interest arbitrage activities. But what is making this "emerging market carry trade" so unique is not the interest rates, but the fact that investors are shorting two of the world's core currencies: the dollar and the euro.

Consider the strategy outlined in Exhibit C—a Euro-Indian rupee carry trade. An investor borrows EUR20 million at an incredibly low rate (again, because of ECB strategy to stimulate the sluggish European economy), say 1.00% per annum or 0.50% for 180 days. The EUR20 million are then exchanged for Indian rupees (INR), the current spot rate at the start of 2012 being a dramatic low of INR67.4 = EUR1.00. The resulting INR1,348,000,000 are put into an interest-bearing deposit with any of a number of Indian banks attempting to attract capital. The rate of interest offered, 2.50% (1.25% for 180 days), is not particularly high, but is greater than that available in the dollar, euro, or even yen markets. The account value at the end of 180 days, INR1,364,850,000, is then returned to euros at the spot rate of INR68.00 = EUR1.00, but at a loss. Although

---

EXHIBIT B   **The Aussie-Yen Carry Trade**

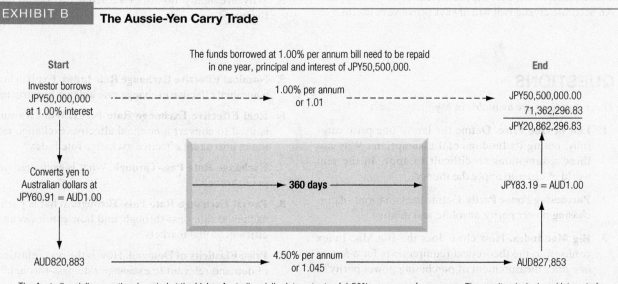

The Australian dollars are then invested at the higher Australian dollar interest rate of 4.50% per annum for one year. The result, principal, and interest of AUD 827,853, is then converted back to Japanese yen at the current spot rate. With luck, talent, or both, the result is profitable.

EXHIBIT C   **The Euro-Rupee Carry Trade**

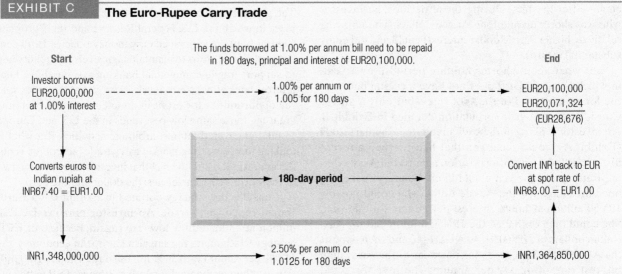

The Indian rupees are invested at 2.50% per annum, 1.25% for 180 days. The result, principal and interest of INR1,364,850,000, is converted back to euros at the spot rate in the market in 180 days of INR68.00 = EUR1.00. Unfortunately, the small movement in the spot rate has eliminated the interest arbitrage profits.

the rupee had not moved much, it had moved enough to eliminate the arbitrage profits.

Carry trade activity is often described in the global press as if it is *easy* or *riskless profit*. It is not. As in the case of the euro-rupee just described, the combination of interest rates and exchange rates is subject to a volatile global marketplace, which does indeed have a lot of moving parts. An accurate crystal ball will always prove very useful.

Mini-Case Questions

1. Why are interest rates so low in the traditional core markets of USD and EUR?
2. What makes this "emerging market carry trade" so different from traditional forms of uncovered interest arbitrage?
3. Why are many investors shorting the dollar and the euro?

QUESTIONS

These questions are available in MyFinanceLab.

1. **Law of One Price.** Define the law of one price carefully, noting its fundamental assumptions. Why are these assumptions so difficult to apply in the real world in order to apply the theory?

2. **Purchasing Power Parity.** Define the two forms of purchasing power parity, absolute and relative.

3. **Big Mac Index.** How close does the Big Mac Index conform to the theoretical requirements for a *law of one price* measurement of purchasing power parity?

4. **Undervaluation and Purchasing Power Parity.** According to the theory of purchasing power parity, what should happen to a currency that is undervalued?

5. **Nominal Effective Exchange Rate Index.** Explain how a nominal effective exchange rate index is constructed.

6. **Real Effective Exchange Rate Index.** What formula is used to convert a nominal effective exchange rate index into a real effective exchange rate index?

7. **Exchange Rate Pass-Through.** What is exchange rate pass-through?

8. **Partial Exchange Rate Pass-Through.** What is partial exchange rate pass-through, and how can it occur in efficient global markets?

9. **Price Elasticity of Demand.** How is the price elasticity of demand relevant to exchange rate pass-through?

10. **The Fisher Effect.** Define the Fisher effect. To what extent do empirical tests confirm that the Fisher effect exists in practice?

11. **Approximate Form of Fisher Effect.** Why is the approximate form of the Fisher Effect frequently used instead of the precise formulation? Does this introduce significant analysis error?

12. **The International Fisher Effect.** Define the international Fisher effect. To what extent do empirical tests confirm that the international Fisher effect exists in practice?

13. **Interest Rate Parity.** Define interest rate parity. What is the relationship between interest rate parity and forward rates?

14. **Covered Interest Arbitrage.** Define the terms *covered interest arbitrage* and *uncovered interest arbitrage*. What is the difference between these two transactions?

15. **Uncovered Interest Arbitrage.** Explain what expectations an investor or speculator would need to undertake an uncovered interest arbitrage investment?

16. **Forward Rate Calculation.** If someone you were working with argued that the current forward rate quoted on a currency pair is the market's expectation of where the future spot rate will end up, what would you say?

17. **Forward Rate as an Unbiased Predictor.** Some forecasters believe that foreign exchange markets for the major floating currencies are "efficient," and forward exchange rates are unbiased predictors of future spot exchange rates. What is meant by "unbiased predictor" in terms of the reliability of the forward rate in estimating future spot exchange rates?

18. **Transaction Costs.** If transaction costs for undertaking covered or uncovered interest arbitrage were large, how do you think it would influence arbitrage activity?

19. **Carry Trade.** The term carry trade is used quite frequently in the business press. What does it mean, and what conditions and expectations do investors need to hold to undertake carry trade transactions?

20. **Market Efficiency.** Many academics and professionals have tested the foreign exchange and interest rate markets to determine their efficiency. What have they concluded?

PROBLEMS

These problems are available in MyFinanceLab.

1. **Pulau Penang Island Resort.** Theresa Nunn is planning a 30-day vacation on Pulau Penang, Malaysia, one year from now. The present charge for a luxury suite plus meals in Malaysian ringgit (RM) is RM1,045/day.

The Malaysian ringgit presently trades at RM3.1350/\$. She determines that the dollar cost today for a 30-day stay would be \$10,000. The hotel informs her that any increase in its room charges will be limited to any increase in the Malaysian cost of living. Malaysian inflation is expected to be 2.75% per annum, while U.S. inflation is expected to be 1.25%.
   a. How many dollars might Theresa expect to need one year hence to pay for her 30-day vacation?
   b. By what percent will the dollar cost have gone up? Why?

2. **Argentine Tears.** The Argentine peso was fixed through a currency board at Ps1.00/\$ throughout the 1990s. In January 2002, the Argentine peso was floated. On January 29, 2003, it was trading at Ps3.20/\$. During that one-year period, Argentina's inflation rate was 20% on an annualized basis. Inflation in the United States during that same period was 2.2% annualized.
   a. What should have been the exchange rate in January 2003 if PPP held?
   b. By what percentage was the Argentine peso undervalued on an annualized basis?
   c. What were the probable causes of undervaluation?

3. **Derek Tosh and Yen-Dollar Parity.** Derek Tosh is attempting to determine whether U.S./Japanese financial conditions are at parity. The current spot rate is a flat ¥89.00/\$, while the 360-day forward rate is ¥84.90/\$. Forecast inflation is 1.100% for Japan, and 5.900% for the United States. The 360-day euroyen deposit rate is 4.700%, and the 360-day eurodollar deposit rate is 9.500%.
   a. Diagram and calculate whether international parity conditions hold between Japan and the United States.
   b. Find the forecasted change in the Japanese yen/U.S. dollar (¥/\$) exchange rate one year from now.

4. **Sydney to Phoenix.** Terry Lamoreaux owns homes in Sydney, Australia, and Phoenix, United States. He travels between the two cities at least twice a year. Because of his frequent trips, he wants to buy some new high-quality luggage. He has done his research and has decided to purchase a Briggs and Riley three-piece luggage set. There are retail stores in Phoenix and Sydney. Terry was a finance major and wants to use purchasing power parity to determine if he is paying the same price regardless of where he makes his purchase.
   a. If the price of the three-piece luggage set in Phoenix is \$850 and the price of the same three-piece set in Sydney is A\$930, using purchasing power parity,

is the price of the luggage truly equal if the spot rate is A$1.0941/$?

b. If the price of the luggage remains the same in Phoenix one year from now, determine the price of the luggage in Sydney in one year's time if PPP holds true. The U.S. inflation rate is 1.15% and the Australian inflation rate is 3.13%.

5. **Starbucks in Croatia.** Starbucks opened its first store in Zagreb, Croatia in October 2010. In Zagreb, the price of a tall vanilla latte is 25.70 Croatian kunas (kn or HRK). In New York City, the price of a tall vanilla latte is $2.65. The exchange rate between Croatian kunas and U.S. dollars is kn5.6288 = $1. According to purchasing power parity, is the Croatian kuna overvalued or undervalued?

6. **Corolla Exports and Pass-Through.** Assume that the export price of a Toyota Corolla from Osaka, Japan is ¥2,150,000. The exchange rate is ¥87.60/$. The forecast rate of inflation in the United States is 2.2% per year and in Japan it is 0.0% per year. Use this data to answer the following questions on exchange rate pass-through.

a. What was the export price for the Corolla at the beginning of the year expressed in U.S. dollars?

b. Assuming purchasing power parity holds, what should be the exchange rate at the end of the year?

c. Assuming 100% exchange rate pass-through, what will be the dollar price of a Corolla at the end of the year?

d. Assuming 75% exchange rate pass-through, what will be the dollar price of a Corolla at the end of the year?

7. **Takeshi Kamada—CIA Japan (A).** Takeshi Kamada, a foreign exchange trader at Credit Suisse (Tokyo), is exploring covered interest arbitrage possibilities. He wants to invest $5,000,000 or its yen equivalent, in a covered interest arbitrage between U.S. dollars and Japanese yen. He faced the following exchange rate and interest rate quotes. Is CIA profit possible? If so, how?

Arbitrage funds available	$5,000,000
Spot rate (¥/$)	118.60
180-day forward rate (¥/$)	117.80
180-day U.S. dollar interest rate	4.800%
180-day Japanese yen interest rate	3.400%

8. **Takeshi Kamada—UIA Japan (B).** Takeshi Kamada, Credit Suisse (Tokyo), observes that the ¥/$ spot rate has been holding steady, and that both dollar and yen interest rates have remained relatively fixed over the past week. Takeshi wonders if he should try an uncovered interest arbitrage (UIA) and thereby save the cost of forward cover. Many of Takeshi's research associates—and their computer models—are predicting the spot rate to remain close to ¥118.00/$ for the coming 180 days. Using the same data as in Problem 7, analyze the UIA potential.

9. **Copenhagen Covered (A).** Heidi Høi Jensen, a foreign exchange trader at JPMorgan Chase, can invest $5 million, or the foreign currency equivalent of the bank's short-term funds, in a covered interest arbitrage with Denmark. Using the following quotes, can Heidi make a covered interest arbitrage (CIA) profit?

Arbitrage funds available	$5,000,000
Spot exchange rate (kr/$)	6.1720
3-month forward rate (kr/$)	6.1980
U.S. dollar 3-month interest rate	3.000%
Danish kroner 3-month interest rate	5.000%

10. **Copenhagen Covered (B).** Heidi Høi Jensen is now evaluating the arbitrage profit potential in the same market after interest rates change. (Note that any time the difference in interest rates does not exactly equal the forward premium, it must be possible to make a CIA profit one way or another.)

Arbitrage funds available	$5,000,000
Spot exchange rate (kr/$)	6.1720
3-month forward rate (kr/$)	6.1980
U.S. dollar 3-month interest rate	4.000%
Danish kroner 3-month interest rate	5.000%

11. **Copenhagen Covered (C).** Heidi Høi Jensen is again evaluating the arbitrage profit potential in the same market after another change in interest rates. (Remember that any time the difference in interest rates does not exactly equal the forward premium, it must be possible to make a CIA profit one way or another.)

Arbitrage funds available	$5,000,000
Spot exchange rate (kr/$)	6.1720
3-month forward rate (kr/$)	6.1980
U.S. dollar 3-month interest rate	3.000%
Danish kroner 3-month interest rate	6.000%

12. **Casper Landsten—CIA (A).** Casper Landsten is a foreign exchange trader for a bank in New York. He has $1 million (or its Swiss franc equivalent) for a short-term money market investment and wonders whether he should invest in U.S. dollars for three months or

make a CIA investment in the Swiss franc. He faces the following quotes:

Arbitrage funds available	$1,000,000
Spot exchange rate (SFr/$)	1.2810
3-month forward rate (SFr/$)	1.2740
U.S. dollar 3-month interest rate	4.800%
Swiss franc 3-month interest rate	3.200%

13. Casper Landsten—UIA (B). Casper Landsten, using the same values and assumptions as in Problem 12, decides to seek the full 4.800% return available in U.S. dollars by not covering his forward dollar receipts— an uncovered interest arbitrage (UIA) transaction. Assess this decision.

14. Casper Landsten—Thirty Days Later. One month after the events described in Problems 12 and 13, Casper Landsten once again has $1 million (or its Swiss franc equivalent) to invest for three months. He now faces the following rates. Should he again enter into a covered interest arbitrage (CIA) investment?

Arbitrage funds available	$1,000,000
Spot exchange rate (SFr/$)	1.3392
3-month forward rate (SFr/$)	1.3286
U.S. dollar 3-month interest rate	4.750%
Swiss franc 3-month interest rate	3.625%

15. Statoil of Norway's Arbitrage. Statoil, the national oil company of Norway, is a large, sophisticated, and active participant in both the currency and petrochemical markets. Although it is a Norwegian company, because it operates within the global oil market, it considers the U.S. dollar, rather than the Norwegian krone, as its functional currency. Ari Karlsen is a currency trader for Statoil and has immediate use of either $3 million (or the Norwegian krone equivalent). He is faced with the following market rates and wonders whether he can make some arbitrage profits in the coming 90 days.

Arbitrage funds available	$3,000,000
Spot exchange rate (Nok/$)	6.0312
3-month forward rate (Nok/$)	6.0186
U.S. dollar 3-month interest rate	5.000%
Norwegian krone 3-month interest rate	4.450%

16. Separated by the Atlantic. Separated by more than 3,000 nautical miles and five time zones, money and foreign exchange markets in both London and New York are very efficient. The following information has been collected from the respective areas:

Assumptions	London	New York
Spot exchange rate ($/€)	1.3264	1.3264
1-year Treasury bill rate	3.900%	4.500%
Expected inflation rate	Unknown	1.250%

a. What do the financial markets suggest for inflation in Europe next year?
b. Estimate today's 1-year forward exchange rate between the dollar and the euro?

17. Chamonix Chateau Rentals. You are planning a ski vacation to Mt. Blanc in Chamonix, France, one year from now. You are negotiating the rental of a chateau. The chateau's owner wishes to preserve his real income against both inflation and exchange rate changes, and so the present weekly rent of €9,800 (Christmas season) will be adjusted upward or downward for any change in the French cost of living between now and then. You are basing your budgeting on purchasing power parity (PPP). French inflation is expected to average 3.5% for the coming year, while U.S. dollar inflation is expected to be 2.5%. The current spot rate is $1.3620/€. What should you budget as the U.S. dollar cost of the 1-week rental?

Spot exchange rate ($/€)	$1.3620
Expected U.S. inflation for coming year	2.500%
Expected French inflation for coming year	3.500%
Current chateau nominal weekly rent (€)	9,800.00

18. East Asiatic Company—Thailand. The East Asiatic Company (EAC), a Danish company with subsidiaries throughout Asia, has been funding its Bangkok subsidiary primarily with U.S. dollar debt because of the cost and availability of dollar capital as opposed to Thai baht-denominated (B) debt. The treasurer of EAC-Thailand is considering a 1-year bank loan for $250,000. The current spot rate is B32.06/$, and the dollar-based interest is 6.75% for the 1-year period. 1-year loans are 12.00% in baht.
a. Assuming expected inflation rates for the coming year of 4.3% and 1.25% in Thailand and the United States, respectively, according to purchase power parity, what would be the effective cost of funds in Thai baht terms?
b. If EAC's foreign exchange advisers believe strongly that the Thai government wants to push the value of the baht down against the dollar by 5% over the coming year (to promote its export competitiveness in dollar markets), what might be the effective cost of funds in baht terms?
c. If EAC could borrow Thai baht at 13% per annum, would this be cheaper than either part (a) or part (b)?

19. Maltese Falcon. Imagine that the mythical solid gold falcon, initially intended as a tribute by the Knights of Malta to the King of Spain in appreciation for his gift of the island of Malta to the order in 1530, has recently been recovered. The falcon is 14 inches high and solid gold, weighing approximately 48 pounds. Assume that gold prices have risen to $440/ounce, primarily as a result of increasing political tensions. The falcon is currently held by a private investor in Istanbul, who is actively negotiating with the Maltese government on its purchase and prospective return to its island home. The sale and payment are to take place one year from now, and the parties are negotiating over the price and currency of payment. The investor has decided, in a show of goodwill, to base the sales price only on the falcon's specie value—its gold value.

The current spot exchange rate is 0.39 Maltese lira (ML) per 1.00 U.S. dollar. Maltese inflation is expected to be about 8.5% for the coming year, while U.S. inflation, on the heels of a double-dip recession, is expected to come in at only 1.5%. If the investor bases value in the U.S. dollar, would he be better off receiving Maltese lira in one year (assuming purchasing power parity) or receiving a guaranteed dollar payment (assuming a gold price of $420 per ounce one year from now).

20. Malaysian Risk. Clayton Moore is the manager of an international money market fund managed out of London. Unlike many money funds that guarantee their investors a near risk-free investment with variable interest earnings, Clayton Moore's fund is a very aggressive fund that searches out relatively high interest earnings around the globe, but at some risk. The fund is pound-denominated. Clayton is currently evaluating a rather interesting opportunity in Malaysia. Since the Asian Crisis of 1997, the Malaysian government enforced a number of currency and capital restrictions to protect and preserve the value of the Malaysian ringgit.

The ringgit was fixed to the U.S. dollar at RM3.80/$ for seven years. In 2005, the Malaysian government allowed the currency to float against several major currencies. The current spot rate today is RM3.13485/$. Local currency time deposits of 180-day maturities are earning 8.900% per annum. The London eurocurrency market for pounds is yielding 4.200% per annum on similar 180-day maturities. The current spot rate on the British pound is $1.5820/£, and the 180-day forward rate is $1.5561/£.

21. The Beer Standard. In 1999, The Economist reported the creation of an index, or standard, for the evaluation of African currency values using the local prices of beer. Beer, instead of Big Macs, was chosen as the product for comparison because McDonald's had not penetrated the African continent beyond South Africa, and beer met most of the same product and market characteristics required for the construction of a proper currency index. Investec, a South African investment banking firm, has replicated the process of creating a measure of purchasing power parity (PPP) like that of the Big Mac Index of *The Economist*, for Africa.

The index compares the cost of a 375-milliliter bottle of clear lager beer across Sub-Saharan Africa. As a measure of PPP, the beer needs to be relatively homogeneous in quality across countries, and must possess substantial elements of local manufacturing, inputs, distribution, and service in order to actually provide a measure of relative purchasing power. The beer is first priced in local currency (purchased in the taverns of the locals, and not in the high-priced tourist centers). The price is then converted to South African rand and the rand-price compared to the local currency price as one measure of whether the local currency is undervalued or overvalued versus the South African rand. Use the data in the table below and complete the calculation of whether the individual currencies are undervalued or overvalued.

Problem 21.

| | | **Beer Prices** | | | | |
Country	Beer	Local Currency	Price in Currency	Price in Rand	Implied PPP Rate	Spot Rate
South Africa	Castle	Rand	2.30	—	—	—
Botswana	Castle	Pula	2.20	2.94	0.96	0.75
Ghana	Star	Cedi	1,200.00	3.17	521.74	379.10
Kenya	Tusker	Shilling	41.25	4.02	17.93	10.27
Malawi	Carlsberg	Kwacha	18.50	2.66	8.04	6.96
Mauritius	Phoenix	Rupee	15.00	3.72	6.52	4.03
Namibia	Windhoek	N$	2.50	2.50	1.09	1.00
Zambia	Castle	Kwacha	1,200.00	3.52	521.74	340.68
Zimbabwe	Castle	Z$	9.00	1.46	3.91	6.15

Problem 23.

Calendar Year	2001	2002	2003	2004	2005	2006
Kalina Price (rubles)	260,000					
Russian inflation (forecast)		14.0%	12.0%	11.0%	8.0%	8.0%
U.S. inflation (forecast)		2.5%	3.0%	3.0%	3.0%	3.0%
Exchange rate (rubles = USD1.00)	30.00					

22. **Grupo Bimbo (Mexico).** Grupo Bimbo, headquartered in Mexico City, is one of the largest bakery companies in the world. On January 1st, when the spot exchange rate is Ps10.80/$, the company borrows $25.0 million from a New York bank for one year at 6.80% interest (Mexican banks had quoted 9.60% for an equivalent loan in pesos). During the year, U.S. inflation is 2% and Mexican inflation is 4%. At the end of the year the firm repays the dollar loan.

a. If Bimbo expected the spot rate at the end of one year to be that equal to purchasing power parity, what would be the cost to Bimbo of its dollar loan in peso-denominated interest?

b. What is the real interest cost (adjusted for inflation) to Bimbo, in peso-denominated terms, of borrowing the dollars for one year, again assuming purchasing power parity ?

c. If the actual spot rate at the end of the year turned out to be Ps9.60/$, what was the actual peso-denominated interest cost of the loan?

23. **AvtoVAZ of Russia's Kalina Export Pricing Analysis.** AvtoVAZ OAO, a leading auto manufacturer in Russia, was launching a new automobile model in 2001, and is in the midst of completing a complete pricing analysis of the car for sales in Russia and export. The new car, the Kalina, would be initially priced at Rubles 260,000 in Russia, and if exported, $8,666.67 in U.S. dollars at the current spot rate of Rubles 30 = $1.00. AvtoVAZ intends to raise the price domestically with the rate of Russian inflation over time, but is worried about how that compares to the export price given U.S. dollar inflation and the future exchange rate. Use the data table above to answer the pricing analysis questions.

a. If the domestic price of the Kalina increases with the rate of inflation, what would its price be over the 2002–2006 period?

b. Assuming that the forecasts of U.S. and Russian inflation prove accurate, what would the value of the ruble be over the coming years if its value versus the dollar followed purchasing power parity?

c. If the export price of the Kalina were set using the purchasing power parity forecast of the ruble-dollar exchange rate, what would the export price be over the 2002–2006 period?

d. How would the Kalina's export price evolve over time if it followed Russian inflation and the exchange rate of the ruble versus the dollar remained relatively constant over this period of time?

e. Vlad, one of the newly hired pricing strategists, believes that prices of automobiles in both domestic and export markets will both increase with the rate of inflation, and that the ruble/dollar exchange rate will remain fixed. What would this imply or forecast for the future export price of the Kalina?

f. If you were AvtoVAZ, what would you hope would happen to the ruble's value versus the dollar over time given your desire to export the Kalina? Now if you combined that "hope" with some assumptions about the competition—other automobile sales prices in dollar markets over time—how might your strategy evolve?

g. So what did the Russian ruble end up doing over the 2001–2006 period?

INTERNET EXERCISES

1. **Big Mac Index Updated.** Use *The Economist's* Web site to find the latest edition of the Big Mac Index of currency overvaluation and undervaluation. (You will need to do a search for "Big Mac Currencies.") Create a worksheet to compare how the British pound, the euro, the Swiss franc, and the Canadian dollar have changed from the version presented in this chapter.

The Economist   www.economist.com/markets-data

2. **Purchasing Power Parity Statistics.** The Organization for Economic Cooperation and Development (OECD) publishes detailed measures of prices and purchasing power for its member countries. Go to the OECD's Web site and download the spreadsheet file with the historical data for purchasing power for the member countries.

OECD   www.oecd.org/std/prices-ppp/

3. **International Interest Rates.** A number of Web sites publish current interest rates by currency and maturity. Use the *Financial Times* Web site listed here to isolate the interest rate differentials between the U.S. dollar, the British pound, and the euro for all maturities up to and including one year.

Financial Times   markets.ft.com/RESEARCH/ Markets/Interest-Rates

Data Listed by the *Financial Times*:

- ■ International money rates (bank call rates for major currency deposits)

- ■ Money rates (LIBOR and CD rates, etc.)

- ■ 10-year spreads (individual country spreads versus the euro and U.S. 10-year treasuries). Check which countries actually have lower 10-year government bond rates than the United States and the euro. Probably Switzerland and Japan.

- ■ Benchmark government bonds (a sampling of representative government issuances by major countries and recent price movements). Note which countries are showing longer maturity benchmark rates.

- ■ Emerging market bonds (government issuances, Brady bonds, etc.)

- ■ Eurozone rates (miscellaneous bond rates for assorted European-based companies; includes debt ratings)

4. **World Bank's International Comparison Program.** The World Bank has an ongoing research program that focuses on the relative purchasing power of 107 different economies globally. Download the latest data tables and highlight which economies seem to be showing the greatest growth in recent years in relative purchasing power. Search the Internet for the World Bank's ICP program site.

CHAPTER 6 **APPENDIX**
An Algebraic Primer to International Parity Conditions

The following is a purely algebraic presentation of the parity conditions explained in this chapter. It is offered to provide those who wish for additional theoretical detail and definition ready access to the step-by-step derivation of the various conditions.

The Law of One Price

The *law of one price* refers to the state in which, in the presence of free trade, perfect substitutability of goods, and costless transactions, the equilibrium exchange rate between two currencies is determined by the ratio of the price of any commodity I denominated in two different currencies. For example,

$$S_t = \frac{P_{i,t}^{\$}}{P_{i,t}^{\text{SF}}},$$

where $P_i^{\$}$ and P_i^{SF} refer to the prices of the same commodity i, at time t, denominated in U.S. dollars and Swiss francs, respectively. The spot exchange rate, S_t, is simply the ratio of the two currency prices.

Purchasing Power Parity

The more general form in which the exchange rate is determined by the ratio of two price indexes is termed the absolute version of *purchasing power parity* (PPP). Each price index reflects the currency cost of the identical "basket" of goods across countries. The exchange rate that equates purchasing power for the identical collection of goods is then stated as follows:

$$S_t = \frac{P_t^{\$}}{P_t^{\text{SF}}},$$

where $P_t^{\$}$ and P_t^{SF} are the price index values in U.S. dollars and Swiss francs at time t, respectively. If $\pi^{\$}$ and π^{SF} represent the rate of inflation in each country, the spot exchange rate at time $t + 1$ would be

$$S_{t+1} = \frac{P_t^{\$}(1 + \pi^{\$})}{P_t^{\text{SF}}(1 + \pi^{\text{SF}})} = S_t \left[\frac{(1 + \pi^{\$})}{(1 + \pi^{\text{SF}})} \right].$$

The change from period t to $t + 1$ is then

$$\frac{S_{t+1}}{S_t} = \frac{\dfrac{P_t^{\$}(1 + \pi^{\$})}{P_t^{\text{SF}}(1 + \pi^{\text{SF}})}}{\dfrac{P_t^{\$}}{P_t^{\text{SF}}}} = \frac{S_t \left[\dfrac{(1 + \pi^{\$})}{(1 + \pi^{\text{SF}})} \right]}{S_t} = \frac{(1 + \pi^{\$})}{(1 + \pi^{\text{SF}})}.$$

177

Isolating the percentage change in the spot exchange rate between periods t to $t + 1$ is then

$$\frac{S_{t+1} - S_t}{S_t} = \frac{S_t\left[\frac{(1 + \pi^\$)}{(1 + \pi^{SF})}\right] - S_t}{S_t} = \frac{(1 + \pi^\$) - (1 + \pi^{SF})}{(1 + \pi^{SF})}.$$

This equation is often approximated by dropping the denominator of the right-hand side if it is considered to be relatively small. It is then stated as

$$\frac{S_{t+1} - S_t}{S_t} = (1 + \pi^\$) - (1 + \pi^{SF}) = \pi^\$ - \pi^{SF}.$$

Forward Rates

The *forward exchange rate* is that contractual rate that is available to private agents through banking institutions and other financial intermediaries who deal in foreign currencies and debt instruments. The annualized percentage difference between the forward rate and the spot rate is termed the *forward premium*,

$$f^{SF} = \left[\frac{F_{t,t+1} - S_t}{S_t}\right] \times \left[\frac{360}{n_{t,t+1}}\right].$$

where f^{SF} is the forward premium on the Swiss franc, $F_{t,t+1}$ is the forward rate contracted at time t for delivery at time $t + 1$, S_t is the current spot rate, and $n_{t,t+1}$ is the number of days between the contract date (t) and the delivery date ($t + 1$).

Covered Interest Arbitrage (CIA) and Interest Rate Parity (IRP)

The process of *covered interest arbitrage* is when an investor exchanges domestic currency for foreign currency in the spot market, invests that currency in an interest-bearing instrument, and signs a forward contract to "lock in" a future exchange rate at which to convert the foreign currency proceeds (gross) back to domestic currency. The net return on CIA is

$$\text{Net Return} = \left[\frac{(1 + i^{SF})F_{t,t+1}}{S_t}\right] - (1 + i^\$),$$

where S_t and $F_{t,t+1}$ are the spot and forward rates (\$/SF), i^{SF} is the nominal interest rate (or yield) on a Swiss franc-denominated monetary instrument, and $i^\$$ is the nominal return on a similar dollar-denominated instrument.

If they possess exactly equal rates of return—that is, if CIA results in zero riskless profit—*interest rate parity* (IRP) holds, and appears as

$$(1 + i^\$) = \left[\frac{(1 + i^{SF})F_{t,t+1}}{S_t}\right]$$

or alternatively as

$$\frac{(1 + i^\$)}{(1 + i^{SF})} = \frac{F_{t,t+1}}{S_t}.$$

If the percent difference of both sides of this equation is found (the percentage difference between the spot and forward rate is the *forward premium*, then the relationship between the forward premium and relative interest rate differentials is

$$\frac{F_{t,t+1} - S_t}{S_t} = f^{\text{SF}} = \frac{i^\$ - i^{\text{SF}}}{1 + i^{\text{SF}}}.$$

If these values are not equal (thus, the markets are not in equilibrium), there exists a potential for riskless profit. The market will then be driven back to equilibrium through CIA by agents attempting to exploit such arbitrage potential, until CIA yields no positive return.

Fisher Effect

The *Fisher effect* states that all nominal interest rates can be decomposed into an implied real rate of interest (return) and an expected rate of inflation:

$$i^\$ = [(1 + r^\$)(1 + \pi^\$)] - 1,$$

where $r^\$$ is the real rate of return and $\pi^\$$ is the expected rate of inflation for dollar-denominated assets. The subcomponents are then identifiable:

$$i^\$ = r^\$ + \pi^\$ + r^\$\pi^\$.$$

As with PPP, there is an approximation of this function that has gained wide acceptance. The cross-product term of $r^\$\pi^\$$ is often very small and therefore dropped altogether:

$$i^\$ = r^\$ + \pi^\$.$$

International Fisher Effect

The *international Fisher effect* is the extension of this domestic interest rate relationship to the international currency markets. If capital, by way of covered interest arbitrage (CIA), attempts to find higher rates of return internationally resulting from current interest rate differentials, the real rates of return between currencies are equalized (e.g., $r^\$ = r^{\text{SF}}$):

$$\frac{S_{t+1} - S_t}{S_t} = \frac{(1 + i^\$) - (1 + i^{\text{SF}})}{(1 + i^{\text{SF}})} = \frac{i^\$ - i^{\text{SF}}}{(1 + i^{\text{SF}})}.$$

If the nominal interest rates are then decomposed into their respective real and expected inflation components, the percentage change in the spot exchange rate is

$$\frac{S_{t+1} - S_t}{S_t} = \frac{(r^\$ + \pi^\$ + r^\$\pi^\$) - (r^{\text{SF}} + \pi^{\text{SF}} + r^{\text{SF}}\pi^{\text{SF}})}{1 + r^{\text{SF}} + \pi^{\text{SF}} + r^{\text{SF}}\pi^{\text{SF}}}.$$

The international Fisher effect has a number of additional implications, if the following requirements are met: (1) capital markets can be freely entered and exited; (2) capital markets possess investment opportunities that are acceptable substitutes; and (3) market agents have complete and equal information regarding these possibilities.

Given these conditions, international arbitragers are capable of exploiting all potential riskless profit opportunities, until real rates of return between markets are equalized ($r^\$ = r^{\text{SF}}$).

Thus, the expected rate of change in the spot exchange rate reduces to the differential in the expected rates of inflation:

$$\frac{S_{t+1} - S_t}{S_t} = \frac{\pi^\$ + r^\$\pi^\$ - \pi^{SF} - r^{SF}\pi^{SF}}{1 + r^{SF} + \pi^{SF} + r^{SF}\pi^{SF}}.$$

If the approximation forms are combined (through the elimination of the denominator and the elimination of the interactive terms of r and π), the change in the spot rate is simply

$$\frac{S_{t+1} - S_t}{S_t} = \pi^\$ - \pi^{SF}.$$

Note the similarity (identical in equation form) of the approximate form of the international Fisher effect to purchasing power parity, discussed previously (the only potential difference is that between *ex post* and *ex ante* (expected) inflation.

CHAPTER 7

Foreign Currency Derivatives: Futures and Options

Unless derivatives contracts are collateralized or guaranteed, their ultimate value also depends on the creditworthiness of the counterparties to them. In the meantime, though, before a contract is settled, the counterparties record profits and losses—often huge in amount—in their current earnings statements without so much as a penny changing hands. The range of derivatives contracts is limited only by the imagination of man (or sometimes, so it seems, madmen).

—Warren Buffett, *Berkshire Hathaway Annual Report*, 2002.

LEARNING OBJECTIVES

- Explain how foreign currency futures are quoted, valued, and used for speculation purposes
- Explore the buying and writing of foreign currency options in terms of risk and return
- Examine how foreign currency option values change with exchange rate movements and over time
- Analyze how foreign currency option values change with price component changes

Financial management of the multinational enterprise in the twenty-first century will certainly include the use of *financial derivatives*. These derivatives, so named because their values are derived from an underlying asset like a stock or a currency, are powerful tools used in business today for two very distinct management objectives, *speculation* and *hedging*. The financial manager of an MNE may purchase financial derivatives in order to take positions in the expectation of profit—*speculation*—or may use these instruments to reduce the risks associated with the everyday management of corporate cash flow—*hedging*. Before these financial instruments can be used effectively, however, the financial manager must understand certain basics about their structure and pricing.

In this chapter, we cover the primary foreign currency financial derivatives used today in multinational finance. Here we focus on the fundamentals of their valuation and use for speculative purposes; Chapter 9 will describe how these foreign currency derivatives can be used to hedge commercial transactions. The Mini-Case at the end of this chapter, *KiKos and the South Korean Won*, illustrates how currency options can be combined to form rather complex products—even for their buyers.

A word of caution—of reservation—before proceeding further. Financial derivatives are powerful tools in the hands of careful and competent financial managers. They can also be destructive devices when used recklessly and carelessly. The history of finance is littered with cases in which financial managers—either intentionally or unintentionally—took huge positions resulting in significant losses for their companies, and occasionally, their outright collapse. In the right hands and with the proper controls, however, derivatives may provide management with opportunities to enhance and protect corporate financial performance. User beware.

Foreign Currency Futures

A *foreign currency futures contract* is an alternative to a forward contract that calls for future delivery of a standard amount of foreign exchange at a fixed time, place, and price. It is similar to futures contracts that exist for commodities (hogs, cattle, lumber, etc.), interest-bearing deposits, and gold.

Most world money centers have established foreign currency futures markets. In the United States, the most important market for foreign currency futures is the *International Monetary Market* (IMM) of Chicago, a division of the Chicago Mercantile Exchange.

Contract Specifications

Contract specifications are established by the exchange on which futures are traded. Using the Chicago IMM as an example, the major features of standardized futures trading can be illustrated by the Mexican peso futures traded on the Chicago Mercantile Exchange (CME), as shown in Exhibit 7.1.

Each futures contract is for 500,000 Mexican pesos. This is the notional principal. Trading in each currency must be done in an even multiple of currency units. The method of stating exchange rates is in American terms, the U.S. dollar cost (price) of a foreign currency (unit), $/MXN, where the CME is mixing the old dollar symbol with the ISO 4217 code for the peso, MXN. In Exhibit 7.1 this is U.S. dollars per Mexican peso. Contracts mature on the third Wednesday of January, March, April, June, July, September, October, or December. Contracts may be traded through the second business day prior to the Wednesday on which they mature. Unless holidays interfere, the last trading day is the Monday preceding the maturity date.

EXHIBIT 7.1   Mexican Peso (CME) (MXN 500,000; $ per MXN)

Maturity	Open	High	Low	Settle	Change	Lifetime High	Lifetime Low	Open Interest
Mar	0.10953	0.10988	0.10930	0.10958	. . .	0.11000	0.09770	34,481.00
June	0.10790	0.10795	0.10778	0.10773	. . .	0.10800	0.09730	3,405.00
Sept	0.10615	0.10615	0.10610	0.10573	. . .	0.10615	0.09930	1,481.00

All contracts are for 500,000 Mexican pesos. "Open" means the opening price on the day. "High" means the high price on the day. "Low" indicates the lowest price on the day. "Settle" is the closing price on the day. "Change" indicates the change in the settle price from the previous day's close. "High" and "Low" to the right of "Change" indicate the highest and lowest prices this specific contract (as defined by its maturity) has experienced over its trading history. "Open Interest" indicates the number of contracts outstanding.

One of the defining characteristics of futures is the requirement that the purchaser deposit a sum as an initial *margin* or *collateral*. This requirement is similar to requiring a performance bond, and it can be met by a letter of credit from a bank, Treasury bills, or cash. In addition, a maintenance margin is required. The value of the contract is marked to market daily, and all changes in value are paid in cash daily. *Marked-to-market* means that the value of the contract is revalued using the closing price for the day. The amount to be paid is called the variation margin.

Only about 5% of all futures contracts are settled by the physical delivery of foreign exchange between buyer and seller. More often, buyers and sellers offset their original position prior to delivery date by taking an opposite position. That is, an investor will normally close out a futures position by selling a futures contract for the same delivery date. The complete buy/sell or sell/buy is called a "round turn."

Customers pay a commission to their broker to execute a round turn and a single price is quoted. This practice differs from that of the interbank market, where dealers quote a bid and an offer and do not charge a commission. All contracts are agreements between the client and the exchange clearinghouse, rather than between the two clients involved. Consequently, clients need not worry that a specific counterparty in the market will fail to honor an agreement, termed *counterparty risk*. The clearinghouse is owned and guaranteed by all members of the exchange.

Using Foreign Currency Futures

Any investor wishing to speculate on the movement of the Mexican peso versus the U.S. dollar could pursue one of the following futures strategies. Keep in mind the principle of a futures contract: A speculator who buys a futures contract is locking in the price at which she must buy that currency on the specified future date. A speculator who sells a futures contract is locking in the price at which she must sell that currency on that future date.

Short Positions. If Amber McClain, a speculator working for International Currency Traders, believes that the Mexican peso will fall in value versus the U.S. dollar by March, she could sell a March futures contract, taking a short position. By selling a March contract, Amber locks in the right to sell 500,000 Mexican pesos at a set price. If the price of the peso falls by the maturity date as she expects, Amber has a contract to sell pesos at a price above their current price on the spot market. Hence, she makes a profit.

Using the quotes on Mexican peso (MXN) futures in Exhibit 7.1, Amber sells one March futures contract for 500,000 pesos at the closing price, termed the settle price, of $.10958/MXN. The value of her position at maturity—at the expiration of the futures contract in March—is then

$$\text{Value at maturity (Short position)} = -\text{Notional principal} \times (\text{Spot} - \text{Futures}).$$

Note that the short position is entered into the valuation as a negative notional principal. If the spot exchange rate at maturity is $.09500/MXN, the value of her position on settlement is

$$\text{Value} = -\text{MXN500,000} \times (\$.09500/\text{MXN} - \$.10958/\text{MXN}) = \$7,290.$$

Amber's expectation proved correct; the Mexican peso fell in value versus the U.S. dollar. We could say, "Amber ends up buying at $.09500 and sells at $.10958 per peso."

All that was really required of Amber to speculate on the Mexican peso's value was for her to form an opinion on the Mexican peso's future exchange value versus the U.S. dollar. In this case, she opined that it would fall in value by the March maturity date of the futures contract.

Long Positions. If Amber McClain expected the peso to rise in value versus the dollar in the near term, she could take a long position, by buying a March future on the Mexican peso. Buying a March future means that Amber is locking in the price at which she must buy Mexican pesos at the future's maturity date. Amber's futures contract at maturity would have the following value:

$$\text{Value at maturity (Long position)} = \text{Notional principal} \times (\text{Spot} - \text{Futures}).$$

Again using the March settle price on Mexican peso futures in Exhibit 7.1, $.10958/MXN, if the spot exchange rate at maturity is $.1100/MXN, Amber has indeed guessed right. The value of her position on settlement is then

$$\text{Value} = \text{MXN500,000} \times (\$.11000/\text{MXN} - \$.10958/\text{MXN}) = \$210.$$

In this case, Amber makes a profit in a matter of months of $210 on the single futures contract. We could say, "Amber buys at $.10958 and sells at $.11000 per peso."

But what happens if Amber's expectation about the future value of the Mexican peso proves wrong? For example, if the Mexican government announces that the rate of inflation in Mexico has suddenly risen dramatically, and the peso falls to $.08000/MXN by the March maturity date, the value of Amber's futures contract on settlement is

$$\text{Value} = \text{MXN500,000} \times (\$.08000/\text{MXN} - \$.10958/\text{MXN}) = (\$14,790).$$

In this case, Amber McClain suffers a speculative loss. Futures contracts could obviously be used in combinations to form a variety of more complex positions. When we combine contracts, valuation is fairly straightforward and additive in character.

Foreign Currency Futures versus Forward Contracts

Foreign currency futures contracts differ from forward contracts in a number of important ways. Individuals find futures contracts useful for speculation because they usually do not have access to forward contracts. For businesses, futures contracts are often considered inefficient and burdensome because the futures position is marked to market on a daily basis over the life of the contract. Although this does not require the business to pay or receive cash daily, it does result in more frequent margin calls from its financial service providers than the business typically wants.

Currency Options

A *foreign currency option* is a contract that gives the option purchaser (the buyer) the right, but not the obligation, to buy or sell a given amount of foreign exchange at a fixed price per unit for a specified time period (until the maturity date). A key phrase in this definition is "but not the obligation," which means that the owner of an option possesses a valuable choice.

In many ways, buying an option is like buying a ticket to a benefit concert. The buyer has the right to attend the concert, but is not obliged to. The buyer of the concert ticket risks nothing more than what she pays for the ticket. Similarly, the buyer of an option cannot lose more than what he pays for the option. If the buyer of the ticket decides later not to attend the concert, prior to the day of the concert, the ticket can be sold to someone else who wishes to go.

Option Fundamentals

There are two basic types of options, *calls* and *puts*. A call is an option to buy foreign currency, and a put is an option to sell foreign currency. The buyer of an option is termed the *holder*, while the seller of an option is referred to as the *writer* or *grantor*.

Every option has three different price elements: 1) the *exercise price* or *strike price*, the exchange rate at which the foreign currency can be purchased (*call*) or sold (*put*); 2) the *premium*, which is the cost, price, or value of the option itself; and 3) the underlying or actual spot exchange rate in the market.

An *American option* gives the buyer the right to exercise the option at any time between the date of writing and the expiration or maturity date. A *European option* can be exercised only on its expiration date, not before. Nevertheless, American and European options are priced almost the same because the option holder would normally sell the option itself before maturity. The option would then still have some "time value" above its "intrinsic value" if exercised (explained later in this chapter).

The premium or option price is the cost of the option, usually paid in advance by the buyer to the seller. In the *over-the-counter (OTC) market* (options offered by banks), premiums are quoted as a percentage of the transaction amount. Premiums on exchange-traded options are quoted as a domestic currency amount per unit of foreign currency.

An option whose exercise price is the same as the spot price of the underlying currency is said to be *at-the-money* (ATM). An option that would be profitable, excluding the cost of the premium, if exercised immediately is said to be *in-the-money* (ITM). An option that would not be profitable, again excluding the cost of the premium, if exercised immediately is referred to as *out-of-the-money* (OTM).

Foreign Currency Options Markets

In the past three decades, the use of foreign currency options as a hedging tool and for speculative purposes has blossomed into a major foreign exchange activity. A number of banks in the United States and other capital markets offer flexible foreign currency options on transactions of $1 million or more. The bank market, or *over-the-counter market* as it is called, offers custom-tailored options on all major trading currencies for any period up to one year, and in some cases, two to three years.

The Philadelphia Stock Exchange introduced trading in standardized foreign currency option contracts in the United States in 1982. The Chicago Mercantile Exchange and other exchanges in the U.S. and abroad have followed suit. Exchange-traded contracts are particularly appealing to speculators and individuals who do not normally have access to the over-the-counter market. Banks also trade on the exchanges because it is one of several ways they can offset the risk of options they may have transacted with clients or other banks.

Increased use of foreign currency options is a reflection of the explosive growth in the use of other kinds of options and the resultant improvements in option pricing models. The original option-pricing model developed by Fischer Black and Myron Scholes in 1973 has been expanded, adapted, and commercialized in hundreds of forms since that time. One wonders if Black and Scholes truly appreciated what a monster they may have created!

Options on the Over-the-Counter Market. Over-the-counter (OTC) options are most frequently written by banks for U.S. dollars against British pounds sterling, Canadian dollars, Japanese yen, Swiss francs, or the euro, but increasingly are available for nearly every major traded currency. As *Global Finance in Practice 7.1* notes, the rise of the Chinese renminbi in recent years has sparked its own option growth.

The main advantage of OTC options is that they are tailored to the specific needs of the firm. Financial institutions are willing to write or buy options that vary by amount (*notional principal*), strike price, and maturity. Although the OTC markets were relatively illiquid in the early years, these markets have grown to such proportions that liquidity is now quite good. On the other hand, the buyer must assess the writing bank's ability to fulfill the option contract. The financial risk associated with the counterparty (counterparty risk) is an ever-present issue in international markets as a result of the increasing use of financial contracts like options and swaps. Exchange-traded options are more the territory of individuals and financial institutions than of business firms.

GLOBAL FINANCE IN PRACTICE 7.1

Euro-Renminbi (EUR-RMB) Options Growth

Daily trading volume in Chinese renminbi currency options boomed in 2014 and 2015 for a variety of reasons. First and foremost was the growth in trade and the increasing settlement of cross-border transactions with Chinese firms in Chinese RMB. Although in the past this trade was dominated by the U.S. dollar, the Chinese government and Chinese firms are in the process of shifting more of the currency transaction burden to external counterparties.

The RMB cross-rate against the euro has seen rapid growth specifically. Many European companies have been pushed to settle more and more transactions in RMB, but they wish to do so without going through the U.S. dollar, which has been the more common practice—going USD/RMB and then EUR/USD.

Options, in addition to forwards, have seen growing demand and liquidity because of the People's Bank of China's management of the RMB versus the USD (although this changed in 2014 as the RMB fell versus the dollar on occasion). For many years this focus on the dollar created a one-sided movement, as the RMB slowly and steadily was revalued against the dollar. But the euro's movements against the RMB were not one-sided, creating more and more demand—and more interest—for currency options between the two.

If an investor wishes to purchase an option in the OTC market, the investor will normally place a call to the currency option desk of a major money center bank, specify the currencies, maturity, strike rate(s), and ask for an *indication*, a bid-offer quote. The bank will normally take a few minutes to a few hours to price the option and return the call.

Options on Organized Exchanges. Options on the physical (underlying) currency are traded on a number of organized exchanges worldwide, including the Philadelphia Stock Exchange (PHLX) and the Chicago Mercantile Exchange. Exchange-traded options are settled through a clearinghouse, so that buyers do not deal directly with sellers. The clearinghouse is the counterparty to every option contract and it guarantees fulfillment. Clearinghouse obligations are in turn the obligation of all members of the exchange, including a large number of banks. In the case of the Philadelphia Stock Exchange, clearinghouse services are provided by the *Options Clearing Corporation* (OCC).

Currency Option Quotations and Prices

Typical quotes in *The Wall Street Journal* for options on Swiss francs are shown in Exhibit 7.2. The Journal's quotes refer to transactions completed on the Philadelphia Stock Exchange

EXHIBIT 7.2   **Swiss Franc Option Quotations (U.S. cents/SF)**

Option and Underlying	Strike Price	Calls – Last			Puts – Last		
		Aug	Sep	Dec	Aug	Sep	Dec
58.51	56.0	—	—	2.76	0.04	0.22	1.16
58.51	56.5	—	—	—	0.06	0.30	—
58.51	57.0	1.13	—	1.74	0.10	0.38	1.27
58.51	57.5	0.75	—	—	0.17	0.55	—
58.51	58.0	0.71	1.05	1.28	0.27	0.89	1.81
58.51	58.5	0.50	—	—	0.50	0.99	—
58.51	59.0	0.30	0.66	1.21	0.90	1.36	—
58.51	59.5	0.15	0.40	—	2.32	—	—
58.51	60.0	—	0.31	—	2.32	2.62	3.30

Each option = 62,500 Swiss francs. The August, September, and December listings are the option maturities or expiration dates.

on the previous day. Although a multitude of strike prices and expiration dates are quoted (shown in the exhibit), not all were actually traded the previous trading day, and in that case no premium price is shown. Currency option strike prices and premiums on the U.S. dollar are typically quoted as direct quotations on the U.S. dollar and indirect quotations on the foreign currency ($/SF, $/¥, etc.).

Buyer of a Call

Options differ from all other types of financial instruments in the patterns of risk they produce. The option owner, the holder, has the choice of exercising the option or allowing it to expire unused. The owner will exercise it only when exercising is profitable, which means only when the option is in the money. In the case of a call option, as the spot price of the underlying currency moves up, the holder has the possibility of unlimited profit. On the down side, however, the holder can abandon the option and walk away with a loss never greater than the premium paid.

Exhibit 7.2 illustrates the three different prices that characterize any foreign currency option. The three prices that characterize an "August 58.5 call option" (highlighted in Exhibit 7.2) are the following:

1. **Spot rate.** "Option and Underlying" in the exhibit means that 58.51 cents, or $0.5851, was the spot dollar price of one Swiss franc at the close of trading on the preceding day.

2. **Exercise price.** The exercise price, or "Strike Price" in the exhibit, means the price per franc that must be paid if the option is exercised. The August call option on francs of 58.5 means $0.5850/SF. Exhibit 7.2 lists nine different strike prices, ranging from $0.5600/SF to $0.6000/SF, although more were available on that date than are listed.

3. **Premium.** The premium is the cost or price of the option. The price of the August 58.5 call option on Swiss francs was 0.50 U.S. cents per franc, or $0.0050/SF. There was no trading of the September and December 58.5 call on that day. The premium is the market value of the option, and therefore the terms premium, cost, price, and value are all interchangeable when referring to an option.

The August 58.5 call option premium is 0.50 cents per franc, and in this case, the August 58.5 put's premium is also 0.50 cents per franc. Since one option contract on the Philadelphia Stock Exchange consists of 62,500 francs, the total cost of one option contract for the call (or put in this case) is SF62,500 × $0.0050/SF = $312.50.

Hans Schmidt is a currency speculator in Zurich. The position of Hans as a buyer of a call is illustrated in Exhibit 7.3. Assume he purchases the August call option on Swiss francs described previously, the one with a strike price of $0.585, and a premium of $0.005/SF. The vertical axis measures profit or loss for the option buyer at each of several different spot prices for the franc up to the time of maturity.

At all spot rates below the strike price of $0.585, Hans would choose not to exercise his option. This is obvious because at a spot rate of $0.580 for example, he would prefer to buy a Swiss franc for $0.580 on the spot market, rather than exercising his option to buy a franc at $0.585. If the spot rate were to remain at $0.580 or below until August when the option expired, Hans would not exercise the option. His total loss would be limited to only what he paid for the option, the $0.005/SF purchase price. Regardless of how far the spot rate was to fall, his loss would be limited to the original $0.005/SF cost.

Alternatively, at all spot rates above the strike price of $0.585, Hans would exercise the option, paying only the strike price for each Swiss franc. For example, if the spot rate were $0.595 per franc at maturity, he would exercise his call option, buying Swiss francs for $0.585 each instead of purchasing them on the spot market at $0.595 each. He could sell the Swiss francs immediately in the spot market for $0.595 each, pocketing a gross profit of $0.010/SF, or a net profit of $0.005/SF after deducting the original cost of the option of $0.005/SF.

| EXHIBIT 7.3 | Profit and Loss for the Buyer of a Call Option |

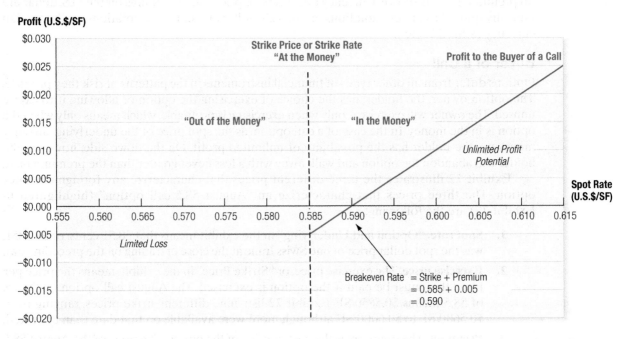

The buyer of a call option on has unlimited profit potential (*in the money*), and limited loss potential, the amount of the premium (*out of the money*).

Hans' profit, if the spot rate is greater than the strike price, with strike price $0.585, a premium of $0.005, and a spot rate of $0.595, is

$$\text{Profit} = \text{Spot Rate} - (\text{Strike Price} + \text{Premium})$$
$$= \$0.595/\text{SF} - (\$0.585/\text{SF} + \$0.005/\text{SF})$$
$$= \$0.005/\text{SF}$$

More likely, Hans would realize the profit through executing an offsetting contract on the options exchange rather than taking delivery of the currency. Because the dollar price of a franc could rise to an infinite level (off the upper right-hand side of Exhibit 7.3), maximum profit is unlimited. The buyer of a call option thus possesses an attractive combination of outcomes: limited loss and unlimited profit potential.

Note that break-even price of $0.590/SF is the price at which Hans neither gains nor loses on exercising the option. The premium cost of $0.005, combined with the cost of exercising the option of $0.585, is exactly equal to the proceeds from selling the francs in the spot market at $0.590. Hans will still exercise the call option at the break-even price. This is because by exercising it he at least recoups the premium paid for the option. At any spot price above the exercise price but below the break-even price, the gross profit earned on exercising the option and selling the underlying currency covers part (but not all) of the premium cost.

Writer of a Call

The position of the writer (seller) of the same call option is illustrated in Exhibit 7.4. If the option expires when the spot price of the underlying currency is below the exercise price of

| EXHIBIT 7.4 | Profit and Loss for the Writer of a Call Option |

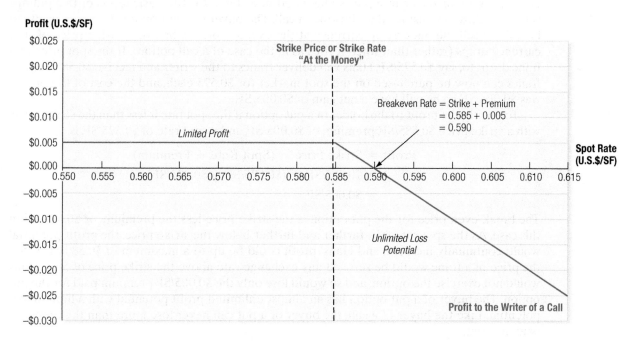

The writer of a call option on has unlimited loss potential and limited profit potential, the amount of the premium.

$0.585, the option holder does not exercise. What the holder loses, the writer gains. The writer keeps as profit the entire premium paid of $0.005/SF. Above the exercise price of 58.5, the writer of the call must deliver the underlying currency for $0.585/SF at a time when the value of the franc is above $0.585. If the writer wrote the option "naked," that is, without owning the currency, then the writer will need to buy the currency at spot and, in this scenario, take the loss. The amount of such a loss is unlimited and increases as the price of the underlying currency rises.

Once again, what the holder gains, the writer loses, and vice versa. Even if the writer already owns the currency, the writer will experience an opportunity loss, surrendering against the option the same currency that could have been sold for more in the open market. For example, the profit to the writer of a call option of strike price $0.585, premium $0.005, a spot rate of $0.595/SF is

$$\text{Profit} = \text{Premium} - (\text{Spot Rate} - \text{Strike Price})$$
$$= \$0.005/\text{SF} - (\$0.595/\text{SF} - \$0.585/\text{SF})$$
$$= \$0.005/\text{SF}$$

but only if the spot rate is greater than or equal to the strike rate. At spot rates less than the strike price, the option will expire worthless and the writer of the call option will keep the premium earned. The maximum profit that the writer of the call option can make is limited to the premium. The writer of a call option has a rather unattractive combination of potential outcomes: limited profit potential and unlimited loss potential—but there are ways to limit such losses through other offsetting techniques that we will discuss later in this chapter.

Buyer of a Put

Hans' position as buyer of a put is illustrated in Exhibit 7.5. The basic terms of this put are similar to those we just used to illustrate a call. The buyer of a put option, however, wants to be able to sell the underlying currency at the exercise price when the market price of that currency drops (rather than when it rises as in the case of a call option). If the spot price of a franc drops to, say, $0.575/SF, Hans will deliver francs to the writer and receive $0.585/SF. The francs can now be purchased on the spot market for $0.575 each, and the cost of the option was $0.005/SF, so he will have a net gain of $0.005/SF.

Explicitly, the profit to the holder of a put option if the spot rate is less than the strike price, with a strike price $0.585/SF, premium of $0.005/SF, and a spot rate of $0.575/SF, is

$$\text{Profit} = \text{Strike Price} - (\text{Spot Rate} + \text{Premium})$$
$$= \$0.585/SF - (\$0.575/SF + \$0.005/SF)$$
$$= \$0.005/SF$$

The break-even price for the put option is the strike price less the premium, or $0.580/SF in this case. As the spot rate falls further and further below the strike price, the profit potential would continually increase, and Hans' profit could be up to a maximum of $0.580/SF, when the price of a franc would be zero. At any exchange rate above the strike price of 58.5, Hans would not exercise the option, and so would lose only the $0.005/SF premium paid for the put option. The buyer of a put option has an almost unlimited profit potential with a limited loss potential. Like the buyer of a call, the buyer of a put can never lose more than the premium paid up front.

EXHIBIT 7.5 **Profit and Loss for the Buyer of a Put Option**

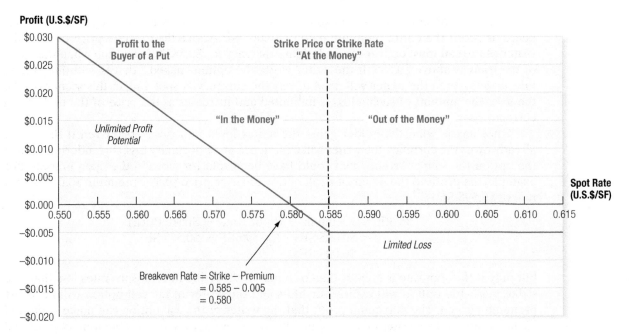

The buyer of a put option on has unlimited profit potential (*in the money*), and limited loss potential, the amount of the premium (*out of the money*).

Writer of a Put

The position of the writer who sold the put to Hans is shown in Exhibit 7.6. Note the symmetry of profit/loss, strike price, and break-even prices between the buyer and the writer of the put. If the spot price of francs drops below $0.585 per franc, Hans will exercise the option. Below a price of $0.585 per franc, the writer will lose more than the premium received from writing the option ($0.005/SF), falling below breakeven. Between $0.580/SF and $0.585/SF the writer will lose part, but not all, of the premium received. If the spot price is above $0.585/SF, Hans will not exercise the option, and the option writer will pocket the entire premium of $0.005/SF.

The profit (loss) earned by the writer of a $0.585 strike price put, premium $0.005, at a spot rate of $0.575, is

$$\text{Profit (loss)} = \text{Premium} - (\text{Strike Price} - \text{Spot Rate})$$
$$= \$0.005/\text{SF} - (\$0.585/\text{SF} - \$0.575/\text{SF})$$
$$= \$0.005/\text{SF}$$

but only for spot rates that are less than or equal to the strike price. At spot rates greater than the strike price, the option expires out-of-the-money and the writer keeps the premium. The writer of the put option has the same combination of outcomes available to the writer of a call: limited profit potential and loss potential.

Global Finance in Practice 7.2 describes one of the largest, and most successful, currency option speculations ever made, that by Andrew Krieger against the New Zealand kiwi. We should all be so *good*.

EXHIBIT 7.6 **Profit and Loss for the Writer of a Put Option**

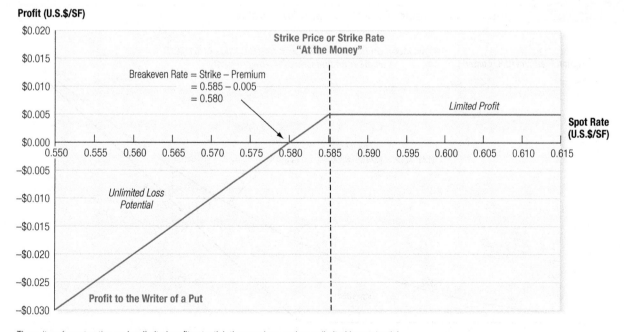

The writer of a put option on has limited profit potential, the premium, and an unlimited loss potential.

The New Zealand Kiwi, Key, and Krieger

In 1987 Andrew Krieger was a 31-year-old currency trader for Bankers Trust of New York (BT). Following the U.S. stock market crash in October 1987, the world's currency markets moved rapidly to exit the dollar. Many of the world's other currencies—including small ones that were in stable, open, industrialized markets, like that of New Zealand—became the subject of interest. As the world's currency traders dumped dollars and bought kiwis, the value of the kiwi rose sharply.

Krieger believed that the markets were overreacting. He took a short position on the kiwi, betting on its eventual fall. And he did so in a big way, combining spot, forward, and options positions. (Krieger supposedly had approval for positions rising to nearly $700 million in size, while all other BT traders were restricted to $50 million.) Krieger, on behalf of BT, is purported to have shorted 200 million kiwi—more than the entire New Zealand money supply at the time. His view proved correct. The kiwi fell, and Krieger was able to earn millions in currency gains for BT. Ironically, only months later, Krieger resigned from BT when annual bonuses were announced and he was reportedly awarded only $3 million on the more than $300 million profit.

Eventually, the New Zealand central bank lodged complaints with BT, in which the CEO at the time, Charles S. Sanford Jr., seemingly added insult to injury when he reportedly remarked "We didn't take too big a position for Bankers Trust, but we may have taken too big a position for that market."

Option Pricing and Valuation

Exhibit 7.7 illustrates the profit/loss profile of a European-style call option on British pounds. The call option allows the holder to buy British pounds at a strike price of $1.70/£. It has a 90-day maturity. The value of this call option is actually the sum of two components:

$$\text{Total Value (premium)} = \text{Intrinsic Value} \times \text{Time Value}$$

The pricing of any currency option combines six elements. For example, this European-style call option on British pounds has a premium of $0.033/£ (3.3 cents per pound) at a spot

EXHIBIT 7.7   **Option Intrinsic Value, Time Value, and Total Value**

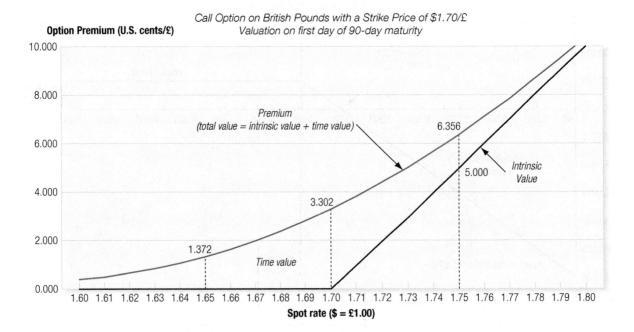

Call Option on British Pounds with a Strike Price of $1.70/£
Valuation on first day of 90-day maturity

rate of $1.70/£. This premium is calculated using the following assumptions: a spot rate of $1.70/£, a 90-day maturity, a $1.70/£ forward rate, both U.S. dollar and British pound interest rates of 8.00% per annum, and an option volatility for the 90-day period of 10.00% per annum.

Intrinsic value is the financial gain if the option is exercised immediately. It is shown by the solid line in Exhibit 7.7, which is zero until it reaches the strike price, then rises linearly (1 cent for each 1-cent increase in the spot rate). Intrinsic value will be zero when the option is *out-of-the-money*—that is, when the strike price is above the market price—as no gain can be derived from exercising the option. When the spot rate rises above the strike price, the intrinsic value becomes positive because the option is always worth at least this value if exercised. On the date of maturity, an option will have a value equal to its intrinsic value (zero time remaining means zero time value).

Exhibit 7.7 (graphically) and Exhibit 7.8 (table) illustrate all three value elements of the $1.70/£ strike 90-day call option on British pounds across a spectrum of spot rates. When the spot rate is $1.75/£, the option is *in-the-money*, and has positive time (¢1.356/£) and intrinsic values (¢5.000/£). When the spot rate is $1.70/£—the same as the option strike rate, the option is *at-the-money*, has no intrinsic value but does have time value (¢3.302/£). When the spot rate is $1.65/£, the option is *out-of-the-money*, has no intrinsic value but does have a time value (¢1.372/£).

The *time value* of an option exists because the price of the underlying currency, the spot rate, can potentially move further and further into the money before the option's expiration. Time value is shown in Exhibit 7.7 as the area between the total value of the option and its intrinsic value. An investor will pay something today for an out-of-the-money option (i.e., zero intrinsic value) on the chance that the spot rate will move far enough before maturity to move the option in-the-money. Consequently, the price of an option is always somewhat greater than its intrinsic value, since there is always some chance—some might say "hope everlasting"—that the intrinsic value will rise by the expiration date.

An investor will pay something today for an out-of-the-money option (i.e., zero intrinsic value) on the chance that the spot rate will move far enough before maturity to move the option in-the-money. Consequently, the price of an option is always somewhat greater than its intrinsic value, since there is always some chance that the intrinsic value will rise between the present and the expiration date.

Advanced Topic: Currency Option Pricing Sensitivity

If currency options are to be used effectively, either for the purposes of speculation or risk management (covered in the coming chapters), the individual trader needs to know how option values—premiums—react to their various components. The following section will analyze these six basic sensitivities:

1. The impact of changing forward rates
2. The impact of changing spot rates

EXHIBIT 7.8	Call Option Premiums: Intrinsic Value and Time Value Components							
Strike Rate ($/£)	Spot Rate ($/£)	Money	Call Premium (U.S. cents/£)	=	Intrinsic Value (U.S. cents/£)	+	Time Value (U.S. cents/£)	Option Delta (0 to 1)
1.70	1.75	In-the-money (ITM)	6.37	=	5.00	+	1.37	0.71
1.70	1.70	At-the-money (ATM)	3.30	=	0.00	+	3.30	0.50
1.70	1.65	Out-of-the-money (OTM)	1.37	=	0.00	+	1.37	0.28

3. The impact of time to maturity
4. The impact of changing volatility
5. The impact of changing interest differentials
6. The impact of alternative option strike prices

Forward Rate Sensitivity

Although rarely noted, standard foreign currency options are priced around the forward rate because the current spot rate and both the domestic and foreign interest rates (home currency and foreign currency rates) are included in the option premium calculation.

Recall from Chapter 4 that the forward rate is calculated from the current spot rate and the two subject currency interest rates for the desired maturity. For example, the 90-day forward rate for the call option on British pounds described above is calculated as follows:

$$F_{90} = \$1.70/\pounds \times \frac{\left[1 + \left(0.08 \times \dfrac{90}{360}\right)\right]}{\left[1 + \left(0.08 \times \dfrac{90}{360}\right)\right]} = \$1.70/\pounds.$$

Regardless of the specific strike rate chosen and priced, the forward rate is central to valuation. The option-pricing formula calculates a subjective probability distribution centered on the forward rate. This approach does not mean that the market expects the forward rate to be equal to the future spot rate, it is simply a result of the arbitrage-pricing structure of options.

The forward rate focus also provides helpful information for the trader managing a position. When the market prices a foreign currency option, it does so without any bullish or bearish sentiment on the direction of the foreign currency's value relative to the domestic currency. If the trader has specific expectations about the future spot rate's direction, those expectations can be put to work. A trader will not be inherently betting against the market. In a following section we will also describe how a change in the interest differential between currencies, the theoretical foundation of forward rates, also alters the value of the option.

Spot Rate Sensitivity (*delta*)

The call option on British pounds depicted in Exhibit 7.8 possesses a premium that exceeds the intrinsic value of the option over the entire range of spot rates surrounding the strike rate. As long as the option has time remaining before expiration, the option will possess this time value element. This characteristic is one of the primary reasons why an American-style option, which can be exercised on any day up to and including the expiration date, is seldom actually exercised prior to expiration. If the option holder wishes to liquidate it for its value, then it would normally be sold and not exercised, so any remaining time value can also be captured by the holder. If the current spot rate falls on the side of the option's strike price—which would induce the option holder to exercise the option upon expiration—the option also has an intrinsic value. The call option illustrated in Exhibit 7.7 is *in-the-money* (ITM) at spot rates to the right of the strike rate of $1.70/£, *at-the-money* (ATM) at $1.70/£, and *out-of-the-money* (OTM) at spot rates less than $1.70/£.

The vertical distance between the market value and the intrinsic value of a call option on pounds is greatest at a spot rate of $1.70/£. At $1.70/£ the spot rate equals the strike price (at-the-money). This premium of 3.30 cents per pound consists entirely of time value. In fact, the value of any option that is currently out-of-the-money (OTM) is made up entirely of time value. The further the option's strike price is out-of-the-money, the lower the value or premium of the option. This is because the market believes the probability of this option actually moving into the exercise range prior to expiration is significantly less than for an option that is already at-the-money. If the spot rate were to fall to $1.68/£, the option premium falls to 2.39 cents/£—again, entirely time value. If the spot rate were to rise above the strike rate to $1.72/£, the premium rises to 4.39 cents/£. In this case the premium represents an intrinsic value of 2.00 cents ($1.72/£ − $1.70/£) plus a time value element of 2.39 cents. Note the symmetry of time value premiums (2.39 cents) to the left and to the right of the strike rate.

The symmetry of option valuation about the strike rate can be observed by decomposing the option premiums into their respective intrinsic and time values. Exhibit 7.8 illustrates how varying the current spot rate by ±$0.05 about the strike rate of $1.70/£ alters each option's intrinsic and time values.

The sensitivity of the option premium to a small change in the spot exchange rate is called the *delta*. For example, the delta of the $1.70/£ call option, when the spot rate changes from $1.70/£ to $1.71/£, is simply the change in the premium divided by the change in the spot rate:

$$\text{Delta} = \frac{\Delta \text{ premium}}{\Delta \text{ spot rate}} = \frac{\$0.038/£ - \$0.033/£}{\$1.71/£ - \$1.70/£} = 0.5$$

If the delta of the specific option is known, it is easy to determine how the option's value will change as the spot rate changes. If the spot rate changes by one cent ($0.01/£), given a delta of 0.5, the option premium would change by 0.5 × $0.01, or $0.005. If the initial premium was $0.033/£, and the spot rate increased by 1 cent (from $1.70/£ to $1.71/£), the new option premium would be $0.033 + $0.005 = $0.038/£. Delta varies between +1 and 0 for a call option and −1 and 0 for a put option.

Traders in options categorize individual options by their delta rather than in-the-money, at-the-money, or out-of-the-money. As an option moves further in-the-money, delta rises toward 1.0. As an option moves further out-of-the-money, delta falls toward zero. Note that the out-of-the-money option in Exhibit 7.8 has a delta of only 0.28.[1]

Rule of Thumb: *The higher the delta (deltas of .7, or .8 and up are considered high) the greater the probability of the option expiring in-the-money.*

Time to Maturity: Value and Deterioration (*theta*)

Option values increase with the length of time to maturity. The expected change in the option premium from a small change in the time to expiration is termed *theta*.

Theta is calculated as the change in the option premium over the change in time. If the $1.70/£ call option were to age 1 day from its initial 90-day maturity, the theta of the call option would be the difference in the two premiums, 3.30 cents/£ and 3.28 cents/£ (assuming a spot rate of $1.70/£):

$$\text{theta} = \frac{\Delta \text{ premium}}{\Delta \text{ time}} = \frac{\text{cents } 3.30/£ - \text{cents } 3.28/£}{90 - 89} = 0.02.$$

Theta is based not on a linear relationship with time, but rather the square root of time. Option premiums deteriorate at an increasing rate as they approach expiration. In fact, the majority of the option premium—depending on the individual option—is lost in the final 30 days prior to expiration.

---

[1] The expected change in the option's delta resulting from a small change in the spot rate is termed *gamma*. It is often used as a measure of the stability of a specific option's *delta*. *Gamma* is utilized in the construction of more sophisticated hedging strategies, which focus on *deltas* (delta-neutral strategies).

This exponential relationship between option premium and time is seen in the ratio of option values between the three-month and the one-month at-the-money maturities. The ratio for the at-the-money call option is not 3 to 1 (holding all other components constant), but rather 1.73 times the price.

$$\frac{\text{Premium of 3 month}}{\text{Premium of 1 month}} = \frac{\sqrt{3}}{\sqrt{1}} = \frac{1.73}{1.00} = 1.73$$

The rapid deterioration of option values in the last days prior to expiration is seen by once again calculating the theta of the $1.70/£ call option, but now as its remaining maturity moves from 15 days to 14 days:

$$\text{theta} = \frac{\Delta \text{ premium}}{\Delta \text{ time}} = \frac{\text{cents } 1.37/£ - \text{cents } 1.32/£}{15 - 14} = 0.05.$$

A decrease of one day in the time to maturity now reduces the option premium by 0.05 cents/£, rather than only 0.02 cents/£ as it did when the maturity was 90 days.

The implications of time value deterioration for traders are quite significant. A trader purchasing an option with only one or two months until expiration will see rapid deterioration of the option's value. If the trader were to then sell the option, it would have a significantly smaller market value in the periods immediately following its purchase. At the same time, however, a trader who is buying options of longer maturities will pay more, but not proportionately more, for the longer maturity option. A six-month option's premium is approximately 2.45 times more expensive than the one-month, while the 12-month option would be only 3.46 times more expensive than the one-month. This implies that two three-month options do not equal one six-month option.

> **Rule of Thumb:** *A trader will normally find longer-maturity options better values, giving the trader the ability to alter an option position without suffering significant time value deterioration.*

Sensitivity to Volatility (*lambda*)

There are few words in the financial field more used and abused than *volatility*. Option *volatility* is defined as the standard deviation of daily percentage changes in the underlying exchange rate. Volatility is important to option value because of an exchange rate's perceived likelihood to move either into or out of the range in which the option would be exercised. If the exchange rate's volatility is rising, and therefore the risk of the option being exercised is increasing, the option premium would be increasing.

Volatility is stated in percent per annum. For example, an option may be described as having a 12.6% annual volatility. The percentage change for a single day can be found as follows:

$$\frac{12.6\%}{\sqrt{365}} = \frac{12.6\%}{19.105} = 0.66\% \text{ daily volatility.}$$

For our $1.70/£ call option, an increase in annual volatility of 1 percentage point—for example from 10.0% to 11.0%—will increase the option premium from $0.033/£ to $0.036/£. The marginal change in the option premium is equal to the change in the option premium itself divided by the change in volatility:

$$\frac{\Delta \text{ premium}}{\Delta \text{ volatility}} = \frac{\$0.036/£ - \$0.033/£}{0.11 - 0.10} = 0.3.$$

The primary problem with volatility is that it is unobservable; it is the only input into the option pricing formula that is determined subjectively by the trader pricing the option. No single correct method for its calculation exists. The problem is one of forecasting; historical volatility is not necessarily an accurate predictor of the future volatility of the exchange rate's movement, yet there is little to go on except history.

Volatility is viewed three ways: historic, where the volatility is drawn from a recent period of time; forward-looking, where the historic volatility is altered to reflect expectations about the future period over which the option will exist; and implied, where the volatility is backed out of the market price of the option.

■ **Historic volatility.**   Historic volatility is normally measured as the percentage movement in the spot rate on a daily, 6-, or 12-hour basis over the previous 10, 30, or even 90 days.

■ **Forward-looking volatility.**   Alternatively, an option trader may adjust recent historic volatilities for expected market swings or events, either upward or downward.

If option traders believe that the immediate future will be the same as the recent past, the historic volatility will equal the forward-looking volatility. If, however, the future period is expected to experience greater or lesser volatility, the historic measure must be altered for option pricing.

■ **Implied volatility.**   Implied volatility is equivalent to having the answers to the test; implied volatilities are calculated by being backed out of the market option premium values traded. Since volatility is the only unobservable element of the option premium price, after all other components are accounted for, the residual value of volatility implied by the price is found.

Selected implied volatilities for a number of currency pairs are listed in Exhibit 7.9. The exhibit clearly illustrates that option volatilities vary considerably across currencies, and that the relationship between volatility and maturity (time to expiration) does not move just one direction. For example, the first exchange rate quoted, the US$/euro cross-rate, initially falls from 8.1% volatility at one week to 7.4% for the 1-month and 2-month maturities, and then rises to 9.3% for the 3-year maturity.

Because volatilities are the only judgmental component that the option writer contributes, they play a critical role in the pricing of options. All currency pairs have historical

EXHIBIT 7.9   **Foreign Currency Implied Volatilities (percent)**

Currency (cross)	Symbol	1 week	1 month	2 month	3 month	6 month	1 year	2 year	3 year
European euro	EUR	8.1	7.4	7.4	7.4	7.8	8.5	9.0	9.3
Japanese yen	JPY	12.3	11.4	11.1	11.0	11.0	11.2	11.8	12.7
Swiss franc	CHF	8.9	8.4	8.4	8.4	8.9	9.5	9.8	9.9
British pound	GBP	7.7	7.3	7.2	7.1	7.3	7.5	7.9	8.2
Canadian dollar	CAD	6.4	6.4	6.3	6.4	6.7	7.1	7.4	7.6
Australian dollar	AUD	11.2	10.7	10.5	10.3	10.4	10.6	10.8	11.0
British pound/euro	GBPEUR	6.7	6.4	6.5	6.4	6.8	7.3	7.6	7.8
Euro/Japanese yen	EURJPY	11.6	11.1	11.2	11.3	11.8	12.6	13.4	14.1

Source: Federal Reserve Bank of New York.

Note: These implied volatility rates are averages of mid-level rates for bid and ask "at-money quotations" on selected currencies at 11 A.M. on the last business day of the month, September 30, 2013.

series that contribute to the formation of the expectations of option writers. But in the end, the truly talented option writers are those with the intuition and insight to price the future effectively.

Like all futures markets, option volatilities react instantaneously and negatively to unsettling economic and political events (or rumor). A doubling of volatility for an at-the-money option will result in an equivalent doubling of the option's price. Most currency option traders focus their activities on predicting movements of currency volatility in the short run, because short-run movements move price the most. For example, option volatilities rose significantly in the months preceding the Persian Gulf War, in September 1992 when the European Monetary System was in crisis, in 1997 after the onset of the Asian financial crisis, in the days following the terrorist attacks on the United States in September 2001, and in the months following the onset of the global financial crisis in September 2008. In all instances option volatilities for major cross-currency combinations rose to nearly 20% for extended periods. As a result, premium costs rose by corresponding amounts.

> **Rule of Thumb:** *Traders who believe volatilities will fall significantly in the near-term will sell (write) options now, hoping to buy them back for a profit immediately after volatilities fall causing option premiums to fall.*

Sensitivity to Changing Interest Rate Differentials (*rho* and *phi*)

At the start of this section we pointed out that currency option prices and values are focused on the forward rate. The forward rate is in turn based on the theory of Interest Rate Parity discussed previously in Chapter 6. Interest rate changes in either currency will alter the forward rate, which in turn will alter the option's premium or value. The expected change in the option premium from a small change in the domestic interest rate (home currency) is term *rho*. The expected change in the option premium from a small change in the foreign interest rate (foreign currency) is termed *phi*.

Continuing with our numerical example, an increase in the U.S. dollar interest rate from 8.0% to 9.0% *increases* the ATM call option premium on British pounds from \$0.033/£ to \$0.035/£. This is a *rho* value of positive 0.2.

$$\text{rho} = \frac{\Delta \text{ premium}}{\Delta \text{ U.S. dollar interest rate}} = \frac{\$0.035/£ - \$0.033/£}{9.0\% - 8.0\%} = 0.2.$$

A similar 1% increase in the foreign interest rate, the pound sterling rate in this case, *reduces* the option value (premium) from \$0.033/£ to \$0.031/£. The *phi* for this call option premium is therefore a negative 0.2.

$$\text{phi} = \frac{\Delta \text{ premium}}{\Delta \text{ foreign interest rate}} = \frac{\$0.031/£ - \$0.033/£}{9.0\% - 8.0\%} = -0.2.$$

For example, throughout the 1990s U.S. dollar (domestic currency) interest rates were substantially lower than pound sterling (foreign currency) interest rates. This meant that the pound consistently sold forward at a discount versus the U.S. dollar. If this interest differential were to widen (either from U.S. interest rates falling or foreign currency interest rates rising, or some combination of both), the pound would sell forward at a larger discount. An increase in the forward discount is the same as a decrease in the forward rate (in U.S. dollars per unit of foreign currency). The option premium condition above states that the premium must increase as interest rate differentials increase (assuming spot rates remain unchanged).

For the option trader, an expectation on the differential between interest rates can obviously help in the evaluation of where the option value is headed. For example, when foreign interest rates are higher than domestic interest rates, the foreign currency sells forward at a discount. This results in relatively lower call option premiums (and lower put option premiums).

> **Rule of Thumb:** *A trader who is purchasing a call option on foreign currency should do so before the domestic interest rate rises. This will allow the trader to purchase the option before its price increases.*

Alternative Strike Prices and Option Premiums

The sixth and final element that is important in option valuation (but, thankfully, has no Greek alias) is the selection of the actual strike price. Although we have conducted all of our sensitivity analysis using the strike price of $1.70/£ (a forward-at-the-money strike rate), a firm purchasing an option in the over-the-counter market may choose its own strike rate. Options with strike rates that are already in-the-money will have both intrinsic and time value elements. Options with strike rates that are out-of-the-money will have only a time value component.

Exhibit 7.10 briefly summarizes the various "Greek" elements and impacts discussed in the previous sections. The option premium is one of the most complex concepts in financial theory, and the application of option pricing to exchange rates does not make it any simpler. Only with a considerable amount of time and effort can the individual be expected to attain a "second-sense" in the management of currency option positions.

Prudence in Practice

In the following chapters we will illustrate how derivatives can be used to reduce the risks associated with the conduct of multinational financial management. It is critical, however, that the user of any financial tool or technique—including financial derivatives—follow sound principles and practices. Many a firm has been ruined as a result of the misuse of derivatives. A word to the wise: Do not fall victim to what many refer to as the *gambler's dilemma—confusing luck with talent.*

Major corporate financial disasters related to financial derivatives continue to be a problem in global business. As is the case with so many issues in modern society, technology is not at fault, rather human error in its use.

EXHIBIT 7.10   **Summary of Option Premium Components**

Greek	Definition	Interpretation
Delta	Expected change in the option premium for a small change in the spot rate	The higher the delta, the more likely the option will move in-the-money
Theta	Expected change in the option premium for a small change in time to expiration	Premiums are relatively insensitive until the final 30 or so days
Lambda	Expected change in the option premium for a small change in volatility	Premiums rise with increases in volatility
Rho	Expected change in the option premium for a small change in the domestic interest rate	Increases in domestic interest rates cause increasing call option premiums
Phi	Expected change in the option premium for a small change in the foreign interest rate	Increases in foreign interest rates cause decreasing call option premiums

SUMMARY POINTS

- Foreign currency futures contracts are standardized forward contracts. Unlike forward contracts, however, trading occurs on the floor of an organized exchange rather than between banks and customers. Futures also require collateral and are normally settled through the purchase of an offsetting position.

- Corporate financial managers typically prefer foreign currency forwards over futures out of simplicity of use and position maintenance. Financial speculators typically prefer foreign currency futures over forwards because of the liquidity of the futures markets.

- Foreign currency options are financial contracts that give the holder the right, but not the obligation, to buy (in the case of calls) or sell (in the case of puts) a specified amount of foreign exchange at a predetermined price on or before a specified maturity date.

- The use of a currency option as a speculative device for the buyer of an option arises from the fact that an option gains in value as the underlying currency rises (for calls) or falls (for puts). The amount of loss when the underlying currency moves opposite to the desired direction is limited to the option premium.

- The use of a currency option as a speculative device for the writer (seller) of an option arises from the option premium. If the option—either a put or call—expires out-of-the-money (valueless), the writer of the option has earned, and retains, the entire premium.

- Speculation is an attempt to profit by trading on expectations about prices in the future. In the foreign exchange market, one speculates by taking a position in a foreign currency and then closing that position afterward; a profit results only if the rate moves in the direction that the speculator expected.

- Currency option valuation, the determination of the option's premium, is a complex combination of the current spot rate, the specific strike rate, the forward rate (which itself is dependent on the current spot rate and interest differentials), currency volatility, and time to maturity.

- The total value of an option is the sum of its intrinsic value and time value. Intrinsic value depends on the relationship between the option's strike price and the current spot rate at any single point in time, whereas time value estimates how intrinsic value may change—for the better—prior to maturity.

MINI-CASE

KiKos and the South Korean Won[2]

That possibility arises from a fundamental tenet of international law that is not written down in any law book: In extremis, the locals win.

— "Bad Trades, Except in Korea," by Floyd Norris, *The New York Times*, April 2, 2009

South Korean exporters in 2006, 2007, and into 2008 were not particularly happy with exchange rate trends. The South Korean won (KRW) had been appreciating, slowly but steadily, for years against the U.S. dollar. This was a major problem for Korean manufacturers, as much of their sales was exports to buyers paying in U.S. dollars. As the dollar continued to weaken, each dollar resulted in fewer and fewer Korean won—and nearly all of their costs were in Korean won. Korean banks, in an effort to service these

hedging needs, became the sale and promotion of *Knock-In Knock-Out option agreements (KiKos)*.

Knock-In Knock-Outs (KiKos)

Many South Korean manufacturers had suffered falling margins on sales for years. Already operating in highly competitive markets, the appreciation of the won had cut further and further into their margins after currency settlement. As seen in Exhibit A, the won had traded in a narrow range for years. But that was little comfort as the difference between KRW1,000 and KRW 930 to the dollar was a big chunk of margin.

South Korean banks had started promoting KiKos as a way of managing this currency risk. The *Knock-In Knock-Out* (KiKo) was a complex option structure, which

EXHIBIT A   **South Korean Won's Steady Appreciation**

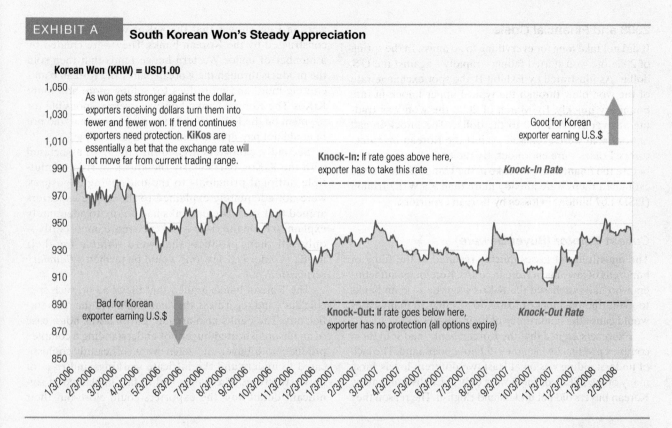

Korean Won (KRW) = USD1.00

As won gets stronger against the dollar, exporters receiving dollars turn them into fewer and fewer won. If trend continues exporters need protection. **KiKos** are essentially a bet that the exchange rate will not move far from current trading range.

Knock-In: If rate goes above here, exporter has to take this rate   *Knock-In Rate*

Good for Korean exporter earning U.S.$

Bad for Korean exporter earning U.S.$

Knock-Out: If rate goes below here, exporter has no protection (all options expire)   *Knock-Out Rate*

combined the sale of call options on the KRW (the knock-in component) and the purchase of put options on the USD (the knock-out component). These structures then established the trading range seen in Exhibit A that the banks and exporters believed that the won would stay within. In one case the bank salesman told a Korean manufacturer "we are 99% sure that the Korean won will continue to stay within this trading range for the year."[3]

But that was not the entirety of the KiKo structure. The bottom of the range, essentially a *protective put* on the dollar, assured the exporter of being able to sell dollars at a set rate if the won did indeed continue to appreciate. This strike rate was set close-in to the current market and was therefore quite expensive. In order to finance that purchase the sale of calls on the knock-in rate was a *multiple* (sometimes call the *turbo* feature) meaning that the exporter sold call options on a multiple, sometimes two or three times, the amount of the currency exposure. The exporters were "over-hedged." This multiple yielded higher earnings on the call options that financed the purchased puts and provided added funds to be contributed to the

final KiKo feature. This final feature was that the KiKo assured the exporter a single "better-than-market-rate" on the exchange of dollars for won as long as the exchange rate stayed within the bounds. Thus, the combined structure allowed the South Korean exporters to continue to exchange dollars for won at a rate like *KRW* 980 = *USD* when the spot market rate might have only been KRW 910.

This was not, however, a "locked-in rate." The exchange rate had to stay within the upper and lower bounds to reap the higher "guaranteed" exchange rate. If the spot rate moved dramatically below the knock-out rate, the knock-out feature would cancel the agreement. This was particularly troublesome because this was the very range in which the exporters needed protection. On the upper side, the knock-in feature, if the spot rate moved above the knock-in rate the exporter was required to deliver the dollars to the bank at that specific rate, although movement in this direction was actually in the exporter's favor. And the potential costs of the knock-in position were essentially unlimited, as a multiple of the exposure had been sold, putting the exporter into a purely speculative position.

[3]"KIKO Hedges Slay Korean Exporters, Threaten Banks," Bomi Lim, *Bloomberg BusinessWeek*, October 17, 2008.

2008 and Financial Crisis

It did not take long for everything to go amiss. In the spring of 2008 the won started falling—rapidly—against the U.S. dollar. As illustrated by Exhibit B, the spot exchange rate of the won blew through the typical upper knock-in rate boundary quickly. By March of 2008 the won was trading at over KRW 1,000 to the dollar. The knock-in call options sold were exercised against the Korean manufacturers. Losses were enormous. By the end of August, days before the financial crisis broke in the United States, it was estimated there were already more than KRW 1.7 trillion (USD 1.67 billion) in losses by Korean exporters.

Caveat Emptor (Buyer Beware)

The magnitude of losses quickly resulted in the filing of hundreds of lawsuits in Korean courts. Korean manufacturers who had purchased the KiKos sued the Korean banks to avoid the payment of losses, losses that in many cases would cause the bankruptcy of their businesses.

Exporters argued that the Korean banks had sold them complex products, which they did not understand. The lack of understanding was on at least two different levels. First, many of the KiKo contracts were only in English, and many Korean buyers did not understand English. The reason they

were in English was that the KiKos were not originally constructed by the Korean banks. They were created by a number of major Western hedge funds that then sold the products through the Korean banks, the Korean banks earning more and more fees for selling more and more KiKos. The Korean banks, however, were responsible for payment on the KiKos; if the exporting companies did not or could not pay-up, the banks would have to pay.

Secondly, exporters argued that the risks associated with the KiKos, particularly the knock-in risks of multiple notional principals to the underlying exposures, were not adequately explained to them. The exporters argued that the Korean banks had a *duty* to adequately explain to them the risks—and even more importantly—only sell them products that were *suitable* for their needs. (Under U.S. law this would be termed a *fiduciary responsibility*.)

The Korean banks argued that they had no such specific duty, and regardless, they had explained the risks sufficiently. The banks also argued that this was not a case of an unsophisticated buyer not understanding a complex product; both buyer and seller were sufficiently sophisticated to understand the intricate workings and risks of these structures. The banks had in fact explained in significant detail how the exporters could close-out their

EXHIBIT B	South Korean Won's Fall and the Knock-In

Korean Won (KRW) = USD1.00

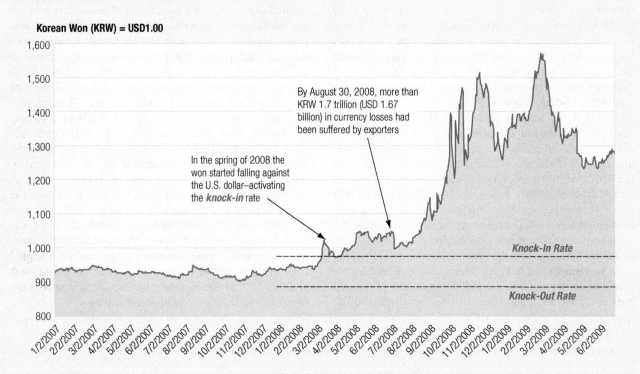

In the spring of 2008 the won started falling against the U.S. dollar—activating the *knock-in* rate

By August 30, 2008, more than KRW 1.7 trillion (USD 1.67 billion) in currency losses had been suffered by exporters

Knock-In Rate

Knock-Out Rate

positions and then limit the losses, but the exporters had chosen not to do so.

In the end the Korean courts found in favor of the exporters in some cases, in favor of the banks in others. One principle that the courts followed was that the exporters found themselves in "changed circumstances" in which the change in the spot exchange rate was *unforeseeable*, and *the losses resulting—too great*. But some firms, for example GM Daewoo, lost $1.11 billion. Some Korean banks suffered significant losses as well, and may have in fact helped transmit the financial crisis of 2008 from the United States and the European Union to many of the world's emerging markets.[4]

Mini-Case Questions

1. What were the expectations—and the fears—of the South Korean exporting firms that purchase the KiKos?
2. What is the responsibility of a bank that is offering and promoting these derivative products to its customers? Does it have some duty to protect their interests? Who do you think was at fault in this case?
3. If you were a consultant advising firms on their use of foreign currency derivative products, what lessons would you draw from this case, and how would you communicate that to your clients?

QUESTIONS

These questions are available in MyFinanceLab.

1. **Foreign Currency Futures.** What is a foreign currency future?

2. **Futures Terminology.** Explain the meaning and probable significance for international business of the following contract specifications: a) notional principal; b) margin; and c) marked-to-market

3. **Long and a Short.** How can foreign currency futures be used to speculate on the exchange rate movements, and what role do long and short positions play in that speculation?

4. **Futures and Forwards.** How do foreign currency futures and foreign currency forwards compare?

5. **Puts and Calls.** Define a put and call on the British pound sterling.

6. **Options versus Futures.** Explain the difference between foreign currency options and futures and when either might be most appropriately used.

7. **Call Option Contract.** Suppose that exchange-traded American call options on pounds sterling with a strike price of 1.460 and a maturity of next March are now quoted at 3.67. What does this mean to a potential buyer?

8. **Premiums, Prices & Costs.** What is the difference between the price of an option, the value of an option, the premium on an option, and the cost of a foreign currency option?

9. **Three Prices.** What are the three different prices or "rates" integral to every foreign currency option contract?

10. **Writing Options.** Why would anyone write an option, knowing that the gain from receiving the option premium is fixed but the loss, if the underlying price goes in the wrong direction, can be extremely large?

11. **Decision Prices.** Once an option has been purchased, only two prices or rates are part of the holder's decision-making process. Which two and why?

12. **Option Cash Flows and Time.** The cash flows associated with a call option on euros by a U.S. dollar-based investor occur at different points in time. What are they and how much does the time element matter?

13. **Option Valuation.** The value of an option is stated to be the sum of its intrinsic value and its time value. Explain what is meant by these terms.

14. **Time Value Deterioration.** An option's value declines over time, but it does not do so evenly. Explain what that means for option valuation.

15. **Option Values and Money.** Options are often described as in-the-money, at-the-money, or out-of-the-money. What does that mean and how is it determined?

16. **Option Pricing and the Forward Rate.** What is the relationship or link between the forward rate and the foreign currency option premium?

17. **Option Deltas.** What is an option delta? How does it change when the option is in-the-money, at-the-money, or out-of-the-money?

18. **Historic versus Implied Volatility.** What is the difference between a historic volatility and an implied volatility?

[4]"Exotic Derivatives Losses in Emerging Markets: Questions of Suitability, Concerns for Stability," by Randall Dodd, International Monetary Fund, IMF Working Paper WP/09, July 2009.

PROBLEMS

These problems are available in MyFinanceLab.

1. **Saguaro Funds.** Tony Begay, a currency trader for Chicago-based Saguaro Funds, uses futures quotes on the British pound (£) to speculate on the value of the pound. Use the futures quotes in the table at the bottom of the page to answer the following questions.
   a. If Tony buys 5 June pound futures, and the spot rate at maturity is $1.3980/£, what is the value of her position?
   b. If Tony sells 12 March pound futures, and the spot rate at maturity is $1.4560/£, what is the value of her position?
   c. If Tony buys 3 March pound futures, and the spot rate at maturity is $1.4560/£, what is the value of her position?
   d. If Tony sells 12 June pound futures, and the spot rate at maturity is $1.3980/£, what is the value of her position?

2. **Amber McClain.** Amber McClain, the currency speculator we met in the chapter, sells eight June futures contracts for 500,000 pesos at the closing price quoted in Exhibit 7.1.
   a. What is the value of her position at maturity if the ending spot rate is $0.12000/Ps?
   b. What is the value of her position at maturity if the ending spot rate is $0.09800/Ps?
   c. What is the value of her position at maturity if the ending spot rate is $0.11000/Ps?

3. **Cece Cao in Jakarta.** Cece Cao trades currencies for Sumatra Funds in Jakarta. She focuses nearly all of her time and attention on the U.S. dollar/Singapore dollar ($/S$) cross-rate. The current spot rate is $0.6000/S$. After considerable study, she has concluded that the Singapore dollar will appreciate versus the U.S. dollar in the coming 90 days, probably to about $0.7000/S$. She has the following options on the Singapore dollar to choose from:

Option	Strike Price	Premium
Put on Sing $	$0.6500/S$	$0.00003/S$
Call on Sing $	$0.6500/S$	$0.00046/S$

   a. Should Cece buy a put on Singapore dollars or a call on Singapore dollars?
   b. What is Cece's break-even price on the option purchased in part (a)?
   c. Using your answer from part (a), what is Cece's gross profit and net profit (including premium) if the spot rate at the end of 90 days is indeed $0.7000/S$?
   d. Using your answer from part (a), what is Cece's gross profit and net profit (including premium) if the spot rate at the end of 90 days is $0.8000/S$?

4. **Kapinsky Capital Geneva (A).** Christoph Hoffeman trades currency for Kapinsky Capital of Geneva. Christoph has $10 million to begin with, and he must state all profits at the end of any speculation in U.S. dollars. The spot rate on the euro is $1.3358/€, while the 30-day forward rate is $1.3350/€.
   a. If Christoph believes the euro will continue to rise in value against the U.S. dollar, so that he expects the spot rate to be $1.3600/€ at the end of 30 days, what should he do?
   b. If Christoph believes the euro will depreciate in value against the U.S. dollar, so that he expects the spot rate to be $1.2800/€ at the end of 30 days, what should he do?

5. **Kapinsky Capital Geneva (B).** Christoph Hoffeman of Kapinsky Capital now believes the Swiss franc will appreciate versus the U.S. dollar in the coming 3-month period. He has $100,000 to invest. The current spot rate is $0.5820/SF, the 3-month forward rate is $0.5640/SF, and he expects the spot rates to reach $0.6250/SF in three months.
   a. Calculate Christoph's expected profit assuming a pure spot market speculation strategy.
   b. Calculate Christoph's expected profit assuming he buys or sells SF three months forward.

Problem 1.

British Pound Futures, U.S.$/pound (CME)   **Contract = 62,500 pounds**

Maturity	Open	High	Low	Settle	Change	High	Open Interest
March	1.4246	1.4268	1.4214	1.4228	0.0032	1.4700	25,605
June	1.4164	1.4188	1.4146	1.4162	0.0030	1.4550	809

6. **Peleh's Puts.** Peleh writes a put option on Japanese yen with a strike price of $0.008000/¥ (¥125.00/$) at a premium of 0.0080¢ per yen and with an expiration date six months from now. The option is for ¥12,500,000. What is Peleh's profit or loss at maturity if the ending spot rates are ¥110/$, ¥115/$, ¥120/$, ¥125/$, ¥130/$, ¥135/$, and ¥140/$?

7. **Chavez S.A.** Chavez S.A., a Venezuelan company, wishes to borrow $8,000,000 for eight weeks. A rate of 6.250% per annum is quoted by potential lenders in New York, Great Britain, and Switzerland, using, respectively, international, British, and the Swiss-eurobond definitions of interest (day count conventions). From which source should Chavez borrow?

8. **Cachita Haynes at Vatic Capital.** Cachita Haynes works as a currency speculator for Vatic Capital of Los Angeles. Her latest speculative position is to profit from her expectation that the U.S. dollar will rise significantly against the Japanese yen. The current spot rate is ¥120.00/$. She must choose between the following 90-day options on the Japanese yen:

Option	Strike Price	Premium
Put on yen	¥125/$	$0.00003/S$
Call on yen	¥125/$	$0.00046/S$

a. Should Cachita buy a put on yen or a call on yen?
b. What is Cachita's break-even price on the option purchased in part (a)?
c. Using your answer from part (a), what is Cachita's gross profit and net profit (including premium) if the spot rate at the end of 90 days is ¥140/$?

9. **Calling All Euros.** Assume a call option on euros is written with a strike price of $1.2500/€ at a premium of 3.80¢ per euro ($0.0380/€) and with an expiration date three months from now. The option is for 100,000. Calculate your profit or loss should you exercise before maturity at a time when the euro is traded spot at strike prices beginning at $1.10/€, rising to $1.40/€ in increments of $0.05.

10. **Arthur Doyle at Baker Street.** Arthur Doyle is a currency trader for Baker Street, a private investment house in London. Baker Street's clients are a collection of wealthy private investors who, with a minimum stake of £250,000 each, wish to speculate on the movement of currencies. The investors expect annual returns in excess of 25%. Although officed in London, all accounts and expectations are based in U.S. dollars.

Arthur is convinced that the British pound will slide significantly—possibly to $1.3200/£—in the coming 30 to 60 days. The current spot rate is $1.4260/£. Arthur wishes to buy a put on pounds, which will yield the 25% return expected by his investors. Which of the following put options would you recommend he purchase? Prove your choice is the preferable combination of strike price, maturity, and up-front premium expense.

Strike Price	Maturity	Premium
$1.36/£	30 days	$0.00081/£
$1.34/£	30 days	$0.00021/£
$1.32/£	30 days	$0.00004/£
$1.36/£	60 days	$0.00333/£
$1.34/£	60 days	$0.00150/£
$1.32/£	60 days	$0.00060/£

11. **Calandra Panagakos at CIBC.** Calandra Panagakos works for CIBC Currency Funds in Toronto. Calandra is something of a contrarian—as opposed to most of the forecasts, she believes the Canadian dollar (C$) will appreciate versus the U.S. dollar over the coming 90 days. The current spot rate is $0.6750/C$. Calandra may choose between the following options on the Canadian dollar.

Option	Strike Price	Premium
Put on C$	$0.7000	$0.00003/S$
Call on C$	$0.7000	$0.00049/S$

a. Should Calandra buy a put on Canadian dollars or a call on Canadian dollars?
b. What is Calandra's break-even price on the option purchased in part (a)?
c. Using your answer from part (a), what is Calandra's gross profit and net profit (including premium) if the spot rate at the end of 90 days is indeed $0.7600?
d. Using your answer from part (a), what is Calandra's gross profit and net profit (including premium) if the spot rate at the end of 90 days is $0.8250?

Pricing Your Own Options

An Excel workbook entitled *FX Option Pricing* is downloadable from this book's Web site. The workbook has five spreadsheets constructed for pricing currency options for the following five currency pairs (dollar/euro shown here): U.S. dollar/euro, U.S. dollar/Japanese yen, euro/Japanese yen, U.S. dollar/British pound, and euro/British pound. Use the appropriate spreadsheet from the workbook to answer Problems 12–16.

12. **U.S. Dollar/Euro.** The table above indicates that a 1-year call option on euros at a strike rate of $1.25/€ will cost the buyer $0.0632/€, or 4.99%. But that assumed a volatility of 12.000% when the spot rate was $1.2674/€. What would that same call option cost if the volatility was reduced to 10.500% when the spot rate fell to $1.2480/€?

13. **U.S. Dollar/Japanese Yen.** What would be the premium expense, in home currency, for a Japanese firm to purchase an option to sell 750,000 U.S. dollars, assuming the initial values listed in the FX Option Pricing workbook?

14. **Euro/Japanese Yen.** A French firm is expecting to receive JPY10.4 million in 90 days as a result of an export sale to a Japanese semiconductor firm. What will it cost, in total, to purchase an option to sell the yen at €0.0072/JPY?

15. **U.S. Dollar/British Pound.** Assuming the same initial values for the dollar/pound cross rate in the FX Option Pricing workbook, how much more would a call option on pounds be if the maturity was doubled from 90 to 180 days? What percentage increase is this for twice the length of maturity?

16. **Euro/British Pound.** How would the call option premium change on the right to buy pounds with euros if the euro interest rate changed to 4.000% from the initial values listed in the FX Option Pricing workbook?

INTERNET EXERCISES

1. **Financial Derivatives and the ISDA.** The International Swaps and Derivatives Association (ISDA) publishes a wealth of information about financial derivatives, their valuation and their use, in addition to providing master documents for their contractual use between parties. Use the following ISDA Internet site to find the definitions to 31 basic financial derivative questions and terms:

    ISDA   www.isda.org/educat/faqs.html

2. **Risk Management of Financial Derivatives.** If you think this book is long, take a look at the freely downloadable U.S. Comptroller of the Currency's handbook

on risk management related to the care and use of financial derivatives!

Comptroller   www.occ.gov/publications/
of the   publications-by-type/
Currency   comptrollers-handbook/deriv.pdf

3. **Option Pricing.** OzForex Foreign Exchange Services is a private firm with an enormously powerful foreign currency derivative-enabled Web site. Use the following site to evaluate the various "Greeks" related to currency option pricing.

    OzForex   www.ozforex.com.au/forex-tools/tools/
              fx-options-calculator

4. **Garman-Kohlhagen Option Formulation.** For those brave of heart and quantitatively adept, check out the following Internet site's detailed presentation of the Garman-Kohlhagen option formulation used widely in business and finance today.

    Riskglossary.com   www.riskglossary.com/link/
                       garman_kohlhagen_1983.htm

5. **Chicago Mercantile Exchange.** The Chicago Mercantile Exchange trades futures and options on a variety of currencies, including the Brazilian real. Navigate to FX under the Trading tab on the following site to evaluate the uses of these currency derivatives:

    Chicago   www.cmegroup.com
    Mercantile
    Exchange

6. **Implied Currency Volatilities.** The single unobservable variable in currency option pricing is the volatility, since volatility inputs are the expected standard deviation of the daily spot rate for the coming period of the option's maturity. Use the New York Federal Reserve's Web site to obtain current implied currency volatilities for major trading cross-rate pairs.

    Federal Reserve   www.ny.frb.org/markets/
    Bank of New York impliedvolatility.html

7. **Montreal Exchange.** The Montreal Exchange is a Canadian exchange devoted to the support of financial derivatives in Canada. Use its Web site to view the latest on MV volatility—the volatility of the Montreal Exchange Index itself—in recent trading hours and days.

    Montreal Exchange   www.m-x.ca/
                        marc_options_en.php

CHAPTER 7 **APPENDIX**
Currency Option Pricing Theory

The foreign currency option model presented here, the European style option, is the result of the work of Black and Scholes (1972), Cox and Ross (1976), Cox, Ross, and Rubinstein (1979), Garman and Kohlhagen (1983), and Bodurtha and Courtadon (1987). Although we do not explain the theoretical derivation of the following option-pricing model, the original model derived by Black and Scholes is based on the formation of a riskless hedged portfolio composed of a long position in the security, asset, or currency, and a European call option. The solution to this model's expected return yields the option *premium*.

The basic theoretical model for the pricing of a European call option is:

$$C = e^{-r_f T} SN(d_1) - E_e^{-r_d T} N(d_2)$$

where

C	premium on a European call
e	continuous time discounting
S	spot exchange rate ($/foreign currency)
E	exercise or strike rate
T	time to expiration
N	cumulative normal distribution function
r_f	foreign interest rate
r_d	domestic interest rate
σ	standard deviation of asset price (volatility)
ln	natural logarithm

The two density functions, d_1 and d_2, are defined:

$$d_1 = \frac{\ln\left(\dfrac{S}{E}\right) + \left(r_d - r_f + \dfrac{\sigma^2}{2}\right)T}{\sigma\sqrt{T}}$$

and:

$$d_2 = d_1 - \sigma\sqrt{T}.$$

This expression can be rearranged so the premium on a European call option is written in terms of the forward rate:

$$C = e^{-r_f T} FN(d_1) - e^{-r_d T} EN(d_2)$$

where the spot rate and foreign interest rate have been replaced with the forward rate, F, and both the first and second terms are discounted over continuous time, e. If we now slightly simplify, we find that the option premium is the present value of the difference between two cumulative normal density functions.

$$C = [FN(d_1) - EN(d_2)]e^{-r_d T}.$$

The two density functions are now defined:

$$d_1 = \frac{\ln\left(\dfrac{F}{E}\right) + \left(\dfrac{\sigma^2}{2}\right)T}{\sigma\sqrt{T}}$$

and

$$d_2 = d_1 - \sigma\sqrt{T}.$$

Solving each of these equations for d_1 and d_2 allows the determination of the European call option premium. The premium for a European put option, P, is similarly derived:

$$P = [F(N(d_1) - 1) - E(N(d_2) - 1)]e^{-r_d T}.$$

The European Call Option: Numerical Example

The actual calculation of the option premium is not as complex as it appears from the preceding set of equations. Assuming the following basic exchange rate and interest rate values, computation of the option premium is relatively straightforward.

Spot rate	= \$1.7000/£
90-day Forward	= \$1.7000/£
Strike rate	= \$1.7000/£
U.S. dollar interest rate	= 8.00% (per annum)
Pound sterling interest rate	= 8.00% (per annum)
Time (days)	= 90
Std. Dev. (volatility)	= 10.00%
e (infinite discounting)	= 2.71828

The value of the two density functions are first derived:

$$d_1 = \frac{\ln\left(\dfrac{F}{E}\right) + \left(\dfrac{\sigma^2}{2}\right)T}{\sigma\sqrt{T}} = \frac{\ln\left(\dfrac{1.7000}{1.7000}\right) + \left(\dfrac{0.1000^2}{2}\right)\dfrac{90}{365}}{0.1000\sqrt{\dfrac{90}{365}}} = 0.025$$

and

$$d_2 = 0.025 - 0.1000\sqrt{\frac{90}{365}} = -0.025.$$

The values of d_1 and d_2 are then found in a cumulative normal probability table,

$$N(d_1) = N(0.025) = 0.51; \quad N(d_2) = N(-0.025) = 0.49.$$

The premium of the European call with a "forward-at-the-money" strike rate is

$$C = [(1.7000)(0.51) - (1.7000)(0.49)]2.71828^{-0.08(90/365)} = \$0.033/£.$$

This is the call option *premium, price, value* or *cost*.

CHAPTER 8

Interest Rate Risk and Swaps

Confidence in markets and institutions, it's a lot like oxygen. When you have it, you don't even think about it. Indispensable. You can go years without thinking about it. When it's gone for five minutes, it's the only thing you think about. The confidence has been sucked out of the credit markets and institutions.

—Warren Buffett, October 1, 2008.

LEARNING OBJECTIVES

- Explain interest rate fundamentals, including basic floating rates of interest and fixed rates of interest
- Define corporate interest rate risk and demonstrate how to manage it
- Explore the use of interest rate futures and forward rate agreements in the management of interest rate risk
- Examine the use of interest rate swaps to manage the interest rate risks of multinational firms
- Detail how cross-currency swaps may be used to manage both foreign currency and interest rate risks in multinational financial management

All firms—domestic or multinational, small or large, leveraged or unleveraged—are sensitive to changes in interest rates. Although a variety of interest rate risks exist, this book focuses on the financial management of the nonfinancial (nonbank) multinational firm. The international financial marketplace in which these multinational firms operate is largely defined by both interest rates and exchange rates, and those theoretical linkages were established in Chapter 6 on parity relationships. We now turn to the interest rate structures and challenges of firms operating in a multi-currency interest rate world.

The first part of the chapter, interest rate foundations, details the various reference rates and floating rates that all multinationals deal in. The chapter then turns to the government-corporate interest rate relationships that define the costs and availability of capital. The third section focuses on the various forms of interest rate risk confronting multinational firms. The fourth and final section details how a variety of financial derivatives, including interest rate swaps, can be used to manage these interest rate risks. The Mini-Case at the end of the chapter, *Argentina and the Vulture Funds*, illustrates the risks a sovereign nation faces in attempting to recover from excessive debt and default.

Interest Rate Foundations

We begin our discussion of interest rates with some definitions. A *reference rate*—for example, U.S. dollar LIBOR—is the rate of interest used in a standardized quotation, loan agreement, or financial derivative valuation. LIBOR, the London Interbank Offered Rate, is by far the most widely used and quoted reference rate.

As defined by the British Bankers Association (BBA), LIBOR is quoted for overnight, 1 week, 1 month, 2 months, etc., through 12-month maturities. Of these, 1-month, 3-month, and 6-month LIBOR are the most significant maturities due to their widespread use in various loan and derivative agreements. The dollar and euro are the most widely used currencies, although the BBA also calculates the Japanese yen LIBOR, and other currency LIBOR rates at the same time in London from samples of banks. As *Global Finance in Practice 8.1* describes, LIBOR itself has been the subject of controversy in recent years.

The interbank interest rate market is not, however, confined to London. Most major domestic financial centers construct their own interbank offered rates for local loan agreement purposes. These rates include PIBOR (Paris Interbank Offered Rate), MIBOR (Madrid Interbank Offered Rate), SIBOR (Singapore Interbank Offered Rate), and FIBOR (Frankfurt Interbank Offered Rate), to name but a few.

GLOBAL FINANCE IN PRACTICE 8.1

The Trouble with LIBOR

"The idea that my word is my Libor is dead."
—Mervyn King, Bank of England Governor.

No single interest rate is more fundamental to the operation of the global financial markets than the London Interbank Offered Rate (LIBOR). But beginning as early as 2007, a number of participants in the interbank market on both sides of the Atlantic suspected that there was trouble with LIBOR.

LIBOR is published under the auspices of the British Bankers Association (BBA). Each day, a panel of 16 major multinational banks is requested to submit estimated borrowing rates in the unsecured interbank market, which are then collected, massaged, and published in three steps.

Step 1. The banks on the LIBOR panels must submit their estimated borrowing rates by 11:10 A.M. London time. The submissions are made directly to Thomson Reuters, which executes the process on behalf of the BBA.

Step 2. Thomson Reuters discards the lowest 25% and highest 25% of interest rates submitted. It then calculates an average rate by maturity and currency using the remaining 50% of borrowing rate quotes.

Step 3. The BBA publishes the day's LIBOR rates 20 minutes later, by 11:30 A.M. London time.

This same process is used to publish LIBOR for multiple currencies across a comprehensive set of maturities.

The 3-month and 6-month maturities are the most significant maturities due to their widespread use in various loan and derivative agreements, with the dollar and the euro being the most widely used currencies.

One problem with LIBOR is the origin of the rates submitted by banks. First, rates are based on "estimated borrowing rates" to avoid reporting only actual transactions, as many banks may not conduct actual transactions in all maturities and currencies each day. As a result, the origin of the rate submitted by each bank becomes, to some degree, discretionary.

Secondly, banks—specifically money-market and derivative traders within the banks—have a number of interests that may be impacted by borrowing costs reported by the bank that day. One such example can be found in the concerns of banks in the interbank market in September 2008, when the credit crisis was in full-bloom. When a bank reported that it was being charged a higher rate by other banks, it was effectively self-reporting the market's assessment that it was increasingly risky. In the words of one analyst, it was akin "to hanging a sign around one's neck that I am carrying a contagious disease." Market analysts are now estimating that many of the banks in the LIBOR panel were reporting borrowing rates that were anywhere from 30 to 40 basis points lower than actual rates throughout the financial crisis. Court documents continue to shed light on the depth of the market's manipulation, although it is not really known to what degree attempts at manipulation have been successful.

Exhibit 8.1 shows 3-month USD-LIBOR over the past 30 years. It has obviously moved over large ranges, from over 11% in the late 1980s to near-zero percent in the post 2008 financial crisis years. The most recent period—that in which the U.S. Federal Reserve has pursued quantitative easing (QE)—in which the Federal Reserve has continued to pump liquidity into the financial system, has clearly kept 3-month LIBOR at very low levels.

But as we noted in Chapter 6 on international parity conditions, interest rates are currency specific. Exhibit 8.2 makes that very apparent as it presents 3-month LIBOR rates for five of the world's largest financial markets—the U.S. dollar, the euro, the Swiss franc, British pound, and Japanese yen. The true volatility of short-term interest rates is clear here, in that in just the past 15 years these major loan rate foundations have bounced from near 7.00% to near or—in some cases recently—below zero percent.

Corporate Cost of Debt

Individual borrowers—whether they are governments or companies—possess their own individual *credit quality*, the market's assessment of their ability to repay debt in a timely manner. These credit assessments result in the assignments of designations of differences in the *cost and access to capital*. This means individual organizational borrowers not only pay different rates to borrow (different interest rates), but also have access to different amounts of capital or debt.

The cost of debt for any individual borrower will therefore possess two components, the *risk-free rate of interest* ($k_{US}^{\$}$), plus a *credit risk premium* ($RPM_{Rating}^{\$}$) reflecting the assessed

---

EXHIBIT 8.1   **U.S. Dollar 3-month LIBOR (monthly, 1986–2014)**

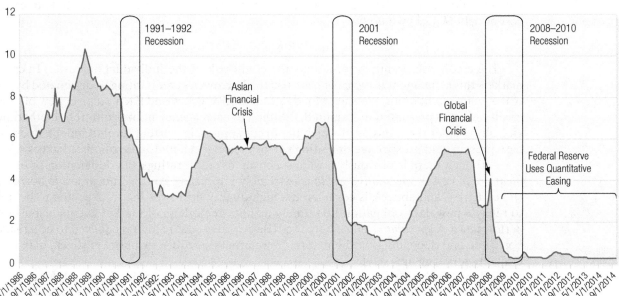

Source: LIBOR data from the Federal Reserve Economic Data (FRED), Federal Reserve Bank of St. Louis.

EXHIBIT 8.2   **3-month LIBOR for Select Currencies (Daily, Jan. 1999–Jan. 2015)**

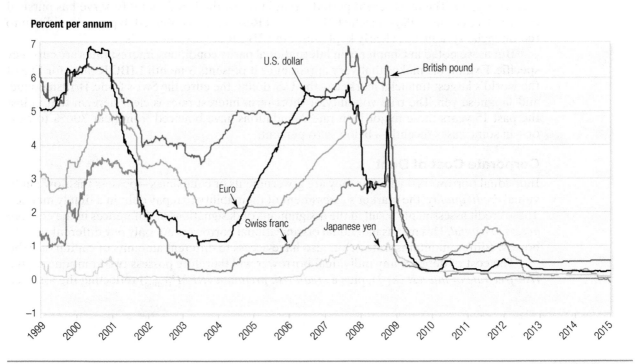

credit quality of the individual borrower. For an individual borrower in the United States, the cost of debt $(k_{\text{Debt}}^{\$})$ would be:

$$k_{\text{Debt}}^{\$} = k_{\text{US}}^{\$} + RPM^{\$}_{\text{Rating}}$$

The credit risk premium represents the credit risk of the individual borrower. In credit markets this assignment is typically based on the borrower's credit rating as designated by one of the major credit rating agencies, Moody's, Standard & Poors, and Fitch. An overview of those credit ratings is presented in Exhibit 8.3. Although each agency utilizes different methodologies, all consider the industry in which the firm operates, its current level of indebtedness, its past, present, and prospective operating performance, and a multitude of other factors.

Although there is obviously a wide spectrum of credit ratings, the designation of *investment grade* versus *speculative grade* is extremely important. An *investment grade* borrower (Baa3, BBB–, and above), is considered a high-quality borrower that is expected to be able to repay a new debt obligation in a timely manner regardless of market events or business performance. A *speculative grade* borrower (Ba1 or BB+ and below) is believed to be a riskier borrower, and depending on the nature of a market downturn or business shock, may have difficulty servicing new debt.

Exhibit 8.3 also illustrates how the cost of debt changes with credit quality. At this time the U.S. Treasury was paying 1.74% for funds for a 5-year period. The average single-A borrower ("A") paid 2.35% per annum for funds for a 5-year period at this time, 0.61% above the U.S. Treasury. Note that the costs of credit quality—*credit spreads*—are quite minor for borrowers of investment grade. Speculative grade borrowers, however, are charged a hefty premium in

| EXHIBIT 8.3 | Credit Ratings and Cost of Funds |

Investment Grade	Moody's	S&P	Fitch	5-year Average Rate	Spread Over Treasury*
Prime	Aaa	AAA	AAA	1.92%	0.18%
High grade	Aa1	AA+	AA+		
	Aa2	AA	AA	2.24%	0.50%
	Aa3	AA−	AA−		
Upper medium grade	A1	A+	A+		
	A2	A	A	2.35%	0.61%
	A3	A−	A−		
Lower medium grade	Baa1	BBB+	BBB+		
	Baa2	BBB	BBB	2.81%	1.07%
	Baa3	BBB−	BBB−		
Speculative Grade					
Speculative Grade	Ba1	BB+	BB+		
	Ba2	BB	BB	4.69%	2.95%
	Ba3	BB−	BB−		
Highly speculative	B1	B+	B+		
	B2	B	B	7.01%	5.27%
	B3	B−	B−		
Substantial risks	Caa1	CCC+	CCC	8.56%	6.82%
Extremely speculative	Caa2	CCC			
Default imminent	Caa3	CCC−			
In default	C	C, D	DDD, DD, D		

*These are long-term credit ratings. Rates quoted are for October 28, 2014, all for 5-year maturities. The 5-year U.S. Treasury rate was 1.74%.

the market. For example the average single-B ("B") borrower paid 7.01% per annum for a 5-year maturity, a full 5.27% above U.S. Treasuries.

The cost of debt for corporate borrowers also changes over maturity. Exhibit 8.4 graphically presents the full range of maturities of the same credit ratings and costs presented in the previous exhibit. Once again, it is the U.S. Treasury *yield curve*, the U.S. government's cost of funds over varying maturities, which establishes the base rates at which all corporate credits are then priced. Note that AAA-rated firms (currently there are only three—ExxonMobil, Microsoft, Johnson & Johnson) pay very little more than the U.S. government to borrow money. The bulk of the larger firms operating in the U.S. today, in the S&P500, are either A, BBB, or BB in rating. The U.S. Treasury yield curve is quite flat, but still upward sloping indicating that short-term funds are cheaper than long-term funds. (We have limited the graphic to 10 years in maturity. U.S. Treasuries actually extend out to 30 years.)

Every country with an established financial system will have some version of this same government yield curve plus corporate borrowing spread structure. In addition to the U.S., the largest in the world would be those of the European Union, the United Kingdom, and Japan.

| EXHIBIT 8.4 | U.S. Corporate Credit Spreads (October 28, 2014) |

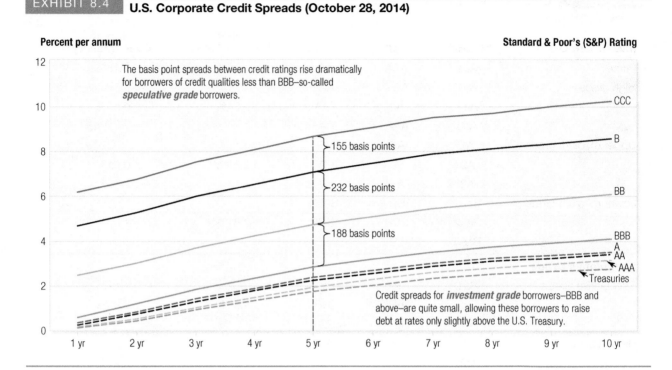

Percent per annum   **Standard & Poor's (S&P) Rating**

The basis point spreads between credit ratings rise dramatically for borrowers of credit qualities less than BBB–so-called *speculative grade* borrowers.

155 basis points

232 basis points

188 basis points

Credit spreads for *investment grade* borrowers–BBB and above–are quite small, allowing these borrowers to raise debt at rates only slightly above the U.S. Treasury.

CCC · B · BB · BBB · A · AA · AAA · Treasuries

As we will see later in this chapter, however, different countries—and currencies—possess very different costs of funds.

Credit Risk and Repricing Risk

For a corporate borrower, it is especially important to distinguish between *credit risk* and *repricing risk*. *Credit risk*, sometimes termed *roll-over risk*, is the possibility that a borrower's creditworthiness at the time of renewing a credit—its credit rating—is reclassified by the lender. This can result in changing fees, changing interest rates, altered credit line commitments, or even denial. *Repricing risk* is the risk of changes in interest rates charged (earned) at the time a financial contract's rate is reset. A borrower that is renewing a credit will face current market conditions on the base rate used for financing, a true floating-rate.

Sovereign Debt

Debt issued by governments—*sovereign debt*—is historically considered debt of the highest quality, higher than that of non-government borrowers within that same country. This quality preference stems from the ability of a government to tax its people and, if need be, print more money. Although the first may cause significant economic harm in the form of unemployment, and the second significant financial harm in the form of inflation, they are both tools available to the sovereign. The government therefore has the ability to service its own debt, one way or another, when that debt is denominated in its own currency. In certain sovereign combinations, like the EU, these fundamentals may be altered, as described in *Global Finance in Practice 8.2* on European sovereign debt.

A government, typically through its central banking authority, also conducts its own monetary policy. That policy will, in combination with economic conditions of growth and inflation,

European Sovereign Debt

The European Union is a complex organism compared to the customary structure of fiscal and monetary policy institutions described in the typical Economics 101 course. With the adoption of a common currency, the EU members participating in the euro gave up exclusive rights over the ability to print money (to service debt). Because it is a common currency, no one EU member has the right to simply print more euros—that is the policy realm of the European Central Bank (ECB). The members of the EU do have relative freedom to set their own fiscal policies—government spending, taxation, and the creation of government surpluses or deficits.

Following the onslaught of the global financial crisis in 2008–2009, a number of EU member countries suffered significant economic crises. Part of their economic woes included growing fears over their ability to service their outstanding sovereign debt. As seen below, these market fears and concerns drove the cost of their funds on the international marketplace very high, showing a dramatic separation of sovereign debt costs.

European Sovereign Debt Interest Costs

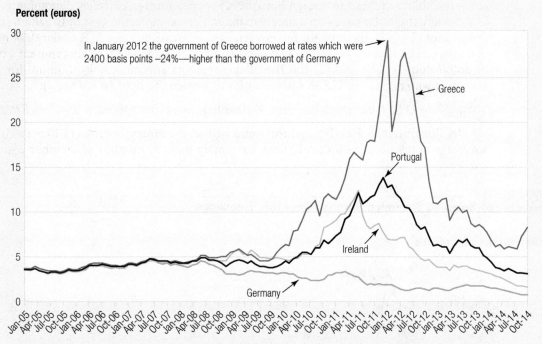

Source: Long-term interest rate statistics for EU member states, European Central Bank, www.ecb.int/stats/money/long. 10-year maturities.

determine the entire structure of its own interest rates of all maturities. Depending on the depth and breadth of domestic financial markets—the sophistication of the domestic financial marketplace—those maturities may be very short or very long. A large industrial country like the United States or Japan may issue its own debt instruments, (borrow money) at maturities as long as 30 years or longer. All denominated in its own currency, and saleable on the global market to buyers, domestic and foreign. The U.S. for example has financed a large portion of its government debt by selling U.S. Treasuries to investors all over the world, to private individuals, organizations, and even governments.

Domestic interest rates are in domestic currency, and as discussed in Chapter 6 on international parity conditions, interest rates are themselves currency specific. A direct comparison of interest rates across countries is only truly economically possible if the rates have all been converted to one currency (as is the case when looking at various opportunities for uncovered interest arbitrage), or when different countries raise debt in a common currency (say the U.S. dollar), or if exchange rates *never* change.

Sovereign Spreads

Many emerging market country governments often raise debt capital in the international markets, and they do so typically in one of the world's most widely traded currencies like the U.S. dollar, European euro, or Japanese yen. Exhibit 8.5 provides a comparison of what several sovereign borrowers have had to pay for U.S. dollar funds over and above that of the U.S. Treasury, the *U.S. dollar sovereign spread*, over the last two decades.

What Exhibit 8.5 details is the global financial market's assessment of *sovereign credit risk*—the ability of these sovereign borrowers to repay foreign currency denominated debt—U.S. dollar debt in this case—in a timely manner. For example, the cost of Brazilian sovereign dollar debt ($k_{Brazil}^{\$}$), the cost for the government of Brazil to raise U.S. dollar debt on global markets, can be decomposed into two basic components: (1) the U.S. government's own cost of dollar debt ($k_{US}^{\$}$); and (2) the Brazilian sovereign spread, a risk-premium for a dollar borrower who must earn U.S. dollars in order to service the debt ($RPM_{Brazil}^{\$}$):

$$k_{Brazil}^{\$} = \text{U.S. Treasury dollar rate} + \text{Brazilian sovereign spread} = k_{US}^{\$} + RPM_{Brazil}^{\$}$$

As illustrated in Exhibit 8.4, for some of these country borrowers like Pakistan, the sovereign spread has periodically been extremely high. A number of member countries of

EXHIBIT 8.5   **Selected Sovereign Spreads Over U.S. Treasuries**

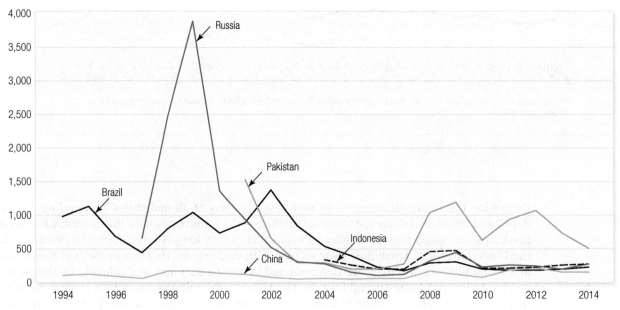

Percent paid over U.S. Treasuries

the European Union—particularly Greece, Portugal, and Ireland—have themselves struggled with recessionary economies and rising costs of debt in recent years. In early 2015, Russia was downgraded to "speculative" status by the major credit rating agencies. This downgrade, a result of economic deterioration associated with sanctions by Western nations (related to Ukraine) and the fall in the price of oil (that provides more than 50% of Russian government revenues), reduced Russia's access to capital.

Interest Rate Risk

The single largest interest rate risk of the nonfinancial firm is debt service. The debt structure of the MNE will possess differing maturities of debt, different interest rate structures (such as fixed versus floating-rate), and different currencies of denomination. Therefore the management of this debt portfolio can be quite complex, and in all cases, important.

The second most prevalent source of interest rate risk for the MNE lies in its holdings of interest-sensitive securities. Unlike debt, which is recorded on the right-hand-side of the firm's balance sheet (a liability), the marketable securities portfolio of the firm appears on the left-hand-side (an asset). Marketable securities represent potential earnings or interest inflows to the firm. Ever-increasing competitive pressures have pushed financial managers to tighten their management of interest rates on both the left and right sides of the firm's balance sheet.

International interest rate calculations are also a concern of any firm borrowing or investing. Interest rate calculations differ by the number of days used in the period's calculation and in the definition of how many days there are in a year (for financial purposes). Exhibit 8.6 illustrates three different examples of how different calculation methodologies result in different 1-month payments of interest on a $10 million loan, 5.500% per annum interest, for an exact period of 28 days.

The first example shown, International Practice, uses a 28-day month in a 360-day financial year. The result is an interest payment of $42,777.78:

$$0.055 \times \$10,000,000 \times (28/360) = \$42,777.78$$

If, however, the interest rate calculation had utilized the Swiss (Eurobond) Practice of a standard 30-day calculation, the interest cost for the same 1-month period would have been $45,833.33, a substantial $3,055.56 higher. Clearly, calculation methods matter.

EXHIBIT 8.6   **International Interest Rate Calculations**

International interest rate calculations differ by the number of days used in the period's calculation and their definition of how many days there are in a year (for financial purposes). The following example highlights how the different methods result in different 1-month payments of interest on a $10 million loan, 5.500% per annum interest, for an exact period of 28 days.

			$10 million @ 5.500% per annum	
Practice	**Day Count in Period**	**Days/Year**	**Days Used**	**Interest Payment**
International	Exact number of days	360	28	$42,777.78
British	Exact number of days	365	28	$42,191.78
Swiss (Eurobond)	Assumed 30 days/month	360	30	$45,833.33

Debt Structures and Strategies

Consider the three different bank loan structures being considered by U.S. medical equipment manufacturer, MedStat. Each is intended to provide $10 million in financing for a 3-year period.

Strategy 1: Borrow $1 million for three years at a fixed rate of interest.

Strategy 2: Borrow $1 million for three years at a floating rate, LIBOR + 2%, LIBOR to be reset annually.

Strategy 3: Borrow $1 million for one year at a fixed rate, then renew the credit annually.

Although the lowest cost of funds is always a major selection criterion, it is not the only one. If MedStat chooses Strategy 1, it assures itself of the funds for the full three years at a known interest rate. It has maximized the predictability of cash flows for the debt obligation. What it has sacrificed to some degree is the ability to enjoy a lower interest rate in the event that interest rates fall over the period. Of course, it has also eliminated the risk that it would face higher rates if interest rates rose over the period.

Strategy 2 offers what Strategy 1 did not—flexibility. It too assures MedStat of full funding for the 3-year period. This eliminates *credit risk*. *Repricing risk* is, however, alive and well in Strategy 2. If LIBOR changes dramatically by the second or third year, the LIBOR rate change is passed through completely to MedStat the borrower. The spread, however, remains fixed (reflecting the credit standing that has been locked in for the full three years). Flexibility comes at a cost in this case, the risk that interest rates can go up as well as down.

Strategy 3 offers MedStat more flexibility and more risk. First, the firm is borrowing at the shorter end of the yield curve. If the yield curve is positively sloped, as is commonly the case in major industrial markets, the base interest rate should be lower. But the short end of the yield curve is also the more volatile. It responds to short-term events in a much more pronounced fashion than longer-term rates. The strategy also exposes the firm to the possibility that its credit rating may change dramatically by the time for credit renewal, for better or worse. Noting that credit ratings in general attempt to establish whether a firm can meet its debt-service obligations under worsening economic conditions, firms that are highly creditworthy (investment-rated grades) may view Strategy 3 as a more relevant alternative than do firms of lower quality (speculative grades).

Although the previous example gives only a partial picture of the complexity of funding decisions and choices within the firm, it demonstrates the many ways credit risks and repricing risks are inextricably intertwined. The expression "interest rate exposure" is a complex concept, and the proper measurement of the exposure prior to its management is critical. We now proceed to describe the interest rate risk of the most common form of corporate debt, floating-rate loans.

MedStat's Floating-Rate Loans

Floating-rate loans are a widely used source of debt for firms worldwide. They are also the source of the single largest and most frequently observed corporate interest rate exposure. Exhibit 8.7 depicts the costs and cash flows for a 3-year floating rate loan taken out by MedStat. The loan of US$10 million will be serviced with annual interest payments and total principal repayment at the end of the 3-year period.

The loan is priced at U.S. dollar LIBOR + 1.250% (note that the cost of money, interest, is often referred to as price). The LIBOR base will be reset each year on an agreed-upon date (say two days prior to payment). Whereas the LIBOR component is truly floating, the spread of 1.250% is actually a fixed component of the interest payment, which is known with certainty for the life of the loan.

EXHIBIT 8.7   **MedStat's Floating-Rate Loan**

The expected interest rates and cash flows associated with a 3-year $10 million floating-rate loan with annual payments. MedStat pays an intitation (origination) fee of 1.500% of principal up-front.

Loan Interest	Year 0	Year 1	Year 2	Year 3
LIBOR (floating)		5.000%	5.000%	5.000%
Credit spread (fixed)		1.250%	1.250%	1.250%
Total interest payable		6.250%	6.250%	6.250%
Principal Payments				
Loan principal	$10,000,000			
Origination fees 1.50%	(150,000)			
Loan proceeds	$9,850,000			
Principal repayment				($10,000,000)
Interest Cash Flows				
LIBOR (floating)		($500,000)	($500,000)	($500,000)
Credit spread (fixed)		(125,000)	(125,000)	(125,000)
Total interest payable		($625,000)	($625,000)	($625,000)
Total loan cash flows	$9,850,000	($625,000)	($625,000)	($10,625,000)
All-in-Cost (AIC) or IRR	*6.820%*			

Note: The effective cost of funds (before-tax) for MedStat—the *all-in-cost* (AIC)—is calculated by finding the *internal rate of return* (IRR) of the total cash flows associated with the loan and its repayment. The AIC of the original loan agreement, without fees, would be 6.250%.

MedStat will not know the actual interest cost of the loan until the loan has been completely repaid. Caitlin Kelly, the CFO of MedStat, may forecast what LIBOR will be for the life of the loan, but she will not know with certainty until all payments have been completed. This uncertainty is not only an interest rate risk, but it is also an actual cash flow risk associated with the interest payment. (A fixed interest rate loan has interest rate risk, in this case opportunity cost, which does not put actual cash flows at risk.)

Exhibit 8.7 illustrates the cash flows and *all-in-costs* (AIC) of the floating-rate loan. The AIC is found by calculating the internal rate of return (IRR) of the total cash flow stream, including proceeds up-front and repayment over time. This baseline analysis assumes that LIBOR remains at 5.000% for the life of the loan. Including the up-front loan origination fees of 1.50%, the AIC to MedStat is 6.820% (or 6.250% without fees). But this is only hypothetical, as MedStat and its bank both assume that over time LIBOR will change. Which direction it will change, and by how much per year, is of course unknown. The loan's LIBOR component, but not the credit spread, creates a debt-service cash flow risk over time for MedStat.

If MedStat Corporation decided, after it had taken out the loan, that it wished to manage the interest rate risk associated with the loan agreement, it would have a number of management alternatives:

- **Refinancing.** MedStat could go back to its lender and restructure and refinance the entire agreement. This is not always possible and it is often expensive.

- **Forward rate agreements (FRAs).** MedStat could lock in the future interest rate payments with *forward rate agreements* (FRAs), interest rate contracts similar to foreign exchange forward contracts.

- **Interest rate futures.** Interest rate futures have gained substantial acceptance in the corporate sector. MedStat could lock in the future interest rate payments by taking an interest rate futures position.

- **Interest rate swaps.** MedStat could enter into an additional agreement with a bank or swap dealer in which it exchanged—*swapped*—future cash flows in such a way that the interest rate payments on the floating-rate loan would become fixed.

The following section details how the three interest rate derivative management solutions (all above but *refinancing*) work and might be utilized by a corporate borrower.

Interest Rate Futures and FRAs

Like foreign currency, there are a multitude of interest rate-based financial derivatives. We first describe *interest rate futures* and *forward rate agreements* (FRAs) before moving on to the interest rate swap.

Interest Rate Futures

Unlike foreign currency futures, *interest rate futures* are relatively widely used by financial managers and treasurers of nonfinancial companies. Their popularity stems from the high liquidity of the interest rate futures markets, their simplicity in use, and the rather standardized interest rate exposures most firms possess. The two most widely used futures contracts are the eurodollar futures traded on the Chicago Mercantile Exchange (CME) and the U.S. Treasury Bond Futures of the Chicago Board of Trade (CBOT). To illustrate the use of futures for managing interest rate risks, we will focus on the 3-month eurodollar futures contracts. Exhibit 8.8 presents eurodollar futures for two years (they actually trade 10 years into the future).

The yield of a futures contract is calculated from the settlement price, which is the closing price for that trading day. For example, a financial manager examining the eurodollar quotes in Exhibit 8.8 for a March 2011 contract would see that the settlement price on the previous day was 94.76, an annual yield of 5.24%:

$$\text{Yield} = (100.00 - 94.76) = 5.24\%$$

Since each contract is for a 3-month period (quarter) and a notional principal of $1 million, each basis point is actually worth $2,500 ($0.01 \times \$1,000,000 \times 90/360$).

EXHIBIT 8.8   **Eurodollar Futures Prices**

Maturity	Open	High	Low	Settle	Yield	Open Interest
June 10	94.99	95.01	94.98	95.01	4.99	455,763
Sept	94.87	94.97	94.87	94.96	5.04	535,932
Dec	94.60	94.70	94.60	94.68	5.32	367,036
March 11	94.67	94.77	94.66	94.76	5.24	299,993
June	94.55	94.68	94.54	94.63	5.37	208,949
Sept	94.43	94.54	94.43	94.53	5.47	168,961
Dec	94.27	94.38	94.27	94.36	5.64	130,824

Typical presentation by *The Wall Street Journal*. Only regular quarterly maturities shown. All contracts are for $1 million; points of 100%. Open interest is number of contracts outstanding.

If a financial manager were interested in hedging a floating-rate interest payment due in March 2011, she would need to sell a future, to take a short position. This strategy is referred to as a short position because the manager is selling something she does not own (as in shorting common stock). If interest rates rise by March, as the manager fears, the futures price will fall and she will be able to close the position at a profit. This profit will roughly offset the losses associated with rising interest payments on her debt. If the manager is wrong, however, and interest rates actually fall by the maturity date, causing the futures price to rise, she will suffer a loss that will wipe out the "savings" derived from making a lower floating-rate interest payment than she expected. So by selling the March 2011 futures contract, the manager locks in an interest rate of 5.24%.

Obviously, interest rate futures positions could be, and are on a regular basis, purchased purely for speculative purposes. Although that is not the focus of the managerial context here, the example shows how any speculator with a directional view on interest rates could take positions in expectations of profit.

As mentioned previously, the most common interest rate exposure of the nonfinancial firm is interest payable on debt. Such exposure is not, however, the only interest rate risk. As more and more firms manage their entire balance sheet, the interest earnings from the left-hand-side are under increasing scrutiny. If financial managers are expected to earn higher interest on interest-bearing securities (marketable securities), they may well find a second use for the interest rate futures market—to lock in future interest earnings. Exhibit 8.9 provides an overview of these two basic interest rate exposures and management strategies.

Forward Rate Agreements

A *forward rate agreement* (FRA) is an interbank-traded contract to buy or sell interest rate payments on a notional principal. These contracts are settled in cash. The buyer of an FRA obtains the right to lock in an interest rate for a desired term that begins at a future date. The contract specifies that the seller of the FRA will pay the buyer the increased interest expense on a nominal sum (the notional principal) of money if interest rates rise above the agreed rate, but the buyer will pay the seller the differential interest expense if interest rates fall below the agreed rate. Maturities available are typically 1, 3, 6, 9, and 12 months.

Like foreign currency forward contracts, FRAs are useful on individual exposures. They are contractual commitments of the firm that allow little flexibility to enjoy favorable movements, such as when LIBOR is falling as described in the previous section. Firms also use FRAs if they plan to invest in securities at future dates but fear that interest rates might fall prior to the investment date. Because of the limited maturities and currencies available, however, FRAs are not widely used outside the largest industrial economies and currencies.

EXHIBIT 8.9   **Interest Rate Futures Strategies for Common Exposures**

Exposure or Position	Futures Action	Interest Rates	Position Outcome
Paying interest on future date	Sell a Futures (short position)	If rates go up	Futures price falls; short earns a profit
		If rates go down	Futures price rises; short earns a loss
Earning interest on future date	Buy a Futures (long position)	If rates go up	Futures price falls; long earns a loss
		If rates go down	Futures price rises; long earns a profit

Interest Rate Swaps

Swaps are contractual agreements to exchange or swap a series of cash flows. These cash flows are most commonly the interest payments associated with debt service, the payments associated with fixed-rate and floating-rate debt obligations.

Swap Structures

The following three key concepts clarify the differences between swap agreements:

1. If the agreement is for one party to swap its fixed interest rate payment for the floating interest rate payments of another, it is termed an *interest rate swap*, sometimes referred to as a *plain-vanilla swap*.

2. If the agreement is to swap currencies of debt service, for example Swiss franc interest payments in exchange for U.S. dollar interest payments, it is termed a *currency swap* or *cross-currency swap*.

3. A single swap may combine elements of both interest rate swaps and currency swaps. For example, a swap agreement may swap fixed-rate dollar payments for floating-rate euro payments.

In all cases, the swap serves to alter the firm's cash flow obligations, as in changing floating-rate payments into fixed-rate payments associated with an existing debt obligation. The swap itself is not a source of capital, but rather an alteration of the cash flows associated with payment.

The two parties may have various motivations for entering into the agreement. For example, a very common position is as follows: A corporate borrower of good credit standing has existing floating-rate debt service payments. The borrower, after reviewing current market conditions and forming expectations about the future, may conclude that interest rates are about to rise. In order to protect the firm against rising debt-service payments, the company's treasury may enter into a swap agreement to pay fixed/receive floating. This means the firm will now make fixed interest rate payments and receive from the swap counterparty floating interest rate payments. The floating-rate payments that the firm receives are used to service the debt obligation of the firm, so the firm on a net basis is now making fixed interest rate payments. Using derivatives it has synthetically changed floating-rate debt into fixed-rate debt. It has done so without going through the costs and intricacies of refinancing existing debt obligations.

The cash flows of an interest rate swap are interest rates applied to a set amount of capital (notional principal). For this reason, these cash flows are also referred to as *coupon swaps*. Firms entering into interest rate swaps set the notional principal so that the cash flows resulting from the interest rate swap cover their interest rate management needs.

Interest rate swaps are contractual commitments between a firm and a swap dealer and are completely independent of the interest rate exposure of the firm. That is, the firm may enter into a swap for any reason it sees fit and then swap a notional principal that is less than, equal to, or even greater than the total position being managed. For example, a firm with a variety of floating-rate loans on its books may, if it wishes, enter into interest rate swaps for only 70% of the existing principal. The size of the swap notional principal is a choice purely in the hands of corporate management, and is not confined to the size of existing floating-rate loan obligations. It should also be noted that the interest rate swap market is filling a gap in market efficiency. If all firms had free and equal access to capital markets, regardless of interest rate structure or currency of denomination, the swap market would most likely not exist. The fact that the swap market not only exists but also flourishes and provides benefits to all parties is in some ways the proverbial "free lunch."

Illustrative Case: MedStat's Floating-Rate Debt

MedStat is a U.S.-based firm with a $40 million floating-rate bank loan. The company is finishing the first two years of the loan agreement (it is the end of the third quarter of 2017), with three years remaining. The loan is priced at LIBOR + 1.250%, the 3-month LIBOR rate plus a 1.250% credit risk premium. The recent movements of LIBOR and MedStat's floating-rate debt are depicted in Exhibit 8.10.

As seen in Exhibit 8.10, LIBOR has started moving upward in the past year. MedStat's management is now worried that interest rates will continue to rise and the company's interest costs rise with them. Management is considering entering into a *pay-fixed receive-floating* plain-vanilla interest rate swap. MedStat's New York bank has quoted it a fixed-rate payment of 3.850% in exchange for LIBOR. The *notional principal* of the swap, the base amount for calculating the interest cash flows, is something MedStat must choose. In this case they decide to enter into a notional principal equal to the full amount of the floating-rate loan, $40 million.

The interest rates of the proposed swap, when combined with MedStat's current floating-rate debt obligations, would appear as follows:

Debt/Swap Component	Floating	Fixed
Floating-rate loan	(LIBOR)	(1.250%)
Swap (pay fixed, receive floating)	LIBOR	(3.850%)
Floating-rate loan after swap	—	(5.100%)

EXHIBIT 8.10   **MedStat Considers a Plain Vanilla Interest Rate Swap**

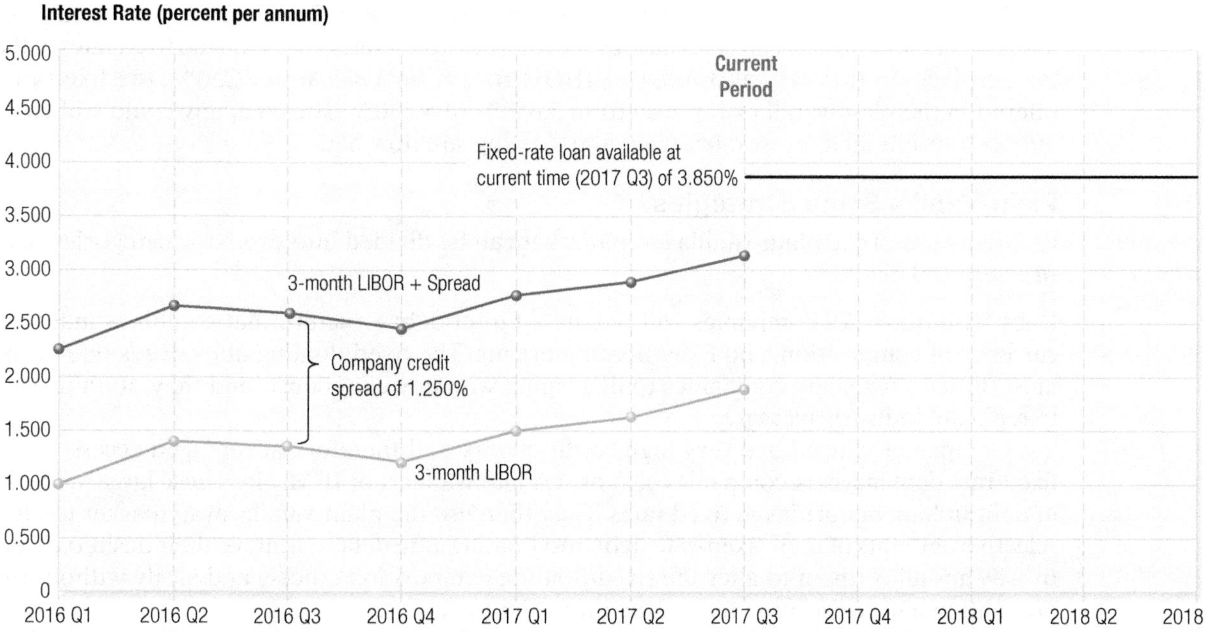

MedStat has $40 million in floating rate loans, paying a floating interest rate of LIBOR + 1.250%. Over the past year LIBOR has been trending upward. Company management is now considering swapping its floating-rate debt for fixed-rate payments—a *pay fixed, receive floating*—plain vanilla interest rate swap. If it swaps now it can lock-in a fixed rate payment of 3.850% (pay fixed component) in exchange for LIBOR (received floating component).

MedStat will now receive a floating-rate payment from the bank of LIBOR, which is then used to pay the LIBOR component on its floating-rate loan. What then remains is for MedStat to pay the fixed-rate spread on the loan, the 1.250% spread, plus the fixed-rate payment of the swap, 3.850%.

The fixed rate quoted to MedStat is based on a corporate issuer of AA quality for a 3-year maturity, which is the length of time the swap needs in order to cover the floating-rate loan. The fixed rates available in the swap market will therefore always reflect the current government and corporate yield curves in the appropriate currency market, in this case the U.S. dollar.

Note that the swap agreement swaps only the floating-rate component, LIBOR, and does not swap or deal with the credit spread in any way. This is because of two principles: (1) the swap market does not wish to deal with the credit risk of any individual borrower, only the core fixed and floating-rate foundations; and (2) the fixed-rate credit spread is indeed a fixed-rate component, and does not change over the life of the loan. The swap market is intended only for the true floating-rate component, and therefore the final combined fixed-rate obligation of MedStat after the swap is purely a fixed-rate payment at a combined rate of 5.100%.

The plain-vanilla interest rate swap is a very cheap and effective method of altering the cash flows associated with debt. They allow a firm to alter the interest rate associated with any debt obligation without suffering the costs (time and money) of repayment and refinancing.

Alternative Futures

Whether MedStat's swap to pay fixed proves to be the right strategy depends on what happens in the coming series of quarters. If LIBOR increases over the coming two or three quarters, but only marginally, then the decision to swap may not turn out to be the best one. If, however, LIBOR rises significantly, the swap may save MedStat sizable interest expenses.

It is also important to consider what might have happened if MedStat had decided not to execute the swap. Exhibit 8.11 provides one potential future—that LIBOR continues to rise— and MedStat's interest cost management options worsen. Both floating rates and fixed rates available are now higher. This is because as short-term interest rates rise, so do the fixed rates in the marketplace. In this case as 3-month LIBOR rose from 1.885% to 2.250%, the fixed rate offered in the plain-vanilla swap rose from 3.850% to 4.200%. The company could still enter into a swap, but all interest rates, fixed and floating, are now higher.

Plain-Vanilla Swap Strategies

Company use of the plain-vanilla swap market can be divided into two basic categories, *debt structure* and *debt cost*.

Debt Structure. All companies will pursue a target debt structure that combines maturity, currency of composition, and fixed/floating pricing. The fixed/floating objective is one of the most difficult for many companies to determine with any confidence, and they often just try to replicate industry averages.

Companies which have very high credit quality and therefore advantaged access to the fixed-rate debt markets, companies of A or AA like Walmart or IBM, often raise large amounts of debt in long maturities at fixed rates. They then use the plain-vanilla swap market to alter selective amounts of their fixed-rate debt into floating-rate debt to achieve their desired objective. Swaps allow them to alter the fixed/floating composition quickly and easily without the origination and registration fees of the direct debt markets.

Companies of lower credit quality, sometimes those of less than investment grade, often find the fixed-rate debt market not open to them. Getting fixed-rate debt is either impossible or too costly. They will generally raise their debt at floating rates and then periodically evaluate whether the plain-vanilla swap market offers any attractive alternatives to swap from paying-floating to paying-fixed.

EXHIBIT 8.11   **MedStat's Deteriorating Choices as LIBOR Rises**

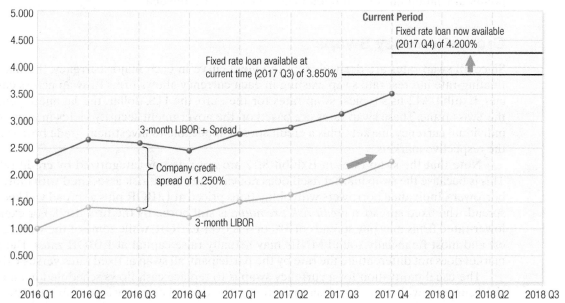

Interest Rate (percent per annum)

Current Period
Fixed rate loan now available
(2017 Q4) of 4.200%

Fixed rate loan available at
current time (2017 Q3) of 3.850%

3-month LIBOR + Spread

Company credit
spread of 1.250%

3-month LIBOR

If MedStat had decided not to execute the swap, and LIBOR continued to rise, MedStat would be left with increasingly poor choices. If MedStat decided that it would now enter into the pay fixed receive floating swap, the fixed rate now available to the company is now higher, 4.200%, rather than the previous quote of 3.850%. This is because as short-term floating rates of interest like LIBOR rise, so do the fixed rates of interest in the marketplace.

The plain-vanilla swap market is, of course, also frequently used by firms to adjust their fixed/floating debt structure for changing interest rate expectations. This was the case of MedStat described previously. Expectations of rising interest rates led the company to use plain-vanilla swaps to swap out of floating-rate payments into fixed-rate payments. During the 2009–2014 period, with interest rates often hitting historical lows in U.S. dollar and European euro debt markets, many firms used the swap market frequently to swap increasingly into fixed-rate obligations.

Debt Cost. All firms are always interested in opportunities to lower the cost of their debt. The plain-vanilla swap market is one highly accessible and low-cost method of doing so.

For an example of the second use, to achieve cheaper funding costs, we can again use the example of MedStat. Assume that MedStat regularly explored opportunities in the debt and swap markets. In our previous example, in the third quarter of 2017 MedStat discovered that it could swap $40 million of its existing debt for an all-in-fixed-rate cost (swap fixed-rate plus remaining credit spread) of 5.100%. At that same time, banks may have offered the company 3-year fixed-rate loans of the same size at a fixed rate of 5.20% or 5.30%. MedStat could, if it wished, swap floating-rate debt for fixed-rate debt to lock in cheaper fixed-rate debt.

These lower costs, achieved through the plain-vanilla swap market, may simply reflect short-term market imperfections and inefficiencies or the comparative advantage some borrowers have in selected markets via selective financial service providers. The savings may be large—30, 40, or even 50 basis points on occasion—or quite small. It is up to the management of the firm and its corporate treasury to determine how much savings is needed to make it worth spending the time and effort needed to execute the swaps. Banks promote the swap market and will regularly market deals to corporate treasuries. A corporate treasurer once remarked to the author that "unless the proposed structure or deal can save me 15 or 20 basis points, at a minimum, do not bother calling me to push the deal."

Cross-Currency Swaps

Since all swap rates are derived from the yield curve in each major currency, the fixed-to-floating-rate interest rate swap existing in each currency allows firms to swap across currencies. Exhibit 8.12 lists typical swap rates for the euro, the U.S. dollar, the Japanese yen, and the Swiss franc. These swap rates are based on the government security yields in each of the individual currency markets, plus a credit spread applicable to investment grade borrowers in the respective markets.

Note that the swap rates in Exhibit 8.12 are not rated or categorized by credit ratings. This is because the swap market itself does not carry the credit risk associated with individual borrowers. Individual borrowers with obligations priced at LIBOR plus a spread will keep the spread. The fixed spread, a *credit risk premium*, is still borne by the firm itself. For example, lower-rated firms may pay spreads of 3% or 4% over LIBOR, while some of the world's largest and most financially sound MNEs may actually raise capital at LIBOR rates. The swap market does not differentiate the rate by the participant; all swap at fixed rates versus LIBOR.

The usual motivation for a currency swap is to replace cash flows scheduled in an undesired currency with flows in a desired currency. The desired currency is probably the currency

EXHIBIT 8.12 | **Interest Rate Swap Quotes (December 31, 2014)**

Years	Euro € Bid	Euro € Ask	£ Sterling Bid	£ Sterling Ask	Swiss franc Bid	Swiss franc Ask	U.S. dollar Bid	U.S. dollar Ask	Japanese yen Bid	Japanese yen Ask
1	0.14	0.18	0.63	0.66	−0.14	−0.08	0.42	0.45	0.11	0.17
2	0.16	0.20	0.91	0.95	−0.18	−0.10	0.86	0.89	0.11	0.17
3	0.20	0.24	1.11	1.15	−0.14	−0.06	1.26	1.29	0.13	0.19
4	0.26	0.30	1.28	1.33	−0.07	0.01	1.55	1.58	0.15	0.21
5	0.34	0.38	1.42	1.47	0.02	0.10	1.75	1.78	0.19	0.25
6	0.42	0.46	1.53	1.58	0.11	0.19	1.90	1.93	0.24	0.30
7	0.51	0.55	1.62	1.67	0.21	0.29	2.02	2.05	0.30	0.36
8	0.60	0.64	1.69	1.74	0.30	0.38	2.11	2.10	0.36	0.42
9	0.70	0.74	1.76	1.81	0.39	0.47	2.19	2.22	0.42	0.48
10	0.79	0.83	1.82	1.87	0.47	0.55	2.26	2.29	0.49	0.55
12	0.95	0.99	1.91	1.98	0.59	0.69	2.37	2.40	0.61	0.69
15	1.12	1.16	2.02	2.11	0.75	0.85	2.48	2.51	0.82	0.90
20	1.30	1.34	2.12	2.25	0.95	1.05	2.59	2.62	1.09	1.17
25	1.39	1.43	2.15	2.28	1.06	1.16	2.64	2.67	1.22	1.30
30	1.44	1.48	2.17	2.30	1.11	1.21	2.67	2.70	1.29	1.37

LIBOR

Typical presentation by the *Financial Times*. Bid and ask spreads as of close of London business. US$ is quoted against 3-month LIBOR; Japanese yen against 6-month LIBOR; Euro and Swiss franc against 6-month LIBOR.

in which the firm's future operating revenues will be generated. Firms often raise capital in currencies in which they do not possess significant revenues or other natural cash flows. The reason they do so is cost; specific firms may find capital costs in specific currencies attractively priced to them under special conditions. Having raised the capital, however, the firm may wish to swap its repayment into a currency in which it has future operating revenues.

The utility of the currency swap market to an MNE is significant. An MNE wishing to swap a 10-year fixed 2.29% U.S. dollar cash flow stream could swap to 0.83% fixed in euros, 1.87% fixed in British pounds, 0.55% fixed in Swiss francs, or 0.55% fixed in Japanese yen. It could swap from fixed dollars not only to fixed rates, but also to floating LIBOR rates in the various currencies as well. Any of which could be arranged with the swap dealer/bank in a matter of only hours and at a fraction of the transaction costs and fees of actually borrowing in those currencies.

MedStat Corporation: Swapping Floating Dollars into Fixed-Rate British Pounds

We return to MedStat Corporation to demonstrate how to use a cross-currency swap. After raising $10 million in floating-rate financing, and subsequently swapping into fixed-rate payments, MedStat decides that it would prefer to make its debt service payments in British pounds. MedStat had recently signed a sales contract with a British buyer that will be paying pounds to MedStat over the next 3-year period. This would be a natural inflow of British pounds for the coming three years, and MedStat wishes to match the currency of denomination of the cash flows through a cross-currency swap.

MedStat enters into a 3-year *pay British pounds and receive U.S. dollars* cross-currency swap. Both interest rates are fixed. MedStat will pay 1.15% (*ask* rate) fixed British pound interest, and receive 1.26% (*bid* rate) fixed U.S. dollars. These swap rates are taken from Exhibit 8.12.

As illustrated in Exhibit 8.13, the 3-year currency swap entered into by MedStat is different from the plain-vanilla interest rate swap in two important ways:

1. The spot exchange rate in effect on the date of the agreement establishes what the notional principal is in the *target currency*. The *target currency* is the currency MedStat is swapping into, in this case the British pound. The $10,000,000 notional principal converts to a £6,410,256 notional principal. This is the principal used to establish the actual cash flows MedStat is committing to making (1.15% × £6,410,256 = £73,718).

2. The notional principal itself is part of the swap agreement. In plain-vanilla interest rate swaps, both interest payment cash flows were based on the same notional principal (in the same currency). Hence, there was no need to include the principal in the agreement. In a cross-currency swap, however, because the notional principals are denominated in two different currencies, and the exchange rate between those two currencies may change over time, the notional principals are part of the swap agreement.

At the time of the swap's inception, both positions have the same net present value. MedStat's swap commits it to three future cash payments in British pounds in return for receiving three payments in U.S. dollars. The payments are set. Accounting practices will require MedStat to regularly track and value its position, mark-to-market the swap, on the basis of current exchange rates and interest rates. If after the swap is initiated, the British pound appreciates versus the dollar, and MedStat is paying pounds, MedStat will record a loss on the swap for accounting purposes. (Similarly, the swap dealer's side of the transaction will record a gain.) At the same time, if interest rates in British pound markets rise, and MedStat's swap commits it to a fixed rate of 1.15%, then a gain will result from the interest component of the swap's value. In short, gains and losses on the swap, at least for accounting purposes, will persist throughout the swap's life. The currency swaps described here are *non-amortizing swaps*, where the swap

EXHIBIT 8.13   **MedStat's Cross-Currency Swap**

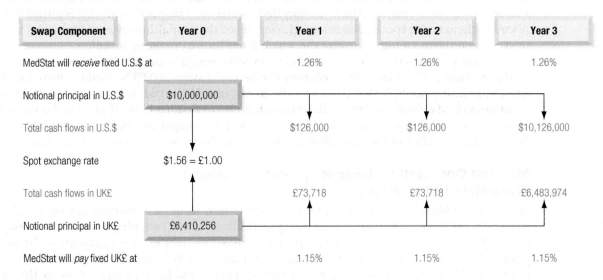

Swap Component	Year 0	Year 1	Year 2	Year 3
MedStat will *receive* fixed U.S.$ at		1.26%	1.26%	1.26%
Notional principal in U.S.$	$10,000,000			
Total cash flows in U.S.$		$126,000	$126,000	$10,126,000
Spot exchange rate	$1.56 = £1.00			
Total cash flows in UK£		£73,718	£73,718	£6,483,974
Notional principal in UK£	£6,410,256			
MedStat will *pay* fixed UK£ at		1.15%	1.15%	1.15%

Note: The U.S. dollar receive fixed rate is the three-year bid rate from Exhibit 8.12. The British pound pay fixed rate is the three-year ask rate from Exhibit 8.12. All rates are per annum, with annual payments and full principal repayment at maturity.

parties pay the entire principal at maturity, rather than over the life of the swap agreement. This is relatively standard practice in the market.

MedStat Corporation: Unwinding Swaps

As with all original loan agreements, it may happen that at some future date the partners to a swap may wish to terminate the agreement before it matures. If, for example, after one year MedStat Corporation's British sales contract is terminated, MedStat will no longer need the swap as part of its hedging program. MedStat could terminate or *unwind* the swap with the swap dealer.

Unwinding a currency swap requires the discounting of the remaining cash flows under the swap agreement at current interest rates, then converting the target currency (British pounds in this case) back to the home currency of the firm (U.S. dollars for MedStat). If MedStat has two payments remaining on the swap agreement (an interest-only payment, and a principal and interest payment), and the 2-year fixed rate of interest for pounds is now 1.50%, the present value of MedStat's commitment in British pounds is:

$$PV(\pounds) = \frac{\pounds 73,717.95}{(1.015)^1} + \frac{\pounds 6,483,974.36}{(1.015)^2} = \pounds 6,366,374.41$$

At the same time, the present value of the remaining cash flows on the dollar side of the swap is determined using the current 2-year fixed dollar interest rate, which is now 1.40%:

$$PV(\$) = \frac{\$126,000.00}{(1.04)^1} + \frac{\$10,126,000.00}{(1.04)^2} = \$9,972,577.21$$

MedStat's currency swap, if unwound at this time, would yield a present value of net inflows (what it receives under the swap) of $9,972,577.21 and a present value of outflows (what it pays under the swap) of £6,366,374.41. If the exchange rate is now $1.65/£, the net settlement of this currency swap will be:

$$\text{Settlement} = \$9,972,577.21 - (£6,366,374.41 \times \$1.65/£) = -\$531,940.56.$$

MedStat must therefore make a cash payment to the swap dealer of $531,941 to *unwind* the swap. MedStat's cash loss on this swap resulted largely from the appreciation of the British pound against the U.S. dollar (the interest rates both rose marginally). Since MedStat had promised to pay in the currency that is now stronger in value—the pound—unwinding the swap is costly. (If for example the exchange rate was still $1.56/£, MedStat would have closed out the position with a gain of $41,033.) It is important to remember, however, that the swap was entered into as a hedge of MedStat's long British pound position; it was not meant as a financial investment. One infamous unwinding required courts of law, that of the swaps Procter & Gamble purchased from Bankers Trust, as described in *Global Finance in Practice 8.3*.

Procter & Gamble and Bankers Trust

In 1994, Procter & Gamble (P&G) announced that it had incurred a $157 million pre-tax loss from closing out an interest rate swap transaction it had entered into with Bankers Trust. The loss would result in a $102 million after-tax charge to third-quarter earnings. According to P&G, this swap was a highly complex and speculative transaction that was counter to P&G's policy of conservatively managing their debt portfolio. Derivatives like these are dangerous and we were badly burned. We won't let this happen again. We are seriously considering our legal options relative to Bankers Trust, the financial institution that designed these swaps and brought them to us.

P&G wanted to continue to make floating-rate payments in return for receiving fixed-rate payments in order to maintain a balance between its fixed and floating-rate debt obligations. The swap would essentially transform existing fixed-rate debt obligations held by P&G into floating-rate obligations at a quite attractive rate. According to Bankers Trust, P&G stated that it was confident that interest rates would not rise significantly over the next year (although interest rates were at that time at historically low levels). P&G wanted to achieve the same favorable floating-rate as under another swap which had just matured, CP minus 40 basis points. But it did not want to incur significant risk.

Bankers Trust sold P&G a "5/30 Year Linked Swap," a 5-year swap structure with a notional principal of $200 million. Bankers Trust would pay P&G a fixed rate of 5.30% on a semi-annual basis. P&G in turn would pay Bankers Trust commercial paper (CP) less 75 basis points for the first six months of the swap agreement, and thereafter, CP less 75 basis points plus an additional spread (the "Spread") which could never be less than zero, and which would be fixed at the end of the first six months. The Spread was calculated as follows:

$$\text{Spread} = \frac{98.5 \times \dfrac{(5\text{-year Treasury yield})}{5.78\%} - \dfrac{30\text{-year } 6.25\%}{\text{Treasury price}}}{100}$$

The Spread formula was actually a speculative play on the entire U.S. Treasury yield curve. If 5-year Treasuries stayed roughly where they were at, the spread stayed near zero. However, because the Spread could increase geometrically with interest rate increases, rather than the customary arithmetic increases of standard interest rate movements, the instrument was considered highly leveraged.

The Spread formula was also extraordinarily sensitive to rising 5-year Treasury note yields. Each 1% increase in 5-year yields would increase P&G's payments under the leveraged swap by more than 17% of notional principal per year (CP plus 1,700 basis points), while each 1% decline in long bond prices would cost P&G 1% of notional principal. P&G had interpreted it as 0.17%, not 17%. A number of analysts noted the rather peculiar way in which the Spread was expressed. The division by 100 gave a number less than one, which appeared as a fraction of 1%.

Counterparty Risk

Counterparty risk is the potential exposure any individual firm bears that the second party to any financial contract will be unable to fulfill its obligations under the contract's specifications. Concern over counterparty risk periodically rises, usually associated with large and well-publicized derivative and swap defaults. The rapid growth in the currency and interest rate financial derivatives markets has actually been accompanied by a surprisingly low default rate to date, particularly in a global market that is unregulated in principle.

Counterparty risk has long been one of the major factors that favor the use of exchange-traded rather than over-the-counter derivatives. Most exchanges, like the Philadelphia Stock Exchange for currency options or the Chicago Mercantile Exchange for Eurodollar futures, are themselves the counterparty to all transactions. This allows all firms a high degree of confidence in being able to buy or sell exchange-traded products quickly and with little concern over the credit quality of the exchange itself. Financial exchanges typically require a small fee from all traders on the exchanges to pay for insurance funds created expressly for the purpose of protecting all parties. Over-the-counter products, however, are direct credit exposures to the firm because the contract is generally between the buying firm and the selling financial institution. Most financial derivatives in today's world financial centers are sold or brokered only by the largest and soundest financial institutions. This structure does not mean, however, that firms can enter continuing agreements with these institutions without some degree of real financial risk and concern.

A firm entering into a currency or interest rate swap agreement retains ultimate responsibility for the timely servicing of its own debt obligations. Although a swap agreement may constitute a contract to exchange U.S. dollar payments for euro payments, the firm that actually holds the dollar debt is still legally responsible for payment. The original debt remains on the borrower's books. In the event that a swap counterparty does not make the payment as agreed, the firm legally holding the debt is still responsible for debt service. In the event of such a failure, the euro payments would cease by the right of offset, and the losses associated with the failed swap would be mitigated.

The real exposure of an interest or currency swap is not the total notional principal, but the mark-to-market values of differentials in interest or currency interest payments (replacement cost) since the inception of the swap agreement. It is similar to the change in swap value discovered by unwinding a swap described previously. This amount is typically only 2% to 3% of the notional principal.

SUMMARY POINTS

- The single largest interest rate risk of the nonfinancial firm is debt service. The debt structure of the MNE will possess differing maturities of debt, different interest rate structures (such as fixed versus floating rate), and different currencies of denomination.

- The increasing volatility of world interest rates, combined with the increasing use of short-term and variable-rate debt by firms worldwide, has led many firms to actively manage their interest rate risks.

- The primary sources of interest rate risk to a multinational nonfinancial firm are short-term borrowing, short-term investing, and long-term debt service.

- The techniques and instruments used in interest rate risk management in many ways resemble those used in currency risk management. The primary instruments used for interest rate risk management include forward rate agreements (FRAs), forward swaps, interest rate futures, and interest rate swaps.

- The interest rate and currency swap markets allow firms that have limited access to specific currencies and interest rate structures to gain access at relatively low costs. This in turn allows these firms to manage their currency and interest rate risks more effectively.

- A cross-currency interest rate swap allows a firm to alter both the currency of denomination of cash flows in debt service and the fixed-to-floating or floating-to-fixed interest rate structure.

Argentina and the Vulture Funds

Argentina's default on its foreign currency denominated sovereign debt in 2001 had proved to be a never-ending nightmare. Now, in June 2014, 13 years after the default, the U.S. Supreme Court had confirmed a lower-court ruling, which would force Argentina to consider defaulting on its international debt obligations once again. But the story was a tangled one, which included investors all over the world, international financial law, courts in New York State and the European Union, and a battle between hedge funds and so-called *vulture funds*—funds that purchased distressed sovereign debt at low prices and then pursued full repayment through litigation. Time was running out.

The Default

Argentina's currency and economy had nearly collapsed in 1999. The rising sovereign debt obligations of the Argentine government, debt denominated in U.S. dollars and European euros, could not be serviced by the faltering economy. In late December 2001, Argentina officially defaulted on its foreign debt: $81.8 billion in private debt, $6.3 billion to the Paris Club; and $9.5 billion to the International Monetary Fund (IMF).

The debt had been originally issued in 1994 and registered under New York State governing law. New York law was specifically chosen because Argentina had become known as a *serial defaulter*. The bonds were issued under a specific structure, a Fiscal Agency Agreement (hence FAA Bonds), which would require all debt service be made through escrow accounts in New York.

The normal process following default is for the debtor to enter into talks with its creditors to *restructure* its debt obligations. Reaching consensus on sovereign debt restructuring, however, may be difficult because there is no international statutory regime for sovereign default similar to domestic bankruptcy codes. This leaves three options: (1) a collective solution; or (2) a voluntary exchange of old debt for restructured obligations; or (3) litigation.

The first option, the collective solution, is usually accomplished via the use of *collective action clauses* (CCAs) that impose similar reorganization terms on all creditors once a specific percentage of creditors, 75% to 90%, have agreed to the restructuring terms. These CCAs prevent a small number of creditors—*holdouts*—from blocking restructuring. Unfortunately, the Argentine bonds did not have collective action clauses.

Given the absence of a collective action clause, Argentina was left with the second option, a voluntary exchange of old debt for new debt. Debt restructuring itself normally includes four key elements: (1) a reduction in the principal of the obligation; (2) a reduction in the interest rate; (3) an extension of the obligation's maturity; and (4) capitalization of missed interest payments. The total result is a reduction in the net present value of the debt obligation, the so-called *haircut*, which may range anywhere from 30% to 70%.

A second common clause included within sovereign debt, the *pari passu* clause (Latin for "equal steps," and is read as "equal among equals"), calls for all creditors to be treated equally, assuring that private or individual deals are not made in preference to some creditors over others. Argentina's sovereign debt did include a *pari passu* clause (Argentine Fiscal Agency Agreement (1994), Clause 1©):

> *"The Securities will constitute [. . .] direct, unconditional, unsecured and unsubordinated obligations of the Republic and shall at all times rank pari passu and without any preference among themselves. The payment obligations of the Republic under the Securities shall at all times rank at least equally with all its other present and future unsecured and unsubordinated External Indebtedness (as defined in this Agreement)."*

Restructuring

Following the December 2001 default, Argentina initiated restructuring discussions with its creditors. From the very beginning it took a hard line. Nearly every dimension of the proposed restructuring was in debate, but the 70% haircut Argentina proposed was the single biggest problem. What followed, as seen in Exhibit A, was a long and twisted path toward debt resolution. It is important to note that many buyers of these bonds in the secondary markets were not original investors—they were intentionally buying defaulted debt.

After three years of contentious talks and two unsuccessful proposals, Argentina moved forward with a unilateral reorganization offer to all creditors in early 2005. In preparation for the offer, Argentina passed the Lock

EXHIBIT A   **Argentine Sovereign Bond Price and Default**

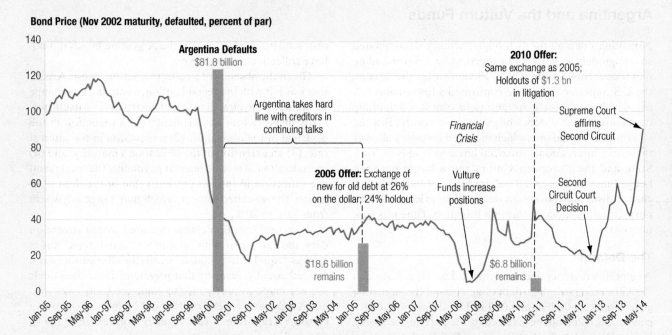

Bond Price (Nov 2002 maturity, defaulted, percent of par)

Law, *Ley cerrojo*. The Lock Law prohibited Argentina from making any arrangement to pay the unexchanged bonds: *"The national State shall be prohibited from conducting any type of in-court, out-of-court or private settlement with respect to the FAA [Fiscal Agency Agreement] bonds."* The Lock Law was meant to provide additional incentive for all creditors to exchange old debt for new debt—immediately.

Characterized as a "take it or leave it" offer, creditors were offered 26% to 30% of the net present value of the original face value of bond obligations, and as a result of the Lock Law, a one-time only offer. The offer was accepted by 75% of Argentina's creditors, reducing outstanding private debt from $81.8 billion to $18.6 billion. The offer was executed by exchanging the original bonds for new Argentina bonds.

The following year Argentina repaid the $9.5 billion in debt owed to the IMF. In 2010, in an attempt to eliminate the remaining outstanding private debt, Argentina temporarily suspended the Lock Law to allow the same bond exchange terms be offered once again to private debt holders. This second offer reduced the outstanding private debt principal to $8.6 billion. A full 92% of all original creditors had now exchanged the original debt for the reduced

value new debt instruments. But a number of holdouts still refused the exchange and instead pursued litigation.

In the prospectus associated with the 2005 and 2010 exchange offers (Exhibit B details both offers), Argentina made it very clear that it did not intend to ever make further payments on the original FAA bonds:

[FAA Bonds] that are in default and that are not tendered may remain in default indefinitely . . . Argentina does not expect to resume payments on any [FAA Bonds] that remain outstanding following the expiration of the exchange offer . . . there can be no assurance that [holders of unexchanged FAA Bonds] will receive any future payments or be able to collect through litigation . . .

Dueling Hedge Funds

Distressed sovereign debt is not a rarity, and so it is not surprising that a number of hedge funds have made the buying and selling of publicly traded distressed debt a business line. There was, however, a fundamental difference between hedge fund investing in corporate distressed debt and sovereign distressed debt. A fund purchasing a substantial portion of distressed corporate debt may become integrally

The Terms of the 2010 Argentina Bond Exchange

Bond Characteristic	Retail Investors	Institutional Investors
Bond Type	Par Bond (pays face value)	Discount Bond (66.3% reduction from face value)
Amount	Upto $2.0 billion	Upto $16.3 billion
Maturity Date	Dec-38	31-Dec-33
Annual Interest Rate	2.5%–5.25% increasing over time	8.28%
Past Due Interest (PDI)	Cash payment	Separate 2017 global bonds @ 8.75%
Bank Commission	0.40%	0.40%

Source: Securities and Exchange Commission, Amendment #5 to Argentina 18-K, April 19, 2010, and Prospectus Directive, April 27, 2010, pp. 11, 33–42, and 106–112.

involved in turning around the company; that was not the case with distressed sovereign debt.

Gramercy. One sovereign debt investor of note was Gramercy. In Gramercy's promotion of its distressed debt fund shown in Exhibit C, it highlighted its investments in the debt of Argentina. Note that the stated *target return* greatly exceeds the *yield to maturity* (the rate of return expected by a prospective investor at issuance who holds the security to maturity). The higher target return is based on the highly discounted purchase price of the securities relative to what Gramercy hopes/expects to sell the securities at a future date. Gramercy reportedly held $400 million in Argentine bonds in 2012 and still in 2014.

In addition to its purchase of distressed sovereign debt, Gramercy also aggressively protected its investments on the downside through the use of *credit default swaps (CDSs)*. The credit default swap is a derivative contract that derives its value from the credit quality and performance of any specific asset. A CDS is a way to bet whether a specific security would either fail to pay on time or fail to pay at all.

In some cases it provided insurance against the possibility that a borrower might not pay. In other instances, it was a way in which a speculator could bet against the increasingly risky security. This was a result of the fact that, as opposed to traditional insurance where an owner of the asset purchased insurance for their asset, an investor in CDSs need not own the asset (like your neighbor purchasing fire insurance on your house).

Gramercy's Chief Investment Officer was Robert Koenigsberger, formerly the manager of the sovereign debt restructuring team at Lehman Brothers. Koenigsberger had acted as an advisor to Argentina in its arguments in U.S. and European courts. It was in fact Gramercy that had persuaded Argentina to reopen the 2005 restructuring negotiations in 2010.

As Gramercy championed the 2010 restructuring offering, it argued that the markets were mis-pricing Argentine debt on the basis of government debt as percentage of GDP. Argentina's credit default swap (CDS) spread was a full 700 basis points—7.00%—above U.S. Treasuries, although Argentina's debt/GDP ratio was a moderate 46.4%. At the same time Brazil had a CDS spread of 119

EXHIBIT C **Gramercy's Holdings of Argentine Debt**

Security	Maturity	Yield to Maturity	Average Entry Price	Target Exit Price	Target Return
Argentina Par 2.5% (USD)	Dec 2038	11.82%	$32.67	$59.00	88.25%
Argentina Bonar 7% (USD)	Apr 2017	14.42%	$66.20	$100.00	61.63%
Argentina Discount 8.28% (USD)	Dec 2033	12.81%	$65.79	$107.50	75.99%

Source: Abstracted from "Gramercy Distressed Opportunity Fund II," September 2012, Current Investments, p. 12.

basis points with a 61.7% debt/GDP ratio, and Turkey a 150 basis point CDS spread with a 49.0% debt/GDP ratio. If Gramercy was correct, and if the market "corrected its error," the prices of the Argentine bonds would rise dramatically.

Elliot. A second hedge fund, Elliot Management Corp, had waged war on Argentina and its debt for years. Elliot, led by Paul Singer, a 68-year-old billionaire and a high-profile supporter of the U.S. Republican Party, was one of the chief litigants against Argentina. Singer had been called the father of *vulture funds*, and had used this same investment strategy in Peru, the Democratic Republic of Congo, and Panama, in recent years.

Elliot's fund, NML Capital Ltd., had first invested in Argentine bonds before the 2001 default, but had purchased most of its holdings as late as 2008, in the midst of the global financial crisis, at rock-bottom prices. (One story reported that Elliot paid $48.7 million for $832 million in bonds, $0.06 on the dollar.) It now claimed it was due $2.5 billion. As the lead holdout, Elliot had refused the offered exchange in 2005 or 2010. Elliot was known for hard-knuckle tactics, actually having detained an Argentine naval training vessel, the *ARA Libertad*, in a Ghanaian port for more than two months in an attempt to attach collateral.

NML Ltd. v Republic of Argentina

Because the distressed debt was under the jurisdiction of New York State law, specifically the Fiscal Agency Agreement (FAA), the case was eventually heard in U.S. District Court. The FAA stipulated that the repayments on the bonds were to be made by Argentina through a trustee based in New York, giving U.S. courts jurisdiction. On October 25, 2012, Judge Thomas P. Griesa for the United States District Court for the Southern District of New York, in NML Capital, Ltd. v. The Republic of Argentina, found in favor of the plaintiff:

> . . . *the judgements of the district court (1) granting summary judgement to plaintiffs on their claims for breach of the Equal Treatment Provision and (2) ordering Argentina to make "Ratable Payments" to plaintiffs concurrent with or in advance of its payments to holders of the 2005 and 2010 restructured debt are affirmed.*

The impact of the court's decision was dramatic. Argentina immediately announced that it would not honor the court's decision. One month later Argentina lost its [first] appeal, with Judge Griesa instructing Argentina to move forward quickly to comply with the Court's judgement. Again, Argentina refused to comply. Fitch, one of the three major global sovereign credit rating services, now downgraded Argentina's credit rating (long-term foreign currency) from B to CC, noted that "a default by Argentina is probable."

The markets closely followed the case. As illustrated by Exhibit D, the value of the outstanding Argentine exchange bonds plummeted in the days and months following the rulings. For example, the Argentine 2017 Global Bond (exchange bond from 2005) carrying an 8.75% coupon fell 9.9% in the one day following the court judgement, and a cumulative 24% in the following week.

In subsequent hearings Judge Griesa cautioned Argentina on altering the payment processes on the bonds. This was in response to Argentina's newest strategy to circumvent the U.S. courts by processing payments to the

---

EXHIBIT D   **District Court Ruling Impact on Exchange Bond Values**

Date	Republic Global Bond (Exchange), 8.75%, 2017			Republic Global Bond (Exchange 2005), 8.28%, 2033		
	Price	Change	Spread vs. U.S. Treasury	Price	Change	Chg in Spread vs. U.S. Treasury
October 25, 2012	$100.053		3.04%	$80.428		
October 26, 2012	$90.157	−9.9%		$72.125	−10.3%	1.5%
November 2, 2012	$76.483	−23.6%	7.79%	$61.278	−23.8%	3.74%

Source: Petition for a Write of Certiorari, Supreme Court of the United States, No. 13, Exchange Bondholder Group v. NNL Capital., Ltd., February 21, 2014.

exchanged bondholders through financial institutions outside the United States. The FAA bonds and their governing law expressly required processing through New York financial institutions.

On appeal Argentina argued that the District Court had misconstrued the *pari passu* clause, but the Second Circuit court was not persuaded. The court noted that the combination of the issuance and service on the exchange bonds, without making equal payments to the holdouts, and simultaneously stating under the Lock Law that the holdouts would not ever be paid, was in effect ranking or subordinating the original debt. The court went on to note the particularly critical role the *pari passu* clause plays:

> *When sovereigns default they do not enter bankruptcy proceedings where the legal rank of debt determines the order in which creditors will be paid. Instead, sovereigns can choose for themselves the order in which creditors will be paid. In this context, the [Pari Passu Clause] prevents Argentina as payor from discriminating against the FAA bonds in favor of other unsubordinated, foreign bonds.*

To its credit, Argentina had made substantial efforts at repairing its relationship with the outside world. In late May it had committed to repaying the $9.7 billion owed the Paris Club, and had agreed to pay Repsol of Spain, $5 billion in Argentine bonds for its seizure of its Argentine subsidiary earlier in 2013.

On Monday June 25, 2014, Argentina deposited $832 million in a New York bank in preparation for payment of interest on its exchange bonds. This would be in direct conflict with the ruling of the courts. According to Argentina's economic minister: "Complying with a ruling doesn't exempt us from honouring our obligations. Argentina will meet its obligations, will pay its debt, will honour its promises." A full-page official communique was placed in the *Financial Times* explaining the country's position. But the court order prevented the banks from dispensing payments on the restructured debt unless holdout creditors were also paid. Argentina argued that because it had deposited the money in the transfer accounts for payment, the bondholders had been paid. The courts disagreed.

What Now?

Argentina's response to the latest U.S. court rulings was outrage. Within days Argentina announced a plan to swap existing bonds governed by U.S. law for debt issued under Argentine law. Although unprecedented in sovereign debt markets, the move had been anticipated by market analysts. If investors were willing to undertake the exchange they would receive some of the highest interest yields in the world—foregoing default—but they would give up all legal rights and protection provided under U.S. law.

Judge Griesa released a statement that any attempt by Argentina to proceed with payments on its restructured debt, without settling with holdouts first, was illegal. Argentina continued to argue that this was impossible. According to the Argentine Finance Minister, if Argentina settled with the holdouts, the exchange bondholders could possibly sue for equal treatment—the RUFO clause (Rights Upon Future Offer), and total claims could reach $120 billion. The RUFO clause, contained in all of the exchange bonds issued in 2005 and 2010, guaranteed exchange bondholders the same rights and payments that might possibly be provided the holdout bondholders. Given the renewed currency crises the country was currently suffering, Argentina's hard currency reserves were estimated at only $30 billion, inadequate to threaten invoking RUFO interests.

On June 30, 2014, Argentina failed to make payments on its outstanding exchange bonds. The country now had a single 30-day grace period before entering into selective default. In early July, representatives of the Argentine Finance Ministry and the Elliot Group met to see if they could find an acceptable solution.

Mini-Case Questions

1. What is the role played by legal clauses such as the *collective action clause* and the *pari passu* clause in sovereign debt issuances?
2. What is the difference between a typical hedge fund investing in sovereign debt (even distressed sovereign debt) and the so-called *vulture funds*?
3. If you were appointed a mediator by the court to find a solution, what options or alternatives would you suggest for resolving the crisis?

QUESTIONS

These questions are available in MyFinanceLab.

1. **Reference Rates.** What is an interest "reference rate" and how is it used to set rates for individual borrowers?

2. **My Word is My LIBOR.** Why has LIBOR played such a central role in international business and financial contracts? Why has this been questioned in recent debates over its value reported?

3. **Credit Risk Premium.** What is a credit risk premium?

4. **Credit and Repricing Risk.** From the point of view of a borrowing corporation, what are credit and repricing risks? Explain steps a company might take to minimize both.

5. **Credit Spreads.** What is a credit spread? What credit rating changes have the most profound impact on the credit spread paid by corporate borrowers?

6. **Investment Grade versus Speculative Grade.** What do the general categories of investment grade and speculative grade represent?

7. **Sovereign Debt.** What is sovereign debt? What specific characteristic of sovereign debt constitutes the greatest risk to a sovereign issuer?

8. **Floating Rate Loan Risk.** Why do borrowers of lower credit quality often find their access limited to floating-rate loans?

9. **Interest Rate Futures.** What is an interest rate future? How can they be used to reduce interest rate risk by a borrower?

10. **Interest Rate Futures Strategies.** What would be the preferred strategy for a borrower paying interest on a future date if he expected interest rates to rise?

11. **Forward rate agreement.** How can a business firm that has borrowed on a floating-rate basis use a forward rate agreement to reduce interest rate risk?

12. **Plain Vanilla.** What is a plain-vanilla interest rate swap? Are swaps a significant source of capital for multinational firms?

13. **Swaps and Credit Quality.** If interest rate swaps are not the cost of government borrowing, what credit quality do they represent?

14. **LIBOR Flat.** Why do fixed-for-floating interest rate swaps never swap the credit spread component on a floating-rate loan?

15. **Debt Structure Swap Strategies.** How can interest rate swaps be used by a multinational firm to manage its debt structure?

16. **Cost-Based Swap Strategies.** How do corporate borrowers use interest rate or cross-currency swaps to reduce the costs of their debt?

17. **Cross-Currency Swaps.** Why would one company with interest payments due in pounds sterling want to swap those payments for interest payments due in U.S. dollars?

18. **Value Swings in Cross-Currency Swaps.** Why are there significantly larger swings in the value of a cross-currency swap than there is in a plain-vanilla interest rate swap?

19. **Unwinding a Swap.** How does a company cancel or unwind a swap?

20. **Counterparty risk.** How does organized exchange trading in swaps remove any risk that the counterparty in a swap agreement will not complete the agreement?

PROBLEMS

These problems are available in MyFinanceLab.

1. **U.S. Treasury Bill Auction Rates—March 2009.** The interest yields on U.S. Treasury securities in early 2009 fell to very low levels as a result of the combined events surrounding the global financial crisis. Calculate the simple and annualized yields for the 3-month and 6-month Treasury bills auctioned on March 9, 2009, listed here.

	3-Month T-Bill	6-Month T-Bill
Treasury bill, face value	$10,000.00	$10,000.00
Price at sale	$9,993.93	$9,976.74
Discount	$6.07	$23.26

2. **Credit Crisis, 2008.** During financial crises, short-term interest rates will often change quickly (typically up) as indications that markets are under severe stress. The interest rates shown in the table below are for selected dates in September–October 2008. Different publications define the TED Spread in different ways, but one measure is the differential between the overnight LIBOR interest rate and the 3-month U.S. Treasury bill rate.

Date	Overnight USD LIBOR	3-Month U.S. Treasury	TED Spread
9/8/08	2.15%	1.70%	_____
9/9/08	2.14%	1.65%	_____
9/10/08	2.13%	1.65%	_____
9/11/08	2.14%	1.60%	_____
9/12/08	2.15%	1.49%	_____
9/15/08	3.11%	0.83%	_____
9/16/08	6.44%	0.79%	_____
9/17/08	5.03%	0.04%	_____
9/18/08	3.84%	0.07%	_____
9/19/08	3.25%	0.97%	_____
9/22/08	2.97%	0.85%	_____
9/23/08	2.95%	0.81%	_____
9/24/08	2.69%	0.45%	_____
9/25/08	2.56%	0.72%	_____
9/26/08	2.31%	0.85%	_____
9/29/08	2.57%	0.41%	_____
9/30/08	6.88%	0.89%	_____
10/1/08	3.79%	0.81%	_____
10/2/08	2.68%	0.60%	_____
10/3/08	2.00%	0.48%	_____
10/6/08	2.37%	0.48%	_____
10/7/08	3.94%	0.79%	_____
10/8/08	5.38%	0.65%	_____
10/9/08	5.09%	0.55%	_____
10/10/08	2.47%	0.18%	_____
10/13/08	2.47%	0.18%	_____
10/14/08	2.18%	0.27%	_____
10/15/08	2.14%	0.20%	_____
10/16/08	1.94%	0.44%	_____
10/17/08	1.67%	0.79%	_____

a. Calculate the spread between the two market rates shown here in September and October 2008.

b. On what date is the spread the narrowest? The widest?

c. When the spread widens dramatically, presumably demonstrating some form of financial anxiety or crisis, which of the rates moves the most and why?

3. **Underwater Mortgages.** Bernie Madeoff pays $240,000 for a new four-bedroom 2,400-square-foot home outside Tonopah, Nevada. He plans to make a 20% down payment, but is having trouble deciding whether he wants a 15-year fixed rate (6.400%) or a 30-year fixed rate (6.875%) mortgage.

a. What is the monthly payment for both the 15- and 30-year mortgages, assuming a fully amortizing loan of equal payments for the life of the mortgage? Use a spreadsheet calculator for the payments.

b. Assume that instead of making a 20% down payment, he makes a 10% down payment, and finances the remainder at 7.125% fixed interest for 15 years. What is his monthly payment?

c. Assume that the home's total value falls by 25%. If the homeowner is able to sell the house, but now at the new home value, what would be his gain or loss on the home and mortgage assuming all of the mortgage principal remains? Use the same assumptions as in part a.

4. **Botany Bay Corporation.** Botany Bay Corporation of Australia seeks to borrow US$30,000,000 in the eurodollar market. Funding is needed for two years. Investigation leads to three possibilities. Compare the alternatives and make a recommendation.

1. Botany Bay could borrow the US$30,000,000 for two years at a fixed 5% rate of interest.

2. Botany Bay could borrow the US$30,000,000 at LIBOR + 1.5%. LIBOR is currently 3.5%, and the rate would be reset every six months.

3. Botany Bay could borrow the US$30,000,000 for one year only at 4.5%. At the end of the first year Botany Bay Corporation would have to negotiate for a new 1-year loan.

5. **Chrysler LLC.** Chrysler LLC, the now privately held company sold off by DaimlerChrysler, must pay floating rate interest three months from now. It wants to lock in these interest payments by buying an interest rate futures contract. Interest rate futures for three months from now settled at 93.07, for a yield of 6.93% per annum.

a. If the floating interest rate three months from now is 6.00%, what did Chrysler gain or lose?

b. If the floating interest rate three months from now is 8.00%, what did Chrysler gain or lose?

6. **CB Solutions.** Heather O'Reilly, the treasurer of CB Solutions, believes interest rates are going to rise, so she wants to swap her future floating rate interest payments for fixed rates. Presently, she is paying per annum on $5,000,000 of debt for the next two years, with payments due semiannually. LIBOR is currently

Problem 7.

Loan		Payments	1	2	3	4
Principal	$100	Interest	(10.00)	(7.85)	(5.48)	(2.87)
Interest rate	0.10	Principal	(21.55)	(23.70)	(26.07)	(28.68)
Maturity (years)	4.0	Total	(31.55)	(31.55)	(31.55)	(31.55)

4.00% per annum. Heather has just made an interest payment today, so the next payment is due six months from today.

Heather finds that she can swap her current floating rate payments for fixed payments of 7.00% per annum. (CB Solution's weighted average cost of capital is 12%, which Heather calculates to be 6% per 6-month period, compounded semiannually).

a. If LIBOR rises at the rate of 50 basis points per 6-month period starting tomorrow, how much does Heather save or cost her company by making this swap?

b. If LIBOR falls at the rate of 25 basis points per 6-month period, starting tomorrow, how much does Heather save or cost her company by making this swap?

7. **Negotiating the Rate.** A sovereign borrower is considering a $100 million loan for a 4-year maturity. It will be an amortizing loan, meaning that the interest and principal payments will total, annually, to a constant amount over the maturity of the loan. There is, however, a debate over the appropriate interest rate. The borrower believes the appropriate rate for its current creditstanding in the market today is 10%, but a number of the international banks with which it is negotiating are arguing that it is most likely 12%, at the minimum 11%. What impact do these different interest rates have on the prospective annual payments?

8. **Saharan Debt Negotiations.** The country of Sahara is negotiating a new loan agreement with a consortium of international banks. Both sides have a tentative agreement on the principal—$220 million. But there are still wide differences of opinion on the final interest rate

and maturity. The banks would like a shorter loan, four years in length, while Sahara would prefer a long maturity of six years. The banks also believe the interest rate will need to be 12.250% per annum, but Sahara believes that is too high, arguing for 11.750%.

a. What would the annual amortizing loan payments be for the bank consortium's proposal?

b. What would the annual amortizing loan payments be for Sahara's loan preferences?

c. How much would annual payments drop on the bank consortium's proposal if the same loan was stretched out from four to six years?

9. **Delos Debt Renegotiations (A).** Delos borrowed €80 million two years ago. The loan agreement, an amortizing loan, was for six years at 8.625% interest per annum. Delos has successfully completed two years of debt-service, but now wishes to renegotiate the terms of the loan with the lender to reduce its annual payments.

a. What were Delos's annual principal and interest payments under the original loan agreement?

b. After two years debt service, how much of the principal is still outstanding?

c. If the loan were restructured to extend another two years, what would the annual payments—principal and interest—be? Is this a significant reduction from the original agreement's annual payments?

10. **Delos Debt Renegotiations (B).** Delos is continuing to renegotiate its prior loan agreement (€80 million for six years at 8.625% per annum), two years into the agreement. Delos is now facing serious tax revenue shortfalls, and fears for its ability to service its debt obligations. So it has decided to get more

Problem 8.

Loan	0	Payments	1	2	3	4	5	6
Principal	$220	Interest	(26.950)	(23.650)	(19.946)	(15.788)	(11.120)	(5.881)
Interest rate	12.250%	Principal	(26.939)	(30.239)	(33.943)	(38.101)	(42.769)	(48.008)
Maturity (years)	6.0	Total	(53.889)	(53.889)	(53.889)	(53.889)	(53.889)	(53.889)

aggressive, and has gone back to its lenders with a request for a *haircut*, a reduction in the remaining loan amount. The banks have, so far, only agreed to restructure the loan agreement for another two years (new loan of six years on the remaining principal balance) but at an interest rate a full 200 basis points higher, 10.625%.

a. If Delos accepts the current bank proposal of the remaining principal for six years (extending the loan an additional two years since two of the original six years have already passed), but at the new interest rate, what are its annual payments going to be? How much relief does this provide Delos on annual debt-service?

b. Delos's demands for a *haircut* are based on getting the new annual debt service payments down. If Delos does agree to the new loan terms, what size of *haircut* should it try and get from its lenders to get its payments down to €10 million per year?

11. **Raid Gauloises.** Raid Gauloises is a rapidly growing French sporting goods and adventure racing outfitter. The company has decided to borrow €20,000,000 via a euro-euro floating rate loan for four years. Raid must decide between two competing loan offers from two of its banks.

Banque de Paris has offered the 4-year debt with an up-front initiation fee of 1.8%. Banque de Sorbonne, however, has offered a higher spread, but with no loan initiation fees up front, for the same term and principal. Both banks reset the interest rate at the end of each year.

Euro-LIBOR is currently 4.00%. Raid's economist forecasts that LIBOR will rise by 0.5 percentage points each year. Banque de Sorbonne, however, officially forecasts euro-LIBOR to begin trending upward at the rate of 0.25 percentage points per year. Raid Gauloises' cost of capital is 11%. Which loan proposal do you recommend for Raid Gauloises?

12. **Firenza Motors.** Firenza Motors of Italy recently took out a 4-year €5 million loan on a floating rate basis. It is now worried, however, about rising interest costs. Although it had initially believed interest rates in the eurozone would be trending downward when taking

out the loan, recent economic indicators show growing inflationary pressures. Analysts are predicting that the European Central Bank will slow monetary growth driving interest rates up.

Firenza is now considering whether to seek some protection against a rise in euro-LIBOR, and is considering a forward rate agreement (FRA) with an insurance company. According to the agreement, Firenza would pay to the insurance company at the end of each year the difference between its initial interest cost at (6.50%) and any fall in interest cost due to a fall in LIBOR. Conversely, the insurance company would pay to Firenza 70% of the difference between Firenza's initial interest cost and any increase in interest costs caused by a rise in LIBOR.

Purchase of the floating rate agreement will cost €100,000, paid at the time of the initial loan. What are Firenza's annual financing costs now if LIBOR rises and if LIBOR falls? Firenza uses 12% as its weighted average cost of capital. Do you recommend that Firenza purchase the FRA?

13. **Lluvia and Paraguas.** Lluvia Manufacturing and Paraguas Products both seek funding at the lowest possible cost. Lluvia would prefer the flexibility of floating rate borrowing, while Paraguas wants the security of fixed rate borrowing. Lluvia is the more creditworthy company. They face the following rate structure. Lluvia, with the better credit rating, has lower borrowing costs in both types of borrowing.

Lluvia wants floating rate debt, so it could borrow at LIBOR + 1%. However, it could borrow fixed at 8% and swap for floating rate debt. Paraguas wants fixed rate debt, so it could borrow fixed at 12%. However, it could borrow floating at LIBOR + 2% and swap for fixed rate debt. What should they do?

14. **Ganado's Cross-Currency Swap: SFr for US$.** Ganado Corporation entered into a 3-year cross-currency interest rate swap to receive U.S. dollars and pay Swiss francs. Ganado, however, decided to unwind the swap after one year—thereby having two years left on the settlement costs of unwinding the swap after one year. Repeat the calculations for unwinding, but assume that the rates shown below now apply.

Problem 14.

Assumptions	Values	Swap Rates	3-Year Bid	3-Year Ask
Notional principal	$10,000,000	Original: U.S. dollar	5.56%	5.59%
Original spot exchange rate, SFr/$	1.5000	Original: Swiss francs	1.93%	2.01%
New (1 year later) spot exchange rate, $/euro	5.5560			
New fixed U.S. dollar interest	5.20%			
New fixed Swiss franc interest	2.20%			

15. **Ganado's Cross-Currency Swap: Yen for Euros.** Use the table of swap rates in the chapter (Exhibit 8.12), and assume Ganado enters into a swap agreement to receive euros and pay Japanese yen, on a notional principal of €5,000,000. The spot exchange rate at the time of the swap is ¥104/€.
   a. Calculate all principal and interest payments, in both euros and Swiss francs, for the life of the swap agreement.
   b. Assume that one year into the swap agreement Ganado decides it wants to unwind the swap agreement and settle it in euros. Assuming that a 2-year fixed rate of interest on the Japanese yen is now 0.80%, and a 2-year fixed rate of interest on the euro is now 3.60%, and the spot rate of exchange is now ¥114/€, what is the net present value of the swap agreement? Who pays whom what?

16. **Falcor.** Falcor is the U.S.-based automotive parts supplier that was spun-off from General Motors in 2000. With annual sales of over $26 billion, the company has expanded its markets far beyond traditional automobile manufacturers in the pursuit of a more diversified sales base. As part of the general diversification effort, the company wishes to diversify the currency of denomination of its debt portfolio as well. Assume Falcor enters into a $50 million 7-year cross-currency interest rate swap to do just that—pay euros and receive dollars. Using the data in Exhibit 8.12, solve the following:
   a. Calculate all principal and interest payments in both currencies for the life of the swap.
   b. Assume that three years later Falcor decides to unwind the swap agreement. If 4-year fixed rates of interest in euros have now risen to 5.35%, 4-year fixed rate dollars have fallen to 4.40%, and the current spot exchange rate is $1.02/€, what is the net present value of the swap agreement? Explain the payment obligations of the two parties precisely.

INTERNET EXERCISES

1. **Financial Derivatives and the ISDA.** The International Swaps and Derivatives Association (ISDA) publishes a wealth of information about financial derivatives, their valuation and their use, in addition to providing master documents for their contractual use between parties. Use the following ISDA Internet site to find the definitions to 31 basic financial derivative questions and terms:

   ISDA   www.isda.org/educat/faqs.html

2. **Risk Management of Financial Derivatives.** If you think this book is long, take a look at the freely downloadable U.S. Comptroller of the Currency's handbook on risk management related to the care and use of financial derivatives!

   | Comptroller of the Currency | www.occ.gov/publications/ publications-by-type/comptrollers-handbook/deriv.pdf |

3. **Option Pricing.** OzForex Foreign Exchange Services is a private firm with an enormously powerful foreign currency derivative-enabled Web site. Use the following site to evaluate the various "Greeks" related to currency option pricing.

   OzForex   www.ozforex.com.au/forex-tools/ tools/fx-options-calculator

4. **Garman-Kohlhagen Option Formulation.** For those brave of heart and quantitatively adept, check out the following Internet site's detailed presentation of the Garman-Kohlhagen option formulation used widely in business and finance today.

   Riskglossary.com   www.riskglossary.com/link/ garman_-kohlhagen_1983.htm

5. **Chicago Mercantile Exchange.** The Chicago Mercantile Exchange trades futures and options on a variety of currencies, including the Brazilian real. Navigate to FX under the Trading tab on the following site to evaluate the uses of these currency derivatives:

   Chicago Mercantile Exchange   www.cmegroup.com

6. **Implied Currency Volatilities.** The single unobservable variable in currency option pricing is the volatility, since volatility inputs are the expected standard deviation of the daily spot rate for the coming period of the option's maturity. Use the New York Federal Reserve's Web site to obtain current implied currency volatilities for major trading crossrate pairs.

   Federal Reserve Bank of New York   www.ny.frb.org/markets/ impliedvolatility.html

7. **Montreal Exchange.** The Montreal Exchange is a Canadian exchange devoted to the support of financial derivatives in Canada. Use its Web site to view the latest on MV volatility, the volatility of the Montreal Exchange Index itself in recent trading hours and days.

   Montreal Exchange   www.m-x.ca/ marc_options_en.php

CHAPTER 9

Foreign Exchange Rate Determination

The herd instinct among forecasters makes sheep look like independent thinkers.

—Edgar R. Fiedler.

LEARNING OBJECTIVES

- Explore the three major theoretical approaches to exchange rate determination
- Detail how direct and indirect foreign exchange market intervention is conducted by central banks
- Analyze the primary causes of exchange rate disequilibrium in emerging market currencies
- Observe how forecasters combine technical analysis with the three major theoretical approaches to forecasting exchange rates

What determines the exchange rate between currencies? This has proven to be a very difficult question to answer. Companies and agents need foreign currency for buying imports, or they may earn foreign currency by exporting. Investors need foreign currency to invest in interest-bearing instruments in foreign markets, such as fixed-income securities (bonds), shares in publicly traded companies, or other new types of hybrid instruments. Tourists, migrant workers, speculators on currency movements—all of these economic agents buy and sell and supply and demand currencies every day. This chapter offers a basic theoretical framework with which to organize these elements, forces, and principles.

Chapter 6 described the international parity conditions that integrate exchange rates with inflation and interest rates and provided a theoretical framework for both the global financial markets and the management of international financial business. Chapter 3 provided a detailed analysis of how an individual country's international economic activity, its balance of payments, can impact exchange rates. This chapter builds on those discussions of exchange rate determination schools of thought and looks at a third school of thought—the asset market approach. The chapter then turns to government intervention in the foreign exchange market. In the third and final section, we discuss a number of approaches to foreign exchange forecasting in practice. The chapter concludes with the Mini-Case entitled *Russian Ruble Roulette*—a study in how the Russian ruble's value has evolved in just the past three-year period from internal change and external forces.

Exchange Rate Determination: The Theoretical Thread

There are basically three views of the exchange rate. The first takes the exchange rate as the relative price of monies (the monetary approach); the second, as the relative price of goods (the purchasing-power-parity approach); and the third, the relative price of bonds.

—Rudiger Dornbusch, "Exchange Rate Economics: Where Do We Stand?,"
Brookings Papers on Economic Activity 1, 1980, pp. 143–194.

Professor Dornbusch's tripartite categorization of exchange rate theory is a good starting point, but in some ways not robust enough—in our humble opinion—to capture the multitude of theories and approaches. So, in the spirit of both tradition and completeness, we have amended Dornbusch's three categories with several additional streams of thought in the following discussion.

Exhibit 9.1 provides an overview of the three major schools of thought—*theoretical determinants*—of exchange rates. The exhibit is organized by the three major schools of thought (*parity conditions, balance of payments approach, asset market approach*) and second by the individual drivers within those approaches. At first glance, the idea that there are three different theories may appear daunting, but it is important to remember that these are not competing theories, but rather *complementary* theories. Without the depth and breadth of the various approaches combined, our ability to capture the complexity of the global market for currencies is lost.

In addition to the three schools of thought described in Exhibit 9.1, note that two other institutional dimensions are considered—whether the country possesses the *capital markets* and *banking systems* needed to drive and discover value. Finally, note that most *determinants*

EXHIBIT 9.1   **The Determinants of Foreign Exchange Rates**

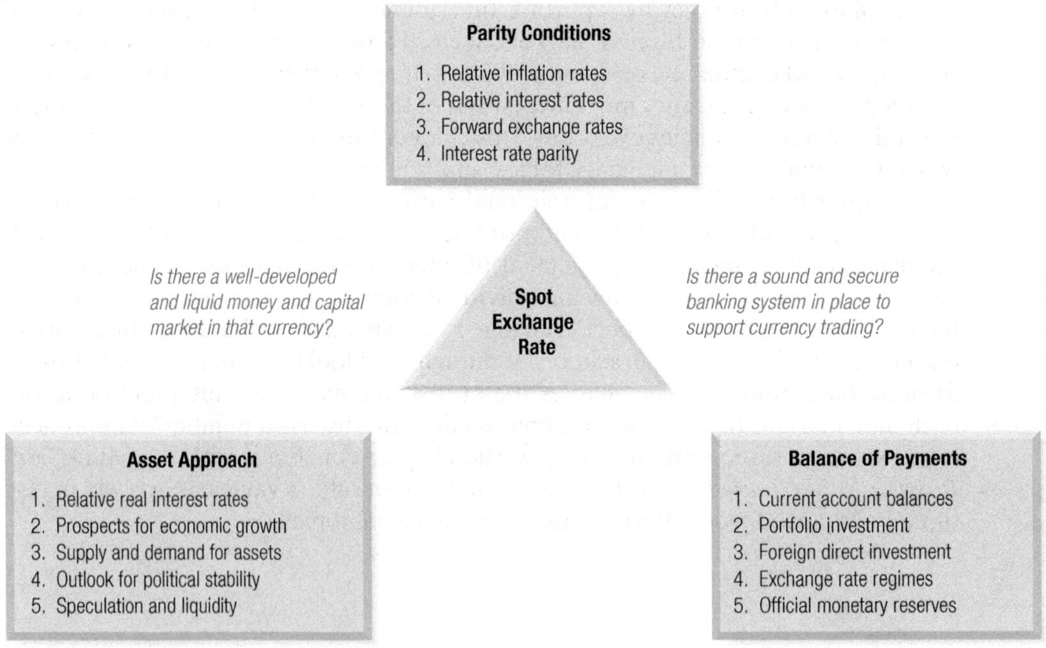

Parity Conditions

1. Relative inflation rates
2. Relative interest rates
3. Forward exchange rates
4. Interest rate parity

Is there a well-developed and liquid money and capital market in that currency?

Spot Exchange Rate

Is there a sound and secure banking system in place to support currency trading?

Asset Approach

1. Relative real interest rates
2. Prospects for economic growth
3. Supply and demand for assets
4. Outlook for political stability
5. Speculation and liquidity

Balance of Payments

1. Current account balances
2. Portfolio investment
3. Foreign direct investment
4. Exchange rate regimes
5. Official monetary reserves

of the spot exchange rate are also affected by *changes* in the spot rate. In other words, they are not only linked but also mutually determined.

Purchasing Power Parity Approaches

> *Under the skin of an international economist lies a deep-seated belief in some variant of the PPP theory of the exchange rate.*
>
> —Paul Krugman, 1976.

The most widely accepted of all exchange rate determination theories, the theory of purchasing power parity (PPP) states that the long-run equilibrium exchange rate is determined by the ratio of domestic prices relative to foreign prices, as explained in Chapter 6. The *purchasing power parity (PPP) approach* is both the oldest and most widely followed of the exchange rate theories, and most theories of exchange rate determination have PPP elements embedded within their frameworks.

There are a number of different versions of PPP: the Law of One Price, Absolute Purchasing Power Parity, and Relative Purchasing Power Parity (all were discussed in detail in Chapter 6). The latter of the three theories, Relative Purchasing Power Parity, is thought to be the most consistently relevant to possibly explaining what drives exchange rate values. In essence, it states that changes in relative prices between countries drive the change in exchange rates over time.

If, for example, the current spot exchange rate between the Japanese yen and U.S. dollar was ¥90.00 = $1.00, and Japanese and U.S. prices were to change at 2% and 1% over the coming period, respectively, the spot exchange rate next period would be ¥90.89/$.

$$S_{t+1} = S_t \times \frac{1 + \Delta \text{ in Japanese prices}}{1 + \Delta \text{ in U.S. prices}} = ¥90.00/\$ \times \frac{1.02}{1.01} = ¥90.89/\$.$$

Although PPP seems to possess a core element of common sense, it has proven to be quite poor at forecasting exchange rates (at least in the short to medium term). The problems are both theoretical and empirical. The theoretical problems lie primarily with its basic assumption that the only thing that matters is relative price changes. Yet many currency supply and demand forces are driven by other forces, including investment incentives, economic growth, and political change. The empirical issues are primarily in deciding which measures or indexes of prices to use across countries, in addition to the ability to provide a "predicted change in prices" with the chosen indexes.

Balance of Payments (Flows) Approach

After PPP, the most frequently used theoretical approach to exchange rate determination is probably that involving the supply and demand for currencies in the foreign exchange market. These exchange rate flows reflect current account and financial account transactions recorded in a nation's balance of payments, as described in Chapter 3. The basic *balance of payments approach* argues that the equilibrium exchange rate is found when the net inflow (outflow) of foreign exchange arising from current account activities matches the net outflow (inflow) of foreign exchange arising from financial account activities.

The balance of payments approach continues to enjoy widespread appeal, as balance of payments transactions are among the most frequently captured and reported of international economic activity. Trade surpluses and deficits, current account growth in service activity, and, recently, the growth and significance of international capital flows continue to fuel this theoretical fire.

Criticisms of the balance of payments approach arise from the theory's emphasis on flows of currency and capital rather than on stocks of money or financial assets. Relative stocks of money or financial assets play no role in exchange rate determination in this theory, a weakness explored in the following discussion of monetary and asset market approaches. Curiously, while the balance of payments approach is largely dismissed by the academic community, the practitioner public-market participants, including currency traders themselves, still rely on variations of this theory for much of their decision making.

Monetary Approach

The *monetary approach* in its simplest form states that the exchange rate is determined by the supply and demand for national monetary stocks, as well as the expected future levels and rates of growth of monetary stocks. Other financial assets, such as bonds, are not considered relevant for exchange rate determination as both domestic and foreign bonds are viewed as perfect substitutes. It is *all* about money stocks.

The monetary approach focuses on changes in the supply and demand for money as the primary determinant of inflation. Changes in relative inflation rates in turn are expected to alter exchange rates through a purchasing power parity effect. The monetary approach then assumes that prices are flexible in the short run as well as the long run, so that the transmission mechanism of inflationary pressure is immediate in impact.

A weakness of monetary models of exchange rate determination is that real economic activity is relegated to a role in which it only influences exchange rates through changes in the demand for money. The monetary approach is also criticized for its omission of a number of factors that are generally agreed upon by area experts as important to exchange rate determination, including (1) the failure of PPP to hold in the short to medium term; (2) money demand appears to be relatively unstable over time; and (3) the level of economic activity and the money supply appear to be interdependent, not independent.

Asset Market Approach (Relative Price of Bonds)

The *asset market approach*, sometimes called the *relative price of bonds* or *portfolio balance approach*, argues that exchange rates are determined by the supply and demand for financial assets of a wide variety. Shifts in the supply and demand for financial assets alter exchange rates. Changes in monetary and fiscal policy alter expected returns and perceived relative risks of financial assets, which in turn alter rates.

Many of the macroeconomic theoretical developments in recent years focused on how monetary and fiscal policy changes altered the relative perceptions of return and risk to the stocks of financial assets driving exchange rate changes. The frequently cited works of Mundell-Fleming are in this genre. Theories of currency substitution, the ability of individual and commercial investors to alter the composition of their portfolios, follow the same basic premises of the portfolio balance and rebalance framework.

Unfortunately, for all of the good work and research over the past 50 years, the ability to forecast exchange rate values in the short term to long term is—in the words of the authors below—*sorry*. Although academics and practitioners alike agree that in the long run fundamental principles such as purchasing power and external balances drive currency values, none of the fundamental theories have proven to be very useful in the short to medium term.

> ... *the case for macroeconomic determinants of exchange rates is in a sorry state [The] results indicate that no model based on such standard fundamentals like money supplies, real income, interest rates, inflation rates and current account balances will ever succeed in explaining or predicting a high percentage of the variation in the exchange rate, at least at short- or medium-term frequencies.*
>
> —Jeffrey A. Frankel and Andrew K. Rose, "A Survey of Empirical Research on Nominal Exchange Rates," NBER Working Paper No. 4865, 1994.

Technical Analysis

The forecasting inadequacies of fundamental theories has led to the growth and popularity of *technical analysis*, the belief that the study of past price behavior provides insights into future price movements. The primary feature of technical analysis is the assumption that exchange rates, or for that matter all market-driven prices, follow trends. And those trends may be analyzed and projected to provide insights into short-term and medium-term price movements in the future. *Global Finance in Practice 9.1* illustrates one example of a simplified technical analysis of the Japanese yen–U.S. dollar crossrate.

Most theories of technical analysis differentiate fair value from market value. Fair value is the true long-term value that the price will eventually retain. The market value is subject to a multitude of changes and behaviors arising from widespread market participant perceptions and beliefs at the time.

The Asset Market Approach to Forecasting

The asset market approach assumes that whether foreigners are willing to hold claims in monetary form depends on an extensive set of investment considerations or drivers. These drivers, as previously depicted in Exhibit 9.1, include the following elements:

- Relative real interest rates are a major consideration for investors in foreign bonds and short-term money market instruments.

- Prospects for economic growth and profitability are an important determinant of cross-border equity investment in both securities and foreign direct investment.

GLOBAL FINANCE IN PRACTICE 9.1

Technical Analysis of the JPY/USD Rate (January 2011– February 2014)

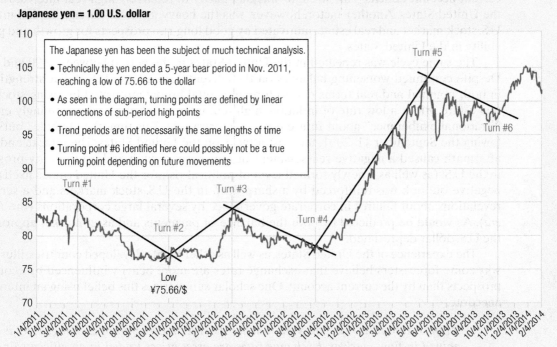

Japanese yen = 1.00 U.S. dollar

The Japanese yen has been the subject of much technical analysis.

- Technically the yen ended a 5-year bear period in Nov. 2011, reaching a low of 75.66 to the dollar
- As seen in the diagram, turning points are defined by linear connections of sub-period high points
- Trend periods are not necessarily the same lengths of time
- Turning point #6 identified here could possibly not be a true turning point depending on future movements

Turn #1
Turn #2
Turn #3
Turn #4
Turn #5
Turn #6

Low
¥75.66/$

- Capital market liquidity is particularly important to foreign institutional investors. Cross-border investors are not only interested in the ease of buying assets, but also in the ease of selling those assets quickly for fair market value if desired.

- A country's economic and social infrastructure are important indicators of its ability to survive unexpected external shocks and to prosper in a rapidly changing world economic environment.

- Political safety is exceptionally important to both foreign portfolio and direct investors. The outlook for political safety is usually reflected in political risk premiums for a country's securities and for purposes of evaluating foreign direct investment in that country.

- The credibility of corporate governance practices is important to cross-border portfolio investors. Poor corporate governance practices can reduce foreign investors' influence and cause subsequent loss of the firm's focus on shareholder wealth objectives.

- Contagion is defined as the spread of a crisis in one country to its neighboring countries and other countries with similar characteristics—at least in the eyes of cross-border investors. Contagion can cause an "innocent" country to experience capital flight with a resulting depreciation of its currency.

- Speculation can either cause a foreign exchange crisis or make an existing crisis worse. We will observe this effect through the three illustrative cases—the Asian crisis, Russian crisis, and Argentine crisis—discussed later in this chapter.

Foreign investors are willing to hold securities and undertake foreign direct investment in highly developed countries based primarily on relative real interest rates and the outlook for economic growth and profitability. All the other drivers described in Exhibit 9.1 are assumed to be satisfied.

For example, during the 1981–1985 period, the U.S. dollar strengthened despite growing current account deficits. This strength was due partly to relatively high real interest rates in the United States. Another factor, however, was the heavy inflow of foreign capital into the U.S. stock market and real estate, motivated by good long-run prospects for growth and profitability in the United States.

The same cycle was repeated in the United States in the period between 1990 and 2000. Despite continued worsening balances on the current account, the U.S. dollar strengthened in both nominal and real terms due to foreign capital inflow motivated by rising stock and real estate prices, a low rate of inflation, high real interest returns, and a seemingly endless "irrational exuberance" about future economic prospects. This time the "bubble" burst following the September 11, 2001, terrorist attacks on the United States. The attacks and their aftermath caused a negative reassessment of long-term growth and profitability prospects in the U.S. (as well as a newly formed level of political risk for the United States itself). This negative outlook was reinforced by a sharp drop in the U.S. stock markets and a series of revelations about failures in corporate governance by several large corporations (the *Enron era*). As would be predicted by both the balance of payments and asset market approaches, the U.S. dollar depreciated.

The experience of the United States, as well as other highly developed countries, illustrates why some forecasters believe that exchange rates are more heavily influenced by economic prospects than by the current account. One scholar summarizes this belief using an interesting anecdote.

Many economists reject the view that the short-term behavior of exchange rates is determined in flow markets. Exchange rates are asset prices traded in an efficient financial market. Indeed, an exchange rate is the relative price of two currencies and therefore is determined by the willingness to hold each currency. Like other asset prices, the exchange rate is determined by expectations about the future, not current trade flows.

A parallel with other asset prices may illustrate the approach. Let's consider the stock price of a winery traded on the Bordeaux stock exchange. A frost in late spring results in a poor harvest, in terms of both quantity and quality. After the harvest the wine is finally sold, and the income is much less than the previous year. On the day of the final sale there is no reason for the stock price to be influenced by this flow. First, the poor income has already been discounted for several months in the winery stock price. Second, the stock price is affected by future, in addition to current, prospects. The stock price is based on expectations of future earnings, and the major cause for a change in stock price is a revision of these expectations.

A similar reasoning applies to exchange rates: Contemporaneous international flows should have little effect on exchange rates to the extent they have already been expected. Only news about future economic prospects will affect exchange rates. Since economic expectations are potentially volatile and influenced by many variables, especially variables of a political nature, the short-run behavior of exchange rates is volatile.

—Bruno Solnik, *International Investments*, 3rd Edition, Reading, MA: Addison Wesley, 1996, p. 58. Reprinted with permission of Pearson Education, Inc.

The asset market approach to forecasting is also applicable to emerging markets. In this case, however, a number of additional variables contribute to exchange rate determination. These variables are, as described previously, illiquid capital markets, weak economic and social infrastructure, political instability, corporate governance, contagion effects, and speculation. These variables will be illustrated in the section detailing major currency crises later in this chapter.

Currency Market Intervention

A fundamental problem with exchange rates is that no commonly accepted method exists to estimate the effectiveness of official intervention into foreign exchange markets. Many interrelated factors affect the exchange rate at any given time, and no quantitative model exists that is able to provide the magnitude of any causal relationship between intervention and an exchange rate when so many interdependent variables are acting simultaneously.

—"Japan's Currency Intervention: Policy Issues," Dick K. Nanto, *CRS Report to Congress (United States)*, July 13, 2007, CRS-7.

The value of a country's currency is of significant interest to an individual government's economic and political policies and objectives. Those interests sometimes extend beyond the individual country, but may actually reflect some form of collective country interest. Although many countries have moved from fixed exchange rate values long ago, the governments and central bank authorities of the multitude of floating rate currencies still privately and publicly profess what value their currency "should hold" in their eyes, regardless of whether the market for that currency agrees at that time. *Foreign currency intervention*—the active management, manipulation, or intervention in the market's valuation of a country's currency—is a component of currency valuation and forecast that cannot be overlooked.

Motivations for Intervention

There is a long-standing saying that "what worries bankers is inflation, but what worries elected officials is unemployment." The principle is actually quite useful in understanding the various motives for currency market intervention. Depending upon whether a country's central bank is an independent institution (e.g., the U.S. Federal Reserve), or a subsidiary of its elected government (as the Bank of England was for many years), the bank's policies may either fight inflation or fight slow economic growth, but rarely can do both.

Historically, a primary motive for a government to pursue currency value change was to keep the country's currency cheap so that foreign buyers would find its exports cheap. This policy, long referred to as "beggar-thy-neighbor," has given rise to numerous competitive devaluations over the years. It has not, however, fallen out of fashion. The slow economic growth and continuing employment problems in many countries in 2012, 2013, and 2014 led some governments, the United States and the European Union being prime examples, to strive to hold their currency values down.

Alternatively, the fall in the value of the domestic currency will sharply reduce the purchasing power of its people. If the economy is forced, for a variety of reasons, to continue to purchase imported products (e.g., petroleum imports because of no domestic substitute), a currency devaluation or depreciation may prove highly inflationary—and in the extreme, impoverish the country's people (as in the case of Venezuela).

It is frequently noted that most countries would like to see stable exchange rates and to avoid the entanglements associated with manipulating currency values. Unfortunately, that would also imply that they are also happy with the current exchange rate's impact on country-level competitiveness. One must look no further than the continuing highly public debate between the U.S. and China over the value of the yuan. The U.S. believes the yuan is under-valued, making Chinese exports to the U.S. overly cheap, which in turn, results in a growing current account deficit for the United States and current account surplus for China.

The International Monetary Fund, as one of its basic principles (Article IV), encourages members to avoid pursuing "currency manipulation" to gain competitive advantages over other members. The IMF defines manipulation as "protracted large-scale intervention in one direction in the exchange market." It seems, however, that many governments often choose to ignore the IMF's advice.

Intervention Methods

There are many ways in which an individual or collective set of governments and central banks can alter the value of their currencies. It should be noted, however, that the methods of market intervention used are very much determined by the size of the country's economy, the magnitude of global trading in its currency, and the depth and breadth of development in its domestic financial markets. A short list of the intervention methods would include direct intervention, indirect intervention, and capital controls.

Direct Intervention. This is the active buying and selling of the domestic currency against foreign currencies. This traditionally required a central bank to act like any other trader in the currency market—albeit a big one. If the goal were to increase the value of the domestic currency, the central bank would purchase its own currency using its foreign exchange reserves, at least to the acceptable limits that it could endure depleting its reserves.

If the goal were to decrease the value of its currency—to fight an appreciation of its currency's value on the foreign exchange market—it would sell its own currency in exchange for foreign currency, typically a major hard currency like the dollar and euro. Although there are no physical limits to its ability to sell its own currency (it could theoretically continue to "print money" endlessly), central banks are cautious in the degree to which they may potentially change their monetary supplies through intervention.

Direct intervention was the primary method used for many years, but beginning in the 1970s, the world's currency markets grew enough that any individual player, even a central bank, could find itself with insufficient resources to move the market. As one trader stated a number of years ago, "We at the bank found ourselves little more than a grain of sand on the beach of the market."

One solution to the market size challenge has been the occasional use of coordinated intervention, in which several major countries, or a collective such as the G8 of industrialized countries, agree that a specific currency's value is out of alignment with their collective interests. In that situation, the countries may work collectively to intervene and push a currency's value in a desired direction. The September 1985 Plaza Agreement, an agreement signed at the Plaza Hotel in New York City by the members of the Group of Ten, was one such coordinated intervention agreement. The members, collectively, had concluded that currency values had become too volatile or too extreme in movement for sound economic policy management. The problem with coordinated intervention is, of course, achieving agreement between nations. This has proven to be a major sticking point in the principle's use.

Indirect Intervention. This is the alteration of economic or financial fundamentals that are thought to be drivers of capital to flow in and out of specific currencies. This was a logical development for market manipulation given the growth in size of the global currency markets relative to the financial resources of central banks.

The most obvious and widely used factor here is interest rates. Following the financial principles outlined in the previous discussion of parity conditions, higher real rates of interest attract capital. If a central bank wishes to "defend its currency" for example, it might follow a restrictive monetary policy, which would drive real rates of interest up. The method is therefore no longer limited to the quantity of foreign exchange reserves held by the country. Instead, it is limited only by the country's willingness to suffer the domestic impacts of higher real interest rates in order to attract capital inflows and therefore drive up the demand for its currency.

Alternatively, in a country wishing for its currency to fall in value, particularly when confronted with a continual appreciation of its value against major trading partner currencies, the central bank may work to lower real interest rates, reducing the returns to capital.

Because indirect intervention uses tools of monetary policy, a fundamental dimension of economic policy, the magnitude and extent of impacts may reach far beyond currency value. Overly stimulating economic activity, or increasing money supply growth beyond real economic activity, may prove inflationary. The use of such broad-based tools like interest rates to manipulate currency values requires a determination of importance, which in some cases may involve a choice to pursue international economic goals at the expense of domestic economic policy goals.

Capital Controls. This is the restriction of access to foreign currency by government. This involves limiting the ability to exchange domestic currency for foreign currency. When access and exchange is permitted, trading takes place only with official designees of the government or central bank, and only at dictated exchange rates.

Often, governments will limit access to foreign currencies to commercial trade: for example, allowing access to hard currency for the purchase of imports only. Access for investment purposes—particularly for short-term portfolios where investors are moving in and out of interest-bearing accounts and purchasing or selling securities or other funds—is often prohibited or limited. The Chinese regulation of access and trading of the Chinese yuan is a prime example of the use of capital controls over currency value. In addition to the government's setting the daily rate of exchange, access to the exchange is limited by a difficult and timely bureaucratic process for approval, and limited to commercial trade transactions.

Failure. It is also important to remember that intervention may—and often does—fail. The Turkish currency crisis of 2014 is a classic example of a drastic indirect intervention that ultimately only slowed the rate of capital flight and currency collapse. Turkey had enjoyed some degree of currency stability throughout 2012 and 2013, but the Turkish economy (one of the so-called "Fragile Five" countries, along with South Africa, India, Indonesia, and Brazil) suffered a widening current account deficit and rising inflation in late 2013. With the increasing anxieties in emerging markets in the fourth quarter of 2013 over the U.S. Federal Reserve's announcement that it would be slowing its bond purchasing (the *Taper Program*, essentially a tighter monetary policy), capital began exiting Turkey. The lira came under increasing downward pressure.

Turkey, however, was also under a great deal of domestic political strife, as the president of Turkey believed that the central bank should be stimulating the Turkish economy by lowering interest rates. Lower rates provided additional incentive for short-term capital flight. Pressures intensified in early January 2014, resulting in a sudden increase in the Turkish one-week bank repurchase interest rate (or *repo rate*) from 4.5% to 10.0%. Although the first few hours indicated some relief with the lira returning to a slightly stronger value versus the dollar (and euro), within days it was trading weaker once again. Indirect intervention in this case had not only proven a failure, but the attempted cure may in the end have worsened the economy.

Understanding the motivations and methods for currency market intervention is critical to any analysis of the determination of future exchange rates. And although it is often impossible to determine, in the end, whether intervention was successful, either direct or indirect in form, it is ever-present. Governments will always try to protect their currencies during periods of weakness. In the end, it may depend on both luck and talent. *Global Finance in Practice 9.2* provides a short list of possible best practices for effective intervention.

The debate was renewed in September 2010 when Japan intervened in the foreign exchange markets for the first time in nearly six years. In an attempt to slow the appreciating yen, Japan reportedly bought nearly 20 billion U.S. dollars in exchange. Finance Ministry officials had stated publicly that 82 yen per dollar was probably the limit of their tolerance for yen appreciation—and their tolerance was being tested.

Rules of Thumb for Effective Intervention

A variety of factors, features, and tactics, according to many currency traders, influence the effectiveness of an intervention effort.

- **Don't Lean into the Wind.** Markets that are moving significantly in one direction, like the strengthening of the Japanese yen in the fall of 2010, are very tough to turn. Termed "leaning into the wind," intervention during a strong market movement will most likely result in a very expensive failure. Currency traders argue that central banks should time their intervention very carefully, choosing moments when trading volumes are light and direction nearly flat. Don't lean into the wind, read it.

- **Coordinate Timing and Activity.** Use traders or associates in a variety of geographic markets and trading centers, possibly other central banks, if possible. The markets are much more likely to be influenced if they believe the intervention activity is reflecting a grass-roots movement, and not the singular activity of a single trading entity or bank.

- **Use Good News.** Particularly when trying to quell a currency fall, time the intervention to coincide with positive economic, financial, or business news closely associated with a country's currency market. Traders often argue that "markets wish to celebrate good news," and currencies may be no different.

- **Don't Be Cheap—Overwhelm Them.** Traders fear missing the moment, and a large, coordinated, well-timed intervention can make them fear they are leaning in the wrong direction. A successful intervention is in many ways a battle of psychology; play on insecurities. If it appears the intervention is gradually having the desired impact, throw ever-increasing assets into the battle. Don't get cheap.

As illustrated in Exhibit 9.2, the Bank of Japan intervened on September 13 as the yen approached 82 yen per dollar. (The Bank of Japan is independent in its ability to conduct Japanese monetary policy, but as the organizational subsidiary of the Japanese Ministry of Finance, it must conduct foreign exchange operations on behalf of the Japanese government.) Japanese officials reportedly notified authorities in both the United States and the European Union of their activity, but noted that they had not asked for permission or support.

The intervention resulted in public outcry from Beijing to Washington to London over the "new era of currency intervention." Although market intervention is always looked down upon by free market proponents, the move by Japan was seen as particularly frustrating as it came at a time when the United States was continuing to pressure China to revalue its currency, the renminbi. As noted by economist Nouriel Roubini, "We are in a world where everyone wants a weak currency," a marketplace in which all countries are looking to stimulate their domestic economies through exceptionally low interest rates and corresponding weak currency values—"a global race to the bottom."

Ironically, as illustrated in Exhibit 9.2, it appears that the intervention was largely unsuccessful. When the Bank of Japan started buying dollars in an appreciating yen market—the so-called "leaning into the wind" strategy—it was hoping to either stop the appreciation, change the direction of the spot rate movement, or both. In either pursuit, it appears to have failed. As one analyst commented, it turned out to be a "short-term fix to a long-term problem." Although the yen spiked downward (more yen per dollar) for a few days, it returned once again to an appreciating path within a week.

EXHIBIT 9.2   **Intervention and the Japanese Yen, 2010**

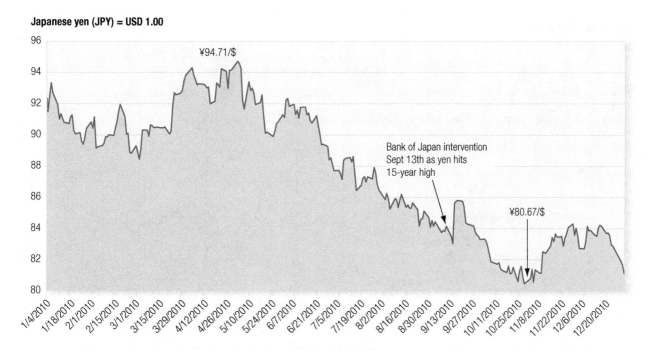

Japanese yen (JPY) = USD 1.00

Japan's frequent interventions, described in Exhibit 9.3, have been the subject of much study. In an August 2005 study by the IMF, it was noted that between 1991 and 2005, the Bank of Japan had intervened on 340 days, while the U.S. Federal Reserve intervened on 22 days and the European Central Bank intervened on only 4 days (since its inception in 1998). Although the IMF has never found Japanese intervention to be officially "currency manipulation," an analysis by Takatoshi Ito in 2004 concluded that there was on average a one-yen-per-dollar change in market rates, roughly 1%, as a result of Japanese intervention over time.

> *There is no historical case in which [yen] selling intervention succeeded in immediately stopping the preexisting long-term uptrend in the Japanese yen.*
>
> —Tohru Sasaki, Currency Strategist, JPMorgan.

Japan's interventions are not, however, a lone example of attempted market manipulation. Even the Swiss National Bank, a bastion of market and instrument stability, repeatedly intervened in 2009 to stop the appreciation of the Swiss franc against both the dollar and the euro.

Organizations of economic union and integration present extreme examples of currency market intervention. As described in Chapter 2, the launch of the euro in 1999 followed two decades of economic, monetary, and currency coordination and intervention. This system, the European Monetary System or EMS, used an elaborate system of bilateral responsibility, in which both country governments were committed to maintaining the parity exchange rates. Their commitment extended to both types of intervention, direct and indirect. *Global Finance in Practice 9.3* describes the EMS system, the so-called "snake in a tunnel."

EXHIBIT 9.3   The History of Japanese Intervention

Japanese yen (JPY) = USD1.00

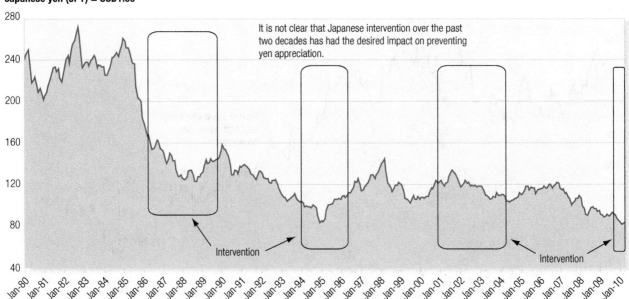

The European Monetary System's "Snake in a Tunnel"

The European Monetary System (EMS), in use between 1979 and the launch of the euro in 1999, established a *central rate* (or *parity rate*) of exchange between two currencies, which was to be the target long-term exchange rate. The exchange rate was then allowed to trade freely as long as it stayed within a band of ±2.25% of that rate. If or when the rate crossed the *upper* or *lower intervention rates* (shown in the figure), *both* countries were required to begin intervention efforts to drive the exchange rate back toward the *central rate* trading zone.

One of the basic tenets of the system that drove much of its success was the *bilateral responsibility* it established between the countries in maintaining the exchange rate. If, for example, the German mark (Deutsche mark, DM) was appreciating against the French franc (FF), and the market rate crossed the *intervention rate* threshhold, both governments were required to undertake *intervention*, either *direct intervention* (buying and selling their currencies in the market) or *indirect intervention* (such as changing interest rates), or both, to maintain trading about the central rate.

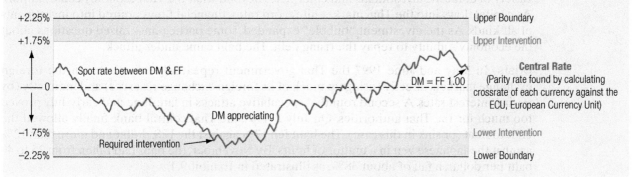

Disequilibrium: Exchange Rates in Emerging Markets

Although the three different schools of thought on exchange rate determination described earlier (parity conditions, balance of payments approach, and asset approach) make understanding exchange rates appear to be straightforward, that is rarely the case. The large and liquid capital and currency markets follow many of the principles outlined so far relatively well in the medium to long term. The smaller and less liquid markets, however, frequently demonstrate behaviors that seemingly contradict theory. The problem lies not in the theory, but in the relevance of the assumptions underlying the theory. An analysis of the emerging market crises illustrates a number of these seeming contradictions.

After a number of years of relative global economic tranquility, beginning in the second half of the 1990s, a series of currency crises shook all emerging markets. The Asian crisis of July 1997 and the fall of the Argentine peso in 2002 demonstrate an array of emerging market economic failures, each with its own complex causes and unknown outlooks. These crises also illustrated the growing problem of capital flight and short-run international speculation in currency and securities markets. We will use each of the individual crises to focus on a specific dimension of the causes and consequences.

The Asian Crisis of 1997

At a 1998 conference sponsored by the Milken Institute, a speaker noted that the world's preoccupation with the economic problems of Indonesia was incomprehensible because "the total gross domestic product of Indonesia is roughly the size of North Carolina." The following speaker observed, however, that the last time he had checked, "North Carolina did not have a population of 220 million people."

The roots of the Asian currency crisis extended from a fundamental change in the economics of the region, the transition of many Asian nations from being net exporters to net importers. Starting as early as 1990 in Thailand, the rapidly expanding economies of the Far East began importing more than they exported, requiring major net capital inflows to support their currencies. As long as the capital continued to flow in—capital for manufacturing plants, dam projects, infrastructure development, and even real estate speculation—the pegged exchange rates of the region could be maintained. When the investment capital inflows stopped, however, crisis was inevitable.

The most visible roots of the crisis were in the excesses of capital inflows into Thailand. With rapid economic growth and rising profits forming the backdrop, Thai firms, banks, and finance companies had ready access to capital on the international markets, finding U.S. dollar debt cheap offshore. Thai banks continued to raise capital internationally, extending credit to a variety of domestic investments and enterprises beyond what the Thai economy could support. As capital flows into the Thai market hit record rates, financial flows poured into investments of all kinds. As the investment "bubble" expanded, some participants raised questions about the economy's ability to repay the rising debt. The baht came under attack.

Crisis. In May and June 1997, the Thai government repeatedly intervened in the foreign exchange markets directly (using up much of its foreign exchange reserves) and indirectly (by raising interest rates). A second round of speculative attacks in late June and early July proved too much for the Thai authorities. On July 2, 1997, the Thai central bank finally allowed the baht to float (or sink in this case). The baht fell 17% against the U.S. dollar and more than 12% against the Japanese yen in a matter of hours. By November, the baht had fallen from 25 to 40 baht per dollar, a fall of about 38%, as illustrated in Exhibit 9.4.

EXHIBIT 9.4   **The Thai Baht and the Asian Crisis**

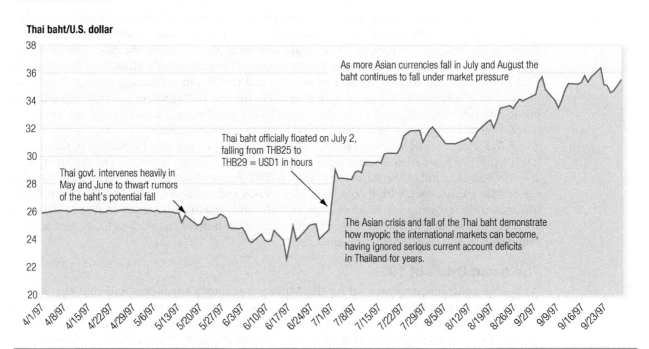

Within days, in Asia's own version of the *tequila effect*, a number of neighboring Asian nations, some with and some without similar characteristics to Thailand, came under speculative attack by currency traders and capital markets. The Philippine peso, the Malaysian ringgit, and the Indonesian rupiah all fell in the months following the July baht devaluation. In late October 1997, Taiwan caught the markets off balance with a surprise competitive devaluation of 15%. The Taiwanese devaluation seemed only to renew the momentum of the crisis. Although the Hong Kong dollar survived (at great expense to its foreign exchange reserves), the Korean won (KRW) was not so lucky. In November 1997, the historically stable won also fell victim, falling from 900 Korean won per dollar to more than 1100. The only currency that had not fallen besides the Hong Kong dollar was the Chinese renminbi, which was not freely convertible at the time.

Causal Complexities. The Asian economic crisis—for it was more than just a currency collapse—had many roots besides traditional balance of payments difficulties. The causes were different in each country, yet there were specific underlying similarities, which allow comparison: corporate socialism, corporate governance, and banking stability.

■ The rapidly growing export-led countries of Asia had known only stability. Because of the influence of government and politics in the business arena, even in the event of failure, it was believed that government would not allow firms to fail, workers to lose their jobs, or banks to close. Practices that had persisted for decades without challenge, such as lifetime employment, were now no longer sustainable.

■ Many firms operating within the Far Eastern business environments were largely controlled either by families or by groups related to the governing party or body of the country. This tendency has been labeled cronyism. Cronyism means that the interests of minority stockholders and creditors are often secondary at best to the primary motivations of corporate management.

■ The banking sector had fallen behind. Bank regulatory structures and markets had been deregulated nearly without exception across the globe. The central role played by banks in the conduct of business had largely been ignored. As firms across Asia collapsed, government coffers were emptied and banks failed. Without banks, the "plumbing" of business conduct was shut down.

In the aftermath, the international speculator and philanthropist George Soros was the object of much criticism for being the cause of the crisis because of massive speculation by his and other hedge funds. Soros, however, was likely only the messenger. *Global Finance in Practice 9.4* details the Soros debate.

The Argentine Crisis of 2002

Now, most Argentines are blaming corrupt politicians and foreign devils for their ills. But few are looking inward, at mainstream societal concepts such as viveza criolla, an Argentine cultural quirk that applauds anyone sly enough to get away with a fast one. It is one reason behind massive tax evasion here: One of every three Argentines does so—and many like to brag about it.

—"Once-Haughty Nation's Swagger Loses Its Currency," Anthony Faiola,
The Washington Post, March 13, 2002.

Argentina's economic ups and downs have historically been tied to the health of the Argentine peso. South America's southernmost resident—which oftentimes considered itself more European than Latin American—had been wracked by hyperinflation, international indebtedness,

GLOBAL FINANCE IN PRACTICE 9.4

Was George Soros to Blame for the Asian Crisis?

For Thailand to blame Mr. Soros for its plight is rather like condemning an undertaker for burying a suicide.
— *The Economist*, August 2, 1997, p. 57.

In the weeks following the start of the Asian Crisis in July 1997, officials from a number of countries, including Thailand and Malaysia, blamed the international financier George Soros for causing the crisis. Particularly vocal was the Prime Minister of Malaysia, Dr. Mahathir Mohamad, who repeatedly implied that Soros had a political agenda associated with Burma's prospect of joining the Association of Southeast Asian Nations (ASEAN). Mahathir noted in a number of public speeches that Soros might have been making a political statement, and not just speculating against currency values. Mahathir argued that the poor people of Malaysia, Thailand, the Philippines, and Indonesia would pay a great price for Soros's attacks on Asian currencies.

George Soros is probably the most famous currency speculator (and possibly the most successful) in global history. Admittedly responsible for much of the European financial crisis of 1992 and the fall of the French franc in 1993, he once again was the recipient of critical attention following the fall of the Thai baht and Malaysian ringgit.

Nine years later, in 2006, Mahathir and Soros met for the first time. Mahathir apologized and withdrew his previous accusations. In Soros's book published in 1998, *The Crisis of Global Capitalism: Open Society Endangered*, Soros explained that his fund had shorted the Thai baht and Malaysian ringgit (signed agreements to deliver the currency to other buyers at future dates) beginning in early 1997. He argued this meant that later in the spring, when his fund attempted to cover their positions, they were buyers of the currencies, not sellers, and therefore were on the "good side" and were in effect helping to support the currency values as the fund moved to realize its profits. Unfortunately, as many currency analysts have argued, the large short positions formed early in 1997 were clear signals in the market (word does move rapidly in currency markets) that Soros Funds expected the baht and ringgit to fall.

and economic collapse in the 1980s. By 1991, the people of Argentina had had enough. Economic reform was a common goal of the Argentine people. They were not interested in quick fixes, but lasting change and a stable future. They nearly got it.

In 1991, the Argentine peso had been pegged to the U.S. dollar at a one-to-one rate of exchange. The policy was a radical departure from traditional methods of fixing the rate of a currency's value. Argentina adopted a *currency board*, a structure—rather than merely a commitment—to limit the growth of money in the economy. Under a currency board, the central bank may increase the money supply in the banking system only with increases in its holdings of hard currency reserves. The reserves were, in this case, U.S. dollars. By removing the ability of government to expand the rate of growth of the money supply, Argentina believed it was eliminating the source of inflation that had devastated its standard of living. It was both a recipe for conservative and prudent financial management, and a decision to eliminate the power of politicians, elected and unelected, to exercise judgment both good and bad. It was an automatic and unbendable rule.

This "cure" was a restrictive monetary policy that slowed economic growth. The country's unemployment rate rose to double-digit levels in 1994 and stayed there. The real GDP growth rate settled into recession in late 1998, and the economy continued to shrink through 2000. Argentine banks allowed depositors to hold their money in either pesos or dollars. This was intended to provide a market-based discipline to the banking and political systems, and to demonstrate the government's unwavering commitment to maintaining the peso's value parity with the dollar. Although intended to build confidence in the system, in the end it proved disastrous to the Argentine banking system.

Economic Crisis. The 1998 recession proved to be unending. Three-and-a-half years later, Argentina was still in recession. By 2001, crisis conditions had revealed three very important underlying problems with Argentina's economy: (1) The Argentine peso was overvalued; (2) The currency board regime had eliminated monetary policy alternatives for macroeconomic policy; and (3) The Argentine government budget deficit was out of control. Inflation had not been eliminated, and the world's markets were watching.

Most of the major economies of South America now slid into recession. With slowing economic activity, imports fell. Most South American currencies now fell against the U.S. dollar, but because the Argentine peso remained pegged to the dollar, Argentine exports grew increasingly overpriced. The sluggish economic growth in Argentina warranted expansionary economic policies, but the currency board's basic premise was that the money supply to the financial system could not be expanded any further or faster than the ability of the economy to capture dollar reserves—eliminating monetary policy.

Government spending was not slowing, however. As the unemployment rate grew higher, as poverty and social unrest grew—both in Buenos Aires, the civil center of Argentina, and in the outer provinces—government was faced with pressure to close the economic and social gaps. Government spending continued to increase, but tax receipts did not. Argentina turned to the international markets to aid in the financing of its government's deficit spending. The total foreign debt of the country began rising dramatically. Only a number of IMF capital injections prevented the total foreign debt of the country from skyrocketing. By the end of the 1990s, however, total foreign debt had doubled, and the economy's earning power had not.

As economic conditions continued to deteriorate, banks suffered increasing runs. Depositors, fearing that the peso would be devalued, lined up to withdraw their money—both Argentine peso and U.S. dollar cash balances. Pesos were converted to dollars, once again adding fuel to the growing fire of currency crisis. The government, fearing that the increasing financial drain on banks would cause their collapse, closed the banks. Consumers, unable to withdraw more than $250 per week, were instructed to use debit cards and credit cards to make purchases and to conduct the everyday transactions required by society.

Devaluation. On Sunday, January 6, 2002, in the first act of his presidency, President Eduardo Duhalde devalued the peso from 1.00 Argentine peso per U.S. dollar to 1.40. But the economic pain continued. Two weeks after the devaluation, the banks were still closed. On February 3, 2002, the Argentine government announced that the peso would be floated, as shown in Exhibit 9.5. The government would no longer attempt to fix or manage its value to any specific level, allowing the market to find or set the exchange rate.

The lessons to be drawn from the Argentine story are somewhat complex. From the very beginning, Argentina and the IMF both knew the currency board system was a risky strategy, but given its long and disastrous experience with exchange rates, one that had been deemed worth taking. In the end, however, despite best efforts, the use of such a radically strict exchange rate system in which government had given up nearly all control over its sovereign monetary system, proved unsustainable.

| EXHIBIT 9.5 | **The Collapse of the Argentine Peso** |

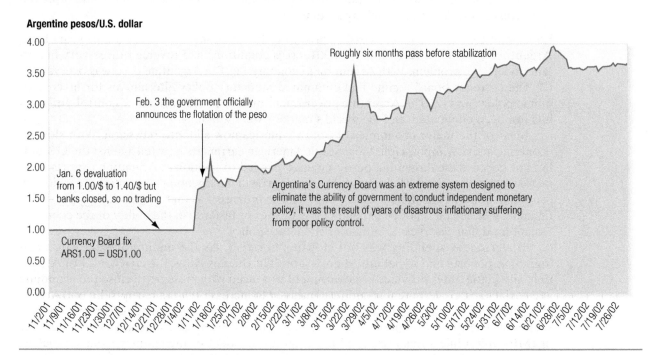

Argentine pesos/U.S. dollar

Roughly six months pass before stabilization

Feb. 3 the government officially
announces the flotation of the peso

Jan. 6 devaluation
from 1.00/$ to 1.40/$ but
banks closed, so no trading

Argentina's Currency Board was an extreme system designed to
eliminate the ability of government to conduct independent monetary
policy. It was the result of years of disastrous inflationary suffering
from poor policy control.

Currency Board fix
ARS1.00 = USD1.00

Forecasting in Practice

There are numerous foreign exchange forecasting services, many of which are provided by banks and independent consultants. In addition, some multinational firms have their own in-house forecasting capabilities. Predictions can be based on econometric models, technical analysis, intuition, and a certain measure of gall.

Exhibit 9.6 summarizes the various forecasting periods, regimes, and most widely followed methodologies. (Remember, if we authors could predict the movement of exchange rates with regularity, we surely wouldn't write books.) Whether any of the forecasting services are worth their cost depends both on the motive for forecasting as well as the required accuracy. For example, long-run forecasts may be motivated by a multinational firm's desire to initiate a foreign investment in Japan, or perhaps to raise long-term funds in Japanese yen. Or a portfolio manager may be considering diversifying for the long term in Japanese securities. The longer the time horizon of the forecast, the more inaccurate, but also the less critical the forecast is likely to be.

Short-term forecasts are typically motivated by a desire to hedge a receivable, payable, or dividend for a period of perhaps three months. In this case, the long-run economic fundamentals may not be as important as technical factors in the marketplace, government intervention, news, and passing whims of traders and investors. Accuracy of the forecast is critical, since most of the exchange rate changes are relatively small even though the day-to-day volatility may be high.

Forecasting services normally undertake fundamental economic analysis for long-term forecasts, and some base their short-term forecasts on the same basic model. Others base their short-term forecasts on technical analysis similar to that conducted in security analysis.

| EXHIBIT 9.6 | The Collapse of the Argentine Peso |

Forecast Period	Regime	Recommended Forecast Methods
SHORT-RUN	*Fixed-Rate*	1. Assume the fixed rate is maintained 2. Indications of stress on fixed rate? 3. Capital controls; black market rates 4. Indicators of government's capability to maintain fixed-rate? 5. Changes in official foreign currency reserves

	Floating-Rate	1. Technical methods which capture trend
		2. Forward rates as forecasts
		  (a) <30 assume a random walk
		  (b) 30–90 days, forward rates
		3. 90–360 days, combine trend with fundamental analysis
		4. Fundamental analysis of inflationary concerns
		5. Government declarations and agreements regarding exchange rate goals
		6. Cooperative agreements with other countries

LONG-RUN	*Fixed-Rate*	1. Fundamental analysis 2. BOP management 3. Ability to control domestic inflation 4. Ability to generate hard currency reserves to use for intervention 5. Ability to run trade surpluses
	Floating-Rate	1. Focus on inflationary fundamentals and PPP 2. Indicators of general economic health such as economic growth and stability 3. Technical analysis of long-term trends; new research indicates possibility of long-term technical "waves"

They attempt to correlate exchange rate changes with various other variables, regardless of whether there is any economic rationale for the correlation.

The chances of these forecasts being consistently useful or profitable depend on whether one believes the foreign exchange market is efficient. The more efficient the market is, the more likely it is that exchange rates are "random walks," with past price behavior providing no clues to the future. The less efficient the foreign exchange market is, the better the chance that forecasters may get lucky and find a key relationship that holds, at least for the short run. If the relationship is consistent, however, others will soon discover it and the market will become efficient again with respect to that piece of information.

Technical Analysis

Technical analysts, traditionally referred to as chartists, focus on price and volume data to determine past trends that are expected to continue into the future. The single most important element of technical analysis is that future exchange rates are based on the current exchange rate. Exchange rate movements, similar to equity price movements, can be subdivided into three periods: (1) day-to-day movement, which is seemingly random; (2) short-term movements, ranging from several days to trends lasting several months; and (3) long-term movements, characterized by up and down long-term trends. Long-term technical analysis has gained new popularity as a result of recent research into the possibility that long-term "waves" in currency movements exist under floating exchange rates.

The longer the time horizon of the forecast, the more inaccurate the forecast is likely to be. Whereas forecasting for the long run must depend on economic fundamentals of exchange rate determination, many of the forecast needs of the firm are short to medium term in their time horizon and can be addressed with less theoretical approaches. Time series techniques infer no theory or causality but simply predict future values from the recent past. Forecasters freely mix fundamental and technical analysis, presumably because forecasting is like playing horseshoes—getting close counts. *Global Finance in Practice 9.5* provides a short analysis of how accurate one prestigious forecaster was over a three-year period.

Cross-Rate Consistency in Forecasting

International financial managers must often forecast their home currency exchange rates for the set of countries in which the firm operates, not only to decide whether to hedge or to make

---

GLOBAL FINANCE IN PRACTICE 9.5

JPMorgan Chase Forecast of the Dollar/Euro

There are many different foreign exchange forecasting services and service providers. JPMorgan Chase (JPMC) is one of the most prestigious and widely used. A review of JPMC's forecasting accuracy for the U.S. dollar/euro spot exchange rate ($/€) for the 2002 to 2005 period, in 90-day increments, is presented in the exhibit. The graph shows the actual spot exchange rate for the period and JPMC's forecast for the spot exchange rate for the same period.

There is good news and there is bad news. The good news is that JPMC hit the actual spot rate dead-on in both May and November 2002. The bad news is that after that, they missed. Somewhat worrisome is when the forecast got the direction wrong. For example, in February 2004, JPMC had

forecast the spot rate to move from the current rate of $1.27/€ to $1.32/€, but in fact, the dollar had appreciated dramatically in the following three-month period to close at $1.19/€. This was in fact a massive difference. The lesson learned is probably that regardless of how professional and prestigious a forecaster may be, and how accurate they may have been in the past, forecasting the future—by anyone for anything—is challenging to say the least.

*This analysis uses exchange rate data as published in the print edition of *The Economist*, appearing quarterly. The source of the exchange rate forecasts, as noted in *The Economist*, is JPMorgan Chase.

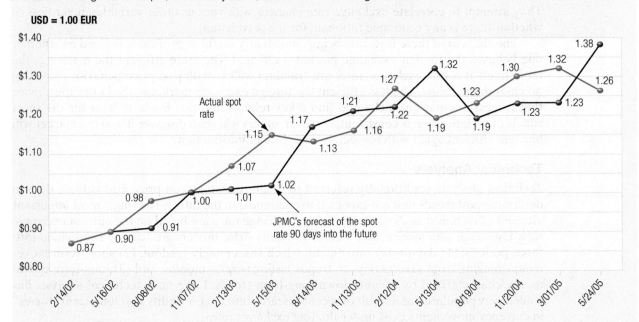

an investment, but also as part of preparing multi-country operating budgets in the home country's currency. These are the operating budgets against which the performance of foreign subsidiary managers will be judged. Checking cross-rate consistency—the reasonableness of the cross rates implicit in individual forecasts—acts as a reality check.

Forecasting: What to Think?

Obviously, with the variety of theories and practices, forecasting exchange rates into the future is a daunting task. Here is a synthesis of our thoughts and experience:

- It appears, from decades of theoretical and empirical studies, that exchange rates do adhere to the fundamental principles and theories outlined in the previous sections. Fundamentals do apply in the long term. There is, therefore, something of a fundamental equilibrium path for a currency's value.

- It also seems that in the short term, a variety of random events, institutional frictions, and technical factors may cause currency values to deviate significantly from their long-term fundamental path. This is sometimes referred to as noise. Clearly, therefore, we might expect deviations from the long-term path not only to occur, but also to occur with some regularity and relative longevity.

Exhibit 9.7 illustrates this synthesis of forecasting thought. The long-term equilibrium path of the currency—although relatively well-defined in retrospect—is not always apparent in the short term. The exchange rate itself may deviate in something of a cycle or wave about the long-term path.

| EXHIBIT 9.7 | **Short-Term Noise versus Long-Term Trends** |

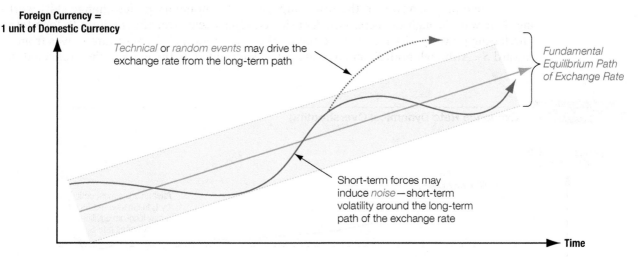

If market participants have *stabilizing expectations*, when forces drive the currency's value below the long-term fundamental equilibrium path, they will buy the currency driving its value back toward the fundamental equilibrium path. If market participants have *destabilizing expectations*, and forces drive the currency's value away from the fundamental path, participants may not move immediately or in significant volume to push the currency's value back toward the fundamental equilibrium path for an extended period of time (or possibly establish a new long-term fundamental path).

If market participants agree on the general long-term path and possess stabilizing expectations, the currency's value will periodically return to the long-term path. It is critical, however, that when the currency's value rises above the long-term path, most market participants see it as being overvalued and respond by selling the currency—causing its price to fall. Similarly, when the currency's value falls below the long-term path, market participants respond by buying the currency—driving its value up. This is what is meant by stabilizing expectations: Market participants continually respond to deviations from the long-term path by buying or selling to drive the currency back to the long-term path.

If, for some reason, the market becomes unstable, as illustrated by the dotted deviation path in Exhibit 9.7, the exchange rate may move significantly away from the long-term path for longer periods of time. Causes of these destabilizing markets—weak infrastructure (such as the banking system) and political or social events that dictate economic behaviors—are often the actions of speculators and inefficient markets.

Exchange Rate Dynamics: Making Sense of Market Movements

Although the various theories surrounding exchange rate determination are clear and sound, it may appear on a day-to-day basis that the currency markets do not pay much attention to the theories—they don't read the books! The difficulty is in understanding which fundamentals are driving markets at which points in time.

One example of this relative confusion over exchange rate dynamics is the phenomenon known as overshooting. Assume that the current spot rate between the dollar and the euro, as illustrated in Exhibit 9.8, is S_0. The U.S. Federal Reserve announces an expansionary monetary policy that cuts U.S. dollar interest rates. If euro-denominated interest rates remain unchanged, the new spot rate expected by the exchange markets based on interest differentials is S_1. This immediate change in the exchange rate is typical of how the markets react to news, distinct economic and political events that are observable. The immediate change in the value of the dollar-euro is therefore based on interest differentials.

As time passes, however, the price impacts of the monetary policy change start working their way through the economy. As price changes occur over the medium to long term, purchasing power parity forces drive the market dynamics, and the spot rate moves from S_1 toward S_2. Although both S_1 and S_2 were rates determined by the market, they reflected the

EXHIBIT 9.8   **Exchange Rate Dynamics: Overshooting**

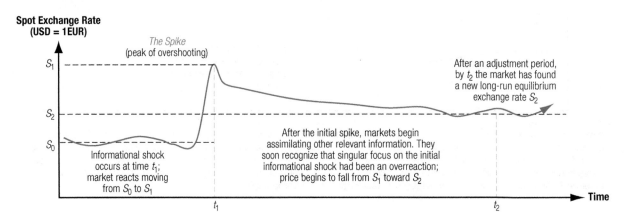

dominance of different theoretical principles. As a result, the initial lower value of the dollar of S_1 is described as overshooting the longer-term equilibrium value of S_2.

This is, of course, only one possible series of events and market reactions. Currency markets are subject to new news every hour of every day, making it very difficult to forecast exchange rate movements in short periods of time. In the longer term, as shown in Exhibit 9.8, the markets do customarily return to fundamentals of exchange rate determination.

SUMMARY POINTS

- There are three major schools of thought to explaining the economic determinants of exchange rates: parity conditions, the balance of payments approach, and the asset market approach.

- The recurrence of exchange rate crises demonstrates not only how sensitive currency values continue to be to economic fundamentals, but also how vulnerable many emerging market currencies are.

- Foreign exchange market intervention may be conducted via direct intervention, buying and selling the country's own currency, or indirect intervention, by changing the motivations and rules for capital to move into or out of a country and its currency.

- Many emerging market currencies periodically experience fundamental exchange rate disequilibrium. In the past, the most frequent cause of disequilibrium was hyperinflation, but today the most frequently

experienced challenge is the large and rapid inflow and outflow of non-current account capital.

- Exchange rate forecasting is part of global business. All businesses of all kinds must form some expectation of what the future holds.

- Short-term forecasting of exchange rates in practice focus on time series trends and current spot rates. Longer-term forecasting requires a return to the basic analysis of exchange rate fundamentals such as BOP, relative inflation and interest rates, and the long-run properties of purchasing power parity.

- In the short term, a variety of random events, institutional frictions, and technical factors may cause currency values to deviate significantly from their long-term fundamental path. In the long term, it does appear that exchange rates follow a fundamental equilibrium path, one consistent with the fundamental theories of exchange rate determination.

MINI-CASE

Russian Ruble Roulette[1]

The Russian ruble has experienced a multitude of regime shifts since the opening of the Russian economy under Perestroika in 1991.[2] After a number of years of a highly controlled official exchange rate accompanied by tight capital controls, the 1998 economic crisis prompted a movement to a heavily managed float. Using both direct intervention and indirect intervention (interest rate policy), the ruble held surprisingly steady until 2008. But all of that stopped in 2008 when the global credit crisis, which started in the United States, spread to Russia. As illustrated by Exhibit A, the impact on the value of the ruble proved disastrous.

Russian Crisis 1998

In an effort to protect the value of the ruble, the Bank of Russia spent $200 billion—a full one-third of its foreign exchange reserves—throughout 2008 and into 2009. Although the market began to calm in early 2009, the Bank decided to introduce a more flexible exchange rate regime for the management of the ruble.

The new system was a *dual-currency floating rate band* for the ruble. A *dual-currency basket* was formed from two currencies, the U.S. dollar (55%) and the euro (45%), for

[2]There is no established English spelling for the Russian currency—the *rouble* or the *ruble*. There is a journalistic tradition that most North American publications use *ruble*, while European organizations favor *rouble*, as does the *Oxford English Dictionary*.

The Russian Ruble: 1995–2015

the calculation of the central ruble rate. Around this basket rate, a *neutral zone* was established in which no currency intervention would be undertaken. This initial neutral zone was 1.00 Ruble versus the basket. Around the neutral zone a set of *operational band boundaries* were established; an *upper band* and *lower band* for intervention purposes.

If the ruble remained in the *neutral zone*, no intervention would be made. If, however, the ruble's value hit either *operational band*, the Bank of Russia would intervene by buying rubles (upper band) or selling rubles (lower band) to stabilize its value. The Bank was allowed a maximum of $700 million per day in purchases of rubles. Once hitting that limit, the Bank was to move the band(s) in increments of 5 kopecks (100 kopecks = 1.00 ruble) per day.[3]

As illustrated in Exhibit B, the ruble continued to slide (appreciating) against the basket throughout 2009 and into 2010. The dual-currency band was continually adjusted—downward—in an effort to put a "moving floor" underneath the currency. Finally, in late-2010, the ruble stabilized.

As part of its continual program to allow the ruble to grow as a global currency, the distance between the upper and lower bands has been repeatedly increased over time. Starting with a *floating band* (the spread between the upper and lower band in Exhibit B) that was only 2 rubles per basket value, the band was expanded to eventually reach 7 rubles.

The ruble's new-found relative stability was rewarded in October 2013 when the Bank of Russia announced that it was expanding the neutral "no-intervention" zone from 1 ruble to 3.1 rubles. This was followed by an announcement in January 2014 that the Bank would begin moving to end daily intervention, with a plan to end all intervention sometime in 2015. (Daily intervention in recent months had averaged about $60 million, a relatively small amount given history.) If, however, the ruble did hit either of its bands, the Bank did acknowledge that it was prepared to reenter the market to preserve stability.

The impetus for moving to a freely-floating ruble was to both allow the changes in the currency's value to "absorb" global economic changes, and allow the central bank to increase its focus on controlling inflationary forces. Russian inflation has been stubbornly high in recent years, and with the U.S. Federal Reserve announcing that it would be slowing/stopping its loose money policy in the wake of the financial crisis of 2008–2009, inflationary pressures were sure to continue. But then the regime shift plan began to unravel.

[3]The daily foreign exchange intervention limit has been adjusted downward a number of times since the dual-currency band was instituted. In January 2014 the limit had contracted to $350 million per day.

EXHIBIT B   **Russian Ruble Floating Band** (January 2009–January 2014)

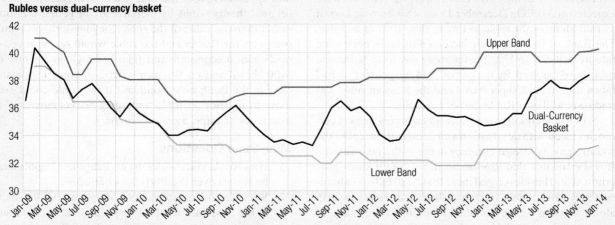

Rubles versus dual-currency basket

Dual-currency basket calculated by authors (monthly average exchange rates).

2014: Western Sanctions and the Price of Oil

"External shocks" is a phrase few central bankers ever want to hear. In the spring of 2014, however, that was exactly what the Russian ruble experienced. In March 2014 the European Union, the United States, and a host of other Western industrial countries imposed political and economic sanctions on Russia in opposition to its aggressive activities in Eastern Ukraine and its annexation of Crimea and Sevastopol. This had an immediate impact on restricting Russian exports, as well as shutting down a number of major foreign direct investment projects in Russia.

But the external shocks were not limited to sanctions. Beginning in the summer of 2014 the price of crude oil (Brent Blend crude is the predominant world price of oil) started falling. Oil was Russia's primary export; the country, the government, Russian business, all relied heavily on oil and gas export earnings to fund their economy. The pressures on the ruble—which many termed a *commodity currency* because Russia relied so heavily on oil—intensified.

As illustrated in Exhibit C, the ruble's fall began to accelerate in the fall of 2014. Sanctions were starting to impose real costs, the price of oil was falling faster and faster, and capital

EXHIBIT C   **The Russian Ruble, Sanctions, and the Price of Oil** (January 2014–January 2015)

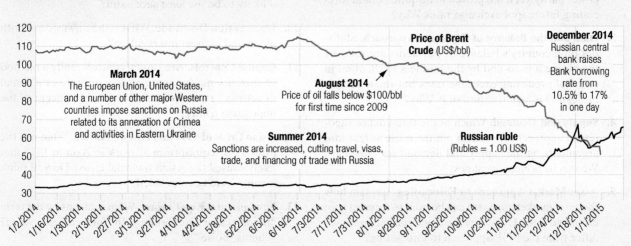

began to flee Russia. By December the Russian central bank estimated that more than $130 billion in capital had already left Russia, and another $120 billion in capital outflows were expected in 2015. On December 15, on what became known as Red Monday, the ruble lost more than 10% of its value. The Bank of Russia quickly increased its bank borrowing rate from 10.5% to 17%, the following day to 18%, but the ruble barely slowed. By the end of 2014 the price of oil had fallen to roughly $50 per barrel and the Russian ruble was trading at well over Rubles 60 per U.S. dollar.

Now a new concern rose which all emerging market currencies faced with devaluation and depreciation: could the country pay its foreign currency debts in the near future? It was estimated that Russian borrowers of all kinds, government and business, faced more than $120 billion in hard-currency foreign debt (largely dollars and euros) in 2015 alone. Russian businesses of all kinds, including some of the world's largest oil companies, were now restricted from borrowing internationally. So they borrowed domestically, pumping out ruble-denominated debt at a breakneck pace. What would that mean for borrowers and debt-holders in the coming years?

The Bank of Russia's plan to implement a long-term currency strategy, which had been put in place back in 2009, now appeared to be something of a train wreck. The Bank's original theory—that by increasingly targeting inflation rather than the value of the ruble, the long-term economic prospects for Russia and the ruble would be improved—made sound economic and financial logic in a world of $100/bbl oil and no Western sanctions. Many inside and outside the Bank now wondered if the ruble could ever move from being a simple "emerging market currency" to a currency of value, a *reserve currency*.

Mini-Case Questions

1. How would you classify the exchange rate regime used by Russia over the 1991–2014 period?
2. What did the establishment of *operational bands* do to the expectations of ruble speculators? Would these expectations be stabilizing or destabilizing in your opinion?
3. Would Western sanctions alone been devastating to the ruble's value, or was it the plummeting price of oil that had the larger impact?

QUESTIONS

These questions are available in MyFinanceLab.

1. **Exchange Rate Determination.** What are the three basic theoretical approaches to exchange rate determination?

2. **PPP Inadequacy.** The most widely accepted theory of foreign exchange rate determination is purchasing power parity, yet it has proven to be quite poor at forecasting future spot exchange rates. Why?

3. **Data and the Balance of Payments Approach.** Statistics on a country's balance of payments are used by the business press and by the business itself, often in terms of predicting exchange rates, but the academic profession is highly critical of it. Why?

4. **Supply and Demand.** Which of the three major theoretical approaches seems to put the most weight into arguments on the supply and demand for currency? What is its primary weakness?

5. **Asset Market Approach to Forecasting.** Explain how the asset market approach can be used to forecast spot exchange rates. How does the asset market approach differ from the BOP approach to forecasting?

6. **Technical Analysis.** Explain how technical analysis can be used to forecast exchange rates.

7. **Intervention.** What is foreign currency intervention? How is it accomplished?

8. **Intervention Motivation.** Why do governments and central banks intervene in the foreign exchange markets? If markets are efficient, why not let them determine the value of a currency?

9. **Direct Intervention Usefulness.** When is direct intervention likely to be the most successful? And when is it likely to be the least successful?

10. **Intervention Downside.** What is the downside of both direct and indirect intervention?

11. **Capital Controls.** Are capital controls really a method of currency market intervention, or more of a denial of activity? How does this fit with the concept of the impossible trinity?

12. **Asian Crisis of 1997 and Disequilibrium.** What was the primary disequilibrium at work in Asia in 1997 that likely caused the Asian financial crisis? Do you think it could have been avoided?

13. **Fundamental Equilibrium.** What is meant by the term "fundamental equilibrium path" for a currency value? What is "noise"?

14. **Argentina's Failure.** What was the basis of the Argentine Currency Board, and why did it fail, in 2002?

15. **Term Forecasting.** What are the major differences between short-term and long-term forecasts for a fixed exchange rate versus a floating exchange rate?

16. **Exchange Rate Dynamics.** Explain the meaning of "overshooting"? What causes it and how is it corrected?

17. **Foreign Currency Speculation.** The emerging market crises of 1997–2002 were worsened because of rampant speculation. Do speculators cause such crisis or do they simply respond to market signals of weakness? How can a government manage foreign exchange speculation?

18. **Cross-Rate Consistency in Forecasting.** Explain the meaning of "cross-rate consistency" as used by MNEs. How do MNEs use a check of cross-rate consistency in practice?

19. **Stabilizing versus Destabilizing Expectations.** Define stabilizing and destabilizing expectations, and describe how they play a role in the long-term determination of exchange rates.

20. **Currency Forecasting Services.** Many multinational firms use forecasting services regularly. If forecasting is essentially "foretelling the future," and that is theoretically impossible, why would these firms spend money on these services?

PROBLEMS

These problems are available in MyFinanceLab.

1. **Ecuadorian Hyper-Inflation.** The Ecuadorian sucre (S) suffered from hyper-inflationary forces throughout 1999. Its value moved from S5,000/\$ to S25,000/\$. What was the percentage change in its value?

2. **Canadian Loonie.** The Canadian dollar's value against the U.S. dollar has seen some significant changes over recent history. Use the following graph of the C\$/US\$ exchange rate for the 30-year period between 1980 and end-of-year 2010 to estimate the percentage change in the Canadian dollar's value (affectionately known as the "loonie") versus the dollar for the following periods.
   a. January 1980–January 1986
   b. January 1986–October 1991
   c. October 1991–December 2001
   d. October 2001–April 2011
   e. April 2011–January 2015

3. **Mexico's Cada Seis Años.** Mexico was famous—or infamous—for many years for having two things every six years (cada seis años in Spanish): a presidential election and a currency devaluation. This was the case in 1976, 1982, 1988, and 1994. In its last devaluation on December 20, 1994, the value of the Mexican peso (Ps) was officially changed from Ps3.30/\$ to Ps5.50/\$. What was the percentage devaluation?

4. **Turkish Lira Devaluation.** The Turkish lira (TL) was officially devalued by the Turkish government in February 2001 during a severe political and economic crisis. The Turkish government announced on February 21 that the lira would be devalued by 20%. The spot exchange rate on February 20 was TL68,000/\$.
   a. What was the exchange rate after devaluation?
   b. What was percentage change after falling to TL100,000/\$?

Problem 2.

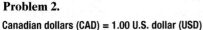 Canadian dollars (CAD) = 1.00 U.S. dollar (USD)

 January 1980–January 2015 (monthly average)

Problem 5.

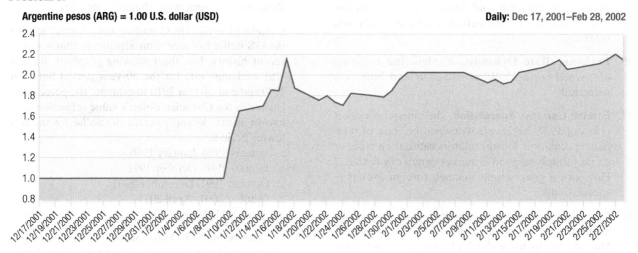

Argentine pesos (ARG) = 1.00 U.S. dollar (USD)   **Daily:** Dec 17, 2001–Feb 28, 2002

5. **Argentine Peso' Anguish.** As illustrated in the graph, the Argentine peso moved from its fixed exchange rate of Ps1.00/$ to over Ps2.00/$ in a matter of days in early January 2002. After a brief period of high volatility, the peso's value appeared to settle down into a range varying between 2.0 and 2.5 pesos per dollar. If you were forecasting the Argentine peso further into the future, to March 30, 2002, how would you use the information in the graphic—the value of the peso freely-floating in the weeks following devaluation—to forecast its future value?

6. **Brokedown Palace.** The Thai baht (THB) was devalued by the Thai government from THB25/$ to THB29/$ on July 2, 1997. What was the percentage devaluation of the baht?

7. **Brazilian Reais Carnival.** The Brazilian reais' (BRL or R$) value was BRL1.80 to USD1.00 on Thursday January 24, 2008, then plunged in value to BRL2.39 to USD1.00 on January 26, 2009. What was the percentage change in its value?

Russo-Swiss Cross—2015.

Switzerland has long served as a cornerstone of banking conservatism and a safe port in a world of currency storms. Swiss banks had long been used by investors from all over the world as financially secure institutions that would preserve the depositor's wealth with confidentiality. A large part of the security offered was the Swiss franc itself. Russian citizens of wealth had used Swiss banks heavily in recent years as a place where they could shelter their capital from politics, both inside Russia and the outside (EU and US) world. But beginning in the fall of 2014 the Russian ruble's value began to slide against the Swiss franc, threatening their wealth. Use data in the exhibit and the table below to answer Problems 8 and 9.

Exchange Rates	Nov. 7, 2013	Nov. 7, 2014	Dec. 4, 2014	Dec. 16, 2014	Dec. 24, 2014	Jan. 16, 2015
Russian rubles per Swiss franc	35.286	48.252	56.249	70.285	55.362	76.639
Russian rubles per US dollar	32.408	46.730	54.416	67.509	54.619	65.071
US dollars per Swiss franc	1.0888	1.0326	1.0337	1.0411	1.0136	1.1778

Problem 8 and 9.

Russian Ruble Tumbles Against Swiss Franc (RUB = 1.00 CHF, Nov. 2014–Jan. 2015)

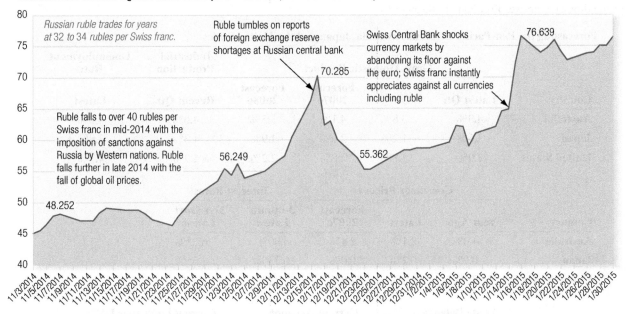

8. **Mikhail Khodorkovsky's Dilemma.** Mikhail Khodorkovsky was one of the infamous Russian oligarchs, accumulating billions of dollars in wealth in the mid-1990s with the fall of the Soviet Union. But in 2003 he had been imprisoned by the Russian state for a decade. Upon his release from prison in 2013 he had taken up residence in Switzerland—with his money.

    In November 2014 Mikhail held a portfolio of USD200 million and CHF150 million in Swiss banks, in addition to accounts in Russia still holding RUB1.2 billion. Using the exchange rate table, answer the following:

    a. What is the value of Mikhail's portfolio as measured in Russian rubles?

    b. What is the value of Mikhail's portfolio as measured in Swiss francs?

    c. What is the value of Mikhail's portfolio as measured in U.S. dollars?

    d. Which currency demonstrated the greatest fluctuations in total value over the six dates?

9. **Trepak–The Russian Dance.** Calculate the percentage change in the value of the ruble for the three different crossrates shown above for the six dates. Did the ruble fall further against the U.S. dollar or the Swiss franc?

10. **BP and Rosneft 2015.** BP (UK) and Rosneft (Russia) had severed a long-term joint venture in 2013, with

Rosneft buying BP out with $55 billion in cash and a 20% interest (equity interest) in Rosneft itself. Rosneft financed a large part of the buyout by borrowing heavily. The following year, in July 2014, BP received a dividend on its ownership interest in Rosneft of RUB24 billion.

    But Rosneft's performance had been declining, as was the Russian ruble. The winter of 2014–2015 in Europe was a relatively mild one, and Europe's purchases of Rosneft's natural gas had fallen as had the price of natural gas. Rosneft's total sales were down, and the ruble had clearly fallen dramatically (table above). And to add debt to injury, Rosneft was due to make a payment of USD19.5 billion in 2015 on its debt from the BP buyout.

    a. Assuming a spot rate of RUB34.78 = USD1.00 in July 2014, how much was the dividend paid to BP in U.S. dollars?

    b. If Rosneft were to pay the same dividend to BP in July 2015, and the spot rate at that time was RUB75 = 1.00USD, what would BP receive in U.S. dollars?

    c. If the combination of Western sanctions against Russia and lower global oil prices truly sent the Russian economy into recession, and the spot rate was RUB75 = 1.00USD in July 2015, what might BP's dividend be in July 2015?

Forecasting the Pan-Pacific Pyramid

Use the table below containing economic, financial, and business indicators from October 20, 2007, issue of *The Economist* (print edition) to answer Problems 11 through 16.

Forecasting the Pan-Pacific Pyramid: Australia, Japan and the United States

Country	Gross Domestic Product				Industrial Production	Unemployment Rate
	Latest Qtr	Qtr*	Forecast 2007e	Forecast 2008e	Recent Qtr	Latest
Australia	4.3%	3.8%	4.1%	3.5%	4.6%	4.2%
Japan	1.6%	−1.2%	2.0%	1.9%	4.3%	3.8%
United States	1.9%	3.8%	2.0%	2.2%	1.9%	4.7%

Country	Consumer Prices			Interest Rates	
	Year Ago	Latest	Forecast 2007e	3-month Latest	1-yr Govt Latest
Australia	4.0%	2.1%	2.4%	6.90%	6.23%
Japan	0.9%	−0.2%	0.0%	0.73%	1.65%
United States	2.1%	2.8%	2.8%	4.72%	4.54%

Country	Trade Balance	Current Account		Current Units (per US$)	
	Last 12 mos (billion $)	Last 12 mos (billion $)	Forecast 07 (% of GDP)	Oct 17th	Year Ago
Australia	−13.0	−$47.0	−5.7%	1.12	1.33
Japan	98.1	$197.5	4.6%	117	119
United States	−810.7	−$793.2	−5.6%	1.00	1.00

Source: Data abstracted from *The Economist*, October 20, 2007, print edition. Unless otherwise noted, percentages are percentage changes over one year. Rec Qtr = recent quarter. Values for 2007e are estimates or forecasts.

11. **Current Spot Rates.** What are the current spot exchange rates for the following cross rates?
   a. Japanese yen/U.S. dollar exchange rate
   b. Japanese yen/Australian dollar exchange rate
   c. Australian dollar/U.S. dollar exchange rate

12. **Purchasing Power Parity Forecasts.** Assuming purchasing power parity, and assuming that the forecasted change in consumer prices is a good proxy of predicted inflation, forecast the following cross rates:
   a. Japanese yen/U.S. dollar in one year
   b. Japanese yen/Australian dollar in one year
   c. Australian dollar/U.S. dollar in one year

13. **International Fischer Forecasts.** Assuming International Fischer applies to the coming year, forecast the following future spot exchange rates using the government bond rates for the respective country currencies:
   a. Japanese yen/U.S. dollar in one year
   b. Japanese yen/Australian dollar in one year
   c. Australian dollar/U.S. dollar in one year

14. **Implied Real Interest Rates.** If the nominal interest rate is the government bond rate, and the current change in consumer prices is used as expected inflation, calculate the implied "real" rates of interest by currency.
   a. Australian dollar "real" rate
   b. Japanese yen "real" rate
   c. U.S. dollar "real" rate

15. **Forward Rates.** Using the spot rates and 3-month interest rates above, calculate the 90-day forward rates for:
   a. Japanese yen/U.S. dollar exchange rate
   b. Japanese yen/Australian dollar exchange rate
   c. Australian dollar/U.S. dollar exchange rate

16. **Real Economic Activity and Misery.** Calculate the country's Misery Index (unemployment + inflation) and then use it like interest differentials to forecast the future spot exchange rate, one year into the future.
   a. Japanese yen/U.S. dollar exchange rate in one year

b. Japanese yen/Australian dollar exchange rate in one year

c. Australian dollar/U.S. dollar exchange rate in one year

17. **Yen-Euro Cross.** The Japanese yen-euro cross rate is one of the more significant currency values for global trade and commerce. The graph below shows this cross rate from when the euro was launched in January 1999 through January of 2015. Estimate the change in the value of the yen over the following four trend periods.

a. Jan 1999–Sept 2000
b. Sep 2000–Sept 2008
c. Sept 2008–Sept 2012
d. Sept 2012–Jan 2015

Japanese yen (JPY) = 1.00 European euro (EUR)   Jan 1, 1999–Jan 31, 2015

INTERNET EXERCISES

1. **Financial Forecast Center.** The Financial Forecast Center offers a variety of exchange rate and interest rate forecasts for business policy use.

Financial Forecast Center	www.forecasts.org/index.htm

2. **IMF Exchange Rates.** The IMF's exchange rate data is available online

IMF Exchange rates	www.imf.org/external/np/fin/ert/ GUI/Pages/CountryDataBase .aspx

3. **Recent Economic and Financial Data.** Use the following Web sites to obtain recent economic and financial data used for all approaches to forecasting presented in this chapter.

Economist.com	www.economist.com/markets-data
FT.com	www.ft.com
EconEdLink	www.econedlink.org/ economic-resources/ focus-on-economic-data.php

4. **OzForex Weekly Comment.** The OzForex Foreign Exchange Services Web site provides a weekly commentary on major political and economic factors and events that move current markets. Using their Web site, see what they expect to happen in the coming week on the three major global currencies—the dollar, yen, and euro.

OzForex	www.ozforex.com.au/ news-commentary/weekly

5. **Exchange Rates, Interest Rates, and Global Markets.** The magnitude of market data can seem overwhelming on occasion. Use the following Bloomberg markets page to organize your mind and your global data.

Bloomberg Financial News	www.bloomberg.com/markets

6. **Banque Canada and the Canadian Dollar Forward Market.** Use the following Web site to find the latest spot and forward quotes of the Canadian dollar against the Bahamian dollar and the Brazilian real.

Banque Canada	www.bankofcanada.ca/rates/ exchange/

Foreign Exchange Exposure

CHAPTER 10

Transaction Exposure

There are two times in a man's life when he should not speculate: when he can't afford it and when he can.

—"Following the Equator, Pudd'nhead Wilson's New Calendar," Mark Twain.

LEARNING OBJECTIVES

- Distinguish between the three major foreign exchange exposures experienced by firms
- Analyze the pros and cons of hedging foreign exchange transaction exposure
- Examine the alternatives available to a firm for managing a large and significant transaction exposure
- Evaluate the institutional practices and concerns of conducting foreign exchange risk management
- Explore advanced dimensions of foreign currency hedging

Foreign exchange exposure is a measure of the potential for a firm's profitability, net cash flow, and market value to change because of a change in exchange rates. An important task of the financial manager is to measure foreign exchange exposure and to manage it so as to maximize the profitability, net cash flow, and market value of the firm. This chapter provides an in-depth discussion of transaction exposure, which is the first category of two main accounting exposures. The following chapters focus on translation exposure, which is the second category of accounting exposures, and operating exposure. The chapter concludes with a Mini-Case, *China Noah Corporation*, examining what a Chinese firm's currency hedging practices.

Types of Foreign Exchange Exposure

What happens to a firm when foreign exchange rates change? There are two distinct categories of foreign exchange exposure for the firm, those that are based in accounting and those that arise from economic competitiveness. Accounting exposures, specifically described as *transaction exposure* and *translation exposure*, arise from contracts and accounts being denominated in foreign currency. The economic exposure, which we will describe as *operating exposure*, is the potential change in the value of the firm from its changing global competitiveness as determined by exchange rates. Exhibit 10.1 shows schematically the three main types of foreign exchange exposure: transaction, translation, and operating:

- *Transaction exposure* measures changes in the value of outstanding financial obligations incurred prior to a change in exchange rates but not due to be settled until after the exchange rates change. Thus, it deals with changes in cash flows that result from existing contractual obligations.

EXHIBIT 10.1   **The Foreign Exchange Exposures of the Firm**

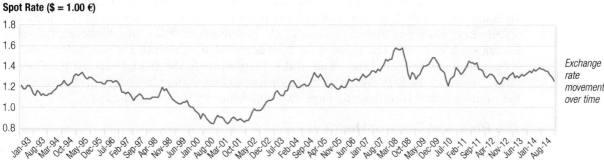

Realized Exposures

Transaction Exposure
Changes in the recorded value of identifiable transactions of the firm like receivables and payables. Results in *realized* foreign exchange gains and losses in income and taxes.

Short-term to medium-term to long-term change

Economic/Operating Exposure
Changes in the expected future cash flows of the firm from unexpected changes in exchange rates. The firm's future cash flows are changed from *realized* changes in its own sales, earnings, and cash flows, as well as changes in competitor responses to exchange rates over time.

→ Time

Unrealized Exposures

Translation Exposure
Changes in the periodic consolidated value of the firm; results in no change in cash flow or global tax liabilities–*unrealized*–changes only the consolidated financial results reported to the market (if publicly traded). Often labeled *Accounting Exposure*.

Spot Rate ($ = 1.00 €)

Exchange rate movement over time

- *Translation exposure* is the potential for accounting-derived changes in owner's equity to occur because of the need to "translate" foreign currency financial statements of foreign subsidiaries into a single reporting currency to prepare worldwide consolidated financial statements.

- *Operating exposure*—also called *economic exposure*, *competitive exposure*, or *strategic exposure*—measures the change in the present value of the firm resulting from any change in future operating cash flows of the firm caused by an unexpected change in exchange rates. The change in value depends on the effect of the exchange rate change on future sales volume, prices, and costs.

Transaction exposure and operating exposure both exist because of unexpected changes in future cash flows. However, while transaction exposure is concerned with future cash flows already contracted for, operating exposure focuses on expected (not yet contracted for) future cash flows that might change because a change in exchange rates has altered international competitiveness.

Why Hedge?

MNEs possess a multitude of cash flows that are sensitive to changes in exchange rates, interest rates, and commodity prices. Chapters 10, 11, and 12 focus exclusively on the sensitivity of the individual firm's value and of its future cash flows to changes in exchange rates. We begin by exploring the question of whether exchange rate risk should or should not be managed.

Hedging Defined

Many firms attempt to manage their currency exposures through *hedging*. Hedging requires a firm to take a position—an asset, a contract, or a derivative—the value of which will rise or fall in a manner that counters the fall or rise in value of an existing position—the exposure. Hedging protects the owner of the existing asset from loss. However, it also eliminates any gain from an increase in the value of the asset hedged. The question remains: What is to be gained by the firm from hedging?

According to financial theory, the value of a firm is the net present value of all expected future cash flows. The fact that these cash flows are *expected* emphasizes that nothing about the future is certain. If the reporting currency value of many of these cash flows is altered by exchange rate changes, a firm that hedges its currency exposures reduces the variance in the value of its future expected cash flows. *Currency risk* can then be defined as the variance in expected cash flows arising from unexpected changes in exchange rates.

Exhibit 10.2 illustrates the distribution of expected net cash flows of the individual firm. Hedging these cash flows narrows the distribution of the cash flows about the mean of the distribution. Currency hedging reduces risk. Reduction of risk is not, however, the same as adding value or return. The value of the firm depicted in Exhibit 10.2 would be increased only if hedging actually shifted the mean of the distribution to the right. In fact, if hedging is not "free," meaning the firm must expend resources to hedge, then hedging will add value only if the rightward shift is sufficiently large to compensate for the cost of hedging.

The Pros and Cons of Hedging

Is a reduction in the variability of cash flows sufficient reason for currency risk management?

EXHIBIT 10.2   **Hedging's Impact on the Expected Cash Flows of the Firm**

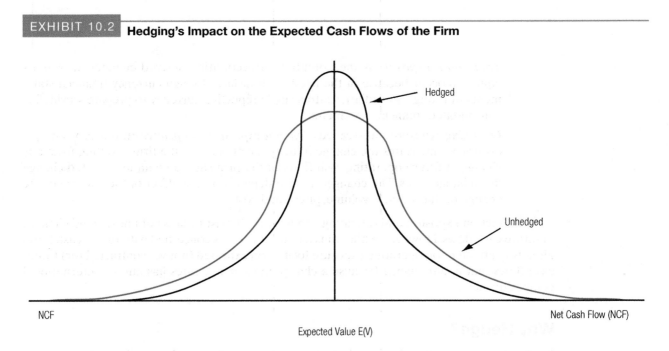

Hedged

Unhedged

NCF

Net Cash Flow (NCF)

Expected Value E(V)

Hedging reduces the variability of expected cash flows about the mean of the distribution.
This reduction of distribution variance is a reduction of risk.

Pros. Proponents of hedging cite the following arguments:

- Reduction in risk of future cash flows improves the planning capability of the firm. If the firm can more accurately predict future cash flows, it may be able to undertake specific investments or activities that it might not otherwise consider.

- Reduction of risk in future cash flows reduces the likelihood that the firm's cash flows will fall below a level sufficient to make debt service payments required for continued operation. This minimum cash flow level, often referred to as the *point of financial distress*, lies to the left of the center of the distribution of expected cash flows. Hedging reduces the likelihood that the firm's cash flows will fall to this level.

- Management has a comparative advantage over the individual shareholder in knowing the actual currency risk of the firm. Regardless of the level of disclosure provided by the firm to the public, management always possesses an advantage in the depth and breadth of knowledge concerning the real risks.

- Markets are usually in disequilibrium because of structural and institutional imperfections, as well as unexpected external shocks (such as an oil crisis or war). Management is in a better position than shareholders to recognize disequilibrium conditions and to take advantage of single opportunities to enhance firm's value through *selective hedging*—hedging only exceptional exposures or the occasional use of hedging when management has a definite expectation of the direction of exchange rates.

Cons. Opponents of hedging commonly make the following arguments:

- Shareholders are more capable of diversifying currency risk than is the management of the firm. If stockholders do not wish to accept the currency risk of any specific firm, they can diversify their portfolios to manage the risk in a way that satisfies their individual preferences and risk tolerance.

- Currency hedging does not increase the expected cash flows of the firm. Currency risk management does, however, consume firm resources and so reduces cash flow. The impact on value is a combination of the reduction of cash flow (which lowers value) and the reduction in variance (which increases value).

- Management often conducts hedging activities that benefit management at the expense of the shareholders. The field of finance called *agency theory* frequently argues that management is generally more risk-averse than are shareholders.

- Managers cannot outguess the market. If and when markets are in equilibrium with respect to parity conditions, the expected net present value of hedging should be zero.

- Management's motivation to reduce variability is sometimes for accounting reasons. Management may believe that it will be criticized more severely for incurring foreign exchange losses than for incurring even higher cash costs by hedging. Foreign exchange losses appear in the income statement as a highly visible separate line item or as a footnote, but the higher costs of protection through hedging are buried in operating or interest expenses.

- Efficient market theorists believe that investors can see through the "accounting veil" and therefore have already factored the foreign exchange effect into a firm's market valuation. Hedging would only add cost.

Every individual firm in the ends decides whether it wishes to hedge, for what purpose, and how. But as illustrated by *Global Finance in Practice 10.1*, this often results in even more questions and more doubts.

Hedging and the German Automobile Industry

The leading automakers in Germany have long been some of the world's biggest advocates of currency hedging. Companies like BMW, Mercedes, Porsche—and Porsche's owner Volkswagen—have aggressively hedged their foreign currency earnings for years in response to their structural exposure: while they manufacture in the eurozone, they increasingly rely on sales in dollar, yen, or other foreign (non-euro) currency markets.

How individual companies hedge, however, differs dramatically. Some companies, like BMW, state clearly that they "hedge to protect earnings," but that they do not speculate. Others, like Porsche and Volkswagen in the past, have

sometimes generated more than 40% of their earnings from their "hedges."

Hedges that earn money continue to pose difficulties for regulators, auditors, and investors worldwide. How a hedge is defined, and whether a hedge should only "cost" but not "profit," has delayed the implementation of many new regulatory efforts in the United States and Europe in the post-2008 financial crisis era. If a publicly traded company—for example an automaker—can consistently earn profits from hedging, is its core competency automobile manufacturing and assembly, or hedging/speculating on exchange rate movements?

Measurement of Transaction Exposure

Transaction exposure measures gains or losses that arise from the settlement of existing financial obligations whose terms are stated in a foreign currency. Transaction exposure arises from any of the following:

1. Purchasing or selling on credit—on *open account*—goods or services when prices are stated in foreign currencies
2. Borrowing or lending funds when repayment is to be made in a foreign currency
3. Being a party to an unperformed foreign exchange forward contract
4. Otherwise acquiring assets or incurring liabilities denominated in foreign currencies

The most common example of transaction exposure arises when a firm has a receivable or payable denominated in a foreign currency. Exhibit 10.3 demonstrates how this exposure is born. The total transaction exposure consists of quotation, backlog, and billing exposures.

EXHIBIT 10.3 The Life Span of a Transaction Exposure

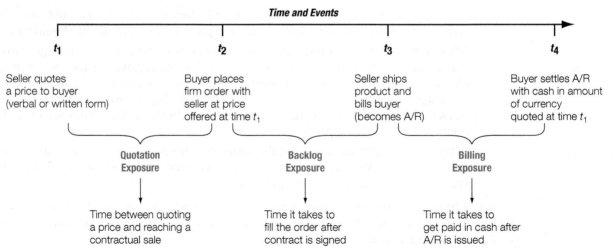

Time and Events

t_1 — Seller quotes a price to buyer (verbal or written form)

t_2 — Buyer places firm order with seller at price offered at time t_1

t_3 — Seller ships product and bills buyer (becomes A/R)

t_4 — Buyer settles A/R with cash in amount of currency quoted at time t_1

Quotation Exposure — Time between quoting a price and reaching a contractual sale

Backlog Exposure — Time it takes to fill the order after contract is signed

Billing Exposure — Time it takes to get paid in cash after A/R is issued

A transaction exposure is created at the first moment the seller quotes a price in foreign currency terms to a potential buyer (t_1). The quote can be either verbal, as in a telephone quote, or as a written bid or a printed price list. This is *quotation exposure.* When the order is placed (t_2), the potential exposure created at the time of the quotation (t_1) is converted into actual exposure, called *backlog exposure*, because the product has not yet been shipped or billed. Backlog exposure lasts until the goods are shipped and billed (t_3), at which time it becomes *billing exposure,* which persists until payment is received by the seller (t_4).

Purchasing or Selling on Open Account. Suppose that Ganado Corporation, a U.S. firm, sells merchandise on open account to a Belgian buyer for €1,800,000, with payment to be made in 60 days. The spot exchange rate on the date of the sale is $1.1200/€, and the seller expects to exchange the euros for €1,800,000 × $1.12/€ = $2,016,000 when payment is received. The $2,016,000 is the value of the sale that is posted to the firm's books. Accounting practices stipulate that the foreign currency transaction be listed at the spot exchange rate in effect on the date of the transaction.

Transaction exposure arises because of the risk that Ganado will receive something other than the $2,016,000 expected and booked. For example, if the euro weakens to $1.1000/€ when payment is received, the U.S. seller will receive only €1,800,000 × $1.100/€ or $1,980,00, some $36,000 less than what was expected at the time of sale.

Transaction settlement: €1,800,000 × $1.1000/€	= $1,980,000
Transaction booked: €1,800,000 × $1.1200/€	= $2,016,000
Foreign exchange gain (loss) on sale	= ($36,000)

If the euro should strengthen to $1.3000/€, however, Ganado receives $2,340,000, an increase of $324,000 over the amount expected. Thus, Ganado's exposure is the chance of either a loss or a gain on the resulting dollar settlement versus the amount at which the sale was booked.

This U.S. seller might have avoided transaction exposure by invoicing the Belgian buyer in dollars. Of course, if the U.S. company attempted to sell only in dollars, it might not have obtained the sale in the first place. Even if the Belgian buyer agrees to pay in dollars, transaction exposure is not eliminated. Instead, the exposure is transferred to the Belgian buyer, whose dollar account payable has an unknown cost 60 days hence.

Borrowing or Lending. A second example of transaction exposure arises when funds are borrowed or loaned, and the amount involved is denominated in a foreign currency. For example, in 1994, PepsiCo's largest bottler outside of the United States was the Mexican company, Grupo Embotellador de Mexico (Gemex). In mid-December 1994, Gemex had U.S. dollar debt of $264 million. At that time, Mexico's new peso ("Ps") was traded at Ps3.45/$, a pegged rate that had been maintained with minor variations since January 1, 1993, when the new currency unit had been created. On December 22, 1994, the peso was allowed to float because of economic and political events within Mexico, and in one day it sank to Ps4.65/$. For most of the following January it traded in a range near Ps5.50/$.

Dollar debt in mid-December 1994: US$264,000,000 × Ps 3.45/US$ =	Ps910,800,000
Dollar debt in mid-January 1995: US$264,000,000 × Ps 5.50/US$ =	Ps1,452,000,000
Dollar debt increase measured in Mexican pesos	Ps541,200,000

The number of pesos needed to repay the dollar debt increased by 59%! In U.S. dollar terms, the drop in the value of the peso meant that Gemex needed the peso-equivalent of an additional $98,400,000 to repay its debt.

Unperformed Foreign Exchange Contracts. When a firm enters into a forward exchange contract, it deliberately creates transaction exposure. This risk is usually incurred to hedge an existing transaction exposure. For example, a U.S. firm might want to offset an existing obligation to purchase ¥100 million to pay for an import from Japan in 90 days. One way to offset this payment is to purchase ¥100 million in the forward market today for delivery in 90 days. In this manner any change in value of the Japanese yen relative to the dollar is neutralized. Thus, the potential transaction loss (or gain) on the account payable is offset by the transaction gain (or loss) on the forward contract.

Contractual Hedges. Foreign exchange transaction exposure can be managed by contractual, operating, and financial hedges. The main contractual hedges employ the forward, money, futures, and options markets. Operating hedges utilize operating cash flows—cash flows originating from the operating activities of the firm—and include risk-sharing agreements and leads and lags in payment strategies. Financial hedges utilize financing cash flows—cash flows originating from the financing activities of the firm—and include specific types of debt and foreign currency derivatives, such as swaps. Operating and financing hedges will be described in greater detail in later chapters.

The term *natural hedge* refers to an offsetting operating cash flow, a payable arising from the conduct of business. A financial hedge refers to either an offsetting debt obligation (such as a loan) or some type of financial derivative such as an interest rate swap. Care should be taken to distinguish hedges in the same way finance distinguishes cash flows—operating from financing. The following case illustrates how contractual hedging techniques may be used to protect against transaction exposure.

Ganado's Transaction Exposure

Maria Gonzalez is the chief financial officer of Ganado. She has just concluded negotiations for the sale of a turbine generator to Regency, a British firm, for £1,000,000. This single sale is quite large in relation to Ganado's present business. Ganado has no other current foreign customers, so the currency risk of this sale is of particular concern. The sale is made in March with payment due three months later in June. Exhibit 10.4 summarizes the financial and market information Maria has collected for the analysis of her currency exposure problem. The unknown—the transaction exposure—is the actual realized value of the receivable in U.S. dollars at the end of 90 days.

Ganado operates on relatively narrow margins. Although Maria and Ganado would be very happy if the pound appreciated versus the dollar, concerns center on the possibility that the pound will fall. When Ganado had priced and budgeted this contract, it had set a very slim minimum acceptable margin at a sales price of $1,700,000; Ganado wanted the deal for both financial and strategic purposes. The budget rate, the lowest acceptable dollar per pound exchange rate, was therefore established at $1.70/£. Any exchange rate below this budget rate would result in Ganado realizing no profit on the deal.

Four alternatives are available to Ganado to manage the exposure: (1) remain unhedged; (2) hedge in the forward market; (3) hedge in the money market; or (4) hedge in the options market.

Unhedged Position

Maria may decide to accept the transaction risk. If she believes the foreign exchange advisor, she expects to receive £1,000,000 × $1.76 = $1,760,000 in three months. However, that amount is at risk. If the pound should fall to, say, $1.65/£, she will receive only $1,650,000. Exchange risk is not one sided, however; if the transaction is left uncovered and the pound strengthens even more than forecast, Ganado will receive considerably more than $1,760,000.

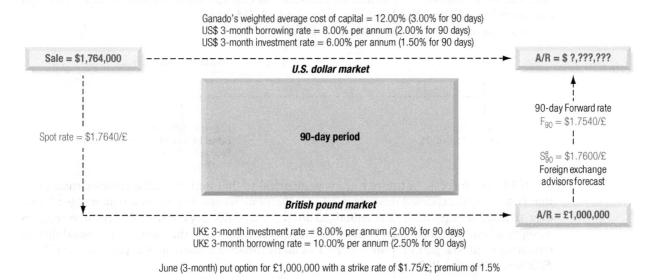

EXHIBIT 10.4   **Ganado's Transaction Exposure**

Ganado's weighted average cost of capital = 12.00% (3.00% for 90 days)
US$ 3-month borrowing rate = 8.00% per annum (2.00% for 90 days)
US$ 3-month investment rate = 6.00% per annum (1.50% for 90 days)

Sale = $1,764,000 - → A/R = $?,???,???

U.S. dollar market

90-day period

90-day Forward rate
F_{90} = $1.7540/£

S_{90}^e = $1.7600/£
Foreign exchange
advisors forecast

Spot rate = $1.7640/£

British pound market

- → A/R = £1,000,000

UK£ 3-month investment rate = 8.00% per annum (2.00% for 90 days)
UK£ 3-month borrowing rate = 10.00% per annum (2.50% for 90 days)

June (3-month) put option for £1,000,000 with a strike rate of $1.75/£; premium of 1.5%

The essence of an unhedged approach is as follows:

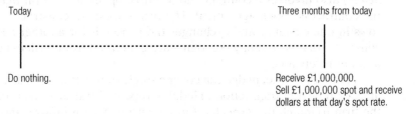

Today

Three months from today

Do nothing.

Receive £1,000,000.
Sell £1,000,000 spot and receive
dollars at that day's spot rate.

Forward Market Hedge

A *forward hedge* involves a forward (or futures) contract and a source of funds to fulfill that contract. The forward contract is entered into at the time the transaction exposure is created. In Ganado's case, that would be in March, when the sale to Regency was booked as an account receivable.

When a foreign currency denominated sale such as this is made, it is booked at the spot rate of exchange existing on the booking date. In this case, the spot rate on the date of sale was $1.7640/£, so the receivable was booked as $1,764,000. Funds to fulfill the forward contract will be available in June, when Regency pays £1,000,000 to Ganado. If funds to fulfill the forward contract are on hand or are due because of a business operation, the hedge is considered covered, perfect, or square, because no residual foreign exchange risk exists. Funds on hand or to be received are matched by funds to be paid.

In some situations, funds to fulfill the forward exchange contract are not already available or due to be received later, but must be purchased in the spot market at some future date. Such a hedge is open or uncovered. It involves considerable risk because the hedger must take a chance on purchasing foreign exchange at an uncertain future spot rate in order to fulfill the forward contract. Purchase of such funds at a later date is referred to as covering.

Should Ganado wish to hedge its transaction exposure with a forward, it will sell £1,000,000 forward today at the 3-month forward rate of $1.7540/£. This is a covered transaction in which the firm no longer has any foreign exchange risk. In three months the firm will receive £1,000,000 from the British buyer, deliver that sum to the bank against its forward sale, and receive $1,754,000. This would be recorded on Ganado's income statement as a foreign exchange loss of $10,000 ($1,764,000 as booked, $1,754,000 as settled).

The essence of a forward hedge is as follows:

Today

Sell £1,000,000
forward @ $1.7540/£.

Three months from today

Receive £1,000,000.
Deliver £1,000,000 against forward sale.
Receive $1,754,000.

If Maria's forecast of future rates was identical to that implicit in the forward quotation, that is, $1.7540/£, expected receipts would be the same whether or not the firm hedges. However, realized receipts under the unhedged alternative could vary considerably from the certain receipts when the transaction is hedged. Never underestimate the value of predictability of outcomes (and 90 nights of sound sleep). But many things can interrupt sleep, as seen in *Global Finance in Practice 10.2*.

Money Market Hedge (Balance Sheet Hedge)

Like a forward market hedge, a *money market hedge* (also commonly called a *balance sheet hedge*) also involves a contract and a source of funds to fulfill that contract. In this instance, the contract is a loan agreement. The firm seeking to construct a money market hedge borrows in one currency and exchanges the proceeds for another currency. Funds to fulfill the contract—that is, to repay the loan—are generated from business operations, in this case, the account receivable.

A money market hedge can cover a single transaction, such as Ganado's £1,000,000 receivable, or repeated transactions. Hedging repeated transactions is called matching. It requires the firm to match the expected foreign currency cash inflows and outflows by currency and maturity. For example, if Ganado had numerous sales denominated in pounds to British customers over a long period of time, then it would have somewhat predictable U.K. pound cash inflows. The appropriate money market hedge technique in that case would be to borrow

Currency Losses at Greenpeace

Foreign currency losses are not limited to multinational companies in search of profits in the global marketplace. Stuff happens—to everyone. In 2014 Greenpeace, the home of the Rainbow Warrior, announced that it had suffered a foreign exchange loss of €3.8 million on unauthorized trades. In a July 14, 2014, press release, Greenpeace explained and apologized:

The losses are a result of a serious error of judgment by an employee in our International Finance Unit acting beyond the limits of their authority and without following proper procedures. Greenpeace International entered *into contracts to buy foreign currency at a fixed exchange rate while the euro was gaining in strength. This resulted in a loss of 3.8 million euros against a range of other currencies.*

Although it does sound as if the individual trader was not authorized to make the forward contract purchases (Greenpeace has not released any further detail), the purchase of euros forward to try to protect the organization against a rising euro sounds more like losses related to hedging rather than speculation.

U.K. pounds in an amount matching the typical size and maturity of expected pound inflows. Then, if the pound depreciated or appreciated, the foreign exchange effect on cash inflows in pounds would be offset by the effect on cash outflows in pounds from repaying the pound loan plus interest.

The structure of a money market hedge resembles that of a forward hedge. The difference is that the cost of the money market hedge is determined by different interest rates than the interest rates used in the formation of the forward rate. The difference in interest rates facing a private firm borrowing in two separate country markets may be different from the difference in risk-free government bill rates or eurocurrency interest rates in these same markets. In efficient markets interest rate parity should ensure that these costs are nearly the same, but not all markets are efficient at all times.

To hedge in the money market, Maria will borrow pounds in London at once, immediately convert the borrowed pounds into dollars, and repay the pound loan in three months with the proceeds from the sale of the generator. She will need to borrow just enough to repay both the principal and interest with the sale proceeds. The borrowing interest rate will be 10% per annum, or 2.5% for three months. Therefore, the amount to borrow now for repayment in three months is

$$\frac{£1,000,000}{1 + 0.025} = £975,610.$$

Maria would borrow £975,610 now, and in three months repay that amount plus £24,390 of interest with the account receivable. Ganado would exchange the £975,610 loan proceeds for dollars at the current spot exchange rate of $1.7640/£, receiving $1,720,976 at once.

The money market hedge, if selected by Ganado, creates a pound-denominated liability—the pound loan—to offset the pound-denominated asset—the account receivable. The money market hedge works as a hedge by matching assets and liabilities according to their currency of denomination. Using a simple T-account illustrating Ganado's balance sheet, the loan in British pounds is seen to offset the pound-denominated account receivable:

Assets		**Liabilities and Net Worth**	
Account receivable	£1,000,000	Bank loan (principal)	£975,610
		Interest payable	24,390
	£1,000,000		£1,000,000

The loan acts as a *balance sheet hedge* against the pound-denominated account receivable.

To compare the forward hedge with the money market hedge, one must analyze how Ganado's loan proceeds will be utilized for the next three months. Remember that the loan proceeds are received today, but the forward contract proceeds are received in three months. For comparison purposes, one must either calculate the future value of the loan proceeds or the present value of the forward contract proceeds. Since the primary uncertainty here is the dollar value in three months, we will use future value here.

As both the forward contract proceeds and the loan proceeds are relatively certain, it is possible to make a clear choice between the two alternatives based on the one that yields the higher dollar receipts. This result, in turn, depends on the assumed rate of investment or use of the loan proceeds.

At least three logical choices exist for an assumed investment rate for the loan proceeds for the next three months. First, if Ganado is cash rich, the loan proceeds might be invested in U.S. dollar money market instruments that yield 6% per annum. Second, Maria might simply use the pound loan proceeds to pay down dollar loans that currently cost Ganado 8% per annum. Third, Maria might invest the loan proceeds in the general operations of the firm, in which case the cost of capital of 12% per annum would be the appropriate rate. The field of finance generally uses the company's cost of capital to move capital forward and backward in time, and we will therefore use the WACC of 12% (3% for the 90-day period here) to calculate the future value of proceeds under the money market hedge:

$$\$1,720,976 \times 1.03 = \$1,772,605$$

A break-even rate can now be calculated between the forward hedge and the money market hedge. Assume that r is the unknown 3-month investment rate (expressed as a decimal) that would equalize the proceeds from the forward and money market hedges. We have

$$(\text{Loan proceeds}) \times (1 + \text{rate}) = (\text{forward proceeds})$$
$$\$1,720,976 \times (1 + r) = \$1,754,000$$
$$r = 0.0192$$

One can convert this 3-month (90 days) investment rate to an annual whole percentage equivalent, assuming a 360-day financial year, as follows:

$$0.0192 \times \frac{360}{90} \times 100 = 7.68\%$$

In other words, if Maria Gonzalez can invest the loan proceeds at a rate higher than 7.68% per annum, she would prefer the money market hedge. If she can only invest at a rate lower than 7.68%, she would prefer the forward hedge.

The essence of a money market hedge is as follows:

The money market hedge therefore results in cash received up-front (at the start of the period), which can then be carried forward in time for comparison with the other hedging alternatives.

Options Market Hedge

Maria Gonzalez could also cover her £1,000,000 exposure by purchasing a put option. This technique — an *option hedge* — allows her to speculate on the upside potential for appreciation of the pound while limiting downside risk to a known amount. Maria could purchase from her bank a 3-month put option on £1,000,000 at an at-the-money (ATM) strike price of $1.75/£ with a premium cost of 1.50%. The cost of the option — the premium — is

$$(\text{Size of option}) \times (\text{premium}) \times (\text{spot rate}) = \text{cost of option},$$
$$£1,000,000 \quad \times \quad 0.015 \quad \times \quad \$1.7640 \quad = \quad \$26,460.$$

Because we are using future value to compare the various hedging alternatives, it is necessary to project the premium cost of the option forward three months. We will use the cost of capital of 12% per annum or 3% per quarter. Therefore the premium cost of the put option as of June would be $26,460(1.03) = $27,254. This is equal to $0.0273 per pound ($27,254 ÷ £1,000,000).

When the £1,000,000 is received in June, the value in dollars depends on the spot rate at that time. The upside potential is unlimited, the same as in the unhedged alternative. At any exchange rate above $1.75/£, Ganado would allow its option to expire unexercised and would exchange the pounds for dollars at the spot rate. If the expected rate of $1.76/£ materializes, Ganado would exchange the £1,000,000 in the spot market for $1,760,000. Net proceeds would be $1,760,000 minus the $27,254 cost of the option, or $1,732,746.

In contrast to the unhedged alternative, downside risk is limited with an option. If the pound depreciates below $1.75/£, Maria would exercise her option to sell (put) £1,000,000 at $1.75/£, receiving $1,750,000 gross, but $1,722,746 net of the $27,254 cost of the option. Although this downside result is worse than the downside of either the forward or money market hedges, the upside potential is unlimited.

The essence of the at-the-money (ATM) put option market hedge is as follows:

Today

Buy put option to
sell pounds @ $1.75/£.
Pay $26,460 for put option.

Three months from today

Receive £1,000,000.
Either deliver £1,000,000 against put,
receiving $1,750,000; or sell £1,000,000
spot if current spot rate is > $1.75/£.

We can calculate a trading range for the pound that defines the break-even points for the option compared with the other strategies. The upper bound of the range is determined by comparison with the forward rate. The pound must appreciate enough above the $1.7540 forward rate to cover the $0.0273/£ cost of the option. Therefore, the break-even upside spot price of the pound must be $1.7540 + $0.0273 = $1.7813. If the spot pound appreciates above $1.7813, proceeds under the option strategy will be greater than under the forward hedge. If the spot pound ends up below $1.7813, the forward hedge would have been superior in retrospect.

The lower bound of the range is determined by the unhedged strategy. If the spot price falls below $1.75/£, Maria will exercise her put and sell the proceeds at $1.75/£. The net proceeds will be $1.75/£ less than the $0.0273 cost of the option, or $1.7227/£. If the spot rate falls below $1.7227/£, the net proceeds from exercising the option will be greater than the net proceeds from selling the unhedged pounds in the spot market. At any spot rate above $1.7227/£, the spot proceeds from remaining unhedged will be greater.

Foreign currency options have a variety of hedging uses. A put option is useful to construction firms and exporters when they must submit a fixed price bid in a foreign currency without knowing until some later date whether their bid is successful. Similarly, a call option is useful to hedge a bid for a foreign firm if a potential future foreign currency payment may be required. In either case, if the bid is rejected, the loss is limited to the cost of the option.

Comparison of Alternatives

Exhibit 10.5 shows the value of Ganado's £1,000,000 account receivable over a range of possible ending spot exchange rates and hedging alternatives. This exhibit makes it clear that the firm's view of likely exchange rate changes aids in the hedging choice as follows:

- If the exchange rate is expected to move against Ganado, to the left of $1.76/£, the money market hedge is clearly the preferred alternative with a guaranteed value of $1,772,605.

- If the exchange rate is expected to move in Ganado's favor, to the right of $1.76/£, then the preferred alternative is less clearcut, lying between remaining unhedged, the money market hedge, or the put option.

| EXHIBIT 10.5 | **Ganado's A/R Transaction Exposure Hedging Alternatives** |

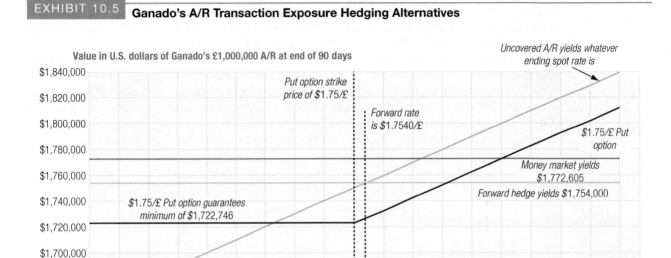

Remaining unhedged is most likely an unacceptable choice. If Maria's expectations regarding the future spot rate prove to be wrong, and the spot rate falls below $1.70/£, she will not reach her budget rate. The put option offers a unique alternative. If the exchange rate moves in Ganado's favor, the put option offers nearly the same upside potential as the unhedged alternative except for the up-front costs. If, however, the exchange rate moves against Ganado, the put option limits the downside risk to $1,722,746.

Strategy Choice and Outcome

So how should Maria Gonzalez choose among the alternative hedging strategies? She must select on the basis of two decision criteria: (1) the *risk tolerance* of Ganado, as expressed in its stated policies; and (2) her own view, or expectation of the direction (and distance) the exchange rate will move over the coming 90-day period.

Ganado's *risk tolerance* is a combination of management's philosophy toward transaction exposure and the specific goals of treasury activities. Many firms believe that currency risk is simply a part of doing business internationally, and therefore, begin their analysis from an unhedged baseline. Other firms, however, view currency risk as unacceptable, and either begin their analysis from a full forward contract cover baseline, or simply mandate that all transaction exposures be fully covered by forward contracts regardless of the value of other hedging alternatives. The treasury in most firms operates as a *cost* or *service center* for the firm. On the other hand, if the treasury operates as a *profit center*, it might tolerate taking more risk.

The final choice between hedges—if Maria Gonzalez does expect the pound to appreciate—combines the firm's risk tolerance, its view, and its confidence in its view. Transaction exposure management with contractual hedges requires managerial judgment. *Global Finance in Practice 10.3* describes how hedging choices may also be influenced by profitability concerns and forward premiums.

Forward Rates and the Cost of Hedging

Some multinational firms measure the cost of hedging as the "total cash flow expenses of the hedge" as a percentage of the initial booked foreign currency transaction. They define the "total cash flow expense of the hedge" as any cash expenses for purchase (e.g., option premium paid up-front, including the time value of money) plus any difference in the final cash flow settlement versus the booked transaction.

If a firm were using forwards, there is no up-front cost, so the total cash flow expense is simply the difference between the forward settlement and the booked transaction (using this definition of hedging expense). This is the forward premium. But the size of the forward premium has sometimes motivated firms to avoid using forward contracts.

Assume a U.S.-based firm has a GBP1 million one-year receivable. The current spot rate is USD1.6000 = GBP1.00. If U.S. dollar and British pound interest rates were 2.00% and 4.00%, respectively, the forward rate would be USD1.5692. This is a forward premium of −1.923% (the pound is selling forward at a 1.923% discount versus the dollar), and in this firm's view, the cost of hedging the transaction is then 1.923%.

However, if British pound interest rates were significantly higher, say 8.00%, then the one year forward rate would be USD1.5111, a forward premium of −5.556%. Some multinationals see using a forward in this case, in which more than 5.5% of the transaction's settlement is "lost" to hedging as too expensive. The definition of "too expensive" must be based on the philosophy of the individual firm and its risk tolerance for currency risk, but fundamentals of financial theory would argue that the two cases are not truly different. However, in business, depending on how pricing was conducted, a loss of 5.56% on the sale settlement could destroy much of the net margin on the sale.

Management of an Account Payable

The management of an account payable, where the firm would be required to make a foreign currency payment at a future date, is similar but not identical to the management of an account receivable. If Ganado had a £1,000,000 account payable due in 90 days, the hedging choices would appear as follows:

Remain Unhedged. Ganado could wait 90 days, exchange dollars for pounds at that time, and make its payment. If Ganado expects the spot rate in 90 days to be $1.7600/£, the payment would be expected to cost $1,760,000. This amount is, however, uncertain; the spot exchange rate in 90 days could be very different from that expected.

Forward Market Hedge. Ganado could buy £1,000,000 forward, locking in a rate of $1.7540/£, and a total dollar cost of $1,754,000. This is $6,000 less than the expected cost of remaining unhedged, and therefore clearly preferable to the first alternative.

Money Market Hedge. The money market hedge is distinctly different for a payable as opposed to a receivable. To implement a money market hedge in this case, Ganado would exchange U.S. dollars spot and invest them for 90 days in a pound-denominated interest-bearing account. The principal and interest in British pounds at the end of the 90-day period would be used to pay the £1,000,000 account payable.

In order to assure that the principal and interest exactly equal the £1,000,000 due in 90 days, Ganado would discount the £1,000,000 by the pound investment interest rate of 8% for 90 days in order to determine the pounds needed today:

$$\frac{£1,000,000}{1 + \left(.08 \times \dfrac{90}{360}\right)} = £980,392.16.$$

This £980,392.16 needed today would require $1,729,411.77 at the current spot rate of $1.7640/£:

$$£980,392.16 \times \$1.7640/£ = \$1,729,411.77.$$

Finally, in order to compare the money market hedge outcome with the other hedging alternatives, the $1,729,411.77 cost today must be carried forward 90 days to the same future date as the other hedge choices. If the current dollar cost is carried forward at Ganado's WACC of 12%, the total cost of the money market hedge is $1,781,294.12. This is higher than the forward hedge and therefore unattractive.

$$\$1,729,411.77 \times \left[1 + \left(.12 \times \frac{90}{360} \right) \right] = \$1,781,294.12.$$

Option Hedge. Ganado could cover its £1,000,000 account payable by purchasing a call option on £1,000,000. A June call option on British pounds with a near at-the-money strike price of $1.75/£ would cost 1.5% (premium) or

$$£1,000,000 \times 0.015 \times \$1.7640/£ = \$26,460.$$

This premium, regardless of whether the call option is exercised or not, will be paid up-front. Its value, carried forward 90 days at the WACC of 12%, would raise its end of period cost to $27,254. If the spot rate in 90 days is less than $1.75/£, the option would be allowed to expire and the £1,000,000 for the payable would be purchased on the spot market. The total cost of the call option hedge, if the option is not exercised, is theoretically smaller than any other alternative (with the exception of remaining unhedged, because the option premium is still paid and lost). If the spot rate in 90 days exceeds $1.75/£, the call option would be exercised. The total cost of the call option hedge, if exercised, is as follows:

Exercise call option (£1,000,000 × $1.75/£)	$1,750,000
Call option premium (carried forward 90 days)	27,254
Total maximum expense of call option hedge	$1,777,254

EXHIBIT 10.6   Ganado's A/P Transaction Exposure Hedging Alternatives

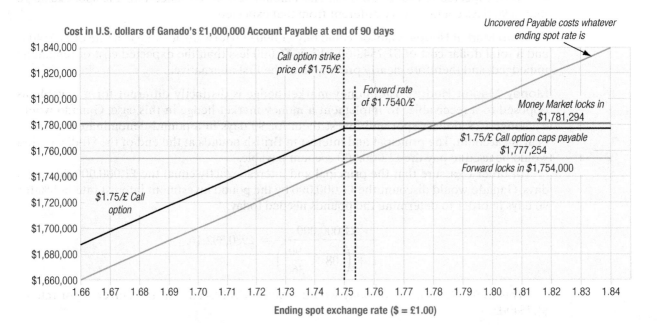

Payable Hedging Strategy Choice. The four hedging methods of managing a £1,000,000 account payable for Ganado are summarized in Exhibit 10.6. The costs of the forward hedge and money market hedge are certain. The cost using the call option hedge is calculated as a maximum, and the cost of remaining unhedged is highly uncertain.

As with Ganado's account receivable, the final hedging choice depends on the confidence of Maria's exchange rate expectations, and her willingness to bear risk. The forward hedge provides the lowest cost of making the account payable payment that is certain. If the dollar strengthens against the pound, ending up at a spot rate less than $1.75/£, the call option could potentially be the lowest cost hedge. Given an expected spot rate of $1.76/£, however, the forward hedge appears to be the preferred alternative.

Risk Management in Practice

There are as many different approaches to exposure management as there are firms. A variety of surveys of corporate risk management practices in recent years in the United States, the United Kingdom, Finland, Australia, and Germany, indicate no real consensus exists regarding the best approach. The following is our attempt to assimilate the basic results of these surveys and combine them with our own personal experiences.

Which Goals?

The treasury function of most private firms, the group typically responsible for transaction exposure management, is usually considered a cost center. It is not expected to add profit to the firm's bottom line (which is not the same thing as saying it is not expected to add value to the firm). Currency risk managers are expected to err on the conservative side when managing the firm's money.

Which Exposures?

Transaction exposures exist before they are actually booked as foreign currency-denominated receivables and payables. However, many firms do not allow the hedging of quotation exposure or backlog exposure as a matter of policy. The reasoning is straightforward: until the transaction exists on the accounting books of the firm, the probability of the exposure actually occurring is considered to be less than 100%. Conservative hedging policies dictate that contractual hedges be placed only on existing exposures.

Which Contractual Hedges?

As might be expected, transaction exposure management programs are generally divided along an "option-line," those that use options and those that do not. Firms that do not use currency options rely almost exclusively on forward contracts and money market hedges. *Global Finance in Practice 10.4* demonstrates how market condition may change firm hedging choices.

Many MNEs have established rather rigid transaction exposure risk management policies, which mandate proportional hedging. These policies generally require the use of forward contract hedges on a percentage (e.g., 50, 60, or 70%) of existing transaction exposures. As the maturity of the exposures lengthens, the percentage forward-cover required decreases. The remaining portion of the exposure is then selectively hedged on the basis of the firm's risk tolerance, view of exchange rate movements, and confidence level. Although rarely acknowledged by the firms themselves, selective hedging is essentially speculation. A significant question remains as to whether a firm or a financial manager can consistently predict the future direction of exchange rates.

The Credit Crisis and Option Volatilities in 2009

The global credit crisis had a number of lasting impacts on corporate foreign exchange hedging practices in late 2008 and early 2009. Currency volatilities rose to some of the highest levels seen in years, and stayed there. This caused option premiums to rise so dramatically that many companies were much more selective in their use of currency options in their risk management programs.

The dollar-euro volatility was a prime example. As recently as July 2007, the implied volatility for the most widely traded currency cross was below 7% for maturities from one week to three years. By October 31, 2008, the 1-month implied volatility had reached 29%. Although this was seemingly the peak, 1-month implied volatilities were still over 20% on January 30, 2009.

This makes options very expensive. For example, the premium on a 1-month call option on the euro with a strike rate forward-at-the-money at the end of January 2009 rose from $0.0096/€ to $0.0286/€ when volatility is 20%, not 7%. For a notional principal of €1 million, that is an increase in price from $9,600 to $28,600. That will put a hole in any treasury department's budget.

An increasing number of firms, however, are actively hedging not only backlog exposures, but also selectively hedging quotation and anticipated exposures. Anticipated exposures are transactions for which there are—at present—no contracts or agreements between parties, but are anticipated on the basis of historical trends and continuing business relationships. Although this may appear to be overly speculative on the part of these firms, it may be that hedging expected foreign-currency payables and receivables for future periods is the most conservative approach to protect the firm's future operating revenues.

Advanced Topics in Hedging

There are other theoretical dimensions to currency hedging that are not often considered in actual industry practice, including the *optimal hedge ratio*, *hedge symmetry*, *hedge effectiveness*, and *hedge timing*.

Hedge Ratio

Transaction exposure is an uncertainty in the value of an asset, such as the value of a specific amount of foreign currency, which may be recognized or realized at a future point in time. In our example in this chapter, Ganado expected to receive £1,000,000 in 90 days, but does not know for certain what that £1,000,000 will be worth in U.S. dollars at that time (the spot exchange rate in 90 days).

The objective of *currency hedging* is to minimize the change in the value of the exposed asset or cash flow from a change in exchange rates. *Hedging* is accomplished by combining the *exposed asset* with a *hedge asset* to create a two-asset portfolio in which the two assets react in relatively equal but opposite directions to an exchange rate change. Once formed, the most common objective of hedging is to construct a hedge that will result in a total change in value of the two-asset portfolio (Δ Portfolio Value)—if perfect—of zero.

$$\Delta \text{Portfolio Value} = \Delta \text{Spot} + \Delta \text{Hedge} = 0.$$

A traditional forward hedge forms a two-asset portfolio, combining the spot exposure with forward cover. The value of the two-asset portfolio is then the sum of the foreign currency amount at the current spot rate (the exposure), with the hedge amount sold forward at the forward rate.

$$\text{Two-Asset Portfolio} = [(\text{Exposure} - \text{Hedge amount}) \times \text{Spot}] + [\text{Hedge amount} \times \text{Forward rate}].$$

For example, if Ganado hedged 100% of its £1,000,000 account receivable with a forward contract at time $t = 90$ (90 days until settlement), assuming a spot rate of $1.7640/£ and a 90-day forward rate of $1.7540/£, this two-asset portfolio would be:

$$V_t = [(£1,000,000 - £1,000,000) \times \$1.7540/£] + [£1,000,000 \times \$1.7540/£] = \$1,754,000.$$

Note that when there is a full forward cover, there is no uncovered exposure remaining. The variance in the terminal value of this two-asset portfolio with respect to the spot exchange rate over the following 90-day period is zero. Its value is set and certain. Also note that if the spot rate and the forward rate were exactly equal (which they are not here), the total position would be termed a *perfect hedge*.

If, however, Maria Gonzalez at Ganado decided to *selectively hedge* the exposure, covering less than 100% of the exposure, the value of the two-asset portfolio would change with the spot exchange rate. The change in value could be either up or down. In this case, Maria Gonzalez would need to follow a methodology for determining what proportion, β, of the exposure, X_t, to cover (so βX_t is the amount of the exposure covered). Now the two-asset portfolio is written:

$$V_t = [(X_t - \beta X_t) \times S_t] + [\beta X_t \times F_t].$$

where the hedge ratio, β, is defined

$$\beta = \frac{\text{Value of currency hedge}}{\text{Value of currency exposure}}$$

If the entire exposure was covered as in Ganado's example above, that is a *hedge ratio* of 1.0 or 100%. The hedge ratio, β, is the percentage of an individual exposure's nominal amount covered by a financial instrument such as a forward contract or currency option.

Hedge Symmetry

Some hedges can be constructed to result in no change in value to any and all exchange rate changes. The hedge is constructed so that whatever spot value is lost as a result of adverse exchange rate movements (ΔSpot), that value is replaced by an equal but opposite change in the value of the hedge asset, (ΔHedge). The commonly used 100% forward contract cover is such a hedge. For example in the case of Ganado, if the entire £1,000,000 account receivable is sold forward, Ganado is assured of the same dollar proceeds at the end of the 90-day period regardless of which direction the exchange rate moves over the exposure period.

But changes in the underlying spot exchange rate need not only result in losses; gains from exchange rate changes are equally possible. In the case of Ganado, if the dollar were to weaken against the pound over the 90-day period, the dollar value of the account receivable would go up. Ganado may choose to construct a hedge, which would minimize the losses in the combined two-asset portfolio (minimize negative ΔValue), and also allow positive changes in value (positive ΔValue) from exchange rate changes. A hedge constructed using a foreign currency option would be pursuing this additional hedging objective. For Ganado, this would be the purchase of a put option on the pound to protect against value losses, and also allow Ganado to possibly reap value increases in the event the exchange rate moved in its favor.

Hedge Effectiveness

The effectiveness of a hedge is determined to what degree the change in spot asset's value is correlated with the equal but opposite change in the hedge asset's value to a change in the underlying spot exchange rate. In currency markets, spot and futures rates are nearly—but not precisely—perfectly correlated. This less-than-perfect correlation is termed *basis risk*.

Hedge Timing

The hedger must also determine the timing of the hedge objective. Does the hedger wish to protect the value of the exposed asset only at the time of its maturity or settlement, or at various points in time over the life of the exposure? For example in the case of Ganado, the various hedging alternatives explored in the problem analysis—the forward, money market, and purchased option hedges—were all constructed and evaluated for the dollar value of the combined hedge portfolio only at the *end* of the 90-day period. In some cases, however, Ganado might wish to protect the value of the exposed asset prior to maturity, for example, at the end of a financial reporting period prior to the actual maturity of the exposure.

SUMMARY POINTS

- MNEs encounter three types of currency exposure: transaction exposure, translation exposure, and operating exposure.

- Transaction exposure measures gains or losses that arise from the settlement of financial obligations whose terms are stated in a foreign currency.

- Considerable theoretical debate exists as to whether firms should hedge currency risk. Theoretically, hedging reduces the variability of the cash flows to the firm. It does not increase the cash flows to the firm. In fact, the costs of hedging may potentially lower them.

- Transaction exposure can be managed by contractual techniques and certain operating strategies. Contractual hedging techniques include forward, futures, money market, and option hedges.

- The choice of which contractual hedge to use depends on the individual firm's currency risk tolerance and its expectation of the probable movement of exchange rates over the transaction exposure period.

- Risk management in practice requires a firm's treasury to identify its goals, choose which contractual hedges it wishes to use, and decide what proportion of the currency exposure should be hedged.

MINI-CASE

China Noah Corporation[1]

China's voracious consumer appetites are already reaching into every corner of Indonesia. The increasing weight of China in every market is a global trend, but growing Chinese, as well as Indian, demand is making an especially big impact in Indonesia. Nick Cashmore of the Jakarta office of CLSA, an investment bank, has coined a new term to describe this symbiotic relationship: "Chindonesia."

—"Special Report on Indonesia: More Than a Single Swallow," *The Economist*, September 10, 2009.

In early 2010, Mr. Savio Chow, CFO of China Noah Corporation (Noah), was concerned about the foreign exchange exposure his company could be creating by shifting much of its procurement of wood to Indonesia. Noah was a leading floorboard manufacturer in China that purchased more than USD100 million in lumber annually, primarily from local wood suppliers in China. But now Mr. Chow planned to shift a large portion of his raw material procurement to Indonesian suppliers in light of the abundant wood resources in Indonesia and the increasingly tight wood supply market in China. Chow knew he needed an explicit strategy for managing the currency exposure.

China Noah

Noah, a private company owned by its founding family, was one of the largest floorboard producers in China. The company was established in 1982 by the current chairman, Mr. Se Hok Pan, a Macau resident. Most of the company's senior management team had been with the company since inception.

Noah's primary product was solid wood flooring, which used 100% natural wood cut into floorboards, sanded, and

[1]Copyright © 2014 Thunderbird, School of Global Management. All rights reserved. This case was prepared by Liangqin Xiao and Yan Ying under the direction of Professor Michael H. Moffett for the purpose of classroom discussion only, and not to indicate either effective or ineffective management.

protected with a layer of gloss. Rapid Chinese economic growth, together with the rising living standards and the emphasis on environmental conservation in China, had created a consumer preference for timber products for both households and offices. Besides being natural, wood products were considered beneficial for both mental and physical health. Noah operated five flooring manufacturing plants and a distributor/retail network of over 1,500 outlet stores across China.

As shown in Exhibit A, Noah had grown rapidly in recent years, with sales growing from CNY986 million in 2006 to CNY1,603 million in 2009 (approximately USD200 million at the current spot rate of CNY6.92=USD1.00). Net profit had risen from CNY115 million to CNY187 million (USD27 million) in the same period. Mr. Chow was a planner, and as is also illustrated by Exhibit A, he and Noah were expecting sales to grow at an annual average rate of 20% for the coming five years. Noah's return on sales was expected to be good this year at 13.5%. But if Chow's forecasts were accurate, they would plummet to 3.7% by 2015.

Supply Chain

One of the key characteristics of the floorboard industry is that wood makes up the vast majority of all raw material and direct cost. In the past three years Noah had spent between CNY60 and CNY65 on wood purchasing for every square meter of floorboard manufactured. This meant wood was almost 90% of cost of goods sold. Given the competitiveness of the floorboard industry, the ability to control and potentially lower wood cost was the dominant driver of corporate profitability.

Noah had never owned any forests of its own, buying wood from Chinese forest owners or lumber traders. Chinese wood prices had long been quite cheap by global standards, partly as a result of a large-scale illegal logging industry. But wood supplies had now tightened dramatically as forest resources became increasingly scarce due to China's shift toward environmental protection, and this tightening supply was sending wood prices upward.

The World Wildlife Fund estimated that domestic wood supplies met only half of the country's current timber

| EXHIBIT A | China Noah's Consolidated Statement of Income (actual and forecast, million Chinese yuan) |

(CNY million)	2007	2008	2009	2010e	2011e	2012e	2013e	2014e	2015e
Sales revenue	1,290.4	1,394.6	1,602.7	1,923.2	2,307.9	2,769.5	3,323.4	3,988.0	4,785.6
Cost of goods sold	(849.4)	(943.4)	(1,110.0)	(1,294.0)	(1,610.3)	(2,000.7)	(2,491.1)	(3,096.8)	(3,848.2)
Gross profit	441.0	451.2	492.7	629.3	697.6	768.8	832.2	891.2	937.4
Gross margin	*34.2%*	*32.4%*	*30.7%*	*32.7%*	*30.2%*	*27.8%*	*25.0%*	*22.3%*	*19.6%*
Selling expense	(216.0)	(208.0)	(201.8)	(242.3)	(290.8)	(349.0)	(418.7)	(502.5)	(603.0)
G&A expense	(19.6)	(20.0)	(20.1)	(24.1)	(28.9)	(34.7)	(41.7)	(50.0)	(60.0)
EBITDA	205.7	223.6	271.1	362.8	377.9	385.1	371.8	338.7	274.4
EBITDA margin	*15.9%*	*16.0%*	*16.9%*	*18.9%*	*16.4%*	*13.9%*	*11.2%*	*8.5%*	*5.7%*
Depreciation	(40.3)	(45.3)	(49.4)	(57.5)	(60.8)	(64.0)	(67.3)	(70.5)	(73.7)
EBIT	165.6	178.4	221.9	305.3	317.1	321.1	304.5	268.2	200.7
EBIT margin	*12.8%*	*12.8%*	*13.8%*	*15.9%*	*13.7%*	*11.6%*	*9.2%*	*6.7%*	*4.2%*
Interest expense	(7.1)	(12.0)	(15.1)	(15.9)	(13.9)	(11.2)	(7.7)	(4.4)	(2.2)
EBT	158.5	166.4	206.8	289.4	303.2	309.9	296.8	263.8	198.5
Income tax	(8.4)	(18.0)	(20.0)	(28.9)	(30.3)	(31.0)	(29.7)	(26.4)	(19.9)
Net income	150.1	148.4	186.8	260.5	272.9	278.9	267.1	237.5	178.7
Return on sales	*11.6%*	*10.8%*	*11.7%*	*13.5%*	*11.8%*	*10.1%*	*8.0%*	*6.0%*	*3.7%*

Assumes sales growth of 20% per year. Estimated costs assume INR 1344 = 1.00 RMB. Projected selling expenses assumed 12.6% of sales, G&A expenses at 1.3% of sales, and income tax expenses at 10% of EBT. Cost of goods sold assumptions for 2010e–2015e are based on Exhibit C, which follows.

consumption, and a variety of price forecasts had quite honestly frightened Chow. For example, Morgan Stanley was forecasting Chinese wood prices to rise by 15% to 20% over the coming five years. Major Chinese floorboard producers, including Noah, were now looking at countries like Brazil, Russia, and Indonesia for more sustainable, legal, and cheaper sources of wood.

Noah's Indonesia Deal

Over the past few months Chow had been pursuing a number of Indonesian wood supplier deals to replace a portion of its Chinese sourcing. Preliminary price quotes were encouraging, prices of CNY62.6/m^2 coming in roughly 8% cheaper than current Chinese prices. The previous week he had presented a potential Indonesian supplier's term sheet (Exhibit B) to Noah's board of directors. The term sheet was based on 30% Indonesian sourcing of the total 17.2 million square meters of flooring Noah expected to sell in 2010.

Chow wished to move quickly to try to control—and possibly reduce—Noah's wood costs for the current year and possibly for years to come. The current price quote from the consortia of Indonesian wood producers was 84,090 Indonesian rupiah per square meter (IDR/m^2), which translated into a price in Chinese yuan per square meter (CNY/m^2) of 62.6.

$$\text{Price}^{\text{CNY}} = \frac{\text{IDR84,080/m}^2}{\text{IDR1,344/CNY}} = \text{CNY62.6/m}^2.$$

At CNY 62.6, this was a 7.7% discount to the current Chinese price of 67.8 for the same wood. Since Chinese prices were expected to rise 4% to 5% per year for the foreseeable future, but the Indonesian consortia was willing to contractually limit annual price increases to just 4% per annum, the discount might increase if the IDR/CNY exchange rate remained the same.

Chow expected Noah's production to more than double over the next five years, from 17.2 million square meters in

2010 to 42.8 million in 2015, as shown in Exhibit C. If he sourced 30% of Noah's wood from Indonesia in 2010, and then increased that proportion 10% per year, Indonesia would account for roughly half of Noah's wood sourcing by 2015.

Indonesian Growth

Indonesia's forests covered 60% of the country. In recent years, the country's high population and rapid industrialization had already led to serious environmental issues, including large-scale deforestation, and like that in China, much of it illegal. That said, Indonesia was rapidly emerging as an important exporter of wood.

In terms of macroeconomics, Indonesia had been less affected by the recent global recession, in comparison to its neighbors. Statistics indicated that Indonesia's GDP grew by 4.5% in 2009, and that it was expected to grow by nearly 7% per year over the coming decade. Indonesia could soon move to economic parity with the BRICs (Brazil, Russia, India, and China). Stable political conditions, despite the 2009 elections and strong domestic demand, could deliver that growth.

Foreign Currency Risk

Indonesia had been one of countries hardest hit by the Asian financial crisis of 1997–1998. Against the U.S. dollar, the Indonesian rupiah dropped from about IDR2,600/USD to a low point of IDR14,000/USD, its economy shrinking a shocking 14%—although it did rebound in the following years. The rupiah had since stabilized in the IDR8,000/USD to IDR10,000/USD range.

As illustrated by Exhibit D, the rupiah had traded in a relatively narrow range of IDR1,000/CNY to IDR1,400/CNY over the past 10 years, with the exception of the recent global credit crisis. Because the Indonesian rupiah was a free-floating currency and the Chinese renminbi a highly controlled and managed currency, crisis had always hit the rupiah much harder. Since Noah was considering

---

EXHIBIT B   **Term Sheet from an Indonesian Wood Consortium**

Buyer	China Noah Corporation
Seller	An Indonesian wood supply consortium
Quantity	5.16 million m^2, 30% of Noah's wood production in 2010
Unit price	IDR84,090/m^2 (equivalent to CNY62.6/m^2, exchanged at spot rate)
Total payment	IDR433,840 million
Payment schedule	The payment must be settled in Indonesian rupiah (IDR) in 6 months.

EXHIBIT C	China Noah Corporation's Cost of Goods Sold Composition

CNY/m² Floorboard	2007	2008	2009	2010e	2011e	2012e	2013e	2014e	2015e
Wood Cost									
Chinese wood cost	59.9	61.8	65.2	67.8	70.5	73.3	76.3	79.3	82.5
Percent Chinese	100%	100%	100%	70%	67%	64%	60%	56%	52%
Indonesian wood cost (IDR/m²):				84,090	87,454	90,952	94,590	98,373	102,308
Cost in CNY/m²				62.6	65.1	67.7	70.4	73.2	76.1
Percent Indonesian	0%	0%	0%	30%	33%	36%	40%	44%	48%
Weighted wood cost	59.9	61.8	65.2	66.2	68.7	71.3	73.9	76.6	79.4
Other Cost									
Packaging	2.9	2.9	3.0	3.1	3.2	3.3	3.4	3.5	3.6
Utilities	0.4	0.4	0.4	0.4	0.4	0.4	0.5	0.5	0.5
Labor	1.0	1.0	1.1	1.1	1.2	1.2	1.2	1.3	1.3
Delivery	1.4	1.4	1.5	1.5	1.6	1.6	1.7	1.7	1.8
Sand	0.9	0.9	0.8	0.8	0.8	0.8	0.8	0.9	0.9
Other	2.0	2.0	2.0	2.1	2.1	2.2	2.3	2.3	2.4
Total COGS (CNY/m²)	68.5	70.4	74.0	75.2	78.0	80.8	83.8	86.8	89.9
Flooring Output (milliom m²)	12.4	13.4	15.0	17.2	20.6	24.8	29.7	35.7	42.8
Wood/COGS	87.4%	87.8%	88.1%	88.0%	88.1%	88.2%	88.2%	88.3%	88.3%
Total COGS (million CNY)	849.4	943.4	1,110.0	1,294.0	1,610.3	2,000.7	2,491.1	3,096.8	3,848.2

Assumes INR 1344 = 1.00 CNY for 2010–2015; flooring output growth at 20%, Chinese wood prices increasing 4% per year, Indonesian wood prices increasing 4% per year, percentage increase in Indonesian sourcing 10% per year from 30% in 2010.

a fundamental change in its wood-sourcing strategy and structure, the exchange rate between the rupiah and the yuan over the long-term was considered critical.

Hedging Foreign Exchange Exposure

The Indonesia wood sourcing contract would expose Noah to exchange rate risk over a series of 6-month periods (March 2010, September 2010, March 2011, etc.). Chow, having little experience in managing exchange rate risk, had obtained some detailed advice from Noah's financial advisors, Morgan Stanley.

Morgan Stanley had noted that the Chinese government was under constant pressure from many countries, including the United States, to revalue the yuan. Unlike the yuan, the rupiah floated in value, although its value often tracked closely against the U.S. dollar. If the Chinese yuan was indeed revalued against the U.S. dollar, and the Indonesian rupiah tracked the dollar "down," a weaker rupiah could result.

Chow's first step in considering hedging alternatives was to collect currency and derivative quotes on the Indonesian rupiah/Chinese yuan spot rate into the immediate future.

Spot Rate Forecast. Morgan Stanley's forecast of the IDR/CNY spot rate through 2015 showed a rupiah that slowly appreciated against the yuan over the coming five years, as illustrated in Exhibit E.

Forward Rates. Chow had also requested forward rate quotes from several of its bankers. An average of their quotes is also presented in Exhibit E. Unlike the spot rate forecast, the forward rate quotes, based on interest rate differentials, locked in a rapidly discounted rupiah against the yuan out over the same 5-year period.

Currency Options. Forward hedging would eliminate Noah's downside risk, but would also eliminate any opportunity to benefit from an even weaker Indonesian rupiah—if that

Indonesian Rupiah to Chinese Yuan Spot Rate (monthly)

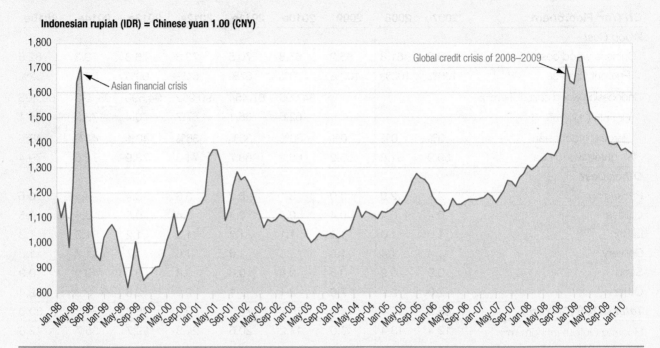

Indonesian rupiah (IDR) = Chinese yuan 1.00 (CNY)

were to occur. Given Chow's limited experience with foreign exchange derivatives, currency options made him nervous. But, as he told his controller, he was determined to consider all appropriate techniques available.

Exhibit F lists possible option positions for Noah with varying strike prices, based on currently available market data. The quotes in Exhibit F are of course only for the first 6-month payment; longer maturities

Forecast and Forward Rates on the IDR/CNY Spot Rate

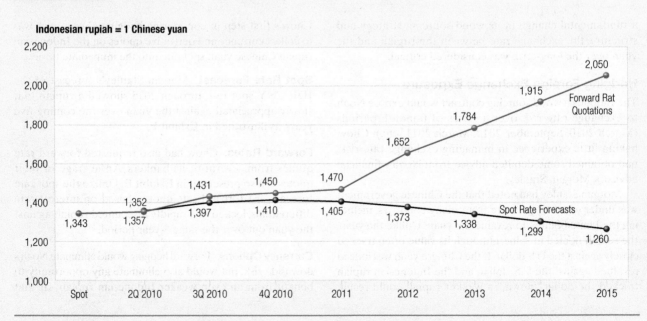

Indonesian rupiah = 1 Chinese yuan

EXHIBIT F	Currency Option Strike Rates and Premiums	

Strike Rate (IDR/CNY)	CNY Put Option Premium (per CNY)	CNY Call Option Premium (per CNY)
1300	2.82%	30.59%
1350	3.20%	28.16%
1400	3.73%	25.89%
1450	4.41%	23.77%
1500	5.23%	21.79%
1550	6.19%	19.94%
1600	7.32%	18.27%
1650	8.58%	16.73%
1700	9.98%	15.33%
1750	11.50%	14.05%
1800	13.13%	12.88%

Note: Quotes are for options of a 1 million CNY notional principal

would be needed for the hedging of future rupiah exposures.

Money Market Hedging. Since Noah's foreign exchange exposure was a payable, Indonesian rupiah at 6-month intervals into the future, money market hedging would entail depositing funds now into Indonesian rupiah-denominated accounts bearing rupiah interest. Indonesian interest rates were consistently higher than comparable rates in China, where rates were subject to government regulations and restrictions. As illustrated in Exhibit G, the 6-month deposit rate in China on CNY was currently

EXHIBIT G	Indonesian and Chinese Deposit Rates of Interest

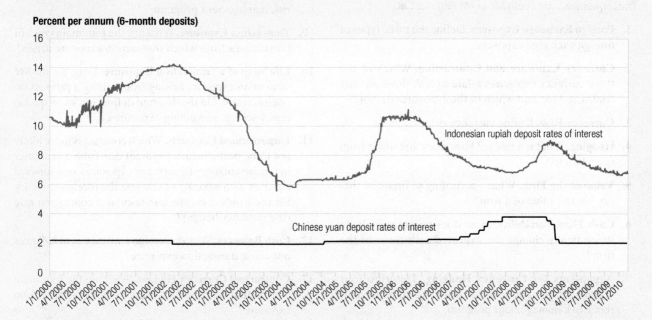

Percent per annum (6-month deposits)

Indonesian rupiah deposit rates of interest

Chinese yuan deposit rates of interest

1.98%, while the same rate in Indonesia was a hefty 6.74%.

Currency Adjustment Clauses. Chow was encouraged to also consider a *Currency Adjustment Clause* or CAC as his banker termed it. Noah's bankers argued that if the Indonesian sourcing was to become a long-term partnership between Noah and the Indonesian consortia, then a CAC would basically allow the two parties to share the currency risks, up or down. The rationale for the CAC was similar to a profit/risk-sharing program, where buyers/sellers initially agree to lock in a local currency price on a settlement denominated in a foreign currency—in this case the Indonesian rupiah purchase price. As long as the exchange rate stays within some defined boundary around a central foreign exchange rate, say $\pm5\%$ around the current spot rate of IDR1344/CNY, the rupiah price would remain fixed. If, however, the spot rate at the time of payment had moved beyond the $\pm5\%$ boundary, the two parties could share the difference between the current spot rate and the original central rate.

Mr. Chow estimated that a 4% fluctuation around the spot FX rate would be a reasonable benchmark to trigger the profit/risk sharing. In this sense, a possible CAC for Noah will initially lock in its payment obligation at CNY62.6/M2. Once the exchange rate movement exceeded the $\pm4\%$ boundary, the CAC would call for an automatic price recalculation by predetermined methods such as using the mid-point between the spot exchange rate and the

exchange rate on the settlement date. Mr. Chow thought that the Indonesian suppliers were likely to respond positively to a CAC because of the many forecasts that the IDR was likely to fall against the CNY.

Time was running short. China Noah's Board was waiting on a currency management strategy proposal from Mr. Chow.

Mini-Case Questions

1. What is the business reason for China Noah's potential currency exposure? Does the company really need to subject itself to substantial exchange rate risk? Is the risk "material" to China Noah? Do you think China Noah should hedge?

2. How does China Noah's profitability (using return on sales as the primary metric) change depending on whether the IDR/CNY exchange rate follows (a) forecast spot rates, (b) forward rate quotes, or (c) fixed rate baseline assumption?

3. Assuming Noah made 6-month payments on its wood purchases from Indonesia, what is the schedule of foreign currency amounts over time?

4. What would be your outlook on the future direction of the Indonesian rupiah and the Chinese renminbi? Should this influence the hedging approach used by Noah?

5. Which of the hedging choices would you recommend?

QUESTIONS

These questions are available in MyFinanceLab.

1. **Foreign Exchange Exposure.** Define the three types of foreign exchange exposure.

2. **Currency Exposure and Contracting.** Which of the three currency exposures relate to cash flows already contracted for, and which of the exposures do not?

3. **Currency Risk.** Define currency risk.

4. **Hedging.** What is a hedge? How does that differ from speculation?

5. **Value of the Firm.** What—according to financial theory—is the value of a firm?

6. **Cash Flow Variability.** How does currency hedging theoretically change the expected cash flows of the firm?

7. **Arguments for Currency Hedging.** Describe four arguments in favor of a firm pursuing an active currency risk management program.

8. **Arguments against Currency Hedging.** Describe six arguments against a firm pursuing an active currency risk management program.

9. **Transaction Exposure.** What are the four main types of transactions from which transaction exposure arises?

10. **Life Span of a Transaction Exposure.** Diagram the life span of an exposure arising from selling a product on open account. On the diagram define and show quotation, backlog, and billing exposures.

11. **Unperformed Contracts.** Which contract is more likely not to be performed: a payment due from a customer in foreign currency (a currency exposure), or a forward contract with a bank to exchange the foreign currency for the firm's domestic currency at a contracted rate (the currency hedge)?

12. **Cash Balances.** Why do foreign currency cash balances not cause transaction exposure?

13. **Contractual Currency Hedges.** What are the four main contractual instruments used to hedge transaction exposure?

14. **Money Market Hedges.** How does a money market hedge differ for an account receivable versus that of an account payable? Is it really a meaningful difference?

15. **Balance Sheet Hedging.** What is the difference between a balance sheet hedge, a financing hedge, and a money market hedge?

16. **Forward versus Money Market Hedging.** Theoretically, shouldn't forward contract hedges and money market hedges have the same identical outcome? Don't they both use the same three specific inputs—the initial spot rate, the domestic cost of funds, and the foreign cost of funds?

17. **Foreign Currency Option Premia.** Why do many firms object to paying for foreign currency option hedges? Do firms pay for forward contract hedges? How do forwards and options differ if at all?

18. **Decision Criteria.** Ultimately, a treasurer must choose among alternative strategies to manage transaction exposure. Explain the two main decision criteria that must be used.

19. **Risk Management Hedging Practices.** According to surveys of corporate practices, which currency exposures do most firms regularly hedge?

20. **Hedge Ratio.** What is the hedge ratio? Why would the hedge ratio ever be less than one?

PROBLEMS

These problems are available in MyFinanceLab.

1. **BioTron Medical, Inc.** Brent Bush, CFO of a medical device distributor, BioTron Medical, Inc., was approached by a Japanese customer, Numata, with a proposal to pay cash (in yen) for its typical orders of ¥12,500,000 every other month if it were given a 4.5% discount. Numata's current terms are 30 days with no discounts. Using the following quotes and estimated cost of capital for Numata, Bush will compare the proposal with covering yen payments with forward contracts. Should Brent Bush accept Numata's proposal?

Spot rate:	¥111.40/$
30-day forward rate:	¥111.00/$
90-day forward rate:	¥110.40/$
180-day forward rate:	¥109.20/$
Numata's WACC	8.850%
BioTron's WACC	9.200%

2. **Bobcat Company.** Bobcat Company, U.S.-based manufacturer of industrial equipment, just purchased a Korean company that produces plastic nuts and bolts for heavy equipment. The purchase price was W7,500 million. W1,000 million has already been paid, and the remaining W6,500 million is due in six months. The current spot rate is W1,110/$, and the 6-month forward rate is W1,175/$. The 6-month Korean won interest rate is 16% per annum, the 6-month U.S. dollar rate is 4% per annum. Bobcat can invest at these interest rates, or borrow at 2% per annum above those rates. A 6-month call option on won with a W1,200/$ strike rate has a 3.0% premium, while the 6-month put option at the same strike rate has a 2.4% premium.

Bobcat can invest at the rates given above, or borrow at 2% per annum above those rates. Bobcat's weighted average cost of capital is 10%. Compare alternate ways that Bobcat might deal with its foreign exchange exposure. What do you recommend and why?

3. **Siam Cement.** Siam Cement, the Bangkok-based cement manufacturer, suffered enormous losses with the coming of the Asian crisis in 1997. The company had been pursuing a very aggressive growth strategy in the mid-1990s, taking on massive quantities of foreign-currency-denominated debt (primarily U.S. dollars). When the Thai baht (B) was devalued from its pegged rate of B25.0/$ in July 1997, Siam's interest payments alone were over $900 million on its outstanding dollar debt (with an average interest rate of 8.40% on its U.S. dollar debt at that time). Assuming Siam Cement took out $50 million in debt in June 1997 at 8.40% interest, and had to repay it in one year when the spot exchange rate had stabilized at B42.0/$, what was the foreign exchange loss incurred on the transaction?

4. **P&G India.** Procter and Gamble's affiliate in India, P&G India, procures much of its toiletries product line from a Japanese company. Because of the shortage of working capital in India, payment terms by Indian importers are typically 180 days or longer. P&G India wishes to hedge an 8.5 million Japanese yen payable. Although options are not available on the Indian rupee (Rs), forward rates are available against the yen. Additionally, a common practice in India is for companies like P&G India to work with a currency agent who will, in this case, lock in the current spot exchange rate in exchange for a 4.85% fee. Using the following exchange rate and interest rate data, recommend a hedging strategy.

Spot rate:	¥120.60/$
180-day forward rate	¥2.400/Rps
Expected spot, 180 days	¥2.6000
180-day Indian rupee investing rate	8.000%
180-day Japanese yen investing rate	1.500%
Currency agent's exchange rate	4.850%
P&G India's cost of capital	12.000%

5. **Elan Pharmaceuticals.** Elan Pharmaceuticals, a U.S.-based multinational pharmaceutical company, is evaluating an export sale of its cholesterol-reduction drug with a prospective Indonesian distributor. The purchase would be for 1,650 million Indonesian rupiah (Rp), which at the current spot exchange rate of Rp9,450/$, translates into nearly $175,000. Although not a big sale by company standards, company policy dictates that sales must be settled for at least a minimum gross margin, in this case, a cash settlement of $168,000. The current 90-day forward rate is Rp9,950/$. Although this rate appeared unattractive, Elan had to contact several major banks before even finding a forward quote on the rupiah. The consensus of currency forecasters at the moment, however, is that the rupiah will hold relatively steady, possibly falling to Rp9,400/$ over the coming 90 to 120 days. Analyze the prospective sale and make a hedging recommendation.

6. **Embraer of Brazil.** Embraer of Brazil is one of the two leading global manufacturers of regional jets (Bombardier of Canada is the other). Regional jets are smaller than the traditional civilian airliners produced by Airbus and Boeing, seating between 50 and 100 people on average. Embraer has concluded an agreement with a regional U.S. airline to produce and deliver four aircraft one year from now for $80 million.

   Although Embraer will be paid in U.S. dollars, it also possesses a currency exposure of inputs—it must pay foreign suppliers $20 million for inputs one year from now (but they will be delivering the subcomponents throughout the year). The current spot rate on the Brazilian real (R$) is R$1.8240/$, but it has been steadily appreciating against the U.S. dollar over the past three years. Forward contracts are difficult to acquire and are considered expensive. Citibank Brasil has not explicitly provided Embraer a forward rate quote, but has stated that it will probably be pricing a forward off the current 4.00% U.S. dollar eurocurrency rate and the 10.50% Brazilian government bond rate. Advise Embraer on its currency exposure.

7. **Krystal.** Krystal is a U.S.-based company that manufactures, sells, and installs water purification equipment.

On April 11, the company sold a system to the City of Nagasaki, Japan, for installation in Nagasaki's famous Glover Gardens (where Puccini's Madame Butterfly waited for the return of Lt. Pinkerton). The sale was priced in yen at ¥20,000,000, with payment due in three months.

Spot exchange rate:	¥118.255/$ (closing mid-rates)
1-month forward rate:	¥117.760/$, a 5.04% per annum premium
3-month forward:	¥116.830/$, a 4.88% per annum premium
1-year forward:	¥112.450/$ a 5.16% per annum premium

Money Rates	United States	Japan	Differential
One month	4.8750%	0.09375%	4.78125%
Three months	4.9375%	0.09375%	4.84375%
Twelve months	5.1875%	0.31250%	4.87500%

Note that the interest rate differentials vary slightly from the forward discounts on the yen because of time differences for the quotes. The spot ¥118.255/$, for example, is a mid-point range. On April 11, the spot yen traded in London from ¥118.30/$ to ¥117.550/$. Aquatech's Japanese competitors are currently borrowing yen from Japanese banks at a spread of two percentage points above the Japanese money rate. Krystal's weighted average cost of capital is 16%, and the company wishes to protect the dollar value of this receivable.

3-month options are available from Kyushu Bank: call option on ¥20,000,000 at exercise price of ¥118.00/$: a 1% premium; or a put option on ¥20,000,000, at exercise price of ¥118.00/$: a 3% premium.

a. What are the costs and benefits of alternative hedges? Which would you recommend, and why?

b. What is the break-even reinvestment rate when comparing forward and money market alternatives?

8. **Caribou River.** Caribou River, Ltd., a Canadian manufacturer of raincoats, does not selectively hedge its transaction exposure. Instead, if the date of the transaction is known with certainty, all foreign currency-denominated cash flows must utilize the following mandatory forward cover formula:

Mandatory Forward Cover	0–90 days	91–180 days	180 days
Paying the points forward	75%	60%	50%
Receiving the points forward	100%	90%	50%

Caribou expects to receive multiple payments in Danish kroner over the next year. DKr3,000,000 is due in 90 days; DKr2,000,000 is due in 180 days; and DKr1,000,000 is due in one year. Using the following spot and forward exchange rates, what would be the amount of forward cover required by company policy for each period?

Spot rate, Dkr/C$	4.70
3-month forward rate, Dkr/C$	4.71
6-month forward rate, Dkr/C$	4.72
12-month forward rate, Dkr/C$	4.74

9. **Pupule Travel.** Pupule Travel, a Honolulu, Hawaii-based 100% privately owned travel company, has signed an agreement to acquire a 50% ownership share of Taichung Travel, a Taiwan-based privately owned travel agency specializing in servicing inbound customers from the United States and Canada. The acquisition price is 7 million Taiwan dollars (T$7,000,000) payable in cash in three months.

Thomas Carson, Pupule Travel's owner, believes the Taiwan dollar will either remain stable or decline slightly over the next three months. At the present spot rate of T$35/$, the amount of cash required is only $200,000, but even this relatively modest amount will need to be borrowed personally by Thomas Carson. Taiwanese interest-bearing deposits by nonresidents are regulated by the government, and are currently set at 1.5% per year. He has a credit line with Bank of Hawaii for $200,000 with a current borrowing interest rate of 8% per year. He does not believe that he can calculate a credible weighted average cost of capital since he has no stock outstanding and his competitors are all also privately held. Since the acquisition would use up all his available credit, he wonders if he should hedge this transaction exposure. He has the following quotes from the Bank of Hawaii:

Spot rate (T$/$)	35.00
3-month forward rate (T$/$)	32.40
3-month Taiwan dollar deposit rate	1.500%
3-month dollar borrowing rate	6.500%
3-month call option on T$	not available

Analyze the costs and risks of each alternative, and then make a recommendation as to which alternative Thomas Carson should choose.

10. **Mattel Toys.** Mattel is a U.S.-based company whose sales are roughly two-thirds in dollars (Asia and the Americas) and one-third in euros (Europe). In September, Mattel delivers a large shipment of toys (primarily Barbies and Hot Wheels) to a major distributor in Antwerp. The receivable, €30 million, is due in 90 days, standard terms for the toy industry in Europe. Mattel's treasury team has collected the following currency and market quotes. The company's foreign exchange advisors believe the euro will be at about $1.4200/€ in 90 days. Mattel's management does not use currency options in currency risk management activities. Advise Mattel as to which hedging alternative is probably preferable.

Current spot rate ($/€)	$1.4158
Credit Suisse 90-day forward rate ($/€)	$1.4172
Barclays 90-day forward rate ($/€)	$1.4195
Mattel Toys WACC ($)	9.600%
90-day eurodollar interest rate	4.000%
90-day euro interest rate	3.885%
90-day eurodollar borrowing rate	5.000%
90-day euro borrowing rate	5.000%

11. **Chronos Time Pieces.** Chronos Time Pieces of Boston exports watches to many countries, selling in local currencies to stores and distributors. Chronos prides itself on being financially conservative. At least 70% of each individual transaction exposure is hedged, mostly in the forward market, but occasionally with options. Chronos' foreign exchange policy is such that the 70% hedge may be increased up to a 120% hedge if devaluation or depreciation appears imminent. Chronos has just shipped to its major North American distributor. It has issued a 90-day invoice to its buyer for €1,560,000. The current spot rate is $1.2224/€, the 90-day forward rate is $1.2270/€. Chronos' treasurer, Manny Hernandez, has a very good track record in predicting exchange rate movements. He currently believes the euro will weaken against the dollar in the coming 90 to 120 days, possibly to around $1.16/€.

a. Evaluate the hedging alternatives for Chronos if Manny is right (Case 1: $1.16/€) and if Manny is wrong (Case 2: $1.26/€). What do you recommend?
b. What does it mean to hedge 120% of a transaction exposure?
c. What would be considered the most conservative transaction exposure management policy by a firm? How does Chronos compare?

12. **Farah Jeans.** Farah Jeans of San Antonio, Texas, is completing a new assembly plant near Guatemala City. A final construction payment of Q8,400,000 is due in six months. ("Q" is the symbol for Guatemalan quetzals.) Lucky 13 uses 20% per annum as its weighted

Problem 13.

Date	Event	Spot Rate ($/£)	Forward Rate ($/£)	Days Forward
February 1	Price quotation for Pegg	1.7850	1.7771	210
March 1	Contract signed for sale	1.7465	1.7381	180
	Contract amount, pounds	£1,000,000		
June 1	Product shipped to Pegg	1.7689	1.7602	90
August 1	Product received by Pegg	1.7840	1.7811	30
September 1	Pegg makes payment	1.7290	—	—

average cost of capital. Today's foreign exchange and interest rate quotations are as follows:

Construction payment due in 6 months (A/P, quetzals)	8,400,000
Present spot rate (quetzals/$)	7.0000
6-month forward rate (quetzals/$)	7.1000
Guatemalan 6-month interest rate (per annum)	14.000%
U.S. dollar 6-month interest rate (per annum)	6.000%
Farah's weighted average cost of capital (WACC)	20.000%

Farah's treasury manager, concerned about the Guatemalan economy, wonders if Lucky 13 should be hedging its foreign exchange risk. The manager's own forecast is as follows:

Expected spot rate in 6-months (quetzals/$)

Highest expected rate (reflecting a significant devaluation)	8.0000
Expected rate	7.3000
Lowest expected rate (reflecting a strengthening of the quetzal)	6.4000

What realistic alternatives are available to Farah for making payments? Which method would you select and why?

13. **Burton Manufacturing.** Jason Stedman is the director of finance for Burton Manufacturing, a U.S.-based manufacturer of handheld computer systems for inventory management. Burton's system combines a low-cost active tag that is attached to inventory items (the tag emits an extremely low-grade radio frequency) with custom-designed hardware and software that tracks the low-grade emissions for inventory control. Burton has completed the sale of an inventory management system to a British firm, Pegg Metropolitan (U.K.), for a total payment of £1,000,000.

The exchange rates shown at the top of this page were available to Burton on the dates shown, corresponding to the events of this specific export sale. Assume each month is 30 days.

a. What will be the amount of foreign exchange gain (loss) upon settlement?

b. If Jason hedges the exposure with a forward contract, what will be the net foreign exchange gain (loss) on settlement?

14. **Micca Metals, Inc.** Micca Metals, Inc. is a specialty materials and metals company located in Detroit, Michigan. The company specializes in specific precious metals and materials that are used in a variety of pigment applications in many industries including cosmetics, appliances, and a variety of high tinsel metal fabricating equipment. Micca just purchased a shipment of phosphates from Morocco for 6,000,000 dirhams, payable in six months.

6-month call options on 6,000,000 dirhams at an exercise price of 10.00 dirhams per dollar are available from Bank Al-Maghrub at a premium of 2%. 6-month put options on 6,000,000 dirhams at an exercise price of 10.00 dirhams per dollar are available at a premium of 3%. Compare and contrast alternative ways that Micca might hedge its foreign exchange transaction exposure. What is your recommendation?

Assumption	Value
Shipment of phosphates from Morocco, Moroccan dirhams	6,000,000
Micca's cost of capital (WACC)	14.000%
Spot exchange rate, dirhams/$	10.00
6-month forward rate, dirhams/$	10.40

15. **Maria Gonzalez and Ganado.** Ganado—the U.S.-based company discussed in this chapter—has concluded another large sale of telecommunications equipment to Regency (U.K.). Total payment of £3,000,000 is

due in 90 days. Maria Gonzalez has also learned that Ganado will only be able to borrow in the United Kingdom at 14% per annum (due to credit concerns of the British banks). Given the following exchange rates and interest rates, what transaction exposure hedge is now in Ganado's best interest?

Assumption	Value	
90-day A/R in pounds	£3,000,000.00	
Spot rate, US$ per pound ($/£)	$1.7620	
90-day forward rate, US$ per pound ($/£)	$1.7550	
3-month U.S. dollar investment rate	6.000%	
3-month U.S. dollar borrowing rate	8.000%	
3-month U.K. investment interest rate	8.000%	
3-month U.K. borrowing interest rate	14.000%	
Ganado's WACC	12.000%	
Expected spot rate in 90 days ($/£)	$1.7850	
Put options on the British pound:	Strike rate ($/£)	Premium
	$1.75	1.500%
	$1.71	1.000%

16. **Larkin Hydraulics.** On May 1, Larkin Hydraulics, a wholly owned subsidiary of Caterpillar (U.S.), sold a 12-megawatt compression turbine to Rebecke-Terwilleger Company of the Netherlands for €4,000,000, payable as €2,000,000 on August 1 and €2,000,000 on November 1. Larkin derived its price quote of €4,000,000 on April 1 by dividing its normal U.S. dollar sales price of $4.320,000 by the then current spot rate of $1.0800/€.

By the time the order was received and booked on May 1, the euro had strengthened to $1.1000/€, so the sale was in fact worth €4,000,000 × $1.1000/€ = $4,400,000. Larkin had already gained an extra $80,000 from favorable exchange rate movements. Nevertheless, Larkin's director of finance now wondered if the firm should hedge against a reversal of the recent trend of the euro. Four approaches were possible:

1. Hedge in the forward market: The 3-month forward exchange quote was $1.1060/€ and the 6-month forward quote was $1.1130/€.

2. Hedge in the money market: Larkin could borrow euros from the Frankfurt branch of its U.S. bank at 8.00% per annum.

3. Hedge with foreign currency options: August put options were available at strike price of $1.1000/€ for a premium of 2.0% per contract, and November put options were available at $1.1000/€ for a premium of 1.2%. August call options at $1.1000/€ could be purchased for a premium of 3.0%, and November call options at $1.1000/€ were available at a 2.6% premium.

4. Do nothing: Larkin could wait until the sales proceeds were received in August and November, hope the recent strengthening of the euro would continue, and sell the euros received for dollars in the spot market.

Larkin estimates the cost of equity capital to be 12% per annum. As a small firm, Larkin Hydraulics is unable to raise funds with long-term debt. U.S. T-bills yield 3.6% per annum. What should Larkin do?

17. **Navarro's Intra-Company Hedging.** Navarro was a U.S.-based multinational company that manufactured and distributed specialty materials for soundproofing construction. It had recently established a new European subsidiary in Barcelona, Spain, and was now in the process of establishing operating rules for transactions between the U.S. parent company and the Barcelona subsidiary. Ignacio Lopez was International Treasurer for Navarro, and was leading the effort at establishing commercial policies for the new subsidiary.

Navarro's first shipment of product to Spain was upcoming. The first shipment would carry an intracompany invoice amount of $500,000. The company was now trying to decide whether to invoice the Spanish subsidiary in U.S. dollars or European euros, and in turn, whether the resulting transaction exposure should be hedged. Ignacio's idea was to take a recent historical period of exchange rate quotes and movements and simulate the invoicing and hedging alternatives available to Navarro to try to characterize the choices.

Ignacio looked at the 90-day period that had ended the previous Friday (standard intracompany payment terms for transcontinental transactions was 90 days). The quarter had opened with a spot rate of $1.0640/€, with the 90-day forward rate quoted at $1.0615/€ the same day. The quarter had closed with a spot rate of $1.0980/€.

a. Which unit would have suffered the gain (loss) on currency exchange if intracompany sales were invoiced in U.S. dollars ($), assuming both completely unhedged and fully hedged?

b. Which unit would have suffered the gain (loss) on currency exchange if intracompany sales were invoiced in euros (€), assuming both completely unhedged and fully hedged?

18. **Korean Airlines.** Korean Airlines (KAL) has just signed a contract with Boeing to purchase two new 747-400's for a total of $60,000,000, with payment in two equal tranches. The first tranche of $30,000,000 has just been paid. The next $30,000,000 is due three months from today. KAL currently has excess cash of 25,000,000,000 won in a Seoul bank, and it is from these funds that KAL plans to make its next payment.

The current spot rate is W800/$, and permission has been obtained for a forward rate (90 days), W794/$. The 90-day eurodollar interest rate is 6.000%, while the 90-day Korean won deposit rate (there is no eurowon rate) is 5.000%. KAL can borrow in Korea at 6.250%, and can probably borrow in the U.S. dollar market at 9.375%.

A three-month call option on dollars in the over-the-counter market, for a strike price of W790/$ sells at a premium of 2.9%, payable at the time the option is purchased. A 90-day put option on dollars, also at a strike price of W790/$, sells at a premium of 1.9% (assuming a 12% volatility). KAL's foreign exchange advisory service forecasts the spot rate in three months to be W792/$.

How should KAL plan to make the payment to Boeing if KAL's goal is to maximize the amount of won cash left in the bank at the end of the 3-month period? Make a recommendation and defend it.

INTERNET EXERCISES

1. **Current Volatilities.** You wish to price your own options, but you need current volatilities on the euro, British pound, and Japanese yen. Using the following Web sites, collect spot rates and volatilities in order to price forward at-the-money put options for your option pricing analysis.

| Federal Reserve Bank of New York | www.newyorkfed.org/markets/foreignex.html |
| RatesFX.com | www.ratesfx.com/ |

2. **Hedging Objectives.** All multinational companies will state the goals and objectives of their currency risk management activities in their annual reports. Beginning with the following firms, collect samples of corporate "why hedge?" discussions for a contrast and comparison discussion.

Nestlè	www.nestle.com
Disney	www.disney.com
Nokia	www.nokia.com
BP	www.bp.com

3. **Changing Translation Practices: FASB.** The Financial Accounting Standards Board promulgates standard practices for the reporting of financial results by companies in the United States. It also, however, often leads the way in the development of new practices and emerging issues around the world. One major issue today is the valuation and reporting of financial derivatives and derivative agreements by firms. Use the FASB's home page and the Web pages of several of the major accounting firms and other interest groups around the world to see current proposed accounting standards and the current state of reaction to the proposed standards.

| FASB home page | raw.rutgers.edu/ |
| Treasury Management of NY | www.tmany.org/ |

Ganado, the same U.S.-based firm used throughout the chapter, still possesses a long £1,000,000 exposure—an account receivable—to be settled in 90 days. Exhibit 10A.1 summarizes the assumptions, exposure, and traditional option alternatives to be used throughout this appendix. The firm believes that the exchange rate will move in its favor over the 90-day period (the British pound will appreciate versus the U.S. dollar). Despite having this directional view or currency expectation, the firm wishes downside protection in the event the pound were to depreciate instead.

The exposure management zones that are of most interest to the firm are the two opposing triangles formed by the uncovered and forward rate profiles. The firm would like to retain all potential area in the upper-right triangle, but minimize its own potential exposure to the bottom-left triangle. The put option's "kinked-profile" is consistent with what the firm wishes if it believes the pound will appreciate.

EXHIBIT 10A.1 **Ganado's A/R Exposure and Put Option Hedges**

		Put Option	Strike Rate	Premium
Spot rate	$1.4790/£			
90-day forward rate	$1.4700/£	Forward ATM	$1.47	$0.0318/£
90-day euro-$ interest rate	3.250%	Out-of-the-money put	$1.44	$0.0188/£
90-day euro-£ interest rate	5.720%			
90-day $/£ volatility	11.000%			

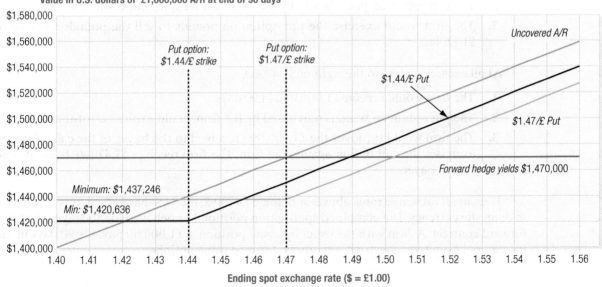

Value in U.S. dollars of £1,000,000 A/R at end of 90 days

The firm could consider any number of different put option strike prices, depending on what minimum assured value—degree of self-insurance—the firm is willing to accept. Exhibit 10A.1 illustrates two different put option alternatives: a forward-ATM put of strike price $1.4700/£, and a forward-OTM put with strike price $1.4400/£. Because foreign currency options are actually priced about the forward rate (see Chapter 8), not the spot rate, the correct specification of whether an option, put or call, is ITM, ATM, or OTM is in reference to the same maturity forward rate. The forward-OTM put provides protection at lower cost, but also at a lower level of protection.

The Synthetic Forward

At a forward rate of $1.4700/£, the proceeds of the forward contract in 90 days will yield $1,470,000. A second alternative for the firm would be to construct a *synthetic forward* using options. The synthetic forward requires the firm to combine two options, of equal size and maturity, both with strike rates at the forward rate:

1. Buy a put option on £ bought at a strike price of $1.4700/£, paying a premium of $0.0318/£

2. Sell a call option on £ at a strike price of $1.4700/£, earning a premium of $0.0318/£

The purchase of the put option requires a premium payment, and the sale of the call option earns the firm a premium payment. If both options are struck at the forward rate (*forward-ATM*), the premiums should be identical and the net premium payment should have a value of zero.

Exhibit 10A.2 illustrates the uncovered position, the basic forward rate hedge, and the individual profiles of the put and call options for the possible construction of a synthetic forward. The outcome of the combined position is easily confirmed by simply tracing what would happen at all exchange rates to the left of $1.4700/£, and what would happen to the right of $1.4700/£.

At all exchange rates to the left of $1.4700/£:

1. The firm would receive £1,000,000 in 90 days.

2. The call option on pounds sold by the firm would expire out-of-the-money.

3. The firm would exercise the put option on pounds to sell the pounds received at $1.4700/£.

At all exchange rates to the right of $1.4700/£:

1. The firm would receive £1,000,000 in 90 days.

2. The put option on pounds purchased by the firm would expire out-of-the-money.

3. The firm would turnover the £1,000,000 received to the buyer of the call, who now exercises the call option against the firm. The firm receives $1.4700/£ from the call option buyer.

Thus, at all exchange rates above or below $1.4700/£, the U.S.-based firm nets $1,470,000 in domestic currency. The combined spot-option position has behaved identically to that of a forward contract. A firm with the exact opposite position, a £1,000,00 payable 90 days in the future, could similarly construct a synthetic forward using options.[2]

---

[2] A U.S.-based firm possessing a future foreign currency denominated payment of £1 million could construct a synthetic forward hedge by (1) buying a call option on £1 million at a strike price of $1.4700/£; and (2) selling a put option on £1 million at the same strike price of $1.4700/£, when 1.47 is the forward rate.

EXHIBIT 10A.2 **Ganado's Synthetic Forward A/R Exposure Hedge**

Spot rate	$1.4790/£				
90-day forward rate	$1.4700/£	Put Option		Strike Rates	Premium
90-day euro-$ interest rate	3.250%	Forward ATM put		$1.47	$0.0318/£
90-day euro-£ interest rate	5.720%	Forward ATM call		$1.44	$0.0188/£
90-day $/£ volatility	11.000%				

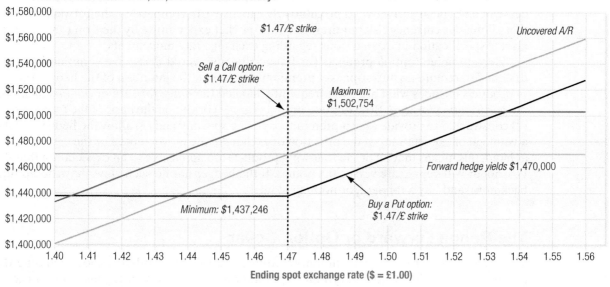

But why would a firm undertake this relatively complex position in order to simply create a forward contract? The answer is found by looking at the option premiums earned and paid. We have assumed that the option strike prices used were precisely forward-ATM rates, and the resulting option premiums paid and earned were exactly equal. But this need not be the case. If the option strike prices (remember that they must be identical for both options, bought and sold) are not precisely on the forward-ATM, the two premiums may differ by a slight amount. The net premium position may then end up as a net premium earning or a net premium payment. If positive, this amount would be added to the proceeds from the receivable to result in a higher total dollar value received than through the use of a traditional forward contract. A second possibility is that the firm, by querying a number of different financial service providers offering options, finds attractive pricing that "beats" the forward. Although this means that theoretically the options market is out of equilibrium, it happens quite frequently.

Second-Generation Currency Risk Management Products

Second-generation risk management products are constructed from the two basic derivatives used throughout this book: the forward and the option. We will subdivide them into two groups: (1) the zero-premium option products, which focus on pricing in and around the forward rate and (2) the exotic option products (for want of a better name), which focus on alternative pricing targets. Although all of the following derivatives are sold as financial products by risk management firms, we will present each as the construction of the position from common building blocks, or LEGO®s, as they have been termed, used in traditional currency risk management, forwards and options. As a group, they are collectively referred to as complex options.

Zero-Premium Option Products

The primary problem with the use of options for risk management in the eyes of the firms is the up-front premium payment. Although the premium payment is only a portion of the total payoff profile of the hedge, many firms view the expenditure of substantial funds for the purchase of a financial derivative as prohibitively expensive. In comparison, the forward contract that eliminates currency risk requires no out-of-pocket expenditure by the firm (and requires no real specification of expectations regarding exchange rate movements).

Zero-premium option products (or financially engineered derivative combinations) are designed to require no out-of-pocket premium payment at the initiation of the hedge. This set of products includes what are most frequently labeled the *range forward* or *option collar* and the *participating forward*. Both of these products are (1) priced on the basis of the forward rate; (2) constructed to provide a zero-premium payment up-front; and (3) allow the hedger to take advantage of expectations of the direction of exchange rate movements. For the case problem at hand, this means that all of the following products are applicable to an expectation that the U.S. dollar will depreciate versus the pound. If the hedger has no such view, they should turn back now (and buy a forward, or nothing at all)!

The Range Forward or Option Collar

The basic *range forward* has been marketed under a variety of other names, including the *option collar*, *flexible forward*, *cylinder option*, *option fence* or simply *fence*, *mini-max*, or *zero-cost tunnel*. Regardless of which alias it trades under, it is constructed via two steps:

1. Buying a put option with a strike rate below the forward rate, for the full amount of the long currency exposure (100% coverage)

2. Selling a call option with a strike rate above the forward rate, for the full amount of the long currency exposure (100% coverage), with the same maturity as the purchased put

The hedger chooses one side of the "range" or spread, normally the downside (put strike rate), which then dictates the strike rate at which the call option will be sold. The call option must be chosen at an equal distance from the forward rate as the put option strike price from the forward rate. The distance from the forward rate for the two strike prices should be calculated in percentage, as in ±3% from the forward rate.

If the hedger believes there is a significant possibility that the currency will move in the firm's favor, and by a sizable degree, the put floor rate may be set relatively low in order for the ceiling to be higher or further out from the forward rate and still enjoy a zero net premium. How far down the downside protection is set is a difficult issue for the firm to determine. Often the firm's treasurer will determine at what bottom exchange rate the firm would be able to recover the minimum necessary margin on the business underlying the cash flow exposure, sometimes called the budget rate.

Exhibit 10A.3 illustrates the final outcome of a range forward constructed by buying a put with strike price $1.4500/£, paying a premium of $0.0226/£, with selling a call option with strike price $1.4900/£, earning a premium of $0.0231/£. The hedger has bounded the range over which

EXHIBIT 10A.3	**Ganado's Range Forward A/R Exposure Hedge**

Spot rate	$1.4790/£			
90-day forward rate	$1.4700/£	**Put Option**	**Strike Rates**	**Premium**
90-day euro-$ interest rate	3.250%	Out-of-the-money put	$1.45	$0.0226/£
90-day euro-£ interest rate	5.720%	Out-of-the-money call	$1.49	$0.0231/£
90-day $/£ volatility	11.000%			

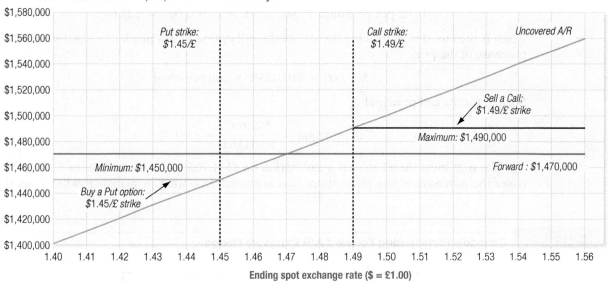

Value in U.S. dollars of £1,000,000 A/R at end of 90 days

the firm's A/R value moves as an uncovered position, with a put option floor and a sold call option ceiling. Although the put and call option premiums are in this case not identical, they are close enough to result in a near zero net premium (a premium expense of $500 in this case):

$$\text{Net premium} = (\$0.0226/£ - \$0.0231/£) \times £1,000,000 = -\$500$$

The benefits of the combined position are readily observable, given that the put option premium alone amounts to $22,600. If the strike rates of the options are selected independently of the desire for an exact zero net premium up front (still bracketing the forward rate), it is termed an *option collar* or *cylinder option*.

The Participating Forward

The *participating forward*, also called a *zero-cost ratio option* and *forward participation agreement*, is an option combination that allows the hedger to share in—or *participate*—potential upside movements while providing option-based downside protection—all at a zero net premium. The participating forward is constructed via two steps:

1. Buying a put option with a strike price below the forward rate, for the full amount of the long currency exposure (100% coverage)

2. Selling a call option with a strike price that is the same as the put option, for a portion of the total currency exposure (less than 100% coverage)

Similar to the range forward, the buyer of a participating forward will choose the put option strike rate first. Because the call option strike rate is the same as the put, all that remains is to determine the participation rate, the proportion of the exposure sold as a call option.

Exhibit 10A.4 illustrates the construction of a participating forward for the chapter problem. The firm first chooses the put option protection level, in this case $1.4500/£, with a premium of $0.0226/£. A call option sold with the same strike rate of $1.4500/£ would earn the firm $0.0425/£. The call premium is substantially higher than the put premium because the call option is already in-the-money (ITM). The firm's objective is to sell a call option only on the number of pounds needed to fund the purchase of the put option. The total put option premium is

$$\text{Total put premium} = \$0.0226/£ \times £1,000,000 = \$22,600,$$

which is then used to determine the size of the call option that is needed to exactly offset the purchase of the put:

$$\$22,600 = \$0.0425/£ \times \text{call principal}$$

Solving for the call principal:

$$\text{Call principal} = \frac{\$22,600}{\$0.0425/£} = £531,765.$$

The firm must therefore sell a call option on £531,765 with a strike rate of $1.4500/£ to cover the purchase of the put option. This mismatch in option principals is what gives the

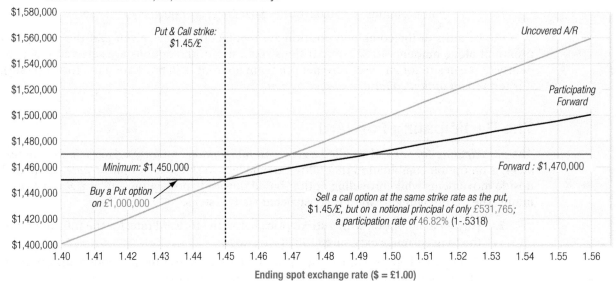

EXHIBIT 10A.4   Ganado's Participating Forward A/R Exposure Hedge

Instruments	Strike Rate	Premium	Notional Principal
Buy a put	$1.4500/£	$0.0226/£	£1,000,000
Sell a call	$1.4500/£	$0.0425/£	£531,765

Value in U.S. dollars of £1,000,000 A/R at end of 90 days

Put & Call strike: $1.45/£

Uncovered A/R

Participating Forward

Minimum: $1,450,000

Forward : $1,470,000

Buy a Put option on £1,000,000

Sell a call option at the same strike rate as the put, $1.45/£, but on a notional principal of only £531,765; a participation rate of 46.82% (1-.5318)

Ending spot exchange rate ($ = £1.00)

participating forward its unique shape. The ratio of option premiums, as well as the ratio of option principals, is termed the percent cover:

$$\text{Percent cover} = \frac{\$0.026/£}{\$0.0425/£} = \frac{£531,765}{£1,000,000} = 0.5318 = 53.18\%.$$

The *participation rate* is the residual percentage of the exposure that is not covered by the sale of the call option. For example, if the percent cover is 53.18%, the participation rate would be 1—the percent cover, or 46.82%. This means that for all favorable exchange rate movements (those above \$1.4500/£), the hedger would "participate" or enjoy 46.8% of the differential. However, like all option-based hedges, downside exposure is bounded by the put option strike rate.

The expectations of the buyer are similar to the range forward; only the degree of foreign currency bullishness is greater. For the participating forward to be superior in outcome to the range forward, it is necessary for the exchange rate to move further in the favorable direction.

Ratio Spreads

One of the older methods of obtaining a zero-premium option combination, and one of the most dangerous from a hedger's perspective, is the ratio spread. This structure leaves the hedger with a large uncovered exposure.

Let us assume that Ganado decides that it wishes to establish a floor level of protection by purchasing a \$1.4700/£ put option (forward-ATM) at a cost of \$0.0318/£ (total cost of \$31,800). This is a substantial outlay of up-front capital for the option premium, and the firm's risk management division has no budget funding for this magnitude of expenditures. The firm, feeling strongly that the dollar will depreciate against the pound, decides to "finance" the purchase of the put with the sale of an OTM call option. The firm reviews market conditions and considers a number of call option strike prices that are significantly OTM, strike prices of \$1.5200/£, \$1.5400/£, or further out.

It is decided that the \$1.5400/£ call option, with a premium of \$0.0089/£, is to be written and sold to earn the premium and finance the put purchase. However, because the premium on the OTM call is so much smaller than the forward-ATM put premium, the size of the call option written must be larger. The firm determines the amount of the call by solving the simple problem of premium equivalency as follows:

$$\text{Cost of put premium} = \text{Earnings call premium.}$$

Substituting in the put and call option premiums yields

$$\$0.0318/£ \times £1,000,000 = \$0.0089/£ \times £ \text{ call.}$$

Solving for the size of the call option to be written as follows:

$$\frac{\$31,800}{\$0.0089/£} = £3,573,034.$$

The reason that this strategy is called a *ratio spread* is that the final position, call option size to put option size, is a ratio greater than 1 (in this case, £3,573,034 ÷ £1,000,000, or a ratio of about 3.57).

The risk to the firm in the use of a ratio spread is, however, dramatic. Although unlikely, it is possible that the spot rate could move far enough by the end of the period to put the call options written in-the-money. That would mean in our example that the firm would have written an uncovered call on £2,573,034, the call option notional principal less the exposure (£3,573,034 − £1,000,000). The loss potential in covering this position is unlimited.

An alternative form of the ratio spread is the *calendar spread*. The calendar spread would combine the 90-day put option with the sale of an OTM call option with a maturity that is longer; for example, 120 or 180 days. The longer maturity of the call option written earns the firm larger premium earnings requiring a smaller "ratio." As a number of firms using this strategy have learned the hard way, however, if the expectations of the hedger prove incorrect, and the spot rate moves past the strike price of the call option written, the firm is faced with delivering a foreign currency that it does not have. In this example, if the spot rate moved above $1.5400/£, the firm would have to cover a position of £2,573,034.

The Average Rate Option

These options are normally classified as "path-dependent" currency options because their values depend on averages of spot rates over some pre-specified period of time. Here we describe two examples of path-dependent options, the *average rate option* and the *average strike option*:

1. *Average rate option* (ARO), also known as an *Asian Option*, sets the option strike rate up front, and is exercised at maturity if the average spot rate over the period (as observed by scheduled sampling) is less than the preset option strike rate.

2. *Average strike option* (ASO) establishes the option strike rate as the average of the spot rate experienced over the option's life, and is exercised if the strike rate is greater than the end of period spot rate.

The *average rate option* is difficult to depict because its value depends not on the ending spot rate, but rather the path the spot rate takes over its specified life span. For example, an average rate option with strike price $1.4700/£ would have a premium of only $0.0186/£. The average rate would be calculated by weekly observations (12 full weeks, the first observation occurring 13 days from purchase) of the spot rate. Numerous different averages or paths of spot rate movement obviously exist. A few different scenarios aid in understanding how the ARO differs in valuation.

1. The spot rate moves very little over the first 70 to 80 days of the period, with a sudden movement in the spot rate below $1.4700/£ in the days prior to expiration. Although the final spot rate is below $1.4700/£, the average for the period is above $1.4700, so the option cannot be exercised. The receivable is exchanged at the spot rate (below $1.4700/£) and the cost of the option premium is still incurred.

2. The dollar slowly and steadily depreciates versus the pound, the rate rising from $1.4790/£ to $1.48, $1.49, and on up. At the end of the 90 days the option expires out of the money, the receivable is exchanged at the favorable spot rate, and the firm has enjoyed average rate option protection at substantially lower premium expense.

A variation on the average rate is the *lookback option*, with strike and without strike. A *lookback option* with strike is a European-style option with a preset strike rate that on maturity is valued versus the highest or lowest spot rate reached over the option life. A lookback option without strike is typically a European-style option that sets the strike rate at maturity as the lowest exchange rate achieved over the period for a call option, or the highest exchange rate experienced over the period for a put option, and is exercised on the basis of this strike rate versus the ending spot rate.

A variety of differing average rate currency option products are sold by financial institutions, each having a distinct payoff structure. Because of the intricacy of the path-dependent option's value, care must be taken in the use of these instruments. As is always the case with more and more complex financial derivatives, *caveat emptor*.

CHAPTER 11

Translation Exposure

What gets measured gets managed.

—Anonymous

LEARNING OBJECTIVES

- Examine how the process of consolidation of a multinational firm's financial results creates translation exposure
- Illustrate both the theoretical and practical differences between the two primary methods of translating or remeasuring foreign currency-denominated financial statements
- Understand how translation can potentially alter the value of a multinational firm
- Explore the costs, benefits, and effectiveness of managing translation exposure

Translation exposure, the second category of accounting exposures, arises because financial statements of foreign subsidiaries—which are stated in foreign currency—must be restated in the parent's reporting currency so that the firm can prepare consolidated financial statements. Foreign subsidiaries of U.S. companies, for example, must restate foreign currency-denominated financial statements into U.S. dollars so that the foreign values can be added to the parent's U.S. dollar-denominated balance sheet and income statement. Using our example U.S. firm, Ganado, this is shown conceptually in Exhibit 11.1. This accounting process is called *translation*. *Translation exposure* is the potential for an increase or decrease in the parent's net worth and reported net income that is caused by a change in exchange rates since the last translation.

Although the main purpose of translation is to prepare consolidated financial statements, translated statements are also used by management to assess the performance of foreign subsidiaries. While such assessment by management might be performed using the local currency statements, restatement of all subsidiary statements into the single "common denominator" of one currency facilitates management comparison. This chapter reviews the predominate methods used in translation today, and concludes with the Mini-Case, *McDonald's, Hoover Hedges, and Cross-Currency Swaps*, illustrating how one major multinational manages its investment and translation risks.

EXHIBIT 11.1	**Ganado's Cross-Border Investments and Consolidation**

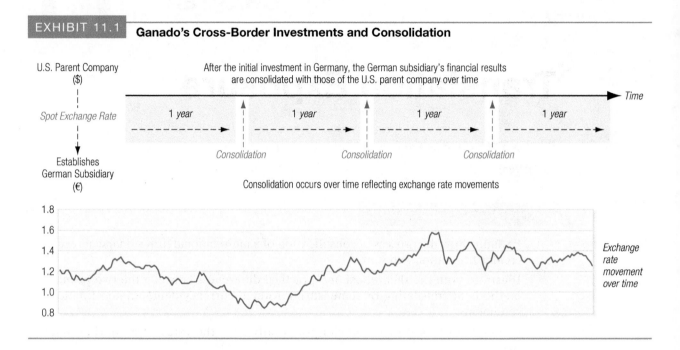

Overview of Translation

There are two financial statements for each subsidiary that must be translated for consolidation: the *income statement* and the *balance sheet*. Statements of cash flow are not translated from the foreign subsidiaries. The consolidated statement of cash flow is constructed from the consolidated statement of income and consolidated balance sheet. Because the consolidated results for any multinational firm are constructed from all of its subsidiary operations, including foreign subsidiaries, the possibility of a change in consolidated net income or consolidated net worth from period to period, as a result of a change in exchange rates, is high.

For any individual financial statement, internally, if the same exchange rate were used to remeasure each and every line item on the individual statement—the *income statement* and *balance sheet*—there would be no imbalances resulting from the remeasurement. But if a different exchange rate were used for different line items on an individual statement, an imbalance would result. Different exchange rates are used in remeasuring different line items because translation principles are a complex compromise between historical and current values. The question, then, is what is to be done with the imbalance?

Subsidiary Characterization

Most countries specify the translation method to be used by a foreign subsidiary based on its business operations. For example, a foreign subsidiary's business can be categorized as either an integrated foreign entity or a self-sustaining foreign entity. An *integrated foreign entity* is one that operates as an extension of the parent company, with cash flows and general business lines that are highly interrelated with those of the parent. A *self-sustaining foreign entity* is one that operates in the local economic environment independent of the parent company. The differentiation is important to the logic of translation. A foreign subsidiary should be valued principally in terms of the currency that is the basis of its economic viability.

It is not unusual for a single company to have both types of foreign subsidiaries, integrated and self-sustaining. For example, a U.S.-based manufacturer, which produces subassemblies in the United States that are then shipped to a Spanish subsidiary for finishing and resale in the European Union, would likely characterize the Spanish subsidiary as an integrated foreign entity. The dominant currency of economic operation is likely the U.S. dollar. That same U.S. parent may also own an agricultural marketing business in Venezuela that has few cash flows or operations related to the U.S. parent company or U.S. dollar. The Venezuelan subsidiary may source all inputs and sell all products in Venezuelan bolivar. Because the Venezuelan subsidiary's operations are independent of its parent, and its functional currency is the Venezuelan bolivar, it would be classified as a self-sustaining foreign entity.

Functional Currency

A foreign subsidiary's *functional currency* is the currency of the primary economic environment in which the subsidiary operates and in which it generates cash flows. In other words, it is the dominant currency used by that foreign subsidiary in its day-to-day operations. It is important to note that the geographic location of a foreign subsidiary and its functional currency may be different. The Singapore subsidiary of a U.S. firm may find that its functional currency is the U.S. dollar (integrated subsidiary), the Singapore dollar (self-sustaining subsidiary), or a third currency such as the British pound (also a self-sustaining subsidiary).

The United States, rather than distinguishing a foreign subsidiary as either integrated or self-sustaining, requires that the functional currency of the subsidiary be determined. Management must evaluate the nature and purpose of each of its individual foreign subsidiaries to determine the appropriate functional currency for each. If a foreign subsidiary of a U.S.-based company is determined to have the U.S. dollar as its functional currency, it is essentially an extension of the parent company (equivalent to the integrated foreign entity designation used by most countries). If, however, the functional currency of the foreign subsidiary is determined to be different from the U.S. dollar, the subsidiary is considered a separate entity from the parent (equivalent to the self-sustaining entity designation).

Translation Methods

Two basic methods for translation are employed worldwide: the *current rate method* and the *temporal method*. Regardless of which method is employed, a translation method must not only designate at what exchange rate individual balance sheet and income statement items are remeasured, but also designate where any imbalance is to be recorded, either in current income or in an equity reserve account in the balance sheet.

Current Rate Method

The *current rate method* is the most prevalent in the world today. Under this method, all financial statement line items are translated at the "current" exchange rate with few exceptions.

- **Assets and liabilities.**   All assets and liabilities are translated at the current rate of exchange; that is, at the rate of exchange in effect on the balance sheet date.
- **Income statement items.**   All items, including depreciation and cost of goods sold, are translated at either the actual exchange rate on the dates the various revenues, expenses, gains, and losses were incurred or at an appropriately weighted average exchange rate for the period.
- **Distributions.**   Dividends paid are translated at the exchange rate in effect on the date of payment.

- **Equity items.** Common stock and paid-in capital accounts are translated at historical rates. Year-end retained earnings consist of the original year-beginning retained earnings plus or minus any income or loss for the year.

Gains or losses caused by translation adjustments are not included in the calculation of consolidated net income. Rather, translation gains or losses are reported separately and accumulated in a separate equity reserve account (on the consolidated balance sheet) with a title such as "cumulative translation adjustment" (CTA), but it depends on the country. If a foreign subsidiary is later sold or liquidated, translation gains or losses of past years accumulated in the CTA account are reported as one component of the total gain or loss on sale or liquidation. The total gain or loss is reported as part of the net income or loss for the period in which the sale or liquidation occurs.

Temporal Method

Under the *temporal method*, specific assets and liabilities are translated at exchange rates consistent with the timing of the item's creation. The temporal method assumes that a number of individual line item assets, such as inventory and net plant and equipment, are restated regularly to reflect market value. If these items were not restated, but were instead carried at historical cost, the temporal method becomes the monetary/nonmonetary method of translation, a form of translation that is still used by a number of countries today. Line items include the following:

- **Monetary assets** (primarily cash, marketable securities, accounts receivable, and long-term receivables) **and monetary liabilities** (primarily current liabilities and long-term debt). These are translated at current exchange rates. Nonmonetary assets and liabilities (primarily inventory and fixed assets) are translated at historical rates.

- **Income statement items.** These are translated at the average exchange rate for the period, except for items such as depreciation and cost of goods sold that are directly associated with nonmonetary assets or liabilities. These accounts are translated at their historical rate.

- **Distributions.** Dividends paid are translated at the exchange rate in effect on the date of payment.

- **Equity items.** Common stock and paid-in capital accounts are translated at historical rates. Year-end retained earnings consist of the original year-beginning retained earnings plus or minus any income or loss for the year, plus or minus any imbalance from translation.

Under the temporal method, gains or losses resulting from remeasurement are carried directly to current consolidated income, and not to equity reserves. Hence, foreign exchange gains and losses arising from the translation process do introduce volatility to consolidated earnings.

U.S. Translation Procedures

The United States differentiates foreign subsidiaries based on functional currency, not subsidiary characterization. A note on terminology: Under U.S. accounting and translation practices, use of the current rate method is termed "translation" while use of the temporal method is termed "remeasurement." The primary principles of U.S. translation are summarized as follows:

- If the financial statements of the foreign subsidiary of a U.S. company are maintained in U.S. dollars, translation is not required.

- If the financial statements of the foreign subsidiary are maintained in the local currency and the local currency is the functional currency, they are translated by the current rate method.

- If the financial statements of the foreign subsidiary are maintained in the local currency and the U.S. dollar is the functional currency, they are remeasured by the temporal method.

- If the financial statements of foreign subsidiaries are in the local currency and neither the local currency nor the dollar is the functional currency, then the statements must first be remeasured into the functional currency by the temporal method, and then translated into dollars by the current rate method.

- U.S. translation practices, summarized in Exhibit 11.2, have a special provision for translating statements of foreign subsidiaries operating in hyperinflation countries. These are countries where cumulative inflation has been 100% or more over a three-year period. In this case, the subsidiary must use the temporal method.

A final note: The selection of the functional currency is determined by the economic realities of the subsidiary's operations, and is not a discretionary management decision on preferred procedures or elective outcomes. Since many U.S.-based multinationals have numerous foreign subsidiaries, some dollar-functional and some foreign currency-functional, currency gains and losses may be passing through both current consolidated income and/or accruing in equity reserves.

International Translation Practices

Many of the world's largest industrial countries use International Accounting Standards Committee (IASC), and therefore the same basic translation procedure. A foreign subsidiary is an integrated foreign entity or a self-sustaining foreign entity; integrated foreign entities are typically remeasured using the temporal method (or some slight variation thereof); and self-sustaining foreign entities are translated at the current rate method, also termed the closing-rate method.

EXHIBIT 11.2   **Flow Chart for U.S. Translation Practices**

Purpose: *Foreign currency financial statements must be translated into U.S. dollars*

If the financial statements of the foreign subsidiary are expressed in a foreign currency, the following determinations need to be made.

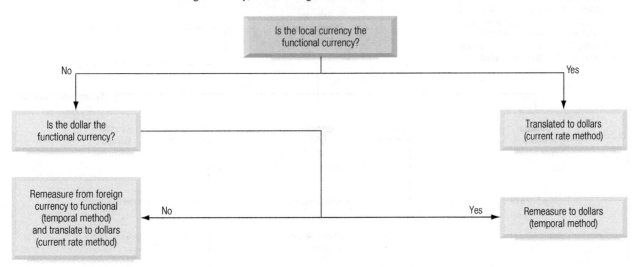

* The term "remeasure" means to translate, as to change the unit of measure, from a foreign currency to the functional currency.

Ganado Corporation's Translation Exposure

Ganado Corporation, first introduced in Chapter 1 and shown in Exhibit 11.3, is a U.S.-based corporation with a U.S. business unit as well as foreign subsidiaries in both Europe and China. The company is publicly traded and its shares are traded on the New York Stock Exchange (NYSE).

Each subsidiary of Ganado—the United States, Europe, and China—will have its own set of financial statements. Each set of financials will be constructed in the local currency (renminbi, dollar, euro), but the subsidiary income statements and balance sheets will also be translated into U.S. dollars, the reporting currency of the company for consolidation and reporting. As a U.S.-based corporation whose shares are traded on the NYSE, Ganado will report all of its final results in U.S. dollars.

Translation Exposure: Income

Ganado Corporation's sales and earnings by operating unit for 2009 and 2010 are described in Exhibit 11.4.

- **Consolidated sales.**   For 2010, the company generated $300 million in sales in its U.S. unit, $158.4 million in its European subsidiary (€120 million at $1.32/€), and $89.6 million in its Chinese subsidiary (Rmb600 million at Rmb6.70/$). Total global sales for 2010 were $548.0 million. This constituted sales growth of 2.8% over 2009.

- **Consolidated earnings.**   The company's earnings (profits) fell in 2010, dropping to $53.1 million from $53.2 million in 2009. Although not a large fall, Wall Street would not react favorably to a fall in consolidated earnings.

EXHIBIT 11.3   **Ganado Corporation: A U.S. Multinational**

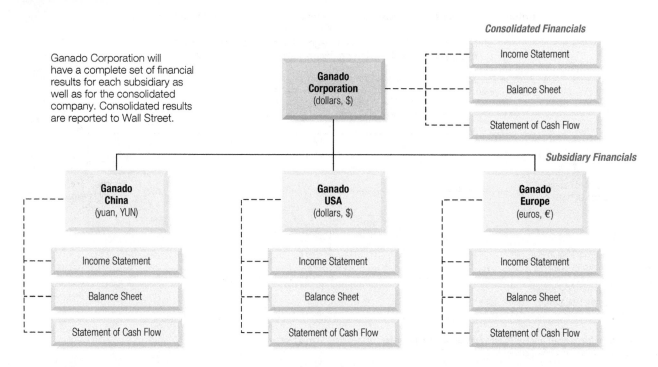

EXHIBIT 11.4	Ganado Corporation, Selected Financial Results, 2009–2010

	Sales (millions, local currency)			Average Exchange Rate ($/€ and YUN/$)			Sales (millions of US$)		
	2009	2010	% Change	2009	2010	% Change	2009	2010	% Change
United States	$280	$300	7.1%	—	—		$280.0	$300.0	7.1%
Europe	€118	€120	1.7%	1.4000	1.3200	−5.71%	$165.2	$158.4	−4.1%
China	YUN 600	YUN 600	0.0%	6.8300	6.7000	1.94%	$87.8	$89.6	1.9%
Total							$533.0	$548.0	2.8%

	Earnings (millions, local currency)			Average Exchange Rate ($/€ and YUN/$)			Earnings (millions of US$)		
	2009	2010	% Change	2009	2010	% Change	2009	2010	% Change
United States	$28.2	$28.6	1.4%	—	—		$28.2	$28.6	1.4%
Europe	€10.4	€10.5	1.0%	1.4000	1.3200	−5.71%	$14.6	$13.9	−4.8%
China	YUN 71.4	YUN 71.4	0.0%	6.8300	6.7000	1.94%	$10.5	$10.7	1.9%
Total							$53.2	$53.1	−0.2%

A closer look at the sales and earnings by country, however, yields some interesting insights. Sales and earnings in the U.S. unit rose, sales growing 7.1% and earnings growing 1.4%. Since the U.S. unit makes up more than half of the total company's sales and profits, this is very important. The Chinese subsidiary's sales and earnings were identical in 2009 and 2010 when measured in local currency, Chinese renminbi. The Chinese renminbi, however, was revalued against the U.S. dollar by the Chinese government, from Rmb6.83/$ to Rmb6.70/$. The result was an increase in the dollar value of both Chinese sales and profits.

The European subsidiary's financial results are even more striking. Sales and earnings in Europe in euros grew from 2009 to 2010. Sales grew 1.7% while earnings increased 1.0%. But the euro depreciated against the dollar, falling from $1.40/€ to $1.32/€. This depreciation of 5.7% resulted in the financial results of European operations falling in dollar terms. As a result, Ganado's consolidated earnings, as reported dollars, fell in 2010. One can imagine the discussion and debate within Ganado, and among the analysts who follow the firm, over the fall in earnings reported to Wall Street.

Translation Exposure: Balance Sheet

Let us continue the example of Ganado, focusing here on the balance sheet of its European subsidiary. We will illustrate translation by both the temporal method and the current rate method, to show the arbitrary nature of a translation gain or loss. The functional currency of Ganado Europe is the euro, and the reporting currency of its parent, Ganado Corporation, is the U.S. dollar.

Our analysis assumes that plant and equipment and long-term debt were acquired, and common stock issued, by Ganado Europe sometime in the past when the exchange rate was $1.2760/€. Inventory currently on hand was purchased or manufactured during the immediately prior quarter when the average exchange rate was $1.2180/€. At the close of business on Monday, December 31, 2010, the current spot exchange rate was $1.2000/€. When business reopened on January 3, 2011, after the New Year holiday, the euro had dropped in value versus the dollar to $1.0000/€.

Current Rate Method. Exhibit 11.5 illustrates translation loss using the current rate method. Assets and liabilities on the pre-depreciation balance sheet are translated at the current exchange rate of $1.2000/€. Capital stock is translated at the historical rate of $1.2760/€, and retained earnings are translated at a composite rate that is equivalent to having each past year's addition to retained earnings translated at the exchange rate in effect that year.

The sum of retained earnings and the CTA account must "balance" the liabilities and net worth section of the balance sheet with the asset side. For this example, we have assumed the two amounts used for the December 31 balance sheet. As shown in Exhibit 11.5, the "just before depreciation" dollar translation reports an accumulated translation loss from prior periods of $136,800. This balance is the cumulative gain or loss from translating euro statements into dollars in prior years.

After the depreciation, Ganado Corporation translates assets and liabilities at the new exchange rate of $1.0000/€. Equity accounts, including retained earnings, are translated just as they were before depreciation, and as a result, the cumulative translation loss increases to $1,736,800. The increase of $1,600,000 in this account (from a cumulative loss of $136,800 to a new cumulative loss of $1,736,800) is the translation loss measured by the current rate method.

This translation loss is a decrease in equity, measured in the parent's reporting currency, of "net exposed assets." An exposed asset is an asset whose value drops with the depreciation of the functional currency and rises with an appreciation of that currency. Net exposed assets in this context are exposed assets minus exposed liabilities. Net exposed assets are positive ("long") if exposed assets exceed exposed liabilities. They are negative ("short") if exposed assets are less than exposed liabilities.

EXHIBIT 11.5   Ganado Europe's Translation Loss after Depreciation of the Euro: Current Rate Method

		December 31, 2010		January 2, 2011	
Assets	**In Euros (€)**	**Exchange Rate (US$/euro)**	**Translated Accounts (US$)**	**Exchange Rate (US$/euro)**	**Translated Accounts (US$)**
Cash	1,600,000	1.2000	$ 1,920,000	1.0000	$ 1,600,000
Accounts receivable	3,200,000	1.2000	3,840,000	1.0000	3,200,000
Inventory	2,400,000	1.2000	2,880,000	1.0000	2,400,000
Net plant & equipment	4,800,000	1.2000	5,760,000	1.0000	4,800,000
Total	12,000,000		$14,400,000		$12,000,000
Liabilities & Net Worth					
Accounts payable	800,000	1.2000	$ 960,000	1.0000	$800,000
Short-term bank debt	1,600,000	1.2000	1,920,000	1.0000	1,600,000
Long-term debt	1,600,000	1.2000	1,920,000	1.0000	1,600,000
Common stock	1,800,000	1.2760	2,296,800	1.2760	2,296,800
Retained earnings	6,200,000	1.2000 (a)	7,440,000	1.2000 (b)	7,440,000
Translation adjustment (CTA)	—		$ (136,800)		$ (1,736,800)
Total	12,000,000		$14,400,000		$12,000,000

(a) Dollar retained earnings before depreciation are the cumulative sum of additions to retained earnings of all prior years, translated at exchange rates in each year.
(b) Translated into dollars at the same rate as before depreciation of the euro.

Temporal Method. Translation of the same accounts under the temporal method shows the arbitrary nature of any gain or loss from translation. This is illustrated in Exhibit 11.6. Monetary assets and monetary liabilities in the pre-depreciation euro balance sheet are translated at the current rate of exchange, but other assets and the equity accounts are translated at their historic rates. For Ganado Europe, the historical rate for inventory differs from that for net plant and equipment because inventory was acquired more recently.

Under the temporal method, translation losses are not accumulated in a separate equity account but passed directly through each quarter's income statement. Thus, in the dollar balance sheet translated before depreciation, retained earnings were the cumulative result of earnings from all prior years translated at historical rates in effect each year, plus translation gains or losses from all prior years. In Exhibit 11.6, no translation loss appears in the pre-depreciation dollar balance sheet because any losses would have been closed to retained earnings.

The effect of the depreciation is to create an immediate translation loss of $160,000. This amount is shown as a separate line item in Exhibit 11.6 to focus attention on it for this example. Under the temporal method, this translation loss of $160,000 would pass through the income statement, reducing reported net income and reducing retained earnings. Ending retained earnings would, in fact, be $7,711,200 minus $160,000, or $7,551,200. Whether gains and losses pass through the income statement under the temporal method depends upon the country.

EXHIBIT 11.6   **Ganado Europe's Translation Loss after Depreciation of the Euro: Temporal Method**

		December 31, 2010		January 2, 2011	
Assets	**In Euros (€)**	**Exchange Rate (US$/euro)**	**Translated Accounts (US$)**	**Exchange Rate (US$/euro)**	**Translated Accounts (US$)**
Cash	1,600,000	1.2000	$ 1,920,000	1.0000	$ 1,600,000
Accounts receivable	3,200,000	1.2000	3,840,000	1.0000	3,200,000
Inventory	2,400,000	1.2180	2,923,200	1.2180	2,923,200
Net plant & equipment	4,800,000	1.2760	6,124,800	1.2760	6,124,800
Total	12,000,000		$14,808,000		$13,848,000
Liabilities & Net Worth					
Accounts payable	800,000	1.2000	$960,000	1.0000	$800,000
Short-term bank debt	1,600,000	1.2000	1,920,000	1.0000	1,600,000
Long-term debt	1,600,000	1.2000	1,920,000	1.0000	1,600,000
Common stock	1,800,000	1.2760	2,296,800	1.2760	2,296,800
Retained earnings	6,200,000	1.2437 (a)	7,711,200	1.2437 (b)	7,711,200
Translation gain (loss)	—			(c)	$ (160,000)
Total	12,000,000		$14,808,000		$13,848,000

(a) Dollar retained earnings before depreciation are the cumulative sum of additions to retained earnings of all prior years, translated at exchange rates in each year.
(b) Translated into dollars at the same rate as before depreciation of the euro.
(c) Under the temporal method, the translation loss of $160,000 would be closed into retained earnings through the income statement rather than left as a separate line item as shown here. Ending retained earnings would actually be $7,711,200 − $160,000 = $7,551,200.

Foreign Subsidiary Valuation

The value contribution of a subsidiary of a multinational firm to the firm as a whole is a topic of increasing debate in global financial management. Most multinational companies report the earnings contribution of foreign operations either individually or by region when they are significant to the total earnings of the consolidated firm. Changes in the value of a subsidiary as a result of the change in an exchange rate can be decomposed into those changes specific to the income and the assets of the subsidiary.

Subsidiary Earnings

The earnings of the subsidiary, once remeasured into the home currency of the parent company, contributes directly to the consolidated income of the firm. An exchange rate change results in fluctuations in the value of the subsidiary's income to the global corporation. If the individual subsidiary in question constitutes a relatively significant or material component of consolidated income, the multinational firm's reported income (and earnings per share, EPS) may be seen to change purely as a result of translation.

Subsidiary Assets

Changes in the reporting currency value of the net assets of the subsidiary are passed into consolidated income or equity. If the foreign subsidiary was designated as "dollar functional," remeasurement results in a transaction exposure, which is passed through current consolidated income. If the foreign subsidiary was designated as "local currency functional," translation results in a translation adjustment and is reported in consolidated equity as a translation adjustment. It does not alter reported consolidated net income.

In the case of Ganado, the translation gain or loss is larger under the current rate method because inventory and net property, plant, and equipment, as well as all monetary assets, are deemed exposed. When net exposed assets are larger, gains or losses from translation are also larger. If management expects a foreign currency to depreciate, it could minimize translation exposure by reducing net exposed assets. If management anticipates an appreciation of the foreign currency, it should increase net exposed assets to benefit from a gain.

Depending on the accounting method, management might select different assets and liabilities for reduction or increase. Thus, "real" decisions about investing and financing might be dictated by which accounting technique is used, when in fact, accounting impacts should be neutral.

As illustrated in *Global Finance in Practice 11.1*, transaction, translation, and operating exposures can become intertwined in the valuation of business units—in this case, the valuation of a foreign subsidiary.

Managing Translation Exposure

> *"Covering P&L translation risk is more complex to hedge and therefore not done by corporates to the same extent as transactional risk,"* says Francois Masquelier, chairman of the Association of Corporate Treasurers of Luxembourg. *"Of course, reported earnings can have positive or negative effects depending on what the currency does vis-à-vis your functional currency. If you have losses in the US then it can reduce those losses (when USD is weaker versus EUR), but if you have profit it can reduce that contribution to the earnings before interest, tax, depreciation and amortisation and therefore your net profit."*
>
> —"Translation risk hits corporate earnings," *FX Week*, 09 May 2014.

The main technique to minimize translation exposure is called a balance sheet hedge. At times, some firms have attempted to hedge translation exposure in the forward market. Such action amounts to speculating in the forward market in the hope that a cash profit will be realized to offset the noncash loss from translation. Success depends on a precise prediction of future exchange rates, for such a hedge will not work over a range of possible future spot rates. In addition, the profit from the forward "hedge" (i.e., speculation) is taxable, but the translation loss does not reduce taxable income.

Balance Sheet Hedge

A balance sheet hedge requires an equal amount of exposed foreign currency assets and liabilities on a firm's consolidated balance sheet. If this can be achieved for each foreign currency, net translation exposure will be zero. A change in exchange rates will change the value of exposed liabilities in an equal amount but in a direction opposite to the change in value of exposed assets. If a firm translates by the temporal method, a zero net exposed position is called "monetary balance." Complete monetary balance cannot be achieved under the current rate method because total assets would have to be matched by an equal amount of debt, but the equity section of the balance sheet must still be translated at historic exchange rates.

The cost of a balance sheet hedge depends on relative borrowing costs. If foreign currency borrowing costs, after adjusting for foreign exchange risk, are higher than parent currency borrowing costs, the balance sheet hedge is costly, and vice versa. Normal operations, however, already require decisions about the magnitude and currency denomination of specific balance sheet accounts. Thus, balance sheet hedges are a compromise in which the denomination of balance sheet accounts is altered, perhaps at a cost in terms of interest expense or operating efficiency, in order to achieve some degree of foreign exchange protection.

To achieve a balance sheet hedge, Ganado Corporation must either (1) reduce exposed euro assets without simultaneously reducing euro liabilities, or (2) increase euro liabilities without simultaneously increasing euro assets. One way to achieve this is to exchange existing euro cash for dollars. If Ganado Europe does not have large euro cash balances, it can borrow euros and exchange the borrowed euros for dollars. Another subsidiary could also borrow euros and exchange them for dollars. That is, the essence of the hedge is for the parent or any of its subsidiaries to create euro debt and exchange the proceeds for dollars.

Current Rate Method. Under the current rate method, Ganado should borrow as much as €8,000,000. The initial effect of this first step is to increase both an exposed asset (cash) and an exposed liability (notes payable) on the balance sheet of Ganado Europe, with no immediate effect on net exposed assets. The required follow-up step can take two forms: (1) Ganado Europe could exchange the acquired euros for U.S. dollars and hold those dollars itself, or (2) it could transfer the borrowed euros to Ganado Corporation, perhaps as a euro dividend or as repayment of intracompany debt. Ganado Corporation could then exchange the euros for dollars. In some countries, local monetary authorities will not allow their currency to be freely exchanged.

An alternative would be for Ganado Corporation or a sister subsidiary to borrow the euros, thus keeping the euro debt entirely off Ganado's books. However, the second step is still essential to eliminate euro exposure; the borrowing entity must exchange the euros for dollars or other unexposed assets. Any such borrowing should be coordinated with all other euro borrowings to avoid the possibility that one subsidiary is borrowing euros to reduce translation exposure at the same time as another subsidiary is repaying euro debt. (Note that euros can be "borrowed," by simply delaying repayment of existing euro debt; the goal is to increase euro debt, not to borrow in a literal sense.)

Temporal Method. If translation is by the temporal method, the much smaller amount of only €800,000 need be borrowed. As before, Ganado Europe could use the proceeds of the loan to acquire U.S. dollars. However, Ganado Europe could also use the proceeds to acquire inventory or fixed assets in Europe. Under the temporal method, these assets are not regarded as exposed and do not drop in dollar value when the euro depreciates.

When Is a Balance Sheet Hedge Justified?

If a firm's subsidiary is using the local currency as the functional currency, the following circumstances could justify when to use a balance sheet hedge:

- The foreign subsidiary is about to be liquidated, so that value of its CTA would be realized.
- The firm has debt covenants or bank agreements that state the firm's debt/equity ratios will be maintained within specific limits.
- Management is evaluated based on certain income statement and balance sheet measures that are affected by translation losses or gains.
- The foreign subsidiary is operating in a hyperinflationary environment.

If a firm is using the parent's home currency as the functional currency of the foreign subsidiary, all transaction gains/losses are passed through to the income statement. Hedging this consolidated income to reduce its variability may be important to investors and bond rating agencies. In the end, accounting exposure is a topic of great concern and complex choices for all multinationals. As demonstrated by *Global Finance in Practice 11.2*, despite the best of intentions and structures, business itself may dictate hedging outcomes.

GLOBAL FINANCE IN PRACTICE 11.2

When Business Dictates Hedging Results

GM Asia, a regional subsidiary of GM Corporation, U.S., held major corporate interests in a variety of countries and companies, including Daewoo Auto of South Korea. GM had acquired control of Daewoo's automobile operations in 2001. The following years had been very good for the Daewoo unit, and by 2009, GM Daewoo was selling automobile components and vehicles to more than 100 countries.

Daewoo's success meant that it had expected sales (receivables) from buyers all over the world. What was even more remarkable was that the global automobile industry now used the U.S. dollar more than ever as its currency of contract for cross-border transactions. This meant that Daewoo did not really have dozens of foreign currencies to manage, just one, the U.S. dollar. So Daewoo of Korea had, in late 2007 and early 2008, entered into a series of forward exchange contracts. These currency contracts locked in the Korean won value of the many dollar-denominated receivables the

company expected to receive from international automobile sales in the coming year. In the eyes of many, this was a conservative and responsible currency hedging policy; that is, until the global financial crisis and the following global collapse of automobile sales.

The problem for Daewoo was not that the Korean won per U.S. dollar exchange rate had moved dramatically; it had not. The problem was that Daewoo's sales, like all other automobile industry participants, had collapsed. The sales had not taken place, and therefore the underlying exposures, the expected receivables in dollars by Daewoo, had not happened. But GM still had to contractually deliver on the forward contracts. It would cost GM Daewoo Won 2,300 billion. GM's Daewoo unit was now broke, its equity wiped out by currency hedging gone bad. GM Asia needed money, quickly, and selling interests in its highly successful Chinese and Indian businesses was the only solution.

SUMMARY POINTS

- Translation exposure results from translating foreign currency-denominated statements of foreign subsidiaries into the parent's reporting currency to prepare consolidated financial statements.

- A foreign subsidiary's functional currency is the currency of the primary economic environment in which the subsidiary operates and in which it generates cash flows. In other words, it is the dominant currency used by that foreign subsidiary in its day-to-day operations.

- Technical aspects of translation include questions about when to recognize gains or losses, the distinction between functional and reporting currency, and the treatment of subsidiaries in hyperinflation countries.

- Translation gains and losses can be quite different from operating gains and losses, not only in magnitude but also in sign. Management may need to determine which is of greater significance prior to deciding which exposure is to be managed first.

- The main technique for managing translation exposure is a balance sheet hedge. This calls for having an equal amount of exposed foreign currency assets and liabilities.

- Even if management chooses to follow an active policy of hedging translation exposure, it is nearly impossible to offset both transaction and translation exposure simultaneously. If forced to choose, most managers will protect against transaction losses because they impact consolidated earnings.

McDonald's, Hoover Hedges, and Cross-Currency Swaps[1]

McDonald's Corporation (NYSE: MCD) is one of the world's most well known and valuable brands. But as McDonald's has grown and expanded globally, so have the investment risks associated with is investment in more than 100 countries. Like most multinational firms, it considers its equity investment in foreign affiliates capital at risk—risk of loss, nationalization, and currency valuation. McDonald's has been quite innovative in its hedging of these combined currency risks over time, finding new ways to construct old solutions—*Hoover Hedges*—but doing so with cross-currency swaps.

Hoover Hedges

A multinational firm that establishes a foreign subsidiary puts capital at risk, a long-time fundamental of international business. Financially, when the parent company creates and invests in a foreign subsidiary it creates an asset, its foreign investment in a foreign subsidiary, which corresponds to the equity investment on the balance sheet of the foreign subsidiary. But the equity investment in the foreign subsidiary is now in local currency, the currency of the foreign business environment. If this is the predominant currency of this subsidiary's business, it is termed the *functional currency* of the subsidiary. Going forward, as the exchange rate between the two country currencies changes, the parent company's equity investment is subject to foreign exchange risk.

Many multinationals have attempted to hedge this equity investment exposure with what can be described as a *balance sheet hedge*. Since the parent company possesses a long-term asset in the foreign currency, the company tries to hedge this by creating a matching long-term liability in the same currency. A long-term loan in the currency of the foreign subsidiary has typically been used. The loan itself is often structured as a bullet repayment loan, in which interest payments are made over time but the entire principal is due in a single final payment at maturity. In this way, the principal on the long-term loan acts as a match to the long-term equity investment.

These hedges are typically referred to as *Hoover Hedges* following the court case of Hoover Company (a vacuum cleaner manufacturer) versus the U.S. Internal Revenue Service[2]. The primary issue in the case was whether the gains and losses from short sales in foreign currency that the Hoover Company used as hedges were to be considered ordinary losses, business expenses, or capital losses and gains, for tax purposes. Although borrowing in the local currency is frequently used, there are a number of other potential hedges of equity investments including short sales and the use of traditional foreign currency derivatives like forward contracts and currency options.

McDonald's Business Forms

McDonald's has structured its business in a variety of different ways depending on marketplace. In the United States the company has utilized a franchising structure where it awards a franchise to a private investor. That investor then has exclusive rights over the sale and distribution of McDonald's products and services within the designated franchise zone. McDonald's corporation will own the land and building, but the franchisee is responsible for the investment in all equipment and furnishings required for the restaurant under the franchise agreement—*from the paint-in*—as they describe it. This structure allows McDonald's to expand with a lower level of capital investment (the franchisee is investing a significant portion), and at the same time create a financial incentive for the franchisee to remain focused and committed to the restaurant's success and profitability. In return McDonald's earns a royalty from the franchise's sales, typically 5% to 5.5% of sales.

Alternatively, in markets in which the company wishes more direct control, and is willing to make substantially larger capital investments itself, it uses the more common form of direct ownership. Although having to put up all the capital needed for the establishment of the business, it gains more direct control over operations. Much of McDonald's international expansion has been structured under this more common direct ownership approach, but at the risk of substantial amounts of capital as the company sought to gain a major presence in a growing number of countries.

The British Subsidiary and Currency Exposure

In the United Kingdom McDonald's owns the majority of its restaurants. These investments create three different British pound-denominated currency exposures for the parent company.

1. The British subsidiary has equity capital, which is a British pound-denominated asset of the parent company.

2. The parent company provides intracompany debt in the form of a four-year loan. The loan is denominated in British pounds, and carries a fixed rate of interest.

3. The British subsidiary pays a fixed percentage of gross sales in royalties to the parent company. This too is pound-denominated.

An additional technical detail further complicates the situation. When the parent company makes an intracompany loan to the British subsidiary, it must *designate*—according to U.S. accounting and tax law practices—whether the loan is considered to be "permanently invested" in that country. Although on the surface it seems illogical to consider four years permanent, the loan itself could simply be continually rolled over by the parent company and never actually be repaid.

If the loan was not considered permanent, the foreign exchange gains and losses related to the loan flow directly to the parent company's income statement, according to Financial Accounting Standard #52, the primary standard for U.S. foreign currency reporting. If, however, the loan is designated as permanent, the foreign exchange gains and losses related to the intracompany loan would flow only to the cumulative translation adjustment account (CTA), a segment of consolidated equity on the company's consolidated balance sheet. To date, McDonald's has chosen to designate the loan as permanent. The functional currency of the British subsidiary for consolidation purposes is the local currency, the British pound.

Cross-Currency Swap Hedging

Anka Gopi is an assistant manager in Treasury—and a McDonald's shareholder. She is currently reviewing the existing hedging strategy employed by McDonald's against the pound exposures.

McDonald's has been hedging the rather complex British pound exposure by entering into a cross-currency U.S. dollar—British pound sterling cross-currency swap. The current swap is a seven-year swap to receive dollars and pay pounds. Like all cross-currency swaps, the agreement requires McDonald's (U.S.) to make regular pound-denominated interest payments and a bullet principal repayment (notional principal) at the end of the swap agreement.

Exhibit A provides a brief map of how the cross-currency swap strategy works. The cross-currency swap

---

[2]*The Hoover Company, Petitioner v. Commissioner of Internal Revenue, Respondent*, 72 T.C. 206 (1979). United States Tax Court, Filed April 24, 1979.

EXHIBIT A	McDonald's Cross-Currency Swap Strategy for the U.K.

Because the British subsidiary makes all payments to the U.S. parent company in British pounds, McDonald's U.S. is long British pounds. By entering into a swap to *pay pounds* (£) and *receive dollars* ($), the swap creates an outflow of £ serviced by the $ inflows. But the cross-currency swap has one additional major feature useful to McDonald's: the cross-currency swap has a large principal which is outstanding (bullet repayment) which acts as *a counterweight–a match*–to the long-term investment in the U.K. subsidiary.

serves as a hedge of both the regular royalty and interest payments in British pounds made to the U.S. parent, and the outstanding swap notional principal in British pounds serves as a hedge of the equity investment by McDonald's U.S. in the British subsidiary. According to accounting practice, a company may elect to take the interest associated with a foreign currency-denominated loan and carry that directly to the parent company's consolidated income. This had been done in the past and McDonald's had benefitted from the inclusion.

Issues for Discussion

One of Anka's concerns is that under FAS #133, Accounting for Derivative Instruments and Hedging Activities, the firm has to mark-to-market the entire cross-currency swap position, including principal, and carry this to *other comprehensive income* (OCI). This has proven a bit troublesome in the past because cross-currency swaps are subject to so much volatility in value when marked-to-market, a direct result of the large notional principal bullet repayment feature they typically carry.

Anka wondered how important OCI was to investors. OCI was a measure of "below the line income," income required under U.S. GAAP and reported in the footnotes to the financial statements. It was below net income (and therefore below earnings and earnings per share as reported to the markets), and included a variety of adjustments arising from consolidated equity (such as these gains and losses associated with hedging instruments and positions).

Anka Gopi wished to reconsider the current hedging strategy. She begins by listing the pros and cons of the current strategy, comparing these to alternative strategies, and then deciding what if anything should be done about it at this time.

Mini-Case Questions

1. How does the cross-currency swap effectively hedge the three primary exposures McDonald's has relative to its British subsidiary?
2. How does the cross-currency swap hedge the long-term equity position in the foreign subsidiary?
3. To what degree, if at all, Should Anka—and McDonald's—worry about OCI?

QUESTIONS

These questions are available in MyFinanceLab.

1. **Translation.** What does the word translation mean? Why is translation exposure called an accounting exposure?

2. **Causation.** What activity gives rise to translation exposure?

3. **Converting Financial Assets.** In the context of preparing consolidated financial statements, are the words translate and convert synonyms?

4. **Subsidiary Characterization.** What is the difference between a self-sustaining foreign subsidiary and an integrated foreign subsidiary?

5. **Functional Currency.** What is a functional currency? What do you think a "non-functional currency" would be?

6. **Functional Currency Designation.** Can or should a company change the functional currency designation of a foreign subsidiary from year to year? If so, when would it be justified?

7. **Translation Methods.** What are the two basic methods for translation used globally?

8. **Current Versus Historical.** One of the major differences between translation methods is which balance sheet components are translated at which exchange rates, current or historical. Why would accounting practices ever use historical exchange rates?

9. **Translating Assets.** What are the major differences in translating assets between the current rate method and the temporal method?

10. **Translating Liabilities.** What are the major differences in translating liabilities between the current rate method and the temporal method?

11. **Earnings or Equity.** Where do you think that most companies would prefer currency translation imbalances or adjustments to go—earnings or consolidated equity? Why?

12. **Translation Exposure Management.** What are the primary options firms have to manage translation exposure?

13. **Accounting or Cash Flow.** A U.S.-based multinational company generates more than 80% of its profits (earnings) outside the U.S. in the euro zone and Japan, and both the euro and the yen fall significantly in value versus the dollar as occurred in the second half of 2014. Is the impact on the firm only accounting or does it alter cash flow, or both?

14. **Balance Sheet Hedge Justification.** When is a balance sheet hedge justified?

15. **Realization and Recognition.** When would a multinational firm, if ever, realize and recognize the cumulative translation losses recorded over time associated with a subsidiary?

16. **Tax Obligations.** How does translation alter the global tax liabilities of a firm? If a multinational firm's consolidated earnings increase as a result of consolidation and translation, what is the impact on tax liabilities?

17. **Hyperinflation.** What is hyperinflation and what are the consequences for translating foreign financial statements in countries experiencing hyperinflation?

18. **Transaction Versus Translation Losses.** What are the main differences between losses from transaction exposure and translation exposure?

PROBLEMS

These problems are available in MyFinanceLab.

1. **Ganado Europe (A).** Using facts in the chapter for Ganado Europe, assume the exchange rate on January 2, 2011, in Exhibit 11.5 dropped in value from $1.2000/€ to $0.9000/€ (rather than to $1.0000/€). Recalculate Ganado Europe's translated balance sheet for January 2, 2011, with the new exchange rate using the current rate method.
   a. What is the amount of translation gain or loss?
   b. Where should it appear in the financial statements?

2. **Ganado Europe (B).** Using facts in the chapter for Ganado Europe, assume as in Problem 1 that the exchange rate on January 2, 2011, in Exhibit 11.6 dropped in value from $1.2000/€ to $0.9000/€ (rather than to $1.0000/€). Recalculate Ganado Europe's translated balance sheet for January 2, 2011, with the new exchange rate using the temporal rate method.
   a. What is the amount of translation gain or loss?
   b. Where should it appear in the financial statements?
   c. Why does the translation loss or gain under the temporal method differ from the loss or gain under the current rate method?

3. **Ganado Europe (C).** Using facts in the chapter for Ganado Europe, assume the exchange rate on January 2, 2011, in Exhibit 11.5 appreciated from $1.2000/€ to $1.500/€. Calculate Ganado Europe's translated balance sheet for January 2, 2011, with the new exchange rate using the current rate method.
   a. What is the amount of translation gain or loss?
   b. Where should it appear in the financial statements?

4. **Ganado Europe (D).** Using facts in the chapter for Ganado Europe, assume as in Problem 3 that the exchange rate on January 2, 2011, in Exhibit 11.6 appreciated from $1.2000/€ to $1.5000/€. Calculate Ganado Europe's translated balance sheet for January 2, 2011, with the new exchange rate using the temporal method.
   a. What is the amount of translation gain or loss?
   b. Where should it appear in the financial statements?

5. **Tristan Narvaja, S.A. (A).** Tristan Narvaja, S.A., is the Uruguayan subsidiary of a U.S. manufacturing company. Its balance sheet for January 1 follows. The January 1 exchange rate between the U.S. dollar and the peso Uruguayo ($U) is $U20/$. Determine Tristan Narvaja's contribution to the translation exposure of its parent on January 1, using the current rate method.

Balance Sheet (thousands of pesos Uruguayo, $U)

Assets		Liabilities & Net Worth	
Cash	$U60,000	Current liabilities	$U30,000
Accounts receivable	120,000	Long-term debt	90,000
Inventory	120,000	Capital stock	300,000
Net plant & equipment	240,000	Retained earnings	120,000
	$U540,000		$U540,000

a. Determine Tristan Narvaja contribution to the translation exposure of its parent on January 1st, using the current rate method.
b. Calculate Tristan Narvaja contribution to its parent's translation loss if the exchange rate on December 31st is $U20/US$. Assume all peso Uruguayo accounts remain as they were at the beginning of the year.

6. **Tristan Narvaja, S.A. (B).** Using the same balance sheet as in Problem 5, calculate Tristan Narvaja's contribution to its parent's translation loss if the exchange rate on December 31 is $U22/$. Assume all peso accounts remain as they were at the beginning of the year.

7. **Tristan Narvaja, S.A. (C).** Calculate Tristan Narvaja's contribution to its parent's translation gain or loss using the current rate method if the exchange rate on December 31 is $U12/$. Assume all peso accounts remain as they were at the beginning of the year.

8. **Bangkok Instruments, Ltd. (A).** Bangkok Instruments, Ltd., the Thai subsidiary of a U.S. corporation, is a seismic instrument manufacturer. Bangkok Instruments manufactures instruments primarily for the oil and gas industry globally, though with recent commodity price increases of all kinds—including copper—its business has begun to grow rapidly. Sales are primarily to multinational companies based in the United States and Europe. Bangkok Instruments' balance sheet in thousands of Thai baht (B) as of March 31 is as follows:

Bangkok Instruments, Ltd.
Balance Sheet, March 1, thousands of Thai bahts

Assets		Liabilities and Net Worth	
Cash	B24,000	Accounts payable	B18,000
Accounts receivable	36,000	Bank loans	60,000
Inventory	48,000	Common stock	18,000
Net plant & equipment	60,000	Retained earnings	72,000
	B168,000		B168,000

Exchange rates for translating Bangkok Instruments' balance sheet into U.S. dollars are:

B40.00/$ April 1st exchange rate after 25% devaluation.

B30.00/$ March 31st exchange rate, before 25% devaluation. All inventory was acquired at this rate.

B20.00/$ Historic exchange rate at which plant and equipment were acquired.

The Thai baht dropped in value from B30/$ to B40/$ between March 31 and April 1. Assuming no change in balance sheet accounts between these two days, calculate the gain or loss from translation by both the current rate method and the temporal method. Explain the translation gain or loss in terms of changes in the value of exposed accounts.

9. **Bangkok Instruments, Ltd. (B).** Using the original data provided for Bangkok Instruments, assume that the Thai baht appreciated in value from B30/$ to B25/$ between March 31 and April 1. Assuming no change in balance sheet accounts between those two days, calculate the gain or loss from translation by both the current rate method and the temporal method. Explain the translation gain or loss in terms of changes in the value of exposed accounts.

10. **Cairo Ingot, Ltd.** Cairo Ingot, Ltd., is the Egyptian subsidiary of Trans-Mediterranean Aluminum, a British multinational that fashions automobile engine blocks from aluminum. Trans-Mediterranean's home reporting currency is the British pound. Cairo Ingot's December 31 balance sheet is shown below. At the date of this balance sheet the exchange rate between Egyptian pounds and British pounds sterling was £E5.50/UK£.

Assets		Liabilities and Net Worth	
Cash	£E 16,500,000	Accounts payable	£E 24,750,000
Accounts receivable	33,000,000	Long-term debt	49,500,000
Inventory	49,500,000	Invested capital	90,750,000
Net plant and equipment	66,000,000		
	£E165,000,000		£E165,000,000

What is Cairo Ingot's contribution to the translation exposure of Trans-Mediterranean on December 31, using the current rate method? Calculate the translation exposure loss to Trans-Mediterranean if the exchange rate at the end of the following quarter is £E6.00/£. Assume all balance sheet accounts are the same at the end of the quarter as they were at the beginning.

INTERNET EXERCISES

1. **Foreign Source Income.** If you are a citizen of the United States, and you receive income from outside the U.S.—foreign source income—how must you report this income? Use the following Internal Revenue Service Web site to determine current reporting practices for tax purposes.

U.S. Internal Revenue Service	www.irs.gov/Individuals/ International-Taxpayers/ Foreign-Currency-and -Currency-Exchange-Rates

2. **Translation in the United Kingdom.** What are the current practices and procedures for translation of financial statements in the United Kingdom? Use the following Web site to start your research.

Institute of Chartered Accountants in England and Wales	www.icaew.com/en/technical/ financial-reporting/uk-gaap/ uk-gaap-standards/

3. **Changing Translation Practices: FASB.** The Financial Accounting Standards Board (FASB) promulgates standard practices for the reporting of financial results by companies in the United States. It also, however, often leads the way in the development of new practices and emerging issues around the world. One major issue today is the valuation and reporting of financial derivatives and derivative agreements by firms. Use the FASB and Treasury Management Association Web pages to see current proposed accounting standards and the current state of reaction to the proposed standards.

FASB home page	raw.rutgers.edu/raw/fasb/
Treasury Management	www.tma.org/Association

4. **Yearly Average Exchange Rates.** When translating foreign currency values into U.S. dollar values for individual reporting purposes in the United States, which average exchange rates should you use? Use the following Web site to find the current average rates.

U.S. Internal Revenue Service	www.irs.gov/Individuals/ International-Taxpayers/ Yearly-Average-Currency -Exchange-Rates

CHAPTER 12

Operating Exposure

Coyote is always waiting. And Coyote is always hungry.
—Navajo Folk Saying.

LEARNING OBJECTIVES

- Examine how operating exposure arises in a multinational firm through unexpected changes in corporate cash flows
- Analyze how to measure operating exposure's impact on a business unit through the sequence of volume, price, cost, and other key variable changes
- Evaluate strategic alternatives to managing operating exposure
- Detail the proactive policies firms use in managing operating exposure

This chapter examines the *economic exposure* of a firm over time, what we term *operating exposure*. *Operating exposure*, also referred to as *competitive exposure* or *strategic exposure*, measures changes in the present value of a firm resulting from changes in future operating cash flows caused by unexpected changes in exchange rates. Operating exposure analysis assesses the impact of changing exchange rates on a firm's own operations over subsequent months and years and on its competitive position vis-à-vis other firms. The goal is to identify strategic moves or operating techniques the firm might wish to adopt to enhance its value in the face of unexpected exchange rate changes.

Operating exposure and transaction exposure are related in that they both deal with future cash flows. They differ in terms of which cash flows management considers and why those cash flows change when exchange rates change. We begin by revisiting the structure of our firm, Ganado Corporation, and how its structure dictates its likely operating exposure. The chapter continues with a series of strategies and structures used in the management of operating exposure, and concludes with a Mini-Case, *Toyota's European Operating Exposure.*

A Multinational's Operating Exposure

The structure and operations of a multinational company determine the nature of its operating exposure. Ganado Corporation's basic structure and currencies of operation are described in Exhibit 12.1. As a U.S.-based publicly traded company, ultimately all financial metrics and values have to be consolidated and expressed in U.S. dollars. That accounting exposure of the firm—translation exposure—was described in Chapter 10. Operationally, however, the functional currencies of the individual subsidiaries in combination determine the overall operating exposure of the firm in total.

EXHIBIT 12.1   **Ganado Corporation: Structure and Operations**

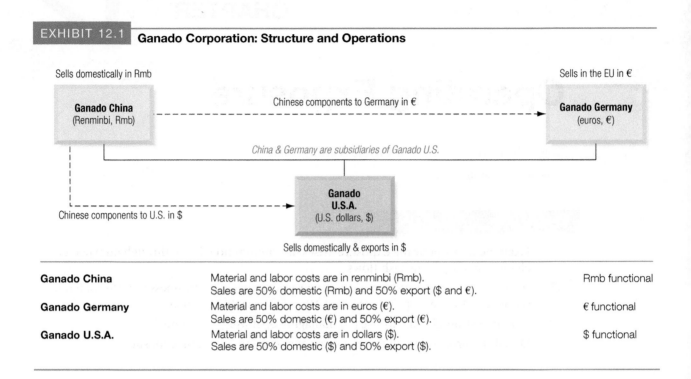

Ganado China	Material and labor costs are in renminbi (Rmb). Sales are 50% domestic (Rmb) and 50% export ($ and €).	Rmb functional
Ganado Germany	Material and labor costs are in euros (€). Sales are 50% domestic (€) and 50% export (€).	€ functional
Ganado U.S.A.	Material and labor costs are in dollars ($). Sales are 50% domestic ($) and 50% export ($).	$ functional

The operating exposure of any individual business or business unit is the net of cash inflows and outflows by currency, and how that compares to other companies competing in the same markets. Accounts receivable are the cash flow proceeds from sales, and accounts payable are all ongoing operating costs associated with the purchase of labor, materials, and other inputs. The net result—in general—is in essence the lifeblood of any business and the source of value created by the firm over time.

For example, Ganado Germany sells locally and it exports, but all sales are invoiced in euros. All operating cash inflows are therefore in its home currency, the euro. On the cost side, labor costs are local and in euros, as well as many of its material input purchases being local and in euros. Ganado Germany also purchases components from Ganado China, but those too are invoiced in euros. Ganado Germany is clearly euro-functional, with all cash inflows and outflows in euros.

Ganado Corporation U.S. is similar in structure to Ganado Germany. All cash inflows from sales, domestic and international, are in U.S. dollars. All costs, labor, and materials, sourced domestically and internationally, are invoiced in U.S. dollars. This includes purchases from Ganado China. Ganado U.S. is, therefore, obviously dollar functional.

Ganado China is more complex. Cash outflows, labor and materials, are all domestic and paid in Chinese renminbi. Cash inflows, however, are generated across three different currencies as the company sells locally in renminbi, as well as exporting to both Germany in euros and the United States in dollars. On net, although having some cash inflows in both dollars and euros, the dominant currency cash flow is the renminbi.

Static versus Dynamic Operating Exposure

Measuring the operating exposure of a firm like Ganado requires forecasting and analyzing all the firm's future individual transaction exposures together with the future exposures of all the firm's competitors and potential competitors worldwide. Exchange rate changes in the short term affect current and immediate contracts, generally termed transactions. But over the longer term, as prices change and competitors react, the more fundamental economic and competitive drivers of the business may alter all cash flows of all units. A simple example will clarify the point.

Assume Ganado's three business units are roughly equal in size. In 2012, the dollar starts depreciating in the market against the euro. At the same time, the Chinese government continues the gradual revaluation of the renminbi. The operating exposure of each individual business unit then needs to be examined statically (transaction exposures) and dynamically (future business transactions not yet contracted for).

- **Ganado China.** Sales in U.S. dollars will result in fewer renminbi proceeds in the immediate period. Sales in euros may stay roughly the same in renminbi proceeds depending on the relative movement of the Rmb against the euro. General profitability will fall in the short run. In the longer term, depending on the markets for its products and the nature of competition, it may need to raise the price at which it sells its export products, even to its U.S. parent company.
- **Ganado Germany.** Since this business unit's cash inflows and outflows are all in euros, there is no immediate transaction exposure or change. It may suffer some rising input costs in the future if Ganado China does indeed eventually push through price increases of component sales. Profitability is unaffected in the short term.
- **Ganado U.S.** Like Ganado Germany, Ganado U.S. has all local currency cash inflows and outflows. A fall in the value of the dollar will have no immediate impact (transaction exposure), but may change over the medium to long term as input costs from China may rise over time as the Chinese subsidiary tries to regain prior profit margins. But, like Germany, short-term profitability is unaffected.

The net result for Ganado is possibly a fall in the total profitability of the firm in the short term, primarily from the fall in profits of the Chinese subsidiary; that is, the short-term transaction/operating exposure impact. The fall in the dollar in the short term, however, is likely to have a positive impact on translation exposure, as profits and earnings in renminbi and euros translate into more dollars. Wall Street prefers returns sooner rather than later.

Operating and Financing Cash Flows

The cash flows of the MNE can be divided into operating cash flows and financing cash flows. Operating cash flows for Ganado arise from intercompany (between unrelated companies) and intracompany (between units of the same company) receivables and payables, rent and lease payments for the use of facilities and equipment, royalty and license fees for the use of technology and intellectual property, and assorted management fees for services provided.

Financing cash flows are payments for the use of intercompany and intracompany loans (principal and interest) and stockholder equity (new equity investments and dividends). Each of these cash flows can occur at different time intervals, in different amounts, and in different currencies of denomination, and each has a different predictability of occurrence. We summarize cash flow possibilities in Exhibit 12.2 for Ganado China and Ganado U.S.

Expected versus Unexpected Changes in Cash Flow

Operating exposure is far more important for the long-run health of a business than changes caused by transaction or translation exposure. However, operating exposure is inevitably

EXHIBIT 12.2   Financial and Operating Cash Flows Between Parent and Subsidiary

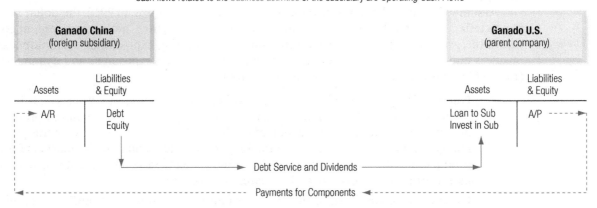

Cash flows related to the *financing* of the subsidiary are *Financial Cash Flows*
Cash flows related to the *business activities* of the subsidiary are *Operating Cash Flows*

subjective because it depends on estimates of future cash flow changes over an arbitrary time horizon. Thus, it does not spring from the accounting process but rather from operating analysis. Planning for operating exposure is a total management responsibility depending upon the interaction of strategies in finance, marketing, purchasing, and production.

An expected change in foreign exchange rates is not included in the definition of operating exposure, because both management and investors should have factored this information into their evaluation of anticipated operating results and market value. This "expected change" arises from differing perspectives as follows:

- From a management perspective, budgeted financial statements already reflect information about the effect of an expected change in exchange rates.

- From a debt service perspective, expected cash flow to amortize debt should already reflect the international Fisher effect. The level of expected interest and principal repayment should be a function of expected exchange rates rather than existing spot rates.

- From an investor's perspective, if the foreign exchange market is efficient, information about expected changes in exchange rates should be widely known and thus reflected in a firm's market value. Only unexpected changes in exchange rates, or an inefficient foreign exchange market, should cause market value to change.

- From a broader macroeconomic perspective, operating exposure is not just the sensitivity of a firm's future cash flows to unexpected changes in foreign exchange rates, but also its sensitivity to other key macroeconomic variables. This factor has been labeled as macroeconomic uncertainty.

We explore this further in *Global Finance in Practice 12.1*. Chapter 6 described the parity relationships among exchange rates, interest rates, and inflation rates. However, these variables are often in disequilibrium with one another. Therefore, unexpected changes in interest rates and inflation rates could also have a simultaneous but differential impact on future cash flows.

Expecting the Devaluation—Ford and Venezuela

Key to the understanding of operating exposure is that expected change in foreign exchange rates is not included in the firm's operating exposure. The assumption is that the market has already taken this value change into account. But is that assumption a sound one?

Consider the case of Ford Motor Company. In December 2013, Ford was very open and public about what it expected to happen to the Venezuelan currency—further devaluation—and what that would mean for Ford's financial results. In filings with the Securities and Exchange Commission (SEC), Ford reported that it had $802 million in investments in Venezuela, that it expected the Venezuelan bolivar to fall from 6.3 to the dollar to 12, and that it could end up suffering a $350 million financial loss as a result. The company was speaking from some experience. Earlier in the year it had lost $186 million when Venezuela devalued the bolivar to 6.3 from 4.3 per dollar.

Measuring Operating Exposure

An unexpected change in exchange rates impacts a firm's expected cash flows at four levels, depending on the time horizon used, as summarized in Exhibit 12.3.

Short Run. The first level impact is on expected cash flows in the one-year operating budget. The gain or loss depends on the currency of denomination of expected cash flows. These are both existing transaction exposures and anticipated exposures. The currency of denomination cannot be changed for existing obligations, or even for implied obligations such as purchase or sales commitments. Apart from real or implied obligations, in the short run it is difficult to change sales prices or renegotiate factor costs. Therefore, realized cash flows will differ from those expected in the budget. However, as time passes, prices and costs can be changed to reflect the new competitive realities caused by a change in exchange rates.

Medium Run: Equilibrium. The second level impact is on expected medium-run cash flows, such as those expressed in two- to five-year budgets, assuming parity conditions hold among foreign exchange rates, national inflation rates, and national interest rates. Under equilibrium conditions, the firm should be able to adjust prices and factor costs over time to maintain the expected level of cash flows. In this case, the currency of denomination of expected cash flows is not as important as the countries in which cash flows originate. National monetary, fiscal, and balance of payments policies determine whether equilibrium conditions will exist and whether firms will be allowed to adjust prices and costs.

EXHIBIT 12.3   Operating Exposure's Phases of Adjustment and Response

Phase	Time	Price Changes	Volume Changes	Structural Changes
Short Run	Less than 1 year	Prices are fixed/contracted	Volumes are contracted	No competitive market changes
Medium Run: Equilibrium	2 to 5 years	Complete pass-through of exchange rate changes	Volumes begin a partial response to prices	Existing competitors begin partial responses
Medium Run: Disequilibrium	2 to 5 years	Partial pass-through of exchange rate changes	Volumes begin a partial response to prices	Existing competitors begin partial responses
Long Run	More than 5 years	Completely flexible	Completely flexible	Threat of new entrants and changing competitor responses

If equilibrium exists continuously, and a firm is free to adjust its prices and costs to maintain its expected competitive position, its operating exposure may be zero. Its expected cash flows would be realized and therefore its market value unchanged since the exchange rate change was anticipated. However, it is also possible that equilibrium conditions exist but the firm is unwilling or unable to adjust operations to the new competitive environment. In such a case, the firm would experience operating exposure because its realized cash flows would differ from expected cash flows. As a result, its market value might also be altered.

Medium Run: Disequilibrium. The third level impact is on expected medium-run cash flows assuming disequilibrium conditions. In this case, the firm may not be able to adjust prices and costs to reflect the new competitive realities caused by a change in exchange rates. The primary problem may be the reactions of existing competitors. The firm's realized cash flows will differ from its expected cash flows. The firm's market value may change because of the unanticipated results.

Long Run. The fourth level impact is on expected long-run cash flows, meaning those beyond five years. At this strategic level a firm's cash flows will be influenced by the reactions of both existing competitors and potential competitors—possible new entrants—to exchange rate changes under disequilibrium conditions. In fact, all firms that are subject to international competition, whether they are purely domestic or multinational, are exposed to foreign exchange operating exposure in the long run whenever foreign exchange markets are not continuously in equilibrium.

Measuring Operating Exposure: Ganado Germany

Exhibit 12.4 presents the dilemma facing Ganado as a result of an unexpected change in the value of the euro, the currency of economic consequence for the German subsidiary. Ganado derives much of its reported profits—the earnings and earnings per share (EPS) as reported

---

EXHIBIT 12.4   **Ganado and Ganado Germany**

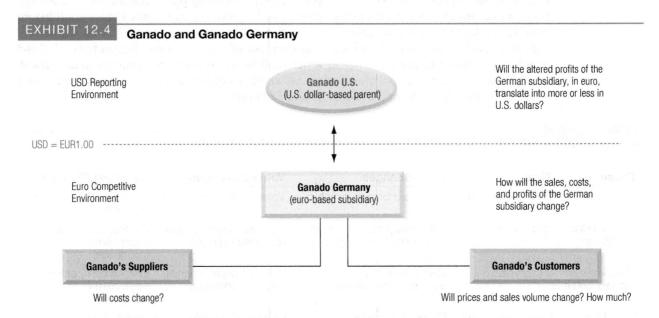

An unexpected depreciation in the value of the euro alters both the competitiveness of the subsidiary and the financial results which are consolidated with the parent company.

to Wall Street—from its European subsidiary. If the euro were to unexpectedly fall in value, how would the value of Ganado Germany's business change?

Value, in the world of finance, is generated by operating cash flow. If Ganado wished to attempt to measure the operating exposure of Ganado Germany to an unexpected exchange rate change, it would do so by evaluating the likely impact of that exchange rate on the operating cash flows of Ganado Germany. Specifically, how would prices, costs, and volume sales change? How would competitors and their respective prices, costs, and volumes change? The following section illustrates how those very values might respond in the short run and medium run to a fall in the value of the euro against Ganado's home currency, the dollar.

The Base Case

Ganado Germany manufactures in Germany, sells domestically, and exports; and all sales are invoiced in euros. Exhibit 12.5 summarizes the current baseline forecast for Ganado Germany income and operating cash flows for the 2014–2018 period (assume it is currently 2013). Sales

EXHIBIT 12.5 Ganado Germany's Valuation: Baseline Analysis

Assumptions	2014	2015	2016	2017	2018
Sales volume (units)	1,000,000	1,000,000	1,000,000	1,000,000	1,000,000
Sales price per unit	€12.80	€12.80	€12.80	€12.80	€12.80
Direct cost per unit	€9.60	€9.60	€9.60	€9.60	€9.60
German corporate tax rate	29.5%	29.5%	29.5%	29.5%	29.5%
Exchange rate ($/€)	1.2000	1.2000	1.2000	1.2000	1.2000

Income Statement	2014	2015	2016	2017	2018
Sales revenue	€12,800,000	€12,800,000	€12,800,000	€12,800,000	€12,800,000
Direct cost of goods sold	−9,600,000	−9,600,000	−9,600,000	−9,600,000	−9,600,000
Cash operating expenses (fixed)	−890,000	−890,000	−890,000	−890,000	−890,000
Depreciation	−600,000	−600,000	−600,000	−600,000	−600,000
Pretax profit	€1,710,000	€1,710,000	€1,710,000	€1,710,000	€1,710,000
Income tax expense	−504,450	−504,450	−504,450	−504,450	−504,450
Net income	€1,205,550	€1,205,550	€1,205,550	€1,205,550	€1,205,550

Cash Flows for Valuation					
Net income	€1,205,550	€1,205,550	€1,205,550	€1,205,550	€1,205,550
Add back depreciation	600,000	600,000	600,000	600,000	600,000
Changes in net working capital	0	0	0	0	0
Free cash flow for valuation, in euros	€1,805,550	€1,805,550	€1,805,550	€1,805,550	€1,805,550
Cash flow from operations, in dollars	$2,166,660	$2,166,660	$2,166,660	$2,166,660	$2,166,660
Present Value @ 15%	$7,262,980				

Notes: We assume, to simplify the analysis, that Ganado Germany has no debt and therefore no interest expenses. We also assume there are no additional capital expenditures required over the five years shown. We also assume no terminal value; Ganado is valued on its coming expected five years of cash flow only. Net working capital requirements (accounts receivable + inventory − accounts payable) require no additions in the base case due to constant sales. In subsequent scenarios it is assumed receivables are maintained at 45 days of sales, inventory at 10 days of cost of goods sold, and accounts payable at 38 days of sales.

volume is assumed to be a constant 1 million units per year, with a per unit sales price of €12.80 and a per unit direct cost of €9.60. The corporate income tax rate in Germany is 29.5%, and the exchange rate is $1.20/€.

These assumptions generate sales of €12,800,000, and €1,205,550 in net income. Adding net income to depreciation and changes in net working capital (which are zero in the base case) generates €1,805,550 or $2,166,660 in operating cash flow at $1.20/€. Ganado's management values its subsidiaries by finding the present value of these total free cash flows over the coming five-year period, in U.S. dollars, assuming a 15% discount rate. The baseline analysis finds a present value of Ganado Germany of $7,262,980.

On January 1, 2014, before any commercial activity begins, the euro unexpectedly drops from $1.2000/€ to $1.0000/€. Operating exposure depends on whether an unexpected change in exchange rates causes unanticipated changes in sales volume, sales prices, or operating costs.

Following a euro depreciation, Ganado Germany might choose to maintain its domestic sales prices constant in euro terms, or it might try to raise domestic prices because competing imports are now priced higher in Europe. The firm might choose to keep export prices constant in terms of foreign currencies, in terms of euros, or somewhere in between (partial pass-through). The strategy undertaken depends to a large measure on management's opinion about the price elasticity of demand, which would also include management's assessment of competitor response. On the cost side, Ganado Germany might raise prices because of more expensive imported raw material or components, or perhaps because all domestic prices in Germany have risen and labor is now demanding higher wages to compensate for domestic inflation.

Ganado Germany's domestic sales and costs might also be partly determined by the effect of the euro depreciation on demand. To the extent that the depreciation, by making prices of German goods initially more competitive, stimulates purchases of European goods in import-competing sectors of the economy as well as exports of German goods, German national income should increase. This assumes that the favorable effect of a euro depreciation on comparative prices is not immediately offset by higher domestic inflation. Thus, Ganado Germany might be able to sell more goods domestically because of price and income effects and internationally because of price effects.

To illustrate the effect of various post-depreciation scenarios on Ganado Germany's operating exposure, consider four simple cases.

Case 1: Depreciation (all variables remain constant)

Case 2: Increase in sales volume (other variables remain constant)

Case 3: Increase in sales price (other variables remain constant)

Case 4: Sales price, cost, and volume increase

To calculate the changes in value under each of the scenarios, we will use the same five-year horizon for any change in cash flow induced by the change in the dollar/euro exchange rate.

Case 1: Depreciation—All Variables Remain Constant

Assume that in the five years ahead no changes occur in sales volume, sales price, or operating costs. Profits for the coming year in euros will be as expected, and cash flow from operations will still be €1,805,550. There is no change in NWC because all results in euros remain the same. The exchange rate change, however, means that operating cash flows measured in U.S. dollars decline to $1,805,550. The present value of this series of operating cash flows is $6,052,483, a fall in Ganado Germany's value—when measured in U.S. dollars—of $1,210,497.

Case 2: Volume Increases—Other Variables Remain Constant

Assume that, following the depreciation in the euro, sales within Europe increase by 40%, to 1,400,000 units (assume all other variables remain constant). The depreciation has now made German-made telecom components more competitive with imports. Additionally, export volume increases because German-made components are now cheaper in countries whose currencies have not weakened. The sales price is kept constant in euro terms because management of Ganado Germany has not observed any change in local German operating costs and because it sees an opportunity to increase market share.

Ganado Germany's net income rises to €2,107,950, and operating cash flows the first year rise to €2,504,553, after a one-time increase in net working capital of €203,397 (using a portion of the increased cash flows). Operating cash flow is €2,707,950 per year for the following four years. The present value of Ganado Germany has risen by $1,637,621 over baseline to $8,900,601.

Case 3: Sales Price Increases—Other Variables Remain Constant

Assume the euro sales price is raised from €12.80 to €15.36 per unit to maintain the same U.S. dollar-equivalent price (the change offsets the depreciation of the euro) and that all other variables remain constant.

	Before	After
Price in euro	€12.80	€15.36
Exchange rate	$1.20/€	$1.00/€
Price in US$	$15.36	$15.36

Also assume that volume remains constant (the baseline 1,000,000 units) in spite of this price increase; that is, customers expect to pay the same dollar-equivalent price, and local costs do not change.

Ganado Germany is now better off following the depreciation than it was before because the sales price, which is pegged to the international price level, increased. And volume did not drop. Net income rises to €3,010,350 per year, with operating cash flow rising to €3,561,254 in 2014 (after a working capital increase of €49,096) and €3,610,350 per year in the following four years. Ganado Germany has now increased in value to $12,059,761.

Case 4: Price, Cost, and Volume Increases

The final case we examine, illustrated in Exhibit 12.6, is a combination of possible outcomes. Price increases by 10% to €14.08, direct cost per unit increases by 5% to €10.00, and volume rises by 10% to 1,100,000 units. Revenues rise by more than costs, and net income for Ganado Germany rises to €2,113,590. Operating cash flow rises to €2,623,683 in 2014 (after NWC increase), and €2,713,590 for each of the following four years. Ganado Germany's present value is now $9,018,195.

Other Possibilities

If any portion of sales revenues were incurred in other currencies, the situation would be different. Ganado Germany might leave the foreign sales price unchanged, in effect raising the euro-equivalent price. Alternatively, it might leave the euro-equivalent price unchanged, thus lowering the foreign sales price in an attempt to gain volume. Of course, it could also position itself between these two extremes. Depending on elasticities and the proportion of foreign to domestic sales, total sales revenue might rise or fall.

EXHIBIT 12.6	**Ganado Germany: Case 4—Sales Price, Volume, and Costs Increase**				

Assumptions	2014	2015	2016	2017	2018
Sales volume (units)	1,100,000	1,100,000	1,100,000	1,100,000	1,100,000
Sales price per unit	€14.08	€14.08	€14.08	€14.08	€14.08
Direct cost per unit	€10.00	€10.00	€10.00	€10.00	€10.00
German corporate tax rate	29.5%	29.5%	29.5%	29.5%	29.5%
Exchange rate ($/€)	1.0000	1.0000	1.0000	1.0000	1.0000

Income Statement	2014	2015	2016	2017	2018
Sales revenue	€15,488,000	€15,488,000	€15,488,000	€15,488,000	€15,488,000
Direct cost of goods sold	−11,000,000	−11,000,000	−11,000,000	−11,000,000	−11,000,000
Cash operating expenses (fixed)	−890,000	−890,000	−890,000	−890,000	−890,000
Depreciation	−600,000	−600,000	−600,000	−600,000	−600,000
Pretax profit	€2,998,000	€2,998,000	€2,998,000	€2,998,000	€2,998,000
Income tax expense	−884,410	−884,410	−884,410	−884,410	−884,410
Net income	€2,113,590	€2,113,590	€2,113,590	€2,113,590	€2,113,590

Cash Flows for Valuation					
Net income	€2,113,590	€2,113,590	€2,113,590	€2,113,590	€2,113,590
Add back depreciation	600,000	600,000	600,000	600,000	600,000
Changes in net working capital	−89,907	0	0	0	0
Free cash flow for valuation, in euros	€2,623,683	€2,713,590	€2,713,590	€2,713,590	€2,713,590
Cash flow from operations, in dollars	$2,623,683	$2,713,590	$2,713,590	$2,713,590	$2,713,590
Present Value @ 15%	$9,018,195				

If some or all raw material or components were imported and paid for in hard currencies, euro operating costs would increase after the depreciation of the euro. Another possibility is that local (not imported) euro costs would rise after a depreciation.

Measurement of Loss

Exhibit 12.7 summarizes the change in Ganado Germany's value across our small set of simple cases given an instantaneous and permanent change in the value of the euro from $1.20/€ to $1.00/€. These cases estimate Ganado Germany's operating exposure by measuring the change in the subsidiary's value as measured by the present value of its operating cash flows over the coming five-year period.

In Case 1, in which the euro depreciates (all variables remain constant), Ganado's German subsidiary's value falls by the percent change in the exchange rate, −16.7%. In Case 2, in which volume increased by 40% as a result of increasing price competitiveness, the German subsidiary's value increased 22.5%. In Case 3, in which the change in the exchange rate was completely passed through to a higher sales price, resulting in a massive 66% increase in subsidiary value. The final case, Case 4, combined increases in all three income drivers. The resulting change in subsidiary valuation of 24.2%, may be creeping toward a "realistic outcome," but there are obviously an infinite number of possibilities, which subsidiary management should be able to narrow. In the end, although the measurement of operating exposure

| EXHIBIT 12.7 | Summary of Ganado Germany Value Changes to Depreciation of the Euro |

Case	Exchange Rate	Price	Volume	Cost	Valuation	Change in Value	Percent Change in Value
Baseline	$1.20/€	€12.80	1,000,000	€9.60	$7,262,980	—	
1: No variable changes	$1.00/€	€12.80	1,000,000	€9.60	$6,052,483	($1,210,497)	−16.7%
2: Volume increases	$1.00/€	€12.80	1,400,000	€9.60	$8,900,601	$1,637,621	22.5%
3: Sales price increases	$1.00/€	€15.60	1,000,000	€9.60	$12,059,761	$4,796,781	66.0%
4: Price, cost, volume increase	$1.00/€	€14.08	1,100,000	€10.00	$9,018,195	$1,755,215	24.2%

is indeed difficult, it is not impossible—and may be worth the time and effort—in progressive financial management.

Strategic Management of Operating Exposure

The objective of managing both operating and transaction exposure is to anticipate and influence the effect of unexpected changes in exchange rates on a firm's future cash flows, rather than merely hoping for the best. To meet this objective, management can diversify the firm's operating and financing base. Management can also change the firm's operating and financing policies if it is concerned. *Global Finance in Practice 12.2* highlights one of the challenges to management awareness—fixed exchange rates.

GLOBAL FINANCE IN PRACTICE 12.2

Do Fixed Exchange Rates Increase Corporate Currency Risk in Emerging Markets?

It has long been argued that when firms know the exchange rate cannot or will not change, they will conduct their business as if currency exposure—at least against the major currency(s) to which their home currency is fixed—will not occur. As one study of currency risk in India noted, "These results support the hypothesis that pegged exchange rates induce moral hazard and increase financial fragility."

Moral hazard is the concept that a party—an agent, an individual, or a firm—will take on more risk when it either knows or believes that a second party will handle, accommodate, or insure the negative repercussions of the firm's risk-taking decisions. In other words, a firm may take more risk when it knows that someone else will pick up the tab. In a fixed or managed exchange rate regime, that "someone else" is represented by the central bank, which tells all those undertaking cross-currency contractual obligations and exposures that the exchange rate will not change.

Although there is still scant research on this specific practice for most of the emerging markets, it could prove to be a significant issue in the years to come, as many emerging markets become the object of major new international capital flows—the so-called globalization of finance. If commercial firms in those markets are not aware of the risk that the country itself may be taking by opening the door to international capital flows, both in and out of the country, and the impact they may have on the country's exchange rate, those firms may be in for a wild ride in the immediate years to come.

Sources: "Does the currency regime shape unhedged currency exposure?," by Ila Patnaik and Ajay Shah, *Journal of International Money and Finance*, 29, 2010, pp. 760–769. See also "Moral Hazard, Financial Crises, and the Choice of Exchange Rate Regimes," Apanard Angkinand and Thomas Willett, June 2006; and "Exchange-Rate Regimes for Emerging Markets: Moral Hazard and International Borrowing," by Ronald I. McKinnon and Huw Pill, *Oxford Review of Economic Policy*, Vol. 15, No. 3, 1999.

The key to managing operating exposure at the strategic level is for management to recognize a disequilibrium in parity conditions when it occurs and to be pre-positioned to react most appropriately. This task can best be accomplished if a firm diversifies internationally both its operating and its financing bases. Diversifying operations means diversifying sales, location of production facilities, and raw material sources. Diversifying the financing base means raising funds in more than one capital market and in more than one currency.

A diversification strategy permits the firm to react either actively or passively, depending on management's risk preference, to opportunities presented by disequilibrium conditions in the foreign exchange, capital, and product markets. Such a strategy does not require management to predict disequilibrium but only to recognize it when it occurs. It does require management to consider how competitors are pre-positioned with respect to their own operating exposures. This knowledge should reveal which firms would be helped or hurt competitively by alternative disequilibrium scenarios.

Diversifying Operations

Diversification of operations is one structural strategy to pre-positioning the firm for managing operating exposure. Consider the case in which purchasing power parity is temporarily in disequilibrium. Although the disequilibrium may have been unpredictable, management can often recognize its symptoms as soon as they occur. For example, management might notice a change in comparative costs in the firm's plants located in different countries. It might also observe changed profit margins or sales volume in one area compared to another, depending on price and income elasticities of demand and competitors' reactions.

Recognizing a temporary change in worldwide competitive conditions permits management to make changes in operating strategies. Management might make marginal shifts in sourcing raw materials, components, or finished products. If spare capacity exists, production runs can be lengthened in one country and reduced in another. The marketing effort can be strengthened in export markets where the firm's products have become more price competitive because of the disequilibrium condition. The challenge of course is to know when the change is temporary or semi-permanent, as *Global Finance in Practice 12.3* describes.

Even if management does not actively alter normal operations when exchange rates change, the firm should experience some beneficial portfolio effects. The variability of its cash flows is probably reduced by international diversification of its production, sourcing, and sales because exchange rate changes under disequilibrium conditions are likely to increase the firm's competitiveness in some markets while reducing it in others. In that case, operating exposure would be neutralized.

In contrast to the internationally diversified MNE, a purely domestic firm might be subject to the full impact of foreign exchange operating exposure even though it does not have foreign currency cash flows. For example, it could experience intense import competition in its domestic market from competing firms producing in countries with undervalued currencies.

A purely domestic firm does not have the option to react to an international disequilibrium condition in the same manner as an MNE. In fact, a purely domestic firm will not be positioned to recognize that a disequilibrium exists because it lacks comparative data from its own internal sources. By the time external data are available, it is often too late to react. Even if a domestic firm recognizes the disequilibrium, it cannot quickly shift production and sales into foreign markets in which it has had no previous presence.

Constraints exist that may limit the feasibility of diversifying production locations. The technology of a particular industry may require large economies of scale. For example,

The United Kingdom and Europe: Trans-Channel Currency Shifts

The United Kingdom's largest trading partner is the European Union, and although the two have been heavily intertwined for many years, the U.K. has not joined the euro. Keeping a separate currency, the British pound, and the associated ability to define its own monetary policy and currency has been a fundamental pillar of British pride and independence. But that independence has come at a price—what might be termed trans-channel (think English Channel) currency shifts.

The past 20 years have seen at least three different currency eras of relative strength between the pound and the euro. Prior to the launch of the euro there was a relatively "weak pound" period. But in 1996 there was a seismic shift—roughly from GBP0.80 = EUR1.00 to GBP0.65—and this shift persisted for more than a decade. During this period all things British became relatively more expensive on the Continent. British export prices were decidedly less competitive, while European exports to the U.K. gained at the U.K.'s expense. As has been the case so many times around the globe, this basic terms of trade shift altered fundamental national economies.

The Trans-Channel Eras of Currency Shift

British pounds (GBP) = 1.00 European euro (EUR)

With the global financial crisis of Fall 2008, the pound weakened dramatically against the euro, settling around 0.85 GBP = 1.00 EUR. In 2014, however, it started to strengthen once again

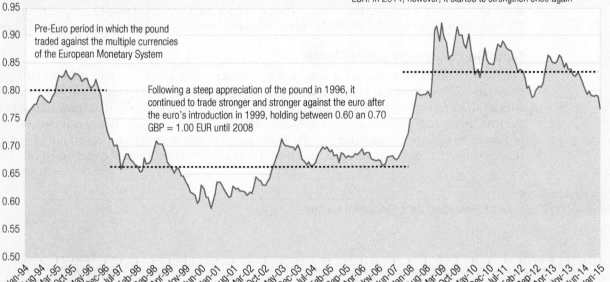

Pre-Euro period in which the pound traded against the multiple currencies of the European Monetary System

Following a steep appreciation of the pound in 1996, it continued to trade stronger and stronger against the euro after the euro's introduction in 1999, holding between 0.60 an 0.70 GBP = 1.00 EUR until 2008

But the tectonic plates moved once again in 2008, with a new trans-channel currency shift moving the British pound weaker once again. Now the pound weakened from GBP0.65 = EUR1.00 to a more volatile, but seemingly medium to long-term fundamental path around GBP0.85 = EUR1.00. British products and services—those priced in British pounds—were once again highly affordable to customers on the continent. How long this new era will last is anyone's guess, but it has already persisted for seven years. The ways in which individual multinational firms on either side of the Channel have managed this periodic trans-channel currency shift is operating exposure management at work.

high-tech firms, such as Intel, prefer to locate in places where they have easy access to high-tech suppliers, a highly educated workforce, and one or more leading universities. Their R&D efforts are closely tied to initial production and sales activities.

Diversifying Financing

If a firm diversifies its financing sources, it will be pre-positioned to take advantage of temporary deviations from the international Fisher effect. If interest rate differentials do not equal expected changes in exchange rates, opportunities to lower a firm's cost of capital will exist. However, to be able to switch financing sources, a firm must already be well known in the international investment community, with banking contacts firmly established. Again, this is not typically an option for a domestic firm.

As we will demonstrate in Chapter 13, diversifying sources of financing, regardless of the currency of denomination, can lower a firm's cost of capital and increase its availability of capital. The ability to source capital from outside of a segmented market is especially important for firms resident in emerging markets.

Proactive Management of Operating Exposure

Operating and transaction exposures can be partially managed by adopting operating or financing policies that offset anticipated foreign exchange exposures. Five of the most commonly employed proactive policies are (1) matching currency cash flows; (2) risk-sharing agreements; (3) back-to-back or parallel loans; (4) cross-currency swaps, and (5) contractual approaches.

Matching Currency Cash Flows

One way to offset an anticipated continuous long-term exposure to a particular currency is to acquire debt denominated in that currency. Exhibit 12.8 depicts the exposure of a U.S. firm with continuing export sales to Canada. In order to compete effectively in Canadian markets, the firm invoices all export sales in Canadian dollars. This policy results in a continuing receipt of Canadian dollars month after month. If the export sales are part of a continuing supplier relationship, the long Canadian dollar position is relatively predictable and constant. This

EXHIBIT 12.8   **Debt Financing as a Financial Hedge**

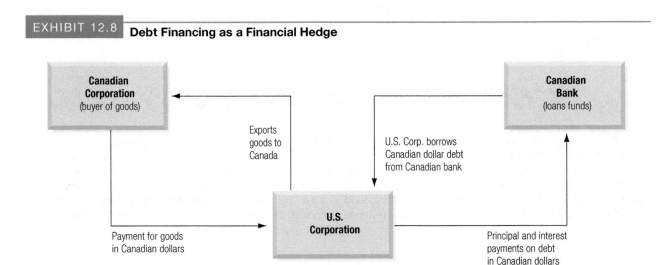

Exposure: The sale of goods to Canada creates a foreign
currency exposure from the inflow of Canadian dollars.

Hedge: The Canadian dollar debt payments act as a financial
hedge by requiring debt service, an outflow of Canadian dollars.

endless series of transaction exposures could of course be continually hedged with forward contracts or other contractual hedges, as discussed in Chapter 9.

But what if the firm sought out a continual use—an outflow—for its continual inflow of Canadian dollars? If the U.S. firm were to acquire part of its debt-capital in the Canadian dollar markets, it could use the relatively predictable Canadian dollar cash inflows from export sales to service the principal and interest payments on Canadian dollar debt and be cash flow matched. The U.S.-based firm has hedged an operational cash inflow by creating a financial cash outflow, and so it does not have to actively manage the exposure with contractual financial instruments such as forward contracts. This form of hedging, sometimes referred to as matching, is effective in eliminating currency exposure when the exposure cash flow is relatively constant and predictable over time.

The list of potential matching strategies is nearly endless. A second alternative would be for the U.S. firm to seek out potential suppliers of materials or components in Canada as a substitute for U.S. or other foreign firms. The firm would then possess both an operational Canadian dollar cash inflow—a *receivable*—and a Canadian dollar operational cash outflow—a *payable*. If the cash flows were roughly the same in magnitude and timing, the strategy would be a *natural hedge*, the term "natural" referring to operating-based activities.

A third alternative, often referred to as *currency switching*, would be to pay foreign suppliers with Canadian dollars. For example, if the U.S. firm imported components from Mexico, the Mexican firms themselves might welcome payment in Canadian dollars because they are short Canadian dollars in their multinational cash flow network.

Risk-Sharing Agreements

An alternative arrangement for managing a long-term cash flow exposure between firms with a continuing buyer-supplier relationship is risk-sharing. *Risk-sharing* is a contractual arrangement in which the buyer and seller agree to "share" or split currency movement impacts on payments between them. If the two firms are interested in a long-term relationship based on product quality and supplier reliability, and not on the whims of the currency markets, a cooperative agreement to share the burden of currency risk may be in order.

If Ford's North American operations import automotive parts from Mazda (Japan) every month, year after year, major swings in exchange rates can benefit one party at the expense of the other. (Ford is a major stockholder of Mazda, but it does not exert control over its operations. Therefore, the risk-sharing agreement is particularly appropriate; transactions between the two are both intercompany and intracompany in nature. A risk-sharing agreement solidifies the partnership.) One potential solution would be for Ford and Mazda to agree that all purchases by Ford will be made in Japanese yen at the current exchange rate, as long as the spot rate on the date of invoice is between, say, ¥115/$ and ¥125/$. If the exchange rate is between these values on the payment dates, Ford agrees to accept whatever transaction exposure exists (because it is paying in a foreign currency). If, however, the exchange rate falls outside this range on the payment date, Ford and Mazda will share the difference equally.

For example, Ford has an account payable of ¥25,000,000 for the month of March. If the spot rate on the date of invoice is ¥110/$, the Japanese yen would have appreciated versus the dollar, causing Ford's costs of purchasing automotive parts to rise. Since this rate falls outside the contractual range, Mazda would agree to accept a total payment in Japanese yen that would result from a difference of ¥5/$ (i.e., ¥115 − ¥110). Ford's payment would be as follows:

$$\left[\frac{¥25,000,000}{¥115.00/\$ - \left(\frac{¥5.00/\$}{2} \right)} \right] = \frac{¥25,000,000}{¥112.50/\$} = \$222,222.22.$$

At a spot rate of ¥110/$, Ford's costs for March would be $227,272.73 without the risk-sharing agreement. With the agreement, however, Ford's payment is calculated using an exchange rate of ¥112.50/$, a payment of $222,222.22. The risk-sharing agreement constitutes a savings for Ford of $5,050.51 (this "savings" is a reduction in a cost increase, not a true cost reduction). Both parties therefore incur costs and benefits from exchange rate movements outside the specified band. Note that the movement could just as easily have been in Mazda's favor if the spot rate had moved to ¥130/$.

The risk-sharing arrangement is intended to smooth the impact on both parties of volatile and unpredictable exchange rate movements. Of course, a sustained appreciation of one currency versus the other would require the negotiation of a new sharing agreement, but the ultimate goal of the agreement is to alleviate currency pressures on the continuing business relationship.

Risk-sharing agreements like these have been in use for nearly 50 years on world markets. They became something of a rarity during the 1960s when exchange rates were relatively stable under the Bretton Woods Agreement. But with the return to floating exchange rates in the 1970s, firms with long-term customer-supplier relationships across borders have returned to some old ways of maintaining mutually beneficial long-term trade. *Global Finance in Practice 12.4* describes how one U.S.-based firm, Harley-Davidson, has used risk-sharing.

GLOBAL FINANCE IN PRACTICE 12.4

Hedging Hogs: Risk-Sharing at Harley-Davidson

Harley-Davidson (U.S.) is representative of a company with centralized manufacturing (all in the United States with costs in U.S. dollars) and global sales (predominantly in U.S. dollars, European euros, Australian dollars, and Japanese yen). Dealerships in individual companies therefore purchase from Harley and sell into the local market in local currency. The foreign dealerships need to be assured of stable product costs—in local currency and at a predictable purchase price of the hogs from Harley (U.S.)—in order to offer stable and competitive prices in-country. The figure below illustrates how a risk-sharing structure can be devised by Harley to provide the desired stability to its Australian dealerships.

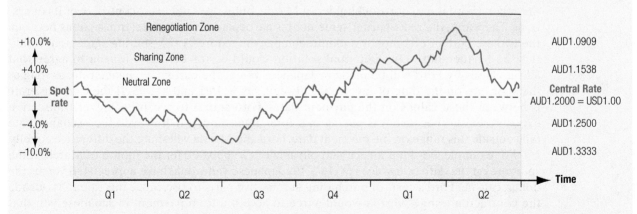

As long as the spot rate during the period stays within the *Neutral Zone*, Harley would be responsible for managing the currency exposures that it would incur on the corporate level. If the spot rate moves into the *Sharing Zone* during the period, the exchange rate used in pricing is adjusted to "share" the change equally. If the spot rate moves beyond the *Sharing Zone* into the *Renegotiation Zone* during the period, Harley will work with foreign distributors to establish a new central rate moving forward.

Back-to-Back or Parallel Loans

A *back-to-back loan*, also referred to as a *parallel loan* or *credit swap*, occurs when two business firms in separate countries arrange to borrow each other's currency for a specific period of time. At an agreed terminal date they return the borrowed currencies. The operation is conducted outside the foreign exchange markets, although spot quotations may be used as the reference point for determining the amount of funds to be swapped. Such a swap creates a covered hedge against exchange loss, since each company, on its own books, borrows the same currency it repays. Back-to-back loans are also used at a time of actual or anticipated legal limitations on the transfer of investment funds to or from either country.

The structure of a typical back-to-back loan is illustrated in Exhibit 12.9. A British parent firm that wants to invest funds in its Dutch subsidiary locates a Dutch parent firm that wants to invest funds in the United Kingdom. Avoiding the exchange markets entirely, the British parent lends pounds to the Dutch subsidiary in the United Kingdom, while the Dutch parent lends euros to the British subsidiary in the Netherlands. The two loans would be for equal values at the current spot rate and for a specified maturity. At maturity, the two separate loans would each be repaid to the original lender, again without any need to use the foreign exchange markets. Neither loan carries any foreign exchange risk, and neither loan normally needs the approval of any governmental body regulating the availability of foreign exchange for investment purposes.

Parent company guarantees are not needed on the back-to-back loans because each loan carries the right of offset in the event of default of the other loan. A further agreement can provide for maintenance of principal parity in case of changes in the spot rate between the two countries. For example, if the pound dropped by more than, say, 6% for as long as 30 days, the British parent might have to advance additional pounds to the Dutch subsidiary to bring the principal value of the two loans back to parity. A similar provision would protect the British if

---

EXHIBIT 12.9   **Back-to-Back Loans for Currency Hedging**

1. British firm wishes to invest funds in its Dutch subsidiary

2. British firm identifies a Dutch firm wishing to invest funds in its British subsidiary

Indirect Financing

British Parent Firm

Dutch Parent Firm

Direct loan in pounds

Direct loan in euros

Dutch Firm's British Subsidiary

British Firm's Dutch Subsidiary

3. British firm loans British pounds directly to the Dutch firm's British subsidiary

4. British firm's Dutch subsidiary loans euros from the Dutch parent

The back-to-back loan provides a method for parent-subsidiary cross-border financing without incurring direct currency exposure.

the euro should weaken. Although this parity provision might lead to changes in the amount of home currency each party must lend during the period of the agreement, it does not increase foreign exchange risk, because at maturity all loans are repaid in the same currency loaned.

There are two fundamental impediments to widespread use of the back-to-back loan. First, it is difficult for a firm to find a partner, termed a *counterparty*, for the currency, amount, and timing desired. Second, a risk exists that one of the parties will fail to return the borrowed funds at the designated maturity—although this risk is minimized because each party to the loan has, in effect, 100% collateral, albeit in a different currency. These disadvantages have led to the rapid development and wide use of the cross-currency swap.

Cross-Currency Swaps

A *cross-currency swap* resembles a back-to-back loan except that it does not appear on a firm's balance sheet. As detailed in Chapter 8, the term swap is used in a variety of ways in international finance, and care should be used to identify the exact use in a specific case. In a *currency swap*, a firm and a swap dealer (or swap bank) agree to exchange an equivalent amount of two different currencies for a specified period of time. Currency swaps can be negotiated for a wide range of maturities up to 30 years in some cases. The swap dealer or swap bank acts as a middleman in setting up the swap agreement.

A typical currency swap first requires two firms to borrow funds in markets and currencies in which they are well known. For example, a Japanese firm would typically borrow yen on a regular basis in its home market. If, however, the Japanese firm were exporting to the United States and earning U.S. dollars, it might wish to construct a matching cash flow hedge that would allow it to use the U.S. dollars earned to make regular debt-service payments on U.S. dollar debt. If, however, the Japanese firm is not well known in the U.S. financial markets, it may have no ready access to U.S. dollar debt.

One way in which this Japanese firm could, in effect, borrow dollars, is to participate in a cross-currency swap as seen in Exhibit 12.10. The firm could swap its yen-denominated debt

---

EXHIBIT 12.10   **Using Cross-Currency Swaps**

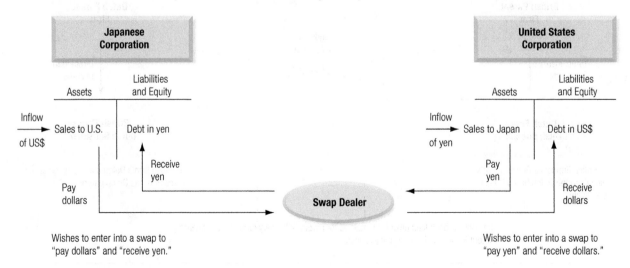

Both the Japanese corporation and the U.S. corporation would like to enter into a cross-currency swap which would allow them to use foreign currency cash inflows to service debt.

service payments with another firm that has U.S. dollar-debt service payments. This swap would have the Japanese firm "paying dollars" and "receiving yen." The Japanese firm would then have dollar debt service without actually borrowing U.S. dollars. Simultaneously, a U.S. corporation would actually be entering into a cross-currency swap in the opposite direction — "paying yen" and "receiving dollars." The swap dealer takes the role of a middleman. Swap dealers arrange most swaps on a "blind basis," meaning that the initiating firm does not know who is on the other side of the swap arrangement — the counterparty. The initiating firm views the dealer or bank as its counterparty. Because the swap markets are dominated by the major money center banks worldwide, the counterparty risk is acceptable. Because the swap dealer's business is arranging swaps, the dealer can generally arrange for the currency, amount, and timing of the desired swap.

Accountants in the United States treat the currency swap as a foreign exchange transaction rather than as debt and they treat the obligation to reverse the swap at some later date as a forward exchange contract. Forward exchange contracts can be matched against assets, but they are entered in a firm's footnotes rather than as balance sheet items. The result is that both translation and operating exposures are avoided, and neither a long-term receivable nor a long-term debt is created on the balance sheet.

Contractual Approaches: Hedging the Unhedgeable

Some MNEs now attempt to hedge their operating exposure with contractual strategies. A number of firms like Merck (U.S.) have undertaken long-term currency option positions — hedges designed to offset lost earnings from adverse exchange rate changes. This hedging of what many of these firms refer to as strategic exposure or competitive exposure seems to fly in the face of traditional theory.

The ability of firms to hedge the "unhedgeable" is dependent upon predictability: (1) the predictability of the firm's future cash flows, and (2) the predictability of the firm's competitor's responses to exchange rate changes. Although the management of many firms may believe they are capable of predicting their own cash flows, in practice few feel capable of accurately predicting competitor response. Many firms still find timely measurement of exposure challenging.

Merck is an example of a firm whose management feels capable of both. The company possesses relatively predictable long-run revenue streams due to the product-niche nature of the pharmaceuticals industry. As a U.S.-based exporter to foreign markets, markets in which sales levels by product are relatively predictable and prices are often regulated by government, Merck can accurately predict net long-term cash flows in foreign currencies five and ten years into the future. Merck has a relatively undiversified operating structure, and it is highly centralized in terms of where research, development, and production costs are located. Merck's managers feel the company has no real alternatives but contractual hedging if it is to weather long-term unexpected exchange rate changes. Merck has purchased over-the-counter (OTC) long-term put options on foreign currencies versus the U.S. dollar as insurance against potential lost earnings from exchange rate changes.

A significant question remains as to the true effectiveness of hedging operating exposure with contractual hedges. The fact remains that even after feared exchange rate movements and put option position payoffs have occurred, the firm is competitively disadvantaged. The capital outlay required for the purchase of such sizeable put option positions is capital not used for the potential diversification of operations, which in the long run might have more effectively maintained the firm's global market share and international competitiveness.

SUMMARY POINTS

- Operating exposure measures the change in value of the firm that results from changes in future operating cash flows caused by an unexpected change in exchange rates.

- Operating strategies for the management of operating exposure emphasize the structuring of firm operations in order to create matching streams of cash flows by currency.

- The objective of operating exposure management is to anticipate and influence the effect of unexpected changes in exchange rates on a firm's future cash flow, rather than being forced into passive reaction to such changes.

- Proactive policies include matching currency cash flows, currency risk-sharing clauses, back-to-back loan structures, and cross-currency swap agreements.

- Contractual approaches (i.e., options and forwards) have occasionally been used to hedge operating exposure but are costly and possibly ineffectual.

MINI-CASE

Toyota's European Operating Exposure[1]

It was January 2002, and Toyota Motor Europe Manufacturing (TMEM) had a problem. More specifically, Mr. Toyoda Shuhei, the new President of TMEM, had a problem. He was on his way to Toyota Motor Company's (Japan) corporate offices outside Tokyo to explain the continuing losses of the European manufacturing and sales operations. The CEO of Toyota Motor Company, Mr. Hiroshi Okuda, was expecting a proposal from Mr. Shuhei to reduce and eventually eliminate the European losses. The situation was intense given that TMEM was the only major Toyota subsidiary suffering losses.

Toyota and Auto Manufacturing

Toyota Motor Company was the number one automobile manufacturer in Japan, the third largest manufacturer in the world by unit sales (5.5 million units or one auto every six seconds), but number eight in sales in Continental Europe. The global automobile manufacturing industry had been experiencing, like many industries, continued consolidation in recent years as margins were squeezed, economies of scale and scope pursued, and global sales slowed.

Toyota was no different. It had continued to rationalize its manufacturing along regional lines. Toyota had continued to increase the amount of local manufacturing in North America. In 2001, over 60% of Toyota's North American sales were locally manufactured. But Toyota's European sales were nowhere close to this yet. Most of Toyota's automobile and truck manufacturing for Europe was still done in Japan. In 2001, only 24% of the autos sold in Europe were manufactured in Europe (including the United Kingdom). The remainder was imported from Japan (see Exhibit A).

Toyota Motor Europe sold 634,000 automobiles in 2000. This was the second largest foreign market for Toyota, second only to North America. TMEM expected significant growth in European sales, and was planning to expand European manufacturing and sales to 800,000 units by 2005. But for fiscal 2001, the unit reported operating losses of ¥9.897 billion ($82.5 million at ¥120/$). TMEM had three assembly plants in the United Kingdom, one plant in Turkey, and one plant in Portugal. In November 2000, Toyota Motor Europe announced publicly that it would not generate positive profits for the next two years due to the weakness of the euro.

Toyota had recently introduced a new model to the European market, the Yaris, which was proving very successful. The Yaris, a super-small vehicle with a 1,000cc engine, had sold more than 180,000 units in 2000. Although the Yaris had been specifically designed for the European market, the decision had been made early on to manufacture it in Japan.

Currency Exposure

The primary source of the continuing operating losses suffered by TMEM was the falling value of the euro. Over the recent two-year period, the euro had fallen in value

EXHIBIT A **Toyota's European Currency Operating Structure**

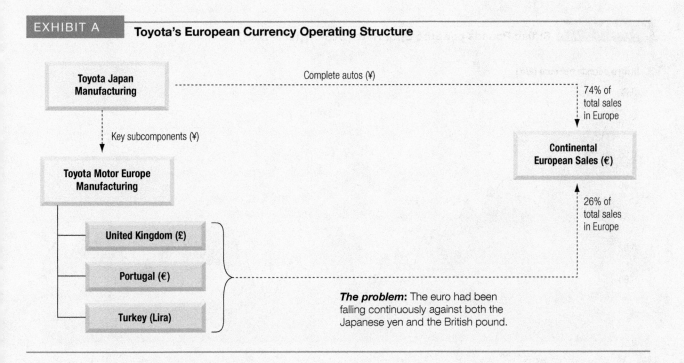

EXHIBIT B **Japanese yen per Euro Spot Rate (weekly, 1999–2001)**

against both the Japanese yen and the British pound. As demonstrated in Exhibit A, the cost base for most of the autos sold within the Continental European market was the Japanese yen. Exhibit B illustrates the slide of the euro against the Japanese yen.

As the yen rose against the euro, costs increased significantly when measured in euros. If Toyota wished to preserve its price competitiveness in the European market, it had to absorb most of the exchange rate changes and suffer reduced margins on both completed cars and key

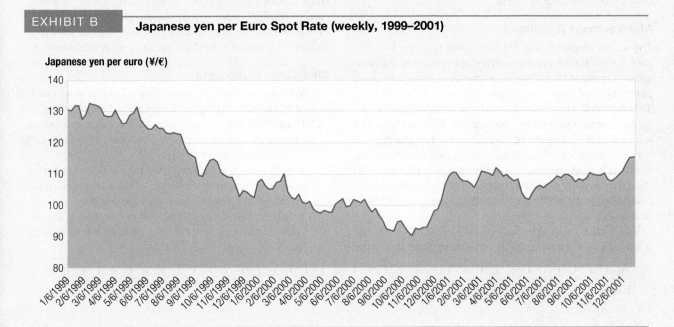

EXHIBIT C   **British Pounds per Euro Spot Rate (weekly, 1999–2001)**

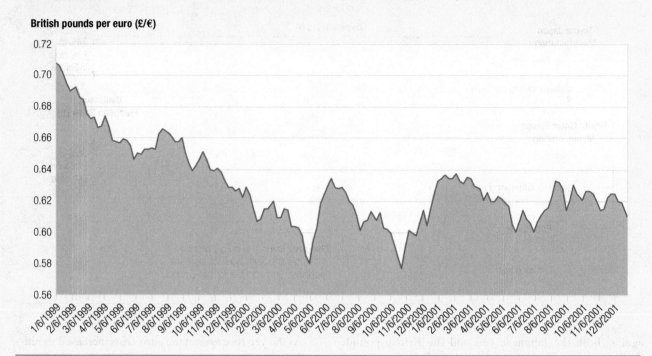

British pounds per euro (£/€)

subcomponents shipped to its European manufacturing centers. Deciding to manufacture the Yaris in Japan had only exacerbated the situation.

Management Response

Toyota management was not sitting passively by. In 2001, they had initiated some assembly operations in Valenciennes, France. Although accounting for a relatively small percentage of total European sales as of January 2002, Toyota planned to continue to expand its European capacity and capabilities and to source about 25% of European sales by 2004. Relocation of Yaris assembly to Valenciennes was scheduled for 2002. The continuing problem, however, was that it was an *assembly* facility, meaning that many of the expensive value-added components of the autos being assembled were still based in either Japan or the United Kingdom.

Mr. Shuhei, with the approval of Mr. Okuda, had also initiated a local sourcing and procurement program for the United Kingdom manufacturing operations. TMEM wished to decrease the number of key components imported from Toyota Japan in order to reduce the currency exposure of the U.K. unit. But again, the continuing problem of the British pound's value against the euro, as shown in Exhibit C, reduced the effectiveness of even this solution.

Mini-Case Questions

1. Why do you think Toyota waited so long to move much of its manufacturing for European sales to Europe?

2. If the British pound were to join the European Monetary Union would the problem be resolved? How likely do you think this is?

3. If you were Mr. Shuhei, how would you categorize your problems and solutions? What was a short-term problem? What was a long-term problem?

4. What measures would you recommend that Toyota Europe take to resolve the continuing operating losses?

QUESTIONS

These questions are available in MyFinanceLab.

1. **Exposure Definitions.** Define operating exposure, economic exposure, and competitive exposure. Can you provide any insights into what may be behind the use of the different terms?

2. **Operating Exposure versus Translation Exposure.** What do you see as the primary difference between operating exposure and translation exposure? Would this have the same meaning to a private firm as a publicly traded firm?

3. **Unexpected Exchange Rate Changes.** Why do unexpected exchange rate changes contribute to operating exposure, but expected exchange rate changes do not?

4. **Time Horizon.** Explain the time horizons used to analyze and measure unexpected changes in exchange rates.

5. **Static versus Dynamic.** What are examples of static exposures versus dynamic exposures?

6. **Operating versus Financing Cash Flows.** According to financial theory, which is more important to the value of the firm, financing or operating cash flows?

7. **Macroeconomic Uncertainty.** Explain how the concept of macroeconomic uncertainty expands the scope of analyzing operating exposure.

8. **Strategic Response.** The objective of both operating and transaction exposure management is to anticipate and influence the effect of unexpected changes in exchange rates on a firm's future cash flows. What strategic alternative policies exist to enable management to manage these exposures?

9. **Managing Operating Exposure.** The key to managing operating exposure at the strategic level is for management to recognize a disequilibrium in parity conditions when it occurs and to be pre-positioned to react most appropriately. How can this task best be accomplished?

10. **Diversification.** How can a multinational firm diversify operations? How can it diversify its financing? Do you believe these are effective ways of managing operating exposure?

11. **Proactive Management.** Operating exposures can be partially managed by adopting operating or financing policies that offset anticipated foreign exchange exposures. What are four of the most commonly employed proactive policies?

12. **Matching Currency Exposure.** Explain how matching currency cash flows can offset operating exposure.

13. **Risk Sharing.** An alternative arrangement for managing operating exposure between firms with a continuing buyer-supplier relationship is risk sharing. Explain how risk sharing works.

14. **Back-to-Back Loans.** Explain how back-to-back loans can hedge foreign exchange operating exposure. Would firms have any specific worries about their partner in a back-to-back loan arrangement?

15. **Currency Swaps.** Explain how currency swaps can hedge foreign exchange operating exposure. What are the accounting advantages of currency swaps?

16. **Hedging the Unhedgeable.** How do some firms attempt to hedge their long-term operation exposure with contractual hedges? What assumptions do they make in order to justify contractual hedging of their operating exposure? How effective is such contractual hedging in your opinion?

PROBLEMS

These problems are available in MyFinanceLab.

1. **Mauna Loa Macadamia.** Mauna Loa Macadamia, a macadamia nut subsidiary of Hershey's with plantations on the slopes of its namesake volcano in Hilo, Hawaii, exports macadamia nuts worldwide. The Japanese market is its biggest export market, with average annual sales invoiced in yen to Japanese customers of ¥1,200,000,000. At the present exchange rate of ¥125/$, this is equivalent to $9,600,000. Sales are relatively equally distributed throughout the year. They show up as a ¥250,00,000 account receivable on Mauna Loa's balance sheet. Credit terms to each customer allow for 60 days before payment is due. Monthly cash collections are typically ¥100,000,000.

   Mauna Loa would like to hedge its yen receipts, but it has too many customers and transactions to make it practical to sell each receivable forward. It does not want to use options because they are considered to be too expensive for this particular purpose. Therefore, they have decided to use a "matching" hedge by borrowing yen.
   a. How much should Mauna Loa borrow in yen?
   b. What should be the terms of payment on the yen loan?

2. **Acuña Leather Goods.** DeMagistris Fashion Company, based in New York City, imports leather coats

from Acuña Leather Goods, a reliable and longtime supplier, based in Buenos Aires, Argentina. Payment is in Argentine pesos. When the peso lost its parity with the U.S. dollar in January 2002, it collapsed in value to Ps4.0/$ by October 2002. The outlook was for a further decline in the peso's value. Since both DeMagistris and Acuña wanted to continue their longtime relationship, they agreed on a risk-sharing arrangement. As long as the spot rate on the date of an invoice is between Ps3.5/$ and Ps4.5/$, DeMagistris will pay based on the spot rate. If the exchange rate falls outside this range, they will share the difference equally with Acuña Leather Goods. The risk-sharing agreement will last for six months, at which time the exchange rate limits will be reevaluated. DeMagistris contracts to import leather coats from Acuña for Ps8,000,000 or $2,000,000 at the current spot rate of Ps4.0/$ during the next six months.

a. If the exchange rate changes immediately to Ps6.00/$, what will be the dollar cost of six months of imports to DeMagistris?

b. At Ps6.00/$, what will be the peso export sales of Acuña Leather Goods to DeMagistris Fashion Company?

3. **Manitowoc Crane (A).** Manitowoc Crane (U.S.) exports heavy crane equipment to several Chinese dock facilities. Sales are currently 10,000 units per year at the yuan equivalent of $24,000 each. The Chinese yuan (renminbi) has been trading at Yuan8.20/$, but a Hong Kong advisory service predicts the renminbi will drop in value next week to Yuan9.00/$, after which it will remain unchanged for at least a decade. Accepting this forecast as given, Manitowoc Crane faces a pricing decision in the face of the impending devaluation. It may either (1) maintain the same yuan price and in effect sell for fewer dollars, in which case Chinese volume will not change; or (2) maintain the same dollar price, raise the yuan price in China to offset the devaluation, and experience a 10% drop in unit volume. Direct costs are 75% of the U.S. sales price.

a. What would be the short-run (one-year) impact of each pricing strategy?

b. Which do you recommend?

4. **Manitowoc Crane (B).** Assume the same facts as in Problem 3. Additionally, financial management believes that if it maintains the same yuan sales price, volume will increase at 12% per annum for eight years. Dollar costs will not change. At the end of 10 years, Manitowoc's' patent expires and it will no longer export to China. After the yuan is devalued to Yuan9.20/$, no further devaluations are expected.

If Manitowoc Crane raises the yuan price so as to maintain its dollar price, volume will increase at only 1% per annum for eight years, starting from the lower initial base of 9,000 units. Again, dollar costs will not change, and at the end of eight years Manitowoc Crane will stop exporting to China. Manitowoc's weighted average cost of capital is 10%. Given these considerations, what should be Manitowoc's pricing policy?

5. **MacLoren Automotive.** MacLoren Automotive manufactures British sports cars, a number of which are exported to New Zealand for payment in pounds sterling. The distributor sells the sports cars in New Zealand for New Zealand dollars. The New Zealand distributor is unable to carry all of the foreign exchange risk, and would not sell MacLoren models unless MacLoren could share some of the foreign exchange risk.

MacLoren has agreed that sales for a given model year will initially be priced at a "base" spot rate between the New Zealand dollar and pound sterling set to be the spot mid-rate at the beginning of that model year. As long as the actual exchange rate is within ±5% of that base rate, payment will be made in pounds sterling. That is, the New Zealand distributor assumes all foreign exchange risk. However, if the spot rate at time of shipment falls outside of this ±5% range, MacLoren will share equally (i.e., 50/50) the difference between the actual spot rate and the base rate. For the current model year the base rate is NZ$1.6400/£.

a. What are the outside ranges within which the New Zealand importer must pay at the then current spot rate?

b. If MacLoren ships 10 sports cars to the New Zealand distributor at a time when the spot exchange rate is NZ$1.7000/£, and each car has an invoice cost £32,000, what will be the cost to the distributor in New Zealand dollars? How many pounds will MacLoren receive, and how does this compare with MacLoren's expected sales receipt of £32,000 per car?

c. If MacLoren Automotive ships the same 10 cars to New Zealand at a time when the spot exchange rate is NZ$1.6500/£, how many New Zealand dollars will the distributor pay? How many pounds will MacLoren Automotive receive?

d. Does a risk-sharing agreement such as this one shift the currency exposure from one party of the transaction to the other?

e. Why is such a risk-sharing agreement of benefit to MacLoren? Why is it of benefit to the New Zealand distributor?

Problem 8.
(a)

Date	1Q 2007	2Q 2007	3Q 2007	4Q 2007	1Q 2008	2Q 2008
Price (millions of pounds, £)	£22.50	£22.50	£22.50	£22.50	£22.50	£22.50
Spot rate (euro = 1.00 pound)	1.4918	1.4733	1.4696	1.4107	1.3198	1.2617
Price (millions of euros, €)	€33.57					

Date	3Q 2008	4Q 2008	1Q 2009	2Q 2009	3Q 2009	4Q 2009
Price (millions of pounds, £)	£22.50	£22.50	£22.50	£22.50	£22.50	£22.50
Spot rate (euro = 1.00 pound)	1.2590	1.1924	1.1017	1.1375	1.1467	1.1066
Price (millions of euros, €)						

6. **Ganado Germany—All Domestic Competitors.** Using the Ganado Germany analysis in Exhibits 12.5 and 12.6 where the euro depreciates, how would prices, costs, and volumes change if Ganado Germany was operating in a mature mostly-domestic market with major domestic competitors

7. **Ganado Germany—All Foreign Competitors.** Ganado Germany is now competing in a number of international (export) markets, growth markets, in which most of its competitors are foreign. Now how would you expect Ganado Germany's operating exposure to respond to the depreciation of the euro?

8. **Rolls-Royce Turbine Engines.** Rolls-Royce is struggling with its pricing strategy with a number of its major customers in Continental Europe, particularly Airbus. Since Rolls-Royce is a British company with most manufacturing of the Airbus engines in the United Kingdom, costs are predominantly denominated in British pounds. But in the period shown in table (a) at the top of this page, 2007–2009, the pound steadily weakened against the euro. Rolls-Royce has traditionally denominated its sales contracts with Airbus in Airbus' home currency, the euro. After completing the table answer the following questions:

a. Assuming each Rolls-Royce engine marketed to Airbus is initially priced at £22.5 million each, how has the price of that engine changed over the period shown when priced in euros at the current spot rate?
b. What is the cumulative percentage change in the price of the engine in euros for the three-year period?
c. If the price elasticity of demand for Rolls-Royce turbine sales to Airbus is relatively inelastic, and the price of the engine in British pounds never changes over the period, what does this price change mean for Rolls-Royce's total sales revenue on sales to Airbus of this engine?
d. Compare the prices and volumes for the first quarter of each of the three years shown in table (b) below. Who has benefitted the most from the exchange rate changes?

9. **Hurte-Paroxysm Products, Inc. (A).** Hurte-Paroxysm Products, Inc. (HP) of the United States, exports computer printers to Brazil, whose currency, the reais (R$) has been trading at R$3.40/US$. Exports to Brazil are currently 50,000 printers per year at the reais equivalent of $200 each. A strong rumor exists that the reais will be devalued to R$4.00/$ within two weeks by the Brazilian government. Should the devaluation take place, the reais is expected to remain unchanged for another decade.

Problem 8.
(b)

	1Q 2007	1Q 2008	1Q 2009	% Chg
Price (in millions of £)	£22.50	£22.50	£22.50	
Spot rate (€/£)	1.4918	1.3198	1.1017	
Price (in millions of €)	€33.57	€29.70	€24.79	
Sales volume (engines)	200	220	240	20.0%
Total cost to Airbus (millions of €)	€6,713.10	€6,533.01	€5,949.18	
Total revenue to RR (millions of £)	£4,500.00	£4,950.00	£5,400.00	

Accepting this forecast as given, HP faces a pricing decision that must be made before any actual devaluation: HP may either (1) maintain the same reais price and in effect sell for fewer dollars, in which case Brazilian volume will not change, or (2) maintain the same dollar price, raise the reais price in Brazil to compensate for the devaluation, and experience a 20% drop in volume. Direct costs in the United States are 60% of the U.S. sales price. What would be the short-run (one-year) implication of each pricing strategy? Which do you recommend?

10. **Hurte-Paroxysm Products, Inc. (B).** Assume the same facts as in Problem 9. HP also believes that if it maintains the same price in Brazilian reais as a permanent policy, volume will increase at 10% per annum for six years, costs will not change. At the end of six years, HP's patent expires and it will no longer export to Brazil. After the reais is devalued to R$4.00/US$, no further devaluation is expected. If HP raises the price in reais so as to maintain its dollar price, volume will increase at only 4% per annum for six years, starting from the lower initial base of 40,000 units. Again, dollar costs will not change, and at the end of six years, HP will stop exporting to Brazil. HP's weighted average cost of capital is 12%. Given these considerations, what do you recommend for HP's pricing policy? Justify your recommendation.

INTERNET EXERCISES

1. **Operating Exposure: Recent Examples.** Using the following major periodicals as starting points, find a current example of a firm with a substantial operating exposure problem. To aid in your search, you might focus on businesses that have major operations in countries with recent currency crises, either through depreciation or major home currency appreciation.

Financial Times	www.ft.com/
The Economist	www.economist.com/
The Wall Street Journal	www.wsj.com/

2. **SEC Edgar Files.** To analyze an individual firm's operating exposure more carefully, it is necessary to have more detailed information available than in the normal annual report. Choose a specific firm with substantial international operations, for example, Coca-Cola or PepsiCo, and search the Security and Exchange Commission's Edgar Files for more detailed financial reports of their international operations.

Search SEC EDGAR Archives	www.sec.gov/cgi-bin/ srch-edgar

Financing the Global Firm

CHAPTER 13

The Global Cost and Availability of Capital

Capital must be propelled by self-interest; it cannot be enticed by benevolence.

— Walter Bagehot, 1826–1877.

LEARNING OBJECTIVES

- Explore the evolution of how corporate strategy and financial globalization may align
- Examine how the cost of capital in the capital asset pricing model changes for the multinational
- Evaluate the effect of market liquidity and segmentation on the cost of capital
- Compare the weighted average cost of capital for an MNE with its domestic counterpart

How can firms tap global capital markets for the purpose of minimizing their cost of capital and maximizing capital's availability? Why should they do so? Is global capital cheaper? This chapter explores these questions, concluding with the Mini-Case, *Novo Industri A/S (Novo)*, which details one of the most influential corporate financial strategies ever executed.

Financial Globalization and Strategy

Global integration of capital markets has given many firms access to new and cheaper sources of funds, beyond those available in their home markets. These firms can then accept more long-term projects and invest more in capital improvements and expansion. If a firm is located in a country with illiquid and/or segmented capital markets, it can achieve this lower global cost and greater availability of capital through a properly designed and implemented strategy. The dimensions of the cost and availability of capital are presented in Exhibit 13.1.

A firm that must source its long-term debt and equity in a highly illiquid domestic securities market will probably have a relatively high cost of capital and will face limited availability of such capital, which in turn will lower its competitiveness both internationally and vis-à-vis foreign firms entering its home market. This category of firms includes both firms resident in emerging countries, where the capital market remains undeveloped, and firms too small to gain access to their own national securities markets. Many family-owned firms find themselves in this category because they choose not to utilize securities markets to source their long-term capital needs.

EXHIBIT 13.1   **Dimensions of the Cost and Availability of Capital Strategy**

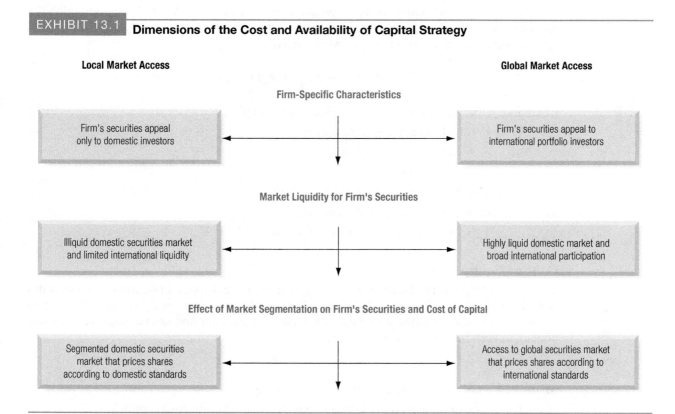

Firms resident in industrial countries with small capital markets often source their long-term debt and equity at home in these partially liquid domestic securities markets. The firms' cost and availability of capital is better than that of firms in countries with illiquid capital markets. However, if these firms can tap the highly liquid global markets, they can also strengthen their competitive advantage in sourcing capital.

Firms resident in countries with segmented capital markets must devise a strategy to escape dependence on that market for their long-term debt and equity needs. A national capital market is segmented if the required rate of return on securities in that market differs from the required rate of return on securities of comparable expected return and risk traded in other securities markets. Capital markets become segmented because of such factors as excessive regulatory control, perceived political risk, anticipated foreign exchange risk, lack of transparency, asymmetric availability of information, cronyism, insider trading, and many other market imperfections. Firms constrained by any of these conditions must develop a strategy to escape their own limited capital markets and source some of their long-term capital abroad.

Cost of Capital

A domestic firm normally finds its cost of capital by evaluating where and from whom it will raise its capital. The cost will obviously differ on the mix of investors interested in the firm, investors willing and able to buy its equity shares, and the debt available to the firm, raised from the domestic bank and debt market.

The firm then calculates its *weighted average cost of capital* (WACC) by combining the cost of equity with the cost of debt in proportion to the relative weight of each in the firm's optimal long-term financial structure. More specifically,

$$k_{WACC} = k_e \frac{E}{V} + k_d(1 - t)\frac{D}{V}$$

where k_{WACC} = weighted average after-tax cost of capital

k_e = risk-adjusted cost of equity

k_d = before-tax cost of debt

t = marginal tax rate

E = market value of the firm's equity

D = market value of the firm's debt

V = market value of the firm's securities $(D + E)$

Cost of Equity

The most widely accepted and used method of calculating the cost of equity for a firm today is the *capital asset pricing model* (CAPM). CAPM defines the cost of equity to be the sum of a risk-free interest component and a firm-specific spread, over and above that risk-free component, as seen in the following formula:

$$k_e = k_{rf} + \beta_j(k_m - k_{rf})$$

where k_e = expected (required) rate of return on equity

k_{rf} = rate of interest on risk-free bonds (Treasury bonds, for example)

β_j = coefficient of *systematic risk* for the firm (beta)

k_m = expected (required) rate of return on the market portfolio of stocks

The key component of CAPM is *beta* (β_j), the measure of systematic risk. *Systematic risk* is a measure of how the firm's returns vary with those of the market in which it trades. Beta is calculated as a function of the total variability of expected returns of the firm's stock relative to the market index (k_m) and the degree to which the variability of expected returns of the firm is correlated to the expected returns on the market index. More formally,

$$\beta_j = \frac{\rho_{jm}\sigma_j}{\sigma_m}$$

where β_j(beta) = measure of systematic risk for security j

ρ(rho) = correlation between security j and the market

σ_j(sigma) = standard deviation of the return on firm j

σ_m(sigma) = standard deviation of the market return

Beta will have a value of less than 1.0 if the firm's returns are less volatile than the market, 1.0 if the firm's returns are the same as the market, or greater than 1.0 if its returns are more volatile—or risky—than the market. CAPM analysis assumes that the required return estimated is an indication of what more is necessary to keep an investor's capital invested in the equity considered. If the equity's return does not reach the expected return, CAPM assumes that individual investors will liquidate their holdings.

CAPM's biggest challenge is that, for a beta to be most useful, it should be an indicator of the future rather than the past. A prospective investor is interested in how the individual firm's returns will vary in the coming periods. Unfortunately, since the future is not known, the beta used in any firm's estimate of equity cost is based on evidence from the recent past.

Cost of Debt

Firms acquire debt in either the form of loans from commercial banks—the most common form of debt—or as securities sold to the debt markets, such as instruments like notes and bonds. The normal procedure for measuring the cost of debt requires a forecast of interest rates for the next few years, the proportions of various classes of debt the firm expects to use, and the corporate income tax rate. The interest costs of the different debt components are then averaged according to their proportion in the debt structure. This before-tax average, k_d, is then adjusted for corporate income taxes by multiplying it by the expression $(1 - \text{tax rate})$, to obtain $k_d(1 - t)$, the weighted average after-tax cost of debt.

The weighted average cost of capital is normally used as the risk-adjusted discount rate whenever a firm's new projects are in the same general risk class as its existing projects. On the other hand, a project-specific required rate of return should be used as the discount rate if a new project differs from existing projects in business or financial risk.

International Portfolio Theory and Diversification

The potential benefits to companies from raising capital on global markets are based on international portfolio theory, the benefits of international diversification. We briefly review these principles before examining the costs and capacities for raising capital in the global market.

Portfolio Risk Reduction

The risk of a portfolio is measured by the ratio of the variance of the portfolio's return relative to the variance of the market return. This is the *beta* of the portfolio. As an investor increases the number of securities in a portfolio, the portfolio's risk declines rapidly at first, and then asymptotically approaches the level of *systematic* risk of the market. The total risk of any portfolio is therefore composed of *systematic risk* (the market) and *unsystematic risk* (the individual securities). Increasing the number of securities in the portfolio reduces the unsystematic risk component but leaves the systematic risk component unchanged. A fully diversified domestic portfolio would have a beta of 1.0. This is standard—domestic—financial theory.

Exhibit 13.2 illustrates the incremental gains of diversifying both domestically and internationally. The lower line in Exhibit 13.2 (portfolio of international stocks) represents a portfolio in which foreign securities have been added. It has the same overall risk shape as the U.S. stock portfolio, but it has a lower portfolio beta. This means that the international portfolio's market risk is lower than that of a domestic portfolio. This situation arises because the returns on the foreign stocks are not perfectly correlated with U.S. stocks.

Foreign Exchange Risk

The foreign exchange risks of a portfolio, whether it is a securities portfolio or the general portfolio of activities of the MNE, are reduced through international diversification. The construction of internationally diversified portfolios is both the same as and different from creating a traditional domestic portfolio. Internationally diversified portfolios are the same in principle because the investor is attempting to combine assets that are less than perfectly correlated, reducing the total risk of the portfolio. In addition, by adding assets from outside the home market—assets that previously were not available to be averaged into the portfolio's expected returns and risks—the investor is tapping into a larger pool of potential investments.

But international portfolio construction is also different in that when the investor acquires assets or securities from outside the investor's host-country market, the investor may also be

EXHIBIT 13.2   **Market Liquidity, Segmentation, and the Marginal Cost of Capital**

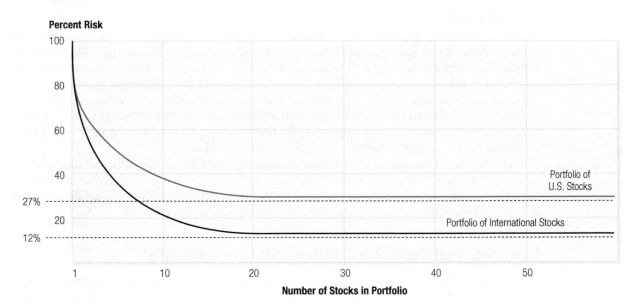

When the portfolio is diversified internationally, the portfolio's beta—the level of systematic risk which cannot be diversified away—is lowered.

acquiring a *foreign currency-denominated asset*[1]. Thus, the investor has actually acquired two additional assets—the currency of denomination and the asset subsequently purchased with the currency—one asset in principle, but two in expected returns and risks.

Japanese Equity Example. A numerical example can illustrate the difficulties associated with international portfolio diversification and currency risk. A U.S.-based investor takes $1,000,000 on January 1, and invests in shares traded on the Tokyo Stock Exchange (TSE). The spot exchange rate on January 1 is ¥130.00/$. The $1 million therefore yields ¥130,000,000. The investor uses ¥130,000,000 to acquire shares on the Tokyo Stock Exchange at ¥20,000 per share, acquiring 6,500 shares, and holds the shares for one year.

At the end of one year the investor sells the 6,500 shares at the market price, which is now ¥25,000 per share; the shares have risen ¥5,000 per share in price. The 6,500 shares at ¥25,000 per share yield proceeds of ¥162,500,000.

The Japanese yen are then changed back into the investor's home currency, the U.S. dollar, at the spot rate of ¥125.00/$ now in effect. This results in total U.S. dollar proceeds of $1,300,000.00. The total return on the investment is then

$$\frac{\$1,300,000 - \$1,000,000}{\$1,000,000} = 30.00\%$$

[1]This is not always the case. For example, many U.S.-based investors routinely purchase and hold Eurodollar bonds on the secondary market only (illegal during primary issuance), which would not pose currency risk to the U.S.-based investor for they are denominated in the investor's home currency.

The total U.S. dollar return is actually a combination of the return on the Japanese yen (which in this case was positive) and the return on the shares listed on the Tokyo Stock Exchange (which was also positive). This value is expressed by isolating the percentage change in the share price (r^{shares}) in combination with the percentage change in the currency value ($r^{¥/\$}$):

$$R^\$ = [(1 + r^{¥/\$})(1 + r^{shares, ¥})] - 1$$

In this case, the value of the Japanese yen, in the eyes of a U.S.-based investor, rose 4.00% (from ¥130/\$ to ¥125/\$—see percentage change calculation in Chapter 5), while the shares traded on the Tokyo Stock Exchange rose 25.00%. The total investment return in U.S. dollars is therefore

$$R^\$ = [(1 + .0400)(1 + .2500)] - 1 = .3000 \text{ or } 30.00\%$$

Obviously, the risk associated with international diversification, when it includes currency risk, is inherently more complex than that of domestic investments. You should also see, however, that the presence of currency risk may alter the correlations associated with securities in different countries and currencies, providing new portfolio composition and diversification possibilities. In conclusion:

▨ International diversification benefits induce investors to demand foreign securities (the so-called "buy-side").

▨ If the addition of a foreign security to the portfolio of the investor aids in the reduction of risk for a given level of return, or if it increases the expected return for a given level of risk, then the security adds value to the portfolio.

▨ Any security that adds value will be in demand by investors. Given the limits of the potential supply of securities, increased demand will bid up the price of the security, resulting in a lower cost of capital for the firm. The firm issuing the security, the "sell-side," is therefore able to raise capital at a lower cost.

International CAPM (ICAPM)

The traditional form of CAPM, the domestic CAPM discussed earlier, assumes the firm's equity trades in a purely domestic market. The beta and market risk premium ($k_m - k_{rf}$) used in the cost of equity calculation, therefore, were based on a purely domestic market of securities and choices. But what if globalization has opened up the global markets, integrating them, and allowing investors to choose among stocks of a global portfolio?

International CAPM (ICAPM) assumes that there is a global market in which the firm's equity trades, and estimates of the firm's beta, β_j^g and the market risk premium, ($k_m^g - k_{rf}^g$), must then reflect this global portfolio.

$$k_e^{global} = k_{rf}^g + \beta_j^g (k_m^g - k_{rf}^g)$$

The value of the risk-free rate, k_{rf}^g, may not change (so that $k_{rf}^g = k_{rf}$), as a U.S. Treasury note may be the risk-free rate for a U.S.-based investor regardless of the domestic or international portfolio. The market return, k_m^g, will change, reflecting average expected global market returns for the coming periods. The firm's beta, β_j^g, will most assuredly change, as it will now reflect the expected variations against a greater global portfolio. How that beta will change, however, depends.

Sample Calculation: Ganado's Cost of Capital

Maria Gonzalez, Ganado's Chief Financial Officer, wants to calculate the company's weighted average cost of capital in both forms, the traditional CAPM and also ICAPM.

Maria assumes the risk-free rate of interest (k_{rf}) as 4%, using the U.S. government 10-year Treasury bond rate. The expected rate of return of the market portfolio (k_m) is assumed to be 9%, the expected rate of return on the market portfolio held by a well-diversified domestic investor. Ganado's estimate of its own systematic risk—its beta—against the domestic portfolio is 1.2. Ganado's cost of equity (sometimes termed the *price of equity*) is then

$$k_e = k_{rf} + \beta(k_m - k_{rf}) = 4.00\% + 1.2(9.00\% - 4.00\%) = 10.00\%.$$

Ganado's cost of debt (k_d) the before tax cost of debt estimated by observing the current yield on Ganado's outstanding bonds combined with bank debt, is 8%. Using 35% as the corporate income tax rate for the United States, Ganado's after-tax cost of debt is then

$$k_d(1 - t) = 8.00(1 - 0.35) = 8.00(0.65) = 5.20\%.$$

Ganado's long-term capital structure is 60% equity (E/V) and 40% debt (D/V), where V is Ganado's total market value. Ganado's weighted average cost of capital k_{WACC} is then

$$k_{WACC} = k_e\frac{E}{V} + k_d(1 - t)\frac{D}{V} = 10.00\%(.60) + 5.20\%(.40) = 8.08\%.$$

This is Ganado's cost of capital using the traditional domestic CAPM estimate of the cost of equity.

But Maria Gonzalez wonders if this is the proper approach for Ganado. As Ganado has globalized its own business activities, the investor base that owns Ganado's shares has also globally diversified. Ganado's shares are now listed in London and Tokyo, in addition to their home listing on the New York Stock Exchange. Over 40% of Ganado's stock is now held in the globally diversified portfolios of foreign investors (outside the United States).

A second calculation of Ganado's cost of equity, this time using the ICAPM, yields different results. Ganado's beta, when calculated against a larger global equity market index, which includes these foreign markets and their investors, is a lower 0.90. The expected market return of 8.00% for a larger globally integrated equity market is lower as well. The ICAPM cost of equity is a much lower value of 7.60%.

$$k_e^{global} = k_{rf}^g + \beta_j^g(k_m^g - k_{rf}^g) = 4.00\% + 0.90(8.00\% - 4.00\%) = 7.60\%$$

Maria now recalculates Ganado's WACC using the ICAPM estimate of equity costs, assuming the same debt and equity proportions and the same cost of current debt. Ganado's WACC is now estimated at a lower cost of 6.64%.

$$k_{WACC}^{ICAPM} = k_e^{global}\frac{E}{V} + k_d(1 - t)\frac{D}{V} = 7.60\%(.60) + 5.20\%(.40) = 6.64\%$$

Maria believes that this is a more appropriate estimate of Ganado's cost of capital. It is fully competitive with Ganado's main rivals in the telecommunications hardware industry segment worldwide, which are mainly headquartered in the United States, the United Kingdom, Canada, Finland, Sweden, Germany, Japan, and the Netherlands. The key to Ganado's favorable global cost and availability of capital going forward is its ability to attract and hold the international portfolio investors that own its stock.

ICAPM Considerations

In theory, the primary distinction in the estimation of the cost of equity for an individual firm using an internationalized version of the CAPM is the definition of the "market" and a recalculation of the firm's beta for that market. The three basic components of the CAPM model must then be reconsidered.

Nestlé, the Swiss-based multinational firm that produces and distributes a variety of confectionary products, serves as an excellent example of how the international investor may view the global cost of capital differently from a domestic investor, and what that means for Nestlé's estimate of its own cost of equity[2]. The numerical example for Nestlé is summarized in Exhibit 13.3.

In the case of Nestlé, a prospective Swiss investor might assume a risk-free return in Swiss francs of 3.3%—the rate of return on an index of Swiss government bond issues. That same Swiss investor might also assume an expected market return in Swiss francs of 10.2%—an average return on a portfolio of Swiss equities, the *Financial Times* Swiss index. Assuming a risk-free rate of 3.30%, an expected market return of 10.2%, and a $\beta_{\text{Nestlé}}$ of 0.885, a Swiss investor would expect Nestlé to yield 9.4065% for the coming year.

$$k_e^{\text{Nestlé}} = k_{\text{RF}} + (K_{\text{M}} - k_{\text{RF}})\beta_{\text{Nestlé}} = 3.3 + (10.2 - 3.3)\,0.885 = 9.4065\%.$$

But what if Swiss investors held internationally diversified portfolios instead? Both the expected market return and the beta estimate for Nestlé itself would be defined and determined differently. For the same period as before, a global portfolio index such as the *Financial Times* index in Swiss francs (FTA-Swiss) would show a market return of 13.7% (as opposed to the domestic Swiss index return of 10.2%). In addition, a beta for Nestlé estimated on Nestlé's returns versus the global portfolio index would be much smaller, 0.585 (as opposed to the 0.885

EXHIBIT 13.3   **The Cost of Equity for Nestlé of Switzerland**

Nestlé's estimate of its cost of equity will depend upon whether a Swiss investor is thought to hold a domestic portfolio of equity securities or a global portfolio.

Domestic Portfolio for Swiss Investor	Global Portfolio for Swiss Investor
k_{RF} = 3.3% (Swiss bond index yield)	k_{RF} = 3.3% (Swiss bond index yield)
k_{M} = 10.2% (Swiss market portfolio in SF)	k_{M} = 13.7% (*Financial Times* Global index in SF)
$\beta_{\text{Nestlé}}$ = 0.885 (Nestlé versus Swiss market portfolio)	$\beta_{\text{Nestlé}}$ = 0.585 (Nestlé versus FTA-Swiss index)

$$k_{\text{Nestlé}} = k_{\text{RF}} + \beta_{\text{Nestlé}}(k_{\text{M}} - k_{\text{RF}})$$

Required return on Nestlé:	Required return on Nestlé:
$k_e^{\text{Nestlé}}$ = 9.4065%	$k_e^{\text{Nestlé}}$ = 9.3840%

Source: All values are taken from René Stulz, "The Cost of Capital in Internationally Integrated Markets: The Case of Nestlé," *European Financial Management*, Vol. 1, No. 1, March 1995, pp. 11–22.

---

[2]René Stulz, "The Cost of Capital in Internationally Integrated Markets: The Case of Nestlé," *European Financial Management*, Vol. 1, No. 1, March 1995, pp. 11–22.

found previously). An internationally diversified Swiss investor would expect the following return on Nestlé:

$$k_e^{\text{Nestlé}} = k_{\text{RF}} + (k_{\text{M}} - k_{\text{RF}})\beta_{\text{Nestlé}} = 3.3 + (13.7 - 3.3)\, 0.585 = 9.3840\%.$$

Admittedly, this is not a lot of difference in the end. However, given the magnitude of change in both the values of the market return average and the beta for the firm, it is obvious that the result could easily have varied by several hundred basis points. The proper construction of the investor's portfolio and the proper portrayal of the investor's perceptions of risk and opportunity cost are clearly important to identifying the global cost of a company's equity capital. In the end, it all depends on the specific case—the firm, the country-market, and the *global portfolio*.

We follow the practice here of describing the internationally diversified portfolio as the global portfolio rather than the *world portfolio*. The distinction is important. The *world portfolio* is an index of all securities in the world. However, even with the increasing trend of deregulation and financial integration, a number of securities markets still remain segmented or restricted in their access. Those securities actually available to an investor are the *global portfolio*.

There are, in fact, a multitude of different proposed formulations for calculating the international cost of capital. The problems with both formulation and data expand dramatically as the analysis is extended to rapidly developing or emerging markets. Harvey (2005) serves as a first place to start if you wish to expand your reading and research.[3]

Global Betas

International portfolio theory typically concludes that adding international securities to a domestic portfolio will reduce the portfolio's risks. Although this idea is fundamental to much of international financial theory, it still depends on individual firms in individual markets. Nestlé's beta went down when calculated using a global portfolio of equities, but that may not always be the case. Depending on the firm, its business line, the country it calls home, and the industry domestically and globally in which it competes, the global beta may go up or down.

One company often noted by researchers is Petrobrás, the national oil company of Brazil. Although government controlled, the company is publicly traded. Its shares are listed in São Paulo and New York. It operates in a global oil market in which prices and values are set in U.S. dollars. As a result, its domestic or home beta has been estimated at 1.3, but its global beta higher, at 1.7[4]. This is only one example of many.

Although it seems obvious to some that the returns to the individual firm should become less correlated to those of the market as the market is redefined ever-larger, it turns out to be more of a case of empirical analysis, not preconceived notions of correlation and covariance.

Equity Risk Premiums

In practice, calculating a firm's equity risk premium is much more controversial. Although the capital asset pricing model (CAPM) has now become widely accepted in global business as the preferred method of calculating the cost of equity for a firm, there is rising debate over what numerical values should be used in its application, especially the *equity risk premium*.

---

[3]"12 Ways to Calculate the International Cost of Capital," Campbell R. Harvey, Duke University, unpublished, October 14, 2005.

[4]*The Real Cost of Capital*, Tim Ogier, John Rugman, and Lucinda Spicer, Financial Times Prentice Hall, Pearson Publishing, 2005, p. 139.

EXHIBIT 13.4	**Alternative Estimates of Cost of Equity for a Hypothetical U.S. Firm Assuming** $\beta = 1$ **and** $k_{rf} = 4\%$		
Source	**Equity Risk Premium** $(k_m - k_{rf})$	**Cost of Equity** $k_{rf} + \beta\,(k_m - k_{rf})$	**Differential**
Ibbotson	8.800%	12.800%	3.800%
Finance textbooks	8.500%	12.500%	3.500%
Investor surveys	7.100%	11.100%	2.100%
Dimson, et al.	5.000%	9.000%	Baseline

Source: Equity risk premium quotes from "Stockmarket Valuations: Great Expectations," *The Economist*, January 31, 2002.

The equity risk premium is the average annual return of the market expected by investors over and above riskless debt, the term $(k_m - k_{rf})$.

The field of finance does agree that a cost of equity calculation should be forward looking, meaning that the inputs to the equation should represent what is expected to happen over the relevant future time horizon. As is typically the case, however, practitioners use historical evidence as the basis for their forward-looking projections. The current debate begins with a debate over what has happened in the past.

In a large study completed in 2001 by Dimson, Marsh, and Stanton (updated in 2003), the authors estimated the equity risk premium in 16 different developed countries for the 1900–2002 period. The study found significant differences in equity returns over bill and bond returns (proxies for the risk-free rate) over time by country. For example, Italy was found to have had the highest equity risk premium, 10.3%, followed by Germany with 9.4% and Japan at 9.3%. Denmark had the lowest at 3.8%.

The debate over which equity risk premium to use in practice was highlighted in this same study by looking at what equity risk premiums are being recommended for the United States by a variety of different sources. As illustrated in Exhibit 13.4, a hypothetical firm with a beta of 1.0 (estimated market risk equal to that of the market) might have a cost of equity as low as 9.000% and as high as 12.800% using this set of alternative values. Note that here the authors used geometric returns, not arithmetic returns. Fernandez and del Campo (2010), in their annual survey of market risk premiums used by analysts and academics, found the average risk premium used by U.S. and Canadian analysts is 5.1%, European analysts 5.0%, and British analysts 5.6%.

How important is it for a company to accurately predict its cost of equity? The corporation must annually determine which potential investments it will accept and reject due to its limited capital resources. If the company is not accurately estimating its cost of equity, and therefore its general cost of capital, it will not be accurately estimating the net present value of potential investments.

The Demand for Foreign Securities:
The Role of International Portfolio Investors

Gradual deregulation of equity markets during the past three decades not only elicited increased competition from domestic players but also opened up markets to foreign competitors. International portfolio investment and the cross-listing of equity shares on foreign markets have become commonplace.

What motivates portfolio investors to purchase and hold foreign securities in their portfolios? The answer lies in an understanding of "domestic" portfolio theory and how it has been extended to handle the possibility of global portfolios. More specifically, it requires an understanding of the principles of portfolio risk reduction, portfolio rate of return, and foreign currency risk.

Both domestic and international portfolio managers are asset allocators. Their objective is to maximize a portfolio's rate of return for a given level of risk, or to minimize risk for a given rate of return. International portfolio managers can choose from a larger bundle of assets than portfolio managers limited to domestic-only asset allocations. As a result, internationally diversified portfolios often have a higher expected rate of return, and they nearly always have a lower level of portfolio risk, since national securities markets are imperfectly correlated with one another.

Portfolio asset allocation can be accomplished along many dimensions depending on the investment objective of the portfolio manager. For example, portfolios can be diversified according to the type of securities. They can be composed of stocks only or bonds only or a combination of both. They also can be diversified by industry or by size of capitalization (small-cap, mid-cap, and large-cap stock portfolios).

For our purposes, the most relevant dimensions are diversification by country, geographic region, stage of development, or a combination of these (global). An example of diversification by country is the Korea Fund. It was at one time the only vehicle allowing foreign investors to hold South Korean securities, but foreign ownership restrictions have more recently been liberalized. A typical regional diversification would be one of the many Asian funds. These performed exceptionally well until the "bubble" burst in Japan and Southeast Asia during the second half of the 1990s. Portfolios composed of emerging market securities are examples of diversification by stage of development. They are composed of securities from different countries, geographic regions, and stage of development.

The Link Between Cost and Availability of Capital

Ganado's weighted average cost of capital (WACC) was calculated assuming that equity and debt capital would always be available at the same required rate of return even if Ganado's capital budget expands. This is a reasonable assumption considering Ganado's excellent access through the NYSE to international portfolio investors in global capital markets. It is a bad assumption, however, for firms resident in illiquid or segmented capital markets, small domestic firms, and family-owned firms resident in any capital market. We will now examine how market liquidity and market segmentation can affect a firm's cost of capital.

Improving Market Liquidity

Although no consensus exists about the definition of market liquidity, we can observe market liquidity by noting the degree to which a firm can issue a new security without depressing the existing market price. In the domestic case, an underlying assumption is that total availability of capital to a firm at any time is determined by supply and demand in the domestic capital markets.

A firm should always expand its capital budget by raising funds in the same proportion as its optimal financial structure. As its budget expands in absolute terms, however, its marginal cost of capital will eventually increase. In other words, a firm can only tap the capital market for some limited amount in the short run before suppliers of capital balk at providing further funds, even if the same optimal financial structure is preserved. In the long run, this may not be a limitation, depending on market liquidity.

In the multinational case, a firm is able to improve market liquidity by raising funds in the euromarkets (money, bond, and equity), by selling security issues abroad, and by tapping local capital markets through foreign subsidiaries. Such activities should logically expand the capacity of an MNE to raise funds in the short run over what might have been raised if the firm were limited to its home capital market. This situation assumes that the firm preserves its optimal financial structure.

Market Segmentation

If all capital markets are fully integrated, securities of comparable expected return and risk should have the same required rate of return in each national market after adjusting for foreign exchange risk and political risk. This definition applies to both equity and debt, although it often happens that one or the other may be more integrated than its counterpart.

Capital market segmentation is a financial market imperfection caused mainly by government constraints, institutional practices, and investor perceptions. The following are the most important imperfections:

- Asymmetric information between domestic and foreign-based investors
- Lack of transparency
- High securities transaction costs
- Foreign exchange risks
- Political risks
- Corporate governance differences
- Regulatory barriers

Market imperfections do not necessarily imply that national securities markets are inefficient. A national securities market can be efficient in a domestic context and yet segmented in an international context. According to finance theory, a market is efficient if security prices in that market reflect all available relevant information and adjust quickly to any new relevant information. Therefore, the price of an individual security reflects its "intrinsic value," and any price fluctuations will be "random walks" around this value. Market efficiency assumes that transaction costs are low, that many participants are in the market, and that these participants have sufficient financial strength to move security prices. Empirical tests of market efficiency show that most major national markets are reasonably efficient.

An efficient national securities market might very well correctly price all securities traded in that market on the basis of information available to the investors who participate in that market. However, if that market is segmented, foreign investors would not be participants.

Availability of capital depends on whether a firm can gain liquidity for its debt and equity securities and a price for those securities based on international rather than national standards. In practice, this means that the firm must define a strategy to attract international portfolio investors and thereby escape the constraints of its own illiquid or segmented national market.

The Effect of Market Liquidity and Segmentation

The degree to which capital markets are illiquid or segmented has an important influence on a firm's marginal cost of capital and thus on its weighted average cost of capital. The marginal cost of capital is the weighted average cost of the next currency unit raised. This is illustrated in Exhibit 13.5, which shows the transition from a domestic to a global marginal cost of capital.

Exhibit 13.5 shows that the MNE has a given marginal return on capital at different budget levels, represented in the line MRR. This demand is determined by ranking potential projects according to net present value or internal rate of return. Percentage rate of return to both users

Market Liquidity, Segmentation, and the Marginal Cost of Capital

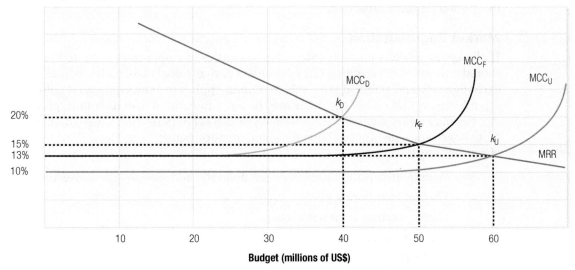

and suppliers of capital is shown on the vertical scale. If the firm is limited to raising funds in its domestic market, the line MCC_D shows the marginal domestic cost of capital (vertical axis) at various budget levels (horizontal axis). Remember that the firm continues to maintain the same debt ratio as it expands its budget, so that financial risk does not change. The optimal budget in the domestic case is $40 million, where the marginal return on capital (MRR) just equals the marginal cost of capital (MCC_D). At this budget, the marginal domestic cost of capital, k_D, would be equal to 20%.

If the MNE has access to additional sources of capital outside an illiquid domestic capital market, the marginal cost of capital should shift to the right (the line MCC_F). In other words, foreign markets can be tapped for long-term funds at times when the domestic market is saturated because of heavy use by other borrowers or equity issuers, or when it is unable to absorb another issue of the MNE in the short run. Exhibit 13.5 shows that by a tap of foreign capital markets the firm has reduced its marginal international cost of capital, to 15%, even while it raises an additional $10 million. This statement assumes that about $20 million is raised abroad, since only about $30 million could be raised domestically at a 15% marginal cost of capital.

If the MNE is located in a capital market that is both illiquid and segmented, the line represents the decreased marginal cost of capital if it gains access to other equity markets. As a result of the combined effects of greater availability of capital and international pricing of the firm's securities, the marginal cost of capital, k_U, declines to 13% and the optimal capital budget climbs to $60 million.

Most of the tests of market segmentation suffer from the usual problem for models—namely, the need to abstract from reality in order to have a testable model. In our opinion, a realistic test would be to observe what happens to a single security's price when, after it has been traded only in a domestic market, it is "discovered" by foreign investors, and is then traded in a foreign market. Arbitrage should keep the market price equal in both markets. However, if during the transition we observe a significant change in the security's price, uncorrelated with price movements in either of the underlying securities markets, we can infer that the domestic market was segmented.

In academic circles, tests based on case studies are often considered to be "casual empiricism," since no theory or model exists to explain what is being observed. Nevertheless, something may be learned from such cases, just as scientists learn from observing nature in an uncontrolled environment. Furthermore, case studies that preserve real-world complications may illustrate specific kinds of barriers to market integration and ways in which they might be overcome.

Unfortunately, few case studies have been documented in which a firm has "escaped" from a segmented capital market. In practice, escape usually means being listed on a foreign stock market such as New York or London, and/or selling securities in foreign capital markets. We will explore one firm's escape from a segmented market with a discussion of Novo in the Mini-Case at the end of the chapter.

Globalization of Securities Markets

During the 1980s, numerous Nordic and other European firms cross-listed on major foreign exchanges such as London and New York. They placed equity and debt issues in major securities markets. In most cases, they were successful in lowering their WACC and increasing its availability. This is the subject of this chapter's Mini-Case.

During the 1990s, national restrictions on cross-border portfolio investment were gradually eased under pressure from the Organization for Economic Cooperation and Development (OECD), a consortium of most of the world's most industrialized countries. Liberalization of European securities markets was accelerated because of the European Union's efforts to develop a single European market without barriers. Emerging nation markets followed suit, as did the former Eastern Bloc countries after the breakup of the Soviet Union. Emerging national markets have often been motivated by the need to source foreign capital to finance large-scale privatization.

Now, market segmentation has been significantly reduced, although the liquidity of individual national markets remains limited. Most observers believe that for better or for worse, we have achieved a global market for securities. The good news is that many firms have been assisted to become MNEs because they now have access to a global cost and availability of capital. The bad news is that the correlation among securities markets has increased, thereby reducing, but not eliminating, the benefits of international portfolio diversification. Globalization of securities markets has also led to more volatility and speculative behavior, as shown by the emerging market crises of the 1995–2001 period, and the 2008–2009 global credit crisis.

Corporate Governance and the Cost of Capital. Would global investors be willing to pay a premium for a share in a good corporate governance company? A recent study of Norwegian and Swedish firms measured the impact of foreign board membership (Anglo-American) on firm value. They summarized their findings as follows:[5]

> Using a sample of firms with headquarters in Norway or Sweden the study indicates a significantly higher value for firms that have outsider Anglo-American board member(s), after a variety of firm-specific and corporate governance related factors have been controlled for. We argue that this superior performance reflects the fact that these companies have successfully broken away from a partly segmented domestic capital market by "importing" an Anglo-American corporate governance system. Such an "import" signals a willingness on the part of the firm to expose itself to improved corporate governance and enhances its reputation in the financial market.

[5]Lars Oxelheim and Trond Randøy, "The impact of foreign board membership on firm value," *Journal of Banking and Finance*, Vol. 27, No. 12, 2003, p. 2369.

Strategic Alliances

Strategic alliances are normally formed by firms that expect to gain synergies from joint efforts. For example, allied firms might share the cost of developing technology or pursue complementary marketing activities. These firms might gain economies of scale or scope or a variety of other commercial advantages. However, one synergy that may sometimes be overlooked is the possibility for a financially strong firm to help a financially weak firm to lower its cost of capital by providing attractively priced equity or debt financing.

The Cost of Capital for MNEs Compared to Domestic Firms

Is the weighted average cost of capital for MNEs higher or lower than for their domestic counterparts? The answer is a function of the marginal cost of capital, the relative after-tax cost of debt, the optimal debt ratio, and the relative cost of equity.

Availability of Capital

Earlier in this chapter, we saw that international availability of capital to MNEs, or to other large firms that can attract international portfolio investors, may allow them to lower their cost of equity and debt compared with most domestic firms. In addition, international availability permits an MNE to maintain its desired debt ratio, even when significant amounts of new funds must be raised. In other words, an MNE's marginal cost of capital is constant for considerable ranges of its capital budget. This statement is not true for most domestic firms. They must either rely on internally generated funds or borrow in the short and medium term from commercial banks.

Financial Structure, Systematic Risk, and the Cost of Capital for MNEs

Theoretically, MNEs should be in a better position than their domestic counterparts to support higher debt ratios because their cash flows are diversified internationally. The probability of a firm's covering fixed charges under varying conditions in product, financial, and foreign exchange markets should improve if the variability of its cash flows is minimized.

By diversifying cash flows internationally, the MNE might be able to achieve the same kind of reduction in cash flow variability as portfolio investors receive from diversifying their security holdings internationally. The same argument applies to cash flow diversification—namely that returns are not perfectly correlated between countries. For example, in 2000 Japan was in recession, but the United States was experiencing rapid growth. Therefore, we might have expected returns, on either a cash flow or an earnings basis, to be depressed in Japan and favorable in the United States. An MNE with operations located in both these countries could rely on its strong U.S. cash inflow to cover debt obligations, even if its Japanese subsidiary produced weak net cash inflows.

Despite the theoretical elegance of this hypothesis, empirical studies have come to the opposite conclusion.[6] Despite the favorable effect of international diversification of cash flows, bankruptcy risk was only about the same for MNEs as for domestic firms. However, MNEs faced higher agency costs, political risk, foreign exchange risk, and asymmetric information. These have been identified as the factors leading to lower debt ratios and even a higher cost of long-term debt for MNEs. Domestic firms rely much more heavily on short and intermediate debt, which lie at the low cost end of the yield curve.

---

[6]Kwang Chul Lee, and Chuck C.Y. Kwok, "Multinational Corporations vs. Domestic Corporations: International Environmental Factors and Determinants of Capital Structure," *Journal of International Business Studies*, Summer 1988, pp. 195–217.

Even more surprising, one study found that MNEs have a higher level of systematic risk than their domestic counterparts.[7] The same factors caused this phenomenon as caused the lower debt ratios for MNEs. The study concluded that the increased standard deviation of cash flows from internationalization more than offset the lower correlation from diversification.

As we stated earlier, the systematic risk term, β_j, is defined as

$$\beta_j = \frac{\rho_{jm}\sigma_j}{\sigma_m}$$

where ρ_{jm} is the correlation coefficient between security j and the market; σ_j is the standard deviation of the return on firm j; and σ_m is the standard deviation of the market return. The MNE's systematic risk could increase if the decrease in the correlation coefficient, ρ_{jm}, due to international diversification, is more than offset by an increase in σ_j, the standard deviation due to the aforementioned risk factors. This conclusion is consistent with the observation that many MNEs use a higher hurdle rate to discount expected foreign project cash flows. In essence, they are accepting projects they consider to be riskier than domestic projects, thus potentially skewing upward their perceived systematic risk. At the least, MNEs need to earn a higher rate of return than their domestic equivalents in order to maintain their market value.

Other studies have found that internationalization actually allows emerging market MNEs to carry a higher level of debt and lower their systematic risk. This occurs because the emerging market MNEs are investing in more stable economies abroad, a strategy that lowers their operating, financial, foreign exchange, and political risks. The reduction in risk more than offsets increased agency costs and allows the firms to enjoy higher leverage and lower systematic risk than their U.S.-based MNE counterparts.

The Riddle: Is the Cost of Capital Higher for MNEs?

The riddle is that the MNE is supposed to have a lower marginal cost of capital (MCC) than a domestic firm because of the MNE's access to a global cost and availability of capital. On the other hand, the empirical studies we mentioned show that the MNE's weighted average cost of capital (WACC) is actually higher than for a comparable domestic firm because of agency costs, foreign exchange risk, political risk, asymmetric information, and other complexities of foreign operations.

The answer to this riddle lies in the link between the cost of capital, its availability, and the opportunity set of projects. As the opportunity set of projects increases, eventually the firm needs to increase its capital budget to the point where its marginal cost of capital is increasing. The optimal capital budget would still be at the point where the rising marginal cost of capital equals the declining rate of return on the opportunity set of projects. However, this would be at a higher weighted average cost of capital than would have occurred for a lower level of the optimal capital budget.

To illustrate this linkage, Exhibit 13.6 shows the marginal cost of capital given different optimal capital budgets. Assume that there are two different demand schedules based on the opportunity set of projects for both the multinational enterprise (MNE) and domestic counterpart (DC).

The line MRR$_{DC}$ depicts a modest set of potential projects. It intersects the line MCC$_{MNE}$ at 15% and a \$100 million budget level. It intersects the MCC$_{DC}$ at 10% and a \$140 million budget level. At these low budget levels the MCC$_{MNE}$ has a higher MCC and probably

[7] David M. Reeb, Chuck C.Y. Kwok, and H. Young Back, "Systematic Risk of the Multinational Corporation," *Journal of International Business Studies*, Second Quarter 1998, pp. 263–279.

EXHIBIT 13.6 **The Cost of Capital for MNE and Domestic Counterpart Compared**

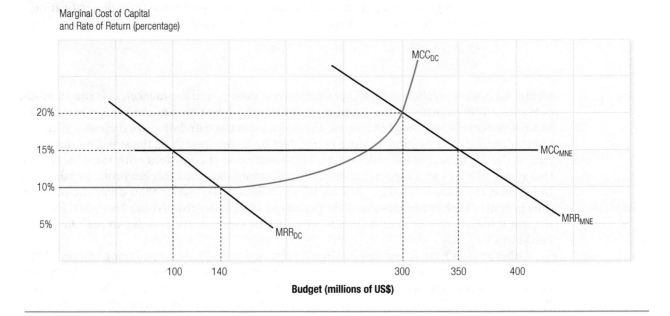

weighted average cost of capital than its domestic counterpart (DC), as discovered in the recent empirical studies.

The line MRR_{MNE} depicts a more ambitious set of projects for both the MNE and its domestic counterpart. It intersects the line MCC_{MNE} still at 15% and a $350 million budget. However, it intersects the MCC_{DC} at 20% and a budget level of $300 million. At these higher budget levels, the MCC_{MNE} has a lower MCC and probably weighted average cost of capital than its domestic counterpart, as predicted earlier in this chapter. In order to generalize this conclusion, we would need to know under what conditions a domestic firm would be willing to undertake the optimal capital budget despite its increasing the firm's marginal cost of capital. At some point the MNE might also have an optimal capital budget at the point where its MCC is rising.

Empirical studies show that neither mature domestic firms nor MNEs are typically willing to assume the higher agency costs or bankruptcy risk associated with higher MCCs and capital budgets. In fact, most mature firms demonstrate some degree of corporate wealth maximizing behavior. They are somewhat risk averse and tend to avoid returning to the market to raise fresh equity. They prefer to limit their capital budgets to what can be financed with free cash flows. Indeed, they have a so-called pecking order that determines the priority of which sources of funds they will tap and in what order. This behavior motivates shareholders to monitor management more closely. They tie management's compensation to stock performance (options). They may also require other types of contractual arrangements that are collectively part of agency costs.

In conclusion, if both MNEs and domestic firms do actually limit their capital budgets to what can be financed without increasing their MCC, then the empirical findings that MNEs have higher WACC stands. If the domestic firm has such good growth opportunities that it chooses to undertake growth despite an increasing marginal cost of capital, then the MNE would have a lower WACC. Exhibit 13.7 summarizes these conclusions.

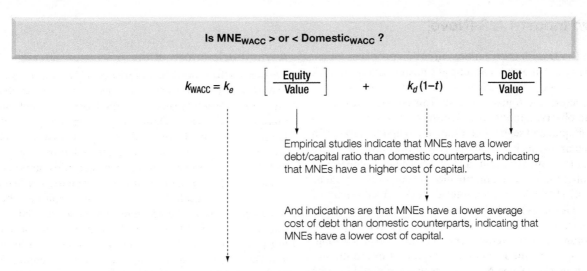

EXHIBIT 13.7   **Do MNEs Have a Higher or Lower Cost of Capital Than Their Domestic Counterparts?**

Is MNE$_{WACC}$ > or < Domestic$_{WACC}$?

$$k_{WACC} = k_e \left[\frac{\text{Equity}}{\text{Value}} \right] + k_d(1-t) \left[\frac{\text{Debt}}{\text{Value}} \right]$$

Empirical studies indicate that MNEs have a lower debt/capital ratio than domestic counterparts, indicating that MNEs have a higher cost of capital.

And indications are that MNEs have a lower average cost of debt than domestic counterparts, indicating that MNEs have a lower cost of capital.

The cost of equity required by investors is higher for multinational firms than for domestic firms. Possible explanations are higher levels of *political risk*, *foreign exchange risk*, and higher *agency costs* of doing business in a multinational managerial environment. However, at relatively high levels of the optimal capital budget, the MNE would have a lower cost of capital.

SUMMARY POINTS

■ Gaining access to global capital markets should allow a firm to lower its cost of capital. This can be achieved by increasing the market liquidity of its shares and by escaping from segmentation of its home capital market.

■ The cost and availability of capital is directly linked to the degree of market liquidity and segmentation. Firms having access to markets with high liquidity and a low level of segmentation should have a lower cost of capital and greater ability to raise new capital.

■ A firm is able to increase its market liquidity by raising debt in the euromarket—by selling security issues in individual national capital markets and as euroequities—and tapping local capital markets through foreign subsidiaries. Increased market liquidity causes the marginal cost of capital line to "flatten out to the right." This results in the firm being able to raise more capital at a lower marginal cost.

■ A national capital market is segmented if the required rate of return on securities in that market differs from the required rate of return on securities of comparable expected return and risk that are traded on other national securities markets.

■ Capital market segmentation is a financial market imperfection caused by government constraints and investor perceptions. Segmentation results in a higher cost of capital and less availability of capital.

■ If a firm is resident in a segmented capital market, it can still escape from this market by sourcing its debt and equity abroad. The result should be a lower marginal cost of capital, improved liquidity for its shares, and a larger capital budget.

■ Whether or not MNEs have a lower cost of capital than their domestic counterparts depends on their optimal financial structures, systematic risk, availability of capital, and the optimal capital budget.

Novo Industri A/S (Novo)[8]

Novo is a Danish multinational firm that produces industrial enzymes and pharmaceuticals (mostly insulin). In 1977, Novo's management decided to "internationalize" its capital structure and sources of funds. This decision was based on the observation that the Danish securities market was both illiquid and segmented from other capital markets. In particular, the lack of availability and high cost of equity capital in Denmark resulted in Novo having a higher cost of capital than its main multinational competitors, such as Eli Lilly (U.S.), Miles Laboratories (U.S.—a subsidiary of Bayer, Germany), and Gist Brocades (the Netherlands).

Apart from the cost of capital, Novo's projected growth opportunities signaled the eventual need to raise new long-term capital beyond what could be raised in the illiquid Danish market. Since Novo is a technology leader in its specialties, planned capital investments in plant, equipment, and research could not be postponed until internal financing from cash flow became available. Novo's competitors would preempt any markets not served by Novo.

Even if an equity issue of the size required could have been raised in Denmark, the required rate of return would have been unacceptably high. For example, Novo's price/earnings ratio was typically around 5; that of its foreign competitors was well over 10. Yet Novo's business and financial risk appeared to be about equal to that of its competitors. A price/earnings ratio of 5 appeared appropriate for Novo only within a domestic Danish context when compared with other domestic firms of comparable business and financial risk.

If Denmark's securities markets were integrated with world markets, one would expect foreign investors to rush in and buy "undervalued" Danish securities. In that case, firms like Novo would enjoy an international cost of capital comparable to that of its foreign competitors. Strangely enough, no Danish governmental restrictions existed that would have prevented foreign investors from holding Danish securities. Therefore, one must look for investor perception as the main cause of market segmentation in Denmark at that time.

At least six characteristics of the Danish equity market were responsible for market segmentation: (1) asymmetric information base of Danish and foreign investors, (2) taxation, (3) alternative sets of feasible portfolios, (4) financial risk, (5) foreign exchange risk, and (6) political risk.

Asymmetric Information

Certain institutional characteristics of Denmark caused Danish and foreign investors to lack information about one another's equity securities. The most important information barrier was a Danish regulation that prohibited Danish investors from holding foreign private sector securities. Therefore, Danish investors had no incentive to follow developments in foreign securities markets or to factor such information into their evaluation of Danish securities. As a result, Danish securities might have been priced correctly in the efficient market sense relative to one another, considering the Danish information base, but priced incorrectly considering the combined foreign and Danish information base. Another detrimental effect of this regulation was that foreign securities firms did not locate offices or personnel in Denmark, since they had no product to sell there. Lack of a physical presence in Denmark reduced the ability of foreign security analysts to follow Danish securities.

A second information barrier was that there were too few Danish security analysts following Danish securities. Only one professional Danish securities analysis service was published (Børsinformation), and that was in the Danish language. A few Danish institutional investors employed in-house analysts, but their findings were not available to the public. Almost no foreign security analysts followed Danish securities because they had no product to sell and the Danish market was too small (small-country bias).

Other information barriers included language and accounting principles. Naturally, financial information was normally published in the Danish language using Danish accounting principles. A few firms, such as Novo, published English versions, but almost none used U.S. or British accounting principles or attempted to show any reconciliation with such principles.

Taxation

Danish taxation policy had all but eliminated investment in common stock by individuals. Until a tax law change in July 1981, capital gains on shares held for over two years were taxed at a 50% rate. Shares held for less than two

[8]This is a condensed version of Arthur Stonehill and Kåre B. Dullum, *Internationalizing the Cost of Capital in Theory and Practice: The Novo Experience and National Policy Implications* (Copenhagen: Nyt Nordisk Forlag Arnold Busck, 1982; and New York: Wiley, 1982). Reprinted with permission.

years, or for "speculative" purposes, were taxed at personal income tax rates, with the top marginal rate being 75%. In contrast, capital gains on bonds were tax-free. This situation resulted in bonds being issued at deep discounts because the redemption at par at maturity was considered a capital gain. Thus, most individual investors held bonds rather than stocks. This factor reduced the liquidity of the stock market and increased the required rate of return on stocks if they were to compete with bonds.

Feasible Portfolios

Because of the prohibition on foreign security ownership, Danish investors had a very limited set of securities from which to choose a portfolio. In practice, Danish institutional portfolios were composed of Danish stocks, government bonds, and mortgage bonds. Since Danish stock price movements are closely correlated with each other, Danish portfolios possessed a rather high level of systematic risk. In addition, government policy had been to provide a relatively high real rate of return on government bonds after adjusting for inflation. The net result of taxation policies on individuals, and attractive real yields on government bonds was that required rates of return on stocks were relatively high by international standards.

From a portfolio perspective, Danish stocks provided an opportunity for foreign investors to diversify internationally. If Danish stock price movements were not closely correlated with world stock price movements, inclusion of Danish stocks in foreign portfolios would reduce those portfolios' systematic risk. Furthermore, foreign investors were not subject to the high Danish income tax rates, due to protections provided by tax treaties that typically limit foreign investor tax rates to 15% on dividends and capital gains. As a result of the international diversification potential, foreign investors might have required a lower rate of return on Danish stocks than the rate required by Danish investors, other things being equal. However, other things were not equal because foreign investors perceived Danish stocks to carry more financial, foreign exchange, and political risk than their own domestic securities.

Financial, Foreign Exchange, and Political Risks

Financial leverage utilized by Danish firms was relatively high by U.S. and U.K. standards but not abnormal for Scandinavia, Germany, Italy, or Japan. In addition, most of the debt was short term with variable interest rates. The way in which foreign investors viewed financial risk in Danish firms depended on what norms they followed in their home countries. We know from Novo's experience in tapping the eurobond market in 1978, that Morgan Grenfell, Novo's British investment banker, advised Novo to maintain a debt ratio (debt/total capitalization) closer to 50% rather than the traditional Danish 65% to 70%.

Foreign investors in Danish securities are subject to foreign exchange risk. Whether this factor is a plus or minus depends on the investor's home currency, perceptions about the future strength of the Danish krone, and its impact on a firm's operating exposure. Through personal contacts with foreign investors and bankers, Novo's management did not believe foreign exchange risk was a factor in Novo's stock price because its operations were perceived as being well diversified internationally. Over 90% of its sales were to customers located outside of Denmark.

With respect to political risk, Denmark was perceived as a stable Western democracy but with the potential to cause periodic problems for foreign investors. In particular, Denmark's national debt was regarded as too high for comfort, although this judgment had not yet shown up in the form of risk premiums on Denmark's eurocurrency syndicated loans.

The Road to Globalization

Although Novo's management in 1977 wished to escape from the shackles of Denmark's segmented and illiquid capital market, many barriers had to be overcome. It is worthwhile to explore some of these obstacles, because they typify the barriers faced by other firms from segmented markets that wish to internationalize their capital sources.

Closing the Information Gap. Novo had been a family-owned firm from its founding in the 1920s by the two Pedersen brothers. Then in 1974, it went public and listed its "B" shares on the Copenhagen Stock Exchange. The "A" shares were held by the Novo Foundation, and these shares were sufficient to maintain voting control. However, Novo was essentially unknown in investment circles outside of Denmark. To overcome this disparity in the information base, Novo increased the level of its financial and technical disclosure in both Danish and English versions.

The information gap was further closed when Morgan Grenfell successfully organized a syndicate to underwrite and sell a $20 million convertible eurobond issue for Novo in 1978. In connection with this offering, Novo listed its shares on the London Stock Exchange to facilitate conversion and to gain visibility. These twin actions were the key to dissolving the information barrier and, of course, they also raised a large amount of long-term capital on favorable terms, which would have been unavailable in Denmark.

Despite the favorable impact of the eurobond issue on availability of capital, Novo's cost of capital actually increased when Danish investors reacted negatively to the potential dilution effect of the conversion right. During 1979, Novo's share price declined from around Dkr300 per share to around Dkr220 per share.

The Biotechnology Boom. During 1979, a fortuitous event occurred. Biotechnology began to attract the interest of the U.S. investment community, with several sensationally oversubscribed stock issues by such start-up firms as Genentech and Cetus. Thanks to the aforementioned domestic information gap, Danish investors were unaware of these events and continued to value Novo at a low price/earnings ratio of 5, compared with over 10 for its established competitors and 30 or more for these new potential competitors.

In order to profile itself as a biotechnology firm with a proven track record, Novo organized a seminar in New York City on April 30, 1980. Soon after the seminar a few sophisticated individual U.S. investors began buying Novo's shares and convertibles through the London Stock Exchange. Danish investors were only too happy to supply this foreign demand. Therefore, despite relatively strong demand from U.S. and British investors, Novo's share price increased only gradually, climbing back to the Dkr300 level by midsummer. However, during the following months, foreign interest began to snowball, and by the end of 1980 Novo's stock price had reached the Dkr600 level. Moreover, foreign investors had increased their proportion of share ownership from virtually nothing to around 30%. Novo's price/earnings ratio had risen to around 16, which was now in line with that of its international competitors but not with the Danish market. At this point one must conclude that Novo had succeeded in internationalizing its cost of capital. Other Danish securities remained locked in a segmented capital market.

Directed Share Issue in the United States. During the first half of 1981, under the guidance of Goldman Sachs and with the assistance of Morgan Grenfell and Copenhagen Handelsbank, Novo prepared a prospectus for SEC registration of a U.S. share offering and eventual listing on the New York Stock Exchange. The main barriers encountered in this effort, which would have general applicability, were connected with preparing financial statements that could be reconciled with U.S. accounting principles and the higher level of disclosure required by the SEC. In particular, industry segment reporting was a problem both from a disclosure perspective and an accounting perspective because the accounting data were not available internally in that format. As it turned out, the investment barriers in the U.S. were relatively tractable, although expensive and time consuming to overcome.

The more serious barriers were caused by a variety of institutional and governmental regulations in Denmark. The latter were never designed so that firms could issue shares at market value, since Danish firms typically issued stock at par value with preemptive rights. By this time, however, Novo's share price, driven by continued foreign buying, was so high that virtually nobody in Denmark thought it was worth the price that foreigners were willing to pay. In fact, prior to the time of the share issue in July 1981, Novo's share price had risen to over Dkr1500, before settling down to a level around Dkr1400. Foreign ownership had increased to over 50% of Novo's shares outstanding!

Stock Market Reactions. One final piece of evidence on market segmentation can be gleaned from the way Danish and foreign investors reacted to the announcement of the proposed $61 million U.S. share issue on May 29, 1981. Novo's share price dropped 156 points the next trading day in Copenhagen, equal to about 10% of its market value. As soon as trading started in New York, the stock price immediately recovered all of its loss. The Copenhagen reaction was typical for an illiquid market. Investors worried about the dilution effect of the new share issue, because it would increase the number of shares outstanding by about 8%. They did not believe that Novo could invest the new funds at a rate of return that would not dilute future earnings per share. They also feared that the U.S. shares would eventually flow back to Copenhagen if biotechnology lost its glitter.

The U.S. reaction to the announcement of the new share issue was consistent with what one would expect in a liquid and integrated market. U.S. investors viewed the new issue as creating additional demand for the shares as Novo became more visible due to the selling efforts of a large aggressive syndicate. Furthermore, the marketing effort was directed at institutional investors who were previously underrepresented among Novo's U.S. investors. They had been underrepresented because U.S. institutional investors want to be assured of a liquid market in a stock in order to be able to get out, if desired, without depressing the share price. The wide distribution effected by the new issue, plus SEC registration and a New York Stock Exchange listing, all added up to more liquidity and a global cost of capital.

Effect on Novo's Weighted Average Cost of Capital. During most of 1981 and the years thereafter, Novo's share price was driven by international portfolio investors transacting on the New York, London, and Copenhagen stock exchanges. This reduced Novo's weighted average cost of capital and lowered its marginal cost of capital. Novo's systematic risk was reduced from its previous level, which was determined by nondiversified (internationally)

Danish institutional investors and the Novo Foundation. However, its appropriate debt ratio level was also reduced to match the standards expected by international portfolio investors trading in the United States, United Kingdom, and other important markets. In essence, the U.S. dollar became Novo's functional currency when being evaluated by international investors. Theoretically, its revised weighted average cost of capital should have become a new reference hurdle rate when evaluating new capital investments in Denmark or abroad.

Other firms that follow Novo's strategy are also likely to have their weighted average cost of capital become a function of the requirements of international portfolio investors. Firms resident in some of the emerging market countries have already experienced "dollarization" of trade and financing for working capital. This phenomenon might be extended to long-term financing and the weighted average cost of capital.

The Novo experience can be a model for other firms wishing to escape from segmented and illiquid home equity markets. In particular, MNEs based in emerging markets often face barriers and lack of visibility similar to what Novo faced. They could benefit by following Novo's proactive strategy employed to attract international portfolio investors. However, a word of caution is advised. Novo had an excellent operating track record and a very strong worldwide market niche in two important industry sectors, insulin and industrial enzymes. This record continues to attract investors in Denmark and abroad. Other companies aspiring to achieve similar results would also need to have such a favorable track record to attract foreign investors.

Globalization of Securities Markets. During the 1980s, numerous other Nordic and other European firms followed Novo's example. They cross-listed on major foreign exchanges such as London and New York. They placed equity and debt issues in major securities markets. In most cases, they were successful in lowering their WACC and increasing its availability.

During the 1980s and 1990s, national restrictions on cross-border portfolio investment were gradually eased under pressure from the Organization for Economic Cooperation and Development (OECD), a consortium of most of the world's most industrialized countries. Liberalization of European securities markets was accelerated because of the European Union's efforts to develop a single European market without barriers. Emerging nation markets followed suit, as did the former Eastern Bloc countries after the breakup of the Soviet Union. Emerging national markets have often been motivated by the need to source foreign capital to finance large-scale privatization.

Now, market segmentation has been significantly reduced, although the liquidity of individual national markets remains limited. Most observers believe that for better or for worse, we have achieved a global market for securities. The good news is that many firms have been assisted to become MNEs because they now have access to a global cost and availability of capital. The bad news is that the correlation among securities markets has increased, thereby reducing, but not eliminating, the benefits of international portfolio diversification. Globalization of securities markets has also led to more volatility and speculative behavior as shown by the emerging market crises of the 1995–2001 period.

Mini-Case Questions

1. What were the impacts on Novo as a result of operating in a segmented market?
2. What were the primary causes of the market segmentation?
3. Ultimately, what actions did Novo take to escape its segmented market?

QUESTIONS

These questions are available in MyFinanceLab.

1. **Segmented Market.** What are the most common challenges a firm resident in a segmented market faces in regards to its access to capital?

2. **Dimensions of Capital.** Global integration has given many firms access to new and cheaper sources of funds beyond those available in their home markets. What are the dimensions of a strategy to capture this lower cost and greater availability of capital?

3. **Cost of Capital Benefits.** What are the benefits of achieving a lower cost and greater availability of capital?

4. **Equity Cost and Risk.** What are the classifications used in defining risk in the estimation of a firm's cost of equity?

5. **Equity Risk Premiums.** What is an equity risk premium? For an equity risk premium to be truly useful, what need it do?

6. **Portfolio Investors.** Both domestic and international portfolio managers are asset allocators. What is their portfolio management objective?

7. **International Portfolio Management.** What is the main advantage that international portfolio managers have compared to portfolio managers limited to domestic-only asset allocation?

8. **International CAPM.** What are the fundamental distinctions that the international CAPM tries to capture, which traditional domestic CAPM does not?

9. **Dimensions of Asset Allocation.** Portfolio asset allocation can be accomplished along many dimensions depending on the investment objective of the portfolio manager. Identify the various dimensions.

10. **Market Liquidity.** What is meant by the term market liquidity? What are the main disadvantages for a firm to be located in an illiquid market?

11. **Market Segmentation.** What is market segmentation, and what are the six main causes of market segmentation?

12. **Market Liquidity.** What is the effect of market liquidity and segmentation on a firm's cost of capital?

13. **Emerging Markets.** Firms located in illiquid and segmented emerging markets would benefit from nationalizing their own cost of capital. What do they need to do, and what conditions must exist for their efforts to succeed?

14. **Cost of Capital for MNEs.** Do multinational firms have a higher or lower cost of capital than their domestic counterparts? Is this surprising?

15. **Multinational Use of Debt.** Do multinational firms use relatively more or less debt than their domestic counterparts? Why?

16. **Multinationals and Beta.** Do multinational firms have higher/lower betas than their domestic counterparts?

17. **The "Riddle."** What is the riddle?

18. **Emerging Market Listings.** Why might emerging market multinationals list their shares abroad?

PROBLEMS

These problems are available in MyFinanceLab.

1. **Ganado's Cost of Capital.** Maria Gonzalez now estimates the risk-free rate to be 3.60%, the company's credit risk premium is 4.40%, the domestic beta is estimated at 1.05, the international beta is estimated at 0.85, and the company's capital structure is now 30% debt. All other values remain the same as those

presented in this chapter in "Sample Calculation: Ganado's Cost of Capital." For both the domestic CAPM and ICAPM, calculate the following:
a. Ganado's cost of equity
b. Ganado's cost of debt
c. Ganado's weighted average cost of capital

2. **Ganado and Equity Risk Premiums.** Using the original weighted average cost of capital data for Ganado used in the chapter in "Sample Calculation: Ganado's Cost of Capital," calculate both the CAPM and ICAPM weighted average costs of capital for the following equity risk premium estimates.
a. 8.00%
b. 7.00%
c. 5.00%
d. 4.00%

3. **Thunderhorse Oil.** Thunderhorse Oil is a U.S. oil company. Its current cost of debt is 7%, and the 10-year U.S. Treasury yield, the proxy for the risk-free rate of interest, is 3%. The expected return on the market portfolio is 8%. The company's effective tax rate is 39%. Its optimal capital structure is 60% debt and 40% equity.
a. If Thunderhorse's beta is estimated at 1.1, what is Thunderhorse's weighted average cost of capital?
b. If Thunderhorse's beta is estimated at 0.8, significantly lower because of the continuing profit prospects in the global energy sector, what is Thunderhorse's weighted average cost of capital?

4. **Nestlé of Switzerland Revisited.** Nestlé of Switzerland is revisiting its cost of equity analysis in 2014. As a result of extraordinary actions by the Swiss Central Bank, the Swiss bond index yield (10-year maturity) has dropped to a record low of 0.520%. The Swiss equity markets have been averaging 8.400% returns, while the *Financial Times* global equity market returns, indexed back to Swiss francs, is at 8.820%. Nestlé's corporate treasury staff has estimated the company's domestic beta at 0.825, but its global beta (against the larger global equity market portfolio) at .515.
a. What is Nestlé's cost of equity based on the domestic portfolio of a Swiss investor?
b. What is Nestlé's cost of equity based on a global portfolio for a Swiss investor?

5. **Corcovado Pharmaceuticals.** Corcovado Pharmaceutical's cost of debt is 7%. The risk-free rate of interest is 3%. The expected return on the market portfolio is 8%. After effective taxes, Corcovado's effective tax rate is 25%. Its optimal capital structure is 60% debt and 40% equity.

a. If Corcovado's beta is estimated at 1.1, what is its weighted average cost of capital?

b. If Corcovado's beta is estimated at 0.8, significantly lower because of the continuing profit prospects in the global energy sector, what is its weighted average cost of capital?

6. **WestGas Conveyance, Inc.** WestGas Conveyance, Inc., is a large U.S. natural gas pipeline company that wants to raise $120 million to finance expansion. WestGas wants a capital structure that is 50% debt and 50% equity. Its corporate combined federal and state income tax rate is 40%. WestGas finds that it can finance in the domestic U.S. capital market at the rates listed below. Both debt and equity would have to be sold in multiples of $20 million, and these cost figures show the component costs, each, of debt and equity if raised half by equity and half by debt.

A London bank advises WestGas that U.S. dollars could be raised in Europe at the following costs, also in multiples of $20 million, while maintaining the 50/50 capital structure.

Each increment of cost would be influenced by the total amount of capital raised. That is, if WestGas first borrowed $20 million in the European market at 6% and matched this with an additional $20 million of equity, additional debt beyond this amount would cost 12% in the United States and 10% in Europe. The same relationship holds for equity financing.

a. Calculate the lowest average cost of capital for each increment of $40 million of new capital, where WestGas raises $20 million in the equity market and an additional $20 in the debt market at the same time.

b. If WestGas plans an expansion of only $60 million, how should that expansion be financed? What will be the weighted average cost of capital for the expansion?

7. **Kashmiri's Cost of Capital.** Kashmiri is the largest and most successful specialty goods company based in Bangalore, India. It has not yet entered the North American marketplace, but is considering establishing both manufacturing and distribution facilities in the United States through a wholly owned subsidiary. It has approached two different investment banking advisors, Goldman Sachs and Bank of New York, for estimates of what its costs of capital would be several years into the future when it planned to list its American subsidiary on a U.S. stock exchange. Using the following assumptions by the two different advisors, calculate the prospective costs of debt, equity, and the WACC for Kashmiri (U.S.):

Assumptions	Symbol	Goldman Sachs	Bank of New York
Estimate of correlation between security and market	β	0.90	0.85
Estimate of standard deviation of Kashmiri's returns	ρ_{jm}	24.0%	30.0%
Estimate of standard deviation of market's return	σ_j	18.0%	22.0%
Risk-free rate of interest	k_{rf}	3.0%	3.0%
Estimate of Kashmiri's cost of debt in U.S. market	k_d	7.5%	7.8%
Estimate of market return, forward-looking	k_m	9.0%	12.0%
Corporate tax rate	t	35.0%	35.0%
Proportion of debt	D/V	35%	40%
Proportion of equity	E/V	65%	60%

8. **Cargill's Cost of Capital.** Cargill is generally considered to be the largest privately held company in the world. Headquartered in Minneapolis, Minnesota, the company has been averaging sales of over $113 billion per year over the past five-year period. Although the company does not have publicly traded shares, it is still extremely important for it to calculate its weighted average cost of capital properly in order to make rational decisions on new investment proposals. Assuming a risk-free rate of 4.50%, an effective tax rate of 48%, and a market risk premium of 5.50%, estimate the weighted average cost of capital first for Companies A and B, and then make a "guesstimate" of what you believe a comparable WACC would be for Cargill.

	Company A	Company B	Cargill
Company sales	$10.5 billion	$45 billion	$113 billion
Company's beta	0.83	0.68	??
Credit rating	AA	A	AA
Weighted average cost of debt	6.885%	7.125%	6.820%
Debt to total capital	34%	41%	28%
International sales/Sales	11%	34%	54%

9. **The Tombs.** You have joined your friends at the local watering hole, The Tombs, for your weekly debate on international finance. The topic this week is whether the cost of equity can ever be cheaper than the cost of debt. The group has chosen Brazil in the mid-1990s as the subject of the debate. One of the group members has torn a table of data out of a book (shown below), which is then the subject of the analysis.

Larry argues, "It's all about expected versus delivered. You can talk about what equity investors expect, but they often find that what is delivered for years at a time is so small — even sometimes negative — that in effect, the cost of equity is cheaper than the cost of debt."

Moe interrupts, "But you're missing the point. The cost of capital is what the investor requires in compensation for the risk taken going into the investment. If he doesn't end up getting it, and that was happening here, then he pulls his capital out and walks."

Curly is the theoretician. "Ladies, this is not about empirical results; it is about the fundamental concept of risk-adjusted returns. An investor in equities knows he will reap returns only after all compensation has been made to debt providers. He is therefore always subject to a higher level of risk to his return than debt instruments, and as the capital asset pricing model states, equity investors set their expected returns as a risk-adjusted factor over and above the returns to risk-free instruments." At this point, Larry and Moe simply stare at Curly — pause — and order more beer. Using the Brazilian data presented, comment on this week's debate at The Tombs.

Brazilian Economic Performance	1995	1996	1997	1998	1999
Inflation rate (IPC)	23.20%	10.00%	4.80%	1.00%	10.50%
Bank lending rate	53.10%	27.10%	24.70%	29.20%	30.70%
Exchange rate (reais/$)	0.972	1.039	1.117	1.207	1.700
Equity returns (São Paulo Bovespa)	16.0%	28.0%	30.2%	33.5%	151.9%

Genedak-Hogan

Use the table below to answer Problems 10 through 12. Genedak-Hogan is an American conglomerate that is actively debating the impacts of international diversification of its operations on its capital structure and cost of capital. The firm is planning on reducing consolidated debt after diversification.

Assumptions	Symbol	Before Diversification	After Diversification
Correlation between G-H and the market	ρ_{jm}	0.88	0.76
Standard deviation of G-H's returns	σ_j	28.0%	26.0%
Standard deviation of market's returns	σ_m	18.0%	18.0%
Risk-free rate of interest	k_{rf}	3.0%	3.0%
Additional equity risk premium for internationalization	RPM	0.0%	3.0%
Estimate of G-H's cost of debt in U.S. market	k_d	7.2%	7.0%
Market risk premium	$k_m - k_{rf}$	5.5%	5.5%
Corporate tax rate	t	35.0%	35.0%
Proportion of debt	D/V	38%	32%
Proportion of equity	E/V	62%	68%

10. **Genedak-Hogan Cost of Equity.** Senior management at Genedak-Hogan are actively debating the implications of diversification on its cost of equity. All agree that the company's returns will be less correlated with the reference market return in the future, the financial advisors believe that the market will assess an additional 3.0% risk premium for "going international" to the basic CAPM cost of equity. Calculate Genedak-Hogan's cost of equity before and after international diversification of its operations, with and without the hypothetical additional risk premium, and comment on the discussion.

11. **Genedak-Hogan's WACC.** Calculate the weighted average cost of capital for Genedak-Hogan before and after international diversification.
    a. Did the reduction in debt costs reduce the firm's weighted average cost of capital? How would you describe the impact of international diversification on its costs of capital?
    b. Adding the hypothetical risk premium to the cost of equity introduced in Problem 10 (an added 3.0% to the cost of equity because of international diversification), what is the firm's WACC?

12. **Genedak-Hogan's WACC and Effective Tax Rate.** Many MNEs have greater ability to control and reduce their effective tax rates when expanding international operations. If Genedak-Hogan was able to reduce its consolidated effective tax rate from 35% to 32%, what would be the impact on its WACC?

INTERNET EXERCISES

1. **International Diversification via Mutual Funds.** All major mutual fund companies now offer a variety of internationally diversified mutual funds. The degree of international composition across funds, however, differs significantly. Use the Web sites listed, and others of interest, to
   a. Distinguish between international funds, global funds, worldwide funds, and overseas funds
   b. Determine how international funds have been performing, in U.S. dollar terms, relative to mutual funds offering purely domestic portfolios

Fidelity	www.fidelity.com
T. Rowe Price	www.troweprice.com
Merrill Lynch	www.ml.com
Kemper	www.kempercorporation.com

2. **Novo Industri.** Novo Industri A/S merged with Nordisk Gentofte in 1989. Nordisk Gentofte was Novo's main European competitor. The combined company, now called Novo Nordisk, has become the leading producer of insulin worldwide. Its main competitor is Eli Lilly of the United States. Using standard investor information, and the Web sites for Novo Nordisk and Eli Lilly, determine if, during the most recent five years, Novo Nordisk has maintained a cost of capital competitive with Eli Lilly. In particular, examine the P/E ratios, share prices, debt ratios, and betas. Try to calculate each firm's actual cost of capital.

Novo Nordisk	www.novonordisk.com
Eli Lilly and Company	www.lilly.com
BigCharts.com	www.bigcharts.com
Yahoo! Finance	www.finance.yahoo.com

3. **Cost of Capital Calculator.** Ibbotson and Associates, a unit of Morningstar, is one of the leading providers of quantitative estimates of the cost of capital across markets. Use the Ibbotson and Associates Web site—specifically the Cost of Capital Center—to prepare an overview of the major theoretical approaches and numerical estimates for cross-border costs of capital by Ibbotson and Associates.

Ibbotson and Associates	corporate.morningstar.com

CHAPTER 14

Raising Equity and Debt Globally

Do what you will, the capital is at hazard. All that can be required of a trustee to invest, is, that he shall conduct himself faithfully and exercise a sound discretion. He is to observe how men of prudence, discretion, and intelligence manage their own affairs, not in regard to speculation, but in regard to the permanent disposition of their funds, considering the probable income, as well as the probable safety of the capital to be invested.

—*Prudent Man Rule*, Justice Samuel Putnam, 1830.

LEARNING OBJECTIVES

- Design a strategy to source capital equity globally
- Examine the potential differences in the optimal financial structure of the multinational firm compared to that of the domestic firm
- Describe the various financial instruments that can be used to source equity in the global equity markets
- Understand the different forms of foreign listings—depositary receipts—in U.S. markets
- Analyze the unique role private placement enjoys in raising global capital
- Evaluate the different goals and considerations relevant to a firm pursuing foreign equity listing and issuance
- Explore the different structures that can be used to source debt globally

Chapter 13 analyzed why gaining access to global capital markets should lower a firm's cost of capital, increase its access to capital, and improve the liquidity of its shares by overcoming market segmentation. A firm pursuing this lofty goal, particularly a firm from a segmented or emerging market, must first design a financial strategy that will attract international investors. This involves choosing among alternative paths to access global capital markets.

This chapter focuses on firms that reside in less liquid, segmented, or emerging markets. They are the ones that need to tap liquid and unsegmented markets in order to attain the global cost and availability of capital. Firms resident in large and highly industrialized countries already have access to their own domestic, liquid, and unsegmented markets. Although they too source equity and debt abroad, it is unlikely to have as significant an impact on their cost and availability of capital. In fact, for these firms, sourcing funds abroad is often motivated solely by the need to fund large foreign acquisitions rather than to fund existing operations.

This chapter begins with the design of a financial strategy to source both equity and debt globally. It then analyzes the optimal financial structure for an MNE and its subsidiaries, one that minimizes its cost of capital. We then explore the alternative paths that a firm may follow in raising capital in global markets. The chapter concludes with the Mini-Case, *Petrobrás of Brazil and the Cost of Capital*, which examines how the international markets discriminate in their treatment of multinational firms by home and industry.

Designing a Strategy to Source Capital Globally

Designing a capital sourcing strategy requires management to agree upon a long-run financial objective and then choose among the various alternative paths to get there. Exhibit 14.1 is a visual presentation of alternative paths to the ultimate objective of attaining a global cost and availability of capital.

Normally, the choice of paths and implementation is aided by an early appointment of an investment bank as official advisor to the firm. Investment bankers are in touch with the potential foreign investors and their current requirements. They can also help navigate the various

EXHIBIT 14.1   **Alternative Paths to Globalize the Cost and Availability of Capital**

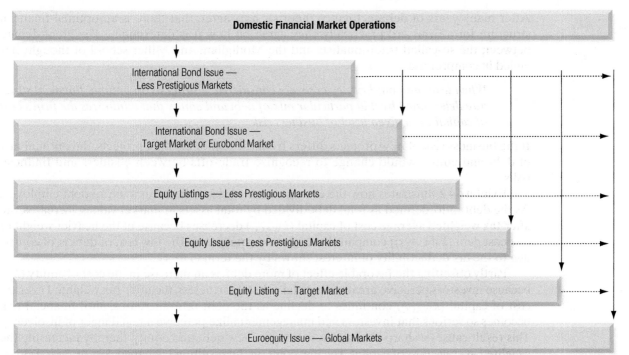

Source: Oxelhiem, Stonehill, Randøy, Vikkula, Dullum, and Modén, *Corporate Strategies in Internationalizing the Cost of Capital*, Copenhagen: Copenhagen Business School Press, 1998, p. 119.

institutional requirements and barriers that must be satisfied. Their services include advising if, when, and where a cross-listing should be initiated. They usually prepare the required prospectus if an equity or debt issue is desired, help to price the issue, and maintain an aftermarket to prevent the share price from falling below its initial price.

Most firms raise their initial capital in their own domestic market (see Exhibit 14.1). Next, they are tempted to skip all the intermediate steps and drop to the bottom line, a euroequity issue in global markets. This is the time when a good investment bank advisor will offer a "reality check." Most firms that have only raised capital in their own domestic market are not sufficiently well known to attract foreign investors. Remember from Chapter 12 that Novo was advised by its investment bankers to start with a convertible eurobond issue and simultaneously cross-list their shares and their bonds in London. This was despite the fact that Novo had an outstanding track record of financial and business performance.

Exhibit 14.1 shows that most firms should start sourcing abroad with an international bond issue. It could be placed on a less prestigious foreign market. This could be followed by an international bond issue in a target market or in the eurobond market. The next step might be to cross-list and issue equity in one of the less prestigious markets in order to attract the attention of international investors. The next step could then be to cross-list shares on a highly liquid prestigious foreign stock exchange such as London (LSE), NYSE, Euronext, or NASDAQ. The ultimate step would be to place a directed equity issue in a prestigious target market or a euroequity issue in global equity markets.

Optimal Financial Structure

After many years of debate, finance theorists now agree that there is an optimal financial structure for a firm, and practically, they agree on how it is determined. The great debate between the so-called traditionalists and the Modigliani and Miller school of thought has ended in compromise:

> *When taxes and bankruptcy costs are considered, a firm has an optimal financial structure determined by that particular mix of debt and equity that minimizes the firm's cost of capital for a given level of business risk.*

If the business risk of new projects differs from the risk of existing projects, the optimal mix of debt and equity would change to recognize trade-offs between business and financial risks.

Exhibit 14.2 illustrates how the cost of capital varies with the amount of debt employed. As the debt ratio, defined as total debt divided by total assets at market values, increases, the after-tax weighted average cost of capital (k_{WACC}) decreases because of the heavier weight of low-cost debt [$k_d(1 - t)$] compared to high-cost equity (k_e). The low cost of debt is, of course, due to the tax deductibility of interest shown by the term $(1 - t)$.

Partly offsetting the favorable effect of more debt, is an increase in the cost of equity (k_e), because investors perceive greater financial risk. Nevertheless, the after-tax weighted average cost of capital (k_{WACC}) continues to decline as the debt ratio increases, until financial risk becomes so serious that investors and management alike perceive a real danger of insolvency. This result causes a sharp increase in the cost of new debt and equity, thereby increasing the weighted average cost of capital. The low point on the resulting U-shaped cost of capital curve, 14% in Exhibit 14.2, defines the debt ratio range in which the cost of capital is minimized.

Most theorists believe that the low point is actually a rather broad flat area encompassing a wide range of debt ratios, 30% to 60% in Exhibit 14.2, where little difference exists in the cost of capital. They also generally agree that, at least in the United States, the range of the

EXHIBIT 14.2 **The Cost of Capital and Financial Structure**

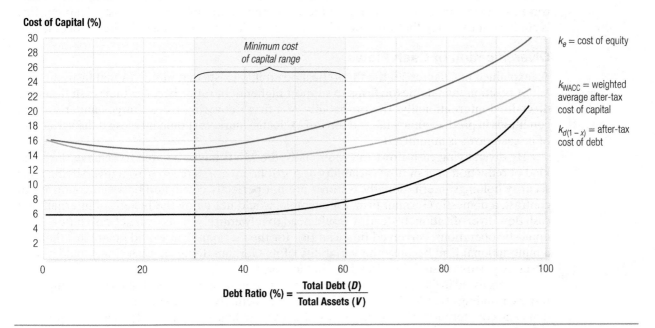

flat area and the location of a particular firm's debt ratio within that range are determined by such variables as (1) the industry in which it competes; (2) volatility of its sales and operating income; and (3) the collateral value of its assets.

Optimal Financial Structure and the Multinational

The domestic theory of optimal financial structures needs to be modified by four more variables in order to accommodate the case of the multinational enterprise. These variables are (1) availability of capital; (2) diversification of cash flows; (3) foreign exchange risk; and (4) expectations of international portfolio investors.[1]

Availability of Capital

Chapter 13 demonstrated that access to capital in global markets allows an MNE to lower its cost of equity and debt compared with most domestic firms. It also permits an MNE to maintain its desired debt ratio, even when significant amounts of new funds must be raised. In other words, a multinational firm's marginal cost of capital is constant for considerable ranges of its capital budget. This statement is not true for most small domestic firms because they do not have access to the national equity or debt markets. They must either rely on internally generated funds or borrow for the short and medium terms from commercial banks.

Multinational firms domiciled in countries that have illiquid capital markets are in almost the same situation as small domestic firms unless they have gained a global cost and availability

[1]An excellent recent study on the practical dimensions of optimal capital structure can be found in "An Empirical Model of Optimal Capital Structure," Jules H. Binsbergen, John R. Graham, and Jie Yang, *Journal of Applied Corporate Finance*, Vol. 23, No. 4, Fall 2011, pp. 34–59.

of capital. They must rely on internally generated funds and bank borrowing. If they need to raise significant amounts of new funds to finance growth opportunities, they may need to borrow more than would be optimal from the viewpoint of minimizing their cost of capital. This is equivalent to saying that their marginal cost of capital is increasing at higher budget levels.

Diversification of Cash Flows

As explained in Chapter 13, the theoretical possibility exists that multinational firms are in a better position than domestic firms to support higher debt ratios because their cash flows are diversified internationally. The probability of a firm's covering fixed charges under varying conditions in product, financial, and foreign exchange markets should increase if the variability of its cash flows is minimized.

By diversifying cash flows internationally, the MNE might be able to achieve the same kind of reduction in cash flow variability as portfolio investors receive from diversifying their security holdings internationally. Returns are not perfectly correlated between countries. In contrast, a domestic German firm, for example, would not enjoy the benefit of international cash flow diversification. Instead, it would need to rely entirely on its own net cash inflow from domestic operations. Perceived financial risk for the German firm would be greater than for a multinational firm because the variability of its German domestic cash flows could not be offset by positive cash flows elsewhere in the world.

As discussed in Chapter 13, the diversification argument has been challenged by empirical research findings that MNEs in the United States actually have lower debt ratios than their domestic counterparts. The agency costs of debt were higher for the MNEs, as were political risks, foreign exchange risks, and asymmetric information.

Foreign Exchange Risk and the Cost of Debt

When a firm issues foreign currency-denominated debt, its effective cost equals the after-tax cost of repaying the principal and interest in terms of the firm's own currency. This amount includes the nominal cost of principal and interest in foreign currency terms, adjusted for any foreign exchange gains or losses.

For example, if a U.S.-based firm borrows SF1,500,000 for one year at 5.00% interest, and during the year the Swiss franc appreciates from an initial rate of SF1.5000/\$ to SF1.4400/\$, what is the dollar cost of this debt $(k_d^\$)$? The dollar proceeds of the initial borrowing are calculated at the current spot rate of SF1.5000/\$:

$$\frac{\text{SF1,500,000}}{\text{SF1.5000/\$}} = \$1,000,000.$$

At the end of one year the U.S.-based firm is responsible for repaying the SF1,500,000 principal plus 5.00% interest, or a total of SF1,575,000. This repayment, however, must be made at an ending spot rate of SF1.4400/\$:

$$\frac{\text{SF1,500,000} \times 1.05}{\text{SF1.4400/\$}} = \$1,093,750.$$

The actual dollar cost of the loan's repayment is not the nominal 5.00% paid in Swiss franc interest, but 9.375%:

$$\left[\frac{\$1,093,750}{\$1,000,000}\right] - 1 = 0.09375 \approx 9.375\%$$

The dollar cost is higher than expected due to appreciation of the Swiss franc against the U.S. dollar. This total home-currency cost is actually the result of the combined percentage cost of debt and percentage change in the foreign currency's value. We can find the total cost of borrowing Swiss francs by a U.S.-dollar based firm, $k_d^\$$, by multiplying one plus the Swiss franc interest expense, k_d^{SF}, by one plus the percentage change in the SF/$ exchange rate, s:

$$k_d^\$ = [(1 + k_d^{SF}) \times (1 + s)] - 1,$$

where $k_d^{SF} = 5.00\%$ and $s = 4.1667\%$. The percentage change in the value of the Swiss franc versus the U.S. dollar, when the home currency is the U.S. dollar, is

$$\frac{S_1 - S_2}{S_2} \times 100 = \frac{SF1.5000/\$ - SF1.4400/\$}{SF1.4400/\$} \times 100 = +4.1667\%.$$

The total expense, combining the nominal interest rate and the percentage change in the exchange rate, is

$$k_d^\$ = [(1 + .0500) \times (1 + .041667)] - 1 = .09375 \approx 9.375\%.$$

The total percentage cost of capital is 9.375%, not simply the foreign currency interest payment of 5%. The after-tax cost of this Swiss franc denominated debt, when the U.S. income tax rate is 34%, is

$$k_d^\$(1 - t) = 9.375\% \times 0.66 = 6.1875\%.$$

The firm would report the added 4.1667% cost of this debt in terms of U.S. dollars as a foreign exchange transaction loss, and it would be deductible for tax purposes.

Expectations of International Portfolio Investors

Chapter 13 highlighted the fact that the key to gaining a global cost and availability of capital is attracting and retaining international portfolio investors. Those investors' expectations for a firm's debt ratio and overall financial structures are based on global norms that have developed over the past 30 years. Because a large proportion of international portfolio investors and based in the most liquid and unsegmented capital markets, such as the United States and the United Kingdom, their expectations tend to predominate and override individual national norms. Therefore, regardless of other factors, if a firm wants to raise capital in global markets, it must adopt global norms that are close to the U.S. and U.K. norms. Debt ratios up to 60% appear to be acceptable. Higher debt ratios are more difficult to sell to international portfolio investors.

Raising Equity Globally

Once a multinational firm has established its financial strategy and considered its *desired* and *target capital structure*, it then proceeds to raise capital outside of its domestic market—both debt and equity—using a variety of capital raising paths and instruments.

Exhibit 14.3 describes three key critical elements to understanding the issues that any firm must confront when seeking to raise equity capital. Although the business press does not often make a clear distinction, there is a fundamental distinction between an *equity issuance* and an *equity listing*. A firm seeking to raise equity capital is ultimately in search of an issuance—the IPO or SPO described in Exhibit 14.3. This generates cash proceeds to be used for funding and executing the business. But often issuances must be preceded by listings, in which the shares are traded on an exchange and, therefore, in a specific country market, gaining name recognition, visibility, and hopefully preparing the market for an issuance.

That said, an issuance need not be public. A firm, public or private, can place an issue with private investors, a *private placement*. (Note that *private placement* may refer to either equity or debt.) Private placements can take a variety of different forms, and the intent of investors

---

EXHIBIT 14.3   **Equity Avenues, Activities, and Attributes**

Equity Issuance

- *Initial Public Offering* (IPO)—the initial sale of shares to the public of a private company. IPOs raise capital and typically use underwriters.
- *Seasoned Public Offering* (SPO)—a subsequent sale of additional shares in the publicly traded company, raising additional equity capital.
- *Euroequity*—the initial sale of shares in two or more markets and countries simultaneously.
- *Directed Issue*—the sale of shares by a publicly traded company to a specific target investor or market, public or private, often in a different country.

Equity Listing

- Shares of a publicly traded firm are listed for purchase or sale on an exchange. An investment banking firm is typically retained to make a market in the shares.
- *Cross-listing* is the listing of a company's shares on an exchange in a different country market. It is intended to expand the potential market for the firm's shares to a larger universe of investors.
- *Depositary receipt* (DR)—a certificate of ownership in the shares of a company issued by a bank, representing a claim on underlying foreign securities. In the United States they are termed *American Depositary Receipts* (ADRs), and when sold globally, *Global Depositary Receipts* (GDRs).

Private Placement

- The sale of a security (equity or debt) to a private investor. The private investors are typically institutions such as pension funds, insurance companies, or high net-worth private entities.
- *SEC Rule 144A private placement sales* are sales of securities to *qualified institutional buyers* (QIBs) in the United States without SEC registration. QIBs are nonbank firms that own and invest in $100 million or more on a discretionary basis.
- *Private Equity*—equity investments in firms by large limited partnerships, institutional investors, or wealthy private investors, with the intention of taking the subject firms private, revitalizing their businesses, and then selling them publicly or privately in one to five years.

---

may be passive (e.g., Rule 144A investors) or active (e.g., private equity, where the investor intends to control and change the firm).

Publicly traded companies, in addition to raising equity capital, are also in pursuit of greater market visibility and reaching ever-larger potential investor audiences. The expectation is that the growing investor audience will result in higher share prices over time—increasing the returns to owners. Privately held companies are more singular in their objective: to raise greater quantities of equity at the lowest possible cost—privately. As discussed in Chapter 4, ownership trends in the industrialized markets have tended toward more private ownership, while many multinational firms from emerging market countries have shown growing interest in going public.

Exhibit 14.4 provides an overview of the four major equity alternatives available to multinational firms today. A firm wishing to raise equity capital outside of its home market may take a *public pathway* or a *private* one. The *public pathway* includes a *directed public share issue* or a *euroequity issue*. Alternatively, and one that has been used with greater frequency over the past decade, is a *private pathway*—private placements, private equity, or a private share sale under strategic alliance.

Initial Public Offering (IPO)

A private firm initiates public ownership of the company through an *initial public offering*, or IPO. Most IPOs begin with the organization of an underwriting and syndication group

EXHIBIT 14.4   **Equity Alternatives in the Global Market**

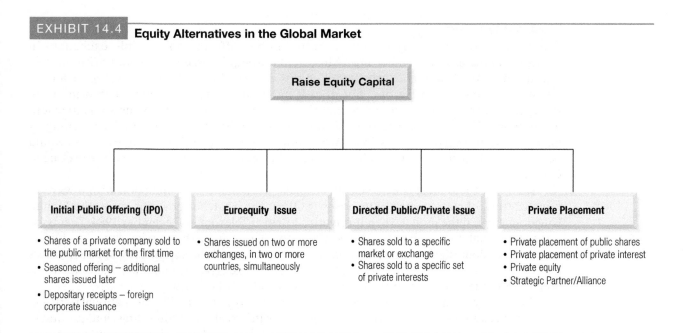

comprised of investment banking service providers. This group then assists the company in preparing the regulatory filings and disclosures required, depending on the country and stock exchange the firm is using. The firm will, in the months preceding the IPO date, publish a *prospectus*. The prospectus will provide a description of the company's history, business, operating and financing results, associated business, financial or political risks, and the company's business plan for the future, all to aid prospective buyers in their assessment of the firm.

The initial issuance of shares by a company typically represents somewhere between 15% and 25% of the ownership in the firm (although a number in recent years have been as little as 6% to 8%). The company may follow the IPO with additional share sales called *seasoned offerings* or *follow-on offerings* (FOs) in which more of the firm's ownership is sold in the public market. The total shares or proportion of shares traded in the public market is often referred to as the *public float* or *free float*.

Once a firm has "gone public," it is open to a considerably higher level of public scrutiny. This scrutiny arises from the detailed public disclosures and financial filings it must make periodically as required by government security regulators and individual stock exchanges. This continuous disclosure is not trivial in either cost or competitive implications. Public firm financial disclosures can be seen as divulging a tremendous amount of information that customers, suppliers, partners, and competitors may use in their relationship with the firm. Private firms have a distinct competitive advantage in this arena.[2]

An added distinction about the publicly traded firm's shares is that they only raise capital for the firm upon issuance. Although the daily rise and fall of share prices drives the returns to the owners of those shares, that daily price movement does not change the capital of the company.

---

[2]A publicly traded firm like Walmart will produce hundreds of pages of operational details, financial results, and management discussion on a quarterly basis. That is in comparison to large private firms like Cargill or Koch, where finding a full single page of financial results would be an achievement.

Euroequity Issue

A *euroequity* or *euroequity issue* is an initial public offering on multiple exchanges in multiple countries at the same time. Almost all euroequity issues are underwritten by an international syndicate. The term "euro" in this context does not imply that the issuers or investors are located in Europe, nor does it mean the shares are denominated in euros. It is a generic term for international securities issues originating and being sold anywhere in the world. The euroequity seeks to raise more capital in its issuance by reaching as many different investors as possible. Two examples of high-profile euroequity issues would be those of British Telecommunications and the famous Italian luxury goods producer, Gucci.

The largest and most spectacular issues have been made in conjunction with a wave of privatizations of state-owned enterprises (SOEs). The Thatcher government in the United Kingdom created the model when it privatized British Telecom in December 1984. That issue was so large that it was necessary and desirable to sell *tranches* to foreign investors in addition to the sale to domestic investors. (A *tranche* is an allocation of shares, typically to underwriters that are expected to sell to investors in their designated geographic markets.) The objective is both to raise the funds and to ensure post-issue worldwide liquidity.

Euroequity privatization issues have been particularly popular with international portfolio investors because most of the firms are very large, with excellent credit ratings and profitable quasi-government monopolies at the time of privatization. The British privatization model has been so successful that numerous others have followed like the Deutsche Telecom initial public offering of $13 billion in 1996.

State-owned enterprises (SOEs) — government-owned firms from emerging markets — have successfully implemented large-scale privatization programs with these foreign tranches. Telefonos de Mexico, the giant Mexican telephone company, completed a $2 billion euroequity issue in 1991 and has continued to have an extremely liquid listing on the NYSE.

One of the largest euroequity offerings by a firm resident in an illiquid market was the 1993 sale of $3 billion in shares by YPF Sociedad Anónima, Argentina's state-owned oil company. About 75% of its shares were placed in tranches outside of Argentina, with 46% in the United States alone. Its underwriting syndicate represented a virtual "who's who" of the world's leading investment banks.

Directed Public Share Issues

A *directed public share issue* or *directed issue* is defined as one that is targeted at investors in a single country and underwritten in whole or in part by investment institutions from that country. The issue may or may not be denominated in the currency of the target market and is typically combined with a cross-listing on a stock exchange in the target market.[3]

A directed issue might be motivated by a need to fund acquisitions or major capital investments in a target foreign market. This is an especially important source of equity for firms that reside in smaller capital markets and that have outgrown that market.

Nycomed, a small but well-respected Norwegian pharmaceutical firm, was an example of this type of motivation for a directed issue combined with cross-listing. Its commercial strategy for growth was to leverage its sophisticated knowledge of certain market niches and technologies within the pharmaceutical field by acquiring other promising firms — primarily firms in Europe and the United States — that possessed relevant technologies, personnel, or

---

[3]The share issue by Novo in 1981(Chapter 12) was a good example of a successful directed share issue that both improved the liquidity of Novo's shares and lowered its cost of capital.

The Planned Directed Equity Issue of PA Resources of Sweden

One example of the use of directed public share issues was the 2005 issuance of PA Resources (PAR.ST), a Swedish oil and gas reserve acquisition and development firm. First listed on the Oslo, Norway, stock exchange in 2001, PAR announced in 2005 a potential private placement of up to 7 million shares that were specifically directed at Norwegian and international investors (non-U.S. investors). The proceeds of the issuance were expected to partially fund the development of recent oil and gas reserve acquisitions made by the company in the North Sea and Tunisia.

The directed issue was reportedly heavily oversubscribed following the announcement. Like many directed issuances outside the United States the offer expressly stated that the securities would not be offered or sold in the U.S., as the issue had not and would not be registered in the U.S. under the U.S. Securities Act of 1933.

market niches. The acquisitions were paid for partly with cash and partly with shares. The company funded its acquisition strategy by selling two directed issues abroad. In 1989 it cross-listed on the London Stock Exchange (LSE) and raised $100 million in equity from foreign investors. Nycomed followed its LSE listing and issuance with a cross-listing and issuance on the NYSE, raising another $75 million from U.S. investors. *Global Finance in Practice 14.1* offers another example of a directed issue, in this case, a publicly traded firm in Sweden and Norway issuing a euroequity to partially fund the development of a recent oil property acquisition.

Depositary Receipts

Depositary receipts (DRs) are negotiable certificates issued by a bank to represent the underlying shares of stock that are held in trust at a foreign custodian bank. *Global depositary receipts* (GDRs) refer to certificates traded outside of the United States, and *American depositary receipts* (ADRs) refer to certificates traded in the United States and denominated in U.S. dollars. For a company that is incorporated outside the United States and that wants to be listed on a U.S. stock exchange, the primary way of doing so is through an ADR program. For a company incorporated anywhere in the world that wants to be listed in any foreign market, this is done via a GDR program.

ADRs are sold, registered, and transferred in the U.S. in the same manner as any share of stock, with each ADR representing either a multiple or portion of the underlying foreign share. This multiple/portion allows ADRs to carry a price per share appropriate for the U.S. market (typically under $20 per share), even if the price of the foreign share is inappropriate when converted to U.S. dollars directly. A number of ADRs, like the ADR of Telefonos de Mexico (TelMex) of Mexico shown in Exhibit 14.5, have been some of the most active shares on U.S. exchanges for many years.

The first ADR program was created for a British company, Selfridges Provincial Stores Limited, a famous British retailer, in 1927. Created by J.P. Morgan, the shares were listed on the New York Curb Exchange, which in later years was transformed into the American Stock Exchange. As with many financial innovations, depositary receipts were created to defeat a regulatory restriction. In this case, the British government had prohibited British companies from registering their shares on foreign markets without British transfer agents. Depositary receipts, in essence, create a synthetic share abroad, and therefore do not require actual registration of shares outside Britain.

EXHIBIT 14.5   **TelMex's American Depositary Receipt (Sample)**

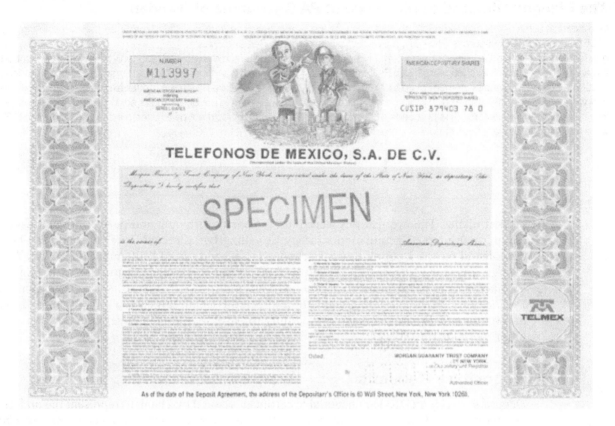

ADR Mechanics

Exhibit 14.6 illustrates the issuance process of a DR program, in this case a U.S.-based investor purchasing shares in a publicly traded Brazilian company—an American depositary receipt or *ADR program*:

1. The U.S. investor instructs his broker to make a purchase of shares in the publicly traded Brazilian company.

2. The U.S. broker contacts a local broker in Brazil (either through the broker's international offices or directly), placing the order.

3. The Brazilian broker purchases the desired ordinary shares and delivers them to a custodian bank in Brazil.

4. The U.S. broker converts the U.S. dollars received from the investor into Brazilian reais to pay the Brazilian broker for the shares purchased.

5. On the same day that the shares are delivered to the Brazilian custodian bank, the custodian notifies the U.S. depositary bank of their deposit.

6. Upon notification, the U.S. depositary bank issues and delivers DRs for the Brazilian company shares to the U.S. broker.

7. The U.S. broker then delivers the DRs to the U.S. investor.

EXHIBIT 14.6 **The Structural Execution of ADRs**

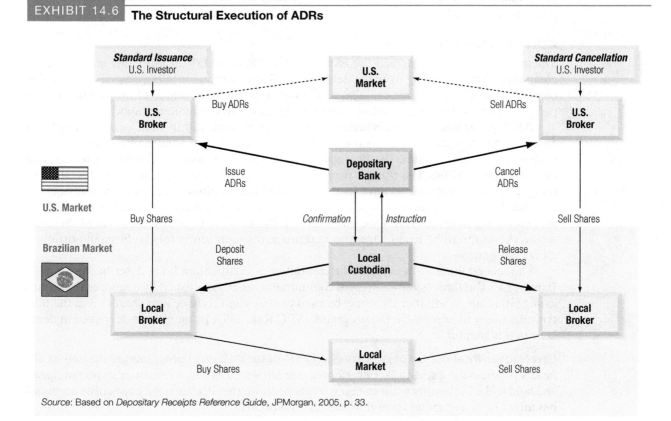

Source: Based on *Depositary Receipts Reference Guide*, JPMorgan, 2005, p. 33.

The DRs are now held and tradable like any other common stock in the United States. In addition to the process just described, it is possible for the U.S. broker to obtain the DRs for the U.S. investor by purchasing existing DRs, not requiring a new issuance. Exhibit 14.6 also describes the alternative process mechanics of a sale or cancellation of ADRs.

Once the ADRs are created, they are tradable in the U.S. market like any other U.S. security. ADRs can be sold to other U.S. investors by simply transferring them from the existing ADR holder (the seller) to another DR holder (the buyer). This is termed *intra-market trading*. This transaction would be settled in the same manner as any other U.S. transaction, with settlement in U.S. dollars on the third business day after the trade date and typically using the depository trust company (DTC). Intra-market trading accounts for nearly 95% of all DR trading today.

ADRs can be exchanged for the underlying foreign shares, or vice versa, so arbitrage keeps foreign and U.S. prices of any given share the same after adjusting for transfer costs. For example, investor demand in one market will cause a price rise there, which will cause an arbitrage rise in the price on the other market even when investors there are not as bullish on the stock.

ADRs convey certain technical advantages to U.S. shareholders. Dividends paid by a foreign firm are passed to its custodial bank and then to the bank that issued the ADR. The issuing bank exchanges the foreign currency dividends for U.S. dollars and sends the dollar dividend to the ADR holders. ADRs are in registered form, rather than in bearer form. Transfer of ownership occurs in the United States in accordance with U.S. laws and procedures. Normally, trading costs are lower than when buying or selling the underlying shares in their home market, and settlement is faster. Withholding taxes is simpler because it is handled by the depositary bank.

ADR Program Structures

The previous section described the issuance of a DR (an ADR in this case) on a Brazilian company's shares resulting from the desire of a U.S.-based investor to buy shares in a Brazilian company. But DR programs can also be viewed from the perspective of the Brazilian company—as part of its financial strategy to reach investors in the United States.

ADR programs differ in whether they are sponsored and in their certification level. *Sponsored ADRs* are created at the request of a foreign firm wanting its shares listed or traded in the United States. The firm applies to the U.S. SEC and a U.S. bank for registration and issuance of ADRs. The foreign firm pays all costs of creating such sponsored ADRs. If a foreign firm does not seek to have its shares listed in the United States but U.S. investors are interested, a U.S. securities firm may initiate creation of the ADRs—an *unsponsored ADR program*. Unsponsored ADRs are still required by the SEC to obtain approval of the firms whose shares are to be listed. Unsponsored programs represent a relatively small portion of all DR programs.

The second dimension of ADR differentiation is certification level, described in detail in Exhibit 14.7. The three general levels of commitment are distinguished by degree of disclosure, listing alternatives, whether they may be used to raise capital (issue new shares), and the time typically taken to implement the programs. (SEC Rule 144A programs are described in detail later in this chapter.)

Level I (over-the-counter or *pink sheets*) DR Programs. Level I programs are the easiest and fastest programs to execute. A Level I program allows the foreign securities to be purchased and held by U.S. investors without being registered with the SEC. It is the least costly approach but might have a minimal favorable impact on liquidity.

Level II DR Programs. Level II applies to firms that want to list existing shares on a U.S. stock exchange. They must meet the full registration requirements of the SEC and the rules of the specific exchange. This also means reconciling their financial accounts with those used under U.S. GAAP, raising the cost considerably.

EXHIBIT 14.7   American Depositary Receipt (ADR) Programs by Level

Type	Description	Degree of Disclosure	Listing Alternatives	Ability to Raise Capital	Implementation Timetable
Level I	Over-the-Counter ADR Program	None: home country standards apply	Over-the-counter (OTC)	—	6 weeks
Level I GDR	Rule 144A/Reg. S GDR Program	None	Not listed	Yes, available only to QIBs	3 weeks
Level II	U.S.-Listed ADR Program	Detailed Sarbanes Oxley	U.S. stock exchange listings	—	13 weeks
Level II GDR	Rule 144A/Reg. S GDR Program	None	DIFX	None	2 weeks
Level III	U.S.-Listed ADR Program	Rigorous Sarbanes Oxley	U.S. stock exchange listings	Yes, public offering	14 weeks
Level III GDR	Rule 144A/Reg. S GDR Program	EU Prospectus Directive and/or U.S. Rule 144A	London, Luxembourg, U.S. Portal	Yes, available to QIBs	2 weeks

Level III DR Programs. Level III applies to the sale of a new equity issued in the United States—raising equity capital. It requires full registration with the SEC and an elaborate stock prospectus. This is the most expensive alternative, but is the most fruitful for foreign firms wishing to raise capital in the world's largest capital markets and possibly generate greater returns for all shareholders.

DR Markets Today: Who, What, and Where

The rapid growth in emerging markets in recent years has been partly a result of the ability of companies from these countries to both list their shares and issue new shares on global equity markets. Their desire to access greater pools of affordable capital, as well as the desire for many of their owners to monetize existing value, has led to an influx of emerging market companies into the DR market.

The Who. The *Who* of global DR programs today is a mix of major multinationals from all over the world, but in recent years participation has shifted back toward industrial country companies. For example, in 2013 the largest issues came from established multinationals like BP, Vodafone, Royal Dutch Shell, and Nestlé, but also included Lukoil and Gazprom of Russia and Taiwan Semiconductor Manufacturing of Taiwan. The oil and gas sector was clearly the largest in both 2012 and 2013, but followed closely by pharmaceutical and telecommunications firms. It's also important to note that in recent years, as illustrated by Exhibit 14.8, the market has clearly been in decline.

The What. The *What* of the global DR market today is a fairly even split between IPO and follow-on offerings or FOs (additional offerings of equity shares post-IPO). It does appear that IPOs continue to make up the majority of DR equity-raising activity.

EXHIBIT 14.8   **Equity Capital Raised Through Depositary Receipts**

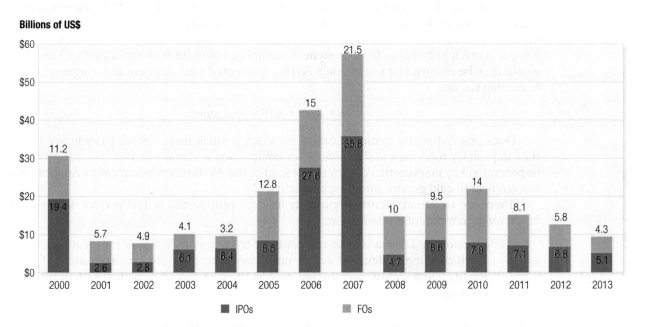

Billions of US$

Year	IPOs	FOs
2000	19.4	11.2
2001	2.6	5.7
2002	2.8	4.9
2003	6.1	4.1
2004	6.4	3.2
2005	8.5	12.8
2006	27.6	15
2007	35.8	21.5
2008	4.7	10
2009	8.6	9.5
2010	7.9	14
2011	7.1	8.1
2012	6.8	5.8
2013	5.1	4.3

Source: "Depositary Receipts, Year in Review 2013," JPMorgan, p. 5. Data derived by JPMorgan from other depositary banks, Bloomberg, and stock exchanges, 2014.

The Where. Given the dominance of emerging market companies in DR markets today, it is not surprising that the *Where* of the DR market is dominated by New York and London. By the end of 2013 there were more than 2,300 sponsored DR programs from more than 86 countries. Of those 2,300, just over half were U.S. programs (ADRs), with the remainder being GDR programs split between the London and Luxembourg stock exchanges.

Even more important than the number of programs participating in the DR markets is the capital that has been raised by companies via DR programs globally. Exhibit 14.8 distinguishes between equity capital raised through initial equity share offerings (IPOs) and seasoned offerings (follow-on). The DR market has periodically proved very fruitful as an avenue for raising capital. It is also obvious which years have been better for equity issuances—years like 2000 and 2006–2007.

Global Registered Shares (GRS)

A *global registered share* (GRS) is a share of equity that is traded across borders and markets without conversion, where one share on the home exchange equals one share on the foreign exchange. The identical share is listed on different stock exchanges, but listed in the currency of the exchange. GRSs can theoretically be traded "with the sun"—following markets as they open and close around the globe and around the clock. The shares are traded electronically, eliminating the specialized forms and depositaries required by share issuances like DRs.

The differences between GRSs and GDRs can be seen in the following example. Assume a German multinational has shares listed on the Frankfurt Stock Exchange, and those shares are currently trading at €4.00 per share. If the current spot rate is \$1.20/€, those same shares would be listed on the NYSE at \$4.80 per share.

$$€4.00 \times \$1.20/€ = \$4.80$$

This would be a standard GRS. But \$4.80 per share is an extremely low share price for the NYSE and the U.S. equity market.

If, however, the German firm's shares were listed in New York as ADRs, they would be converted to a value that was strategically priced for the target market—the United States. Strategic pricing in the U.S. means having share prices that are generally between \$10 and \$20 per share, a price range long-thought to maximize buyer interest and liquidity. The ADR would then be constructed so that each ADR represented four shares in the company on the home market, or:

$$\$4.80 \times 4 = \$19.20 \text{ per share}$$

Does this distinction matter? Clearly the GRS is much more similar to ordinary shares than depositary receipts, and it allows easier comparison and analysis. But if target pricing is important in key markets like that of the U.S., then the ADR offers better opportunities for a foreign firm to gain greater presence and activity.[4]

There are two fundamental arguments used by proponents of GRSs over ADRs, both based on pure forces of globalization:

1. Investors and markets alike will continue to grow in their desire for securities, which are increasingly identical across markets—taking on the characteristics of commodity—like securities, changing only by the currency of denomination of the local exchange.

---

[4] GRSs are not a new innovation, as they are identical to the structure used for cross-border trading of Canadian equities in the United States for many years. More than 70 Canadian firms are listed on the NYSE-Euronext. Of course, one could argue that has been facilitated by near-parity of the U.S. and Canadian dollar for years as well.

2. Regulations governing security trading across country markets will continue to converge toward a common set of global principles, eliminating the need for securities customized for local market attributes or requirements.

Other potential distinctions include the possibility of retaining all voting rights (GRSs do, by definition, while some ADRs may not) and the general principle that ADRs are designed for one singular cultural and legal environment—the United States. All argument aside, at least to date, the GRS has not replaced the ADR or GDR.

Private Placement

Raising equity through *private placement* is increasingly common across the globe. Publicly traded and private firms alike raise private equity capital on occasion. A private placement is the sale of a security to a small set of qualified institutional buyers. The investors are traditionally insurance companies and investment companies. Since the securities are not registered for sale to the public, investors have typically followed a "buy and hold" policy. In the case of debt, terms are often custom designed on a negotiated basis. Private placement markets now exist in most countries.

SEC Rule 144A

In 1990, the SEC approved Rule 144A. It permits *qualified institutional buyers* (QIBs) to trade privately placed securities without the previous holding period restrictions and without requiring SEC registration.

A QIB is an entity (except a bank or a savings and loan) that owns and invests on a discretionary basis $100 million in securities of non-affiliates. Banks and savings and loans must meet this test but also must have a minimum net worth of $25 million. The SEC has estimated that about 4,000 QIBs exist, mainly investment advisors, investment companies, insurance companies, pension funds, and charitable institutions. Simultaneously, the SEC modified its regulations to permit foreign issuers to tap the U.S. private placement market through an SEC Rule 144A issue, also without SEC registration. A trading system called PORTAL was established to support the distribution of primary issues and to create a liquid secondary market for these issues.

Since SEC registration has been identified as the main barrier to foreign firms wishing to raise funds in the United States, SEC Rule 144A placements are proving attractive to foreign issuers of both equity and debt securities. Atlas Copco, the Swedish multinational engineering firm, was the first foreign firm to take advantage of SEC Rule 144A. It raised $49 million in the United States through an ADR equity placement as part of its larger $214 million euroequity issue in 1990. Since then, several billion dollars have been raised each year by foreign issuers with private equity placements in the United States. However, it does not appear that such placements have a favorable effect on either liquidity or stock price.

Private Equity Funds

Private equity funds are usually limited partnerships of institutional and wealthy investors, such as college endowment funds, that raise capital in the most liquid capital markets. They are best known for buying control of publicly owned firms, taking them private, improving management, and then reselling them after one to three years. They are resold in a variety of ways including selling the firms to other firms, to other private equity funds, or by taking them public once again. The private equity funds themselves are frequently very large, but may also utilize a large amount of debt to fund their takeovers. These "alternatives" as they are called, demand fees of 2% of assets plus 20% of profits. Equity funds have had some highly visible successes.

Many mature family-owned firms resident in emerging markets are unlikely to qualify for a global cost and availability of capital even if they follow the strategy suggested in this chapter. Although they might be consistently profitable and growing, they are still too small, too invisible to foreign investors, lacking in managerial depth, and unable to fund the up-front costs of a globalization strategy. For these firms, private equity funds may be a solution.

Private equity funds differ from traditional venture capital funds. The latter usually operate mainly in highly developed countries. They typically invest in start-up firms with the goal of exiting the investment with an initial public offering (IPO) placed in those same highly liquid markets. Very little venture capital is available in emerging markets, partly because it would be difficult to exit with an IPO in an illiquid market. The same exiting problem faces the private equity funds, but they appear to have a longer time horizon. They invest in already mature and profitable companies. They are content with growing companies through better management and mergers with other firms.

Foreign Equity Listing and Issuance

According to the alternative equity pathways in the global market illustrated earlier in Exhibit 14.1, a firm needs to choose one or more stock market on which to cross-list its shares and sell new equity. Just where to go depends mainly on the firm's specific motives and the willingness of the host stock market to accept the firm. By cross-listing and selling its shares on a foreign exchange, a firm typically tries to accomplish one or more of the following objectives:

- Improve the liquidity of its shares and support a liquid secondary market for new equity issues in foreign markets

- Increase its share price by overcoming mispricing in a segmented and illiquid home capital market

- Increase the firm's visibility and acceptance to its customers, suppliers, creditors, and host governments

- Establish a liquid secondary market for shares used to acquire other firms in the host market and to compensate local management and employees of foreign subsidiaries[5]

Improving Liquidity

Quite often foreign investors have acquired a firm's shares through normal brokerage channels, even though the shares are not listed in the investor's home market or are not traded in the investor's preferred currency. Cross-listing is a way to encourage such investors to continue to hold and trade these shares, thus marginally improving secondary market liquidity. This is usually done through ADRs.

Firms domiciled in countries with small illiquid capital markets often outgrow those markets and are forced to raise new equity abroad. Listing on a stock exchange in the market in which these funds are to be raised is typically required by the underwriters to ensure post-issue liquidity in the shares.

The first section of this chapter suggested that firms start by cross-listing in a less liquid market, followed by an equity issue in that market (see Exhibit 14.1). In order to maximize liquidity, however, the firm ideally should cross-list and issue equity in a more liquid market and eventually offer a global equity issue.

---

[5] A recent example of this trading expansion opportunity is Kosmos Energy. Following the company's IPO in the United States in May 2011 (NYSE: KOS), the company listed its shares on the Ghanaian Stock Exchange. Ghana was the country in which the oil company had made its major discoveries and generated nearly all of its income.

In order to maximize liquidity, it is desirable to cross-list and/or sell equity in the most liquid markets. Stock markets have, however, been subject to two major forces in recent years, which are changing their very behavior and liquidity—*demutualization* and *diversification*.

Demutualization is the ongoing process by which the small controlling seat owners on a number of exchanges have been giving up their exclusive powers. As a result, the actual ownership of the exchanges has become increasingly public. *Diversification* represents the growing diversity of both products (derivatives, currencies, etc.) and foreign companies/shares being listed. This has increased the activities and profitability of many exchanges while simultaneously offering a more global mix for reduced cost and increased service.

Stock Exchanges. With respect to stock exchanges, New York and London are clearly the most liquid. The recent merger of the New York Stock Exchange (NYSE) and Euronext, which itself was a merger of stock exchanges in Amsterdam, Brussels, and Paris, has extended the NYSE's lead over both the NASDAQ (New York) and the London Stock Exchange (LSE). Tokyo has declined a bit over the past 20 years in terms of trading value globally, as many foreign firms chose to delist from the Tokyo exchange. Few foreign firms remain cross-listed now in Tokyo. Deutsche Börse (Germany) has a fairly liquid market for domestic shares but a much lower level of liquidity for trading foreign shares. On the other hand, it is an appropriate target market for firms resident in the European Union, especially those that have adopted the euro. It is also used as a supplementary cross-listing location for firms that are already cross-listed on the LSE, NYSE, or NASDAQ.

Why are New York and London so dominant? They offer what global financial firms are looking for: plenty of skilled people, ready access to capital, good infrastructure, attractive regulatory and tax environments, and low levels of corruption. Location and the use of English, increasingly acknowledged as the language of global finance, are also important factors.

Electronic Trading. Most exchanges have moved heavily into electronic trading in recent years. In fact, the U.S. stock market is now a network of 50 different venues connected by an electronic system of published quotes and sales prices. This shift to electronic trading has had broad-reaching effects. For example, the role of the *specialist* on the floor of the NYSE has been greatly reduced with a corresponding reduction in employment by specialist firms. Specialists are no longer responsible for ensuring an orderly movement for their stocks, but they are still important in making more liquid markets for the less-traded shares. The same fate has reduced the importance of market makers on the London Stock Exchange (LSE).

Electronic trading has allowed hedge funds and other high-frequency traders to dominate the market. High-frequency traders now account for 60% of daily volumes. Conversely, volume controlled by the NYSE fell from 80% in 2005 to 25% in 2010. Trades are executed immediately by computer. Spreads between buy and sell orders are now in decimal points as low as a penny a share instead of an eighth of a point. Liquidity has greatly increased but so has the risk of unexpected swings in prices. For example, on May 6, 2010, the Dow Jones Average fell 9.2% at one point but eventually recovered by the end of the day. During that single day of trading, nineteen billion shares were bought and sold.

Promoting Shares and Share Prices

Although cross-listing and equity issuance can occur together, their impacts are separable and significant in and of themselves.

Cross-Listing. Does merely cross-listing on a foreign stock exchange have a favorable impact on share prices? It depends on the degree to which markets are segmented.

If a firm's home capital market is segmented, the firm could theoretically benefit by cross-listing in a foreign market if that market values the firm or its industry more than does the home market. This was certainly the situation experienced by Novo when it listed on the NYSE in 1981 (see Chapter 12). However, most capital markets are becoming more integrated with global markets. Even emerging markets are less segmented than they were just a few years ago.

Equity Issuance. It is well known that the combined impact of a new equity issue undertaken simultaneously with a cross-listing has a more favorable impact on stock price than cross-listing alone. This occurs because the new issue creates an instantly enlarged shareholder base. Marketing efforts by the underwriters prior to the issue engender higher levels of visibility. Post-issue efforts by the underwriters to support at least the initial offering price also reduce investor risk.

Increasing Visibility and Political Acceptance

MNEs list in markets where they have substantial physical operations. Commercial objectives are to enhance corporate image, advertise trademarks and products, get better local press coverage, and become more familiar with the local financial community in order to raise working capital locally.

Political objectives might include the need to meet local ownership requirements for a multinational firm's foreign joint venture. Local ownership of the parent firm's shares might provide a forum for publicizing the firm's activities and how they support the host country.

Establish Liquid Secondary Markets

The establishment of a local liquid market for the firm's equity may aid in financing acquisitions and in the creation of stock-based management compensation programs for subsidiaries.

Funding Growth by Acquisitions. Firms that follow a strategy of growth by acquisition are always looking for creative alternatives to cash for funding these acquisitions. Offering their shares as partial payment is considerably more attractive if those shares have a liquid secondary market. In that case, the target's shareholders have an easy way to convert their acquired shares to cash if they prefer cash to a share swap. However, a share swap is often attractive as a tax-free exchange.

Compensating Management and Employees. If an MNE wishes to use stock options and share purchase compensation plans as a component of the compensation scheme for local management and employees, local listing on a liquid secondary market would enhance the perceived value of such plans. It should reduce transaction and foreign exchange costs for the local beneficiaries.

Barriers to Cross-Listing and Selling Equity Abroad

Although a firm may decide to cross-list and/or sell equity abroad, certain barriers exist. The most serious barriers are the future commitment to providing full and transparent disclosure of operating results and balance sheets as well as a continuous program of investor relations.

The Commitment to Disclosure and Investor Relations. A decision to cross-list must be balanced against the implied increased commitment to full disclosure and a continuing investor relations program. For firms resident in the Anglo-American markets, listing abroad might not appear to be much of a barrier. For example, the SEC's disclosure rules for listing in the United States are so stringent and costly that any other market's rules are mere child's play. Reversing the logic, however, non-U.S. firms must consider disclosure requirements carefully before cross-listing in the United States. Not only are the disclosure requirements breathtaking, timely quarterly information is also required by U.S. regulators and investors. As a result, the foreign firm must maintain a costly continuous investor relations program for its U.S. shareholders, including frequent "road shows" and the time-consuming personal involvement of top management.

Disclosure Is a Double-Edged Sword. The U.S. school of thought presumes that the world-wide trend toward more comprehensive, more transparent, and more standardized financial disclosure of operating results and financial positions will have the desirable effect of lowering the cost of equity capital. As we observed in 2002 and 2008, lack of full and accurate disclosure and poor transparency contributed to the U.S. stock market decline as investors fled to safer securities such as U.S. government bonds. This action increased the equity cost of capital for all firms.

The opposing school of thought—the other edge of the sword—is that the U.S. level of required disclosure is an onerous, costly burden. It discourages many potential listers, and thereby narrows the choice of securities available to U.S. investors at reasonable transaction costs.

Raising Debt Globally

The international debt markets offer the borrower a variety of different maturities, repayment structures, and currencies of denomination. The markets and their many different instruments vary by source of funding, pricing structure, maturity, and subordination or linkage to other debt and equity instruments.

Exhibit 14.9 provides an overview of the three basic categories described in the following sections, along with their primary components as issued or traded in the international debt markets today. As shown in the exhibit, the three major sources of debt funding on the international markets are the international bank loans and syndicated credits, euronote market, and international bond market.

EXHIBIT 14.9   **International Debt Markets and Instruments**

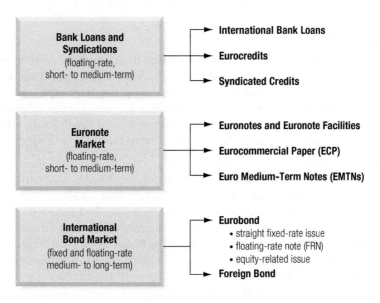

Bank Loans and Syndications

International Bank Loans. *International bank loans* have traditionally been sourced in the eurocurrency loan markets. Eurodollar bank loans are also called "eurodollar credits" or simply "eurocredits." The latter title is broader because it encompasses nondollar loans in the eurocurrency loan market. The key factor attracting both depositors and borrowers to the eurocurrency loan market is the narrow interest rate spread within that market. The difference between deposit and loan rates is often less than 1%.

Eurocredits. *Eurocredits* are bank loans to MNEs, sovereign governments, international institutions, and banks denominated in eurocurrencies and extended by banks in countries other than the country in whose currency the loan is denominated. The basic borrowing interest rate for eurocredits has long been tied to the London Interbank Offered Rate (LIBOR), which is the deposit rate applicable to interbank loans within London. Eurocredits are lent for both short- and medium-term maturities, with maturities for six months or less regarded as routine. Most eurocredits are for a fixed term with no provision for early repayment.

Syndicated Credits. The syndication of loans has enabled banks to spread the risk of large loans among a number of banks. Syndication is particularly important because many large MNEs need credit in excess of a single bank's loan limit. A *syndicated bank credit* is arranged by a lead bank on behalf of its client. Before finalizing the loan agreement, the lead bank seeks the participation of a group of banks, with each participant providing a portion of the total funds needed. The lead bank will work with the borrower to determine the amount of the total credit, the floating-rate base and spread over the base rate, maturity, and fee structure for managing the participating banks. There are two elements to the periodic expenses of the syndicated credit:

1. Actual interest expense of the loan, normally stated as a spread in basis points over a variable-rate base such as LIBOR

2. Commitment fees paid on any unused portions of the credit—the spread paid over LIBOR by the borrower is considered the risk premium, reflecting the general business and financial risk applicable to the borrower's repayment capability

Euronote Market

The *euronote market* is the collective term used to describe short- to medium-term debt instruments sourced in the eurocurrency markets. Although a multitude of differentiated financial products exists, they can be divided into two major groups—*underwritten facilities* and *nonunderwritten facilities*. *Underwritten facilities* are used for the sale of euronotes in a number of different forms. *Nonunderwritten facilities* are used for the sale and distribution of euro-commercial paper (ECP) and euro medium-term notes (EMTNs).

Euronotes and Euronote Facilities. A major development in international money markets was the establishment of underwriting facilities for the sale of short-term, negotiable, promissory notes—*euronotes*. Among the facilities for their issuance were *revolving underwriting facilities* (rufs), *note issuance facilities* (nifs), and *standby note issuance facilities* (snifs). These facilities were provided by international investment and commercial banks. The euronote was a substantially cheaper source of short-term funds than were syndicated loans because the securitized and underwritten form allowed the ready establishment of liquid secondary markets, allowing the notes to be placed directly with the investing public. The banks received substantial fees initially for their underwriting and placement services.

Eurocommercial Paper (ECP). *Eurocommercial paper* (ECP), like commercial paper issued in domestic markets around the world, is a short-term debt obligation (nonunderwritten) of a corporation or bank. Maturities are typically one, three, and six months. The paper is sold normally at a discount or occasionally with a stated coupon. Although the market is capable of supporting issues in any major currency, over 90% of issues outstanding are denominated in U.S. dollars.

Euro Medium-Term Notes (EMTNs). The *euro medium-term note* (EMTN) market effectively bridges the maturity gap between ECP and the longer-term and less flexible international bond. Although many of these notes were initially underwritten, most EMTNs are now nonunderwritten.

The rapid initial growth of the EMTN market followed directly on the heels of the same basic instrument that began in the U.S. domestic market when the U.S. SEC instituted *SEC Rule #415*, allowing companies to obtain *shelf registrations* for debt issues. Once such a registration was obtained, the corporation could issue notes on a continuous basis without the need to obtain new registrations for each additional issue. This, in turn, allowed a firm to sell short- and medium-term notes through a much cheaper and more flexible issuance facility than ordinary bonds.

The EMTN's basic characteristics are similar to those of a bond, with principal, maturity, coupon structures, and rates being comparable. The EMTN's typical maturities range from as little as nine months to a maximum of 10 years. Coupons are typically paid semiannually, and coupon rates are comparable to similar bond issues. The EMTN does, however, have three unique characteristics: (1) the EMTN is a facility, allowing continuous issuance over a period of time, unlike a bond issue that is essentially sold all at once; (2) because EMTNs are sold continuously, in order to make debt service (coupon redemption) manageable, coupons are paid on set calendar dates regardless of the date of issuance; (3) EMTNs are issued in relatively small denominations, from $2 million to $5 million, making medium-term debt acquisition much more flexible than the large minimums customarily needed in the international bond markets.

International Bond Market

The international bond market sports a rich array of innovative instruments created by imaginative investment bankers who are unfettered by the usual controls and regulations governing domestic capital markets. Indeed, the international bond market rivals the international banking market in terms of the quantity and cost of funds provided to international borrowers. All international bonds fall within two generic classifications, eurobonds and foreign bonds. The distinction between categories is based on whether the borrower is a domestic or a foreign resident, and whether the issue is denominated in the local currency or a foreign currency.

Eurobonds. A *Eurobond* is underwritten by an international syndicate of banks and other securities firms, and is sold exclusively in countries other than the country in whose currency the issue is denominated. For example, a bond issued by a firm resident in the United States, denominated in U.S. dollars, and sold to investors in Europe and Japan (but not to investors in the United States), is a eurobond.

Eurobonds are issued by MNEs, large domestic corporations, sovereign governments, governmental enterprises, and international institutions. They are offered simultaneously in a number of different national capital markets, but not in the capital market or to residents of the country in whose currency the bond is denominated. Almost all eurobonds are in bearer form with call provisions (the ability of the issuer to call the bond in prior to maturity) and sinking funds (required accumulations of funds by the firms to assure repayment of the obligation).

The syndicate that offers a new issue of eurobonds might be composed of underwriters from a number of countries, including European banks, foreign branches of U.S. banks, banks from offshore financial centers, investment and merchant banks, and nonbank securities firms. There are three types of eurobond issues:

- **The Straight Fixed-Rate Issue.**   The *straight fixed-rate issue* is structured like most domestic bonds, with a fixed coupon, set maturity date, and full principal repayment upon final maturity. Coupons are normally paid annually, rather than semiannually, primarily because the bonds are bearer bonds and annual coupon redemption is more convenient for the holders.

- **The Floating-Rate Note.**   The *floating-rate note* (FRN) normally pays a semiannual coupon that is determined using a variable-rate base. A typical coupon would be set at some fixed spread over LIBOR. This structure, like most variable-rate interest-bearing instruments, was designed to allow investors to shift more of the interest-rate risk of a financial investment to the borrower. Although many FRNs have fixed maturities, in recent years many issues are perpetuities, with no principal repayment, taking on the characteristics of equity.

- **The Equity-Related Issue.**   The *equity-related international bond* resembles the straight fixed-rate issue in practically all price and payment characteristics, with the added feature that it is convertible to stock prior to maturity at a specified price per share (or alternatively, number of shares per bond). The borrower is able to issue debt with lower coupon payments due to the added value of the equity conversion feature.

Foreign Bonds. A *foreign bond* is underwritten by a syndicate composed of members from a single country, sold principally within that country, and denominated in the currency of that country. The issuer, however, is from another country. A bond issued by a firm resident in Sweden, denominated in U.S. dollars, and sold in the United States to U.S. investors by U.S. investment bankers, is a foreign bond. Foreign bonds have nicknames: foreign bonds sold in the United States are *Yankee bonds*; foreign bonds sold in Japan are *Samurai bonds*; and foreign bonds sold in the United Kingdom are *Bulldogs*.

Unique Characteristics of Eurobond Markets

Although the eurobond market evolved at about the same time as the eurodollar market, the two markets exist for different reasons, and each could exist independently of the other. The eurobond market owes its existence to several unique factors: the absence of regulatory interference, less stringent disclosure practices, favorable tax treatment, and ratings.

Absence of Regulatory Interference. National governments often impose tight controls on foreign issuers of securities denominated in the local currency and sold within their national boundaries. However, governments in general have less stringent limitations for securities denominated in foreign currencies and sold within their markets to holders of those foreign currencies. In effect, eurobond sales fall outside the regulatory domain of any single nation.

Less Stringent Disclosure. Disclosure requirements in the eurobond market are much less stringent than those of the U.S. Securities and Exchange Commission (SEC) for sales within the United States. U.S. firms often find that the registration costs of a eurobond offering are less than those of a domestic issue and that less time is needed to bring a new issue to market. Non-U.S. firms often prefer eurodollar bonds over bonds sold within the United States because they do not wish to undergo the costs and disclosure needed to register with the

SEC. However, the SEC has relaxed disclosure requirements for certain private placements (Rule #144A), which has improved the attractiveness of the U.S. domestic bond and equity markets.

Favorable Tax Treatment. Eurobonds offer tax anonymity and flexibility. Interest paid on eurobonds is generally not subject to an income withholding tax. As one might expect, eurobond interest is not always reported to tax authorities. Eurobonds are usually issued in bearer form, meaning that the name and country of residence of the owner is not on the certificate. To receive interest, the bearer cuts an interest coupon from the bond and turns it in at a banking institution listed on the issue as a paying agent. European investors are accustomed to the privacy provided by bearer bonds and are very reluctant to purchase registered bonds, which require holders to reveal their names before they receive interest. It follows, then, that bearer bond status is often tied to tax avoidance.

Access to debt capital is obviously impacted by everything from the legal and tax environments to basic societal norms. Indeed, even religion plays a part in the use and availability of debt capital. *Global Finance in Practice 14.2* illustrates one area rarely seen by Westerners, *Islamic finance.*

GLOBAL FINANCE IN PRACTICE 14.2

Islamic Finance

Muslims, the followers of Islam, now make up roughly one-fourth of the world's population. The countries of the world that are predominantly Muslim create roughly 10% of global GDP and comprise a large share of the emerging marketplace. Islamic law speaks to many dimensions of the individual and organizational behaviors for its practitioners—including business. Islamic finance, the specific area of our interest, imposes a number of restrictions on Muslims, which have a dramatic impact on the funding and structure of Muslim businesses.

The Islamic form of finance is as old as the religion of Islam itself. The basis for all Islamic finance lies in the principles of the Sharia, or Islamic Law, which is taken from the Qur'an. Observance of these principles precipitates restrictions on business and finance practices as follows:

- Making money from money is not permissible
- Earning interest is prohibited
- Profit and loss should be shared
- Speculation (gambling) is prohibited
- Investments should support only halal activities

For the conduct of business, the key to understanding the Sharia prohibition on earning interest is to understand that profitability from traditional Western investments arises from the returns associated with carrying risk. For example, a traditional Western bank may extend a loan to a business. It is

agreed that the bank will receive its principal and interest in return regardless of the ultimate profitability of the business (the borrower). In fact, the debt is paid off before returns to equity occur. Similarly, an individual who deposits money in a Western bank will receive interest earnings on that deposit regardless of the profitability of the bank and of the bank's associated investments.

Under Sharia law, however, an Islamic bank cannot pay interest to depositors. Therefore, the depositors in an Islamic bank are, in effect, shareholders (much like credit unions in the West), and the returns they receive are a function of the profitability of the bank's investments. Their returns cannot be fixed or guaranteed, because that would break the principle of profit and loss being shared.

Recently, however, a number of Islamic banking institutions have opened in Europe and North America. A Muslim now can enter into a sequence of purchases that allows him to purchase a home without departing from Islamic principles. The buyer selects the property, which is then purchased by an Islamic bank. The bank in turn resells the house to the prospective buyer at a higher price. The buyer is allowed to pay off the purchase over a series of years. Although the difference in purchase prices is, by Western thinking, implicit interest, this structure does conform to Sharia law. Unfortunately, in both the United States and the United Kingdom, this "implicit interest" is not a tax-deductible expense for the homeowner as interest would be.

Ratings. Rating agencies, such as Moody's and Standard and Poor's (S&P), provide ratings for selected international bonds for a fee. Moody's ratings for international bonds imply the same creditworthiness as for domestic bonds of U.S. issuers. Moody's limits its evaluation to the issuer's ability to obtain the necessary currency to repay the issue according to the original terms of the bond. The agency excludes any assessment of risk to the investor caused by changing exchange rates.

Moody's rates international bonds at the request of the issuer. Based on supporting financial statements and other material obtained from the issuer, it makes a preliminary rating and then informs the issuer who has an opportunity to comment. After Moody's determines its final rating, the issuer may decide not to have the rating published. Consequently, a disproportionately large number of published international ratings fall into the highest categories, since issuers that receive a lower rating do not allow publication.

Purchasers of eurobonds do not rely only on bond-rating services or on detailed analyses of financial statements. The general reputation of the issuing corporation and its underwriters has been a major factor in obtaining favorable terms. For this reason, larger and better-known MNEs, state enterprises, and sovereign governments are able to obtain the lowest interest rates. Firms whose names are better known to the general public, possibly because they manufacture consumer goods, are often believed to have an advantage over equally qualified firms whose products are less widely known.

SUMMARY POINTS

- Designing a capital sourcing strategy requires management to design a long-run financial strategy. The firm must then choose among the various alternative paths to achieve its goals, including where to cross-list its shares, and where to issue new equity, and in what form.

- A multinational firm's marginal cost of capital is constant for considerable ranges of its capital budget. This statement is not true for most small domestic firms.

- By diversifying cash flows internationally, the MNE may be able to achieve the same kind of reduction in cash flow variability that portfolio investors receive from diversifying their portfolios internationally.

- When a firm issues foreign currency-denominated debt, its effective cost equals the after-tax cost of repaying the principal and interest in terms of the firm's own currency. This amount includes the nominal cost of principal and interest in foreign currency terms, adjusted for any foreign exchange gains or losses.

- There is a variety of different equity pathways that firms may choose between when pursuing global sources of equity, including euroequity issues, direct foreign issuances, depositary receipt programs, and private placements.

- Depositary receipt programs, either American or global, provide an extremely effective way for firms from outside of the established industrial country markets to improve the liquidity of their existing shares, or issue new shares.

- Private placement is a growing segment of the market, allowing firms from emerging markets to raise capital in the largest of capital markets with limited disclosure and cost.

- The international debt markets offer the borrower a variety of different maturities, repayment structures, and currencies of denomination. The markets and their many different instruments vary by source of funding, pricing structure, maturity, and subordination or linkage to other debt and equity instruments.

- Eurocurrency markets serve two valuable purposes: (1) eurocurrency deposits are an efficient and convenient money market device for holding excess corporate liquidity, and (2) the eurocurrency market is a major source of short-term bank loans to finance corporate working capital needs, including the financing of imports and exports.

Petrobrás of Brazil and the Cost of Capital[6]

The national oil company of Brazil, Petrobrás, suffered from an ailment common in emerging markets—a high and uncompetitive cost of capital. Despite being widely considered the global leader in deepwater technology (the ability to drill and develop oil and gas fields more than a mile below the ocean's surface), unless it could devise a strategy to lower its cost of capital, it would be unable to exploit its true organizational competitive advantage.

Many market analysts argued that the Brazilian company should follow the strategy employed by a number of Mexican companies and buy its way out of its dilemma. If Petrobrás were to acquire one of the many independent North American oil and gas companies, it might transform itself from being wholly "Brazilian" to partially "American" in the eyes of capital markets, and possibly lower its weighted average cost of capital (WACC) to between 6% and 8%.

Petróleo Brasileiro S.A. (Petrobrás) was an integrated oil and gas company founded in 1954 by the Brazilian government as the national oil company of Brazil. The company was listed publicly in São Paulo in 1997 and on the New York Stock Exchange (NYSE: PBR) in 2000. Despite the equity listings, the Brazilian government continued to be the controlling shareholder, with 33% of the total capital and 55% of the voting shares. As the national oil company of Brazil, the company's singular purpose was the reduction of Brazil's dependency on imported oil. A side effect of this focus, however, had been a lack of international diversification. Many of the company's critics argued that being both Brazilian and undiversified internationally resulted in an uncompetitive cost of capital.

Need for Diversification

Petrobrás in 2002 was the largest company in Brazil, and the largest publicly traded oil company in Latin America. It was not, however, international in its operations. This inherent lack of international diversification was apparent to international investors, who assigned the company the same country risk factors and premiums they did to all other Brazilian companies. The result was a cost of capital in 2002, as seen in Exhibit A, that was 6% higher than the other firms shown.

Petrobrás embarked on a globalization strategy, with several major transactions heading up the process. In December 2001, Repsol-YPF of Argentina and Petrobrás concluded an exchange of operating assets valued at $500 million. In the exchange, Petrobrás received 99% interest in the Eg3 S.A. service station chain, while Repsol-YPF gained a 30% stake in a refinery, a 10% stake in an offshore oil field, and a fuel resale right to 230 service stations in Brazil. The agreement included an eight-year guarantee against currency risks.

In October 2002, Petrobrás purchased Perez Companc (Pecom) of Argentina. Pecom had quickly come into play following the Argentine financial crisis in January 2002. Although Pecom had significant international reserves and production capability, the combined forces of a devalued Argentine peso, a largely dollar-denominated debt portfolio, and a multitude of Argentine government regulations that hindered its ability to hold and leverage hard currency resources, the company had moved quickly to find a buyer to refund its financial structure. Petrobrás took advantage of the opportunity. Pecom's ownership had been split between its original controlling family owners and their foundation, 58.6%, and public flotation of the remaining 41.4%. Petrobrás had purchased the controlling interest, the full 58.6% interest, outright from the family.

Over the next three years, Petrobrás focused on restructuring much of its debt (and the debt it had acquired via the Pecom acquisition) and investing in its own growth. But progress toward revitalizing its financial structure came slowly, and by 2005 there was renewed discussion of a new equity issuance to increase the firm's equity capital.[7] But at what cost? What was the company's cost of capital?

[7]By 2005, the company's financial strategy was showing significant diversification. Total corporate funding was well-balanced: bonds, $4 billion; BNDES (bonds issued under the auspices of a Brazilian economic development agency), $3 billion; project finance, $5 billion; other, $4 billion.

EXHIBIT A   **Petrobrás' Uncompetitive Cost of Capital**

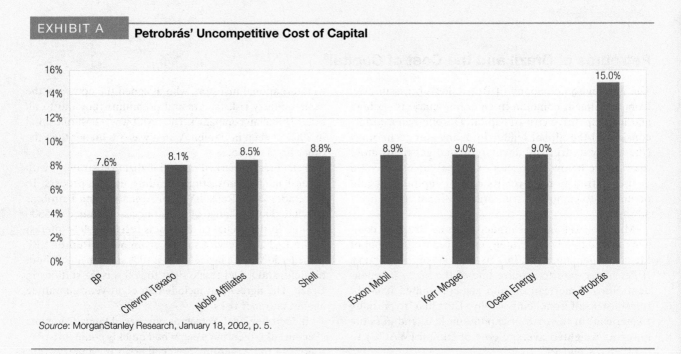

Source: MorganStanley Research, January 18, 2002, p. 5.

Country Risk

Exhibit A presented the cost of capital of a number of major oil and gas companies across the world, including Petrobrás in 2002. This comparison could occur only if all capital costs were calculated in a common currency, in this case, the U.S. dollar. The global oil and gas markets had long been considered "dollar-denominated," and any company operating in these markets, regardless of where it actually operated in the world, was considered to have the dollar as its functional currency. Once that company listed its shares in a U.S. equity market, the dollarization of its capital costs became even more accepted.

But what was the cost of capital—in dollar terms—for a Brazilian business? Brazil has a long history of bouts with high inflation, economic instability, and currency devaluations and depreciations (depending on the regime de jure). One of the leading indicators of the global market's opinion of Brazilian country risk was the sovereign spread, the additional yield or cost of dollar funds that the Brazilian government had to pay on global markets over and above that which the U.S. Treasury paid to borrow dollar funds. As illustrated in Exhibit B, the Brazilian sovereign spread had been both

high and volatile over the past decade.[8] The spread was sometimes as low as 400 basis points (4.0%), as in recent years, or as high as 2,400 basis points (24%), during the 2002 financial crisis in which the real was first devalued then floated. And that was merely the cost of debt for the government of Brazil. How was this sovereign spread reflected in the cost of debt and equity for a Brazilian company like Petrobrás?

One approach to the estimation of Petrobrás' cost of debt in U.S. dollar terms ($k_d^\$$) was to build it up: the government of Brazil's cost of dollar funds adjusted for a private corporate credit spread.

$$k_d^\$ = \text{U.S. Treasury} + \text{Brazilian} + \text{Petrobrás}$$
$$\text{risk-free rate} \quad \text{sovereign} \quad \text{credit}$$
$$\text{spread} \quad \text{spread}$$

$$k_d^\$ = 4.000\% + 4.000\% + 1.000\% = 9.000\%$$

If the U.S. Treasury risk-free rate was estimated using the Treasury 10-year bond rate (yield), a base rate in August 2005 could be 4.0%. The Brazilian sovereign spread, as seen in Exhibit B, appeared to be 400 basis points, or an additional 4.0%. Even if Petrobrás' credit spread was only

---

[8] The measure of sovereign spread presented in Exhibit B is that calculated by JPMorgan in its *Emerging Market Bond Index Plus* (EMBI+) index. This is the most widely used measure of country risk by practitioners.

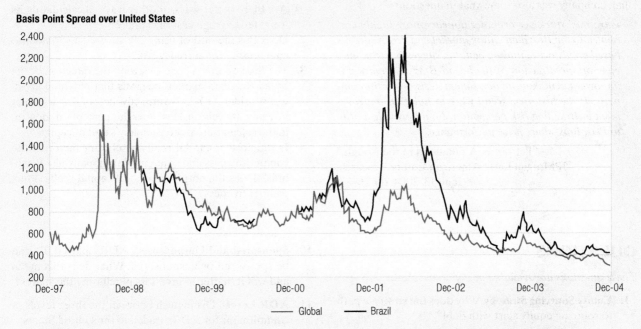

EXHIBIT B   **The Brazilian Sovereign Spread**

Basis Point Spread over United States

Source: JPMorgan's EMBI+ Spread, as quoted by Latin Focus, www.latin-focus.com/latinfocus/countries/brazilbisprd.htm, August 2005.

1.0%, the company's current cost of dollar debt would be 9%. This cost was clearly higher than the cost of debt for most of the world's oil majors who were probably paying only 5% on average for debt in late 2005.

Petrobrás' cost of equity would be similarly affected by the country risk-adjusted risk-free rate of interest. Using a simple expression of the Capital Asset Pricing Model (CAPM) to estimate the company's cost of equity capital in dollar terms ($k_e^{\$}$):

$$k_e^{\$} = \text{risk-free rate} \\ +(\beta_{\text{Petrobrás}} \times \text{market risk premium}) = 8.000\% \\ +(1.10 \times 5.500\%) = 14.05\%$$

This calculation assumed the same risk-free rate as used in the cost of debt previously, with a beta (NYSE basis) of 1.10 and a market risk premium of 5.500%. Even with these relatively conservative assumptions (many would argue that the company's beta was actually higher or lower, and that the market risk premium was 6.0% or higher), the company's cost of equity was 14%.

$$\text{WACC} = (\text{debt/capital} \times k_d^{\$} \times (1 - \text{tax rate})) \\ +(\text{equity/capital} \times k_e^{\$})$$

Assuming a long-term target capital structure of one-third debt and two-thirds equity, and an effective corporate tax rate of 28% (after special tax concessions, surcharges, and incentives for the Brazilian oil and gas industry), Petrobrás' WACC was estimated at a little over 11.5%:

$$\text{WACC} = (0.333 \times 9.000\% \times 0.72) \\ +(0.667 \times 14.050\%) = 11.529\%.$$

So, after all of the efforts to internationally diversify the firm and internationalize its cost of capital, why was Petrobrás' cost of capital still so much higher than its global counterparts? Not only was the company's weighted average cost of capital high compared to other major global players, this was the same high cost of capital used as the basic discount rate in evaluating many potential investments and acquisitions.

A number of the investment banking firms that covered Petrobrás noted that the company's share price had shown a very high correlation with the EMBI+ sovereign spread for Brazil (shown in Exhibit B), hovering around 0.84 for a number of years. Similarly, Petrobrás' share price was also historically correlated—inversely—with the Brazilian reais/U.S. dollar exchange rate. This correlation had averaged

−0.88 over the 2000–2004 period. Finally, the question of whether Petrobrás was considered an oil company or a Brazilian company was also somewhat in question:

Petrobrás' stock performance appears more highly correlated to the Brazilian equity market and credit spreads based on historical trading patterns, suggesting that one's view on the direction of the broad Brazilian market is important in making an investment decision on the company. If the historical trend were to hold, an improvement in Brazilian risk perception should provide a fillip to Petrobrás' share price performance.

— "Petrobrás: A Diamond in the Rough,"
JPMorgan Latin American Equity Research,
June 18, 2004, pp. 26–27.

Mini-Case Questions

1. Why do you think Petrobrás' cost of capital is so high? Are there better ways, or other ways, of calculating its weighted average cost of capital?

2. Does this method of using the sovereign spread also compensate for currency risk?

3. The final quote on "one's view on the direction of the broad Brazilian market" suggests that potential investors consider the relative attractiveness of Brazil in their investment decision. How does this perception show up in the calculation of the company's cost of capital?

4. Is the cost of capital really a relevant factor in the competitiveness and strategy of a company like Petrobrás? Does the corporate cost of capital really affect competitiveness?

QUESTIONS

These questions are available in MyFinanceLab.

1. **Equity Sourcing Strategy.** Why does the strategic path to sourcing equity start with debt?

2. **Optimal Financial Structure.** If the cost of debt is less than the cost of equity, why doesn't the firm's cost of capital continue to decrease with the use of more and more debt?

3. **Multinationals and Cash Flow Diversification.** How does the multinational's ability to diversify its cash flows alter its ability to use greater amounts of debt?

4. **Foreign Currency-Denominated Debt.** How does borrowing in a foreign currency change the risk associated with debt?

5. **Three Keys to Global Equity.** What are the three key elements related to raising equity capital in the global marketplace?

6. **Global Equity Alternatives.** What are the alternative structures available for raising equity capital on the global market?

7. **Directed Public Issues.** What is a directed public issue? What is the purpose of this kind of international equity issuance?

8. **Depositary Receipts.** What is a depositary receipt? Give examples of equity shares listed and issued in foreign equity markets in this form?

9. **GDRs, ADRs, and GRSs.** What is the difference between a GDR, ADR, and GRS? How are these differences significant?

10. **Sponsored and Unsponsored.** ADRs and GDRs can be sponsored or unsponsored. What does this mean and will it matter to investors purchasing the shares?

11. **ADR Levels.** Distinguish between the three levels of commitment for ADRs traded in the United States.

12. **IPOs and FOs.** What is the significance of IPOs versus FOs?

13. **Foreign Equity Listing and Issuance.** Give five reasons why a firm might cross-list and sell its shares on a very liquid stock exchange.

14. **Cross-Listing Abroad.** What are the main reasons for firms to cross-list abroad?

15. **Barriers to Cross-Listing.** What are the main barriers to cross-listing abroad?

16. **Private Placement.** What is a private placement? What are the comparative pros and cons of private placement versus a pubic issue?

17. **Private Equity.** What is private equity and how do private equity funds differ from traditional venture capital firms?

18. **Bank Loans versus Securitized Debt.** For multinational corporations, what is the advantage of securitized debt instruments sold on a market versus bank borrowing?

19. **International Debt Instruments.** What are the primary alternative instruments available for raising debt on the international marketplace?

20. **Eurobond versus Foreign Bonds.** What is the difference between a eurobond and a foreign bond and why do two types of international bonds exist?

21. **Funding Foreign Subsidiaries.** What are the primary methods of funding foreign subsidiaries, and how do host government concerns affect those choices?

22. **Local Norms.** Should foreign subsidiaries of multinational firms conform to the capital structure norms of the host country or to the norms of their parent's country?

23. **Internal Financing of Foreign Subsidiaries.** What is the difference between "internal" financing and "external" financing for a subsidiary?

24. **External Financing of Foreign Subsidiaries.** What are the primary alternatives for the external financing of a foreign subsidiary?

PROBLEMS

These problems are available in MyFinanceLab.

1. **Copper Mountain Group (U.S.).** The Copper Mountain Group, a private equity firm headquartered in Boulder, Colorado, borrows £5,000,000 for one year at 7.375% interest.
   a. What is the dollar cost of this debt if the pound depreciates from $2.0260/£ to $1.9460/£ over the year?
   b. What is the dollar cost of this debt if the pound appreciates from $2.0260/£ to $2.1640/£ over the year?

2. **Foreign Exchange Risk and the Cost of Borrowing Swiss Francs.** The chapter demonstrated that a firm borrowing in a foreign currency could potentially end up paying a very different effective rate of interest than what it expected. Using the same baseline values of a debt principal of SF1.5 million, a one year period, an initial spot rate of SF1.5000/$, a 5.000% cost of debt, and a 34% tax rate, what is the effective cost of debt for one year for a U.S. dollar-based company if the exchange rate at the end of the period was:
   a. SF1.5000/$ c. SF1.3860/$
   b. SF1.4400/$ d. SF1.6240/$

3. **McDougan Associates (U.S.).** McDougan Associates, a U.S.-based investment partnership, borrows €80,000,000 at a time when the exchange rate is $1.3460/€. The entire principal is to be repaid in three years, and interest is 6.250% per annum, paid annually in euros. The euro is expected to depreciate vis-à-vis the dollar at 3% per annum. What is the effective cost of this loan for McDougan?

4. **Morning Star Air (China).** Morning Star Air, headquartered in Kunming, China, needs US$25,000,000

for one year to finance working capital. The airline has two alternatives for borrowing:
   a. Borrow US$25,000,000 in Eurodollars in London at 7.250% per annum
   b. Borrow HK$39,000,000 in Hong Kong at 7.00% per annum, and exchange these Hong Kong dollars at the present exchange rate of HK$7.8/US$ for U.S. dollars.

At what ending exchange rate would Morning Star Air be indifferent between borrowing U.S. dollars and borrowing Hong Kong dollars?

5. **Pantheon Capital, S.A.** If Pantheon Capital, S.A. is raising funds via a euro-medium-term note with the following characteristics, how much in dollars will Pantheon receive for each $1,000 note sold?

Coupon rate: 8.00% payable semiannually on June 30 and December 31
Date of issuance: February 28, 2011
Maturity: August 31, 2011

6. **Westminster Insurance Company.** Westminster Insurance Company plans to sell $2,000,000 of eurocommercial paper with a 60-day maturity and discounted to yield 4.60% per annum. What will be the immediate proceeds to Westminster Insurance?

7. **Sunrise Manufacturing, Inc.** Sunrise Manufacturing, Inc., a U.S. multinational company, has the following debt components in its consolidated capital section. Sunrise's finance staff estimates their cost of equity to be 20%. Current exchange rates are also listed below. Income taxes are 30% around the world after allowing for credits. Calculate Sunrise's weighted average cost of capital. Are any assumptions implicit in your calculation?

Assumption	Value
Tax rate	30.00%
10-year eurobonds (euros)	6,000,000
20-year yen bonds (yen)	750,000,000
Spot rate ($/euro)	1.2400
Spot rate ($/pound)	1.8600
Spot rate (yen/$)	109.00

8. **Petrol Ibérico.** Petrol Ibérico, a European gas company, is borrowing US$650,000,000 via a syndicated eurocredit for six years at 80 basis points over LIBOR. LIBOR for the loan will be reset every six months. The funds will be provided by a syndicate of eight leading investment bankers, which will charge up-front fees totaling 1.2% of the principal amount. What is the effective interest cost for the first year if LIBOR is 4.00% for the first six months and 4.20% for the second six months.

9. **Adamantine Architectonics.** Adamantine Architectonics consists of a U.S. parent and wholly owned subsidiaries in Malaysia (A-Malaysia) and Mexico (A-Mexico). Selected portions of their non-consolidated balance sheets, translated into U.S. dollars, are shown in the table below. What are the debt and equity proportions in Adamantine's consolidated balance sheet?

A-Malaysia (accounts in ringgits)		A-Mexico (accounts in pesos)	
Long-term debt	RM11,400,000	Long-term debt	PS20,000,000
Shareholders' equity	RM15,200,000	Shareholders' equity	PS60,000,000

Adamantine Architectonics
(Nonconsolidated Balance Sheet—Selected Items Only)

Investment in subsidiaries		Parent long-term debt	$12,000,000
In A-Malaysia	$4,000,000	Common stock	5,000,000
In A-Mexico	6,000,000	Retained earnings	20,000,000
Current exchange rates:			
Malaysia	RM3.80/$		
Mexico	PS10/$		

Petrobrás of Brazil: Estimating its Weighted Average Cost of Capital

Petrobrás Petróleo Brasileiro S.A. or Petrobras is the national oil company of Brazil. It is publicly traded, but the government of Brazil holds the controlling share. It is the largest company in the Southern Hemisphere by market capitalization and the largest in all of Latin America. As an oil company, the primary product of its production has a price set on global markets—the price of oil—and much of its business is conducted the global currency of oil, the U.S. dollar. Problems 10–15 examine a variety of different financial institutions' attempts to estimate the company's cost of capital.

10. **JPMorgan.** JPMorgan's Latin American Equity Research department produced the following WACC calculation for Petrobrás of Brazil versus Lukoil of Russia in their June 18, 2004, report. Evaluate the methodology and assumptions used in the calculation. Assume a 28% tax rate for both companies.

	Petrobrás	Lukoil
Risk-free rate	4.8%	4.8%
Sovereign risk	7.0%	3.0%
Equity risk premium	4.5%	5.7%
Market cost of equity	16.3%	13.5%
Beta (relevered)	0.87	1.04
Cost of debt	8.4%	6.8%
Debt/capital ratio	0.333	0.475
WACC	14.7%	12.3%

11. **UNIBANCO.** UNIBANCO estimated the weighted average cost of capital for Petrobrás to be 13.2% in Brazilian reais in August of 2004. Evaluate the methodology and assumptions used in the calculation.

Risk-free rate	4.5%	Cost of debt (after-tax)	5.7%
Beta	0.99	Tax rate	34%
Market premium	6.0%	Debt/total capital	40%
Country risk premium	5.5%	WACC (R$)	13.2%
Cost of equity (US$)	15.9%		

12. **Citigroup SmithBarney (Dollar).** Citigroup regularly performs a U.S. dollar-based discount cash flow (DCF) valuation of Petrobrás in its coverage. That DCF analysis requires the use of a discount rate, which they base on the company's weighted average cost of capital. Evaluate the methodology and assumptions used in the 2003 Actual and 2004 Estimates of Petrobrás' WACC shown in the table on the next page.

Problem 12.

Capital Cost Components	July 28, 2005		March 8, 2005	
	2003A	2004E	2003A	2004E
Risk-free rate	9.400%	9.400%	9.000%	9.000%
Levered beta	1.07	1.09	1.08	1.10
Risk premium	5.500%	5.500%	5.500%	5.500%
Cost of equity	15.285%	15.395%	14.940%	15.050%
Cost of debt	8.400%	8.400%	9.000%	9.000%
Tax rate	28.500%	27.100%	28.500%	27.100%
Cost of debt, after-tax	6.006%	6.124%	6.435%	6.561%
Debt/capital ratio	32.700%	32.400%	32.700%	32.400%
Equity/capital ratio	67.300%	67.600%	67.300%	67.600%
WACC	12.20%	12.30%	12.10%	12.30%

13. **Citigroup SmithBarney (Reais).** In a report dated June 17, 2003, Citigroup SmithBarney calculated a WACC for Petrobrás denominated in Brazilian reais (R$). Evaluate the methodology and assumptions used in this cost of capital calculation.

Risk-free rate (Brazilian C-Bond)	9.90%
Petrobrás levered beta	1.40
Market risk premium	5.50%
Cost of equity	17.60%
Cost of debt	10.00%
Brazilian corporate tax rate	34.00%
Long-term debt ratio (% of capital)	50.60%
WACC (R$)	12.00%

14. **BBVA Investment Bank.** BBVA utilized a rather innovative approach to dealing with both country and currency risk in their December 20, 2004, report on Petrobrás. Evaluate the methodology and assumptions used in this cost of capital calculation.

Cost of Capital Component	2003 Estimate	2004 Estimate
U.S. 10-year risk-free rate (in US$)	4.10%	4.40%
Country risk premium (in US$)	6.00%	4.00%
Petrobrás premium "adjustment"	1.00%	1.00%
Petrobrás risk-free rate (in US$)	9.10%	7.40%
Market risk premium (in US$)	6.00%	6.00%
Petrobrás beta	0.80	0.80

Cost of Capital Component	2003 Estimate	2004 Estimate
Cost of equity (in US$)	13.90%	12.20%
Projected 10-year currency devaluation	2.50%	2.50%
Cost of equity (in R$)	16.75%	14.44%
Petrobrás cost of debt after-tax (in R$)	5.50%	5.50%
Long-term equity ratio (% of capital)	69%	72%
Long-term debt ratio (% of capital)	31%	28%
WACC (in R$)	13.30%	12.00%

15. **Petrobrás' WACC Comparison.** The various estimates of the cost of capital for Petrobrás of Brazil appear to be very different, but are they? Reorganize your answers to Problems 10–14 into those costs of capital that are in U.S. dollars versus Brazilian reais. Use the estimates for 2004 as the basis of comparison.

16. **Grupo Modelo S.A.B. de C.V.** Grupo Modelo, a brewery out of Mexico that exports such well-known varieties as Corona, Modelo, and Pacifico, is Mexican by incorporation. However, the company evaluates all business results, including financing costs, in U.S. dollars. The company needs to borrow $10,000,000 or the foreign currency equivalent for four years. For all issues, interest is payable once per year, at the end of the year. Available alternatives are as follows:
a. Sell Japanese yen bonds at par yielding 3% per annum. The current exchange rate is ¥106/$, and the yen is expected to strengthen against the dollar by 2% per annum.

b. Sell euro-denominated bonds at par yielding 7% per annum. The current exchange rate is $1.1960/€, and the euro is expected to weaken against the dollar by 2% per annum.

c. Sell U.S. dollar bonds at par yielding 5% per annum.

Which course of action do you recommend Grupo Modelo take and why?

INTERNET EXERCISES

1. **Global Equities.** Bloomberg provides extensive coverage of the global equity markets 24 hours a day. Using the Bloomberg site listed here, note how different the performance indices are on the same equity markets at the same point in time all around the world.

Bloomberg   www.bloomberg.com/markets/stocks/world-indexes/

2. **JPMorgan and Bank of New York Mellon.** JPMorgan and Bank of New York Mellon provide up to

the minute performance of American Depositary Receipts in the U.S. marketplace. The site highlights the high-performing equities of the day.

a. Prepare a briefing for senior management in your firm encouraging them to consider internationally diversifying the firm's liquid asset portfolio with ADRs.

b. Identify whether the ADR program level (I, II, III, 144A) has any significance to which securities you believe the firm should consider.

JPMorgan ADRs   www.adr.com

Bank of New York Mellon   www.adrbnymellon.com

3. **London Stock Exchange.** The London Stock Exchange (LSE) lists many different global depositary receipts among its active equities. Use the LSE's Internet site to track the performance of the largest GDRs active today.

London Stock Exchange   www.londonstockexchange.com/traders-and-brokers/security-types/gdrs/gdrs.htm

Financial Structure of Foreign Subsidiaries

If we accept the theory that minimizing the cost of capital for a given level of business risk and capital budget is an objective that should be implemented from the perspective of the consolidated MNE, then the financial structure of each subsidiary is relevant only to the extent that it affects this overall goal. In other words, an individual subsidiary does not really have an independent cost of capital. Therefore, its financial structure should not be based on the objective of minimizing its cost of capital.

Financial structure norms for firms vary widely from one country to another but vary less for firms domiciled in the same country. This statement is the conclusion of a long line of empirical studies that have investigated the question of what factors drive financial structure. Most of these international studies concluded that country-specific environmental variables are key determinants of debt ratios. These variables include historical development, taxation, corporate governance, bank influence, existence of a viable corporate bond market, attitude toward risk, government regulation, availability of capital, and agency costs, to name a few.

Local Norms

Within the constraint of minimizing its consolidated worldwide cost of capital, should an MNE take differing country debt ratio norms into consideration when determining its desired debt ratio for foreign subsidiaries? For definition purposes the debt considered here should include only funds borrowed from sources outside the MNE. This debt would include local and foreign currency loans as well as eurocurrency loans.

The reason for this definition is that parent loans to foreign subsidiaries are often regarded as equivalent to equity investment both by host country and by investing firms. A parent loan is usually subordinated to other debt and does not create the same threat of insolvency as an external loan. Furthermore, the choice of debt or equity investment is often considered arbitrary (by some) and subject to negotiation between host country and parent firm.

The main advantages of a finance structure for foreign subsidiaries that conforms to local debt norms are as follows:

- A localized financial structure reduces criticism of foreign subsidiaries that have been operating with too high a proportion of debt (judged by local standards), often resulting in the accusation that they are not contributing a fair share of risk capital to the host country.

- A localized financial structure helps management evaluate return on equity investment relative to local competitors in the same industry.

- In economies where interest rates are relatively high because of a scarcity of capital, the high cost of local funds reminds management that return on assets needs to exceed the local price of capital.

The main disadvantages of localized financial structures are as follows:

- An MNE is expected to have a comparative advantage over local firms in overcoming imperfections in national capital markets through better availability of capital and the ability to diversify risk.

- If each foreign subsidiary of an MNE localizes its financial structure, the resulting consolidated balance sheet might show a financial structure that does not conform to any particular country's norm.

- The debt ratio of a foreign subsidiary is only cosmetic, because lenders ultimately look to the parent and its consolidated worldwide cash flow as the source of repayment.

In our opinion, a compromise position is possible. Both multinational and domestic firms should try to minimize their overall weighted average cost of capital for a given level of business risk and capital budget, as finance theory suggests. However, if debt is available to a foreign subsidiary at equal cost to that which could be raised elsewhere, after adjusting for foreign exchange risk, then localizing the foreign subsidiary's financial structure should incur no cost penalty and would also enjoy the advantages listed above.

Financing the Foreign Subsidiary

In addition to choosing an appropriate financial structure for foreign subsidiaries, financial managers of multinational firms need to choose among alternative sources of funds—internal and external to the multinational—with which to finance foreign subsidiaries.

Ideally, the choice among the sources of funds should minimize the cost of external funds after adjusting for foreign exchange risk. The firm should choose internal sources in order to minimize worldwide taxes and political risk, while ensuring that managerial motivation in the foreign subsidiaries is geared toward minimizing the firm's consolidated worldwide cost of capital, rather than the subsidiary's cost of capital.

Internal Sources of Funding

Exhibit 14A.1 provides an overview of the internal sources of financing for foreign subsidiaries. In general, although a minimum amount of equity capital from the parent company is required, multinationals often strive to minimize the amount of equity in foreign subsidiaries in order to limit risks of losing that capital. Equity investment can take the form of either cash or real goods (machinery, equipment, inventory, etc.).

While debt is the preferable form of subsidiary financing, access to local host country debt is limited in the early stages of a foreign subsidiary's life. Without a history of proven operational capability and debt service capability, the foreign subsidiary may need to acquire its debt from the parent company or from unrelated parties with a parental guarantee (after operations have been initiated). Once the operational and financial capabilities of the subsidiary have been established, it may then actually enjoy preferred access to debt locally.

EXHIBIT 14A.1   **Internal Financing of the Foreign Subsidiary**

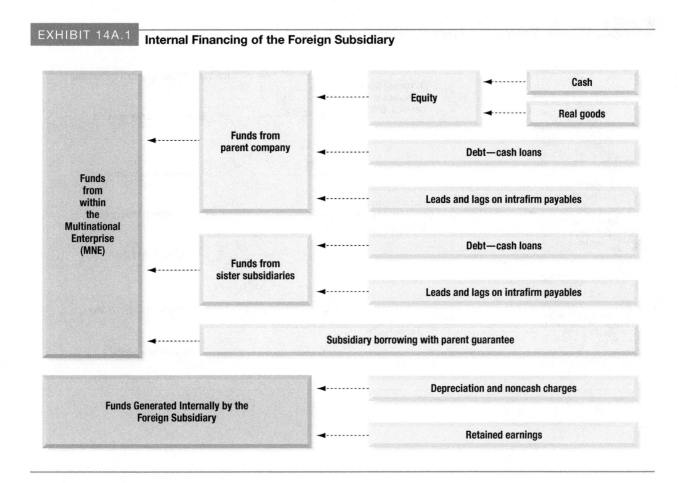

External Sources of Funding

Exhibit 14A.2 provides an overview of the sources of foreign subsidiary financing external to the MNE. The sources are first decomposed into three categories: (1) debt from the parent's country; (2) debt from countries outside the parent's country; and (3) local equity.

Debt acquired from external parties in the parent's country reflects the lenders' familiarity with and confidence in the parent company itself, although the parent is in this case not providing explicit guarantees for the repayment of the debt. Local currency debt is particularly valuable to the foreign subsidiary that has substantial local currency cash inflows arising from its business activities. In the case of some emerging markets, however, local currency debt is in short supply for all borrowers, local or foreign.

EXHIBIT 14A.2   **External Financing of the Foreign Subsidiary**

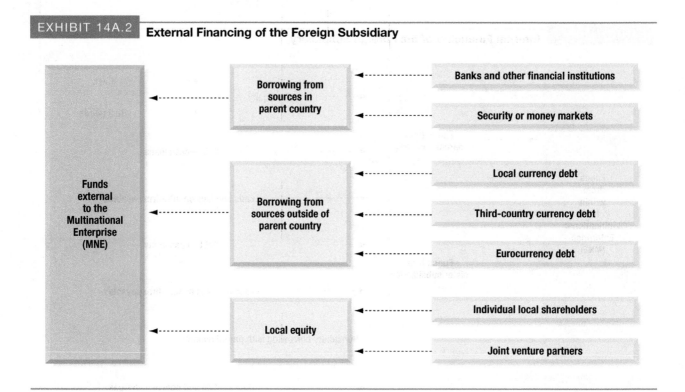

CHAPTER 15

Multinational Tax Management

Over and over again courts have said that there is nothing sinister in so arranging one's affairs as to keep taxes as low as possible. Everybody does so, rich and poor, and all do right, for nobody owes any public duty to pay more than the law demands: taxes are enforced extractions, not voluntary contributions. To demand more in the name of morals is mere cant.

—Judge Learned Hand, Commissioner v. Newman, 159 F.2d 848 (CA-2, 1947).

LEARNING OBJECTIVES

■ Identify the differences between tax systems employed by government around the globe and how they affect multinational enterprises

■ Explore the unique characteristics of multinational tax management

■ Illustrate the use of offshore international financial centers for multinational tax reduction

■ Analyze how Google has creatively repositioned its global profits for tax management purposes

■ Examine the process and motivations for re-incorporating outside the United States, the so-called *corporate inversion*

Tax planning for multinational operations is an extremely complex but vitally important aspect of international business. To plan effectively, MNEs must understand not only the intricacies of their own operations worldwide, but also the different structures and interpretations of tax liabilities across countries. The primary objective of multinational tax planning is the minimization of the firm's worldwide tax burden. This objective, however, must not be pursued without full recognition that decision-making within the firm must always be based on the economic fundamentals of the firm's line of business, and not on convoluted policies undertaken purely for the reduction of tax liability. As evident from previous chapters, taxes have a major impact on corporate net income and cash flow through their influence on foreign investment decisions, financial structure, cost of capital, foreign exchange management, and financial control.

This chapter provides an overview of how taxes are applied to MNEs globally; how the United States taxes the global earnings of U.S.-based multinationals; and how U.S.-based MNEs manage their global tax liabilities. We do this in four parts. The first section acquaints the reader with the overall international tax environment. The second part examines how a multinational firm manages its global tax liabilities. Although we use U.S. taxes as illustrations,

421

our intention is not to make this chapter or this book U.S.-centric; most U.S. practices have close parallels in other countries, modified to fit their specific national systems. The third section of the chapter examines the use of tax-haven subsidiaries and international offshore financial centers. This is followed by an illustrative case about Google, which explores how profit positioning is used to reduce global taxes. The fifth and final part explores the rapidly expanding area of corporate inversions. The chapter concludes with the Mini-Case, *Apple's Global iTax Strategy*.

Tax Principles

The sections that follow explain the most important aspects of the international tax environments and specific features that affect MNEs. Before we explain the specifics of multinational taxation in practice, however, it is necessary to introduce two fundamental tax principles: tax morality and tax neutrality.

Tax Morality

The MNE faces not only a morass of foreign taxes, but also an ethical question. In many countries, taxpayers—corporate or individual—do not voluntarily comply with the tax laws. This is termed *tax morality*. Smaller domestic firms and individuals are the chief violators. The MNE must decide whether to follow a practice of full disclosure to tax authorities or adopt the philosophy of "when in Rome, do as the Romans do." Given the local prominence of most foreign subsidiaries, most MNEs follow the full disclosure practice. Some firms, however, believe that their competitive position would be eroded if they did not avoid taxes to the same extent as their domestic competitors. There is obviously no prescriptive answer to the problem, since business ethics are partly a function of cultural heritage and historical development.

Some countries have imposed what seem to be arbitrary punitive tax penalties on MNEs for violations of local tax laws. Property or wealth tax assessments are sometimes perceived by the foreign firm to be excessively large when compared with those levied on locally owned firms. The problem then is how to respond to tax penalties that are punitive or discriminatory. Other countries have sought to make the question of tax morality compliance more difficult by increasing the transparency associated with tax payment.

Tax Neutrality

When a government decides to levy a tax, it must consider not only the potential revenue from the tax and how efficiently it can be collected, but also the effect the proposed tax can have on private economic behavior. For example, the U.S. government's policy on taxation of foreign-source income does not have as its sole objective the raising of revenue; rather it has multiple objectives, including the following:

- Neutralizing tax incentives that might favor (or disfavor) U.S. private investment in developed countries
- Providing an incentive for U.S. private investment in developing countries
- Improving the U.S. balance of payments by removing the advantages of artificial tax havens and encouraging repatriation of funds
- Raising revenue

The ideal tax should not only raise revenue efficiently but also have as few negative effects on economic behavior as possible. Some theorists argue that the ideal tax should be completely neutral in its effect on private decisions and completely equitable among taxpayers. This is *tax neutrality*. However, other theorists claim that national policy objectives such as balance of payments or investment in developing countries should be encouraged through an active tax incentive policy, as opposed to requiring taxes to be neutral and equitable. Most tax systems compromise between these two viewpoints.

One way to view neutrality is to require that the burden of taxation on each dollar, euro, pound, or yen of profit earned in home-country operations by an MNE be equal to the burden of taxation on each currency-equivalent of profit earned by the same firm in its foreign operations. This is called *domestic tax neutrality*. A second way to view neutrality is to require that the tax burden on each foreign subsidiary of the firm be equal to the tax burden on its competitors in the same country. This is called *foreign tax neutrality*. The later interpretation is often supported by MNEs because it focuses more on the competitiveness of the individual firm in individual country markets.

Tax neutrality is not to be confused with *tax equity*. In theory, an equitable tax is one that imposes the same total tax burden on all taxpayers who are similarly situated and located in the same tax jurisdiction. In the case of foreign investment income, the U.S. Treasury argues that since the United States uses the nationality principle to claim tax jurisdiction, U.S.-owned foreign subsidiaries are in the same tax jurisdiction as U.S. domestic subsidiaries. Therefore, a dollar earned in foreign operations should be taxed at the same rate and paid at the same time as a dollar earned in domestic operations.

National Tax Environments

Despite the fundamental objectives of national tax authorities, it is widely agreed that taxes do affect economic decisions made by MNEs. Tax treaties between nations, and differential tax structures, rates, and practices all result in a less than level playing field for the MNEs competing on world markets. Different countries use different categorizations of income (e.g., distributed versus undistributed profits), use different tax rates, and have radically different tax regimes, all of which drive different global tax management strategies by multinational firms.

Nations structure their tax systems along two basic approaches: the *worldwide approach* or the *territorial approach*. Both approaches are attempts to determine which firms, foreign or domestic by incorporation, or which incomes, foreign or domestic in origin, are subject to the taxation of host-country tax authorities.

Worldwide Approach. The *worldwide approach*, also referred to as the *residential approach* or *national approach*, levies taxes on the income earned by firms that are incorporated in the host country, regardless of where the income was earned (domestically or abroad). An MNE earning income both at home and abroad would therefore find its worldwide income taxed by its host-country tax authorities.

For example, the United States taxes the income earned by firms based in the U.S. regardless of whether the income earned by the firm is domestically sourced or foreign sourced. In the U.S., ordinary foreign-sourced income is taxed only as remitted to the parent firm. As with all questions of tax, however, numerous conditions and exceptions exist. The primary problem is that this does not address the income earned by foreign firms operating within the United States. Countries like the U.S. then apply the principle of territorial taxation to foreign firms within their legal jurisdiction, taxing all income earned by foreign firms in the United States.

Territorial Approach. The *territorial approach*, also termed the *source approach*, focuses on the income earned by firms within the legal jurisdiction of the host country, not on the country of firm incorporation. Countries like Germany, which follow the territorial approach, apply taxes equally to foreign or domestic firms on income earned within the country, but in principle not on income earned outside the country. The territorial approach, like the worldwide approach, results in a major gap in coverage if resident firms earn income outside the country, but are not taxed by the country in which the profits are earned (operations in a so-called *tax haven*). In this case, tax authorities extend tax coverage to income earned abroad if it is not currently covered by foreign tax jurisdictions. Once again, a mix of the two tax approaches is necessary for full coverage of income.

As illustrated by Exhibit 15.1, 28 of 34 countries within the Organization for Economic Cooperation and Development (OECD) currently utilize a territorial tax system. The predominance of territorial systems has grown rapidly, as more than half of these same OECD countries used worldwide systems only 10 years ago. In 2009 alone, both Japan and the United Kingdom switched from worldwide to territorial systems.

Tax Deferral. If the worldwide approach to international taxation were followed to the letter, it would end the tax-deferral privilege for many MNEs. Foreign subsidiaries of MNEs pay host-country corporate income taxes, but many parent countries defer claiming additional income taxes on that foreign-source income until it is remitted to the parent firm—this is called *tax deferral*. For example, U.S. corporate income taxes on some types of foreign-source income of U.S.-owned subsidiaries incorporated abroad are deferred until the earnings are remitted to the U.S. parent. However, the ability to defer corporate income taxes is highly restricted and has been the subject of many of the tax law changes in the past three decades.

The deferral privilege has been challenged a number of times in recent U.S. presidential elections. A number of different candidates have argued that tax deferrals create an incentive for outsourcing abroad—so-called *offshoring*—of certain manufacturing and service activities by U.S. firms. The added concern to the potential loss of American jobs was the potential reduction in tax collections in the United States, enlarging the already sizeable U.S. government fiscal deficit.

Tax Treaties

A network of bilateral tax treaties, many of which are modeled after one proposed by the OECD, provides a means of reducing double taxation. Tax treaties normally define whether taxes are to be imposed on income earned in one country by the nationals of another, and if

---

EXHIBIT 15.1   **Tax Regimes of the OECD 34 Countries**

Territorial Taxation				Worldwide Taxation
Australia	France	Netherlands	Sweden	Chile
Austria	Germany	New Zealand	Switzerland	Ireland
Belgium	Greece	Norway	Turkey	Israel
Canada	Hungary	Poland	United Kingdom	Korea (South)
Czech Republic	Iceland	Portugal		Mexico
Denmark	Italy	Slovakia		United States
Estonia	Japan	Slovenia		
Finland	Luxembourg	Spain		

Source: Data drawn from *Evolution of Territorial Tax Systems in the OECD*, PWC, April 2, 2013.

so, how. Tax treaties are bilateral, with the two signatories specifying what rates are applicable to which types of income between the two countries.

Individual bilateral tax jurisdictions as specified through tax treaties are particularly important for firms that are primarily exporting to another country rather than doing business in that country via a "permanent establishment" (for example, a manufacturing plant). A firm that only exports would not want any of its other worldwide income taxed by the importing country. Tax treaties define what is a "permanent establishment" and what constitutes a limited presence for tax purposes. Tax treaties also typically result in reduced withholding tax rates between the two signatory countries, the negotiation of the treaty itself serving as a forum for opening and expanding business relationships between the two countries.

Tax Types

Taxes are classified as *direct taxes* if they are applied directly to income or *indirect taxes* if they are based on some other measurable performance characteristic of the firm. Exhibit 15.2 illustrates the wide range of corporate income tax rates and indirect rates across the world today.

Income Tax. Most governments rely on income taxes, both personal and corporate, for their primary revenue source. Corporate income tax rates differ widely globally, and take a variety of different forms. Some countries, for example, impose different corporate tax rates on distributed income (often lower) versus undistributed income (often higher), in an attempt to motivate companies to distribute greater portions of their income to their owners. Corporate income taxes—*statutory tax rates*—have been declining for more than a decade. That decline, however, now seems to have leveled off. Many governments amongst the industrialized countries are now under pressure to increase their tax revenues, leading to a variety of changes in indirect taxation.

These differences reflect a rapidly changing global tax environment. Corporate income taxes have been falling rapidly and widely over the past two decades. On average, the non-OECD countries have relatively lower rates on average. The highly industrialized world, for better or worse, has been reluctant to reduce corporate income tax rates as aggressively as many emerging market nations. Corporate income tax rates, like any burden on the profitability of commercial enterprise, has become a competitive element used by many countries to attempt to promote inward investment from abroad. And if corporate tax rates are indeed an element of competition, as illustrated in Exhibit 15.3, the United States is losing this competitive battle, with the highest corporate tax rate among the 30 countries shown. In 2011, for the first time in the past 50 years, the global average corporate income tax rate fell below 23%.

Withholding Tax. Passive income (such as dividends, interest, and royalties) earned by a resident of one country within the tax jurisdiction of a second country, are normally subject to a *withholding tax* in the second country. The reason for the institution of withholding taxes is actually quite simple: governments recognize that most international investors will not file a tax return in each country in which they invest. The government, therefore, wishes to ensure that a minimum tax payment is received. As the term "withholding" implies, taxes are withheld by the corporation from the payment made to the investor, and those withheld taxes are then turned over to government authorities. Withholding taxes is a major subject of bilateral tax treaties and generally range between 0 and 25%.

Value-Added Tax. One type of tax that has achieved great prominence is the *value-added tax*. The value-added tax is a type of national sales tax collected at each stage of production or sale of consumption goods in proportion to the value added during that stage. In general,

EXHIBIT 15.2   **Corporate and Indirect Tax Rates for Selected Countries**

Country	Corporate	Indirect	Country	Corporate	Indirect	Country	Corporate	Indirect
Afghanistan	20%	na	Ghana	25%	17.5%	Papua New Guinea	30%	10%
Albania	15%	20%	Gibraltar	10%	na	Paraguay	10%	10%
Algeria	19%	17%	Greece	26%	23%	Peru	30%	18%
Angola	35%	10%	Guatemala	28%	12%	Philippines	30%	12%
Argentina	35%	21%	Guernsey	0%	na	Poland	19%	23%
Armenia	20%	20%	Honduras	30%	15%	Portugal	23%	23%
Aruba	28%	1.5%	Hong Kong	16.5%	na	Qatar	10%	na
Australia	30%	10%	Hungary	19%	27%	Romania	16%	24%
Austria	25%	20%	Iceland	20%	25.5%	Russia	20%	18%
Bahamas	0%	na	India	33.99%	**	St. Maarten	34.5%	5%
Bahrain	0%	na	Indonesia	25%	10%	Samoa	27%	15%
Bangladesh	27.5%	15.0%	Iran	15%	na	Saudi Arabia	20%	na
Barbados	25%	17.5%	Ireland	12.5%	23%	Serbia	15%	20%
Belarus	18%	20%	Isle of Man	0%	20%	Sierra Leone	30%	15%
Belgium	33.99%	21%	Israel	26.5%	18%	Singapore	17%	7%
Bermuda	0%	na	Italy	31.4%	22%	Slovak Republic	22%	20%
Bolivia	25%	13%	Jamaica	25%	16.5%	Slovenia	17%	22%
Bonaire, St. Eustatius, Saba	0%	8%	Japan	35.64%	8%	South Africa	28%	14%
			Jersey	0%	5%	South Korea	24.2%	10%
Bosnia and Herzegovina	10%	17%	Jordan	14%	16%	Spain	30%	21%
			Kazakhstan	20%	12%	Sri Lanka	28%	12%
Botswana	22%	12%	Kenya	30%	16%	Sudan	35%	17%
Brazil	34%	*	Kuwait	15%	na	Sweden	22%	25%
Bulgaria	10%	20%	Latvia	15%	21%	Switzerland	17.92%	8%
Cambodia	20%	10%	Lebanon	15%	10%	Syria	22%	na
Canada	26.5%	**	Libya	20%	na	Taiwan	17%	5%
Cayman Islands	0%	na	Liechtenstein	12.5%	8%	Tanzania	30%	18%
Chile	20%	19%	Lithuania	15%	21%	Thailand	20%	7%
China	25%	**	Luxembourg	29.22%	15%	Trinidad & Tobago	25%	15%
Colombia	25%	16%	Macau	12%	na			
Costa Rica	30%	13%	Macedonia	10%	18%	Tunisia	25%	18%
Croatia	20%	25%	Malawai	30%	16.5%	Turkey	20%	18%
Curacao	27.5%	6%	Malaysia	25%	10%	Uganda	30%	18%
Cyprus	12.5%	19%	Malta	35%	18%	Ukraine	18%	20%
Czech Republic	19%	21%	Mauritius	15%	15%	United Arab Emirates	55%	na
Denmark	24.5%	25%	Mexico	30%	16%			
Dominican Republic	28%	18%	Montenegro	9%	19%	United Kingdom	21%	20%
			Morocco	30%	20%			
Ecuador	22%	12%	Mozambique	32%	17%	United States	40%	**
Egypt	25%	10%	Namibia	33%	15%	Uruguay	25%	22%
El Salvador	30%	13%	Netherlands	25%	21%	Vanuatu	0%	12.5%
Estonia	21%	20%	New Zealand	28%	15%	Venezuela	34%	12%
Fiji	20%	15%	Nigeria	30%	5%	Vietnam	22%	10%
Finland	20%	24%	Norway	27%	25%	Yemen	20%	5%
France	33.33%	20%	Oman	12%	na	Zambia	35%	16%
Georgia	15%	18%	Pakistan	34%	17%	Zimbabwe	25.75%	15%
Germany	29.58%	19%	Panama	25%	7%			

Source: *KPMG's Corporate and Indirect Tax Rate Survey,* 2014. *Bonaire, St. Eustatius, and Saba are grouped together as per ISO standards.
**Countries impose indirect taxes at also state, provincial, municipal (or similar) level, making determination of a specific rate case specific.

EXHIBIT 15.3 | Comparative OECD Corporate Tax Rates, 2014

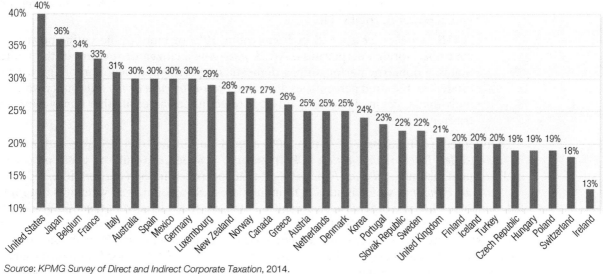

Source: KPMG Survey of Direct and Indirect Corporate Taxation, 2014.

production goods such as plant and equipment have not been subject to the value-added tax. Certain necessities—such as medicines and other health-related expenses, education and religious activities—are usually exempt or taxed at lower rates.

The value-added tax has been adopted as the main source of revenue from indirect taxation by all members of the EU, most non-EU countries in Europe, a number of Latin American countries, Canada, and scattered other countries.

Other National Taxes. There are other national taxes, which vary in importance from country to country. The turnover tax (tax on the purchase or sale of securities in some countries) and higher taxes on undistributed profits (a higher income tax rate on retained earnings of firms) are examples of other national taxes. Property and inheritance taxes, also termed transfer taxes, are imposed in a variety of ways to achieve intended social redistribution of income and wealth as much as to raise revenue. There are other "red-tape charges" for public services that are in reality user taxes. Sometimes foreign exchange purchases or sales are in effect hidden taxes inasmuch as the government earns revenue rather than just regulating imports and exports for balance of payments reasons.

Controlled Foreign Corporations

The rule that U.S. shareholders do not pay U.S. taxes on foreign-source income until that income is remitted to the United States was amended in 1962 by the creation of special *Subpart F income*. The revision was designed to prevent the use of arrangements between operating companies and base companies located in tax havens as a means of deferring U.S. taxes and to encourage greater repatriation of foreign incomes. The Tax Reform Act of 1986 retained the concept of Subpart F income but made a number of changes that expanded categories of income subject to taxation, reduced exceptions, and raised or lowered thresholds.

Several definitions are needed to understand Subpart F income:

- A *controlled foreign corporation* (CFC) is any foreign corporation in which U.S. shareholders, including corporate parents, own more than 50% of the combined voting power or total value.

- A U.S. *shareholder* is a U.S. person owning 10% or more of the voting power of a controlled foreign corporation. A U.S. *person* is a citizen or resident of the United States, a domestic partnership, a domestic corporation, or any nonforeign trust or estate. The required percentages are based on *constructive ownership*, under which an individual is deemed to own shares registered in the names of other family members, trusts, and so on.

Under these definitions, a more-than-50% owned "subsidiary" of a U.S. corporation would be a *controlled foreign corporation* (CFC), and the U.S. parent would be taxed on certain undistributed income—*Subpart F income*—of that controlled foreign corporation.

Subpart F income, which is subject to immediate U.S. taxation even when it is not remitted, is income of a type otherwise easily shifted offshore to avoid current taxation. It includes (1) *passive income* received by the foreign corporation such as dividends, interest, rents, royalties, net foreign currency gains, net commodities gains, and income from the sale of non-income-producing property; (2) income from the insurance of U.S. risks; (3) financial service income; (4) shipping income; (5) oil-related income; and (6) certain related-party sales and service income.

Subpart F, restated, provides that if a foreign corporation is considered to be a controlled foreign corporation, each U.S. shareholder owning 10% or more of that CFC must include the shareholder's *pro rated* share of the CFC's Subpart F income in the shareholder's gross income. Thus, Subpart F income is subject to current U.S. taxation at the shareholder level even though not remitted to the United States. Exhibit 15.4 illustrates the mechanics of how a U.S.

EXHIBIT 15.4   **U.S. Taxation of Foreign Source Income and Subpart F**

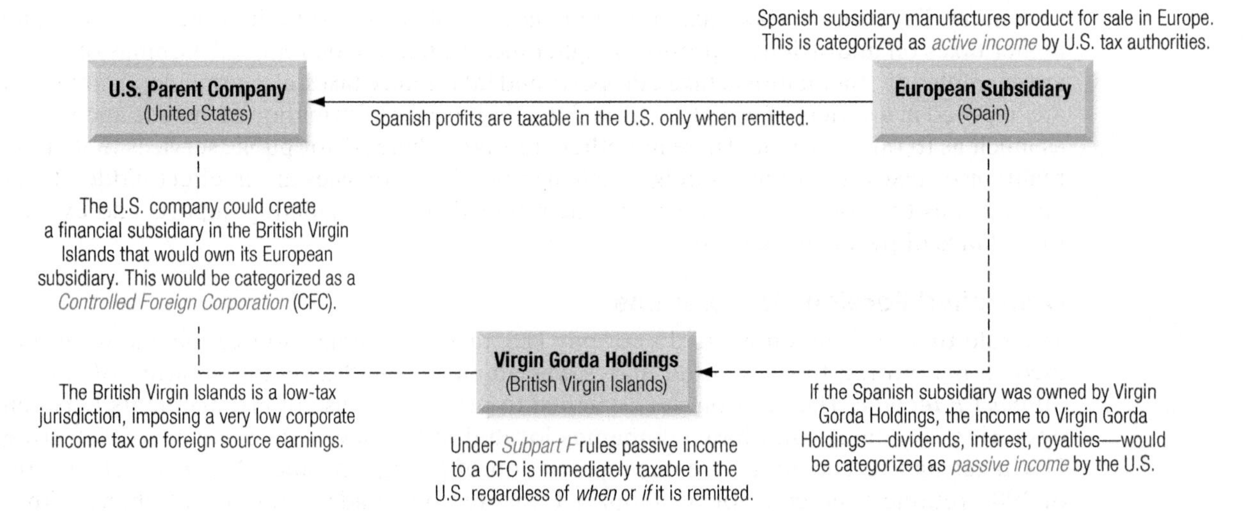

Spanish subsidiary manufactures product for sale in Europe. This is categorized as *active income* by U.S. tax authorities.

U.S. Parent Company (United States)

Spanish profits are taxable in the U.S. only when remitted.

European Subsidiary (Spain)

The U.S. company could create a financial subsidiary in the British Virgin Islands that would own its European subsidiary. This would be categorized as a *Controlled Foreign Corporation* (CFC).

Virgin Gorda Holdings (British Virgin Islands)

The British Virgin Islands is a low-tax jurisdiction, imposing a very low corporate income tax on foreign source earnings.

Under *Subpart F* rules passive income to a CFC is immediately taxable in the U.S. regardless of *when* or *if* it is remitted.

If the Spanish subsidiary was owned by Virgin Gorda Holdings, the income to Virgin Gorda Holdings—dividends, interest, royalties—would be categorized as *passive income* by the U.S.

corporation attempting to use a financial subsidiary in the British Virgin Islands (one which has no economic function other than to own for tax purposes) would be treated under Subpart F income principles.

Multinational Tax Management

The operational goal of the multinational enterprise is the maximization of consolidated after-tax income. This requires the MNE to minimize its effective global tax burden. Since multinational firms incorporated in a worldwide tax country like the United States are taxed on their worldwide income, and not just their income in-country (territorial taxation), they will devise and pursue tax structures and strategies to minimize tax payments across all countries in which they operate. This section will focus on those strategies, and specifically on those employed by U.S. multinational firms.

The following tax structures and strategies are not illegal, but rather may be considered extremely aggressive efforts to reduce tax liabilities—*tax avoidance*. Whereas illegal activities are termed *tax evasion*, *tax avoidance* is used to describe extremely aggressive strategies and structures used by business to reduce taxes far below what most governments expect. This latter category would include the use of offshore tax havens. The question that remains, however, is whether the MNEs are also pursuing the nonfinancial interests or responsibilities of the firm equitably or ethically.

There are a variety of different strategies, structures, and practices which are used in tax avoidance. Most methods are premised on shifting taxable profits to low tax environments while minimizing taxable income in higher tax jurisdictions. We will focus our discussion on five international tax management practices: *allocation of debt, foreign tax credits, transfer pricing, cross-crediting*, and *check-the-box*.

Allocation of Debt and Earnings Stripping

A multinational firm may allocate debt differently across its various foreign subsidiaries to reduce tax liabilities in high tax environments. Units in high tax environments may be assigned very high debt obligations in an attempt to maximize the interest deductibility provisions offered in that country. Often termed *earnings stripping*, this method is typically limited by host government requirements for minimum equity capitalizations—*thin capitalization rules*.

The United States defines *thin capitalization* as a debt-to-equity ratio above 1.5 to 1.0, with net interest exceeding 50% of adjusted taxable income (taxable income plus interest plus depreciation). Interest expenses that exceed 50% that are paid to a related corporation are not deductible toward U.S. tax liabilities.

Foreign Tax Credits and Deferral

To prevent double taxation of the same income, most countries grant a *foreign tax credit* for income taxes paid to the host country. Countries differ on how they calculate the foreign tax credit and what kinds of limitations they place on the total amount claimed. Normally foreign tax credits are also available for withholding taxes paid to other countries on dividends, royalties, interest, and other income remitted to the parent. The value-added tax and other sales taxes are not eligible for a foreign tax credit but are typically deductible from pre-tax income as an expense.

A *tax credit* is a direct reduction of taxes that would otherwise be due and payable. It differs from a *deductible expense*, which is an expense used to reduce taxable income before the tax rate is applied. A $100 tax credit reduces taxes payable by the full $100, whereas a $100 deductible expense reduces taxable income by $100 and taxes payable by $100 × t, where t is the tax rate. Tax credits are more valuable on a dollar-for-dollar basis than are deductible expenses.

If there were no credits for foreign taxes paid, sequential taxation by the host government and then by the home government would result in a very high cumulative tax rate. For example, assume the wholly owned foreign subsidiary of an MNE earns $10,000 before local income taxes and pays a dividend equal to all of its after-tax income. The host-country income tax rate is 30%, and the home country of the parent tax rate is 35%, assuming no withholding taxes. Total taxation with and without tax credits is shown in Exhibit 15.5.

If tax credits are not allowed, sequential levying of both a 30% host-country tax and then a 35% home-country tax on the income that remains results in an effective 54.5% tax as a percentage of the original before tax income, a cumulative rate that would make many MNEs uncompetitive with local firms. The effect of allowing tax credits is to limit total taxation on the original before-tax income to no more than the highest single rate among jurisdictions. In the case depicted in Exhibit 15.5, the effective overall tax rate of 35% with foreign tax credits is equivalent to the higher tax rate of the home country (and is the tax rate payable if the income had been earned at home, domestically sourced income). The $500 of additional home-country tax under the tax credit system in Exhibit 15.5 is the amount needed to bring total taxation ($3,000 already paid plus the additional $500) up to but not beyond 35% of the original $10,000 of before-tax foreign income.

The problem, however, is that if this company repatriates the profits of its foreign businesses to the parent company, it owes more taxes. Period. If it leaves those profits in that foreign country, it enjoys what is referred to as *deferral*—it is able to defer incurring additional parent-country taxes on the foreign-source income until it does repatriate those earnings. As shown in *Global Finance in Practice 15.1*, this has motivated some countries like the U.S. to try periodically to provide tax incentives for repatriating profits.

EXHIBIT 15.5	**Foreign Tax Credits**		
		Without Foreign Tax Credits	**With Foreign Tax Credits**
Before-tax foreign income		$10,000	$10,000
Less foreign tax @ 30%		−3,000	−3,000
Available to parent and paid as dividend		$ 7,000	$ 7,000
Less additional parent-country tax at 35%		−2,450	—
Less incremental tax (after credits)		—	−500
Profit after all taxes		$ 4,550	$ 6,500
Total taxes, both jurisdictions		$ 5,450	$ 3,500
Effective overall tax rate (total taxes paid ÷ foreign income)		54.5%	35.0%

GLOBAL FINANCE IN PRACTICE 15.1

Offshore Profits and Dividend Repatriation

It is estimated that U.S.-based multinationals have one trillion dollars in un-repatriated profits offshore. Repatriating those profits, given the relatively higher effective corporate income tax rate in the U.S. compared to many other countries, would trigger significant additional tax charges in the U.S. In an effort to facilitate repatriation of those profits in 2004, the U.S. government passed the Homeland Investment Act of 2004. The Act provided a window of opportunity in 2005 in which profits could be repatriated with only an additional tax obligation of 5.25%.

The temporary tax law change clearly had the desired impact of stimulating the repatriation of profits, as illustrated in the exhibit. Dividend repatriations skyrocketed in 2005 to over $360 billion from $60 billion the previous year. After the temporary tax revision expired, repatriated dividends returned to trend.

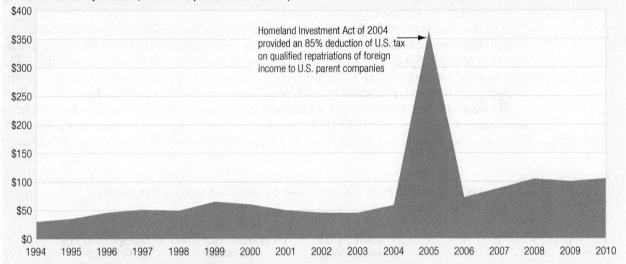

U.S. Dividend Repatriations, 1994–2010 (billions of U.S. dollars)

Homeland Investment Act of 2004 provided an 85% deduction of U.S. tax on qualified repatriations of foreign income to U.S. parent companies

Source: Bureau of Economic Analysis, Joint Committee on Taxation, Congressional Research Service.

The apparent political motivation for the original tax holiday was jobs creation in the United States by U.S. companies. However, evidence indicates that U.S. parent companies used the repatriated profits for a variety of purposes, such as returning money to shareholders via dividends and share repurchases, but not for creating new jobs. As discussions get underway regarding another tax holiday aimed at repatriating over one trillion dollars in corporate profits held offshore, these same debates are rising once again.

Transfer Pricing

The pricing of goods, services, and technology transferred to a foreign subsidiary from an affiliated company—*transfer pricing*—is the first and foremost method of transferring funds out of a foreign subsidiary. These costs enter directly into the cost of goods sold component of the subsidiary's income statement. This is a particularly sensitive problem for MNEs. Even purely domestic firms find it difficult to reach agreement on the best method for setting prices on transactions between related units. In the multinational case, managers must balance conflicting considerations. These include fund positioning and income taxes.

Fund Positioning Effect. A parent firm wishing to transfer funds out of a particular country can charge higher prices on goods sold to its subsidiary in that country—to the degree that government regulations allow. A foreign subsidiary can be financed by the reverse technique, a lowering of transfer prices. Payment by the subsidiary for imports from its parent or sister subsidiary transfers funds out of the subsidiary. A higher transfer price permits funds to be accumulated in the selling country. Multiple sourcing of component parts on a worldwide basis allows the act of switching between suppliers from within the corporate family to function as a device to transfer funds.

Income Tax Effect. A major consideration in setting a transfer price is the *income tax effect*. The firm's global profits can be influenced when transfer prices are set to minimize taxable income in a country with a high income tax rate and to maximize taxable income in a country with a low income tax rate. A parent wishing to reduce the taxable profits of a subsidiary in a high-tax environment may set transfer prices at a higher point to increase the costs of the subsidiary, thereby reducing taxable income.

The income tax effect is illustrated in the hypothetical example presented in Exhibit 15.6. Ganado Europe is operating in a relatively high-tax environment, assuming German corporate income taxes of 45%. Ganado U.S. is in a significantly lower tax environment, assuming a U.S. corporate income tax rate of 35%, motivating Ganado to charge Ganado Europe a higher transfer price on goods produced in the United States and sold to Ganado Europe.

If Ganado Corporation adopts a high-markup policy by "selling" its merchandise at an intracompany sales price of $1,700,000, the same $800,000 of pre-tax consolidated income is allocated more heavily to low-tax Ganado U.S. and less heavily to high-tax Ganado Europe.

EXHIBIT 15.6   Effect of Low versus High Transfer Price on Ganado Europe's Net Income
(thousands of U.S. dollars)

	Ganado U.S. (subsidiary)	Ganado Europe (subsidiary)	Europe and U.S. Combined
Low-Markup Policy			
Sales	$1,400	$2,000	$2,000
Less cost of goods sold*	(1,000)	(1,400)	(1,000)
Gross profit	$ 400	$ 600	$1,000
Less operating expenses	(100)	(100)	(200)
Taxable income	$ 300	$ 500	$ 800
Less income taxes	35% (105)	45% (225)	(330)
Net income	$ 195	$ 275	$ 470
High-Markup Policy			
Sales	$1,700	$2,000	$2,000
Less cost of goods sold*	(1,000)	(1,700)	(1,000)
Gross profit	$ 700	$ 300	$1,000
Less operating expenses	(100)	(100)	(200)
Taxable income	$ 600	$ 200	$ 800
Less income taxes	35% (210)	45% (90)	(300)
Net income	$ 390	$ 110	$ 500

*Ganado U.S. sales price becomes cost of goods sold for Ganado Europe.

(Note that it is Ganado Corporation, the corporate parent that must adopt a transfer pricing policy that directly alters the profitability of each of the individual subsidiaries, not the subsidiary itself.) As a consequence, total taxes drop by $30,000 and consolidated net income increases by $30,000 to $500,000. All while total sales remain constant.

Ganado would naturally prefer the high-markup policy for sales from the United States to Europe. Needless to say, government tax authorities are aware of the potential income distortion from transfer price manipulation. A variety of regulations exist on the reasonableness of transfer prices, including fees and royalties as well as prices set for merchandise. Government taxing authorities obviously have the right to reject transfer prices deemed inappropriate.

U.S. IRS regulations provide three methods to establish arm's length prices: *comparable uncontrolled prices*, *resale prices*, and *cost-plus calculations*. All three of these methods are recommended for use in member countries by the OECD Committee on Fiscal Affairs. In some cases, combinations of these three methods are used.

Section 482 of the U.S. Internal Revenue Code is typical of laws covering transfer prices. Under this authority, the IRS can reallocate gross income, deductions, credits, or allowances between related corporations in order to prevent tax evasion or to reflect more clearly a proper allocation of income. The burden of proof is on the taxpaying firm to show that the IRS has been arbitrary or unreasonable in reallocating income. This "guilty until proved innocent" approach means that MNEs must keep good documentation of the logic and costs behind their transfer prices. The "correct price" according to the guidelines is the one that reflects an arm's length price; a sale of the same goods or service to a comparable unrelated customer.

Managerial Incentives and Evaluation. When a firm is organized with decentralized profit centers, transfer pricing between centers can disrupt evaluation of managerial performance. This problem is not unique to MNEs; it is also a controversial issue in the "centralization versus decentralization" debate in domestic circles. In the domestic case, however, a modicum of coordination at the corporate level can alleviate some of the distortion that occurs when any profit center suboptimizes its profit for the corporate good. Also, in most domestic cases, the company can file a single (for that country) consolidated tax return, so the issue of cost allocation between affiliates is not critical from a tax-payment point of view.

For the multinational, coordination is often hindered by longer and less-efficient channels of communication, the need to consider the unique variables that influence international pricing, and separate taxation. Even with the best of intent, a manager in one country finds it difficult to know what is best for the firm as a whole when buying at a negotiated price from related companies in another country. If corporate headquarters establishes transfer prices and sourcing alternatives, one of the main advantages of a decentralized profit center system disappears: local management loses the incentive to act for its own benefit.

Exhibit 15.6 illustrated a transfer pricing example where an increase in the transfer price led to a worldwide income gain: Ganado Corporation's income rose by $195,000 (from $195,000 to $390,000) while Ganado Europe's income fell by only $165,000 (from $275,000 to $110,000), for a net gain of $30,000. Should the managers of the European subsidiary lose their bonuses (or their jobs) because of their "sub-par" performance? Bonuses are usually determined by a company-wide formula based in part on the profitability of individual subsidiaries, but in this case, Ganado Europe "sacrificed" for the greater good of the whole. Arbitrarily changing transfer prices can create measurement problems.

Transferring profit from high-tax Ganado Europe to low-tax Ganado Corporation in the U.S. changes a multitude of cash flows and performance metrics for one or both companies:

- Import tariffs paid (importer only) and hence profit levels
- Measurements of foreign exchange exposure, such as the amount of net exposed assets, because of changes in amounts of cash and receivables

- Liquidity tests, such as the current ratio, receivables turnover, and inventory turnover
- Operating efficiency, as measured by the ratio of gross profit to either sales or to total assets
- Income tax payments
- Profitability, as measured by the ratio of net income to either sales or capital invested
- Dividend payout ratio, in that a constant dividend will show as a varied payout ratio as net income changes (alternatively, if the payout ratio is kept constant, the amount of dividend is changed by a change in transfer price)
- Internal growth rate, as measured by the ratio of retained earnings to existing ownership equity

Effect on Joint-Venture Partners. Joint ventures pose a special problem in transfer pricing, because serving the interest of local stockholders by maximizing local profit may be suboptimal from the overall viewpoint of the MNE. Often, the conflicting interests are irreconcilable. Indeed, the local joint venture partner could be viewed as a potential "Trojan horse" if they complain to local authorities about the transfer pricing policy.

Cross-Crediting

One of the most valuable management methods available to companies in a worldwide tax system like that of the United States is the ability to *cross-credit* foreign tax credits with foreign tax deficits in the same period. If a U.S. multinational remits profits from two different countries, one in a high-tax environment (relative to the U.S.) and the other in low-tax environment (relative to the U.S.), if the income is from one of the two major "baskets" of foreign source income (active or passive), the excess foreign tax credits from one can be cross-credited against the foreign tax deficits of the other.

Exhibit 15.7 summarizes how our U.S.-based multinational, Ganado, may manage dividend remittances from two of its foreign subsidiaries using cross-crediting. Ganado's dividend remittances from its two foreign subsidiaries create two different and offsetting tax credit positions.

- Because corporate income tax rates in Germany (40%) are higher than those in the United States (35%), dividends remitted to the U.S. parent from Ganado Germany result in *excess foreign tax credits*. Any applicable withholding taxes on dividends between Germany and the U.S. only increase the amount of the excess credit.
- Because corporate income tax rates in Brazil (25%) are lower than those in the United States (35%), dividends remitted to the U.S. parent from Ganado Brazil result in *deficit foreign tax credits*. If there are withholding taxes applied to the dividends by Brazil on remittances to the United States, this will reduce the size of the deficit, but not eliminate it.

Ganado's management would like to manage the two dividend remittances in order to match the deficits with the credits. The most straightforward method of doing this would be to adjust the amount of dividend distributed from each foreign subsidiary so that, after all applicable income and withholding taxes have been applied, Ganado's excess foreign tax credits from Ganado Germany exactly match the excess foreign tax deficits from Ganado Brazil. There are a number of other methods of managing the global tax liabilities of Ganado, so-called repositioning of funds, where firms strive to structure global operations to record their profits in a low-tax environment, as shown in the Mini-Case on Apple at the end of this chapter.

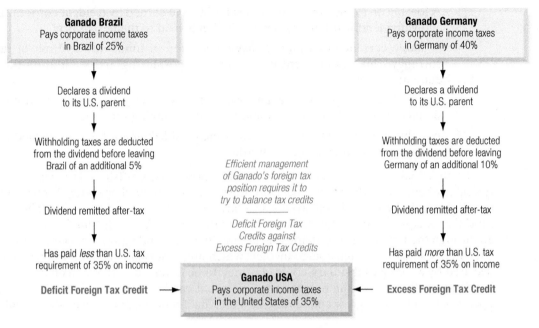

EXHIBIT 15.7   **Ganado's Cross-Crediting of Foreign Tax Credits**

Ganado Brazil
Pays corporate income taxes
in Brazil of 25%

Declares a dividend
to its U.S. parent

Withholding taxes are deducted
from the dividend before leaving
Brazil of an additional 5%

Dividend remitted after-tax

Has paid *less* than U.S. tax
requirement of 35% on income

Deficit Foreign Tax Credit

*Efficient management
of Ganado's foreign tax
position requires it to
try to balance tax credits*

*Deficit Foreign Tax
Credits against
Excess Foreign Tax Credits*

Ganado Germany
Pays corporate income taxes
in Germany of 40%

Declares a dividend
to its U.S. parent

Withholding taxes are deducted
from the dividend before leaving
Germany of an additional 10%

Dividend remitted after-tax

Has paid *more* than U.S. tax
requirement of 35% on income

Excess Foreign Tax Credit

Ganado USA
Pays corporate income taxes
in the United States of 35%

Note: Ganado pays taxes to the U.S. government separately on domestic-source income and foreign-source income.

Check-the-Box and Hybrid Entities

The U.S. Treasury's attempt to stop the repositioning of profits by U.S.-based multinationals in low-tax jurisdictions took a major step backward in 1997 when Treasury introduced what is called *check-the-box* subsidiary characterization. In an attempt to allow simplification of taxation, the U.S. Treasury changed its required filing practices to allow multinational firms to categorize subsidiaries for taxation purposes by simply "checking-the-box" on a single form.

One of the box choices offered, a *disregarded entity*, allowed the unit to "disappear" for tax purposes, as its results would be consolidated with those of its parent company. These combined units are termed *hybrid entities*. In the end, it allowed U.S. multinationals that have tiered ownership of offshore units to once again begin repositioning profits in low-tax environments and gain essentially permanent deferral for those earnings. In 2007, the U.S. Treasury codified this process in what is now referred to as the *look-through-rules* on this tax treatment of disregarded entities. The Mini-Case at the end of this chapter on Apple Computer's use of this structure provides additional detail.

Tax Havens and International Offshore Financial Centers

Many MNEs have foreign subsidiaries that act as tax havens for corporate funds awaiting reinvestment or repatriation. Tax-haven subsidiaries, categorically referred to as International Offshore Financial Centers, are partially a result of tax-deferral features on earned foreign income allowed by some of the parent countries. Tax-haven subsidiaries are typically established in a country that can meet the following requirements:

- A low tax on foreign investment or sales income earned by resident corporations and a low dividend withholding tax on dividends paid to the parent firm.

- A stable currency that allows easy conversion of funds into and out of the local currency. This requirement can be met by permitting and facilitating the use of eurocurrencies.

- The facilities to support financial services; for example, good communications, professional qualified office workers, and reputable banking services.

- A stable government that encourages the establishment of foreign-owned financial and service facilities within its borders.

Exhibit 15.8 provides a map of most of the world's major offshore financial centers. The typical tax-haven subsidiary owns the common stock of its related operating foreign subsidiaries. There might be several tax-haven subsidiaries scattered around the world. The tax-haven subsidiary's equity is typically 100% owned by the parent firm. All transfers of funds might go through the tax-haven subsidiaries, including dividends and equity financing. Thus, the parent country's tax on foreign-source income, which might normally be paid when a dividend is declared by a foreign subsidiary, could continue to be deferred until the tax-haven subsidiary itself pays a dividend to the parent firm. This event can be postponed indefinitely if foreign operations continue to grow and require new internal financing from the tax-haven subsidiary.

EXHIBIT 15.8   **International Offshore Financial Centers**

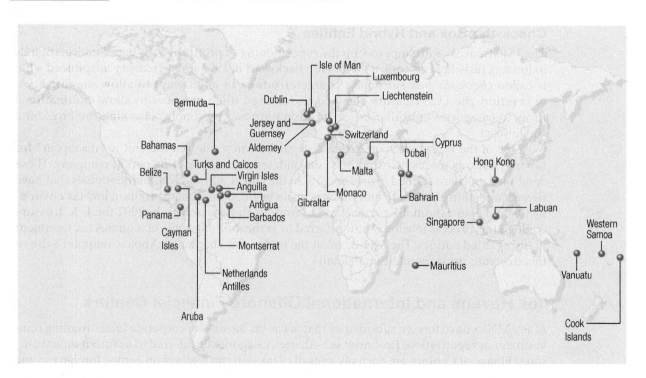

EXHIBIT 15.9   **The Activities of Offshore Financial Centers**

Offshore financial centers provide financial management services to foreign users in exchange for foreign exchange earnings. There are several comparative advantages for clients, including very low tax rates, minimal administrative formalities, and confidentiality and discretion. This environment allows wealthy international clients to minimize potential tax liability while protecting income and assets from political, fiscal, and legal risks. There are many vehicles through which offshore financial services can be provided. They include the following:

- Offshore banking, which can handle foreign exchange operations for corporations or banks. These operations are not subject to capital, corporate, capital gains, dividend, or interest taxes or to exchange controls.

- International business corporations, which are often tax-exempt, limited-liability companies used to operate businesses or raise capital through issuing shares, bonds, or other instruments.

- Offshore insurance companies, which are established to minimize taxes and manage risk.

- Asset management and protection, which allows individuals and corporations in countries with fragile banking systems or unstable political regimes to keep assets offshore to protect against the collapse of domestic currencies and banks.

- Tax planning, which means multinationals may route transactions through offshore centers to minimize taxes through transfer pricing. Individuals can make use of favorable tax regimes offered by offshore centers through trusts and foundations.

The tax concessions and secrecy offered by offshore financial centers can be used for many legitimate purposes, but they have also been used for illegitimate ends, including money laundering and tax evasion.

Thus, MNEs are able to operate a corporate pool of funds for foreign operations without having to repatriate foreign earnings through the parent country's tax machine.

For U.S. MNEs, the tax-deferral privilege enjoyed by foreign subsidiaries (it is considered a privilege because they do not pay tax on the foreign income until they remit dividends back to the parent company) was not originally a tax loophole. On the contrary, it was granted by the U.S. government to allow U.S. firms to expand overseas and to place those firms on par with foreign competitors, which also enjoy similar types of tax deferral and export subsidies of one type or another. Exhibit 15.9 provides a categorization of the primary activities of offshore financial centers.

Unfortunately, some firms distorted the original intent of *tax deferral* into *tax evasion*. Transfer prices on goods and services bought from or sold to foreign affiliates were artificially rigged to leave all the income from the transaction in the tax-haven subsidiary. This manipulation was accomplished by routing the legal title to the goods or services through the tax-haven subsidiary, even though physically the goods or services never entered the tax-haven country. Needless to say, tax authorities of both exporting and importing countries were dismayed by the lack of taxable income in such transactions.

One purpose of the U.S. Internal Revenue Act of 1962 was to eliminate the tax advantages of these "paper" foreign corporations without destroying the tax-deferral privilege for those foreign manufacturing and sales subsidiaries that were established for business and economic motives rather than tax motives. Although the tax motive has been removed, some firms have found these subsidiaries useful as finance control centers for foreign operations.

Google: An Illustrative Case of Profit Repositioning

> *It's called capitalism.* —Eric Schmidt, Chairman, Google, 2012.

Google, the dominant Internet search engine famous for encouraging all employees to *Do no evil* in the company code of conduct, has been the subject of much scrutiny over its global tax strategy in recent years. It is representative of the challenges for all companies and all governments in an increasingly complex world of digital commerce in which it is often difficult to determine where a taxable event occurred or where a taxable activity was performed.

Google's offshore tax strategy, the *Double-Irish-Dutch Sandwich*, is based on repositioning the ownership of much of its intellectual property to a subsidiary in a low-tax environment like Ireland (see Exhibit 15.10), and then establishing high transfer prices on various forms of services and overheads to other units, positioning most of the profits in the near-zero tax environment of Bermuda. The company negotiated for years with U.S. tax authorities, eventually gaining consent in an *advanced pricing agreement*. The agreement, as yet undisclosed, established allowable transfer prices and practices between the various Google-owned units used to minimize global taxes.

What is core to Google's structure shown in Exhibit 15.10 is known as *permanent establishment* (PE). *Permanent establishment* rules allow firms such as Google to fix a tax base in a low-tax country like Ireland, while generating lots of business in a country where tax rates are higher, like France. Companies in principle are taxed not on "where they do business," but on "where they finalize their business deals with customers"—the country or jurisdiction where the final contract is signed. In the case of Google, that means most sales throughout the European Union are finalized in Ireland. It is estimated that 75% of the top 50 U.S. software, Internet, and computer hardware companies use similar PE structures that help them avoid taxes.

Basis Erosion and Profit Shifting (BEPS)

The increasingly aggressive structures and strategies used by multinational firms all over the world—not just by U.S. multinationals—to avoid or delay paying taxes has led to a call by

---

EXHIBIT 15.10   Google's Global Tax Structure and U.K. Sales

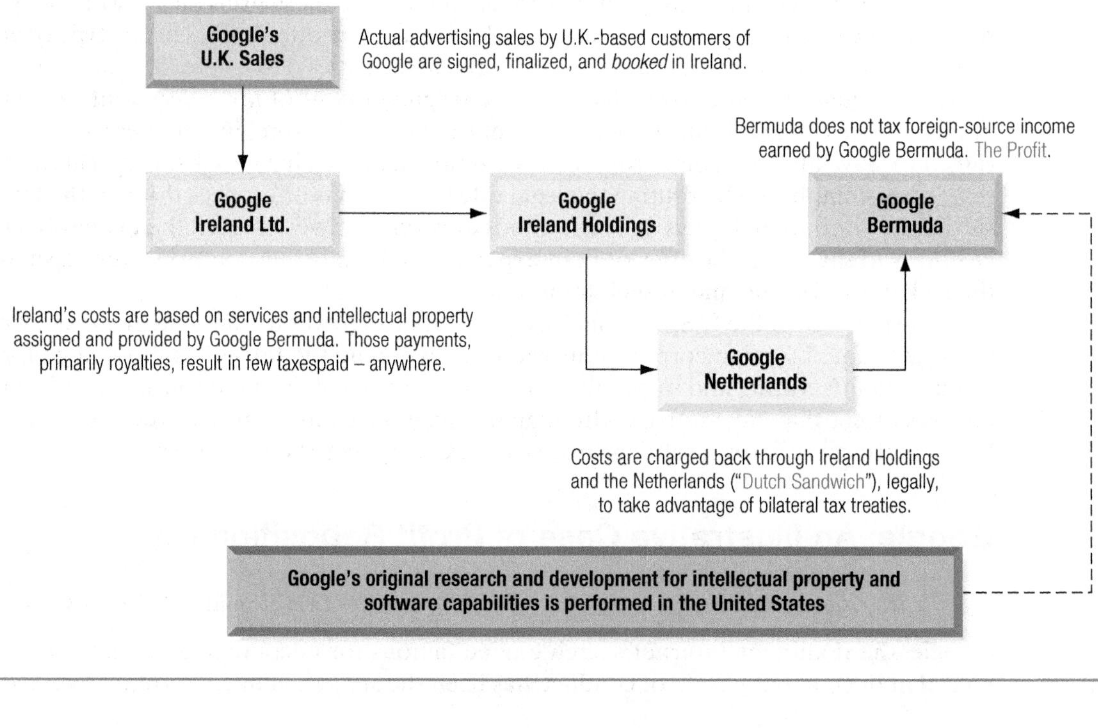

Google's U.K. Sales — Actual advertising sales by U.K.-based customers of Google are signed, finalized, and *booked* in Ireland.

Bermuda does not tax foreign-source income earned by Google Bermuda. The Profit.

Google Ireland Ltd. → Google Ireland Holdings → Google Bermuda

Ireland's costs are based on services and intellectual property assigned and provided by Google Bermuda. Those payments, primarily royalties, result in few taxes paid – anywhere.

Google Netherlands

Costs are charged back through Ireland Holdings and the Netherlands ("Dutch Sandwich"), legally, to take advantage of bilateral tax treaties.

Google's original research and development for intellectual property and software capabilities is performed in the United States

the finance ministers of the G20 in conjunction with the OECD to create an action plan to stop *basis erosion and profit shifting* (BEPS). Interestingly, this is not an effort to stop illegal activity—for most of the repositioning of profits and reduction in tax liabilities in total is legal—but to explore new initiatives to change tax laws and practices globally to reassert taxing powers.

The debate that has raged in recent years about Google, Apple, and many other multinational companies is that they are generating massive profits around the world and often paying few corporate income taxes—anywhere to anyone. All the while many other multinationals continue to pay effective tax rates in the high 20s to low 30s (percentage of consolidated pre-tax income). If that is indeed the case, it is not a level playing field, and one that many traditional manufacturers, who cannot move their products and property digitally around the globe, feel is biased against them.

Corporate Inversion

A *corporate inversion* is the changing of a company's country of incorporation. Its purpose is to reduce its effective global tax liabilities by re-incorporating in a lower-tax jurisdiction, typically a country using a territorial tax regime. Although the company's operations may be completely unchanged, and its corporate headquarters may remain in the original country of incorporation, it will now have a new corporate home, and its old country of incorporation will now be only one of many countries in which the firm operates foreign subsidiaries.

The typical transaction is one in which a U.S. corporation and its foreign subsidiary in another country, like Bermuda, exchange ownership shares. As a result of the exchange, the U.S. company becomes the U.S. subsidiary of a Bermudian corporation. There is no change in the control of the corporation, only its place of incorporation. This is colloquially referred to as a *naked inversion*. Such inversions enjoyed a short period of popularity in the late 1990s and early 2000s (Ingersoll Rand, Tyco, and Foster Wheeler among others successfully reincorporated outside the U.S.), but after one particularly tumultuous attempt—that of Stanley Works in 2002, which was not completed—the U.S. passed the American Jobs Creation Act (AJCA) of 2004 which altered the rules governing inversions in two important ways:

First, if the new foreign parent corporation is still 80% or more owned and controlled by the former parent company's stockholders, the company would continue to be treated as a domestic or U.S. incorporated company. This means that it would continue to be taxed on its worldwide income, and its U.S. "subsidiary" would be treated as its effective parent company. This is referred to as the *80% rule*.

Second, if the new foreign parent corporation is still at least 60% controlled by former parent company shareholders, but less than 80%, the new company is not allowed any tax credits on gains (toll taxes) on the legal transfer of assets from the old company to the new. This made many naked inversions financially unattractive even when there was a substantial change in ownership structure.

Today there are three basic types of corporate inversions in use: (1) the substantial business presence; (2) merger with a larger foreign firm; and (3) merger with a smaller foreign firm.

Substantial Business Presence

The AJCA elements addressing corporate inversions are specifically structured to stop corporate inversions undertaken purely for tax reduction purposes. The Act, however, does not hinder inversions when the re-incorporation takes place in a country in which the company does indeed have a "substantial business presence," defined as the company having 25% of its assets, income, and employees in the country to which it is moving. This will, therefore, rarely include a traditional tax haven.

Merger with Larger Foreign Firm

The second major form of inversion is when a U.S. company is merged with a large foreign firm and the new combined entity is incorporated in the foreign country. The added stipulation is that the previous U.S. ownership must have a minority position (less than 50% ownership) in the new combined entity. One high-profile example of this was the combination of two major deep-water oil drilling firms, Pride (U.S.) with Ensco (U.K.) in 2011.

Merger with Smaller Foreign Firm

The third form of corporate inversion is when a U.S. corporation merges with a smaller foreign firm, often one incorporated in Ireland, the United Kingdom, or Luxembourg. The control of the newly created company remains with the previous U.S. stockholders, so it does not fulfill the 80% rule. However, because the new corporate home is still often a low-tax jurisdiction, the ability of the U.S. to impose worldwide tax principles to the new firm is limited. Combining Easton (U.S.) and Cooper Industries (Ireland) in 2012, thereby creating a new Irish corporation, is one such example.

The rapid evolution of corporate inversions for U.S.-based multinationals over the past 20 years has heightened the awareness of the relatively high U.S. corporate tax rates, including its worldwide regime, and corporate concerns over global competitiveness. That more inversions are resulting in movements of newly merged incorporations to other major developed countries like Ireland, and not to tax havens like Bermuda, the Cayman Islands, or the Bahamas, has increased tensions over worldwide treaty shopping and political debate over a corporate race to the bottom of the corporate tax environment. The merger between Applied Materials (U.S.) and Tokyo Electron (Japan) in 2013, in which the two merged and reincorporated in the Netherlands, is one such complex example. The continuing concern is reflected in the words of Senator Charles E. Grassley, ranking member of the U.S. Senate Committee on Finance who noted in 2002, "These expropriations aren't illegal. But they're sure immoral."

SUMMARY POINTS

- Nations structure their tax systems along one of two basic approaches: the worldwide approach or the territorial approach. Both approaches are attempts to determine which firms and which incomes, foreign or domestic by incorporation and origin, are subject to the taxation of host-country tax authorities.

- Tax treaties normally define whether taxes are to be imposed on income earned in one country by the nationals of another, and if so, how. Tax treaties are bilateral, with the two signatories specifying what rates are applicable to which types of income between the two countries.

- Transfer pricing is the pricing of goods, services, and technology between related companies. High- or low-transfer prices have an effect on income taxes, fund positioning, managerial incentives and evaluation, and joint venture partners.

- The U.S. differentiates foreign source income from domestic source income. Each is taxed separately, and tax deficits/credits in one category may not be used against deficits/credits in the other category.

- If a U.S.-based MNE receives income from a foreign country that imposes higher corporate income taxes than does the United States, total creditable taxes will exceed U.S. taxes on that foreign income, resulting in excess foreign tax credits.

- MNEs have foreign subsidiaries that act as tax havens for corporate funds awaiting reinvestment or repatriation. Tax havens are typically located in countries that have a low corporate tax rate, a stable currency, facilities to support financial services, and a stable government.

- Many U.S.-based companies have used corporate inversions in an effort to reduce their effective tax rates by re-incorporating offshore in lower tax environments. An alternative to a corporate inversion is acquiring a firm incorporated within one of these low-tax environments, and then adopting its incorporation for the new combined company.

Apple's Global iTax Strategy[1]

Apple does not use tax gimmicks. Apple does not move its intellectual property into offshore tax havens and use it to sell products back into the U.S. in order to avoid U.S. tax; it does not use revolving loans from foreign subsidiaries to fund its domestic operations; it does not hold money on a Caribbean island; and it does not have a bank account in the Cayman Islands. Apple has substantial foreign cash because it sells the majority of its products outside the U.S.

—Apple CEO Tim Cook, in testimony before the U.S. Senate Permanent Subcommittee on Investigations, 2013.

Life—or at least your public reputation—is difficult to manage when you are possibly the world's largest and most profitable company, and you are constantly criticized for not paying enough in taxes. Such is the angst at Apple. Apple has engineered one of the more aggressive tax-saving global tax strategies in global business.

Global Operations

Apple is in many ways organized like any other large multinational company. It is headquartered in Cupertino, California, and is a U.S. incorporated company—Apple, Inc. It is here, as noted in Exhibit A, that essentially all of its global research and development is conducted, and therefore its intellectual property created. Although Apple does use contract manufacturers for most of its product construction and assembly (primarily in China), it does manufacture all of its A5 series of microprocessors—the self-described *engine* of Apple products—at its

Apple's International Product Value Stream

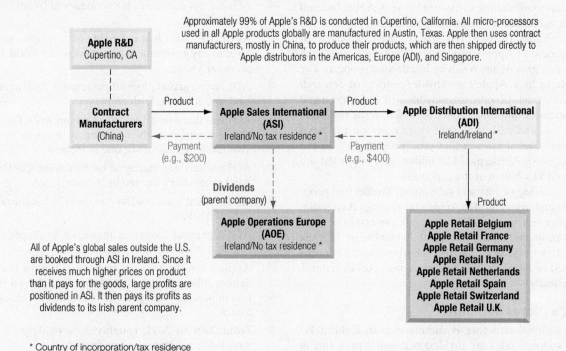

Approximately 99% of Apple's R&D is conducted in Cupertino, California. All micro-processors used in all Apple products globally are manufactured in Austin, Texas. Apple then uses contract manufacturers, mostly in China, to produce their products, which are then shipped directly to Apple distributors in the Americas, Europe (ADI), and Singapore.

All of Apple's global sales outside the U.S. are booked through ASI in Ireland. Since it receives much higher prices on product than it pays for the goods, large profits are positioned in ASI. It then pays its profits as dividends to its Irish parent company.

* Country of incorporation/tax residence

---

production facility in Austin, Texas. Manufactured final product is then shipped directly to Apple's distribution centers globally.

All of Apple's sales outside the Americas are booked through its Irish subsidiary ASI, Apple Sales International. ASI purchases the product from the contract manufacturers, taking title to the goods, and then resells to its international distribution company, ADI, Apple Distribution International. As demonstrated in Exhibit A, ASI then enjoys large profitability on the resale. ASI, as is typical of much of global commerce today, may take legal title to the goods but the goods never physically pass through Ireland, being shipped directly from Chinese manufacturers to the distribution centers for in-country sales, like all of the retail companies listed for Europe in Exhibit A. ASI then pays out all of its profits, in the form of dividends, to Apple Operations Europe (AOE), its parent company, also in Ireland.

Apple's tax management is then based on a series of structures it has established, beginning with the principle of a *cost sharing agreement* on the development and ownership of intellectual property.

Cost Sharing

Apple has a *cost sharing agreement* between Apple Inc. and ASI, an agreement between the parent company and the Irish subsidiary. The two units agree to share in the cost of development of Apple's products, and in turn to share the economic rights of any resulting intellectual property. For example, in 2011, Apple's worldwide spending on research and development (approximately 95% of which occurred in Cupertino, California) totaled $2.4 billion. The two units, Apple Inc. and ASI, then split these costs on the basis of Apple's global sales in that year, roughly 40% in the Americas (Apple Inc. paid $1.0 billion) and 60% offshore (ASI paid $1.4 billion of the expenses).

This sharing of cost and subsequent intellectual property ownership is central to Apple's tax strategy. As a result, the profits accruing to ASI based on its ownership of intellectual property are not immediately taxable by U.S. tax authorities because of its Irish incorporation. Theoretically, ASI or AOE should then be paying taxes in Ireland. Theoretically.

Apple's Global Structure

Apple's global structure is summarized in Exhibit B.[2] Apple's global sales are divided between Apple Inc. in the United States and Apple Sales International (ASI) of Ireland. ASI is responsible for the sale of all Apple products in Europe, the Middle East, Asia, Africa, India, and the Pacific.

The key to understanding both the structure and function of Apple's tax strategy is the incorporation of its major foreign affiliate holding companies—AOI, AOE, ASI, and ADI, and Apple Retail Europe—in Ireland. Ireland has a low (by global standards) statutory corporate income tax rate of 12%. Apple, however, has negotiated a lower rate with the Irish government, at just less than 2%, since 2003.[3] This is accomplished, according to Apple, by the way the Irish government has chosen to calculate Apple's taxable income.

As illustrated in Exhibit B, Apple Operations International (AOI) is the single legal entity which owns and controls all of Apple's activities outside the Americas. AOI itself is testimony to the global and digital structure of global enterprise management today.

- AOI was incorporated in Ireland in 1980. Apple, however, has been unable to locate any documents that explain *why* Ireland was chosen as its place of incorporation.

- AOI has not declared a tax residency in Ireland or any other country.

- As of 2013, AOI had not paid any corporate income taxes to any government anywhere in the world, in the previous 5 years.

- AOI has no actual physical presence in Dublin or Ireland, and it has no Irish employees.

- AOI has three directors, all of whom work for other Apple companies while serving as directors of AOI. Two reside in California, one in Ireland.

- AOI's assets are managed by Braeburn Capital, an Apple subsidiary located in Nevada, U.S.A.

- AOI's actual asset holdings are held in bank accounts in New York.

- AOI's general ledger is managed at Apple's U.S. shared service center in Austin, Texas.

- Apple's tax director, in testimony before the U.S. Senate subcommittee, stated that he believed AOI's functions were managed and controlled in the United States.

- From 2009 to 2011, roughly 30% of Apple's total worldwide net income came from AOI.

[2]Memorandum of the Permanent Subcommittee on Investigations, Re: *Offshore Profit Shifting and the U.S. Tax Code—Part 2 (Apple Inc.)*, U.S. Senate, May 21, 2013.

[3]This is based on data presented by Apple in 2013, most of which focused on the 2008–2011 period.

EXHIBIT B   **Apple's Global Organizational Structure**

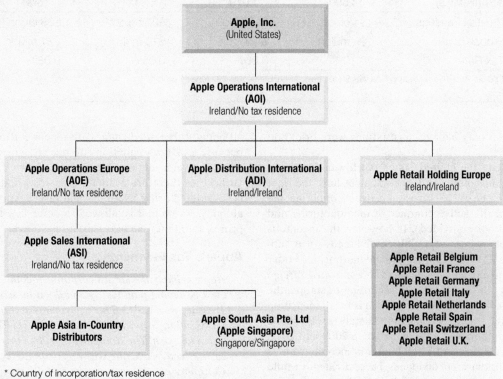

* Country of incorporation/tax residence
Source: Permanent Subcommittee on Investigations, May 2013.

AOI is clearly a legal entity of a digital construction, but its lack of a tax residence anywhere on earth is obviously curious.[4]

Tax Residency

Under Irish law, Irish tax residency requires that the company be either managed or controlled in Ireland. Obviously, by the details reported above, AOI is not. The United States requirements for tax residency require that the business be incorporated in the United States, which AOI is not. So, in Apple's opinion, AOI has no tax residency anywhere on earth, and the company has therefore never established its tax residency. Apple executives, when asked in a U.S. Senate subcommittee hearing whether AOI was actually managed and controlled from the U.S., answered that ". . . it had not determined the answer to that question."

Apple Sales International (ASI), like AOI, has no tax residency anywhere. Exhibit C indicates that ASI paid just $10 million (million with an "m") in taxes on more than $22 billion (billion with a "b") in pre-tax earnings in 2011. That is an effective tax rate of less than 0.05%. For the 2009–2011 period, ASI had pre-tax earnings of $38 billion and paid a total of $21 million in taxes. Curiously, although ASI is not a tax resident of Ireland, it has filed corporate tax returns in Ireland, which is why we have these numbers.

Whether businesses like AOI or ASI are really, in any functional form, separate from Apple Inc. continues to be debated. In 2008, Apple Inc., ASI, and AOE signed an amended cost sharing agreement. The signatory for AOE was Apple's Treasurer. The signatory for ASI was Tim Cook, Apple's COO. The signatory for AOE was Apple's

---

[4]Apple reorganized its Irish employees in 2012. Apple's 2,455 employees were redistributed across five different Irish business units, the majority of which were now assigned to ADI. ASI's employment in Ireland expanded from zero to 250 employees.

EXHIBIT C	Global Taxes Paid by Apple Sales International (ASI)			
ASI (Ireland)	**2009**	**2010**	**2011**	**Total**
Pre-Tax Earnings	$4 billion	$12 billion	$22 billion	$38 billion
Global Tax	$4 million	$7 million	$10 million	$21 million
Tax Rate	0.10%	0.06%	0.05%	0.06%

Source: Apple Consolidating Financial Statements, APL-PSI-000130-232 (sealed exhibit).

CFO. One can only believe negotiations were brief and efficient.

According to the United States' worldwide tax structure, foreign subsidiaries of U.S. companies have the right to defer payment of U.S. taxes on *active income* (income derived from the active conduct of manufacturing and sales and services provided). If, however, the income is *passive*—such as interest, royalties, dividends—it is subject to immediate taxation by U.S. tax authorities under Subpart F income rules as applied to *controlled foreign corporations* (CFCs). Therefore, according to statute, the income earned by ASI and AOI, which is by all indication passive income, should become immediately taxable in the U.S. by tax authorities.[5] In fact, between 2009 and 2012, AOI reportedly received $29.9 billion in income from its lower-tier subsidiaries in dividends. These dividends would ordinarily be immediately taxable by U.S. tax authorities according to Subpart F statutes.

Apple, however, has avoided this tax exposure through the use of what is known as *check-the-box*. In 1996, in an attempt to simplify the U.S. tax code, the Treasury Department adopted a new practice that allowed companies to "check-the-box" on a tax form to describe a foreign corporate entity (like ASI or AOE) for tax purposes as irrelevant—a so-called *disregarded entity*. This simplified the tax filings for multinational companies dramatically. This disregarded entity status allowed U.S.-based multinationals like Apple to set up high-volume profitability subsidiaries in low tax jurisdictions such as Ireland or Luxembourg. For Apple, that meant that all of the companies shown in Exhibit B below Apple Operations International (AOI),

all being *disregarded entities*, disappear for tax purposes because U.S. tax regulations do not recognize payments made between units within a single entity. The U.S. tax authorities therefore evaluate only AOI, and its income is considered *active* as it buys and resells Apple products globally. As such, it is allowed to defer U.S. taxes on its profits until that time of repatriation—if ever.[6]

Apple's Tax Payments

Apple is likely the largest corporate income tax payer in the US, having paid nearly $6 billion in taxes to the US Treasury in FY2012. These payments account for $1 in every $40 in corporate income tax the US Treasury collected last year. The Company's FY2012 total US federal cash effective tax rate was approximately 30.5%. The Company expects to pay over $7 billion in taxes to the US Treasury in its current fiscal year. In accordance with US law, Apple pays US corporate income taxes on the profits earned from its sales in the US and on the investment income of its Controlled Foreign Corporations ("CFCs"), including the investment earnings of its Irish subsidiary, Apple Operations International ("AOI").

—Testimony of Apple Inc. before the
Permanent Subcommittee on Investigations,
U.S. Senate, May 21, 2013.

Apple does indeed pay a lot of taxes. As illustrated by Exhibit D, in 2011 alone, according to its 10-K filing, Apple had total taxes payable for the year of $8.283 billion. Yet its effective tax rate in the United States was a combined 22.4% Federal plus State, and its effective tax

---

[5]Theoretically, Apple had two types of income that should have been subject to Subpart F statutes: 1) foreign base company sales income (FBCS), the sales income that Apple had assigned to ASI in Ireland for no reason but consolidation and positioning of profits; and 2) foreign personal holding company income (FPHC), the passive income earned in Ireland from dividends, royalties, fees, and interest.

[6]After the passage of check-the-box legislation U.S. tax authorities nearly immediately realized their mistake. Tax authorities later admitted that the statutory changes were meant for corporations operating within the territorial boundaries of the United States, and they had not even evaluated how they might affect international operations and tax liability. Although tax authorities have repeatedly tried to get check-the-box repealed, they have been unable to, primarily because of political opposition.

EXHIBIT D	Apple's Provisions for U.S. Taxes—2011			
Tax Liability	**Federal**	**State**	**Foreign**	**Total**
Current	$ 3,884	$ 762	$ 769	
Deferred	2,998	37	(167)	
Net	$ 6,882	$ 799	$ 602	$8,283
Effective Tax Rate	20.1%	2.3%	1.8%	24.2%

Source: Apple 2011 Annual Report (Form 10-K).

rate on profits earned outside the U.S. was 1.8%. Given that the U.S. statutory corporate income tax rate alone is 35%, Apple's strategy and structure seems to be working to reduce its taxes—globally.

One of the principles underlying worldwide taxation and deferral under the U.S. tax system is that company operations and profits earned in other countries are paying taxes in those countries. The deferral provision, however, was suspended with the creation of Subpart F income statutes in 1962 to deter the use of tax havens to position offshore profits and permanently defer paying corporate income taxes. At this point one must conclude that the complexity of global business combined with changes made to the U.S. tax code over the past 50 years have undermined what was originally intended.

Apple supports comprehensive reform of the US corporate tax system. The Company supports a dramatic simplification of the corporate tax system that is revenue neutral, eliminates all tax expenditures, lowers tax rates and implements a reasonable tax on foreign earnings that allows free movement of capital back to the US. Apple believes such comprehensive reform would stimulate economic growth. Apple supports this plan even though it would likely result in Apple paying more US corporate tax.

—Testimony of Apple Inc. before the
Permanent Subcommittee on Investigations,
U.S. Senate, May 21, 2013.

Mini-Case Questions

1. What is the single most important element of Apple's global tax strategy?
2. Why do most of Apple's businesses in Ireland not have a country of tax residence?
3. Why does Apple Operations International (AOI)—the Irish subsidiary that captures most of Apple's global profits outside the Americas—not pay taxes to Ireland or the United States?

QUESTIONS

These questions are available in MyFinanceLab.

1. **Primary Objective.** What is the primary objective of multinational tax planning?

2. **Tax Morality.** What is meant by the term "tax morality"? If for example, your company has a subsidiary in Russia where some believe tax evasion is a fine art, should you comply with Russian tax laws or violate the laws as do your local competitors?

3. **Tax Neutrality.** What is tax neutrality? What is the difference between domestic neutrality and foreign neutrality?

4. **Worldwide versus Territorial.** What is the difference between the worldwide and territorial approaches to taxation?

5. **Direct or Indirect.** What is the difference between a direct tax and an indirect tax?

6. **Tax Deferral.** What is meant by tax deferral in the U.S. system of taxation? What is the deferral privilege?

7. **Value-Added Tax.** What is a value-added tax, and how does it differ from an income tax?

8. **Withholding Tax.** What is a withholding tax and why do governments impose them?

9. **Tax Treaty.** What is usually included within a tax treaty?

10. **Active versus Passive.** What do the terms active and passive mean in the context of U.S. taxation of foreign source income?

11. **Tax Types.** Taxes are classified based on whether they are applied directly to income, called direct taxes, or to some other measurable performance characteristic of the firm, called indirect taxes. Identify each of the following as a "direct tax," an "indirect tax," or something else:
    a. Corporate income tax paid by a Japanese subsidiary on its operating income
    b. Royalties paid to Saudi Arabia for oil extracted and shipped to world markets
    c. Interest received by a U.S. parent on bank deposits held in London
    d. Interest received by a U.S. parent on a loan to a subsidiary in Mexico
    e. Principal repayment received by U.S. parent from Belgium on a loan to a wholly owned subsidiary in Belgium
    f. Excise tax paid on cigarettes manufactured and sold within the United States
    g. Property taxes paid on the corporate headquarters building in Seattle
    h. A direct contribution to the International Committee of the Red Cross for refugee relief
    i. Deferred income tax, shown as a deduction on the U.S. parent's consolidated income tax
    j. Withholding taxes withheld by Germany on dividends paid to a United Kingdom parent corporation

12. **Foreign Tax Credit.** What is a foreign tax credit? Why do countries give credit for taxes paid on foreign source income?

13. **Earnings Stripping.** What is earnings stripping, and what are some examples of how multinational firms pursue it?

14. **Controlled Foreign Corporation.** What is a controlled foreign corporation and what is its significance in global tax management?

15. **Transfer Pricing.** What is a transfer price and can a government regulate it? What difficulties and motives does a parent multinational firm face in setting transfer prices?

16. **Fund Positioning.** What is fund positioning?

17. **Income Tax Effect.** What is the income tax effect, and how may a multinational firm alter transfer prices as a result of the income tax effect?

18. **Correct Pricing.** What is Section 482 of the U.S. Internal Revenue Code and what guidelines does it recommend when setting transfer prices?

19. **Cross-Crediting.** Define cross-crediting and explain why it may or may not be consistent with a worldwide tax regime.

20. **Check-the-Box.** Explain how the check-the-box regulatory change altered the effectiveness of Subpart F income regulations.

21. **Measuring Managerial Performance.** What role does transfer pricing have within multinational companies when measuring management performance? How can transfer pricing practices within a firm conflict with performance measurement?

22. **Tax Haven Subsidiary.** What is a tax haven? Is it the same thing as an international offshore financial center? What is the purpose of a multinational creating and operating a financial subsidiary in a tax haven?

23. **Corporate Inversion.** What is a corporate inversion, and why do many U.S. corporations want to pursue it although it is highly criticized by public and private parties alike?

PROBLEMS

These problems are available in MyFinanceLab.

1. **Avon's Foreign-Source Income.** Avon is a U.S.-based direct seller of a wide array of products. Avon markets leading beauty, fashion, and home products in more than 100 countries. As part of the training in its corporate treasury offices, it has its interns build a spreadsheet analysis of the following hypothetical subsidiary earnings/distribution analysis. Use the tax analysis presented in Exhibit 15.7 for your basic structure.

Baseline Values	Case 1	Case 2
a. Foreign corporate income tax rate	28%	45%
b. U.S. corporate income tax rate	35%	35%
c. Foreign dividend withholding tax rate	15%	0%
d. U.S. ownership in foreign firm	100%	100%
e. Dividend payout rate of foreign firm	100%	100%

a. What is the total tax payment, foreign and domestic combined, for this income?
b. What is the effective tax rate paid on this income by the U.S.-based parent company?
c. What would be the total tax payment and effective tax rate if the foreign corporate tax rate was 45% and there were no withholding taxes on dividends?
d. What would be the total tax payment and effective tax rate if the income was earned by a branch of the U.S. corporation?

2. **Pacific Jewel Airlines (Hong Kong).** Pacific Jewel Airlines is a U.S.-based air freight firm with a wholly owned subsidiary in Hong Kong. The subsidiary, Jewel Hong Kong, has just completed a long-term planning report for the parent company in San Francisco, in which it has estimated the following expected earnings and payout rates for the years 2011–2014.

Jewel Hong Kong Income Items (millions US$)

	2011	2012	2013	2014
Earnings before interest and taxes (EBIT)	8,000	10,000	12,000	14,000
Less interest expenses	(800)	(1,000)	(1,200)	(1,400)
Earnings before taxes (EBT)	7,200	9,000	10,800	12,600

The current Hong Kong corporate tax rate on this category of income is 16.5%. Hong Kong imposes no withholding taxes on dividends remitted to U.S. investors (per the Hong Kong–United States bilateral tax treaty). The U.S. corporate income tax rate is 35%. The parent company wants to repatriate 75% of net income as dividends annually.

a. Calculate the net income available for distribution by the Hong Kong subsidiary for the years 2011–2014.
b. What is the expected amount of the dividend to be remitted to the U.S. parent each year?
c. After estimating the theoretical U.S. tax liability on the expected dividend (what is often termed *gross-up* in the U.S.), what is the total dividend after tax, including all Hong Kong and U.S. taxes, expected each year?
d. What is the effective tax rate on this foreign-sourced income per year?

3. **Kraftstoff of Germany.** Kraftstoff is a German-based company that manufactures electronic fuel-injection carburetor assemblies for several large automobile companies in Germany, including Mercedes, BMW, and Opel. The firm, like many firms in Germany today, is revising its financial policies in line with the increasing degree of disclosure required by firms if they wish to list their shares publicly in or out of Germany. The company's earnings before tax (EBT) is €483,500,000.

Kraftstoff's primary problem is that the German corporate income tax code applies a different income tax rate to income depending on whether it is retained (45%) or distributed to stockholders (30%).

a. If Kraftstoff planned to distribute 50% of its net income, what would be its total net income and total corporate tax bills?
b. If Kraftstoff was attempting to choose between a 40% and 60% payout rate to stockholders, what arguments and values would management use in order to convince stockholders which of the two payouts is in everyone's best interest?

4. **Gamboa's Tax Averaging.** Gamboa, Incorporated, is a relatively new U.S.-based retailer of specialty fruits and vegetables. The firm is vertically integrated with fruit and vegetable-sourcing subsidiaries in Central America, and distribution outlets throughout the southeastern and northeastern regions of the United States. Gamboa's two Central American subsidiaries are in Belize and Costa Rica.

Maria Gamboa, the daughter of the firm's founder, is being groomed to take over the firm's financial management in the near future. Like many firms of Gamboa's size, it has not possessed a very high degree of sophistication in financial management simply out of time and cost considerations. Maria, however, has recently finished her MBA and is now attempting to put some specialized knowledge of U.S. taxation practices to work to save Gamboa money. Her first concern is *tax averaging* for foreign tax liabilities arising from the two Central American subsidiaries.

Costa Rican operations are slightly more profitable than Belize, which is particularly good since Costa Rica is a relatively low-tax country. Costa Rican corporate taxes are a flat 30%, and there are no withholding taxes imposed on dividends paid by foreign firms with operations there. Belize has a higher corporate income tax rate, 40%, and imposes a 10% withholding tax on all dividends distributed to foreign investors. The current U.S. corporate income tax rate is 35%.

	Belize	Costa Rica
Earnings before taxes	$1,000,000	$1,500,000
Corporate income tax rate	40%	30%
Dividend withholding tax rate	10%	0%

a. If Maria Gamboa assumes a 50% payout rate from each subsidiary, what are the additional taxes due on foreign-sourced income from Belize and Costa Rica individually? How much in additional U.S. taxes would be due if Maria averaged the tax credits/liabilities of the two units?

b. Keeping the payout rate from the Belize subsidiary at 50%, how should Maria change the payout rate of the Costa Rican subsidiary in order to most efficiently manage her total foreign tax bill?

c. What is the minimum effective tax rate that Maria can achieve on her foreign sourced income?

Chinglish Dirk

Use the following company case to answer Problems 5–7.

Chinglish Dirk Company (Hong Kong) exports razor blades to its wholly owned parent company, Torrington Edge (Great Britain). Hong Kong tax rates are 16% and British tax rates are 30%. Chinglish calculates its profit per container as follows (all values in British pounds).

Constructing Transfer (Sales) Price per Unit	Chinglish Dirk (British pounds)	Torrington Edge (British pounds)
Direct costs	£10,000	£16,100
Overhead	4,000	1,000
Total costs	£14,000	£17,100
Desired markup	2,100	2,565
Transfer price (sales price)	£16,100	£19,665
Income Statement		
Sales price	£16,100,000	£19,665,000
Less total costs	(14,000,000)	(17,100,000)
Taxable income	£2,100,000	£2,565,000
Less taxes	(336,000)	(769,500)
Profit, after-tax	£1,764,000	£1,795,500

5. **Chinglish Dirk (A).** Corporate management of Torrington Edge is considering repositioning profits within the multinational company. What happens to the profits of Chinglish Dirk and Torrington Edge, and the consolidated results of both, if the markup at Chinglish was increased to 20% and the markup at Torrington was reduced to 10%? What is the impact of this repositioning on consolidated tax payments?

6. **Chinglish Dirk (B).** Encouraged by the results from the previous problem's analysis, corporate management of Torrington Edge wishes to continue to reposition profit in Hong Kong. It is, however, facing two constraints. First, the final sales price in Great Britain must be £20,000 or less to remain competitive. Secondly, the British tax authorities—in working with Torrington Edge's cost accounting staff—has established a maximum transfer price allowed (from Hong Kong) of £17,800. What combination of markups do you recommend for Torrington Edge to institute? What is the impact of this repositioning on consolidated profits on after-tax and total tax payments?

7. **Chinglish Dirk (C).** Not to leave any potential tax repositioning opportunities unexplored, Torrington Edge wants to combine the components of Problems 4 and 5 with a redistribution of overhead costs. If overhead costs could be reallocated between the two units, but still total £5,000 per unit, and maintain a minimum of £1,750 per unit in Hong Kong, what is the impact of this repositioning on consolidated profits after-tax and total tax payments?

INTERNET EXERCISES

1. **Global Taxes.** Web sites like TaxWorld.org provide detailed insights into the conduct of business and the associated tax and accounting requirements of doing business in a variety of countries.

   International Tax Resources   www.taxworld.org/OtherSites/International/international.htm

2. **International Taxpayer.** The United States Internal Revenue Service (IRS) provides detailed support and document requirements for international taxpayers. Use the IRS site to find the legal rules and regulations and definitions for international residents tax liabilities when earning income and profits in the United States.

   U.S. IRS Taxpayer   www.irs.gov/businesses/small/international/index.html

3. **Official Government Tax Authorities.** Tax laws are constantly changing, and an MNE's tax planning and management processes must therefore include a continual updating of tax practices by country. Use the following government tax sites to address specific issues related to those countries:

   Hong Kong's ownership change to China   www.gov.hk/en/business/taxes/profittax/

   Ireland's international financial services center   www.revenue.ie

4. **Tax Practices for International Business.** Many of the major accounting firms provide online information and advisory services for international business activities as related to tax and accounting practices. Use the following Web sites to find current information on tax law changes and practices.

Ernst and Young	www.ey.com/tax/
Deloitte & Touche	www.deloitte.com/view/en_US/us/Services/tax/index.htm
KPMG	www.kpmg.com
Price Waterhouse Coopers	www.pwc.com/us/en/tax-services/index.jhtml
Ernst & Young	www.eyi.com

CHAPTER 16

International Trade Finance

What the wise man does in the beginning, the fool does in the end.

—Niccolò Machiavelli.

LEARNING OBJECTIVES

- Learn how international trade alters both the supply chain and general value chain of the firm
- Discover the key elements of an import or export business transaction system
- Explore how the three key documents in import/export combine to finance both the transaction and to manage its risks
- Describe the variety of government programs to help finance exports
- Examine the major trade financing alternatives
- Evaluate the use of forfaiting for medium- to long-term trade financing

The purpose of this chapter is to explain how international trade—exports and imports—is financed. The content is of direct practical relevance to both domestic firms—that simply import and export—and to multinational firms—that trade with related and unrelated entities.

The chapter opens with an explanation of the types of trade relationships that exist and a discussion of the trade dilemma: exporters want to be paid before they export and importers do not want to pay until they receive the goods. The next section then describes the key elements of an international trade transaction. This is followed by an exploration of the three key trade documents—the letter of credit, draft, and bill of lading—and how they are used to manage the various risks of international import and export. The fourth section of the chapter describes government export financing programs, followed by a detailed examination of alternative trade financing vehicles and instruments. The sixth and final section explores the use of forfaiting for the financing of long-term receivables. The Mini-Case at the end of the chapter, *Crosswell International and Brazil*, illustrates how an export requires the integration of management, marketing, and finance.

The Trade Relationship

As we saw in Chapter 1, the first significant global activity by a domestic firm is the importing and exporting of goods and services. The purpose of this chapter is to analyze the *international trade phase* for a domestic firm, which begins to import goods and services from foreign suppliers and to export to foreign buyers. In the case of Ganado, this trade phase began with suppliers from Mexico and buyers from Canada.

Trade financing shares a number of common characteristics with the traditional value chain activities conducted by all firms. All companies must search out suppliers for the many goods and services required as inputs to their own goods production or service provision processes. Ganado's Purchasing and Procurement Department must determine whether each potential supplier is capable of producing the product to required quality specifications and in a timely and reliable manner, and whether the supplier will work with Ganado in the ongoing process of product and process improvement for continued competitiveness. All must be at an acceptable price and payment terms. As illustrated in Exhibit 16.1, this same series of issues applies to potential customers, as their continued business is equally critical to Ganado's operations and success.

Understanding the nature of the relationship between the exporter and the importer is critical to understanding the methods for import-export financing utilized in industry. Exhibit 16.2 provides an overview of the three categories of import/export relationships: *unaffiliated unknown*, *unaffiliated known*, and *affiliated*.

- A foreign importer with which Ganado has not previously conducted business would be considered *unaffiliated unknown*. In this case, the two parties would need to enter into a detailed sales contract, outlining the specific responsibilities and expectations of the business agreement. Ganado would also need to seek out protection against the possibility that the importer would not make payment in full in a timely fashion.

- A foreign importer with which Ganado has previously conducted business successfully would be considered *unaffiliated known*. In this case, the two parties may still enter into a detailed sales contract, but specific terms and shipments or provisions of services may be significantly looser in definition. Depending on the depth of the relationship, Ganado may seek some third-party protection against noncompletion or conduct the business on an open account basis.

- A foreign importer, which is a subsidiary business unit of Ganado, such as Ganado Brazil, would be an *affiliated* party (sometimes referred to as *intrafirm trade*). Because both businesses are part of the same MNE, the most common practice would be to conduct the trade transaction without a contract or protection against nonpayment. This is not, however, always the case. In a variety of international business situations it may still be in Ganado's best interest to detail the conditions for the business transaction, and to possibly protect against any political or country-based interruption to the completion of the trade transaction.

EXHIBIT 16.1   Financing Trade: The Flow of Goods and Funds

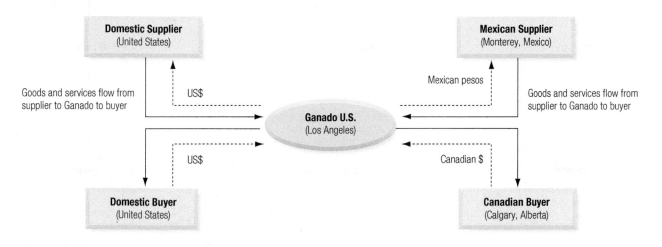

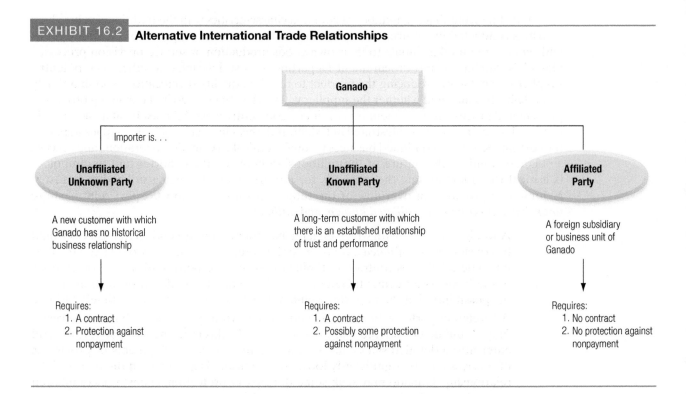

EXHIBIT 16.2   **Alternative International Trade Relationships**

International trade must work around a fundamental dilemma. Imagine an importer and an exporter who would like to do business with one another. Because of the distance between the two, it is not possible to simultaneously hand over goods with one hand and accept payment with the other. The importer would prefer the arrangement at the top of Exhibit 16.3, while the exporter's preference is shown at the bottom.

EXHIBIT 16.3   **The Mechanics of Import and Export**

EXHIBIT 16.4 **The Bank as the Import/Export Intermediary**

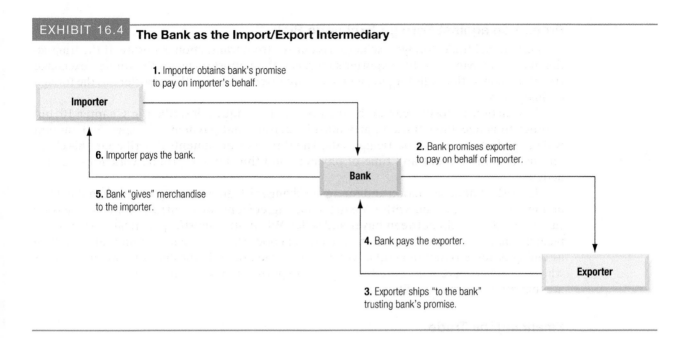

The fundamental dilemma of being reluctant to trust a stranger in a foreign land is resolved by using a highly respected bank as intermediary. A greatly simplified view is described in Exhibit 16.4. In this simplified view, the importer obtains the bank's promise to pay on its behalf, knowing that the exporter will trust the bank. The bank's promise to pay is called a *letter of credit* (L/C). The exporter ships the merchandise to the importer's country. Title to the merchandise is given to the bank on a document called a *bill of lading* (B/L). The exporter asks the bank to pay for the goods, and the bank does so. The document requesting payment is called a *sight draft*. The bank, having paid for the goods, now passes title to the importer, whom the bank trusts. At that time or later, depending on their agreement, the importer reimburses the bank.

Financial managers of MNEs must understand these three basic documents, because their firms will often trade with unaffiliated parties, and also because the system of documentation provides a source of short-term capital that can be drawn upon even when shipments are to sister subsidiaries.

Benefits of the System

The three key documents and their interaction will be discussed in detail later in this chapter. They constitute a system developed and modified over centuries to protect both importer and exporter from the risk of noncompletion and foreign exchange risk, as well as to provide a means of financing.

Protection against Risk of Noncompletion

As stated above, once importer and exporter agree on terms, the seller usually prefers to maintain legal title to the goods until paid, or at least until assured of payment. The buyer, however, will be reluctant to pay before receiving the goods, or at least before receiving title to them. Each wants assurance that the other party will complete its portion of the transaction. The letter of credit, bill of lading, and sight draft are part of a system carefully constructed to determine who bears the financial loss if one of the parties defaults at any time.

Protection against Foreign Exchange Risk

In international trade, foreign exchange risk arises from transaction exposure. If the transaction requires payment in the exporter's currency, the importer carries the foreign exchange risk. If the transaction calls for payment in the importer's currency, the exporter has the foreign exchange risk.

Transaction exposure can be hedged by the techniques described in Chapter 10, but in order to hedge, the exposed party must be certain that payment of a specified amount will be made on or near a particular date. The three key documents described in this chapter ensure both amount and time of payment and thus lay the groundwork for effective hedging.

The risk of noncompletion and foreign exchange risk are most important when the international trade is episodic, with no outstanding agreement for recurring shipments and no sustained relationship between buyer and seller. When the import/export relationship is of a recurring nature, as in the case of manufactured goods shipped weekly or monthly to a final assembly or retail outlet in another country, and when the relationship is between countries whose currencies are considered strong, the exporter may well bill the importer on open account after a normal credit check.

Financing the Trade

Most international trade involves a time lag during which funds are tied up while the merchandise is in transit. Once the risks of noncompletion and of exchange rate changes are disposed of, banks are willing to finance goods in transit. A bank can finance goods in transit, as well as goods held for sale, based on the key documents, without exposing itself to questions about the quality of merchandise or aspects of shipment.

Noncompletion Risks

In order to understand the risks associated with international trade transactions, it is helpful to understand the sequence of events in any such transaction. Exhibit 16.5 illustrates, in principle, the series of events associated with a single export transaction.

From a financial management perspective, the two primary risks associated with an international trade transaction are currency risk (discussed previously in Chapter 10) and *risk of noncompletion*. Exhibit 16.5 illustrates the traditional business problem of credit management: the exporter quotes a price, finalizes a contract, and ships the goods, losing physical control over the goods based on trust of the buyer or the promise of a bank to pay based on documents presented. The risk of default on the part of the importer—*risk of noncompletion*—is present as soon as the financing period begins, as depicted in Exhibit 16.5.

In many cases, the initial task of analyzing the creditworthiness of foreign customers is similar to procedures for analyzing domestic customers. If Ganado has had no experience with a foreign customer but the customer is a large, well-known firm in its home country, Ganado may simply ask for a bank credit report on that firm. Ganado may also talk to other firms that have had dealings with the foreign customer. If these investigations show the foreign customer (and country) to be completely trustworthy, Ganado would likely ship to them on open account, with a credit limit, just as they would for a domestic customer. This is the least costly method of handling exports because there are no heavy documentation or bank charges. However, before a regular trading relationship has been established with a new or unknown firm, Ganado must face the possibility of nonpayment for its exports or noncompletion of its imports. The risk of nonpayment can be eliminated through the use of a letter of credit issued by a creditworthy bank.

EXHIBIT 16.5   **The Trade Transaction Time Line and Structure**

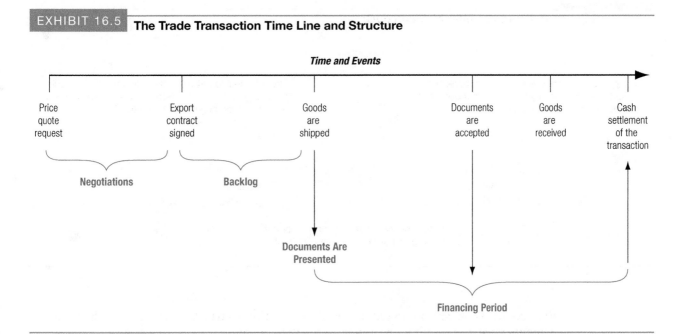

Key Documents

The three key documents described in the following pages—the letter of credit, draft, and bill of lading—constitute a system developed and modified over centuries to protect both importer and exporter from the risk of noncompletion of the trade transaction as well as to provide a means of financing. These three key trade documents are part of a carefully constructed system to determine who bears the financial loss if one of the parties defaults at any time.

Letter of Credit (L/C)

A *letter of credit* (L/C) is a document issued by a bank at the request of an importer (the applicant/buyer) by which the bank promises to pay an exporter (the beneficiary of the letter) upon presentation of documents specified in the L/C. An L/C reduces the risk of noncompletion, because the bank agrees to pay against documents rather than actual merchandise. The relationship between the three parties can be seen in Exhibit 16.6.

A beneficiary (exporter) and an applicant (importer) agree on a transaction and the importer then applies to its local bank for the issuance of an L/C. The importer's bank issues an L/C and cuts a sales contract based on its assessment of the importer's creditworthiness, or the bank might require a cash deposit or other collateral from the importer in advance. The importer's bank will want to know the type of transaction, the amount of money involved, and what documents must accompany the draft that will be drawn against the L/C.

If the importer's bank is satisfied with the credit standing of the applicant, it will issue an L/C guaranteeing to pay for the merchandise if shipped in accordance with the instructions and conditions contained in the L/C.

The essence of an L/C is the promise of the issuing bank to pay *against specified documents* that must accompany any draft drawn against the credit. The L/C is not a guarantee of the underlying commercial transaction. Indeed, the L/C is a separate transaction from any sales or

EXHIBIT 16.6   **Parties to a Letter of Credit (L/C)**

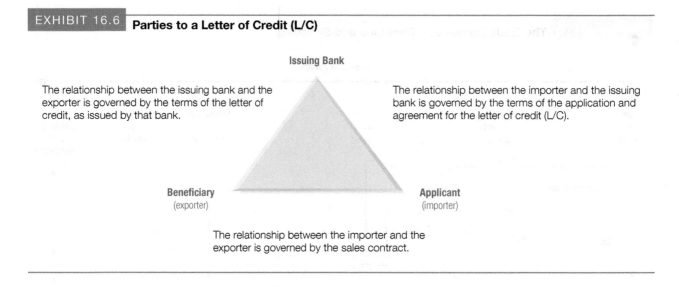

Issuing Bank

The relationship between the issuing bank and the exporter is governed by the terms of the letter of credit, as issued by that bank.

The relationship between the importer and the issuing bank is governed by the terms of the application and agreement for the letter of credit (L/C).

Beneficiary
(exporter)

Applicant
(importer)

The relationship between the importer and the exporter is governed by the sales contract.

other contracts on which it might be based. To constitute a true L/C transaction, the following elements must be present with respect to the issuing bank:

1. The issuing bank must receive a fee or other valid business consideration for issuing the L/C.
2. The bank's L/C must contain a specified expiration date or a definite maturity.
3. The bank's commitment must have a stated maximum amount of money.
4. The bank's obligation to pay must arise only on the presentation of specific documents, and the bank must not be called on to determine disputed questions of fact or law.
5. The bank's customer must have an unqualified obligation to reimburse the bank on the same condition as the bank has paid.

Commercial letters of credit are also classified based on whether they are revocable and confirmed.

- **Irrevocable versus Revocable.**   An irrevocable L/C obligates the issuing bank to honor drafts drawn in compliance with the credit and can be neither canceled nor modified without the consent of all parties, including in particular the beneficiary (exporter). A revocable L/C can be canceled or amended at any time before payment; it is intended to serve as a means of arranging payment but not as a guarantee of payment.

- **Confirmed versus Unconfirmed.**   A confirmed L/C is issued by one bank and can be confirmed by another bank, in which case the confirming bank can honor drafts drawn in compliance with the L/C. An unconfirmed L/C is the obligation only of the issuing bank. An exporter is likely to want a foreign bank's L/C confirmed by a domestic bank when the exporter has doubts about the foreign bank's ability to pay. Such doubts can arise when the exporter is unsure of the financial standing of the foreign bank, or if political or economic conditions in the foreign country are unstable. The essence of an L/C is shown in Exhibit 16.7.

Most commercial letters of credit are *documentary*, meaning that certain documents must be included with drafts drawn under their terms. Required documents usually include a bill

| EXHIBIT 16.7 | **Essence of a Letter of Credit (L/C)** |

Bank of the East, Ltd.
[Name of Issuing Bank]

Date: September 18, 2012
L/C Number 123456

Bank of the East, Ltd. hereby issues this irrevocable documentary Letter of Credit to Jones Company *[name of exporter]* for US$500,000, payable 90 days after sight by a draft drawn against Bank of the East, Ltd., in accordance with Letter of Credit number 123456.

The draft is to be accompanied by the following documents:
1. Commercial invoice in triplicate
2. Packing list
3. Clean on board order bill of lading
4. Insurance documents, paid for by buyer

At maturity Bank of the East, Ltd. will pay the face amount of the draft to the bearer of that draft.

Authorized Signature

---

of lading (discussed in more detail later in the chapter), a commercial invoice, and any of the following: consular invoice, insurance certificate or policy, and packing list.

Advantages and Disadvantages of Letters of Credit. The primary advantage of an L/C is that it reduces risk—the exporter can sell against a bank's promise to pay rather than against the promise of a commercial firm. The exporter is also in a more secure position as to the availability of foreign exchange to pay for the sale, since banks are more likely to be aware of foreign exchange conditions and rules than is the importing firm itself. If the importing country should change its foreign exchange rules during the course of a transaction, the government is likely to allow already outstanding bank letters of credit to be honored for fear of throwing its own domestic banks into international disrepute. Of course, if the L/C is confirmed by a bank in the exporter's country, the exporter avoids any problem of blocked foreign exchange.

An exporter may find that an order backed by an irrevocable L/C will facilitate obtaining pre-export financing in the home country. If the exporter's reputation for delivery is good, a local bank may lend funds to process and prepare the merchandise for shipment. Once the merchandise is shipped in compliance with the terms and conditions of the L/C, payment for the business transaction is made and funds will be generated to repay the pre-export loan.

Another advantage of an L/C to the importer is that the importer need not pay out funds until the documents have arrived at a local port or airfield and unless all conditions stated in the credit have been fulfilled.

The main disadvantages of L/Cs are the fees charged by the importer's bank for issuing its L/C and the possibility that the L/C reduces the importer's borrowing line of credit with its bank. It may, in fact, be a competitive disadvantage for the exporter to demand automatically an L/C from an importer, especially if the importer has a good credit record and there is no concern regarding the economic or political conditions of the importer's country.

In balance, though, the value of the L/C has been well established since the beginning of commerce, as detailed in *Global Finance in Practice 16.1.*

GLOBAL FINANCE IN PRACTICE 16.1

Florence—The Birthplace of Trade Financing

Merchant banking for international trade largely began in a landlocked city, Florence, Italy. In the late 13th and early 14th century as commerce grew throughout Europe and the Mediterranean, banking began to develop in both Venice and Florence.

It was a time in which commerce was still in its infancy, with the Catholic Church prohibiting many aspects of commerce, including the loaning of money in return for interest—*usury*. Although usury has come to mean the illegal activity of charging excessive rates of interest, the term originally referred to charging interest of any kind.

The *florin* is a small gold coin first minted in Florence in 1252. Named after the city, the florin flourished as a means of transacting trade across Europe in the following century. Merchants conducted their trade on a bench—a *banco*—which

gave rise to the term for the safe place in which to keep one's money.

But the coins were heavy, and if a merchant were traveling from one city or country to another to conduct trade, the weight was substantial, as was the chance of being robbed. So the merchants created the first financial derivative, a draft on the *banco*—a *letter of exchange*—which could be carried from one city to another and was recognized as a credit for florins on account at their home banco. Payment was guaranteed within three months. Of course, with the creation of banks came the first failures—bankruptcies.

From the very beginning, whether it was the loaning of money, the validity of a letter of exchange, or even the value of a currency, all were instruments or activities that involved risk, or *risque* in the Italian of the time.

Draft

A *draft*, sometimes called a *bill of exchange* (B/E), is the instrument normally used in international commerce to effect payment. A draft is simply an order written by an exporter (seller) instructing an importer (buyer) or its agent to pay a specified amount of money at a specified time. Thus, it is the exporter's formal demand for payment from the importer.

The person or business initiating the draft is known as the *maker* (also known as the *drawer* or *originator*). Normally, maker is the exporter who sells and ships the merchandise. The party to whom the draft is addressed is the *drawee*. The drawee is asked to honor the draft, that is, to pay the amount requested according to the stated terms. In commercial transactions, the drawee is either the buyer, in which case the draft is called a *trade draft*, or the buyer's bank, in which case the draft is called a *bank draft*. Bank drafts are usually drawn according to the terms of an L/C. A draft may be drawn as a bearer instrument, or it may designate a person to whom payment is to be made. This person, known as the payee, may be the drawer itself or it may be some other party such as the drawer's bank.

Negotiable Instruments. If properly drawn, drafts can become *negotiable instruments*. As such, they provide a convenient instrument for financing the international movement of the merchandise. To become a *negotiable instrument*, a draft must conform to the following requirements (Uniform Commercial Code, Section 3104(1)):

1. It must be in writing and signed by the maker or drawer.
2. It must contain an unconditional promise or order to pay a definite sum of money.
3. It must be payable on demand or at a fixed or determinable future date.
4. It must be payable to order or to bearer.

If a draft is drawn in conformity with the above requirements, a person receiving it with proper endorsements becomes a "holder in due course." This is a privileged legal status that enables the holder to receive payment despite any personal disagreements between drawee and maker because of controversy over the underlying transaction. If the drawee dishonors the draft, payment must be made to any holder in due course by any prior endorser or by the maker. This clear definition of the rights of parties who hold a negotiable instrument as a holder in due course has contributed significantly to the widespread acceptance of various forms of drafts, including personal checks.

Types of Drafts. Drafts are of two types: *sight drafts* and *time drafts*. A *sight draft* is payable on presentation to the drawee; the drawee must pay at once or dishonor the draft. A *time draft*, also called a *usance draft*, allows a delay in payment. It is presented to the drawee, who accepts it by writing or stamping a notice of acceptance on its face. Once accepted, the time draft becomes a promise to pay by the accepting party (the buyer). When a time draft is drawn on and accepted by a bank, it becomes a bankers' acceptance; when drawn on and accepted by a business firm, it becomes a *trade acceptance* (T/A).

The time period of a draft is referred to as its *tenor*. To qualify as a negotiable instrument, and so be attractive to a holder in due course, a draft must be payable on a fixed or determinable future date. For example, "60 days after sight" is a fixed date, which is established precisely at the time the draft is accepted. However, payment "on arrival of goods" is not determinable since the date of arrival cannot be known in advance. Indeed, there is no assurance that the goods will arrive at all.

Bankers' Acceptances. When a draft is accepted by a bank, it becomes a *bankers' acceptance*. As such it is the unconditional promise of that bank to make payment on the draft when it matures. In quality, the bankers' acceptance is practically identical to a marketable bank certificate of deposit (CD). The holder of a bankers' acceptance need not wait until maturity to liquidate the investment, but may sell the acceptance in the money market, where constant trading in such instruments occurs. The amount of the discount depends entirely on the credit rating of the bank that signs the acceptance, or another bank that reconfirmed the bankers' acceptance, for a fee. The total cost or *all-in cost* of using a bankers' acceptance compared to other short-term financing instruments is analyzed later in this chapter.

Bill of Lading (B/L)

The third key document for financing international trade is the *bill of lading* (B/L). The bill of lading is issued to the exporter by a common carrier transporting the merchandise. It serves three purposes: as a receipt, a contract, and a document of title.

As a receipt, the bill of lading indicates that the carrier has received the merchandise described on the face of the document. The carrier is not responsible for ascertaining that the containers hold what is alleged to be their contents, so descriptions of merchandise on bills of lading are usually short and simple. If shipping charges are paid in advance, the bill of lading will usually be stamped "freight paid" or "freight prepaid." If merchandise is shipped collect—a less common procedure internationally than domestically—the carrier maintains a lien on the goods until freight is paid.

As a contract, the bill of lading indicates the obligation of the carrier to provide certain transportation in return for certain charges. Common carriers cannot disclaim responsibility for their negligence by inserting special clauses in a bill of lading. The bill of lading may specify alternative ports in the event that delivery cannot be made to the designated port, or it may specify that the goods will be returned to the exporter at the exporter's expense.

As a document of title, the bill of lading is used to obtain payment or a written promise of payment before the merchandise is released to the importer. The bill of lading can also function as collateral against which funds may be advanced to the exporter by its local bank prior to or during shipment and before final payment by the importer.

The bill of lading is typically made payable to the order of the exporter, who thus retains title to the goods after they have been handed to the carrier. Title to the merchandise remains with the exporter until payment is received, at which time the exporter endorses the bill of lading (which is negotiable) in blank (making it a bearer instrument) or to the party making the payment, usually a bank. The most common procedure would be for payment to be advanced against a documentary draft accompanied by the endorsed order bill of lading. After paying the draft, the exporter's bank forwards the documents through bank clearing channels to the bank of the importer. The importer's bank, in turn, releases the documents to the importer after payment (sight drafts); after acceptance (time drafts addressed to the importer and marked D/A); or after payment terms have been agreed upon (drafts drawn on the importer's bank under provisions of an L/C).

Documentation in a Typical Trade Transaction

Although a trade transaction could conceivably be handled in many ways, we shall now turn to a hypothetical example that illustrates the interaction of the various documents. Assume that Ganado U.S. receives an order from a Canadian buyer. For Ganado, this will be an export financed under an L/C requiring a bill of lading, with the exporter collecting via a time draft that was accepted by the Canadian buyer's bank. Such a transaction proceeds as follows, illustrated in Exhibit 16.8.

1. The Canadian buyer (the Importer in Exhibit 16.8) places an order with Ganado (the Exporter in Exhibit 16.8), asking if Ganado is willing to ship under an L/C.

2. Ganado agrees to ship under an L/C and specifies relevant information such as prices and terms.

3. The Canadian buyer applies to its bank, Northland Bank (Bank I in Exhibit 16.8), for an L/C to be issued in favor of Ganado for the merchandise it wishes to buy.

4. Northland Bank issues the L/C in favor of Ganado and sends it to Ganado's bank, Southland Bank (Bank X in Exhibit 16.8).

5. Southland Bank advises Ganado of the opening of an L/C in Ganado's favor. Southland Bank may or may not confirm the L/C to add its own guarantee to the document.

6. Ganado ships the goods to the Canadian buyer.

7. Ganado prepares a time draft and presents it to its bank, Southland Bank. The draft is drawn (i.e., addressed to) Northland Bank in accordance with Northland Bank's L/C and accompanied by other documents as required, including the bill of lading. Ganado endorses the bill of lading in blank (making it a bearer instrument) so that title to the goods goes with the holder of the documents—Southland Bank at this point in the transaction.

8. Southland Bank presents the draft and documents to Northland Bank for acceptance. Northland Bank accepts the draft by stamping and signing it (making it a bankers' acceptance), takes possession of the documents, and promises to pay the now-accepted draft at maturity—say, 60 days.

9. Northland Bank returns the accepted draft to Southland Bank. Alternatively, Southland Bank might ask Northland Bank to accept and discount the draft. Should this occur, Northland Bank would remit the cash less a discount fee rather than return the accepted draft to Southland Bank.

10. Southland Bank, having received back the accepted draft, now a bankers' acceptance, may choose between several alternatives. Southland Bank may sell the acceptance in the open market at a discount to an investor, typically a corporation or financial institution with with excess cash it wants to invest for a short period of time. Southland Bank may also hold the acceptance in its own portfolio.

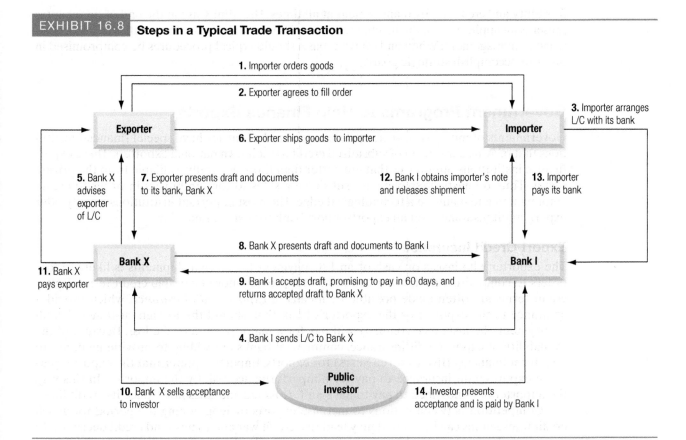

EXHIBIT 16.8   **Steps in a Typical Trade Transaction**

1. Importer orders goods

2. Exporter agrees to fill order

3. Importer arranges L/C with its bank

Exporter   **Importer**

6. Exporter ships goods to importer

5. Bank X advises exporter of L/C

7. Exporter presents draft and documents to its bank, Bank X

12. Bank I obtains importer's note and releases shipment

13. Importer pays its bank

8. Bank X presents draft and documents to Bank I

Bank X   **Bank I**

11. Bank X pays exporter

9. Bank I accepts draft, promising to pay in 60 days, and returns accepted draft to Bank X

4. Bank I sends L/C to Bank X

Public Investor

10. Bank X sells acceptance to investor

14. Investor presents acceptance and is paid by Bank I

11. If Southland Bank discounted the acceptance with Northland Bank (mentioned in Step 9) or discounted it in the local money market, Southland Bank will transfer the proceeds less any fees and discount to Ganado. Another possibility would be for Ganado itself to take possession of the acceptance, hold it for 60 days, and present it for collection. Normally, however, exporters prefer to receive the discounted cash value of the acceptance at once rather than wait for the acceptance to mature and receive a slightly greater amount of cash at a later date.

12. Northland Bank notifies the Canadian buyer of the arrival of the documents. The Canadian buyer signs a note or makes some other agreed upon plan to pay Northland Bank for the merchandise in 60 days, Northland Bank releases the underlying documents so that the Canadian buyer can obtain physical possession of the shipment at once.

13. After 60 days, Northland Bank receives funds from the Canadian buyer to pay the maturing acceptance.

14. On the same day, the 60th day after acceptance, the holder of the matured acceptance presents it for payment and receives its face value. The holder may present it directly to Northland Bank, or return it to Southland Bank and have Southland Bank collect it through normal banking channels.

Although this is a typical transaction involving an L/C, few international trade transactions are probably ever truly typical. Business, and more specifically international business, requires flexibility and creativity by management at all times. The Mini-Case at the end of this chapter presents an application of the mechanics of a real business situation. The result is a classic challenge to management: When and on what basis should typical procedures be compromised in order to accomplish strategic goals?

Government Programs to Help Finance Exports

Governments of most export-oriented industrialized countries have special financial institutions that provide some form of subsidized credit to their own national exporters. These export finance institutions offer terms that are better than those generally available from the private sector. Thus, domestic taxpayers are subsidizing sales to foreign buyers in order to create employment and maintain a technological edge. The most important institutions usually offer export credit insurance and an export-import bank for export financing.

Export Credit Insurance

The exporter who insists on cash or an L/C payment for foreign shipments is likely to lose orders to competitors from other countries that provide more favorable credit terms. Better credit terms are often made possible by means of *export credit insurance*, which provides assurance to the exporter or the exporter's bank that, should the foreign customer default on payment, the insurance company will pay for a major portion of the loss. Because of the availability of export credit insurance, commercial banks are willing to provide medium- to long-term financing (five to seven years) for exports. Importers prefer that the exporter purchase export credit insurance to pay for nonperformance risk by the importer. In this way, the importer does not need to pay to have an L/C issued and does not reduce its credit line.

Competition between nations to increase exports by lengthening the period for which credit transactions can be insured may lead to a credit war and to unsound credit decisions. To prevent such an unhealthy development, a number of leading trading nations joined together in 1934 to create the Berne Union (officially, the Union d'Assureurs des Credits Internationaux) for the purpose of establishing a voluntary international understanding on export credit terms. The Berne Union recommends maximum credit terms for many items including, for example, heavy capital goods (five years), light capital goods (three years), and consumer durable goods (one year).

In the United States, export credit insurance is provided by the *Foreign Credit Insurance Association* (FCIA). This is an unincorporated association of private commercial insurance companies operating in cooperation with the Export-Import Bank (discussed in the next section). The FCIA provides policies protecting U.S. exporters against the risk of nonpayment by foreign debtors as a result of commercial and political risks. Losses due to commercial risk are those that result from the insolvency or protracted payment default of the buyer. Political losses arise from actions of governments beyond the control of buyer or seller.

Export-Import Bank and Export Financing

The *Export-Import Bank (Eximbank)* is another independent agency of the U.S. government, established in 1934 to stimulate and facilitate the foreign trade of the United States. Interestingly, the Eximbank was originally created primarily to facilitate exports to the Soviet Union. In 1945, the Eximbank was re-chartered "to aid in financing and to facilitate exports and imports and the exchange of commodities between the United States and any foreign country or the agencies or nationals thereof."

The Eximbank facilitates the financing of U.S. exports through various loan guarantee and insurance programs. The Eximbank guarantees repayment of medium-term (181 days to five years) and long-term (five years to ten years) export loans extended by U.S. banks to foreign borrowers. The Eximbank's medium- and long-term, direct-lending operation is based on participation with private sources of funds. Essentially, the Eximbank lends dollars to borrowers outside the United States for the purchase of U.S. goods and services. Proceeds of such loans are paid to U.S. suppliers. The loans themselves are repaid with interest in dollars to the Eximbank. The Eximbank requires private participation in these direct loans in order to: (1) ensure that it complements rather than competes with private sources of export financing; (2) spread its resources more broadly; and (3) ensure that private financial institutions will continue to provide export credit.

The Eximbank also guarantees lease transactions, finances the costs involved in the preparation by U.S. firms of engineering, planning, and feasibility studies for non-U.S. clients on large capital projects; and supplies counseling for exporters, banks, or others needing help in finding financing for U.S. goods.

Trade Financing Alternatives

In order to finance international trade receivables, firms use the same financing instruments that they use for domestic trade receivables, plus a few specialized instruments that are only available for financing international trade. Exhibit 16.9 identifies the main short-term financing instruments and their costs.

Bankers' Acceptances

Bankers' acceptances, described earlier in this chapter, can be used to finance both domestic and international trade receivables. Exhibit 16.9 shows that bankers' acceptances earn a yield comparable to other money market instruments, especially marketable bank certificates of deposit. However, the all-in cost to a firm of creating and discounting a bankers' acceptance also depends upon the commission charged by the bank that accepts the firm's draft.

The first owner of the bankers' acceptance created from an international trade transaction will be the exporter, who receives the accepted draft back after the bank has stamped it "accepted." The exporter may hold the acceptance until maturity and then collect. On an acceptance of, say, $100,000 for three months the exporter would receive the face amount less the bank's acceptance commission of 1.5% per annum:

Face amount of the acceptance	$100,000
Less 1.5% per annum commission for three months	− 375 (.015 × 3/12 × $100,000)
Amount received by exporter in three months	$ 99,625

EXHIBIT 16.9   **Instruments for Financing Short-Term Domestic and International Trade Receivables**

Instrument	Cost or Yield for 3-Month Maturity
Bankers' acceptances*	1.14% yield annualized
Trade acceptances*	1.17% yield annualized
Factoring	Variable rate but much higher cost than bank credit lines
Securitization	Variable rate but competitive with bank credit lines
Bank credit lines	4.25% plus points (fewer points if covered by export credit insurance)
Commercial paper*	1.15% yield annualized

*These instruments compete with 3-month marketable bank time certificates of deposit that yield 1.17%.

Alternatively, the exporter may "discount"—that is, sell at a reduced price—the acceptance to its bank in order to receive funds at once. The exporter will then receive the face amount of the acceptance less both the acceptance fee and the going market rate of discount for bankers' acceptances. If the discount rate were 1.14% per annum as shown in Exhibit 16.9, the exporter would receive the following:

Face amount of the acceptance	$100,000
Less 1.5% per annum commission for three months	− 375 (0.015 × 3/12 × $100,000)
Less 1.14% per annum discount rate for three months	− 285 (0.0114 × 3/12 × $100,000)
Amount received by exporter at once	$ 99,340

Therefore, the annualized all-in cost of financing this bankers' acceptance is as follows:

$$\frac{\text{Commission} + \text{Discount}}{\text{Proceeds}} \times \frac{360}{90} = \frac{\$375 + \$285}{\$99,340} \times \frac{360}{90} = 0.0266 \text{ or } 2.66\%.$$

The discounting bank may hold the acceptance in its own portfolio, earning for itself the 1.14% per annum discount rate, or the acceptance may be resold in the acceptance market to portfolio investors. Investors buying bankers' acceptances provide the funds that finance the transaction.

Trade Acceptances

Trade acceptances are similar to bankers' acceptances except that the accepting entity is a commercial firm, like General Motors Acceptance Corporation (GMAC), rather than a bank. The cost of a trade acceptance depends on the credit rating of the accepting firm plus the commission it charges. Like bankers' acceptances, trade acceptances are sold at a discount to banks and other investors at a rate that is competitive with other money market instruments (see Exhibit 16.9).

Factoring

Specialized firms, known as factors, purchase receivables at a discount on either a non-recourse or recourse basis. Non-recourse means that the factor assumes the credit, political, and foreign exchange risk of the receivables it purchases. Recourse means that the factor can give back receivables that are not collectable. Since the factor must bear the cost and risk of assessing the creditworthiness of each receivable, the cost of factoring is usually quite high. It is more than borrowing at the prime rate plus points.

The all-in cost of factoring non-recourse receivables is similar in structure to acceptances. The factor charges a commission to cover the non-recourse risk, typically 1.5%–2.5%, plus interest deducted as a discount from the initial proceeds. On the other hand, the firm selling the non-recourse receivables avoids the cost of determining the creditworthiness of its customers. It also does not have to show debt borrowed to finance these receivables on its balance sheet. Furthermore, the firm avoids both foreign exchange and political risk on these non-recourse receivables. *Global Finance in Practice 16.2* provides an example of the costs.

Securitization

The securitization of export receivables for financing trade is an attractive supplement to bankers' acceptance financing and factoring. A firm can securitize its export receivables by selling them to a legal entity established to create marketable securities based on a package of

GLOBAL FINANCE IN PRACTICE 16.2

Factoring in Practice

A U.S.-based manufacturer that may have suffered significant losses during first the global credit crisis and the following global recession is cash-short. Sales, profits, and cash flows have fallen. The company is now struggling to service its high levels of debt. It does, however, have a number of new sales agreements. It is considering factoring one of its biggest new sales, a sale for $5 million to a Japanese company. The receivable is due in 90 days. After contacting a factoring agent, it is quoted the following numbers.

Face amount of receivable	$5,000,000
Non-recourse fee (1.5%)	−75,000

Factoring fee	
(2.5% per month × 3 months)	−375,000
Net proceeds on sale (received now)	$4,550,000

If the company wishes to factor its receivable it will net $4.55 million, 91% of the face amount. Although this may at first sight appear expensive, the firm would net the proceeds in cash up-front, not having to wait 90 days for payment. And it would not be responsible for collecting on the receivable. If the firm were able to "factor-in" the cost of factoring in the initial sale, all the better. Alternatively, it might offer a discount for cash paid in the first 10 days after shipment.

individual export receivables. An advantage of this technique is to remove the export receivables from the exporter's balance sheet because they have been sold without recourse.

The receivables are normally sold at a discount. The size of the discount depends on four factors:

1. The historic collection risk of the exporter
2. The cost of credit insurance
3. The cost of securing the desirable cash flow stream to the investors
4. The size of the financing and services fees

Securitization is more cost effective if there is a large value of transactions with a known credit history and default probability. A large exporter could establish its own securitization entity. While the initial setup cost is high, the entity can be used on an ongoing basis. Alternatively, smaller exporters could use a common securitization entity provided by a financial institution, thereby saving the expensive setup costs.

Bank Credit Lines

A firm's bank credit line can typically be used to finance, up to a fixed upper limit, say 80%, of accounts receivable. Export receivables can be eligible for inclusion in bank credit line financing. However, credit information on foreign customers may be more difficult to collect and assess. If a firm covers its export receivables with export credit insurance, it can greatly reduce the credit risk of those receivables. This insurance enables the bank credit line to cover more export receivables and lower the interest rate for that coverage. Of course, any foreign exchange risk must be handled by the transaction exposure techniques described in Chapter 10.

The cost of using a bank credit line is usually the prime rate of interest plus points to reflect a particular firm's credit risk. As usual, 100 points is equal to 1%. In the United States, borrowers are also expected to maintain a compensating deposit balance at the lending institution. In Europe and many other places, lending is done on an overdraft basis. An overdraft agreement allows a firm to overdraw its bank account up to the limit of its credit line. Interest at prime plus points is based only on the amount of overdraft borrowed. In either case, the all-in cost of bank borrowing using a credit line is higher than acceptance financing as shown in Exhibit 16.9.

Commercial Paper

A firm can issue commercial paper—unsecured promissory notes—to fund its short-term financing needs, including both domestic and export receivables. However, it is only the large well-known firms with favorable credit ratings that have access to either the domestic or euro commercial paper market. As shown in Exhibit 16.9, commercial paper interest rates lie at the low end of the yield curve.

Forfaiting: Medium- and Long-Term Financing

Forfaiting is a specialized technique to eliminate the risk of nonpayment by importers in instances where the importing firm and/or its government is perceived by the exporter to be too risky for open account credit. The name of the technique comes from the French *à forfait*, a term that implies "to forfeit or surrender a right."

Role of the Forfaiter

The essence of forfaiting is the non-recourse sale by an exporter of bank-guaranteed promissory notes, bills of exchange, or similar documents received from an importer in another country. The exporter receives cash at the time of the transaction by selling the notes or bills at a discount from their face value to a specialized finance firm called a forfaiter. The forfaiter arranges the entire operation prior to the execution of the transaction. Although the exporting firm is responsible for the quality of delivered goods, it receives a clear and unconditional cash payment at the time of the transaction. All political and commercial risk of nonpayment by the importer is carried by the guaranteeing bank. Small exporters, who trust their clients to pay, find the forfaiting technique invaluable because it eases cash flow problems.

During the Soviet era, expertise in the technique was centered in German and Austrian banks, which used forfaiting to finance sales of capital equipment to eastern European, "Soviet Bloc," countries. British, Scandinavian, Italian, Spanish, and French exporters have now adopted the technique, but U.S. and Canadian exporters are reported to be slow to use forfaiting, possibly because they are suspicious of its simplicity and lack of complex documentation. Nevertheless, some American firms now specialize in the technique, and the Association of Forfaiters in the Americas (AFIA) has more than 20 members. Major export destinations financed via the forfaiting technique are Asia, Eastern Europe, the Middle East, and Latin America.

A Typical Forfaiting Transaction

A typical forfaiting transaction involves five parties, as shown in Exhibit 16.10. The steps in the process are as follows:

Step 1: Agreement. Importer and exporter agree on a series of imports to be paid for over a period of time, typically three to five years. However, periods as long as 10 years and as short as 180 days have been financed using the technique. The normal minimum size for a transaction is $100,000. The importer agrees to make periodic payments, often against progress on delivery or completion of a project.

Step 2: Commitment. The forfaiter promises to finance the transaction at a fixed discount rate, with payment to be made when the exporter delivers to the forfaiter the appropriate promissory notes or other specified paper. The agreed-upon discount rate is based on the cost of funds in the euromarket, usually on LIBOR for the average life of the transaction, plus a margin over LIBOR to reflect the perceived risk in the deal. This risk premium is influenced

EXHIBIT 16.10 **Typical Forfaiting Transaction**

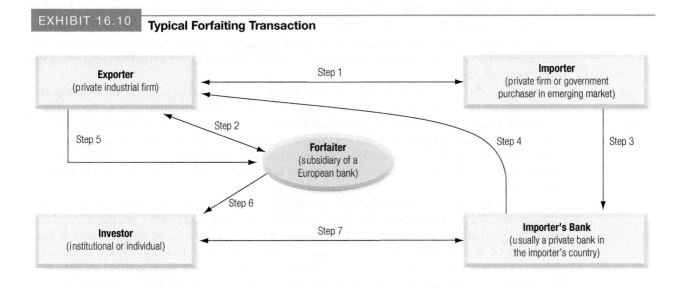

by the size and tenor of the deal, country risk, and the quality of the guarantor institution. On a five-year deal, for example, with 10 semiannual payments, the rate used would be based on the 2.25-year LIBOR rate. This discount rate is normally added to the invoice value of the transaction so that the cost of financing is ultimately borne by the importer. The forfaiter charges an additional commitment fee of from 0.5% per annum to as high as 6.0% per annum from the date of its commitment to finance until receipt of the actual discount paper issued in accordance with the finance contract. This fee is also normally added to the invoice cost and passed on to the importer.

Step 3: Aval or Guarantee. The importer obligates itself to pay for its purchases by issuing a series of promissory notes, usually maturing every six or twelve months, against progress on delivery or completion of the project. These promissory notes are first delivered to the importer's bank where they are endorsed (that is, guaranteed) by that bank. In Europe, this unconditional guarantee is referred to as an *aval*, which translates into English as "backing." At this point, the importer's bank becomes the primary obligor in the eyes of all subsequent holders of the notes. The bank's *aval* or guarantee must be irrevocable, unconditional, divisible, and assignable. Because U.S. banks do not issue avals, U.S. transactions are guaranteed by a standby letter of credit (L/C), which is functionally similar to an aval but more cumbersome. For example, L/Cs can normally be transferred only once.

Step 4: Delivery of Notes. The now-endorsed promissory notes are delivered to the exporter.

Step 5: Discounting. The exporter endorses the notes "without recourse" and discounts them with the forfaiter, receiving the agreed-upon proceeds. Proceeds are usually received two days after the documents are presented. By endorsing the notes "without recourse," the exporter frees itself from any liability for future payment on the notes and thus receives the discounted proceeds without having to worry about any further payment difficulties.

Step 6: Investment. The forfaiting bank either holds the notes until full maturity as an investment or endorses and rediscounts them in the international money market. Such subsequent sale by the forfaiter is usually without recourse. The major rediscount markets are in London and Switzerland, plus New York for notes issued in conjunction with Latin American business.

Step 7: Maturity. At maturity, the investor holding the notes presents them for collection to the importer or to the importer's bank. The promise of the importer's bank is what gives the documents their value.

In effect, the forfaiter functions as both a money market firm (e.g., a lender of short-term financing) and a specialist in packaging financial deals involving country risk. As a money market firm, the forfaiter divides the discounted notes into appropriately sized packages and resells them to various investors having different maturity preferences. As a country risk specialist, the forfaiter assesses the risk that the notes will eventually be paid by the importer or the importer's bank and puts together a deal that satisfies the needs of both exporter and importer.

Success of the forfaiting technique springs from the belief that the aval—the guarantee of a commercial bank—can be depended on. Although commercial banks are the normal and preferred guarantors, guarantees by government banks or government ministries of finance are accepted in some cases. On occasion, large commercial enterprises have been accepted as debtors without a bank guarantee. An additional aspect of the technique is that the endorsing bank's aval is perceived to be an "off balance sheet" obligation—the debt is presumably not considered by others in assessing the financial structure of the commercial banks.

SUMMARY POINTS

- International trade takes place between three categories of relationships: unaffiliated unknown parties, unaffiliated known parties, and affiliated parties.

- International trade transactions between affiliated parties typically do not require contractual arrangements or external financing. Trade transactions between unaffiliated parties typically require contracts and some type of external financing, such as that available through letters of credit.

- The basic procedure of financing international trade rests on the interrelationship between three key documents: the letter of credit (L/C), the bill of lading, and the draft.

- In the L/C, the bank substitutes its credit for that of the importer and promises to pay if certain documents are submitted to the bank. The exporter may now rely on the promise of the bank rather than on the promise of the importer.

- The exporter typically ships with a bill of lading, attaches the bill of lading to a draft that orders payment from the importer's bank, and presents these documents, plus any of a number of additional documents, through its own bank to the importer's bank.

- If the documents are in order, the importer's bank either pays the draft (a sight draft) or accepts the draft (a time draft). In the latter case, the bank promises to pay in the future. At this step, the importer's bank acquires title to the merchandise through the bill of lading and releases the merchandise to the importer.

- The total costs to an exporter of entering a foreign market include the transaction costs of the trade financing, the import and export duties and tariffs applied by exporting and importing nations, and the costs of foreign market penetration.

- Trade financing uses the same financing instruments as does domestic receivables financing, plus some specialized instruments that are only available for financing international trade. A popular instrument for short-term financing is a bankers' acceptance.

- Other short-term financing instruments with a domestic counterpart are trade acceptances, factoring, securitization, bank credit lines (usually covered by export credit insurance), and commercial paper.

Crosswell International and Brazil[1]

Crosswell International is a U.S.-based manufacturer and distributor of health care products, including children's diapers. Crosswell has been approached by Leonardo Sousa, the president of Material Hospitalar, a distributor of health care products throughout Brazil. Sousa is interested in distributing Crosswell's major diaper product, *Precious Diapers*, but only if an acceptable arrangement regarding pricing and payment terms can be reached.

Exporting to Brazil

Crosswell's manager for export operations, Geoff Mathieux, followed up the preliminary discussions by putting together an estimate of export costs and pricing for discussion purposes with Sousa. Crosswell needs to know all of the costs and pricing assumptions for the entire supply and value chain as it reaches the consumer. Mathieux believes it critical that any arrangement that Crosswell enters into results in a price to consumers in the Brazilian marketplace that is both fair to all parties involved and competitive, given the market niche Crosswell hopes to penetrate. This first cut on pricing *Precious Diapers* into Brazil is presented in Exhibit A.

Crosswell proposes to sell the basic diaper line to the Brazilian distributor for $34.00 per case, *FAS (free alongside ship)* Miami docks. This means that the seller, Crosswell, agrees to cover all costs associated with getting the diapers to the Miami docks. The costs of loading the diapers onto the ship, of the actual shipping (freight), and of the associated documents is $4.32 per case. The running subtotal, $38.32 per case, is termed *CFR (cost and freight)*. Finally, the insurance expenses related to the potential loss of the goods while in transit to final port of destination, export insurance, are $0.86 per case. The total *CIF (cost, insurance, and freight)* is $39.18 per case, or 97.95 Brazilian real per case, assuming an exchange rate of 2.50 Brazilian real (R$) per U.S. dollar ($). In summary, the CIF cost of R$97.95 is the price charged by the exporter to the importer on arrival in Brazil, and is calculated as follows:

$$CIF = FAS + freight + export\ insurance$$
$$= (\$34.00 + \$4.32 + \$0.86) \times R\$2.50/\$$$
$$= R\$97.95.$$

The actual cost to the distributor in getting the diapers through the port and customs warehouses must also be calculated in terms of what Leonardo Sousa's costs are in reality. The various fees and taxes detailed in Exhibit A

raise the fully landed cost of the *Precious Diapers* to R$107.63 per case. The distributor would now bear storage and inventory costs totaling R$8.33 per case, which would bring the costs to R$115.96. The distributor then adds a margin for distribution services of 20% (R$23.19), raising the price as sold to the final retailer to R$139.15 per case.

Finally, the retailer (a supermarket or other retailer of consumer health care products) would include its expenses, taxes, and markup to reach the final shelf price to the customer of R$245.48 per case. This final retail price estimate now allows both Crosswell and Material Hospitalar to evaluate the price competitiveness of the *Precious Ultra-Thin Diaper* in the Brazilian marketplace, and provides a basis for further negotiations between the two parties.

The *Precious Ultra-Thin Diaper* will be shipped via container. Each container will hold 968 cases of diapers. The costs and prices in Exhibit A are calculated on a per case basis, although some costs and fees are assessed by container.

Mathieux provides the export price quotation shown in Exhibit A, an outline of a potential representation agreement (for Sousa to represent Crosswell's product lines in the Brazilian marketplace), and payment and credit terms to Leonardo Sousa. Crosswell's payment and credit terms are that Sousa either pay in full in cash in advance, or remit a confirmed irrevocable documentary L/C with a time draft specifying a tenor of 60 days.

Crosswell also requests from Sousa financial statements, banking references, foreign commercial references, descriptions of regional sales forces, and sales forecasts for the Precious Diaper line. These last requests allow Crosswell to assess Material Hospitalar's ability to be a dependable, creditworthy, and capable long-term partner and representative of the firm in the Brazilian marketplace. The discussions that follow focus on finding acceptable common ground between the two parties and on working to increase the competitiveness of the Precious Diaper product line in the Brazilian marketplace.

Crosswell's Proposal

The proposed sale by Crosswell to Material Hospitalar, at least in the initial shipment, is for 10 containers of 968 cases of diapers at $39.18 per case, CIF Brazil, payable in U.S. dollars. This is a total invoice amount of $379,262.40. Payment terms are that a confirmed L/C will be required of Material Hospitalar on a U.S. bank. The payment will

EXHIBIT A	Export Pricing for the Precious Diaper Line to Brazil

The *Precious Ultra-Thin Diaper* will be shipped via container. Each container will hold 968 cases of diapers. The costs and prices below are calculated on a per case basis, although some costs and fees are assessed per container.

Exports Costs & Pricing to Brazil	Per Case	Rates & Calculation
FAS price per case, Miami	$34.00	
Freight, loading & documentation	4.32	$4,180 per container/968 = $4.32
CFR price per case, Brazilian port (Santos)	$38.32	
Export insurance	0.86	2.25% of CIF
CIF to Brazilian port	$39.18	
CIF to Brazilian port, in Brazilian real	R$97.95	2.50 Real/US$ × $39.18
Brazilian Importation Costs		
Import duties	1.96	2.00% of CIF
Merchant marine renovation fee	2.70	25.00% of freight
Port storage fees	1.27	1.30% of CIF
Port handling fees	0.01	R$12 per container
Additional handling fees	0.26	20.00% of storage & handling
Customs brokerage fees	1.96	2.00% of CIF
Import license fee	0.05	R$50 per container
Local transportation charges	1.47	1.50% of CIF
Total cost to distributor in real	R$107.63	
Distributor's Costs & Pricing		
Storage cost	1.47	1.50% of CIF × months
Cost of financing diaper inventory	6.86	7.00% of CIF × months
Distributor's margin	23.19	20.00% of Price + storage + financing
Price to retailer in real	R$139.15	
Brazilian Retailer Costs & Pricing		
Industrial product tax (IPT)	20.87	15.00% of price to retailer
Mercantile circulation services tax (MCS)	28.80	18.00% of price + IPT
Retailer costs and markup	56.65	30.00% of price + IPT + MCS
Price to consumer in real	R$245.48	

Diaper Prices to Consumers	Diapers per Case	Price per Diaper
Small size	352	R$0.70
Medium size	256	R$0.96
Large size	192	R$1.28

be based on a time draft of 60 days, presentation to the bank for acceptance with other documents on the date of shipment. Both the exporter and the exporter's bank will expect payment from the importer or the importer's bank 60 days from this date of shipment.

What Should Crosswell Expect?

Assuming Material Hospitalar acquires the L/C and it is confirmed by Crosswell's bank in the United States, Crosswell will ship the goods after the initial agreement, say 15 days, as illustrated in Exhibit B.

| EXHIBIT B | **Export Payment Terms on Crosswell's Export to Brazil** |

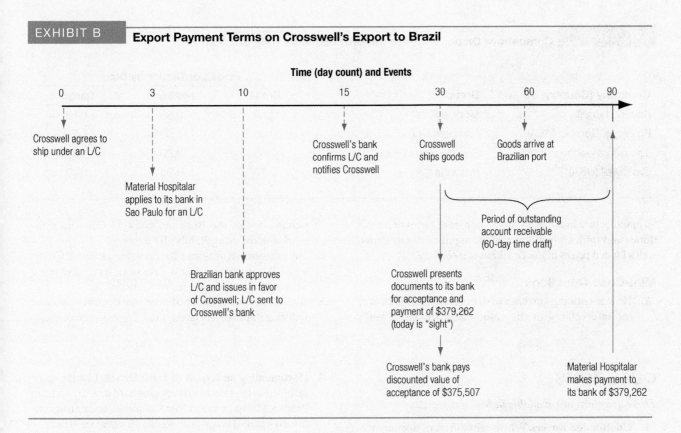

Simultaneous with the shipment, when Crosswell has lost physical control over the goods, Crosswell will present the bill of lading (acquired at the time of shipment) with the other needed documents to its bank requesting payment. Because the export is under a confirmed L/C, assuming all documents are in order, Crosswell's bank will give Crosswell two choices:

1. Wait the full time period of the time draft of 60 days and receive the entire payment in full ($379,262.40).

2. Receive the discounted value of this amount today. The discounted amount, assuming U.S. dollar interest rate of 6.00% per annum (1.00% per 60 days):

$$\frac{\$379,262.40}{(1 + 0.01)} = \frac{\$379,262.40}{1.01} = \$375,507.33.$$

Because the invoice is denominated in U.S. dollars, Crosswell need not worry about currency value changes (currency risk). And because its bank has confirmed the L/C, it is protected against changes or deteriorations in Material Hospitalar's ability to pay on the future date.

What Should Material Hospitalar Expect?

Material Hospitalar will receive the goods on or before day 60. It will then move the goods through its distribution system to retailers. Depending on the payment terms between Material Hospitalar and its buyers (retailers), it could either receive cash or terms for payment for the goods. Because Material Hospitalar purchased the goods via the 60-day time draft and an L/C from its Brazilian bank, total payment of $379,262.40 is due on day 90 (shipment and presentation of documents was on 30 + 60 day time draft) to the Brazilian bank. Material Hospitalar, because it is a Brazilian-based company and has agreed to make payment in U.S. dollars (foreign currency), carries the currency risk of the transaction.

Crosswell/Material Hospitalar's Concern

The concern the two companies hold, however, is that the total price to the consumer in Brazil, R$245.48 per case, or R$0.70/diaper (small size), is too high. The major competitors in the Brazilian market for premium quality diapers, Kenko do Brasil (Japan), Johnson and Johnson (U.S.), and Procter and Gamble (U.S.), are cheaper (see Exhibit C). The competitors

EXHIBIT C   Competitive Diaper Prices in the Brazilian Market (in Brazilian real)

| Company (Country) | Brand | Price per Diaper by Size | | |
		Small	Medium	Large
Kenko (Japan)	Monica Plus	0.68	0.85	1.18
Procter & Gamble (USA)	Pampers Uni	0.65	0.80	1.08
Johnson & Johnson (USA)	Sempre Seca Plus	0.65	0.80	1.08
Crosswell (USA)	Precious	0.70	0.96	1.40

all manufacture in-country, thus avoiding the series of import duties and tariffs, which have added significantly to Crosswell's landed prices in the Brazilian marketplace.

Mini-Case Questions

1. How are pricing, currency of denomination, and financing interrelated in the value-chain for Crosswell's penetration of the Brazilian market? Can you summarize them using Exhibit B?

2. How important is Sousa to the value-chain of Crosswell? What worries might Crosswell have regarding Sousa's ability to fulfill his obligations?

3. If Crosswell is to penetrate the market, some way of reducing its prices will be required. What do you suggest?

QUESTIONS

These questions are available in MyFinanceLab.

1. **Unaffiliated Buyers.** Why might different documentation be used for an export to a non-affiliated foreign buyer who is a new customer as compared to an export to a non-affiliated foreign buyer to whom the exporter has been selling for many years?

2. **Affiliated Buyers.** For what reason might an exporter use standard international trade documentation (letter of credit, draft, order bill of lading) on an intrafirm export to its parent or sister subsidiary?

3. **Related Party Trade.** What reasons can you give for the observation that intrafirm trade is now greater than trade between non-affiliated exporters and importers?

4. **Documents.** Explain the difference between a letter of credit (L/C) and a draft. How are they linked?

5. **Risks.** What is the major difference between "currency risk" and "risk of noncompletion"? How are these risks handled in a typical international trade transaction?

6. **Letter of Credit.** Identify each party to a letter of credit (L/C) and indicate its responsibility.

7. **Confirmed Letter of Credit.** Why would an exporter insist on a confirmed letter of credit?

8. **Documenting an Export of Hard Drives.** List the steps involved in the export of computer hard disk drives from Penang, Malaysia, to San Jose, California, using an unconfirmed letter of credit authorizing payment on sight.

9. **Documenting an Export of Lumber from Portland to Yokohama.** List the steps involved in the export of lumber from Portland, Oregon, to Yokohama, Japan, using a confirmed letter of credit, payment to be made in 120 days.

10. **Governmentally Supplied Credit.** Various governments have established agencies to insure against nonpayment for exports and/or to provide export credit. This shifts credit risk away from private banks and to the citizen taxpayers of the country whose government created and backs the agency. Why would such an arrangement be of benefit to the citizens of that country?

PROBLEMS

These problems are available in MyFinanceLab.

1. **Nikken Microsystems (A).** Assume Nikken Microsystems has sold Internet servers to Telecom España for €700,000. Payment is due in three months and will be made with a trade acceptance from Telecom España Acceptance. The acceptance fee is 1.0% per annum of the face amount of the note. This acceptance will be sold at a 4% per annum discount. What is the annualized percentage all-in cost in euros of this method of trade financing?

2. **Nikken Microsystems (B).** Assume that Nikken Microsystems prefers to receive U.S. dollars rather than euros for the trade transaction described in Problem 1. It is considering two alternatives: (1) sell the acceptance for euros at once and convert the euros immediately to U.S. dollars at the spot rate of exchange of $1.00/€ or (2) hold the euro acceptance until maturity but at the start sell the expected euro proceeds forward for dollars at the 3-month forward rate of $1.02/€.

   a. What are the U.S. dollar net proceeds received at once from the discounted trade acceptance in alternative 1?

   b. What are the U.S. dollar net proceeds received in three months in alternative 2?

   c. What is the break-even investment rate that would equalize the net U.S. dollar proceeds from both alternatives?

   d. Which alternative should Nikken Microsystems choose?

3. **Motoguzzie (A).** Motoguzzie exports large-engine motorcycles (greater than 700cc) to Australia and invoices its customers in U.S. dollars. Sydney Wholesale Imports has purchased $3,000,000 of merchandise from Motoguzzie, with payment due in six months. The payment will be made with a bankers' acceptance issued by Charter Bank of Sydney at a fee of 1.75% per annum. Motoguzzie has a weighted average cost of capital of 10%. If Motoguzzie holds this acceptance to maturity, what is its annualized percentage all-in cost?

4. **Motoguzzie (B).** Assuming the facts in Problem 3, Bank of America is now willing to buy Motoguzzie's bankers' acceptance for a discount of 6% per annum. What would be Motoguzzie's annualized percentage all-in cost of financing its $3,000,000 Australian receivable?

5. **Nakatomi Toyota.** Nakatomi Toyota buys its cars from Toyota Motors (U.S.), and sells them to U.S. customers. One of its customers is EcoHire, a car rental firm that buys cars from Nakatomi Toyota at a wholesale price. Final payment is due to Nakatomi Toyota in six months. EcoHire has bought $200,000 worth of cars from Nakatomi, with a cash down payment of $40,000 and the balance due in six months without any interest charged as a sales incentive. Nakatomi Toyota will have the EcoHire receivable accepted by Alliance Acceptance for a 2% fee, and then sell it at a 3% per annum discount to Wells Fargo Bank.

   a. What is the annualized percentage all-in cost to Nakatomi Toyota?

   b. What are Nakatomi's net cash proceeds, including the cash down payment?

6. **Forfaiting at Umaru Oil (Nigeria).** Umaru Oil of Nigeria has purchased $1,000,000 of oil drilling equipment from Gunslinger Drilling of Houston, Texas. Umaru Oil must pay for this purchase over the next five years at a rate of $200,000 per year due on March 1 of each year.

   Bank of Zurich, a Swiss forfaiter, has agreed to buy the five notes of $200,000 each at a discount. The discount rate would be approximately 8% per annum based on the expected 3-year LIBOR rate plus 200 basis points, paid by Umaru Oil. Bank of Zurich would also charge Umaru Oil an additional commitment fee of 2% per annum from the date of its commitment to finance until receipt of the actual discounted notes issued in accordance with the financing contract. The $200,000 promissory notes will come due on March 1 in successive years.

   The promissory notes issued by Umaru Oil will be endorsed by their bank, Lagos City Bank, for a 1% fee and delivered to Gunslinger Drilling. At this point, Gunslinger Drilling will endorse the notes without recourse and discount them with the forfaiter, Bank of Zurich, receiving the full $200,000 principal amount. Bank of Zurich will sell the notes by rediscounting them to investors in the international money market without recourse. At maturity, the investors holding the notes will present them for collection at Lagos City Bank. If Lagos City Bank defaults on payment, the investors will collect on the notes from Bank of Zurich.

   a. What is the annualized percentage all-in cost to Umaru Oil of financing the first $200,000 note due March 1, 2011?

   b. What might motivate Umaru Oil to use this relatively expensive alternative for financing?

7. **Sunny Coast Enterprises (A).** Sunny Coast Enterprises has sold a combination of films and DVDs to Hong Kong Media Incorporated for US$100,000, with payment due in six months. Sunny Coast Enterprises has the following alternatives for financing this receivable: (1) Use its bank credit line. Interest would be at the prime rate of 5% plus 150 basis points per annum. Sunny Coast Enterprises would need to maintain a compensating balance of 20% of the loan's face amount. No interest will be paid on the compensating balance by the bank. (2) Use its bank credit line but purchase export credit insurance for a 1% fee. Because of the reduced risk, the bank interest rate would be reduced to 5% per annum without any points.

   a. What are the annualized percentage all-in costs of each alternative?

   b. What are the advantages and disadvantages of each alternative?

   c. Which alternative would you recommend?

8. **Sunny Coast Enterprises (B).** Sunny Coast Enterprises has been approached by a factor that offers to purchase the Hong Kong Media Imports receivable at a 16% per annum discount plus a 2% charge for a non-recourse clause.
   a. What is the annualized percentage all-in cost of this factoring alternative?
   b. What are the advantages and disadvantages of the factoring alternative compared to the alternatives in Problem 7?

9. **Whatchamacallit Sports (A).** Whatchamacallit Sports (Whatchamacallit) is considering bidding to sell $100,000 of ski equipment to Phang Family Enterprises of Seoul, Korea. Payment would be due in six months. Since Whatchamacallit cannot find good credit information on Phang, Whatchamacallit wants to protect its credit risk. It is considering the following financing solution.

   Phang's bank issues a letter of credit on behalf of Phang and agrees to accept Whatchamacallit's draft for $100,000 due in six months. The acceptance fee would cost Whatchamacallit $500, plus reduce Phang's available credit line by $100,000. The bankers' acceptance note of $100,000 would be sold at a 2% per annum discount in the money market. What is the annualized percentage all-in cost to Whatchamacallit of this bankers' acceptance financing?

10. **Whatchamacallit Sports (B).** Whatchamacallit could also buy export credit insurance from FCIA for a 1.5% premium. It finances the $100,000 receivable from Phang from its credit line at 6% per annum interest. No compensating bank balance would be required.
    a. What is Whatchamacallit's annualized percentage all-in cost of financing?
    b. What are Phang's costs?
    c. What are the advantages and disadvantages of this alternative compared to the bankers' acceptance financing in Problem 9? Which alternative would you recommend?

11. **Inca Breweries of Peru.** Inca Breweries of Lima, Peru, has received an order for 10,000 cartons of beer from Alicante Importers of Alicante, Spain. The beer will be exported to Spain under the terms of a letter of credit issued by a Madrid bank on behalf of Alicante Importers. The letter of credit specifies that the face value of the shipment, $720,000 U.S. dollars, will be paid 90 days after the Madrid bank accepts a draft drawn by Inca Breweries in accordance with the terms of the letter of credit.

    The current discount rate on 3-month bankers' acceptance is 8% per annum, and Inca Breweries estimates its weighted average cost of capital to be 20% per annum. The commission for selling a bankers' acceptance in the discount market is 1.2% of the face amount.

    How much cash will Inca Breweries receive from the sale if it holds the acceptance until maturity? Do you recommend that Inca Breweries hold the acceptance until maturity or discount it at once in the U.S. bankers' acceptance market?

12. **Swishing Shoe Company.** Swishing Shoe Company of Durham, North Carolina, has received an order for 50,000 cartons of athletic shoes from Southampton Footware, Ltd., of England, payment to be in British pounds sterling. The shoes will be shipped to Southampton Footware under the terms of a letter of credit issued by a London bank on behalf of Southampton Footware. The letter of credit specifies that the face value of the shipment, £400,000, will be paid 120 days after the London bank accepts a draft drawn by Southampton Footware in accordance with the terms of the letter of credit.

    The current discount rate in London on 120-day bankers' acceptances is 12% per annum, and Southampton Footware estimates its weighted average cost of capital to be 18% per annum. The commission for selling a bankers' acceptance in the discount market is 2.0% of the face amount.
    a. Would Swishing Shoe Company gain by holding the acceptance to maturity, as compared to discounting the bankers' acceptance at once?
    b. Does Swishing Shoe Company incur any other risks in this transaction?

13. **Going Abroad.** Assume that Great Britain charges an import duty of 10% on shoes imported into the United Kingdom. Swishing Shoe Company, in Problem 12, discovers that it can manufacture shoes in Ireland and import them into Britain free of any import duty. What factors should Swishing consider in deciding to continue to export shoes from North Carolina versus manufacture them in Ireland?

INTERNET EXERCISES

1. **Letter of Credit Services.** Commercial banks worldwide provide a variety of services to aid in the financing of foreign trade. Contact any of the many major multinational banks (a few are listed below) and determine what types of letter of credit services and other trade financing services they are able to provide.

Bank of America	www.bankamerica.com
Barclays	www.barclays.com
Deutsche Bank	www.deutschebank.com
Union Bank of Switzerland	www.unionbank.com
Swiss Bank Corporation	www.swissbank.com

2. **The Balanced World.** The Balanced World Web site is the equivalent of a social networking site for those interested in discussing a multitude of financial issues in greater depth and breadth. There is no limit to breadth of topics in finance and financial management that are posted and discussed.

The Balanced World   www.thebalancedworld.com

Foreign Investments and Investment Analysis

CHAPTER 17

Foreign Direct Investment and Political Risk

In this world, shipmates, Sin that pays its way can travel freely, and without passport; whereas Virtue, if a pauper, is stopped at all frontiers.

—Herman Melville, Chapter 9, The Sermon, in *Moby Dick*, 1851.

LEARNING OBJECTIVES

- Demonstrate how key competitive advantages support a strategy to sustain direct foreign investment
- Show how the OLI Paradigm provides a theoretical foundation for the globalization process
- Explore the motivations and factors that determine where multinationals invest abroad
- Compare and contrast the modes of foreign investment
- Illustrate the new forces driving corporate competition in emerging markets
- Evaluate the various factors that may be used to predict when and where political risks will arise
- Describe the forms of transfer risk and how multinationals may mitigate these blockages
- Learn how to evaluate the cultural and institutional factors often leading to firm-specific political risk
- Examine the unique complexities of global specific risk

The strategic decision to undertake *foreign direct investment* (FDI), and thus become an MNE, starts with a self-evaluation. This self-evaluation combines a series of questions including the nature of the firm's competitive advantage, what business form and commensurate risks the firm should use and accept upon entry, and what political risks—both in the macro and micro context—the firm will be facing. This chapter explores this sequence of self-evaluation, as well as methods for both measuring and managing the multitude of risks taken in foreign investment. As part of this international exploration, an illustrative case on corporate competition from the emerging markets, highlighting how many of tomorrow's most competitive MNEs may be arising from the emerging markets themselves. The chapter concludes with the Mini-Case, *Strategic Portfolio Theory, Black Swans, and [Avoiding] Being the Turkey*, a discussion of modern investment and portfolio theory in a world of growing and unknown risks.

Sustaining and Transferring Competitive Advantage

In deciding whether to invest abroad, management must first determine whether the firm has a sustainable competitive advantage that enables it to compete effectively in the home market. The competitive advantage must be firm-specific, transferable, and powerful enough to compensate the firm for the potential disadvantages of operating abroad (foreign exchange risks, political risks, and increased agency costs).

Based on observations of firms that have successfully invested abroad, we can conclude that some of the competitive advantages enjoyed by MNEs are (1) economies of scale and scope arising from their large size; (2) managerial and marketing expertise; (3) advanced technology owing to their heavy emphasis on research; (4) financial strength; (5) differentiated products; and sometimes (6) competitiveness of their home markets.

Economies of Scale and Scope

Economies of scale and scope can be developed in production, marketing, finance, research and development, transportation, and purchasing. All of these areas have significant competitive advantages of being large, whether size is due to international or domestic operations. Production economies can come from the use of large-scale automated plant and equipment or from an ability to rationalize production through global specialization.

For example, some automobile manufacturers, such as Ford, rationalize manufacturing by producing engines in one country, transmissions in another, and bodies in another and assembling still elsewhere, with the location often being dictated by comparative advantage. Marketing economies occur when firms are large enough to use the most efficient advertising media to create global brand identification, as well as to establish global distribution, warehousing, and servicing systems. Financial economies derive from access to the full range of financial instruments and sources of funds, such as the euroequity and eurobond markets. In-house research and development programs are typically restricted to large firms because of the minimum-size threshold for establishing a laboratory and scientific staff. Transportation economies accrue to firms that can ship in carload or shipload lots. Purchasing economies come from quantity discounts and market power.

Managerial and Marketing Expertise

Managerial expertise includes skill in managing large industrial organizations from both a human and a technical viewpoint. It also encompasses knowledge of modern analytical techniques and their application in functional areas of business. Managerial expertise can be developed through prior experience in foreign markets. In most empirical studies, multinational firms have been observed to export to a market before establishing a production facility there. Likewise, they have prior experience sourcing raw materials and human capital in other foreign countries either through imports, licensing, or FDI. In this manner, the MNEs can partially overcome the supposed superior local knowledge of host-country firms.

Advanced Technology

Advanced technology includes both scientific and engineering skills. It is not limited to MNEs, but firms in the most industrialized countries have had an advantage in terms of access to continuing new technology spin-offs from the military and space programs. Empirical studies have supported the importance of technology as a characteristic of MNEs.

Financial Strength

Companies demonstrate financial strength by achieving and maintaining a global cost and availability of capital. This is a critical competitive cost variable that enables them to fund FDI and other foreign activities. MNEs that are resident in liquid and unsegmented capital markets are normally blessed with this attribute. However, MNEs that are resident in small industrial or emerging market countries can still follow a proactive strategy of seeking foreign portfolio and corporate investors.

Small- and medium-size firms often lack the characteristics that attract foreign (and maybe domestic) investors. They are too small or unattractive to achieve a global cost of capital. This limits their ability to fund FDI, and their higher marginal cost of capital reduces the number of foreign projects that can generate the higher required rate of return.

Differentiated Products

Firms create their own firm-specific advantages by producing and marketing differentiated products. Such products originate from research-based innovations or heavy marketing expenditures to gain brand identification. Furthermore, the research and marketing process continues to produce a steady stream of new differentiated products. It is difficult and costly for competitors to copy such products, and they always face a time lag if they try. Having developed differentiated products for the domestic home market, the firm may decide to market them worldwide, a decision consistent with the desire to maximize return on heavy research and marketing expenditures.

Competitiveness of the Home Market

A strongly competitive home market can sharpen a firm's competitive advantage relative to firms located in less competitive home markets. This phenomenon is known as the "competitive advantage of nations," a concept originated by Michael Porter of Harvard, and summarized in Exhibit 17.1.

EXHIBIT 17.1   **Determinants of National Competitive Advantage: Porter's Demand**

A firm's competitiveness can be significantly strengthened based on its having competed in a highly competitive home market. Home country competitive advantage must be based on at least one of four critical components.

Source: Based on concepts described by Michael Porter in "The Competitive Advantage of Nations," *Harvard Business Review*, March–April 1990.

A firm's success in competing in a particular industry depends partly on the availability of factors of production (land, labor, capital, and technology) appropriate for that industry. Countries that are either naturally endowed with the appropriate factors or able to create them will probably spawn firms that are both competitive at home and potentially so abroad. For example, a well-educated workforce in the home market creates a competitive advantage for firms in certain high-tech industries. Firms facing sophisticated and demanding customers in the home market are able to hone their marketing, production, and quality control skills. Japan is such a market.

Firms in industries that are surrounded by a critical mass of related industries and suppliers will be more competitive because of this supporting cast. For example, electronic firms located in centers of excellence, such as in the San Francisco Bay area, are surrounded by efficient, creative suppliers who enjoy access to educational institutions at the forefront of knowledge.

A competitive home market forces firms to fine-tune their operational and control strategies for their specific industry and country environment. Japanese firms learned how to organize to implement their famous just-in-time inventory control system. One key was to use numerous subcontractors and suppliers that were encouraged to locate near the final assembly plants.

In some cases, host-country markets have not been large or competitive, but MNEs located there have nevertheless developed global niche markets served by foreign subsidiaries. Global competition in oligopolistic industries substitutes for domestic competition. For example, a number of MNEs resident in Scandinavia, Switzerland, and the Netherlands fall into this category. They include Novo Nordisk (Denmark), Norske Hydro (Norway), Nokia (Finland), L.M. Ericsson (Sweden), Astra (Sweden), ABB (Sweden/Switzerland), Roche Holding (Switzerland), Royal Dutch Shell (the Netherlands), Unilever (the Netherlands), and Philips (the Netherlands).

Emerging market countries have also spawned aspiring global MNEs in niche markets even though they lack competitive home-country markets. Some of these are traditional exporters in natural resource fields such as oil, agriculture, and minerals, but they are in transition to becoming MNEs. They typically start with foreign sales subsidiaries, joint ventures, and strategic alliances. Examples are Petrobrás (Brazil), YPF (Argentina), and Cemex (Mexico). Another category is firms that have been recently privatized in the telecommunications industry. Examples are Telefonos de Mexico and Telebrás (Brazil). Still others started as electronic component manufacturers but are making the transition to manufacturing abroad. Examples are Samsung Electronics (Korea) and Acer Computer (Taiwan).

The OLI Paradigm and Internationalization

The OLI Paradigm (Buckley and Casson, 1976; Dunning, 1977) is an attempt to create an overall framework to explain why MNEs choose FDI rather than serve foreign markets through alternative modes such as licensing, joint ventures, strategic alliances, management contracts, and exporting.[1]

The OLI Paradigm states first that a firm must first have some competitive advantage in its home market—"O" for ownership advantages—that can be transferred abroad if the firm

[1]Peter J. Buckley and Mark Casson, *The Future of the Multinational Enterprise*, London: McMillan, 1976; and John H. Dunning, "Trade Location of Economic Activity and the MNE: A Search for an Eclectic Approach," in *The International Allocation of Economic Activity*, Bertil Ohlin, Per-Ove Hesselborn, and Per Magnus Wijkman, eds., New York: Holmes and Meier, 1977, pp. 395–418.

is to be successful in foreign direct investment. Second, the firm must be attracted by specific characteristics of the foreign market—"L" for location advantages—that will allow it to exploit its competitive advantages in that market. Third, the firm will maintain its competitive position by attempting to control the entire value chain in its industry—"I" for internationalization advantages. This leads it to foreign direct investment rather than licensing or *outsourcing*.

Ownership Advantages

As described earlier, a firm must have competitive advantages in its home market. These must be firm-specific, not easily copied, and in a form that allows them to be transferred to foreign subsidiaries. For example, economies of scale and financial strength are not necessarily firm-specific because they can be achieved by many other firms. Certain kinds of technology can be purchased, licensed, or copied. Even differentiated products can lose their advantage to slightly altered versions, given enough marketing effort and the right price.

Location Advantages

These factors are typically market imperfections or genuine comparative advantages that attract FDI to particular locations. These factors might include a low-cost but productive labor force, unique sources of raw materials, a large domestic market, defensive investments to counter other competitors, or centers of technological excellence.

Internationalization Advantages

According to the theory, the key ingredient for maintaining a firm-specific competitive advantage is possession of proprietary information and control of the human capital that can generate new information through expertise in research. Needless to say, once again, large research-intensive firms are most likely to fit this description.

Minimizing transactions costs is the key factor in determining the success of an inter-nationalization strategy. Wholly owned FDI reduces the agency costs that arise from asymmetric information, lack of trust, and the need to monitor foreign partners, suppliers, and financial institutions. Self-financing eliminates the need to observe specific debt covenants on foreign subsidiaries that are financed locally or by joint venture partners. If a multinational firm has a low global cost and high availability of capital, why share it with joint venture partners, distributors, licensees, and local banks, all of which probably have a higher cost of capital?

The Financial Strategy

Financial strategies are directly related to the OLI Paradigm in explaining FDI, as shown in Exhibit 17.2. Proactive financial strategies can be formulated in advance by the MNE's financial managers. These include strategies necessary to gain an advantage from lower global cost and greater availability of capital. Other proactive financial strategies are negotiating financial subsidies and/or reduced taxation to increase free cash flows, reducing financial agency costs through FDI, and reducing operating and transaction exposure through FDI.

Reactive financial strategies, as illustrated in Exhibit 17.2, depend on discovering market imperfections. For example, the MNE can exploit misaligned exchange rates and stock prices. It also needs to react to capital controls that prevent the free movement of funds and react to opportunities to minimize worldwide taxation.

| EXHIBIT 17.2 | **Finance Factors and the OLI Paradigm** |

	Proactive Financial Strategies	**Reactive Financial Strategies**
Ownership Advantages	• Competitive sourcing of capital globally • Strategic cross-listing • Accounting & disclosure transparency • Maintaining financial relationships • Maintaining competitive credit rating	
Location Advantages	• Competitive sourcing of capital globally • Maintaining competitive credit rating • Negotiating tax & financial subsidies	• Exploiting exchange rates • Exploiting stock prices • Reacting to capital controls • Minimizing taxation
Internationalization Advantages	• Maintaining competitive credit rating • Reducing agency costs through FDI	• Minimizing taxation

Source: Constructed by authors based on "On the Treatment of Finance-Specific Factors Within the OLI Paradigm," by Lars Oxelheim, Arthur Stonehill, and Trond Randøy, *International Business Review* 10, 2001, pp. 381–398.

Deciding Where to Invest

The decision about where to invest abroad for the first time is not the same as the decision about where to reinvest abroad. This decision is influenced by behavioral factors. A firm learns from its first few investments abroad and what it learns influences subsequent investments.

In theory, a firm should identify its competitive advantages. Then it should search worldwide for market imperfections and comparative advantage until it finds a country where it expects to enjoy a competitive advantage large enough to generate a risk-adjusted return above the firm's *hurdle rate*, the minimum acceptable rate of return on new investments.

In practice, firms have been observed to follow a sequential search pattern as described in the behavioral theory of the firm. Human rationality is bounded by one's ability to gather and process all the information that would be needed to make a perfectly rational decision based on all the facts. This observation lies behind two behavioral theories of FDI described next—the behavioral approach and the international network theory.

The Behavioral Approach to FDI

The behavioral approach to analyzing the FDI decision is typified by the so-called Swedish School of economists.[2] The Swedish School has rather successfully explained not just the initial

[2]John Johansen, and F. Wiedersheim-Paul, "The Internationalization of the Firm: Four Swedish Case Studies," *Journal of Management Studies*, Vol. 12, No. 3, 1975; and John Johansen and Jan Erik Vahlne, "The Internationalization of the Firm: A Model of Knowledge Development and Increasing Foreign Market Commitments," *Journal of International Business Studies*, Vol. 8, No. 1, 1977.

decision to invest abroad but also later decisions to reinvest elsewhere and to change the structure of a firm's international involvement over time. Based on the internationalization process of a sample of Swedish MNEs, the economists observed that these firms tended to invest first in countries that were not too far distant in psychic terms. Close *psychic distance* defined countries with a cultural, legal, and institutional environment similar to Sweden's, such as Norway, Denmark, Finland, Germany, and the United Kingdom. The initial investments were modest in size to minimize the risk of an uncertain foreign environment. As the Swedish firms learned from their initial investments, they became willing to take greater risks with respect to both the psychic distance of the countries and the size of the investments.

MNEs in a Network Perspective

As the Swedish MNEs grew and matured, so did the nature of their international involvement, what is often termed the *network perspective*. Today, each MNE is perceived as being a member of an international network, with nodes based in each of the foreign subsidiaries, as well as the parent firm itself. Centralized (hierarchical) control has given way to decentralized (heterarchical) control. Foreign subsidiaries compete with each other and with the parent for expanded resource commitments, thus influencing the strategy and reinvestment decisions.[3]

Many of these MNEs have become political coalitions with competing internal and external networks. Each subsidiary (and the parent) is embedded in its host country's network of suppliers and customers. It is also a member of a worldwide network based on its industry. Finally, it is a member of an organizational network under the nominal control of the parent firm. Complicating matters still further is the possibility that the parent itself may have evolved into a *transnational firm*, one that is owned by a coalition of investors located in different countries.

Modes of Foreign Investment

The globalization process includes a sequence of decisions regarding where production is to occur, who is to own or control intellectual property, and who is to own the actual production facilities. Exhibit 17.3 provides a roadmap to explain this FDI sequence.

Exporting versus Production Abroad

There are several advantages to limiting a firm's activities to exports. Exporting has none of the unique risks facing FDI, joint ventures, strategic alliances, and licensing. Political risks are minimal. Agency costs, such as monitoring and evaluating foreign units, are avoided. The amount of front-end investment is typically lower than in other modes of foreign involvement. Foreign exchange risks remain, however. The fact that a significant share of exports (and imports) is executed between MNEs and their foreign subsidiaries and affiliates further reduces the risk of exports compared to other modes of involvement.

There are also disadvantages of limiting a firm's activities to exports. A firm is not able to internalize and exploit the results of its research and development as effectively as if it invested directly. The firm also risks losing markets to imitators and global competitors that might be more cost efficient in production abroad and distribution. As these firms capture foreign markets, they might become so strong that they can export into the domestic exporter's own market. Remember that defensive FDI is often motivated by the need to prevent this kind of predatory behavior as well as to preempt foreign markets before competitors can get started.

[3]Mats Forsgren, *Managing the Internationalization Process: The Swedish Case*, London: Routledge, 1989.

EXHIBIT 17.3   **The FDI Sequence: Foreign Presence and Investment**

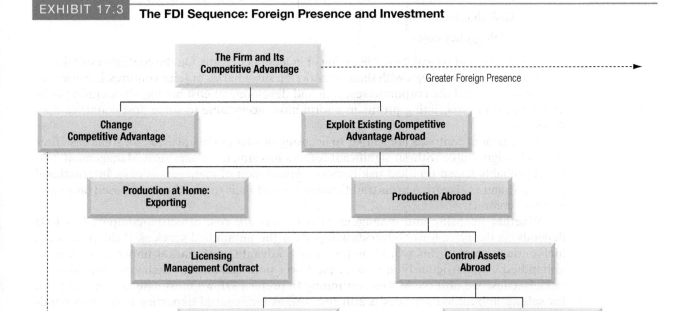

Source: Adapted from Gunter Dufey and R. Mirus, "Foreign Direct Investment: Theory and Strategic Considerations,"
unpublished, University of Michigan, 1985. Reprinted with permission from the authors. All rights reserved.

Licensing and Management Contracts

Licensing is a popular method for domestic firms to profit from foreign markets without the need to commit sizable funds. Since the foreign producer is typically wholly owned locally, political risk is minimized. In recent years, a number of host countries have demanded that MNEs sell their services in pieces ("unbundled form") rather than only through FDI. Such countries would like their local firms to purchase managerial expertise and knowledge of product and factor markets through management contracts, and purchase technology through licensing agreements.

The main disadvantage of licensing is that license fees are likely to be lower than FDI profits, although the return on the marginal investment might be higher. Other disadvantages include the following:

- Possible loss of quality control
- Establishment of a potential competitor in third-country markets
- Possible improvement of the technology by the local licensee, which then enters the firm's home market

- Possible loss of opportunity to enter the licensee's market with FDI later
- Risk that technology will be stolen
- High agency costs

MNEs have not typically used licensing of independent firms. On the contrary, most licensing arrangements have been with their own foreign subsidiaries or joint ventures. License fees are a way to spread the corporate research and development cost among all operating units and a means of repatriating profits in a form more acceptable to some host countries than dividends.

Management contracts are similar to licensing insofar as they provide for some cash flow from a foreign source without significant foreign investment or exposure. Management contracts probably lessen political risk because repatriation of managers is easy. International consulting and engineering firms traditionally conduct their foreign business based on a management contract.

Whether licensing and management contracts are cost effective compared to FDI depends on the price host countries will pay for the unbundled services. If the price were high enough, many firms would prefer to take advantage of market imperfections in an unbundled way, particularly in view of the lower political, foreign exchange, and business risks. Because we observe MNEs continuing to prefer FDI, we must assume that the price for selling unbundled services is still too low, as managerial expertise is often dependent on a delicate mix of organizational support factors that cannot be transferred abroad efficiently.

Joint Venture versus Wholly Owned Subsidiary

A *joint venture* (JV) is defined here as shared ownership in a foreign business. A foreign business unit that is partially owned by the parent company is typically termed a *foreign affiliate*. A foreign business unit that is 50% or more owned (and therefore controlled) by the parent company is typically designated a foreign subsidiary. A JV would therefore typically be described as a foreign affiliate, but not a foreign subsidiary.

A joint venture between an MNE and a host-country partner is a viable strategy if, and only if, the MNE finds the right partner. Some of the obvious advantages of having a compatible local partner are as follows:

- The local partner understands the customs, mores, and institutions of the local environment. An MNE might need years to acquire such knowledge on its own with a 100%-owned greenfield subsidiary. (Greenfield investments are started with a clean slate, having no prior history of development.)
- The local partner can provide competent management, not just at the top but also at the middle levels of management.
- If the host country requires that foreign firms share ownership with local firms or investors, 100% foreign ownership is not a realistic alternative to a joint venture.
- The local partner's contacts and reputation enhance access to the host-country's capital markets.
- The local partner may possess technology that is appropriate for the local environment or perhaps can be used worldwide.
- The public image of a firm that is partially locally owned may improve its sales possibilities if the purpose of the investment is to serve the local market.

Despite this impressive list of advantages, joint ventures are not as common as 100%—owned foreign subsidiaries because MNEs fear interference by the local partner in certain critical decision areas. Indeed, what is optimal from the viewpoint of the local venture may be suboptimal for the multinational operation as a whole. The most important potential conflicts or difficulties are these:

- Political risk is increased rather than reduced if the wrong partner is chosen. The local partner must be credible and ethical or the venture is worse off for being a joint venture.

- Local and foreign partners may have divergent views about the need for cash dividends, or about the desirability of growth financed from retained earnings versus new financing.

- Transfer pricing on products or components bought from or sold to related companies creates a potential for conflict of interest.

- Control of financing is another problem area. An MNE cannot justify its use of cheap or available funds raised in one country to finance joint venture operations in another country.

- Ability of a firm to rationalize production on a worldwide basis can be jeopardized if such rationalization would act to the disadvantage of local joint venture partners.

- Financial disclosure of local results might be necessary with locally traded shares, whereas if the firm is wholly owned from abroad such disclosure is not needed. Disclosure gives nondisclosing competitors an advantage in setting strategy.

Valuation of equity shares is difficult. How much should the local partner pay for its share? What is the value of contributed technology, or of contributed land in a country where all land is state owned? It is highly unlikely that foreign and host-country partners have similar opportunity costs of capital, expectations about the required rate of return, or similar perceptions of appropriate premiums for business, foreign exchange, and political risks. Insofar as the venture is a component of the portfolio of each investor, its contribution to portfolio return and variance may be quite different for each.

Strategic Alliances

The term strategic alliance conveys different meanings to different observers. In one form of cross-border strategic alliance, two firms exchange a share of ownership with one another. A strategic alliance can be a takeover defense if the prime purpose is for a firm to place some of its stock in stable and friendly hands. If that is all that occurs, it is just another form of portfolio investment.

In a more comprehensive strategic alliance, in addition to exchanging stock, the partners establish a separate joint venture to develop and manufacture a product or service. Numerous examples of such strategic alliances can be found in the automotive, electronics, telecommunications, and aircraft industries. Such alliances are particularly suited to high-tech industries where the cost of research and development is high and timely introduction of improvements is important.

A third level of cooperation might include joint marketing and servicing agreements in which each partner represents the other in certain markets. Some observers believe such arrangements begin to resemble the cartels prevalent in the 1920s and 1930s. Because they reduce competition, cartels have been banned by international agreements and by many national laws.

Illustrative Case: Corporate Competition from the Emerging Markets

> *BCG [Boston Consulting Group] argues . . . they have managed to resolve three trade-offs that are usually associated with corporate growth: of volume against margin; rapid expansion against low leverage (debt); and growth against dividends. On average the challengers have increased their sales three times faster than their established global peers since 2005. Yet they have also reduced their debt-to-equity ratio by three percentage points and achieved a higher ratio of dividends to share price in every year but one.*
>
> — "Nipping at Their Heels: Firms from the Developing World Are Rapidly Catching Up with Their Old-World Competitors," *The Economist*, January 22, 2011, p. 80.

Leadership in all companies, public and private, new and old, start-ups and maturing, have all heard the same threat in recent years: the emerging market competitors are coming. Despite the threat, there have been other forces at work that would prevent their advancing—or advancing too fast: the ability to raise sufficient capital at a reasonable cost; the ability to reach the larger and more profitable markets; the competition in markets that value name recognition and brand identity; and global reach. But a number of market prognosticators—the gurus and consultants—are now contending that these new competitors are already here.

One such analysis was recently published by BCG, the Boston Consulting Group.[4] BCG labels these firms the *global challengers*, companies based in rapidly developing economies that are "shaking up" the established economic order. Their list of 100 global companies, most of which are from Brazil, Russia, India, China, and Mexico, are all innovative and aggressive, but have also proven to be financially fit.

The value created by these firms for their shareholders is very convincing. The *total shareholder return* (TSR) for the global challengers between 2005 and 2009 was 22%; the same TSR for their global peers, public companies in comparable business lines from the industrialized economies, was a mere 5%. These firms have, according to BCG, been able to achieve these results by resolving three classic trade-offs confronting emerging players. These strategic trade-offs turn out to be uniquely financial in character.

The Three Trade-Offs

The three trade-offs could also be characterized as three financial dimensions of competitiveness—the market, the financing, and the offered return.

Trade-Off #1: Volume versus Margin. Traditional business thinking assumes that large-scale, large-market sales, like that of Walmart, require incredibly low prices, which in turn impose low margin returns to the scale competitors. Higher margin products and services are usually reserved for specialty market segments, which may be much more expensive to service, but are found justifiable by the higher prices and higher margins they offer. BCG argues that the global challengers have been able to have both volume and margin, relying on exceptionally low direct costs of materials and labor, combined with the latest in technology and execution found in the developed country markets.

Trade-Off #2: Rapid Expansion versus Low Leverage. One of the key advantages always held by the world's largest companies is their preferred access to capital. The advantages

---

[4]"Companies on the Move, Rising Stars from Rapidly Developing Economies Are Reshaping Global Industries," Boston Consulting Group, January 2011

afforded companies in large market economies—capitalist economies—is access to plentiful and affordable capital. Companies arising from the emerging markets have often been held back in their expansion efforts, not having the capital to exercise their ambitions. Only after gaining access to the world's largest capital markets, providers of both debt and equity, can these firms pose a serious threat beyond their immediate country market or region. In the past, access meant higher levels of debt and the associated risks and burdens of higher leverage.

But the global challengers have again fought off the trade-off, finding ways to increase both equity and debt in proportion, and therefore to grow without taking on a riskier financial structure. The obvious solution has been to gain increasing access to affordable equity, often in London and New York.

Trade-Off #3: Growth versus Dividends. Financial theory has always emphasized the critical distinctions between what opportunities and threats growth firms and value firms offer investors. Growth firms are typically smaller firms, start-ups, companies with unique business models based on new technologies or services. They have enormous upside potential, but need more time, more experience, more breadth, and most importantly, more capital. Investors in these companies know the risks are high, and as a result, accept those risks in focusing on prospective returns from capital gains, not dividend distributions. Investors also know that these firms, often very small firms, will show large share price movements quickly with commensurate business developments. For that, the firm needs to be nimble, quick, and not laden with debt.

Value companies—a polite term for mature or older, larger, well-established global competitors—are of a size in which new business developments, new markets, or new technologies, are rarely large enough to move share prices significantly and quickly. Investors in these companies, according to agency theory, do not "trust management" to take sufficient risks to generate returns. Therefore, they prefer the firm to bear some artificial financial burdens to assure diligence. Those financial burdens are typically higher levels of debt and growing distributions of profit as dividends. Both elements serve as financial disciplines, requiring management to maintain watchfulness over costs and cash flows to service debt, and generate sufficient profitability over time to supply dividends.

The global challengers have arguably thwarted this trade-off as well, paying dividends at growing rates and similar dividend yields to more mature firms with stronger and sustained cash flows. This may actually be the easiest of the three to accomplish given their already substantial sizes and strong profitability.

Continuing Questions

If these global challengers can defeat these traditional financial trade-offs, can they overcome the corporate strategic challenges that so many firms from so many markets flailed against before them? As *The Economist* notes, "All this is impressive, but it seems implausible that these trade-offs have been 'resolved.'?"[5]

Many emerging market analysts and rapidly developing economy analysts argue that these firms not only understand emerging markets, but also they have demonstrated sustained innovation and remained financially healthy. Others argue that these three factors are likely to be more simultaneous than causal. It is clear, however, that most of these new global players are arising from large underdeveloped and underserved markets—markets that are providing large bases for their rapid development.

One strategy being rapidly deployed by many of these firms is the use of strategic partnerships, joint ventures, or share swap agreements.[6] In each of these forms, the companies are

---

[5]"Nipping at Their Heels: Firms from the Developing World Are Rapidly Catching Up with Their Old-World Competitors," *The Economist*, January 22, 2011, p. 80.

gaining a competitive reach, a global partner, and access to technology and markets without major growth on their part. Despite the use of these partnerships, this does not directly address the continuing debate as to whether firms can grow as successfully into different businesses in different markets—diversified global conglomerates. Although a strategy employed in the past, it is one not followed frequently today.

Predicting Political Risk

How can multinational firms anticipate political risks such as government regulations that, from the firm's perspective, are discriminatory or wealth depriving? First, the firm must be able to define and classify the political risks it may face.

Defining and Classifying Political Risk

In order for an MNE to identify, measure, and manage its political risks, it needs to define and classify these risks. At the macro level, firms attempt to assess a host country's political stability and attitude toward foreign investors. At the micro level, firms analyze whether their firm-specific activities are likely to conflict with host-country goals as evidenced by existing regulations. The most difficult task, however, is to anticipate changes in host-country goal priorities, new regulations to implement reordered priorities, and the likely impact of such changes on the firm's operations.

Exhibit 17.4 further classifies the political risks facing MNEs as being firm-specific, country-specific, or global-specific.

EXHIBIT 17.4   Classification of Political Risks

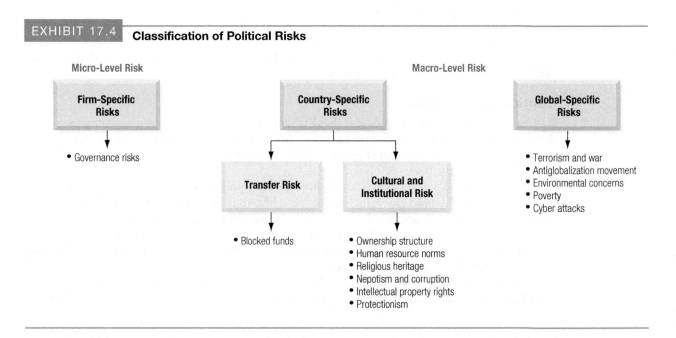

- *Firm-specific risks*, also known as *micro risks*, are those that affect the MNE at the project or corporate level. Governance risk due to goal conflict between an MNE and its host government is the main political firm-specific risk.
- *Country-specific risks*, also known as *macro risks*, are those that affect the MNE at the project or corporate level but originate at the country level. The two main political risk categories at the country level are transfer risk and cultural and institutional risks. Cultural and institutional risks spring from ownership structure, human resource norms, religious heritage, nepotism and corruption, intellectual property rights, and protectionism.
- *Global-specific risks* are those that affect the MNE at the project or corporate level but originate at the global level. Examples are terrorism, the antiglobalization movement, environmental concerns, poverty, and cyber attacks.

This method of classification differs sharply from the traditional method that classifies risks according to the disciplines of economics, finance, political science, sociology, and law. We prefer our classification system because it is easier to relate the identified political risks to existing and recommended strategies to manage these risks.

Predicting Firm-Specific Risk (Micro Risk)

From the viewpoint of a multinational firm, assessing the political stability of a host country is only the first step in predicting firm-specific risk, since the real objective is to anticipate the effect of political changes on activities of a specific firm. Indeed, different foreign firms operating within the same country may have very different degrees of vulnerability to changes in host-country policy or regulations. One does not expect a Kentucky Fried Chicken franchise to experience the same risk as a Ford manufacturing plant.

The need for firm-specific analyses of political risk has led to a demand for "tailor-made" studies undertaken in-house by professional risk analysts. This demand is heightened by the observation that outside professional analysts rarely even agree on the degree of macro-political risk that exists in any set of countries.

In-house political risk analysts relate the macro risk attributes of specific countries to the particular characteristics and vulnerabilities of their client firms. Mineral extractive firms, manufacturing firms, multinational banks, private insurance carriers, and worldwide hotel chains are all exposed in fundamentally different ways to politically inspired restrictions. Even with the best possible firm-specific analysis, MNEs cannot be sure that the political or economic situation will not change. Thus, it is necessary to plan protective steps in advance to minimize the risk of damage from unanticipated changes.

Predicting Country-Specific Risk (Macro Risk)

Macro political risk analysis is still an emerging field of study. Political scientists in academia, industry, and government study *country risk* for the benefit of multinational firms, government foreign policy decision makers, and defense planners.

Political risk studies usually include an analysis of the historical stability of the country in question, evidence of present turmoil or dissatisfaction, indications of economic stability, and trends in cultural and religious activities. Data are usually assembled by reading local newspapers, monitoring radio and television broadcasts, reading publications from diplomatic sources, tapping the knowledge of outstanding expert consultants, contacting other businesspeople who have had recent experience in the host country, and finally conducting on-site visits.

Despite this impressive list of activities, the prediction track record of business firms, the diplomatic service, and the military has been spotty at best. When one analyzes trends, whether in politics or economics, the tendency is to predict an extension of the same trends into the future. It is a rare forecaster who is able to predict a cataclysmic change in direction. Who predicted the overthrow of Ferdinand Marcos in the Philippines? Indeed, who predicted the collapse of communism in the Soviet Union and the Eastern European satellites? Who predicted the fall of President Suharto in Indonesia in 1998? As illustrated by *Global Finance in Practice 17.1*, the 2011 public protests in Egypt serve as one corporate reminder of risk and the reaction of markets to perceived vulnerability.

Despite the difficulty of predicting country risk, the MNE must still attempt to do so in order to prepare itself for the unknown. A number of institutional services provide country risk ratings on a regular basis.

Predicting Global-Specific Risk (Macro Risk)

Predicting global-specific risk is even more difficult than the other two types of political risk. Nobody predicted the surprise attacks on the World Trade Center and the Pentagon in the United States on September 11, 2001. On the other hand, the aftermath of this attack, that is, the war on global terrorism, increased U.S. homeland security, and the destruction of part of the

---

GLOBAL FINANCE IN PRACTICE 17.1

Apache Takes a Hit from Egyptian Protests

The January and February 2011 protests in Egypt took billions of dollars of value away from Apache Corporation (NYSE: APA). The U.S.-based oil exploration and production company has significant holdings and operations in Egypt, and the political turmoil that engulfed the country in early 2011 caused the investment public to start dumping Apache's shares. Although actual oil and gas production was not disrupted during this period, Apache did evacuate all expatriate workers from Egypt. Egypt made up roughly 30% of Apache's revenue in 2011, 26% of total production, and 13% of its estimated proved reserves of oil and gas.

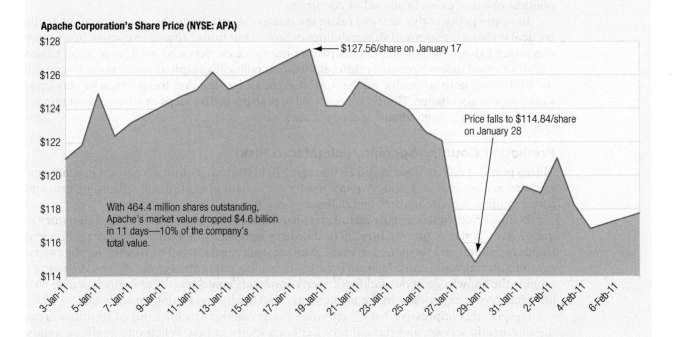

Apache Corporation's Share Price (NYSE: APA)

$127.56/share on January 17

Price falls to $114.84/share on January 28

With 464.4 million shares outstanding, Apache's market value dropped $4.6 billion in 11 days—10% of the company's total value.

terrorist network in Afghanistan was predictable. Nevertheless, we have come to expect future surprise terrorist attacks. U.S.-based MNEs are particularly exposed not only to Al-Qaeda but also to other unpredictable interest groups willing to use terror or mob action to promote such diverse causes as antiglobalization, environmental protection, and even anarchy. Since there is a great need to predict terrorism, we can expect to see a number of new indices, similar to country-specific indices, but devoted to ranking different types of terrorist threats, their locations, and potential targets.

Firm-Specific Political Risk: Governance Risk

The firm-specific political risks that confront MNEs include *foreign exchange risks* and *governance risks*. The various business and foreign exchange risks were detailed in Chapters 10, 11, and 12. We focus our discussion here on governance risks.

Governance risk is measured by the MNE's ability to exercise effective control over its operations within a country's legal and political environment. For an MNE, however, governance is a subject similar in structure to consolidated profitability—it must be addressed for the individual business unit and subsidiary, as well as for the MNE as a whole.

The most important type of governance risk for the MNE on the subsidiary level arises from a goal conflict between bona fide objectives of host governments and the private firms operating within their spheres of influence. Governments are normally responsive to a constituency consisting of their citizens. Firms are responsive to a constituency consisting of their owners and other stakeholders. The valid needs of these two separate sets of constituents need not be the same, but governments set the rules. Consequently, governments impose constraints on the activities of private firms as part of their normal administrative function.

Historically, conflicts between objectives of MNEs and host governments have arisen over such issues as the firm's impact on economic development, perceived infringement on national sovereignty, foreign control of key industries, sharing or nonsharing of ownership and control with local interests, impact on a host country's balance of payments, influence on the foreign exchange value of its currency, control over export markets, use of domestic versus foreign executives and workers, and exploitation of national resources. Attitudes about conflicts are often colored by views about free enterprise versus state socialism, the degree of nationalism or internationalism present, or the place of religious views in determining appropriate economic and financial behavior.

The best approach to goal conflict management is to anticipate problems and negotiate understandings ahead of time. Different cultures apply different ethics to the question of honoring prior "contracts," especially when they were negotiated with a previous administration. Nevertheless, prenegotiation of all conceivable areas of conflict provides a better basis for a successful future for both parties than does overlooking the possibility that divergent objectives will evolve over time. Prenegotiation often includes negotiating investment agreements, buying investment insurance and guarantees, and designing risk-reducing operating strategies to be used after the foreign investment decision has been made.

Investment Agreements

An *investment agreement* spells out specific rights and responsibilities of both the foreign firm and the host government. The presence of MNEs is as often sought by development-seeking host governments, just as a particular foreign location may be sought by an MNE. All parties have alternatives and so bargaining is appropriate. An investment agreement should spell out policies on financial and managerial issues, including the following:

- The basis on which fund flows, such as dividends, management fees, royalties, patent fees, and loan repayments, may be remitted
- The basis for setting transfer prices

- The right to export to third-country markets
- Obligations to build, or fund, social and economic overhead projects, such as schools, hospitals, and retirement systems
- Methods of taxation, including the rate, the type of taxation, and means by which the rate base is determined
- Access to host-country capital markets, particularly for long-term borrowing
- Permission for 100% foreign ownership versus required local ownership (joint venture) participation
- Price controls, if any, applicable to sales in the host-country markets
- Requirements for local sourcing versus import of raw materials and components
- Permission to use expatriate managerial and technical personnel, and to bring them and their personal possessions into the country free of exorbitant charges or import duties
- Provision for arbitration of disputes
- Provision for planned divestment, should such be required, indicating how the going concern will be valued and to whom it will be sold

Investment Insurance and Guarantees: OPIC

MNEs can sometimes transfer political risk to a host-country public agency through an investment insurance and guarantee program. Many developed countries have such programs to protect investments by their nationals in developing countries.

For example, in the United States the investment insurance and guarantee program is managed by the government-owned *Overseas Private Investment Corporation* (OPIC). An OPIC's purpose is to mobilize and facilitate the participation of U.S. private capital and skills in the economic and social progress of less-developed friendly countries and areas, thereby complementing the developmental assistance of the United States. An OPIC offers insurance coverage for four separate types of political risk, which have their own specific definitions for insurance purposes:

1. Inconvertibility is the risk that the investor will not be able to convert profits, royalties, fees, or other income, as well as the original capital invested, into dollars.
2. *Expropriation* is the risk that the host government takes a specific step that for one year prevents the investor or the foreign subsidiary from exercising effective control over use of the property.
3. War, revolution, insurrection, and civil strife coverage applies primarily to the damage of physical property of the insured, although in some cases inability of a foreign subsidiary to repay a loan because of a war may be covered.
4. Business income coverage provides compensation for loss of business income resulting from events of political violence that directly cause damage to the assets of a foreign enterprise.

Operating Strategies after the FDI Decision

Although an investment agreement creates obligations on the part of both foreign investor and host government, conditions change and agreements are often revised in the light of such changes. The changed conditions may be economic, or they may be the result of political changes within the host government. The firm that sticks rigidly to the legal interpretation of its original agreement may well find that the host government first applies pressure in areas not covered by the agreement and then possibly reinterprets the agreement to conform to the political reality of that country.

Most MNEs, in their own self-interest, follow a policy of adapting to changing host-country priorities whenever possible. The essence of such adaptation is anticipating host-country priorities and ensuring that the activities of the firm are of continued value to the host country. Such an approach assumes the host government acts rationally in seeking its country's self-interest and is based on the idea that the firm should initiate reductions in goal conflict. Future bargaining position can be enhanced by careful consideration of policies in production, logistics, marketing, finance, organization, and personnel.

Local Sourcing

Host governments may require foreign firms to purchase raw material and components locally as a way to maximize value-added benefits and to increase local employment. From the viewpoint of the foreign firm trying to adapt to host-country goals, local sourcing reduces political risk, albeit at a trade-off with other factors. Local strikes or other turmoil may shut down the operation and such issues as quality control, high local prices, and unreliable delivery schedules become important. Often, through local sourcing, the MNE lowers political risk only by increasing its financial and commercial risk.

Facility Location

Production facilities may be located to minimize risk. The natural location of different stages of production may be resource-oriented, footloose, or market-oriented. Oil, for instance, is drilled in and around the Persian Gulf, Russia, Venezuela, and Indonesia. No choice exists for where this activity takes place. Refining is footloose; a refining facility can be moved easily to another location or country. Whenever possible, oil companies have built refineries in politically safe countries, such as Western Europe, or small islands (such as Singapore or Curaçao), even though costs might be reduced by refining nearer the oil fields. They have traded reduced political risk and financial exposure for higher transportation and refining costs.

Control

Control—of transportation, technology, markets, brands and trademarks—is key to the management of a multitude of political risks.

Transportation. Control of transportation has been an important means to reduce political risk. Oil pipelines that cross national frontiers, oil tankers, ore carriers, refrigerated ships, and railroads have all been controlled at times to influence the bargaining power of both nations and companies.

Technology. Control of key patents and processes is a viable way to reduce political risk. If a host country cannot operate a plant because it does not have technicians capable of running the process, or of keeping up with changing technology, abrogation of an investment agreement with a foreign firm is unlikely. Control of technology works best when the foreign firm is steadily improving its technology.

Markets. Control of markets is a common strategy to enhance a firm's bargaining position. As effective as the OPEC cartel was in raising the price received for crude oil by its member countries in the 1970s, marketing was still controlled by the international oil companies. OPEC's need for the oil companies limited the degree to which its members could dictate terms.

Control of export markets for manufactured goods is also a source of leverage in dealings between MNEs and host governments. The MNE would prefer to serve world markets from sources of its own choosing, basing the decision on considerations of production cost, transportation, tariff barriers, political risk exposure, and competition. The selling pattern that maximizes long-run profits from the viewpoint of the worldwide firm rarely maximizes exports, or value added, from the perspective of the host countries. Some will argue that if the same plants were owned by local nationals and were not part of a worldwide integrated system, more goods would be exported by the host country. The contrary argument is that self-contained local firms might never obtain foreign market share because they lack economies of scale on the production side and are unable to market in foreign countries.

Brand Name/Trademark. Control of a brand name or trademark can have an effect almost identical to that of controlling technology. It gives the MNE a monopoly on something that may or may not have substantive value but quite likely represents value in the eyes of consumers. Ability to market under a world brand name is valuable for local firms and may represent an important bargaining chip for maintaining an investment.

Thin Equity Base

Foreign subsidiaries can be financed with a thin equity base and a large proportion of local debt. If the debt is borrowed from locally owned banks, host-government actions that weaken the financial viability of the firm also endanger local creditors.

Multiple-Source Borrowing

If the firm must finance with foreign source debt, it may borrow from banks in a number of countries rather than just from host-country banks. If, for example, debt is owed to banks in Tokyo, Frankfurt, London, and New York, nationals in a number of foreign countries have a vested interest in keeping the borrowing subsidiary financially strong. If the multinational is U.S.-owned, a fallout between the United States and the host government is less likely to cause the local government to move against the firm if it also owes funds to these other countries.

Country-Specific Risk: Transfer Risk

Country-specific risks affect all firms, domestic and foreign, that are resident in a host country. Exhibit 17.5 presents a taxonomy of most of the contemporary political risks that emanate from a specific country location. The main country-specific political risks are *transfer risk*, *cultural risk*, and *institutional risk*. This section focuses on transfer risk.

Blocked Funds

Transfer risk is the risk of limitations on the MNE's ability to transfer funds into and out of a host country without restrictions. When a government runs short of foreign exchange and cannot obtain additional funds through borrowing or attracting new foreign investment, it usually limits transfers of foreign exchange out of the country, a restriction known as *blocked funds*. In theory, such limitations do not discriminate against foreign-owned firms because they apply to everyone; in practice, though, foreign firms have more at stake because of their foreign ownership.

Depending on the size of a foreign exchange shortage, the host government might simply require approval of all transfers of funds abroad, thus reserving the right to set a priority on the use of scarce foreign exchange. In very severe cases, the government might make its currency nonconvertible into other currencies, thereby fully blocking transfers of funds abroad. In between these positions are policies that restrict the size and timing of dividends, debt amortization, royalties, and service fees.

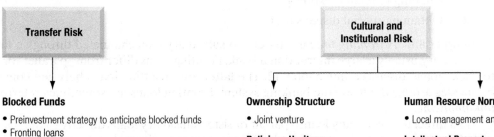

EXHIBIT 17.5   **Management Strategies for Country-Specific Risks**

Transfer Risk

Cultural and Institutional Risk

Blocked Funds
- Preinvestment strategy to anticipate blocked funds
- Fronting loans
- Creating unrelated exports
- Obtaining special dispensation
- Forced reinvestment

Ownership Structure
- Joint venture

Religious Heritage
- Understand and respect host country religious heritage

Nepotism and Corruption
- Disclose bribery policy to both employees and clients
- Retain a local legal adviser

Human Resource Norms
- Local management and staffing

Intellectual Property
- Legal action in host country courts
- Support worldwide treaty to protect intellectual property rights

Protectionism
- Support government actions to create regional markets

MNEs can react to the potential for blocked funds at three stages:

1. Prior to investing, a firm can analyze the effect of blocked funds on expected return on investment, the desired local financial structure, and optimal links with subsidiaries.

2. During operations, a firm can attempt to move funds through a variety of repositioning techniques.

3. Funds that cannot be moved must be reinvested in the local country in a manner that avoids deterioration in their real value because of inflation or exchange depreciation.

Preinvestment Strategy to Anticipate Blocked Funds

Management can consider blocked funds in their capital budgeting analysis. Temporary blockage of funds normally reduces the expected net present value and internal rate of return on a proposed investment. Whether the investment should nevertheless be undertaken depends on whether the expected rate of return, even with blocked funds, exceeds the required rate of return on investments of the same risk class. Preinvestment analysis also includes the potential to minimize the effect of blocked funds by financing with local borrowing instead of parent equity, swap agreements, and other techniques to reduce local currency exposure and thus the need to repatriate funds. Sourcing and sales links with subsidiaries can be predetermined to maximize the potential for moving blocked funds.

Moving Blocked Funds

What can a multinational firm do to transfer funds out of countries having exchange or remittance restrictions? At least six popular strategies are used:

1. Providing alternative conduits for repatriating funds
2. Transferring pricing goods and services between related units of the MNE

3. Leading and lagging payments

4. Signing fronting loans

5. Creating unrelated exports

6. Obtaining special dispensation

Fronting Loans. A *fronting loan* is a parent-to-subsidiary loan channeled through a financial intermediary, usually a large international bank. Fronting loans differ from "parallel" or "back-to-back" loans, discussed in Chapter 12. The latter are offsetting loans between commercial businesses arranged outside the banking system. Fronting loans are sometimes referred to as *link financing*.

In a direct intracompany loan, a parent or sister subsidiary loans directly to the borrowing subsidiary, and later, the borrowing subsidiary repays the principal and interest. In a fronting loan, by contrast, the "lending" parent or subsidiary deposits funds in, say, a London bank, and that bank loans the same amount to the borrowing subsidiary in the host country. From the London bank's point of view, the loan is risk-free, because the bank has 100% collateral in the form of the parent's deposit. In effect, the bank "fronts" for the parent—hence the name. Interest paid by the borrowing subsidiary to the bank is usually slightly higher than the rate paid by the bank to the parent, allowing the bank a margin for expenses and profit.

The bank chosen for the fronting loan is usually in a neutral country, away from both the lender's and the borrower's legal jurisdiction. Use of fronting loans increases the chances for repayment should political turmoil occur between the home and host countries. Government authorities are more likely to allow a local subsidiary to repay a loan to a large international bank in a neutral country than to allow the same subsidiary to repay a loan directly to its parent. To stop payment to the international bank would hurt the international credit image of the country, whereas to stop payment to the parent corporation would have minimal impact on that image and might even provide some domestic political advantage.

Creating Unrelated Exports. Another approach to blocked funds, which benefits both the subsidiary and host country, is the creation of unrelated exports. Because the main reason for stringent exchange controls is usually a host country's persistent inability to earn hard currencies, anything an MNE can do to create new exports from the host country helps the situation and provides a potential means to transfer funds out. Some new exports can often be created from present productive capacity with little or no additional investment, especially if they are in product lines related to existing operations. Other new exports may require reinvestment or new funds, although if the funds reinvested consist of those already blocked, little is lost in the way of opportunity costs.

Special Dispensation. If all else fails and the multinational firm is investing in an industry that is important to the economic development of the host country, the firm may bargain for special dispensation to repatriate some portion of the funds that otherwise would be blocked. Firms in "desirable" industries such as telecommunications, semiconductor manufacturing, instrumentation, pharmaceuticals, or other research and high-tech industries may receive preference over firms in mature industries. The amount of preference received often depends on the relative bargaining strength of the two parties.

Self-Fulfilling Prophecies. In seeking "escape routes" for blocked funds—or for that matter in trying to position funds through any of the techniques discussed in this chapter—the MNE may increase political risk and cause a change from partial blockage to full blockage. The possibility of such a self-fulfilling cycle exists any time a firm takes action that, no matter how legal, thwarts the underlying intent of politically motivated controls. In the statehouses of the world, as in the editorial offices of the local press and TV, MNEs and their subsidiaries are always potential scapegoats.

Forced Reinvestment. If funds are indeed blocked from transfer into foreign exchange, they are by definition "reinvested." Under such a situation, the firm must find local opportunities that will maximize the rate of return for a given acceptable level of risk.

If blockage is expected to be temporary, the most obvious alternative is to invest in local money market instruments. Unfortunately, in many countries, such instruments are not available in sufficient quantity or with adequate liquidity. In some cases, government Treasury bills, bank deposits, and other short-term instruments have yields that are kept artificially low relative to local rates of inflation or probable changes in exchange rates. Thus, the firm often loses real value during the period of blockage.

If short-term portfolio investments, such as bonds, bank time deposits, or direct loans to other companies, are not possible, investment in additional production facilities may be the only alternative. Often, this investment is what the host country is seeking by its exchange controls, even if exchange controls is counterproductive to additional foreign investment. Examples of forced direct reinvestment can be cited for Peru, where an airline invested in hotels and in maintenance facilities for other airlines; for Turkey, where a fish canning company constructed a plant to manufacture cans needed for packing the catch.

If investment opportunities in additional production facilities are not available, funds may simply be used to acquire other assets expected to increase in value with local inflation. Typical purchases might be land, office buildings, or commodities that are exported to global markets. Even inventory stockpiling might be a reasonable investment, given the low opportunity cost of the blocked funds.

Country-Specific Risk: Cultural and Institutional Risk

When investing in some of the emerging markets, MNEs that are resident in the most industrialized countries face serious risks because of cultural and institutional differences. Many such differences include the following:

- Differences in allowable ownership structures
- Differences in human resource norms
- Differences in religious heritage
- Nepotism and corruption in the host country
- Protection of intellectual property rights
- Protectionism
- Legal liabilities

Ownership Structure

Historically, many countries have required that MNEs share ownership of their foreign subsidiaries with local firms or citizens. Thus, joint ventures were the only way an MNE could operate in some host countries. Prominent countries that used to require majority local ownership were Japan, Mexico, China, India, and Korea. This requirement has been eliminated or modified in more recent years by these countries and most others. However, firms in certain industries are still either excluded from ownership completely or must accept being a minority owner. These industries are typically related to national defense, agriculture, banking, or other sectors that are deemed critical for the host nation.

Human Resource Norms

MNEs are often required by host countries to employ a certain proportion of host-country citizens rather than staffing mainly with foreign expatriates. It is often very difficult to fire local employees due to host-country labor laws and union contracts. This lack of flexibility to downsize in response to business cycles affects both MNEs and their local competitors. It also qualifies as a country-specific risk. Cultural differences can also inhibit an MNE's staffing policies. For example, it is somewhat difficult for a woman manager to be accepted by local employees and managers in selected Middle Eastern countries.

Religious Heritage

MNEs in some cases may encounter host country environments in which the political attitudes, and therefore business attitudes, of the state are intertwined with religious beliefs that conflict with MNE business practices. Despite religious differences, however, MNEs have operated successfully in emerging markets, especially in extractive and natural resource industries, such as oil, natural gas, minerals, and forest products. The main MNE strategy is to understand and respect the host country's religious traditions.

Nepotism and Corruption

MNEs must deal with endemic *nepotism* and corruption in a number of important foreign investment locations. Indonesia was famous for nepotism and corruption under the now-deposed Suharto government. Nigeria, Kenya, Uganda, and a number of other African countries have a history of nepotism and corruption after they threw out their colonial governments after World War II. China and Russia have recently launched well-publicized crackdowns on those evils.

Bribery is not limited to emerging markets. It is also a problem in even the most industrialized countries, including the United States and Japan. In fact, the United States has an anti-bribery law that would imprison any U.S. business executive found guilty of bribing a foreign government official. This law was passed in reaction to a U.S. aircraft manufacturer's attempt to bribe a senior Japanese government official.

MNEs are caught in a dilemma. Should they employ bribery if their local competitors use this strategy? The nearly universal response is, absolutely not. Most MNEs have a set of principles and practices that they follow in the execution of their business globally. These principles and ethics are universal, and not situation specific. Regardless of what competitors may or may not do, the individual MNE must follow its own set of principles—even if it means losing the business.

Intellectual Property Rights

Rogue businesses in some host countries have historically infringed on the intellectual property rights of both MNEs and individuals. *Intellectual property rights* grant the exclusive use of patented technology and copyrighted creative materials. Examples of patented technology are unique manufactured products, processing techniques, and prescription pharmaceutical drugs. Examples of copyrighted creative materials are software programs, educational materials (textbooks), and entertainment products (e.g., music, film, art).

The agreement on Trade-Related Aspects of Intellectual Property Rights (TRIPS) to protect intellectual property rights has been ratified by most major countries. It remains to be seen whether host governments are strong enough to enforce their official efforts to stamp out intellectual piracy. Complicating this task is the thin line that exists between the real item being protected and generic versions of the same item.

Protectionism

Protectionism is defined as the attempt by a national government to protect certain of its designated industries from foreign competition through such methods as tariffs or other import restrictions. Protected industries are usually related to defense, agriculture, and "infant" industries.

Defense. Even though the U.S. is a vocal proponent of open markets, a foreign firm proposing to buy a U.S. critical defense supplier would not be welcome. The same attitude exists in many other countries, such as France, which has always wished to maintain an independent defense capability.

Agriculture. Agriculture is another sensitive industry. No MNE would be foolish enough as to attempt to buy agricultural properties, such as rice operations, in Japan. Japan has worked diligently to maintain an independent ability to feed its own population. Agriculture is the typical "Mother Earth" industry that most countries want to protect for their own citizens.

Infant Industries. The traditional protectionist argument is that newly emerging "infant" industries need protection from foreign competition until they are firmly established. The infant industry argument is usually directed at limiting imports but not necessarily MNEs. In fact, most host countries encourage MNEs to establish operations in new industries that do not presently exist in the host country. Sometimes the host country offers foreign MNEs "infant industry" status for a limited number of years. This status could lead to tax subsidies, construction of infrastructure, employee training, and other aids to help the MNE get started. Host countries are especially interested in attracting MNEs that promise to export, either to their own foreign subsidiaries elsewhere or to unrelated parties.

Tariff and Non-Tariff Barriers. The traditional methods for countries to implement protectionist barriers were through tariff and non-tariff regulations. A multitude of international negotiations and treaties have greatly reduced the general level of tariffs over the past decades. However, many non-tariff barriers remain. These non-tariff barriers restrict imports by something other than a financial cost and are often difficult to identify because they are promulgated as health, safety, or sanitation requirements.

Strategies to Manage Protectionism. MNEs have only a very limited ability to overcome host country protectionism. However, MNEs do enthusiastically support efforts to reduce protectionism by joining in regional markets. The best examples of regional markets are the European Union (EU), the North American Free Trade Association (NAFTA), and the Latin American Free Trade Association (MERCOSUR). Among the objectives of regional markets are elimination of internal trade barriers, such as tariffs and non-tariff barriers, as well as the free movement of citizens for employment purposes.

Legal Liabilities

Despite good intentions, MNEs are often confronted with unexpected legal liabilities. *Global Finance in Practice 17.2* illustrates why Hospira, a U.S.-based pharmaceutical manufacturer, decided to cancel an FDI project in Italy as a result of potential legal and associated financial liabilities.

Drugs, Public Policy, and the Death Penalty

Foreign direct investment can be a very tricky thing. Just ask Hospira, a U.S.-based pharmaceutical manufacturer. Hospira, of Lake Forest, Illinois (U.S.), stopped manufacturing of Pentothal (sodium thiopental) in North Carolina in the United States in mid-2009. It intended to shift all production to Italy.

But Hospira quickly found that the Italian government required assurance that any product produced in Italy would not be used in capital punishment procedures in the United States. Hospira, however, after extensive discussions with its wholesalers and distributors could not be certain that product, once entering into the sales and distribution system in the U.S., could not be diverted to departments of corrections for us in capital punishment.

Hospira concluded that it could not afford the risk that Italian authorities would hold it liable for the product if it was so diverted, and therefore decided to exit the market.

The news was met with dismay by the medical industry. Pentothal, at one time a widely used anesthetic, is today only used in a variety of special cases. The drug is preferred in specific cases because it does not cause blood pressure to drop severely, including the care of the elderly, patients with heart disease, or expecting mothers requiring emerging C-sections in which the possibility of low blood pressure is threatening. Second-best solutions would now have to be good enough.

Global-Specific Risk

Global-specific risks faced by MNEs have come to the forefront in recent years. Exhibit 17.6 summarizes some of these risks, and strategies that can be used to manage them. The most visible recent risk was, of course, the attack by terrorists on the twin towers of the World Trade Center in New York on September 11, 2001. Many MNEs had major operations in the World Trade Center and suffered heavy casualties among their employees. In addition to terrorism, other global-specific risks include the antiglobalization movement, environmental concerns, poverty in emerging markets, and cyber attacks on computer information systems.

EXHIBIT 17.6 **Management Strategies for Global-Specific Risks**

Terrorism and War
- Support government efforts to fight terrorism and war
- Crisis planning
- Cross-border supply chain integration

Antiglobalization
- Support government efforts to reduce trade barriers
- Recognize that MNEs are the targets

Environmental Concerns
- Show sensitivity to environmental concerns
- Support government efforts to maintain a level playing field for pollution controls

Poverty
- Provide stable, relatively well-paying jobs
- Establish the strictest of occupational safety standards

Cyber Attacks
- No effective strategy except Internet security efforts
- Support government anti-cyber attack efforts

MNE movement toward multiple primary objectives:
Profitability, Sustainable Development, Corporate Social Responsibility

Terrorism and War

Although the World Trade Center attack and its aftermath, the war in Afghanistan, have affected nearly everyone worldwide, many other acts of terrorism have been committed in recent years. More terrorist acts are expected to occur in the future. Particularly exposed are the foreign subsidiaries of MNEs and their employees. As mentioned earlier, foreign subsidiaries are especially exposed to war, ethnic strife, and terrorism because they are symbols of their respective parent countries.

Crisis Planning. No MNE has the tools to avert terrorism. Hedging, diversification, insurance, and the like are not suited to the task. Therefore, MNEs must depend on governments to fight terrorism and protect their foreign subsidiaries (and now even the parent firm). In return, governments expect financial, material, and verbal support from MNEs to support antiterrorist legislation and proactive initiatives to destroy terrorist cells wherever they exist.

MNEs can be subject to damage by being in harm's way. Nearly every year one or more host countries experience some form of ethnic strife, outright war with other countries, or terrorism. It seems that foreign MNEs are often singled out as symbols of "oppression" because they represent their parent country, especially if it is the United States.

Cross-Border Supply Chain Integration. The drive to increase efficiency in manufacturing has driven many MNEs to adopt just-in-time (JIT) near-zero inventory systems. Focusing on so-called inventory velocity, the speed at which inventory moves through a manufacturing process, arriving only as needed and not before, has allowed these MNEs to generate increasing profits and cash flows with less capital being bottled-up in the production cycle itself. This finely tuned supply chain system, however, is subject to significant political risk if the supply chain itself extends across borders.

Supply Chain Interruptions. Consider the cases of Dell Computer, Ford Motor Company, Apple Computer, Herman Miller, and The Limited in the days following the terrorist attacks of September 11, 2001. An immediate result of the attacks was the grounding of all aircraft into or out of the United States. Similarly, the land (Mexico and Canada) and sea borders of the United States were also shut down and not reopened for several days in some specific sites. Ford Motor Company shut down five of its manufacturing plants in the days following September 11 because of inadequate inventories of critical automotive inputs supplied from Canada.

Dell Computer, with one of the most highly acclaimed and admired virtually integrated supply chains, depends on computer parts and subassembly suppliers and manufacturers in both Mexico and Canada to fulfill its everyday assembly and sales needs. In recent years, Dell carried less than three days sales (total inventory—by cost of goods sold). Suppliers are integrated electronically with Dell's order fulfillment system, and deliver required components and subassemblies as sales demands require. But with the closure of borders and grounding of air freight, the company was literally brought to a near standstill because of its supply chain's reliance on the ability to treat business units and suppliers in different countries as if they were all part of a single seamless political unit.

As a result of these newly learned lessons, many MNEs are now evaluating the degree of exposure their own supply chains possess in regard to cross-border stoppages or other cross-border political events. These companies are not, however, about to abandon JIT. It is estimated that many U.S. companies alone have saved more than $1 billion a year in inventory carrying costs by using JIT methods over the past decade. This substantial benefit is now being weighed against the costs and risks associated with the post-September 11 supply chain interruptions. To avoid suffering a similar fate in the future, manufacturers, retailers, and suppliers are now employing a range of tactics:

- **Inventory Management.** Manufacturers and assemblers are now considering carrying more buffer inventory in order to hedge against supply and production-line disruptions. Retailers, meanwhile, should think about the timing and frequency of their replenishment. Rather than stocking up across the board, companies are focusing on the most critical parts to the product or service, and those components, which are uniquely available from international sources.

- **Sourcing.** Manufacturers are now being more selective about where the critical inputs to their products come from. Although sourcing strategies will have to vary by location (those involving Mexico for example will differ dramatically from Canada), firms are attempting to work more closely with existing suppliers to minimize cross-border exposures and reduce the potential costs with future stoppages.

- **Transportation.** Retailers and manufacturers alike are reassessing their cross-border shipping arrangements. Although the mode of transportation employed is a function of value, volume, and weight, many firms are now reassessing whether higher costs for faster shipment balance out the more tenuous delivery under airline stoppages from labor, terrorist, or even bankruptcy disruptions in the future.

Antiglobalization Movement

During the past decade, there has been a growing negative reaction by some groups to reduced trade barriers and efforts to create regional markets, particularly to NAFTA and the European Union. NAFTA has been vigorously opposed by those sectors of the labor movement that could lose jobs to Mexico. Opposition within the European Union centers on loss of cultural identity, dilution of individual national control as new members are admitted, over-centralization of power in a large bureaucracy in Brussels, and the loss of independent monetary policies including the definition of their own individual currencies.

The antiglobalization movement has become more visible following riots in Seattle during the 2001 annual meeting of the World Trade Organization. However, antiglobalization forces were not solely responsible for these riots, or for subsequent riots in Quebec and Prague in 2001. Other disaffected groups, such as environmentalists and even anarchists, joined in to make their causes more visible. MNEs do not have the tools to combat antiglobalism. Indeed they are blamed for fostering the problem in the first place. Once again, MNEs must rely on governments and crisis planning to manage these risks.

Environmental Concerns

MNEs have been accused of "exporting" their environmental problems to other countries. The accusation is that MNEs, frustrated by pollution controls in their home country, have relocated these activities to countries with weaker pollution controls. Another accusation is that MNEs contribute to the problem of global warming. However, that accusation applies to all firms in all countries. It is based on the manufacturing methods employed by specific industries and on consumers' desire for certain products, such as large automobiles and sport utility vehicles that are not fuel efficient.

Once again, solving environmental problems is dependent on governments passing legislation and implementing pollution control standards. In 2001, a treaty attempting to reduce global warming was ratified by most nations, with the notable exception of the United States. However, the United States has promised to combat global warming using its own strategies. The United States objected to provisions in the worldwide treaty that allowed emerging nations to follow less restrictive standards, while the economic burden would fall on the most industrialized countries, particularly the United States.

Poverty

MNEs have located foreign subsidiaries in countries plagued by extremely uneven income distribution. At one end of the spectrum is an elite class of well-educated, well-connected, and productive persons. At the other end is a very large class of persons living at or below the poverty level. They lack education, social and economic infrastructure, and political power.

MNEs might be contributing to this disparity by employing the elite class to manage their operations. On the other hand, MNEs are creating relatively stable and well-paying jobs for those who would be otherwise unemployed and living below the poverty level. Despite being accused of supporting "sweat shop" conditions, MNEs usually compare favorably to their local competitors.

Cyber Attacks

The rapid growth of the Internet has fostered a whole new category of digital terrorists that aspire to disrupt institutions and organizations of all kinds. Although a domestic problem initially, cyber attacks have been international in their source and structure now for over a decade. MNEs can face costly cyber attacks by disaffected persons with a grudge because of their visibility and the complexity of their internal information systems, as Sony learned in December 2014.

At this time, we know of no uniquely international strategies that MNEs can use to combat cyber attacks. MNEs are using the same strategies to manage foreign cyber attacks as they use for domestic attacks. Once again, they must rely on governments to control cyber attacks.

SUMMARY POINTS

- In order to invest abroad a firm must have a sustainable competitive advantage in the home market. This must be strong enough and transferable enough to overcome the disadvantages of operating abroad.

- Competitive advantages stem from economies of scale and scope arising from large size; managerial and marketing expertise; superior technology; financial strength; differentiated products; and competitiveness of the home market.

- The *OLI Paradigm* is an attempt to create an overall framework to explain why MNEs choose FDI rather than serve foreign markets through alternative modes, such as licensing, joint ventures, strategic alliances, management contracts, and exporting.

- Political risks can be defined on three levels: firm-specific, country-specific, or global-specific.

- The main country-specific risks are transfer risk, known as blocked funds, and certain cultural and institutional risks.

- Cultural and institutional risks emanate from host-country policies with respect to ownership structure, human resource norms, religious heritage, nepotism and corruption, intellectual property rights, protectionism, and legal liabilities.

- Managing cultural and institutional risks requires the MNE to understand the differences, take legal actions in host-country courts, support worldwide treaties to protect intellectual property rights, and support government efforts to create regional markets.

- MNEs should adopt a crisis plan to protect its employees and property against global-specific risks. However, the main reliance remains on governments to protect its citizens and firms from these global-specific threats.

Strategic Portfolio Theory, Black Swans, and [Avoiding] Being the Turkey[7]

Put another way, investors are more volatile than investments. Economic reality governs the returns earned by our businesses, and Black Swans are unlikely. But emotions and perceptions—the swings of hope, greed, and fear among the participants in our financial system—govern the returns earned in our markets. Emotional factors magnify or minimize this central core of economic reality, and Black Swans can appear at any time.

—John C. Bogle, founder of The Vanguard Group[8]

Modern Portfolio Theory (MPT), like all theories, has been subject to much criticism. Most of that criticism is focused either on the failings of the theory's fundamental assumptions, or in the way the theory and its assumptions have been applied. Ultimately, it has been accused of failing to predict the major financial crises of our time, like Black Monday in 1987, the credit crisis in the U.S. in 2008, or the current financial crisis over sovereign debt in Europe.

Criticisms of Modern Portfolio Theory

Modern portfolio theory was the creation of Harry Markowitz in which he applied principles of linear programming to the creation of asset portfolios.[9] Markowitz demonstrated that an investor could reduce the standard deviation of portfolio returns by combining assets that were less than perfectly correlated in their returns. The theory assumes that all investors have access to the same information at the same point in time. It assumes all investors are rational and risk averse, and will take on additional risk only if compensated by higher expected returns. It assumes all investors are similarly rational, although different investors will have different trade-offs between risk and return based on their own risk aversion characteristics. The usual measure of risk used in portfolio theory is the standard deviation of returns, assuming a normal distribution of returns over time.

As one would expect, the criticisms of portfolio theory are pointed at each and every assumption behind the theory. For example, the field of behavioral economics argues that investors are not necessarily rational—that in some cases, gamblers buy risk. All investors do not have access to the same information, that insider trading persists, that some investors are biased, that some investors regularly beat the market through market timing. Even the mathematics comes under attack, as to whether standard deviations are the appropriate measure of risk to minimize, or whether the standard normal distribution is appropriate.

Many of the major stock market collapses in recent history, like that of Black Monday's crash on October 19, 1987, when the Dow fell 23%, were "missed" by the purveyors of portfolio strategy. Statistical studies of markets and their returns over time often show returns that are not normally distributed, but are subject to greater deviation from the mean than traditional normal distributions—evidence of so-called *fat tails*. Much of the work of Benoit Mandelbrot, the father of fractal geometry, revolved around the possibility that financial markets exhibited fat tail distributions. Mandelbrot's analysis in fact showed that the Black Monday event was a 20-sigma event, one which, according to *normal distributions* (*bell curve* or Gaussian model), was so improbable as not likely to ever occur.[10] And if something has not happened in the past, portfolio theory assumes it cannot happen in the future. Yet it did.

The argument and criticism that has been deployed with the greatest traction seems to be that portfolio theory is typically executed using historical data—the numbers from the past—assuming a distribution that the data does not fit.

Any attempts to refine the tools of modern portfolio theory by relaxing the bell curve assumptions, or by "fudging" and adding the occasional "jumps" will not be sufficient. We live in a world primarily driven by random jumps, and tools designed for random walks address the wrong problem. It would be like tinkering with models of gases in an attempt to characterise them as solids and call them "a good approximation"[11]

---

[7] Copyright © 2012 Thunderbird School of Global Management. All rights reserved. This case was prepared by Professor Michael H. Moffett for the purpose of classroom discussion only and not to indicate either effective or ineffective management.

[8] "Black Monday and Black Swans," remarks by John C. Bogle, Founder and Former Chief Executive, The Vanguard Group, before the Risk Management Association, Boca Raton, Florida, October 11, 2007, p.6.

[9] H.M. Markowitz, "Portfolio Selection," *Journal of Finance*, Vol. 7, March 1952, pp. 77–91.

[10] For a more detailed discussion see Richard L. Hudson and Benoit B. Mandelbrot, *The (Mis)Behavior of Markets: A Fractal View of Risk, Ruin, and Reward*, Basic Books, 2004.

[11] Benoit Mandelbrot and Nassim Taleb, "A focus on the exceptions that prove the rule," *The Financial Times*, March 23, 2006.

Black Swan Theory

Nassim Nicholas Taleb published a book in 2001 entitled *Fooled by Randomness* in which he introduced the analogy of the black swan.[12] The argument is quite simple: prior to the discovery of Australia and the existence of black swans, all swans were thought to be white. Black swans did not exist because no one had ever seen one. But that did not mean they did not exist. Taleb then applied this premise to financial markets, arguing that simply because a specific event had never occurred did not mean it couldn't.

Taleb argued that a black swan event is characterized by three fundamentals, in order, *Rarity*, *Extremeness*, and *Retrospective Predictability*:

1. *Rarity*: The event is a shock or surprise to the observer.
2. *Extremeness*: The event has a major impact.
3. *Retrospective Predictability*: After the event has occurred, the event is rationalized by hindsight, and found to have been predictable.

Although the third argument is a characteristic of human intellectual nature, it is the first element that is fundamental to the debate. If an event has not been recorded, does that mean it cannot occur? Portfolio theory is a mathematical analysis of provided inputs. Its outcomes are no better than its inputs. The theory itself does not predict price movements. It simply allows the identification of portfolios in which the risk is at a minimum for an expected level of return.

Taleb does not argue that he has some secret ability to predict the future when historical data cannot. Rather, he argues that investors should structure their portfolios, their investments, to protect against the extremes, the *improbable events* rather than the *probable* ones. He argues for what many call "investment humility," to acknowledge that the world we live in is not always the one we think we live in, and to understand how much we will never understand.

What Drives the Improbable?

So what causes the "random jumps" noted by Mandelbrot and Taleb? A number of investment theorists, including John Maynard Keynes and John Bogle, have argued over the past century that equity returns are driven by two fundamental forces, *enterprise* (economic or business returns over time) and *speculation* (the psychology or emotions of the individuals in the market). First Keynes and then Bogle eventually concluded that *speculation* would win out over *enterprise*, very much akin to arguing that hope will win out over logic.

This is what Bogle means in the opening quotation when he states that "investors are more volatile than investments."

All agree that the behavior of the *speculators* (in the words of Keynes) or the *jumps* of market returns (in the words of Mandelbrot) are largely unpredictable. And all that one can do is try and protect against the unpredictable by building more robust systems and portfolios than can hopefully withstand the improbable. But most also agree that the "jumps" are exceedingly rare, and depending upon the holding period, the market may return to more fundamental values, if given the time. But the event does indeed have a lasting impact.

Portfolio theory remains a valuable tool. It allows investors to gain approximate values over the risk and expected returns they are likely to see in their total positions. But it is fraught with failings, although it is not at all clear what would be better. Even some of history's greatest market timers have noted that they have followed modern portfolio theory, but have used subjective inputs on what is to be expected in the future. Even Harry Markowitz—on the last page of the same article that started it all, *Portfolio Selection*—noted that ". . . in the selection of securities we must have procedures for finding reasonable ρ_i and σ_{ij}. These procedures, I believe, should combine statistical techniques and the judgment of practical men."

But the judgement of practical men is—well—difficult to validate. We need predictions of what may come, even if it is generally based on the past. Kenneth Arrow, a famed economist and Nobel Prize winner relayed the following story of how during the World War II a group of statisticians were tasked with forecasting weather patterns.

> *The statisticians subjected these forecasts to verification and found they differed in no way from chance. The forecasters themselves were convinced and requested that the forecasts be discontinued. The reply read approximately like this: "The Commanding General is well aware that the forecasts are no good. However, he needs them for planning purposes."*[13]

Taleb, in a recent edition of *Black Swan Theory*, notes that the event is a surprise to the specific observer, and that what is a surprise to the turkey is not a surprise to the butcher. The challenge is "to avoid being the turkey."

Mini-Case Questions

1. What are the primary assumptions behind modern portfolio theory?
2. What do many of MPT's critics believe are the fundamental problems with the theory?
3. How would you suggest MPT be used in investing your own money?

---

[12] Nassim Nicholas Taleb, *Fooled by Randomness: The Hidden Role of Chance in Life and in the Markets*, Random House, 2001. He later extended his premise to many other fields beyond finance in *The Black Swan: The Impact of the Highly Improbable*, Random House, 2007.

[13] "I Know a Hawk from a Handsaw," *Emminent Economists: Their Life Philosophies*, edited by Michael Szenberg, Cambridge University Press, 1992, p. 47.

QUESTIONS

These questions are available in MyFinanceLab.

1. **Evolving into Multinationalism.** As a firm evolves from purely domestic into a true multinational enterprise, it must consider (1) its competitive advantages, (2) its production location, (3) the type of control it wants to have over any foreign operations, and (4) how much monetary capital to invest abroad. Explain how each of these considerations is important to the success of foreign operations.

2. **Market Imperfections.** MNEs strive to take advantage of market imperfections in national markets for products, factors of production, and financial assets. Large international firms are better able to exploit such imperfections. What are their main competitive advantages?

3. **Competitive Advantage.** In deciding whether to invest abroad, management must first determine whether the firm has a sustainable competitive advantage that enables it to compete effectively in the home market. What are the necessary characteristics of this competitive advantage?

4. **Economies of Scale and Scope.** Explain briefly how economies of scale and scope can be developed in production, marketing, finance, research and development, transportation, and purchasing.

5. **Competitiveness of the Home Market.** A strongly competitive home market can sharpen a firm's competitive advantage relative to firms located in less competitive markets. Explain what is meant by the "competitive advantage of nations."

6. **OLI Paradigm.** The OLI Paradigm is an attempt to create an overall framework to explain why MNEs choose FDI rather than serve foreign markets through alternative modes. Explain what is meant by the "O," the "L," and the "I" of the paradigm.

7. **Financial Links to OLI.** Financial strategies are directly related to the OLI Paradigm.
   a. Explain how proactive financial strategies are related to OLI.
   b. Explain how reactive financial strategies are related to OLI.

8. **Where to Invest.** The decision about where to invest abroad is influenced by behavioral factors.
   a. Explain the behavioral approach to FDI.
   b. Explain the international network theory explanation of FDI.

9. **Exporting versus Producing Abroad.** What are the advantages and disadvantages of limiting a firm's activities to exporting compared to producing abroad?

10. **Licensing and Management Contracts versus Producing Abroad.** What are the advantages and disadvantages of licensing and management contracts compared to producing abroad?

11. **Joint Venture versus Wholly Owned Production Subsidiary.** What are the advantages and disadvantages of forming a joint venture to serve a foreign market compared to serving that market with a wholly owned production subsidiary?

12. **Greenfield Investment versus Acquisition.** What are the advantages and disadvantages of serving a foreign market through a greenfield foreign direct investment compared to an acquisition of a local firm in the target market?

13. **Cross-Border Strategic Alliance.** The term "cross-border strategic alliance" conveys different meanings to different observers. What are the meanings?

14. **Governance Risk.** Answer the following questions:
   a. What is meant by the term "governance risk?"
   b. What is the most important type of governance risk?

15. **Investment Agreement.** An investment agreement spells out specific rights and responsibilities of both the foreign firm and the host government. What are the main financial policies that should be included in an investment agreement?

16. **Investment Insurance and Guarantees: OPIC.** Answer the following questions:
   a. What is OPIC?
   b. What types of political risks can OPIC insure against?

17. **Operating Strategies after the FDI Decision.** The following operating strategies, among others, are expected to reduce damage from political risk. Explain each and how it reduces damage.
   a. Local sourcing
   b. Facility location
   c. Control of technology
   d. Thin equity base
   e. Multiple-source borrowing

18. **Country-Specific Risk.** Define the following terms:
   a. Transfer risk
   b. Blocked funds

19. **Blocked Funds.** Explain the strategies used by an MNE to counter blocked funds.

20. **Cultural and Institutional Risks.** Identify and explain the main types of cultural and institutional risks, except protectionism.

21. **Strategies to Manage Cultural and Institutional Risks.** Explain the strategies that an MNE can use to manage each of the cultural and institutional risks that you identified in Question 20, except protectionism.

22. **Protectionism Defined.** Respond to the following:
    a. Define protectionism and identify the industries that are typically protected.
    b. Explain the "infant industry" argument for protectionism.

23. **Managing Protectionism.** Answer the following questions:
    a. What are the traditional methods for countries to implement protectionism?
    b. What are some typical non-tariff barriers to trade?
    c. How can MNEs overcome host country protectionism?

24. **Global-Specific Risks.** What are the main types of political risks that are global in origin?

25. **Managing Global-Specific Risks.** What are the main strategies used by MNEs to manage the global-specific risks you have identified in Question 24?

26. **U.S. Antibribery Law.** The United States has a law prohibiting U.S. firms from bribing foreign officials and business persons, even in countries where bribery is a normal practice. Some U.S. firms claim this places the United States at a disadvantage compared to host-country firms and other foreign firms that are not hampered by such a law. Discuss the ethics and practicality of the U.S. antibribery law.

INTERNET EXERCISES

1. **Global Corruption Report.** Transparency International (TI) is considered by many to be the leading nongovernmental anticorruption organization in the world today. Recently, it has introduced its own annual survey analyzing current developments, identifying ongoing challenges, and offering potential solutions to individuals and organizations. One dimension of this analysis is the Bribe Payers Index. Visit TI's Web site to view the latest edition of the Bribe Payers Index.

Corruption Index	www.transparency.org/policy_research/surveys_indices/cpi
Bribe Payers Index	www.transparency.org/policy_research/surveys_indices/bpi

2. **Sovereign Credit Ratings Criteria.** The evaluation of credit risk and all other relevant risks associated with the multitude of borrowers on world debt markets requires a structured approach to international risk assessment. Use Standard and Poor's criteria, described in depth on their Web page, to differentiate the various risks (local currency risk, default risk, currency risk, transfer risk, etc.) contained in major sovereign ratings worldwide. (You may need to complete a free login for this site.)

Standard and Poor's	www.standardandpoors.com/ratings/sovereigns/ratings-list/en/

3. **Milken Capital Access Index.** The Milken Institute's Capital Access Index (CAI) is one of the most recent informational indices that aids in the evaluation of how accessible world capital markets are to MNEs and governments of many emerging market countries. According to the CAI, which countries have seen the largest deterioration in their access to capital in the last two years?

Milken Institute	www.milken-inst.org

4. **Overseas Private Investment Corporation.** The Overseas Private Investment Corporation (OPIC) provides long-term political risk insurance and limited recourse project financing aid to U.S.-based firms investing abroad. Using the organization's Web page, answer the following questions:
    a. Exactly what types of risk will OPIC insure against?
    b. What financial limits and restrictions are there on this insurance protection?
    c. How should a project be structured to aid in its approval for OPIC coverage?

Overseas Private Investment Corp.	www.opic.gov

5. **Political Risk and Emerging Markets.** Check the World Bank's political risk insurance blog for current issues and topics in emerging markets.

Political Insurance Blog	blogs.worldbank.org/miga/category/tags/political-risk-insurance

CHAPTER 18

Multinational Capital Budgeting and Cross-Border Acquisitions

When it comes to finances, remember that there are no withholding taxes on the wages of sin.

—Mae West (1892–1980), *Mae West on Sex, Health and ESP*, 1975.

LEARNING OBJECTIVES

- Extend the domestic capital budgeting analysis to evaluate a greenfield foreign project
- Distinguish between the project viewpoint and the parent viewpoint of a potential foreign investment
- Adjust the capital budgeting analysis of a foreign project for risk
- Examine the use of project finance to fund and evaluate large global projects
- Introduce the principles of cross-border mergers and acquisitions

This chapter describes in detail the issues and principles related to the investment in real productive assets in foreign countries, generally referred to as *multinational capital budgeting*. The chapter first describes the complexities of budgeting for a foreign project. Second, we describe the insights gained by valuing a project from both the *project's viewpoint* and the *parent's viewpoint* using an illustrative case involving an investment by Cemex of Mexico in Indonesia. This illustrative case also explores *real option analysis*. Next, the use of *project financing* today is discussed, and the final section describes the stages involved in affecting cross-border acquisitions. The chapter concludes with the Mini-Case, *Elan and Royalty Pharma*, about a hostile takeover (acquisition) attempt that played out in the summer of 2013.

Although the original decision to undertake an investment in a particular foreign country may be determined by a mix of strategic, behavioral, and economic factors, the specific project should be justified—as should all reinvestment decisions—by traditional financial analysis. For example, a production efficiency opportunity may exist for a U.S. firm to invest abroad, but the type of plant, mix of labor and capital, kinds of equipment, method of financing, and other project variables must be analyzed with traditional discounted cash flow analysis. The firm must also consider the impact of the proposed foreign project on consolidated earnings, cash flows from subsidiaries in other countries, and on the market value of the parent firm.

Multinational capital budgeting for a foreign project uses the same theoretical framework as domestic capital budgeting—with a few very important differences. The basic steps are as follows:

- Identify the initial capital invested or put at risk.

- Estimate cash flows to be derived from the project over time, including an estimate of the terminal or salvage value of the investment.

- Identify the appropriate discount rate for determining the present value of the expected cash flows.

- Use traditional capital budgeting methods, such as net present value (NPV) and internal rate of return (IRR), to assess and rank potential projects.

Complexities of Budgeting for a Foreign Project

Capital budgeting for a foreign project is considerably more complex than the domestic case. Two broad categories of factor contribute to this greater complexity, cash flows and managerial expectations.

Cash Flows

- Parent cash flows must be distinguished from project cash flows. Each of these two types of flows contributes to a different view of value.

- Parent cash flows often depend on the form of financing. Thus, we cannot clearly separate cash flows from financing decisions, as we can in domestic capital budgeting.

- Additional cash flows generated by a new investment in one foreign subsidiary may be in part or in whole taken away from another subsidiary, with the net result that the project is favorable from a single subsidiary's point of view but contributes nothing to worldwide cash flows.

- The parent must explicitly recognize remittance of funds because of differing tax systems, legal and political constraints on the movement of funds, local business norms, and differences in the way financial markets and institutions function.

- An array of nonfinancial payments can generate cash flows from subsidiaries to the parent, including payment of license fees and payments for imports from the parent.

Management Expectations

- Managers must anticipate differing rates of national inflation because of their potential to cause changes in competitive position, and thus changes in cash flows over a period of time.

- Managers must keep the possibility of unanticipated foreign exchange rate changes in mind because of possible direct effects on the value of local cash flows, as well as indirect effects on the competitive position of the foreign subsidiary.

- Use of segmented national capital markets may create an opportunity for financial gains or may lead to additional financial costs.

- Use of host-government subsidized loans complicates both capital structure and the parent's ability to determine an appropriate weighted average cost of capital for discounting purposes.

- Managers must evaluate political risk because political events can drastically reduce the value or availability of expected cash flows.

- Terminal value is more difficult to estimate because potential purchasers from the host, parent, or third countries, or from the private or public sector, may have widely divergent perspectives on the value to them of acquiring the project.

Since the same theoretical capital budgeting framework is used to choose among competing foreign and domestic projects, it is critical that we have a common standard. Thus, all foreign complexities must be quantified as modifications to either expected cash flow or the rate of discount. Although in practice many firms make such modifications arbitrarily, readily available information, theoretical deduction, or just plain common sense can be used to make less arbitrary and more reasonable choices.

Project versus Parent Valuation

Consider a foreign direct investment like that illustrated in Exhibit 18.1. A U.S. multinational invests capital in a foreign project in a foreign country, the results of which—if they occur—are generated over time. Similar to any investment, domestically or internationally, the return on the investment is based on the outcomes to the parent company. Given that the initial investment is in the parent's own or home currency, the U.S. dollar as shown here, then those returns over time need to be denominated in that same currency for evaluation purposes.

A strong theoretical argument exists in favor of analyzing any foreign project from the *viewpoint of the parent*. Cash flows to the parent are ultimately the basis for dividends to stockholders, reinvestment elsewhere in the world, repayment of corporate-wide debt, and other purposes that affect the firm's many interest groups. However, since most of a project's

---

EXHIBIT 18.1   **Multinational Capital Budgeting: Project and Parent Viewpoints**

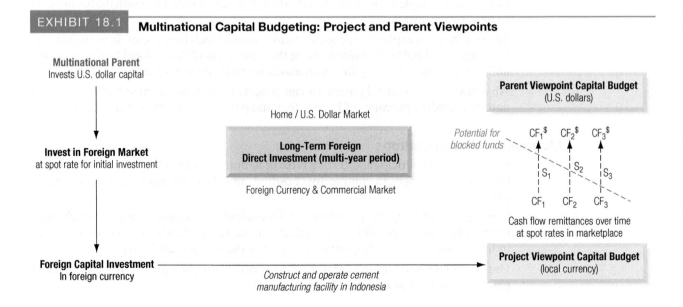

cash flows to its parent or sister subsidiaries are financial cash flows rather than operating cash flows, the parent viewpoint violates a cardinal concept of capital budgeting, namely, that financial cash flows should not be mixed with operating cash flows. Often the difference is not important because the two are almost identical, but in some instances a sharp divergence in these cash flows will exist. For example, funds that are permanently blocked from repatriation, or "forcibly reinvested," are not available for dividends to the stockholders or for repayment of parent debt. Therefore, shareholders will not perceive the blocked earnings as contributing to the value of the firm, and creditors will not count on them in calculating interest coverage ratios and other metrics of debt service capability.

Evaluation of a project from the local viewpoint—the *project viewpoint*—serves a number of useful purposes as well. In evaluating a foreign project's performance relative to the potential of a competing project in the same host country, we must pay attention to the project's local return. Almost any project should at least be able to earn a cash return equal to the yield available on host government bonds with a maturity equal to the project's economic life, if a free market exists for such bonds. Host-government bonds ordinarily reflect the local risk-free rate of return, including a premium equal to the expected rate of inflation. If a project cannot earn more than such a bond yield, the parent firm should buy host government bonds rather than invest in a riskier project.

Multinational firms should invest only if they can earn a risk-adjusted return greater than locally based competitors can earn on the same project. If they are unable to earn superior returns on foreign projects, their stockholders would be better off buying shares in local firms, where possible, and letting those companies carry out the local projects. Apart from these theoretical arguments, surveys over the past 40 years show that in practice MNEs continue to evaluate foreign investments from both the parent and project viewpoint.

The attention paid to project returns in various surveys may reflect emphasis on maximizing reported earnings per share as a corporate financial goal of publicly traded companies. It is not clear that privately held firms place the same emphasis on consolidated results, given that few public investors ever see their financial results. Consolidation practices, including translation as described in Chapter 11, remeasure foreign project cash flows, earnings, and assets *as if* they are "returned" to the parent company. And as long as foreign earnings are not blocked, they can be consolidated with the earnings of both the remaining subsidiaries and the parent.[1] Even in the case of temporarily blocked funds, some of the most mature MNEs do not necessarily eliminate a project from financial consideration. They take a very long-run view of world business opportunities.

If reinvestment opportunities in the country where funds are blocked are at least equal to the parent firm's required rate of return (after adjusting for anticipated exchange rate changes), temporary blockage of transfer may have little practical effect on the capital budgeting outcome, because future project cash flows will be increased by the returns on forced reinvestment. Since large multinationals hold a portfolio of domestic and foreign projects, corporate liquidity is not impaired if a few projects have blocked funds; alternate sources of funds are available to meet all planned uses of funds. Furthermore, a long-run historical perspective on blocked funds does indeed lend support to the belief that funds are almost never permanently blocked. However, waiting for the release of such funds can be frustrating, and sometimes the blocked funds lose value while blocked because of inflation or unexpected exchange rate deterioration, even though they have been reinvested in the host country to protect at least part of their value in real terms.

---

[1] U.S. firms must consolidate foreign subsidiaries that are over 50% owned. If a firm is owned between 20% and 49% by a parent, it is called an *affiliate. Affiliates* are consolidated with the parent owner on a pro rata basis. Subsidiaries less than 20% owned are normally carried as unconsolidated investments.

In conclusion, most firms appear to evaluate foreign projects from both parent and project viewpoints. The parent's viewpoint gives results closer to the traditional meaning of net present value in capital budgeting theoretically, but as we will demonstrate, possibly not in practice. Project valuation provides a closer approximation of the effect on consolidated earnings per share, which all surveys indicate is of major concern to practicing managers. To illustrate the foreign complexities of multinational capital budgeting, we analyze a hypothetical market-seeking foreign direct investment by Cemex in Indonesia.

Illustrative Case: Cemex Enters Indonesia[2]

Cementos Mexicanos, Cemex, is considering the construction of a cement manufacturing facility on the Indonesian island of Sumatra. The project, Semen Indonesia (the Indonesian word for "cement" is semen), would be a wholly owned *greenfield investment* with a total installed capacity of 20 million metric tonnes per year (mmt/y). Although that is large by Asian production standards, Cemex believes that its latest cement manufacturing technology would be most efficiently utilized with a production facility of this scale.

Cemex has three driving reasons for the project: (1) the firm wishes to initiate a productive presence of its own in Southeast Asia, a relatively new market for Cemex; (2) the long-term prospects for Asian infrastructure development and growth appear very good over the longer term; and (3) there are positive prospects for Indonesia to act as a produce-for-export site as a result of the depreciation of the Indonesian rupiah (IDR or Rp) in recent years.

Cemex, the world's third-largest cement manufacturer, is an MNE headquartered in an emerging market but competing in a global arena. The firm competes in the global marketplace for both market share and capital. The international cement market, like markets in other commodities such as oil, is a dollar-based market. For this reason, and for comparisons against its major competitors in both Germany and Switzerland, Cemex considers the U.S. dollar its functional currency.

Cemex's shares are listed in both Mexico City and New York (OTC: CMXSY). The firm has successfully raised capital—both debt and equity—outside Mexico in U.S. dollars. Its investor base is increasingly global, with the U.S. share turnover rising rapidly as a percentage of total trading. As a result, its cost and availability of capital are internationalized and dominated by U.S. dollar investors. Ultimately, the Semen Indonesia project will be evaluated—in both cash flows and capital cost—in U.S. dollars.

Overview

The first step in analyzing Cemex's potential investment in Indonesia is to construct a set of *pro forma* financial statements for Semen Indonesia, all in Indonesian rupiah (IDR). The next step is to create two capital budgets, the project viewpoint and parent viewpoint. Semen Indonesia will take only one year to build the plant, with actual operations commencing in year 1. The Indonesian government has only recently deregulated the heavier industries to allow foreign ownership.

All of the following analysis is conducted assuming that purchasing power parity (PPP) holds for the rupiah to dollar exchange rate for the life of the Indonesian project. This is a standard financial assumption made by Cemex for its foreign investments. Thus, if we assume an initial spot rate of Rp10,000/\$, and Indonesian and U.S. inflation rates of 30% and 3% per annum, respectively, for the life of the project, forecasted spot exchange rates follow the usual

---

[2] Cemex is a real company. However, the greenfield investment described here is hypothetical.

PPP calculation. For example, the forecasted exchange rate for year 1 of the project would be as follows:

$$\text{Spot rate (year 1)} = \text{Rp10,000/US\$} \times \frac{1 + .30}{1 + .03} = \text{Rp12,621/US\$}$$

The financial statements shown in Exhibits 18.2 through 18.5 are based on these assumptions.

Capital Investment. Although the cost of building new cement manufacturing capacity anywhere in the industrial countries is now estimated at roughly \$150/tonne of installed capacity, Cemex believed that it could build a state-of-the-art production and shipment facility in Sumatra at roughly \$110/tonne (see Exhibit 18.2). Assuming a 20 million metric ton per year (mmt/y) capacity, and a year 0 average exchange rate of Rp10,000/\$, this cost will constitute an investment of Rp22 trillion (\$2.2 billion). This figure includes an investment of Rp17.6 trillion in plant and equipment, giving rise to an annual depreciation charge of Rp1.76 trillion if we assume a 10-year straight-line depreciation schedule. The relatively short depreciation schedule is one of the policies of the Indonesian tax authorities meant to attract foreign investment.

Financing. This massive investment would be financed with 50% equity, all from Cemex, and 50% debt—75% from Cemex and 25% from a bank consortium arranged by the Indonesian government. Cemex's own U.S. dollar-based weighted average cost of capital (WACC) was currently estimated at 11.98%. The WACC for the project itself on a local Indonesian level in rupiah terms was estimated at 33.257%. The details of this calculation are discussed later in this chapter.

The cost of the U.S. dollar-denominated loan is stated in rupiah terms assuming purchasing power parity and U.S. dollar and Indonesian inflation rates of 3% and 30% per annum, respectively, throughout the subject period. The explicit debt structures, including repayment schedules, are presented in Exhibit 18.3. The loan arranged by the Indonesian government, part of the government's economic development incentive program, is an eight-year loan, in rupiah, at 35% annual interest, fully amortizing. The interest payments are fully deductible against corporate tax liabilities.

The majority of the debt, however, is being provided by the parent company, Cemex. After raising the capital from its financing subsidiary, Cemex will re-lend the capital to Semen Indonesia. The loan is denominated in U.S. dollars, five years maturity, with an annual interest rate of 10%. Because the debt will have to be repaid from the rupiah earnings of the Indonesian enterprise, the pro forma financial statements are constructed so that the expected costs of servicing the dollar debt are included in the firm's pro forma income statement. The dollar loan, if the rupiah follows the purchasing power parity forecast, will have an effective interest expense in rupiah terms of 38.835% before taxes. We find this rate by determining the internal rate of return of repaying the dollar loan in full in rupiah (see Exhibit 18.3).

The loan by Cemex to the Indonesian subsidiary is denominated in U.S. dollars. Therefore, the loan will have to be repaid in U.S. dollars, not rupiah. At the time of the loan agreement, the spot exchange rate is Rp10,000/\$. This is the assumption used in calculating the "scheduled" repaying of principal and interest in rupiah. The rupiah, however, is expected to depreciate in line with purchasing power parity. As it is repaid, the "actual" exchange rate will therefore give rise to a foreign exchange loss as it takes more and more rupiah to acquire U.S. dollars for debt service, both principal and interest. The foreign exchange losses on this debt service will be recognized on the Indonesian income statement.

Revenues. Given the current existing cement manufacturing in Indonesia, and its currently depressed state as a result of the Asian crisis, all sales are based on export. The 20 mmt/y

| EXHIBIT 18.2 | Investment and Financing of the Semen Indonesia Project (in 000s) |

Invesment

Invesment		Financing	
Average exchange rate, Rp/$	10,000	Equity	11,000,000,000
Cost of installed capacity ($/tonne)	$110	Debt:	11,000,000,000
Installed capacity	20,000	Rupiah debt	2,750,000,000
Investment in US$	$2,200,000	US$ debt in rupiah	8,250,000,000
Investment in rupiah	22,000,000,000	Total	22,000,000,000
Percentage of investment in plant & equip	80%		
Plant and equipment (000s Rp)	17,600,000,000	Note: US$ debt principal	$825,000
Depreciation of capital equipment (years)	10.00		
Annual depreciation (millions)	(1,760,000)		

Costs of Capital: Cemex (Mexico)

Risk-free rate	6.000%	Cemex beta	1.50
Credit premium	2.000%	Equity risk premium	7.000%
Cost of debt	8.000%	Cost of equity	16.500%
Corporate income tax rate	35.000%	Percent equity	60.0%
Cost of debt after-tax	5.200%	**WACC**	**11.980%**
Percent debt	40.0%		

Cost of Capital: Semen Indonesia (Indonesia)

Risk-free rate	33.000%	Semen Indonesia beta	1.000
Credit premium	2.000%	Equity risk premium	6.000%
Cost of rupiah debt	35.000%	Cost of equity	40.000%
Indonesia corporate income tax rate	30.000%	Percent equity	50.0%
Cost of US$ debt, after-tax	5.200%	**WACC**	**33.257%**
Cost of US$ debt (rupiah equivalent)	38.835%		
Cost of US$ debt, after-tax (rupiah eq)	27.184%		
Percent debt	50.0%		

The cost of the US$ loan is stated in rupiah terms assuming purchasing power parity and U.S. dollar and Indionesian inflation rates of 3% and 30% per annum, respectively, throughout the subject period.

Semen Indonesia (Rp)	Amount	Financing Proportion	Cost	After-tax Cost	Component Cost
Rupiah loan	2,750,000,000	12.5%	35.000%	24.500%	3.063%
Cemex loan	8,250,000,000	37.5%	38.835%	27.184%	10.194%
Total debt	11,000,000,000	50.0%			
Equity	11,000,000,000	50.0%	40.000%	40.000%	20.000%
Total financing	22,000,000,000	100.0%		**WACC**	**33.257%**

facility is expected to operate at only 40% capacity (producing 8 million metric tonnes). Cement produced will be sold in the export market at $58/tonne (delivered). Note also that, at least for the conservative baseline analysis, we assume no increase in the price received over time.

EXHIBIT 18.3	Semen Indonesia's Debt Service Schedules and Foreign Exchange Gains/Losses					
Spot rate (Rp/$)	10,000	12,621	15,930	20,106	25,376	32,028
Project Year	**0**	**1**	**2**	**3**	**4**	**5**
Indonesian loan @ 35% for 8 years (millions of rupiah)						
Loan principal	2,750,000					
Interest payment		(962,500)	(928,921)	(883,590)	(822,393)	(739,777)
Principal payment		(95,939)	(129,518)	(174,849)	(236,046)	(318,662)
Total payment		(1,058,439)	(1,058,439)	(1,058,439)	(1,058,439)	(1,058,439)
Cemex loan @ 10% for 5 years (millions of U.S. dollars)						
Loan principal	825					
Interest payment		($82.50)	($68.99)	($54.12)	($37.77)	($19.78)
Principal payment		($135.13)	($148.65)	($163.51)	($179.86)	($197.85)
Total payment		($217.63)	($217.63)	($217.63)	($217.63)	($217.63)
Cemex loan converted to Rp at scheduled and current spot rates (millions of Rp):						
Scheduled at Rp10,000/$:						
Interest payment		(825,000)	(689,867)	(541,221)	(377,710)	(197,848)
Principal payment		(1,351,329)	(1,486,462)	(1,635,108)	(1,798,619)	(1,978,481)
Total payment		(2,176,329)	(2,176,329)	(2,176,329)	(2,176,329)	(2,176,329)
Actual (at current spot rate):						
Interest payment		(1,041,262)	(1,098,949)	(1,088,160)	(958,480)	(633,669)
Principal payment		(1,705,561)	(2,367,915)	(3,287,494)	(4,564,190)	(6,336,691)
Total payment		(2,746,823)	(3,466,864)	(4,375,654)	(5,522,670)	(6,970,360)
Cash flows in Rp on Cemex loan (millions of Rp):						
Total actual cash flows	8,250,000	(2,746,823)	(3,466,864)	(4,375,654)	(5,522,670)	(6,970,360)
IRR of cash flows	**38.835%**					
Foreign exchange gains (losses) on Cemex loan (millions of Rp):						
Foreign exchange gains (losses) on interest		(216,262)	(409,082)	(546,940)	(580,770)	(435,821)
Foreign exchange gains (losses) on principal		(354,232)	(881,453)	(1,652,385)	(2,765,571)	(4,358,210)
Total foreign exchange losses on debt		(570,494)	(1,290,535)	(2,199,325)	(3,346,341)	(4,794,031)

The loan by Cemex to the Indonesian subsidiary is denominated in U.S. dollars. Therefore, the loan will have to be repaid in U.S. dollars, not rupiah. At the time of the loan agreement, the spot exchange rate is Rp10,000/$. This is the assumption used in calculating the "scheduled" repaying of principal and interest in rupiah. The rupiah, however, is expected to depreciate in line with purchasing power parity. As it is repaid, the "actual" exchange rate will therefore give rise to a foreign exchange loss as it takes more and more rupiah to acquire U.S. dollars for debt service, both principal and interest. The foreign exchange losses on this debt service will be recognized on the Indonesian income statement.

Costs. The cash costs of cement manufacturing (labor, materials, power, etc.) are estimated at Rp115,000 per tonne for year 1, rising at about the rate of inflation, 30% per year. Additional production costs of Rp20,000 per tonne for year 1 are also assumed to rise at the rate of inflation. As a result of all production being exported, loading costs of $2.00/tonne and shipping of $10.00/tonne must also be included. Note that these costs are originally stated in U.S. dollars, and for the purposes of Semen Indonesia's income statement, they must be converted to rupiah terms. This is the case because both shiploading and shipping costs are international services governed by contracts denominated in dollars. As a result, they are expected to rise over time only at the U.S. dollar rate of inflation (3%).

Semen Indonesia's pro forma income statement is illustrated in Exhibit 18.4. This is the typical financial statement measurement of the profitability of any business, whether domestic or international. The baseline analysis assumes a capacity utilization rate of only 40% (year 1),

EXHIBIT 18.4   **Semen Indonesia's Pro Forma Income Statement (millions of rupiah)**

Exchange rate (Rp/US$)	10,000	12,621	15,930	20,106	25,376	32,028
Project Year	**0**	**1**	**2**	**3**	**4**	**5**
Sales volume		8.00	10.00	12.00	12.00	12.00
Sales price (US$)		58.00	58.00	58.00	58.00	58.00
Sales price (Rp)		732,039	923,933	1,166,128	1,471,813	1,857,627
Total revenue		5,856,311	9,239,325	13,993,541	17,661,751	22,291,530
Less cash costs		(920,000)	(1,495,000)	(2,332,200)	(3,031,860)	(3,941,418)
Less other production costs		(160,000)	(260,000)	(405,600)	(527,280)	(685,464)
Less loading costs		(201,942)	(328,155)	(511,922)	(665,499)	(865,149)
Less shipping costs		(1,009,709)	(1,640,777)	(2,559,612)	(3,327,495)	(4,325,744)
Total production costs		(2,291,650)	(3,723,932)	(5,809,334)	(7,552,134)	(9,817,774)
Gross profit		3,564,660	5,515,393	8,184,207	10,109,617	12,473,756
Gross margin		*60.9%*	*59.7%*	*58.5%*	*57.2%*	*56.0%*
Less license fees		(117,126)	(184,787)	(279,871)	(353,235)	(445,831)
Less general & administrative		(468,505)	(831,539)	(1,399,354)	(1,942,793)	(2,674,984)
EBITDA		2,979,029	4,499,067	6,504,982	7,813,589	9,352,941
Less depreciation & amortization		(1,760,000)	(1,760,000)	(1,760,000)	(1,760,000)	(1,760,000)
EBIT		1,219,029	2,739,067	4,744,982	6,053,589	7,592,941
Less interest on Cemex debt		(825,000)	(689,867)	(541,221)	(377,710)	(197,848)
Foreign exchange losses on debt		(570,494)	(1,290,535)	(2,199,325)	(3,346,341)	(4,794,031)
Less interest on local debt		(962,500)	(928,921)	(883,590)	(822,393)	(739,777)
EBT		(1,138,965)	(170,256)	1,120,846	1,507,145	1,861,285
Less income taxes (30%)		—	—	—	(395,631)	(558,386)
Net income		(1,138,965)	(170,256)	1,120,846	1,111,514	1,302,900
Net income (millions of US$)		(90)	(11)	56	44	41
Return on sales		*−19.4%*	*−1.8%*	*8.0%*	*6.3%*	*5.8%*
Dividends distributed		—	—	560,423	555,757	651,450
Retained		(1,138,965)	(170,256)	560,423	555,757	651,450

EBITDA = earnings before interest, taxes, depreciation, and amortization. EBIT = earnings before interest and taxes; EBT = earnings before taxes.

Tax credits resulting from current period losses are carried forward toward next year's tax liabilities. Dividends are not distributed in the first year of operations as a result of losses, and are distributed at a 50% rate in years 2000–2003.

All calculations are exact, but may appear not to add due to reported decimal places. The tax payment for year 3 is zero, and year 4 is less than 30%, as a result of tax loss carry-forwards from previous years.

50% (year 2), and 60% in the following years. Management believes this is necessary since existing in-country cement manufacturers are averaging only 40% of capacity at this time.

Tax credits resulting from current period losses are carried forward toward next year's tax liabilities. Dividends are not distributed in the first year of operations as a result of losses, and are distributed at a 50% rate in years 2–5.

Additional expenses in the pro forma financial analysis include license fees paid by the subsidiary to the parent company of 2.0% of sales, and general and administrative expenses for Indonesian operations of 8.0% per year (and growing an additional 1% per year). Foreign exchange gains and losses are those related to the servicing of the U.S. dollar-denominated debt provided by the parent and are drawn from the bottom of Exhibit 18.3. In summary, the subsidiary operation is expected to begin turning an accounting profit in its fourth year of operations, with profits rising as capacity utilization increases over time.

The loan by Cemex to the Indonesian subsidiary is denominated in U.S. dollars. Therefore, the loan will have to be repaid in U.S. dollars, not rupiah. At the time of the loan agreement, the spot exchange rate is Rp10,000/$. This is the assumption used in calculating the "scheduled" repaying of principal and interest in rupiah. The rupiah, however, is expected to depreciate in line with purchasing power parity. As it is repaid, the "actual" exchange rate will therefore give rise to a foreign exchange loss as it takes more and more rupiah to acquire U.S. dollars for debt service, both principal and interest. The foreign exchange losses on this debt service will be recognized on the Indonesian income statement.

Tax credits resulting from current period losses are carried forward toward next year's tax liabilities. Dividends are not distributed in the first year of operations as a result of losses, and are distributed at a 50% rate in years 2000–2003. All calculations are exact, but may appear not to add due to reported decimal places. The tax payment for year 3 is zero, and year 4 is less than 30%, as a result of tax loss carry-forwards from previous years.

Project Viewpoint Capital Budget

The capital budget for the Semen Indonesia project from a project viewpoint is shown in Exhibit 18.5. We find the net cash flow, *free cash flow* as it is often labeled, by summing EBITDA (earnings before interest, taxes, depreciation, and amortization), recalculated taxes, changes in net working capital (the sum of the net additions to receivables, inventories, and payables necessary to support sales growth), and capital investment.

Note that EBIT, not EBT, is used in the capital budget, which contains both depreciation and interest expense. Depreciation and amortization are noncash expenses of the firm and therefore contribute positive cash flow. Because the capital budget creates cash flows that will be discounted to present value with a discount rate, and the discount rate includes the cost of debt—interest—we do not wish to subtract interest twice. Therefore, taxes are recalculated on the basis of EBIT.[3] The firm's cost of capital used in discounting also includes the deductibility of debt interest in its calculation.

The initial investment of Rp22 trillion is the total capital invested to support these earnings. Although receivables average 50 to 55 days sales outstanding (DSO) and inventories average 65 to 70 DSO, payables and trade credit are also relatively long at 114 DSO in the Indonesian cement industry. Semen Indonesia expects to add approximately 15 net DSO to its investment with sales growth. The remaining elements to complete the project viewpoint's capital budget are the *terminal value* (discussed below) and the *discount rate* of 33.257% (the firm's weighted average cost of capital).

---

[3] This highlights the distinction between an income statement and a capital budget. The project's income statement shows losses the first two years of operations as a result of interest expenses and forecast foreign exchange losses, so it is not expected to pay taxes. But the capital budget, constructed on the basis of EBIT, before these financing and foreign exchange expenses, calculates a positive tax payment.

EXHIBIT 18.5	Semen Indonesia Capital Budget: Project Viewpoint (millions of rupiah)					

Exchange rate (Rp/US$)	10,000	12,621	15,930	20,106	25,376	32,028
Project Year	**0**	**1**	**2**	**3**	**4**	**5**
EBIT		1,219,029	2,739,067	4,744,982	6,053,589	7,592,941
Less recalculated taxes @ 30%		(365,709)	(821,720)	(1,423,495)	(1,816,077)	(2,277,882)
Add back depreciation		1,760,000	1,760,000	1,760,000	1,760,000	1,760,000
Net operating cash flow		2,613,320	3,677,347	5,081,487	5,997,512	7,075,059
Less changes to NWC		(240,670)	(139,028)	(195,379)	(150,748)	(190,265)
Initial investment	(22,000,000)					
Terminal value						21,274,102
Free cash flow (FCF)	(22,000,000)	2,372,650	3,538,319	4,886,109	5,846,764	28,158,896
NPV @ 33.257%	**(7,606,313)**					
IRR	**19.1%**					

NWC = net working capital. NPV = net present value. Discount rate is Semen Indonesia's WACC of 33.257%. IRR = internal rate of return, the rate of discount yielding an NPV of exactly zero. Values in exhibit are exact and are rounded to the nearest million.

Terminal Value. The *terminal value* (TV) of the project represents the continuing value of the cement manufacturing facility in the years after year 5, the last year of the detailed pro forma financial analysis shown in Exhibit 18.5. This value, like all asset values according to financial theory, is the present value of all future free cash flows that the asset is expected to yield. We calculate the TV as the present value of a perpetual *net operating cash flow* (NOCF) generated in the fifth year by Semen Indonesia, the growth rate assumed for that net operating cash flow (g), and the firm's weighted average cost of capital (k_{WACC}):

$$\text{Terminal value} = \frac{NOCF_5 (1 + g)}{k_{WACC} - g} = \frac{7,075,059 (1 + 0)}{.33257 - 0} = \text{Rp21,274,102}$$

or Rp21,274,102 trillion. The assumption that $g = 0$, that is, that net operating cash flows will not grow past year 5 is probably not true, but it is a prudent assumption for Cemex to make when estimating future cash flows. (If Semen Indonesia's business was to continue to grow in-line with the Indonesian economy, g may well be 1% or 2%.) The results of the capital budget from the project viewpoint indicate a negative *net present value* (NPV) and an *internal rate of return* (IRR) of only 19.1% compared to the 33.257% cost of capital. These are the returns the project would yield to a local or Indonesian investor in Indonesian rupiah. The project, from this viewpoint, is not acceptable.

Repatriating Cash Flows to Cemex

Exhibit 18.6 now collects all incremental earnings to Cemex from the prospective investment project in Indonesia. As described in the section, *Project versus Parent Valuation*, a foreign investor's assessment of a project's returns depends on the actual cash flows that are returned to it in its own currency via actual potential cash flow channels. For Cemex, this means that the investment must be analyzed in terms of the actual likely U.S. dollar cash inflows and

EXHIBIT 18.6	Semen Indonesia's Remittance of Income to Parent Company (millions of rupiah and US$)					

Exchange rate (Rp/$)	10,000	12,621	15,930	20,106	25,376	32,028
Project Year	**0**	**1**	**2**	**3**	**4**	**5**
Dividend Remittance						
Dividends paid (Rp)		—	—	560,423	555,757	651,450
Less Indonesian withholding taxes		—	—	(84,063)	(83,364)	(97,717)
Net dividend remitted (Rp)		—	—	476,360	472,393	553,732
Net dividend remitted ($)		—	—	23.69	18.62	17.29
License Fees Remittance						
License fees remitted (Rp)		117,126	184,787	279,871	353,235	445,831
Less Indonesian withholding taxes		(5,856)	(9,239)	(13,994)	(17,662)	(22,292)
Net license fees remitted (Rp)		111,270	175,547	265,877	335,573	423,539
Net license fees remitted ($)		8.82	11.02	13.22	13.22	13.22
Debt Service Remittance						
Promised interest paid ($)		82.50	68.99	54.12	37.77	19.78
Less Indonesian withholding tax @ 10%		(8.25)	(6.90)	(5.41)	(3.78)	(1.98)
Net interest remitted ($)		74.25	62.09	48.71	33.99	17.81
Principal payments remitted ($)		135.13	148.65	163.51	179.86	197.85
Total principal and interest remitted		$209.38	$210.73	$212.22	$213.86	$215.65
Capital Budget: Parent Viewpoint (millions of U.S. dollars)						
Dividends		$0.0	$0.0	$23.7	$18.6	$17.3
License fees		8.8	11.0	13.2	13.2	13.2
Debt service		209.4	210.7	212.2	213.9	215.7
Total earnings		$218.2	$221.8	$249.1	$245.7	$246.2
Initial investment	(1,925.0)					
Terminal value						1,369.1
Net cash flows	($1,925.0)	$218.2	$221.8	$249.1	$245.7	$1,615.3
NPV @ 17.98%	**17.98%**	**(595.6)**				
IRR		**7.21%**				

NPV calculated using a company-determined discount rate of WACC+ foreign investment premium, or 11.98% + 6.00% = 17.98%.

outflows associated with the investment over the life of the project, after-tax, discounted at its appropriate cost of capital.

The *parent viewpoint capital budget* is constructed in two steps:

1. First, we isolate the individual cash flows, cash flows by channel, adjusted for any withholding taxes imposed by the Indonesian government and converted to U.S. dollars. (Statutory withholding taxes on international transfers are set by bilateral tax treaties, but individual firms may negotiate lower rates with governmental tax

authorities. In the case of Semen Indonesia, dividends will be charged a 15% with-holding tax, 10% on interest payments, and 5% license fees.) Mexico does not tax repatriated earnings since they have already been taxed in Indonesia. (The U.S. does levy a contingent tax on repatriated earnings of foreign source income, as discussed in Chapter 16.)

2. The second step, the actual *parent viewpoint capital budget*, combines these U.S. dollar after-tax cash flows with the initial investment to determine the net present value of the proposed Semen Indonesia subsidiary in the eyes (and pocketbook) of Cemex. This is illustrated in Exhibit 18.6, which shows all incremental earnings to Cemex from the prospective investment project. A specific peculiarity of this parent view-point capital budget is that only the capital invested into the project by Cemex itself, $1,925 million, is included in the initial investment (the $1,100 million in equity and the $825 million loan). The Indonesian debt of Rp2.75 billion ($275 million) is not included in the Cemex parent viewpoint capital budget.

Parent Viewpoint Capital Budget

Finally, all cash flow estimates are now constructed to form the parent viewpoint's capital budget, detailed in the bottom of Exhibit 18.6. The cash flows generated by Semen Indonesia from its Indonesian operations, dividends, license fees, debt service, and terminal value are now valued in U.S. dollar terms after-tax.

In order to evaluate the project's cash flows that are returned to the parent company, Cemex must discount these at the corporate cost of capital. Remembering that Cemex considers its functional currency to be the U.S. dollar, it calculates its cost of capital in U.S. dollars. As described in Chapter 13, the customary weighted average cost of capital formula is as follows:

$$k_{WACC} = k_e \frac{E}{V} + k_d(1 - t)\frac{D}{V},$$

where k_e is the risk-adjusted cost of equity, k_d is the before-tax cost of debt, t is the marginal tax rate, E is the market value of the firm's equity, D is the market value of the firm's debt, and V is the total market value of the firm's securities ($E + D$).

$$k_e = k_{rf} + (k_m - k_{rf})\beta_{Cemex} = 6.00\% + (13.00\% - 6.00\%)1.5 = 16.50\%$$

Cemex's cost of equity is calculated using the capital asset pricing model (CAPM): This assumes the risk-adjusted cost of equity (k_e) is based on the risk-free rate of interest (k_{rf}), as measured by the U.S. Treasury intermediate bond yield of 6.00%, the expected rate of return in U.S. equity markets (k_m) is 13.00%, and the measure of Cemex's individual risk relative to the market (β_{Cemex}) is 1.5. The result is a cost of equity—required rate of return on equity investment in Cemex—of 16.50%.

The investment will be funded internally by the parent company, roughly in the same debt/equity proportions as the consolidated firm, 40% debt (D/V) and 60% equity (E/V). The current cost of debt for Cemex is 8.00%, and the effective tax rate is 35%. The cost of equity, when combined with the other components, results in a weighted average cost of capital for Cemex of

$$k_{WACC} = k_e \frac{E}{V} + k_d(1 - t)\frac{D}{V} = (16.50\%)(.60) + (8.00\%)(1 - .35)(.40) = 11.98\%$$

Cemex customarily uses this weighted average cost of capital of 11.98% to discount prospective investment cash flows for project ranking purposes. The Indonesian investment poses a variety of risks, however, which the typical domestic investment does not.

If Cemex were undertaking an investment of the same relative degree of risk as the firm itself, a simple discount rate of 11.980% might be adequate. Cemex, however, generally requires new investments to yield an additional 3% over the cost of capital for domestic investments, and 6% more for international projects (these are company-required spreads, and will differ dramatically across companies). The discount rate for Semen Indonesia's cash flows repatriated to Cemex will therefore be discounted at 11.98% + 6.00%, or 17.98%. The project's baseline analysis indicates a negative NPV with an IRR of 7.21%, which means that it is an unacceptable investment from the parent's viewpoint.

Most corporations require that new investments more than cover the cost of the capital employed in their undertaking. It is therefore not unusual for the firm to require a hurdle rate of 3% to 6% above its cost of capital in order to identify potential investments that will literally add value to stockholder wealth. An NPV of zero means the investment is "acceptable," but NPV values that exceed zero are literally the present value of wealth that is expected to be added to the value of the firm and its shareholders. For foreign projects, as discussed previously, we must adjust for agency costs and foreign exchange risks and costs.

Sensitivity Analysis: Project Viewpoint

So far, the project investigation team has used a set of "most likely" assumptions to forecast rates of return. It is now time to subject the most likely outcome to sensitivity analyses. The same probabilistic techniques are available to test the sensitivity of results to political and foreign exchange risks as are used to test sensitivity to business and financial risks. Many decision makers feel more uncomfortable about the necessity to guess probabilities for unfamiliar political and foreign exchange events than they do about guessing their own more familiar business or financial risks. Therefore, it is more common to test sensitivity to political and foreign exchange risk by simulating what would happen to net present value and earnings under a variety of "what if" scenarios.

Political Risk. *What if Indonesia imposes controls on the payment of dividends or license fees to Cemex?* The impact of blocked funds on the rate of return from Cemex's perspective would depend on when the blockage occurs, what reinvestment opportunities exist for the blocked funds in Indonesia, and when the blocked funds would eventually be released to Cemex. We could simulate various scenarios for blocked funds and rerun the cash flow analysis in Exhibit 18.6 to estimate the effect on Cemex's rate of return.

What if Indonesia should expropriate Semen Indonesia? The effect of expropriation would depend on the following factors:

1. When the expropriation occurs, in terms of number of years after the business began operation
2. How much compensation the Indonesian government will pay, and how long after expropriation the payment will be made
3. How much debt is still outstanding to Indonesian lenders, and whether the parent, Cemex, will have to pay this debt because of its parental guarantee
4. The tax consequences of the expropriation
5. Whether the future cash flows are foregone

Many expropriations eventually result in some form of compensation to the former owners. This compensation can come from a negotiated settlement with the host government or from payment of political risk insurance by the parent government. Negotiating a settlement takes time, and the eventual compensation is sometimes paid in installments over a further period of time. Thus, the present value of the compensation is often much lower than its nominal value. Furthermore, most settlements are based on book value of the firm at the time of expropriation rather than the firm's market value.

The tax consequences of expropriation would depend on the timing and amount of capital loss recognized by Mexico. This loss would usually be based on the uncompensated book value of the Indonesian investment. The problem is that there is often some doubt as to when a write-off is appropriate for tax purposes, particularly if negotiations for a settlement drag on. In some ways, a nice clear expropriation without hope of compensation, such as occurred in Cuba in the early 1960s, is preferred to a slow "bleeding death" in protracted negotiations. The former leads to an earlier use of the tax shield and a one-shot write-off against earnings, whereas the latter tends to depress earnings for years, as legal and other costs continue and no tax shelter is achieved.

Foreign Exchange Risk. The project investigation team assumed that the Indonesian rupiah would depreciate versus the U.S. dollar at the purchasing power parity "rate" (approximately 20.767% per year in the baseline analysis).

What if the rate of rupiah depreciation were greater? Although this event would make the assumed cash flows to Cemex worth less in dollars, operating exposure analysis would be necessary to determine whether the cheaper rupiah made Semen Indonesia more competitive. For example, since Semen Indonesia's exports to Taiwan are denominated in U.S. dollars, a weakening of the rupiah versus the dollar could result in greater rupiah earnings from those export sales. This serves to somewhat offset the imported components that Semen Indonesia purchases from the parent company that are also denominated in U.S. dollars. Semen Indonesia is representative of firms today that have both cash inflows and outflows denominated in foreign currencies, providing a partial natural hedge against currency movements.

What if the rupiah should appreciate against the dollar? The same kind of economic exposure analysis is needed. In this particular case, we might guess that the effect would be positive on both local sales in Indonesia and the value in dollars of dividends and license fees paid to Cemex by Semen Indonesia. Note, however, that an appreciation of the rupiah might lead to more competition within Indonesia from firms in other countries with now lower cost structures, lessening Semen Indonesia's sales.

Sometimes foreign exchange risk and political risks were inseparable, as was the case of Venezuela in 2015 as examined in *Global Finance in Practice 18.1.*

Other Sensitivity Variables. The project rate of return to Cemex would also be sensitive to a change in the assumed terminal value, the capacity utilization rate, the size of the license fee paid by Semen Indonesia, the size of the initial project cost, the amount of working capital financed locally, and the tax rates in Indonesia and Mexico. Since some of these variables are within control of Cemex, it is still possible that the Semen Indonesia project could be improved in its value to the firm and become acceptable.

Sensitivity Analysis: Parent Viewpoint Measurement

When a foreign project is analyzed from the parent's point of view, the additional risk that stems from its "foreign" location can be measured in two ways, *adjusting the discount rates* or *adjusting the cash flows*.

GLOBAL FINANCE IN PRACTICE 18.1

Venezuelan Currency and Capital Controls Force Devaluation of Business

The Venezuelan government's restrictions on access to hard currency have now lasted more than 12 years, and foreign corporate interests have had enough. Throughout 2014 and into 2015 many international investors in Venezuela struggled to run and value their businesses.

Air Canada suspended all flights to Venezuela in March 2014, citing concern over its ability to assure passenger safety in light of ongoing civil protest in the country. Air Canada was also due millions of dollars in back payments for services rendered. International airlines in total claimed that they were owed more than $2 billion in backpayments. Other companies like Avon and Merck wrote down their investments in Venezuela as a result of the continuing fall in the market value of the Venezuelan bolivar. Manufacturing companies like GM continued to struggle to even operate, as restricted access to hard currency prevented them from purchasing critical inputs and components for their products. Factories stopped, layoffs followed.

In February 2015 the Venezuelan government announced once again a "new" currency exchange system. The new system was little different, however, from the old three-tiered system in effect. There is the official exchange rate of roughly 6.3 bolivars to the U.S. dollar. But outside of food and medical purchases, few companies had access to this rate. The second- or middle-tier rate, called SICAD 1, a rate that was offered to select companies, was 12 bolivars. The third-tier rate, SICAD 2, theoretically open to all who needed it, was hovering around 52. A fourth, the black market rate, was trading at 190 bolivars per dollar.

Regardless of the next exchange rate system or next devaluation, multinational firms from all over the world continued to write down their Venezuelan investments. This included Coca Cola (U.S.), Telefonica (Spain) and drugmaker Bayer (Germany). So what was the value of investing or doing business in Venezuela tomorrow?

Adjusting Discount Rates. The first method is to treat all foreign risk as a single problem, by adjusting the discount rate applicable to foreign projects relative to the rate used for domestic projects to reflect the greater foreign exchange risk, political risk, agency costs, asymmetric information, and other uncertainties perceived in foreign operations. However, adjusting the discount rate applied to a foreign project's cash flow to reflect these uncertainties does not penalize net present value in proportion either to the actual amount at risk or to possible variations in the nature of that risk over time. Combining all risks into a single discount rate may thus cause us to discard much information about the uncertainties of the future.

In the case of foreign exchange risk, changes in exchange rates have a potential effect on future cash flows because of operating exposure. The direction of the effect, however, can either decrease or increase net cash inflows, depending on where the products are sold and where inputs are sourced. To increase the discount rate applicable to a foreign project on the assumption that the foreign currency might depreciate more than expected, is to ignore the possible favorable effect of a foreign currency depreciation on the project's competitive position. Increased sales volume might more than offset a lower value of the local currency. Such an increase in the discount rate also ignores the possibility that the foreign currency may appreciate (two-sided risk).

Adjusting Cash Flows. In the second method, we incorporate foreign risks in adjustments to forecasted cash flows of the project. The discount rate for the foreign project is risk-adjusted only for overall business and financial risk, in the same manner as for domestic projects. Simulation-based assessment utilizes scenario development to estimate cash flows to the parent arising from the project over time under different alternative economic futures.

Certainty regarding the quantity and timing of cash flows in a prospective foreign investment is, to quote Shakespeare, "the stuff that dreams are made of." Due to the complexity of economic forces at work in major investment projects, it is paramount that the analyst understand the subjectivity of the forecast cash flows. Humility in analysis is a valuable trait.

Shortcomings of Each. In many cases, however, neither adjusting the discount rate nor adjusting cash flows is optimal. For example, political uncertainties are a threat to the entire investment, not just the annual cash flows. Potential loss depends partly on the terminal value of the unrecovered parent investment, which will vary depending on how the project was financed, whether political risk insurance was obtained, and what investment horizon is contemplated. Furthermore, if the political climate were expected to be unfavorable in the near future, any investment would probably be unacceptable. Political uncertainty usually relates to possible adverse events that might occur in the more distant future, but that cannot be foreseen at the present. Adjusting the discount rate for political risk thus penalizes early cash flows too heavily while not penalizing distant cash flows enough.

Repercussions to the Investor. Apart from anticipated political and foreign exchange risks, MNEs sometimes worry that taking on foreign projects may increase the firm's overall cost of capital because of investors' perceptions of foreign risk. This worry seemed reasonable if a firm had significant investments in Iraq, Iran, Russia, Serbia, or Afghanistan in recent years. However, the argument loses persuasiveness when applied to diversified foreign investments with a heavy balance in the industrial countries of Canada, Western Europe, Australia, Latin America, and Asia where, in fact, the bulk of FDI is located. These countries have a reputation for treating foreign investments by consistent standards, and empirical evidence confirms that a foreign presence in these countries may not increase the cost of capital. In fact, some studies indicate that required returns on foreign projects may even be lower than those for domestic projects.

MNE Practices. Surveys of MNEs over the past 35 years have shown that about half of them adjust the discount rate and half adjust the cash flows. One recent survey indicated a rising use of adjusting discount rates over adjusting cash flows. However, the survey also indicated an increasing use of multifactor methods—discount rate adjustment, cash flow adjustment, real options analysis, and qualitative criteria—in evaluating foreign investments.[4]

Portfolio Risk Measurement

The field of finance has distinguished two different definitions of risk: (1) the risk of the individual security (standard deviation of expected return) and (2) the risk of the individual security as a component of a portfolio (*beta*). A foreign investment undertaken in order to enter a local or regional market—market seeking—will have returns that are more or less correlated with those of the local market. A portfolio-based assessment of the investment's prospects would then seem appropriate. A foreign investment motivated by *resource-seeking* or *production-seeking* objectives may yield returns related to the products or services and markets of the parent company or units located somewhere else in the world and have little to do with local markets. Cemex's proposed investment in Semen Indonesia is both *market-seeking* and *production-seeking* (for export). The decision about which approach is to be used in evaluating prospective foreign investments may be the single most important analytical decision that the MNE makes. An investment's acceptability may change dramatically across criteria.

---

[4] Tom Keck, Eric Levengood, and Al Longield, "Using Discounted Cash Flow Analysis in an International Setting: A Survey of Issues in Modeling the Cost of Capital," *Journal of Applied Corporate Finance*, Vol. 11, No. 3, Fall 1998, pp. 82–99.

For comparisons within the local host country, we should overlook a project's actual financing or parent-influenced debt capacity, since these would probably be different for local investors than they are for a multinational owner. In addition, the risks of the project to local investors might differ from those perceived by a foreign multinational owner because of the opportunities an MNE has to take advantage of market imperfections. Moreover, the local project may be only one out of an internationally diversified portfolio of projects for the multinational owner; if undertaken by local investors it might have to stand alone without international diversification. Since diversification reduces risk, the MNE can require a lower rate of return than is required by local investors.

Thus, the discount rate used locally must be a hypothetical rate based on a judgment as to what independent local investors would probably demand were they to own the business. Consequently, application of the local discount rate to local cash flows provides only a rough measure of the value of the project as a stand-alone local venture, rather than an absolute valuation.

Real Option Analysis

The discounted cash flow (DCF) approach used in the valuation of Semen Indonesia—and capital budgeting and valuation in general—has long had its critics. Investments that have long lives, cash flow returns in later years, or higher levels of risk than those typical of the firm's current business activities are often rejected by traditional DCF financial analysis. More importantly, when MNEs evaluate competitive projects, traditional discounted cash flow analysis is typically unable to capture the strategic options that an individual investment option may offer. This has led to the development of real option analysis. Real option analysis is the application of option theory to capital budgeting decisions.

Real options present a different way of thinking about investment values. At its core, it is a cross between decision-tree analysis and pure option-based valuation. It is particularly useful when analyzing investment projects that will follow very different value paths at decision points in time where management decisions are made regarding project pursuit. This wide range of potential outcomes is at the heart of real option theory. These wide ranges of value are volatilities, the basic element of option pricing theory described previously.

Real option valuation also allows us to analyze a number of managerial decisions, which in practice characterize many major capital investment projects:

- The option to defer
- The option to abandon
- The option to alter capacity
- The option to start up or shut down (switching)

Real option analysis treats cash flows in terms of future value in a positive sense, whereas DCF treats future cash flows negatively (on a discounted basis). Real option analysis is a particularly powerful device when addressing potential investment projects with extremely long life spans or investments that do not commence until future dates. Real option analysis acknowledges the way information is gathered over time to support decision-making. Management learns from both active (searching it out) and passive (observing market conditions) knowledge-gathering and then uses this knowledge to make better decisions.

Project Financing

One of the more unique structures used in international finance is *project finance*, which refers to the arrangement of financing for long-term capital projects, large in scale, long in life, and generally high in risk. This is a very general definition, however, because there are many different forms and structures that fall under this generic heading.

Project finance is not new. Examples of project finance go back centuries, and include many famous early international businesses such as the Dutch East India Company and the British East India Company. These entrepreneurial importers financed their trade ventures to Asia on a voyage-by-voyage basis, with each voyage's financing being like venture capital-investors would be repaid when the shipper returned and the fruits of the Asian marketplace were sold at the docks to Mediterranean and European merchants. If all went well, the individual shareholders of the voyage were paid in full.

Project finance is used widely today in the development of large-scale infrastructure projects in China, India, and many other emerging markets. Although each individual project has unique characteristics, most are highly leveraged transactions, with debt making up more than 60% of the total financing. Equity is a small component of project financing for two reasons: first, the simple scale of the investment project often precludes a single investor or even a collection of private investors from being able to fund it; second, many of these projects involve subjects traditionally funded by governments—such as electrical power generation, dam building, highway construction, energy exploration, production, and distribution.

This level of debt, however, places an enormous burden on cash flow for debt service. Therefore, project financing usually requires a number of additional levels of risk reduction. The lenders involved in these investments must feel secure that they will be repaid; bankers are not by nature entrepreneurs, and do not enjoy entrepreneurial returns from project finance. Project finance has a number of basic properties that are critical to its success.

Separability of the Project from Its Investors

The project is established as an individual legal entity, separate from the legal and financial responsibilities of its individual investors. This not only serves to protect the assets of equity investors, but also it provides a controlled platform upon which creditors can evaluate the risks associated with the singular project, the ability of the project's cash flows to service debt, and to rest assured that the debt service payments will be automatically allocated by and from the project itself (and not from a decision by management within an MNE).

Long-Lived and Capital-Intensive Singular Projects

Not only must the individual project be separable and large in proportion to the financial resources of its owners, but also its business line must be singular in its construction, operation, and size (capacity). The size is set at inception, and is seldom, if ever, changed over the project's life.

Cash Flow Predictability from Third Party Commitments

An oil field or electric power plant produces a homogeneous commodity product that can produce predictable cash flows if third party commitments to take and pay can be established. In addition to revenue predictability, nonfinancial costs of production need to be controlled over time, usually through long-term supplier contracts with price adjustment clauses based on inflation. The predictability of net cash inflows to long-term contracts eliminates much of the individual project's business risk, allowing the financial structure to be heavily debt-financed and still be safe from financial distress.

The predictability of the project's revenue stream is essential in securing project financing. Typical contract provisions that are intended to assure adequate cash flow normally include the following clauses: quantity and quality of the project's output; a pricing formula that enhances the predictability of adequate margin to cover operating costs and debt service payments; a clear statement of the circumstances that permit significant changes in the contract, such as force majeure or adverse business conditions.

Finite Projects with Finite Lives

Even with a longer-term investment, it is critical that the project have a definite ending point at which all debt and equity has been repaid. Because the project is a stand-alone investment in which its cash flows go directly to the servicing of its capital structure and not to reinvestment for growth or other investment alternatives, investors of all kinds need assurances that the project's returns will be attained in a finite period. There is no capital appreciation, only cash flow.

Examples of project finance include some of the largest individual investments undertaken in the past three decades, such as British Petroleum's financing of its interest in the North Sea, and the Trans-Alaska Pipeline. The Trans-Alaska Pipeline was a joint venture between Standard Oil of Ohio, Atlantic Richfield, Exxon, British Petroleum, Mobil Oil, Philips Petroleum, Union Oil, and Amerada Hess. Each of these projects was at or above $1 billion, and represented capital expenditures that no single firm would or could attempt to finance. Yet, through a joint venture arrangement, the higher than normal risk absorbed by the capital employed could be managed.

Cross-Border Mergers and Acquisitions

The drivers of M&A activity, summarized in Exhibit 18.7, are both macro in scope—the global competitive environment—and micro in scope—the variety of industry and firm-level forces and actions driving individual firm value. The primary forces of change in the global competitive environment—technological change, regulatory change, and capital market change—create new business opportunities for MNEs, which they pursue aggressively.

But the global competitive environment is really just the playing field, the ground upon which the individual players compete. MNEs undertake cross-border mergers and acquisitions for a variety of reasons. As shown in Exhibit 18.7, the drivers are strategic responses by MNEs to defend and enhance their global competitiveness.

As opposed to greenfield investment, a *cross-border acquisition* has a number of significant advantages. First and foremost, it is quicker. Greenfield investment frequently requires extended periods of physical construction and organizational development. By acquiring an existing firm, the MNE shortens the time required to gain a presence and facilitate competitive entry into the market. Second, acquisition may be a cost-effective way of gaining competitive advantages, such as technology, brand names valued in the target market, and logistical and distribution advantages, while simultaneously eliminating a local competitor. Third, specific to cross-border acquisitions, international economic, political, and foreign exchange conditions may result in market imperfections, allowing target firms to be undervalued.

Cross-border acquisitions are not, however, without their pitfalls. As with all acquisitions—domestic or cross-border—there are problems of paying too much or suffering excessive financing costs. Melding corporate cultures can be traumatic. Managing the post-acquisition process is frequently characterized by downsizing to gain economies of scale and scope in overhead functions. This results in nonproductive impacts on the firm as individuals attempt to save

---

| EXHIBIT 18.7 | **Driving Forces Behind Cross-Border Acquisition** |

Changes in the Global Business Environment

- Changes in technology
- Changes in regulation
- Changes in capital markets

*Create business opportunities for select firms to both enhance
and defend their competitive positions in global markets*

- Gaining access to strategic proprietary assets
- Gaining market power and dominance
- Achieving synergies in local/global operations and across different industries
- Becoming larger, and then reaping the benefits of size in competition and negotiation
- Diversifying and spreading their risks wider
- Exploiting financial opportunities they may possess and others desire

---

their own jobs. Internationally, additional difficulties arise from host governments intervening in pricing, financing, employment guarantees, market segmentation, and general nationalism and favoritism. In fact, the ability to successfully complete cross-border acquisitions may itself be a test of competency of the MNE when entering emerging markets.

The Cross-Border Acquisition Process

Although the field of finance has sometimes viewed acquisition as mainly an issue of valuation, it is a much more complex and rich process than simply determining what price to pay. As depicted in Exhibit 18.8, the process begins with the strategic drivers discussed in the previous section.

The process of acquiring an enterprise anywhere in the world has three common elements: (1) identification and valuation of the target, (2) execution of the acquisition offer and purchase—the *tender*, and (3) management of the post-acquisition transition.

Stage 1: Identification and Valuation. Identification of potential acquisition targets requires a well-defined corporate strategy and focus.

The identification of the target market typically precedes the identification of the target firm. Entering a highly developed market offers the widest choice of publicly traded firms with relatively well-defined markets and publicly disclosed financial and operational data. In this case, the tender offer is made publicly, although target company management may openly recommend that its shareholders reject the offer. If enough shareholders take the offer, the acquiring company may gain sufficient ownership influence or control to change management. During this rather confrontational process, it is up to the board of the target company to continue to take actions consistent with protecting the rights of shareholders. The board may need to provide rather strong oversight of management during this process to ensure that the acts of management are consistent with protecting and building shareholder value.

Once identification has been completed, the process of valuing the target begins. A variety of valuation techniques are widely used in global business today, each with its own

EXHIBIT 18.8   **The Cross-Border Acquisition Process**

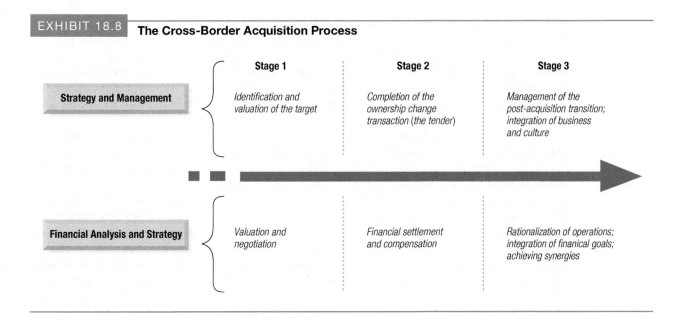

merits. In addition to the fundamental methodologies of discounted cash flow (DCF) and multiples (earnings and cash flows), there are also industry-specific measures that focus on the most significant elements of value in business lines. The completion of various alternative valuations for the target firm aids not only in gaining a more complete picture of what price must be paid to complete the transaction, but also in determining whether the price is attractive.

Stage 2: Execution of the Acquisition. Once an acquisition target has been identified and valued, the process of gaining approval from management and ownership of the target, getting approvals from government regulatory bodies, and finally determining method of compensation—the complete execution of the acquisition strategy—can be time-consuming and complex.

Gaining the approval of the target company has been the highlight of some of the most famous acquisitions in business history. The critical distinction here is whether the acquisition is supported or not by the target company's management.

Although there is probably no "typical transaction," many acquisitions flow relatively smoothly through a friendly process. The acquiring firm will approach the management of the target company and attempt to convince them of the business logic of the acquisition. (Gaining their support is sometimes difficult, but assuring target company management that it will not be replaced is often quite convincing!) If the target's management is supportive, management may then recommend to stockholders that they accept the offer of the acquiring company. One problem that occasionally surfaces at this stage is that influential shareholders may object to the offer, either in principle or based on price, and may therefore feel that management is not taking appropriate steps to protect and build their shareholder value.

The process takes on a very different dynamic when the acquisition is not supported by the target company management—the so-called hostile takeover. The acquiring company may choose to pursue the acquisition without the target's support, and instead go directly to the target shareholders. In this case, the tender offer is made publicly, although target company management may openly recommend that its shareholders reject the offer. If enough share-holders take the offer, the acquiring company may gain sufficient ownership influence or control to change management. During this rather confrontational process, it is up to the board of the target company to continue to take actions consistent with protecting the rights of share-holders. As in Stage 1, the board may need to provide rather strong oversight of management during this process to ensure that the acts of management are consistent with protecting and building shareholder value.

Regulatory approval alone may prove to be a major hurdle in the execution of the deal. An acquisition may be subject to significant regulatory approval if it involves a company in an industry considered fundamental to national security or if there is concern over major concentration and anticompetitive results from consolidation.

The proposed acquisition of Honeywell International (itself the result of a merger of Honeywell U.S. and Allied-Signal U.S.) by General Electric (U.S.) in 2001 was something of a watershed event in the field of regulatory approval. General Electric's acquisition of Honeywell had been approved by management, ownership, and U.S. regulatory bodies when it then sought approval within the European Union. Jack Welch, the charismatic chief executive officer and president of GE did not anticipate the degree of opposition that the merger would face from EU authorities. After a continuing series of demands by the EU that specific businesses within the combined companies be sold off to reduce anticompetitive effects, Welch withdrew the request for acquisition approval, arguing that the liquidations would destroy most of the value-enhancing benefits of the acquisition. The acquisition was canceled. This case may have far-reaching effects on cross-border M&A for years to come, as the power of regulatory authorities within strong economic zones like the EU to block the combination of two MNEs may foretell a change in regulatory strength and breadth.

The last act within this second stage of cross-border acquisition, compensation settlement, is the payment to shareholders of the target company. Shareholders of the target company are typically paid either in shares of the acquiring company or in cash. If a share exchange occurs, the exchange is generally defined by some ratio of acquiring company shares to target company shares (say, two shares of acquirer in exchange for three shares of target), and the stockholder is typically not taxed—the shares of ownership are simply replaced by other shares in a nontaxable transaction.

If cash is paid to the target company shareholder, it is the same as if the shareholder sold the shares on the open market, resulting in a capital gain or loss (a gain, it is hoped, in the case of an acquisition) with tax liabilities. Because of the tax ramifications, shareholders are typically more receptive to share exchanges so that they may choose whether and when tax liabilities will arise.

A variety of factors go into the determination of the type of settlement. The availability of cash, the size of the acquisition, the friendliness of the takeover, and the relative valuations of both acquiring firm and target firm affect the decision. One of the most destructive forces that sometimes arises at this stage is regulatory delay and its impact on the share prices of the two firms. If regulatory body approval drags out over time, the possibility of a drop in share price increases and can change the attractiveness of the share swap.

Stage 3: Post-Acquisition Management. Although the headlines and flash of investment banking activities are typically focused on the valuation and bidding process in an acquisi-tion transaction, post-transaction management is probably the most critical of the three stages in determining an acquisition's success or failure. An acquiring firm can pay too little or too much, but if the post transaction is not managed effectively, the entire return on the investment is squandered. Post-acquisition management is the stage in which the motivations for the transaction must be realized—motivations such as more effective management, synergies arising from the new combination, or the injection of capital at a cost and availability previously out of the reach of the acquisition target, must be effectively implemented after the transaction. The biggest problem, however, is nearly always melding corporate cultures.

The clash of corporate cultures and personalities pose both the biggest risk and the biggest potential gain from cross-border mergers and acquisitions. Although not readily measurable as are price/earnings ratios or share price premiums, in the end, the value is either gained or lost in the hearts and minds of the stakeholders.

Currency Risks in Cross-Border Acquisitions

The pursuit and execution of a cross-border acquisition poses a number of challenging foreign currency risks and exposures for an MNE. As illustrated by Exhibit 18.9, the nature of the currency exposure related to any specific cross-border acquisition evolves as the bidding and negotiating process itself evolves across the bidding, financing, transaction (settlement), and

EXHIBIT 18.9   **Currency Risks in Cross-Border Acquisitions**

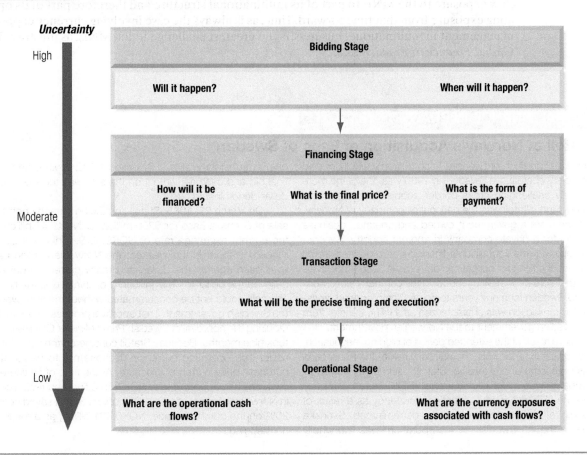

operating stages. The assorted risks, both in the timing and information related to the various stages of a cross-border acquisition, make the management of the currency exposures difficult. As illustrated in Exhibit 18.9, the uncertainty related to the multitude of stages declines over time as stages are completed and contracts and agreements reached.

The initial bid, if denominated in a foreign currency, creates a *contingent foreign currency exposure* for the bidder. This contingent exposure grows in certainty of occurrence over time as negotiations continue, regulatory requests and approvals are gained, and competitive bidders emerge. Although a variety of hedging strategies might be employed, the use of a purchased currency call option is the simplest. The option's notional principal would be for the estimated purchase price, but the maturity, for the sake of conservatism, might possibly be significantly longer than probably needed to allow for extended bidding, regulatory, and negotiation delays.

Once the bidder has successfully won the acquisition, the exposure evolves from a *contingent exposure* to a *transaction exposure*. Although a variety of uncertainties remain as to the exact timing of the transaction settlement, the certainty over the occurrence of the currency exposure is largely eliminated. Some combination of forward contracts and purchased currency options may then be used to manage the currency risks associated with the completion of the cross-border acquisition.

Once consummated, the currency risks and exposures of the cross-border acquisition, now a property and foreign subsidiary of the MNE, changes from being a transaction-based cash flow exposure to the MNE to part of its multinational structure and therefore part of its operating exposure from that time forward. Time, as is always the case involving currency exposure management in multinational business, is the greatest challenge to the MNE, as illustrated by *Global Finance in Practice 18.2*.

GLOBAL FINANCE IN PRACTICE 18.2

Statoil of Norway's Acquisition of Esso of Sweden

Statoil's acquisition of Svenska Esso (Exxon's wholly owned subsidiary operating in Sweden) in 1986 was one of the more uniquely challenging cross-border acquisitions ever completed. First, Statoil was the national oil company of Norway, and therefore a government-owned and operated business bidding for a private company in another country. Second, if completed, the acquisition's financing as proposed would increase the financial obligations of Svenska Esso (debt levels and therefore debt service), reducing the company's tax liabilities to Sweden for many years to come. The proposed cross-border transaction was characterized as a value transfer from the Swedish government to the Norwegian government.

As a result of the extended period of bidding, negotiation, and regulatory approvals, the currency risk of the transaction was both large and extensive. Statoil, being a Norwegian oil company, was a Norwegian kroner (NOK)-based company with the U.S. dollar as its functional currency as a result of the global oil industry being dollar-denominated. Svenska Esso, although Swedish by incorporation, was the wholly

owned subsidiary of a U.S.-based MNE, Exxon, and the final bid and cash settlement on the sale was therefore U.S. dollar-denominated.

On March 26, 1985, Statoil and Exxon agreed upon the sale of Svenska Esso for $260 million, or NOK2.47 billion at the current exchange rate of NOK9.50/$. (This was by all modern standards the weakest the Norwegian krone had ever been against the dollar, and many currency analysts believed the dollar to be significantly overvalued at the time.) The sale could not be consummated without the approval of the Swedish government. That approval process—eventually requiring the approval of Swedish Prime Minister Olaf Palme—took nine months. Because Statoil considered the U.S. dollar as its true operating currency, it chose not to hedge the purchase price currency exposure. At the time of settlement the krone had appreciated to NOK7.65/$, final acquisition cost in Norwegian kroner of NOK1.989 billion. Statoil saved nearly 20% on the purchase price, NOK0.481 billion, as a result of not hedging.

SUMMARY POINTS

- Parent cash flows must be distinguished from project cash flows. Each of these two types of flows contributes to a different view of value.

- Parent cash flows often depend on the form of financing. Thus, cash flows cannot be clearly separated from financing decisions, as is done in domestic capital budgeting.

- Remittance of funds to the parent must be explicitly recognized because of differing tax systems, legal and political constraints on the movement of funds, local business norms, and differences in how financial markets and institutions function.

- When a foreign project is analyzed from the project's point of view, risk analysis focuses on the use of sensitivities, as well as consideration of foreign exchange and political risks associated with the project's execution over time.

- When a foreign project is analyzed from the parent's point of view, the additional risk that stems from its "foreign" location can be measured in at least two ways, adjusting the discount rates or adjusting the cash flows.

- Real option analysis is a different way of thinking about investment values. At its core, it is a cross between decision-tree analysis and pure option-based valuation. It allows us to evaluate the option to defer, the option to abandon, the option to alter size or capacity, and the option to start up or shut down a project.

- Project finance is used widely today in the development of large-scale infrastructure projects in many emerging markets. Although each individual project has unique characteristics, most are highly leveraged transactions, with debt making up more than 60% of the total financing.

- The process of acquiring an enterprise anywhere in the world has three common elements: (1) identification and valuation of the target; (2) completion of the ownership change transaction (the tender); and (3) the management of the post-acquisition transition.

- Cross-border mergers, acquisitions, and strategic alliances, all face similar challenges: They must value the target enterprise on the basis of its projected performance in its market. This process of enterprise valuation combines elements of strategy, management, and finance.

MINI-CASE

Elan and Royalty Pharma[5]

We lived a long time with Elan (ELN). We always appreciated its science and scientists, and, at times, we hated its former management, or whoever caused it to turn from ascending towards becoming a citadel of sciences, especially neurosciences, into an almost bankrupt firm with less everything valuable in it than what was necessary for its survival. What saved it at the time was the emergence of Tysabri, for multiple sclerosis, which we knew it was second to none in treatment of relapsing remitting multiple sclerosis. We were certain that this drug, like Aaron's cane, would swallow up all magicians' staffs.

> —"Biogen Idec Pays Elan $3.25 Billion for Tysabri:
> Do We Leave, Or Stay?," *Seeking Alpha*,
> February 6, 2013.

Elan's shareholders (Elan Corporation, NYSE: ELN) were faced with a difficult choice. Elan's management had made four proposals to shareholders in an attempt to defend itself against a hostile takeover from Royalty Pharma (U.S.), a privately held company. If shareholders voted in favor of any of the four initiatives, it would kill Royalty Pharma's offer. That would allow Elan to stay independent and remain under the control of a management team that had not sparked confidence in recent years. All votes had to be filed by midnight June 16, 2013.

The Players

Elan Corporation was a global biopharmaceutical company headquartered in Dublin, Ireland. Elan focused on the discovery, development and marketing of therapeutic products in neurology including Alzheimer's disease and Parkinson's disease and autoimmune diseases such as multiple sclerosis and Crohn's disease. But over time the company had spun-out, sold-off, or closed most of its business activities. By the spring of 2013, Elan was a company of only two assets: a large pile of cash and a perpetual royalty stream on a leading therapeutic for multiple sclerosis called Tysabri, which it had co-developed with Biogen.

The solution to Elan's problem was the sale of its interest in Tysabri to its partner Biogen. In February 2013 Elan sold its 50% rights in Tysabri to Biogen in return for $3.29 billion in cash and a perpetual royalty stream on Tysabri. Whereas previously Elan earned returns on only its 50% share of Tysabri, the royalty agreement was based on 100% of the asset. The royalty was a step-up rate structure on worldwide sales of 12% in year 1, 18% all subsequent years, plus 25% on all global sales above $2 billion.

The ink had barely dried on Elan's sale agreement in February 2013 when it was approached by a private U.S. firm, Royalty Pharma, about the possible purchase of Elan for $11 per share. Elan acknowledged the proposal publicly, and stated it would consider the proposal along with other strategic options.

Royalty Pharma (RP) is a privately held company (owned by private equity interests) that acquires royalty interests in marketed or late-stage pharmaceutical products. Its business allows the owners of these intellectual products to monetize their interests in order to pursue additional business development opportunities. RP accepts the risk that the price they paid for the asset interest will actually accrue over time. RP owns royalty rights; it does not operate or market.

In March 2013, possibly tired of waiting, RP issued a statement directly to Elan shareholders to encourage them to vote for the proposed acquisition of Elan for $11 per share. At that time, Elan issued a response to RP's statement that characterized the Royalty Pharma proposal as "conditional and opportunistic."

Elan's Defense

Elan's leadership was now under considerable pressure by shareholders to explain why shareholders should not tender their shares to Royalty Pharma. In May, Elan began to detail a collection of initiatives to redefine the company. Going forward, Elan described a series of four complex strategic initiatives that it would pursue to grow and diversify the firm beyond its current two-asset portfolio. Because the company was currently in the offer period of a proposed acquisition, Irish securities laws required that all four of Elan's proposals be approved by shareholders. But from the beginning that appeared difficult given public perception that the initiatives were purely defensive.

Royalty Pharma responded publicly with a letter to Elan's stockholders questioning whether Elan's leadership was really acting in the best interests of the shareholders. It then increased its tender offer to $12.50/share plus a *Contingent Value Right* (CVR). The CVR was a conditional element where all shareholders would receive an additional amount per share in the future—up to an additional $2.50 per share—if Tysabri's future sales reached specific milestone targets. Royalty Pharma's CVR offer required Tysabri sales to hit $2.6 billion by 2015 and $3.1 billion by 2017. Royalty Pharma also made it very clear that if shareholders were to approve any of the Elan's four management proposals, the acquisition offer would lapse.

The Value Debate

Elan, as of May 2013, consisted of $1.787 billion in cash, the Tysabri royalty stream, a few remaining prospective pipeline products, and between $100 and $200 million in annual expenses associated with its business. Elan's leadership wanted to use its cash and its annual royalty earnings to build a new business. Royalty Pharma just wanted to buy Elan, take the cash and royalty stream assets, and shut Elan down.

The valuation debate on Elan revolved around the value of the Tysabri royalty stream. That meant predicting what actual sales were likely to be in the coming decade. Exhibit A presents Royalty Pharma's synopsis of the sales debate, noting that Elan's claims on value have been selectively high, while Royal Pharma has based its latest offer on the Street Consensus numbers.

Predicting royalty earnings on biotechnology products is not all that different than predicting the sales of any product. Pricing, competition, regulation, government policy, changing demographics and conditions—all could change future global sales. That said, there were several more distinct factors of concern.

First, Tysabri was scheduled to go off-patent in 2020 (original patent filing was in 2000). The Street Consensus forecast, the one advocated by Royalty Pharma, predicted Tysabri global sales to peak in that year at $2.74 billion. Sales would slide, but continue, in the following years. Second, competitive products were already entering the market. In the spring, Biogen had finally received FDA approval on an oral treatment for relapsing-remitting forms of multiple sclerosis. It was only one of several new treatments coming to the market. Royalty Pharma had pointed to declining new patient adds over the past two quarters as evidence that aggressive future sales forecasts for Tysabri may be unrealistic—already.

For these and other reasons Royalty Pharma had argued that a conservative sales forecast was critically important for investors to use when deciding whether or not to go with

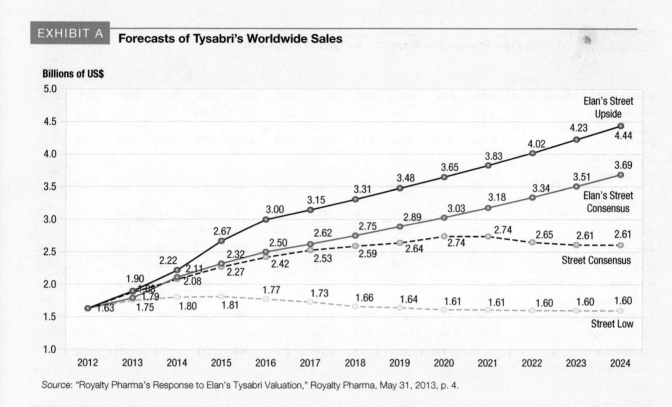

EXHIBIT A   **Forecasts of Tysabri's Worldwide Sales**

Source: "Royalty Pharma's Response to Elan's Tysabri Valuation," Royalty Pharma, May 31, 2013, p. 4.

management or Royalty Pharma's offer. Royalty Pharma's valuation, presented in Exhibit B, used this sales forecast for its baseline analysis. Royalty Pharma's valuation of Elan was based on the following critical assumptions:

- Tysabri's worldwide sales, the top-line of the valuation, were based on the Street Consensus.

- Elan's operating expenses would remain relatively flat, rising at 1% to 2% per year, from $75 million in 2013.

- Elan's net operating losses and Irish incorporation would reduce effective taxes to 1% per year through 2017, rising to Ireland's still relatively low corporate tax rate of 12.5% per year afterward.

- The discount rate would be 7.5% per year up until going off-patent in 2017, and rising to 10% after that.

- Perpetuity value (terminal value) would be based on year 2024's income, discounted at 12%, and assuming an annual growth rate of either −2% or −4% as Tysabri's sales slide into the future.

- There were 518 million shares outstanding as of May 29, 2013, according to Elan's most recent communications.

- Elan's cash total was $1.787 billion, according to Elan's most recent communications.

The result was a base valuation of $10.49 or $10.17 per share, depending on the terminal value decline assumption. As typical of most valuations, the top-line total sales was the single largest driver for all future projected cash flows. The shares outstanding assumption, 518 million shares, reflected the results of a large share repurchase program that Elan had pursued right up to mid-May of 2013. Note that Royalty Pharma expressly decomposed its total valuation into three pieces: (1) the under patent period, (2) the post-patent period, and (3) the perpetuity value. In Royalty Pharma's opinion, the post-patent period represented a significantly higher risk period for actual Tysabri sales.

Market Valuation

Despite the debate over Elan's value, as a publicly traded company, the market made its opinion known every single trading day. On the day prior to receiving the first indication of interest from Royalty Pharma, Elan was trading at $11 per share. (In the days that follow, the market is factoring in what it thinks the effective offer price is from a suitor like Royalty Pharma and the probability of the acquisition occurring.) Elan's share price history for 2013 is shown in Exhibit C.

EXHIBIT 3 **Valuing Elan: Prospective Royalties on Tysabri Plus Cash**

Millions of US$	Rates		Actual 2012	2013	2014	2015	2016	2017	2018	2019	2020	2021	2022	2023	2024
Worldwide Sales			1,631	1,884	2,082	2,266	2,418	2,530	2,591	2,643	2,742	2,744	2,653	2,609	2,611
year-over-year growth				15.5%	10.5%	8.8%	6.7%	4.6%	2.4%	2.0%	3.7%	0.1%	−3.3%	−1.7%	0.1%
Royalties to Elan:															
$0 to $2 billion in sales	18%			151	360	360	360	360	360	360	360	360	360	360	360
Greater than $2 billion	25%				21	67	105	133	148	161	186	186	163	152	153
Total Royalties				151	381	427	465	493	508	521	546	546	523	512	513
Expenses				(75)	(77)	(78)	(80)	(81)	(83)	(84)	(86)	(88)	(90)	(91)	(93)
Pre-tax Income				76	304	349	385	412	425	437	460	458	433	421	420
Less Taxes	1%	12.5%		(1)	(3)	(3)	(4)	(4)	(53)	(55)	(57)	(57)	(54)	(53)	(52)
Net Income				75	300	345	381	407	372	382	402	401	379	369	367
WACC				7.5%	7.5%	7.5%	7.5%	7.5%	7.5%	7.5%	10.0%	10.0%	10.0%	10.0%	10.0%
Discount Factor				1.0000	0.9302	0.8653	0.8050	0.7488	0.6966	0.6480	0.5132	0.4665	0.4241	0.3855	0.3505
PV of Net Income				75	280	299	306	305	259	248	206	187	161	142	129
Perpetuity Value	−2%														2,999
Discount Factor															0.3505
PV of Perpetuity															1,051
NPV (cumulative PV)				$3,647												
Shares Outstanding (millions)				518												
Value per Share				$7.04												
Cash				1,787												
Cash Value per Share				$3.45												
Elan, Total Value per Share				$10.49												

(2%) Perpetuity Growth Rate		Total	Per Share	% of Total		(4%) Perpetuity Growth Rate		Total	Per Share	% of Total
Discounted value 2013–2020		$1,977	$3.82	54.2%		Discounted value 2013–2020		$1,977	$3.82	56.8%
Discounted value 2021–2024		$619	$1.19	17.0%		Discounted value 2021–2024		$619	$1.19	17.8%
Perpetuity value beyond 2024	−2%	$1,051	$2.03	28.8%		Perpetuity value beyond 2024	−4%	$883	$1.70	25.4%
Total Tysabri Value		$3,647	$7.04	100.0%		Total Tysabri Value		$3,479	$6.72	100.0%
Cash		$1,787	$3.45			Cash		$1,787	$3.45	
Total Elan Value		$5,434	$10.49			Total Elan Value		$5,266	$10.17	

Notes: Valuation based on that presented in "Royalty Pharma's Response to Elan's Tysabri Valuation," May 29, 2013, p. 12. Royalties paid for the first 12 months, approximately 2013 in length, are at 12%. Perpetuity value (terminal value) assumes net revenues "grow" at −2% per annum indefinitely and are discounted at 10%. Assumes same 518 million shares outstanding as Elan stated on May 29, 2013. Elan's tax loss carry-forwards reduce the effective tax rate to 1% through 2017; beginning in 2018 the royalty stream is subject to Irish taxation at 12.5%. Royalty Pharma believes that the WACC should rise from 7.50% to 10.0% beginning in 2020 when Tysabri goes off-patent.

EXHIBIT C | **Elan's Share Price (January 1–June 16, 2013)**

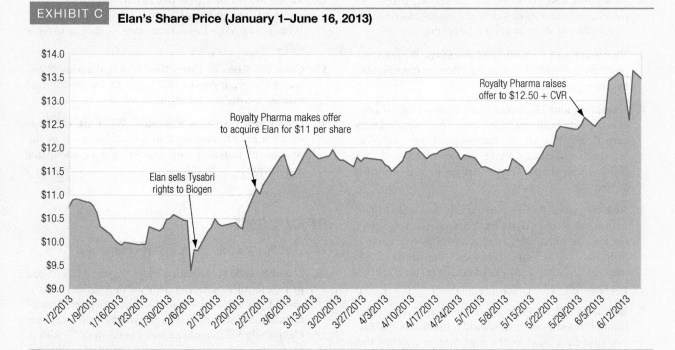

Elan's management had made their case to shareholders. The collection of initiatives that Elan's leadership wished to pursue had to be approved, however, by shareholders. The Extraordinary General Meeting (EGM) of shareholders would be held on Monday, June 17th. At that meeting the results of the shareholder vote (all votes were due by the previous Friday) would be announced.

In the days leading up to the EGM, the battle had become very public, and in the words of one journalist, "quite chippy." In a *Financial Times* editorial, one former Elan board member, Jack Schuler, wrote "I have no confidence that Kelly Martin [Elan's CEO] or the other Elan board members will act in the interests of shareholders. I hope the Elan shareholders realise that their only option is to sell the company to the highest bidder." Elan's current non-executive chairman then responded: "I note that Elan's share price has trebled since Mr. Schuler's departure. The board and management team remain wholly focused on continued value creation and will continue to act in the best interests of our shareholders."

Shareholders had to decide—quickly.

Mini-Case Questions

1. Using the sales forecasts for Tysabri presented in Exhibit A, and using the discounted cash flow model presented in Exhibit B, what do you think Elan is worth?

2. What other considerations do you think should be included in the valuation of Elan?

3. What would be your recommendation to shareholders— to approve management's proposals killing RP's offer—or say "no" to the proposals, probably prompting the acceptance of RP's offer?

QUESTIONS

These questions are available in MyFinanceLab.

1. **Capital Budgeting Theoretical Framework.** Capital budgeting for a foreign project uses the same theoretical framework as domestic capital budgeting. What are the basic steps in domestic capital budgeting?

2. **Foreign Complexities.** Capital budgeting for a foreign project is considerably more complex than the domestic case. What are the factors that add complexity?

3. **Project versus Parent Valuation.** Why should a foreign project be evaluated both from a project and parent viewpoint?

4. **Viewpoint and NPV.** Which viewpoint, project or parent, gives results closer to the traditional meaning of net present value in capital budgeting?

5. **Viewpoint and Consolidated Earnings.** Which viewpoint gives results closer to the effect on consolidated earnings per share?

6. **Operating and Financing Cash Flows.** Capital projects provide both operating cash flows and financial cash flows. Why are operating cash flows preferred for domestic capital budgeting but financial cash flows given major consideration in international projects?

7. **Risk-Adjusted Return.** Should the anticipated internal rate of return (IRR) for a proposed foreign project be compared to (a) alternative home country proposals, (b) returns earned by local companies in the same industry and/or risk class, or (c) both? Justify your answer.

8. **Blocked Cash Flows.** In the evaluation of a potential foreign investment, how should a multinational firm evaluate cash flows in the host foreign country that are blocked from being repatriated to the firm's home country?

9. **Host Country Inflation.** How should an MNE factor host country inflation into its evaluation of an investment proposal?

10. **Cost of Equity.** A foreign subsidiary does not have an independent cost of capital. However, in order to estimate the discount rate for a comparable host-country firm, the analyst should try to calculate a hypothetical cost of capital. How is this done?

11. **Viewpoint Cash Flows.** What are the differences in the cash flows used in a project point of view analysis and a parent point of view analysis?

12. **Foreign Exchange Risk and Capital Budgeting.** How is foreign exchange risk sensitivity factored into the capital budgeting analysis of a foreign project?

13. **Expropriation Risk.** How is expropriation risk factored into the capital budgeting analysis of a foreign project?

14. **Real Option Analysis.** What is real option analysis? How is it a better method of making investment decisions than traditional capital budgeting analysis?

15. **M&A Business Drivers.** What are the primary driving forces that motivate cross-border mergers and acquisitions?

16. **Three Stages of Cross-Border Acquisitions.** What are the three stages of a cross-border acquisition? What are the core financial elements integral to each stage?

17. **Currency Risks in Cross-Border Acquisitions.** What are the currency risks that arise in the process of making a cross-border acquisition?

18. **Contingent Currency Exposure.** What are the largest contingent currency exposures that arise in the process of pursuing and executing a cross-border acquisition?

PROBLEMS

These problems are available in MyFinanceLab.

1. **Carambola de Honduras.** Slinger Wayne, a U.S.-based private equity firm, is trying to determine what it should pay for a tool manufacturing firm in Honduras named Carambola. Slinger Wayne estimates that Carambola will generate a free cash flow of 13 million Honduran lempiras (Lp) next year (2012), and that this free cash flow will continue to grow at a constant rate of 8.0% per annum indefinitely.

   A private equity firm like Slinger Wayne, however, is not interested in owning a company for long, and plans to sell Carambola at the end of three years for approximately 10 times Carambola's free cash flow in that year. The current spot exchange rate is Lp14.80/$, but the Honduran inflation rate is expected to remain at a relatively high rate of 16.0% per annum compared to the U.S. dollar inflation rate of only 2.0% per annum. Slinger Wayne expects to earn at least a 20% annual rate of return on international investments like Carambola.

   a. What is Carambola worth if the Honduran lempira were to remain fixed over the three year investment period?

   b. What is Carambola worth if the Honduran lempira were to change in value over time according to purchasing power parity?

2. **Finisterra, S.A.** Finisterra, S.A., located in the state of Baja California, Mexico, manufactures frozen Mexican food that is popular in the states of California and Arizona (U.S.). In order to be closer to its U.S. market, Finisterra is considering moving some of its manufacturing operations to southern California. Operations in California would begin in year 1 and have the following attributes:

Assumptions	Value
Sales price per unit, year 1 (US$)	$5.00
Sales price increase, per year	3.00%
Initial sales volume, year 1, units	1,000,000
Sales volume increase, per year	10.00%
Production costs per unit, year 1	$4.00
Production cost per unit increase, per year	4.00%
General and administrative expenses per year	$100,000
Depreciation expenses per year	$ 80,000
Finisterra's WACC (pesos)	16.00%
Terminal value discount rate	20.00%

The operations in California will pay 80% of its accounting profit to Finisterra as an annual cash dividend. Mexican taxes are calculated on grossed up dividends from foreign countries, with a credit for host-country taxes already paid. What is the maximum U.S. dollar price Finisterra should offer in year 1 for the investment?

3. **Grenouille Properties.** Grenouille Properties (U.S.) expects to receive cash dividends from a French joint venture over the coming three years. The first dividend, to be paid December 31, 2011, is expected to be €720,000. The dividend is then expected to grow 10.0% per year over the following two years. The current exchange rate (December 30, 2010) is $1.3603/€. Grenouille's weighted average cost of capital is 12%.
   a. What is the present value of the expected euro dividend stream if the euro is expected to appreciate 4.00% per annum against the dollar?
   b. What is the present value of the expected dividend stream if the euro were to depreciate 3.00% per annum against the dollar?

4. **Natural Mosaic.** Natural Mosaic Company (U.S.) is considering investing Rs50,000,000 in India to create a wholly owned tile manufacturing plant to export to the European market. After five years, the subsidiary would be sold to Indian investors for Rs100,000,000. A pro forma income statement for the Indian operation predicts the generation of Rs7,000,000 of annual cash flow, is listed in the following table.

Sales revenue	30,000,000
Less cash operating expenses	(17,000,000)
Gross income	13,000,000
Less depreciation expenses	(1,000,000)
Earnings before interest and taxes	12,000,000
Less Indian taxes at 50%	(6,000,000)
Net income	6,000,000
Add back depreciation	1,000,000
Annual cash flow	7,000,000

The initial investment will be made on December 31, 2011, and cash flows will occur on December 31st of each succeeding year. Annual cash dividends to Natural Mosaic from India will equal 75% of accounting income.

The U.S. corporate tax rate is 40% and the Indian corporate tax rate is 50%. Because the Indian tax rate is greater than the U.S. tax rate, annual dividends paid to Natural Mosaic will not be subject to additional taxes in the United States. There are no capital gains taxes on the final sale. Natural Mosaic uses a weighted average cost of capital of 14% on domestic investments, but will add six percentage points for the Indian investment because of perceived greater risk. Natural Mosaic forecasts the rupee/dollar exchange rate on December 31st for the next six years as listed below.

	R$/$		R$/$
2011	50	2014	62
2012	54	2015	66
2013	58	2016	70

What is the net present value and internal rate of return on this investment?

5. **Doohicky Devices.** Doohickey Devices, Inc., manufactures design components for personal computers. Until the present, manufacturing has been subcontracted to other companies, but for reasons of quality control Doohicky has decided to manufacture itself in Asia. Analysis has narrowed the choice to two possibilities, Penang, Malaysia, and Manila, the Philippines. At the moment only the summary of expected, after-tax, cash flows displayed at the bottom of this page is available. Although most operating outflows would be in Malaysian ringgit or Philippine pesos, some additional U.S. dollar cash outflows would be necessary, as shown in the table at the top of the next page.

The Malaysia ringgit currently trades at RM3.80/$ and the Philippine peso trades at Ps50.00/$. Doohicky expects the Malaysian ringgit to appreciate 2.0% per year against the dollar, and the Philippine peso to depreciate 5.0% per year against the dollar. If the weighted average cost of capital for Doohicky Devices is 14.0%, which project looks most promising?

Problem 5.

Doohicky in Penang (after-tax)	2012	2013	2014	2015	2016	2017
Net ringgit cash flows	(26,000)	8,000	6,800	7,400	9,200	10,000
Dollar cash outflows	—	(100)	(120)	(150)	(150)	—
Doohicky in Manila (after-tax)						
Net peso cash flows	(560,000)	190,000	180,000	200,000	210,000	200,000
Dollar cash outflows	—	(100)	(200)	(300)	(400)	—

Problem 6.

Assumptions	0	1	2	3
Original investment (Czech korunas, K)	250,000,000			
Spot exchange rate (K/$)	32.50	30.00	27.50	25.00
Unit demand		700,000	900,000	1,000,000
Unit sales price		$10.00	$10.30	$10.60
Fixed cash operating expenses		$1,000,000	$1,030,000	$1,060,000
Depreciation		$500,000	$500,000	$500,000
Investment in working capital (K)	100,000,000			

6. **Wenceslas Refining Company.** Privately owned Wenceslas Refining Company is considering investing in the Czech Republic so as to have a refinery source closer to its European customers. The original investment in Czech korunas would amount to K250 million, or $5,000,000 at the current spot rate of K32.50/$, all in fixed assets, which will be depreciated over 10 years by the straight-line method. An additional K100,000,000 will be needed for working capital.

For capital budgeting purposes, Wenceslas assumes sale as a going concern at the end of the third year at a price, after all taxes, equal to the net book value of fixed assets alone (not including working capital). All free cash flow will be repatriated to the United States as soon as possible. In evaluating the venture, the U.S. dollar forecasts are shown in the table above.

Variable manufacturing costs are expected to be 50% of sales. No additional funds need be invested in the U.S. subsidiary during the period under consideration. The Czech Republic imposes no restrictions on repatriation of any funds of any sort. The Czech corporate tax rate is 25% and the United States rate is 40%. Both countries allow a tax credit for taxes paid in other countries. Wenceslas uses 18% as its weighted average cost of capital, and its objective is to maximize present value. Is the investment attractive to Wenceslas Refining?

Hermosa Beach Components (U.S.)

Use the following information and assumptions to answer Problems 7–10.

Hermosa Beach Components, Inc., of California exports 24,000 sets of low-density light bulbs per year to Argentina under an import license that expires in five years. In Argentina, the bulbs are sold for the Argentine peso equivalent of $60 per set. Direct manufacturing costs in the United States and shipping together amount to $40 per set. The market for this type of bulb in Argentina is stable, neither growing nor shrinking, and Hermosa holds the major portion of the market.

The Argentine government has invited Hermosa to open a manufacturing plant so imported bulbs can be replaced by local production. If Hermosa makes the investment, it will operate the plant for five years and then sell the building and equipment to Argentine investors at net book value at the time of sale plus the value of any net working capital. (Net working capital is the amount of current assets less any portion financed by local debt.) Hermosa will be allowed to repatriate all net income and depreciation funds to the United States each year. Hermosa traditionally evaluates all foreign investments in U.S. dollar terms.

Investment. Hermosa's anticipated cash outlay in U.S. dollars in 2012 would be as follows:

Building and equipment	$1,000,000
Net working capital	1,000,000
Total investment	$2,000,000

All investment outlays will be made in 2012, and all operating cash flows will occur at the end of years 2013 through 2017.

Depreciation and Investment Recovery. Building and equipment will be depreciated over five years on a straight-line basis. At the end of the fifth year, the $1,000,000 of net working capital may also be repatriated to the United States, as may the remaining net book value of the plant.

Sales Price of Bulbs. Locally manufactured bulbs will be sold for the Argentine peso equivalent of $60 per set.

Operating Expenses per Set of Bulbs. Material purchases are as follows:

Materials purchased in Argentina (U.S. dollar equivalent)	$20 per set
Materials imported from Hermosa Beach-USA	10 per set
Total variable costs	$30 per set

Transfer Prices. The $10 transfer price per set for raw material sold by the parent consists of $5 of direct and indirect costs incurred in the United States on their manufacture, creating $5 of pre-tax profit to Hermosa Beach.

Taxes. The corporate income tax rate is 40% in both Argentina and the United States (combined federal and state/province). There are no capital gains taxes on the future sale of the Argentine subsidiary, either in Argentina or the United States.

Discount Rate. Hermosa Components uses a 15% discount rate to evaluate all domestic and foreign projects.

7. **Hermosa Components: Baseline Analysis.** Evaluate the proposed investment in Argentina by Hermosa Components (U.S.). Hermosa's management wishes the baseline analysis to be performed in U.S. dollars (and implicitly also assumes the exchange rate remains fixed throughout the life of the project). Create a project viewpoint capital budget and a parent viewpoint capital budget. What do you conclude from your analysis?

8. **Hermosa Components: Revenue Growth Scenario.** As a result of their analysis in Problem 7, Hermosa wishes to explore the implications of being able to grow sales volume by 4% per year. Argentine inflation is expected to average 5% per year, so sales price and material cost increases of 7% and 6% per year, respectively, are thought reasonable. Although material costs in Argentina are expected to rise, U.S.-based costs are not expected to change over the five-year period. Evaluate this scenario for both the project and parent viewpoints. Is the project under this revenue growth scenario acceptable?

9. **Hermosa Components: Revenue Growth and Sales Price Scenario.** In addition to the assumptions employed in Problem 8, Hermosa now wishes to evaluate the prospect of being able to sell the Argentine subsidiary at the end of year 5 at a multiple of the business' earnings in that year. Hermosa believes that a multiple of six is a conservative estimate of the market value of the firm at that time. Evaluate the project and parent viewpoint capital budgets.

10. **Hermosa Components: Revenue Growth, Sales Price, and Currency Risk Scenario.** Melinda Deane, a new analyst at Hermosa and a recent MBA graduate, believes that it is a fundamental error to evaluate the Argentine project's prospective earnings and cash flows in dollars, rather than first estimating their Argentine peso (Ps) value and then converting cash flow returns to the United States in dollars. She believes the correct method is to use the end-of-year spot rate in 2012 of Ps3.50/$ and assume it will change in relation to purchasing power. (She is assuming U.S. inflation to be 1% per annum and Argentine inflation to be 5% per annum). She also believes that Hermosa should use a risk-adjusted discount rate in Argentina that reflects Argentine capital costs (20% is her estimate) and a risk-adjusted discount rate for the parent viewpoint capital budget (18%) on the assumption that international projects in a risky currency environment should require a higher expected return than other lower-risk projects. How do these assumptions and changes alter Hermosa's perspective on the proposed investment?

INTERNET EXERCISES

1. **Capital Projects and the EBRD.** The European Bank for Reconstruction and Development (EBRD) was established to foster market-oriented business development in the former Soviet Bloc. Use the EBRD Web site to determine which projects and companies EBRD is currently undertaking.

European Bank for Reconstruction and Development	www.ebrd.com

2. **Emerging Markets: China.** Long-term investment projects such as electrical power generation require a thorough understanding of all attributes of doing business in that country. China is currently the focus of investment and market penetration strategies of multinational firms worldwide. Using the Web (you might start with the Web sites listed below), build a database on doing business in China, and prepare an update of many of the factors discussed in this chapter.

Ministry of Foreign Trade and Economic Cooperation, PRC	english.mofcom.gov.cn
China Investment Trust and Investment Corporation	www.citic.com/wps/portal/enlimited
ChinaNet Investment Pages	www.chinanet-online.com

3. **BeyondBrics: The *Financial Times*' Emerging Market Hub.** Check the FT's blog on emerging markets for the latest debates and guest editorials.

Financial Times Blog on Emerging Markets	blogs.ft.com/beyond-brics/

Answers to Selected End-of-Chapter Problems

Chapter 1: Multinational Financial Management: Opportunities and Challenges

10. a. $14.77
    b. U.S. = 30.5%, Brazil = 27.1%, Germany = 40.1%, China = 2.4%
    c. 69.5%

13. Appreciation case: +13.9%
    Depreciation case: −13.9%

Chapter 2: The International Monetary System

2. $40,257.65
6. 1.1398
7. −41.82%

Chapter 3: The Balance of Payments

Australia	2000	2001	2002	2003	2004	2005	2006	2007	2008	2009	2010	2011	2012	2013
3.1 What is Australia's balance on goods? (goods exports − goods imports)	−4,813	1,786	−5,431	−15,369	−18,031	−13,372	−9,596	−17,784	−4,915	−4,439	17,479	22,481	−12,186	4,390
3.2 What is Australia's balance on services? (services credit − services debit)	289	−259	−201	−433	−678	542	869	588	−3,098	−1,351	−4,345	−10,244	−12,371	−14,055
3.3 What is Australia's balance on goods and services (balance on goods + balance on services)	−4,524	1,527	−5,632	−15,802	−18,709	−12,830	−8,727	−17,196	−8,013	−5,790	13,134	12,237	−24,557	−9,665
3.4 What is Australia's current account balance? (the sum of the four balances listed above, goods, services, income, and current transfers)	−15,103	−8,721	−17,385	−30,674	−40,066	−41,032	−41,504	−58,031	−47,786	−44,999	−37,177	−44,524	−48,738	−30,777

Chapter 4: Financial Goals and Corporate Governance

1. a. 25.000%
   b. 33.333%
   c. Dividend yield = 8.333%, capital gains = 25.00%, total shareholder return = 33.333%

2. a. 64.23%
   b. 4.19%
   c. 71.12%

Chapter 5: The Foreign Exchange Market

1. a. 4.72
   b. 21,243

10. a. Profit of 26,143.79
    b. Loss of (26,086.96)

Chapter 6: International Parity Conditions

4. a. 1.0941
   b. 1.1155 and 948.19.

7. A CIA profit potential of −0.042% tells Takeshi he should borrow Japanese yen and invest in the higher yielding currency, the U.S. dollar, to earn a CIA profit of 55,000.

Chapter 7: Foreign Currency Derivatives: Futures and Options

1. a. ($49,080.00)
   b. $38,920.00
   c. ($9,080.00)

4. a. Sallie should buy a call on Singapore dollars
   b. $0.65046
   c. Gross profit = $0.05000 Net profit = $0.04954
   d. Gross profit = $0.15000 Net profit = $0.14954

Chapter 8: Interest Rate Risk and Swaps

			3-month	6-month
1.	a.	Discount on sale	$6.07	$23.26
	b.	Simple yield	0.0607%	0.2331%
	c.	Annualized yield	0.2432%	0.4668%
3.	a.	15-year mortgage: $1,662		
		30-year mortgage: $1,261		
	b.	$1,957		
	c.	15-year mortgage: ($12,000)		
		30-year mortgage: ($12,000)		

Chapter 9: Foreign Exchange Rate Determination

1. −80.00%

3. −40.00%

6. −13.79%

Chapter 10: Transaction Exposure

3. Foreign exchange loss of $921,400,000

9. Do nothing: Could be anything
   Forward: $216,049.38
   Money market: $212,190.81
   Forward is preferable choice if bank allows an expanded line

Chapter 11: Translation Exposure

1. a. Translation loss is ($2,400,000)
   b. Loss is accumulated on the consolidated balance sheet, and does not pass through the consolidated income if the subsidiary is foreign currency functional

5. Net exposure is $21,000

Chapter 12: Operating Exposure

1. a. $9,984,000
   b. To be paid out of monthly cash flow

3. Case 1: Same yuan price: $33,913,043
   Case 2: Same dollar price: $54,000,000 (better)

Chapter 13: The Global Cost and Availability of Capital

3. a. 6.550%
   b. 5.950%

4. Domestic: 7.0210%
   Global: 4.7945%

Chapter 14: Raising Equity and Debt Globally

4. 7.7818

8. 4.960%

Chapter 15: Multinational Tax Management

1. Case 1: 38.8%
   Case 2: 45.0%

4. 36.4%

Chapter 16: International Trade Finance

3. 11.765%

5. a. 5.128%
   b. $196,000.00

Chapter 17: Foreign Direct Investment and Political Risk

No quantitative problems in this chapter.

Chapter 18: Multinational Capital Budgeting and Cross-Border Acquisitions

1. a. $7,912,725
   b. $5,587,094
2. Cumulative PV = Ps28,442,771 or $3,555,346

Glossary

Absolute advantage. The ability of an individual party or country to produce more of a product or service with the same inputs as another party. It is therefore possible for a country to have no absolute advantage in any international trade activity. *See also* Comparative advantage.

Absolute purchasing power parity. The theory that the exact rate of exchange between two currencies is found by equalizing the purchasing power of the two currencies.

Accounting exposure. Another name for translation exposure. *See* Translation exposure.

ADR. *See* American Depositary Receipt.

Affiliate. *See* Foreign affiliate.

Affiliated. In business, a close association between two companies. Usually implies a partial but not controlling equity or ownership position by one in the other.

Agency theory. The costs and risks of aligning interests between shareholders of the firm and their agents, management, in the conduct of firm business and strategy. Also referred to as the *agency problem* or *agency issue*.

All-in cost (AIC). The total cost, including interest rate and fees, associated with a loan or debt obligation.

American Depositary Receipt (ADR). A certificate of ownership, issued by a U.S. bank, representing a claim on underlying foreign securities. ADRs may be traded in lieu of trading in the actual underlying shares.

American option. An option that can be exercised at any time up to and including the expiration date.

American terms. Foreign exchange quotations for the U.S. dollar, expressed as the number of U.S. dollars per unit of non-U.S. currency.

Anchor currency. *See* Reserve currency.

Angel investor. An investor who provides capital for small business startups.

Anticipated exposure. A foreign exchange exposure that is believed by management to have a very high likelihood of occurring, but is not yet contractual, and is therefore not yet certain.

Appreciation. In the context of exchange rate changes, a rise in the foreign exchange value of a currency that is pegged to other currencies or to gold. Also called revaluation.

Arbitrage. A trading strategy based on the purchase of a commodity, including foreign exchange, in one market at one price while simultaneously selling it in another market at a more advantageous price, in order to obtain a risk-free profit on the price differential.

Arbitrager. An individual or company that practices arbitrage.

Arm's length price. The price at which a willing buyer and a willing, unrelated seller freely agree to carry out a transaction. In effect, a free market price. Applied by tax authorities in judging the appropriateness of transfer prices between related companies.

Ask. The price at which a dealer is willing to sell foreign exchange, securities, or commodities. Also called offer price.

Asset market approach. A strategy that determines whether foreigners are willing to hold claims in monetary form, depending on an extensive set of investment considerations or drivers.

At-the-money (ATM). An option whose exercise price is the same as the spot price of the underlying currency.

Aval. An endorsement by a third party, acting as guarantor, for the full amount of a debt. The third party (guarantor) commits to cover the payment of the amount of the credit title and its interest in the event the original debtor does not fulfill his or her obligation.

Backlog exposure. The period of time between contract initiation and fulfillment through delivery of services or shipping of goods.

Back-to-back loan. A loan in which two companies in separate countries borrow each other's currency for a specific period of time and repay the other's currency at an agreed maturity. Sometimes the two loans are channeled through an intermediate bank. Back-to-back financing is also called link financing. Also known as a *parallel loan* or a *credit swap*.

Balance of payments (BOP). A financial statement summarizing the flow of goods, services, and investment funds between residents of a given country and residents of the rest of the world.

Balance of trade (BOT). An entry in the balance of payments measuring the difference between the monetary value of merchandise exports and merchandise imports.

Balance on goods and services. A sub-balance within the Current Account of a nation's balance of payments, indicating the net position in the export and import of both goods manufacturing and services trade.

Balance sheet hedge. *See* Money market hedge.

Bank draft. A check for payment drawing upon a bank's own account; a check whose payment is guaranteed by a bank.

Bank for International Settlements (BIS). A bank in Basel, Switzerland, that functions as a bank for European central banks.

Bankers' acceptance. An unconditional promise by a bank to make payment on a draft when it matures. This comes in the form of the bank's endorsement (acceptance) of a draft drawn against that bank in accordance with the terms of a letter of credit issued by the bank.

Barter. International trade conducted by the direct exchange of physical goods, rather than by separate purchases and sales at prices and exchange rates set by a free market.

Base currency. The *base* or *unit currency* is the USD in a currency quotation, such as USD1.0750 = EUR1.00.

Basic balance. In a country's balance of payments, the net of exports and imports of goods and services, unilateral transfers, and long-term capital flows.

Basis erosion and profit shifting (BEPS). The reallocation of corporate profits to lower-tax environments, away from higher-tax environments that are arguably their proper state.

Basis point. One one-hundredth of one percentage point, often used in quotations of spreads between interest rates or to describe changes in yields in securities.

Basis risk. A type of interest rate risk in which the interest rate base is mismatched.

Bearer bond. Corporate or governmental debt in bond form that is not registered to any owner. Possession of the bond implies ownership, and interest is obtained by clipping a coupon attached to the bond. The advantage of the bearer form is easy transfer at the time of a sale, easy use as collateral for a debt, and what some cynics call taxpayer anonymity, meaning that governments find it hard to trace interest payments in order to collect income taxes. Bearer bonds are common in Europe, but are seldom issued any more in the United States. The alternate form to a bearer bond is a registered bond.

Beta. Second letter of the Greek alphabet, used as a statistical measure of risk in the Capital Asset Pricing Model. Beta is the covariance between returns on a given asset and returns on the market portfolio, divided by the variance of returns on the market portfolio.

Bid. The price that a dealer is willing to pay to purchase foreign exchange or a security. Also referred to as the *bid rate*.

Bid-ask spread. The difference between a bid and an ask quotation.

Bill of exchange (B/E). A written order requesting one party (such as an importer) to pay a specified amount of money at a specified time to the writer of the bill. Also called a draft. *See* Sight draft.

Bill of lading (B/L). A contract between a common carrier and a shipper to transport goods to a named destination. The bill of lading is also a receipt for the goods. Bills of lading are usually negotiable, meaning they are made to the order of a particular party and can be endorsed to transfer title to another party.

Billing exposure. The time it takes to get paid in cash after the issuance of an account receivable (A/R).

Black market. An illegal foreign exchange market.

Blocked funds. Funds in one country's currency that may not be exchanged freely for foreign currencies because of exchange controls.

Branch. A foreign operation not incorporated in the host country, in contrast to a subsidiary.

Bretton Woods Agreement. An agreement negotiated at a 1944 international conference and in effect from 1945 to 1971 that established the international monetary system. The conference was held in Bretton Woods, New Hampshire, United States.

BRIC. A frequently used acronym for the four largest emerging market countries—Brazil, Russia, India, and China.

Bridge financing. Short-term financing from a bank, used while a borrower obtains medium- or long-term fixed-rate financing from capital markets.

Bulldog. British pound-denominated bond issued within the United Kingdom by a foreign borrower.

Cable. The U.S. dollar per British pound cross rate.

Call. An option with the right, but not the obligation, to buy foreign exchange or another financial contract at a specified price within a specified time. *See also* Foreign currency option.

Capital account. A section of the balance of payments accounts. Under the revised format of the International Monetary Fund, the capital account measures capital transfers and the acquisition and disposal of nonproduced, nonfinancial assets. Under traditional definitions, still used by many countries, the capital account measures public and private international lending and investment. Most of the traditional definition of the capital account is now incorporated into IMF statements as the financial account.

Capital Asset Pricing Model (CAPM). A theoretical model that relates the return on an asset to its risk, where risk is the contribution of the asset to the volatility of a portfolio. Risk and return are presumed to be determined in competitive and efficient financial markets.

Capital budgeting. The analytical approach used to determine whether investment in long-lived assets or projects is viable. *See also* Multinational capital budgeting.

Capital control. Restrictions, requirements, taxes or prohibitions on the movements of capital across borders as imposed and enforced by governments.

Capital flight. Movement of funds out of a country because of political risk.

Capital gain. A profit or loss arising from the sale of an asset of any kind such as a stock, bond, business, or real estate.

Capital lifecycle. The changing capital needs, in both form, maturity, and amount, which a firm experiences from inception through maturity.

Capital markets. The financial markets of various countries in which various types of long-term debt and/or ownership securities, or claims on those securities, are purchased and sold.

Capital mobility. The degree to which private capital moves freely from country to country in search of the most promising investment opportunities.

Carry trade. The strategy of borrowing in a low interest rate currency to fund investing in higher yielding currencies. Also termed *currency carry trade*, the strategy is speculative in that currency risk is present and not managed or hedged.

Caveat emptor. Latin for "buyer beware."

Certificate of Deposit (CD). A negotiable receipt issued by a bank for funds deposited for a certain period of time. CDs can be purchased or sold prior to their maturity in a secondary market, making them an interest-earning marketable security.

CFR (cost and freight). *See* Cost and freight.

CIF (cost, insurance, and freight). *See* Cost, insurance, and freight.

Classical gold standard. *See* Gold standard.

Clearinghouse. An institution through which financial obligations are cleared by the process of settling the obligations of various members.

Clearinghouse Interbank Payments System (CHIPS). A New York-based computerized clearing system used by banks to settle interbank foreign exchange obligations (mostly U.S. dollars) between members.

Collateral. *See* margin.

Collective action clause (CCA). A contractual clause in a bond or other debt agreement allowing a supermajority of bondholders to agree to a debt restructuring that is legally binding on all bondholders.

Commercial risk. In banking, the likelihood that a foreign debtor will be unable to repay its debts because of business events, as distinct from political ones.

Commodity currency. The currency of a country that is dependent on the export of a certain commodity. For example, the Chilean peso and copper.

Comparative advantage. A theory that everyone gains if each nation specializes in the production of those goods that it produces relatively most efficiently and imports those goods that other countries produce relatively most efficiently. The theory supports free trade arguments.

Competitive exposure. *See* Operating exposure.

Consolidated financial statement. A corporate financial statement in which accounts of a parent company and its subsidiaries are added together to produce a statement which reports the status of the worldwide enterprise as if it were a single corporation. Internal obligations are eliminated in consolidated statements.

Consolidation. In the context of accounting for multinational corporations, the process of preparing a single reporting currency financial statement, which combines financial statements of subsidiaries that are in fact measured in different currencies.

Contagion. The spread of a crisis in one country to its neighboring countries and other countries with similar characteristics—at least in the eyes of cross-border investors.

Contingent foreign currency exposure. The final determination of the exposure is contingent upon another firm's decision, such as a decision to invest or the winning of a business or construction bid.

Contingent Value Right (CVR). A right given to shareholders of an acquired company (or company facing acquisition) that promises them to receive additional cash or shares if a specified event occurs. CVRs are similar to options because they carry an expiration date related to the time in which the contingent event must occur.

Continuous Linked Settlements (CLS). A U.S. financial institution that provides foreign exchange settlement services to members.

Contractual hedge. A foreign currency hedging agreement or contract, typically using a financial derivative such as a forward contract or foreign currency option.

Controlled foreign corporation (CFC). A foreign corporation in which U.S. shareholders own more than 50% of the combined voting power or total value. Under U.S. tax law, U.S. shareholders may be liable for taxes on undistributed earnings of the controlled foreign corporation.

Convertible bond. A bond or other fixed-income security that may be exchanged for a number of shares of common stock.

Convertible currency. A currency that can be exchanged freely for any other currency without government restrictions.

Corporate governance. The relationship among stakeholders used to determine and control the strategic direction and performance of an organization.

Corporate inversion. The reincorporation of a company from a low-tax country environment from a high-tax country environment. Nearly exclusively applies to the United States.

Corporate social responsibility (CSR). A form of corporate self-regulation to pursue business in a legal, ethical manner, and consistent with a series of social norms such as environmental and social sustainability.

Correspondent bank. A bank that holds deposits for and provides services to another bank, located in another geographic area, on a reciprocal basis.

Cost and freight (CFR). Price, as quoted by an exporter, that includes the cost of transportation to the named port of destination.

Cost, insurance, and freight (CIF). Exporter's quoted price including the cost of packaging, freight or carriage, insurance premium, and other charges paid in respect of the goods from the time of loading in the country of export to their arrival at the named port of destination or place of transshipment.

Cost of capital. The cost, expressed as a percentage and on a weighted average basis, of raising equity and debt at current market rates. More commonly referred to as the *weighted average cost of capital*, or WACC.

Counterparty. The opposite party in a double transaction, such as a swap or back-to-back loan, which involves an exchange of financial instruments or obligations now and a reversal of that same transaction at an agreed-upon later date.

Counterparty risk. The potential exposure any individual firm bears that the second party to any financial contract may be unable to fulfill its obligations under the contract's specifications.

Country risk. In banking, the likelihood that unexpected events within a host country will influence a client's or a government's ability to repay a loan. Country risk is often divided into sovereign (political) risk and foreign exchange (currency) risk. *See also* Country-specific risk.

Country-specific risk. Political risk that affects the MNE at the country level, such as transfer risk (blocked funds) and cultural and institutional risks.

Covered interest arbitrage (CIA). The process whereby an investor earns a risk-free profit by (1) borrowing funds in one currency, (2) exchanging those funds in the spot market for a foreign currency, (3) investing the foreign currency at interest rates in a foreign country, (4) selling forward, at the time of original investment, the investment proceeds to be received at maturity, (5) using the proceeds of the forward sale to repay the original loan, and (6) sustaining a remaining profit balance.

Covered transaction. A foreign currency exposure which has been hedged or "covered."

Covering. A transaction in the forward foreign exchange market or money market that protects the value of future cash flows. Covering is another term for hedging. *See* Hedging.

Crawling peg. A foreign exchange rate system in which the exchange rate is adjusted very frequently to reflect prevailing rate of inflation.

Credit risk. The possibility that a borrower's credit worth, at the time of renewing a credit, is reclassified by the lender.

Credit spread. The added interest cost assessed a borrower to compensate the lender or investor for the assessed credit risk of the borrower. The spread is typically based upon the credit rating of the borrower. Also termed *credit risk premium*.

Credit swap. *See* Back-to-back loan.

Crisis planning. The process of educating management and other employees about how to react to various scenarios of violence or other disruptive events.

Cross-border acquisition. A purchase in which one firm acquires another firm located in a different country.

Cross-currency interest rate swap. *See* Currency swap.

Cross-currency swap. *See* Currency swap.

Cross-listing. The listing of shares of common stock on two or more stock exchanges.

Cross rate. An exchange rate between two currencies derived by dividing each currency's exchange rate with a third currency. Colloquially, it is often used to refer to a specific currency pair such as the euro/yen cross rate, as the yen/dollar and dollar/euro are the more common currency quotations.

Crowdfunding. The practice of funding a startup business or enterprise of some kind by raising money in small amounts from a large number of people, typically via the Internet.

Cryptocurrency. A currency created and exchanged using the secure information processes and principles of cryptography. One of the first and most well-known cryptocurrencies is Bitcoin.

Cumulative translation adjustment (CTA) account. An entry in a translated balance sheet in which gains and/or losses from translation have been accumulated over a period of years.

Currency Adjustment Clause (CAC). A contractual arrangement to share or split changes in exchange rates between two parties. Commonly used in long-term supplier contracts.

Currency board. A currency board exists when a country's central bank commits to back its money supply entirely with foreign reserves at all times.

Currency contract period. The period immediately following a change in the value of a currency in which existing contracts do not allow any change in prices—yet.

Currency risk. The variance in expected cash flows arising from unexpected changes in exchange rates.

Currency swap. A transaction in which two counterparties exchange specific amounts of two different currencies at the outset, and then repay over time according to an agreed-upon contract that reflects interest payments and possibly amortization of principal. In a currency swap, the cash flows are similar to those in a spot and forward foreign exchange transaction. *See also* Swap.

Currency switching. Where a firm uses foreign exchange received in the course of business to settle obligations to a third party, often located in a third country.

Currency wars. A state of trade competition in which a country will intentionally devalue its currency in the hope of making its exports more competitive on the global marketplace.

Current account. In the balance of payments, the net flow of goods, services, and unilateral transfers (such as gifts) between a country and all foreign countries.

Current rate method. A method of translating the financial statements of foreign subsidiaries into the parent's reporting currency. All assets and liabilities are translated at the current exchange rate.

D/A. Documents against acceptance. D/A is an international trade term.

Deductible expense. A business expense that is recognized by tax officials as deductible toward the firm's income tax liabilities.

Degree of pass-through. *See* Exchange rate pass-through.

Delta. The change in an option's price divided by the change in the price of the underlying instrument. Hedging strategies are based on delta ratios.

Demand deposit. A bank deposit that can be withdrawn or transferred at any time without notice, in contrast to a time deposit where (theoretically) the bank may require a waiting period before the deposit can be withdrawn. Demand deposits may or may not earn interest. A time deposit is the opposite of a demand deposit.

Depositary receipt (DR). *See* American Depositary Receipt (ADR).

Depreciation. A market-driven change in the value of a currency that results in reduced value or purchasing power.

Derivative. *See* Financial derivative.

Devaluation. The action of a government or central bank authority to drop the spot foreign exchange value of a currency that is pegged to another currency or to gold.

Dim Sum Bond Market. The market for Chinese renminbi (yuan) denominated securities as issued in Hong Kong.

Direct intervention. The purchase or the sale of a country's home currency by its own fiscal or monetary authority in order to influence the value of the domestic currency.

Direct investment. *See* Foreign direct investment (FDI).

Direct quote. The price of a unit of foreign exchange expressed in the home country's currency. The term has meaning only when the home country is specified.

Direct tax. A tax paid directly to the government by the person on whom it is imposed.

Directed issue. *See* Directed public share issue.

Directed public share issue. An issue that is targeted at investors in a single country and underwritten in whole or in part by investment institutions from that country.

Dirty float. A system of floating (i.e., market-determined) exchange rates in which the government intervenes from time to time to influence the foreign exchange value of its currency.

Discount. In the foreign exchange market, the amount by which a currency is cheaper for future delivery than for spot (immediate) delivery. The opposite of discount is premium.

Dividend yield. The current period dividend distribution as a percentage of the beginning of period share price.

Dollarization. The use of the U.S. dollar as the official currency of a country.

Draft. An unconditional written order requesting one party (such as an importer) to pay a specified amount of money at a specified time to the order of the writer of the draft. Also called a bill of exchange. Personal checks are one type of draft.

Dual-currency basket. The use of an index or *basket* composed of two other foreign currencies to benchmark or manage a country's own currency value.

Dutch Disease. A term invented by the *Economist* magazine, referring to the process of a country's currency appreciating in value as a result of the discovery and development of a natural resource like natural gas or oil. The result of the currency appreciation is to make other exports from the country less competitive in the export market. The Dutch reference was the *Economist's* use of the term to explain what happened to the Dutch florin after the discovery of natural gas in the Netherlands in 1959.

Earnings stripping. *See* Basis erosion and profit shifting.

Economic exposure. Another name for operating exposure. *See* Operating exposure.

Effective tax rate. Actual taxes paid as a percentage of actual income before tax.

Efficient market. A market in which all relevant information is already reflected in market prices. The term is most frequently applied to foreign exchange markets and securities markets.

Equity issuance. The issuance to the public market of shares of ownership in a publicly traded company.

Equity listing. The listing of a company's shares on a public stock exchange.

Equity risk premium. The average annual return of the market expected by investors over and above riskless debt.

Euro. A single new currency unit adopted by the 11 participating members of the European Union's European Monetary System in January 1999, replacing their individual currencies. The euro's use has expanded with EU expansion since 1999, totalling 18 participating countries as of 2014.

Eurobank. A bank, or bank department, that bids for time deposits and makes loans in currencies other than that of the country where the bank is located.

Eurobond. A bond originally offered outside the country in whose currency it is denominated. For example, a dollar-denominated bond originally offered for sale to investors outside the United States.

Eurocommercial paper (ECP). Short-term notes (30, 60, 90, 120, 180, 270, and 360 days) sold in international money markets.

Eurocredit. Bank loans to MNEs, sovereign governments, international institutions, and banks denominated in eurocurrencies and extended by banks in countries other than the country in whose currency the loan is denominated.

Eurocurrency. A currency deposited in a bank located in a country other than the country issuing the currency.

Eurodollar. A U.S. dollar deposited in a bank outside the United States. A eurodollar is a type of eurocurrency. Also termed a *Eurodollar deposit*.

Euroequity. A new equity issue that is underwritten and distributed in multiple foreign equity markets, sometimes simultaneously with distribution in the domestic market.

Euronote market. Short- to medium-term debt instruments sold in the eurocurrency market.

European Central Bank (ECB). Conducts monetary policy of the European Monetary Union. Its goal is to safeguard the stability of the euro and minimize inflation.

European Currency Unit (ECU). A composite currency created by the European Monetary System prior to the euro, which was designed to function as a reserve currency numeraire. The ECU was used as the numeraire for denominating a number of financial instruments and obligations.

European Monetary System (EMS). A system of exchange rate and monetary system linkages first established in 1979 between fifteen European countries. The EMS laid the groundwork for the eventual creation of the euro. The EMS has continued to expand its membership over time.

European option. An option that can be exercised only on the day on which it expires.

European terms. Foreign exchange quotations for the U.S. dollar, expressed as the number of non-U.S. currency units per U.S. dollar.

European Union (EU). The official name of the former European Economic Community (EEC) as of January 1, 1994.

Eurozone. The countries that officially use the euro as their currency.

Exchange rate. *See* foreign exchange rate.

Exchange Rate Mechanism (ERM). The means by which members of the EMS formerly maintained their currency exchange rates within an agreed-upon range with respect to the other member currencies.

Exchange rate pass-through. The degree to which the prices of imported and exported goods change as a result of exchange rate changes.

Exercise price. Same as the *strike price*; the agreed upon rate of exchange within an option contract to buy or sell the underlying asset.

Export credit insurance. Provides assurance to the exporter or the exporter's bank that, should the foreign customer default on payment, the insurance company will pay for a major portion of the loss. *See also* Foreign Credit Insurance Association (FCIA).

Export-Import Bank (Eximbank). A U.S. government agency created to finance and otherwise facilitate imports and exports.

Exposed asset. An asset whose value is subject to change as a result of the translation of its value from local currency financial statements to home currency financial statements as a result of financial statement consolidation. The change in value typically results from moving from historical to current exchange rates for translation and remeasurement.

Expropriation. Official government seizure of private property, recognized by international law as the right of any sovereign state provided expropriated owners are given prompt compensation and fair market value in convertible currencies.

Factoring. Specialized firms, known as factors, purchase receivables at a discount on either a non-recourse or recourse basis.

Fair value. The estimated true market value of an item or asset.

FAS (free alongside ship). An international trade term in which the seller's quoted price for goods includes all costs of delivery of the goods alongside a vessel at the port of embarkation.

Fiat currency. Money or currency which derives its value from government regulation or proclamation. Unlike commodity money or *specie*, its value is not tied or based on a precious metal or other physical commodity.

FIBOR. Frankfurt interbank offered rate.

Financial account. A section of the balance of payments accounts. Under the revised format of the International Monetary Fund, the financial account measures long-term financial flows including direct foreign investment, portfolio investments, and other long-term movements. Under the traditional definition, which is still used by many countries, items in the financial account were included in the capital account.

Financial derivative. A financial instrument, such as a futures contract or option, whose value is derived from an underlying asset like a stock or currency.

Financing cash flow. Cash flows originating from financing activities of the firm, including interest payments and dividend distributions.

Firm-specific risk. Political risk that affects the MNE at the project or corporate level. Governance risk due to goal conflict between an MNE and its host government is the main political firm-specific risk.

Fisher Effect. A theory that nominal interest rates in two or more countries should be equal to the required real rate of return to investors plus compensation for the expected amount of inflation in each country.

Fixed exchange rates. Foreign exchange rates tied to the currency of a major country (such as the United States), to gold, or to a basket of currencies such as Special Drawing Rights.

Flexible exchange rates. The opposite of fixed exchange rates. The foreign exchange rate is adjusted periodically by the country's monetary authorities in accordance with their judgment and/or an external set of economic indicators.

Floating exchange rates. Foreign exchange rates determined by demand and supply in an open market that is presumably free of government interference.

Floating-rate note (FRN). Medium-term securities with interest rates pegged to LIBOR and adjusted quarterly or semiannually.

Follow-on offering (FO). Additional offerings of equity shares post-IPO.

Forced delistings. The requirement by a stock exchange for a publicly traded share on that exchange to be delisted from active trading, typically from failure to maintain a minimum level of market capitalization.

Foreign affiliate. A foreign business unit that is less than 50% owned by the parent company.

Foreign bond. A bond issued by a foreign corporation or government for sale in the domestic capital market of another country, and denominated in the currency of that country.

Foreign Credit Insurance Association (FCIA). An unincorporated association of private commercial insurance companies, in cooperation with the Export-Import Bank of the United States, that provides export credit insurance to U.S. firms.

Foreign currency. Any currency other than that used officially for contracts and transactions in the domestic economy.

Foreign currency intervention. Any activity or policy initiative by a government or central bank with the intent of changing a currency value on the open market. They may include *direct intervention*, where the central bank may buy or sell its own currency, or *indirect intervention*, in which it may change interest rates in order to change the attractiveness of domestic currency obligations in the eyes of foreign investors.

Foreign currency option. A financial contract or derivative which guarantees the holder the right to buy or sell a specific amount of foreign currency at a specific rate by a stated expiration or maturity date.

Foreign currency translation. *See* Translation.

Foreign direct investment (FDI). Purchase of physical assets, such as plant and equipment, in a foreign country, to be managed by the parent corporation. FDI is distinguished from foreign portfolio investment.

Foreign exchange broker. An individual or firm that arranges foreign exchange transactions between two parties, but is not itself a principal in the trade. Foreign exchange brokers earn a commission for their efforts.

Foreign exchange dealer (or trader). An individual or firm that buys foreign exchange from one party (at a bid price), and then sells it (at an ask price) to another party. The dealer is a principal in two transactions and profits via the spread between the bid and ask prices.

Foreign exchange intervention. The active entry into the foreign exchange market by buying and selling a currency by an official authority in order to manage or fix the currency's value relative to other traded currencies.

Foreign exchange rate. The price of one country's currency in terms of another currency, or in terms of a commodity such as gold or silver. Also termed *foreign currency exchange rate*. *See also* Exchange rate.

Foreign exchange risk. The likelihood that an unexpected change in exchange rates will alter the home currency value of foreign currency cash payments expected from a foreign source. Also, the likelihood that an unexpected change in exchange rates will alter the amount of home currency needed to repay a debt denominated in a foreign currency.

Foreign subsidiary. A foreign operation incorporated in the host country and owned 50% or more by a parent corporation. Foreign operations that are not incorporated are called branches.

Foreign tax credit. The amount by which a domestic firm may reduce (credit) domestic income taxes for income tax payments to a foreign government.

Foreign tax neutrality. The principle that tax obligations or tax burdens are the same on taxable earnings, regardless of where the earnings were generated, in domestic or foreign markets.

Forfaiting (forfeiting). A technique for arranging non-recourse medium-term export financing, used most frequently to finance imports into Eastern Europe. A third party, usually a specialized financial institution, guaranteeing the financing.

Forward-ATM. The strike rate or exercise price of a foreign exchange derivative set equivalent to the forward exchange rate.

Forward-forward swap. The exchange of two different maturities of forward exchange contracts.

Forward contract. An agreement to exchange currencies of different countries at a specified future date and at a specified forward rate.

Forward discount. The difference between spot and forward rates, expressed as an annual percentage, also known as the *forward premium.*

Forward exchange rate (Forward rate). An exchange rate quoted for settlement at some future date. The rate used in a forward transaction.

Forward hedge. The use of a forward exchange contract to hedge or protect the value of a foreign currency denominated transaction.

Forward premium. *See* Forward discount.

Forward rate. *See* Forward exchange rate.

Forward rate agreement (FRA). An interbank-traded contract to buy or sell interest rate payments on a notional principal.

Forward transaction. An agreed-upon foreign exchange transaction to be settled at a specified future date, often one, two, or three months after the transaction date.

Free alongside ship (FAS). *See* FAS (free alongside ship).

Free cash flow. Operating cash flow less capital expenditures (capex).

Free float. The portion of publicly traded shares of a corporation that are held by public investors as opposed to locked-in stock held by promoters (underwriters), company officers, controlling-interest investors, or government.

Fronting loan. A parent-to-subsidiary loan that is channeled through a financial intermediary such as a large international bank in order to reduce political risk. Presumably government authorities are less likely to prevent a foreign subsidiary repaying an established bank than repaying the subsidiary's corporate parent.

Functional currency. In the context of translating financial statements, the currency of the primary economic environment in which a foreign subsidiary operates and in which it generates cash flows.

Futures, or futures contracts. *See* Interest rate futures.

Gamma. A measure of the sensitivity of an option's delta ratio to small unit changes in the price of the underlying security.

Generally Accepted Accounting Principles (GAAP). Approved accounting principles for U.S. firms, defined by the Financial Accounting Standards Board (FASB).

Global depositary receipt (GDR). Similar to American Depositary Receipts (ADRs), it is a bank certificate issued in multiple countries for shares in a foreign company. Actual company shares are held by a foreign branch of an international bank. The shares are traded as domestic shares, but are offered for sale globally by sponsoring banks.

Global registered shares (GRS). Similar to ordinary shares, global registered shares have the added benefit of being tradable on equity exchanges around the globe in a variety of currencies.

Global reserve currency. *See* Reserve currency.

Global-specific risks. Political risks that originate at the global level, such as terrorism, the antiglobalization movement, environmental concerns, poverty, and cyber attacks.

Gold standard or Gold-exchange standard. A monetary system in which currencies are defined in terms of their gold content, and payment imbalances between countries are settled in gold.

Greenfield investment. An initial investment in a new foreign subsidiary with no predecessor operation in that location. This is in contrast to a new subsidiary created by the purchase of an already existing operation. An investment that starts, conceptually if not literally, with an undeveloped "green field."

Haircut. The percentage of the market value of a financial asset recognized as the collateral value or redeemed value of the asset.

Hard currency. A freely convertible currency that is not expected to depreciate in value in the foreseeable future.

Hedging. Purchasing a contract (including forward foreign exchange) or tangible good that will rise in value and offset a drop in value of another contract or tangible good. Hedges are undertaken to reduce risk by protecting an owner from loss.

Historical exchange rate. In accounting, the exchange rate in effect when an asset or liability was acquired.

Home currency. The currency of a company's incorporation; the currency for financial reporting purposes.

Hoover Hedges. Hedges constructed to protect the value of a long-term investment or loan in a foreign currency.

Hot money. Money that moves internationally from one currency and/or country to another in response to interest rate differences, and moves away immediately when the interest advantage disappears.

Hurdle rate. The required rate of return by a firm on a potential new investment in order to approve accepting the investment. The rate is typically based on the company's current cost of capital, including debt and equity. In some cases the firm will require some premium or additional margin on certain investments above and beyond its cost of capital in the calculation of the hurdle rate (e.g., cost of capital + premium = hurdle rate).

Hyperinflation countries. Countries with a very high rate of inflation. Under United States FASB 52, these are defined as countries where the cumulative three-year inflation amounts to 100% or more.

Impossible trinity. An ideal currency would have exchange rate stability, full financial integration, and monetary independence.

In-the-money (ITM). Circumstance in which an option is profitable, excluding the cost of the premium, if exercised immediately.

Indication. A quotation, typically in the form of a bid rate and ask rate, for a currency or other financial asset.

Indirect intervention. Actions taken by central banks or other monetary authorities to influence the supply and demand for a country's own currency. The most common form of indirect intervention is the alteration of interest rates.

Indirect quote. The price of a unit of a home country's currency expressed in terms of a foreign country's currency.

Initial public offering (IPO). The initial sale of shares of ownership of a company to the general public. The issuing firm raises capital for the conduct of its business and return to its original owners through the IPO.

Integrated foreign entity. An entity that operates as an extension of the parent company, with cash flows and general business lines that are highly interrelated with those of the parent.

Intellectual property rights. Legislation that grants the exclusive use of patented technology and copyrighted creative materials. A worldwide treaty to protect intellectual property rights has been ratified by most major countries, including most recently by China.

Interest rate futures. Exchange-traded agreements calling for future delivery of a standard amount of any good, e.g., foreign exchange, at a fixed time, place, and price.

Interest rate parity (IRP). A theory that the differences in national interest rates for securities of similar risk and maturity should be equal to but opposite in sign (positive or negative) to the forward exchange rate discount or premium for the foreign currency.

Interest rate risk. The risk to the organization arising from interest-bearing debt obligations, either fixed or floating rate obligations. It is typically used to refer to the changing interest rates which a company may incur by borrowing at floating rates of interest.

Interest rate swap. A transaction in which two counterparties exchange interest payment streams of different character (such as floating vs. fixed), based on an underlying notional principal amount.

Internal rate of return (IRR). A capital budgeting approach in which a discount rate is found that matches the present value of expected future cash inflows with the present value of outflows.

International Bank for Reconstruction and Development (IBRD or World Bank). International development bank owned by member nations that makes development loans to member countries.

International CAPM (ICAPM). A strategy in which the primary distinction in the estimation of the cost of equity for an individual firm using an internationalized version of the domestic capital asset pricing model is the definition of the "market" and a recalculation of the firm's beta for that market.

International Fisher effect. A theory that the spot exchange rate should change by an amount equal to the difference in interest rates between two countries.

International Monetary Fund (IMF). An international organization created in 1944 to promote exchange rate stability and provide temporary financing for countries experiencing balance of payments difficulties.

International Monetary Market (IMM). A branch of the Chicago Mercantile Exchange that specializes in trading currency and financial futures contracts.

International monetary system. The structure within which foreign exchange rates are determined, international trade and capital flows are accommodated, and balance of payments adjustments made.

International parity conditions. In the context of international finance, a set of basic economic relationships that provide for equilibrium between spot and forward foreign exchange rates, interest rates, and inflation rates.

International Swaps and Derivatives Association (ISDA). A New York City trade association for over the country (OTC) derivatives. The ISDA maintains the documentation used in most of the financial services trading of financial derivatives used globally.

Intrafirm trade. Trade in goods and services between incorporated units of the same multinational business or enterprise.

Intrinsic value. The financial gain if an option is exercised immediately.

Investment agreement. An agreement that spells out specific rights and responsibilities of both the investing foreign firm and the host government.

Investment grade. A credit rating, typically assigned by Moody's, Standard & Poors, or Fitch, symbolizing the assured ability of a borrower to repay in a timely manner regardless of business or market conditions.

Denoted as BBB- (or equivalent by credit rating agency) or higher.

Islamic finance. Banking or financing activity that is consistent with the principles of *sharia* and Islamic economics.

J-curve. The adjustment path of a country's trade balance following a devaluation or significant depreciation of the country's currency. The path first worsens as a result of existing contracts before improving as a result of more competitive pricing conditions.

Joint venture (JV). A business venture that is owned by two or more entities, often from different countries.

Lag. In the context of leads and lags, payment of a financial obligation later than is expected or required.

Lambda. A measure of the sensitivity of an option premium to a unit change in volatility.

Law of one price. The concept that if an identical product or service can be sold in two different markets, and no restrictions exist on the sale or transportation costs of moving the product between markets, the product's price should be the same in both markets.

Lead. In the context of *leads and lags*, the payment of a financial obligation earlier than is expected or required.

Legal tender. A medium of payment allowed by law or recognized by a legal system to be valid for meeting a financial obligation.

Lender-of-last-resort. The body or institution within an economy which is ultimately capable of preserving the financial survival or viability of individual institutions. Typically the country's central bank.

Letter of credit (L/C). An instrument issued by a bank, in which the bank promises to pay a beneficiary upon presentation of documents specified in the letter.

Link financing. *See* Back-to-back loan or Fronting loan.

Liquid. The ability to exchange an asset for cash at or near its fair market value.

London Interbank Offered Rate (LIBOR). The deposit rate applicable to interbank loans in London. LIBOR is used as the reference rate for many international interest rate transactions.

Long position. A position in which foreign currency assets exceed foreign currency liabilities. The opposite of a long position is a short position.

Maastricht Treaty. A treaty among the 12 European Union countries that specified a plan and timetable for the introduction of a single European currency, to be called the euro.

Macro risk. *See* Country-specific risk.

Macroeconomic uncertainty. Operating exposure's sensitivity to key macroeconomic variables, such as exchange rates, interest rates, and inflation rates.

Managed float. A country allows its currency to trade within a given band of exchange rates.

Margin. A deposit made as security for a financial transaction otherwise financed on credit.

Marked-to-market. The condition in which the value of a futures contract is assigned to market value daily, and all changes in value are paid in cash daily. The value of the contract is revalued using the closing price for the day. The amount to be paid is called the variation margin.

Market capitalization. The total market value of a publicly traded company, calculated as the total number of shares outstanding multiplied by the market-determined price per share.

Market liquidity. The degree to which a firm can issue a new security without depressing the existing market price, as well as the degree to which a change in price of its securities elicits a substantial order flow.

Matching currency cash flows. The strategy of offsetting anticipated continuous long exposure to a particular currency by acquiring debt denominated in that currency.

Merchant bank. A bank that specializes in helping corporations and governments finance by any of a variety of market and/or traditional techniques. European merchant banks are sometimes differentiated from clearing banks, which tend to focus on bank deposits and clearing balances for the majority of the population.

MIBOR. Madrid interbank offered rate.

Micro risk. *See* Firm-specific risk.

Monetary assets or liabilities. Assets in the form of cash or claims to cash (such as accounts receivable), or liabilities payable in cash. Monetary assets minus monetary liabilities are called net monetary assets.

Monetary/nonmonetary method. A method of translating the financial statements of foreign subsidiaries into the parent's reporting currency. All monetary accounts are translated at the current rate, and all nonmonetary accounts are translated at their historical rates. Sometimes called temporal method in the United States.

Money laundering. The process of depositing or inserting illegally generated money or cash into the financial system.

Money market hedge. The use of foreign currency borrowing to reduce transaction or accounting foreign exchange exposure.

Money markets. The financial markets in various countries in which various types of short-term debt instruments, including bank loans, are purchased and sold.

Moral hazard. When an individual or organization takes on more risk than it would normally as a result of the existence or support of a secondary insuring or protecting authority or organization.

Multinational capital budgeting. The financial analysis of foreign investment projects requiring the use of discounted cash flow analysis. Also termed *international capital budgeting* and *capital budgeting of foreign projects*.

Multinational enterprise (MNE). A firm that has operating subsidiaries, branches, or affiliates located in foreign countries.

National approach (to taxes). *See* Worldwide approach (to taxes).

Natural hedge. The use or existence of an offsetting or matching cash flow from firm operating activities to hedge a currency exposure.

Negotiable instrument. A written draft or promissory note, signed by the maker or drawer, that contains an unconditional promise or order to pay a definite sum of money on demand or at a determinable future date, and is payable to order or to bearer. A holder of a negotiable instrument is entitled to payment despite any personal disagreements between the drawee and maker.

Nepotism. The practice of showing favor to relatives over other qualified persons in conferring such benefits as the awarding of contracts, granting of special prices, promotions to various ranks, etc.

Net international investment position (NIIP). The net difference between a country's external financial assets and liabilities as defined by nationality of ownership. A country's external debt includes both its government debt and private debt, and similarly its public and privately held legal residents.

Net operating cash flow (NOCF). The cash generated by the normal operations of the business. It is considered a measure of the value created by the business. It is calculated as the sum of net income, depreciation, and changes in net working capital.

Net present value (NPV). A capital budgeting approach in which the present value of expected future cash inflows is subtracted from the present value of outflows.

Net working capital (NWC). Accounts receivable plus inventories less accounts payable.

Netting. The process of netting intracompany payments in order to reduce the size and frequency of cash and currency exchanges.

Nominal exchange rate. The actual foreign exchange quotation, in contrast to *real exchange rate*, which is adjusted for changes in purchasing power. May be constructed as an index.

Nondeliverable forward (NDF). A forward or futures contract on currencies, settled on the basis of the differential between the contracted forward rate and occurring spot rate, but settled in the currency of the traders. For example, a forward contract on the Chinese yuan that is settled in dollars, not yuan.

North American Free Trade Agreement (NAFTA). A treaty allowing free trade and investment between Canada, the United States, and Mexico.

Note issuance facility (nif). An agreement by which a syndicate of banks indicates a willingness to accept short-term notes from borrowers and resell those notes in the eurocurrency markets. The discount rate is often tied to LIBOR.

Notional principal. The size of a derivative contract, in total currency value, as used in futures contracts, forward contracts, option contracts, or swap agreements.

NPV. *See* Net present value (NPV).

Offer. *See* Ask.

Offer rate. The price of sale, or ask as in *bid-ask* and *bid-offer*.

Official reserves account. Total reserves held by official monetary authorities within the country, such as gold, SDRs, and major currencies.

OLI Paradigm. An attempt to create an overall framework to explain why MNEs choose foreign direct investment rather than serve foreign markets through alternative modes such as licensing, joint ventures, strategic alliances, management contracts, and exporting.

Open account. A sale where goods are shipped and delivered before payment is due or made. Payment is typically made anywhere between 30 and 90 days later, depending on industry and national practices.

Operating cash flows. The primary cash flows generated by a business from the conduct of trade, typically composed of earnings, depreciation and amortization, and changes in net working capital.

Operating exposure. The potential for a change in expected cash flows, and thus in value, of a foreign subsidiary as a result of an unexpected change in exchange rates. Also called economic exposure.

Option. *See* Foreign currency option.

Option collar. The simultaneous purchase of a put option and sale of a call option, or vice versa, resulting in a form of hybrid option.

Order bill of lading. A shipping document through which possession and title to the shipment reside with the owner of the bill.

Out-of-the-money (OTM). An option that would not be profitable, excluding the cost of the premium, if exercised immediately.

Outright forward. *See* Forward rate.

Outright forward transaction. *See* Forward transaction.

Overseas Private Investment Corporation (OPIC). A U.S. government-owned insurance company that insures U.S. corporations against various political risks.

Overshooting. A behavior in financial markets in which a major market adjustment in price changes "overshoots" or surpasses the likely value it will settle at after a longer adjustment period. A market movement akin to an "overreaction."

Over-the-counter (OTC) market. A market for share of stock, options (including foreign currency options), or other financial contracts conducted via electronic

connections between dealers. The over-the-counter market has no physical location or address, and is thus differentiated from organized exchanges that have a physical location where trading takes place.

Panda Bond. The issuance of a yuan-denominated bond in the Chinese market by a foreign borrower.

Parallel loan. Another name for a back-to-back loan, in which two companies in separate countries borrow each other's currency for a specific period of time, and repay the other's currency at an agreed maturity. *See also* Back-to-back loan.

Participating forward. A complex option position which combines a bought put and a sold call option at the same strike price to create a net zero position. Also called zero-cost option and forward participation agreement.

Pass-through or Pass-through period. The time it takes for an exchange rate change to be reflected in market prices of products or services.

Phi. The expected change in an option premium caused by a small change in the foreign interest rate (interest rate for the foreign currency).

PIBOR. Paris interbank offered rate.

Pip. Percentage in point, in reference to an exchange rate fluctuation.

Plain vanilla swap. An interest rate swap agreement exchange fixed interest payments for floating interest payments, all in the same currency.

Points. The smallest units of price change quoted, given a conventional number of digits in which a quotation is stated.

Political risk. The possibility that political events in a particular country will influence the economic well-being of firms in that country.

Portfolio investment. Purchase of foreign stocks and bonds, in contrast to foreign direct investment.

Premium. In a foreign exchange market, the amount by which a currency is more expensive for future delivery than for spot (immediate) delivery. The opposite of premium is discount.

Price currency. The quote currency in a currency price quotation. The euro (EUR) is the price currency in a typical exchange rate quotation on the dollar-euro such as USD1.0750 = EUR1.00.

Price elasticity of demand. From economic theory, the percentage change in the quantity demanded as a result of a one percent change in the product price.

Principal agent problem. *See* Agency theory.

Private equity (PE). Asset ownership in a business that is not publicly traded. Private equity investments are typically made by private equity firms or private equity funds.

Private placement. The sale of a security issue to a small set of qualified institutional buyers.

Project financing. Arrangement of financing for long-term capital projects, large in scale, long in life, and generally high in risk.

Prospectus. A document disclosing the prospective risks and returns associated with the proposed public sale of a security. The prospectus commonly includes material information such as a description of the company's business, financial statements, biographies of officers and directors, detailed information about their compensation, any litigation that is pending, a list of material properties, and any other material information.

Protectionism. A political attitude or policy intended to inhibit or prohibit the import of foreign goods and services. The opposite of free trade policies.

Psychic distance. Firms tend to invest first in countries with a similar cultural, legal, and institutional environment.

Public debt. The debt obligation of a governmental body or sovereign authority.

Purchasing power parity (PPP). A theory that the price of internationally traded commodities should be the same in every country, and hence the exchange rate between the two currencies should be the ratio of prices in the two countries.

Put. An option to sell foreign exchange or financial contracts. *See also* Foreign currency option.

Qualified institutional buyer (QIB). An entity (except a bank or a savings and loan) that owns and invests on a discretionary basis a minimum of $100 million in securities of non-affiliates.

Quota. A limit, mandatory or voluntary, set on the import of a product.

Quotation. In foreign exchange trading, the pair of prices (bid and ask) at which a dealer is willing to buy or sell foreign exchange.

Quotation exposure. The period of time in which a seller has quoted a fixed price in a foreign currency to a potential buyer, but the buyer has yet to agree.

Quote currency. *See* Price currency.

Range forward. A complex option position that combines the purchase of a put option and the sale of a call option with strike prices equidistant from the forward rate. Also called *flexible forward, cylinder option, option fence, mini-max,* and *zero-cost tunnel.*

Real exchange rate. An index of foreign exchange adjusted for relative price-level changes from a base point in time, typically a month or a year. Sometimes referred to as real effective exchange rate, it is used to measure purchasing-power-adjusted changes in exchange rates. Also termed *real effective exchange rate* or *real effective exchange rate index.*

Real option analysis. The application of option theory to capital budgeting decisions.

Reference rate. The rate of interest used in a standardized quotation, loan agreement, or financial derivative valuation.

Registered bond. Corporate or governmental debt in a bond form in which the owner's name appears on the bond and in the issuer's records, and interest payments are made to the owner.

Relative purchasing power parity. A theory that if the spot exchange rate between two countries starts in equilibrium, any change in the differential rate of inflation between them tends to be offset over the long run by an equal but opposite change in the spot exchange rate.

Remittance. A transfer of money or currency from one party to another in payment or gift or saving.

Renminbi (RMB). The alternative official name (the yuan, CNY) of the currency of the People's Republic of China.

Reporting currency. In the context of translating financial statements, the currency in which a parent firm prepares its own financial statements. Usually this is the parent's home currency.

Repositioning of funds. The movement of funds from one currency or country to another. An MNE faces a variety of political, tax, foreign exchange, and liquidity constraints that limit its ability to move funds easily and without cost.

Repricing risk. The risk of changes in interest rates charged or earned at the time a financial contract's rate is reset.

Reserve currency. A currency used by a government or central banking authority as a resource asset or currency to be used in market interventions to alter the market value of the domestic currency.

Residential approach. The levy of taxes against the worldwide income earned by a business by home country tax authorities regardless of where or in which country the income was earned.

Revaluation. A rise in the foreign exchange value of a currency that is pegged to other currencies or to gold. Also called *appreciation*.

Rho. The expected change in an option premium caused by a small change in the domestic interest rate (interest rate for the home currency).

Risk. The likelihood that an actual outcome will differ from an expected outcome. The actual outcome could be better or worse than expected (two-sided risk), although in common practice risk is more often used only in the context of an adverse outcome (one-sided risk). Risk can exist for any number of uncertain future situations, including future spot rates or the results of political events.

Risk-free rate of interest. The return on an asset assumed to possess no possibility of failure to pay. Typically a debt security issued by a government like a U.S. Treasury bill, note, or bond.

Risk-sharing. A contractual arrangement in which the buyer and seller agree to share or split currency movement impacts on payments between them.

Roll-over risk. *See* Credit risk.

Rules of the Game. The basis of exchange rate determination under the international gold standard during most of the 19th and early 20th centuries. All countries agreed informally to follow the rule of buying and selling their currency at a fixed and predetermined price against gold.

Samurai bond. Yen-denominated bond issued within Japan by a foreign borrower.

Sarbanes-Oxley Act. An act passed in 2002 to regulate corporate governance in the United States.

Seasoned offering. *See* Follow-on offering (FO).

SEC Rule 144A. Permits qualified institutional buyers to trade privately placed securities without requiring SEC registration.

SEC Rule 415. Security and Exchange Commission rules which allows shelf registration of security offerings to the public without a separate prospectus for each act of offering. A single prospectus is used for multiple future offerings.

Section 482. The set of U.S. Treasury regulations governing transfer prices.

Securitization. The replacement of nonmarketable loans (such as direct bank loans) with negotiable securities (such as publicly traded marketable notes and bonds), so that the risk can be spread widely among many investors, each of whom can add or subtract the amount of risk carried by buying or selling the marketable security.

Seignorage. The net revenues or proceeds garnered by a government from the printing of its money.

Selective hedging. Hedging only exceptional exposures or the occasional use of hedging when management has a definite expectation of the direction of exchange rates.

Self-sustaining foreign entity. One that operates in the local economic environment independent of the parent company.

Selling short (shorting). The sale of an asset which the seller does not (yet) own. The premise is that the seller believes he will be able to purchase the asset for contract fulfillment at a lower price before sale contract expiration.

Shareholder. An individual or institution that holds legal ownership to a share or stock in a publicly traded company.

Shareholder wealth maximization (SWM). The corporate goal of maximizing the total value of the shareholders' investment in the company.

Shelf registrations. *See* SEC Rule 415.

Short position. *See* Long position.

SIBOR. Singapore interbank offered rate.

Sight draft. A bill of exchange (B/E) that is due on demand; i.e., when presented to the bank. *See also* Bill of exchange (B/E).

Sovereign credit risk. The risk that a host government may unilaterally repudiate its foreign obligations or may prevent local firms from honoring their foreign obligations. Sovereign risk is often regarded as a subset of political risk.

Sovereign debt. The debt obligation or a sovereign or governmental authority or body.

Sovereign spread. The credit spread paid by a sovereign borrower on a major foreign currency-denominated debt obligation. For example, the credit spread paid by the Venezuelan government to borrow U.S. dollars over and above a similar maturity issuance by the U.S. Treasury.

Special Drawing Right (SDR). An international reserve asset, defined by the International Monetary Fund as the value of a weighted basket of five currencies.

Speculation. An attempt to make a profit by trading on expectations about future prices.

Speculative grade. A credit quality that is below BBB, below investment grade. The designation implies a possibility of borrower default in the event of unfavorable economic or business conditions.

Spot rate. The price at which foreign exchange can be purchased (its bid) or sold (its ask) in a spot transaction. *See* Spot transaction.

Spot transaction. A foreign exchange transaction to be settled (paid for) on the second following business day.

Spread. The difference between the bid (buying) quote and the ask (selling) quote.

Stakeholder capitalism model (SCM). Another name for corporate wealth maximization.

State-owned enterprise (SOE). Any organization or business which is owned (in-whole or in-part) and controlled by government, typically created to conduct commercial business activities.

Statutory tax rate. The legally imposed tax rate.

Strategic alliance. A formal relationship, short of a merger or acquisition, between two companies, formed for the purpose of gaining synergies because in some aspect the two companies complement each other.

Strategic exposure. *See* Operating exposure.

Strike price. The agreed upon rate of exchange within an option contract. *See also* Exercise price.

Subpart F income. A type of foreign income, as defined in the U.S. tax code, which under certain conditions is taxed immediately in the United States even though it has not been repatriated to the United States. It is income of a type that is otherwise easily shifted offshore to avoid current taxation.

Swap or Swap transaction. In general it is the simultaneous purchase and sale of foreign exchange or securities, with the purchase executed at once and the sale back to the same party carried out at an agreed-upon price to be completed at a specified future date. Swaps include interest rate swaps, currency swaps, and credit swaps.

Swap rate. A forward foreign exchange quotation expressed in terms of the number of points by which the forward rate differs from the spot rate.

Syndicated loan. A large loan made by a group of banks to a large multinational firm or government. Syndicated loans allow the participating banks to maintain diversification by not lending too much to a single borrower. Also termed a Syndicated bank credit.

Synthetic forward. A complex option position which combines the purchase of a put option and the sale of a call option, or vice versa, both at the forward rate. Theoretically, the combined position should have a net-zero premium.

Systematic risk. In portfolio theory, the risk of the market itself, i.e., risk that cannot be diversified away.

Tariff. A duty or tax on imports that can be levied as a percentage of cost or as a specific amount per unit of import.

Tax deferral. Foreign subsidiaries of MNEs pay host country corporate income taxes, but many parent countries, including the United States, defer claiming additional taxes on that foreign source income until it is remitted to the parent firm.

Tax exposure. The potential for tax liability on a given income stream or on the value of an asset. Usually used in the context of a multinational firm being able to minimize its tax liabilities by locating some portion of operations in a country where the tax liability is minimized.

Tax haven. A country with either no or very low tax rates that uses its tax structure to attract foreign investment or international financial dealings.

Tax morality. The consideration of conduct by an MNE to decide whether to follow a practice of full disclosure to local tax authorities or adopt the philosophy, "When in Rome, do as the Romans do."

Tax neutrality. In domestic tax, the requirement that the burden of taxation on earnings in home country operations by an MNE be equal to the burden of taxation on each currency equivalent of profit earned by the same firm in its foreign operations. Foreign tax neutrality requires that the tax burden on each foreign subsidiary of the firm be equal to the tax burden on its competitors in the same country.

Tax treaties. A network of bilateral treaties that provide a means of reducing double taxation.

Technical analysis. The focus on price and volume data to determine past trends that are expected to continue into the future. Analysts believe that future exchange rates are based on the current exchange rate.

TED Spread. Treasury Eurodollar Spread. The difference, in basis points, between the 3-month interest rate swap index or the 3-month LIBOR interest rate, and the 90-day U.S. Treasury bill rate. It is sometimes used as an indicator of credit crisis or fear over bank credit quality.

Temporal method. In the United States, term for a codification of a translation method essentially similar to the monetary/nonmonetary method.

Tender. To offer for sale or purchase.

Tenor. The length of time of a contract or debt obligation; loan repayment period.

Tequila effect. Term used to describe how the Mexican peso crisis of December 1994 quickly spread to other Latin American currency and equity markets through the contagion effect.

Terminal value (TV). The continuing value of a project or investment beyond the period shown in detail. It represents the present value at a future point in time of all future cash flows assuming a stable perpetual growth rate.

Terms of trade. The weighted average exchange ratio between a nation's export prices and its import prices, used to measure gains from trade. Gains from trade refers to increases in total consumption resulting from production specialization and international trade.

Territorial approach (to taxes). Also called *territorial taxation*. Taxation of income earned by firms within the legal jurisdiction of the host country, not on the country of the firm's incorporation.

Theory of comparative advantage. Based on the concept of *absolute advantage*, in which each country specializes in the production of those goods for which it is uniquely suited, the theory of comparative advantage states that exchange between these countries will result in all parties or countries being better off through specialization and exchange than by attempting to produce all at home.

Theta. The expected change in an option premium caused by a small change in the time to expiration.

Thin capitalization. When a company's capital structure is deemed to be excessively based on debt rather than equity. It is typically used to reduce domestic tax liabilities through interest-based deductions.

Time draft. A draft that allows a delay in payment. It is presented to the drawee, who accepts it by writing a notice of acceptance on its face. Once accepted, the time draft becomes a promise to pay by the accepting party. *See also* Bankers' acceptance.

Total Shareholder Return (TSR). A measure of corporate performance based on the sum of share price appreciation and current dividends.

Trade acceptance (T/A). An international trade term. A bill of exchange drawn directly upon and accepted by an importer or purchaser, rather than a bank, and due at a specified future time.

Tranche. An allocation of shares, typically to underwriters that are expected to sell to investors in their designated geographic markets.

Transaction exposure. The potential for a change in the value of outstanding financial obligations entered into prior to a change in exchange rates but not due to be settled until after the exchange rates change.

Transfer pricing. The setting of prices to be charged by one unit (such as a foreign subsidiary) of a multi-unit corporation to another unit (such as the parent corporation) for goods or services sold between such related units.

Translation. The remeasurement of a financial statement from one currency to another.

Translation exposure. The potential for an accounting-derived change in owners' equity resulting from exchange rate changes and the need to restate financial statements of foreign subsidiaries in the single currency of the parent corporation. *See also* Accounting exposure.

Transnational firm. A company owned by a coalition of investors located in different countries.

Transparency. The degree to which an investor can discern the true activities and value drivers of a company from the disclosures and financial results reported.

Triangular arbitrage. An arbitrage activity of exchanging currency A for currency B for currency C back to currency A to exploit slight disequilibrium in exchange rates.

Triffin Dilemma (or Triffin Paradox). The potential conflict in objectives which may arise between domestic monetary policy and currency policy when a country's currency is used as a reserve currency.

Trilemma of international finance. The difficult but required choice which a government must make between three conflicting international financial system goals: (1) a fixed exchange rate; (2) independent monetary policy; and (3) free mobility of capital.

Turnover tax. A tax based on turnover or sales, and is similar in structure to a VAT, in which taxes may be assessed on intermediate stages of a good's production.

Unaffiliated. An independent third-party.

Unbundling. Dividing cash flows from a subsidiary to a parent into their many separate components, such as royalties, lease payments, dividends, etc., so as to increase the likelihood that some fund flows will be allowed during economically difficult times.

Uncovered interest arbitrage (UIA). The process by which investors borrow in countries and currencies exhibiting relatively low interest rates and convert the proceeds into currencies that offer much higher interest rates. The transaction is "uncovered" because the investor does not sell the higher yielding currency proceeds forward.

Undervalued currency. The status of currency with a current foreign exchange value (i.e., current price in the foreign exchange market) below the worth of that currency. Because "worth" is a subjective concept, undervaluation is a matter of opinion. If the euro has a current market value of $1.20 (i.e., the current exchange rate is $1.20/€) at a time when its "true" value as derived from purchasing power parity or some other method is deemed to be $1.30, the euro is undervalued. The opposite of undervalued is overvalued.

Unit currency. *See* Base currency.

Unsystematic risk. In a portfolio, the amount of risk that can be eliminated by diversification.

Usance draft. *See* Time draft.

Valley of Death. The period in capital raising for a startup firm between seed capital and more formal forms such as angel financing and venture capital. So-named because many business startups fail in this stage as a result of not finding funding sources.

Value date. The date when value is given (i.e., funds are deposited) for foreign exchange transactions between banks.

Value firm. A reference to a larger, older, more mature business that typically demonstrates little movement in its share price.

Value-added tax. A type of national sales tax collected at each stage of production or sale of consumption goods, and levied in proportion to the value added during that stage.

Venture capitalist (VC). An investor or fund that provides capital and funding to early-stage business startups. The startups are typically considered of high potential growth and possibility as a result of unique intellectual property or technology.

Volatility. In connection with options, the standard deviation of daily spot price movement.

Vulture funds. Investment funds which specialize in acquiring debt that is in default and then pursuing legal means to obtain either collateral or full payment via legal means.

Weighted average cost of capital (WACC). The sum of the proportionally weighted costs of different sources of capital, used as the minimum acceptable target return on new investments.

Wire transfer. Electronic transfer of funds.

Working capital management. The management of the net working capital requirements (A/R plus inventories less A/P) of the firm.

World Bank. *See* International Bank for Reconstruction and Development.

Worldwide approach (to taxes). The principle that taxes are levied on the income earned by firms that are incorporated in a host country, regardless of where the income was earned.

Yankee bond. Dollar-denominated bond issued within the United States by a foreign borrower.

Yield to maturity. The rate of interest (discount) that equates future cash flows of a bond, both interest and principal, with the present market price. Yield to maturity is thus the time-adjusted rate of return earned by a bond investor.

Yuan (CNY). The official currency of the People's Republic of China, also termed the renminbi.

Index

Currencies of the World

Country	Currency	ISO-4217 Code	Symbol
Afghanistan	Afghan afghani	AFN	
Albania	Albanian lek	ALL	
Algeria	Algerian dinar	DZD	
American Samoa	see United States		
Andorra	see Spain and France		
Angola	Angolan kwanza	AOA	
Anguilla	East Caribbean dollar	XCD	EC$
Antigua and Barbuda	East Caribbean dollar	XCD	EC$
Argentina	Argentine peso	ARS	
Armenia	Armenian dram	AMD	
Aruba	Aruban florin	AWG	ƒ
Australia	Australian dollar	AUD	$
Austria	European euro	EUR	€
Azerbaijan	Azerbaijani manat	AZN	
Bahamas	Bahamian dollar	BSD	B$
Bahrain	Bahraini dinar	BHD	
Bangladesh	Bangladeshi taka	BDT	
Barbados	Barbadian dollar	BBD	Bds$
Belarus	Belarusian ruble	BYR	Br
Belgium	European euro	EUR	€
Belize	Belize dollar	BZD	BZ$
Benin	West African CFA franc	XOF	CFA
Bermuda	Bermudian dollar	BMD	BD$
Bhutan	Bhutanese ngultrum	BTN	Nu.
Bolivia	Bolivian boliviano	BOB	Bs.
Bosnia-Herzegovina	Bosnia and Herzegovina konvertibilna marka	BAM	KM
Botswana	Botswana pula	BWP	P
Brazil	Brazilian real	BRL	R$
British Indian Ocean Territory	see United Kingdom		
Brunei	Brunei dollar	BND	B$
Bulgaria	Bulgarian lev	BGN	
Burkina Faso	West African CFA franc	XOF	CFA
Burma	see Myanmar		
Burundi	Burundi franc	BIF	FBu
Cambodia	Cambodian riel	KHR	
Cameroon	Central African CFA franc	XAF	CFA
Canada	Canadian dollar	CAD	$
Canton and Enderbury Islands	see Kiribati		
Cape Verde	Cape Verdean escudo	CVE	Esc
Cayman Islands	Cayman Islands dollar	KYD	KY$
Central African Republic	Central African CFA franc	XAF	CFA
Chad	Central African CFA franc	XAF	CFA
Chile	Chilean peso	CLP	$
China	Chinese renminbi	CNY	¥
Christmas Island	see Australia		
Cocos (Keeling) Islands	see Australia		
Colombia	Colombian peso	COP	Col$
Comoros	Comorian franc	KMF	
Congo	Central African CFA franc	XAF	CFA
Congo, Democratic Republic	Congolese franc	CDF	F
Cook Islands	see New Zealand		
Costa Rica	Costa Rican colon	CRC	₡
Côte d'Ivoire	West African CFA franc	XOF	CFA
Croatia	Croatian kuna	HRK	kn
Cuba	Cuban peso	CUC	$
Cyprus	European euro	EUR	€
Czech Republic	Czech koruna	CZK	Kč
Denmark	Danish krone	DKK	Kr
Djibouti	Djiboutian franc	DJF	Fdj
Dominica	East Caribbean dollar	XCD	EC$
Dominican Republic	Dominican peso	DOP	RD$
Dronning Maud Land	see Norway		
East Timor	see Timor-Leste		
Ecuador	uses the U.S. Dollar		
Egypt	Egyptian pound	EGP	£
El Salvador	uses the U.S. Dollar		
Equatorial Guinea	Central African CFA franc	GQE	CFA
Eritrea	Eritrean nakfa	ERN	Nfa
Estonia	Estonian kroon	EEK	KR
Ethiopia	Ethiopian birr	ETB	Br
Faeroe Islands (Føroyar)	see Denmark		
Falkland Islands	Falkland Islands pound	FKP	£
Fiji	Fijian dollar	FJD	FJ$
Finland	European euro	EUR	€
France	European euro	EUR	€
French Guiana	see France		
French Polynesia	CFP franc	XPF	F
Gabon	Central African CFA franc	XAF	CFA
Gambia	Gambian dalasi	GMD	D
Georgia	Georgian lari	GEL	
Germany	European euro	EUR	€
Ghana	Ghanaian cedi	GHS	
Gibraltar	Gibraltar pound	GIP	£
Great Britain	see United Kingdom		
Greece	European euro	EUR	€
Greenland	see Denmark		

Currencies of the World (continued)

Country	Currency	ISO-4217 Code	Symbol
Grenada	East Caribbean dollar	XCD	EC$
Guadeloupe	see France		
Guam	see United States		
Guatemala	Guatemalan quetzal	GTQ	Q
Guernsey	see United Kingdom		
Guinea-Bissau	West African CFA franc	XOF	CFA
Guinea	Guinean franc	GNF	FG
Guyana	Guyanese dollar	GYD	GY$
Haiti	Haitian gourde	HTG	G
Heard and McDonald Islands	see Australia		
Honduras	Honduran lempira	HNL	L
Hong Kong	Hong Kong dollar	HKD	HK$
Hungary	Hungarian forint	HUF	Ft
Iceland	Icelandic króna	ISK	kr
India	Indian rupee	INR	₹
Indonesia	Indonesian rupiah	IDR	Rp
International Monetary Fund	Special Drawing Rights	XDR	SDR
Iran	Iranian rial	IRR	
Iraq	Iraqi dinar	IQD	
Ireland	European euro	EUR	€
Isle of Man	see United Kingdom		
Israel	Israeli new sheqel	ILS	
Italy	European euro	EUR	€
Ivory Coast	see Côte d'Ivoire		
Jamaica	Jamaican dollar	JMD	J$
Japan	Japanese yen	JPY	¥
Jersey	see United Kingdom		
Johnston Island	see United States		
Jordan	Jordanian dinar	JOD	
Kampuchea	see Cambodia		
Kazakhstan	Kazakhstani tenge	KZT	T
Kenya	Kenyan shilling	KES	KSh
Kiribati	see Australia		
Korea, North	North Korean won	KPW	W
Korea, South	South Korean won	KRW	W
Kuwait	Kuwaiti dinar	KWD	
Kyrgyzstan	Kyrgyzstani som	KGS	
Laos	Lao kip	LAK	KN
Latvia	Latvian lats	LVL	Ls
Lebanon	Lebanese lira	LBP	
Lesotho	Lesotho loti	LSL	M
Liberia	Liberian dollar	LRD	L$
Libya	Libyan dinar	LYD	LD
Liechtenstein	uses the Swiss Franc		
Lithuania	Lithuanian litas	LTL	Lt
Luxembourg	European euro	EUR	€
Macau	Macanese pataca	MOP	P
Macedonia (Former Yug. Rep.)	Macedonian denar	MKD	
Madagascar	Malagasy ariary	MGA	FMG
Malawi	Malawian kwacha	MWK	MK
Malaysia	Malaysian ringgit	MYR	RM
Maldives	Maldivian rufiyaa	MVR	Rf
Mali	West African CFA franc	XOF	CFA
Malta	European Euro	EUR	€
Martinique	see France		
Mauritania	Mauritanian ouguiya	MRO	UM
Mauritius	Mauritian rupee	MUR	Rs
Mayotte	see France		
Micronesia	see United States		
Midway Islands	see United States		
Mexico	Mexican peso	MXN	$
Moldova	Moldovan leu	MDL	
Monaco	see France		
Mongolia	Mongolian tugrik	MNT	₮
Montenegro	see Italy		
Montserrat	East Caribbean dollar	XCD	EC$
Morocco	Moroccan dirham	MAD	
Mozambique	Mozambican metical	MZM	MTn
Myanmar	Myanma kyat	MMK	K
Nauru	see Australia		
Namibia	Namibian dollar	NAD	N$
Nepal	Nepalese rupee	NPR	NRs
Netherlands Antilles	Netherlands Antillean gulden	ANG	NAf
Netherlands	European euro	EUR	€
New Caledonia	CFP franc	XPF	F
New Zealand	New Zealand dollar	NZD	NZ$
Nicaragua	Nicaraguan córdoba	NIO	C$
Niger	West African CFA franc	XOF	CFA
Nigeria	Nigerian naira	NGN	₦
Niue	see New Zealand		
Norfolk Island	see Australia		
Northern Mariana Islands	see United States		
Norway	Norwegian krone	NOK	kr
Oman	Omani rial	OMR	
Pakistan	Pakistani rupee	PKR	Rs.
Palau	see United States		